Student's Solutions Manual

to accompany

Precalculus

Sixth Edition

Raymond A. Barnett
Merritt College

Michael Ziegler
Marquette University

Karl Byleen
Marquette University

Prepared by
Fred Safier

McGraw-Hill Higher Education

Boston Burr Ridge, IL Dubuque, IA New York San Francisco St. Louis
Bangkok Bogotá Caracas Kuala Lumpur Lisbon London Madrid Mexico City
Milan Montreal New Delhi Santiago Seoul Singapore Sydney Taipei Toronto

The McGraw·Hill Companies

McGraw-Hill Higher Education

Student's Solutions Manual to accompany
PRECALCULUS, SIXTH EDITION
RAYMOND A. BARNETT, MICHAEL ZIEGLER, AND KARL BYLEEN

Published by McGraw-Hill Higher Education, an imprint of The McGraw-Hill Companies, Inc., 1221 Avenue of the Americas, New York, NY 10020.

1 2 3 4 5 6 7 8 9 0 QPD/QPD 0 9 8 7

ISBN: 978-0-07-291770-3
MHID: 0-07-291770-9

www.mhhe.com

TABLE OF CONTENTS

CHAPTER R

Section R-1

1. True
3. False
5. False
7. True
9. False
11. True
13. 1
15. Division by zero is not defined. Undefined.
17. $0 \div (9 \cdot 8) = 0 \div 72 = 0$
19. $0 \cdot \left(100 + \frac{1}{100}\right) = 0$ since $a \cdot 0 = 0 \cdot a = 0$ for all real numbers a.
21. $\frac{4}{9} + \frac{12}{5} = \frac{4 \cdot 5 + 12 \cdot 9}{9 \cdot 5} = \frac{20 + 108}{45} = \frac{128}{45}$
23. $-\left(\frac{1}{100} + \frac{4}{25}\right) = -\left(\frac{1}{100} + \frac{16}{100}\right) = -\frac{17}{100}$ or $\frac{-17}{100}$
25. $\left(\frac{3}{8}\right)^{-1} + 2^{-1} = \frac{8}{3} + \frac{1}{2} = \frac{8 \cdot 2 + 1 \cdot 3}{3 \cdot 2} = \frac{16 + 3}{6} = \frac{19}{6}$
27. Commutative $(\cdot)$
29. Distributive
31. Inverse $(\cdot)$
33. Inverse (+)
35. Identity (+)
37. Negatives (Theorem 1)
39. The even integers between −3 and 5 are −2, 0, 2, and 4. Hence, the set is written {−2, 0, 2, 4}.
41. The letters in "status" are: s, t, a, u. Hence, the set is written $\{s, t, a, u\}$, or, equivalently, $\{a, s, t, u\}$.
43. Since there are no months starting with B, the set is empty. $\varnothing$
45. (A) The empty set, $\varnothing$, is a subset of every set.
    $\{a\}$ and $\{b\}$ are subsets of S_2.
    S_2 is a subset of itself.
    Thus, there are four subsets of S_2.

    (B) The empty set, $\varnothing$, is a subset of every set.
    $\{a\}$, $\{b\}$, and $\{c\}$ are one-member subsets of S_3.
    $\{b, c\}$, $\{a, c\}$, and $\{a, b\}$ are two-member subsets of S_3.
    S_3 is a subset of itself.
    Thus, there are $1 + 3 + 3 + 1 = 8$ subsets of S_3.

    (C) The empty set, $\varnothing$, is a subset of every set.
    $\{a\}$, $\{b\}$, $\{c\}$, and $\{d\}$ are one-member subsets of S_4.
    $\{a, b\}$, $\{a, c\}$, $\{a, d\}$, $\{b, c\}$, $\{b, d\}$, and $\{c, d\}$ are two-member subsets of S_4.
    $\{b, c, d\}$, $\{a, c, d\}$, $\{a, b, d\}$, and $\{b, c, d\}$ are three-member subsets of S_4.
    S_4 is a subset of itself.
    Thus, there are $1 + 4 + 6 + 4 + 1 = 16$ subsets of S_4.
47. Yes. This restates Zero Property (2). (Theorem 2, Part 2)
49. (A) True.

    (B) False, $\frac{2}{3}$ is an example of a real number that is not irrational.

    (C) True.

51. $\frac{3}{5}$ and -1.43 are two examples of infinitely many.

53. (A) $\{1, \sqrt{144}\}$ (B) $\{-3, 0, 1, \sqrt{144}\}$ (C) $\left\{-3, -\frac{2}{3}, 0, 1, \frac{9}{5}, \sqrt{144}\right\}$ (D) $\{\sqrt{3}\}$

55. (A) 0.888 888…; repeating; repeated digit: 8
(B) 0.272 727…; repeating; repeated digits: 27
(C) 2.236 067 977…; nonrepeating and nonterminating
(D) 1.375; terminating

57. (A) True; commutative property for addition.
(B) False; for example $3 - 5 \neq 5 - 3$.
(C) True; commutative property for multiplication.
(D) False; for example $9 \div 3 \neq 3 \div 9$.

59. F $3 - 8 = -5$ is one of many counterexamples. **61.** T

63. F $\sqrt{2} \cdot \frac{\sqrt{2}}{2} = 1$ is one of many counterexamples. **65.** T

67. (A) List each element of *A*. Follow these with each element of *B* that is not yet listed. {1,2,3,4,6}
(B) List each element of *A* that is also an element of *B* {2,4}.

69.
$$\begin{aligned} \text{Let } c &= 0.090909\ldots \\ \text{Then } 100c &= 9.0909\ldots \\ 100c - c &= (9.0909\ldots) - (0.090909\ldots) \\ 99c &= 9 \\ c &= \frac{9}{99} = \frac{1}{11} \end{aligned}$$

71.
$$\begin{array}{rl} 23 & \\ \underline{12} & \\ 46 & 23 \cdot 2 \\ \underline{230} & 23 \cdot 10 \\ 276 & \end{array}$$

$$\begin{aligned} 23 \cdot 12 &= 23(2 + 10) \\ &= 23 \cdot 2 + 23 \cdot 10 \\ &= 46 + 230 \\ &= 276 \end{aligned}$$

Section R-2

1. 256

3. $\left(\frac{2}{3}\right)^4 = \frac{2^4}{3^4} = \frac{16}{81}$

5. $4^{-4} = \frac{1}{4^4} = \frac{1}{256}$

7. 625

9. $(-3)^{-1} = \frac{1}{-3} = \frac{-1}{3}$

11. $-7^{-2} = -\frac{1}{7^2} = \frac{-1}{49}$

13. 1

15. 10

17. 5

19. $9^{-3/2} = \frac{1}{9^{3/2}} = \frac{1}{\left(9^{1/2}\right)^3} = \frac{1}{3^3} = \frac{1}{27}$

21. Irrational

23. $1^{-3} + 3^{-1} = \frac{1}{1^3} + \frac{1}{3^1} = \frac{1}{1} + \frac{1}{3} = \frac{3+1}{3} = \frac{4}{3}$

25. $\frac{5}{5^4} = \frac{1}{5^3} = \frac{1}{125}$

27. $x^5x^{-2} = x^{5+(-2)} = x^3$

29. $(2y)(3y^2)(5y^4) = 2\cdot3\cdot5yy^2y^4 = 30y^{1+2+4} = 30y^7$

31. $(a^2b^3)^5 = (a^2)^5(b^3)^5 = a^{10}b^{15}$

33. $u^{1/3}u^{5/3} = u^{(1/3)+(5/3)} = u^{6/3} = u^2$

35. $(x^{-3})^{1/6} = x^{-3/6} = x^{-1/2} = \frac{1}{x^{1/2}}$

37. $(49a^4b^{-2})^{1/2} = 49^{1/2}(a^4)^{1/2}(b^{-2})^{1/2} = 7a^2b^{-1} = \frac{7a^2}{b}$

39. $45{,}320{,}000 = 4.5320000. \times 10^7$

7 places left

$= 4.532 \times 10^7$

41. $0.066 = 0.06.6 \times 10^{-2} = 6.6 \times 10^{-2}$

2 places right

negative exponent

43. $0.000\ 000\ 084 = 8.4 \times 10^{-8}$

8 places right

45. $9 \times 10^{-5} = 0.00009. = 0.000\ 09$

5 places left

47. $3.48 \times 10^6 = 3.480\ 000. = 3{,}480{,}000$

6 places right

49. $4.2 \times 10^{-9} = 0.000\ 000\ 004.2 = 0.000\ 000\ 004\ 2$

9 places left

51. $\left(\dfrac{x^4y^{-1}}{x^{-2}y^3}\right)^2 = \dfrac{x^8y^{-2}}{x^{-4}y^6} = x^{8-(-4)}y^{-2-6} = x^{12}y^{-8} = \dfrac{x^{12}}{y^8}$

53. $\left(\dfrac{2x^{-3}y^2}{4xy^{-1}}\right)^{-2} = \left(\dfrac{x^{-3}y^2}{2xy^{-1}}\right)^{-2} = \dfrac{x^6y^{-4}}{2^{-2}x^{-2}y^2} = 2^2x^{6-(-2)}y^{-4-2} = 4x^8y^{-6} = \dfrac{4x^8}{y^6}$

55. $\left(\dfrac{a^{-3}}{b^4}\right)^{1/12} = \dfrac{a^{-3/12}}{b^{4/12}} = \dfrac{a^{-1/4}}{b^{1/3}} = \dfrac{1}{a^{1/4}b^{1/3}}$

57. $\left(\dfrac{4x^{-2}}{y^4}\right)^{-1/2} = \dfrac{4^{-1/2}x^1}{y^{-2}} = \dfrac{xy^2}{4^{1/2}} = \dfrac{xy^2}{2}$

> **Common Error:** $\dfrac{4x^1}{y^{-2}}$. This is wrong; the exponent $(-\frac{1}{2})$ applies to constant as well as variable factors.

59. $\left(\dfrac{8a^{-4}b^3}{27a^2b^{-3}}\right)^{1/3} = \dfrac{8^{1/3}a^{-4/3}b^1}{27^{1/3}a^{2/3}b^{-1}} = \dfrac{2b^{1-(-1)}}{3a^{2/3-(-4/3)}} = \dfrac{2b^2}{3a^2}$

61. $-3(x^3+3)^{-4}(3x^2) = -3 \cdot \dfrac{1}{(x^3+3)^4} \cdot 3x^2$

$= \dfrac{-9x^2}{(x^3+3)^4}$

> **Common Error:** $(x^3 + 3)^{-4} \neq x^{-12} + 3^{-4}$

63. $2^{3^2} = 64$ on the assumption that 2^3^2 is entered, without parentheses.

65. The identity property for multiplication states that for x in R, $x(1) = x$, and 1 is the only element in R with that property. Therefore, if $a^ma^0 = a^m$, since $a^m(1) = a^m$, a^0 should equal 1.

67. $\dfrac{(32.7)(0.000\,000\,008\,42)}{(0.0513)(80{,}700{,}000{,}000)} = \dfrac{(32.7)(8.42 \times 10^{-9})}{(0.0513)(8.07 \times 10^{10})} = 6.65 \times 10^{-17}$

69. $\dfrac{(5{,}760{,}000{,}000)}{(527)(0.000\,007\,09)} = \dfrac{5.76 \times 10^9}{(527)(7.09 \times 10^{-6})} = 1.54 \times 10^{12}$

71. $15^{5/4} = 15^{1.25} = 29.52$

73. $103^{-3/4} = 103^{-0.75} = 0.03093$

75. $2.876^{8/5} = 2.876^{1.6} = 5.421$

77. $(0.000\,000\,077\,35)^{-2/7} = (7.735 \times 10^{-8})^{(-2\div 7)} = 107.6$

79. There are many examples; one simple choice is $x = y = 1$. Then $(x + y)^{1/2} = (1 + 1)^{1/2} = 2^{1/2} = \sqrt{2}$, and $x^{1/2} + y^{1/2} = 1^{1/2} + 1^{1/2} = 1 + 1 = 2$; thus, the left side is not equal to the right side.

81. There are many examples; one simple choice is $x = y = 1$. Then $(x + y)^{1/3} = (1 + 1)^{1/3} = 2^{1/3} = \sqrt[3]{2}$ and $\frac{1}{(x+y)^3} = \frac{1}{(1+1)^3} = \frac{1}{2^3} = \frac{1}{8}$; thus, the left side is not equal to the right side.

83. $(a^{3/n}b^{3/m})^{1/3} = a^{(1/3)(3/n)}b^{(1/3)(3/m)} = a^{1/n}\,b^{1/m}$

85. $(x^{m/4}y^{n/3})^{-12} = x^{-12(m/4)}y^{-12(n/3)} = x^{-3m}\,y^{-4n} = \frac{1}{x^{3m}y^{4n}}$

87. (A) Since $(x^2)^{1/2}$ represents the positive real square root of x^2, it will not equal x if x is negative. So any negative value of x, for example $x = -2$, will make the left side ≠ right side. $[(-2)^2]^{1/2} = 4^{1/2} = 2 \neq -2$.

(B) Similarly any positive (or zero) value of x, for example $x = 2$, will make the left side = right side. $[2^2]^{1/2} = 4^{1/2} = 2 = 2$.

(C) Any value of x will make the left side = right side.
$[2^3]^{1/3} = 8^{1/3} = 2 \quad [(-2)^3]^{1/3} = [-8]^{1/3} = -2 \quad [0^3]^{1/3} = 0^{1/3} = 0$
There is no real value of x such that $(x^3)^{1/3} \neq x$.

89. No. Any negative value of b, for example -4, will make $(b^m)^{1/n}$ not a real number if n is even and m is odd. Thus, $[(-4)^3]^{1/2} = [-64]^{1/2}$ is not a real number.

91. mass of earth in pounds
= mass of earth in grams × number of pounds/gram
$= 6.1 \times 10^{27} \times 2.2 \times 10^{-3}$
$= 13.42 \times 10^{24}$
$= 1.342 \times 10^{25}$
$= 1.3 \times 10^{25}$ pounds to two significant digits

93. 1 operation in 10^{-8} seconds means $1 \div 10^{-8}$ operations/second, that is 10^8 operations in 1 second, that is, 100,000,000 or 100 million. Similarly, 1 operation in 10^{-10} seconds means $1 \div 10^{-10}$, that is 10,000,000,000 or 10 billion operations in 1 second.

Since 1 minute is 60 seconds, multiply each number by 60 to get $60 \times 10^8 = 6 \times 10^9$ or 6,000,000,000 or 6 billion operations in 1 minute and $60 \times 10^{10} = 6 \times 10^{11}$ or 600,000,000,000 or 600 billion operations in 1 minute.

95. One person's share of national debt
= amount of national debt ÷ number of persons = 6,760,000,000,000 ÷ 291,000,000
$= 6.76 \times 10^{12} \div 2.91 \times 10^8$
$= 2.32 \times 10^4$ or $23,200 per person to three significant digits

97. We are to calculate $N = 10x^{3/4}y^{1/4}$ given $x = 256$ units of labor and $y = 81$ units of capital.

$N = 10(256)^{3/4}(81)^{1/4}$
$= 10(256^{1/4})^3(81)^{1/4}$
$= 10(4^3)(3)$
$= 1{,}920$ units of finished product.

99. We are to calculate $d = 0.0212v^{7/3}$ given $v = 70$ miles/hour.

$d = 0.0212(70)^{7/3}$
$= 0.0212(70)^{(7 \div 3)}$
$= 428$ feet (to the nearest foot)

Section R-3

1. $\sqrt[5]{32}$

3. $8\sqrt[3]{x^2}$

5. $4x^{-1/2} = 4\sqrt{x^{-1}}$ or $\dfrac{4}{\sqrt{x}}$

7. $\sqrt[3]{x} - \sqrt[3]{y}$

9. $361^{1/2}$

11. $4xy^{3/5}$

13. $(x^2 + y^2)^{1/3}$

15. -2

17. $-\sqrt{128} = -\sqrt{64 \cdot 2} = -\sqrt{8^2 \cdot 2} - \sqrt{8^2}\sqrt{2} = -8\sqrt{2}$

19. $\sqrt{27} - 5\sqrt{3} = \sqrt{9 \cdot 3} - 5\sqrt{3} = \sqrt{3^2 \cdot 3} - 5\sqrt{3} = \sqrt{3^2}\sqrt{3} - 5\sqrt{3} = 3\sqrt{3} - 5\sqrt{3} = -2\sqrt{3}$

21. $\sqrt[3]{5} - \sqrt[3]{25} + \sqrt[3]{125 \cdot 5} = \sqrt[3]{5} - \sqrt[3]{25} + \sqrt[3]{5^3 \cdot 5} = \sqrt[3]{5} - \sqrt[3]{25} + \sqrt[3]{5^3}\sqrt[3]{5}$
$= \sqrt[3]{5} - \sqrt[3]{25} + 5\sqrt[3]{5} = 6\sqrt[3]{5} - \sqrt[3]{25}$

23. $\sqrt[3]{25}\sqrt[3]{10} = \sqrt[3]{250} = \sqrt[3]{125 \cdot 2} = \sqrt[3]{5^3 \cdot 2} = \sqrt[3]{5^3}\sqrt[3]{2} = 5\sqrt[3]{2}$

25. $\sqrt{9x^8y^4} = \sqrt{9}\sqrt{x^8}\sqrt{y^4} = 3x^4y^2$

27. $\sqrt[4]{16m^4n^8} = \sqrt[4]{16}\sqrt[4]{m^4}\sqrt[4]{n^8} = 2mn^2$

29. $\sqrt[4]{m^2} = \sqrt[2\cdot 2]{m^{2\cdot 1}} = \sqrt[2]{m^1} = \sqrt{m}$

31. $\sqrt[5]{\sqrt[3]{xy}} = \sqrt[15]{xy}$

33. $\sqrt[3]{9x^2}\sqrt[3]{9x} = \sqrt[3]{(9x^2)(9x)} = \sqrt[3]{81x^3} = \sqrt[3]{27x^3 \cdot 3} = \sqrt[3]{27x^3}\sqrt[3]{3} = 3x\sqrt[3]{3}$

35. $\dfrac{1}{2\sqrt{5}} = \dfrac{1\sqrt{5}}{2\sqrt{5}\sqrt{5}} = \dfrac{\sqrt{5}}{2 \cdot 5} = \dfrac{\sqrt{5}}{10}$

37. $\dfrac{1}{\sqrt[3]{7}} = \dfrac{1\sqrt[3]{7^2}}{\sqrt[3]{7}\sqrt[3]{7^2}} = \dfrac{\sqrt[3]{49}}{\sqrt[3]{7^3}} = \dfrac{\sqrt[3]{49}}{7}$

39. $\dfrac{2}{\sqrt{44}} = \dfrac{2}{\sqrt{2^2 \cdot 11}} = \dfrac{2\sqrt{11}}{\sqrt{2^2}\sqrt{11}\sqrt{11}} = \dfrac{2\sqrt{11}}{2 \cdot 11} = \dfrac{\sqrt{11}}{11}$

41. $\dfrac{6x}{\sqrt{3x}} = \dfrac{6x}{\sqrt{3x}}\dfrac{\sqrt{3x}}{\sqrt{3x}} = \dfrac{6x\sqrt{3x}}{3x} = 2\sqrt{3x}$

43. $\dfrac{2}{\sqrt{2} - 1} = \dfrac{2}{(\sqrt{2} - 1)}\dfrac{(\sqrt{2} + 1)}{(\sqrt{2} + 1)} = \dfrac{2(\sqrt{2} + 1)}{2 - 1} = \dfrac{2(\sqrt{2} + 1)}{1} = 2(\sqrt{2} + 1) = 2\sqrt{2} + 2$

45. $\dfrac{\sqrt{2}}{\sqrt{6} + 2} = \dfrac{\sqrt{2}}{(\sqrt{6} + 2)}\dfrac{(\sqrt{6} - 2)}{(\sqrt{6} - 2)} = \dfrac{\sqrt{2}(\sqrt{6} - 2)}{6 - 4} = \dfrac{\sqrt{2}\sqrt{6} - 2\sqrt{2}}{2} = \dfrac{\sqrt{12} - 2\sqrt{2}}{2} = \dfrac{2\sqrt{3} - 2\sqrt{2}}{2}$
$= \dfrac{2(\sqrt{3} - \sqrt{2})}{2} = \sqrt{3} - \sqrt{2}$

47. $x\sqrt[5]{3^6x^7y^{11}} = x\sqrt[5]{(3^5x^5y^{10})(3x^2y)} = x\sqrt[5]{3^5x^5y^{10}}\sqrt[5]{3x^2y} = x(3xy^2)\sqrt[5]{3x^2y} = 3x^2y^2\sqrt[5]{3x^2y}$

49. $\dfrac{\sqrt[4]{32m^7n^9}}{2mn} = \dfrac{\sqrt[4]{16m^4n^8 \cdot 2m^3n}}{2mn} = \dfrac{2mn^2\sqrt[4]{2m^3n}}{2mn} = n\sqrt[4]{2m^3n}$

51. $\sqrt[3]{\sqrt[4]{a^9b^3}} = \sqrt[3\cdot 4]{a^{3\cdot 3}b^{3\cdot 1}} = \sqrt[4]{a^3b}$

53. $\sqrt[3]{2x^2y^4}\,\sqrt[3]{3x^5y} = \sqrt[3]{(2x^2y^4)(3x^5y)} = \sqrt[3]{6x^7y^5} = \sqrt[3]{(x^6y^3)(6xy^2)} = \sqrt[3]{x^6y^3}\,\sqrt[3]{6xy^2} = x^2y\sqrt[3]{6xy^2}$

55. $\dfrac{\sqrt{2m}\sqrt{5}}{\sqrt{20m}} = \dfrac{\sqrt{10m}}{\sqrt{20m}} = \sqrt{\dfrac{10m}{20m}} = \sqrt{\dfrac{1}{2}} = \sqrt{\dfrac{1}{2}\cdot\dfrac{2}{2}} = \sqrt{\dfrac{2}{4}} = \dfrac{\sqrt{2}}{2}$ or $\dfrac{1}{2}\sqrt{2}$

57. $\dfrac{3\sqrt{y}}{2\sqrt{y}-3} = \dfrac{3\sqrt{y}}{(2\sqrt{y}-3)}\dfrac{(2\sqrt{y}+3)}{(2\sqrt{y}+3)} = \dfrac{3\sqrt{y}(2\sqrt{y}+3)}{(2\sqrt{y})^2-(3)^2} = \dfrac{6y+9\sqrt{y}}{4y-9}$

59. $\dfrac{2\sqrt{5}+3\sqrt{2}}{5\sqrt{5}+2\sqrt{2}} = \dfrac{(2\sqrt{5}+3\sqrt{2})}{(5\sqrt{5}+2\sqrt{2})}\dfrac{(5\sqrt{5}-2\sqrt{2})}{(5\sqrt{5}-2\sqrt{2})}$

$= \dfrac{2\sqrt{5}\cdot 5\sqrt{5} - 2\sqrt{5}\cdot 2\sqrt{2} + 3\sqrt{2}\cdot 5\sqrt{5} - 3\sqrt{2}\cdot 2\sqrt{2}}{(5\sqrt{5})^2 + (2\sqrt{2})^2}$

$= \dfrac{(10)(5) - 4\sqrt{10} + 15\sqrt{10} - (6)(2)}{(25)(5) - (4)(2)} = \dfrac{50 + 11\sqrt{10} - 12}{125 - 8} = \dfrac{38 + 11\sqrt{10}}{117}$

61. Here we want to remove the radical terms from the *numerator*.

$\dfrac{\sqrt{t}-\sqrt{x}}{t-x} = \dfrac{(\sqrt{t}-\sqrt{x})}{t-x}\cdot\dfrac{(\sqrt{t}+\sqrt{x})}{\sqrt{t}+\sqrt{x}} = \dfrac{(\sqrt{t})^2-(\sqrt{x})^2}{(t-x)(\sqrt{t}+\sqrt{x})}$

$= \dfrac{\overset{1}{\cancel{t-x}}}{\underset{1}{\cancel{(t-x)}}(\sqrt{t}+\sqrt{x})} = \dfrac{1}{\sqrt{t}+\sqrt{x}}$

63. Here we want to remove the radical terms from the *numerator*.

$\dfrac{\sqrt{x+h}-\sqrt{x}}{h} = \dfrac{(\sqrt{x+h}-\sqrt{x})}{h}\cdot\dfrac{(\sqrt{x+h}+\sqrt{x})}{(\sqrt{x+h}+\sqrt{x})} = \dfrac{(\sqrt{x+h})^2-(\sqrt{x})^2}{h(\sqrt{x+h}+\sqrt{x})} = \dfrac{x+h-x}{h(\sqrt{x+h}+\sqrt{x})}$

$= \dfrac{\overset{1}{\cancel{h}}}{\underset{1}{\cancel{h}}(\sqrt{x+h}+\sqrt{x})} = \dfrac{1}{\sqrt{x+h}+\sqrt{x}}$

65. $\sqrt{0.032\,965} = 0.1816$ **67.** $\sqrt[3]{45.0218} = (45.0218)^{1/3} = 45.0218^{(1\div 3)} = 3.557$

69. $\sqrt[8]{5.477\times 10^{-9}} = (5.477\times 10^{-9})^{(1\div 8)} = 0.09275$

71. $\sqrt[5]{9+\sqrt[5]{9}} = (9 + 9^{(1\div 5)})^{(1\div 5)} = 1.602$

73. $\sqrt{x^2}$ is the principal square root of x^2, that is, the positive (or zero) square root of x^2. The two square roots of x^2 are x and $-x$. $\sqrt{x^2} = -x$ if and only if $-x$ is positive or zero. Hence $\sqrt{x^2} = -x$ if and only if x is negative or zero. $x \le 0$.

75. $\sqrt[3]{x^3}$ is the real third root of x^3. This is always equal to x if x is real. All real numbers.

77. (A) $\sqrt{3}+\sqrt{5} \approx 3.968\ 118\ 785$ (B) $\sqrt{2+\sqrt{3}} + \sqrt{2-\sqrt{3}} \approx 2.449\ 489\ 743$

(C) $1+\sqrt{3} \approx 2.732\ 050\ 808$ (D) $\sqrt[3]{10+6\sqrt{3}} \approx 2.732\ 050\ 808$

(E) $\sqrt{8 + \sqrt{60}} \approx 3.968\ 118\ 785$ (F) $\sqrt{6} \approx 2.449\ 489\ 743$

Thus, (A) and (E), (B) and (F), and (C) and (D) have the same value to 9 decimal places. To show that they are actually equal:
(A) and (E): $(\sqrt{3} + \sqrt{5})^2 = (\sqrt{3})^2 + 2\sqrt{3}\sqrt{5} + (\sqrt{5})^2 = 3 + \sqrt{4}\sqrt{3}\sqrt{5} + 5$
$= 8 + \sqrt{60}$ Therefore, since $\sqrt{3} + \sqrt{5}$ is positive, $\sqrt{3} + \sqrt{5}$ is the positive square root of
$8 + \sqrt{60}$. Thus, $\sqrt{8 + \sqrt{60}} = \sqrt{3} + \sqrt{5}$.

(B) and (F):

$$\begin{aligned}\left(\sqrt{2 + 3} + \sqrt{2 - 3}\right)^2 &= \left(\sqrt{2 + \sqrt{3}}\right)^2 + 2\sqrt{2 + \sqrt{3}}\,\sqrt{2 - \sqrt{3}} + \left(\sqrt{2 - \sqrt{3}}\right)^2 \\ &= 2 + \sqrt{3} + 2\sqrt{(2 + \sqrt{3})(2 - \sqrt{3})} + 2 - \sqrt{3} \\ &= 4 + 2\sqrt{2^2 - (\sqrt{3})^2} \\ &= 4 + 2\sqrt{4 - 3} \\ &= 4 + 2 \cdot 1 \\ &= 6\end{aligned}$$

Therefore, since $\sqrt{2 + \sqrt{3}} + \sqrt{2 - \sqrt{3}}$ is positive, $\sqrt{2 + \sqrt{3}} + \sqrt{2 - \sqrt{3}}$ is the positive square root of 6. Thus, $\sqrt{2 + \sqrt{3}} + \sqrt{2 - \sqrt{3}} = \sqrt{6}$.

(C) and (D):

$$\begin{aligned}(1 + \sqrt{3})^3 &= (1 + \sqrt{3})(1 + \sqrt{3})(1 + \sqrt{3}) \\ &= (1^2 + 2\sqrt{3} + \sqrt{3}^2)(1 + \sqrt{3}) \\ &= (4 + 2\sqrt{3})(1 + \sqrt{3}) \\ &= 4 + 4\sqrt{3} + 2\sqrt{3} + 2\sqrt{3}\sqrt{3} \\ &= 10 + 6\sqrt{3}\end{aligned}$$

Therefore, $1 + \sqrt{3}$ is the real cube root of $10 + 6\sqrt{3}$. Thus, $1 + \sqrt{3} = \sqrt[3]{10 + 6\sqrt{3}}$.

79. $$\frac{1}{\sqrt[3]{a} - \sqrt[3]{b}} = \frac{1}{\left(\sqrt[3]{a} - \sqrt[3]{b}\right)} \frac{\left[(\sqrt[3]{a})^2 + \sqrt[3]{a}\sqrt[3]{b} + (\sqrt[3]{b})^2\right]}{\left[(\sqrt[3]{a})^2 + \sqrt[3]{a}\sqrt[3]{b} + (\sqrt[3]{b})^2\right]}$$

using $(x - y)(x^2 + xy + y^2) = x^3 - y^3$

$$= \frac{1\left[\sqrt[3]{a^2} + \sqrt[3]{ab} + \sqrt[3]{b^2}\right]}{(\sqrt[3]{a})^3 - (\sqrt[3]{b})^3} = \frac{\sqrt[3]{a^2} + \sqrt[3]{ab} + \sqrt[3]{b^2}}{a - b}$$

81. $$\frac{1}{\sqrt{x} - \sqrt{y} + \sqrt{z}} = \frac{1}{\left[(\sqrt{x} - \sqrt{y}) + \sqrt{z}\right]} \frac{\left[(\sqrt{x} - \sqrt{y}) - \sqrt{z}\right]}{\left[(\sqrt{x} - \sqrt{y}) - \sqrt{z}\right]} = \frac{\sqrt{x} - \sqrt{y} - \sqrt{z}}{(\sqrt{x} - \sqrt{y})^2 + (\sqrt{z})^2}$$

$$= \frac{\sqrt{x} - \sqrt{y} - \sqrt{z}}{(\sqrt{x})^2 - 2\sqrt{x}\sqrt{y} + (\sqrt{y})^2 - (\sqrt{z})^2} = \frac{\sqrt{x} - \sqrt{y} - \sqrt{z}}{x - 2\sqrt{xy} + y - z}$$

$$= \frac{\sqrt{x} - \sqrt{y} - \sqrt{z}}{x - 2\sqrt{xy} + y - z} = \frac{\sqrt{x} - \sqrt{y} - \sqrt{z}}{x + y - z - 2\sqrt{xy}}$$

$$= \frac{(\sqrt{x} - \sqrt{y} - \sqrt{z})}{[(x + y - z) - 2\sqrt{xy}]} \frac{[(x + y - z) + 2\sqrt{xy}]}{[(x + y - z) + 2\sqrt{xy}]}$$

$$= \frac{(\sqrt{x} - \sqrt{y} - \sqrt{z})(x + y - z) + 2\sqrt{xy}}{(x + y - z)^2 - (2\sqrt{xy})^2}$$

$$= \frac{(\sqrt{x} - \sqrt{y} - \sqrt{z})(x + y - z) + 2\sqrt{xy}}{(x + y - z)^2 - 4xy}$$

83. $$\frac{\sqrt[3]{x+h}-\sqrt[3]{x}}{h} = \frac{\left(\sqrt[3]{x+h}-\sqrt[3]{x}\right)}{h}\frac{\left[(\sqrt[3]{x+h})^2+\sqrt[3]{x+h}\,\sqrt[3]{x}+(\sqrt[3]{x})^2\right]}{\left[(\sqrt[3]{x+h})^2+\sqrt[3]{x+h}\,\sqrt[3]{x}+(\sqrt[3]{x})^2\right]}$$

$$= \frac{(\sqrt[3]{x+h})^3-(\sqrt[3]{x})^3}{h\left[\sqrt[3]{(x+h)^2}+\sqrt[3]{x(x+h)}+\sqrt[3]{x^2}\right]}$$

Using $(a-b)(a^2+ab+b^2) = a^3-b^3$

$$= \frac{x+h-x}{h\left[\sqrt[3]{(x+h)^2}+\sqrt[3]{x(x+h)}+\sqrt[3]{x^2}\right]}$$

$$= \frac{h}{h\left[\sqrt[3]{(x+h)^2}+\sqrt[3]{x(x+h)}+\sqrt[3]{x^2}\right]}$$

$$= \frac{1}{\sqrt[3]{(x+h)^2}+\sqrt[3]{x(x+h)}+\sqrt[3]{x^2}}$$

85. $\sqrt[kn]{x^{km}} = (x^{km})^{1/kn} = x^{km/kn} = x^{m/n} = \sqrt[n]{x^m}$

87. $$\frac{M_0}{\sqrt{1-\frac{v^2}{c^2}}} = M_0 \div \sqrt{1-\frac{v^2}{c^2}}$$

$$= M_0 \div \sqrt{\frac{c^2-v^2}{c^2}} = M_0 \div \frac{\sqrt{c^2-v^2}}{c}$$

$$= M_0 \cdot \frac{c}{\sqrt{c^2-v^2}}$$

Now rationalize the denominator

$$= M_0 \cdot \frac{c}{\sqrt{c^2-v^2}}\,\frac{\sqrt{c^2-v^2}}{\sqrt{c^2-v^2}}$$

$$= \frac{M_0 c\sqrt{c^2-v^2}}{c^2-v^2}$$

Section R-4

1. 4

3. $(x^4 - 2x^2 + 3) + (x^3 - 1) = x^4 - 2x^2 + 3 + x^3 - 1 = x^4 + x^3 - 2x^2 + 2$
This polynomial has degree 4.

5. $(x^4 - 2x^2 + 3)(x^3 - 1) = x^7 - x^4 - 2x^5 + 2x^2 + 3x^3 - 3$
$= x^7 - 2x^5 - x^4 + 3x^3 + 2x^2 - 3$

7. $(x^4 - 2x^2 + 3) - (x^3 - 1) = x^4 - 2x^2 + 3 - x^3 + 1 = x^4 - x^3 - 2x^2 + 4$

9. Yes. 2 **11.** No **13.** Yes. 3

15. $2(x - 1) + 3(2x - 3) - (4x - 5) = 2x - 2 + 6x - 9 - 4x + 5 = 4x - 6$

17. $2y - 3y[4 - 2(y - 1)] = 2y - 3y[4 - 2y + 2]$
$= 2y - 3y[6 - 2y]$
$= 2y - 18y + 6y^2$
$= 6y^2 - 16y$

19. $m^2 - n^2$

21. $(4t - 3)(t - 2) = 4t^2 - 8t - 3t + 6 = 4t^2 - 11t + 6$

23. $(5y - 1)(3 - 2y) = 15y - 10y^2 - 3 + 2y = -10y^2 + 17y - 3$

25. $(3x + 2y)(x - 3y) = 3x^2 - 9xy + 2xy - 6y^2 = 3x^2 - 7xy - 6y^2$

27. $4m^2 - 49$

29. $(6x - 4y)(5x + 3y) = 30x^2 + 18xy - 20xy - 12y^2 = 30x^2 - 2xy - 12y^2$

31. $9x^2 - 4y^2$

33. $(4x - y)^2 = (4x)^2 - 2(4x)y + y^2 = 16x^2 - 8xy + y^2$

35. $(a + b)(a^2 - ab + b^2) = a^3 - a^2b + ab^2 + a^2b - ab^2 + b^3 = a^3 + b^3$

37.
$$\begin{aligned} 2x - 3\{x + 2[x - (x + 5)] + 1\} &= 2x - 3\{x + 2[x - x - 5] + 1\} \\ &= 2x - 3\{x + 2(-5) + 1\} \\ &= 2x - 3\{x - 10 + 1\} \\ &= 2x - 3\{x - 9\} \\ &= 2x - 3x + 27 \\ &= -x + 27 \end{aligned}$$

39.
$$\begin{aligned} 2\{3[a - 4(1 - a)] - (5 - a)\} &= 2\{3[a - 4 + 4a] - 5 + a\} \\ &= 2\{3[5a - 4] - 5 + a\} \\ &= 2\{15a - 12 - 5 + a\} \\ &= 2\{16a - 17\} \\ &= 32a - 34 \end{aligned}$$

41.
$$\begin{aligned} (2x^2 - 3x + 1)(x^2 + x - 2) &= 2x^4 + 2x^3 - 4x^2 - 3x^3 - 3x^2 + 6x + x^2 + x - 2 \\ &= 2x^4 - x^3 - 6x^2 + 7x - 2 \end{aligned}$$

43.
$$\begin{aligned} (x - 2y)^2(x + 2y)^2 &= [(x - 2y)(x + 2y)]^2 \\ &= [x^2 - 4y^2]^2 \\ &= x^4 - 8x^2y^2 + 16y^4 \end{aligned}$$

45.
$$\begin{aligned} (3u - 2v)^2 - (2u - 3v)(2u + 3v) &= (9u^2 - 12uv + 4v^2) - (4u^2 - 9v^2) \\ &= 9u^2 - 12uv + 4v^2 - 4u^2 + 9v^2 \\ &= 5u^2 - 12uv + 13v^2 \end{aligned}$$

47.
$$\begin{aligned} (z + 2)(z^2 - 2z + 3) + z - 7 &= z^3 - 2z^2 + 3z + 2z^2 - 4z + 6 + z - 7 \\ &= z^3 - 1 \end{aligned}$$

49.
$$\begin{aligned} (2m - n^3) &= (2m - n)^2(2m - n) \\ &= (4m^2 - 4mn + n^2)(2m - n) \\ &= 8m^3 - 4m^2n - 8m^2n + 4mn^2 + 2mn^2 - n^3 \\ &= 8m^3 - 12m^2n + 6mn^2 - n^3 \end{aligned}$$

51. $3(x + h) - 7 - (3x - 7) = 3x + 3h - 7 - 3x + 7 = 3h$

53.
$$\begin{aligned} 2(x + h)^2 - 3(x + h) - (2x^2 - 3x) &= 2(x^2 + 2xh + h^2) - 3x - 3h - 2x^2 + 3x \\ &= 2x^2 + 4xh + 2h^2 - 3x - 3h - 2x^2 + 3x \\ &= 4xh + 2h^2 - 3h \end{aligned}$$

55.
$$\begin{aligned} &2(x + h)^2 - 4(x + h) - 9 - (2x^2 - 4x - 9) \\ &\quad = 2(x^2 + 2xh + h^2) - 4x - 4h - 9 - 2x^2 + 4x + 9 \\ &\quad = 2x^2 + 4xh + 2h^2 - 4x - 4h - 9 - 2x^2 + 4x + 9 \\ &\quad = 4xh + 2h^2 - 4h \end{aligned}$$

57. $(x + h)^3 - 2(x + h)^2 - (x^3 - 2x^2)$

$$= (x + h)^2(x + h) - 2(x + h)^2 - (x^3 - 2x^2)$$
$$= (x^2 + 2xh + h^2)(x + h) - 2(x^2 + 2xh + h^2) - x^3 + 2x^2$$
$$= x^3 + hx^2 + 2xh^2 + 2h^2x + h^2x + h^3 - 2x^2 - 4xh - 2h^2 - x^3 + 2x^2$$
$$= 3hx^2 + 3h^2x + h^3 - 4xh - 2h^2$$

59. The sum of the first two polynomials:
$(3m^2 - 2m + 5) + (4m^2 - m) = 3m^2 - 2m + 5 + 4m^2 - m = 7m^2 - 3m + 5$

The sum of the last two polynomials:
$(3m^2 - 3m - 2) + (m^3 + m^2 + 2) = 3m^2 - 3m - 2 + m^3 + m^2 + 2 = m^3 + 4m^2 - 3m$

Subtract from the sum of the last two polynomials the sum of the first two.

$$(m^3 + 4m^2 - 3m) - (7m^2 - 3m + 5) = m^3 + 4m^2 - 3m - 7m^2 + 3m - 5$$
$$= m^3 - 3m^2 - 5$$

61. The area of the entire region is given by $(a + b)(c + d)$. The sum of the areas of the four small rectangles is given by $ac + ad + bc + bd$. Since the area of the entire region must equal the sum of the areas of the four small rectangles, $(a + b)(c + d) = ac + ad + bc + bd$.

63. The area of the large square is given by c^2. The area of the small square is given by $(a - b)^2$. The area of each of the four triangles is given by $\frac{1}{2}ab$.

Therefore,

$$(a - b)^2 + 4\left(\frac{1}{2}ab\right) = c^2$$

Simplifying, we obtain

$$a^2 - 2ab + b^2 + 2ab = c^2$$
$$a^2 + b^2 = c^2$$

This is the Pythagorean Theorem.

65. $2(x - 2)^3 - (x - 2)^2 - 3(x - 2) - 4$

$$= 2(x - 2)^2(x - 2) - (x - 2)^2 - 3(x - 2) - 4$$
$$= 2(x^2 - 4x + 4)(x - 2) - (x^2 - 4x + 4) - 3(x - 2) - 4$$
$$= 2(x^3 - 2x^2 - 4x^2 + 8x + 4x - 8) - (x^2 - 4x + 4) - 3(x - 2) - 4$$
$$= 2x^3 - 4x^2 - 8x^2 + 16x + 8x - 16 - x^2 + 4x - 4 - 3x + 6 - 4$$
$$= 2x^3 - 13x^2 + 25x - 18$$

67. $-3x\{x[x - x(2 - x)] - (x + 2)(x^2 - 3)\}$

$$= -3x\{x[x - 2x + x^2] - (x^3 - 3x + 2x^2 - 6)\}$$
$$= -3x\{x[-x + x^2] - (x^3 - 3x + 2x^2 - 6)\}$$
$$= -3x\{-x^2 + x^3 - x^3 + 3x - 2x^2 + 6\}$$
$$= -3x\{-3x^2 + 3x + 6\}$$
$$= 9x^3 - 9x^2 - 18x$$

69. One example is given by choosing $a = 1$ and $b = 1$. Then $(a + b)^2 = (1 + 1)^2 = 2^2 = 4$, but $a^2 + b^2 = 1^2 + 1^2 = 1 + 1 = 2$. Thus, in general, $(a + b)^2 \neq a^2 + b^2$. In fact, since $(a + b)^2 = a^2 + 2ab + b^2$, this quantity can only equal $a^2 + b^2$ if $2ab = 0$. By the properties of 0, either $a = 0$ or $b = 0$. In these cases $(a + b)^2$ would equal $a^2 + b^2$, but only in these.

71. The non-zero term with the highest degree in the polynomial of degree m has degree m. This term is not changed during the addition, since there is no term in the polynomial of degree n with the degree $m (m > n)$. It will still be the highest degree term in the sum, so the sum will be a polynomial of degree m.

73. Now the non-zero term with the highest degree in each polynomial has degree m. It is possible for the coefficients of these terms to be equal in absolute value but opposite in sign, in which case they would add to 0, leaving a term of degree less than m as the term with the highest degree in the sum. Otherwise the terms of degree m will combine to give another term of degree m, which will be the highest degree term in the sum. Thus, $(2x^4 + x^2 + 1) + (-2x^4 + x^2 + 1) = 2x^2 + 2$, but $(2x^4 + x^2 + 1) + (2x^4 + 2x^2 + 1) = 4x^4 + 3x^2 + 2$. Summarizing, the degree of the sum may be less than or equal to m.

75. There are three quantities in this problem, perimeter, length, and width. They are related by the perimeter formula $P = 2\ell + 2w$. Since x = length of the rectangle, and the width is 5 meters less than the length, $x - 5$ = width of the rectangle. So $P = 2x + 2(x - 5)$ represents the perimeter of the rectangle. Simplifying: $P = 2x + 2x - 10 = 4x - 10$ (meters)

77. There are several quantities involved in this problem. It is important to keep them distinct by using enough words. We write:

$$\begin{aligned} x &= \text{number of nickels} \\ x - 5 &= \text{number of dimes} \\ (x - 5) + 2 &= \text{number of quarters} \end{aligned}$$

This follows because there are five fewer dimes than nickels $(x - 5)$ and 2 more quarters than dimes (2 more than $x - 5$). Each nickel is worth 5 cents, each dime worth 10 cents, and each quarter worth 25 cents. Hence the value of the nickels is 5 times the number of nickels, the value of the dimes is 10 times the number of dimes, and the value of the quarters is 25 times the number of quarters.

$$\begin{aligned} \text{value of nickels} &= 5x \\ \text{value of dimes} &= 10(x - 5) \\ \text{value of quarters} &= 25[(x - 5) + 2] \end{aligned}$$

$$\begin{aligned} \text{The value of the pile} &= (\text{value of nickels}) + (\text{value of dimes}) \\ &\quad + (\text{value of quarters}) \\ &= 5x + 10(x - 5) + 25[(x - 5) + 2] \end{aligned}$$

Simplifying this expression, we get:

$$\begin{aligned} \text{The value of the pile} &= 5x + 10x - 50 + 25[x - 5 + 2] \\ &= 5x + 10x - 50 + 25[x - 3] \\ &= 5x + 10x - 50 + 25x - 75 \\ &= 40x - 125 \text{ (cents)} \end{aligned}$$

79. The volume of the plastic shell is equal to the volume of the larger sphere $(V = \frac{4}{3}\pi r^3)$ minus the volume of the hole. Since the radius of the hole is x cm and the plastic is 0.3 cm thick, the radius of the larger sphere is $x + 0.3$ cm. Thus, we have

$$\begin{aligned} \begin{pmatrix}\text{Volume of}\\ \text{shell}\end{pmatrix} &= \begin{pmatrix}\text{Volume of}\\ \text{larger sphere}\end{pmatrix} - \begin{pmatrix}\text{Volume of}\\ \text{hole}\end{pmatrix} \\ \text{Volume} &= \tfrac{4}{3}\pi(x + 0.3)^3 - \tfrac{4}{3}\pi x^3 \\ &= \tfrac{4}{3}\pi(x + 0.3)(x + 0.3)^2 - \tfrac{4}{3}\pi x^3 \\ &= \tfrac{4}{3}\pi(x + 0.3)(x^2 + 0.6x + 0.09) - \tfrac{4}{3}\pi x^3 \\ &= \tfrac{4}{3}\pi(x^3 + 0.6x^2 + 0.09x + 0.3x^2 + 0.18x + 0.027) - \tfrac{4}{3}\pi x^3 \\ &= \tfrac{4}{3}\pi(x^3 + 0.9x^2 + 0.27x + 0.027) - \tfrac{4}{3}\pi x^3 \\ &= \tfrac{4}{3}\pi x^3 + 1.2\pi x^2 + 0.36\pi x + 0.036\pi - \tfrac{4}{3}\pi x^3 \\ &= 1.2\pi x^2 + 0.36\pi x + 0.036\pi \text{ (cm}^3\text{)} \end{aligned}$$

Section R-5

1. $2x^2(3x^2 - 4x - 1)$ **3.** $5xy(2x^2 + 4xy - 3y^2)$ **5.** $(x + 1)(5x - 3)$ **7.** $(y - 2z)(2w - x$

9. $x^2 - 2x + 3x - 6 = (x^2 - 2x) + (3x - 6)$
$= x(x - 2) + 3(x - 2)$
$= (x - 2)(x + 3)$

11. $6m^2 + 10m - 3m - 5 = (6m^2 + 10m) - (3m + 5) = 2m(3m + 5) - 1(3m + 5)$
$= (3m + 5)(2m - 1)$

13. $2x^2 - 4xy - 3xy + 6y^2 = (2x^2 - 4xy) - (3xy - 6y^2) = 2x(x - 2y) - 3y(x - 2y)$
$= (x - 2y)(2x - 3y)$

15. $8ac + 3bd - 6bc - 4ad = 8ac - 4ad - 6bc + 3bd$
$= (8ac - 4ad) - (6bc - 3bd)$
$= 4a(2c - d) - 3b(2c - d)$
$= (2c - d)(4a - 3b)$

17. $(2x + 3)(x - 1)$ **19.** Prime

21. $(x + 7y)(x - 2y)$

23. $(2a + 3b)(2a - 3b)$

25. $4x^2 - 20x + 25 = (2x)^2 - 2(2x)(5) + 5^2 = (2x - 5)^2$ **27.** Prime

29. Prime

31. $6x^2 + 48x + 72 = 6(x^2 + 8x + 12) = 6(x + 2)(x + 6)$

33. $2y^3 - 22y^2 + 48y = 2y(y^2 - 11y + 24) = 2y(y - 3)(y - 8)$

35. $16x^2y - 8xy + y = y(16x^2 - 8x + 1) = y[(4x)^2 - 2(4x) + 1] = y(4x - 1)^2$

37. $(3m + 4n)(2m - 3n)$

39. $x^3y - 9xy^3 = xy(x^2 - 9y^2) = xy(x - 3y)(x + 3y)$

41. $3m(m^2 - 2m + 5)$ **43.** $(m + n)(m^2 - mn + n^2)$

45. $8x^3 - 125 = (2x)^3 - 5^3 = (2x - 5)[(2x)^2 + (2x)5 + 5^2]$
$= (2x - 5)(4x^2 + 10x + 25)$

47. $2x(x + 1)^4 + 4x^2(x + 1)^3 = 2x(x + 1)^3[(x + 1) + 2x] = 2x(x + 1)^3(3x + 1)$

49. $6(3x - 5)(2x - 3)^2 + 4(3x - 5)^2(2x - 3) = 2(3x - 5)(2x - 3)[3(2x - 3) + 2(3x - 5)]$
$= 2(3x - 5)(2x - 3)[6x - 9 + 6x - 10]$
$= 2(3x - 5)(2x - 3)(12x - 19)$

51. $5x^4(9 - x)^4 - 4x^5(9 - x)^3 = x^4(9 - x)^3[5(9 - x) - 4x]$
$= x^4(9 - x)^3[45 - 5x - 4x]$
$= x^4(9 - x)^3(45 - 9x)$ or $9x^4(9 - x)^3(5 - x)$

53. $2(x + 1)(x^2 - 5)^2 + 4x(x + 1)^2(x^2 - 5) = 2(x + 1)(x^2 - 5)[x^2 - 5 + 2x(x + 1)]$
$= 2(x + 1)(x^2 - 5)[x^2 - 5 + 2x^2 + 2x]$
$= 2(x + 1)(x^2 - 5)(3x^2 + 2x - 5)$
$= 2(x + 1)(x^2 - 5)(3x + 5)(x - 1)$

55. $(a - b)^2 - 4(c - d)^2 = (a - b)^2 - [2(c - d)]^2$
$= [(a - b) - 2(c - d)][(a - b) + 2(c - d)]$

57. $2am - 3an + 2bm - 3bn = (2am - 3an) + (2bm - 3bn) = a(2m - 3n) + b(2m - 3n)$
$= (2m - 3n)(a + b)$

59. Prime

61. $x^3 - 3x^2 - 9x + 27 = (x^3 - 3x^2) - (9x - 27) = x^2(x - 3) - 9(x - 3) = (x - 3)(x^2 - 9)$
$= (x - 3)(x - 3)(x + 3) = (x - 3)^2(x + 3)$

63. $a^3 - 2a^2 - a + 2 = (a^3 - 2a^2) - (a - 2) = a^2(a - 2) - 1(a - 2) = (a - 2)(a^2 - 1)$
$= (a - 2)(a + 1)(a - 1)$

65. Prime

67. $(x^2 + 2)(x^2 + 4)$

69. $m^4 - n^4 = (m^2)^2 - (n^2)^2 = (m^2 - n^2)(m^2 + n^2) = (m - n)(m + n)(m^2 + n^2)$

71. $y^4 - 11y^2 + 18 = (y^2 - 2)(y^2 - 9) = (y^2 - 2)(y - 3)(y + 3)$

73. $m^2 + 2mn + n^2 - m - n = (m^2 + 2mn + n^2) - (m + n) = (m + n)^2 - 1(m + n)$
$= (m + n)(m + n - 1)$

75. $18a^3 - 8a(x^2 + 8x + 16) = 2a[9a^2 - 4(x^2 + 8x + 16)] = 2a\{(3a)^2 - [2(x + 4)]^2\}$
$= 2a[3a - 2(x + 4)][3a + 2(x + 4)]$

77. $x^4 + 2x^2 + 1 - x^2 = (x^4 + 2x^2 + 1) - x^2 = (x^2 + 1)^2 - x^2$
$= (x^2 + 1 - x)(x^2 + 1 + x) = (x^2 - x + 1)(x^2 + x + 1)$

79. $u^7 - v^7 = (u - v)(u^6 + u^5v + u^4v^2 + u^3v^3 + u^2v^4 + uv^5 + v^6)$
Verifying: The right side =
$(u - v)u^6 + (u - v)u^5v + (u - v)u^4v^2 + (u - v)u^3v^3 + (u - v)u^2v^4 + (u - v)uv^5 + (u - v)v^6$
$= u^7 - u^6v + u^6v - u^5v^2 + u^5v^2 - u^4v^3 + u^4v^3 - u^3v^4 + u^3v^4 - u^2v^5 + u^2v^5 - uv^6 + uv^6 - v^7$
$= u^7 - v^7$

81. 2, 3, 5, 7, 11, 13, 17, 19, 23, 29, 31, 37, 41, 43, 47, 53, 59, 61, 67, 71, 73, 79, 83, 89, 97, 101, 103, 107, 109, 113, 127, 131, 137, 139, 149, 151, 157, 163, 167, 173, 179, 181, 191, 193, 197, 199.

83. (A) If $\sqrt{2} = \frac{a}{b}$, than $2 = \frac{a^2}{b^2}$ and $a^2 = 2b^2$.

(B) Any factor of a must appear in the factorization of a^2 twice for each time it appears in the factorization of a. This means that it appears an even number of times (possibly 0 times).

(C) Any factor of b^2 appears an even number of times by the reasoning of (B). Then the number of times 2 appears must be 1 more than this even number, that is, an odd number of times.

(D) To have two different numbers of appearances of a factor contradicts the uniqueness of a prime factorization.

85. (A) The area of the cardboard can be written as (Original area) - (Removed area), where the original area = $20^2 = 400$ and the removed area consists of 4 squares of area x^2 each; thus,

(Original area) - (Removed area) = $400 - 4x^2$ in expanded form.

In factored form, $400 - 4x^2 = 4(100 - x^2) = 4(10 - x)(10 + x)$.

(B) See figure.

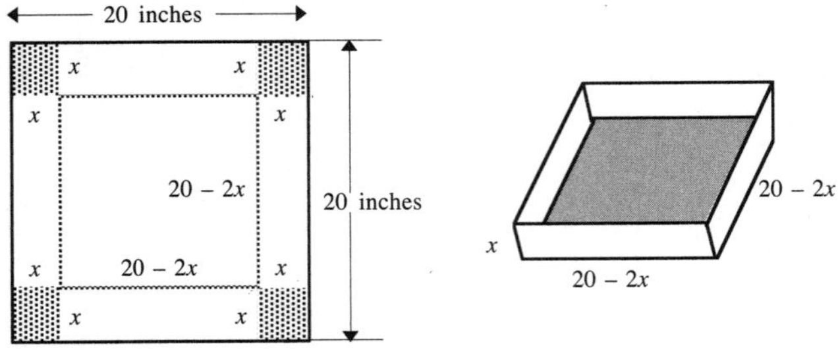

The volume of the box = $\ell wh = x(20 - 2x)(20 - 2x)$

$= 4x(10 - x)(10 - x)$ in factored form

In expanded form, $4x(10 - x)(10 - x) = 4x[10^2 - 2(10)x + x^2]$

$= 4x[100 - 20x + x^2]$

$= 400x - 80x^2 + 4x^3$

Section R-6

1. $\dfrac{42}{105} = \dfrac{21 \cdot 2}{21 \cdot 5} = \dfrac{2}{5}$

3. $\dfrac{x + 1}{x^2 + 3x + 2} = \dfrac{\overset{1}{\cancel{x + 1}}}{\underset{1}{\cancel{(x + 1)}}(x + 2)} = \dfrac{1}{x + 2}$

5. $\dfrac{x^2 - 9}{x^2 + 3x - 18} = \dfrac{\overset{1}{\cancel{(x - 3)}}(x + 3)}{\underset{1}{\cancel{(x - 3)}}(x + 6)} = \dfrac{x + 3}{x + 6}$

7. $\dfrac{3x^2y^3}{x^4y} = \dfrac{x^2y \cdot 3y^2}{x^2y \cdot x^2} = \dfrac{3y^2}{x^2}$

9. $\dfrac{5}{6} - \dfrac{11}{15} = \dfrac{25}{30} - \dfrac{22}{30} = \dfrac{3}{30} = \dfrac{1}{10}$

11. $\left(\dfrac{b^2}{2a} \div \dfrac{b}{a^2}\right) \cdot \dfrac{a}{3b} = \left(\dfrac{b^2}{2a} \cdot \dfrac{a^2}{b}\right) \cdot \dfrac{a}{3b} = \left(\dfrac{\overset{b}{\cancel{b^2}}}{2\underset{1}{\cancel{a}}} \cdot \dfrac{\overset{a}{\cancel{a^2}}}{\underset{1}{\cancel{b}}}\right) \cdot \dfrac{a}{3b}$

$= \dfrac{ba}{2} \cdot \dfrac{a}{3b}$

$= \dfrac{a^2}{6}$

13. $\dfrac{1}{n} - \dfrac{1}{m} = \dfrac{m}{mn} - \dfrac{n}{mn} = \dfrac{m - n}{mn}$

15. $$\frac{x^2 - 1}{x + 2} \div \frac{x + 1}{x^2 - 4} = \frac{x^2 - 1}{x + 2} \cdot \frac{x^2 - 4}{x + 1}$$
$$= \frac{(x - 1)(x + 1)}{x + 2} \cdot \frac{(x - 2)(x + 2)}{x + 1}$$
$$= \frac{(x - 1)\overset{1}{\cancel{(x + 1)}}}{\underset{1}{\cancel{x + 2}}} \cdot \frac{(x - 2)\overset{1}{\cancel{(x + 2)}}}{\underset{1}{\cancel{x + 1}}}$$
$$= (x - 1)(x - 2)$$

17. $$\frac{1}{c} + \frac{1}{b} + \frac{1}{a} = \frac{ab}{abc} + \frac{ac}{abc} + \frac{bc}{abc} = \frac{ab + ac + bc}{abc}$$

19. $$\frac{2a - b}{a^2 - b^2} - \frac{2a + 3b}{a^2 + 2ab + b^2} = \frac{2a - b}{(a + b)(a - b)} - \frac{2a + 3b}{(a + b)(a + b)}$$
$$= \frac{(2a - b)(a + b)}{(a + b)(a + b)(a - b)} - \frac{(2a + 3b)(a - b)}{(a + b)(a + b)(a - b)}$$
$$= \frac{(2a - b)(a + b) - (2a + 3b)(a - b)}{(a + b)^2(a - b)}$$
$$= \frac{2a^2 + ab - b^2 - (2a^2 + ab - 3b^2)}{(a + b)^2(a - b)}$$
$$= \frac{2a^2 + ab - b^2 - 2a^2 - ab + 3b^2}{(a + b)^2(a - b)}$$
$$= \frac{2b^2}{(a + b)^2(a - b)}$$

21. $$m + 2 - \frac{m - 2}{m - 1} = \frac{m + 2}{1} - \frac{m - 2}{m - 1} = \frac{(m + 2)(m - 1)}{m - 1} - \frac{(m - 2)}{m - 1}$$
$$= \frac{(m^2 + m - 2) - (m - 2)}{m - 1} = \frac{m^2 + m - 2 - m + 2}{m - 1}$$
$$= \frac{m^2}{m - 1}$$

23. $$\frac{3}{x - 2} - \frac{2}{2 - x} = \frac{3}{x - 2} - \frac{-2}{x - 2} = \frac{3 + 2}{x - 2} = \frac{5}{x - 2}$$

25. $$\frac{3}{y + 2} + \frac{2}{y - 2} - \frac{4y}{y^2 - 4} = \frac{3(y - 2)}{(y + 2)(y - 2)} + \frac{2(y + 2)}{(y + 2)(y - 2)} - \frac{4y}{(y + 2)(y - 2)}$$
$$= \frac{3(y - 2) + 2(y + 2) - 4y}{(y + 2)(y - 2)} = \frac{3y - 6 + 2y + 4 - 4y}{(y + 2)(y - 2)}$$
$$= \frac{y - 2}{(y + 2)(y - 2)} = \frac{\overset{1}{\cancel{(y - 2)}}}{(y + 2)\underset{1}{\cancel{(y - 2)}}} = \frac{1}{y + 2}$$

27. $$\frac{\frac{x^2}{y^2} - 1}{\frac{x}{y} + 1} = \frac{y^2\left(\frac{x^2}{y^2} - 1\right)}{y^2\left(\frac{x}{y} + 1\right)} = \frac{x^2 - y^2}{xy + y^2} = \frac{(x - y)\overset{1}{\cancel{(x + y)}}}{y\underset{1}{\cancel{(x + y)}}} = \frac{x - y}{y}$$

29. $\dfrac{6x^3(x^2+2)^2 - 2x(x^2+2)^3}{x^4} = \dfrac{2x(x^2+2)^2[3x^2 - (x^2+2)]}{x^4}$

$= \dfrac{2x(x^2+2)^2[3x^2 - x^2 - 2]}{x^4}$

$= \dfrac{2\overset{1}{\cancel{x}}(x^2+2)^2(2x^2-2)}{\underset{x^3}{\cancel{x^4}}}$

$= \dfrac{2(x^2+2)^2(2x^2-2)}{x^3}$

$= \dfrac{2(x^2+2)^2\,2(x^2-1)}{x^3}$

$= \dfrac{4(x^2+2)^2(x+1)(x-1)}{x^3}$

31. $\dfrac{2x(1-3x)^3 + 9x^2(1-3x)^2}{(1-3x)^6} = \dfrac{x(1-3x)^2[2(1-3x) + 9x]}{(1-3x)^6}$

$= \dfrac{x(1-3x)^2[2 - 6x + 9x]}{(1-3x)^6}$

$= \dfrac{x\overset{1}{\cancel{(1-3x)^2}}(2+3x)}{\underset{(1-3x)^4}{\cancel{(1-3x)^6}}}$

$= \dfrac{x(2+3x)}{(1-3x)^4}$

33. $\dfrac{-2x(x+4)^3 - 3(3-x^2)(x+4)^2}{(x+4)^6} = \dfrac{(x+4)^2[-2x(x+4) - 3(3-x^2)]}{(x+4)^6}$

$= \dfrac{(x+4)^2[-2x^2 - 8x - 9 + 3x^2]}{(x+4)^6}$

$= \dfrac{\overset{1}{\cancel{(x+4)^2}}(x^2 - 8x - 9)}{\underset{(x+4)^4}{\cancel{(x+4)^6}}}$

$= \dfrac{x^2 - 8x - 9}{(x+4)^4}$

$= \dfrac{(x+1)(x-9)}{(x+4)^4}$

35. $\dfrac{y}{y^2 - 2y - 8} - \dfrac{2}{y^2 - 5y + 4} + \dfrac{1}{y^2 + y - 2}$

$= \dfrac{y}{(y-4)(y+2)} - \dfrac{2}{(y-4)(y-1)} + \dfrac{1}{(y+2)(y-1)}$

$= \dfrac{y(y-1)}{(y-1)(y-4)(y+2)} - \dfrac{2(y+2)}{(y-4)(y-1)(y+2)} + \dfrac{1(y-4)}{(y-4)(y-1)(y+2)}$

$$= \frac{y^2 - y - 2y - 4 + y - 4}{(y - 4)(y - 1)(y + 2)} = \frac{y^2 - 2y - 8}{(y - 4)(y - 1)(y + 2)} = \frac{\cancel{(y - 4)}\cancel{(y + 2)}}{\cancel{(y - 4)}(y - 1)\cancel{(y + 2)}}$$

$$= \frac{1}{y - 1}$$

37. $$\frac{16 - m^2}{m^2 + 3m - 4} \cdot \frac{m - 1}{m - 4} = \frac{\cancel{(4 - m)}\cancel{(4 + m)}}{\cancel{(m + 4)}\cancel{(m - 1)}} \cdot \frac{\cancel{m - 1}}{\cancel{m - 4}} = \frac{-1}{1} = -1$$

39. $$\frac{x + 7}{ax - bx} + \frac{y + 9}{by - ay} = \frac{x + 7}{x(a - b)} + \frac{y + 9}{y(b - a)} = \frac{y(x + 7)}{xy(a - b)} + \frac{-x(y + 9)}{xy(a - b)}$$

$$= \frac{xy + 7y - xy - 9x}{xy(a - b)} = \frac{7y - 9x}{xy(a - b)}$$

41. $$\frac{x^2 - 16}{2x^2 + 10x + 8} \div \frac{x^2 - 13x + 36}{x^3 + 1} = \frac{x^2 - 16}{2x^2 + 10x + 8} \cdot \frac{x^3 + 1}{x^2 - 13x + 36}$$

$$= \frac{\cancel{(x - 4)}\cancel{(x + 4)}}{2\cancel{(x + 1)}\cancel{(x + 4)}} \cdot \frac{\cancel{(x + 1)}(x^2 - x + 1)}{\cancel{(x - 4)}(x - 9)} = \frac{x^2 - x + 1}{2(x - 9)}$$

43. $$\frac{x^2 - xy}{xy + y^2} \div \left(\frac{x^2 - y^2}{x^2 + 2xy + y^2} \div \frac{x^2 - 2xy + y^2}{x^2y + xy^2}\right)$$

$$= \frac{x^2 - xy}{xy + y^2} \div \left(\frac{x^2 - y^2}{x^2 + 2xy + y^2} \cdot \frac{x^2y + xy^2}{x^2 - 2xy + y^2}\right)$$

$$= \frac{x^2 - xy}{xy + y^2} \div \left(\frac{\cancel{(x - y)}\cancel{(x + y)}}{\cancel{(x + y)}\cancel{(x + y)}} \cdot \frac{xy\cancel{(x + y)}}{\cancel{(x - y)}(x - y)}\right)$$

$$= \frac{x^2 - xy}{xy + y^2} \div \frac{xy}{x - y} = \frac{x^2 - xy}{xy + y^2} \cdot \frac{x - y}{xy}$$

$$= \frac{\cancel{x}(x - y)}{y(x + y)} \cdot \frac{x - y}{\cancel{x}y} = \frac{(x - y)^2}{y^2(x + y)}$$

45. $$\left(\frac{x}{x^2 - 16} - \frac{1}{x + 4}\right) \div \frac{4}{x + 4} = \left(\frac{x}{(x - 4)(x + 4)} - \frac{1}{x + 4}\right) \div \frac{4}{x + 4}$$

$$= \left(\frac{x}{(x - 4)(x + 4)} - \frac{(x - 4)}{(x - 4)(x + 4)}\right) \div \frac{4}{x + 4}$$

$$= \frac{x - x + 4}{(x - 4)(x + 4)} \div \frac{4}{x + 4} = \frac{4}{(x - 4)(x + 4)} \div \frac{4}{x + 4}$$

$$= \frac{\cancel{4}}{(x - 4)\cancel{(x + 4)}} \cdot \frac{\cancel{x + 4}}{\cancel{4}} = \frac{1}{x - 4}$$

47. $$\frac{1 + \frac{2}{x} - \frac{15}{x^2}}{1 + \frac{4}{x} - \frac{5}{x^2}} = \frac{x^2\left(1 + \frac{2}{x} - \frac{15}{x^2}\right)}{x^2\left(1 + \frac{4}{x} - \frac{5}{x^2}\right)} = \frac{x^2 + 2x - 15}{x^2 + 4x - 5} = \frac{\overset{1}{\cancel{(x + 5)}}(x - 3)}{\underset{1}{\cancel{(x + 5)}}(x - 1)} = \frac{x - 3}{x - 1}$$

49. $$\frac{\frac{1}{x + h} - \frac{1}{x}}{h} = \frac{\frac{x}{x(x + h)} - \frac{x + h}{x(x + h)}}{h} = \frac{\frac{x - x - h}{x(x + h)}}{h} = \frac{\frac{-h}{x(x + h)}}{h} = \frac{-h}{x(x + h)} \div h$$

$$= \frac{-h}{x(x + h)} \div \frac{h}{1} = \frac{-\overset{-1}{\cancel{h}}}{x(x + h)} \cdot \frac{1}{\underset{1}{\cancel{h}}} = \frac{-1}{x(x + h)}$$

51. $$\frac{\frac{(x + h)^2}{x + h + 2} - \frac{x^2}{x + 2}}{h} = \frac{\frac{(x + h)^2(x + 2)}{(x + h + 2)(x + 2)} - \frac{x^2(x + h + 2)}{(x + h + 2)(x + 2)}}{h} = \frac{\frac{(x + h)^2(x + 2) - x^2(x + h + 2)}{(x + h + 2)(x + 2)}}{h}$$

$$= \frac{\frac{(x^2 + 2xh + h^2)(x + 2) - x^3 - x^2h - 2x^2}{(x + h + 2)(x + 2)}}{h}$$

$$= \frac{\frac{x^3 + 2x^2h + h^2x + 2x^2 + 4xh + 2h^2 - x^3 - x^2h - 2x^2}{(x + h + 2)(x + 2)}}{h}$$

$$= \frac{\frac{x^2h + h^2x + 4xh + 2h^2}{(x + h + 2)(x + 2)}}{h} = \frac{x^2h + h^2x + 4xh + 2h^2}{(x + h + 2)(x + 2)} \div h$$

$$= \frac{x^2h + h^2x + 4xh + 2h^2}{(x + h + 2)(x + 2)} \div \frac{h}{1} = \frac{\overset{1}{\cancel{h}}(x^2 + hx + 4x + 2h)}{(x + h + 2)(x + 2)} \cdot \frac{1}{\underset{1}{\cancel{h}}}$$

$$= \frac{x^2 + hx + 4x + 2h}{(x + h + 2)(x + 2)}$$

53. (A) The solution is incorrect, because in the first step the quantity 4 has been removed from numerator and denominator. But 4 is a term, and only factors can be cancelled. A correct solution would factor numerator and denominator, then cancel common factors, if any.

(B) $$\frac{x^2 + 5x + 4}{x + 4} = \frac{(x + 1)\overset{1}{\cancel{(x + 4)}}}{\underset{1}{\cancel{x + 4}}} = x + 1$$

55. (A) The solution is incorrect, because in the first step the quantity h has been removed from numerator and denominator. Although h is a factor of the denominator it is not a factor of the numerator as shown and only common factors of numerator and denominator can be cancelled. A correct solution would factor numerator and denominator, revealing that h is actually a factor of the numerator in factored form and allowing a correct cancellation of h.

(B) $$\frac{(x + h)^2 - x^2}{h} = \frac{[(x + h) - x][(x + h) + x]}{h}$$ using the difference of two squares

$$= \frac{[x + h - x][x + h + x]}{h}$$

$$= \frac{\overset{1}{\cancel{h}}[2x + h]}{\underset{1}{\cancel{h}}} = 2x + h$$

57. (A) The solution is incorrect. In adding a quantity to a fractional expression, the quantity cannot simply be placed in the numerator. A correct solution would rewrite $x - 2$ as a fractional expression, first with denominator 1, and then with denominator equal to the LCD of all denominators in the addition.

(B) $$\frac{x^2 - 2x}{x^2 - x - 2} + x - 2 = \frac{x\cancel{(x-2)}^{1}}{(x+1)\cancel{(x-2)}_{1}} + \frac{x-2}{1}$$
$$= \frac{x}{x+1} + \frac{x-2}{1}$$
$$= \frac{x}{x+1} + \frac{(x-2)(x+1)}{x+1}$$
$$= \frac{x + x^2 - x - 2}{x+1}$$
$$= \frac{x^2 - 2}{x - 1}$$

59. (A)(B) The solution is correct.

61. $$\frac{y - \frac{y^2}{y-x}}{1 + \frac{x^2}{y^2 - x^2}} = \frac{(y^2 - x^2)[y - \frac{y^2}{y-x}]}{(y^2 - x^2)[1 + \frac{x^2}{y^2 - x^2}]} = \frac{(y^2 - x^2)y - \cancel{(y^2 - x^2)}^{(y+x)}\frac{y^2}{\cancel{y-x}_{1}}}{y^2 - x^2 + \cancel{(y^2 - x^2)}_{1}\frac{x^2}{\cancel{y^2 - x^2}_{1}}}$$
$$= \frac{y^3 - x^2y - y^3 - xy^2}{y^2 - x^2 + x^2} = \frac{-x^2y - xy^2}{y^2} = \frac{-x\cancel{y}^{1}(x+y)}{\cancel{y^2}_{y}} = \frac{-x(x+y)}{y}$$

63. $$2 - \frac{1}{1 - \frac{2}{a+2}} = 2 - \frac{(a+2)\cdot 1}{(a+2)[1 - \frac{2}{a+2}]} = 2 - \frac{a+2}{a + 2 - \cancel{(a+2)}_{1}\frac{2}{\cancel{a+2}_{1}}} = 2 - \frac{a+2}{a+2-2}$$
$$= 2 - \frac{a+2}{a} = \frac{2}{1} - \frac{a+2}{a} = \frac{2a}{a} - \frac{(a+2)}{a} = \frac{2a - a - 2}{a} = \frac{a-2}{a}$$

65. (A) The multiplicative inverse of x is the unique real number $\frac{1}{x}$ such that $x(\frac{1}{x}) = 1$. Since $\frac{c}{d} \cdot \frac{d}{c} = \frac{cd}{dc} = \frac{cd}{cd} = 1$, $\frac{d}{c}$ is the multiplicative inverse of $\frac{c}{d}$.

(B) By Definition 1 of Section R-1 of the text $a \div b = a(\frac{1}{b}) =$ $a \cdot$ (multiplicative inverse of b).

Hence $\frac{a}{b} \div \frac{c}{d} = \frac{a}{b} \cdot$ (multiplicative inverse of $\frac{c}{d}$)

$= \frac{a}{b} \cdot \frac{d}{c}$ by part A.

CHAPTER R REVIEW

1. (A) Since 3 is an element of {1, 2, 3, 4, 5}, the statement is true. T.

(B) Since 5 is not an element of {4, 1, 2}, the statement is true. T.

(C) Since B is not an element of {1, 2, 3, 4, 5}, the statement is false. F.

(D) Since each element of the set {1, 2, 4} is also an element of the set {1, 2, 3, 4, 5}, the statement is true. T.

(D) Since sets B and C have exactly the same elements, the sets are equal. The statement is false. F.

(E) The statement is false, since not every element of the set {1, 2, 3, 4, 5} is an element of the set {1, 2, 4}. For example, 3 is an element of the first, but not the second. F. (*R*-1)

2. (A) The commutative property (·) states that, in general, $xy = yx$.
Comparing this with
$x(y + z) = ?$
we see that $x(y + z) = (y + z)x$
$(y + z)x$

(B) The associative property (+) states that, in general,
$(x + y) + z = x + (y + z)$
Comparing this with
$? = 2 + (x + y)$
we see that
$(2 + x) + y = 2 + (x + y)$
$(2 + x) + y$

(C) The distributive property states that, in general,
$(x + y)z = xz + yz$
Comparing this with
$(2 + 3)x = ?$
we see that
$(2 + 3)x = 2x + 3x$
$2x + 3x$ (*R-1*)

3. $\left(\frac{2}{3} \cdot \frac{15}{14}\right)^{-1} = \left(\frac{5}{7}\right)^{-1} = \frac{7}{5}$ (*R-1*)

4. $\frac{9}{10} + \frac{5}{12} = \frac{54}{60} + \frac{25}{60} = \frac{79}{60}$ (*R-1*)

5. $-\left(\frac{8}{7} + 2^{-1}\right) = -\left(\frac{8}{7} + \frac{1}{2}\right) = -\left(\frac{16}{14} + \frac{7}{14}\right) = -\left(\frac{23}{14}\right) = -\frac{23}{14}$ (*R-1*)

6. $(4^{-1}9^{-1})^{-1} = \left(\frac{1}{4} \cdot \frac{1}{9}\right)^{-1} = \left(\frac{1}{36}\right)^{-1} = 36$ (*R-1*)

7. 3 (*R-4*)

8. 5 (*R-4*)

9. $(x^3 + 2) + (x^5 - x^3 + 1) = x^3 + 2 + x^5 - x^3 + 1 = x^5 + 3$. This polynomial has degree 5. (*R-4*)

10. $(x^3 + 2)(x^5 - x^3 + 1) = x^8 + 2x^5 - x^6 - 2x^3 + x^3 + 2 = x^8 - x^6 + 2x^5 - x^3 + 2$. This polynomial has degree 8. (*R-4*)

11. See problem 10. $x^8 - x^6 + 2x^5 - x^3 + 2$. (*R-4*)

12. See problem 9. $x^5 + 3$ (*R-4*)

13. 17 (*R-2*) **14.** 6 (*R-2*) **15.** $8^{-2/3} = \frac{1}{8^{2/3}} = \frac{1}{(8^{1/3})^2} = \frac{1}{2^2} = \frac{1}{4}$ (*R-2*)

16. $(-64)^{5/3} = [(-64)^{1/3}]^5 = (-4)^5 = -1024$ (*R-2*)

17. $\left(\frac{9}{16}\right)^{-1/2} = 1 \div \left(\frac{9}{16}\right)^{1/2} = 1 \div \frac{3}{4} = 1 \cdot \frac{4}{3} = \frac{4}{3}$ (*R-2*)

18. $(121^{1/2} + 25^{1/2})^{-3/4} = (11 + 5)^{-3/4} = 16^{-3/4} = \dfrac{1}{16^{3/4}} = \dfrac{1}{(16^{1/4})^3} = \dfrac{1}{2^3} = \dfrac{1}{8}$ (*R*-2)

19. $5x^2 - 3x[4 - 3(x - 2)] = 5x^2 - 3x[4 - 3x + 6]$
$= 5x^2 - 3x[10 - 3x]$
$= 5x^2 - 30x + 9x^2$
$= 14x^2 - 30x$ (*R*-4)

20. $9m^2 - 25n^2$ (*R*-4)

21. $(2x + y)(3x - 4y) = 6x^2 - 8xy + 3xy - 4y^2 = 6x^2 - 5xy - 4y^2$ (*R*-4)

22. $(2a - 3b)^2 = (2a)^2 - 2(2a)(3b) + (3b)^2 = 4a^2 - 12ab + 9b^2$ (*R*-4)

23. $(3x - 2)^2$ (*R*-5)

24. Prime (*R*-5)

25. $6n^3 - 9n^2 - 15n = 3n(2n^2 - 3n - 5) = 3n(2n - 5)(n + 1)$ (*R*-5)

26. $\dfrac{2}{5b} - \dfrac{4}{3a^3} - \dfrac{1}{6a^2b^2} = \dfrac{12a^3b}{30a^3b^2} - \dfrac{40b^2}{30a^3b^2} - \dfrac{5a}{30a^3b^2} = \dfrac{12a^3b - 40b^2 - 5a}{30a^3b^2}$ (*R*-6)

27. $\dfrac{3x}{3x^2 - 12x} + \dfrac{1}{6x} = \dfrac{3x}{3x(x - 4)} + \dfrac{1}{6x} = \dfrac{2(3x)}{6x(x - 4)} + \dfrac{1(x - 4)}{6x(x - 4)} = \dfrac{6x + x - 4}{6x(x - 4)} = \dfrac{7x - 4}{6x(x - 4)}$ (*R*-6)

28. $\dfrac{y - 2}{y^2 - 4y + 4} \div \dfrac{y^2 + 27}{y^2 + 4y + 4} = \dfrac{y - 2}{y^2 - 4y + 4} \cdot \dfrac{y^2 + 4y + 4}{y^2 + 2y}$

$= \dfrac{\overset{1}{\cancel{(y - 2)}}}{\underset{1}{\cancel{(y - 2)}}(y - 2)} \cdot \dfrac{\overset{1}{\cancel{(y + 2)}}(y + 2)}{y\underset{1}{\cancel{(y + 2)}}}$

$= \dfrac{y + 2}{y(y - 2)}$ (*R*-6)

29. $\dfrac{u - \frac{1}{u}}{1 - \frac{1}{u^2}} = \dfrac{u^2\left(u - \frac{1}{u}\right)}{u^2\left(1 - \frac{1}{u^2}\right)} = \dfrac{u^3 - u}{u^2 - 1} = \dfrac{u\overset{1}{\cancel{(u^2 - 1)}}}{\underset{1}{\cancel{u^2 - 1}}} = u$ (*R*-6)

30. $6(xy^3)^5 = 6x^5(y^3)^5 = 6x^5y^{15}$ (*R*-2)

31. $\dfrac{9u^8v^6}{3u^4v^8} = \dfrac{3u^4}{v^2}$ (*R*-2)

32. $(2 \times 10^5)(3 \times 10^{-3}) = (2 \cdot 3) \times 10^{5+(-3)} = 6 \times 10^2$ (*R*-2)

33. $(x^{-3}y^2)^{-2} = (x^{-3})^{-2}(y^2)^{-2} = x^6y^{-4} = \dfrac{x^6}{y^4}$ (*R*-2)

34. $u^{5/3}u^{2/3} = u^{5/3+2/3} = u^{7/3}$ (*R*-2)

35. $(9a^4b^{-2})^{1/2} = 9^{1/2}(a^4)^{1/2}(b^{-2})^{1/2} = 3a^2b^{-1} = \dfrac{3a^2}{b}$ (*R*-3)

36. $3\sqrt[5]{x^2}$ (*R*-3)

37. $-3(xy)^{2/3}$ (*R*-3)

38. $3x\sqrt[3]{x^5y^4} = 3x\sqrt[3]{x^3y^3 \cdot x^2y} = 3x\sqrt[3]{x^3y^3}\sqrt[3]{x^2y} = 3x(xy)\sqrt[3]{x^2y} = 3x^2y\sqrt[3]{x^2y}$ (*R-3*)

39. $\sqrt{2x^2y^5}\sqrt{18x^3y^2} = \sqrt{(2x^2y^5)(18x^3y^2)} = \sqrt{36x^5y^7} = \sqrt{36x^4y^6 \cdot xy} = \sqrt{36x^4y^6}\sqrt{xy}$
$= 6x^2y^3\sqrt{xy}$ (*R-3*)

40. $\frac{6ab}{\sqrt{3a}} = \frac{6ab}{\sqrt{3a}}\frac{\sqrt{3a}}{\sqrt{3a}} = \frac{\overset{2}{\cancel{6a}}b\sqrt{3a}}{\underset{1}{\cancel{3a}}} = 2b\sqrt{3a}$ (*R-3*)

41. $\frac{\sqrt{5}}{3-\sqrt{5}} = \frac{\sqrt{5}}{(3-\sqrt{5})}\frac{(3+\sqrt{5})}{(3+\sqrt{5})} = \frac{\sqrt{5}(3+\sqrt{5})}{3^2-(\sqrt{5})^2} = \frac{\sqrt{5}(3+\sqrt{5})}{9-5} = \frac{\sqrt{5}(3+\sqrt{5})}{4} = \frac{3\sqrt{5}+5}{4}$ (*R-3*)

42. $\sqrt[8]{y^6} = \sqrt[2\cdot 4]{y^{2\cdot 3}} = \sqrt[4]{y^3}$ (*R-3*)

43. The odd integers between -4 and 2 are -3, -1, and 1. Hence the set is written {-3, -1, 1} (*R-1*)

44. Subtraction (*R-1*) **45.** Commutative (+) (*R-1*) **46.** Distributive (*R-1*)

47. Associative (·) (*R-1*) **48.** Negatives (*R-1*) **49.** Identity (+) (*R-1*)

50. (A) T (B) F (*R-1*) **51.** 0 and −3 are two examples of infinitely many. (*R-1*)

52. (A) *a* and *d* (B) None (*R-4*)

53. $(2x - y)(2x + y) - (2x - y)^2 = 4x^2 - y^2 - (4x^2 - 4xy + y^2)$
$= 4x^2 - y^2 - 4x^2 + 4xy - y^2$
$= 4xy - 2y^2$ (*R-4*)

54. This can be simplified directly using the distributive property or:
$(m^2 + 2mn - n^2)(m^2 - 2mn - n^2) = [(m^2 - n^2) + 2mn][(m^2 - n^2) - 2mn]$
$= (m^2 - n^2)^2 - (2mn)^2$
$= m^4 - 2m^2n^2 + n^4 - 4m^2n^2$
$= m^4 - 6m^2n^2 + n^4$ (*R-4*)

55. $5(x + h)^2 - 7(x + h) - (5x^2 - 7x) = 5(x^2 + 2hx + h^2) - 7(x + h) - (5x^2 - 7x)$
$= 5x^2 + 10hx + 5h^2 - 7x - 7h - 5x^2 + 7x$
$= 10hx + 5h^2 - 7h$ (*R-4*)

56. $-2x\{(x^2 + 2)(x - 3) - x[x - x(3 - x)]\}$
$= -2x\{(x^2 + 2)(x - 3) - x[x - 3x + x^2]\}$
$= -2x\{(x^2 + 2)(x - 3) - x[-2x + x^2]\}$
$= -2x\{x^3 - 3x^2 + 2x - 6 + 2x^2 - x^3\}$
$= -2x\{-x^2 + 2x - 6\}$
$= 2x^3 - 4x^2 + 12x$ (*R-4*)

57. $(x - 2y)^3 = (x - 2y)^2(x - 2y)$
$= (x^2 - 4xy + 4y^2)(x - 2y)$
$= x^3 - 2x^2y - 4x^2y + 8xy^2 + 4xy^2 - 8y^3$
$= x^3 - 6x^2y + 12xy^2 - 8y^3$ (*R-4*)

58. $(4x - y)^2 - 9x^2 = (4x - y)^2 - (3x)^2$
$= [(4x - y) - 3x][(4x - y) + 3x]$
$= (x - y)(7x - y)$ (*R*-5)

59. Prime (*R*-5)

60. $3xy(2x^2 + 4xy - 5y^2)$ (*R*-5)

61. $(y - b)^2 - y + b = (y - b)(y - b) - 1(y - b) = (y - b)(y - b - 1)$ (*R*-5)

62. $3x^3 + 24y^3 = 3(x^3 + 8y^3) = 3[x^3 + (2y)^3] = 3(x + 2y)[x^2 - x(2y) + (2y)^2]$
$= 3(x + 2y)(x^2 - 2xy + 4y^2)$ (*R*-5)

63. $y^3 + 2y^2 - 4y - 8 = y^2(y + 2) - 4(y + 2)$
$= (y + 2)(y^2 - 4) = (y + 2)(y + 2)(y - 2)$
$= (y - 2)(y + 2)^2$ (*R*-5)

64. $2x(x - 4)^3 + 3x^2(x - 4)^2 = x(x - 4)^2[2(x - 4) + 3x]$
$= x(x - 4)^2[2x - 8 + 3x]$
$= x(x - 4)^2(5x - 8)$ (*R*-5)

65. $$\frac{3x^2(x + 2)^2 - 2x(x + 2)^3}{x^4} = \frac{x(x + 2)^2[3x - 2(x + 2)]}{x^4}$$
$$= \frac{\overset{1}{\cancel{x}}(x + 2)^2[3x - 2x - 4]}{\underset{x^3}{\cancel{x^4}}}$$
$$= \frac{(x + 2)^2(x - 4)}{x^3}$$ (*R*-6)

66. $$\frac{m - 1}{m^2 - 4m + 4} + \frac{m + 3}{m^2 - 4} + \frac{2}{2 - m}$$
$$= \frac{(m - 1)}{(m - 2)(m - 2)} + \frac{m + 3}{(m - 2)(m + 2)} + \frac{-2}{m - 2}$$
$$= \frac{(m - 1)(m + 2)}{(m - 2)(m - 2)(m + 2)} + \frac{(m + 3)(m - 2)}{(m - 2)(m - 2)(m + 2)} + \frac{-2(m - 2)(m + 2)}{(m - 2)(m - 2)(m + 2)}$$
$$= \frac{(m - 1)(m + 2) + (m + 3)(m - 2) - 2(m - 2)(m + 2)}{(m - 2)^2(m + 2)}$$
$$= \frac{m^2 + m - 2 + m^2 + m - 6 - 2(m^2 - 4)}{(m - 2)^2(m + 2)}$$
$$= \frac{2m^2 + 2m - 8 - 2m^2 + 8}{(m - 2)^2(m + 2)} = \frac{2m}{(m - 2)^2(m + 2)}$$ (*R*-6)

67. $$\frac{y}{x^2} \div \left(\frac{x^2+3x}{2x^2+5x-3} \div \frac{x^3y-x^2y}{2x^2-3x+1}\right) = \frac{y}{x^2} \div \left(\frac{x^2+3x}{2x^2+5x-3} \cdot \frac{2x^2-3x+1}{x^3y-x^2y}\right)$$

$$= \frac{y}{x^2} \div \left(\frac{\cancel{x}\cancel{(x+3)}}{\cancel{(2x-1)}\cancel{(x+3)}} \cdot \frac{\cancel{(2x-1)}\cancel{(x-1)}}{\cancel{x}^2y\cancel{(x-1)}}\right)$$

$$= \frac{y}{x^2} \div \frac{1}{xy}$$

$$= \frac{y}{\cancel{x}^2} \cdot \frac{\cancel{x}y}{1}$$

$$= \frac{y^2}{x} \qquad (R\text{-}6)$$

68. $$\frac{1-\frac{1}{1+\frac{x}{y}}}{1-\frac{1}{1-\frac{x}{y}}} = \frac{1-\frac{y(1)}{y\left(1+\frac{x}{y}\right)}}{1-\frac{y(1)}{y\left(1-\frac{x}{y}\right)}} = \frac{1-\frac{y}{y+x}}{1-\frac{y}{y-x}} = \frac{\frac{y+x}{y+x}-\frac{y}{y+x}}{\frac{y-x}{y-x}-\frac{y}{y-x}}$$

$$= \frac{\frac{x}{y+x}}{\frac{-x}{y-x}} = \frac{x}{y+x} \div \frac{-x}{y-x} = \frac{\cancel{x}}{y+x} \cdot \frac{y-x}{-\cancel{x}} = \frac{-1(y-x)}{y+x} = \frac{x-y}{x+y} \qquad (R\text{-}6)$$

69. $$\frac{a^{-1}-b^{-1}}{ab^{-2}-ba^{-2}} = \frac{\frac{1}{a}-\frac{1}{b}}{\frac{a}{b^2}-\frac{b}{a^2}} = \frac{a^2b^2\left(\frac{1}{a}-\frac{1}{b}\right)}{a^2b^2\left(\frac{a}{b^2}-\frac{b}{a^2}\right)}$$

$$= \frac{ab^2-a^2b}{a^3-b^3} = \frac{ab\cancel{(b-a)}}{\cancel{(a-b)}(a^2+ab+b^2)} = \frac{-ab}{a^2+ab+b^2} \qquad (R\text{-}6,\ R\text{-}2)$$

70. The solution is incorrect. In adding a quantity to a fractional expression, the quantity cannot simply be placed in the numerator. A correct solution would rewrite $x + 2$ as a fractional expression, first with denominator 1, and then with denominator equal to the LCD of all denominators in the expression. Thus,

$$\frac{x^2+2x}{x^2+x-2} + x + 2 = \frac{x(x+2)}{(x-1)(x+2)} + \frac{x+2}{1}$$

$$= \frac{x}{x-1} + \frac{x+2}{1}$$

$$= \frac{x}{x-1} + \frac{(x+2)(x-1)}{x-1}$$

$$= \frac{x+x^2+x-2}{x-1}$$

$$= \frac{x^2+2x-2}{x-1} \qquad (R\text{-}6)$$

71. $$\left(\frac{8u^{-1}}{2^2u^2v^0}\right)^{-2}\left(\frac{u^{-5}}{u^{-3}}\right)^3 = \left(\frac{2^3u^{-1}}{2^2u^2}\right)^{-2}\left(\frac{u^{-5}}{u^{-3}}\right)^3 = (2u^{-3})^{-2}\left(\frac{u^{-5}}{u^{-3}}\right)^3 = 2^{-2}u^6\frac{u^{-15}}{u^{-9}}$$

$$= \frac{1}{2^2}u^6u^{-15-(-9)} = \frac{1}{2^2}u^6u^{-6} = \frac{1}{4} \qquad (R\text{-}2)$$

72. $\dfrac{5^0}{3^2} + \dfrac{3^{-2}}{2^{-2}} = \dfrac{1}{3^2} + \dfrac{2^2}{3^2} = \dfrac{1}{9} + \dfrac{4}{9} = \dfrac{5}{9}$ (R-2)

73. $\left(\dfrac{27x^2y^{-3}}{8x^{-4}y^3}\right)^{1/3} = \left(\dfrac{3^3x^{2-(-4)}y^{-3-3}}{2^3}\right)^{1/3} = \left(\dfrac{3^3x^6y^{-6}}{2^3}\right)^{1/3} = \dfrac{3x^2y^{-2}}{2} = \dfrac{3x^2}{2y^2}$ (R-2)

74. $(a^{-1/3}b^{1/4})(9a^{1/3}b^{-1/2})^{3/2} = a^{-1/3}b^{1/4}9^{3/2}a^{1/2}b^{-3/4}$

$= a^{-1/3+1/2}b^{1/4-3/4}9^{3/2} = 27a^{1/6}b^{-1/2}$

$= \dfrac{27a^{1/6}}{b^{1/2}}$ (R-2)

75. $(x^{1/2} + y^{1/2})^2 = (x^{1/2})^2 + 2(x^{1/2})(y^{1/2}) + (y^{1/2})^2$

$= x + 2x^{1/2}y^{1/2} + y$

Common Error:
$(x^{1/2} + y^{1/2})^2 \neq x + y$.
(Exponents do not distribute over addition.)

(R-2)

76. $(3x^{1/2} - y^{1/2})(2x^{1/2} + 3y^{1/2}) = (3x^{1/2})(2x^{1/2}) + (3x^{1/2})(3y^{1/2}) - (2x^{1/2})(y^{1/2}) - (y^{1/2})(3y^{1/2})$

$= 6x + 9x^{1/2}y^{1/2} - 2x^{1/2}y^{1/2} - 3y$

$= 6x + 7x^{1/2}y^{1/2} - 3y$ (R-2)

77. $\dfrac{0.000\,000\,000\,52}{(1,300)(0.000\,002)} = \dfrac{5.2 \times 10^{-10}}{(1.3 \times 10^3)(2 \times 10^{-6})}$

$= \dfrac{\overset{2}{\cancel{5.2}} \times 10^{-10}}{\underset{1}{\cancel{2.6}} \times 10^{-3}} = 2 \times 10^{-10-(3)}$

$= 2 \times 10^{-7}$ (R-2)

78. $\dfrac{(20,410)(0.000\,003\,477)}{0.000\,000\,022\,09} = \dfrac{(2.041 \times 10^4)(3.477 \times 10^{-6})}{(2.209 \times 10^{-8})} = 3.213 \times 10^6$

Note that the input is given to 4 significant digits, hence we interpret 4 significant digits of the calculator answer. (R-2)

79. 4.434×10^{-5} (R-2)

80. -4.541×10^{-6} (R-2)

81. $82.45^{8/3} = (82.45)^{(8\div3)} = 128,800$ (R-2)

82. $(0.000\ 000\ 419\ 9)^{2/7} = (4.199 \times 10^{-7})^{(2\div7)} = 0.01507$ (R-2)

83. $\sqrt[5]{0.006\,604} = (6.604 \times 10^{-3})^{1/5} = (6.604 \times 10^{-3})^{0.2} = 0.3664$ (R-3)

84. $\sqrt[3]{3 + \sqrt{2}} = (3 + \sqrt{2})^{(1\div3)} = 1.640$ (R-3)

85. $\dfrac{2^{-1/2} - 3^{-1/2}}{2^{-1/3} + 3^{-1/3}} = [2^{(-0.5)} - 3^{(-0.5)}] \div [2^{(-1\div3)} + 3^{(-1\div3)}] = 0.08726$ (R-2)

86. $-2x\sqrt[5]{3^6x^7y^{11}} = -2x\sqrt[5]{3^5x^5y^{10} \cdot 3x^2y} = -2x\sqrt[5]{3^5x^5y^{10}}\,\sqrt[5]{3x^2y}$

$= -2x \cdot 3xy^2\,\sqrt[5]{3x^2y} = -6x^2y^2\,\sqrt[5]{3x^2y}$ (R-3)

87. $\dfrac{2x^2}{\sqrt[3]{4x}} = \dfrac{2x^2}{\sqrt[3]{4x}}\dfrac{\sqrt[3]{2x^2}}{\sqrt[3]{2x^2}} = \dfrac{2x^2\sqrt[3]{2x^2}}{\sqrt[3]{8x^3}} = \dfrac{\overset{x}{\cancel{2x}}{}^2\sqrt[3]{2x^2}}{\underset{1}{\cancel{2x}}} = x\sqrt[3]{2x^2}$ (R-3)

88. $\sqrt[5]{\dfrac{3y^2}{8x^2}} = \sqrt[5]{\dfrac{3y^2}{8x^2}\,\dfrac{4x^3}{4x^3}} = \sqrt[5]{\dfrac{12x^3y^2}{32x^5}} = \dfrac{\sqrt[5]{12x^3y^2}}{\sqrt[5]{32x^5}} = \dfrac{\sqrt[5]{12x^3y^2}}{2x}$ (*R-3*)

89. $\sqrt[9]{8x^6y^{12}} = \sqrt[9]{2^3x^6y^{12}} = \sqrt[3\cdot3]{2^{3\cdot1}x^{3\cdot2}y^{3\cdot4}} = \sqrt[3]{2x^2y^4} = \sqrt[3]{y^3\cdot 2x^2y} = \sqrt[3]{y^3}\,\sqrt[3]{2x^2y} = y\sqrt[3]{2x^2y}$ (*R-3*)

90. $\sqrt{\sqrt[3]{4x^4}} = \sqrt[2\cdot3]{2^{2\cdot1}x^{2\cdot2}} = \sqrt[3]{2x^2}$ (*R-3*)

91. $(2\sqrt{x} - 5\sqrt{y})(\sqrt{x} + \sqrt{y}) = 2\sqrt{x}\sqrt{x} + 2\sqrt{x}\sqrt{y} - 5\sqrt{x}\sqrt{y} - 5\sqrt{y}\sqrt{y}$
$= 2x - 3\sqrt{x}\sqrt{y} - 5y = 2x - 3\sqrt{xy} - 5y$ (*R-3*)

92. $\dfrac{3\sqrt{x}}{2\sqrt{x} - \sqrt{y}} = \dfrac{3\sqrt{x}}{(2\sqrt{x} - \sqrt{y})}\,\dfrac{(2\sqrt{x} + \sqrt{y})}{(2\sqrt{x} + \sqrt{y})} = \dfrac{3\sqrt{x}(2\sqrt{x} + \sqrt{y})}{(2\sqrt{x})^2 - (\sqrt{y})^2}$
$= \dfrac{3\sqrt{x}(2\sqrt{x} + \sqrt{y})}{4x - y} = \dfrac{6x + 3\sqrt{xy}}{4x - y}$ (*R-3*)

93. $\dfrac{2\sqrt{u} - 3\sqrt{v}}{2\sqrt{u} + 3\sqrt{v}} = \dfrac{(2\sqrt{u} - 3\sqrt{v})}{(2\sqrt{u} + 3\sqrt{v})}\,\dfrac{(2\sqrt{u} - 3\sqrt{v})}{(2\sqrt{u} - 3\sqrt{v})} = \dfrac{(2\sqrt{u})^2 - 2(2\sqrt{u})(3\sqrt{v}) + (3\sqrt{v})^2}{(2\sqrt{u})^2 - (3\sqrt{v})^2} = \dfrac{4u - 12\sqrt{uv} + 9v}{4u - 9v}$ (*R-3*)

94. $\dfrac{y^2}{\sqrt{y^2+4} - 2} = \dfrac{y^2}{(\sqrt{y^2+4} - 2)}\,\dfrac{(\sqrt{y^2+4} + 2)}{(\sqrt{y^2+4} + 2)} = \dfrac{y^2(\sqrt{y^2+4} + 2)}{(\sqrt{y^2+4})^2 - (2)^2} = \dfrac{y^2(\sqrt{y^2+4} + 2)}{y^2 + 4 - 4}$
$= \dfrac{\cancel{y^2}(\sqrt{y^2+4} + 2)}{\cancel{y^2}} = \sqrt{y^2+4} + 2$ (*R-3*)

95. $\dfrac{\sqrt{t} - \sqrt{5}}{t - 5} = \dfrac{(\sqrt{t} - \sqrt{5})}{(t - 5)}\,\dfrac{(\sqrt{t} + \sqrt{5})}{\sqrt{t} + \sqrt{5}} = \dfrac{(\sqrt{t})^2 - (\sqrt{5})^2}{(t - 5)(\sqrt{t} + \sqrt{5})} = \dfrac{\overset{1}{\cancel{t - 5}}}{\underset{1}{\cancel{(t - 5)}}(\sqrt{t} + \sqrt{5})} = \dfrac{1}{\sqrt{t} + \sqrt{5}}$ (*R-3*)

96. $\dfrac{4\sqrt{x} - 3}{2\sqrt{x}} = \dfrac{4\sqrt{x}}{2\sqrt{x}} - \dfrac{3}{2\sqrt{x}} = 2 - \dfrac{3}{2x^{1/2}} = 2 - \dfrac{3}{2}x^{-1/2} = 2x^0 - \dfrac{3}{2}x^{-1/2}$ (*R-3*)

97. Let $c = 0.54545454\ldots$. Then

$$\begin{aligned} 100c &= 54.545454\ldots \\ \text{So } (100c - c) &= (54.545454\ldots) - (0.54545454\ldots) \\ &= 54 \\ 99c &= 54 \\ c &= \frac{54}{99} = \frac{6}{11} \end{aligned}$$

The number can be written as the quotient of two integers, so it is rational. (*R-1*)

98. (A) List each element of *M*. Follow these with each element of *N* that is not yet listed. {-4,-3,2,0}, or, in increasing order, {-4,-3,0,2}.

(B) List each element of *M* that is also an element of *N*. {-3,2}. (*R-1*)

99. $x^2 - 4x + 1 = (2 - \sqrt{3})^2 - 4(2 - \sqrt{3}) + 1 = (2)^2 - 2(2)\sqrt{3} + (\sqrt{3})^2 - 8 + 4\sqrt{3} + 1$
$= 4 - 4\sqrt{3} + 3 - 8 + 4\sqrt{3} + 1 = 0$ (*R-3*)

100. $x(2x - 1)(x + 3) = x(2x^2 + 6x - x - 3)$
$= x(2x^2 + 5x - 3)$
$= 2x^3 + 5x^2 - 3x$

$(x - 1)^3 = (x - 1)^2(x - 1)$
$= (x^2 - 2x + 1)(x - 1)$
$= x^3 - x^2 - 2x^2 + 2x + x - 1$
$= x^3 - 3x^2 + 3x - 1$

Therefore,

$x(2x - 1)(x + 3) - (x - 1)^3 = (2x^3 + 5x^2 - 3x) - (x^3 - 3x^2 + 3x - 1)$
$= 2x^3 + 5x^2 - 3x - x^3 + 3x^2 - 3x + 1$
$= x^3 + 8x^2 - 6x + 1$ (*R-4*)

101. $4x(a^2 - 4a + 4) - 9x^3 = x[4(a^2 - 4a + 4) - 9x^2]$
$= x[2^2(a - 2)^2 - (3x)^2]$
$= x[2(a - 2) + 3x][2(a - 2) - 3x]$
$= x(2a - 4 + 3x)(2a - 4 - 3x)$
$= x(2a + 3x - 4)(2a - 3x - 4)$ (*R-5*)

102. (A) $\sqrt{3 + \sqrt{5}} + \sqrt{3 - \sqrt{5}} \approx 3.162\ 277\ 660$

(B) $\sqrt{4 + \sqrt{15}} + \sqrt{4 - \sqrt{15}} \approx 3.162\ 277\ 660$

(C) $\sqrt{10} \approx 3.162\ 277\ 660$

Thus, (A), (B), and (C) all have the same value to 9 decimal places. To show that they are actually equal:

$\left(\sqrt{3 + \sqrt{5}} + \sqrt{3 - \sqrt{5}}\right)^2 = \left(\sqrt{3 + \sqrt{5}}\right)^2 + 2\sqrt{3 + \sqrt{5}}\sqrt{3 - \sqrt{5}} + \left(\sqrt{3 - \sqrt{5}}\right)^2$
$= 3 + \sqrt{5} + 2\sqrt{(3 + \sqrt{5})(3 - \sqrt{5})} + 3 - \sqrt{5}$
$= 6 + 2\sqrt{3^2 - (\sqrt{5})^2}$
$= 6 + 2\sqrt{9 - 5}$
$= 6 + 4$
$= 10$

Similarly,

$\left(\sqrt{4 + \sqrt{15}}\ \sqrt{4 - \sqrt{15}}\right)^2 = \left(\sqrt{4 + \sqrt{15}}\right)^2 + 2\sqrt{4 + \sqrt{15}}\sqrt{4 - \sqrt{15}} + \left(\sqrt{4 - \sqrt{15}}\right)^2$
$= 4 + \sqrt{15} + 2\sqrt{(4 + \sqrt{15})(4 - \sqrt{15})} + 4 - \sqrt{15}$
$= 8 + 2\sqrt{4^2 - (\sqrt{15})^2}$
$= 8 + 2\sqrt{16 - 15}$
$= 8 + 2$
$= 10$

Therefore, since both $\sqrt{3 + \sqrt{5}} + \sqrt{3 - \sqrt{5}}$ and $\sqrt{4 + \sqrt{15}} + \sqrt{4 - \sqrt{15}}$ are positive, they are both equal to the positive square root of 10. Thus,

$$\sqrt{3 + \sqrt{5}} + \sqrt{3 - \sqrt{5}} = \sqrt{4 + \sqrt{15}} + \sqrt{4 - \sqrt{15}} = \sqrt{10}$$ (*R-3*)

103. $\dfrac{8(x - 2)^{-3}(x + 3)^2}{12(x - 2)^{-4}(x + 3)^{-2}} = \dfrac{2(x - 2)^{-3-(-4)}(x + 3)^{2-(-2)}}{3} = \dfrac{2(x - 2)^1(x + 3)^4}{3} = \dfrac{2(x - 2)(x + 3)^4}{3}$
$= \dfrac{2}{3}(x - 2)(x + 3)^4$ (*R-2*)

104. $\left(\dfrac{a^{-2}}{b^{-1}} + \dfrac{b^{-2}}{a^{-1}}\right)^{-1} = \left(\dfrac{b}{a^2} + \dfrac{a}{b^2}\right)^{-1} = \left(\dfrac{b^3}{a^2b^2} + \dfrac{a^3}{a^2b^2}\right)^{-1} = \left(\dfrac{b^3 + a^3}{a^2b^2}\right)^{-1} = \dfrac{a^2b^2}{b^3 + a^3}$ or $\dfrac{a^2b^2}{a^3 + b^3}$ (R-2)

105. $(x^{1/3} - y^{1/3})(x^{2/3} + x^{1/3}y^{1/3} + y^{2/3}) = (x^{1/3})^3 - (y^{1/3})^3 = x - y$ (R-2)

106. $\left(\dfrac{x^{m^2}}{x^{2m-1}}\right)^{1/(m-1)} = (x^{m^2-(2m-1)})^{1/(m-1)} = (x^{m^2-2m+1})^{1/(m-1)} = x^{(m^2-2m+1)/(m-1)} = x^{(m-1)^2/(m-1)}$

$= x^{\cancel{(m-1)^2}/\cancel{m-1}} = x^{m-1}$ (R-2)

107. $\dfrac{1}{1 - \sqrt[3]{x}} = \dfrac{1}{(1 - \sqrt[3]{x})}\dfrac{[1 + \sqrt[3]{x} + (\sqrt[3]{x})^2]}{[1 + \sqrt[3]{x} + (\sqrt[3]{x})^2]} = \dfrac{1 + \sqrt[3]{x} + \sqrt[3]{x^2}}{(1)^3 - (\sqrt[3]{x})^3}$

$= \dfrac{1 + \sqrt[3]{x} + \sqrt[3]{x^2}}{1 - x}$ (R-3)

> **Common Error:**
> $(1 - \sqrt[3]{x})(1 + \sqrt[3]{x}) \neq 1 - x$
> This would not be a correct application of
> $(a - b)(a + b) = a^2 - b^2$

108. $\dfrac{\sqrt[3]{t} - \sqrt[3]{5}}{t - 5} = \dfrac{(\sqrt[3]{t} - \sqrt[3]{5})}{(t - 5)}\dfrac{[(\sqrt[3]{t})^2 + \sqrt[3]{t}\sqrt[3]{5} + (\sqrt[3]{5})^2]}{[(\sqrt[3]{t})^2 + \sqrt[3]{t}\sqrt[3]{5} + (\sqrt[3]{5})^2]} = \dfrac{(\sqrt[3]{t})^3 - (\sqrt[3]{5})^3}{(t - 5)[\sqrt[3]{t^2} + \sqrt[3]{5t} + \sqrt[3]{5^2}]}$

$= \dfrac{\cancel{t-5}}{\cancel{(t-5)}[\sqrt[3]{t^2} + \sqrt[3]{5t} + \sqrt[3]{25}]} = \dfrac{1}{\sqrt[3]{t^2} + \sqrt[3]{5t} + \sqrt[3]{25}}$ (1-7)

109. $\sqrt[(n+1)]{x^{n^2}x^{2n+1}} = \sqrt[(n+1)]{x^{n^2}x^{2n+1}}$

$= (x^{n^2+2n+1})^{1/(n+1)} = x^{(n^2+n+1)/(n+1)} = x^{(n+1)^2/(n+1)} = x^{\cancel{(n+1)^2}/\cancel{(n+1)}} = x^{n+1}$ (R-3)

110. The volume of the concrete wall is equal to the volume of the outer cylinder ($V = \pi r^2 h$) minus the volume of the basin. Since the radius of the basin is x ft and the concrete is 2 ft thick, the radius of the outer cylinder is $x + 2$ ft. Thus, we have

$$\begin{pmatrix}\text{Volume of}\\ \text{concrete}\\ \text{wall}\end{pmatrix} = \begin{pmatrix}\text{Volume of}\\ \text{outer}\\ \text{cylinder}\end{pmatrix} - \begin{pmatrix}\text{Volume}\\ \text{of}\\ \text{basin}\end{pmatrix}$$

$$\begin{aligned}\text{Volume} &= \pi(x + 2)^2\,3 - \pi x^2 3\\ &= \pi(x^2 + 4x + 4)3 - \pi x^2 3\\ &= 3\pi x^2 + 12\pi x + 12\pi - 3\pi x^2\\ &= 12\pi x + 12\pi \text{ (ft}^3)\end{aligned}$$ (R-4)

111. average personal income = total personal income ÷ number of persons
= 9,208,000,000,000 ÷ 291,000,000
= $9.208 \times 10^{12} \div 2.91 \times 10^8$
= 3.16×10^4 or \$31,600 to three significant digits
(R-2)

112. (A) We are to estimate $N = 20x^{1/2}y^{1/2}$ given $x = 1{,}600$ units of capital and $y = 900$ units of labor.

$N = 20(1{,}600)^{1/2}(900)^{1/2} = 20(40)(30) = 24{,}000$ units produced.

(B) Given $x = 3{,}200$ units of capital and $y = 1{,}800$ units of labor, then

$$\begin{aligned} N = 20(3{,}200)^{1/2}(1{,}800)^{1/2} &= 20(2 \cdot 1{,}600)^{1/2}(2 \cdot 900)^{1/2} \\ &= 20 \cdot 2^{1/2}(1{,}600)^{1/2}\, 2^{1/2}(900)^{1/2} \\ &= 20 \cdot 2^{1/2}(40) \cdot 2^{1/2}(30) \\ &= 20(40)(30) \cdot 2^{1/2} \cdot 2^{1/2} \\ &= 24{,}000 \cdot 2^{1} \\ &= 48{,}000 \text{ units produced} \end{aligned}$$

(C) The effect of raising x to $2x$ and y to $2y$ is to replace N by

$$\begin{aligned} 20(2x)^{1/2}(2y)^{1/2} &= 20 \cdot 2^{1/2}x^{1/2}2^{1/2}y^{1/2} \\ &= 20 \cdot 2^{1/2} \cdot 2^{1/2}x^{1/2}y^{1/2} \\ &= 2^{1} \cdot 20x^{1/2}y^{1/2} \\ &= 2N \end{aligned}$$

Thus, the production is doubled at any production level. (*1-6*)

113. $$\frac{1}{\frac{1}{R_1} + \frac{1}{R_2} + \frac{1}{R_3}} = \frac{R_1R_2R_3 \cdot 1}{R_1R_2R_3\left(\frac{1}{R_1} + \frac{1}{R_2} + \frac{1}{R_3}\right)} = \frac{R_1R_2R_3}{R_2R_3 + R_1R_3 + R_1R_2}$$ (*1-4*)

114. (A) The area of the cardboard can be written as (Original area) - (Removed area), where the original area = 16 × 30 = 480 and the removed area consists of 6 squares of area x^2 each; thus

(Original area) - (Removed area) = $480 - 6x^2$ in expanded form.

In factored form, $480 - 6x^2 = 6(80 - x^2)$

(B) See figure.

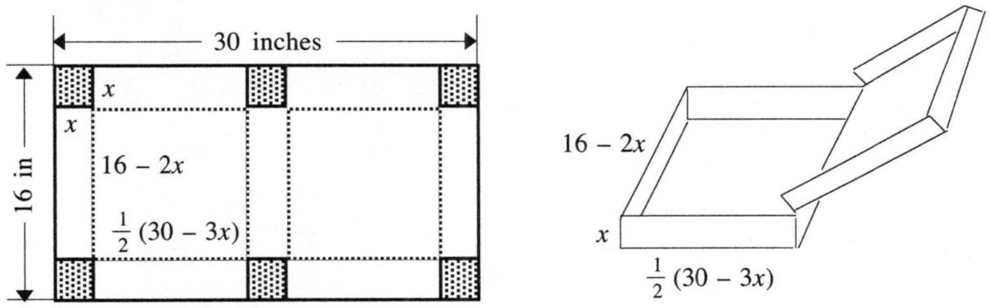

$$\begin{aligned} \text{The volume of the box} = \ell wh &= x(16 - 2x)\frac{1}{2}(30 - 3x) \\ &= x(16 - 2x)(15 - 1.5x) \\ &\quad \text{or} \\ &\quad 3x(8 - x)(10 - x) \quad \text{in factored form.} \end{aligned}$$

In expanded form, $3x(8 - x)(10 - x) = 3x(80 - 18x + x^2)$

CHAPTER 1

Section 1-1

1.
$$\begin{aligned} 4(x+5) &= 6(x-2) \\ 4x+20 &= 6x-12 \\ -2x &= -32 \\ x &= 16 \end{aligned}$$
Solution: 16

3.
$$\begin{aligned} 5+4(w-1) &= 2w+2(w+4) \\ 5+4w-4 &= 2w+2w+4 \\ 4w+1 &= 4w+4 \\ 1 &= 4 \end{aligned}$$
No solution

Common Error:
After line 2, students often write

$$50 - 50 - 1\not{0}^{2}\,\frac{3a-4}{\not{5}} = \cdots$$
$$50 - 6a - 4 = \cdots$$

forgetting to distribute the −2. Put compound numerators in parentheses to avoid this.

5.
$$\begin{aligned} 5 - \frac{3a-4}{5} &= \frac{7-2a}{2} \quad \text{LCD} = 10 \\ 10 \cdot 5 - 10 \cdot \frac{(3a-4)}{5} &= 10 \cdot \frac{(7-2a)}{2} \\ 50 - 2(3a-4) &= 5(7-2a) \\ 50 - 6a + 8 &= 35 - 10a \\ -6a + 58 &= 35 - 10a \\ 4a &= -23 \\ a &= -\frac{23}{4} \text{ or } -5.75 \end{aligned}$$

7.
$$\begin{aligned} \frac{x}{2} + \frac{2x-1}{3} &= \frac{3x+4}{4} \quad \text{LCD} = 12 \\ 12 \cdot \frac{x}{2} + 12 \cdot \frac{(2x-1)}{3} &= 12 \cdot \frac{(3x+4)}{4} \\ 6x + 4(2x-1) &= 3(3x+4) \\ 6x + 8x - 4 &= 9x + 12 \\ 14x - 4 &= 9x + 12 \\ 5x &= 16 \\ x &= \frac{16}{5} \text{ or } 3.2 \end{aligned}$$

9.
$$\begin{aligned} 0.1(t+0.5) + 0.2t &= 0.3(t-0.4) \\ 0.1t + 0.05 + 0.2t &= 0.3t - 0.12 \\ 0.3t + 0.05 &= 0.3t - 0.12 \\ 0.05 &= -0.12 \end{aligned}$$
No solution

11.
$$\begin{aligned} 0.35(s+0.34) + 0.15s &= 0.2s - 1.66 \\ 0.35s + 0.119 + 0.15s &= 0.2s - 1.66 \\ 0.5s + 0.119 &= 0.2s - 1.66 \\ 0.3s &= -1.779 \\ s &= -5.93 \end{aligned}$$
Solution: −5.93

13.
$$\frac{2}{y} + \frac{5}{2} = 4 - \frac{2}{3y}$$
Excluded value: $y \neq 0$ LCD $= 6y$
$$\begin{aligned} 6y \cdot \frac{2}{y} + 6y \cdot \frac{5}{2} &= 6y(4) - 6y \cdot \frac{2}{3y} \\ 12 + 15y &= 24y - 4 \\ -9y &= -16 \\ y &= \frac{16}{9} \text{ or } 1.\overline{7} \end{aligned}$$

15.
$$\frac{z}{z-1} = \frac{1}{z-1} + 2$$
Excluded value: $z \neq 1$ LCD $= z - 1$
$$\begin{aligned} (z-1)\frac{z}{z-1} &= (z-1)\frac{1}{z-1} + 2(z-1) \\ z &= 1 + 2z - 2 \\ -z &= -1 \\ z &= 1 \end{aligned}$$
No solution: 1 is excluded

17.
$$\begin{aligned} \frac{2m}{5} + \frac{m-4}{6} &= \frac{4m+1}{4} - 2 \quad \text{LCD} = 60 \\ 60 \cdot \frac{2m}{5} + 60 \cdot \frac{(m-4)}{6} &= 60 \cdot \frac{(4m+1)}{4} - 60(2) \\ 24m + 10(m-4) &= 15(4m+1) - 120 \\ 24m + 10m - 40 &= 60m + 15 - 120 \\ 34m - 40 &= 60m - 105 \\ -26m &= -65 \\ m &= \frac{65}{26} \\ m &= \frac{5}{2} \text{ or } 2.5 \end{aligned}$$

19. $1 - \frac{x - 3}{x - 2} = \frac{2x - 3}{x - 2}$

Excluded value: $x \neq 2$ LCD = $x - 2$

$$(x - 2)1 - (x - 2)\frac{(x - 3)}{x - 2} = (x - 2)\frac{(2x - 3)}{x - 2}$$

$$x - 2 - (x - 3) = 2x - 3$$

$$x - 2 - x + 3 = 2x - 3$$

$$1 = 2x - 3$$

$$4 = 2x$$

$$2 = x$$

No solution: 2 is excluded

21. $\frac{6}{y + 4} + 1 = \frac{5}{2y + 8}$

Excluded value: $y \neq -4$ LCD $2(y + 4)$

$$2(y + 4)\frac{6}{y + 4} + 2(y + 4)1 = 2(y + 4)\frac{5}{2(y + 4)}$$

$$12 + 2y + 8 = 5$$

$$2y + 20 = 5$$

$$2y = -15$$

$$y = -\frac{15}{2} \text{ or } -7.5$$

23. $\frac{3a - 1}{a^2 + 4a + 4} - \frac{3}{a^2 + 2a} = \frac{3}{a}$

$\frac{3a - 1}{(a + 2)^2} - \frac{3}{a(a + 2)} = \frac{3}{a}$ Excluded values: $a \neq 0, -2$

$$a(a + 2)^2 \frac{(3a - 1)}{(a + 2)^2} - a(a + 2)^2 \frac{3}{a(a + 2)} = a(a + 2)^2 \frac{3}{a}$$

$$a(3a - 1) - (a + 2)3 = (a + 2)^2 3$$

$$3a^2 - a - 3a - 6 = 3(a^2 + 4a + 4)$$

$$3a^2 - 4a - 6 = 3a^2 + 12a + 12$$

$$-16a = 18$$

$$a = -\frac{9}{8} \text{ or } -1.125$$

25.

$$3.142x - 0.4835(x - 4) = 6.795$$

$$3.142x - 0.4835x + 1.934 = 6.795$$

$$2.6585x + 1.934 = 6.795$$

$$2.6585x = 4.861$$

$$x = \frac{4.861}{2.6585}$$

$x = 1.83$ to 3 significant digits

Solution: 1.83

27. $\frac{2.32x}{x - 2} - \frac{3.76}{x} = 2.32$

Excluded values: $x \neq 0, 2$ LCD = $x(x - 2)$

$$x(x - 2)\frac{2.32x}{x - 2} - x(x - 2)\frac{3.76}{x} = 2.32x(x - 2)$$

$$2.32x^2 - 3.76(x - 2) = 2.32x(x - 2)$$

$$2.32x^2 - 3.76x + 7.52 = 2.32x^2 - 4.64x$$

$$-3.76x + 7.52 = -4.64x$$

$$7.52 = -0.88x$$

$$x = -8.55$$

Solution: -8.55

29.
$$\begin{aligned} a_n &= a_1 + (n - 1)d \\ a_1 + (n - 1)d &= a_n \\ (n - 1)d &= a_n - a_1 \\ d &= \frac{a_n - a_1}{n - 1} \end{aligned}$$

31.
$$\begin{aligned} \frac{1}{f} &= \frac{1}{d_1} + \frac{1}{d_2} \quad \text{LCD} = d_1d_2f \\ d_1d_2f\frac{1}{f} &= d_1d_2f\frac{1}{d_1} + d_1d_2f\frac{1}{d_2} \\ d_1d_2 &= d_2f + d_1f \\ d_2f + d_1f &= d_1d_2 \\ (d_2 + d_1)f &= d_1d_2 \\ f &= \frac{d_1d_2}{d_2 + d_1} \end{aligned}$$

33.
$$\begin{aligned} A &= 2ab + 2ac + 2bc \\ 2ab + 2ac + 2bc &= A \\ 2ab + 2ac &= A - 2bc \\ a(2b + 2c) &= A - 2bc \\ a &= \frac{A - 2bc}{2b + 2c} \end{aligned}$$

35.
$$\begin{aligned} y &= \frac{2x - 3}{3x + 5} \\ (3x + 5)y &= 2x - 3 \\ 3xy + 5y &= 2x - 3 \\ 5y + 3 &= 2x - 3xy \\ 5y + 3 &= x(2 - 3y) \\ \frac{5y + 3}{2 - 3y} &= x \\ x &= \frac{5y + 3}{2 - 3y} \end{aligned}$$

37. The "solution" is incorrect. Although 3 is a solution of the two last equations, they are not equivalent to the first equation because both sides have been multiplied by $x - 3$, which is zero when $x = 3$. It is not permitted to multiply both sides of an equation by zero. When $x = 3$, the first equation involves division by zero. Since 3, the only possible solution, is not a solution, the given (first) equation has no solution.

39.
$$\begin{aligned} \frac{x - \frac{1}{x}}{1 + \frac{1}{x}} &= 3 \quad \text{Excl. val.: } x \neq 0 \\ \frac{x\left(x - \frac{1}{x}\right)}{x\left(1 + \frac{1}{x}\right)} &= 3 \\ \frac{x^2 - 1}{x + 1} &= 3 \quad \text{Excl. val.: } x \neq -1 \\ \frac{(x - 1)\overset{1}{\cancel{(x + 1)}}}{\underset{1}{\cancel{x + 1}}} &= 3 \\ x - 1 &= 3 \\ x &= 4 \end{aligned}$$

Solution: 4

41.
$$\begin{aligned} \frac{x + 1 - \frac{2}{x}}{1 - \frac{1}{x}} &= x + 2 \quad \text{Excl. val.: } x \neq 0 \\ \frac{x\left(x + 1 - \frac{2}{x}\right)}{x\left(1 - \frac{1}{x}\right)} &= x + 2 \\ \frac{x^2 + x - 2}{x - 1} &= x + 2 \quad \text{Excl. val.: } x \neq 1 \\ \frac{\overset{1}{\cancel{(x - 1)}}(x + 2)}{\underset{1}{\cancel{x - 1}}} &= x + 2 \\ x + 2 &= x + 2 \end{aligned}$$

Solution: All real numbers except the excluded numbers 0 and 1.

43.

$$y = \frac{a}{1 + \frac{b}{x+c}}$$

$$y = \frac{a(x + c)}{(x + c)\left(1 + \frac{b}{x+c}\right)}$$

$$y = \frac{a(x + c)}{x + c + b}$$

$$y = \frac{ax + ac}{x + c + b}$$

$$y(x + c + b) = ax + ac$$

$$xy + cy + by = ax + ac$$

$$cy + by - ac = ax - xy$$

$$cy + by - ac = x(a - y)$$

$$\frac{cy + by - ac}{a - y} = x$$

$$x = \frac{cy + by - ac}{a - y}$$

45. Let x = the number,
Then 10 less than two thirds the number is one fourth the number

$$\frac{2}{3}x - 10 = \frac{1}{4}x$$

$$\frac{2}{3}x - 10 = \frac{1}{4}x$$

$$12\left(\frac{2}{3}x\right) - 12(10) = 12\left(\frac{1}{4}x\right)$$

$$8x - 120 = 3x$$

$$-120 = -5x$$

$$x = 24$$

The number is 24.

47. Let x = first of the consecutive even numbers
$x + 2$ = second of the numbers
$x + 4$ = third of the numbers
$x + 6$ = fourth of the numbers
first + second + third = 2 more than twice fourth

$$x + x + 2 + x + 4 = 2 + 2(x + 6)$$

$$3x + 6 = 2 + 2x + 12$$

$$3x + 6 = 2x + 14$$

$$x = 8$$

The four consecutive numbers are 8, 10, 12, 14.

49. Let w = width of rectangle
$2w - 3$ = length of rectangle
We use the perimeter formula

$$P = 2a + 2b.$$

$$54 = 2w + 2(2w - 3)$$

$$54 = 2w + 4w - 6$$

$$54 = 6w - 6$$

$$60 = 6w$$

$$10 = w$$

$$17 = 2w - 3$$

dimensions: 17 meters × 10 meters

51. Let P = perimeter of triangle
16 = length of one side
$\frac{2}{7}P$ = length of second side
$\frac{1}{3}P$ = length of third side
We use the perimeter formula

$$P = a + b + c$$

$$P = 16 + \frac{2}{7}P + \frac{1}{3}P$$

$$21P = 21(16) + 21\left(\frac{2}{7}P\right) + 21\left(\frac{1}{3}P\right)$$

$$21P = 336 + 6P + 7P$$

$$21P = 336 + 13P$$

$$8P = 336$$

$$P = 42 \text{ feet}$$

53. Let P = price before discount

$0.20P$ = 20 percent discount on P

Then price before discount - discount = price after discount

$$P - 0.20P = 72$$
$$0.8P = 72$$
$$P = \frac{72}{0.8}$$
$$P = \$90$$

55. Let x = sales of employee

Then $x - 7{,}000$ = sales on which 8% commission is paid

$0.08(x - 7{,}000)$ = (rate of commission) × (sales) = (amount of commission)

$2{,}150 + 0.08(x - 7{,}000)$ = (base salary) + (amount of commission) = earnings

Earnings = 3,170

$$2{,}150 + 0.08(x - 7{,}000) = 3{,}170$$
$$2{,}150 + 0.08x - 560 = 3{,}170$$
$$0.08x + 1{,}590 = 3{,}170$$
$$0.08x = 1{,}580$$
$$x = \frac{1{,}580}{0.08}$$
$$x = \$19{,}750$$

57. (A) We note: The temperature increased 2.5°C for each additional 100 meters of depth. Hence, the temperature increased 25 degrees for each additional kilometer of depth.

Let x = the depth (in kilometers), then $x - 3$ = the depth beyond 3 kilometers.

$25(x - 3)$ = the temperature increase for $x - 3$ kilometers of depth.

T = temperature at 3 kilometers + temperature increase.

$$T = 30 + 25(x - 3)$$

(B) We are to find T when $x = 12$. We use the above relationship as a formula

$$T = 30 + 25(12 - 3)$$
$$= 255^\circ\text{C}$$

(C) We are to find x when $T = 200$. We use the above relationship as an equation.

$$200 = 30 + 25(x - 3)$$
$$200 = 30 + 25x - 75$$
$$200 = -45 + 25x$$
$$245 = 25x$$
$$x = 9.8 \text{ kilometers}$$

59. Let D = distance from earthquake to station

Then $\frac{D}{5}$ = time of primary wave.

$\frac{D}{3}$ = time of secondary wave.

Time difference = time of *slower* secondary wave - time of *faster* primary wave

$$12 = \frac{D}{3} - \frac{D}{5}$$
$$15(12) = 15\left(\frac{D}{3}\right) - 15\left(\frac{D}{5}\right)$$
$$180 = 5D - 3D$$
$$180 = 2D$$
$$D = 90 \text{ miles}$$

Common Error:
Time difference is *not* time of fast wave (short) - time of slow wave (long)

61. We set up the proportion

$$\frac{\text{marked trout in second sample}}{\text{total number in second sample}} = \frac{\text{marked trout in first sample}}{\text{total trout population}}$$

Let x = total population

$$\frac{8}{200} = \frac{200}{x}$$

$$200x\left(\frac{8}{200}\right) = 200x\left(\frac{200}{x}\right)$$

$$8x = 40{,}000$$

$$x = 5{,}000 \text{ trout}$$

63. Let x = amount of distilled water

50 = amount of 30% solution

Then $50 + x$ = amount of 25% solution

acid in 30% solution + acid in distilled water = acid in 25% solution

$$0.3(50) + 0 = 0.25(50 + x)$$

$$0.3(50) = 0.25(50 + x)$$

$$15 = 12.5 + 0.25x$$

$$2.5 = 0.25x$$

$$x = 10 \text{ gallons}$$

65. Let x = amount of 50% solution

5 = amount of distilled water

Then $x - 5$ = amount of 90% solution

acid in 90% solution + acid in distilled water = acid in 50% solution

$$0.9(x - 5) + 0 = 0.5x$$

$$0.9x - 4.5 = 0.5x$$

$$-4.5 = -0.4x$$

$$x = 11.25 \text{ liters}$$

67. Let t = time for both computers to finish the job

Then $t + 1$ = time worked by old computer

t = time worked by new computer

Since the old computer can do 1 job in 5 hours, it works at a rate

(1 job) ÷ (5 hours) = $\frac{1}{5}$ job per hour

Similarly the new computer works at a rate of $\frac{1}{3}$ job per hour.

Part of job completed by old computer in $t + 1$ hours	+	Part of job completed by new computer in t hours	= 1 whole job.
(Rate of old)(time of old)	+	(Rate of new)(Time of new)	= 1
$\frac{1}{5}(t + 1)$	+	$\frac{1}{3}(t)$	= 1

$$15\left(\frac{1}{5}\right)(t + 1) + 15\left(\frac{1}{3}t\right) = 15$$

$$3(t + 1) + 5t = 15$$

$$3t + 3 + 5t = 15$$

$$8t + 3 = 15$$

$$8t = 12$$

$$t = 1.5 \text{ hours}$$

69. Let d = distance flown north

(A) Using $t = \frac{d}{r}$, we note:

$$\begin{aligned}
\text{rate flying north} &= 150 - 30 = 120 \text{ miles per hour}\\
\text{rate flying south} &= 150 + 30 = 180 \text{ miles per hour}\\
\text{time flying north} + \text{time flying south} &= 3 \text{ hours}\\
\frac{d}{120} + \frac{d}{180} &= 3\\
360\frac{d}{120} + 360\frac{d}{180} &= 3(360)\\
3d + 2d &= 1080\\
5d &= 1080\\
d &= 216 \text{ miles}
\end{aligned}$$

(B) We still use the above ideas, except that rate flying north = rate flying south = 150 miles per hour.

$$\begin{aligned}
\frac{d}{150} + \frac{d}{150} &= 3\\
\frac{2d}{150} &= 3\\
\frac{d}{75} &= 3\\
d &= 225 \text{ miles}
\end{aligned}$$

71. Let x = frequency of second note
y = frequency of third note

$$\frac{264}{4} = \frac{x}{5} = \frac{y}{6}$$

$$\frac{264}{4} = \frac{x}{5} \qquad \frac{264}{4} = \frac{y}{6}$$

$$66 = \frac{x}{5} \qquad 66 = \frac{y}{6}$$

$$x = 330 \text{ hertz} \qquad y = 396 \text{ hertz}$$

73. We are to find d when $p = 40$.

$$\begin{aligned}
40 &= -\frac{1}{5}d + 70\\
5(40) &= 5\left(-\frac{1}{5}d\right) + 5(70)\\
200 &= -d + 350\\
-150 &= -d\\
d &= 150 \text{ centimeters}
\end{aligned}$$

75. total height = height in sand + height in water + height in air

$$\begin{aligned}
\text{Let } h &= \text{total height}\\
h &= \frac{1}{5}h + 20 + \frac{2}{3}h\\
15h &= 3h + 300 + 10h\\
15h &= 13h + 300\\
2h &= 300\\
h &= 150 \text{ feet}
\end{aligned}$$

77. The hands will meet when the minute hand has made exactly 1 more revolution than the hour hand.

$$\begin{aligned}
\text{Let } t &= \text{time after 12 o'clock in hours}\\
\text{rate of hour hand} &= \tfrac{1}{12}\text{revolution per hour}\\
\text{rate of minute hand} &= 1 \text{ revolution per hour}\\
\text{distance of minute hand} &= 1 + \text{distance of hour hand}\\
1(t) &= 1 + \tfrac{1}{12}(t)\\
t &= 1 + \tfrac{1}{12}t\\
12t &= 12 + t\\
11t &= 12\\
t &= \tfrac{12}{11} \text{ hours}
\end{aligned}$$

That clock will read $12 + \frac{12}{11}$ o'clock, that is $1\frac{1}{11}$ hours after 12, or 1 hour $\frac{60}{11}$ minutes after 12. This will be $\frac{60}{11}$ minutes, or $5\frac{5}{11}$ minutes after 1 PM.

Section 1-2

1. $-8 \le x \le 7$

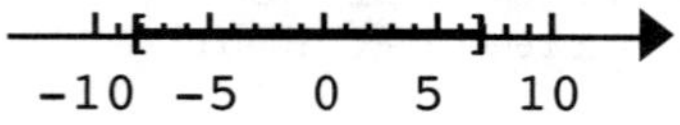

3. $-6 \le x < 6$

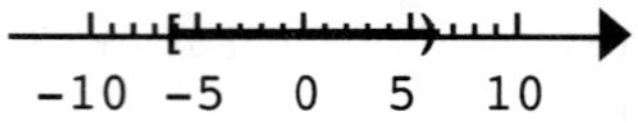

5. $x \ge -6$

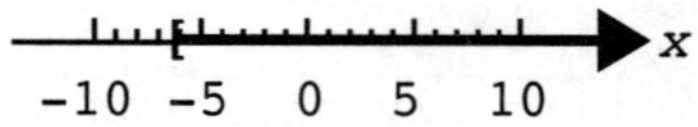

7. $(-2, 6]$

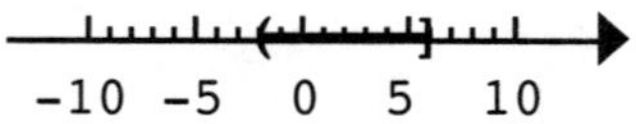

9. $(-7, 8)$

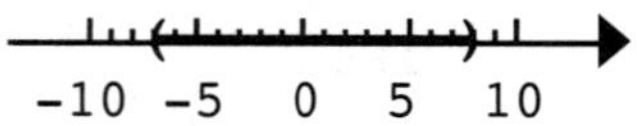

11. $(-\infty, -2]$

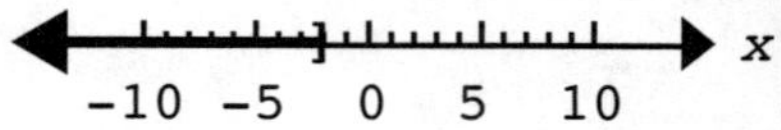

13. $[-7, 2)$; $-7 \le x < 2$

15. $(-\infty, 0]$; $x \le 0$

17. $12 > 6$, $12 + 5 > 6 + 5$

19. $-6 > -8$, $-6 - 3 > -8 - 3$

21. $2 > -1$, $-2(2) < -2(-1)$

23. $2 < 6$, $\frac{2}{2} < \frac{6}{2}$

25.
$$\begin{aligned} 7x - 8 &< 4x + 7 \\ 3x &< 15 \\ x &< 5 \text{ or } (-\infty, 5) \end{aligned}$$

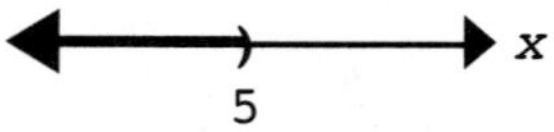

27.
$$\begin{aligned} 3 - x &\ge 5(3 - x) \\ 3 - x &\ge 15 - 5x \\ 4x &\ge 12 \\ x &\ge 3 \text{ or } [3, \infty) \end{aligned}$$

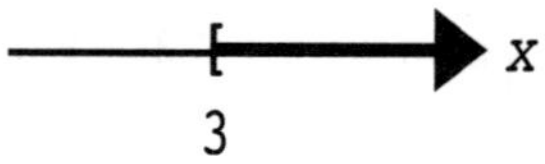

29.
$$\begin{aligned} \frac{N}{-2} &> 4 \\ N &< -8 \text{ or } (-\infty, -8) \end{aligned}$$

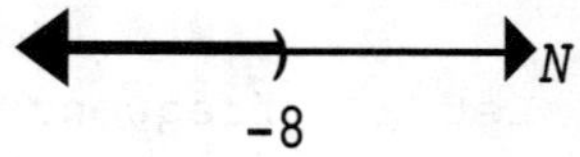

31. $-5t$

$t > 2$ or $(2, \infty)$

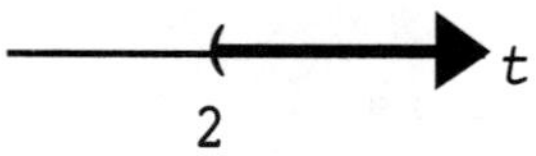

33.
$$\begin{aligned} 3 - m &< 4(m - 3) \\ 3 - m &< 4m - 12 \\ -5m &< -15 \\ m &> 3 \text{ or } (3, \infty) \end{aligned}$$

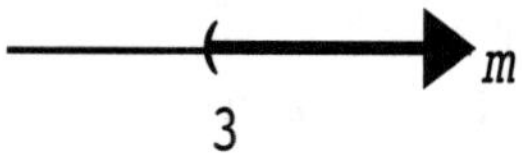

> **Common Error:**
> Neglecting to reverse the order after division by -5.

35.
$$\begin{aligned} -2 - \frac{B}{4} &\le \frac{1 + B}{3} \\ 12\left(-2 - \frac{B}{4}\right) &\le 12\,\frac{(1 + B)}{3} \\ -24 - 3B &\le 4(1 + B) \\ -24 - 3B &\le 4 + 4B \\ -7B &\le 28 \\ B &\ge -4 \text{ or } [-4, \infty) \end{aligned}$$

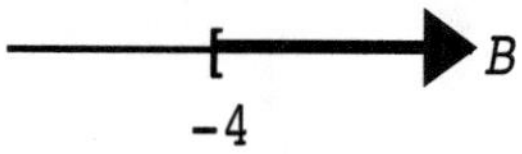

37.
$$\begin{aligned} -4 < 5t + 6 &\le 21 \\ -10 < 5t &\le 15 \\ -2 < t &\le 3 \text{ or } (-2, 3] \end{aligned}$$

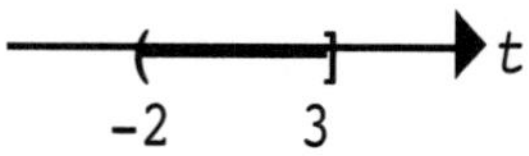

39.

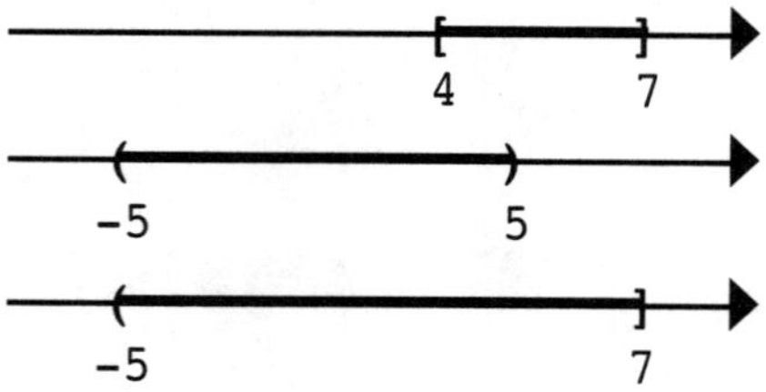

$[4, 7]$

$(-5, 5)$

$(-5, 5) \cup [4, 7] = (-5, 7]$ **$-5 < x \le 7$**

41.

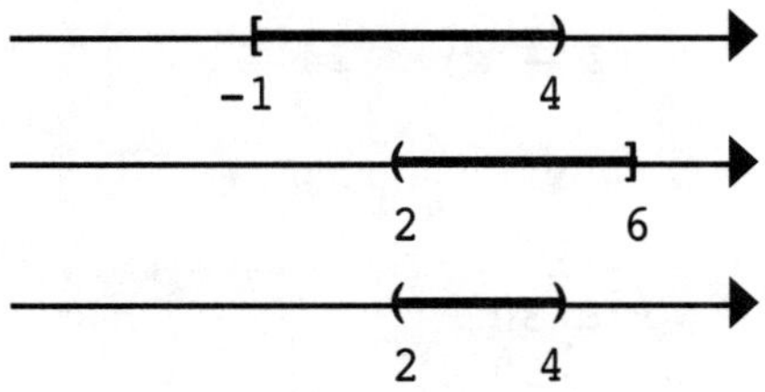

[-1, 4)

(2, 6]

[-1, 4) ∩ (2, 6] = (2, 4) **$2 < x < 4$**

43.

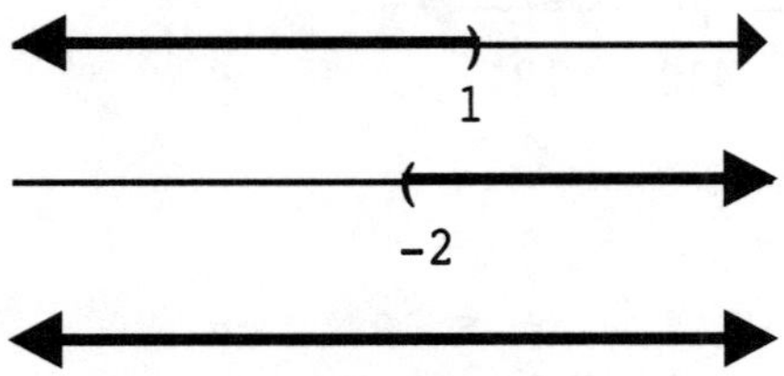

(-∞, 1)

(-2, ∞)

(-∞, 1) ∪ (-2, ∞) = (-∞, ∞) **$-\infty < x < \infty$**

45.

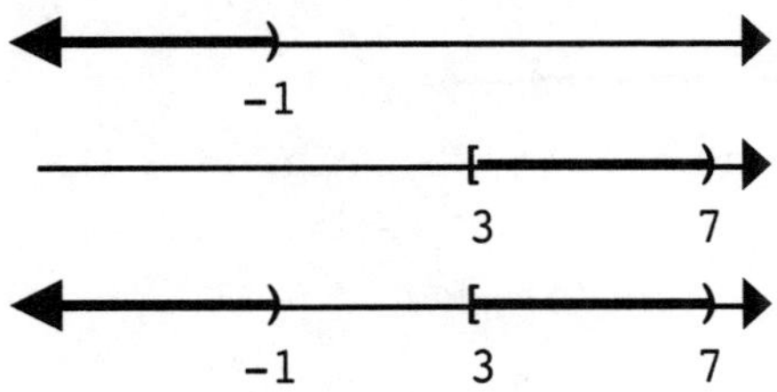

(-∞, -1)

[3, 7)

(-∞, -1) ∪ [3, 7) **$x < -1$ or $3 \le x < 7$**

47.

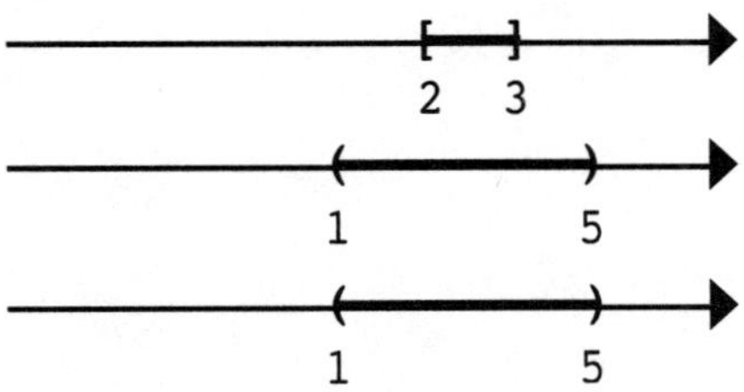

[2, 3]

(1, 5)

[2, 3] ∪ (1, 5) = (1, 5) **$1 < x < 5$**

49.

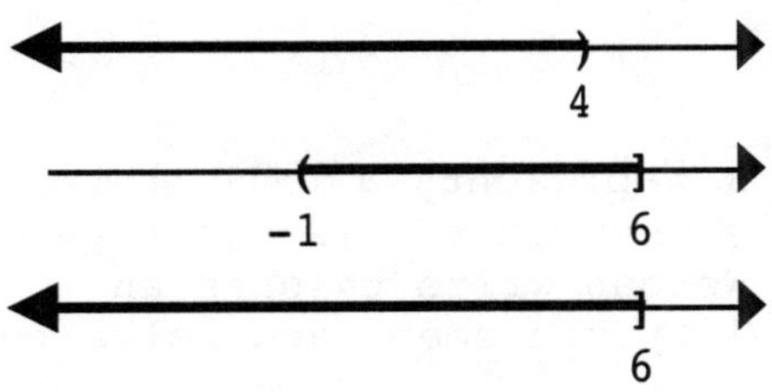

(-∞, 4)

(-1, 6]

(-∞, 4) ∪ (-1, 6] = (-∞, 6] **$x \le 6$**

51.

$$\frac{q}{7} - 3 > \frac{q-4}{3} + 1$$
$$21\left(\frac{q}{7} - 3\right) > 21\left(\frac{(q-4)}{3} + 1\right)$$
$$3q - 63 > 7(q-4) + 21$$
$$3q - 63 > 7q - 28 + 21$$
$$3q - 63 > 7q - 7$$
$$-4q > 56$$
$$q < -14 \text{ or } (-\infty, -14)$$

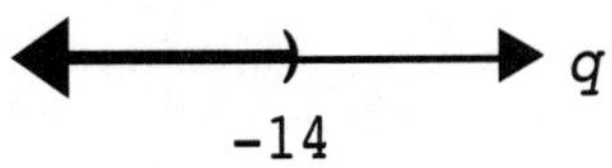

53.

$$\frac{2x}{5} - \frac{1}{2}(x-3) \le \frac{2x}{3} - \frac{3}{10}(x+2) \quad \text{LCD} = 30$$
$$12x - 15(x-3) \le 20x - 9(x+2)$$
$$12x - 15x + 45 \le 20x - 9x - 18$$
$$-3x + 45 \le 11x - 18$$
$$-14x \le -63$$
$$x \ge 4.5 \text{ or } [4.5, \infty)$$

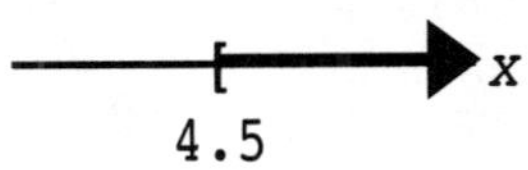

55.
$$-4 \le \frac{9}{5}x + 32 \le 68$$
$$-36 \le \frac{9}{5}x \le 36$$
$$\frac{5}{9}(-36) \le x \le \frac{5}{9}(36)$$
$$-20 \le x \le 20 \text{ or } [-20, 20]$$

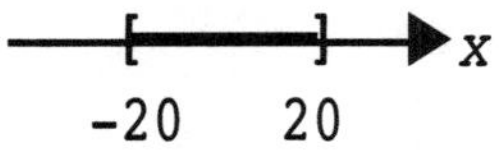

57.
$$-12 < \frac{3}{4}(2 - x) \le 24$$
$$\frac{4}{3}(-12) < 2 - x \le \frac{4}{3}(24)$$
$$-16 < 2 - x \le 32$$
$$-18 < -x \le 30$$
$$18 > x \ge -30$$
$$-30 \le x < 18 \text{ or } [-30, 18)$$

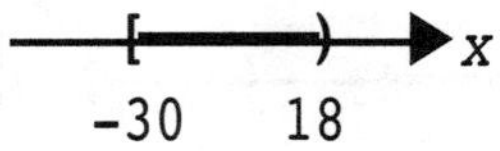

59.
$$16 < 7 - 3x \le 31$$
$$9 < -3x \le 24$$
$$-3 > x \ge -8$$
$$-8 \le x < -3 \text{ or } [-8, -3)$$

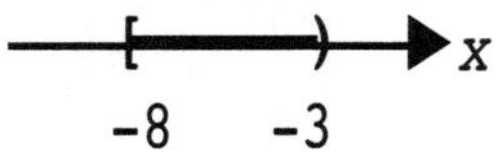

61.
$$-6 < -\frac{2}{5}(1 - x) \le 4$$
$$-30 < -2(1 - x) \le 20$$
$$-30 < -2 + 2x \le 20$$
$$-28 < 2x \le 22$$
$$-14 < x \le 11 \text{ or } (-14, 11]$$

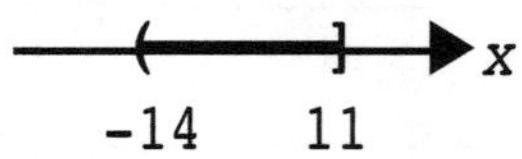

63.
$$0.1(x - 7) < 0.8 - 0.05x$$
$$0.1x - 0.7 < 0.8 - 0.05x$$
$$0.1x + 0.05x - 0.7 < 0.8$$
$$0.15x - 0.7 < 0.8$$
$$0.15x < 0.8 + 0.7$$
$$0.15x < 1.5$$
$$x < 10 \text{ or } (-\infty, 10)$$

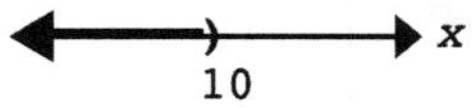

65.
$$0.3x - 2.04 \ge 0.04(x + 1)$$
$$0.3x - 2.04 \ge 0.04x + 0.04$$
$$0.3x - 0.04x - 2.04 \ge 0.04$$
$$0.26x - 2.04 \ge 0.04$$
$$0.26x \ge 0.04 + 2.04$$
$$0.26x \ge 2.08$$
$$x \ge 8 \text{ or } [8, \infty)$$

67. $\sqrt{1 - x}$ represents a real number exactly when $1 - x$ is positive or zero. We can write this as an inequality statement and solve for x.

$$1 - x \ge 0$$
$$-x \ge -1$$
$$x \le 1$$

69. $\sqrt{3x + 5}$ represents a real number exactly when $3x + 5$ is positive or zero. We can write this as an inequality statement and solve for x.

$$3x + 5 \ge 0$$
$$3x \ge -5$$
$$x \ge -\frac{5}{3}$$

71. $\dfrac{1}{\sqrt[4]{2x + 3}}$ represents a real number exactly when $2x + 3$ is positive. (*not* zero).
We can write this as an inequality statement and solve for x.

$$2x + 3 > 0$$
$$2x > -3$$
$$x > -\frac{3}{2}$$

73. (A) For $ab > 0$, ab must be positive, hence a and b must have the same sign.
Either
1. $a > 0$ and $b > 0$ or
2. $a < 0$ and $b < 0$

(B) For $ab < 0$, ab must be negative, hence a and b must have opposite signs.
Either
1. $a > 0$ and $b < 0$ or
2. $a < 0$ and $b > 0$

(C) For $\frac{a}{b} > 0$, $\frac{a}{b}$ must be positive, hence a and b must have the same sign. Answer as in (A).

(D) For $\frac{a}{b} < 0$, $\frac{a}{b}$ must be negative, hence a and b must have opposite signs. Answer as in (B).

75. (A) If $a - b = 1$, then $a = b + 1$. Therefore, a is greater than b. $>$

(B) If $u - v = -2$, then $v = u + 2$. Therefore, u is less than v. $<$

77. If $\frac{b}{a}$ is greater than 1

$\frac{b}{a} > 1$

$a \cdot \frac{b}{a} < a \cdot 1$. (since a is negative.)

$b < a$

$0 < a - b$

$a - b$ is positive

79. (A) F (B) T (C) T

81. If $a < b$, then by definition of $<$, there exists a positive number p such that $a + p = b$. Then, adding c to both sides, we obtain $(a + c) + p = b + c$, where p is positive. Hence, by definition of $<$, we have
$a + c < b + c$

83. (A) If $a < b$, then by definition of $<$, there exists a positive number p such that $a + p = b$. If we multiply both sides of this by the positive number c, we obtain $(a + p)c = bc$, or $ac + pc = bc$, where pc is positive. Hence, by definition of $<$, we have $ac < bc$.

(B) If $a < b$, then by definition of $<$, there exists a positive number p such that $a + p = b$. If we multiply both sides of this by the negative number c, we obtain $(a + p)c = bc$, or $ac + pc = bc$, where pc is negative. Hence, by definition of $<$, we have $ac > bc$.

85. $150 \le T \le 250$

$150 \le 30 + 25(x - 3) \le 250$

$150 \le 30 + 25x - 75 \le 250$

$150 \le 25x - 45 \le 250$

$195 \le 25x \le 295$

$7.8 \le x \le 11.8$

Depth from 7.8 km to 11.8 km.

87. Let x = number of calculators sold
Then Revenue = (price per calculator) × (number of calculators sold) = $63x$
Cost = fixed cost + variable cost
= 650,000 + (Cost per calculator) × (number sold)
= $650{,}000 + 47x$

(A) We want Revenue > Cost

$$63x > 650{,}000 + 47x$$
$$16x > 650{,}000$$
$$x > 40{,}625$$

More than 40,625 calculators must be sold for the company to make a profit.

(B) We want Revenue = Cost

$$63x = 650{,}000 + 47x$$
$$16x = 650{,}000$$
$$x = 40{,}625$$

(C) 40,625 calculators sold represents the break-even point, the boundary between profit and loss.

89. (A) The company might try to increase sales and keep the price the same (see part B). It might try to increase the price and keep the sales the same (see part C). Either of these strategies would need further analysis and implementation that are out of place in a discussion here.

(B) Here the cost has been changed to $650{,}000 + 50.5x$, but the revenue is still $63x$.

Revenue > Cost

$$63x > 650{,}000 + 50.5x$$
$$12.5x > 650{,}000$$
$$x > 52{,}000 \text{ calculators}$$

(C) Let p = the new price. Here the cost is still $650{,}000 + 50.5x$ as in part (B) where x is now known to be 40,625. Thus, cost = 650,000 + 50.5(40,625). The revenue is (price per calculator) × (number of calculators) = $p(40{,}625)$.

Revenue > Cost

$$p(40{,}625) > 650{,}000 + 50.5(40{,}625)$$
$$p > \frac{650{,}000 + 50.5(40{,}625)}{40{,}625}$$
$$p > 66.50$$

The price could be raised by \$3.50 to \$66.50.

91. We want $220 \le W \le 2{,}750$. We are given $W = 110I$. Substituting, we must solve

$$220 \le 110I \le 2{,}750$$
$$2 \le I \le 25 \text{ or } [2, 25].$$

Section 1-3

1. $\sqrt{5}$

3. $|(-6) - (-2)| = |-4| = 4$

5. $|5 - \sqrt{5}| = 5 - \sqrt{5}$ since $5 - \sqrt{5}$ is positive.

7. $|\sqrt{5} - 5| = -(\sqrt{5} - 5) = 5 - \sqrt{5}$ since $\sqrt{5} - 5$ is negative.

9. $d(A,B) = |b - a| = |5 - (-7)| = |12| = 12$

11. $d(A,B) = |b - a| = |-7 - 5| = |-12| = 12$

13. $d(B,0) = |0 - (-4)| = |4| = 4$

15. $d(0,B) = |-4 - 0| = |-4| = 4$

17. $d(B,C) = |5 - (-4)| = |9| = 9$

19. The distance between x and 3 is equal to 4.
$|x - 3| = 4$

21. The distance between m and -2 is equal to 5.
$|m - (-2)| = 5$
$|m + 2| = 5$

23. The distance between x and 3 is less than 5.
$|x - 3| < 5$

25. The distance between p and -2 is more than 6.
$|p - (-2)| > 6$
$|p + 2| > 6$

27. The distance between q and 1 is not less than 2.
$|q - 1| \geq 2$

29. y is 3 units from 5.
$|y - 5| = 3$
$y - 5 = \pm 3$
$y = 5 \pm 3$
$y = 2, 8$

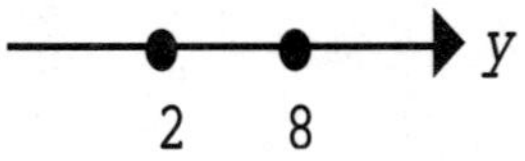

31. y is less than 3 units from 5.
$|y - 5| < 3$
$-3 < y - 5 < 3$
$2 < y < 8$
$(2, 8)$

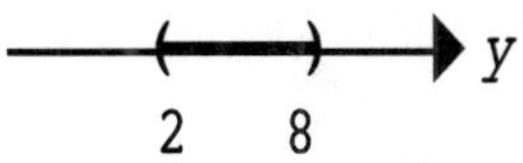

33. y is more than 3 units from 5.
$|y - 5| > 3$
$y - 5 < -3$ or $y - 5 > 3$
$y < 2$ or $y > 8$
$(-\infty, 2) \cup (8, \infty)$

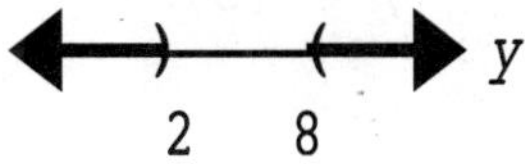

35. $|u - (-8)| = 3$
u is 3 units from -8.
$|u + 8| = 3$
$u + 8 = \pm 3$
$u = -8 \pm 3$
$u = -11$ or -5

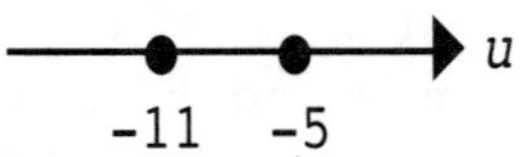

37. $|u - (-8)| \leq 3$
u is no more than 3 units from -8.
$|u + 8| \leq 3$
$-3 \leq u + 8 \leq 3$
$-11 \leq u \leq -5$
$[-11, -5]$

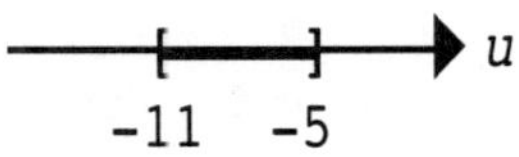

39. $|u - (-8)| \geq 3$
u is at least 3 units from -8.
$|u + 8| \geq 3$
$u + 8 \leq -3$ or $u + 8 \geq 3$
$u \leq -11$ or $u \geq -5$
$(-\infty, -11] \cup [-5, \infty)$

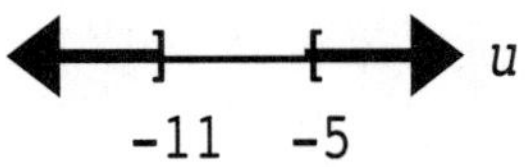

41. $|3x - 7| \leq 4$
$-4 \leq 3x - 7 \leq 4$
$3 \leq 3x \leq 11$
$1 \leq x \leq \frac{11}{3}$
$\left[1, \frac{11}{3}\right]$

43. $|4 - 2t| > 6$
$4 - 2t > 6$ or $4 - 2t < -6$
$-2t > 2$ or $-2t < -10$
$t < -1$ or $t > 5$
$(-\infty, -1) \cup (5, \infty)$

45. $|7m + 11| = 3$
$7m + 11 = 3$ or $7m + 11 = -3$
$7m = -8$ or $7m = -14$
$m = -\frac{8}{7}$ or $m = -2$

47. $\left|\frac{1}{2}w - \frac{3}{4}\right| < 2$
$-2 < \frac{1}{2}w - \frac{3}{4} < 2$
$-8 < 2w - 3 < 8$
$-5 < 2w < 11$
$-\frac{5}{2} < w < \frac{11}{2}$
$-2.5 < w < 5.5$
$(-2.5, 5.5)$

49. $|0.2u + 1.7| \geq 0.5$
$0.2u + 1.7 \geq 0.5$ or $0.2u + 1.7 \leq -0.5$
$0.2u \geq -1.2$ or $0.2u \leq -2.2$
$u \geq -6$ or $u \leq -11$
$[-6, \infty) \cup (-\infty, -11]$

51. $\left|\frac{9}{5}C + 32\right| < 31$

$-31 < \frac{9}{5}C + 32 < 31$

$-63 < \frac{9}{5}C < -1$

$-35 < C < -\frac{5}{9}$

$\left(-35, -\frac{5}{9}\right)$

53. $\sqrt{x^2} < 2$

$|x| < 2$

$-2 < x < 2$

$(-2, 2)$

55. $\sqrt{(1 - 3t)^2} \le 2$

$|1 - 3t| \le 2$

$-2 \le 1 - 3t \le 2$

$-3 \le -3t \le 1$

$1 \ge t \ge -\frac{1}{3}$

$-\frac{1}{3} \le t \le 1$

$\left[-\frac{1}{3}, 1\right]$

57. $\sqrt{(2t - 3)^2} > 3$

$|2t - 3| > 3$

$2t - 3 < -3$ or $2t - 3 > 3$

$2t < 0 \qquad 2t > 6$

$t < 0 \qquad t > 3$

$(-\infty, 0) \cup (3, \infty)$

59. $|2.25 - 1.02x| \le 1.64$

$-1.64 \le 2.25 - 1.02x \le 1.64$

$-3.89 \le -1.02x \le -0.61$

$3.81 \ge x \ge 0.598$

to three significant digits

$.598 \le x \le 3.81$

61. $|21.7 - 11.3x| = 15.2$

$21.7 - 11.3x = 15.2$ or $21.7 - 11.3x = -15.2$

$-11.3x = -6.5$ or $-11.3x = -36.9$

$x = 0.575$ or $x = 3.27$

to three significant digits

63. $0 < |x - 3| < 0.1$

The distance between x and 3 is less than 0.1 but $x \ne 3$.

$-0.1 < x - 3 < 0.1$ except $x \ne 3$

$2.9 < x < 3.1$ but $x \ne 3$

$2.9 < x < 3$ or $3 < x < 3.1$

$(2.9, 3) \approx (3, 3.1)$

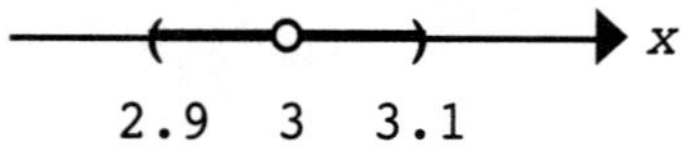

65. $0 < |x - c| < d$

The distance between x and c is less than d but $x \ne c$.

$-d < x - c < d$ except $x \ne c$

$c - d < x < c + d$ but $x \ne c$

$c - d < x < c$ or $c < x < c + d$

$(c - d, c) \approx (c, c + d)$

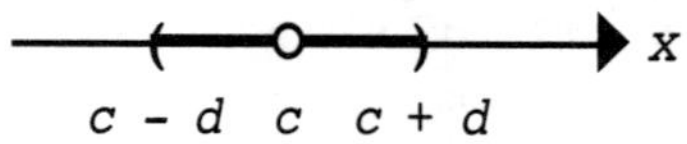

67. We consider two cases for $|x - 2| = 2x - 7$.

Case 1: $x - 2 \ge 0$

For this case, the possible values of x are in the set $x \ge 2$.

Then $|x - 2| = x - 2$

We solve $x - 2 = 2x - 7$

$-x = -5$

$x = 5$ A solution, since 5 is among the possible values of x.

Case 2: $x - 2 < 0$

For this case, the possible values of x are in the set $x < 2$.

Then $|x - 2| = 2 - x$

We solve $2 - x = 2x - 7$

$-3x = -9$

$x = 3$ Not a solution, since 3 is not among the possible values of x.

Solution: $x = 5$

69. We consider two cases for $|3x + 5| = 2x + 6$

Case 1: $3x + 5 \ge 0$

For this case, the possible values of x are in the set $x \ge -\frac{5}{3}$.

Then $|3x + 5| = 3x + 5$.
We solve $3x + 5 = 2x + 6$
$x = 1$ A solution, since 1 is among the possible values of x.

Case 2: $3x + 5 < 0$

For this case, the possible values of x are in the set $x < -\frac{5}{3}$.

Then $|3x + 5| = -(3x + 5)$
We solve $-(3x + 5) = 2x + 6$
$-3x - 5 = 2x + 6$
$-5x = 11$
$x = -2.2$ A solution, since -2.2 is among the possible values of x.

Solution $x = 1, -2.2$

71. We consider four cases for $|x| + |x + 3| = 3$.

Case 1: $x \ge 0$ and $x + 3 \ge 0$, that is, $x \ge 0$ and $x \ge -3$ (or simply $x \ge 0$)
$|x| = x$ and $|x + 3| = x + 3$

Hence:
$$x + x + 3 = 3$$
$$2x + 3 = 3$$
$$x = 0$$
which is a possible value for x in this case.

Case 2: $x < 0$ and $x + 3 \ge 0$, that is, $x < 0$ and $x \ge -3$ (or simply $-3 \le x < 0$)
$|x| = -x$ and $|x + 3| = x + 3$

Hence:
$$-x + x + 3 = 3$$
$$3 = 3$$
This is satisfied by all x, but the condition $-3 \le x < 0$ must be imposed.

Case 3: $x \ge 0$ and $x + 3 < 0$, that is, $x \ge 0$ and $x < -3$. These are mutually contradictory, so no solution is possible in this case.

Case 4: $x < 0$ and $x + 3 \le 0$, that is, $x < 0$ and $x \le -3$ (or simply $x \le -3$)
$|x| = -x$ and $|x + 3| = -(x + 3)$

Hence:
$$-x + -(x + 3) = 3$$
$$-x - x - 3 = 3$$
$$-2x = 6$$
$$x = -3$$
Combining the results of the four cases, $-3 \le x \le 0$ is the solution.

73. We consider four cases for $|2x + 7| - |6 - 3x| = 8$.

Case 1: $2x + 7 \ge 0$ and $6 - 3x \ge 0$, that is, $x \ge -\frac{7}{2}$ and $x \le 2$
(or simply $-\frac{7}{2} \le x \le 2$)
$|2x + 7| = 2x + 7$ and $|6 - 3x| = 6 - 3x$

Hence:
$$2x + 7 - (6 - 3x) = 8$$
$$2x + 7 - 6 + 3x = 8$$
$$5x + 1 = 8$$
$$5x = 7$$
$$x = 1.4$$
which is a possible value for x in this case.

Case 2: $2x + 7 < 0$ and $6 - 3x \geq 0$, that is, $x < -\frac{7}{2}$ and $x \leq 2$ $\left(\text{or simply } x < -\frac{7}{2}\right)$

$|2x + 7| = -(2x + 7)$ and $|6 - 3x| = 6 - 3x$

Hence: $-(2x + 7) - (6 - 3x) = 8$

$$-2x - 7 - 6 + 3x = 8$$
$$x - 13 = 8$$
$$x = 21$$

which is not a possible value for x in this case.

Case 3: $2x + 7 \geq 0$ and $6 - 3x < 0$, that is, $x \geq -\frac{7}{2}$ and $x > 2$ (or simply $x > 2$)

$|2x + 7| = 2x + 7$ $\quad |6 - 3x| = -(6 - 3x)$

Hence: $2x + 7 - [-(6 - 3x)] = 8$

$$2x + 7 + 6 - 3x = 8$$
$$-x + 13 = 8$$
$$-x = -5$$
$$x = 5$$

which is a possible value for x in this case.

Case 4: $2x + 7 < 0$ and $6 - 3x < 0$, that is, $x < -\frac{7}{2}$ and $x > 2$. These are mutually contradictory, so no solution is possible in this case.

Combining the results of the four cases, $x = 1.4$ or $x = 5$ are the solutions.

75. *Case 1:* $x > 0$. Then $|x| = x$. Hence $\frac{x}{|x|} = \frac{x}{x} = 1$.

Case 2: $x = 0$. Then $|x| = 0$. Hence $\frac{x}{|x|}$ is not defined.

Case 3: $x < 0$. Then $|x| = -x$. Hence $\frac{x}{|x|} = \frac{x}{-x} = -1$.

Thus, the possible values of $\frac{x}{|x|}$ are 1 and -1.

77. There are three possible relations between real numbers a and b; either $a = b$, $a > b$, or $a < b$. We examine each case separately.

Case 1: $a = b$

$|b - a| = |0| = 0$;

$|a - b| = |0| = 0$

Case 2: $a > b$

$|b - a| = -(b - a) = a - b$

$|a - b| = a - b$

Case 3: $b > a$

$|b - a| = b - a$

$|a - b| = -(a - b) = b - a$

Thus in all three cases $|b - a| = |a - b|$.

79. If $m < n$, then $m + m < m + n$ (adding m to both sides)
Also, $m + n < n + n$ (adding n to both sides). Hence,

$$m + m < m + n < n + n$$
$$2m < m + n < 2n$$
$$m < \frac{m + n}{2} < n$$

81. *Case 1.* $m > 0$. Then $|m| = m$ $\quad |-m| = -(-m) = m$. Hence $|m| = |-m|$

Case 2. $m < 0$. Then $|m| = -m$ $\quad |-m| = -m$. Hence $|m| = |-m|$

Case 3. $m = 0$. Then $0 = m = -m$, hence $|m| = |-m| = 0$.

83. If $n \neq 0$, $n > 0$ or $n < 0$.

Case 1. $n > 0$. If $m \geq 0$ $|m| = m$, $\frac{m}{n} \geq 0$ $\left|\frac{m}{n}\right| = \frac{m}{n}$ $|n| = n$.

Hence: $\left|\frac{m}{n}\right| = \frac{m}{n} = \frac{|m|}{|n|}$

If $m < 0$ $|m| = -m$, $\frac{m}{n} < 0$ $\left|\frac{m}{n}\right| = -\frac{m}{n}$ $|n| = n$.

Hence: $\left|\frac{m}{n}\right| = -\frac{m}{n} = \frac{-m}{n} = \frac{|m|}{|n|}$

Case 2. $n < 0$. If $m > 0$ $|m| = m$, $\frac{m}{n} < 0$ $\left|\frac{m}{n}\right| = -\frac{m}{n}$ $|n| = -n$

Hence: $\left|\frac{m}{n}\right| = -\frac{m}{n} = \frac{m}{-n} = \frac{|m|}{|n|}$

If $m \leq 0$ $|m| = -m$ $\frac{m}{n} > 0$ $\left|\frac{m}{n}\right| = \frac{m}{n}$ $|n| = -n$

Hence: $\left|\frac{m}{n}\right| = \frac{m}{n} = \frac{-m}{-n} = \frac{|m|}{|n|}$

85. First note that $a \leq b$ is true if $a < b$ or if $a = b$. Hence $a < b$ implies $a \leq b$. Also $a = b$ implies $a \leq b$. Now consider three cases ($m > 0$, $m = 0$, $m < 0$).

Case 1. $m > 0$. Then $|m| = m$. Also $-|m| < 0$

Hence $-|m| < 0 < m = |m|$

$-|m| < m = |m|$

$-|m| \leq m \leq |m|$

Case 2. $m = 0$. Then $-|m| = m = |m| = 0$.

Hence $-|m| \leq m \leq |m|$

Case 3. $m < 0$. Then $|m| = -m$, hence $-|m| = m$. Also $|m| > 0$.

Hence $-|m| = m < 0 < |m|$

$-|m| = m < |m|$

$-|m| \leq m \leq |m|$

87. Consider three cases ($a < b$, $a = b$, $a > b$)

Case 1. $a < b$. Then $\max(a,b) = b$. Also $a - b < 0$, hence $|a - b| = b - a$.

Hence $\frac{1}{2}[a + b + |a - b|] = \frac{1}{2}[a + b + b - a] = \frac{1}{2}(2b) = b$.

Thus $\max(a,b) = b = \frac{1}{2}[a + b + |a - b|]$

Case 2. $a = b$. Then $\max(a,b) = a = b$. Also $a - b = 0$, hence $|a - b| = 0$.

Hence $\frac{1}{2}[a + b + |a - b|] = \frac{1}{2}[a + b] = \frac{1}{2}[2a] = a$ or $\frac{1}{2}[2b] = b$.

Thus $\max(a,b) = a = b = \frac{1}{2}[a + b + |a - b|]$

Case 3. $a > b$. Then $\max(a,b) = a$. Also $a - b > 0$, hence $|a - b| = a - b$

Hence $\frac{1}{2}[a + b + |a - b|] = \frac{1}{2}[a + b + a - b] = \frac{1}{2}(2a) = a$.

Thus $\max(a,b) = a = \frac{1}{2}[a + b + |a - b|]$

89. $\left|\frac{x - 45.4}{3.2}\right| < 1$

$$-1 < \frac{x - 45.4}{3.2} < 1$$

$$-3.2 < x - 45.4 < 3.2$$

$$42.2 < x < 48.6$$

91. The difference between P and 500 has an absolute value of no more than 20.

$|P - 500| \leq 20$

93. The difference between A and 12.436 has an absolute value of less than the error of 0.001.
$|A - 12.436| < 0.001$
$-0.001 < A - 12.436 < 0.001$
$12.435 < A < 12.437$ or, in interval notation, (12.435, 12.437)

95. The difference between N and 2.37 has an absolute value of no more than 0.005.
$|N - 2.37| \le 0.005$.

Then $|x - 2| = x - 2$.
We solve $x - 2 = 2x - 7$
$-x = -5$
$x = 5$ A solution, since 5 is among the possible values of x.

Case 2: $x - 2 < 0$
For this case, the possible values of x are in the set $x < 2$.
Then $|x - 2| = 2 - x$.
We solve $2 - x = 2x - 7$
$-3x = -9$
$x = 3$ Not a solution, since 3 is not among the possible values of x.

Solution: $x = 5$

Section 1-4

1. $2 - 9i = 2 + (-9)i$
(A) real part: 2
(B) imaginary part: $-9i$
(C) conjugate: $2 - (-9)i = 2 + 9i$

3. $-\frac{3}{2} + \frac{5}{6}i$
(A) real part: $-\frac{3}{2}$
(B) imaginary part: $\frac{5}{6}i$
(C) conjugate: $-\frac{3}{2} - \frac{5}{6}i$

5. $6.5 + 2.1i$
(A) real part: 6.5
(B) imaginary part: $2.1i$
(C) conjugate: $6.5 - 2.1i$

7. $i\pi = 0 + \pi i$
(A) real part: 0
(B) imaginary part: πi
(C) conjugate: $0 - \pi i = -\pi i$

9. $4\pi = 4\pi + 0i$
(A) real part: 4π
(B) imaginary part: $0i = 0$
(C) conjugate: $4\pi - 0i = 4\pi$

11. $-5 + i\sqrt{2}$
(A) real part: -5
(B) imaginary part: $i\sqrt{2}$
(C) conjugate: $-5 - i\sqrt{2}$

13. $(3 + 5i) + (2 + 4i) = 3 + 5i + 2 + 4i = 5 + 9i$

15. $(8 - 3i) + (-5 + 6i) = 8 - 3i - 5 + 6i = 3 + 3i$

17. $(9 + 5i) - (6 + 2i) = 9 + 5i - 6 - 2i = 3 + 3i$

19. $(3 - 4i) - (-5 + 6i) = 3 - 4i + 5 - 6i = 8 - 10i$

21. $2 + (3i + 5) = 2 + 3i + 5 = 7 + 3i$

23. $(2i)(4i) = 8i^2 = 8(-1) = -8$

25. $-2i(4 - 6i) = -8i + 12i^2 = -8i + 12(-1) = -12 - 8i$

27. $(1 + 2i)(3 - 4i) = 3 - 4i + 6i - 8i^2 = 3 + 2i - 8(-1) = 3 + 2i + 8 = 11 + 2i$

29. $(3 - i)(4 + i) = 12 + 3i - 4i - i^2 = 12 - i - (-1) = 12 - i + 1 = 13 - i$

31. $(2 + 9i)(2 - 9i) = 4 - 81i^2 = 4 + 81 = 85$ or $85 + 0i$

33. $\frac{1}{2 + 4i} = \frac{1}{(2 + 4i)} \frac{(2 - 4i)}{(2 - 4i)} = \frac{2 - 4i}{4 - 16i^2} = \frac{2 - 4i}{4 + 16} = \frac{2 - 4i}{20} = 0.1 - 0.2i$

35. $\frac{4 + 3i}{1 + 2i} = \frac{(4 + 3i)}{(1 + 2i)} \frac{(1 - 2i)}{(1 - 2i)} = \frac{4 - 5i - 6i^2}{1 - 4i^2} = \frac{4 - 5i + 6}{1 + 4} = \frac{10 - 5i}{5} = 2 - i$

37. $\frac{7 + i}{2 + i} = \frac{(7 + i)}{(2 + i)} \frac{(2 - i)}{(2 - i)} = \frac{14 - 5i - i^2}{4 - i^2} = \frac{14 - 5i + 1}{4 + 1} = \frac{15 - 5i}{5} = 3 - i$

39. $\sqrt{2}\sqrt{8} = \sqrt{16} = 4$

41. $\sqrt{2}\sqrt{-8} = \sqrt{2}\sqrt{8}\sqrt{-1} = \sqrt{2}\sqrt{8}\,i = \sqrt{16}\,i = 4i$

43. $\sqrt{-2}\sqrt{8} = \sqrt{-1}\sqrt{2}\sqrt{8} = i\sqrt{2}\sqrt{8} = i\sqrt{16} = 4i$

45. $\sqrt{-2}\sqrt{-8} = \sqrt{-1}\sqrt{2}\sqrt{-1}\sqrt{8} = i\sqrt{2}\,i\sqrt{8} = i^2 i\sqrt{16} = -1 \cdot 4 = -4$

47. $(2 - \sqrt{-4}) + (5 - \sqrt{-9}) = (2 - i\sqrt{4}) + (5 - i\sqrt{9}) = 2 - 2i + 5 - 3i = 7 - 5i$

49. $(9 - \sqrt{-9}) - (12 - \sqrt{-25}) = (9 - i\sqrt{9}) - (12 - i\sqrt{25})$
$= (9 - 3i) - (12 - 5i) = 9 - 3i - 12 + 5i = -3 + 2i$

51. $(3 - \sqrt{-4})(-2 + \sqrt{-49}) = (3 - i\sqrt{4})(-2 + i\sqrt{49})$
$= (3 - 2i)(-2 + 7i) = -6 + 25i - 14i^2$
$= -6 + 25i + 14 = 8 + 25i$

53. $\frac{5 - \sqrt{-4}}{7} = \frac{5 - i\sqrt{4}}{7} = \frac{5 - 2i}{7} = \frac{5}{7} - \frac{2}{7}i$

55. $\frac{1}{2 - \sqrt{-9}} = \frac{1}{2 - i\sqrt{9}} = \frac{1}{2 - 3i} = \frac{1}{(2 - 3i)} \frac{(2 + 3i)}{(2 + 3i)} = \frac{2 + 3i}{4 - 9i^2}$
$= \frac{2 + 3i}{4 + 9} = \frac{2 + 3i}{13} = \frac{2}{13} + \frac{3}{13}i$

57. $\frac{2}{5i} = \frac{2}{5i} \cdot \frac{i}{i} = \frac{2i}{5i^2} = \frac{2i}{-5} = -\frac{2}{5}i$ or $0 - \frac{2}{5}i$

59. $\frac{1 + 3i}{2i} = \frac{(1 + 3i)}{(2i)} \cdot \frac{i}{i} = \frac{i + 3i^2}{2i^2} = \frac{i - 3}{-2} = \frac{3}{2} - \frac{1}{2}i$

61. $(2 - 3i)^2 - 2(2 - 3i) + 9 = 4 - 12i + 9i^2 - 4 + 6i + 9$
$= 4 - 12i - 9 - 4 + 6i + 9 = -6i$ or $0 - 6i$

63. $x^2 - 2x + 2 = (1 - i)^2 - 2(1 - i) + 2 = 1 - 2i + i^2 - 2 + 2i + 2$
$= 1 - 2i - 1 - 2 + 2i + 2 = 0$ or $0 + 0i$

65. $(2 - 3i)^2 - 2(2 - 3i) + 9 = 4 - 12i + 9i^2 - 4 + 6i + 9$
$= 4 - 12i - 9 - 4 + 6i + 9 = -6i$ or $0 - 6i$

67. $f(x) = x^2 - 2x + 2$

(A) $f(1 + i) = (1 + i)^2 - 2(1 + i) + 2$
$= 1 + 2i + i^2 - 2 - 2i + 2$
$= 1 + 2i - 1 - 2 - 2i + 2$
$= 0$

$f(1 - i) = (1 - i)^2 - 2(1 - i) + 2$
$= 1 - 2i + i^2 - 2 + 2i + 2$
$= 1 - 2i - 1 - 2 + 2i + 2$
$= 0$

Thus, $1 + i$ and $1 - i$ are both zeros of $f(x)$.

(B) There are no real solutions of $x^2 - 2x + 2 = 0$. The only solutions are the complex numbers $1 + i$ and $1 - i$. Therefore there are no real zeros of f. In fact, since $f(x) = (x - 1)^2 + 1$, the graph is a parabola with vertex at (1, 1), and 1 is a minimum value for f. Thus there can be no x intercepts for the graph of f.

69. $(3.17 - 4.08i)(7.14 + 2.76i)$
$= (3.17)(7.14) + (3.17)(2.76i) - (4.08i)(7.14) - (4.08i)(2.76i)$
$= 22.6338 + 8.7492i - 29.1312i - 11.2608i^2$
$= 22.6338 - 20.382i + 11.2608$
$= 33.9 - 20.4i$ to three significant digits

71. $\frac{8.14 + 2.63i}{3.04 + 6.27i} = \frac{(8.14 + 2.63i)}{(3.04 + 6.27i)} \cdot \frac{(3.04 - 6.27i)}{(3.04 - 6.27i)}$
$= \frac{(8.14)(3.04) - (8.14)(6.27i) + (2.63i)(3.04) - (2.63i)(6.27i)}{(3.04)^2 + (6.27)^2}$
$= \frac{24.7456 - 51.0378i + 7.9952i - 16.901i^2}{9.2416 + 39.3129}$
$= \frac{24.7456 - 43.0426i + 16.4901}{48.5545} = \frac{41.2357 - 43.0426i}{48.5545}$
$= 0.849 - 0.886i$ to three significant digits

73. $(2x - 1) + (3y + 2)i = 5 - 4i$
We note: $a + bi = c + di$ if and only if $a = c$ and $b = d$. Thus
$2x - 1 = 5$ and $3y + 2 = -4$
$2x = 6$ $\quad$ $3y = -6$
$x = 3$ $\quad$ $y = -2$

75. $\frac{(1 + x) + (y - 2)i}{1 + i} = 2 - i$
$(1 + i)\frac{(1 + x) + (y - 2)i}{1 + i} = (2 - i)(1 + i)$
$(1 + x) + (y - 2)i = 2 - i + 2i - i^2$
$(1 + x) + (y - 2)i = 3 + i$
We note: $a + bi = c + di$ if and only if $a = c$ and $b = d$. Thus
$1 + x = 3$ and $y - 2 = 1$
$x = 2$ $\quad$ $y = 3$

77. $(2 + i)z + i = 4i$
$(2 + i)z = 3i$
$z = \frac{3i}{2 + i}$
$z = \frac{3i}{2 + i} \cdot \frac{2 - i}{2 - i} = \frac{6i - 3i^2}{4 - i^2} = \frac{6i + 3}{4 + 1} = \frac{3 + 6i}{5}$
$= \frac{3}{5} + \frac{6}{5}i$ or $0.6 + 1.2i$

79. $3iz + (2 - 4i) = (1 + 2i)z - 3i$
$3iz - (1 + 2i)z = -3i - (2 - 4i)$
$(-1 + i)z = -3i - 2 + 4i$
$(-1 + i)z = -2 + i$
$z = \frac{-2 + i}{-1 + i}$
$z = \frac{-2 + i}{-1 + i} \cdot \frac{-1 - i}{-1 - i} = \frac{2 + 2i - i - i^2}{1 - i^2} = \frac{3 + i}{2}$
$= \frac{3}{2} + \frac{1}{2}i$ or $1.5 + 0.5i$

81. The error arises when equating $\sqrt{-1}\sqrt{-1}$ with $\sqrt{(-1)(-1)}$.
For positive real numbers a and b, $\sqrt{-a}\sqrt{-b} \neq \sqrt{(-a)(-b)}$.

83. $(a + bi)(a - bi) = a^2 - b^2i^2 = a^2 + b^2$ or $(a^2 + b^2) + 0i$

85. $(a + bi)(c + di) = ac + adi + bci + bdi^2 = ac + (ad + bc)i - bd$
$= (ac - bd) + (ad + bc)i$

87. $i^{4k} = (i^4)^k = (i^2 \cdot i^2)^k = [(-1)(-1)]^k = 1^k = 1$

89. 1. Definition of addition
2. Commutative property for addition of real numbers.
3. Definition of addition (read from right to left).

91. The product of a complex number and its conjugate is a real number.

$$
\begin{aligned}
z\bar{z} &= (x + yi)(x - yi) \\
&= x^2 - (yi)^2 \\
&= x^2 - y^2i^2 \\
&= x^2 + y^2 \text{ or } \\
&\quad (x^2 + y^2) + 0i.
\end{aligned}
$$

This is a real number.

93. The conjugate of a complex number is equal to the complex number if and only if the number is real. To prove a theorem containing the phrase "if and only if", it is often helpful to prove two parts separately. Thus: $\bar{z} = z$ if z is real; $\bar{z} = z$ only if z is real

Hypothesis: z is real
Conclusion: $\bar{z} = z$
Proof: Assume z is real, then
$z = x + 0i = x$
$\bar{z} = x - 0i = x$
Hence $z = \bar{z}$.

Hypothesis: $\bar{z} = z$
Conclusion: z is real
Proof: Assume $\bar{z} = z$,
that is, $x - yi = x + yi$
Then by the definition of equality
$x = x \quad -y = y$
$-2y = 0$
$y = 0$
Hence $z = x + 0i$, that is, z is real.

95. The conjugate of the sum of two complex numbers is equal to the sum of their conjugates.

$$
\begin{aligned}
\overline{z + w} &= \overline{(x + yi) + (u + vi)} \\
&= \overline{x + yi + u + vi} \\
&= \overline{x + u + (y + v)i} \\
&= (x + u) - (y + v)i \\
&= x + u - yi - vi \\
&= (x - yi) + (u - vi) \\
&= \bar{z} + \bar{w}
\end{aligned}
$$

97. The conjugate of the product of two complex numbers is equal to the product of their conjugates.

$$
\begin{aligned}
\overline{zw} &= \overline{(x + yi)(u + vi)} \\
&= \overline{xu + xvi + yui + yvi^2} \\
&= \overline{xu + (xv + yu)i - yv} \\
&= xu - yv - (xv + yu)i \\
&= xu - xvi - yv - yui \\
&= x(u - vi) - yui + yv(-1) \\
&= x(u - vi) - yui + yvi^2 \\
&= x(u - vi) - yi(u - vi) \\
&= (x - yi)(u - vi) \\
&= \bar{z}\,\bar{w}
\end{aligned}
$$

Section 1-5

1. $2x^2 = 8x$
$2x^2 - 8x = 0$
$2x(x - 4) = 0$
$2x = 0$ or $x - 4 = 0$
$x = 0 \qquad x = 4$

3. $4t^2 + 9 = 12t$
$4t^2 - 12t + 9 = 0$
$(2t - 3)^2 = 0$
$2t - 3 = 0$
$2t = 3$
$t = \frac{3}{2}$ (double root)

5. $3w^2 + 13w = 10$
$3w^2 + 13w - 10 = 0$
$(3w - 2)(w + 5) = 0$
$3w - 2 = 0$ or $w + 5 = 0$
$3w = 2 \qquad w = -5$
$w = \frac{2}{3}$

7. $m^2 - 25 = 0$
$m^2 = 25$
$m = \pm\sqrt{25}$
$m = \pm 5$

9. $c^2 + 9 = 0$
$c^2 = -9$
$c = \pm\sqrt{-9}$
$c = \pm 3i$

11. $4y^2 + 9 = 0$
$4y^2 = -9$
$y^2 = -\frac{9}{4}$
$y = \pm\sqrt{-\frac{9}{4}}$
$y = \pm\frac{i\sqrt{9}}{\sqrt{4}}$
$y = \pm\frac{3i}{2}$

13. $25z^2 - 32 = 0$
$25z^2 = 32$
$z^2 = \frac{32}{25}$
$z = \pm\sqrt{\frac{32}{25}}$
$z = \pm\frac{\sqrt{32}}{\sqrt{25}}$
$z = \pm\frac{4\sqrt{2}}{5}$

15. $(s + 1)^2 = 5$
$s + 1 = \pm\sqrt{5}$
$s = -1 \pm \sqrt{5}$

17. $(n - 3)^2 = -4$
$n - 3 = \pm\sqrt{-4}$
$n - 3 = \pm 2i$
$n = 3 \pm 2i$

19. $x^2 - 2x - 1 = 0$
$a = 1,\ b = -2,\ c = -1$
$b^2 - 4ac = (-2)^2 - 4(1)(-1)$
$= 8$

The discriminant is positive; there are two real roots.

$x = \frac{-b \pm \sqrt{b^2 - 4ac}}{2a}$

$x = \frac{-(-2) \pm \sqrt{8}}{2(1)}$

$x = \frac{2 \pm 2\sqrt{2}}{2}$

$x = 1 \pm \sqrt{2}$

Common Errors:
It is incorrect to cancel this way:
$\frac{\not{2} \pm 2\sqrt{2}}{\not{2}} \neq \pm 2\sqrt{2}$
or this way $\frac{\not{2} \pm 2\sqrt{2}}{\not{2}} \neq 1 \pm 2\sqrt{2}$

21. $x^2 - 2x + 3 = 0$
$a = 1,\ b = -2,\ c = 3$
$b^2 - 4ac = (-2)^2 - 4(1)(3) = -8$
The discriminant is negative; there are no real roots.

$$x = \frac{-b \pm \sqrt{b^2 - 4ac}}{2a}$$
$$x = \frac{-(-2) \pm \sqrt{-8}}{2(1)}$$
$$x = \frac{2 \pm 2i\sqrt{2}}{2}$$
$$x = 1 \pm i\sqrt{2}$$

23. $2t^2 + 8 = 6t$
$2t^2 - 6t + 8 = 0$
$t^2 - 3t + 4 = 0$
$a = 1,\ b = -3,\ c = 4$
$b^2 - 4ac = (-3)^2 - 4(1)(4) = -7$
The discriminant is negative; there are no real roots.

$$t = \frac{-b \pm \sqrt{b^2 - 4ac}}{2a}$$
$$t = \frac{-(-3) \pm \sqrt{-7}}{2(1)}$$
$$t = \frac{3 \pm i\sqrt{7}}{2}$$

25. $2t^2 + 1 = 6t$
$2t^2 - 6t + 1 = 0$
$a = 2,\ b = -6,\ c = 1$
$b^2 - 4ac = (-6)^2 - 4(2)(1) = 28$
The discriminant is positive; there are two real roots.

$$t = \frac{-b \pm \sqrt{b^2 - 4ac}}{2a}$$
$$t = \frac{-(-6) \pm \sqrt{28}}{2(2)}$$
$$t = \frac{6 \pm \sqrt{28}}{4}$$
$$t = \frac{6 \pm 2\sqrt{7}}{4}$$
$$t = \frac{3 \pm \sqrt{7}}{2}$$

27.
$$x^2 - 4x - 1 = 0$$
$$x^2 - 4x = 1$$
$$x^2 - 4x + 4 = 5$$
$$(x - 2)^2 = 5$$
$$x - 2 = \pm\sqrt{5}$$
$$x = 2 \pm \sqrt{5}$$

29.
$$2r^2 + 10r + 11 = 0$$
$$r^2 + 5r + \frac{11}{2} = 0$$
$$r^2 + 5r = -\frac{11}{2}$$
$$r^2 + 5r + \frac{25}{4} = -\frac{11}{2} + \frac{25}{4}$$
$$\left(r + \frac{5}{2}\right)^2 = \frac{3}{4}$$
$$r + \frac{5}{2} = \pm\sqrt{\frac{3}{4}}$$
$$r = -\frac{5}{2} \pm \frac{\sqrt{3}}{\sqrt{4}}$$
$$r = -\frac{5}{2} \pm \frac{\sqrt{3}}{2}$$
$$r = \frac{-5 \pm \sqrt{3}}{2}$$

31.
$$4u^2 + 8u + 15 = 0$$
$$u^2 + 2u + \frac{15}{4} = 0$$
$$u^2 + 2u = -\frac{15}{4}$$
$$u^2 + 2u + 1 = -\frac{15}{4} + 1$$
$$(u + 1)^2 = -\frac{11}{4}$$
$$u + 1 = \pm\sqrt{-\frac{11}{4}}$$
$$u = -1 \pm \frac{i\sqrt{11}}{\sqrt{4}}$$
$$u = -\frac{2}{2} \pm \frac{i\sqrt{11}}{2}$$
$$u = \frac{-2 \pm i\sqrt{11}}{2}$$

33.
$$3w^2 + 4w + 3 = 0$$
$$w^2 + \frac{4}{3}w + 1 = 0$$
$$w^2 + \frac{4}{3}w = -1$$
$$w^2 + \frac{4}{3}w + \frac{4}{9} = -1 + \frac{4}{9}$$
$$\left(w + \frac{2}{3}\right)^2 = -\frac{5}{9}$$
$$w + \frac{2}{3} = \pm\sqrt{-\frac{5}{9}}$$
$$w = -\frac{2}{3} \pm \frac{i\sqrt{5}}{\sqrt{9}}$$
$$w = -\frac{2}{3} \pm \frac{i\sqrt{5}}{3}$$
$$w = \frac{-2 \pm i\sqrt{5}}{3}$$

35.
$$12x^2 + 7x = 10$$
$$12x^2 + 7x - 10 = 0$$
$$(4x + 5)(3x - 2) = 0$$ Polynomial is factorable.
$$4x + 5 = 0 \text{ or } 3x - 2 = 0$$
$$4x = -5 \qquad 3x = 2$$
$$x = -\frac{5}{4} \qquad x = \frac{2}{3}$$

37. $(2y - 3)^2 = 5$ Format for the square root method.
$$2y - 3 = \pm\sqrt{5}$$
$$2y = 3 \pm \sqrt{5}$$
$$y = \frac{3 \pm \sqrt{5}}{2}$$

39.
$$x^2 = 3x + 1$$
$$x^2 - 3x - 1 = 0$$ Polynomial is not factorable, use quadratic formula.
$$x = \frac{-b \pm \sqrt{b^2 - 4ac}}{2a}$$
$$a = 1,\ b = -3,\ c = -1$$
$$x = \frac{-(-3) \pm \sqrt{(-3)^2 - 4(1)(-1)}}{2(1)}$$
$$x = \frac{3 \pm \sqrt{13}}{2}$$

41.
$$7n^2 = -4n$$
$$7n^2 + 4n = 0$$
$$n(7n + 4) = 0$$ Polynomial is factorable.
$$n = 0 \text{ or } 7n + 4 = 0$$
$$7n = -4$$
$$n = -\frac{4}{7}$$

43.
$$1 + \frac{8}{x^2} = \frac{4}{x}$$ Excluded value: $x \neq 0$
$$x^2 + 8 = 4x$$
$$x^2 - 4x + 8 = 0$$ Polynomial is not factorable, use quadratic formula, or complete the square.
$$x^2 - 4x = -8$$
$$x^2 - 4x + 4 = -4$$
$$(x - 2)^2 = -4$$
$$x - 2 = \pm\sqrt{-4}$$
$$x - 2 = \pm i\sqrt{4}$$
$$x - 2 = \pm 2i$$
$$x = 2 \pm 2i$$

45.
$$\frac{24}{10 + m} + 1 = \frac{24}{10 - m}$$ Excluded value: $m \neq -10, 10$: LCD is $(10 + m)(10 - m)$
$$(10 + m)(10 - m)\frac{24}{10 + m} + (10 + m)(10 - m) = (10 + m)(10 - m)\frac{24}{10 - m}$$
$$24(10 - m) + 100 - m^2 = 24(10 + m)$$
$$240 - 24m + 100 - m^2 = 240 + 24m$$
$$340 - 24m - m^2 = 240 + 24m$$
$$0 = m^2 + 48m - 100$$

$$\begin{aligned} m^2 + 48m - 100 &= 0 \quad \text{Polynomial is factorable.} \\ (m + 50)(m - 2) &= 0 \\ m + 50 &= 0 \quad \text{or} \quad m - 2 = 0 \\ m &= -50 \qquad\quad m = 2 \end{aligned}$$

47. $\dfrac{2}{x - 2} = \dfrac{4}{x - 3} - \dfrac{1}{x + 1}$ Excluded values: $x \neq 2, 3, -1$

$$(x - 2)(x - 3)(x + 1)\frac{2}{x - 2} = (x - 2)(x - 3)(x + 1)\frac{4}{x - 3} - (x - 2)(x - 3)(x + 1)\frac{1}{x + 1}$$

$$\begin{aligned} 2(x - 3)(x + 1) &= 4(x - 2)(x + 1) - (x - 2)(x - 3) \\ 2(x^2 - 2x - 3) &= 4(x^2 - x - 2) - (x^2 - 5x + 6) \\ 2x^2 - 4x - 6 &= 4x^2 - 4x - 8 - x^2 + 5x - 6 \\ 2x^2 - 4x - 6 &= 3x^2 + x - 14 \\ 0 &= x^2 + 5x - 8 \\ x^2 + 5x - 8 &= 0 \quad \text{Polynomial is not factorable, use quadratic formula.} \end{aligned}$$

$$x = \frac{-b \pm \sqrt{b^2 - 4ac}}{2a} \quad a = 1,\ b = 5,\ c = -8$$

$$x = \frac{-5 \pm \sqrt{(5)^2 - 4(1)(-8)}}{2(1)} = \frac{-5 \pm \sqrt{57}}{2}$$

49. $\dfrac{x + 2}{x + 3} - \dfrac{x^2}{x^2 - 9} = 1 - \dfrac{x - 1}{3 - x}$ Excluded values: $x \neq 3, -3$

$$(x - 3)(x + 3)\frac{(x + 2)}{x + 3} - (x - 3)(x + 3)\frac{x^2}{x^2 - 9} = (x - 3)(x + 3) - (x - 3)(x + 3)\frac{x - 1}{3 - x}$$

$$\begin{aligned} (x - 3)(x + 2) - x^2 &= x^2 - 9 + (x - 1)(x + 3) \\ x^2 - x - 6 - x^2 &= x^2 - 9 + x^2 + 2x - 3 \\ -x - 6 &= 2x^2 + 2x - 12 \\ 0 &= 2x^2 + 3x - 6 \\ 2x^2 + 3x - 6 &= 0 \quad \text{Polynomial is not factorable, use quadratic formula.} \end{aligned}$$

$$x = \frac{-b \pm \sqrt{b^2 - 4ac}}{2a} \quad a = 2,\ b = 3,\ c = -6$$

$$x = \frac{-3 \pm \sqrt{(3)^2 - 4(2)(-6)}}{2(2)}$$

$$x = \frac{-3 \pm \sqrt{57}}{4}$$

51. According to Theorem 3, Section 1-3, $|3u - 2| = u^2$ is equivalent to $3u - 2 = u^2$ or $3u - 2 = -u^2$. Hence we must solve:

$3u - 2 = u^2$ or $3u - 2 = -u^2$

$0 = u^2 - 3u \neq 2$ $\quad$ $u^2 + 3u - 2 = 0$

$0 = (u - 1)(u - 2)$ $\quad$ $u = \dfrac{-b \pm \sqrt{b^2 - 4ac}}{2a}$ $\quad a = 1,\ b = 3,\ c = -2$

$u - 1 = 0$ or $u - 2 = 0$ $\quad$ $u = \dfrac{-3 \pm \sqrt{(3)^2 - 4(1)(-2)}}{2(1)}$

$u = 1$ $\quad$ $u = 2$

$u = \dfrac{-3 \pm \sqrt{17}}{2}$

53.
$$s = \frac{1}{2}gt^2$$
$$\frac{1}{2}gt^2 = s$$
$$gt^2 = 2s$$
$$t^2 = \frac{2s}{g}$$
$$t = \sqrt{\frac{2s}{g}}$$

55.
$$P = EI - RI^2$$
$$RI^2 - EI + P = 0$$
$$I = \frac{-b \pm \sqrt{b^2 - 4ac}}{2a} \quad a = R,\ b = -E,\ c = P$$
$$I = \frac{-(-E) \pm \sqrt{(-E)^2 - 4(R)(P)}}{2(R)}$$
$$I = \frac{E + \sqrt{E^2 - 4RP}}{2R} \quad \text{(positive square root)}$$

57. $2.07x^2 - 3.79x + 1.34 = 0$
$$x = \frac{-b \pm \sqrt{b^2 - 4ac}}{2a} \quad a = 2.07,\ b = -3.79,\ c = 1.34$$
$$x = \frac{-(-3.79) \pm \sqrt{(-3.79)^2 - 4(2.07)(1.34)}}{2(2.07)} = \frac{3.79 \pm 1.81}{4.14}$$
$$x = 1.35,\ 0.479$$

59. $4.83x^2 + 2.04x - 3.18 = 0$
$$x = \frac{-b \pm \sqrt{b^2 - 4ac}}{2a} \quad a = 4.83,\ b = 2.04,\ c = -3.18$$
$$x = \frac{-2.04 \pm \sqrt{(2.04)^2 - 4(4.83)(-3.18)}}{2(4.83)} = \frac{-2.04 \pm 8.10}{9.66}$$
$$x = -1.05,\ 0.627$$

61. In this problem, $a = 1$, $b = 4$, $c = c$. Thus, the discriminant $b^2 - 4ac = (4)^2 - 4(1)(c) = 16 - 4c$. Hence,
if $16 - 4c > 0$, thus $16 > 4c$ or $c < 4$, there are two distinct real roots.
if $16 - 4c = 0$, thus $c = 4$, there is one real double root,
and if $16 - 4c < 0$, thus $16 < 4c$ or $c > 4$, there are two distinct imaginary roots.

63. $\sqrt{3}x^2 = 8\sqrt{2}x - 4\sqrt{3}$

$\sqrt{3}x^2 - 8\sqrt{2}x + 4\sqrt{3} = 0$

$$x = \frac{-b \pm \sqrt{b^2 - 4ac}}{2a} \quad a = \sqrt{3},\ b = -8\sqrt{2},\ c = 4\sqrt{3}$$

$$x = \frac{-(-8\sqrt{2}) \pm \sqrt{(-8\sqrt{2})^2 - 4(\sqrt{3})(4\sqrt{3})}}{2(\sqrt{3})}$$

$$x = \frac{8\sqrt{2} \pm \sqrt{(64)(2) - (16)(3)}}{2\sqrt{3}}$$

$$x = \frac{8\sqrt{2} \pm \sqrt{80}}{2\sqrt{3}}$$

$$x = \frac{8\sqrt{2} \pm 4\sqrt{5}}{2\sqrt{3}}$$

$$x = \frac{2(4\sqrt{2} \pm 2\sqrt{5})}{2\sqrt{3}} \cdot \frac{\sqrt{3}}{\sqrt{3}}$$

$$x = \frac{4\sqrt{6} \pm 2\sqrt{15}}{3} \text{ or } \frac{4}{3}\sqrt{6} \pm \frac{2}{3}\sqrt{15}$$

65. $x^2 + 2ix = 3$

$x^2 + 2ix - 3 = 0$

$$x = \frac{-b \pm \sqrt{b^2 - 4ac}}{2a} \quad a = 1,\ b = 2i,\ c = -3$$

$$x = \frac{-2i \pm \sqrt{(2i)^2 - 4(1)(-3)}}{2(1)}$$

$$x = \frac{-2i \pm \sqrt{-4 + 12}}{2}$$

$$x = \frac{-2i \pm \sqrt{8}}{2}$$

$$x = \frac{-2i \pm 2\sqrt{2}}{2}$$

$$x = \frac{2(-i \pm \sqrt{2})}{2}$$

$$x = -i \pm \sqrt{2}$$

$$x = \sqrt{2} - i,\ -\sqrt{2} - i$$

67. $x^3 - 1 = 0$

$(x - 1)(x^2 + x + 1) = 0$

$x - 1 = 0$ or $x^2 + x + 1 = 0$

$x = 1 \qquad x = \dfrac{-b \pm \sqrt{b^2 - 4ac}}{2a} \quad a = 1,\ b = 1,\ c = 1$

$$x = \frac{-1 \pm \sqrt{(1)^2 - 4(1)(1)}}{2(1)}$$

$$x = \frac{-1 \pm \sqrt{1 - 4}}{2}$$

$$x = \frac{-1 \pm \sqrt{-3}}{2}$$

$$x = \frac{-1 \pm i\sqrt{3}}{2} \text{ or } -\frac{1}{2} \pm \frac{1}{2}i\sqrt{3}$$

69. If a quadratic equation has two roots, they are $\dfrac{-b \pm \sqrt{b^2 - 4ac}}{2a}$ and $\dfrac{-b - \sqrt{b^2 - 4ac}}{2a}$. If a, b, c are rational, then so are $-b$, $2a$, and $b^2 - 4ac$. Then, *either* $\sqrt{b^2 - 4ac}$ is rational, hence $\dfrac{-b + \sqrt{b^2 - 4ac}}{2a}$ and $\dfrac{-b - \sqrt{b^2 - 4ac}}{2a}$ are both rational, *or*, $\sqrt{b^2 - 4ac}$ is irrational, hence $\dfrac{-b + \sqrt{b^2 - 4ac}}{2a}$ and $\dfrac{-b - \sqrt{b^2 - 4ac}}{2a}$ are both irrational, *or*, $\sqrt{b^2 - 4ac}$ is imaginary, hence $\dfrac{-b + \sqrt{b^2 - 4ac}}{2a}$ and $\dfrac{-b + \sqrt{b^2 - 4ac}}{2a}$ are both imaginary. There is no other possibility; hence, one root cannot be rational while the other is irrational.

71. $r_1 = \dfrac{-b + \sqrt{b^2 - 4ac}}{2a}$ $r_2 = \dfrac{-b - \sqrt{b^2 - 4ac}}{2a}$

$$r_1 r_2 = \frac{(-b + \sqrt{b^2 - 4ac})}{2a}\frac{(-b - \sqrt{b^2 - 4ac})}{2a}$$

$$= \frac{(-b)^2 - (\sqrt{b^2 - 4ac})^2}{4a^2} = \frac{b^2 - (b^2 - 4ac)}{4a^2} = \frac{b^2 - b^2 + 4ac}{4a^2} = \frac{4ac}{4a^2} = \frac{c}{a}$$

73. The ± in front still yields the same two numbers even if a is negative.

75. Let x = one number.
Since their sum is 21,
$21 - x$ = other number
Then, since their product is 104,

$$\begin{aligned} x(21 - x) &= 104 \\ 21x - x^2 &= 104 \\ 0 &= x^2 - 21x + 104 \\ x^2 - 21x + 104 &= 0 \\ (x - 13)(x - 8) &= 0 \\ x - 13 = 0 \text{ or } x - 8 &= 0 \\ x = 13 \quad x &= 8 \end{aligned}$$

The numbers are 8 and 13.

77. Let x = first of the two consecutive even integers.
Then $x + 2$ = second of these integers
Since their product is 168,

$$\begin{aligned} x(x + 2) &= 168 \\ x^2 + 2x &= 168 \\ x^2 + 2x - 168 &= 0 \\ (x - 12)(x + 14) &= 0 \\ x - 12 &= 0 \text{ or } x + 14 = 0 \\ x &= 12 \quad\quad x = -14 \end{aligned}$$

If $x = 12$, the two consecutive positive even integers must be 12 and 14. We discard the other solution, since the numbers must be positive.

79. Let x = amount of increase
Then $x + 4$ = length of new rectangle
$x + 2$ = width of new rectangle
Using $A = ab$, we have
Area of new rectangle = 2 × Area of old rectangle.

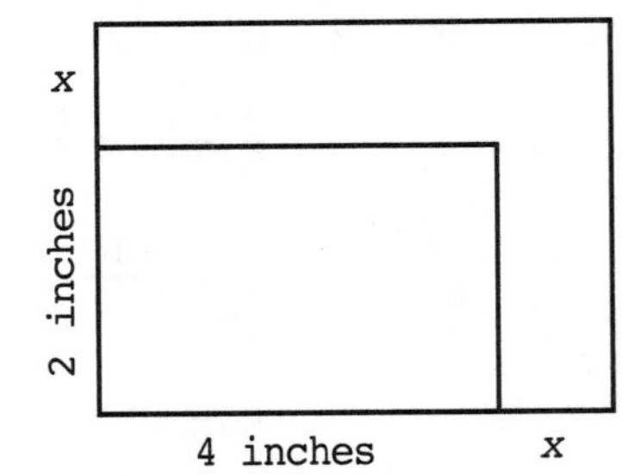

$$\begin{aligned} (x + 4)(x + 2) &= 2(4)(2) \\ x^2 + 6x + 8 &= 16 \\ x^2 + 6x - 8 &= 0 \\ x^2 + 6x &= 8 \\ x^2 + 6x + 9 &= 17 \\ (x + 3)^2 &= 17 \\ x + 3 &= \pm\sqrt{17} \\ x &= -3 \pm \sqrt{17} \end{aligned}$$

We discard the negative solution and take $x = -3 + \sqrt{17} = 1.12$. Then dimensions of new rectangle are $x + 4$ by $x + 2$ or 5.12 by 3.12 inches.

81. If $p = 3$, the demand is given by
$q = 1{,}600 - 200p = 1{,}600 - 200(3) = 1{,}000$ hamburgers.
The revenue is given by
$R = qp = 1{,}000 \cdot 3 = \$3{,}000$

83. The revenue equation is $R = qp = (1{,}600 - 200p)p$. Solve
(A) $2{,}800 = (1{,}600 - 200p)p$

$$200p^2 - 1{,}600p + 2{,}800 = 0$$

$$p^2 - 8p + 14 = 0$$

$$p = \frac{-(-8) \pm \sqrt{(-8)^2 - 4(1)(14)}}{2(1)} = \frac{8 \pm \sqrt{8}}{2}$$

$p = \$2.59$ or $p = \$5.41$

(B) $3,200 = (1,600 - 200p)p$

$$200p^2 - 1,600p + 3,200 = 0$$
$$p^2 - 8p + 16 = 0$$
$$(p - 4)^2 = 0$$
$$p = \$4$$

(C) $3,400 = (1,600 - 200p)p$

$$200p^2 - 1,600p + 3,400 = 0$$
$$p^2 - 8p + 17 = 0$$

Since the discriminant $b^2 - 4ac = (-8)^2 - 4(1)(17) = -4$ is negative, there is no solution.

85. Let r = rate of slow plane.
Then $r + 140$ = rate of fast plane.
After 1 hour $r(1) = r$ = distance traveled by slow plane.
$(r + 140)(1) = r + 140$ = distance traveled by fast plane.

Applying the Pythagorean theorem, we have

$$r^2 + (r + 140)^2 = 260^2$$
$$r^2 + r^2 + 280r + 19,600 = 67,600$$
$$2r^2 + 280r - 48,000 = 0$$
$$r^2 + 140r - 24,000 = 0$$
$$(r + 240)(r - 100) = 0$$
$$r + 240 = 0 \quad \text{or} \quad r - 100 = 0$$
$$r = -240 \qquad r = 100$$

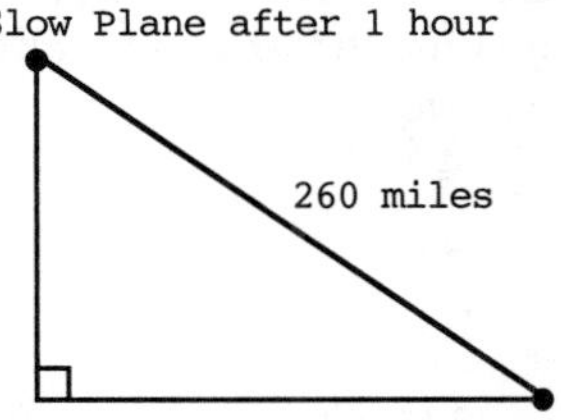

Discarding the negative solution, we have
r = 100 miles per hour = rate of slow plane
$r + 140$ = 240 miles per hour = rate of fast plane.

87. Let t = time for smaller pipe to fill tank alone
$t - 5$ = time for larger pipe to fill tank alone
5 = time for both pipes to fill tank together

Then $\frac{1}{t}$ = rate for smaller pipe

$\frac{1}{t-5}$ = rate for larger pipe

$$\begin{pmatrix}\text{Part of job}\\ \text{completed by}\\ \text{smaller pipe}\end{pmatrix} + \begin{pmatrix}\text{Part of job}\\ \text{completed by}\\ \text{larger pipe}\end{pmatrix} = 1 \text{ whole job}$$

$$\frac{1}{t}(5) + \frac{1}{t-5}(5) = 1$$

$$\frac{5}{t} + \frac{5}{t-5} = 1 \quad \text{Excluded values: } t \neq 0, 5$$

$$t(t-5)\frac{5}{t} + t(t-5) = t(t-5)$$
$$5(t-5) + 5t = t(t-5)$$
$$5t - 25 + 5t = t^2 - 5t$$
$$10t - 25 = t^2 - 5t$$
$$0 = t^2 - 15t + 25$$
$$t^2 - 15t + 25 = 0$$

$$t = \frac{-b \pm \sqrt{b^2 - 4ac}}{2a} \quad a = 1,\ b = -15,\ c = 25$$

$$t = \frac{-(-15) \pm \sqrt{(-15)^2 - 4(1)(25)}}{2(1)} = \frac{15 \pm \sqrt{125}}{2}$$

$$t = 13.09,\ 1.91$$
$$t - 5 = 8.09,\ -3.09$$

Discarding the answer for t which results in a negative answer for $t - 5$, we have 13.09 hours for smaller pipe alone, 8.09 hours for larger pipe alone.

89. Let v = speed of car. Applying the given formula, we have

$$165 = 0.044v^2 + 1.1v$$
$$0 = 0.044v^2 + 1.1v - 165$$
$$0.044v^2 + 1.1v - 165 = 0$$
$$v = \frac{-b \pm \sqrt{b^2 - 4ac}}{2a} \quad a = 0.044,\ b = 1.1,\ c = -165$$
$$v = \frac{-1.1 \pm \sqrt{(1.1)^2 - 4(0.044)(-165)}}{2(0.044)} = \frac{-1.1 \pm 5.5}{0.088}$$
$$v = -75 \text{ or } 50$$

Discarding the negative answer, we have $v = 50$ miles per hour.

91. Let ℓ = length of building

w = width of building.

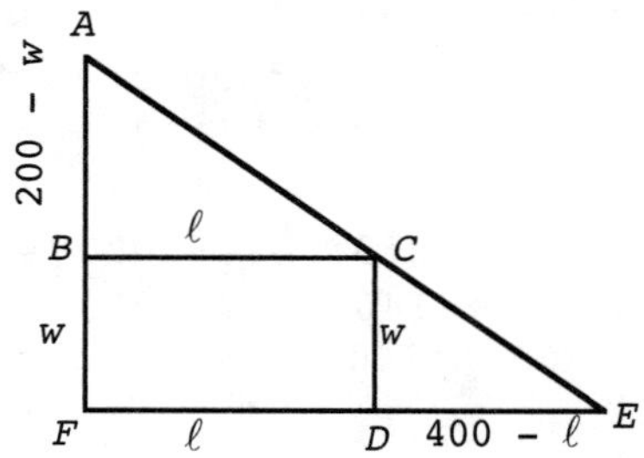

Then, using the hint, in the similar triangles ABC and AFE, we have

$$\frac{\ell}{200 - w} = \frac{400}{200}$$
$$\frac{\ell}{200 - w} = 2$$
$$\ell = 2(200 - w)$$
$$\ell = 400 - 2w$$

Since the cross-sectional area of the building is given as 15,000 ft^2, we have
$\ell w = 15{,}000$

Since the cross-sectional area of the building is given as 15,000 ft^2, we have
$\ell w = 15{,}000$

Substituting, we get,

$$(400 - 2w)w = 15{,}000$$
$$400w - 2w^2 = 15{,}000$$
$$0 = 2w^2 - 400w + 15{,}000$$
$$0 = w^2 - 200w + 7{,}500$$
$$0 = (w - 50)(w - 150)$$

$w - 50 = 0$ or $w - 150 = 0$

$w = 50$ $w = 150$

$\ell = 400 - 2w$ $\ell = 400 - 2w$

$= 400 - 2(50)$ $= 400 - 2(150)$

$= 300$ $= 100$

Thus, there are two solutions: the building is 50 ft wide and 300 ft long or 150 ft wide and 100 ft long.

93. Let x = distance from the warehouse to Factory A. Since the distance from the warehouse to Factory B via Factory A is known (it is the difference in odometer readings: 52937–52846) to be 91 miles, then

$91 - x$ = distance from Factory A to Factory B.

The distance from Factory B to the warehouse is known (it is the difference in odometer readings: 53002–52937) to be 65 miles. Applying the Pythagorean theorem, we have

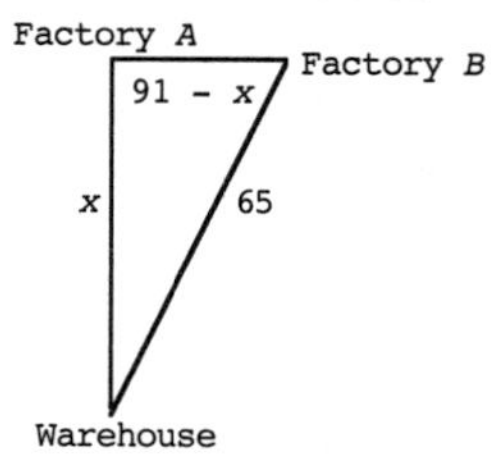

$$x^2 + (91 - x)^2 = 65^2$$
$$x^2 + 8281 - 182x + x^2 = 4225$$
$$2x^2 - 182x + 4056 = 0$$
$$x^2 - 91x + 2028 = 0$$
$$(x - 52)(x - 39) = 0$$

$x - 52 = 0$ or $x - 39 = 0$

$x = 52$ mi $x = 39$ mi

Since we are told that the distance from the warehouse to Factory A was greater than the distance from Factory A to Factory B, we discard the solution $x = 39$, which would lead to $91 - x = 52$ miles, a contradiction. 52 miles.

Section 1-6

1. This statement is true, since $\pm\sqrt{5}$ are the only two solutions of $x^2 = 5$. T

3. This statement is false.
$(\sqrt{x-1} + 1)^2$ is not equal to $x - 1 + 1$ or x, in general.
In fact
$(\sqrt{x-1} + 1)^2 = x - 1 + 2\sqrt{x-1} + 1 = x + 2\sqrt{x-1}$.
This is only equal to x in case $x = 1$. F

5. This statement is false. If $x^3 = 2$ then $x = \sqrt[3]{2}$ or x is equal to one of two non-real complex numbers whose cube is 2. If $x = 8$, x^3 must equal 512. F

7.
$$\begin{aligned} \sqrt{x+2} &= 4 \\ x + 2 &= 16 \\ x &= 14 \end{aligned}$$
Check:
$$\begin{aligned} \sqrt{14+2} &\stackrel{?}{=} 4 \\ \sqrt{16} &\stackrel{?}{=} 4 \\ 4 &\stackrel{\checkmark}{=} 4 \end{aligned}$$
Solution: 14

9.
$$\begin{aligned} \sqrt{3y-2} &= y - 2 \\ 3y - 2 &= y^2 - 4y + 4 \\ 0 &= y^2 - 7y + 6 \\ y^2 - 7y + 6 &= 0 \\ (y-1)(y-6) &= 0 \\ y &= 1, 6 \end{aligned}$$
Check: 1: $\sqrt{3(1)-2} \stackrel{?}{=} 1 - 2$
$\sqrt{1} \stackrel{?}{=} -1$
$1 \neq -1$
Not a solution

6: $\sqrt{3(6)-2} \stackrel{?}{=} 6 - 2$
$\sqrt{16} \stackrel{?}{=} 4$
$4 \stackrel{\checkmark}{=} 4$
A solution

Solution: 6

11.
$$\begin{aligned} \sqrt{5w+6} - w &= 2 \\ \sqrt{5w+6} &= w + 2 \\ 5w + 6 &= w^2 + 4w + 4 \\ 0 &= w^2 - w - 2 \\ w^2 - w - 2 &= 0 \\ (w-2)(w+1) &= 0 \\ w &= 2, -1 \end{aligned}$$

Common Error: $5w + 6 - w^2 = 4$ or $5w + 6 + w^2 = 4$ are not equivalent to the original equation.

Check:
2: $\sqrt{5(2)+6} - 2 \stackrel{?}{=} 2$
$4 - 2 \stackrel{\checkmark}{=} 2$
A solution

−1: $\sqrt{5(-1)+6} - (-1) \stackrel{?}{=} 2$
$1 + 1 \stackrel{\checkmark}{=} 2$
A solution

Solution: 2, −1

13.
$$\begin{aligned} |2x+1| &= x + 2 \\ |2x+1|^2 &= (x+2)^2 \\ 4x^2 + 4x + 1 &= x^2 + 4x + 4 \\ 3x^2 &= 3 \\ x^2 &= 1 \\ x &= 1 \text{ or } x = -1 \end{aligned}$$

Check: $x = 1$
$|2x + 1| = x + 2$
$|2\cdot1 + 1| \stackrel{?}{=} 1 + 2$
$|3| \stackrel{?}{=} 3$
$3 \stackrel{\checkmark}{=} 3$
A solution

$x = -1$
$|2x + 1| = x + 2$
$|2(-1) + 1| \stackrel{?}{=} (-1) + 2$
$|-1| \stackrel{?}{=} 1$
$1 \stackrel{\checkmark}{=} 1$
A solution

Solution: 1, −1

15. $|x - 5| = 7 - 2x$
$|x - 5|^2 = (7 - 2x)^2$
$x^2 - 10x + 25 = 49 - 28x + 4x^2$
$0 = 3x^2 - 18x + 24$
$3(x - 2)(x - 4) = 0$
$x - 2 = 0$ or $x - 4 = 0$
$x = 2 \qquad x = 4$

Check: $x = 2$
$|x - 5| = 7 - 2x$
$|2 - 5| \stackrel{?}{=} 7 - 2 \cdot 2$
$|-3| \stackrel{?}{=} 3$
$3 \stackrel{\checkmark}{=} 3$
A solution

$x = 4$
$|x - 5| = 7 - 2x$
$|4 - 5| \stackrel{?}{=} 7 - 2 \cdot 4$
$|-1| \stackrel{?}{=} -1$
$1 \neq -1$
Not a solution

Solution: 2

17. $|3x - 4| = 2x - 5$
$|3x - 4|^2 = (2x - 5)^2$
$9x^2 - 24x + 16 = 4x^2 - 20x + 25$
$5x^2 - 4x - 9 = 0$
$(5x - 9)(x + 1) = 0$
$5x - 9 = 0$ or $x + 1 = 0$
$x = \frac{9}{5} \qquad x = -1$

Check: $x = \frac{9}{5}$
$|3x - 4| = 2x - 5$
$\left|3\left(\frac{9}{5}\right) - 4\right| \stackrel{?}{=} 2\left(\frac{9}{5}\right) - 5$
$\left|\frac{7}{5}\right| \stackrel{?}{=} -\frac{7}{5}$
$\frac{7}{5} \neq -\frac{7}{5}$
Not a solution

$x = -1$
$|3x - 4| = 2x - 5$
$|3(-1) - 4| \stackrel{?}{=} 2(-1) - 5$
$|-7| \stackrel{?}{=} -7$
$7 \neq -7$
Not a solution

No solution

19. Since $x^5 \neq (x^2)^2$ this is not an equation of quadratic type.

21. If we set $u = t^{2/5}$ then $u^2 = t^{4/5}$ and the equation would become $6u^2 + 11u - 8 = 0$; hence, the equation is of quadratic type.

23. If we set $u = \sqrt{y}$ then $u^2 = y$ and the equation would become $2u^2 + u + 5 = 0$; hence, the equation is of quadratic type.

25. $\sqrt{3t - 2} = 1 - 2\sqrt{t}$
$3t - 2 = 1 - 4\sqrt{t} + 4t$
$-t - 3 = -4\sqrt{t}$
$t^2 + 6t + 9 = 16t$
$t^2 - 10t + 9 = 0$
$(t - 9)(t - 1) = 0$
$t = 9, 1$

Check: 9: $\sqrt{3 \cdot 9 - 2} \stackrel{?}{=} 1 - 2\sqrt{9}$
$\sqrt{25} \stackrel{\checkmark}{=} 1 - 6$
$5 \neq -5$
Not a solution

1: $\sqrt{3 \cdot 9 - 2} \stackrel{?}{=} 1 - 2\sqrt{1}$
$\sqrt{1} \stackrel{?}{=} 1 - 2$
$1 \neq -1$
Not a solution

No solution

27. $m^4 + 2m^2 - 15 = 0$
Let $u = m^2$, then
$u^2 + 2u - 15 = 0$
$(u + 5)(u - 3) = 0$
$u = -5, 3$
$m^2 = -5 \qquad m^2 = 3$
$m = \pm i\sqrt{5} \qquad m = \pm\sqrt{3}$

29. $3x = \sqrt{x^2 - 2}$

$9x^2 = x^2 - 2$

$8x^2 = -2$

$x^2 = -\frac{1}{4}$

$x = \pm\sqrt{-\frac{1}{4}}$

$x = \pm\frac{1}{2}i$

Check: $\frac{1}{2}i$: $3(\frac{1}{2}i) \stackrel{?}{=} \sqrt{(\frac{1}{2}i)^2 - 2}$

$\frac{3}{2}i \stackrel{?}{=} \sqrt{-\frac{1}{4} - 2}$

$\frac{3}{2}i \stackrel{?}{=} \sqrt{-\frac{9}{4}}$

$\frac{3}{2}i \stackrel{\surd}{=} \frac{3}{2}i$

A solution

$-\frac{1}{2}i$: $3(-\frac{1}{2}i) \stackrel{?}{=} \sqrt{(-\frac{1}{2}i)^2 - 2}$

$-\frac{3}{2}i \stackrel{?}{=} \sqrt{-\frac{1}{4} - 2}$

$-\frac{3}{2}i \stackrel{?}{=} \sqrt{-\frac{9}{4}}$

$-\frac{3}{2}i \neq \frac{3}{2}i$

Not a solution

Solution: $\frac{1}{2}i$

31. $2y^{2/3} + 5y^{1/3} - 12 = 0$

Let $u = y^{1/3}$, then

$2u^2 + 5u - 12 = 0$

$(2u - 3)(u + 4) = 0$

$2u - 3 = 0 \qquad u + 4 = 0$

$u = \frac{3}{2} \qquad u = -4$

$y^{1/3} = \frac{3}{2} \qquad y^{1/3} = -4$

$y = \frac{27}{8} \qquad y = -64$

33. $(m^2 - 2m)^2 + 2(m^2 - 2m) = 15$

Let $u = m^2 - 2m$, then

$u^2 + 2u = 15$

$u^2 + 2u - 15 = 0$

$(u + 5)(u - 3) = 0$

$u = -5 \qquad u = 3$

$m^2 - 2m = -5 \qquad m^2 - 2m = 3$

$m^2 - 2m + 1 = -4 \qquad m^2 - 2m - 3 = 0$

$(m - 1)^2 = -4 \qquad (m - 3)(m + 1) = 0$

$m - 1 = \pm 2i \qquad m - 3 = 0 \qquad m + 1 = 0$

$m = 1 \pm 2i \qquad m = 3 \qquad m = -1$

35. $\sqrt{2t + 3} + 2 = \sqrt{t - 2}$

$2t + 3 + 4\sqrt{2t + 3} + 4 = t - 2$

$2t + 7 + 4\sqrt{2t + 3} = t - 2$

$4\sqrt{2t + 3} = -t - 9$

$16(2t + 3) = t^2 + 18t + 81$

$32t + 48 = t^2 + 18t + 81$

$0 = t^2 - 14t + 33$

$0 = (t - 11)(t - 3)$

$t = 11 \qquad t = 3$

Common Error: $2t + 3 + 4 = t - 2$ is not an equivalent equation to the given equation $(\sqrt{2t + 3} + 2)^2 \neq 2t + 3 + 4$

Check: 11: $\sqrt{2(11) + 3} + 2 \stackrel{?}{=} \sqrt{11 - 2}$

$\sqrt{25} + 2 \stackrel{?}{=} \sqrt{9}$

$5 + 2 \neq 3$

Not a solution

3: $\sqrt{2(3) + 3} + 2 \stackrel{?}{=} \sqrt{3 - 2}$

$\sqrt{9} + 2 \stackrel{?}{=} \sqrt{1}$

$3 + 2 \neq 1$

Not a solution

No solution

37. $\sqrt{w + 3} + \sqrt{2 - w} = 3$

$\sqrt{w + 3} = 3 - \sqrt{2 - w}$

$w + 3 = 9 - 6\sqrt{2 - w} + 2 - w$

$w + 3 = 11 - w - 6\sqrt{2 - w}$

$2w - 8 = -6\sqrt{2 - w}$

$w - 4 = -3\sqrt{2 - w}$

$w^2 - 8w + 16 = 9(2 - w)$

$w^2 - 8w + 16 = 18 - 9w$

$w^2 + w - 2 = 0$

$(w + 2)(w - 1) = 0$

$w = -2 \qquad w = 1$

Check: -2: $\sqrt{(-2) + 3} + \sqrt{2 - (-2)} \stackrel{?}{=} 3$

$\sqrt{1} + \sqrt{4} \stackrel{?}{=} 3$

$1 + 2 \stackrel{\surd}{=} 3$

A solution

1: $\sqrt{1 + 3} + \sqrt{2 - 1} \stackrel{?}{=} 3$

$\sqrt{4} + \sqrt{1} \stackrel{\surd}{=} 3$

A solution

Solution: -2, 1

39.
$$\begin{aligned}
\sqrt{8 - z} &= 1 + \sqrt{z + 5}\\
8 - z &= 1 + 2\sqrt{z + 5} + z + 5\\
8 - z &= z + 6 + 2\sqrt{z + 5}\\
2 - 2z &= 2\sqrt{z + 5}\\
1 - z &= \sqrt{z + 5}\\
z^2 - 2z + 1 &= z + 5\\
z^2 - 3z - 4 &= 0\\
(z - 4)(z + 1) &= 0\\
z = 4 \qquad z &= -1
\end{aligned}$$

Check: 4: $\sqrt{8 - 4} \stackrel{?}{=} 1 + \sqrt{4 + 5}$

$\sqrt{4} \stackrel{?}{=} 1 + \sqrt{9}$

$2 \neq 1 + 3$

Not a solution

-1: $\sqrt{8 - (-1)} \stackrel{?}{=} 1 + \sqrt{(-1) + 5}$

$\sqrt{9} \stackrel{?}{=} 1 + \sqrt{4}$

$3 \stackrel{\surd}{=} 1 + 2$

A solution

Solution: -1

41.
$$\begin{aligned}
\sqrt{4x^2 + 12x + 1} - 6x &= 9\\
\sqrt{4x^2 + 12x + 1} &= 6x + 9\\
4x^2 + 12x + 1 &= 36x^2 + 108x + 81\\
0 &= 32x^2 + 96x + 80\\
0 &= 2x^2 + 6x + 5
\end{aligned}$$

$$x = \frac{-b \pm \sqrt{b^2 - 4ac}}{2a} \quad a = 2 \quad b = 6 \quad c = 5$$

$$x = \frac{-6 \pm \sqrt{6^2 - 4(2)(5)}}{2(2)}$$

$$x = \frac{-6 \pm \sqrt{36 - 40}}{4}$$

$$x = \frac{-6 \pm \sqrt{-4}}{4}$$

$$x = \frac{-6 \pm 2i}{4}$$

$$x = -\frac{3}{2} \pm \frac{1}{2}i$$

Check: $-\frac{3}{2} + \frac{1}{2}i$: $\sqrt{4\left(-\frac{3}{2} + \frac{1}{2}i\right)^2 + 12\left(-\frac{3}{2} + \frac{1}{2}i\right) + 1} - 6\left(-\frac{3}{2} + \frac{1}{2}i\right) \stackrel{?}{=} 9$

$$\sqrt{(-3 + i)^2 - 18 + 6i + 1} + 9 - 3i \stackrel{?}{=} 9$$

$$\sqrt{9 - 6i + i^2 - 18 + 6i + 1} + 9 - 3i \stackrel{?}{=} 9$$

$$\sqrt{9 - 6i - 1 - 18 + 6i + 1} + 9 - 3i \stackrel{?}{=} 9$$

$$\sqrt{-9} + 9 - 3i \stackrel{?}{=} 9$$

$$3i + 9 - 3i \stackrel{\surd}{=} 9$$ A solution

$-\frac{3}{2} - \frac{1}{2}i$: $\sqrt{4\left(-\frac{3}{2} - \frac{1}{2}i\right)^2 + 12\left(-\frac{3}{2} - \frac{1}{2}i\right) + 1} - 6\left(-\frac{3}{2} - \frac{1}{2}i\right) \stackrel{?}{=} 9$

$$\sqrt{(-3 - i)^2 - 18 - 6i + 1} + 9 + 3i \stackrel{?}{=} 9$$

$$\sqrt{9 + 6i + i^2 - 18 - 6i + 1} + 9 + 3i \stackrel{?}{=} 9$$

$$\sqrt{9 + 6i - 1 - 18 - 6i + 1} + 9 + 3i \stackrel{?}{=} 9$$

$$\sqrt{-9} + 9 + 3i \stackrel{?}{=} 9$$

$$3i + 9 + 3i \stackrel{?}{=} 9$$

$$9 + 6i \neq 9$$ Not a solution

Solution: $-\frac{3}{2} + \frac{1}{2}i$

43. $y^{-2} - 2y^{-1} + 3 = 0$

Let $u = y^{-1}$, then

$$u^2 - 2u + 3 = 0$$
$$u^2 - 2u = -3$$
$$u^2 - 2u + 1 = -2$$
$$(u - 1)^2 = -2$$
$$u - 1 = \pm i\sqrt{2}$$
$$u = 1 \pm i\sqrt{2}$$
$$y^{-1} = 1 \pm i\sqrt{2}$$
$$\frac{1}{y} = 1 \pm i\sqrt{2}$$
$$y = \frac{1}{1 \pm i\sqrt{2}}$$
$$y = \frac{1}{(1 \pm i\sqrt{2})}\,\frac{(1 \pm i\sqrt{2})}{(1 \pm i\sqrt{2})}$$
$$y = \frac{1 \pm i\sqrt{2}}{1 - (-2)}$$
$$y = \frac{1 \pm i\sqrt{2}}{3}$$

or alternatively, write

$$\frac{1}{y^2} - \frac{2}{y} + 3 = 0 \qquad y \neq 0 \quad \text{LCD} = y^2$$
$$1 - 2y + 3y^2 = 0$$
$$3y^2 - 2y + 1 = 0$$
$$y = \frac{-b \pm \sqrt{b^2 - 4ac}}{2a} \qquad a = 3,\ b = -2,\ c = 1$$
$$y = \frac{-(-2) \pm \sqrt{(-2)^2 - 4(1)(3)}}{2(3)}$$
$$y = \frac{2 \pm \sqrt{-8}}{6}$$
$$y = \frac{2 \pm i2\sqrt{2}}{6}$$
$$y = \frac{1 \pm i\sqrt{2}}{3}$$

45. $2t^{-4} - 5t^{-2} + 2 = 0$

Let $u = t^{-2}$, then

$$2u^2 - 5u + 2 = 0$$
$$(2u - 1)(u - 2) = 0$$
$$2u - 1 = 0 \quad \text{or} \quad u - 2 = 0$$

$$u = \frac{1}{2} \qquad\qquad u = 2$$
$$t^{-2} = \frac{1}{2} \qquad\qquad t^{-2} = 2$$
$$t^2 = 2 \qquad\qquad t^2 = \frac{1}{2}$$
$$t = \pm\sqrt{2} \qquad\qquad t = \pm\sqrt{\frac{1}{2}}$$
$$t = \pm\frac{1}{\sqrt{2}}$$
$$t = \pm\frac{\sqrt{2}}{2}$$

47. $3z^{-1} - 3z^{-1/2} + 1 = 0$

Let $u = z^{-1/2}$, then

$$3u^2 - 3u + 1 = 0$$
$$u = \frac{-b \pm \sqrt{b^2 - 4ac}}{2a} \qquad a = 3,\ b = -3,\ c = 1$$
$$u = \frac{-(-3) \pm \sqrt{(-3)^2 - 4(3)(1)}}{2(3)}$$
$$u = \frac{3 \pm i\sqrt{3}}{6}$$
$$z^{-1/2} = \frac{3 \pm i\sqrt{3}}{6}$$
$$z^{-1} = \left(\frac{3 \pm i\sqrt{3}}{6}\right)^2$$
$$= \frac{9 \pm 2(3)\,i\sqrt{3} + (i\sqrt{3})^2}{36}$$
$$z^{-1} = \frac{9 \pm 6i\sqrt{3} - 3}{36}$$
$$z^{-1} = \frac{6 \pm 6i\sqrt{3}}{36}$$
$$z^{-1} = \frac{1 \pm i\sqrt{3}}{6}$$
$$z = \frac{6}{1 \pm i\sqrt{3}}$$
$$z = \frac{6}{(1 \pm i\sqrt{3})}\,\frac{(1 \pm i\sqrt{3})}{(1 \pm i\sqrt{3})}$$
$$z = \frac{6(1 \pm i\sqrt{3})}{1 - (-3)}$$
$$z = \frac{6(1 \pm i\sqrt{3})}{4} \quad \text{or} \quad \frac{3 \pm 3i\sqrt{3}}{2}$$

49. By squaring: $4m + 8\sqrt{m} - 5 = 0$

$4m - 5 = -8\sqrt{m}$

$16m^2 - 40m + 25 = 64m$

$16m^2 - 104m + 25 = 0$

$(4m - 1)(4m - 25) = 0$

$4m - 1 = 0$ or $4m - 25 = 0$

$4m = 1$ $\quad 4m = 25$

$m = 0.25$ $\quad m = 6.25$

Check: $m = 0.25$

$4(0.25) + 8\sqrt{0.25} - 5 \stackrel{?}{=} 0$

$1 + 8(0.5) - 5 \stackrel{?}{=} 0$

$1 + 4 - 5 \stackrel{\surd}{=} 0$

A solution

$m = 6.25$

$4(6.25) + 8\sqrt{6.25} - 5 \stackrel{?}{=} 0$

$25 + 8(2.5) - 5 \stackrel{?}{=} 0$

$25 + 20 - 5 \stackrel{?}{=} 0$

$40 \neq 0$

Not a solution

By substitution:

$4m + 8\sqrt{m} - 5 = 0$

Let $u = \sqrt{m}$, then

$4u^2 + 8u - 5 = 0$

$(2u + 5)(2u - 1) = 0$

$2u + 5 = 0$ or $2u - 1 = 0$

$u = -2.5$ $\quad u = 0.5$

$\sqrt{w} = -2.5$ $\quad \sqrt{w} = 0.5$

Impossible: $\sqrt{w}$ must be positive

$w = 0.25$

This solution has already been checked.

Solution: 0.25

51. $2w + 3\sqrt{w} = 14$

By squaring:

$2w + 3\sqrt{w} - 14 = 0$

$2w - 14 = -3\sqrt{w}$

$4w^2 - 56w + 196 = 9w$

$4w^2 - 65w + 196 = 0$

$(4w - 49)(w - 4) = 0$

$4w - 49 = 0$ $\quad w - 4 = 0$

$w = \frac{49}{4}$ $\quad w = 4$

Check:

$w = \frac{49}{4}$: $2\left(\frac{49}{4}\right) + 3\sqrt{\frac{49}{4}} \stackrel{?}{=} 14$

$\frac{49}{2} + 3\left(\frac{7}{2}\right) \stackrel{?}{=} 14$

$\frac{70}{2} \neq 14$

Not a solution

$w = 4$: $2(4) + 3\sqrt{4} \stackrel{?}{=} 14$

$8 + 6 \stackrel{\surd}{=} 14$

A solution

Solution: 4

By substitution:

Let $u = \sqrt{w}$, then

$2u^2 + 3u = 14$

$2u^2 + 3u - 14 = 0$

$(2u + 7)(u - 2) = 0$

$2u + 7 = 0$ $\quad u - 2 = 0$

$u = -\frac{7}{2}$ $\quad u = 2$

$\sqrt{w} = -\frac{7}{2}$ $\quad \sqrt{w} = 2$

Impossible: $\sqrt{w}$ must be positive

$w = 4$

Solution: 4

(This has already been checked.)

53. $\sqrt{7 - 2x} - \sqrt{x + 2} = \sqrt{x + 5}$

$$7 - 2x - 2\sqrt{7 - 2x}\sqrt{x + 2} + x + 2 = x + 5$$
$$-2\sqrt{7 - 2x}\sqrt{x + 2} - x + 9 = x + 5$$
$$-2\sqrt{7 - 2x}\sqrt{x + 2} = 2x - 4$$
$$-\sqrt{7 - 2x}\sqrt{x + 2} = x - 2$$
$$(7 - 2x)(x + 2) = x^2 - 4x + 4$$
$$7x + 14 - 2x^2 - 4x = x^2 - 4x + 4$$
$$0 = 3x^2 - 7x - 10$$
$$0 = (3x - 10)(x + 1)$$
$$3x - 10 = 0 \qquad x + 1 = 0$$
$$x = \frac{10}{3} \qquad x = -1$$

Check: $x = \frac{10}{3}$: $\sqrt{7 - 2\left(\frac{10}{3}\right)} - \sqrt{\frac{10}{3} + 2} \stackrel{?}{=} \sqrt{\frac{10}{3} + 5}$

$$\sqrt{\frac{21}{3} - \frac{20}{3}} - \sqrt{\frac{10}{3} + \frac{6}{3}} \stackrel{?}{=} \sqrt{\frac{10}{3} + \frac{15}{3}}$$
$$\sqrt{\frac{1}{3}} - \sqrt{\frac{16}{3}} \stackrel{?}{=} \sqrt{\frac{25}{3}}$$
$$\frac{1}{\sqrt{3}} - \frac{4}{\sqrt{3}} \neq \frac{5}{\sqrt{3}}$$

Not a solution

$x = -1$: $\sqrt{7 - 2(-1)} - \sqrt{(-1) + 2} \stackrel{?}{=} \sqrt{(-1) + 5}$

$$\sqrt{9} - \sqrt{1} \stackrel{?}{=} \sqrt{4}$$
$$3 - 1 \stackrel{\checkmark}{=} 2$$

A solution

Solution: -1

55. $3 + x^{-4} = 5x^{-2} \quad x \neq 0$, LCD $= x^4$

$$3x^4 + 1 = 5x^2$$
$$3x^4 - 5x^2 + 1 = 0$$

Let $u = x^2$, then

$$3u^2 - 5u + 1 = 0$$
$$u = \frac{-b \pm \sqrt{b^2 - 4ac}}{2a} \quad a = 3,\ b = -5,\ c = 1$$
$$u = \frac{-(-5) \pm \sqrt{(-5)^2 - 4(3)(1)}}{2(3)}$$
$$u = \frac{5 \pm \sqrt{13}}{6}$$
$$x^2 = \frac{5 \pm \sqrt{13}}{6}$$
$$x = \pm\sqrt{\frac{5 \pm \sqrt{13}}{6}} \quad \text{(four roots)}$$

57.

$$\begin{aligned} 2\sqrt{x + 5} &= 0.01x + 2.04 \\ 200\sqrt{x + 5} &= x + 204 \\ 40{,}000(x + 5) &= (x + 204)^2 \\ 40{,}000x + 200{,}000 &= x^2 + 408x + 41{,}616 \\ 0 &= x^2 - 39{,}592x - 158{,}384 \end{aligned}$$

Although this is factorable in the integers, one is unlikely to notice this or to detect the factors, so the quadratic formula is used

$$x = \frac{-b \pm \sqrt{b^2 - 4ac}}{2a} \qquad a = 1$$

$$b = -39{,}592$$
$$c = -158{,}384$$

$$x = \frac{-(-39,592) \pm \sqrt{(-39,592)^2 - 4(1)(-158,384)}}{2(1)}$$

$$x = \frac{39,592 \pm 39,600}{2}$$

$x = -4,\ 39{,}596$

Check:

-4: $2\sqrt{-4 + 5} \stackrel{?}{=} 0.01(-4) + 2.04$ $2 \stackrel{\checkmark}{=} 2$

$39{,}596$: $2\sqrt{39,596 + 5} \stackrel{?}{=} 0.01(39{,}596) + 2.04$ $398 \stackrel{\checkmark}{=} 398$

Solution: -4, 39596

59. $2x^{-2/5} - 5x^{-1/5} + 1 = 0$

Let $u = x^{-1/5}$, then

$$2u^2 - 5u + 1 = 0$$

$$u = \frac{-b \pm \sqrt{b^2 - 4ac}}{2a} \qquad a = 2,\ b = -5,\ c = 1$$

$$u = \frac{-(-5) \pm \sqrt{(-5)^2 - 4(2)(1)}}{2(2)}$$

$$u = \frac{5 \pm \sqrt{17}}{4}$$

$$x^{-1/5} = \frac{5 + \sqrt{17}}{4} \qquad x^{-1/5} = \frac{5 - \sqrt{17}}{4}$$

$$x^{1/5} = \frac{4}{5 + \sqrt{17}} \qquad x^{1/5} = \frac{4}{5 - \sqrt{17}}$$

$$x = \left(\frac{4}{5 + \sqrt{17}}\right)^5 \qquad x = \left(\frac{4}{5 - \sqrt{17}}\right)^5$$

$$x \approx 0.016203 \qquad x \approx 1974.98$$

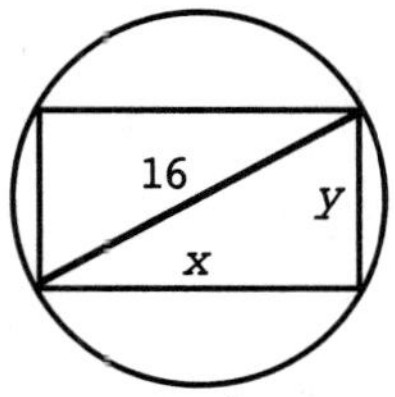

61. Let x = width of cross-section of the beam
y = depth of cross-section of the beam

From the Pythagorean theorem

$$x^2 + y^2 = 16^2$$

Thus, $y = \sqrt{256 - x^2}$

Since the area of the rectangle is given by xy, we have

$$\begin{aligned} xy &= 120 \\ x\sqrt{256 - x^2} &= 120 \\ x^2(256 - x^2) &= 14{,}400 \\ 256x^2 - x^4 &= 14{,}400 \\ -x^4 + 256x^2 - 14{,}400 &= 0 \\ (x^2)^2 - 256x^2 + 14{,}400 &= 0 \\ x^2 &= \frac{-b \pm \sqrt{b^2 - 4ac}}{2a} \quad a = 1,\ b = -256,\ c = 14{,}400 \\ x^2 &= \frac{-(-256) \pm \sqrt{(-256)^2 - 4(1)(14{,}400)}}{2(1)} \\ x^2 &= \frac{256 \pm \sqrt{65{,}536 - 57{,}600}}{2} \\ x^2 &= \frac{256 \pm \sqrt{7{,}936}}{2} \\ x^2 &= 128 \pm \sqrt{1{,}984} \\ x &= \sqrt{128 \pm \sqrt{1{,}984}} \end{aligned}$$

If $x = \sqrt{128 + \sqrt{1{,}984}} \approx 13.1$ then

$$\begin{aligned} y &= \sqrt{256 - x^2} \\ &= \sqrt{256 - (128 + \sqrt{1{,}984})} \\ &= \sqrt{128 - \sqrt{1{,}984}} \approx 9.1 \end{aligned}$$

Thus the dimensions of the rectangle are 13.1 inches by 9.1 inches. Notice that if $x = \sqrt{128 - \sqrt{1{,}984}}$, then $y = \sqrt{128 + \sqrt{1{,}984}}$ and the dimensions are still 13.1 inches by 9.1 inches.

63.

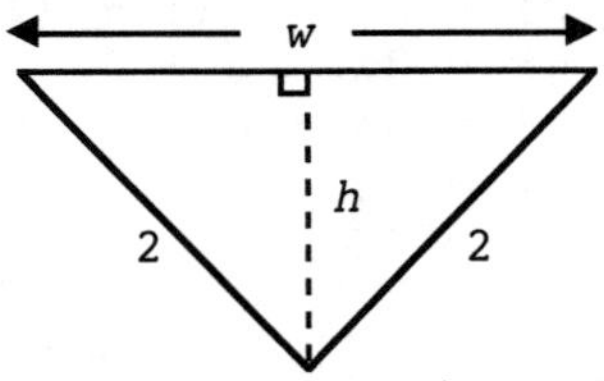

Let w = width of trough
h = altitude of triangular end

Examining the triangular end of the trough sketched above, we see that $h^2 + \left(\frac{1}{2}w\right)^2 = 2^2$. The area of this end, $A = \frac{1}{2}wh$. Since the volume of the trough V is given by $V = A \cdot 6$, we have

$$\begin{aligned} 9 &= 6A \\ 9 &= 6\left(\frac{1}{2}wh\right) \\ 9 &= 3wh \\ 3 &= wh \end{aligned}$$

Since $h^2 = 2^2 - \left(\frac{1}{2}w\right)^2$

$$h^2 = 2^2 - \frac{1}{4}w^2$$

$$h = \sqrt{4 - \frac{1}{4}w^2}$$

Hence we solve

$$3 = w\sqrt{4 - \frac{1}{4}w^2}$$
$$9 = w^2\left(4 - \frac{1}{4}w^2\right)$$
$$9 = 4w^2 - \frac{1}{4}(w^2)^2$$
$$36 = 16w^2 - (w^2)^2$$
$$(w^2)^2 - 16w^2 + 36 = 0$$

$$w^2 = \frac{-b \pm \sqrt{b^2 - 4ac}}{2a} \quad a = 1,\ b = -16,\ c = 36$$
$$w^2 = \frac{-(-16) \pm \sqrt{(-16)^2 - 4(1)(36)}}{2(1)}$$
$$w^2 = \frac{16 \pm \sqrt{256 - 144}}{2}$$
$$w^2 = \frac{16 \pm \sqrt{112}}{2}$$
$$w^2 = 8 \pm 2\sqrt{7}$$
$$w = \sqrt{8 \pm 2\sqrt{7}}$$
$$w = 1.65 \text{ ft or } 3.65 \text{ ft}$$

CHAPTER 1 REVIEW

1.
$$0.05x + 0.25(30 - x) = 3.3$$
$$0.05x + 7.5 - 0.25x = 3.3$$
$$-0.2x + 7.5 = 3.3$$
$$-0.2x = -4.2$$
$$x = \frac{-4.2}{-0.2}$$
$$x = 21$$
(1-1)

2.
$$\frac{5x}{3} - \frac{4 + x}{2} = \frac{x - 2}{4} + 1$$
$$12\frac{5x}{3} - 12\frac{(4 + x)}{2} = 12\frac{(x - 2)}{4} + 12$$
$$20x - 6(4 + x) = 3(x - 2) + 12$$
$$20x - 24 - 6x = 3x - 6 + 12$$
$$14x - 24 = 3x + 6$$
$$11x = 30$$
$$x = \frac{30}{11}$$
(1-1)

3.
$$3(2 - x) - 2 \le 2x - 1$$
$$6 - 3x - 2 \le 2x - 1$$
$$-3x + 4 \le 2x - 1$$
$$-5x \le -5$$
$$x \ge 1$$
$[1, \infty)$

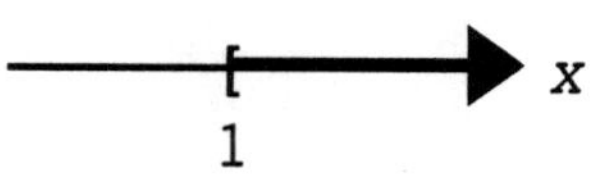

(1-2)

4.
$$|y + 9| < 5$$
$$-5 < y + 9 < 5$$
$$-14 < y < -4$$
$(-14, -4)$

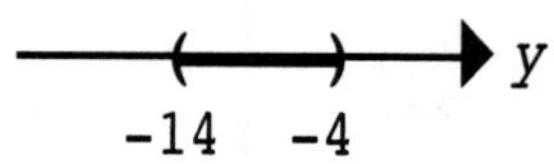

(1-3)

5. $|3 - 2x| \le 5$

$$-5 \le 3 - 2x \le 5$$
$$-8 \le -2x \le 2$$
$$4 \ge x \ge -1$$
$$-1 \le x \le 4$$

$[-1, 4]$ 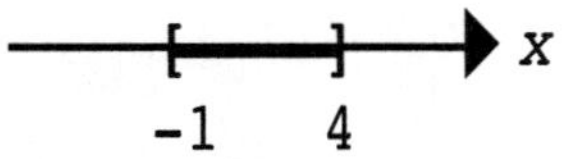

(1-3)

6. (A) $9 - 4i = 9 + (-4)i$
Real part: 9
Imaginary part: $-4i$
Conjugate: $9 - (-4)i = 9 + 4i$

(B) $5i = 0 + 5i$
Real part = 0
Imaginary part: $5i$
Conjugate: $0 + (-5)i = -5i$

(C) $-10 = -10 + 0i$
Real part: -10
Imaginary part: 0
Conjugate: $-10 - 0i = -10$ (1-4)

7. (A) $(-3 + 2i) + (6 - 8i) = -3 + 2i + 6 - 8i = 3 - 6i$

(B) $(3 - 3i)(2 + 3i) = 6 + 3i - 9i^2 = 6 + 3i + 9 = 15 + 3i$

(C) $$\frac{13 - i}{5 - 3i} = \frac{(13 - i)}{(5 - 3i)}\frac{(5 + 3i)}{(5 + 3i)}$$
$$= \frac{65 + 34i - 3i^2}{25 - 9i^2} = \frac{65 + 34i + 3}{25 + 9} = \frac{68 + 34i}{34} = 2 + i$$ (1-4)

8. $2x^2 - 7 = 0$
$$2x^2 = 7$$
$$x^2 = \frac{7}{2}$$
$$x = \pm\sqrt{\frac{7}{2}}$$
$$x = \pm\frac{\sqrt{14}}{2}$$ (1-5)

9. $$2x^2 = 4x$$
$$2x^2 - 4x = 0$$
$$2x(x - 2) = 0$$
$$2x = 0 \quad x - 2 = 0$$
$$x = 0 \quad x = 2$$ (1-5)

10. $$2x^2 = 7x - 3$$
$$2x^2 - 7x + 3 = 0$$
$$(2x - 1)(x - 3) = 0$$
$$2x - 1 = 0 \quad x - 3 = 0$$
$$x = \frac{1}{2} \quad x = 3$$ (1-5)

11. $m^2 + m + 1 = 0$
$$m = \frac{-b \pm \sqrt{b^2 - 4ac}}{2a} \quad a = 1,\ b = 1,\ c = 1$$
$$m = \frac{-1 \pm \sqrt{(1)^2 - 4(1)(1)}}{2(1)}$$
$$m = \frac{-1 \pm \sqrt{-3}}{2}$$
$$m = \frac{-1 \pm i\sqrt{3}}{2}$$
$$m = -\frac{1}{2} \pm \frac{\sqrt{3}}{2}i$$ (1-5)

12. $$y^2 = \frac{3}{2}(y + 1)$$
$$2y^2 = 3(y + 1)$$
$$2y^2 = 3y + 3$$
$$2y^2 - 3y - 3 = 0$$
$$y = \frac{-b \pm \sqrt{b^2 - 4ac}}{2a} \quad a = 2,\ b = -3,\ c = -3$$
$$y = \frac{-(-3) \pm \sqrt{(-3)^2 - 4(2)(-3)}}{2(2)}$$
$$y = \frac{3 \pm \sqrt{33}}{4}$$ (1-5)

13.

$$\begin{aligned} \sqrt{5x-6} - x &= 0 \\ \sqrt{5x-6} &= x \\ 5x - 6 &= x^2 \\ 0 &= x^2 - 5x + 6 \\ x^2 - 5x + 6 &= 0 \\ (x-3)(x-2) &= 0 \\ x &= 2, 3 \end{aligned}$$

Check: $\sqrt{5(2)-6} - 2 \stackrel{?}{=} 0$

$0 \stackrel{\checkmark}{=} 0$

$\sqrt{5(3)-6} - 3 \stackrel{?}{=} 0$

$0 \stackrel{\checkmark}{=} 0$

Solution: 2, 3 *(1-6)*

14. $\sqrt{3-5x}$ represents a real number exactly when $3 - 5x$ is positive or zero. We can write this as an inequality statement and solve for x.

$$\begin{aligned} 3 - 5x &\geq 0 \\ -5x &\geq -3 \\ x &\leq \frac{3}{5} \end{aligned}$$

(1-2)

15.

$$\frac{7}{2-x} = \frac{10-4x}{x^2+3x-10}$$

$$\frac{7}{2-x} = \frac{10-4x}{(x-2)(x+5)} \quad \text{Excluded values: } x \neq 2, -5$$

$$\overset{-1}{\cancel{(x-2)}}(x+5)\frac{7}{\underset{1}{\cancel{2-x}}} = (x-2)(x+5)\frac{10-4x}{(x-2)(x+5)}$$

$$\begin{aligned} -7(x+5) &= 10 - 4x \\ -7x - 35 &= 10 - 4x \\ -3x &= 45 \\ x &= -15 \end{aligned}$$

(1-1)

16.

$$\frac{u-3}{2u-2} = \frac{1}{6} - \frac{1-u}{3u-3}$$

$$\frac{u-3}{2(u-1)} = \frac{1}{6} - \frac{1-u}{3(u-1)} \quad \text{Excluded value: } u \neq 1$$

$$6(u-1)\frac{(u-3)}{2(u-1)} = 6(u-1)\frac{1}{6} - 6(u-1)\frac{(1-u)}{3(u-1)}$$

$$\begin{aligned} 3(u-3) &= u - 1 - 2(1-u) \\ 3u - 9 &= u - 1 - 2 + 2u \\ 3u - 9 &= 3u - 3 \\ -9 &= -3 \end{aligned}$$

No solution *(1-1)*

17.
$$\frac{x+3}{8} \le 5 - \frac{2-x}{3}$$
$$24\frac{(x+3)}{8} \le 120 - 24\frac{(2-x)}{3}$$
$$3(x+3) \le 120 - 8(2-x)$$
$$3x + 9 \le 120 - 16 + 8x$$
$$3x + 9 \le 8x + 104$$
$$-5x \le 95$$
$$x \ge -19$$
$[-19, \infty)$

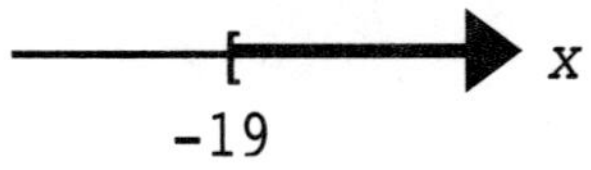

(1-2)

18.
$$|3x - 8| > 2$$
$$3x - 8 < -2 \quad \text{or} \quad 3x - 8 > 2$$
$$3x < 6 \qquad\qquad 3x > 10$$
$$x < 2 \quad \text{or} \quad x > \frac{10}{3}$$
$$(-\infty, 2) \approx \left(\frac{10}{3}, \infty\right)$$

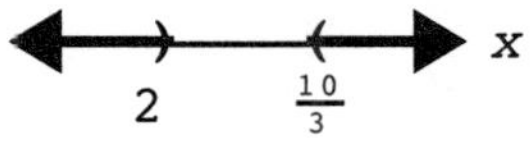

(1-3)

19.
$$\sqrt{(1-2m)^2} \le 3$$
$$|1 - 2m| \le 3$$
$$-3 \le 1 - 2m \le 3$$
$$-4 \le -2m \le 2$$
$$2 \ge m \ge -1$$
$$-1 \le m \le 2$$
$[-1, 2]$

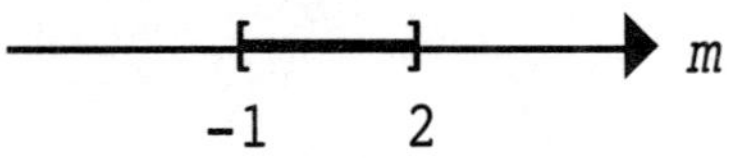

(1-3)

20. (A) $d(A,B) = |-2 - (-8)| = |6| = 6$ (B) $d(B,A) = |-8 - (-2)| = |-6| = 6$ (1-3)

21. (A)
$$(3+i)^2 - 2(3+i) + 3 = 9 + 6i + i^2 - 6 - 2i + 3$$
$$= 9 + 6i - 1 - 6 - 2i + 3$$
$$= 5 + 4i$$
(B) $i^{27} = i^{26}i = (i^2)^{13}i = (-1)^{13}i = (-1)i = -i$ (1-4)

22. (A)
$$(2 - \sqrt{-4}) - (3 - \sqrt{-9}) = (2 - i\sqrt{4}) - (3 - i\sqrt{9}) = (2 - 2i) - (3 - 3i)$$
$$= 2 - 2i - 3 + 3i = -1 + i$$
(B)
$$\frac{2-\sqrt{-1}}{3+\sqrt{-4}} = \frac{2 - i\sqrt{1}}{3 + i\sqrt{4}} = \frac{2-i}{3+2i} = \frac{(2-i)}{(3+2i)}\frac{(3-2i)}{(3-2i)} = \frac{6 - 7i + 2i^2}{9 - 4i^2} = \frac{6 - 7i - 2}{9 + 4}$$
$$= \frac{4-7i}{13} = \frac{4}{13} - \frac{7}{13}i$$
(C)
$$\frac{4+\sqrt{-25}}{\sqrt{-4}} = \frac{4 + i\sqrt{25}}{i\sqrt{4}} = \frac{4+5i}{2i} = \frac{4+5i}{2i}\frac{i}{i} = \frac{4i + 5i^2}{2i^2} = \frac{4i - 5}{-2} = \frac{5}{2} - 2i$$
(1-4)

23.
$$\left(u + \frac{5}{2}\right)^2 = \frac{5}{4}$$
$$u + \frac{5}{2} = \pm\sqrt{\frac{5}{4}}$$
$$u + \frac{5}{2} = \pm\frac{\sqrt{5}}{2}$$
$$u = -\frac{5}{2} \pm \frac{\sqrt{5}}{2}$$
$$u = \frac{-5 \pm \sqrt{5}}{2}$$
(1-5)

24.
$$1 + \frac{3}{u^2} = \frac{2}{u} \quad \text{Excluded value: } u \ne 0$$
$$u^2 + 3 = 2u$$
$$u^2 - 2u = -3$$
$$u^2 - 2u + 1 = -2$$
$$(u-1)^2 = -2$$
$$u - 1 = \pm\sqrt{-2}$$
$$u = 1 \pm \sqrt{-2}$$
$$u = 1 \pm i\sqrt{2}$$
(1-5)

25.

$$\frac{x}{x^2 - x - 6} - \frac{2}{x - 3} = 3$$

$$\frac{x}{(x - 3)(x + 2)} - \frac{2}{x - 3} = 3 \quad \text{Excluded values: } x \neq 3, -2$$

$$(x - 3)(x + 2)\frac{x}{(x - 3)(x + 2)} - (x - 3)(x + 2)\frac{2}{x - 3} = 3(x - 3)(x + 2)$$

$$x - 2(x + 2) = 3(x - 3)(x + 2)$$

$$x - 2x - 4 = 3(x^2 - x - 6)$$

$$-x - 4 = 3x^2 - 3x - 18$$

$$0 = 3x^2 - 2x - 14$$

$$3x^2 - 2x - 14 = 0$$

$$x = \frac{-b \pm \sqrt{b^2 - 4ac}}{2a} \qquad a = 3,\ b = -2,\ c = -14$$

$$x = \frac{-(-2) \pm \sqrt{(-2)^2 - 4(3)(-14)}}{2(3)}$$

$$x = \frac{2 \pm \sqrt{172}}{6}$$

$$x = \frac{2 \pm 2\sqrt{43}}{6}$$

$$x = \frac{1 \pm \sqrt{43}}{3}$$

(*1-5*)

26. $2x^{2/3} - 5x^{1/3} - 12 = 0$

Let $u = x^{1/3}$, then

$$2u^2 - 5u - 12 = 0$$

$$(2u + 3)(u - 4) = 0$$

$$u = -\frac{3}{2}, 4$$

$x^{1/3} = -\frac{3}{2}$ $\qquad$ $x^{1/3} = 4$

$x = -\frac{27}{8}$ $\qquad$ $x = 64$ $\qquad$ (*1-6*)

27. $m^4 + 5m^2 - 36 = 0$

Let $u = m^2$, then

$$u^2 + 5u - 36 = 0$$

$$(u + 9)(u - 4) = 0$$

$$u = -9, 4$$

$m^2 = -9$ $\qquad$ $m^2 = 4$

$m = \pm 3i$ $\qquad$ $m = \pm 2$ $\qquad$ (*1-6*)

28.

$$\sqrt{y - 2} - \sqrt{5y + 1} = -3$$

$$-\sqrt{5y + 1} = -3 - \sqrt{y - 2}$$

$$5y + 1 = 9 + 6\sqrt{y - 2} + y - 2$$

$$5y + 1 = y + 7 + 6\sqrt{y - 2}$$

$$4y - 6 = 6\sqrt{y - 2}$$

$$2y - 3 = 3\sqrt{y - 2}$$

$$4y^2 - 12y + 9 = 9(y - 2)$$

$$4y^2 - 12y + 9 = 9y - 18$$

$$4y^2 - 21y + 27 = 0$$

$$(4y - 9)(y - 3) = 0$$

$$y = \frac{9}{4}, 3$$

> **Common Error:**
> $y - 2 - 5y + 1 = 9$
> is not equivalent to the equation formed by squaring both members of the given equation.

Check:

$$\sqrt{\tfrac{9}{4} - 2} - \sqrt{5\left(\tfrac{9}{4}\right) + 1} \stackrel{?}{=} -3$$

$$\sqrt{\tfrac{1}{4}} - \sqrt{\tfrac{49}{4}} \stackrel{?}{=} -3$$

$$-3 \stackrel{\checkmark}{=} -3$$

$$\sqrt{3 - 2} - \sqrt{5(3) + 1} \stackrel{?}{=} -3$$

$$-3 \stackrel{\checkmark}{=} -3$$

Solution: $\frac{9}{4}$, 3 $\qquad$ (*1-6*)

29.
$$\begin{aligned} 2.15x - 3.73(x - 0.930) &= 6.11x \\ 2.15x - 3.73x + 3.4689 &= 6.11x \\ -1.58x + 3.4689 &= 6.11x \\ 3.4689 &= 7.69x \\ x &= 0.451 \end{aligned}$$
(1-1)

30.
$$\begin{aligned} -1.52 \le 0.770 - 2.04x &\le 5.33 \\ -2.29 \le -2.04x &\le 4.56 \\ 1.12 \ge x &\ge -2.24 \end{aligned}$$
$-2.24 \le x \le 1.12$ or $[-2.24, 1.12]$ (1-2)

31. $|9.71 - 3.62x| > 5.48$

$9.71 - 3.62x > 5.48$ or $9.71 - 3.62x < -5.48$

$-3.62x > -4.23$ $\quad$ $-3.62x < -15.19$

$x < 1.17$ $\quad$ $x > 4.20$ (1-3)

32.
$$\begin{aligned} \frac{3.77 - 8.47i}{6.82 - 7.06i} &= \frac{(3.77 - 8.47i)(6.82 + 7.06i)}{(6.82 - 7.06i)(6.82 + 7.06i)} \\ &= \frac{(3.77)(6.82) + (3.77)(7.06i) - (6.82)(8.47i) - (8.47)(7.06)i^2}{(6.82)^2 + (7.06)^2} \\ &= \frac{25.7114 + 26.6162i - 57.7654i + 59.7982}{46.5124 + 49.8436} \\ &= \frac{85.5096 - 31.1492i}{96.356} \\ &= 0.887 - 0.323i \end{aligned}$$
(1-4)

33. $6.09x^2 + 4.57x - 8.86 = 0$

$$x = \frac{-b \pm \sqrt{b^2 - 4ac}}{2a} \quad a = 6.09,\ b = 4.57,\ c = -8.86$$
$$x = \frac{-4.57 \pm \sqrt{(4.57)^2 - 4(6.09)(-8.86)}}{2(6.09)}$$
$$x = \frac{-4.57 \pm \sqrt{236.7145}}{12.18}$$
$$x = \frac{-4.57 \pm 15.3855}{12.18}$$
$$x = -1.64,\ 0.888$$
(1-5)

34.
$$\begin{aligned} P &= M - Mdt \\ M - Mdt &= P \\ M(1 - dt) &= P \\ M &= \frac{P}{1 - dt} \end{aligned}$$
(1-1)

35.
$$\begin{aligned} P &= EI - RI^2 \\ RI^2 - EI + P &= 0 \end{aligned}$$
$$I = \frac{-b \pm \sqrt{b^2 - 4ac}}{2a} \quad a = R,\ b = -E,\ c = P$$
$$I = \frac{-(-E) \pm \sqrt{(-E)^2 - 4(R)(P)}}{2(R)}$$
$$I = \frac{E \pm \sqrt{E^2 - 4PR}}{2R}$$
(1-5)

36.
$$\begin{aligned} x &= \frac{4y + 5}{2y + 1} \\ x(2y + 1) &= (2y + 1)\frac{4y + 5}{2y + 1} \\ 2xy + x &= 4y + 5 \\ 2xy + x - 4y &= 5 \\ 2xy - 4y &= 5 - x \\ y(2x - 4) &= 5 - x \\ y &= \frac{5 - x}{2x - 4} \end{aligned}$$
(1-1)

37. The original equation can be rewritten as

$$\frac{4}{(x-1)(x-3)} = \frac{3}{(x-1)(x-2)}$$

Thus, $x = 1$ cannot be a solution of this equation. This extraneous solution was introduced when both sides were multiplied by $x - 1$ in the second line. $x = 1$ must be discarded and the only correct solution is $x = -1$. (*1-1*)

38. In this problem, $a = 1$, $b = -6$, $c = c$. Thus, the discriminant $b^2 - 4ac = (-6)^2 - 4(1)(c) = 36 - 4c$. Hence,

if $36 - 4c > 0$, thus $36 > 4c$ or $c < 9$, there are two distinct real roots.
if $36 - 4c = 0$, thus $c = 9$, there is one real double root.
if $36 - 4c < 0$, thus $36 < 4c$ or $c > 9$, there are two distinct imaginary roots.
(*1-5*)

39. The given inequality $a + b < b - a$ is equivalent to, successively,

$$a < -a$$
$$2a < 0$$
$$a < 0$$

Thus its truth is independent of the value of b, and dependent on a being negative. True for all real b and all negative a. (*1-2*)

40. If $a > b$ and b is negative, then $\frac{a}{b} < \frac{b}{b}$, that is, $\frac{a}{b} < 1$, since dividing both sides by b reverses the order of the inequality. $\frac{a}{b}$ is less than 1. (*1-2*)

41.

$$y = \frac{1}{1 - \frac{1}{1-x}}$$
$$y = \frac{1(1-x)}{(1-x)1 - (1-x)\frac{1}{1-x}}$$
$$y = \frac{1-x}{1-x-1}$$
$$y = \frac{1-x}{-x}$$
$$-xy = 1 - x$$
$$x - xy = 1$$
$$x(1 - y) = 1$$
$$x = \frac{1}{1-y}$$

(*1-1*)

42. $0 < |x - 6| < d$ means: the distance between x and 6 is less than d but $x \neq 6$.

$$-d < x - 6 < d \text{ except } x \neq 6$$
$$6 - d < x < 6 + d \text{ but } x \neq 6$$
$$6 - d < x < 6 \text{ or } 6 < x < 6 + d$$
$$(6 - d, 6) \cup (6, 6 + d)$$

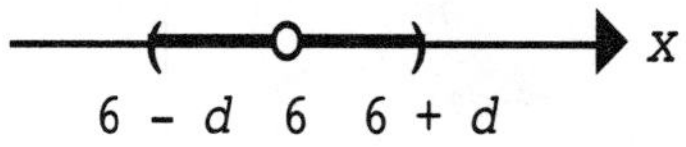

(*1-3*)

43.

$$2x^2 = \sqrt{3}x - \frac{1}{2}$$
$$4x^2 = 2\sqrt{3}x - 1$$
$$4x^2 - 2\sqrt{3}x + 1 = 0$$
$$x = \frac{-b \pm \sqrt{b^2 - 4ac}}{2a} \qquad a = 4,\ b = -2\sqrt{3},\ c = 1$$
$$x = \frac{-(-2\sqrt{3}) \pm \sqrt{(-2\sqrt{3})^2 - 4(4)(1)}}{2(4)}$$
$$x = \frac{2\sqrt{3} \pm \sqrt{-4}}{8}$$
$$x = \frac{2\sqrt{3} \pm 2i}{8}$$
$$x = \frac{\sqrt{3} \pm i}{4} \text{ or } \frac{\sqrt{3}}{4} \pm \frac{1}{4}i$$

(*1-5*)

44. $4 = 8x^{-2} - x^{-4}$

$4x^4 = 8x^2 - 1 \quad x \neq 0 \quad \text{LCD} = x^4$

$4x^4 - 8x^2 + 1 = 0$

Let $u = x^2$, then

$4u^2 - 8u + 1 = 0$

$$u = \frac{-b \pm \sqrt{b^2 - 4ac}}{2a} \quad a = 4,\ b = -8,\ c = 1$$

$$u = \frac{-(-8) \pm \sqrt{(-8)^2 - 4(4)(1)}}{2(4)}$$

$$u = \frac{8 \pm \sqrt{48}}{8}$$

Common Error: It is incorrect to "cancel" the 8's at this point.

$$u = \frac{8 \pm 4\sqrt{3}}{8}$$

$$u = \frac{4(2 \pm \sqrt{3})}{8}$$

$$u = \frac{2 \pm \sqrt{3}}{2}$$

$$x^2 = \frac{2 \pm \sqrt{3}}{2}$$

$$x = \pm\sqrt{\frac{2 \pm \sqrt{3}}{2}} \quad \text{(four real roots)}$$

(1-6)

45. $(a + bi)\left(\frac{a}{a^2 + b^2} - \frac{b}{a^2 + b^2}i\right)$

$$= \frac{(a + bi)}{1}\left(\frac{a}{a^2 + b^2} - \frac{bi}{a^2 + b^2}\right)$$

$$= \frac{a(a + bi)}{a^2 + b^2} - \frac{bi(a + bi)}{a^2 + b^2}$$

$$= \frac{a^2 + abi - abi - b^2i^2}{a^2 + b^2}$$

$$= \frac{a^2 + b^2}{a^2 + b^2} = 1$$

(1-4)

46. Let x = the number

$\frac{1}{x}$ = its reciprocal

Then $x - \frac{1}{x} = \frac{16}{15}$ Excluded value: $x \neq 0$

$15x^2 - 15 = 16x$

$15x^2 - 16x - 15 = 0$

$(5x + 3)(3x - 5) = 0$

$5x + 3 = 0$ or $3x - 5 = 0$

$x = -\frac{3}{5} \qquad x = \frac{5}{3}$

(1-5)

47. (A) $H = 0.7(220 - A)$

(B) We are to find H when $A = 20$.

$H = 0.7(220 - 20)$

$H = 140$ beats per minute.

(C) We are to find A when $H = 126$.

$126 = 0.7(220 - A)$

$126 = 154 - 0.7A$

$-28 = -0.7A$

$A = 40$ years old.

(1-1)

48. Let x = amount of 80% solution
Then $50 - x$ = amount of 30% solution
since 50 = amount of 60% solution

acid in 80% solution + acid in 30% solution = acid in 60% solution

$$0.8(x) + 0.3(50 - x) = 0.6(50)$$
$$0.8x + 15 - 0.3x = 30$$
$$15 + 0.5x = 30$$
$$0.5x = 15$$
$$x = \frac{15}{0.5}$$
x = 30 milliliters of 80% solution
$50 - x$ = 20 milliliters of 30% solution *(1-2)*

49. Let x = the rate of the current
Then $12 - x$ = the rate of the boat upstream
$12 + x$ = the rate of the boat downstream

Solving $d = rt$ for t, we have $t = \frac{d}{r}$. We use this formula, together with

time upstream = 2 + time downstream

$$\text{time upstream} = \frac{\text{distance upstream}}{\text{rate upstream}} = \frac{45}{12 - x}$$

$$\text{time downstream} = \frac{\text{distance downstream}}{\text{rate downstream}} = \frac{45}{12 + x}$$

So, $\frac{45}{12 - x} = 2 + \frac{45}{12 + x}$ Excluded values: $x \neq -12, 12$

$$45(12 + x) = (12 + x)(12 - x)2 + 45(12 - x)$$
$$540 + 45x = 288 - 2x^2 + 540 - 45x$$
$$2x^2 + 90x - 288 = 0$$
$$2(x + 48)(x - 3) = 0$$
$$x = -48, 3$$

Discarding the negative answer, we have rate = 3 miles per hour. *(1-5)*

50. (A) Let x = distance rowed
then $15 - 3 = 12$ km/hr = the rate rowed upstream
$15 + 3 = 18$ km/hr = the rate rowed downstream

Using $t = \frac{d}{r}$ as in the previous problem yields

$$\text{time upstream} = \frac{x}{12}$$
$$\text{time downstream} = \frac{x}{18}$$

So $\frac{x}{12} + \frac{x}{18} = \frac{25}{60}$ LCD = 180

$$\frac{180}{1} \cdot \frac{x}{12} + \frac{180}{1} \cdot \frac{x}{18} = \frac{180}{1} \cdot \frac{25}{60}$$
$$15x + 10x = 75$$
$$25x = 75$$
$$x = 3 \text{ km}$$

(B) Now let x = still-water speed
$x - 3$ = the rate rowed upstream
$x + 3$ = the rate rowed downstream

$$\text{time upstream} = \frac{3}{x-3}$$

$$\text{time downstream} = \frac{3}{x+3}$$

So $\frac{3}{x-3} + \frac{3}{x+3} = \frac{23}{60}$ Excluded values: $x = 3, -3$

$$60(x+3)3 + 60(x-3)3 = 23(x+3)(x-3)$$
$$180x + 540 + 180x - 540 = 23x^2 - 207$$
$$0 = 23x^2 - 360x - 207$$
$$x = \frac{-b \pm \sqrt{b^2 - 4ac}}{2a} \quad a = 23,\ b = -360,\ c = -207$$
$$x = \frac{-(-360) \pm \sqrt{(-360)^2 - 4(23)(-207)}}{2(23)}$$
$$x = 16.2 \text{ or } -0.6$$

Discarding the negative answer, we have $x = 16.2$ km/hr.

(C) Now $18 - 3 = 15$ km/hr = the rate rowed upstream
$18 + 3 = 21$ km/hr = the rate rowed downstream

So $\frac{3}{15} + \frac{3}{21}$ = round trip time
= 0.343 hr
= 0.343 × 60 min
= 20.6 min (*1-1, 1-5*)

51. (A) Apply the given formula with $C = 15$.

$$15 = x^2 - 10x + 31$$
$$0 = x^2 - 10x + 16$$
$$x^2 - 10x + 16 = 0$$
$$(x-8)(x-2) = 0$$
$$x = 2 \text{ or } 8$$

Thus the output could be either 2000 or 8000 units.

(B) Apply the given formula with $C = 6$.

$$6 = x^2 - 10x + 31$$
$$0 = x^2 - 10x + 25$$
$$x^2 - 10x + 25 = 0$$
$$(x-5)^2 = 0$$
$$x = 5$$

Thus the output must be 5000 units. (*1-5*)

52. The break-even points are defined by $C = R$ (cost = revenue). Applying the formulas in this problem and the previous one, we have

$$x^2 - 10x + 31 = 3x$$
$$x^2 - 13x + 31 = 0$$
$$x = \frac{-b \pm \sqrt{b^2 - 4ac}}{2a} \quad a = 1,\ b = -13,\ c = 31$$
$$x = \frac{-(-13) \pm \sqrt{(-13)^2 - 4(1)(31)}}{2(1)}$$
$$x = \frac{13 \pm \sqrt{45}}{2}$$ thousand or approximately 3,146 and 9,854 units (*1-5*)

53. The distance of T from 110 must be no greater than 5. $|T - 110| \le 5$ (*1-3*)

54. Let x = width of page
y = height of page

Then $xy = 480$, thus $y = \frac{480}{x}$.

Since the printed portion is surrounded by margins of 2 cm on each side, we have

$x - 4$ = width of printed portion
$y - 4$ = height of printed portion

Hence

$(x - 4)(y - 4) = 320$, that is

$$(x - 4)\left(\frac{480}{x} - 4\right) = 320$$

Solving this, we obtain:

$$x\left(\frac{480}{x}\right) - 4x - 4\left(\frac{480}{x}\right) + 16 = 320$$
$$480 - 4x - \frac{1,920}{x} + 16 = 320$$
$$-4x - \frac{1,920}{x} = -176 \quad \text{LCD: } x \quad x \neq 0$$
$$-4x^2 - 1,920 = -176x$$
$$0 = 4x^2 - 176x + 1,920$$
$$0 = x^2 - 44x + 480$$
$$0 = (x - 20)(x - 24)$$
$$x - 20 = 0 \quad \text{or} \quad x - 24 = 0$$
$$x = 20 \qquad x = 24$$
$$\frac{480}{x} = 24 \qquad \frac{480}{x} = 20$$

Thus, the dimensions of the page are 20 cm by 24 cm. *(1-5)*

55.

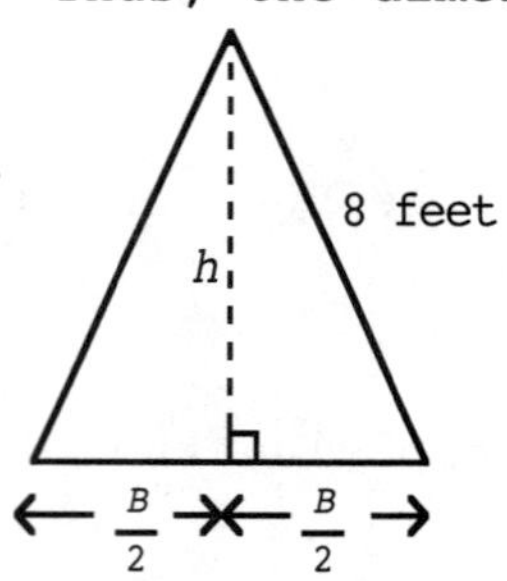

In the isosceles triangle we note:

$$\frac{1}{2}Bh = A = 24$$

Hence

$$h = \frac{48}{B}$$

Applying the Pythagorean theorem, we have

$$h^2 + \left(\frac{B}{2}\right)^2 = 8^2$$
$$\left(\frac{48}{B}\right)^2 + \left(\frac{B}{2}\right)^2 = 8^2$$
$$\frac{2,304}{B^2} + \frac{B^2}{4} = 64$$
$$4B^2\left(\frac{2,304}{B^2}\right) + 4B^2\left(\frac{B^2}{4}\right) = 4B^2(64)$$
$$9,216 + B^4 = 256B^2$$

$$(B^2)^2 - 256B^2 + 9{,}216 = 0$$

$$B^2 = \frac{-b \pm \sqrt{b^2 - 4ac}}{2a} \qquad a = 1,\ b = -256,\ c = 9{,}216$$

$$B^2 = \frac{-(-256) \pm \sqrt{(-256)^2 - 4(1)(9{,}216)}}{2(1)}$$

$$B^2 = \frac{256 \pm \sqrt{65{,}536 - 36{,}864}}{2}$$

$$B^2 = \frac{256 \pm \sqrt{28{,}672}}{2}$$

$$B^2 = 128 \pm 32\sqrt{7}$$

$$B = \sqrt{128 \pm 32\sqrt{7}}$$

$$B = 14.58 \text{ ft or } 6.58 \text{ ft} \qquad (1\text{-}6)$$

CHAPTER 2

Section 2-1

1. The set of all points for which the x coordinate is 0 is the y axis.

3. The set of all points for which the x and y coordinates are negative is quadrant III.

5. The set of all points for which the x coordinate is positive and the y coordinate is negative is quadrant IV.

7. The set of all points for which x is positive, excluding those points for which $y = 0$ (positive x axis), includes quadrants I and IV.

9. The set of all points for which $xy < 0$ includes those points for which the x coordinate is positive and the y coordinate is negative (quadrant IV) and also those points for which the x coordinate is negative and the y coordinate is positive (quadrant II).

11. 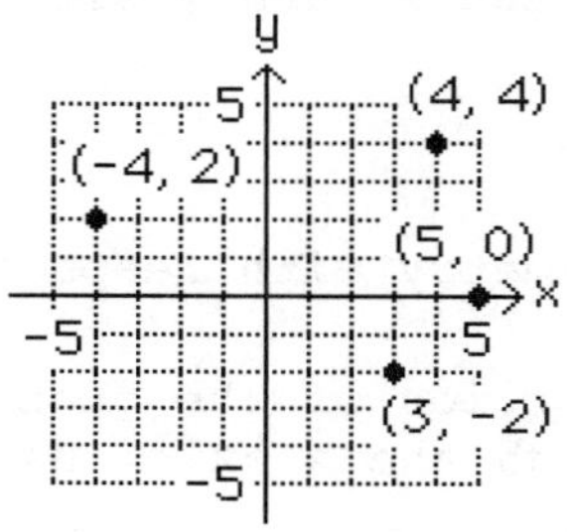

13. 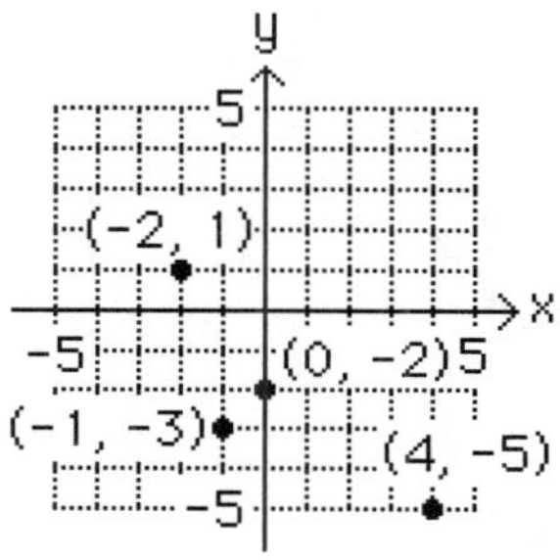

15. Point A has coordinates (2, 4). Its reflection through the y axis is $A'(-2, 4)$.
Point B has coordinates (3, −1). Its reflection through the y axis is $B'(-3, -1)$.
Point C has coordinates (−4, 0). Its reflection through the y axis is $C'(4, 0)$.
Point D has coordinates (−5, 2). Its reflection through the y axis is $D'(5, 2)$.

17. Point A has coordinates (−3, −3). Its reflection through the origin is $A'(3, 3)$.
Point B has coordinates (0, 4). Its reflection through the origin is $B'(0, -4)$.
Point C has coordinates (−3, 2). Its reflection through the origin is $C'(3, -2)$.
Point D has coordinates (5, −1). Its reflection through the origin is $D'(-5, 1)$.

19. $y = 2x - 4$

Test y axis	Test x axis	Test origin
Replace x with $-x$:	Replace y with $-y$:	Replace x with $-x$ and y with $-y$:
$y = 2(-x) - 4$	$-y = 2x - 4$	$-y = 2(-x) - 4$
$y = -2x - 4$	$y = -2x + 4$	$y = 2x + 4$

The graph has none of these symmetries.

x	y
0	-4
2	0
4	4

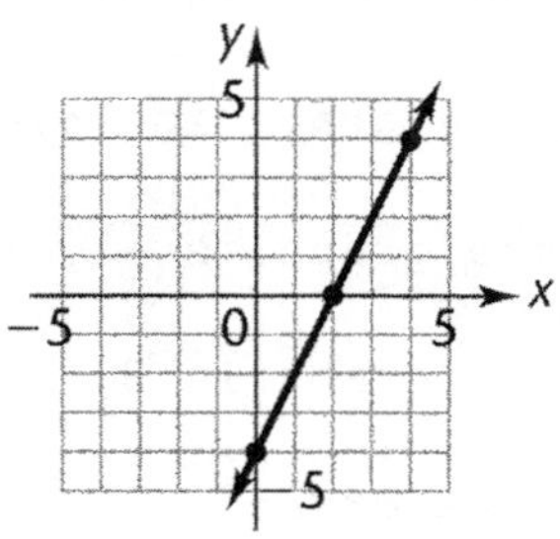

21. $y = \frac{1}{2}x$

Test y axis
Replace x with $-x$:

$y = \frac{1}{2}(-x)$

$y = -\frac{1}{2}x$

Test x axis
Replace y with $-y$:

$-y = \frac{1}{2}x$

$y = -\frac{1}{2}x$

Test origin
Replace x with $-x$ and y with $-y$:

$-y = \frac{1}{2}(-x)$

$y = \frac{1}{2}x$

x	y
0	0
4	2

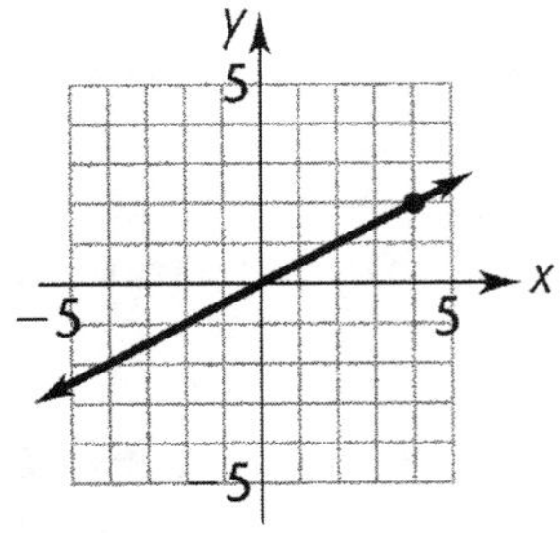

The graph has symmetry with respect to the origin.

We reflect the portion of the graph in quadrant I through the origin, using the origin symmetry.

23. $|y| = x$

Test y axis
Replace x with $-x$:
$|y| = -x$

Test x axis
Replace y with $-y$:
$|-y| = x$
$|y| = x$

Test origin
Replace x with $-x$ and y with $-y$:
$|-y| = -x$
$|y| = -x$

x	y
0	0
4	4

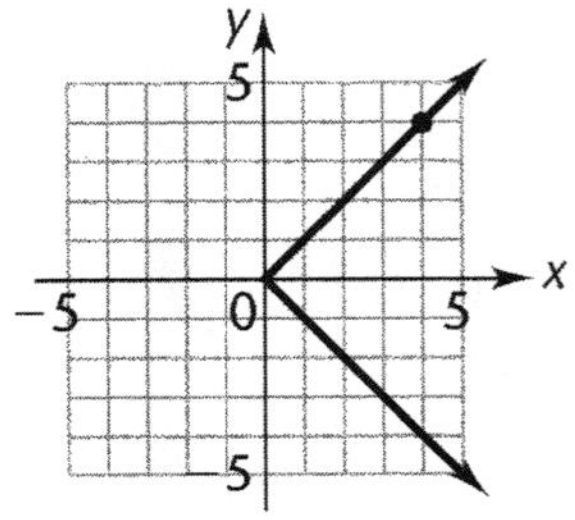

The graph has symmetry with respect to the x axis.

We reflect the portion of the graph where $y \geq 0$ through the x axis, using the x axis symmetry.

25. $|x| = |y|$

Test y axis
Replace x with $-x$:
$|-x| = |y|$
$|x| = |y|$

Test x axis
Replace y with $-y$:
$|x| = |-y|$
$|x| = |y|$

Origin symmetry follows automatically.

x	y
0	0
4	4

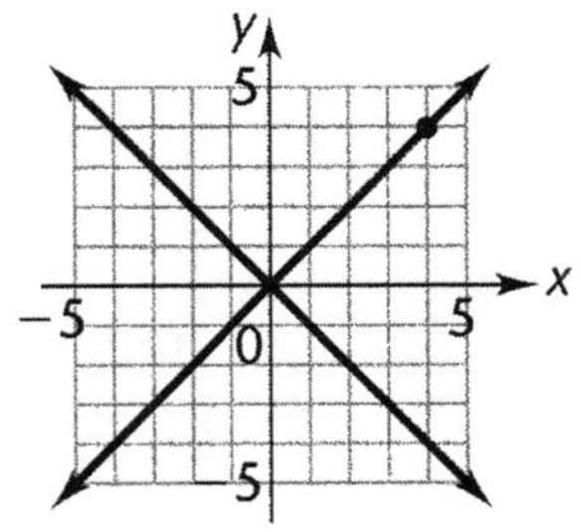

The graph has all three symmetries.

We reflect the portion of the graph in quadrant I through the y axis, and the x axis, and the origin, using all three symmetries.

27. (A) When $x = 8$, the corresponding y value on the graph is 6, to the nearest integer.

(B) When $x = -5$, the corresponding y value on the graph is -5, to the nearest integer.

(C) When $x = 0$, the corresponding y value on the graph is -1, to the nearest integer.

(D) When $y = 6$, the corresponding x value on the graph is 8, to the nearest integer.

(E) When $y = -5$, the corresponding x value on the graph is -5, to the nearest integer.

(F) When $y = 0$, the corresponding x value on the graph is 5, to the nearest integer.

29. (A) When $x = 1$, the corresponding y value on the graph is 6, to the nearest integer.
(B) When $x = -8$, the corresponding y value on the graph is 4, to the nearest integer.
(C) When $x = 0$, the corresponding y value on the graph is 4, to the nearest integer.
(D) When $y = -6$, the corresponding x value on the graph is 8, to the nearest integer.
(E) Three values of x correspond to $y = 4$ on the graph. To the nearest integer they are -8, 0, and 6.
(F) Three values of x correspond to $y = 0$ on the graph. To the nearest integer they are -7, -2, and 7.

31. (A) Plot the points (-2, -2), (0, 0), and (2, 2). There are many possible graphs connecting these points. The simplest is:

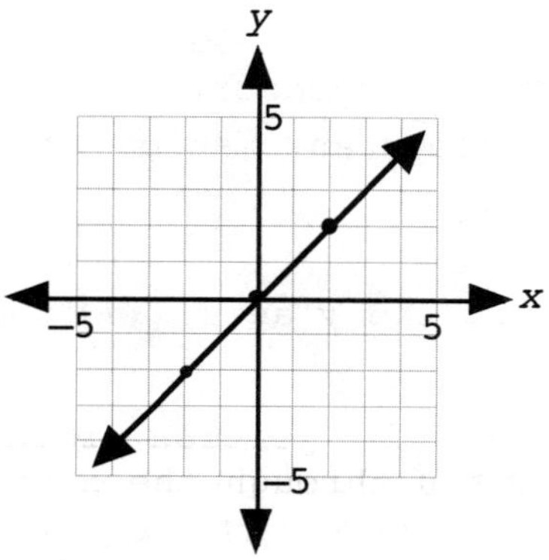

(B) Plot the points (-1, 2), (0, 0), and (1, -2). Again, the simplest graph connecting the points is a line:

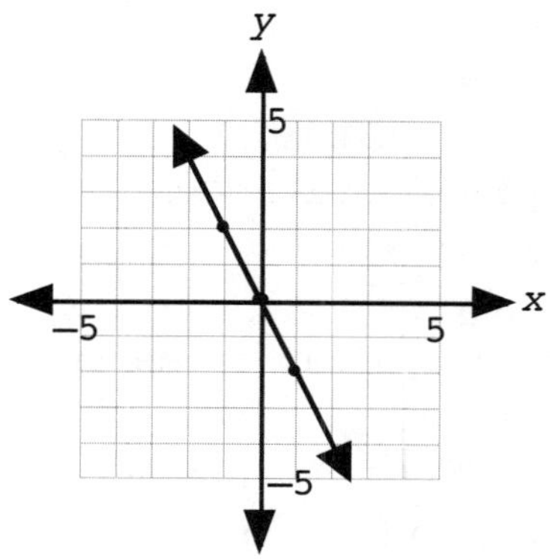

(C)

x	−2	−1	0	1	2
y	−2	2	0	−2	2

Plot the points (-2, -2), (-1, 2), (0, 0), (1, -2), and (2, 2).
The graph of $y = x^3 - 3x$ is shown below.

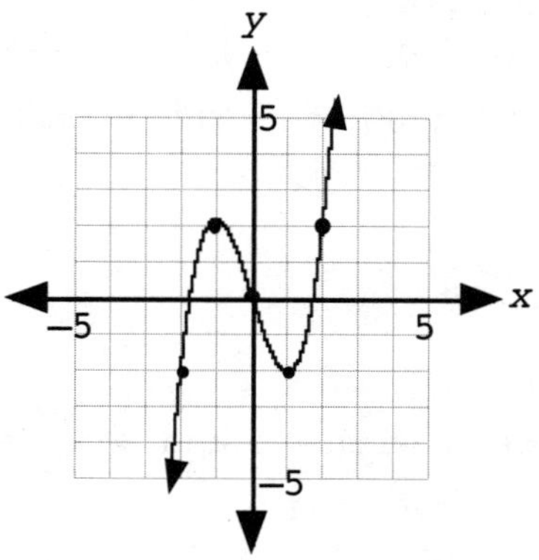

(D) All of the points on the two graphs in (A) and (B) are on the graph of $y = x^3 - 3x$. If you only plot certain points, it's possible to draw an erroneous conclusion about the shape of the graph.

33. (A) Reflect the given graph across the x axis.

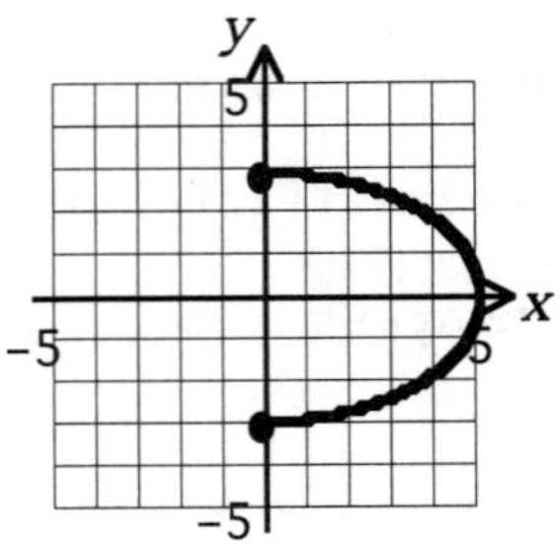

(B) Reflect the given graph across the y axis.

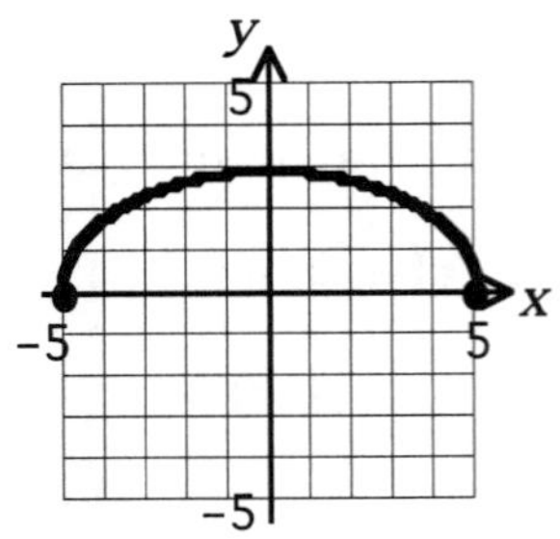

(C) Reflect the given graph through the origin.

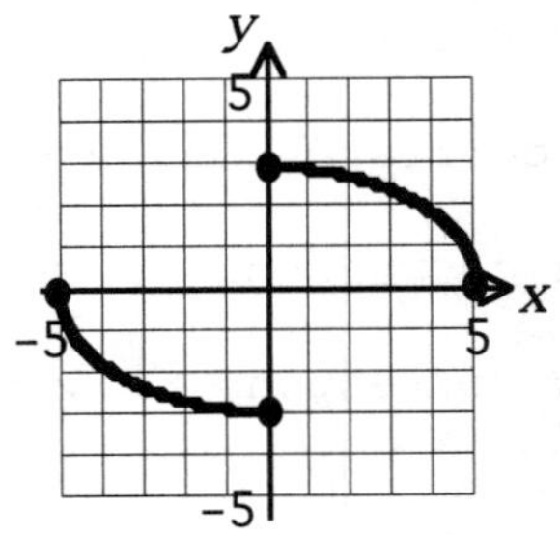

(D) Reflect the given graph across the y axis, then reflect the resulting curve across the x axis.

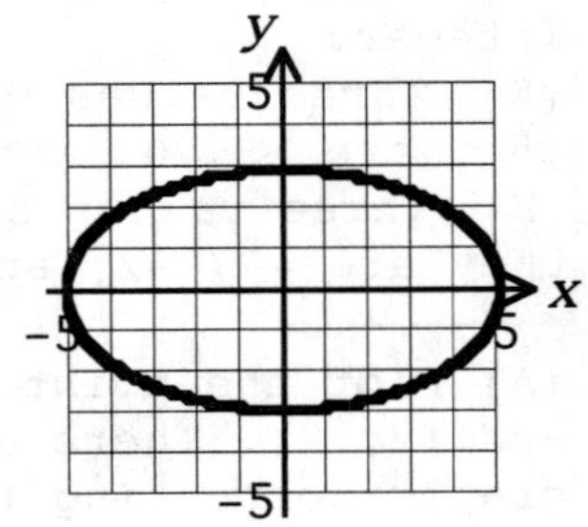

35. $x^2y^2 = 4$

Test y axis
Replace x with $-x$:
$(-x)^2y^2 = 4$
$x^2y^2 = 4$

Test x axis
Replace y with $-y$:
$x^2(-y)^2 = 4$
$x^2y^2 = 4$

Origin symmetry follows automatically.

The graph has symmetry with respect to the x axis, the y axis, and the origin.

37. $x^2 + xy^2 = 3$

Test y axis
Replace x with $-x$:
$(-x)^2 + (-x)y^2 = 3$
$x^2 - xy^2 = 3$

Test x axis
Replace y with $-y$:
$x^2 + x(-y)^2 = 3$
$x^2 + xy^2 = 3$

Test origin
Replace x with $-x$ and y with $-y$:
$(-x)^2 + (-x)(-y)^2 = 3$
$x^2 - xy^2 = 3$

The graph has symmetry with respect to the x axis.

39. $x^2 + 4y + y^2 = 9$

Test y axis
Replace x with $-x$:
$(-x)^2 + 4y + y^2 = 9$
$x^2 + 4y + y^2 = 9$

Test x axis
Replace y with $-y$:
$x^2 + 4(-y) + (-y)^2 = 9$
$x^2 - 4y + y^2 = 9$

Test origin
Replace x with $-x$ and y with $-y$:
$(-x)^2 + 4(-y) + (-y)^2 = 9$
$x^2 - 4y + y^2 = 9$

The graph has symmetry with respect to the y axis.

41. $y^2 = x + 2$

Test y axis
Replace x with $-x$:
$y^2 = -x + 2$

Test x axis
Replace y with $-y$:
$(-y)^2 = x + 2$
$y^2 = x + 2$

Test origin
Replace x with $-x$ and y with $-y$:
$(-y)^2 = (-x) + 2$
$y^2 = -x + 2$

x	y
-2	0
-1	1
2	2

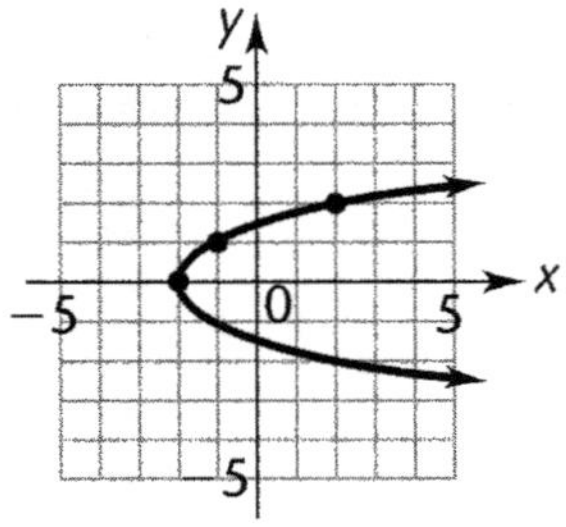

The graph has symmetry with respect to the x axis.
To obtain the portion of the graph for $y \geq 0$, we sketch $y = \sqrt{x + 2}$, $x \geq -2$. We reflect the portion of the graph for $y \geq 0$ across the x axis, using the x axis symmetry.

43. $y = x^2 + 1$

Test y axis
Replace x with $-x$:
$y = (-x)^2 + 1$
$y = x^2 + 1$

Test x axis
Replace y with $-y$:
$(-y) = x^2 + 1$
$y = -x^2 - 1$

Test origin
Replace x with $-x$ and y with $-y$:
$(-y) = (-x)^2 + 1$
$y = -x^2 - 1$

x	y
0	1
1	2
2	5

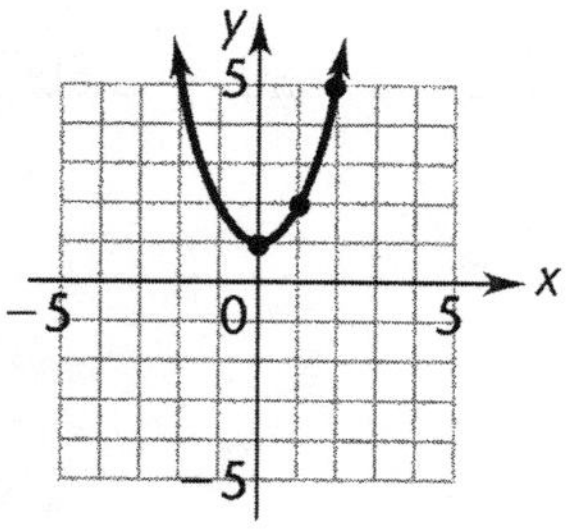

The graph has symmetry with respect to the y axis. We reflect the portion of the graph for $x \geq 0$ across the y axis, using the y axis symmetry.

45. $4y^2 - x^2 = 1$

Test y axis
Replace x with $-x$:
$4y^2 - (-x)^2 = 1$
$4y^2 - x^2 = 1$

Test x axis
Replace y with $-y$:
$4(-y)^2 - x^2 = 1$
$4y^2 - x^2 = 1$

Origin symmetry follows automatically.

The graph has all three symmetries. $y = \pm\frac{1}{2}\sqrt{x^2 + 1}$.

x	y
0	$\frac{1}{2}$
2	$\frac{1}{2}\sqrt{5} \approx 1.1$
4	$\frac{1}{2}\sqrt{17} \approx 2.0$

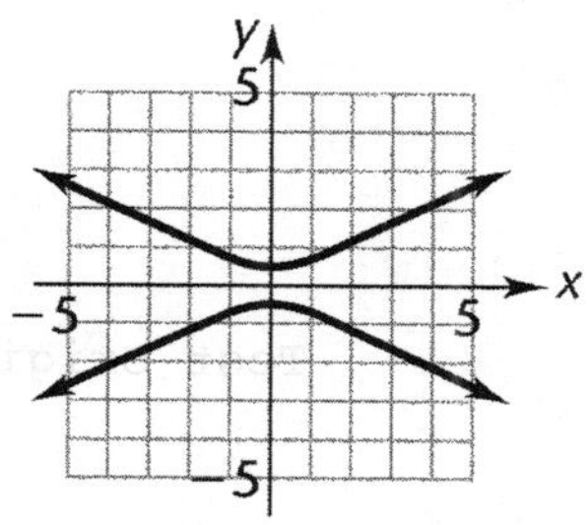

The graph has all three symmetries. To obtain the quadrant I portion of the graph, we sketch $y = \frac{1}{2}\sqrt{x^2 + 1}$, $x \geq 0$. We reflect this graph across the y axis, then reflect everything across the x axis.

47. $y^3 = x$

Test y axis
Replace x with $-x$:
$y^3 = -x$

Test x axis
Replace y with $-y$:
$(-y)^3 = x$
$y^3 = -x$

Test origin
Replace x with $-x$ and y with $-y$:
$(-y)^3 = -x$
$y^3 = x$

x	y
0	0
1	1
8	2

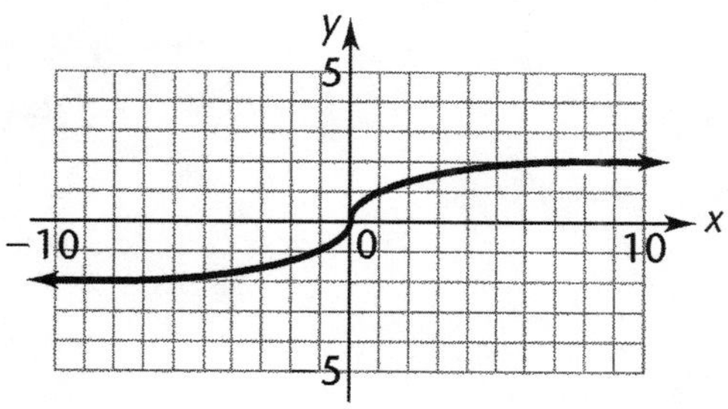

The graph has symmetry with respect to the origin.
We reflect the portion of the graph in quadrant I through the origin, using the origin symmetry.

49. $y = 0.6x^2 - 4.5$

Test y axis
Replace x with $-x$:
$y = 0.6(-x)^2 - 4.5$
$y = 0.6x^2 - 4.5$

Test x axis
Replace y with $-y$:
$-y = 0.6x^2 - 4.5$
$y = -0.6x^2 + 4.5$

Test origin
Replace x with $-x$ and y with $-y$:
$-y = 0.6(-x)^2 - 4.5$
$y = -0.6x^2 + 4.5$

x	y
0	-4.5
1	-3.9
2	-2.1
3	0.9
4	5.1

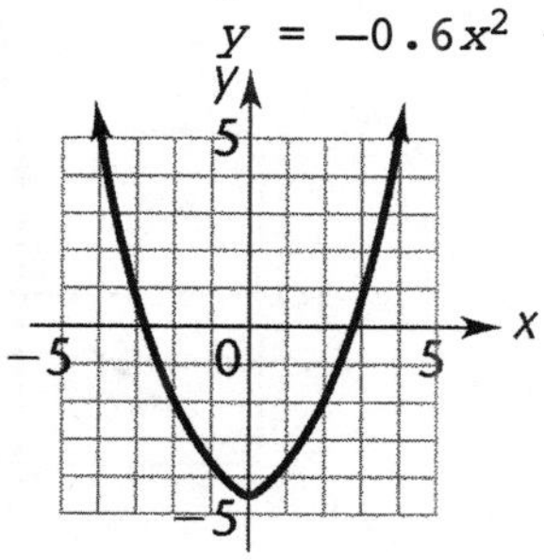

The graph has symmetry with respect to the y axis. We reflect the portion of the graph in quadrant I through the y axis, using the y axis symmetry.

51. $y = x^{2/3}$

Test y axis
Replace x with $-x$:
$y = (-x)^{2/3}$
$y = x^{2/3}$

Test x axis
Replace y with $-y$:
$-y = x^{2/3}$
$y = -x^{2/3}$

Test origin
Replace x with $-x$ and y with $-y$:
$-y = (-x)^{2/3}$
$y = -x^{2/3}$

The graph has symmetry with respect to the y axis.
We reflect the portion of the graph for $x \geq 0$ across the y axis, using the y axis symmetry.

x	y
0	0
1	1
2	$2^{2/3} \approx 1.6$
3	$3^{2/3} \approx 2.1$

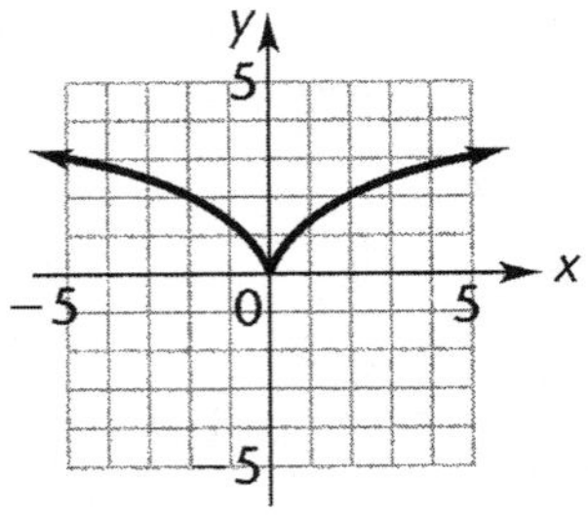

53. (A) and (B)

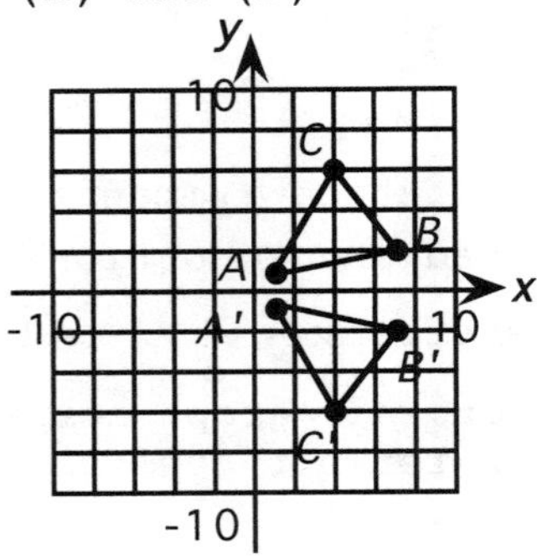

(C) The triangles are mirror images of each other, reflected across the x axis. Changing the sign of the y coordinate reflects the graph across the x axis.

55. (A) and (B)

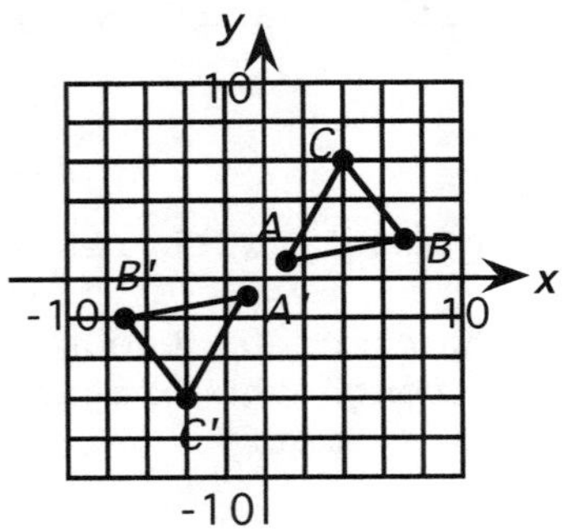

(C) The triangles are mirror images of each other, reflected across the origin. Changing the signs of both coordinates reflects the graph through the origin.

57. $2x^2 + y^2 = 3$
$y^2 = 3 - 2x$
$y = \pm\sqrt{3 - 2x}$

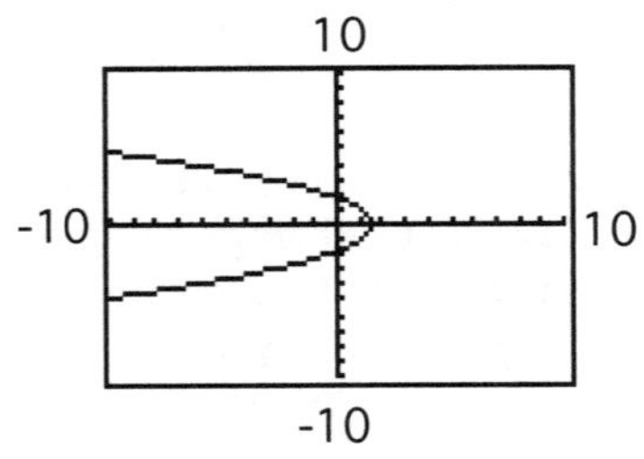

59. $x^2 - (y + 1)^2 = 4$
$-(y + 1)^2 = 4 - x^2$
$(y + 1)^2 = x^2 - 4$
$y + 1 = \pm\sqrt{x^2 - 4}$
$y = -1 \pm \sqrt{x^2 - 4}$

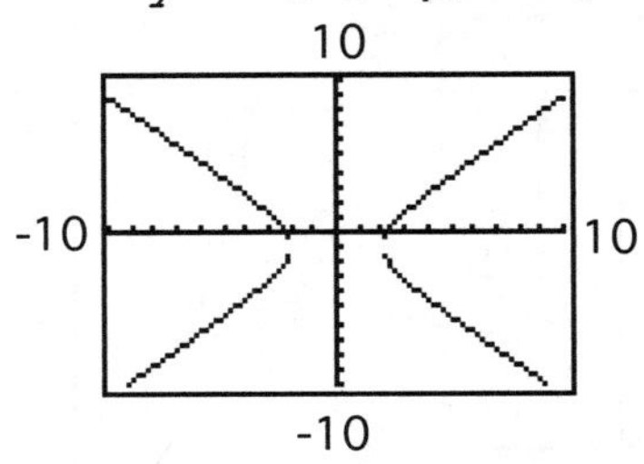

61. $y^3 = |x|$

Test y axis
Replace x with $-x$:
$y^3 = |-x|$
$y^3 = |x|$

Test x axis
Replace y with $-y$:
$(-y)^3 = |x|$
$y^3 = -|x|$

Test origin
Replace x with $-x$ and y with $-y$:
$(-y)^3 = |-x|$
$y^3 = -|x|$

x	y
0	0
1	1
8	2

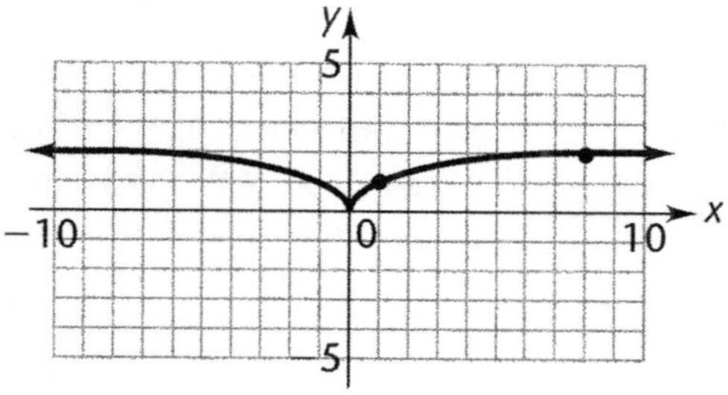

The graph has symmetry with respect to the y axis.
We reflect the portion of the graph for $x \geq 0$ across the y axis using the y axis symmetry.

63. $xy = 1$

Test y axis
Replace x with $-x$:
$(-x)y = 1$
$xy = -1$

Test x axis
Replace y with $-y$:
$x(-y) = 1$
$xy = -1$

Test origin
Replace x with $-x$ and y with $-y$:
$(-x)(-y) = 1$
$xy = 1$

x	y
1	1
2	$\frac{1}{2}$
3	$\frac{1}{3}$
$\frac{1}{2}$	2
$\frac{1}{3}$	3

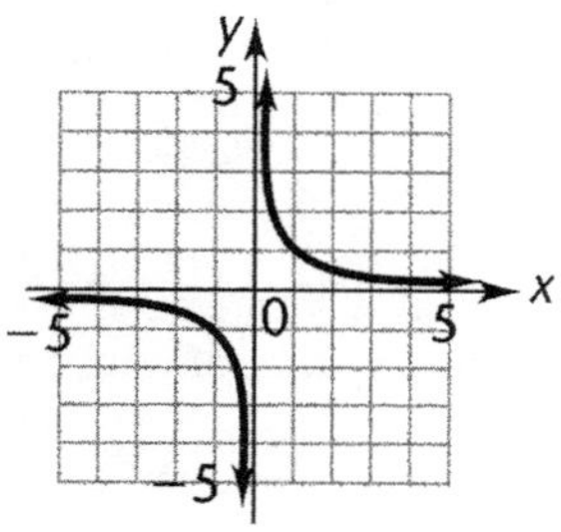

The graph has symmetry with respect to the origin.
We reflect the portion of the graph in quadrant I through the origin, using the origin symmetry.

65. $6x - x^2$

Test y axis
Replace x with $-x$:
$y = 6(-x) - (-x)^2$
$y = -6x - x^2$

Test x axis
Replace y with $-y$:
$-y = 6x - x^2$
$y = -6x + x^2$

Test origin
Replace x with $-x$ and y with $-y$:
$-y = 6(-x) - (-x)^2$
$y = 6x + x^2$

The graph has none of these three symmetries. A larger table of values is needed since we have no symmetry information.

x	y
-1	-7
0	0
1	5
2	8
3	9
4	8
5	5
6	0
7	-7

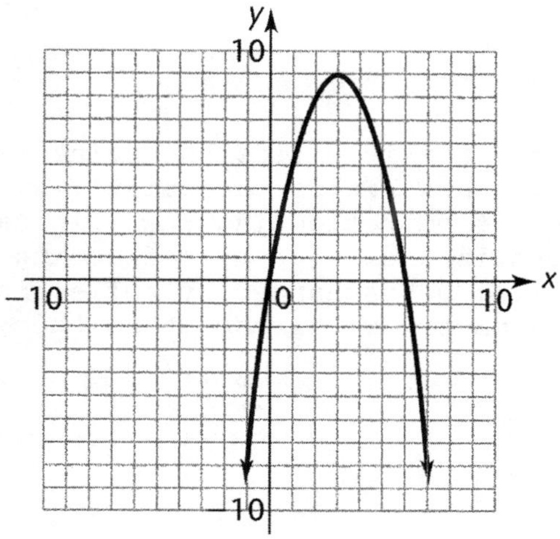

67. $y^2 = |x| + 1$

Test y axis
Replace x with $-x$:
$y^2 = |-x| + 1$
$y^2 = |x| + 1$

Test x axis
Replace y with $-y$:
$(-y^2) = |x| + 1$
$y^2 = |x| + 1$

Origin symmetry follows automatically.

The graph has symmetry with respect to the x axis, the y axis, and the origin.
$y = \pm\sqrt{|x| + 1}$
To obtain the quadrant I portion of this graph, we sketch $y = \sqrt{|x| + 1}$ $x \geq 0$.
We reflect this graph across the y axis, then reflect everything across the x axis.

x	y
0	1
1	$\sqrt{2} \approx 1.4$
3	2

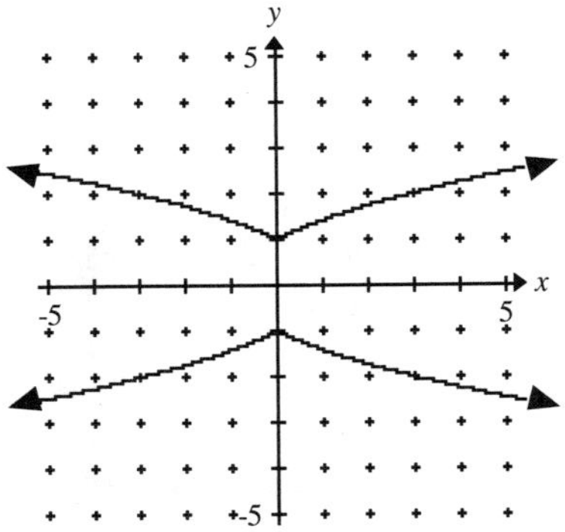

69. $|xy| + 2|y| = 6$

Test y axis
Replace x with $-x$:
$|(-x)y| + 2|y| = 6$
$|xy| + 2|y| = 6$

Test x axis
Replace y with $-y$:
$|x(-y)| + 2|-y| = 6$
$|xy| + 2|y| = 6$

Origin symmetry follows automatically.

x	y
0	3
1	2
4	1

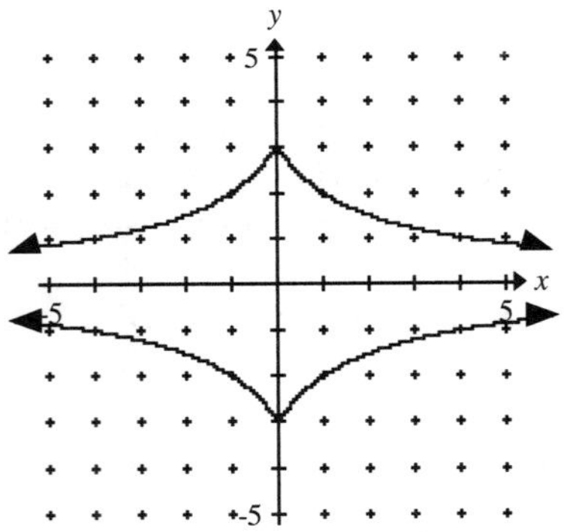

The graph has symmetry with respect to the x axis, the y axis, and the origin. We reflect the portion of the graph in quadrant I across the y axis, then reflect everything across the x axis.

71. Reflecting a point (x, y) across the x axis yields the point $(x, -y)$. Reflecting this point through the origin yields the point $(-x, y)$. This point is the same point that would result from reflecting the original point across the y axis. Therefore, if the graph is unchanged by reflecting across the x axis and across the origin, it will be unchanged by reflecting across the y axis and will necessarily have symmetry with respect to the y axis.

73.

p	$R = (10 - p)p$
5	25
6	24
7	21
8	16
9	9
10	0

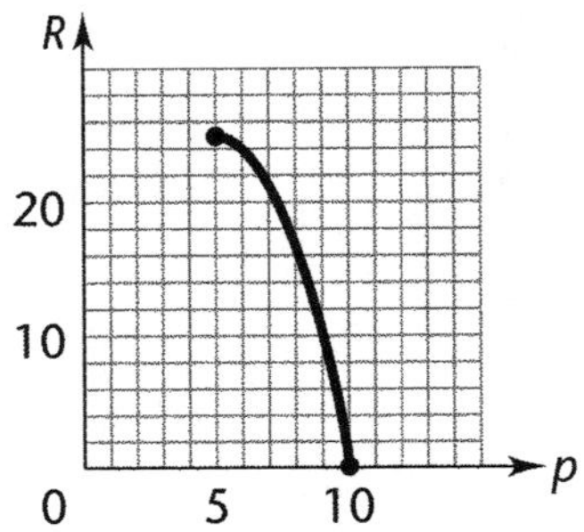

75. (A) \$6.00 on the price scale corresponds to 3,000 cases on the demand scale.

(B) The demand decreases from 3,000 to 2,600 cases, that is, by 400 cases.

(C) The demand increases from 3,000 to 3,600 cases, that is, by 600 cases.

(D) Demand decreases with increasing price and increases with decreasing price. To increase demand from 2,000 to 4,000 cases, a price decrease from \$6.90 to \$5.60 is necessary.

77. (A) 9:00 is halfway from 6 AM to noon, and corresponds to a temperature of 53°.

(B) The highest temperature occurs halfway from noon to 6 PM, at 3 PM. This temperature is 68°.

(C) This temperature occurs at 1 AM, 7 AM, and 11 PM.

79. (A) There is no obvious symmetry. A table of values yields the following approximate values:

x	0	0.5	1	1.5	2
v	0.7	0.6	0.5	0.35	0

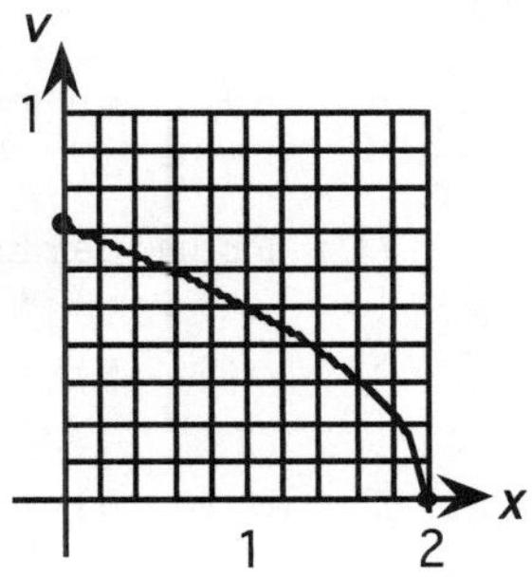

(C) For a displacement of 2 cm, the ball is stationary ($v = 0$). As the vertical displacement approaches 0, the ball gathers speed until $v = 0.5\sqrt{2} \approx 0.7$ m/sec. The velocity carries the ball through the equilibrium position ($x = 0$) to rise again to displacement 2 cm and velocity 0.

Section 2-2

1. $d = \sqrt{(4-1)^2 + (4-0)^2} = \sqrt{9+16} = \sqrt{25} = 5$

Midpoint $= \left(\frac{1+4}{2}, \frac{0+4}{2}\right) = \left(\frac{5}{2}, 2\right)$

3. $d = \sqrt{(5-0)^2 + (10-(-2))^2} = \sqrt{5^2 + 12^2} = \sqrt{25+144} = \sqrt{169} = 13$

Midpoint $= \left(\frac{0+5}{2}, \frac{-2+10}{2}\right) = \left(\frac{5}{2}, 4\right)$

5. $d = \sqrt{(3-(-6))^2 + (4-(-4))^2} = \sqrt{9^2 + 8^2} = \sqrt{81+64} = \sqrt{145}$

Midpoint $= \left(\frac{-6+3}{2}, \frac{-4+4}{2}\right) = \left(\frac{-3}{2}, 0\right)$

7. $d = \sqrt{(-6-(-2))^2 + (-3-(-1))^2} = \sqrt{20} = 2\sqrt{5}$

Midpoint $= \left(\frac{(-6)+(-2)}{2}, \frac{(-3)+(-1)}{2}\right) = (-4, -2)$

9. $(x-0)^2 + (y-0)^2 = 7^2$
$x^2 + y^2 = 49$

11. $(x-2)^2 + (y-3)^2 = 6^2$
$(x-2)^2 + (y-3)^2 = 36$

13. $[x-(-4)]^2 + (y-1)^2 = (\sqrt{7})^2$
$(x+4)^2 + (y-1)^2 = 7$

Common Error: *not* $(x-4)^2$

15. $[x-(-3)]^2 + [y-(-4)]^2 = (\sqrt{2})^2$
$(x+3)^2 + (y+4)^2 = 2$

17. This is a circle with center (0, 0) and radius 2.

$$x^2 + y^2 = 4$$

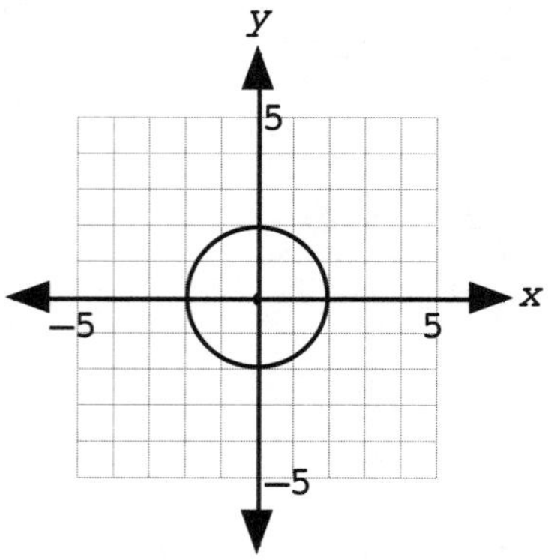

19. This is a circle with center (1, 0) and radius 1.

$$(x - 1)^2 + y^2 = 1$$

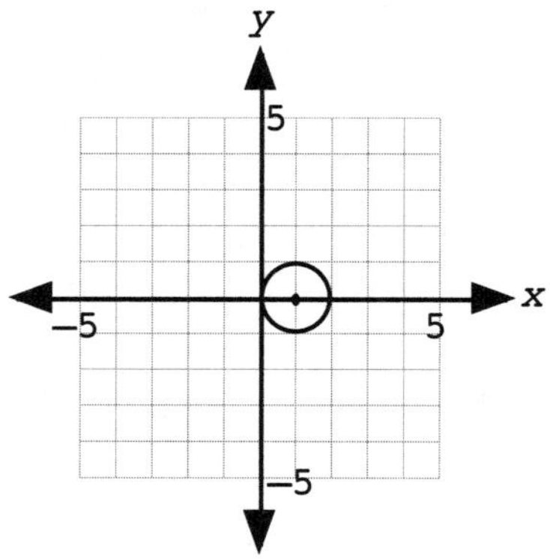

21. This is a circle with center (-2, 1) and radius 3.

$$(x - (-2))^2 + (y - 1)^2 = 9$$
or $(x + 2)^2 + (y - 1)^2 = 9$

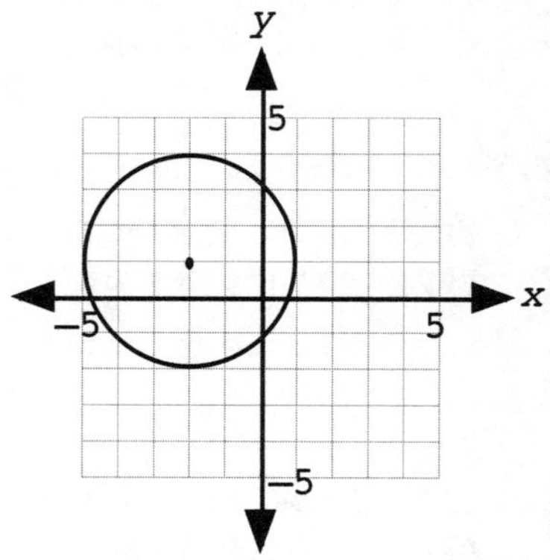

23. (A)
$$-2 = \frac{a_1 + 1}{2}$$
$$-4 = a_1 + 1$$
$$-5 = a_1$$

(B)
$$6 = \frac{a_2 + 3}{2}$$
$$12 = a_2 + 3$$
$$9 = a_2$$

(C) From parts (A) and (B), $A = (-5, 9)$

$$d(A, M) = \sqrt{(-2(-5)^2 + (6 - 9)^2} = \sqrt{3^2 + (-3)^2} = \sqrt{9 + 9} = \sqrt{18}$$

$$d(M, B) = \sqrt{(1 - (-2)^2 + (3 - 6)^2} = \sqrt{3^2 + (-3)^2} = \sqrt{9 + 9} = \sqrt{18}$$

As expected, the distances are the same.

25. The distance formula requires that $\sqrt{[x - (-4)\,]^2 + (7 - 1)^2} = 10$

Solving, we have $\sqrt{(x + 4)^2 + (6)^2} = 10$
$$(x + 4)^2 + (6)^2 = 10^2$$
$$(x + 4)^2 + 36 = 100$$
$$(x + 4)^2 = 64$$
$$x + 4 = \pm 8$$
$$x = -4 \pm 8$$
$$x = -12,\ 4$$

27. The distance formula requires that $\sqrt{[2 - (-1)\,]^2 + (y - 4)^2} = 3$

Solving, we have $\sqrt{(3)^2 + (y - 4)^2} = 3$
$$(3)^2 + (y - 4)^2 = (3)^2$$
$$(y - 4)^2 = 0$$
$$y - 4 = 0$$
$$y = 4$$

29. This is a circle with center (0, 2) and radius 2. That is, the set of all points that are 2 units away from (0, 2).
$$x^2 + (y - 2)^2 = 4$$

31. This is a circle with center (1, 1) and radius 4. That is, the set of all points that are 4 units away from (1, 1).
$$(x - 1)^2 + (y - 1)^2 = 16$$

33. $M = \left(\dfrac{-4.3 + 9.6}{2}, \dfrac{5.2 + (-1.7)}{2}\right) = \left(\dfrac{5.3}{2}, \dfrac{3.5}{2}\right) = (2.65, 1.75)$

$$d(A, M) = \sqrt{(2.65 - (-4.3))^2 + (1.75 - 5.2)^2} = \sqrt{6.95^2 + (-3.45)^2} = \sqrt{48.3025 + 11.9025} = \sqrt{60.205} = 7.76$$

$$d(M, B) = \sqrt{(4.6 - 2.65)^2 + (-1.7 - 1.75)^2} = \sqrt{6.95^2 + (-3.45)^2} = \sqrt{48.3025 + 11.9025} = \sqrt{60.205} = 7.76$$

$$d(A, B) = \sqrt{(9.6 - (-4.3))^2 + (-1.7 - 5.2)^2} = \sqrt{13.9^2 + (-6.9)^2} = \sqrt{193.21 + 47.61} = \sqrt{240.82} = 15.52$$

$$\frac{1}{2}d(A, B) = \frac{1}{2}(15.52) = 7.76$$

35. Write $B = (b_1, b_2)$. -5 is the average of 25 and b_1, so $-5 = \dfrac{25 + b_1}{2}$

$$-10 = 25 + b_1$$
$$-35 = b_1$$

-2 is the average of 10 and b_2, so $-2 = \dfrac{10 + b_2}{2}$

$$-4 = 10 + b_2$$
$$-14 = b_2$$

So $B = (-35, -14)$.

$$d(A, M) = \sqrt{(-5 - 25)^2 + (-2 - 10)^2} = \sqrt{(-30)^2 - (12)^2} = \sqrt{900 + 144} = \sqrt{1,044} = 32.3$$

$$d(M, B) = \sqrt{(-35 - (-5))^2 + (-14 - (-2))^2} = \sqrt{(-30)^2 - (12)^2} = \sqrt{900 + 144} = \sqrt{1,044} = 32.3$$

$$d(A, B) = \sqrt{(-35 - 25)^2 + (-14 - 10)^2} = \sqrt{(-60)^2 + (-24)^2} = \sqrt{3,600 + 576} = \sqrt{4,176} = 64.6$$

$$\frac{1}{2}d(A, B) = \frac{1}{2}(64.6) = 32.3$$

37. Write $A = (a_1, a_2)$. -8 is the average of a_1 and 2, so $-8 = \dfrac{a_1 + 2}{2}$

$$-16 = a_1 + 2$$
$$-18 = a_1$$

-6 is the average of a_2 and 4, so $-6 = \dfrac{a_2 + 4}{2}$

$$-12 = a_2 + 4$$
$$-16 = a_2$$

So $A = (-18, -16)$.

$$d(A, M) = \sqrt{(-8 - (-18))^2 + (-6 - (-16))^2} = \sqrt{10^2 + 10^2} = \sqrt{100 + 100} = \sqrt{200} = 14.14$$

$$d(M, B) = \sqrt{(2 - (-8))^2 + (4 - (-6))^2} = \sqrt{10^2 + 10^2} = \sqrt{100 + 100} = \sqrt{200} = 14.14$$

$$d(A, B) = \sqrt{(2 - (-18))^2 + (4 - (-16))^2} = \sqrt{20^2 + 20^2} = \sqrt{400 + 400} = \sqrt{800} = 28.28$$

$$\frac{1}{2}d(A, B) = \frac{1}{2}(28.28) = 14.14$$

39. $x^2 + (y + 2)^2 = 9$

$(x - 0)^2 + (y - (-2))^2 = 3^2$

Center $(0, -2)$; Radius $= 3$

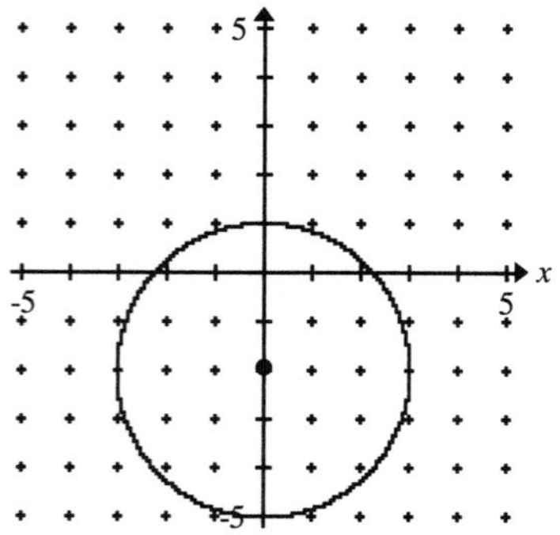

41. $(x + 4)^2 + (y - 2)^2 = 7$

$[x - (-4)]^2 + (y - 2)^2 = (\sqrt{7})^2$

Center $(-4, 2)$; Radius $= \sqrt{7}$

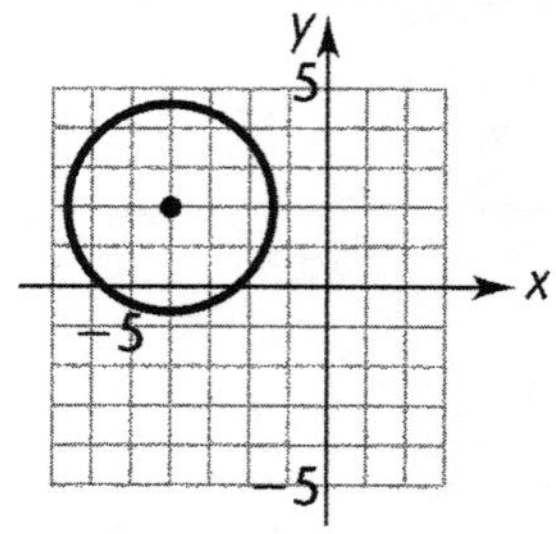

43. $x^2 + 6x + y^2 = 16$

$x^2 + 6x + 9 + y^2 = 16 + 9$

$(x + 3)^2 + y^2 = 25$

Center $(-3, 0)$; Radius $= 5$

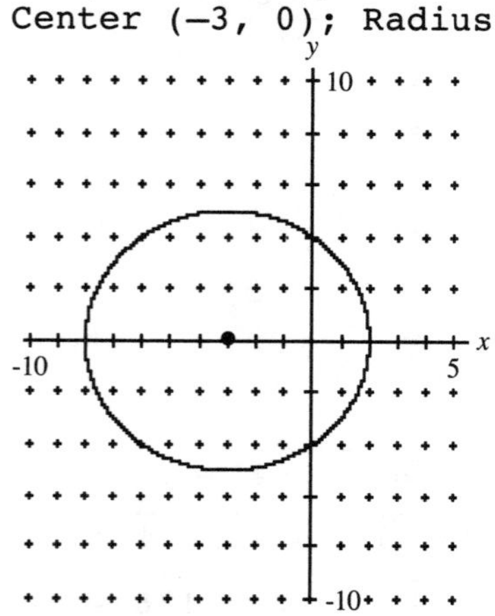

45. $x^2 + y^2 - 6x - 4y = 36$

$x^2 - 6x + y^2 - 4y = 36$

$x^2 - 6x + 9 + y^2 - 4y + 4 = 36 + 9 + 4$

$(x - 3)^2 + (y - 2)^2 = 49$

Center $(3, 2)$; Radius $= 7$

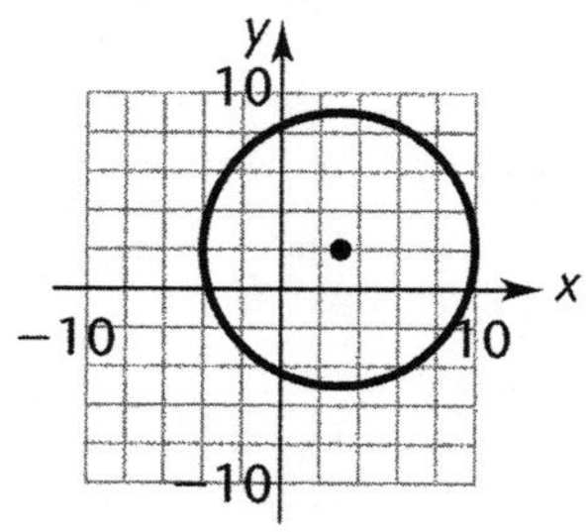

47. $3x^2 + 3y^2 + 24x - 18y + 24 = 0$

$x^2 + y^2 + 8x - 6y + 8 = 0$

$x^2 + 8x + y^2 - 6y = -8$

$x^2 + 8x + 16 + y^2 - 6y + 9 = -8 + 16 + 9$

$(x + 4)^2 + (y - 3)^2 = 17$

Center $(-4, 3)$; Radius $= \sqrt{17}$

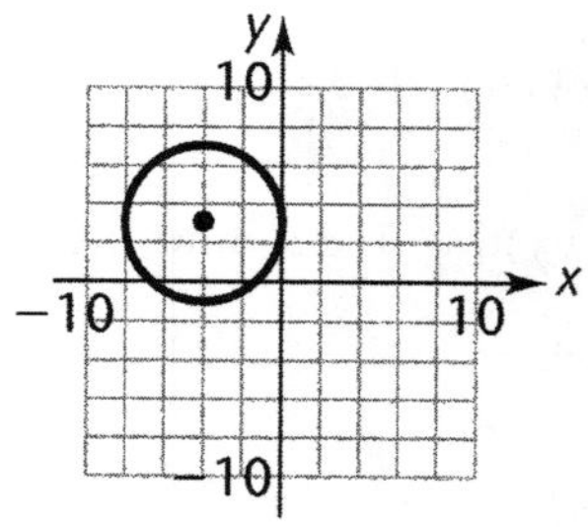

49. $x^2 + y^2 = 3$

$y^2 = 3 - x^2$

$y = \pm\sqrt{3 - x^2}$

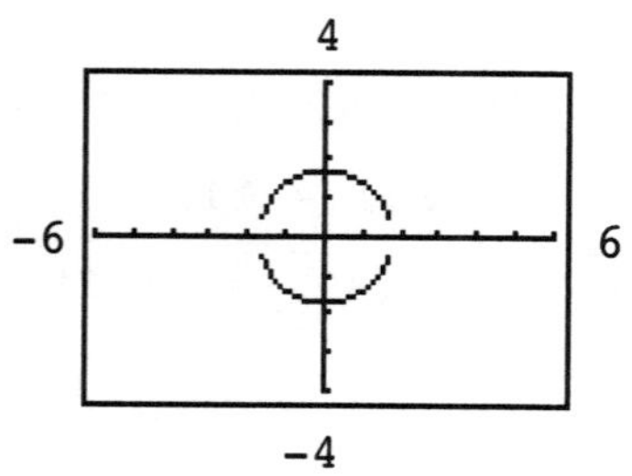

51. $(x + 3)^2 + (y + 1)^2 = 2$

$(y + 1)^2 = 2 - (x + 3)^2$

$y + 1 = \pm\sqrt{2 - (x + 3)^2}$

$y = -1 \pm \sqrt{2 - (x + 3)^2}$

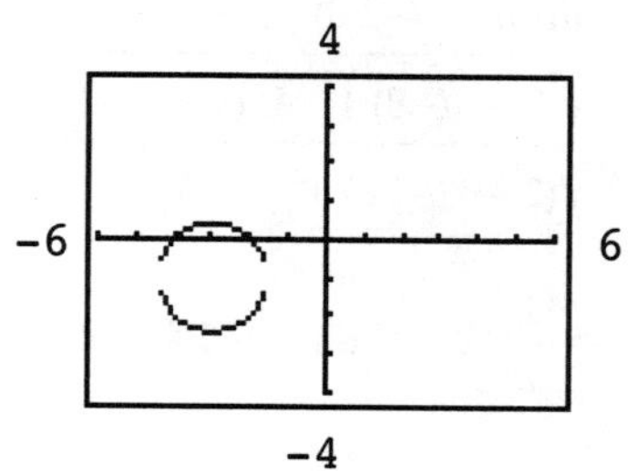

53. Let $A = (-3, 2)$, $B = (1, -2)$, $C = (8, 5)$

$d(A, B) = \sqrt{(1-(-3))^2 + (-2-2)^2} = \sqrt{4^2 + (-4)^2} = \sqrt{16+16} = \sqrt{32}$

$d(B, C) = \sqrt{(8-1)^2 + (5-(-2))^2} = \sqrt{7^2 + 7^2} = \sqrt{49+49} = \sqrt{98}$

$d(A, C) = \sqrt{(8-(-3))^2 + (5-2)^2} = \sqrt{11^2 + 3^2} = \sqrt{121+9} = \sqrt{130}$

Notice that $(d(A, B))^2 + (d(B, C))^2 = 32 + 98 = 130 = (d(A, C))^2$

Since these distances satisfy the Pythagorean Theorem, the three points are vertices of a right triangle. The segment connecting A and C is the hypotenuse (it's the longest side) so we need to find its midpoint.

$$M = \left(\frac{-3+8}{2}, \frac{2+5}{2}\right) = \left(\frac{5}{2}, \frac{7}{2}\right)$$

The vertex opposite the hypotenuse is B

$$d(M, B) = \sqrt{\left(1-\frac{5}{2}\right)^2 + \left(-2-\frac{7}{2}\right)^2} = \sqrt{\left(\frac{-3}{2}\right)^2 + \left(-\frac{-11}{2}\right)^2} = \sqrt{\frac{9}{4} + \frac{121}{4}} = \sqrt{\frac{130}{4}} = \sqrt{32.5}$$

55. Perimeter = sum of lengths of all three sides

$= \sqrt{[1-(-3)]^2 + [(-2)-1]^2} + \sqrt{(4-1)^2 + [3-(-2)]^2} + \sqrt{[4-(-3)]^2 + (3-1)^2}$

$= \sqrt{16+9} + \sqrt{9+25} + \sqrt{49+4}$

$= \sqrt{25} + \sqrt{34} + \sqrt{53}$

$= 18.11$ to two decimal places

57.
$$d(P_1, M) = \sqrt{\left(\frac{x_1+x_2}{2} - x_1\right)^2 + \left(\frac{y_1+y_2}{2} - y_1\right)^2}$$
$$= \sqrt{\left(\frac{x_1+x_2-2x_1}{2}\right)^2 + \left(\frac{y_1+y_2-2y_1}{2}\right)^2}$$
$$= \sqrt{\left(\frac{x_2-x_1}{2}\right)^2 + \left(\frac{y_2-y_1}{2}\right)^2}$$

$$d(M, P_2) = \sqrt{\left(x_2 - \frac{x_1+x_2}{2}\right)^2 + \left(y_1 - \frac{y_1+y_2}{2}\right)^2}$$
$$= \sqrt{\left(\frac{2x_2-x_1-x_2}{2}\right)^2 + \left(\frac{2y_2-y_1-y_2}{2}\right)^2}$$
$$= \sqrt{\left(\frac{x_2-x_1}{2}\right)^2 + \left(\frac{y_2-y_1}{2}\right)^2}$$

$$\frac{1}{2}d(P_1, P_2) = \frac{1}{2}\sqrt{(x_2-x_1)^2 + (y_2-y_1)^2}$$
$$= \sqrt{\frac{1}{4}\left[(x_2-x_1)^2 + (y_2-y_1)^2\right]}$$

Note: The $\frac{1}{2}$ becomes $\frac{1}{4}$ when moved inside the root.

$$= \sqrt{\frac{(x_2-x_1)^2}{4} + \frac{(y_2-y_1)^2}{4}}$$
$$= \sqrt{\left(\frac{x_2-x_1}{2}\right)^2 + \left(\frac{y_2-y_1}{2}\right)^2}$$

All three of these distances are equal.

59. The center of a circle is the midpoint of its diameter. The radius of a circle is half the length of its diameter. Using the midpoint formula, the center of this circle has coordinates given by

$$\left(\frac{x_1 + x_2}{2}, \frac{y_1 + y_2}{2}\right) = \left(\frac{0 + 0}{2}, \frac{5 + (-3)}{2}\right) = (0, 1)$$

The radius of the circle is given by

$$r = \frac{1}{2}\sqrt{(x_2 - x_1)^2 + (y_2 - y_1)^2}$$
$$= \frac{1}{2}\sqrt{(0 - 0)^2 + [5 - (-3)]^2}$$
$$= \frac{1}{2}\sqrt{64} = \frac{1}{2} \cdot 8 = 4$$

Hence the equation of the circle is given by

$$(x - 0)^2 + (y - 1)^2 = 4^2$$
$$x^2 + (y - 1)^2 = 16$$

61. The center of a circle is the midpoint of its diameter. The radius of a circle is half the length of its diameter. Using the midpoint formula, the center of this circle has coordinates given by

$$\left(\frac{x_1 + x_2}{2}, \frac{y_1 + y_2}{2}\right) = \left(\frac{7 + 1}{2}, \frac{-3 + 7}{2}\right) = (4, 2)$$

The radius of the circle is given by

$$r = \frac{1}{2}\sqrt{(x_2 - x_1)^2 + (y_2 - y_1)^2}$$
$$= \frac{1}{2}\sqrt{(1 - 7)^2 + [7 - (-3)]^2}$$
$$= \frac{1}{2}\sqrt{36 + 100} = \frac{1}{2}\sqrt{136} = \frac{1}{2}\sqrt{4 \cdot 34} = \sqrt{34}$$

Hence the equation of the circle is given by

$$(x - 4)^2 + (y - 2)^2 = (\sqrt{34})^2$$
$$(x - 4)^2 + (y - 2)^2 = 34$$

63. The radius of a circle is the distance from the center to any point on the circle. Since the center of this circle is (0, 0) and (3, −9) is a point on the circle, the radius is given by

$$r = \sqrt{(3 - 0)^2 + (-9 - 0)^2} = \sqrt{9 + 81} = \sqrt{90}$$

Hence the equation of the circle is given by

$$x^2 + y^2 = (\sqrt{90})^2$$
$$x^2 + y^2 = 90$$

65. The radius of a circle is the distance from the center to any point on the circle. Since the center of this circle is (2, 2) and (3, -5) is a point on the circle, the radius is given by

$$r = \sqrt{(3 - 2)^2 + [(-5) - 2]^2} = \sqrt{1 + 49} = \sqrt{50}$$

Hence the equation of the circle is given by

$$(x - 2)^2 + (y - 2)^2 = (\sqrt{50})^2$$
$$(x - 2)^2 + (y - 2)^2 = 50$$

67. 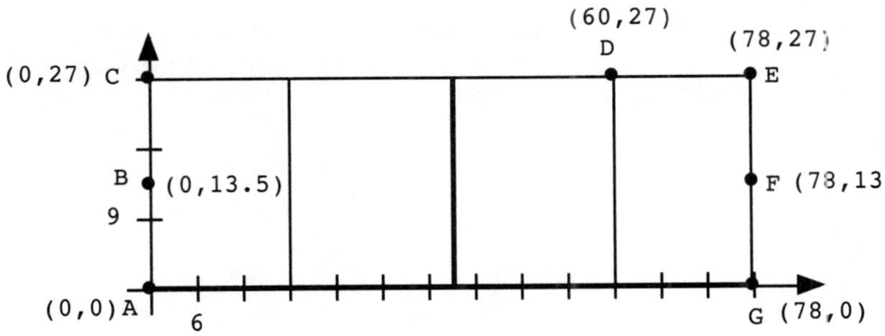

$$d(B, D) = \sqrt{(60-0)^2 + (27-13.5)^2} = \sqrt{60^2 + 13.5^2} = \sqrt{3600 + 182.25}$$
$$= \sqrt{3,782.25} = 66 \text{ ft.}$$

$$d(F, C) = \sqrt{(78-0)^2 + (13.5-27)^2} = \sqrt{78^2 + (-13.5)^2} = \sqrt{6,084 + 182.25}$$
$$= \sqrt{6,266.25} = 79 \text{ ft.}$$

69. Using the hint, we note that $(2, r - 1)$ must satisfy $x^2 + y^2 = r^2$, that is

$$2^2 + (r-1)^2 = r^2$$
$$4 + r^2 - 2r + 1 = r^2$$
$$-2r + 5 = 0$$
$$r = \frac{5}{2} \text{ or } 2.5 \text{ ft.}$$

71. (A) From the drawing, we can write:

$$\begin{pmatrix}\text{Distance from tower}\\ \text{to town } B\end{pmatrix} = 2 \times \begin{pmatrix}\text{Distance from tower}\\ \text{to town } A\end{pmatrix}$$
$$\begin{pmatrix}\text{Distance from } (x,y)\\ \text{to } (36,15)\end{pmatrix} = 2 \times \begin{pmatrix}\text{Distance from } (x,y)\\ \text{to } (0,0)\end{pmatrix}$$
$$\sqrt{(36-x)^2 + (15-y)^2} = 2\sqrt{(0-x)^2 + (0-y)^2}$$
$$\sqrt{(36-x)^2 + (15-y)^2} = 2\sqrt{x^2 + y^2}$$
$$(36-x)^2 + (15-y)^2 = 4(x^2 + y^2)$$
$$1,296 - 72x + x^2 + 225 - 30y + y^2 = 4x^2 + 4y^2$$
$$1,521 = 3x^2 + 3y^2 + 72x + 30y$$
$$507 = x^2 + y^2 + 24x + 10y$$
$$144 + 25 + 507 = x^2 + 24x + 144 + y^2 + 10y + 25$$
$$676 = (x+12)^2 + (y+5)^2$$

The circle has center $(-12, -5)$ and radius 26.

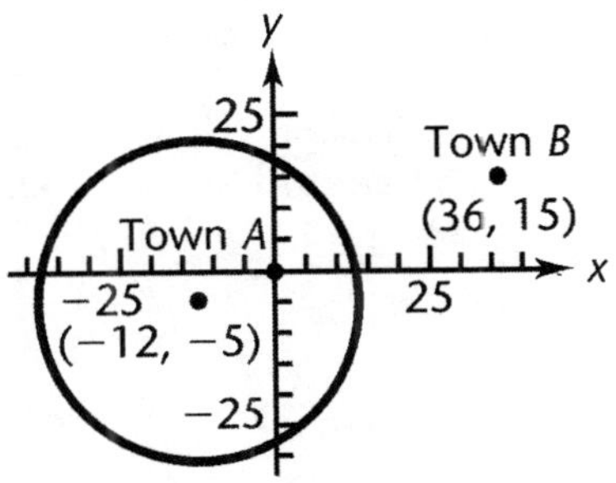

(B) All points due east of Town *A* have *y* coordinate 0 in this coordinate system. The points on the circle for which $y = 0$ are found by substituting $y = 0$ into the equation of the circle and solving for *x*.

$$\begin{aligned} (x + 12)^2 + (y + 5)^2 &= 676 \\ (x + 12)^2 + 25 &= 676 \\ (x + 12)^2 &= 651 \\ x + 12 &= \pm\sqrt{651} \\ x &= -12 \pm \sqrt{651} \end{aligned}$$

Choosing the positive square root so that *x* is greater than -12 (east rather than west) we have $x = -12 + \sqrt{651} \approx 13.5$ miles.

Section 2-3

1.
$$\begin{aligned} 4(x + 5) &= 6(x - 2) \\ 4x + 20 &= 6x - 12 \\ -2x &= -32 \\ x &= 16 \end{aligned}$$
Solution: 16

3.
$$\begin{aligned} 5 + 4(w - 1) &= 2w + 2(w + 4) \\ 5 + 4w - 4 &= 2w + 2w + 4 \\ 4w + 1 &= 4w + 4 \\ 1 &= 4 \end{aligned}$$
No solution

5.
$$\begin{aligned} 5 - \frac{3a - 4}{5} &= \frac{7 - 2a}{2} \quad \text{LCD} = 10 \\ 10 \cdot 5 - 10 \cdot \frac{(3a - 4)}{5} &= 10 \cdot \frac{(7 - 2a)}{2} \\ 50 - 2(3a - 4) &= 5(7 - 2a) \\ 50 - 6a + 8 &= 35 - 10a \\ -6a + 58 &= 35 - 10a \\ 4a &= -23 \\ a &= -\frac{23}{4} \text{ or } -5.75 \end{aligned}$$

> **Common Error:**
> After line 2, students often write
> $50 - \overset{2}{\cancel{10}}\,\frac{3a - 4}{\cancel{5}} = \ldots$
> $50 - 6a - 4 = \ldots$
> forgetting to distribute the -2. Put compound numerators in parentheses to avoid this.

7. The *x* intercept is -2. The *y* intercept is 2. From the point (-2, 0) to the point(0, 2), the value of *y* increases by 2 units as the value of *x* increases by 2 units. Thus slope $= \frac{\text{rise}}{\text{run}} = \frac{2}{2} = 1$.
Equation: $y = mx + b$
$y = 1x + 2$ or $y = x + 2$

9. The *x* intercept is -2. The *y* intercept is -4. From the point (-2, 0) to the point (0, -4) the value of *y* decreases by 4 units as the value of *x* increases by 2 units. Thus, the slope $= \frac{\text{rise}}{\text{run}} = \frac{-4}{2} = -2$.
Equation: $y = mx + b$
$y = -2x + (-4)$ or $y = -2x - 4$

11. The *x* intercept is 3. The *y* intercept is -1. From the point (0, -1) to the point (3, 0) the value of *y* increases by 1 unit as the value of *x* increases by 3 units. Thus, the slope $= \frac{\text{rise}}{\text{run}} = \frac{1}{3}$.
Equation: $y = mx + b$
$y = \frac{1}{3}x + (-1)$ or $y = \frac{1}{3}x - 1$

13. $y = -\frac{3}{5}x + 4$ slope $-\frac{3}{5}$

x	y
0	4
5	1
−5	7

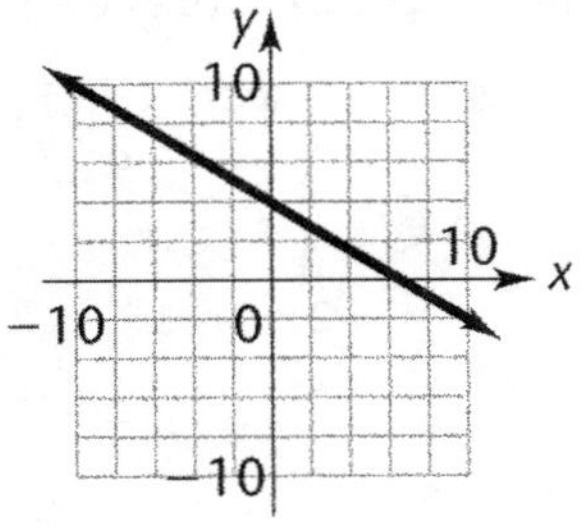

15. $y = -\frac{3}{4}x$ slope $-\frac{3}{4}$

x	y
0	0
4	−3
−4	3

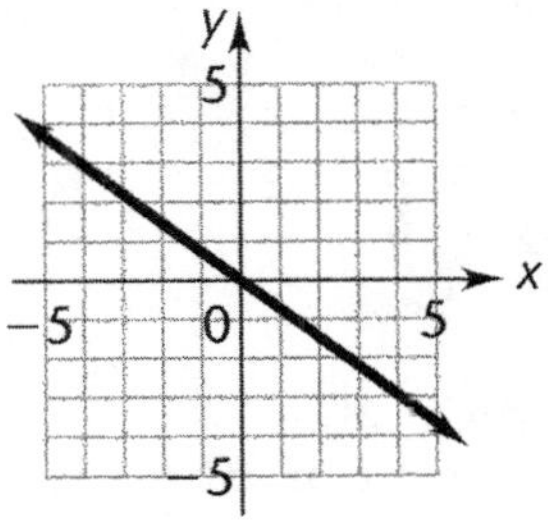

17. $4x + 2y = 0$

$2y = -4x$

$y = -2x$ slope -2

x	y
0	0
1	−2
2	−4

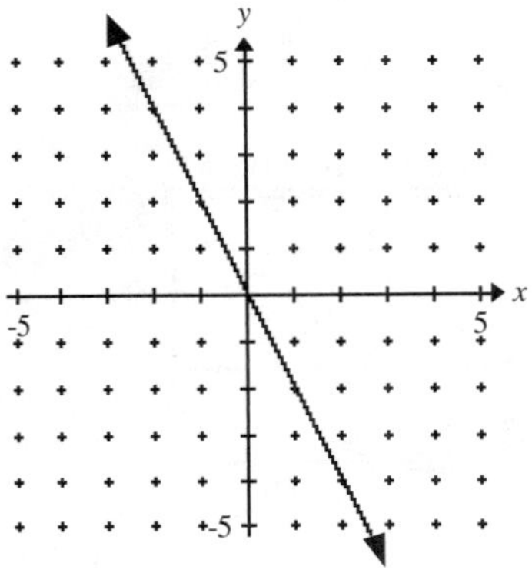

19. $2x - 3y = 15$

$-3y = -2x + 15$

$y = \frac{2}{3}x - 5$ slope $\frac{2}{3}$

x	y
0	−5
3	−3
−3	−7

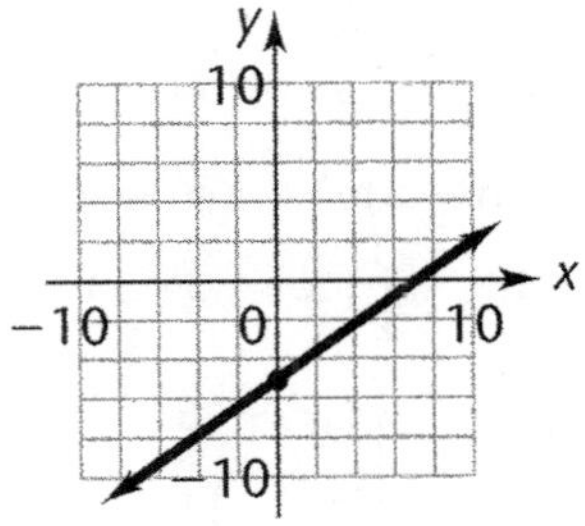

21. $4x - 5y = -24$

$-5y = -4x - 24$

$y = \frac{4}{5}x + \frac{24}{5}$ slope $= \frac{4}{5}$

x	y
−1	4
4	8
9	12

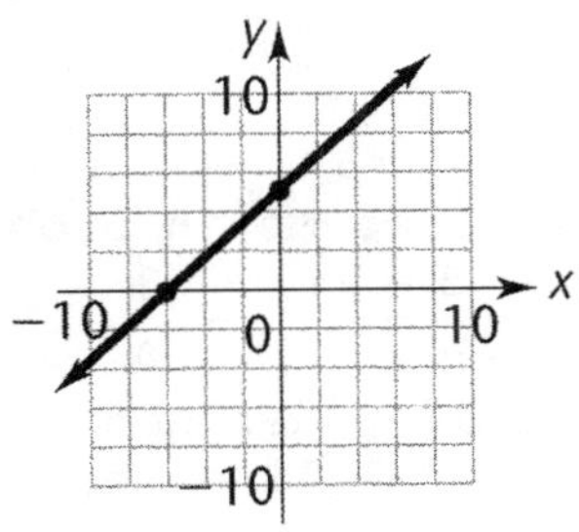

23. $\frac{y}{8} - \frac{x}{4} = 1$

$\frac{y}{8} = \frac{x}{4} + 1$

$y = 2x + 8$ slope 2

x	y
0	8
−4	0
−8	−8

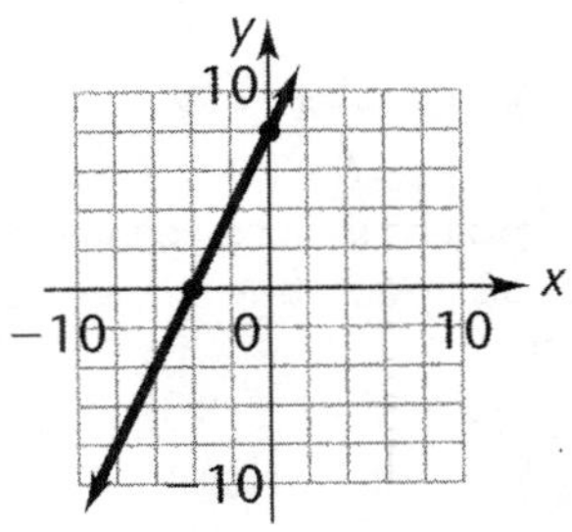

25. $x = -3$ slope not defined vertical line

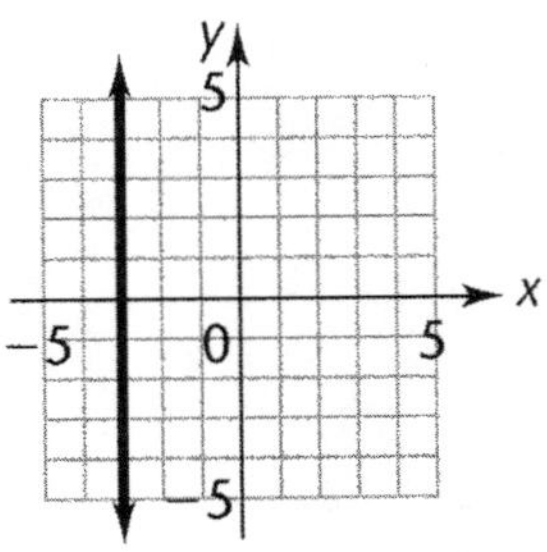

27. $y = 3.5$ slope 0 horizontal line

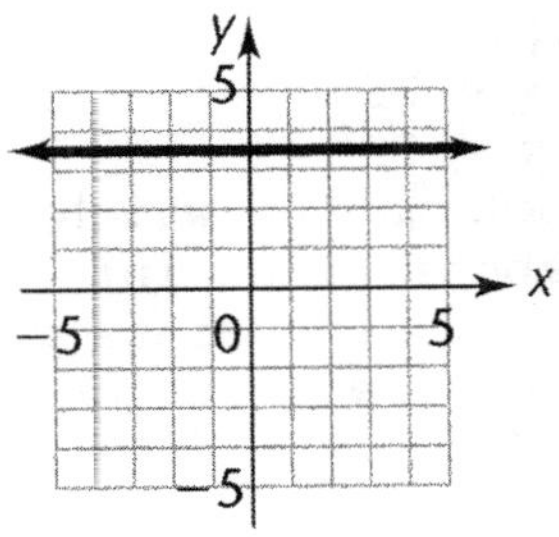

29. Slope and y intercept are given; we use slope-intercept form.

$$y = 1x + 0$$
$$y = x$$
$$0 = x - y$$
$$x - y = 0$$

31. Slope and y intercept are given; we use slope-intercept form.

$$y = -\frac{2}{3}x + (-4)$$
$$y = -\frac{2}{3}x - 4$$
$$3y = -2x - 12$$
$$2x + 3y = -12$$

33. The equation of this horizontal line is $y = -2$.

35. A point and the slope are given; we use point-slope form.

$$y - 3 = -2(x - 0)$$
$$y - 3 = -2x$$
$$y = -2x + 3$$

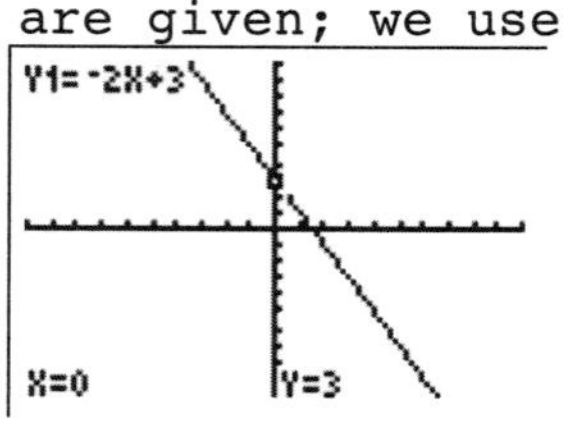

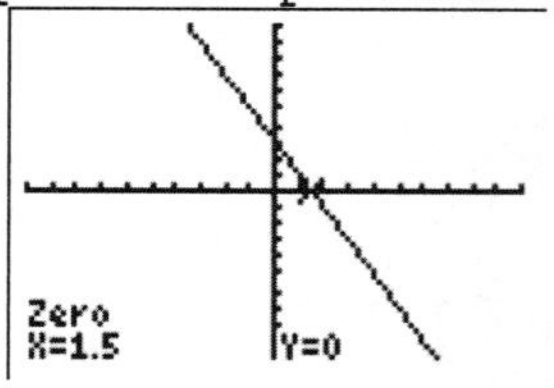

37. A point and the slope are given; we use point-slope form.

$$y - 4 = \frac{3}{2}[x - (-5)]$$
$$y - 4 = \frac{3}{2}[x + 5]$$
$$y - 4 = \frac{3}{2}x + \frac{15}{2}$$
$$y = \frac{3}{2}x + \frac{23}{2}$$

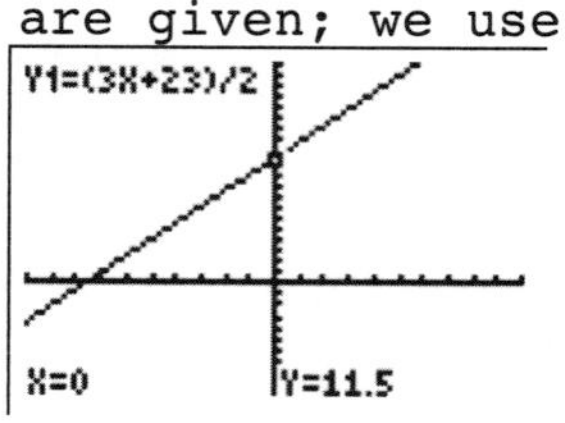

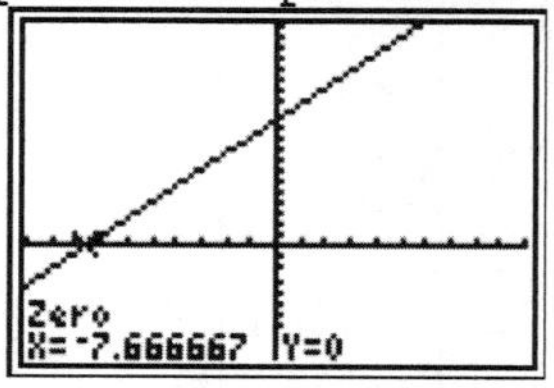

39. A point and the slope are given; we use point-slope form.

$$y - (-3) = -\frac{1}{2}[x - (-2)]$$
$$y + 3 = -\frac{1}{2}[x + 2]$$
$$y + 3 = -\frac{1}{2}x - 1$$
$$y = -\frac{1}{2}x - 4$$

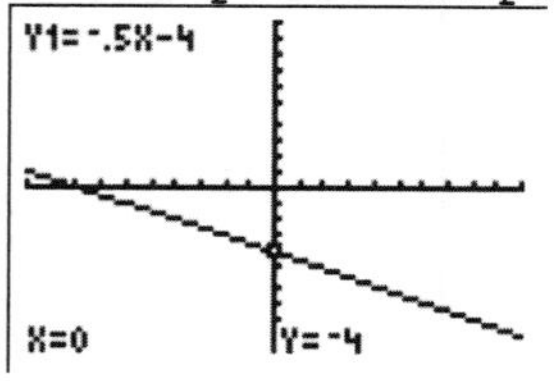

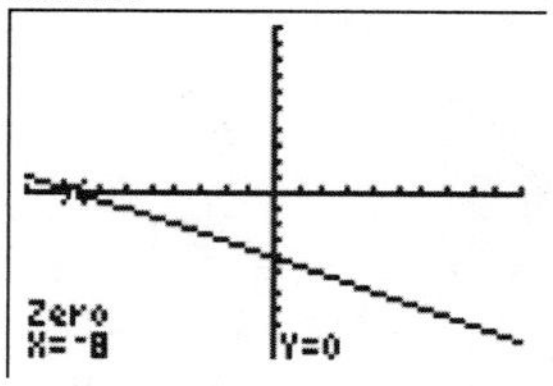

41. Two points are given; we first find the slope, then we use point-slope form.

$$m = \frac{-3 - 5}{4 - 2} = \frac{-8}{2} = -4$$
$$y - 5 = -4(x - 2)$$
$$y - 5 = -4x + 8$$
$$y = -4x + 13$$

or

$$y - (-3) = -4(x - 4)$$
$$y + 3 = -4x + 16$$
$$y = -4x + 13$$

Thus, it does not matter which point is chosen in substituting into the point-slope form; both points must give rise to the same equation.

43. We proceed as in problem 41.

$$m = \frac{5 - 2}{(-3) - (-3)} = \frac{3}{0}$$

Slope is undefined.
A vertical line through $(-3, 2)$ has equation $x = -3$.

45. We proceed as in problem 41.

$$m = \frac{2 - 2}{0 - 4} = \frac{0}{-4} = 0$$
$$y - 2 = 0(x - 0)$$
$$y - 2 = 0$$
$$y = 2$$

47. A point and the slope are given; we use point-slope form.

$$y - 4 = -3(x - 0)$$
$$y - 4 = -3x$$
$$y = -3x + 4$$

Graph:

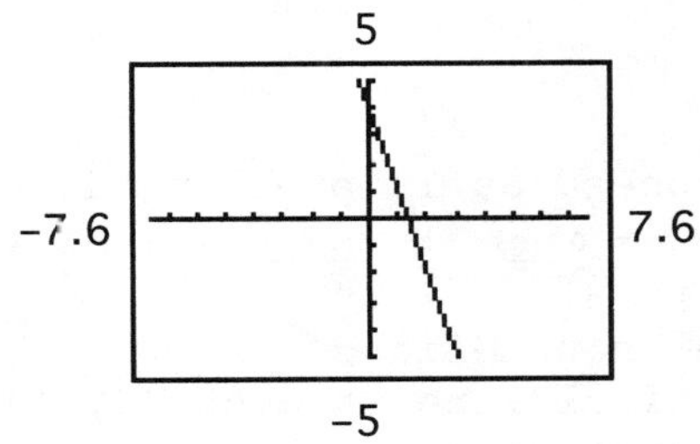

49. A point and the slope are given; we use point-slope form.

$$y - 4 = -\frac{2}{5}[x - (-5)]$$
$$y - 4 = -\frac{2}{5}[x + 5]$$
$$y - 4 = -\frac{2}{5}x - 2$$
$$y = -\frac{2}{5}x + 2$$

Graph:

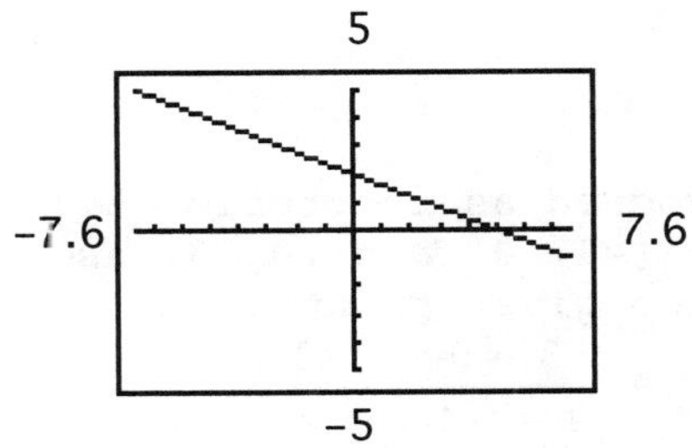

51. Two points are given; we first find slope, then use point-slope form.

$$m = \frac{-2 - 6}{5 - 1} = \frac{-8}{4} = -2$$
$$y - 6 = (-2)(x - 1)$$
$$y - 6 = -2x + 2$$
$$y = -2x + 8$$

or

$$y - (-2) = (-2)(x - 5)$$
$$y + 2 = -2x + 10$$
$$y = -2x + 8$$

Thus, it does not matter which point is chosen in substituting into the point-slope form; both points must give rise to the same equation.

Graph:

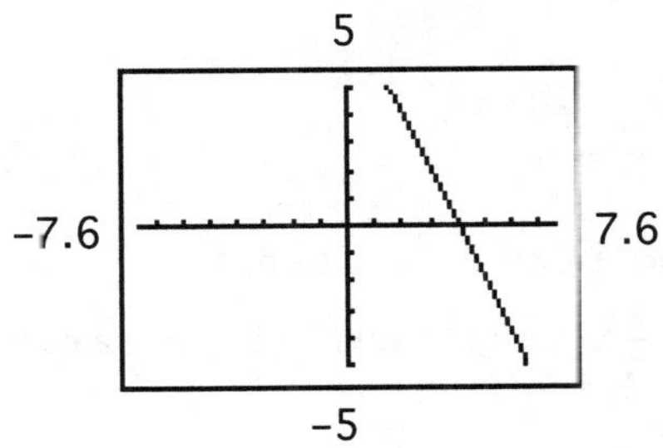

53. We proceed as in problem 51.

$$m = \frac{8 - 0}{-4 - 2} = \frac{8}{-6} = -\frac{4}{3}$$
$$y - 0 = -\frac{4}{3}(x - 2)$$
$$y = -\frac{4}{3}x + \frac{8}{3}$$

Graph:

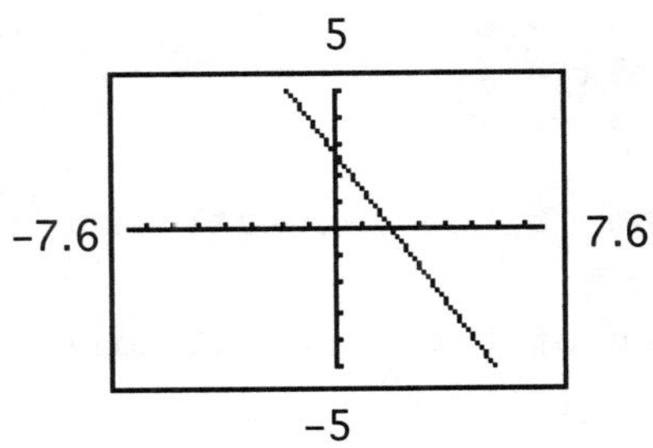

55. We proceed as in problem 51.

$$m = \frac{4 - 4}{5 - (-3)} = \frac{0}{8} = 0$$

$$y - 4 = 0(x - 5)$$
$$y - 4 = 0$$
$$y = 4$$

Graph:

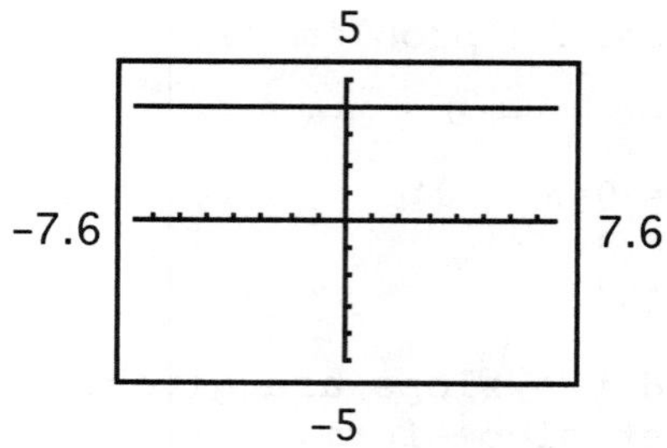

57. We proceed as in problem 51.

$$m = \frac{-3 - 6}{4 - 4} = \frac{-9}{0}$$

slope is undefined.
A vertical line through (4, 6) has equation $x = 4$.

Graph:

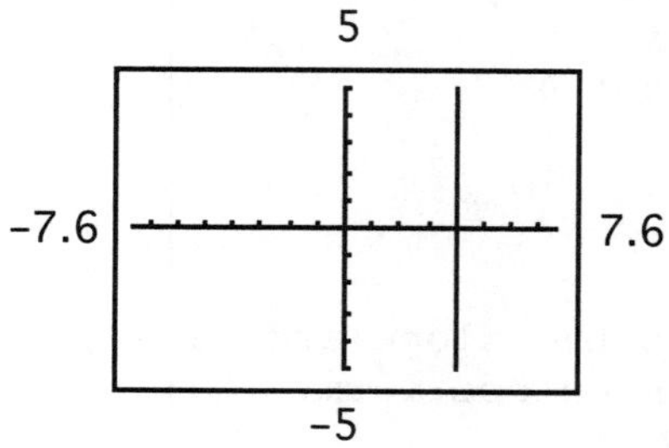

59. We proceed as in problem 51, using (-4, 0) and (0, 3) as the two given points.

$$m = \frac{3 - 0}{0 - (-4)} = \frac{3}{4}$$

$$y - 0 = \frac{3}{4}[x - (-4)]$$

$$y = \frac{3}{4}x + 3$$

Graph:

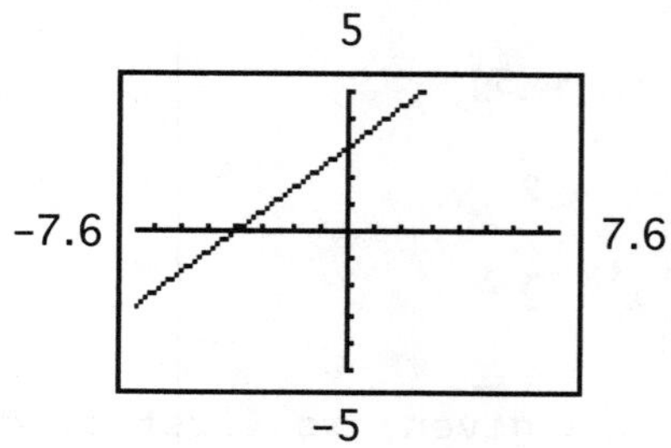

61. A line parallel to $y = 3x - 5$ will have the same slope, namely 3. We now use the point slope form.

$$y - 4 = 3[x - (-3)]$$
$$y - 4 = 3x + 9$$
$$y = 3x + 13$$
$$-3x + y = 13$$
$$3x - y = -13$$

63. A line perpendicular to $y = -\frac{1}{3}x$ will have slope satisfying $-\frac{1}{3}m = -1$, or $m = 3$. We use the point-slope form.

$$y - (-3) = 3(x - 2)$$
$$y + 3 = 3x - 6$$
$$9 = 3x - y$$
$$3x - y = 9$$

65. The equation of the vertical line through (2, 5) is $x = 2$.

67. The equation of the vertical line through (3, -2) is $x = 3$.

69. A line parallel to $3x - 2y = 4$ will have the same slope. The slope of $3x - 2y = 4$, or $2y = 3x - 4$, or $y = \frac{3}{2}x - 2$ is $\frac{3}{2}$. We use the point-slope form,

$$y - 0 = \frac{3}{2}(x - 5)$$
$$y = \frac{3}{2}x - \frac{15}{2}$$
$$\frac{15}{2} = \frac{3}{2}x - y$$
$$15 = 3x - 2y$$
$$3x - 2y = 15$$

Alternatively, we could notice that a line parallel to $3x - 2y = 4$ will have an equation of the form $3x - 2y = C$. Since the required line must contain (5, 0), (5, 0) must satisfy its equation. Therefore, $3(5) - 2(0) = C$. Since $15 = C$, the equation desired is $3x - 2y = 15$.)

71. A line perpendicular to $x + 3y = 9$, which has slope $-\frac{1}{3}$, will have slope satisfying $-\frac{1}{3}m = -1$, or $m = 3$. We use the point-slope form.

$$y - (-4) = 3(x - 0)$$
$$y + 4 = 3x$$
$$4 = 3x - y$$
$$3x - y = 4$$

73. The graphs are straight lines; all have the same y intercept, 2, but their slopes vary as m varies.

75. slope of $AB = \dfrac{-1 - 2}{4 - 0} = -\dfrac{3}{4}$

slope of $DC = \dfrac{-5 - (-2)}{1 - (-3)} = -\dfrac{3}{4}$

Therefore $AB \parallel DC$.

77. slope of $AB = -\dfrac{3}{4}$

slope of $BC = \dfrac{-5 - (-1)}{1 - 4} = \dfrac{4}{3}$

(slope AB)(slope BC) $= \left(-\dfrac{3}{4}\right)\left(\dfrac{4}{3}\right) = -1$

Therefore $AB \perp BC$.

79. midpoint of $AD = \left(\dfrac{0 + (-3)}{2}, \dfrac{2 + (-2)}{2}\right) = \left(-\dfrac{3}{2}, 0\right)$

slope of $AD = \dfrac{-2 - 2}{-3 - 0} = \dfrac{4}{3}$

We require the equation of a line, through the midpoint of AD, which is perpendicular to AD. Its slope will satisfy $\frac{4}{3}m = -1$, or $m = -\frac{3}{4}$. We use the point-slope form.

$$y - 0 = -\frac{3}{4}\left[x - \left(-\frac{3}{2}\right)\right]$$
$$y = -\frac{3}{4}\left(x + \frac{3}{2}\right)$$
$$y = -\frac{3}{4}x - \frac{9}{8}$$
$$8y = -6x - 9$$
$$6x + 8y = -9$$

81. (A) Solve each equation for y in terms of x, then graph.

$$3x + 2y = 6 \qquad 2y = 6 - 3x \qquad y = \frac{6 - 3x}{2}$$

$$3x + 2y = -6 \qquad 2y = -6 - 3x \qquad y = \frac{-6 - 3x}{2}$$

$$3x + 2y = 3 \qquad 2y = 3 - 3x \qquad y = \frac{3 - 3x}{2}$$

$$3x + 2y = -3 \qquad 2y = -3 - 3x \qquad y = \frac{-3 - 3x}{2}$$

x	$y = \frac{6 - 3x}{2}$	$y = \frac{-6 - 3x}{2}$	$y = \frac{3 - 3x}{2}$	$y = \frac{-3 - 3x}{2}$
0	3	-3	$\frac{3}{2}$	$-\frac{3}{2}$
2	0	-6	$-\frac{3}{2}$	$-\frac{9}{2}$
-2	6	0	$\frac{9}{2}$	$\frac{3}{2}$

(B)

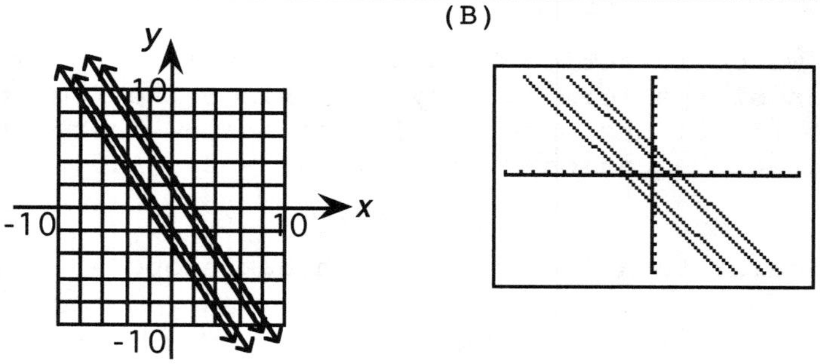

(C) The family of lines all have slope $y = -\frac{A}{B}$ and y intercept $\frac{C}{B}$. The slopes are therefore equal and the lines are parallel.

(D) Solve $Ax + By = C$ for y in terms of x.

$$By = -Ax + C$$

$$y = \frac{-Ax + C}{B}$$

$$y = -\frac{A}{B}x + \frac{C}{B}$$

For fixed A and B and varying C this is a family of lines as described in part (C).

83. Two points are given; we first find the slope, then use the point-slope form.

$$m = \frac{0 - b}{a - 0} = -\frac{b}{a} \quad a \neq 0$$

$$y - b = -\frac{b}{a}(x - 0)$$

$$y - b = -\frac{bx}{a}$$

Divide both sides by b, then $(b \neq 0)$

$$\frac{y - b}{b} = -\frac{x}{a}$$

$$\frac{y}{b} - 1 = -\frac{x}{a}$$

$$\frac{y}{b} = 1 - \frac{x}{a}$$

$$\frac{x}{a} + \frac{y}{b} = 1$$

85. Using the result of problem 83, we have:

$a = 3,\ b = 5$

The equation is

$$\frac{x}{3} + \frac{y}{5} = 1$$

$$5x + 3y = 15$$

87. The circle has center (0, 0). The radius drawn from (0, 0) to the given point (3, 4) has slope given by $M_R = \frac{4-0}{3-0} = \frac{4}{3}$.

Therefore, the slope of the tangent line is given by

$$\left(\frac{4}{3}\right)m = -1 \quad \text{or} \quad m = -\frac{3}{4}$$

We require the equation of a line through (3, 4) with slope $-\frac{3}{4}$. We use the point-slope form.

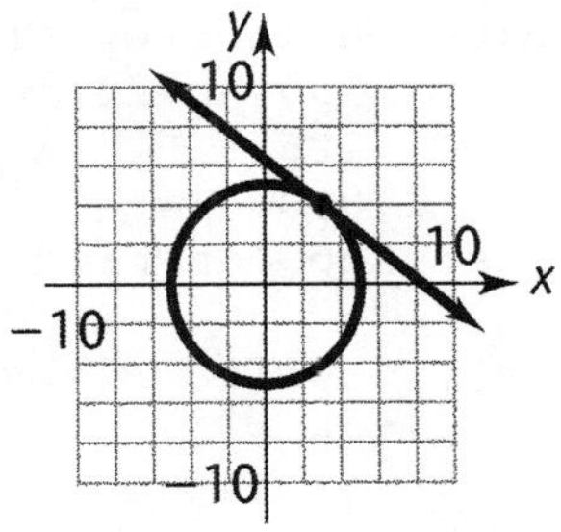

$$y - 4 = -\frac{3}{4}(x - 3)$$

$$y - 4 = -\frac{3}{4}x + \frac{9}{4}$$

$$4y - 16 = -3x + 9$$

$$3x + 4y = 25$$

89. The circle has center (0, 0). The radius drawn from (0, 0) to the given point (5, -5) has slope given by

$$M_R = \frac{-5-0}{5-0} = -1$$

Therefore, the slope of the tangent line is given by

$$(-1)m = -1 \quad \text{or} \quad m = 1$$

We require the equation of a line through (5, -5) with slope 1. We use the point-slope form.

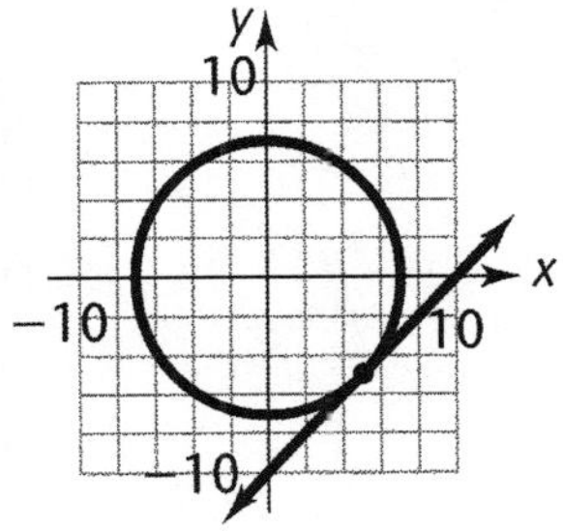

$$y - (-5) = 1(x - 5)$$

$$y + 5 = x - 5$$

$$x - y = 10$$

91. The circle has center (3, -4). The radius drawn from (3, -4) to the given point (8, -16) has slope given by

$$M_R = \frac{-16-(-4)}{8-3} = -\frac{12}{5}$$

Therefore, the slope of the tangent line is given by

$$\left(-\frac{12}{5}\right)m = -1 \quad \text{or} \quad m = \frac{5}{12}$$

We require the equation of a line through (8,-16) with slope $\frac{5}{12}$. We use the point-slope form.

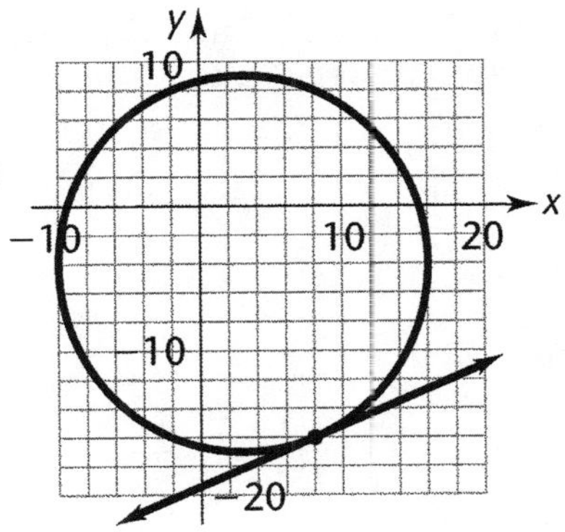

$$y - (-16) = \frac{5}{12}(x - 8)$$

$$y + 16 = \frac{5}{12}x - \frac{40}{12}$$

$$12y + 192 = 5x - 40$$

$$232 = 5x - 12y$$

93. (A)

x	0	5,000	10,000	15,000	20,000	25,000	30,000
$212 - 0.0018x = B$	212	203	194	185	176	167	158

(B) The boiling point drops 9°F for each 5,000 foot increase in altitude.

95. The rental charges are \$25 per day plus \$0.25 per mile driven.

97. Total cost = Fixed cost + variable cost. If x = the number of doughnuts produced, variable cost is $0.12x$ (\$0.12 per doughnut times the number of doughnuts produced.) If $C(x)$ represents total cost,

$C(x) = 124 + 0.12x$

Plug in 250 for $C(x)$:

$$250 = 124 + 0.12x$$
$$126 = 0.12x$$
$$1{,}050 = x$$

The shop can produce 1,050 doughnuts for \$250.

99. (A) We write $s = mw + b$. If $w = 0$, $s = 0$, hence $0 = m \cdot 0 + b$, $b = 0$. If $w = 5$, $s = 2$, hence $2 = m \cdot 5$, $m = 0.4$.

$s = 0.4w$

(B) $s = 0.4(20) = 8$ inches

(C) Solve $3.6 = 0.4w$ to obtain $w = 9$ pounds.

101. (A) We write $F = mC + b$. If $C = 0$, $F = 32$, hence $32 = m \cdot 0 + b$, $b = 32$.

If $C = 100$, $F = 212$, hence $212 = m \cdot 100 + 32$, $100m = 180$, $m = \frac{9}{5}$.

$$F = \frac{9}{5}C + 32$$

(B) If $C = 20$, $F = \frac{9}{5}(20) + 32 = 68°\text{F}$

If $F = 86$, solve

$$86 = \frac{9}{5}C + 32$$
$$54 = \frac{9}{5}C$$
$$C = \frac{5}{9}(54) = 30°\text{C}$$

103. (A) If h is linearly related to t, then we are looking for an equation whose graph passes through $(t_1, h_1) = (9, 23)$ and $(t_2, h_2) = (24, 40)$. We find the slope, and then we use the point-slope form to find the equation.

$$m = \frac{h_2 - h_1}{t_2 - t_1} = \frac{40 - 23}{24 - 9} = \frac{17}{15} \approx 1.13$$
$$h - h_1 = m(t - t_1)$$
$$h - 23 = \frac{17}{15}(t - 9)$$
$$h - 23 = \frac{17}{15}t - \frac{51}{5}$$
$$h = \frac{17}{15}t - \frac{51}{5} + 23$$
$$h = 1.13t + 12.8$$

(B) We are asked for t when $h = 50$.

$$50 = 1.13t + 12.8$$
$$37.2 = 1.13t$$
$$t = 32.9 \text{ hours}$$

105. (A) If L is linearly related to t, then we are looking for an equation whose graph passes through $(t_1, L_1) = (0, 49.2)$ and $(t_2, L_2) = (100, 77.3)$. We find the slope, then we use the point-slope form to find the equation.

$$m = \frac{L_2 - L_1}{t_2 - t_1} = \frac{77.3 - 49.2}{100 - 0} = 0.281$$

$$L - L_1 = m(t - t_1)$$
$$L - 49.2 = 0.281(t - 0)$$
$$L - 49.2 = 0.281t$$
$$L = 0.281t + 49.2$$

(B) We are asked for L when $t = 120$.

$$L = 0.281(120) + 49.2$$
$$L = 82.9 \text{ years}$$

107.

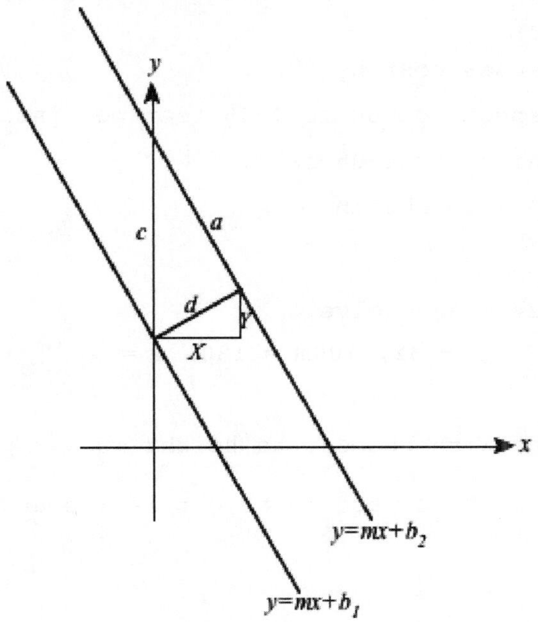

In general, we can show that $d = \dfrac{c}{\sqrt{1 + m^2}}$ as follows:

The Pythagorean Theorem gives:

$$d^2 + a^2 = c^2$$

The two triangles shown are similar, hence corresponding sides are proportional. Thus

$$\frac{a}{d} = \frac{x}{y}$$

The slope of the line segment labeled d is the negative reciprocal of m.

Thus $\dfrac{y}{x} = -\dfrac{1}{m}$

$$\frac{x}{y} = -m$$

It follows that $a = \dfrac{x}{y}d = -md$.

Hence $d^2 + (-md)^2 = c^2$

$$d^2(1 + m^2) = c^2$$

$$d^2 = \frac{c^2}{1 + m^2}$$

$$d = \frac{c}{\sqrt{1 + m^2}}$$

In particular, avenue A is shown to have a rise of -5000 and a run of 4000, hence $m = -\dfrac{5000}{4000} = -1.25$. The equation of avenue A is then (using the slope-intercept form $y = mx + b$) $y = -1.25x + 5000$. Avenue B has the same slope, and y intercept 4000. Substituting in the above formula, with $c = 5000 - 4000 = 1000$, yields

$$d_1 = \frac{1000}{\sqrt{1 + (-1.25)^2}} = 625 \text{ ft.}$$

Section 2-4

1. (A) If cost y is linearly related to the number of golf clubs x_1 then we are looking for an equation whose graph passes through $(x_1, y_1) = (80, 8{,}147)$ and $(x_2, y_2) = (100, 9{,}647)$. We find the slope and then use the point-slope form to find the equation.

$$m = \frac{y_2 - y_1}{x_2 - x_1} = \frac{9{,}647 - 8{,}147}{100 - 80} = 75$$

$$\begin{aligned} y - y_1 &= m(x - x_1) \\ y - 8{,}147 &= 75(x - 80) \\ y - 8{,}147 &= 75x - 6000 \\ y &= 75x + 2{,}147 \end{aligned}$$

(B) The slope of 75 is the rate of change of cost with respect to production, \$75 per golf club.

(C) Increasing production by 1 unit increases cost by \$75.

3. (A) The rate of change of height with respect to DBH is 4.06 feet per inch.

(B) Increasing DBH by 1 inch increases height by 4.06 feet.

(C) Substitute $d = 12$ into $h = 4.06d + 24.1$ to obtain

$$\begin{aligned} h &= 4.06(12) + 24.1 \\ h &= 73 \text{ feet} \end{aligned}$$

(D) Substitute $h = 100$ into $h = 4.06d + 24.1$ and solve.

$$\begin{aligned} 100 &= 4.06d + 24.1 \\ 75.9 &= 4.06d \\ d &= 19 \text{ inches} \end{aligned}$$

5. (A) Robinson: The rate of change of weight with respect to height is 3.7 pounds per inch.
Miller: The rate of change of weight with respect to height is 3 pounds per inch.

(B) 5′6″ = 6 inches over 5 feet
Substitute $h = 6$ into each model.
Robinson: $w = 108 + 3.7(6) = 130.2$ pounds
Miller: $w = 117 + 3.0(6) = 135$ pounds

(C) Substitute $w = 140$ into each model and solve.

Robinson: $$\begin{aligned} 140 &= 108 + 3.7h \\ 32 &= 3.7h \\ h &= 9 \text{ inches, predicting 5′9″.} \end{aligned}$$

Miller: $$\begin{aligned} 140 &= 117 + 3.0h \\ 23 &= 3.0h \\ h &= 8 \text{ inches, predicting 5′8″.} \end{aligned}$$

7. If speed s is linearly related to temperature t, then we are looking for an equation whose graph passes through $(t_1, s_1) = (32, 741)$ and $(t_2, s_2) = (72, 771)$. We find the slope and then use the point-slope form to find the equation.

$$m = \frac{s_2 - s_1}{t_2 - t_1} = \frac{771 - 741}{72 - 42} = 0.75$$

$$\begin{aligned} s - s_1 &= m(t - t_1) \\ s - 741 &= 0.75(t - 32) \\ s - 741 &= 0.75t - 24 \\ s &= 0.75t + 717 \end{aligned}$$

The speed of sound at sea level increases by 0.75 mph for each 1°F change in temperature.

9. If percentage m is linearly related to time t, then we are looking for an equation whose graph passes through $(t_1, m_1) = (0, 28.3)$ and $(t_2, m_2) = (10, 25.2)$. We find the slope and then use the point-slope form to find the equation. M is used for slope to avoid confusion.

$$M = \frac{m_2 - m_1}{t_2 - t_1} = \frac{25.2 - 28.3}{10 - 0} = -0.31$$

$$m - m_1 = M(t - t_1)$$

$$m - 28.3 = -0.31(t - 0)$$

$$m = -0.31t + 28.3$$

To find t when $m = 20$, substitute $m = 20$ and solve.

$$20 = -0.31t + 28.3$$

$$-8.3 = -0.31t$$

$$t = 27$$

27 years after 1990 will be 2017.

11. (A) If value V is linearly related to time t, then we are looking for an equation whose graph passes through $(t_1, V_1) = (0, 142{,}000)$ and $(t_2, V_2) = (10, 67{,}000)$. We find the slope and then use the point-slope form to find the equation.

$$m = \frac{V_2 - V_1}{t_2 - t_1} = \frac{67{,}000 - 142{,}000}{10 - 0} = -7{,}500$$

$$V - V_1 = m(t - t_1)$$

$$V - 142{,}000 = -7{,}500(t - 0)$$

$$V = -7{,}500t + 142{,}000$$

(B) The tractor's value decreases at the rate of \$7,500 per year.

(C) When $t = 6$, $V = -7{,}500(6) + 142{,}000 = \$97{,}000$.

13. (A) If price R is linearly related to cost C, then we are looking for an equation whose graph passes through $(C_1, R_1) = (85, 112)$ and $(C_2, R_2) = (175, 238)$. We find the slope and then use the point-slope form to find the equation.

$$m = \frac{R_2 - R_1}{C_2 - C_1} = \frac{238 - 112}{175 - 85} = 1.4$$

$$R - R_1 = m(C - C_1)$$

$$R - 112 = 1.4(C - 85)$$

$$R - 112 = 1.4C - 119$$

$$R = 1.4C - 7$$

(B) The slope is 1.4. This is the rate of change of retail price with respect to cost.

(C) To find C when $R = 185$, substitute $R = 185$ and solve.

$$185 = 1.4C - 7$$

$$192 = 1.4C$$

$$C = \$137$$

15. (A) If temperature T is linearly related to altitude A, then we are looking for an equation whose graph passes through $(A_1, T_1) = (0, 70)$ and $(A_2, T_2) = (18, -20)$. We find the slope and then use the point-slope form to find the equation.

$$m = \frac{T_2 - T_1}{A_2 - A_1} = \frac{-20 - 70}{18 - 0} = -5$$

$$T - T_1 = m(A - A_1)$$

$$T - 70 = -5(A - 0)$$

$$T = -5A + 70$$

(B) To find A when $T = 0$, substitute $T = 0$ and solve.

$$0 = -5A + 70$$

$A = 14$, that is, 14,000 feet

17. (A) If altitude a is linearly related to time t, then we are looking for an equation whose graph passes through $(t_1, a_1) = (0, 2{,}880)$ and $(t_2, a_2) = (120, 0)$. We find the slope and then use the point-slope form to find the equation.

$$m = \frac{a_2 - a_1}{t_2 - t_1} = \frac{0 - 2{,}880}{120 - 0} = -24$$

$$a - a_1 = m(t - t_1)$$

$$a - 2{,}880 = -24(t - 0)$$

$$a = -24t + 2{,}880$$

(B) Since altitude is decreasing at the rate of 24 feet per second, this is the rate of descent.

19. (A)

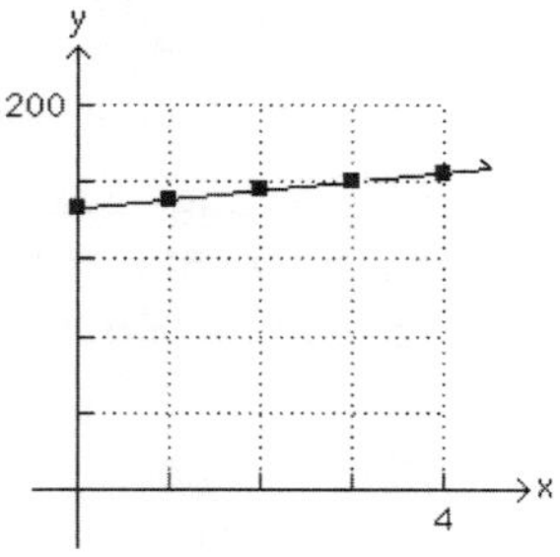

(B) Substitute $x = 10$ into $y = 4.6x + 146$ to obtain $y = 4.6(10) + 146 = 192$. This represents \$192,000.

(C) The average price is increasing at the rate of \$4,600 per year.

21. (A)

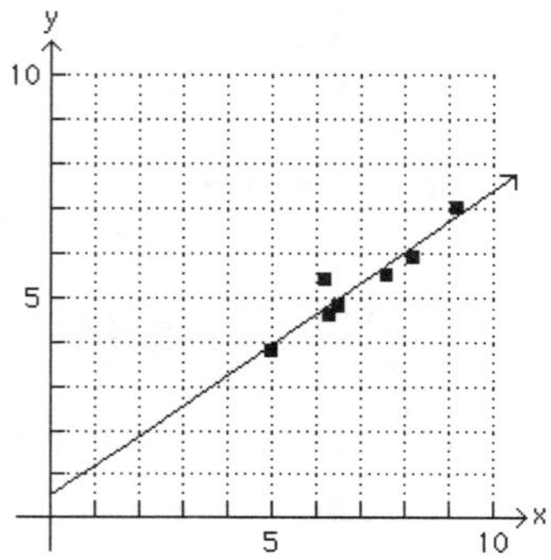

(B) Substitute $x = 4.5$ into $y = 0.69x + 0.5$ to obtain $y = 0.69(4.5) + 0.5 \approx 3.6$ million.

(C) Substitute $x = 5.6$ into $y = 0.69x + 0.5$ to obtain $y = 0.69(5.6) + 0.5 \approx 4.4$ million.

23. A portion of the entered date is shown here along with the results of the linear regression calculations.

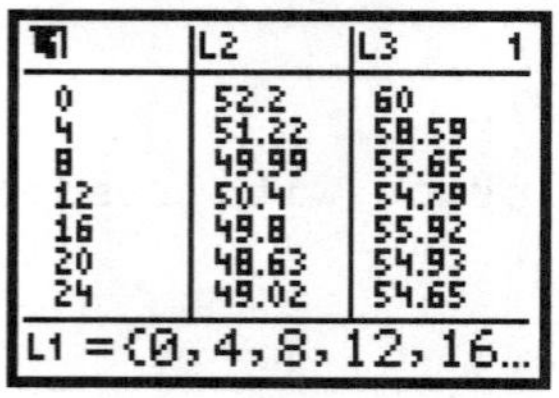

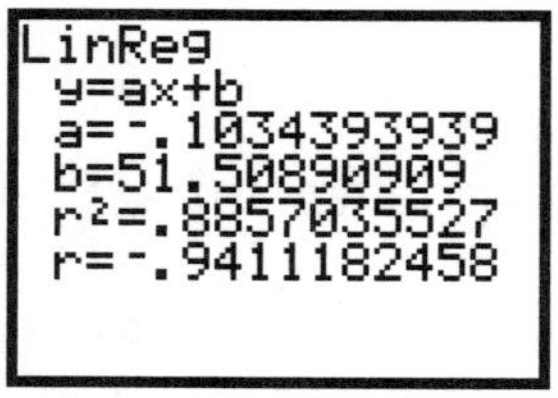

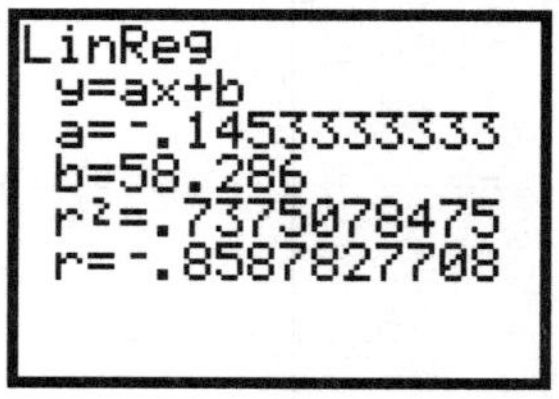

The linear regression model for men's 100-meter freestyle data is seen to be $y = -0.1034x + 51.51$. The linear regression model for women's 100- meter freestyle data is seen to be $y = -0.1453x + 58.29$. A plausible window is shown here, along with the results of the intersection calculation.

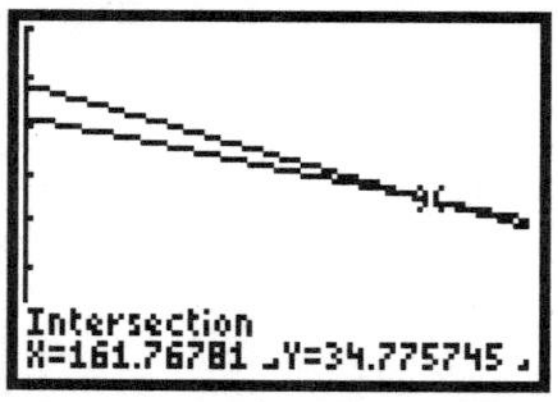

The intersection for $x \approx 162$ implies that the women will catch up with the men in $1968 + 162 = 2130$.

25. Entering the data (note: price is entered as y) and applying the linear regression routine yields the following:

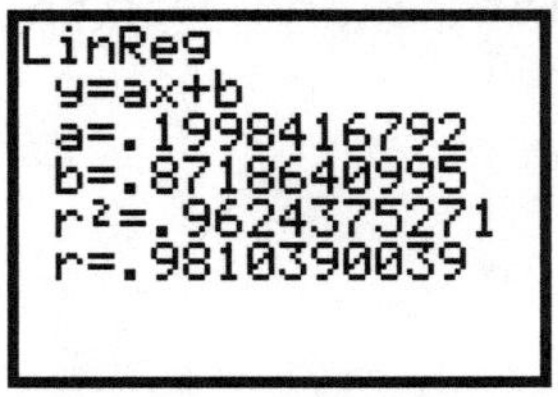

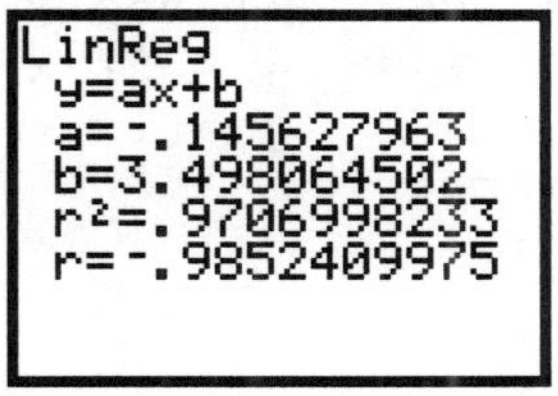

The linear regression model for the price-supply data is seen to be $y = 0.200x + 0.872$. The linear regression model for the price-demand data is seen to be $y = -0.146x + 3.50$. A plausible window is shown here, along with the results of the intersection calculation.

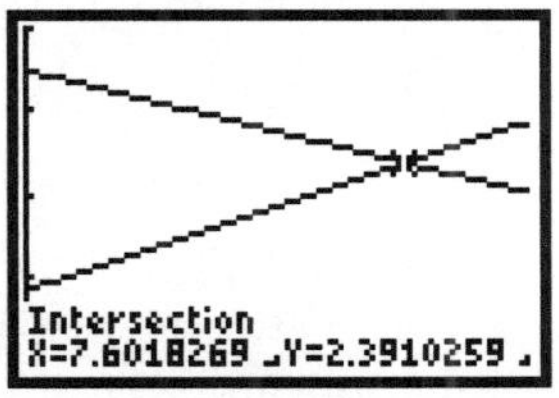

The intersection for $y = 2.39$ implies an equilibrium price of \$2.39.

CHAPTER 2 REVIEW

1.

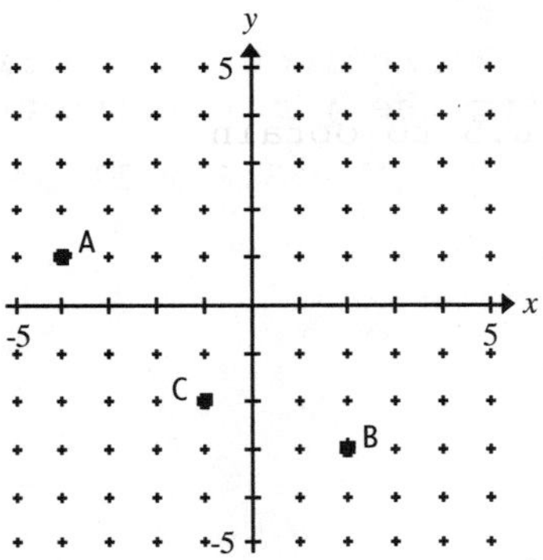

(2, 1)

2.

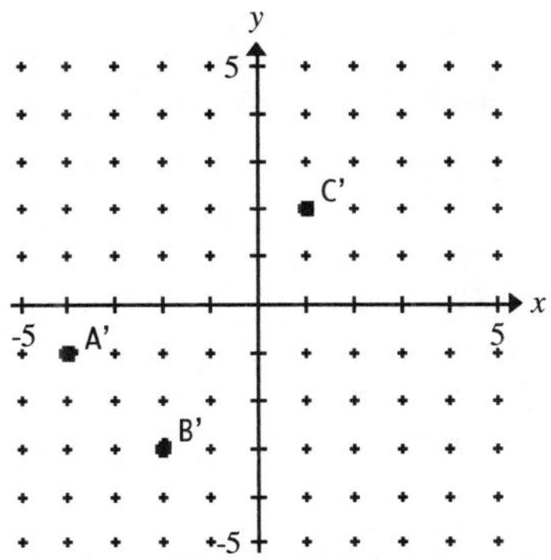

The reflection of A in the x axis is $A'(-4, -1)$.
The reflection of B in the y axis is $B'(-2, -3)$.
The reflection of C in the origin is $C'(1, 2)$.

(2-1)

3. (A) $y = 2x$

Test y axis	Test x axis	Test origin
Replace x with $-x$:	Replace y with $-y$:	Replace x with $-x$ and y with $-y$:
$y = 2(-x)$	$-y = 2x$	$-y = 2(-x)$
$y = -2x$	$y = -2x$	$y = 2x$

The graph has symmetry with respect to the origin. We reflect the portion of the graph in quadrant I through the origin, using the origin symmetry.

x	y
0	0
1	2
2	4

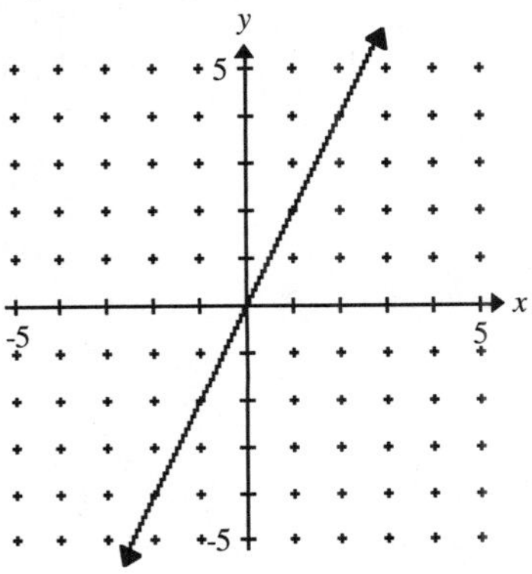

(B) $y = 2x - 1$

Test y axis
Replace x with $-x$:
$y = 2(-x) - 1$
$y = -2x - 1$

Test x axis
Replace y with $-y$:
$-y = 2x - 1$
$y = -2x + 1$

Test origin
Replace x with $-x$ and y with $-y$:
$-y = 2(-x) - 1$
$y = 2x + 1$

The graph has none of these three symmetries.

x	y
0	−1
1	1
−1	−3
2	3
−2	−5

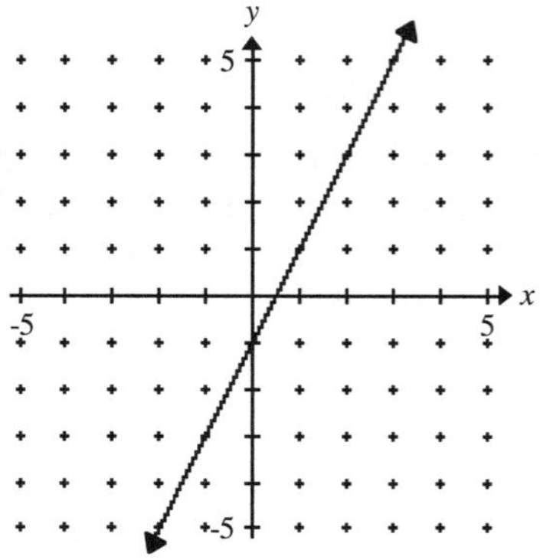

(C) $y = 2|x|$

Test y axis
Replace x with $-x$:
$y = 2|-x|$
$y = 2|x|$

Test x axis
Replace y with $-y$:
$-y = 2|x|$
$y = -2|x|$

Test origin
Replace x with $-x$ and y with $-y$:
$-y = 2|-x|$
$y = -2|x|$

The graph has symmetry with respect to the y axis. We reflect the portion of the graph in quadrant I through the y axis, using the y axis symmetry.

x	y
0	0
1	2
2	4

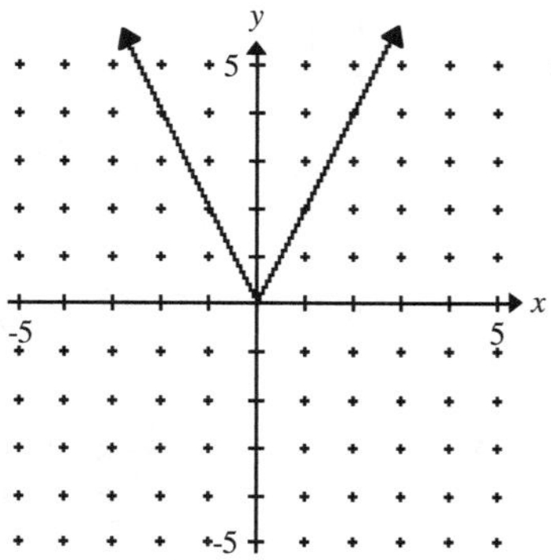

(D) $|y| = 2x$

Test y axis
Replace x with $-x$:
$|y| = 2(-x)$
$|y| = -2x$

Test x axis
Replace y with $-y$:
$|-y| = 2x$
$|y| = 2x$

Test origin
Replace x with $-x$ and y with $-y$:
$|-y| = 2(-x)$
$|y| = -2x$

The graph has symmetry with respect to the x axis. We reflect the portion of the graph in quadrant I through the x axis, using the x axis symmetry.

x	y
0	0
1	2
2	4

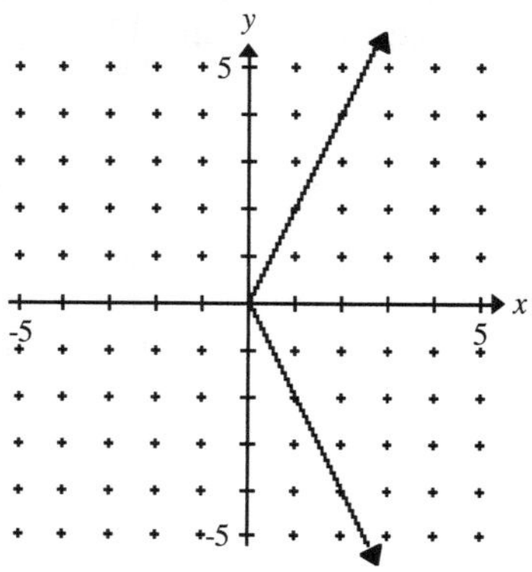

(2-1)

4. (A) −2 (B) −2, 1 (C) −3, 2 *(2-1)*

5. (A) $d(A, B) = \sqrt{[4 - (-2)]^2 + (0 - 3)^2}$
$= \sqrt{36 + 9} = \sqrt{45}$

(B) $m = \frac{0 - 3}{4 - (-2)} = \frac{-3}{6} = -\frac{1}{2}$

(C) The slope m_1 of a line perpendicular to AB must satisfy

$$m_1\left(-\frac{1}{2}\right) = -1.$$

Therefore, $m_1 = 2$. *(2-2, 2-3)*

6. (A) Center at (0, 0) and radius $\sqrt{7}$
$x^2 + y^2 = r^2$
$x^2 + y^2 = (\sqrt{7})^2$
$x^2 + y^2 = 7$

(B) Center at (3, -2) and radius $\sqrt{7}$
$(h, k) = (3, -2)$
$r = \sqrt{7}$
$(x - h)^2 + (y - k)^2 = r^2$
$(x - 3)^2 + [y - (-2)]^2 = (\sqrt{7})^2$
$(x - 3)^2 + (y + 2)^2 = 7$ *(2-2)*

7. $(x + 3)^2 + (y - 2)^2 = 5$
$[x - (-3)]^2 + (y - 2)^2 = (\sqrt{5})^2$
Center: $C(h, k) = (-3, 2)$
Radius: $r = \sqrt{5}$ *(2-2)*

8. (A) −4 is the average of a_1 and 2, so $-4 = \frac{a_1 + 2}{2}$
$-8 = a_1 + 2$
$-10 = a_1$

(B) 3 is the average of a_2 and −5, so $3 = \frac{a_2 + (-5)}{2}$
$6 = a_2 - 5$
$11 = a_2$

(C) $A = (-10, 11)$
$d(A, M) = \sqrt{(-10 - (-4))^2 + (11 - 3)^2} = \sqrt{(-6)^2 + 8^2} = \sqrt{36 + 64} = \sqrt{100} = 10$
$d(M, B) = \sqrt{(-4 - 2)^2 + (3 - (-5))^2} = \sqrt{(-6)^2 + 8^2} = \sqrt{36 + 64} = \sqrt{100} = 10$ *(2-2)*

9. $A = (-1, -2)$, $B = (4, 3)$, $C = (1, 4)$

(A)

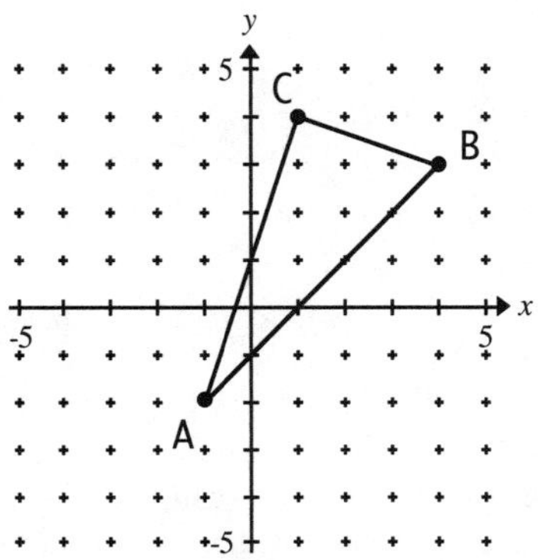

(B) $d(A, B) = \sqrt{(-4 - (-1))^2 + (3 - (-2))^2} = \sqrt{5^2 + 5^2}$
$= \sqrt{25 + 25} = \sqrt{50}$
$d(B, C) = \sqrt{(1 - 4)^2 + (4 - 3)^2} = \sqrt{(-3)^2 + 1^2}$
$= \sqrt{9 + 1} = \sqrt{10}$
$d(A, C) = \sqrt{(1 - (-1))^2 + (4 - (-2))^2} = \sqrt{2^2 + 6^2}$
$= \sqrt{4 + 36} = \sqrt{40} = 2\sqrt{10}$
Perimeter = sum of lengths of all three sides
$= \sqrt{50} + \sqrt{10} + 2\sqrt{10} = 16.56$ to two decimal places

(C) Since $(\sqrt{50})^2 = (\sqrt{10})^2 + (2\sqrt{10})^2$, that is, $50 = 10 + 40$, the triangle is a right triangle.

(D) Midpoint of $AB = \left(\frac{-1 + 4}{2}, \frac{-2 + 3}{2}\right) = (1.5, 0.5)$

Midpoint of $BC = \left(\frac{4 + 1}{2}, \frac{3 + 4}{2}\right) = (2.5, 3.5)$

Midpoint of $AC = \left(\frac{-1 + 1}{2}, \frac{-2 + 4}{2}\right) = (0, 1)$ *(2-3)*

10. The points at two of the vertices of the triangle are $(-4, 3)$ and $(1, 1)$. The line moves 2 units down between these points so the rise is -2. The line moves 5 units to the right between these points so the run is 5. The slope is thus $m = \frac{-2}{5}$.

$$y - 1 = \frac{-2}{5}(x - 1)$$
$$y - 1 = -\frac{2}{5}x + \frac{2}{5}$$
$$5y - 5 = -2x + 2$$
$$2x + 5y = 7$$

(2-3)

11. $3x + 2y = 9$

$$2y = -3x + 9$$
$$y = -\frac{3}{2}x + \frac{9}{2}$$

slope: $-\frac{3}{2}$

x	y
0	$\frac{9}{2}$
3	0
1	3

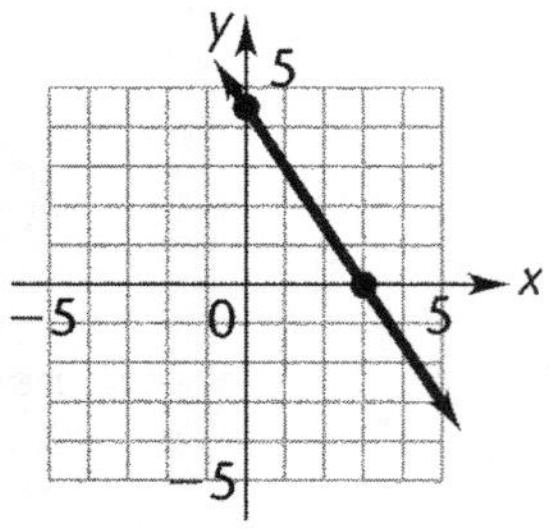

(2-3)

12. The line passes through the two given points, $(6, 0)$ and $(0, 4)$. Thus, its slope is given by

$$m = \frac{0 - 4}{6 - 0} = \frac{-4}{6} = -\frac{2}{3}$$

The equation of the line is, therefore, using the point-slope form,

$$y - 0 = -\frac{2}{3}(x - 6)$$

or $3y = -2(x - 6)$

or $3y = -2x + 12.$

$2x + 3y = 12$ *(2-3)*

13. $y = mx + b \quad m = -\frac{2}{3} \quad b = 2$

$$y = -\frac{2}{3}x + 2$$

(2-3)

14. *vertical:* $x = -3$, slope not defined; *horizontal:* $y = 4$, slope $= 0$ *(2-3)*

15. $y = x^2 - 2$

Test y axis
Replace x with $-x$:
$y = (-x)^2 - 2$
$y = x^2 - 2$

Test x axis
Replace y with $-y$:
$-y = x^2 - 2$
$y = -x^2 + 2$

Test origin
Replace x with $-x$ and y with $-y$:
$-y = (-x)^2 - 2$
$y = -x^2 + 2$

x	y
0	−2
1	−1
2	2

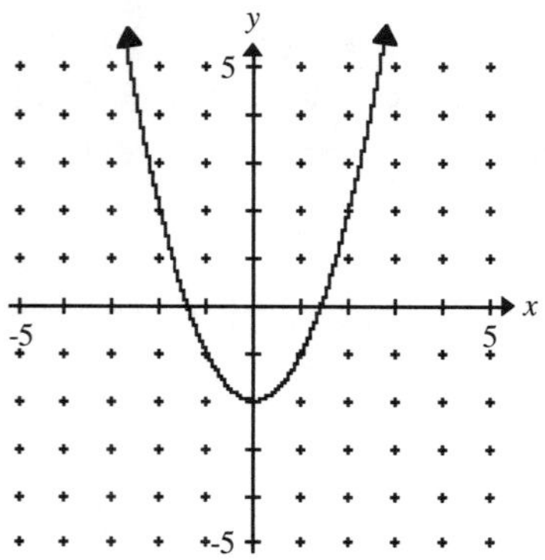

The graph has symmetry with respect to the y axis.

We reflect the portion of the graph in quadrant I through the y axis, using the y axis symmetry.

(2-1)

16. $y^2 = x - 2$

Test y axis
Replace x with $-x$:
$y^2 = -x - 2$

Test x axis
Replace y with $-y$:
$(-y)^2 = x - 2$
$y^2 = x - 2$

Test origin
Replace x with $-x$ and y with $-y$:
$(-y)^2 = -x - 2$
$y^2 = -x - 2$

x	y
2	0
3	±1
6	±2

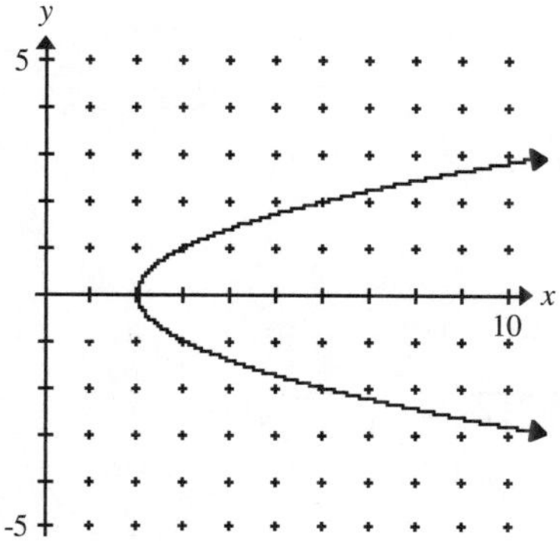

The graph has symmetry with respect to the x axis.

We reflect the portion of the graph in quadrant I through the x axis, using the x axis symmetry.

(2-1)

17. $9y^2 + 4x^2 = 36$

Test y axis
Replace x with $-x$:
$9y^2 + 4(-x)^2 = 36$
$9y^2 + 4x^2 = 36$

Test x axis
Replace y with $-y$:
$9(-y)^2 + 4x^2 = 36$
$9y^2 + 4x^2 = 36$

Origin symmetry follows automatically.

x	y
0	±2
1	$\pm\frac{1}{3}\sqrt{32} \approx 1.9$
2	$\pm\frac{1}{3}\sqrt{20} \approx 1.5$
3	0

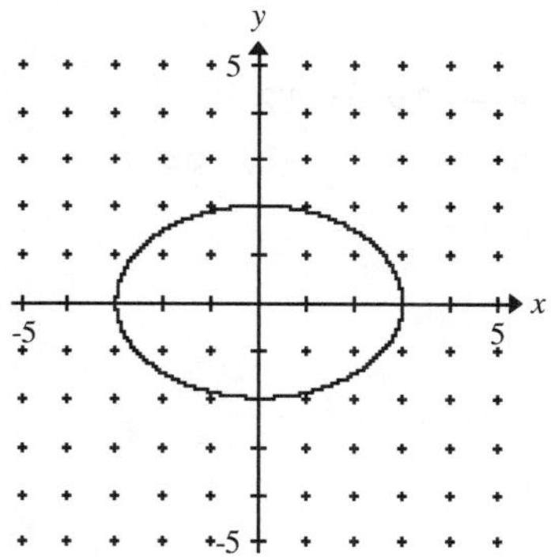

The graph has symmetry with respect to the x axis, the y axis, and the origin.

$y = \pm\frac{1}{3}\sqrt{36 - 4x^2}$

We reflect the portion of the graph of $y = \frac{1}{3}\sqrt{36 - 4x^2}$ in quadrant I through the y axis, then reflect everything through the x axis.

(2-1)

18. $9y^2 - 4x^2 = 36$

Test y axis
Replace x with $-x$:
$9y^2 - 4(-x)^2 = 36$
$9y^2 - 4x^2 = 36$

Test x axis
Replace y with $-y$:
$9(-y)^2 - 4x^2 = 36$
$9y^2 - 4x^2 = 36$

Origin symmetry follows automatically.

x	y
0	±2
1	$\pm\frac{1}{3}\sqrt{40} \approx 2.1$
2	$\pm\frac{1}{3}\sqrt{52} \approx 2.4$
3	$\pm\frac{1}{3}\sqrt{72} \approx 2.8$
4	$\pm\frac{10}{3} \approx 3.3$

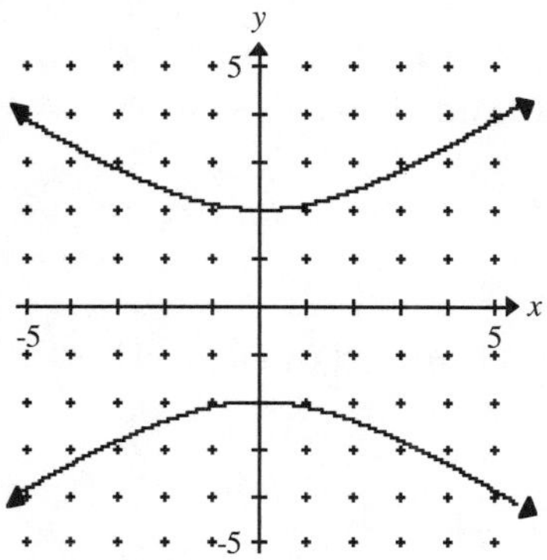

The graph has symmetry with respect to the x axis, the y axis, and the origin.

$y = \pm\frac{1}{3}\sqrt{36 + 4x^2}$

We reflect the portion of the graph of $y = \frac{1}{3}\sqrt{36 + 4x^2}$ in quadrant I through the y axis, then reflect everything through the x axis.

(2-1)

19. The graph is a circle with center (2, 0) and radius 2.

$$(x - h)^2 + (y - k)^2 = r^2$$
$$(x - 2)^2 + (y - 0)^2 = 2^2$$
$$(x - 2)^2 + y^2 = 4$$

(2-2)

20. (A) Since two points are given, we find the slope, then apply the point-slope form.

$$m = \frac{-3 - 3}{0 - (-4)} = \frac{-6}{4} = -\frac{3}{2}$$
$$y - 3 = -\frac{3}{2}[x - (-4)]$$
$$2(y - 3) = -3(x + 4)$$
$$2y - 6 = -3x - 12$$
$$3x + 2y = -6$$

(B) $d(P, Q) = \sqrt{(-3 - 3)^2 + [0 - (-4)]^2} = \sqrt{36 + 16} = \sqrt{52} = 2\sqrt{13}$ (2-2, 2-3)

21. The line $6x + 3y = 5$, or
$3y = -6x + 5$,
or $y = -2x + \frac{5}{3}$, has slope -2.

(A) We require a line through (-2, 1), with slope -2. Applying the point-slope form, we have

$$y - 1 = -2[x - (-2)]$$
$$y - 1 = -2x - 4$$
$$y = -2x - 3$$

(B) We require a line with slope m satisfying $-2m = -1$, or $m = \frac{1}{2}$.

Again applying the point-slope form, we have

$$y - 1 = \frac{1}{2}[x - (-2)]$$
$$y - 1 = \frac{1}{2}x + 1$$
$$y = \frac{1}{2}x + 2$$

(2-3)

22. We are given $C(h, k) = (3, 0)$.
To find r we use the distance formula.
r = distance from the center to (-1, 4)

$$= \sqrt{[(-1) - 3]^2 + (0 - 4)^2}$$
$$= \sqrt{16 + 16}$$
$$= \sqrt{32}$$

Then the equation of the circle is

$$(x - h)^2 + (y - k)^2 = r^2$$
$$(x - 3)^2 + (y - 0)^2 = (\sqrt{32})^2$$
$$(x - 3)^2 + y^2 = 32$$

(2-2)

23.

$$x^2 + y^2 + 4x - 6y = 3$$
$$(x^2 + 4x + ?) + (y^2 - 6y + ?) = 3$$
$$(x^2 + 4x + 4) + (y^2 - 6y + 9) = 3 + 4 + 9$$
$$(x + 2)^2 + (y - 3)^2 = 16$$
$$[x - (-2)]^2 + (y - 3)^2 = 4^2$$

Center: $C(h, k) = C(-2, 3)$
Radius $r = \sqrt{16} = 4$ (2-2)

24. Let (x, y) be a point equidistant from (3, 3) and (6, 0). Then

$$\sqrt{(x - 3)^2 + (y - 3)^2} = \sqrt{(x - 6)^2 + (y - 0)^2}$$
$$(x - 3)^2 + (y - 3)^2 = (x - 6)^2 + y^2$$
$$x^2 - 6x + 9 + y^2 - 6y + 9 = x^2 - 12x + 36 + y^2$$
$$-6x - 6y + 18 = -12x + 36$$
$$6x - 6y = 18$$
$$x - y = 3$$

This is the equation of a line. (2-2, 2-3)

25. If $m = 0$, the equations are reduced to $-y = b$ (a horizontal line) and $x = b$ (a vertical line). In this case, the graphs are perpendicular. Otherwise, $m \neq 0$. Solving for y yields

$$\begin{aligned} mx - y &= b \\ mx &= y + b \\ y &= mx - b \end{aligned} \qquad \begin{aligned} x + my &= b \\ my &= -x + b \\ y &= -\frac{1}{m}x + \frac{b}{m} \end{aligned}$$

The first line has slope m and the second has slope $-\frac{1}{m}$. The graphs are perpendicular in this case also. (2-3)

26.
$$\begin{aligned} x^2 - 4x + y^2 - 2y - 3 &= 0 \\ (x^2 - 4x + ?) + (y^2 - 2y + ?) &= 3 \\ (x^2 - 4x + 4) + (y^2 - 2y + 1) &= 3 + 4 + 1 \\ (x - 2)^2 + (y - 1)^2 &= 8 \\ (x - 2)^2 + (y - 1)^2 &= (\sqrt{8})^2 \end{aligned}$$

Center: (2, 1); radius: $\sqrt{8} = 2\sqrt{2}$ (2-2)

27. The line tangent to the circle is perpendicular to the radius drawn to the point of tangency. The radius drawn to (4, 3) is a line through (4, 3) and (2, 1) and therefore has slope

$$m_1 = \frac{1 - 3}{2 - 4} = 1$$

The tangent line therefore has slope

$m_2 = -\frac{1}{m_1} = -\frac{1}{1} = -1$ and passes through (4, 3). Applying the point-slope form yields

$$\begin{aligned} y - 3 &= (-1)(x - 4) \\ y - 3 &= -x + 4 \\ y &= -x + 7 \end{aligned}$$

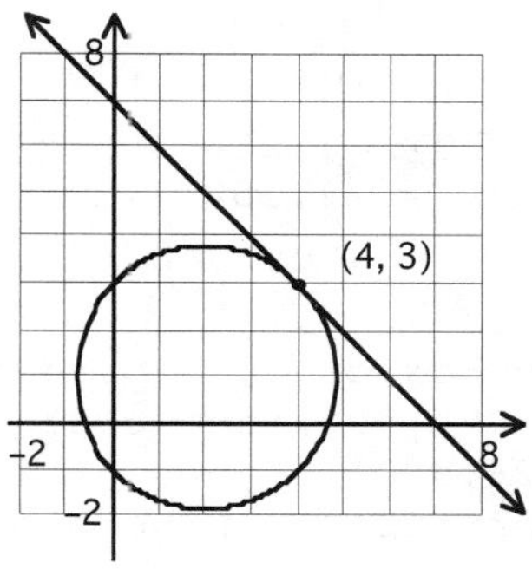

(2-2, 2-3)

28. The radius of a circle is the distance from the center to any point on the circle. Since the center of this circle is(4, –3) and (1, 2) is a point on the circle, the radius is given by

$$r = \sqrt{(1 - 4)^2 + (2 - (-3))^2} = \sqrt{9 + 25} = \sqrt{34}$$

Hence the equation of the circle is given by

$$\begin{aligned} (x - 4)^2 + (y - (-3))^2 &= (\sqrt{34})^2 \\ (x - 4)^2 + (y + 3)^2 &= 34 \end{aligned}$$

(2-2)

29. (A)

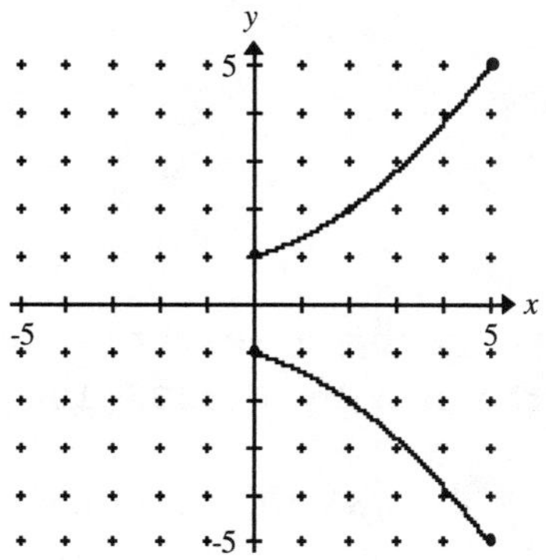

(B)

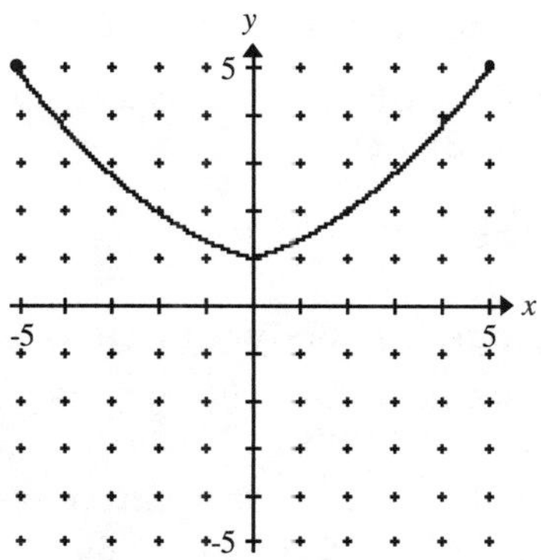

(C)

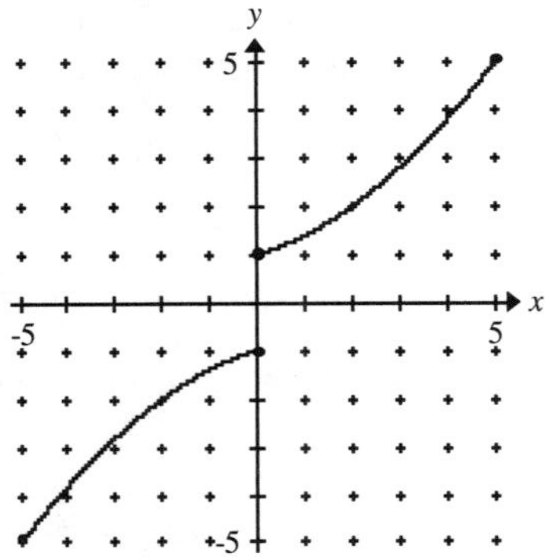

(D)

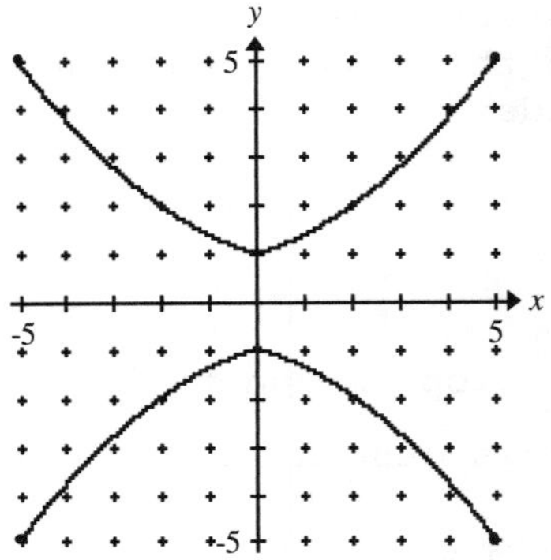

(2-1)

30. $b = 5h$ *(2-3)*

31. $h = 0.25b$ *(2-3)*

32. (A) If value V is linearly related to time t, then we are looking for an equation whose graph passes through $(t_1, V_1) = (0, 12{,}000)$ and $(t_2, V_2) = (8, 2{,}000)$. We find the slope, and then we use the point-slope form to find the equation.

$$m = \frac{V_2 - V_1}{t_2 - t_1} = \frac{2{,}000 - 12{,}000}{8 - 0} = -1{,}250$$

$$V - V_1 = m(t - t_1)$$

$$V - 12{,}000 = -1{,}250(t - 0)$$

$$V = -1{,}250t + 12{,}000$$

(B) Substitute $t = 5$ to obtain $V = -1{,}250(5) + 12{,}000 = \$5{,}750$. *(2-4)*

33. If x is the number of CD's produced, then the variable cost is $5x$. Since the fixed cost is \$24,900, the cost C of producing x CD's is given by

$$C = 5x + 24{,}900$$

To find how many CD's can be produced for \$62,000, substitute $C = 62{,}000$ and solve.

$$62{,}000 = 5x + 24{,}900$$

$$37{,}100 = 5x$$

$$x = 7{,}420 \text{ CD's}$$

(2-4)

34. (A) The rate of change of height with respect to DBH is 2.9.

(B) Increasing DBH by 1 inch increases height by 2.9 feet.

(C) Substitute $d = 3$ into $h = 2.9d + 30.2$ to obtain

$$h = 2.9(3) + 30.2$$

$$h = 38.9$$

$h = 39$ feet to the nearest foot.

(D) Substitute $h = 45$ into $h = 2.9d + 30.2$ and solve

$$45 = 2.9d + 30.2$$

$$14.8 = 2.9d$$

$d = 5$ inches *(2-4)*

35. (A) The rate of change of body surface area with respect to weight is 0.3433.

(B) Increasing the weight by 100 grams increases the BSA by $0.3433(100) = 34.33$ square centimeters.

(C) Substitute wt = 15,000 grams into BSA = 1,321 + 0.3433 wt to obtain BSA = 1,321 + 0.3433(15,000) = 6470.5 cm^2. *(2-4)*

36. In the sketch, we note that the point $(4, r - 2)$ is on the circle with equation $x^2 + y^2 = r^2$, hence $(4, r - 2)$ must satisfy this equation.

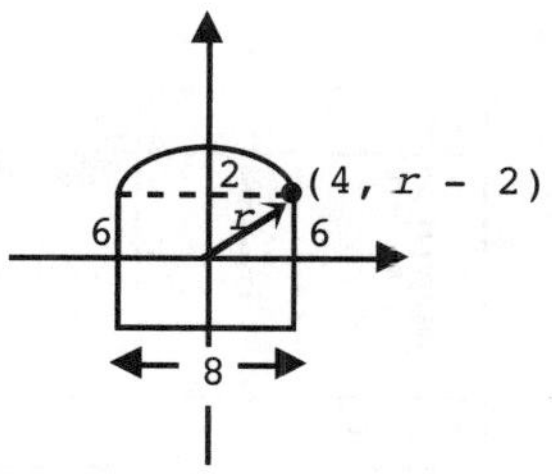

$$4^2 + (r - 2)^2 = r^2$$
$$16 + r^2 - 4r + 4 = r^2$$
$$-4r + 20 = 0$$
$$r = 5 \text{ feet}$$

(2-2)

37. (A) $H = 0.7(220 - A)$

(B) To find H when $A = 20$, substitute $A = 20$ into $H = 0.7(220 - A)$ to obtain $H = 0.7(220 - 20) = 140$ beats per minute.

(C) To find A when $H = 126$, substitute $H = 126$ into $H = 0.7(220 - A)$ and solve.

$$126 = 0.7(220 - A)$$
$$180 = 220 - A$$
$$-40 = -A$$
$$A = 40 \text{ years old.}$$

(2-4)

38. (A) In this example, the independent variable is years since 1900 so the first list under the STAT EDIT screen comes from the top row of Table 1. (Note that x is years after 1900, so the values we input are 12, 32, 52, 72 and 92. The dependent variable is time, so the second list is 5.41, 4.81, 4.51, 4.00, 3.75.)

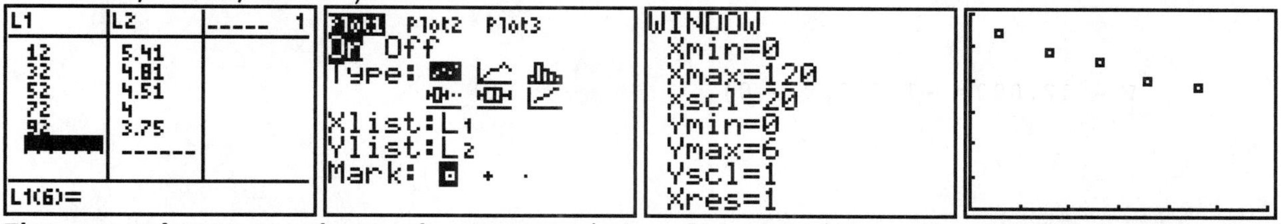

The second screen shows the appropriate plot settings; the third shows a good choice of viewing window; the last is the scatter plot of the data from the table.

To compare the algebraic function, we enter $y_1 = -0.021x + 5.57$ in the equation editor, choose the "ask" option for independent variable in the table setup window, then enter x-values 12, 32, 52, 72, and 92.

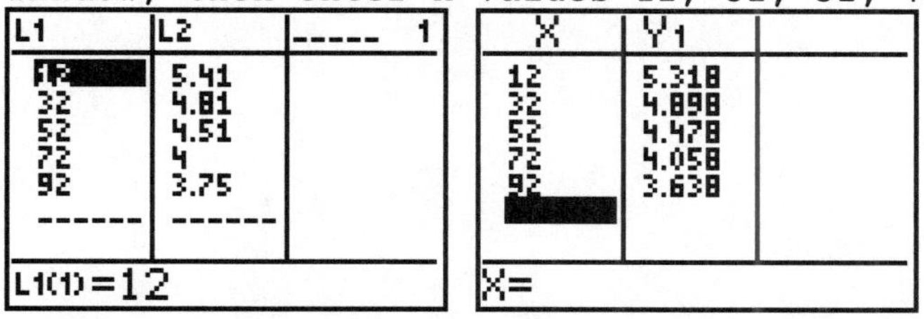

Comparing the two tables, we see that the corresponding times match very closely. Finally, we look at the graphs of both the function and the scatter plot:

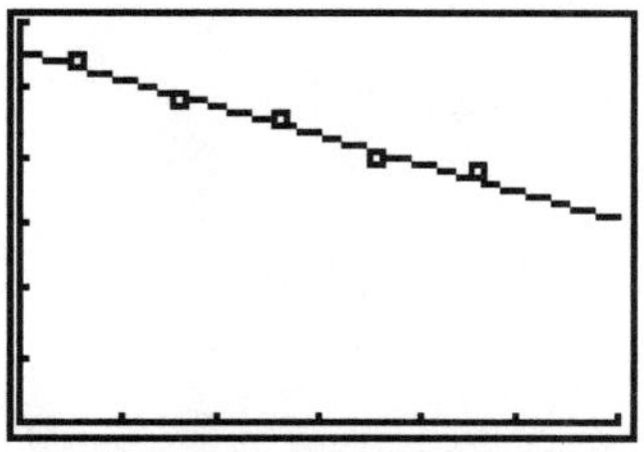

As expected, the graph of the function matches the points from the scatter plot very well.

(B) The year 2008 corresponds to $x = 108$. Substitute $x = 108$ to obtain

$$f(108) = -0.021(108) + 5.57 = 3.302 \text{ minutes}$$

(2-4)

39. (A)

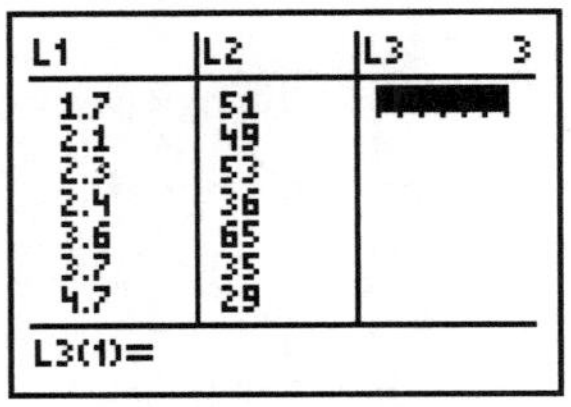

Enter data.

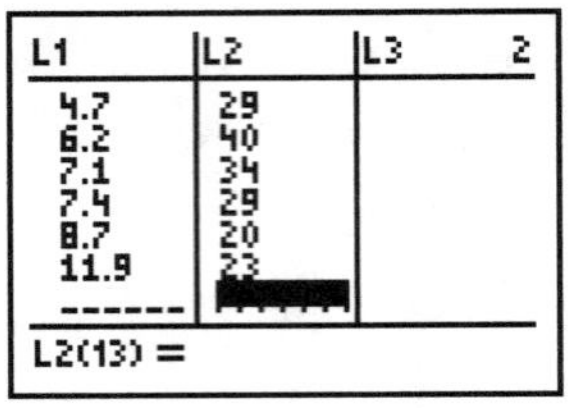

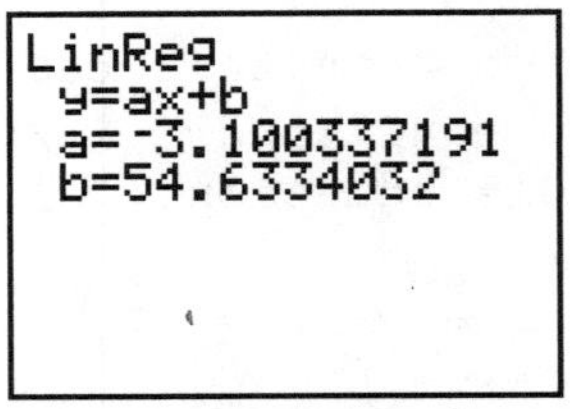

Compute regression equation.

(B)

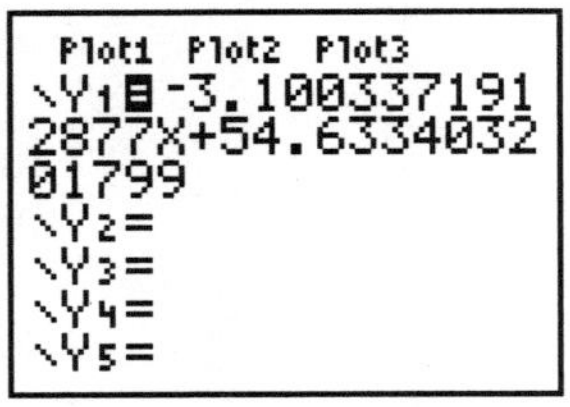

Transfer the regression equation to the equation editor.

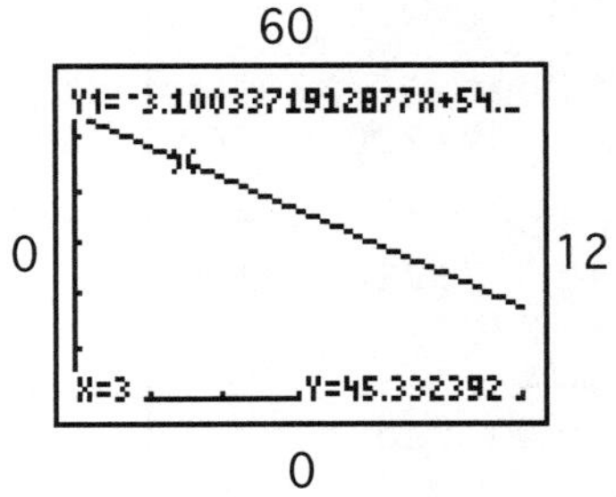

Graph the regression equation and determine the percentage that corresponds to the assessed value of $300.

The model predicts a percentage of 45.33% votes for democrats. (*2-4*)

CHAPTER 3

Section 3-1

1. A function

3. Not a function (two range values correspond to domain values 3 and 5)

5. A function

7. A function; domain = {2, 3, 4, 5}; range = {4, 6, 8, 10}

9. Not a function (two range values correspond to domain values 5 and 10)

11. A function; domain = {0, 1, 2, 3, 4, 5}; range = {1, 2}

13. A function

15. Not a function (fails vertical line test since the y-axis hits the graph 3 times.)

17. Not a function (fails vertical line test since the graph is vertical itself!)

19. (A) This is a function. Each congressional district has only one representative, so all residents correspond to exactly one representative.

(B) This is not a function. Every state has two senators that represent the entire state, so every resident corresponds to two senators. This violates the definition of function.

21. This equation does define y as a function of x. If you plug in any number for x, there is only one choice of y that will then make the equation true.

23. This equation does not define y as a function of x. For many choices of x, there are two values of y that will make the equation true. For example, if $x = 0$ the equation is $y^2 = u$ and y can be either 2 or -2. Since we found a value for x that corresponds to two different values of y, the equation is not a function.

25. $f(x) = 3x - 5$
(A) $f(3) = 3(3) - 5 = 4$
(B) $f(h) = 3h - 5$
(C) $f(3) + f(h) = 4 + (3h - 5) = 3h - 1$
(D) $f(3 + h) = 3(3 + h) - 5 = 9 + 3h - 5 = 3h + 4$

Common Error: $f(3 + h)$ is not the same as $f(3) + h$ or $f(3) + f(h)$.

27. $F(w) = -w^2 + 2w$
(A) $F(4) = -(4)^2 + 2(4) = -16 + 8 = -8$
(B) $F(-4) = -(-4)^2 + 2(-4) = -16 - 8 = -24$
(C) $F(4 + a) = -(4 + a)^2 + 2(4 + a)$
$= -(16 + 8a + a^2) + 8 + 2a$
$= -16 - 8a - a^2 + 8 + 2a$
$= -a^2 - 6a - 8$
(D) $F(2 - a) = -(2 - a)^2 + 2(2 - a)$
$= -(4 - 4a + a^2) + 4 - 2a$
$= -4 + 4a - a^2 + 4 - 2a$
$= -a^2 + 2a$

29. $f(t) = 2 - 3t^2$
(A) $f(-2) = 2 - 3(-2)^2 = 2 - 12 = -10$
(B) $f(-t) = 2 - 3(-t)^2 = 2 - 3t^2$
(C) $-f(t) = -(2 - 3t^2) = -2 + 3t^2$
(D) $-f(-t) = -(2 - 3t^2) = -2 + 3t^2$

31. $F(u) = u^2 - u - 1$
(A) $F(10) = (10)^2 - (10) - 1 = 89$
(B) $F(u^2) = (u^2)^2 - (u^2) - 1 = u^4 - u^2 - 1$
(C) $F(5u) = (5u)^2 - (5u) - 1 = 25u^2 - 5u - 1$
(D) $5F(u) = 5(u^2 - u - 1) = 5u^2 - 5u - 5$

33. When $x = -4$, $y = 7$, so $f(-4) = 7$.

35. When $y = 4$, x can equal -5, -1, or 6.

37. Solving for the dependent variable y, we have
$$y - x^2 = 1$$
$$y = x^2 + 1$$
Since $x^2 + 1$ is a real number for each real number x, the equation defines a function with domain all real numbers.

39. Solving for the dependent variable y, we have
$$2x^3 + y^2 = 4$$
$$y^2 = 4 - 2x^3$$
$$y = \pm\sqrt{4 - 2x^3}$$
Since each positive number has two real square roots, the equation does not define a function. For example, when $x = 0$, $y = \pm\sqrt{4 - 2(0)^3} = \pm\sqrt{4} = \pm 2$.

41. Solving for the dependent variable y, we have
$$x^3 - y = 2$$
$$-y = 2 - x^3$$
$$y = x^3 - 2$$
Since $x^3 - 2$ is a real number for each real number x, the equation defines a function with domain all real numbers.

43. If $2x + |y| = 7$, then $|y| = 2x + 7$. This equation does not define a function. For example, if $x = 0$, then $|y| = 7$, so $y = 7$ or $y = -7$.

45. Solving for the dependent variable y, we have
$$3y + 2|x| = 12$$
$$3y = 12 - 2|x|$$
$$y = \frac{12 - 2|x|}{3}$$
Since $\frac{12 - 2|x|}{3}$ is a real number for each real number x, the equation defines a function with domain all real numbers.

47. If $xy = 0$, then either $x = 0$ or $y = 0$. This equation does not define a function since if $x = 0$, y can be any real number.

49. The domain is the set of all real numbers R, $-\infty < t < \infty$, $(-\infty, \infty)$.

51. Since $f(x)$ is a polynomial, the domain is the set of all real numbers R, $-\infty < x < \infty$, $(-\infty, \infty)$.

53. Since $3u^2 + 4$ is never negative, $\sqrt{3u^2 + 4}$ represents a real number for all replacements of u by real numbers. The domain is R, $-\infty < u < \infty$, $(-\infty, \infty)$.

55. If the denominator of the fraction is zero, the function will be undefined, since division by zero is undefined. For any other values of x, $h(x)$ represents a real number. Solve the equation $4 - x = 0$; $x = 4$. Thus, the domain is all real numbers except 4 or $(-\infty, 4) \cup (4, \infty)$; $x < 4$ or $x > 4$.

57. The square root of a number is defined only if the number is nonnegative. Thus, $\sqrt{t - 4}$ is defined if $t - 4 \geq 0$ and undefined if $t - 4 < 0$. Note that $t - 4 \geq 0$ when $t \geq 4$, so the domain is all real numbers greater than or equal to 4 or $[4, \infty)$.

59. The formula $\sqrt{7 + 3w}$ is defined only if $7 + 3w \ge 0$, since the square root is only defined if the number inside is nonnegative.
We solve the inequality:

$$7 + 3w \ge 0$$
$$3w \ge -7$$
$$w \ge -\frac{7}{3}$$

Thus, the domain is all real numbers greater than or equal to $-\frac{7}{3}$ or $\left[\frac{7}{3}, \infty\right)$.

61. The fraction $\frac{u}{u^2 + 4}$ is defined for any value of u that does <u>not</u> make the denominator zero, so we solve the equation $u^2 + 4 = 0$ to find values that make the function undefined.

$$u^2 + 4 = 0$$
$$u^2 = -4$$

This equation has no solution since the square of a number can't be negative. So there are no values of u that make the fraction undefined and the domain is all real numbers or $(-\infty, \infty)$.

63. The fraction $\frac{v + 2}{v^2 - 16}$ is defined for any value of v that does <u>not</u> make the denominator zero, so we solve the equation $v^2 - 16 = 0$ to find values that make the function undefined.

$$v^2 - 16 = 0$$
$$v^2 = 16$$
$$v = 4, -4$$

Thus, the domain is all real numbers except 4 and -4, or $(-\infty, -4) \cup (-4, 4) \cup (4, \infty)$; $v < -4$, $-4 < v < 4$, or $v > 4$.

65. There are two issues to consider: we need to make certain that $x + 4 \ge 0$ so that the number inside the square root is nonnegative, and we need to avoid x-values that make the denominator zero.

First, $x + 4 \ge 0$ whenever $x \ge -4$. So x must be greater than or equal to -4 to avoid a negative under the root. Also, $x - 1 = 0$ when $x = 1$, so x cannot be 1. The domain is all real numbers greater than or equal to -4 except 1, or $[-4, 1) \cup (1, \infty)$; $-4 \le x < 1$ or $x > 1$.

67. There are two issues to consider: the number inside the square root has to be nonnegative, and the denominator of the fraction has to be nonzero. Since we just have x inside the square root, x has to be greater than or equal to zero. Next, we solve $3 - \sqrt{t} = 0$ to find any t-values that make the denominator zero.

$$3 - \sqrt{t} = 0$$
$$3 = \sqrt{t}$$
$$9 = t$$

Thus, t cannot be 9. The domain is all real numbers greater than or equal to zero except 9, or $[0, 9) \cup (9, \infty)$; $0 \le t < 9$ or $t > 9$.

69. $g(x) = 2x^3 - 5$

71. $G(x) = 2\sqrt{x} - x^2$

73. Function f multiplies the domain element by 2 and subtracts 3 from the result.

75. Function F multiplies the cube of the domain element by 3 and subtracts twice the square root of the domain element from the result.

77.

$$F(s) = 3s + 15$$
$$F(2 + h) = 3(2 + h) + 15$$
$$F(2) = 3(2) + 15$$
$$\frac{F(2 + h) - F(2)}{h} = \frac{[3(2 + h) + 15] - [3(2) + 15]}{h} = \frac{[6 + 3h + 15] - [21]}{h}$$
$$= \frac{3h + 21 - 21}{h} = \frac{3h}{h} = 3$$

79.

$$g(x) = 2 - x^2$$
$$g(3 + h) = 2 - (3 + h)^2$$
$$g(3) = 2 - (3)^2$$
$$\frac{g(3 + h) - g(3)}{h} = \frac{[2 - (3 + h)]^2 - [2 - (3)^2]}{h} = \frac{[2 - 9 - 6h - h^2] - [-7]}{h}$$
$$= \frac{-6h - h^2}{h} = \frac{h(-6 - h)}{h} = -6 - h$$

81.

$$L(w) = -2w^2 + 3w - 1$$
$$L(-2 + h) = -2(-2 + h)^2 + 3(-2 + h) - 1$$
$$L(-2) = -2(-2)^2 + 3(-2) - 1$$
$$\frac{L(-2 + h) - L(-2)}{h} = \frac{[-2(-2 + h)^2 + 3(-2 + h) - 1] - [-2(-2)^2 + 3(-2) - 1]}{h}$$
$$= \frac{[-2(4 - 4h + h^2) - 6 + 3h - 1] - [-15]}{h}$$
$$= \frac{-8 + 8h - 2h^2 - 6 + 3h - 1 + 15}{h}$$
$$= \frac{11h - 2h^2}{h} = \frac{h(11 - 2h)}{h} = 11 - 2h$$

83. (A)

$$f(x) = 4x - 7$$
$$f(x + h) = 4(x + h) - 7$$
$$\frac{f(x + h) - f(x)}{h} = \frac{[4(x + h) - 7] - [4x - 7]}{h}$$
$$= \frac{4x + 4h - 7 - 4x + 7}{h}$$
$$= \frac{4h}{h}$$
$$= 4$$

Common Errors: $f(x + h) \neq f(x) + f(h)$ or $4x - 7 + 4h - 7$

$f(x + h) \neq f(x) + h$ or $4x - 7 + h$

Also note: $-f(x) \neq -4x - 7$

Parentheses must be supplied.

(B)

$$f(x) = 4x - 7$$
$$f(a) = 4a - 7$$
$$\frac{f(x) - f(a)}{x - a} = \frac{(4x - 7) - (4a - 7)}{x - a}$$
$$= \frac{4x - 7 - 4a + 7}{x - a}$$
$$= \frac{4x - 4a}{x - a}$$
$$= \frac{4(x - a)}{x - a}$$
$$= 4$$

85. (A)

$$f(x) = 2x^2 - 4$$
$$f(x + h) = 2(x + h)^2 - 4$$
$$\frac{f(x + h) - f(x)}{h} = \frac{[2(x + h)^2 - 4] - [2x^2 - 4]}{h}$$
$$= \frac{2(x^2 + 2xh + h^2) - 4 - 2x^2 + 4}{h}$$
$$= \frac{2x^2 + 4xh + 2h^2 - 4 - 2x^2 + 4}{h}$$
$$= \frac{4xh + 2h^2}{h}$$
$$= \frac{h(4x + 2h)}{h}$$
$$= 4x + 2h$$

(B)

$$f(x) = 2x^2 - 4$$
$$f(a) = 2a^2 - 4$$
$$\frac{f(x) - f(a)}{x - a} = \frac{(2x^2 - 4) - (2a^2 - 4)}{x - a}$$
$$= \frac{2x^2 - 4 - 2a^2 + 4}{x - a}$$
$$= \frac{2x^2 - 2a^2}{x - a}$$
$$= \frac{2(x - a)(x + a)}{x - a}$$
$$= 2(x + a)$$
$$= 2x + 2a$$

87. (A) $f(x) = -4x^2 + 3x - 2$

$$f(x+h) = -4(x+h)^2 + 3(x+h) - 2$$

$$\frac{f(x+h) - f(x)}{h} = \frac{[-4(x+h)^2 + 3(x+h) - 2] - [-4x^2 + 3x - 2]}{h}$$

$$= \frac{-4(x^2 + 2xh + h^2) + 3(x+h) - 2 + 4x^2 - 3x + 2}{h}$$

$$= \frac{-4x^2 - 8xh - 4h^2 + 3x + 3h - 2 + 4x^2 - 3x + 2}{h}$$

$$= \frac{-8xh - 4h^2 + 3h}{h}$$

$$= \frac{h(-8x - 4h + 3)}{h}$$

$$= -8x - 4h + 3$$

(B) $f(x) = -4x^2 + 3x - 2$

$$f(a) = -4a^2 + 3a - 2$$

$$\frac{f(x) - f(a)}{x - a} = \frac{(-4x^2 + 3x - 2) - (-4a^2 + 3a - 2)}{x - a}$$

$$= \frac{-4x^2 + 3x - 2 + 4a^2 - 3a + 2}{x - a}$$

$$= \frac{-4x^2 + 4a^2 + 3x - 3a}{x - a}$$

$$= \frac{-4(x-a)(x+a) + 3(x-a)}{x - a}$$

$$= \frac{(x-a)[-4(x+a) + 3]}{x - a}$$

$$= -4(x+a) + 3$$

$$= -4x - 4a + 3$$

89. (A) $f(x) = x^3 - 2x$

$$f(x+h) = (x+h)^3 - 2(x+h)$$

$$\frac{f(x+h) - f(x)}{h} = \frac{[(x+h)^3 - 2(x+h)] - [x^3 - 2x]}{h}$$

$$= \frac{x^3 + 3x^2h + 3xh^2 + h^3 - 2x - 2h - x^3 + 2x}{h}$$

$$= \frac{3x^2h + 3xh^2 + h^3 - 2h}{h}$$

$$= \frac{h(3x^2 + 3xh + h^2 - 2)}{h} = 3x^2 + 3xh + h^2 - 2$$

(B) $f(x) = x^3 - 2x$

$$f(a) = a^3 - 2a$$

$$\frac{f(x) - f(a)}{x - a} = \frac{(x^3 - 2x) - (a^3 - 2a)}{x - a}$$

$$= \frac{x^3 - 2x - a^3 + 2a}{x - a}$$

$$= \frac{x^3 - a^3 - 2x + 2a}{x - a}$$

$$= \frac{(x-a)(x^2 + ax + a^2) - 2(x-a)}{x - a}$$

$$= \frac{(x-a)[x^2 + ax + a^2 - 2]}{x - a} = x^2 + ax + a^2 - 2$$

91. (A)
$$f(x) = \sqrt{x+2}$$
$$f(x+h) = \sqrt{x+h+2}$$
$$\frac{f(x+h)-f(x)}{h} = \frac{\sqrt{x+h+2}-\sqrt{x+2}}{h}$$
$$= \frac{\sqrt{x+h+2}-\sqrt{x+2}}{h} \cdot \frac{\sqrt{x+h+2}+\sqrt{x+2}}{\sqrt{x+h+2}+\sqrt{x+2}}$$
$$= \frac{(\sqrt{x+h+2})^2-(\sqrt{x+2})^2}{h(\sqrt{x+h+2}+\sqrt{x+2})}$$
$$= \frac{x+h+2-(x+2)}{h(\sqrt{x+h+2}+\sqrt{x+2})}$$
$$= \frac{h}{h(\sqrt{x+h+2}+\sqrt{x+2})}$$
$$= \frac{1}{\sqrt{x+h+2}+\sqrt{x+2}}$$

(B)
$$f(x) = \sqrt{x+2}$$
$$f(a) = \sqrt{a+2}$$
$$\frac{f(x)-f(a)}{x-a} = \frac{\sqrt{x+2}-\sqrt{a+2}}{x-a}$$
$$= \frac{\sqrt{x+2}-\sqrt{a+2}}{x-a} \cdot \frac{\sqrt{x+2}+\sqrt{a+2}}{\sqrt{x+2}+\sqrt{a+2}}$$
$$= \frac{(\sqrt{x+2})^2-(\sqrt{a+2})^2}{(x-a)(\sqrt{x+2}+\sqrt{a+2})}$$
$$= \frac{x+2-(a+2)}{(x-a)(\sqrt{x+2}+\sqrt{a+2})} = \frac{x-a}{(x-a)(\sqrt{x+2}+\sqrt{a+2})}$$
$$= \frac{1}{\sqrt{x+2}+\sqrt{a+2}}$$

93. (A)
$$f(x) = \frac{4}{x}$$
$$f(x+h) = \frac{4}{x+h}$$
$$\frac{f(x+h)-f(x)}{h} = \frac{\frac{4}{x+h}-\frac{4}{x}}{h}$$
$$= \frac{4x-4(x+h)}{hx(x+h)}$$
$$= \frac{-4h}{hx(x+h)}$$
$$= \frac{-4}{x(x+h)}$$

(B)
$$f(x) = \frac{4}{x}$$
$$f(a) = \frac{4}{a}$$
$$\frac{f(x)-f(a)}{x-a} = \frac{\frac{4}{x}-\frac{4}{a}}{x-a}$$
$$= \frac{4a-4x}{ax(x-a)}$$
$$= \frac{-4(x-a)}{ax(x-a)}$$
$$= \frac{-4}{ax}$$

95. Given w = width and Area = 64, we use $A = \ell w$ to write $\ell = \frac{A}{w} = \frac{64}{w}$.
Then $P = 2w + 2\ell = 2w + 2\left(\frac{64}{w}\right) = 2w + \frac{128}{w}$. Since w must be positive, the domain of $P(w)$ is $w > 0$.

97.

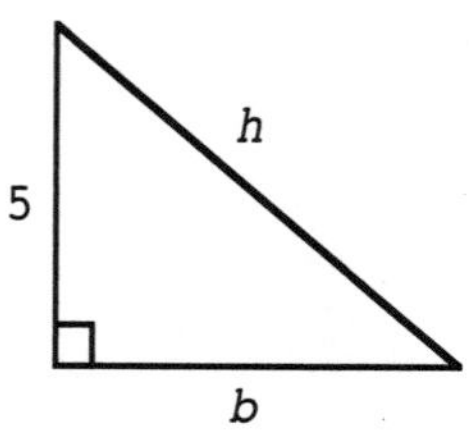

Using the given letters, the Pythagorean theorem gives
$$h^2 = b^2 + 5^2$$
$$h^2 = b^2 + 25$$
$$h = \sqrt{b^2 + 25} \text{ (since } h \text{ is positive)}$$
Since b must be positive, the domain of $h(b)$ is $b > 0$.

99. Daily cost = fixed cost + variable cost

$C(x)$ = \$300 + (\$1.75 per dozen doughnuts) × (number of dozen doughnuts)

$C(x) = 300 + 1.75x$

101. The rental charges are \$20 per day plus \$0.25 per mile driven.

103. (A) $S(0) = 16(0)^2 = 0$; $S(1) = 16(1)^2 = 16$; $S(2) = 16(2)^2 = 16(4) = 64$; $S(3) = 16(3)^2 = 16(9) = 144$ (Note: Remember that the order of operations requires that we apply the exponent first then multiply by 16.)

(B)
$$\frac{S(2+h)-S(2)}{h} = \frac{16(2+h)^2 - 16(2)^2}{h} = \frac{16(4+4h+h^2)-16(4)}{h}$$
$$= \frac{64+64h+16h^2-64}{h} = \frac{64h+16h^2}{h} = \frac{\cancel{h}(64+16h)}{\cancel{h}} = 64+16h$$

(Note: Be careful when evaluating $S(2 + h)$! You need to replace the variable t in the function with $(2 + h)$. $S(2 + h)$ is <u>not</u> the same thing as $S(2) + h$!)

(C) $h = 1$: $64 + 16(1) = 80$; $h = -1$: $64 + 16(-1) = 64 - 16 = 48$
$h = 0.1$: $64 + 16(0.1) = 64 + 1.6 = 65.6$;
$h = -0.1$: $64 + 16(-0.1) = 64 - 1.6 = 62.4$
$h = 0.01$: $64 + 16(0.01) = 64 + 0.16 = 64.16$:
$h = -0.01$: $64 + 16(-0.01) = 64 - 0.16 = 63.84$
$h = 0.001$: $64 + 16(0.001) = 64 + 0.016 = 64.016$
$h = -0.001$: $64 + 16(-0.001) = 64 - 0.016 = 63.984$

(D) The smaller h gets the closer the result is to 64. The numerator of the fraction, $S(2 + h) - S(2)$, is the difference between how far an object has fallen after $2 + h$ seconds and how far it's fallen after 2 seconds. This difference is how far the object falls in the small period of time from 2 to $2 + h$ seconds. When you divide that distance by the time (h), you get the average velocity of the object between 2 and $2 + h$ seconds. Part (C) shows that this average velocity approaches 64 feet per second as h gets smaller.

105.

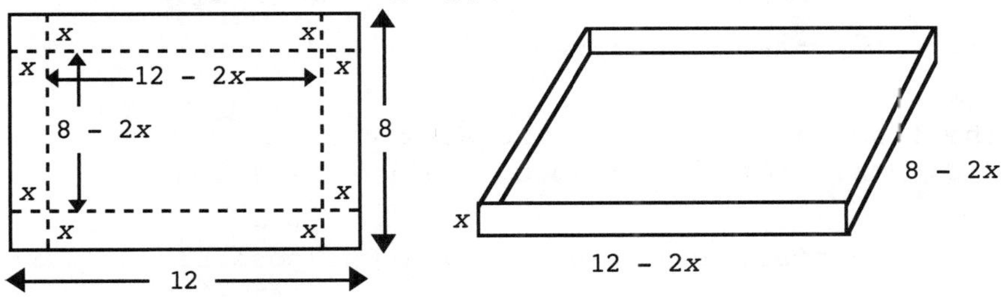

From the above figures it should be clear that
V = length ∞ width ∞ height = $(12 - 2x)(8 - 2x)x$.
Since all distances must be positive, $x > 0$, $8 - 2x > 0$, $12 - 2x > 0$.
Thus, $0 < x$, $4 > x$, $6 > x$, or $0 < x < 4$ (the last condition, $6 > x$, will be automatically satisfied if $x < 4$.) Domain: $0 < x < 4$.

107. From the text diagram, since each pen must have area 50 square feet, we see Area = (length)(width) or 50 = (length)x. Thus, the length of each pen is $\frac{50}{x}$ ft.

The total amount of fencing = 4(width) + 5(length) + 4(width − gate width)

$$F(x) = 4x + 5\left(\frac{50}{x}\right) + 4(x - 3)$$
$$F(x) = 4x + \frac{250}{x} + 4x - 12$$
$$F(x) = 8x + \frac{250}{x} - 12$$

Then $F(4) = 8(4) + \frac{250}{4} - 12 = 82.5$ $\quad F(5) = 8(5) + \frac{250}{5} - 12 = 78$

$F(6) = 8(6) + \frac{250}{6} - 12 = 77.7$ $\quad F(7) = 8(7) + \frac{250}{7} - 12 = 79.7$

109. 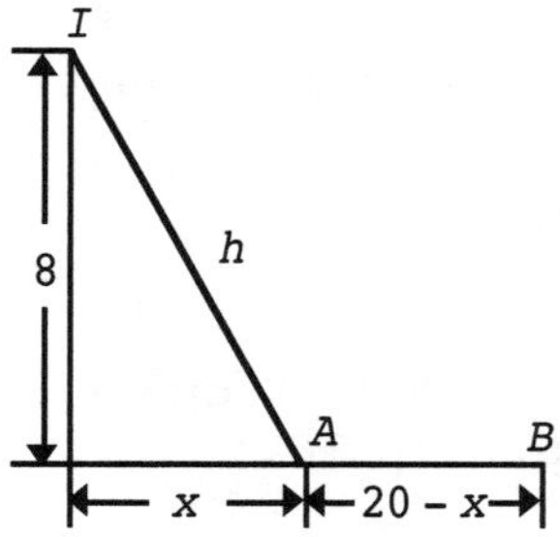

We note that the pipeline consists of the lake section *IA*, and shore section *AB*. The shore section has length $20 - x$. The lake section has length *h*, where $h^2 = 8^2 + x^2$, thus $h = \sqrt{64 + x^2}$.
The cost of all the pipeline =

$$\left(\begin{array}{c}\text{cost of shore}\\ \text{section per mile}\end{array}\right)\left(\begin{array}{c}\text{number of}\\ \text{shore miles}\end{array}\right) + \left(\begin{array}{c}\text{cost of lake}\\ \text{section per mile}\end{array}\right)\left(\begin{array}{c}\text{number of}\\ \text{lake miles}\end{array}\right)$$

$C(x) = 10{,}000(20 - x) + 15{,}000\sqrt{64 + x^2}$

From the diagram we see that *x* must be non-negative, but no more than 20.

Domain: $0 \le x \le 20$

111. Cost = (cost per hour) × number of hours
= (Fuel cost per hour + Fixed cost per hour) × number of hours.
For a 500 mile trip, since $D = rt$, $t = D/r = 500/v$ hours.

Hence cost = $\left(\frac{v^2}{5} + 400\right)\left(\frac{500}{v}\right)$

$$C(v) = \frac{v^2}{5} \cdot \frac{500}{v} + 400 \cdot \frac{500}{v}$$
$$= 100v + \frac{200{,}000}{v}$$

Section 3-2

1. (A) [-4, 4) (B) [-3, 3) (C) 0 (D) 0 (E) [-4, 4) (F) None
(G) None (H) None

3. (A) $(-\infty, \infty)$ (B) $[-4, \infty)$ (C) -3, 1 (D) -3 (E) $[-1, \infty)$ (F) $(-\infty, -1]$
(G) None (H) None

5. (A) $(-\infty, 2) \cup (2, \infty)$ (The function is not defined at $x = 2$.)
(B) $(-\infty, -1) \cup [1, \infty)$ (C) None (D) 1 (E) None (F) $(-\infty, -2] \cup (2, \infty)$
(G) [-2, 2) (H) $x = 2$

7. One possible answer:

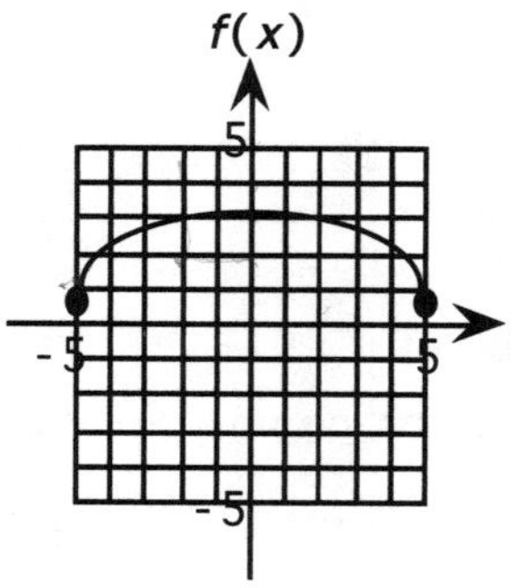

9. One possible answer:

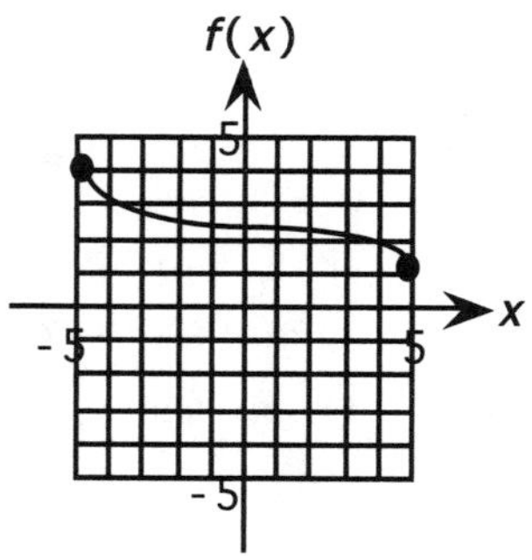

11. One possible answer:

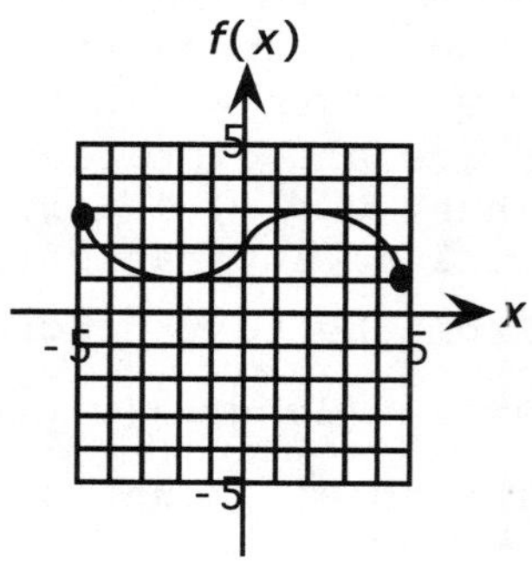

13. $f(x) = 2x + 4$
slope
The *y* intercept is $f(0) = 4$, and the slope is 2. To find the *x* intercept, we solve the equation $f(x) = 0$ for *x*.

$$f(x) = 0$$
$$2x + 4 = 0$$
$$2x = -4$$
$$x = -2$$

The *x* intercept is -2.

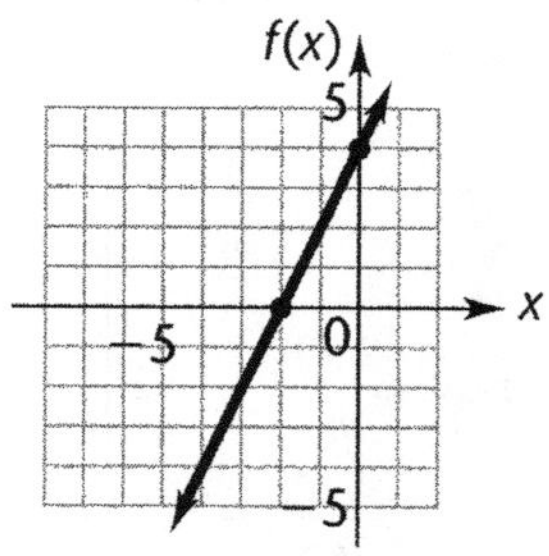

15. $f(x) = -\frac{1}{2}x - \frac{5}{3}$

The y intercept is $f(0) = -\frac{5}{3}$, and the slope is $-\frac{1}{2}$. To find the x intercept, we solve the equation $f(x) = 0$ for x.

$$\begin{aligned} f(x) &= 0 \\ -\frac{1}{2}x - \frac{5}{3} &= 0 \\ -\frac{1}{2}x &= \frac{5}{3} \\ x &= (-2)\frac{5}{3} \\ x &= \frac{-10}{3} \end{aligned}$$

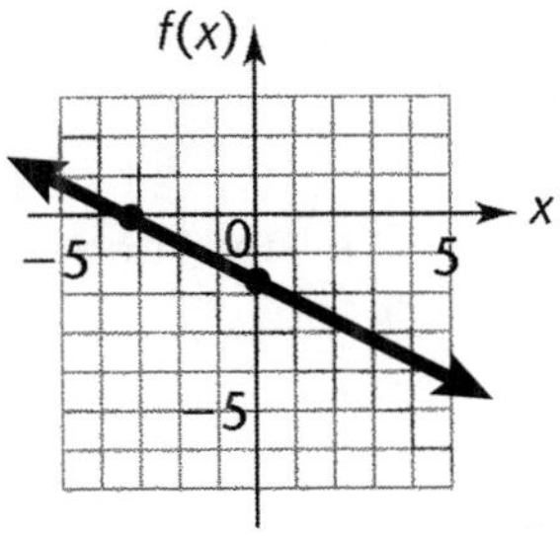

The x intercept is $\frac{-10}{3}$.

17. A linear function must have the form $f(x) = mx + b$. We are given $f(-2) = 7$, hence $7 = f(-2) = m(-2) + b$. Thus $b = 7 + 2m$. Since $f(4) = -2$, we also have $-2 = f(4) = 4m + b$. Substituting, we have

$$\begin{aligned} b &= 7 + 2m \\ -2 &= 4m + b \\ -2 &= 4m + 7 + 2m \\ -2 &= 6m + 7 \\ -9 &= 6m \\ m &= -\frac{3}{2} \\ b &= 7 + 2m = 7 + 2\left(-\frac{3}{2}\right) = 4 \end{aligned}$$

Hence $f(x) = mx + b$ becomes $f(x) = -\frac{3}{2}x + 4$.

19. The rational expression $\frac{3x - 12}{2x + 4}$ is defined everywhere except at the zero of the denominator:

$$\begin{aligned} 2x + 4 &= 0 \\ 2x &= -4 \\ x &= -2 \end{aligned}$$

The domain of f is $\{x \mid x \neq -2\}$.
A rational expression is 0 if and only if the numerator is zero:

$$\begin{aligned} 3x - 12 &= 0 \\ 3x &= 12 \\ x &= 4 \end{aligned}$$

The x intercept of f is 4. The y intercept of f is $f(0) = \frac{3(0) - 12}{2(0) + 4} = -3$.

21. The rational expression $\frac{3x - 2}{4x - 5}$ is defined everywhere except at the zero of the denominator:

$$\begin{aligned} 4x - 5 &= 0 \\ 4x &= 5 \\ x &= \frac{5}{4} \end{aligned}$$

The domain of f is $\left\{x \mid x \neq \frac{5}{4}\right\}$.

A rational expression is 0 if and only if the numerator is zero:

$$3x - 2 = 0$$
$$3x = 2$$
$$x = \frac{2}{3}$$

The x intercept of f is $\frac{2}{3}$. The y intercept of f is $f(0) = \frac{3(0) - 2}{4(0) - 5} = \frac{2}{5}$.

23. The rational expression $\frac{4x}{(x - 2)^2}$ is defined everywhere except at the zero of the denominator:

$$(x - 2)^2 = 0$$
$$x - 2 = 0$$
$$x = 2$$

The domain of f is $\{x \mid x \neq 2\}$.
A rational expression is 0 if and only if the numerator is zero:

$$4x = 0$$
$$x = 0$$

The x intercept of f is 0.

The y intercept of f is $f(0) = \frac{4(0)}{(0 - 2)^2} = 0$.

25. The rational expression $\frac{x^2 - 16}{x^2 - 9}$ is defined everywhere except at the zeros of the denominator:

$$x^2 - 9 = 0$$
$$(x - 3)(x + 3) = 0$$
$$x = 3 \text{ or } -3$$

The domain of f is $\{x \mid x \neq -3, 3\}$.
A rational expression is 0 if and only if the numerator is zero:

$$x^2 - 16 = 0$$
$$(x - 4)(x + 4) = 0$$
$$x = 4 \text{ or } -4$$

The x intercepts of f are ± 4.

The y intercept of f is $f(0) = \frac{0^2 - 16}{0^2 - 9} = \frac{16}{9}$.

27. The rational expression $\frac{x^2 + 7}{x^2 - 25}$ is defined everywhere except at the zeros of the denominator:

$$x^2 - 25 = 0$$
$$(x - 5)(x + 5) = 0$$
$$x = 5 \text{ or } -5$$

The domain of f is $\{x \mid x \neq -5, 5\}$.
A rational expression is 0 if and only if the numerator is zero.
$x^2 + 7 = 0$ has no real solutions, hence f has no x intercept.

The y intercept of f is $f(0) = \frac{0^2 + 7}{0^2 - 25} = -\frac{7}{25}$.

29. (A) For $-1 \le x < 0$, $f(x) = x + 1$, so $f(-1) = -1 + 1 = 0$.
For $0 \le x \le 1$, $f(x) = -x + 1$, so $f(0) = -0 + 1 = 1$ and $f(1) = -1 + 1 = 0$.

(B)

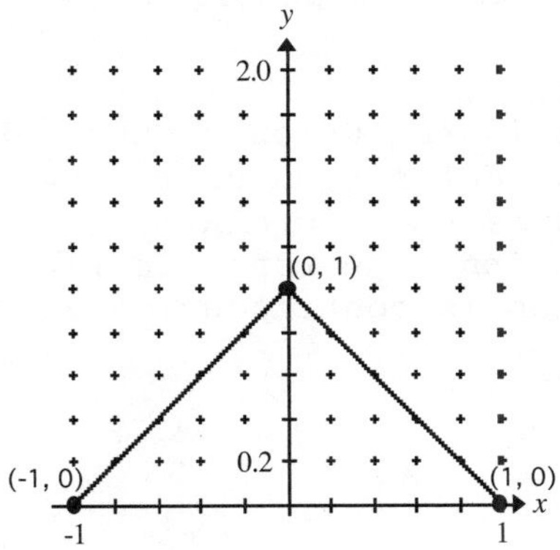

(C) The domain is the union of the intervals used in the definition of f: $[-1, 1]$.
From the graph, the range is $[0, 1]$.
The function is continuous on its domain.

31. (A) For $-3 \le x < -1$, $f(x) = -2$, so $f(-3) = -2$.
$f(-1)$ is not defined.
For $-1 < x \le 2$, $f(x) = 4$, so $f(2) = 4$.

(B)

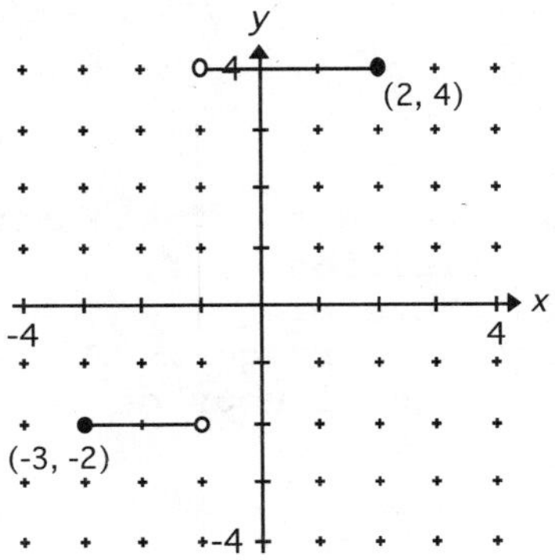

(C) The domain is the union of the intervals used in the definition of f:
$[-3, -1) \cup (-1, 2]$.
From the graph, the range is the set of numbers $\{-2, 4\}$.
The function is discontinuous at $x = -1$.

33. (A) For $x < -1$, $f(x) = x + 2$, so $f(-2) = -2 + 2 = 0$.
$f(-1)$ is not defined.
For $x > -1$, $f(x) = x - 2$, so $f(0) = 0 - 2 = -2$.

(B)

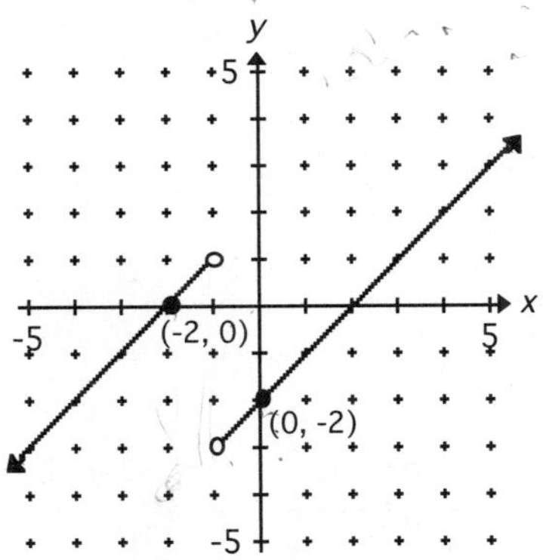

(C) The domain is the union of the intervals used in the definition of f:
$(-\infty, -1) \cup (-1, \infty)$.
From the graph, the range is R.
The function is discontinuous at $x = -1$.

35. (A) For $x < -2$, $f(x) = -2x - 6$, so $f(-3) = -2(-3) - 6 = 0$.
For $-2 \le x < 3$, $f(x) = -2$, so $f(-2) = f(0) = -2$.
For $x \ge 3$, $f(x) = 6x - 20$, so $f(3) = 6 \cdot 3 - 20 = -2$ and $f(4) = 6 \cdot 4 - 20 = 4$.

(B)

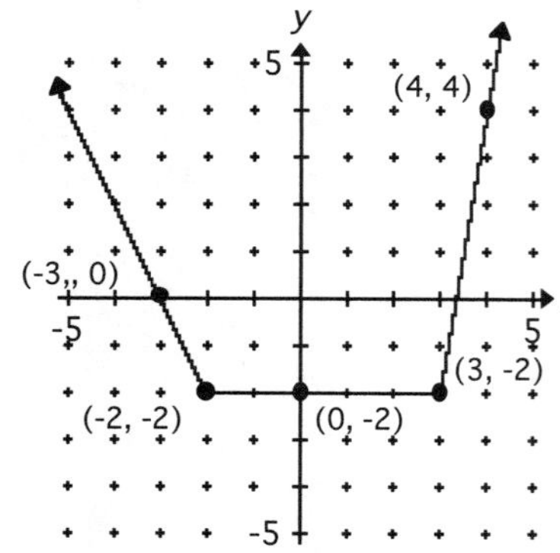

(C) The domain is the union of the intervals used in the definition of f.
The domain is therefore R.
From the graph, the range is $[-2, \infty)$.
The function is continuous on its domain.

37. (A) For $x < -2$, $f(x) = \frac{5}{2}x + 6$, so $f(-3) = \frac{5}{2}(-3) + 6 = -\frac{3}{2}$.
For $-2 \le x \le 3$, $f(x) = 1$, so $f(-2) = f(0) = f(3) = 1$.
For $x > 3$, $f(x) = \frac{3}{2}x - \frac{7}{2}$, so $f(4) = \frac{3}{2}(4) - \frac{7}{2} = \frac{5}{2}$.

(B)

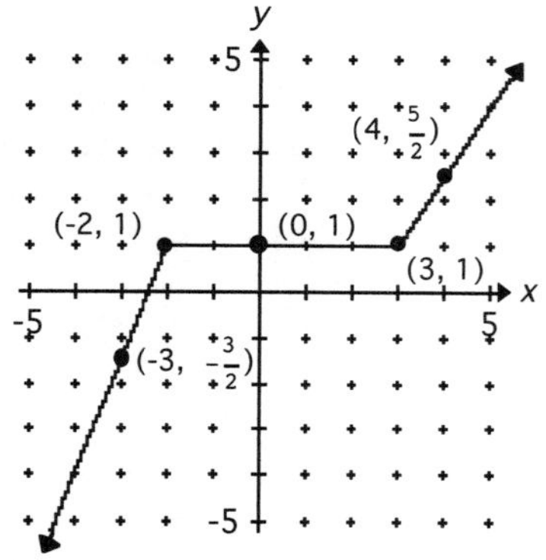

(C) The domain is the union of the intervals used in the definition of f.
The domain is therefore R.
From the graph, the range is R.
The function is continuous on its domain.

39. (A) For $x < 0$, $f(x) = \frac{2}{3}x + 4$, so $f(-1) = \frac{2}{3}(-1) + 4 = \frac{10}{3}$.
$f(0)$ is not defined.
For $0 < x < 2$, $f(x) = -\frac{1}{2}x + 3$, so $f(1) = -\frac{1}{2}(1) + 3 = \frac{5}{2}$.
$f(2)$ is not defined.
For $x > 2$, $f(x) = -\frac{1}{2}x$, so $f(3) = -\frac{1}{2}(3) = -\frac{3}{2}$.

(B)

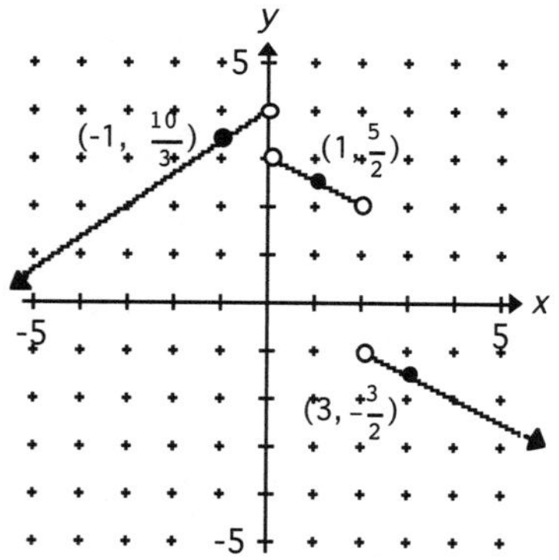

(C) The domain is the union of the intervals used in the definition of f.
$(-\infty, 0) \cup (0, 2) \cup (2, \infty)$.
From the graph, the range is $(-\infty, 4)$.
The function is discontinuous at $x = 0$ and $x = 2$.

41. For $x \le 0$, the graph is a line with slope $\frac{1 - 3}{0 - (-2)} = -1$ and y intercept 1 that is, $y = -x + 1$.

For $x > 0$, the graph is a line with slope $\frac{(-3) - (-1)}{2 - 0} = -1$ and y intercept -1, that is, $y = -x - 1$.

Therefore,

$$f(x) = \begin{cases} -x + 1 & \text{if } x \le 0 \\ -x - 1 & \text{if } x > 0 \end{cases}$$

43. For $x \le -1$, the graph is the horizontal line $y = 3$.

For $x > -1$, the graph is a line with slope $\frac{-3 - 3}{1 - (-1)} = -3$ passing through $(1, -3)$. The point-slope form yields $y - (-3) = -3(x - 1)$ or $y = -3x$.
Therefore,

$$f(x) = \begin{cases} 3 & \text{if } x \le -1 \\ -3x & \text{if } x > -1 \end{cases}$$

An equally valid solution would be

$$f(x) = \begin{cases} 3 & \text{if } x < -1 \\ -3x & \text{if } x \ge -1 \end{cases}$$

45. For $x < -2$, the graph is the horizontal line $y = 3$. The function is not defined at $x = -2$ or at $x = 1$.

For $-2 < x < 1$, the graph is a line with slope $\frac{(-4) - 2}{1 - (-2)} = -2$ which would, if extended, pass through $(-2, 2)$. The point-slope form yields $y - 2 = -2[x - (-2)]$, that is, $y = -2x - 2$.
For $x > 1$, the graph is the horizontal line $y = -1$.
Therefore,

$$f(x) = \begin{cases} 3 & \text{if } x < -2 \\ -2x - 2 & \text{if } -2 < x < 1 \\ -1 & \text{if } x > 1 \end{cases}$$

47. If $x < 0$, then $|x| = -x$ and $f(x) = 1 + |x| = 1 + (-x) = 1 - x$.
If $x \ge 0$, then $|x| = x$ and $f(x) = 1 + x$.
Therefore,

$$f(x) = \begin{cases} 1 - x & \text{if } x < 0 \\ 1 + x & \text{if } x \ge 0 \end{cases}$$

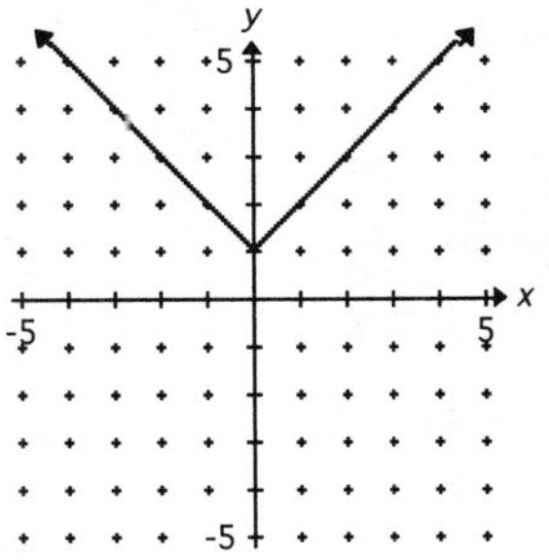

The function is defined for all real numbers; the domain is R. From the graph, the range is $[1, \infty)$. The function is continuous on its domain.

49. If $x - 2 < 0$, that is, $x < 2$, $f(x) = |x - 2| = -(-x - 2) = -x + 2$.
If $x - 2 \ge 0$, that is, $x \ge 2$,
$f(x) = |x - 2| = x - 2$.

Therefore,

$$f(x) = \begin{cases} -x + 2 & \text{if } x < 2 \\ x - 2 & \text{if } x \ge 2 \end{cases}$$

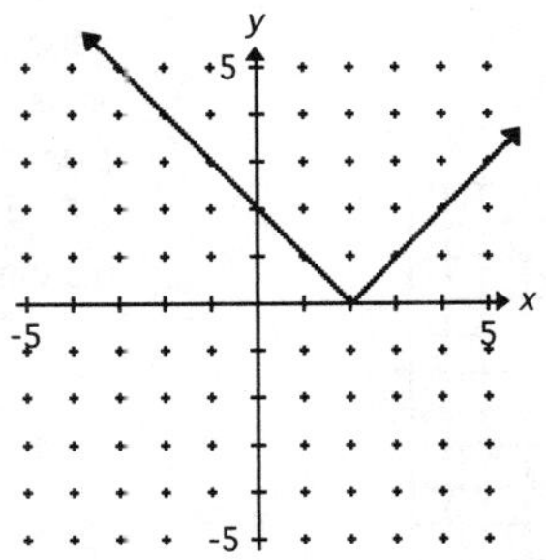

The function is defined for all real numbers; the domain is R. From the graph, the range is $[0, \infty)$. The function is continuous on its domain.

51. If $x + 3 < 0$, that is, $x < -3$, $|x + 3| = -(x + 3)$ and

$$f(x) = \frac{|x+3|}{x+3} = \frac{-(x+3)}{x+3} = -1.$$

If $x + 3 = 0$, that is, $x = -3$, the function is not defined.

If $x + 3 > 0$, that is, $x > -3$, $|x + 3| = x + 3$ and $f(x) = \frac{|x+3|}{x+3} = \frac{x+3}{x+3} = 1.$

Therefore,

$$f(x) = \begin{cases} -1 & \text{if } x < -3 \\ 1 & \text{if } x > -3 \end{cases}$$

The domain is $\{x \mid x \neq -3\}$ or $(-\infty, -3) \cup (-3, \infty)$. The range is the set of numbers $\{-1, 1\}$. The function is discontinuous at $x = -3$.

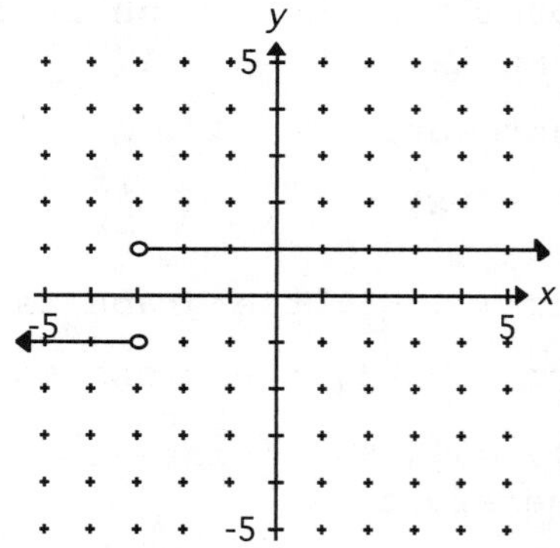

53. If $x < 0$, $|x| = -x$ and $f(x) = |x| + x = -x + x = 0$.

If $x \geq 0$, $|x| = x$ and $f(x) = x + x = 2x$.

Therefore,

$$f(x) = \begin{cases} 0 & \text{if } x < 0 \\ 2x & \text{if } x \geq 0 \end{cases}$$

The function is defined for all real numbers; the domain is R. From the graph, the range is $[0, \infty)$. The function is continuous on its domain.

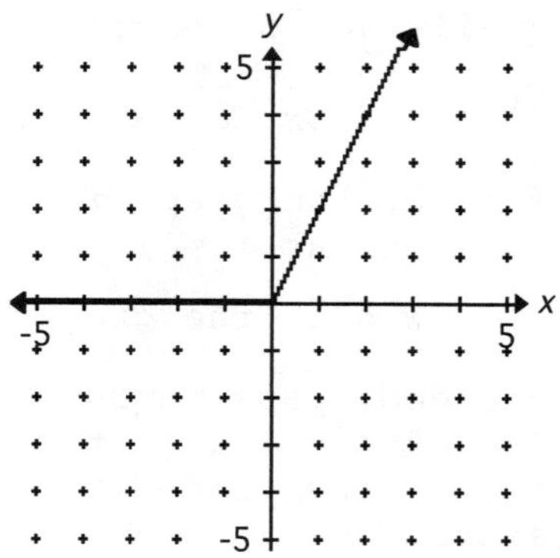

55. If $x - 1 < 0$, that is, $x < 1$, $|x - 1| = -(x - 1)$ and $f(x) = x + \frac{|x-1|}{x-1}$

$= x + \frac{-(x-1)}{x-1} = x - 1.$

If $x - 1 \geq 0$, that is, $x = 1$, the function is not defined.

If $x - 1 > 0$, that is, $x > 1$, $|x - 1| = x - 1$ and $f(x) = x + \frac{|x-1|}{x-1} = x + \frac{x-1}{x-1} = x + 1.$

Therefore,

$$f(x) = \begin{cases} x - 1 & \text{if } x < 1 \\ x + 1 & \text{if } x > 1 \end{cases}$$

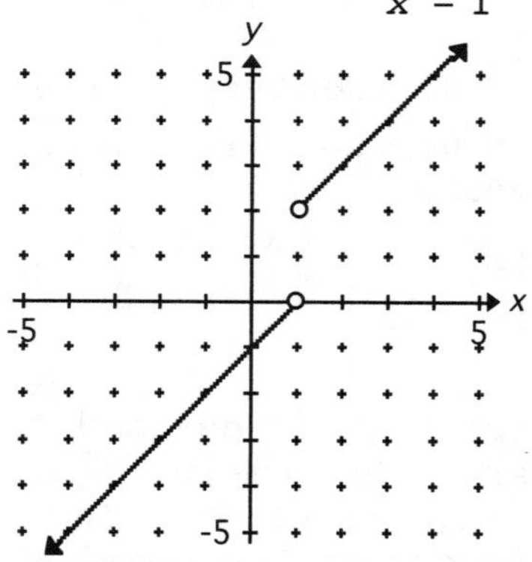

The domain is $\{x \mid x \neq 1\}$ or $(-\infty, 1) \cup (1, \infty)$. From the graph, the range is $(-\infty, 0) \cup (2, \infty)$. The function is discontinuous at $x = 1$.

57. (A) One possible answer:

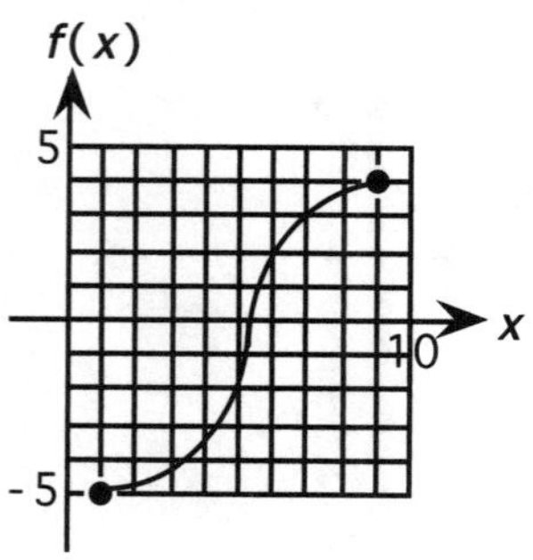

(B) This graph crosses the x axis once. To meet the conditions specified a graph must cross the x axis exactly once. If it crossed more times the function would have to be decreasing somewhere; if it did not cross at all the function would have to be discontinuous somewhere.

59. (A) One possible answer:

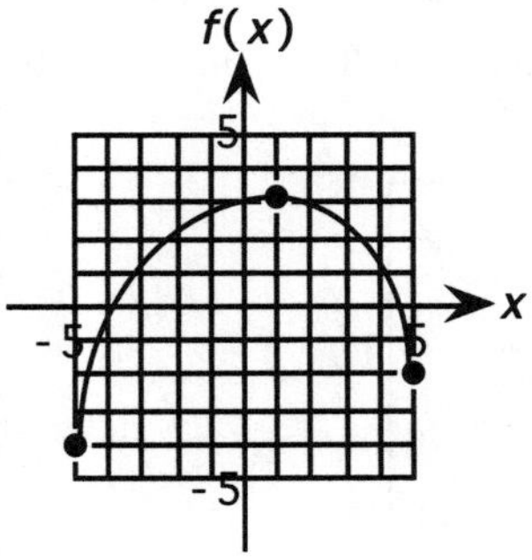

(B) This graph crosses the x axis twice. To meet the conditions specified a graph must cross the x axis at least twice. If it crossed fewer times the function would have to be discontinuous somewhere. However, the graph could cross more times; in fact there is no upper limit on the number of times it can cross the x axis.

61. Graphs of f and g

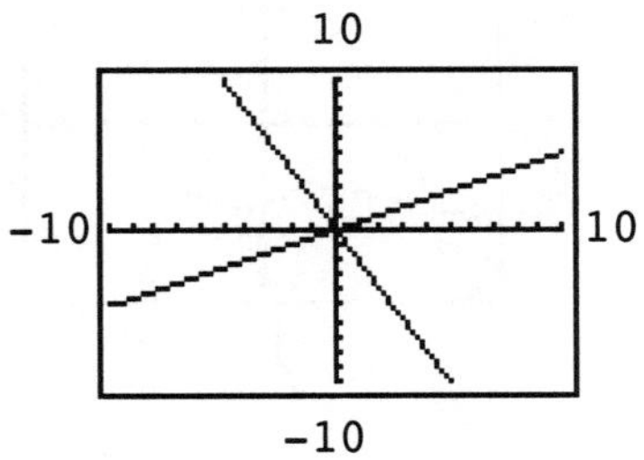

Graph of $m(x) = 0.5[-2x + 0.5x + |-2x - 0.5x|]$
$= 0.5[-1.5x + |-2.5x|]$

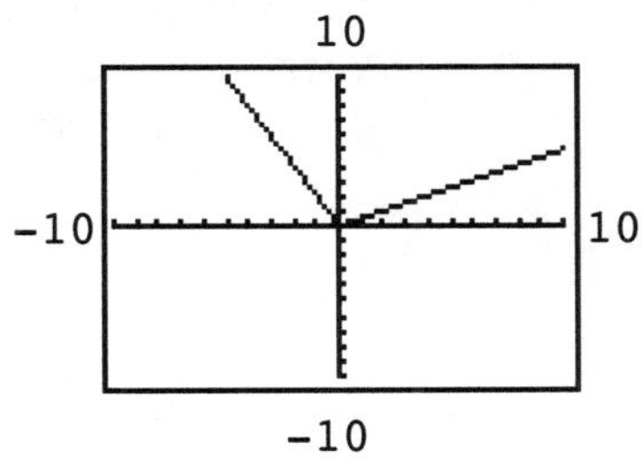

Graph of $n(x) = 0.5[-2x + 0.5x - |-2x - 0.5x|] = 0.5[-1.5x - |-2.5x|]$

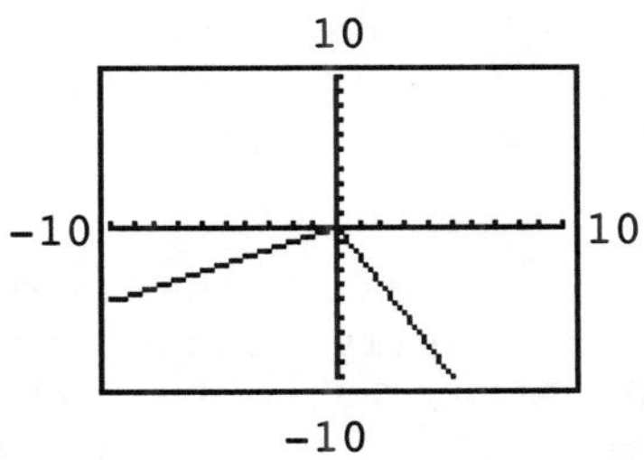

63. Graphs of f and g

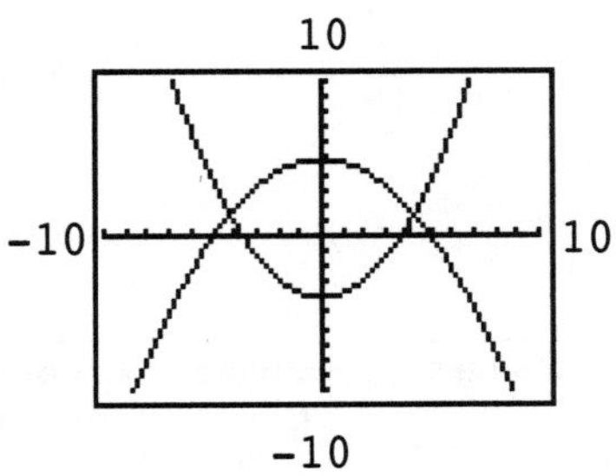

Graph of $m(x) = 0.5[5 - 0.2x^2 + 0.3x^2 - 4$
$+ |5 - 0.2x^2 - (0.3x^2 - 4)|]$
$= 0.5[1 + 0.1x^2 + |9 - 0.5x^2|]$

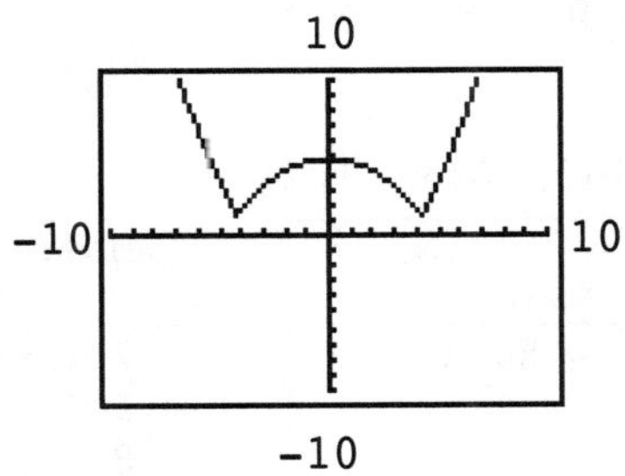

Graph of $n(x) = 0.5[5 - 0.2x^2 + 0.3x^2 - 4$
$- |5 - 0.2x^2 - (0.3x^2 - 4)|]$
$= 0.5[1 + 0.1x^2 - |9 - 0.5x^2|]$

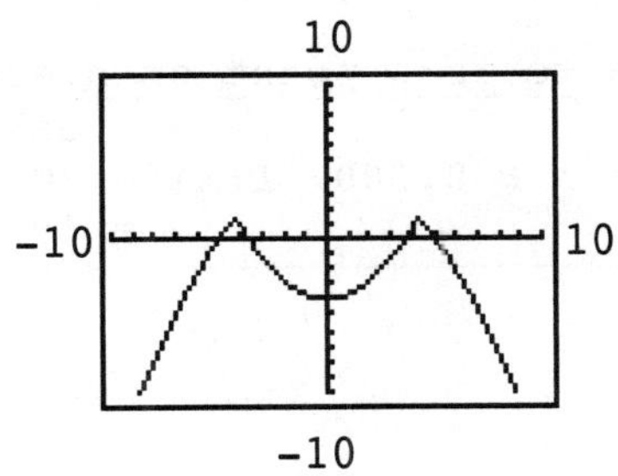

65. Graphs of f and g

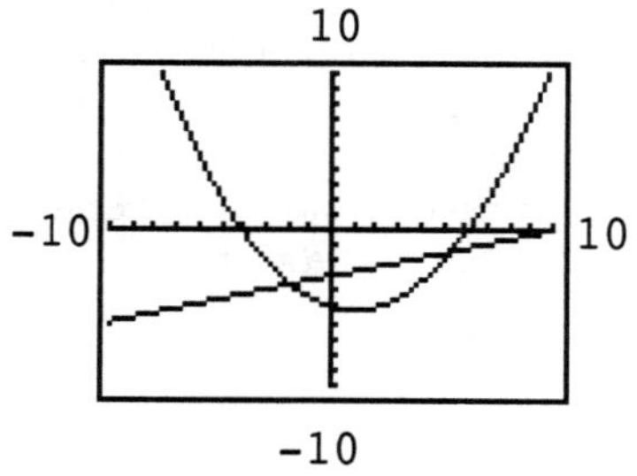

Graph of $m(x) = 0.5[0.2x^2 - 0.4x - 5 + 0.3x - 3 + |0.2x^2 - 0.4x - 5 - (0.3x - 3)|]$

$= 0.5[0.2x^2 - 0.1x - 8 + |0.2x^2 - 0.7x - 2|]$

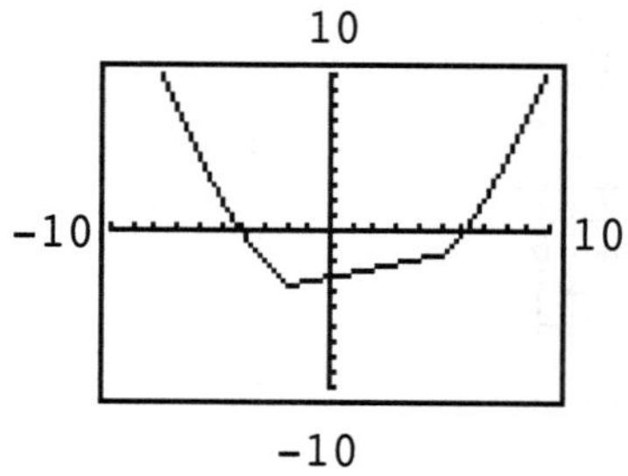

Graph of $n(x) = 0.5[0.2x^2 - 0.4x - 5 + 0.3x - 3 - |0.2x^2 - 0.4x - 5 - (0.3x - 3)|$

$= 0.5[0.2x^2 - 0.1x - 8 - |0.2x^2 - 0.7x - 2|]$

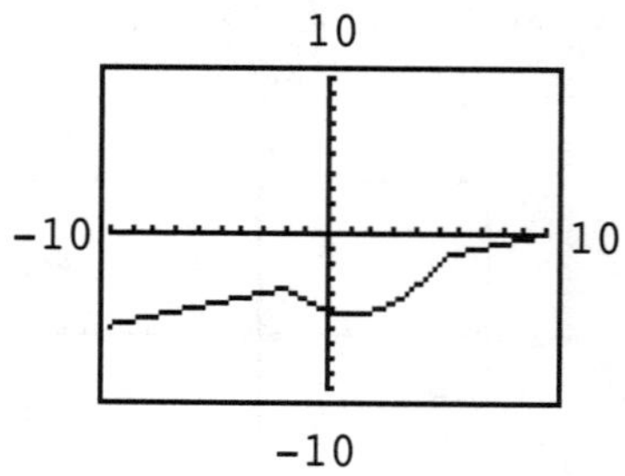

67. The graphs of $m(x)$ show that the value of $m(x)$ is always the larger of the two values for $f(x)$ and $g(x)$. In other words, $m(x) = \max[f(x), g(x)]$. This is confirmed in Section 1-3, Problem 87 where it was shown that
$\max(a, b) = 0.5[a + b + |a - b|]$,
hence $\max[f(x), g(x)] = 0.5[f(x) + g(x) + |f(x) - g(x)|] = m(x)$.

69. Since 100 miles are included, only the daily charge of \$32 applies for mileage between 0 and 100. So if $R(x)$ is the daily cost of rental where x is miles driven, $R(x) = 32$ if $0 \le x \le 100$. After 100 miles, the charge is an extra \$0.16 for each mile: the mileage charge will be 0.16 times the number of miles over 100 which is $x - 100$. So the mileage charge is $0.16(x - 100)$ or $0.16x - 16$. The \$32 charge still applies so when $x \ge 100$ the rental charge is $32 + 0.16x - 16$, or $16 + 0.16x$.

$$R(x) = \begin{cases} 32 & 0 \le x \le 100 \\ 16 + 0.16x & x > 100 \end{cases}$$

71. If $0 \le x \le 3{,}000$, $E(x) = 200$

		Base Salary	+	Commission on Sales Over \$3,000
If \$3,000 < x < 8,000, $E(x)$	=	200	+	$0.04(x - 3000)$
	=	200	+	$0.04x - 120$
	=	80	+	$0.04x$

Common Error:
Commission is not $0.04x$ (4% of sales) not is it $0.04x + 20$ (base salary plus 4% of sales).

There is a point of discontinuity at $x = 8{,}000$.

		Salary	+	Bonus
If $x \ge 8{,}000$. $E(x)$	=	$80 + 0.04x$	+	100
	=	180	+	$0.04x$

Summarizing, $E(x) = \begin{cases} 200 & \text{if } 0 \le x \le 3{,}000 \\ 80 + 0.04x & \text{if } 3{,}000 < x < 8{,}000 \\ 180 + 0.04x & \text{if } 8{,}000 \le x \end{cases}$

$E(5{,}750) = 80 + 0.04(5{,}750) = \$310$
$E(9{,}200) = 180 + 0.04(9{,}200) = \$548$

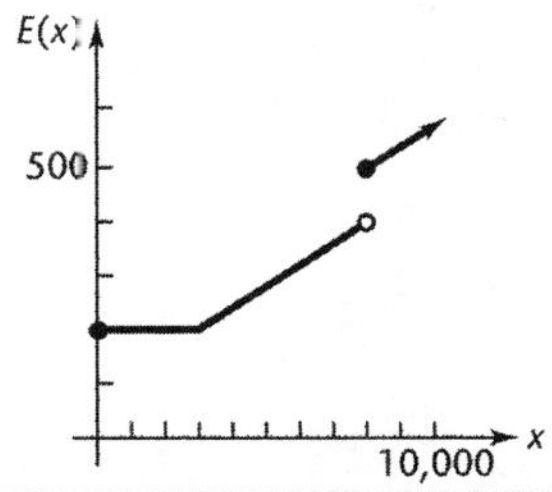

x	$y = 200$	x	$y = 80 + 0.04x$	x	$y = 180 + 0.04x$
0	200	3,000	200	8,000	500
2,000	200	7,000	360	10,000	580

73.

$$\left.\begin{array}{rcrcrcr} f(4) &=& 10[\![0.5 + 0.4]\!] &=& 10(0) &=& 0 \\ f(-4) &=& 10[\![0.5 - 0.4]\!] &=& 10(0) &=& 0 \\ f(6) &=& 10[\![0.5 + 0.6]\!] &=& 10(1) &=& 10 \\ f(-6) &=& 10[\![0.5 - 0.6]\!] &=& 10(-1) &=& -10 \\ f(24) &=& 10[\![0.5 + 2.4]\!] &=& 10(2) &=& 20 \\ f(25) &=& 10[\![0.5 + 2.5]\!] &=& 10(3) &=& 30 \\ f(247) &=& 10[\![0.5 + 24.7]\!] &=& 10(25) &=& 250 \\ f(-243) &=& 10[\![0.5 + 24.3]\!] &=& 10(-24) &=& -240 \\ f(-245) &=& 10[\![0.5 - 24.5]\!] &=& 10(-24) &=& -240 \\ f(-246) &=& 10[\![0.5 - 24.6]\!] &=& 10(-25) &=& -250 \end{array}\right\} f \text{ rounds numbers to the tens place}$$

75. Since $f(x) = [\![10x + 0.5]\!]/10$ rounds numbers to the nearest tenth, (see text example 7) we try $[\![100x + 0.5]\!]/100 = f(x)$ to round to the nearest hundredth.

$f(3.274) = [\![327.9]\!]/100 = 3.27$
$f(7.846) = [\![785.1]\!]/100 = 7.85$
$f(-2.8783) = [\![-287.33]\!]/100 = -2.88$

A few examples suffice to convince us that this is probably correct.
(A proof would be out of place in this book.)
$f(x) = [\![100x + 0.5]\!]/100.$

77. (A) $C(x) = \begin{cases} 15 & 0 < x \le 1 \\ 18 & 1 < x \le 2 \\ 21 & 2 < x \le 3 \\ 24 & 3 < x \le 4 \\ 27 & 4 < x \le 5 \\ 30 & 5 < x \le 6 \end{cases}$

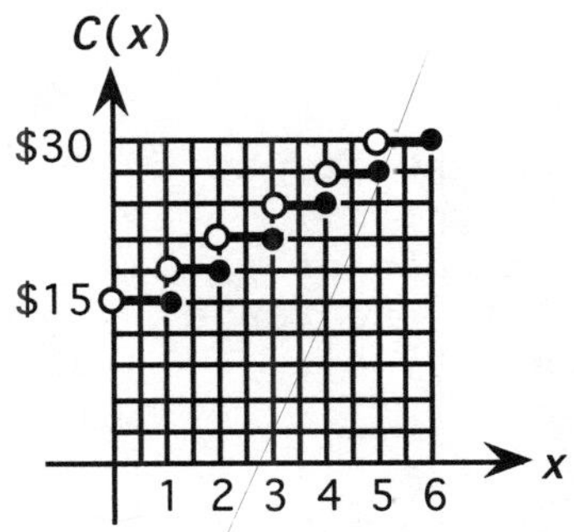

(B) The two functions appear to coincide, for example

$C(3.5) = 24 \quad f(3.5) = 15 + 3[\![3.5]\!] = 15 + 3 \cdot 3 = 24$

However,

$C(1) = 15 \quad f(1) = 15 + 3[\![1]\!] = 15 \cdot 3 \cdot 1 = 18$

The functions are not the same, therefore. In fact, $f(x) \ne C(x)$ at $x = 1, 2, 3, 4, 5, 6$.

79. On the interval [0, 10,000], the tax is $0.03x$.
On the interval (10,000, ∞), the tax is $0.03(10,000) + 0.05(x - 10,000)$ or $300 + 0.05x - 500$, that is, $0.05x - 200$.
Combining the intervals with the above linear expressions, we have

$$T(x) = \begin{cases} 0.03x & \text{if } 0 \le x \le 10,000 \\ 0.05x - 200 & \text{if } x > 10,000 \end{cases}$$

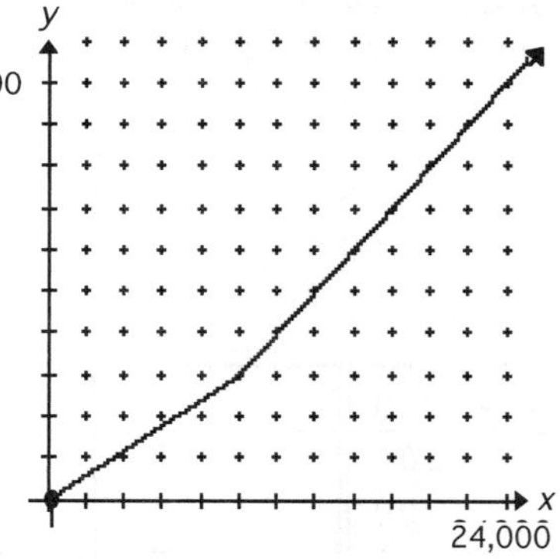

81. On the interval [0, 19,440] the tax is $0.0535x$.
On the interval (19,440, 63,860] the tax is
$1,040.04 + 0.0705(x - 19,440) = 0.0705x - 330.48$
On the interval (63,860, ∞) the tax is
$4,171.65 + 0.0785(x - 63,860) = 0.0785x - 841.36$
Combining the intervals with the above linear expressions, we have

$$T(x) = \begin{cases} 0.0535x & \text{if } 0 \le x \le 19,440 \\ 0.0705x - 330.48 & \text{if } 19,440 < x \le 63,860 \\ 0.0785x - 841.36 & \text{if } x > 63,860 \end{cases}$$

Applying the piecewise definition of T, we have
$T(10,000) = 0.0535(10,000) = \$535$
$T(30,000) = 0.0705(30,000) - 330.48 = \$1,784.52$
$T(100,000) = 0.0785(100,000) - 841.36 = \$7,008.64$

Section 3-3

1. Domain: Since $x \ge 0$, the domain is $[0, \infty)$
Range: Since the range of $f(x) = \sqrt{x}$ is $y \ge 0$, for $h(x) = -\sqrt{x}$, $y \le 0$. Thus, the range of h is $(-\infty, 0]$.

3. Domain: R
Range: Since the range of $f(x) = x^2$ is $y \ge 0$, for $g(x) = -2x^2$, $y \le 0$. Thus, the range of g is $(-\infty, 0]$.

5. Domain: R; Range: R

7. The graph of $y = h(x)$ is the graph of $y = f(x)$ shifted up 2 units. The domain of h is the domain of f, $[-2, 2]$. The range of h is the range of f shifted up 2 units, $[0, 4]$.

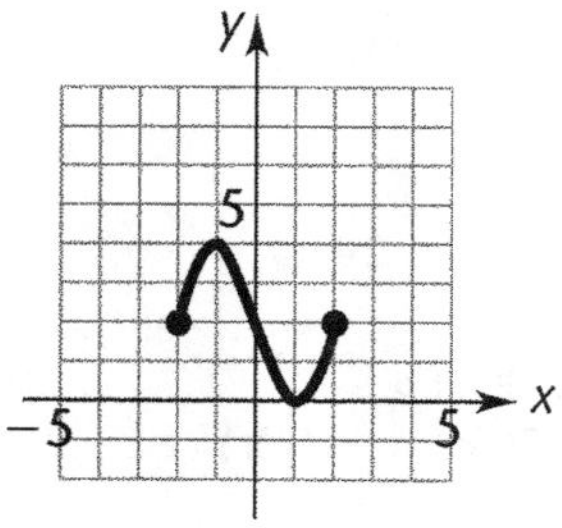

9. The graph of $y = h(x)$ is the graph of $y = g(x)$ shifted up 2 units. The domain of h is the domain of g, $[-2, 2]$. The range of h is the range of g shifted up 2 units, $[1, 3]$.

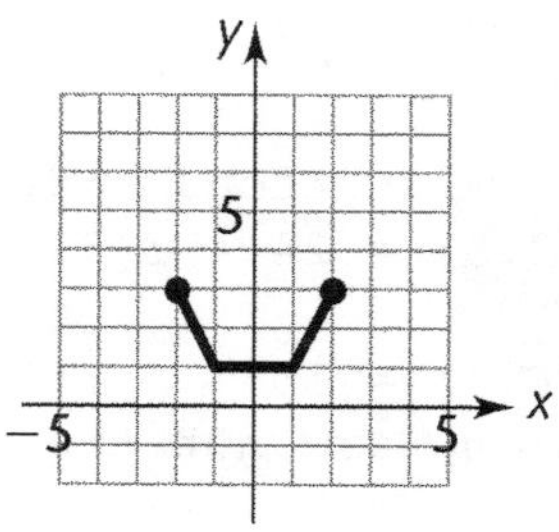

11. The graph of h is the graph of f shifted right 2 units. The domain of h is the domain of f shifted right 2 units, [0, 4]. The range of h is the range of f, [−2, 2].

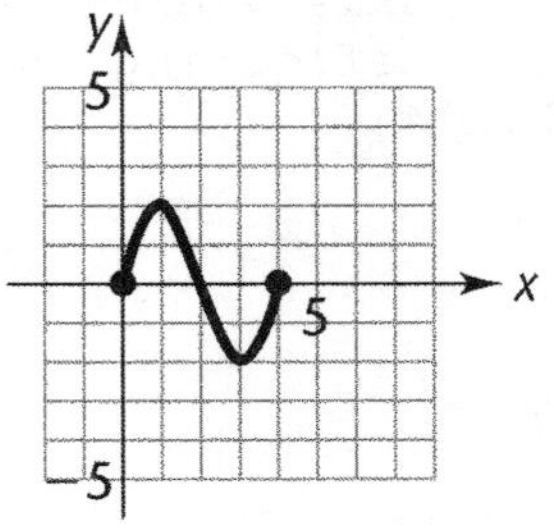

Common Error: $x - 2$ does not indicate shifting left.

13. The graph of h is the graph of g shifted left 2 units. The domain of h is the domain of g shifted left 2 units, [−4, 0]. The range of h is the range of g, [−1, 1].

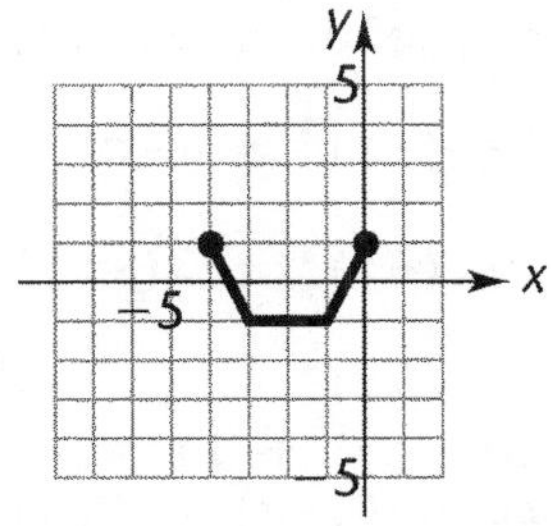

15. The graph of h is the graph of f reflected through the x axis. The domain of h is the domain of f, [−2, 2]. The range of h is the range of f reflected through the x axis, [−2, 2].

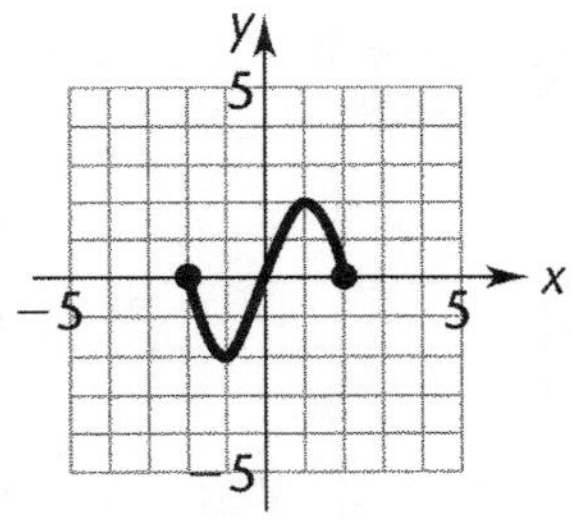

17. The graph of h is the graph of g stretched vertically by a factor of 2. The domain of h is the domain of g, [−2, 2]. The range of f is the range of g stretched vertically by a factor of 2, [−2, 2].

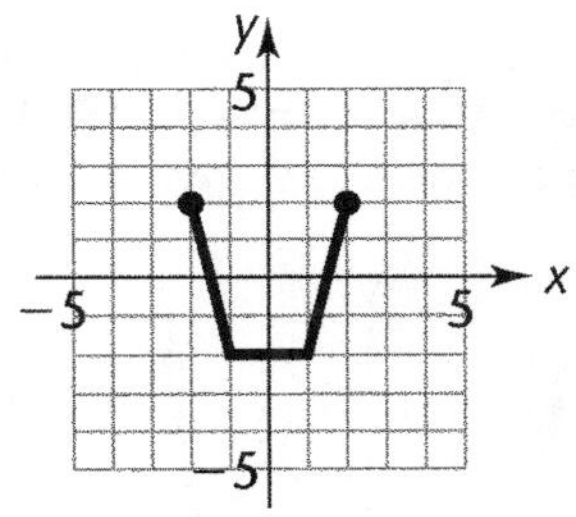

19. The graph of h is the graph of g shrunk horizontally by a factor of $\frac{1}{2}$. The domain of h is the domain of g shrunk horizontally by a factor of $\frac{1}{2}$, [−1, 1]. The range of h is the range of g, [−1, 1].

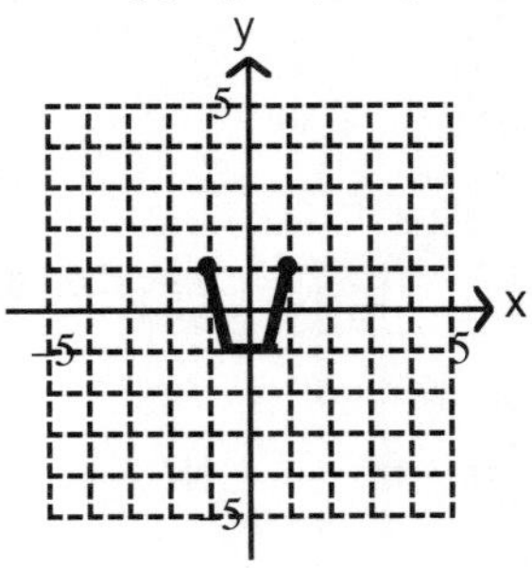

21. The graph of h is the graph of f reflected through the y axis. The domain of h is the domain of f reflected through the y axis, [−2, 2]. The range of h is the range of f, [−2, 2].

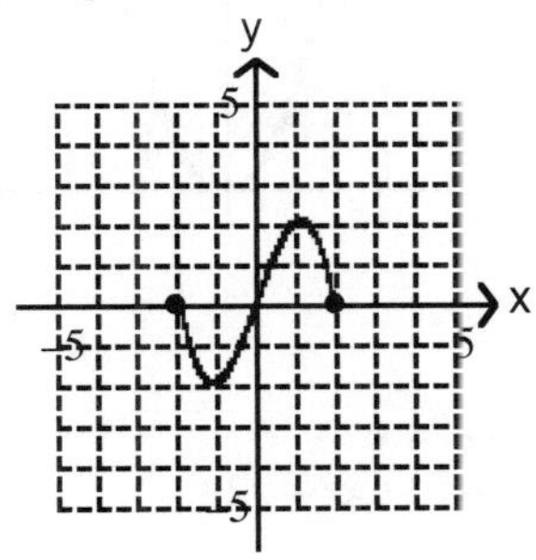

23. The graph of $y = f(x + 3)$ is the graph of $y = f(x)$ shifted left 3 units. Shifting (2, −4) left 3 units yields (−1, −4).

25. The graph of $y = f(x) + 1$ is the graph of $y = f(x)$ shifted up 1 unit. Shifting (2, −4) up 1 unit yields (2, −3).

27. The graph of $y = 3f(x)$ is the graph of $y = f(x)$ stretched vertically by a factor of 3. This multiplies each y coordinate by 3, so (2, −4) would become (2, −12).

29. The graph of $y = f(4x)$ is the graph of $y = f(x)$ shrunk horizontally by a factor of $\frac{1}{4}$. This multiples each x coordinate by $\frac{1}{4}$, so $(2, -4)$ would become $\left(\frac{1}{2}, -4\right)$.

31. The graph of $y = 2 - 3f(-x)$ is the graph of $y = f(x)$ reflected through the y axis, stretched vertically by a factor of 3, reflected through the x axis, and shifted up 2 units. Reflecting $(2, -4)$ through the y axis yields $(-2, -4)$. Stretching vertically by a factor of 3 yields $(-2, -12)$. Reflecting through the x axis yields $(-2, 12)$. Shifting up 2 units yields $(-2, 14)$.

33. $g(-x) = (-x)^3 + (-x) = -x^3 - x = -(x^3 + x) = -g(x)$. Odd

35. $m(-x) = (-x)^4 + 3(-x)^2 = x^4 = 3x^2 = m(x)$. Even

37. $F(-x) = (-x)^5 + 1 = -x^5 + 1$
$-F(x) = -(x^5 + 1) = -x^5 - 1$
Therefore $F(-x) \neq F(x)$. $F(-x) \neq -F(x)$. $F(x)$ is neither even nor odd.

39. $G(-x) = (-x)^4 + 2 = x^4 + 2 = G(x)$. Even

41. $q(-x) = (-x)^2 + (-x) - 3 = x^2 - x - 3$.
$-q(x) = -(x^2 + x - 3) = -x^2 - x + 3$
Therefore $q(-x) \neq q(x)$. $q(-x) \neq -q(x)$. $q(x)$ is neither even nor odd.

43. The graph of $y = x^2$ is stretched vertically by a factor of 4 (**or** shrunk horizontally by a factor of 2.)

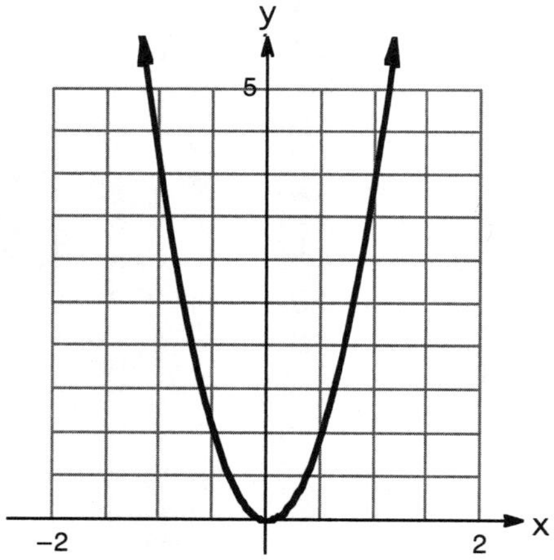

45. The graph of $y = |x|$ is shifted left 2 units.

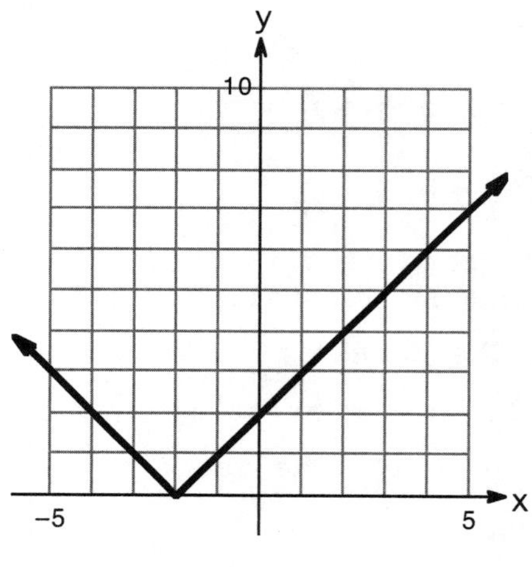

47. $m(x) = -|4x - 8|$
$= -|4(x - 2)|$

The graph of $y = |x|$ is shifted right 2 units, stretched vertically by a factor of 4, and reflected through the x axis.

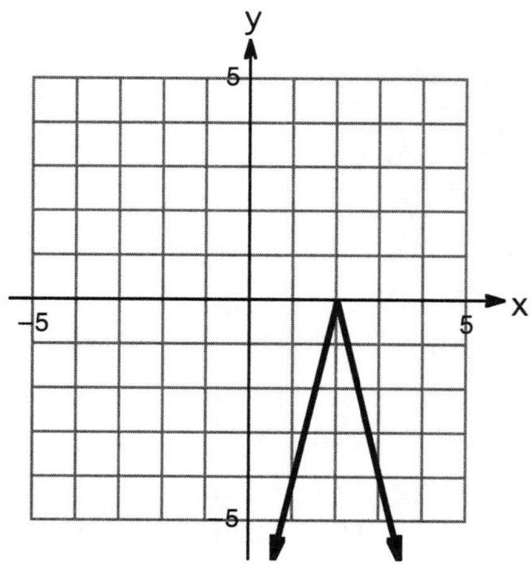

49. The graph of $y = \sqrt{x}$ is reflected through the x axis and shifted up 3 units.

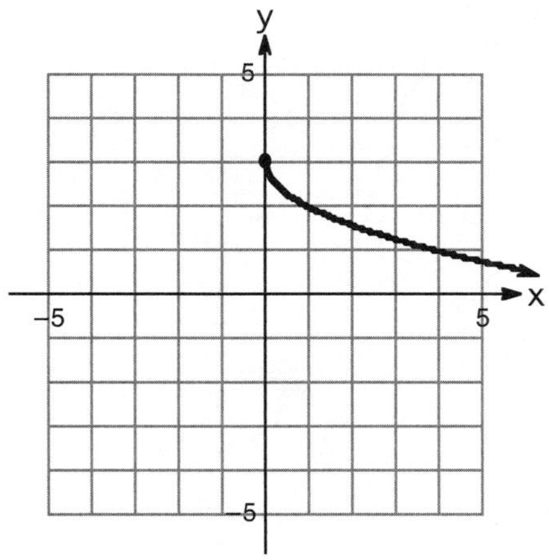

51. The graph of $y = \sqrt{x}$ is shifted right 1 unit, stretched vertically by a factor of 3, and shifted up 2 units.

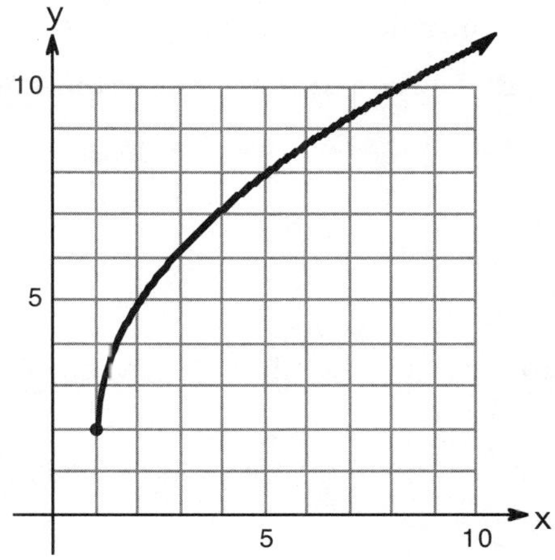

53. The graph of $y = x^2$ is shifted up 3 units.

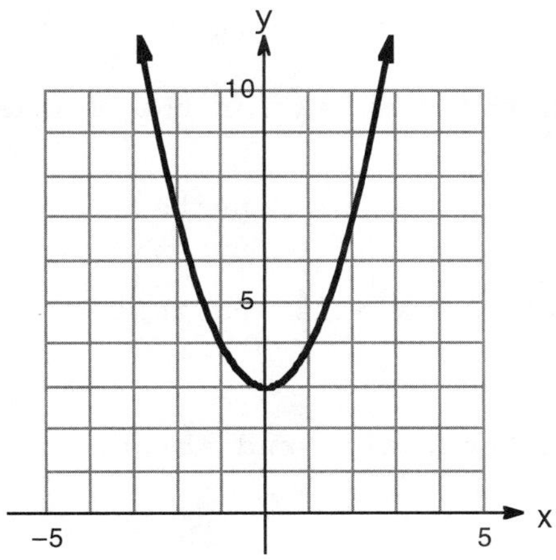

55. The graph of $y = x^2$ is stretched horizontally by a factor of 2, reflected through the x axis, and shifted up 1 unit.

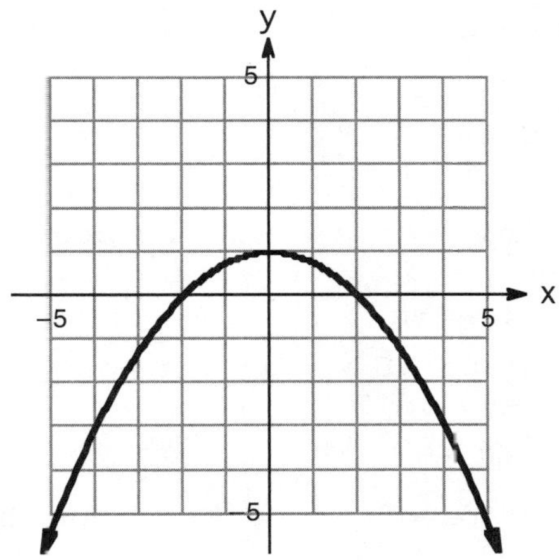

57. The graph of $y = x^3$ is stretched vertically by a factor of 2 and shifted up 1 unit.

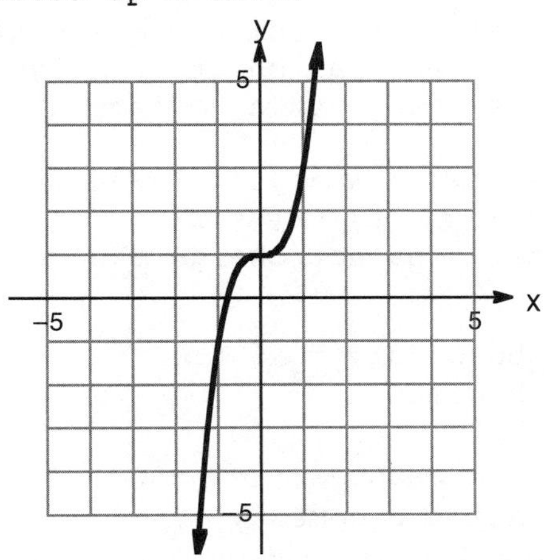

59. The graph of $y = x^2$ is shifted left 2 units.

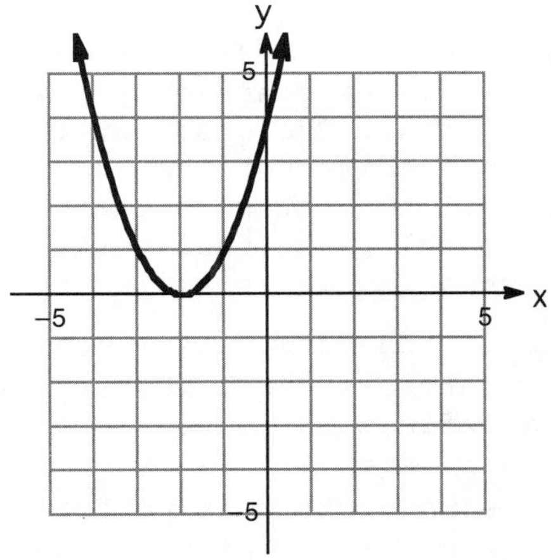

61. The graph of $y = x^2$ is shifted right 2 units, shrunk vertically by a factor of $\frac{1}{2}$, reflected through the x axis, and shifted up 4 units.

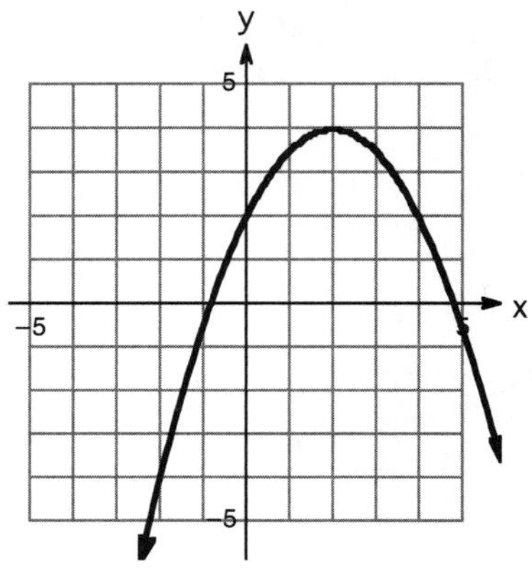

63. The graph of $y = x^2$ has been shifted 2 units right.
$y = (x - 2)^2$

65. The graph of $y = x^3$ has been shifted 2 units down.
$y = x^3 - 2$

67. The graph of $y = |x|$ has been shrunk vertically by a factor of $\frac{1}{4}$.
$y = \frac{1}{4}|x|$ or $y = 0.25|x|$

69. The graph of $y = x^3$ has been reflected through the y axis, (or the x axis).
$y = -x^3$

71. The graph of $y = |x|$ has been shifted left 2 units and up 2 units.
$y = |x + 2| + 2$

73. The graph of $y = \sqrt{x}$ has been reflected through the x axis and shifted up 4 units.
$y = 4 - \sqrt{x}$

75. The graph of $y = x^2$ has been reflected through the x axis and shifted right 1 unit and up 4 units.
$y = 4 - (x - 1)^2$

77. The graph of $y = x^3$ has been shrunk vertically by a factor of $\frac{1}{2}$ and shifted right 3 units and up 1 unit.
$y = \frac{1}{2}(x - 3)^3 + 1$ or $y = 0.5(x - 3)^3 + 1$

79. The graph of $y = f(x - h)$ represents a horizontal shift from the graph of $y = f(x)$. The graph of $y = f(x) + k$ represents a vertical shift from the graph of $y = f(x)$. The graph of $y = f(x - h) + k$ represents both a horizontal and a vertical shift but the order does not matter:
Vertical first then horizontal: $y = f(x) \to y = f(x) + k \to y = f(x - h) + k$
Horizontal first then vertical: $y = f(x) \to y = f(x - h) \to y = f(x - h) + k$
The same result is achieved; reversing the order does not change the result.

81. Consider the graph of $y = x^2$.
If a vertical shift is performed the equation becomes $y = x^2 + k$.
If a reflection is now performed the equation becomes
$y = -(x^2 + k)$ or $y = -x^2 - k$.
If the reflection is performed first the equation becomes $y = -x^2$.
If the vertical shift is now performed the equation becomes $y = -x^2 + k$.
Since $y = -x^2 - k$ and $y = -x^2 + k$ differ (unless $k = 0$), reversing the order changes the result.

83. The graph of $y = f(x - h)$ represents a horizontal shift from the graph of $y = f(x)$. The graph of $y = -f(x)$ represents a reflection of the graph of $y = f(x)$ in the x axis. The graph of $y = -f(x - h)$ represents both a horizontal shift and a reflection but the order does not matter:

Shift first then reflection: $y = f(x) \to y = f(x - h) \to y = -f(x - h)$

Reflection first then shift: $y = f(x) \to y = -f(x) \to y = -f(x - h)$

The same result is achieved; reversing the order does not change the result.

85.

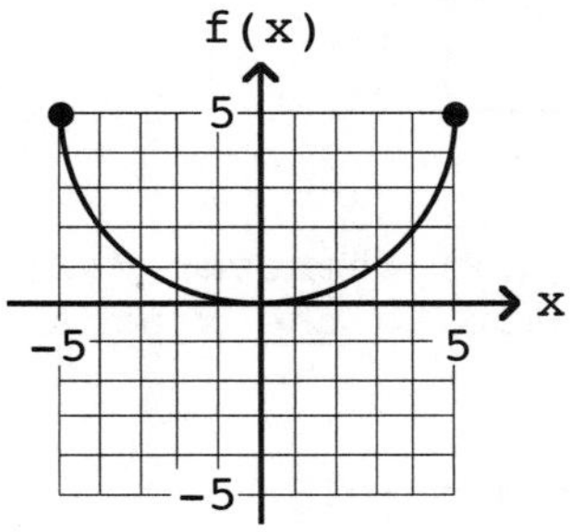

87.

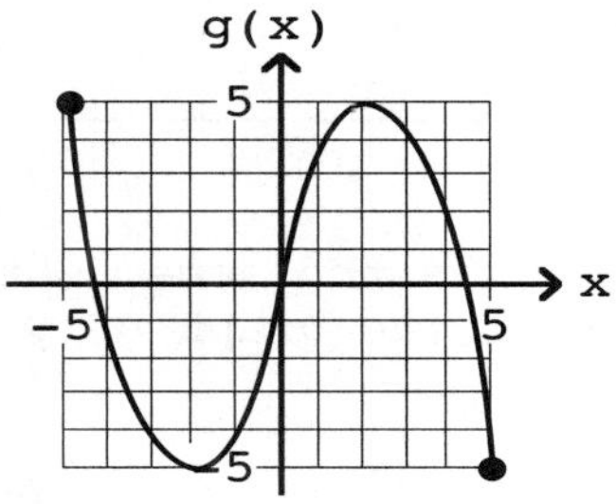

89. (A) $E(x) = \frac{1}{2}[f(x) + f(-x)]$

$E(-x) = \frac{1}{2}[f(-x) + f\{-(-x)\}] = \frac{1}{2}[f(-x) + f(x)] = \frac{1}{2}[f(x) + f(-x)] = E(x)$.

Thus, $E(x)$ is even.

(B) $O(x) = \frac{1}{2}[f(x) - f(-x)]$

$O(-x) = \frac{1}{2}[f(-x) - f\{-(-x)\}] = \frac{1}{2}[f(-x) - f(x)] = -\frac{1}{2}[f(x) - f(-x)] = -O(x)$.

Thus, $O(x)$ is odd.

(C)
$$\begin{aligned} E(x) + O(x) &= \frac{1}{2}[f(x) + f(-x)] + \frac{1}{2}[f(x) - f(-x)] \\ &= \frac{1}{2}f(x) + \frac{1}{2}f(-x) + \frac{1}{2}f(x) - \frac{1}{2}f(-x) \\ &= f(x) \end{aligned}$$

Conclusion: Any function can be written as the sum of two other functions, one even and the other odd.

91. Graph of $f(x)$

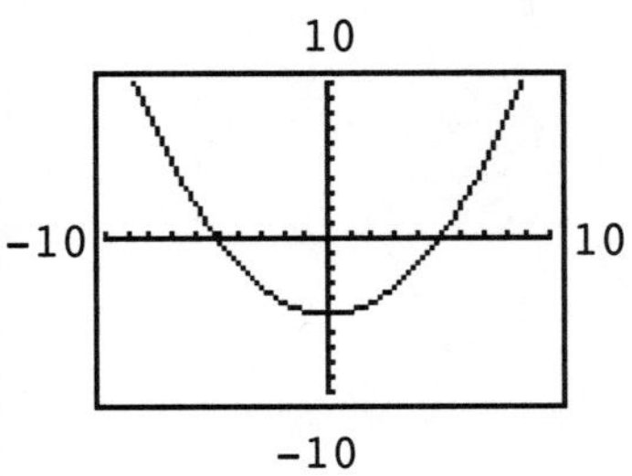

Graph of $|f(x)| = |0.2x^2 - 5|$

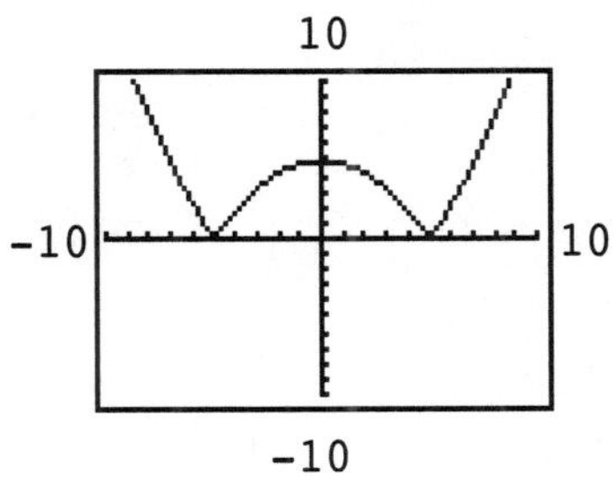

Graph of $-|f(x)| = -|0.2x^2 - 5|$

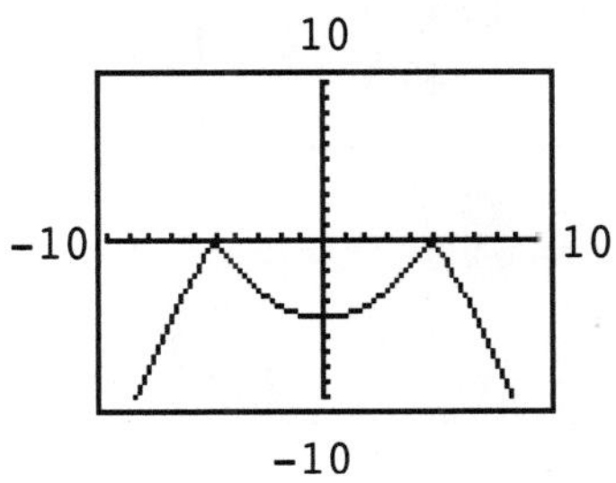

93. Graph of $f(x)$ | Graph of $|f(x)| = |4 - 0.1(x + 2)^3|$ | Graph of $-|f(x)| = -|4 - 0.1(x + 2)^3|$

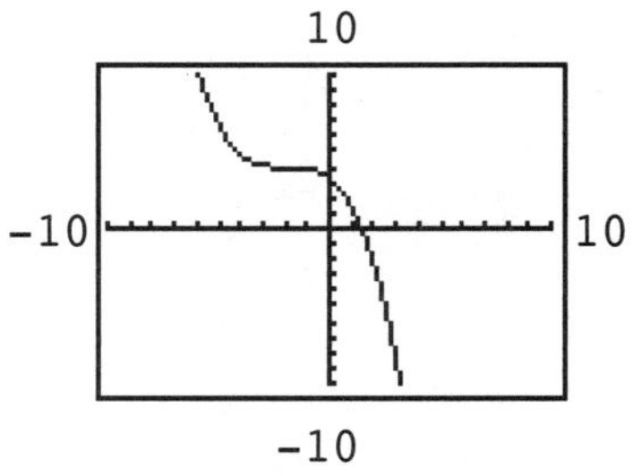

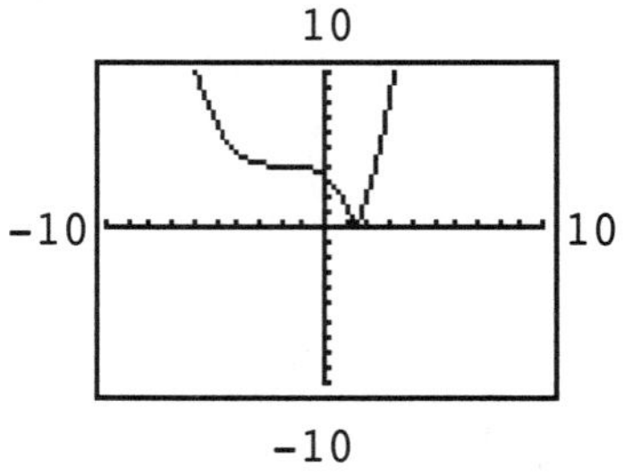

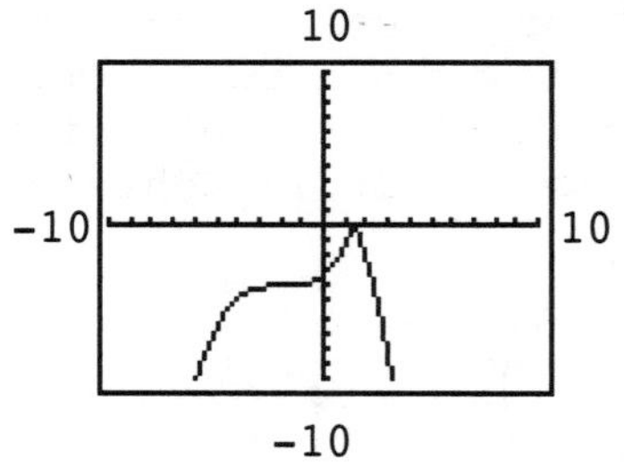

95. The graph of $y = |f(x)|$ is the same as the graph of $y = f(x)$ whenever $f(x) \geq 0$ and is the reflection of the graph of $y = f(x)$ with respect to the x axis whenever $f(x) < 0$.

97. The graph of the function $C(x) = 30{,}000 + f(x)$ is the same as the given graph of the function $f(x)$ shifted up 30,000 units ($).

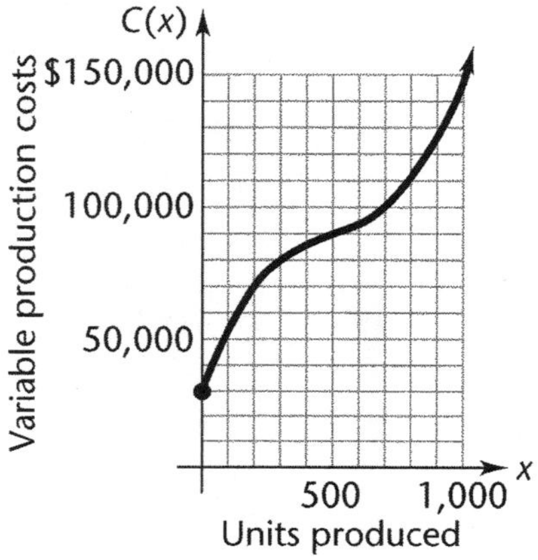

99. $y = 10 + 0.004(x - 10)^3$,
$y = 15 + 0.004(x - 10)^3$,
$y = 20 + 0.004(x - 10)^3$.
Each graph is a vertical shift of the graph of $y = 0.004(x - 10)^3$.

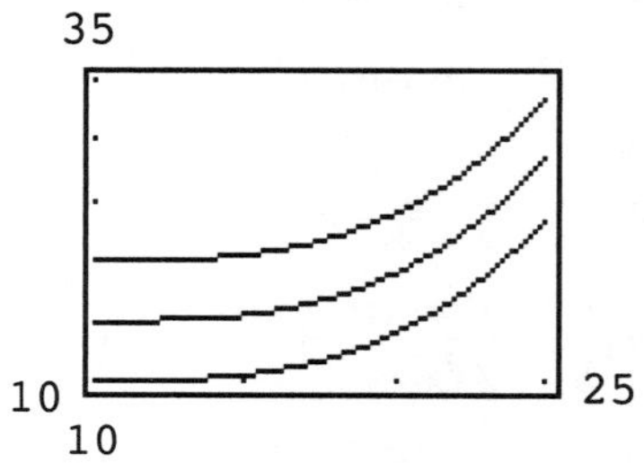

101. $y = \frac{1}{2}x^2 - 2$, $y = \frac{1}{3}x^2 - 3$, $y = \frac{1}{4}x^2 - 4$, $y = \frac{1}{5}x^2 - 5$.
Each graph is a vertical shrinking followed by a vertical shift of the graph of $y = x^2$.

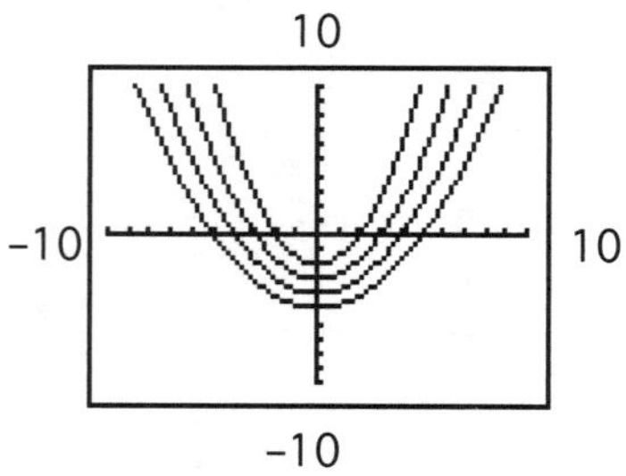

Section 3-4

1.
$$\begin{aligned} f(x) &= x^2 - 4x + 5 \\ &= (x^2 - 4x + ?) + 5 \qquad \tfrac{1}{2}(-4) = -2 \quad (-2)^2 = 4 \\ &= (x^2 - 4x + 4) + 5 - 4 \\ &= (x - 2)^2 + 1 \end{aligned}$$

3. $h(x) = -x^2 - 2x - 1$

$= -(x^2 + 2x + ?) - 1$ $\quad \frac{1}{2}(2) = 1 \quad 1^2 = 1$

$= -(x^2 + 2x + 1) - 1 + 1$

$= -(x + 1)^2$

5. $m(x) = x^2 - 4x + 1$

$= (x^2 - 4x + ?) + 1$ $\quad \frac{1}{2}(-4) = -2 \quad (-2)^2 = 4$

$= (x^2 - 4x + 4) + 1 - 4$

$= (x - 2)^2 - 3$

7. The graph of f is the graph of $y = x^2$ shifted to the right 2 units and up 1 unit.

9. The graph of h is the graph of $y = x^2$ reflected through the x axis and shifted to the left 1 unit.

11. The graph of m is the graph of $y = x^2$ shifted to the right 2 units and down 3 units.

13. The graph of $y = x^2$ has been shifted to the right 2 units. This is the graph of $y = (x - 2)^2$, corresponding to the function k.

15. The graph of $y = x^2$ has been shifted to the right 2 units and down 3 units. This is the graph of $y = (x - 2)^2 - 3$, corresponding to the function m.

17. The graph of $y = x^2$ has been reflected in the x axis and shifted to the left 1 unit, corresponding to the function h.

19. $f(x) = -\frac{1}{4}x^2$ is in vertex form with $a = -\frac{1}{4}$, $h = 0$, $k = 0$. Therefore the graph is a parabola with

x	$f(x)$
0	0
±1	$-\frac{1}{4}$
±2	−1

vertex: (0, 0); axis of symmetry: $x = 0$;
domain: $(-\infty, \infty)$; range: $(-\infty, 0]$; max $f(x) = f(0) = 0$;
increasing on $(-\infty, 0)$; decreasing on $(0, \infty)$.

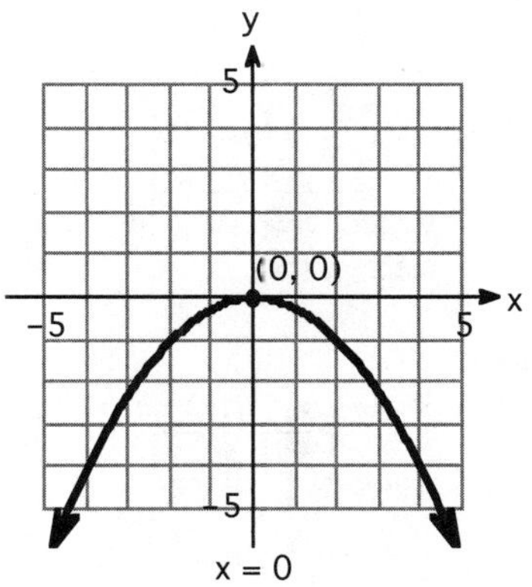

21. $f(x) = \frac{4}{3} - \frac{1}{3}x^2$

$= -\frac{1}{3}x^2 + \frac{4}{3}$

In vertex form, we see $a = -\frac{1}{3}$, $h = 0$, $k = \frac{4}{3}$.

Therefore the graph is a parabola with

x	$f(x)$
0	$\frac{4}{3}$
±1	1
±2	0
±3	$-\frac{5}{3}$

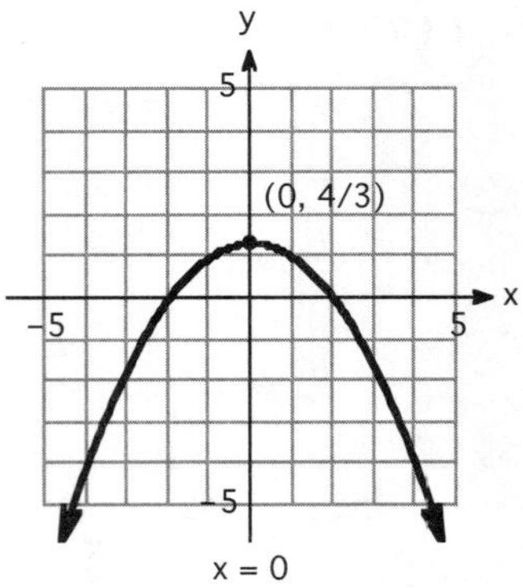

vertex: $\left(0, \frac{4}{3}\right)$; axis of symmetry: $x = 0$; domain: $(-\infty, \infty)$; range: $\left(-\infty, \frac{4}{3}\right]$;

max $f(x) = f(0) = \frac{4}{3}$; increasing on $(-\infty, 0)$; decreasing on $(0, \infty)$.

23. $f(x) = (x + 3)^2$ is in vertex form with $a = 1$, $h = -3$, $k = 0$. Therefore, the graph is a parabola with

x	$f(x)$
−3	0
−4	1
−2	1
−5	4
−1	4

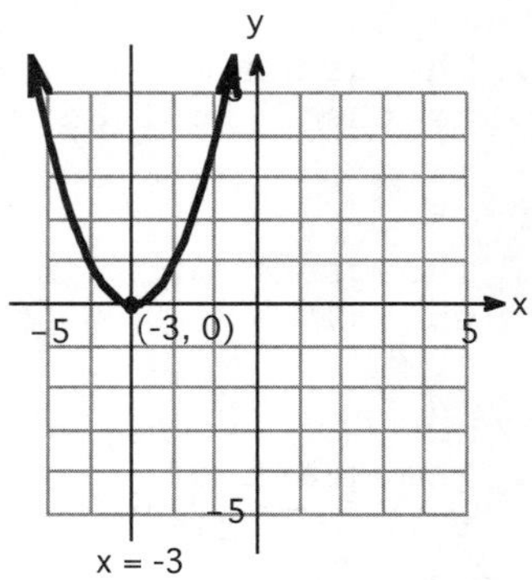

vertex: $(-3, 0)$; axis of symmetry: $x = -3$;
domain: $(-\infty, \infty)$; range: $[0, \infty)$; min $f(x) = f(-3) = 0$;
decreasing on $(-\infty, -3)$; increasing on $(-3, \infty)$

25. $f(x) = 4x - x^2$

$= -(x^2 - 4x + ?)$

$= -(x^2 - 4x + 4) + 4$

$= -1(x - 2)^2 + 4$

In vertex form, we see $a = -1$, $h = 2$, $k = 4$. Therefore, the graph is a parabola with

x	$f(x)$
2	4
1	3
3	3
0	0
4	0

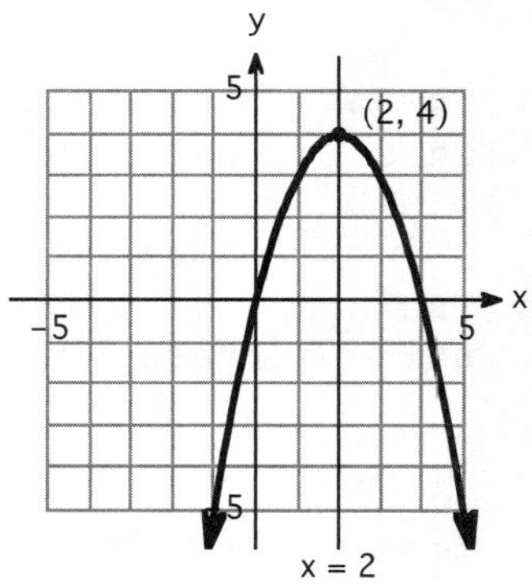

vertex: $(2, 4)$; axis of symmetry: $x = 2$; domain: $(-\infty, \infty)$; range: $(-\infty, 4]$;
max $f(x) = f(2) = 4$; increasing on $(-\infty, 2)$; decreasing on $(2, \infty)$

27. $f(x) = (x - 3)^2 + 2$ is in vertex form with $a = 1$, $h = 3$, $k = 2$. Therefore, the graph is a parabola with

x	$f(x)$
3	2
4	3
2	3
5	6
1	6

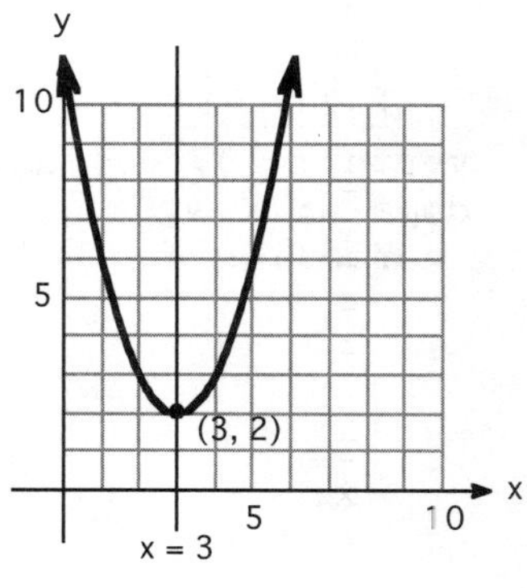

vertex: $(3, 2)$; axis of symmetry: $x = 3$;
domain: $(-\infty, \infty)$; range: $[2, \infty)$; min $f(x) = f(3) = 2$;
decreasing on $(-\infty, 3)$; increasing on $(3, \infty)$

29. $f(x) = -(x + 3)^2 - 2$

$= -1(x + 3)^2 - 2$

In vertex form, we see $a = -1$, $h = -3$, $k = -2$.
Therefore, the graph is a parabola with

x	$f(x)$
−3	−2
−2	−3
−4	−3
−1	−6
−5	−6

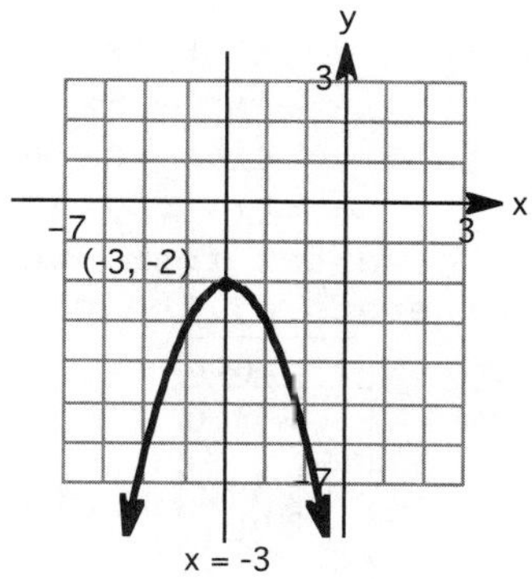

vertex: $(-3, -2)$; axis of symmetry: $x = -3$;
domain: $(-\infty, \infty)$; range: $(-\infty, -2]$; max $f(x) = f(-3) = -2$;
increasing on $(-\infty, -3)$; decreasing on $(-3, \infty)$

31. $f(x) = x^2 - 4x - 5$

$= (x^2 - 4x + ?) - 5$

$= (x^2 - 4x + 4) - 4 - 5$

$= (x - 2)^2 - 9$

In vertex form, we see $a = 1$, $h = 2$, $k = -9$.
Therefore, the graph is a parabola with

x	$f(x)$
2	−9
4	−5
0	−5
6	7
−2	7

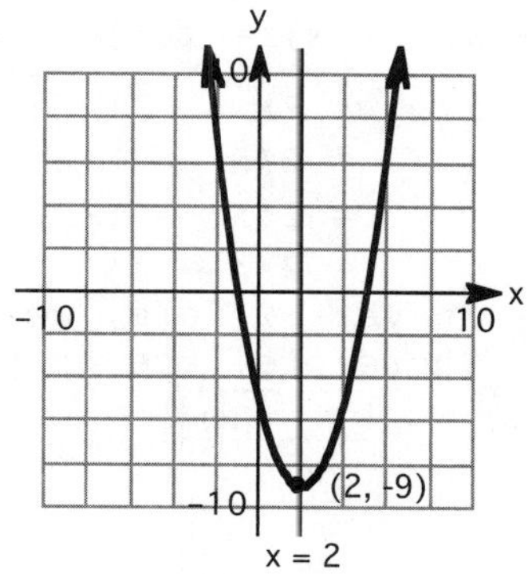

vertex: $(2, -9)$; axis of symmetry: $x = 2$; domain: $(-\infty, \infty)$; range: $[-9, \infty)$;
min $f(x) = f(2) = -9$; decreasing on $(-\infty, 2)$; increasing on $(2, \infty)$

33. $f(x) = -x^2 + 6x$

$= -(x^2 + 6x + ?)$

$= -(x^2 - 6x + 9) + 9$

$= -1(x - 3)^2 + 9$

In vertex form, we see $a = -1$, $h = 3$, $k = 9$.
Therefore, the graph is a parabola with

x	$f(x)$
3	9
2	8
4	8
0	0
6	0

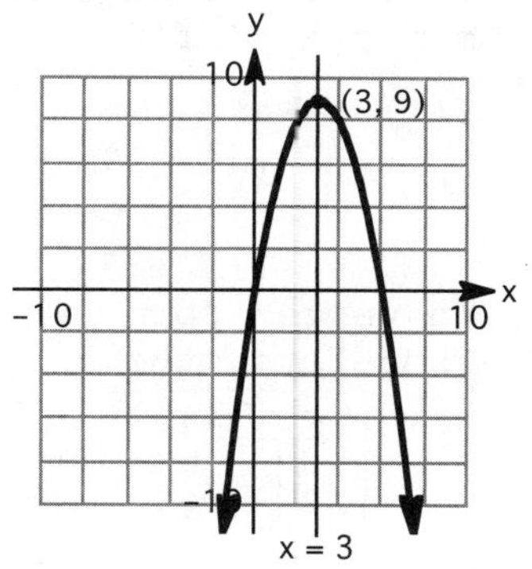

vertex: $(3, 9)$; axis of symmetry: $x = 3$; domain: $(-\infty, \infty)$; range: $(-\infty, 9]$;
max $f(x) = f(3) = 9$; increasing on $(-\infty, 3)$; decreasing on $(3, \infty)$

35. $f(x) = x^2 + 6x + 11$

$= (x^2 + 6x + ?) + 11$

$= (x^2 + 6x + 9) - 9 + 11$

$= (x + 3)^2 + 2$

In vertex form, we see $a = 1$, $h = -3$, $k = 2$.
Therefore, the graph is a parabola with

x	$f(x)$
−3	2
−1	6
−5	6
1	18
−7	18

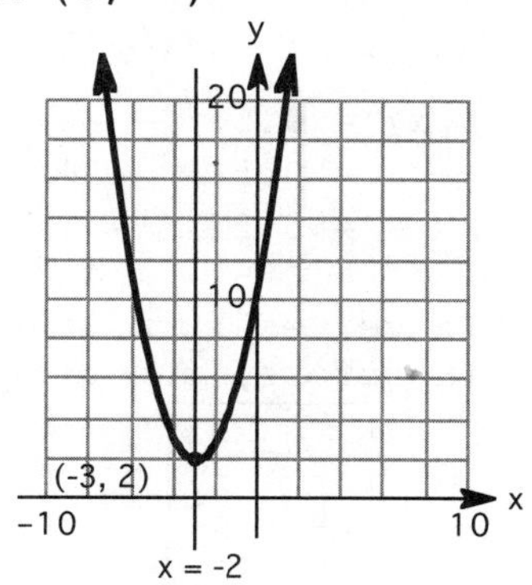

vertex: $(-3, 2)$; axis of symmetry: $x = -2$; domain: $(-\infty, \infty)$; range: $[2, \infty)$;
min $f(x) = f(-3) = 2$; decreasing on $(-\infty, -3)$; increasing on $(-3, \infty)$

37. $f(x) = -x^2 + 6x - 6$
$= -(x^2 - 6x + ?) - 6$
$= -(x^2 - 6x + 9) + 9 - 6$
$= -1(x - 3)^2 + 3$

In vertex form, we see $a = -1$, $h = 3$, $k = 3$. Therefore, the graph is a parabola with

x	$f(x)$
3	3
2	2
4	2
1	−1
5	−1

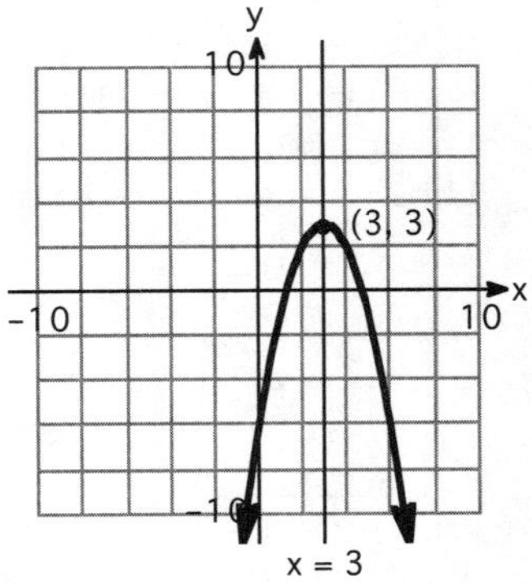

vertex: (3, 3); axis of symmetry: $x = 3$; domain: $(-\infty, \infty)$; range: $(-\infty, 3]$; max $f(x) = f(3) = 3$; increasing on $(-\infty, 3)$; decreasing on $(3, \infty)$

39. $f(x) = 2x^2 - 24x + 90$
$= 2(x^2 - 12x + ?) + 90$
$= 2(x^2 - 12x + 36) - 72 + 90$
$= 2(x - 6)^2 + 18$

In vertex form, we see $a = 2$, $h = 6$, $k = 18$. Therefore, the graph is a parabola with

x	$f(x)$
6	18
5	20
7	20
4	26
8	26

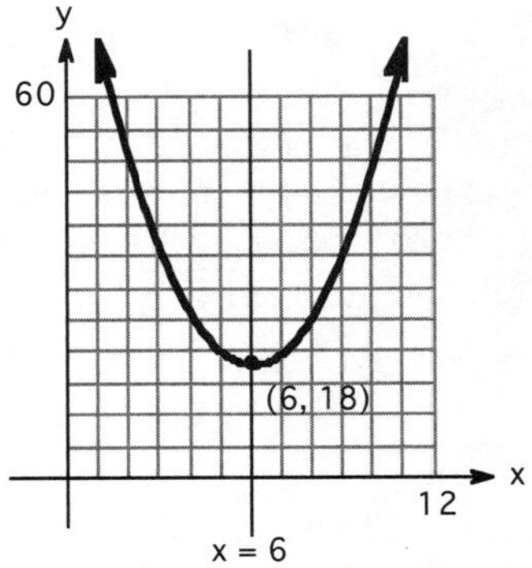

vertex: (6, 18); axis of symmetry: $x = 6$; domain: $(-\infty, \infty)$; range: $[18, \infty)$; min $f(x) = f(6) = 18$; decreasing on $(-\infty, 6)$; increasing on $(6, \infty)$

41. $f(x) = -x^2 - 6x - 4$
$= -(x^2 + 6x + ?) - 4$
$= -(x^2 + 6x + 9) + 9 - 4$
$= -1(x + 3)^2 + 5$

In vertex form, we see $a = -1$, $h = -3$, $k = 5$. Therefore, the graph is a parabola with

x	$f(x)$
−3	5
−2	4
−4	4
0	−4
−6	−4

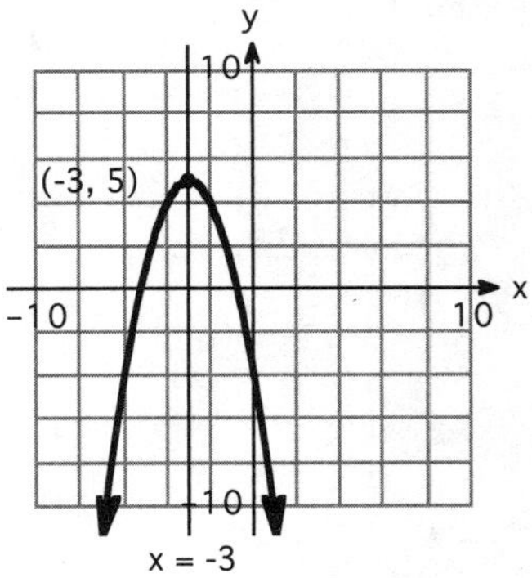

vertex: (-3, 5); axis of symmetry: $x = -3$; domain: $(-\infty, \infty)$; range: $(-\infty, 5]$; max $f(x) = f(-3) = 5$; increasing on $(-\infty, -3)$; decreasing on $(-3, \infty)$

43. $f(x) = 0.5x^2 - 2x - 7$
$= 0.5(x^2 - 4x + ?) - 7$
$= 0.5(x^2 - 4x + 4) - 0.5(4) - 7$
$= 0.5(x - 2)^2 - 9$

In vertex form, we see $a = 0.5$, $h = 2$, $k = -9$. Therefore, the graph is a parabola with

x	$f(x)$
2	−9
0	−7
4	−7
−2	−1
6	−1

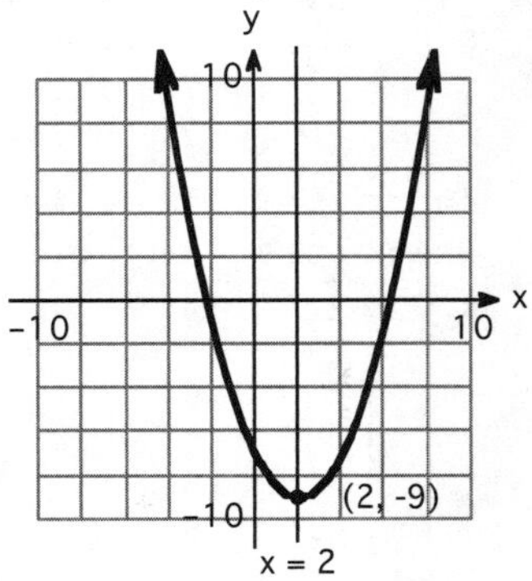

vertex: (2, −9); axis of symmetry: $x = 2$; domain: $(-\infty, \infty)$; range: $[-9, \infty)$; min $f(x) = f(2) = -9$; decreasing on $(-\infty, 2)$; increasing on $(2, \infty)$

45. $f(x) = 0.5x^2 + 2x + 3$
$= 0.5(x^2 + 4x + ?) + 3$
$= 0.5(x^2 + 4x + 4) - 0.5(4) + 3$
$= 0.5(x + 2)^2 + 1$

In vertex form, we see $a = 0.5$, $h = -2$, $k = 1$.

Therefore, the graph is a parabola with

x	$f(x)$
−2	1
0	3
−4	3
2	9
−6	9

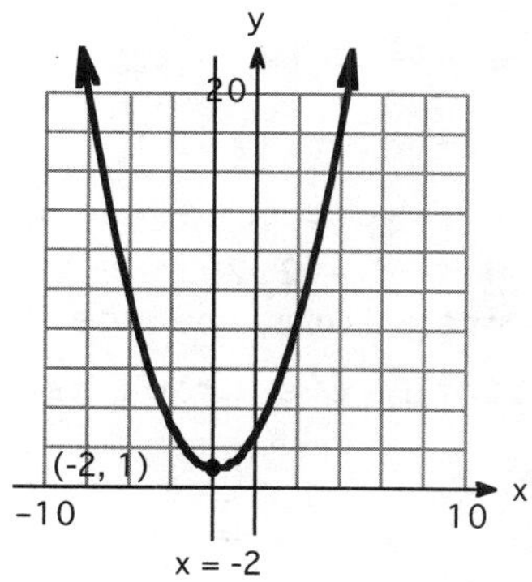

vertex: (−2, 1); axis of symmetry: $x = -2$; domain: $(-\infty, \infty)$; range: $[1, \infty)$; min $f(x) = f(-2) = 1$; decreasing on $(-\infty, -2)$; increasing on $(-2, \infty)$

47. $f(x) = 4x^2 - 12x + 9$

$= 4(x^2 - 3x + ?) + 9 \qquad \frac{1}{2}(-3) = -\frac{3}{2} \qquad \left(-\frac{3}{2}\right)^2 = \frac{9}{4}$

$= 4\left(x^2 - 3x + \frac{9}{4}\right) - 4\left(\frac{9}{4}\right) + 9$

$= 4\left(x - \frac{3}{2}\right)^2 + 0$

In vertex form, we see $a = 4$, $h = \frac{3}{2} = 1.5$, $k = 0$.

Therefore, the graph is a parabola with

x	$f(x)$
1.5	0
2	1
1	1
0	9
3	9

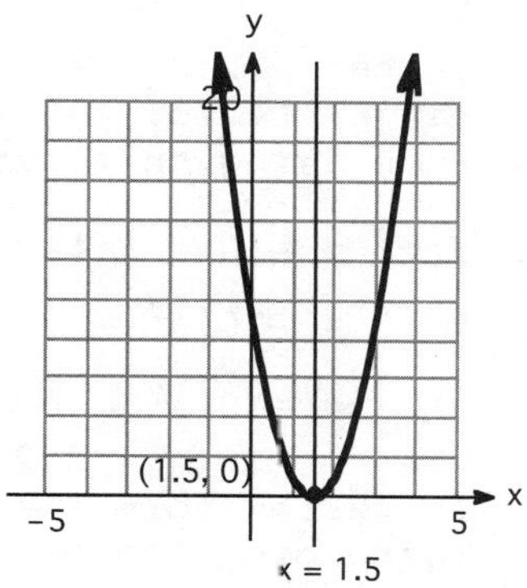

vertex: (1.5, 0); axis of symmetry: $x = 1.5$; domain: $(-\infty, \infty)$; range: $[0, \infty)$; min $f(x) = f(1.5) = 0$; decreasing on $(-\infty, 1.5)$; increasing on $(1.5, \infty)$

49. $f(x) = -2x^2 - 8x - 2$
$= -2(x^2 + 4x + ?) - 2$
$= -2(x^2 + 4x + 4) + 8 - 2$
$= -2(x + 2)^2 + 6$

In vertex form, we see $a = -2$, $h = -2$, $k = 6$.

Therefore, the graph is a parabola with

x	$f(x)$
−2	6
−1	4
−3	4
0	−2
−4	−2

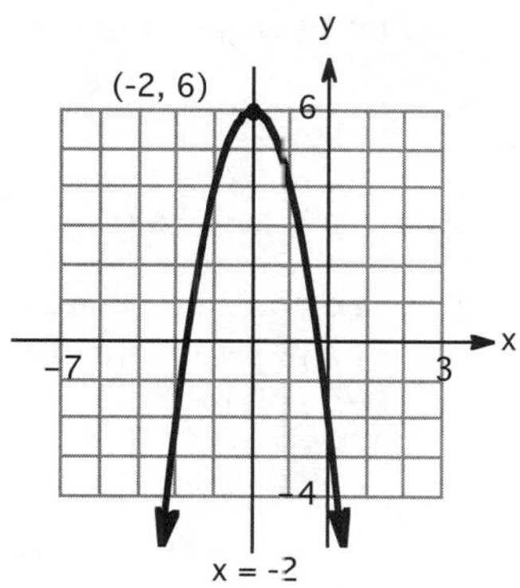

vertex: (−2, 6); axis of symmetry: $x = -2$; domain: $(-\infty, \infty)$; range: $(-\infty, 6]$; max $f(x) = f(-2) = 6$; increasing on $(-\infty, -2)$; decreasing on $(-2, \infty)$

51. $f(x) = 4x^2 - 18x + 25$

$= 4(x^2 - 4.5x + ?) + 25 \qquad \frac{1}{2}(-4.5) = -2.25 \qquad (-2.25)^2 = 5.0625$

$= 4(x^2 - 4.5x + 5.0625) - 4(5.0625) + 25$

$= 4(x - 2.25)^2 + 4.75$

In vertex form, we see $a = 4$, $h = 2.25$, $k = 4.75$.

Therefore, the graph is a parabola with

x	$f(x)$
2.25	4.75
2	5
3	7
1	11
4	17

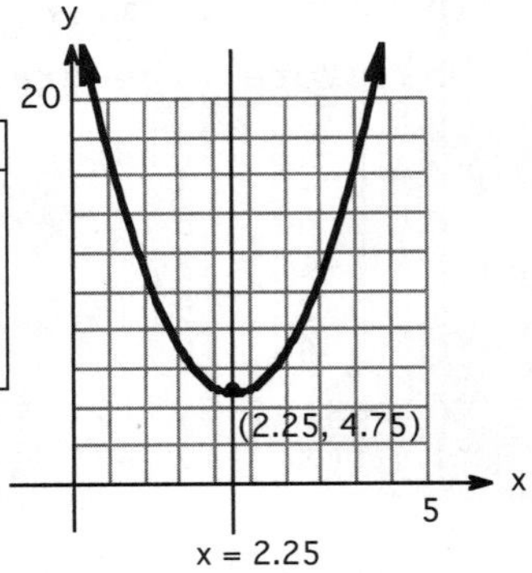

vertex: (2.25, 4.75); axis of symmetry: $x = 2.25$; domain: $(-\infty, \infty)$; range: $[4.75, \infty)$; min $f(x) = f(2.25) = 4.75$; decreasing on $(-\infty, 2.25)$; increasing on $(2.25, \infty)$

53. $f(x) = -10x^2 + 44x + 12$

$= -10(x^2 - 4.4x + ?) + 12 \qquad \frac{1}{2}(-4.4) = -2.2 \qquad (-2.2)^2 = 4.84$

$= -10(x^2 - 4.4x + 4.84) + 10(4.84) + 12$

$= -10(x - 2.2)^2 + 60.4$

In vertex form, we see $a = -10$, $h = 2.2$, $k = 60.4$.

Therefore, the graph is a parabola with

x	$f(x)$
2.2	60.4
2	60
3	54
0	12
5	–18

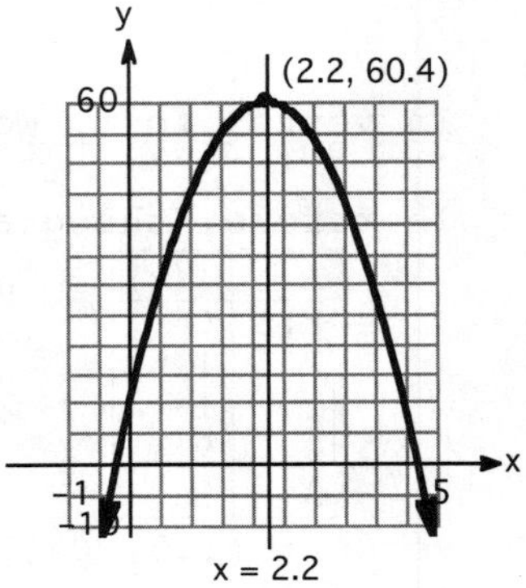

vertex: (2.2, 60.4); axis of symmetry: $x = 2.2$; domain: $(-\infty, \infty)$; range: $(-\infty, 60.4]$; max $f(x) = f(2.2) = 60.4$; increasing on $(-\infty, 2.2)$; decreasing on $(2.2, \infty)$

55. $x^2 < 10 - 3x$

$f(x) = x^2 + 3x - 10 < 0$

$f(x) = (x + 5)(x - 2) < 0$

The zeros of f are –5 and 2. Plotting the graph of $f(x)$, we see that $f(x) < 0$ for $-5 < x < 2$, or $(-5, 2)$.

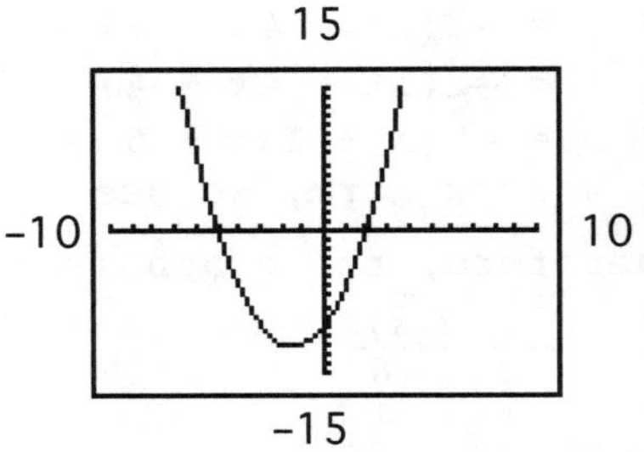

57. $x^2 + 21 > 10x$

$f(x) = x^2 - 10x + 21 > 0$

$f(x) = (x - 3)(x - 7) > 0$

The zeros of f are 3 and 7. Plotting the graph of $f(x)$, we see that $f(x) > 0$ for $x < 3$ and $x > 7$, or $(-\infty, 3) \cup (7, \infty)$.

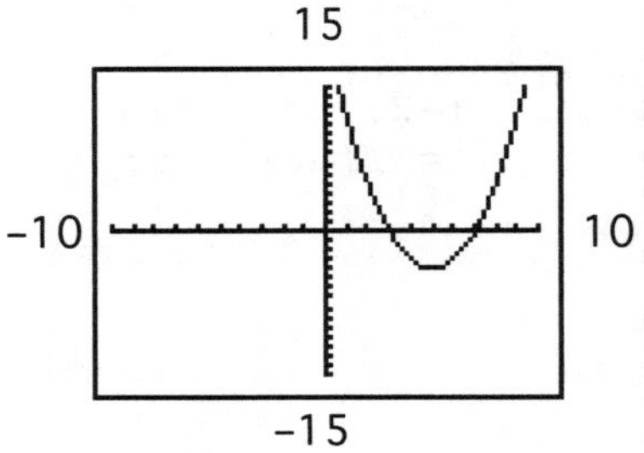

59. $x^2 \le 8x$

$f(x) = x^2 - 8x \le 0$

$f(x) = x(x - 8) \le 0$

The zeros of f are 0 and 8. Plotting the graph of $f(x)$, we see that $f(x) \le 0$ for $0 \le x \le 8$ or (0, 8).

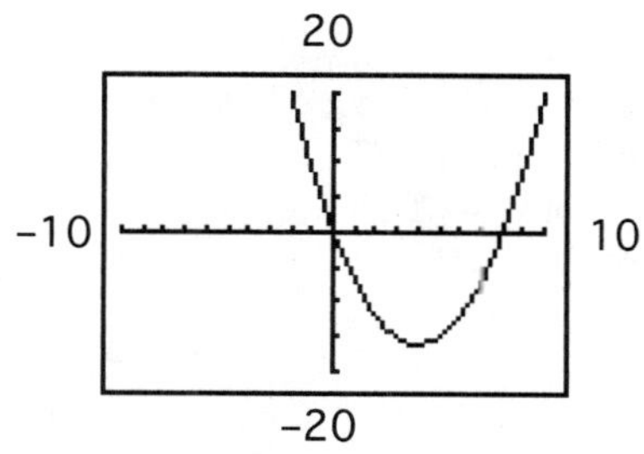

61. $x^2 + 5x \le 0$

$f(x) = x^2 + 5x \le 0$

$f(x) = x(x + 5) \le 0$

The zeros of f are -5 and 0. Plotting the graph of $f(x)$, we see that $f(x) \le 0$ for $-5 \le x \le 0$ or $[-5, 0]$.

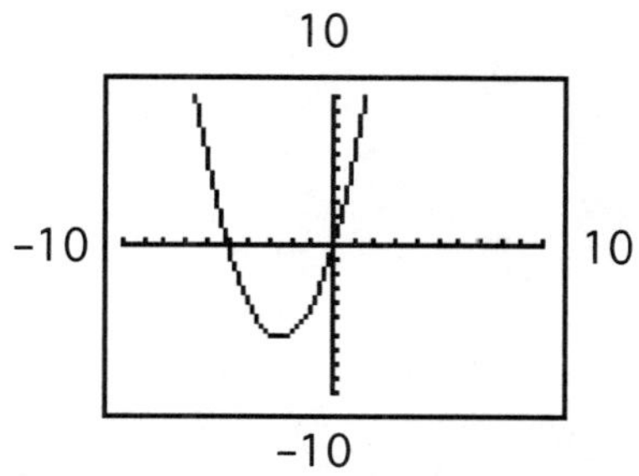

63.

$$x^2 + 1 < 2x$$
$$x^2 - 2x + 1 < 0$$
$$(x - 1)^2 < 0$$

Since the square of no real number is negative, these statements are never true for any real number x. No solution; $\varnothing$ is the solution set.

65. $x^2 < 3x - 3$

$x^2 - 3x + 3 < 0$

We attempt to find all real zeros of the polynomial.

$x^2 - 3x + 3 = 0$

$$x = \frac{-b \pm \sqrt{b^2 - 4ac}}{2a} \qquad a = 1,\ b = -3,\ c = 3$$

$$x = \frac{-(-3) \pm \sqrt{(-3)^2 - 4(1)(3)}}{2(1)}$$

$$x = \frac{3 \pm \sqrt{-3}}{2}$$

The polynomial has no real zeros. Hence the statement is either true for all real x or for no real x. To determine which, we choose a test number, say 0.

$x^2 < 3x - 3$

$0^2 \overset{?}{<} 3(0) - 3$

$0 \overset{?}{<} -3$ False.

The statement is never true for any real number x. No solution (and no graph). $\varnothing$ is the solution set.

67. $x^2 - 1 \geq 4x$

$f(x) = x^2 - 4x - 1 \geq 0$

Find all real zeros of $f(x)$.

$x^2 - 4x - 1 = 0$

$$x = \frac{-b \pm \sqrt{b^2 - 4ac}}{2a} \quad a = 1,\ b = -4,\ c = -1$$

$$x = \frac{-(-4) \pm \sqrt{(-4)^2 - 4(1)(-1)}}{2(1)}$$

$$x = \frac{4 \pm \sqrt{16 + 4}}{2}$$

$$x = \frac{4 \pm \sqrt{20}}{2}$$

$$x = 2 \pm \sqrt{5}$$
$$\approx -0.236,\ 4.236$$

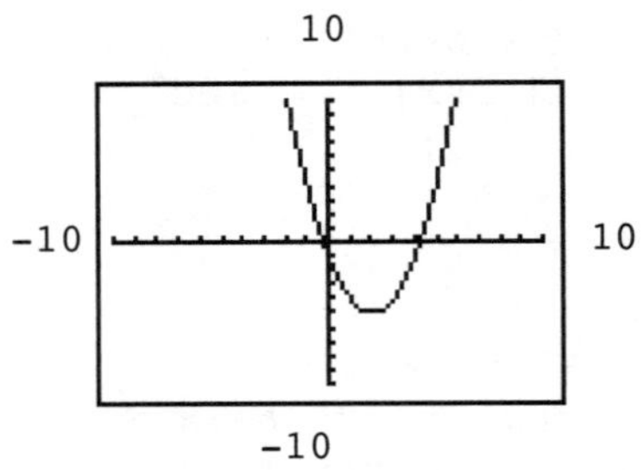

Common Error: $x \neq 2 \pm \sqrt{20}$

Plotting the graph of $f(x)$ we see that $f(x) \geq 0$ for $x \leq 2 - \sqrt{5}$ and $x \geq 2 + \sqrt{5}$, or $(-\infty, 2 - \sqrt{5}] \cup [2 + \sqrt{5}, \infty)$.

69. (A) For $x < 0$, $f(x) = x^2 + 1$,
so $f(-2) = (-2)^2 + 1 = 5$ and
$f(-1) = (-1)^2 + 1 = 2$.

For $x \geq 0$, $f(x) = -x^2 - 1$,
so $f(0) = -(0)^2 - 1 = -1$,
$f(1) = -(1)^2 - 1 = -2$,
$f(2) = -(2)^2 - 1 = -5$

(B)

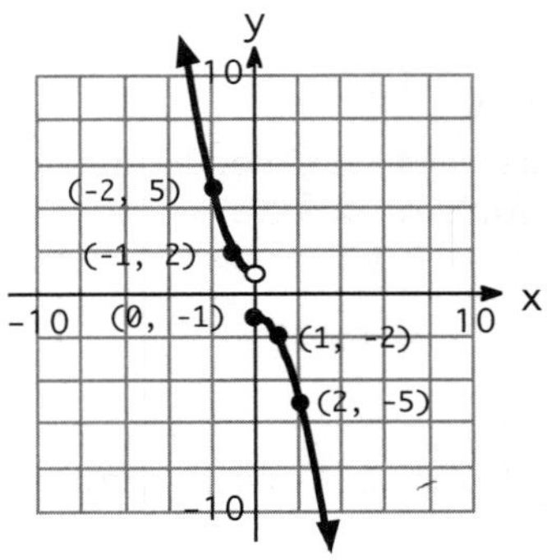

(C) $f(x)$ is defined for all real numbers; the domain of f is $(-\infty, \infty)$.
From the graph, the range of f is $(-\infty, -1] \cup (1, \infty)$.
The function is discontinuous at $x = 0$.

71. (A) For $x \leq 2$, $f(x) = x^2 - 4$,
so $f(-2) = (-2)^2 - 4 = 0$,
$f(0) = 0^2 - 4 = -4$,
and $f(2) = 2^2 - 4 = 0$.

For $x > 2$, $f(x) = -2x + 4$,
so $f(4) = -2(4) + 4 = -4$.

(B)

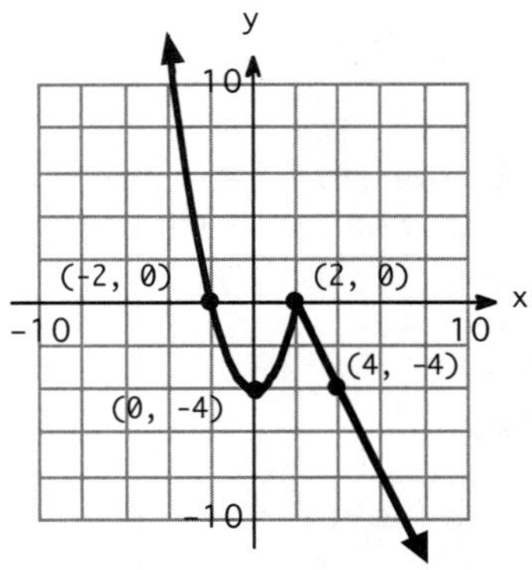

(C) $f(x)$ is defined for all real numbers; the domain of f is $(-\infty, \infty)$.
From the graph, the range is $(-\infty, \infty)$.
The function is continuous on its domain.

73. (A) For $x < -1$, $f(x) = x^2 + 4x$, so
$f(-4) = (-4)^2 + 4(-4) = 0$ and
$f(-2) = (-2)^2 + 4(-2) = -4$.

For $-1 \le x \le 1$, $f(x) = 2x$, so
$f(-1) = 2(-1) = -2$ and
$f(1) = 2(1) = 2$.

For $x > 1$, $f(x) = 4x - x^2$, so
$f(2) = 4(2) - 2^2 = 4$ and
$f(4) = 4(4) - 4^2 = 0$.

(B)

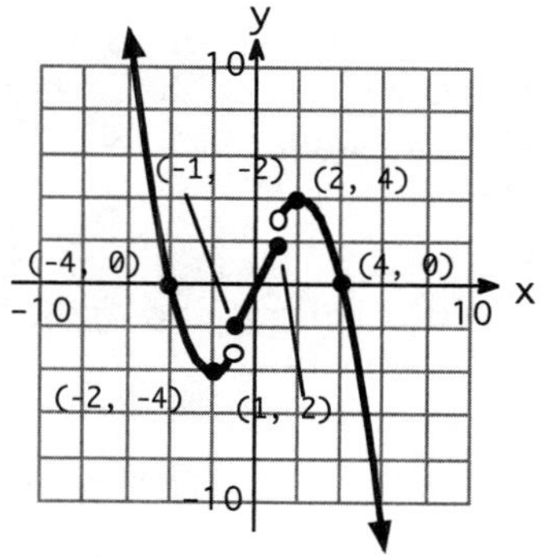

(C) $f(x)$ is defined for all real numbers; the domain of f is $(-\infty, \propto)$.
From the graph, the range is $(-\infty, \infty)$.
The function is discontinuous at $x = -1$ and $x = 1$.

75. The vertex of the parabola is at (1, -4). Therefore the equation must have form
$y = a(x - 1)^2 - 4$
Since the parabola passes through (3, 4), these coordinates must satisfy the equation
$4 = a(3 - 1)^2 - 4$
$8 = 4a$
$a = 2$.
The equation is
$y = 2(x - 1)^2 - 4$
$y = 2(x^2 - 2x + 1) - 4$
$y = 2x^2 - 4x + 2 - 4$
$y = 2x^2 - 4x - 2$

77. The vertex of the parabola is at (-1, 4). Therefore the equation must have form
$y = a(x + 1)^2 + 4$
Since the parabola passes through (1, 2), these coordinates must satisfy the equation.
$2 = a(1 + 1)^2 + 4$
$2 = 4a + 4$
$-2 = 4a$
$a = -0.5$
The equation is
$y = -0.5(x + 1)^2 + 4$
$y = -0.5(x^2 + 2x + 1) + 4$
$y = -0.5x^2 - x - 0.5 + 4$
$y = -0.5x^2 - x + 3.5$

79. Notice that the table does not provide the exact coordinates of the vertex, so we can't tell for certain what they are. We know that $f(-1)$ and $f(3)$ are both zero, so the axis of symmetry is halfway between $x = -1$ and $x = 3$. In other words, the x-coordinate of the vertex is 1; the equation looks like
$f(x) = a(x - 1)^2 + k$.
Plug in $x = -1$:
$f(-1) = a(-1 - 1)^2 + k = a(-2)^2 + k = 4a + k = 0$ (since (-1, 0) is on the graph)
$4a + k = 0$
$k = -49$
Substitute $-4a$ in for k: $f(x) = a(x - 1)^2 - 4a$
Plug in $x = 0$:
$f(0) = a(0 - 1)^2 - 4a = a - 4a = -3a = -3$ (since (0, -3) is on the graph)
$-3a = -3$
$a = 1$
$f(x) = (x - 1)^2 - 4$ or $f(x) = x^2 - 2x - 3$

81. Notice that the table does not provide the exact coordinates of the vertex, so we can't tell for certain what they are. We know that $f(-1)$ and $f(5)$ are equal, so the axis of symmetry is halfway between $x = -1$ and $x = 5$. In other words, the x-coordinate of the vertex is 2; the equation looks like

$f(x) = a(x - 2)^2 + k$

Plug in $x = -1$: $f(-1) = a(-1 - 2)^2 + k = a(-3)^2 + k = 9a + k = 0$
(since $(-1, 0)$ is on the graph)

$9a + k = 0$

$k = -9a$

Substitute $-9a$ in for k: $f(x) = a(x - 2)^2 - 9a$

Plug in $x = 0$: $f(0) = a(0 - 2)^2 - 9a = a(-2)^2 - 9a = 4a - 9a = -5a = 2.5$
(since $(0, 2.5)$ is on the graph)

$-5a = 2.5$

$a = -0.5$

$f(x) = 0.5(x - 2)^2 + 4.5$ or $f(x) = -0.5x^2 + 2x + 2.5$

83. The vertex of the parabola is at $(4, 8)$. Therefore the equation must have form

$y = a(x - 4)^2 + 8$

Since the x intercept is 6, $(6, 0)$ must satisfy the equation

$0 = a(6 - 4)^2 + 8$

$0 = 4a + 8$

$a = -2$

The equation is

$y = -2(x - 4)^2 + 8$

$y = -2(x^2 - 8x + 16) + 8$

$y = -2x^2 + 16x - 32 + 8$

$y = -2x^2 + 16x - 24$

85. The vertex of the parabola is at $(-4, 12)$. Therefore the equation must have form

$y = a(x + 4)^2 + 12$

Since the y intercept is 4, $(0, 4)$ must satisfy the equation.

$4 = a(0 + 4)^2 + 12$

$4 = 16a + 12$

$-8 = 16a$

$a = -0.5$

The equation is

$y = -0.5(x + 4)^2 + 12$

$y = -0.5(x^2 + 8x + 16) + 12$

$y = -0.5x^2 - 4x - 8 + 12$

$y = -0.5x^2 - 4x + 4$

87. The vertex of the parabola is at $(-5, -25)$. Therefore the equation must have form

$y = a(x + 5)^2 - 25$

Since the parabola passes through $(-2, 20)$, these coordinates must satisfy the equation.

$20 = a(-2 + 5)^2 - 25$

$20 = 9a - 25$

$45 = 9a$

$a = 5$

The equation is $y = 5(x + 5)^2 - 25$

$y = 5(x^2 + 10x + 25) - 25$

$y = 5x^2 + 50x + 125 - 25$

$y = 5x^2 + 50x + 100$

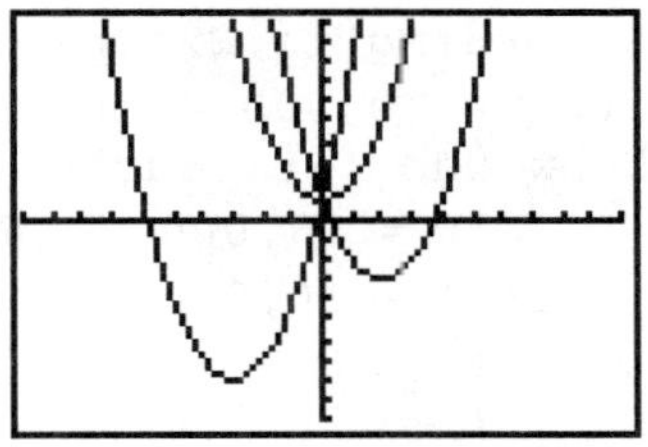

89. The graphs shown are $f(x) = x^2 + 6x + 1$, $f(x) = x^2 + 1$, and $f(x) = x^2 - 4x + 1$. These correspond to $f(x) = x^2 + kx + 1$ with $k = 6, 0$, and -4 respectively.

Note that all have the same shape but a different vertex. In fact, all three are translations of the graph $y = x^2$.

91. The graphs of $f(x) = (x - 1)^2$, $g(x) = (x - 1)^2 + 4$, and $h(x) = (x - 1)^2 - 5$ are shown.

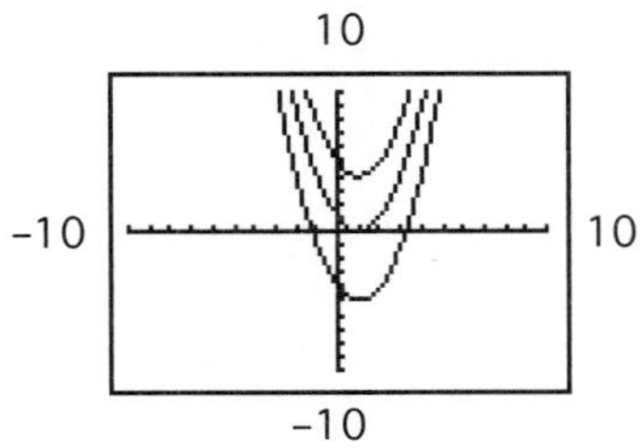

It is clear that $f(x)$ has one x intercept (at $x = 1$), $g(x)$ has no x intercepts and $h(x)$ has two x intercepts.
In general, for $a > 0$, the graph of $f(x) = a(x - h)^2 + k$ can be expected to have no intercepts for $k > 0$, one intercept at $x = h$ for $k = 0$, and two intercepts for $k < 0$.

93. $f(x) = a(x - h)^2 + k$. Since min $f(x) = f(2) = 4$, $h = 2$, $k = 4$, and the vertex is $(2, 4)$. The axis of symmetry is $x = h$ or $x = 2$. Since y is never less than 4, the range is $[4, \infty)$ and there are no x intercepts since y cannot be equal to 0. The domain of f is $(-\infty, \infty)$.

95. Let one number $= x$. Then the other number is $x - 30$. The product is a function of x given by $f(x) = x(x - 30) = x^2 - 30x$. This is a quadratic funtion with $a > 0$, therefore it has a minimum value at the vertex of its graph (a parabola). Completing the square yields

$$f(x) = x^2 - 30x \qquad \frac{1}{2}(-30) = -15 \qquad (-15)^2 = 225$$
$$= x^2 - 30x + 225 - 225$$
$$= (x - 15)^2 - 225$$

Thus the minimum product is -225, when $x = 15$ and $x - 30 = -15$. There is no "highest point" on this parabola and no maximum product.

97. Any correct choice of two points on this line will lead to a linear regression model of $y = 0.5x + 3$. In general, any two points chosen correctly on $y = ax + b$ will lead to a linear regression model of exactly the same equation $y = ax + b$.

99. (A) Since the four sides needing fencing are x, y, $x + 50$, and y, we have $x + y + (x + 50) + y = 250$, on this parabola and no maximum product. Hence, $y = 100 - x$.
Therefore the area
$A(x) = (x + 50)y = (x + 50)(100 - x) = -x^2 + 50x + 5000$.

(Since both x and $100 - x$ must be nonnegative, the domain of $A(x)$ is $0 \le x \le 100$.)

(B) This is a quadratic function with $a < 0$, therefore it has a maximum value at the vertex of its graph (a parabola). Completing the square yields

$$A(x) = -x^2 + 50x + 5{,}000$$
$$= -(x^2 - 50x) + 5{,}000 \qquad \frac{1}{2}(-50) = -25 \qquad (-25)^2 = 625$$
$$= -(x^2 - 50x + 625) + 5000 + 625$$
$$= -(x - 25)^2 + 5{,}625$$

Thus the maximum area is 5,625 when $x = 25$ and $y = 100 - x = 75$.

(C) When $x = 25$, $y = 100 - 25 = 75$. The dimensions of the corral are then $x + 50$ by y, or 75 ft by 75 ft.

101. According to Example 5, the function describing the height of the sandbag is $h(t) = 10{,}000 - 16t^2$ (since the initial height is 10,000 feet). We want to know when it reaches ground level, so plug in zero for $h(t)$, then solve for t.

$$0 = 10{,}000 - 16t^2$$
$$16t^2 = 10{,}000$$
$$t^2 = 625$$
$$t = 25$$

It hits the ground 25 seconds after it's dropped.

103. According to Example 5, the function describing the height of the diver is $h(t) = h_0 - 16t^2$, where h_0 is the initial height in feet. (This initial height is the height of the cliff.) We know that $h(2.5) = 0$ since it takes 2.5 seconds to reach the water; we plug in 2.5 for t and 0 for $h(t)$, which allows us to solve for h_0.

$$0 = h_0 - 16(2.5)^2$$
$$0 = h_0 - 100$$
$$100 = h_0$$

The cliff is 100 feet high.

105. (A) Since $d(t)$ is a quadratic function with maximum value 484 when $t = 5.5$, an equation for $d(t)$ must be of the form

$$d(t) = a(t - 5.5)^2 + 484$$

Since $d(0) = 0$,

$$0 = d(0) = a(0 - 5.5)^2 + 484$$
$$0 = 30.25a + 484$$
$$a = -16$$

Hence
$$d(t) = -16(t - 5.5)^2 + 484$$
$$= -16(t^2 - 11t + 30.25) + 484$$
$$= -16t^2 + 176t - 484 + 484$$
$$= -16t^2 + 176t$$

Since the graph of d must be symmetric with respect to $t = 5.5$, and $d(0) = 0$, $d(11)$ must also equal 0. The distance above the ground will be nonnegative only for values of t between 0 and 11, hence the domain of the function is $0 \le t \le 11$.

(B) Solve $250 = -16t^2 + 176t$ by graphing Y1 = 250 and Y2 = $-16x^2 + 176x$ and applying a built-in routine.

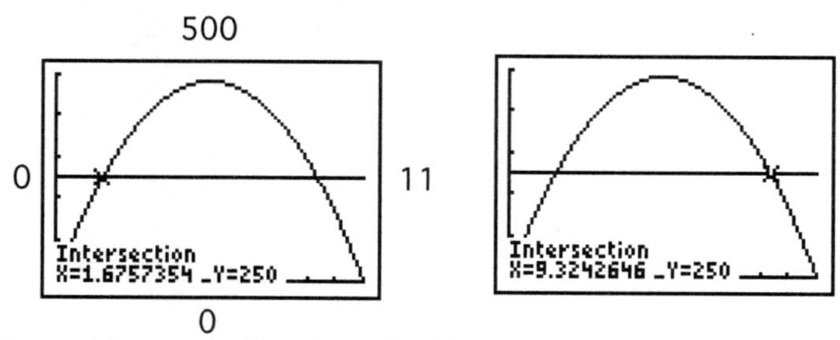

From the graphs $t = 1.68$ sec and $t = 9.32$ sec, to two decimal places.

107. (A) If coordinates are chosen with origin at the center of the base, the parabola is the graph of a quadratic function $h(x)$ with maximum value 14 when $x = 0$. The equation must be of form

$$h(x) = ax^2 + 14$$

Since $h(10) = 0$ (why?)

$$0 = h(10) = a(10)^2 + 14$$
$$0 = 100a + 14$$
$$a = -0.14$$

Hence $h(x) = -0.14x^2 + 14 \quad -10 \le x \le 10$

(B) Suppose the truck were to drive so as to maximize its clearance, that is, in the center of the roadway. Then half its width, or 4 ft, would extend to each side. But if $x = 4$, $h(x) = -0.14(4)^2 + 14 = 11.76$ ft. The arch is only 11.76 feet high, but the truck is 12 feet high. The truck cannot pass through the arch.

(C) From part (B), if $x = 4$, $h(x) = 11.76$ ft is the height of the tallest truck.

109. (A) Label the text figure as follows:

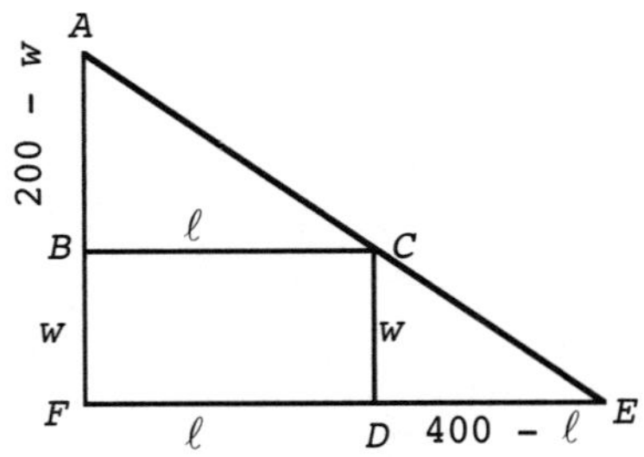

Then, using the hint, in the similar triangles *ABC* and *AFE*, we have

$$\frac{\ell}{200 - w} = \frac{400}{200}$$
$$\frac{\ell}{200 - w} = 2$$
$$\ell = 2(200 - w)$$
$$\ell = 400 - 2w$$

Then $A(w) = \ell w = (400 - 2w)w = 400w - 2w^2$.

Since $w \ge 0$ and $\ell = 400 - 2w \ge 0$, the domain of the function is $0 \le w \le 200$.

(B) Solve $150{,}000 \le 400w - 2w^2$

$$2w^2 - 400w + 15{,}000 \le 0$$
$$w^2 - 200w + 7{,}500 \le 0$$
$$(w - 50)(w - 150) \le 0$$

The solution of this inequality consists of all values of w for which the graph of $y = w^2 - 200w + 7{,}500$ is on or below the w axis. The w intercepts of the graph are 50 and 150.

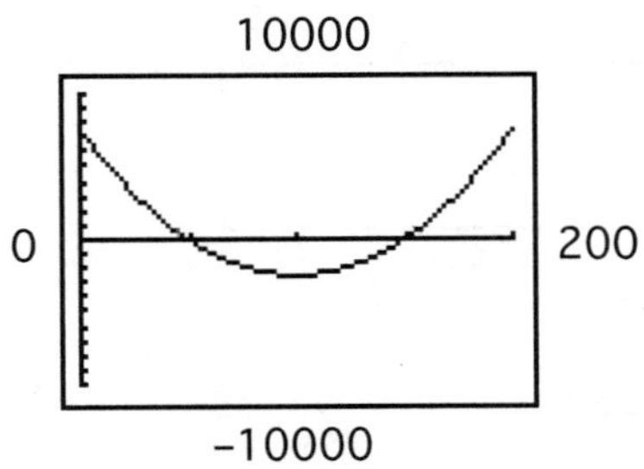

Examining the graph, we see that the solution of the inequality is $50 \le w \le 150$.

(C) The cross-sectional area was found in part (A) to be

$$A(w) = 400w - 2w^2$$

Graphing this and applying a built-in routine to find the maximum, we obtain the graph at the right.

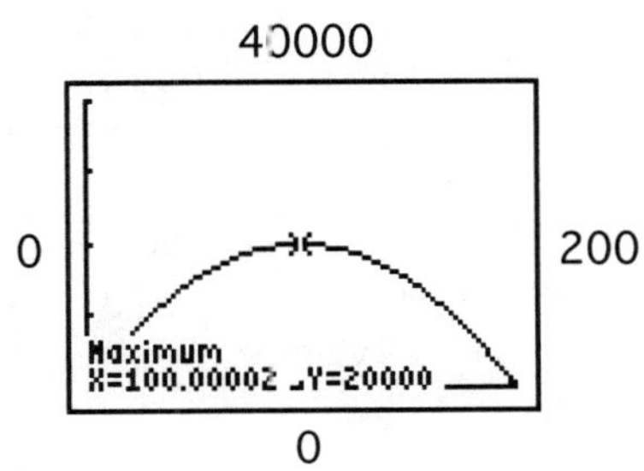

The maximum cross-sectional area of such a building is 20,000 square feet when $w = 100$ ft. A building with area 25,000 square feet is impossible.

111. (A)

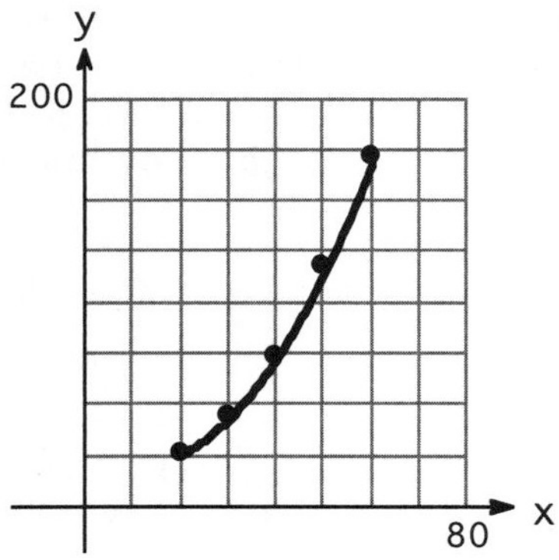

(B) Solve $150 = 0.06x^2 - 1.2x + 26$

$0 = 0.06x^2 - 1.2x - 124$

$$x = \frac{-(-1.2) \pm \sqrt{(-1.2)^2 - 4(0.06)(-124)}}{2(0.06)}$$

$$x = \frac{1.2 \pm 5.59}{0.12}$$

$x = 57$ mph (the negative answer doesn't make sense)

113. (A) Beer consumption in 1960 is given as 0.99.

Solve $0.99 = -0.0006x^2 + 0.03x + 1$

$0 = -0.0006x^2 + 0.03x + 0.01$

$0 = -6x^2 + 300x + 100$ (for convenience)

$$x = \frac{-300 \pm \sqrt{(300)^2 - 4(-6)(100)}}{2(-6)}$$

$x = 50$ (discarding the negative answer)

This represents the year 2010.

(B) Substitute $x = 45$ to obtain

$B(45) = -0.0006(45)^2 + 0.03(45) + 1$
$= 1.14$ gallons

115. A profit will result if $C(x) < R(x)$.

Solve $245 + 1.6x < 10x - 0.04x^2$

$0.04x^2 - 8.4x + 245 < 0$

$f(x) = 0.04x^2 - 8.4x + 245 < 0$

Find the zeros of $f(x)$.

$$x = \frac{-(-8.4) \pm \sqrt{(-8.4)^2 - 4(0.04)(245)}}{2(0.04)}$$

$x = 35$ or 175

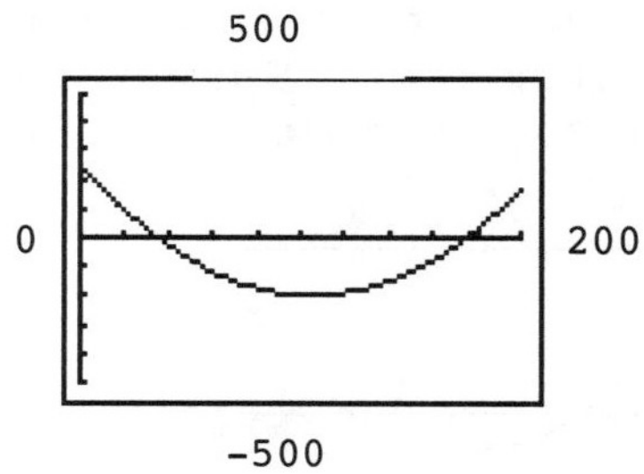

Plotting the graph of $f(x)$ we see that $f(x) < 0$ for $35 < x < 175$. The break-even points are therefore $(35, R(35)) = (35, 301)$ and $(175, R(175)) = (175, 525)$.

117. The revenue function is $R(x) = xd(x) = x(9.3 - 0.15x)$ or $R(x) = -0.15x^2 + 9.3x$.

$R(x)$ has a maximum value at the vertex of this parabola, which is given by

$$x = -\frac{b}{2a} = -\frac{9.3}{2(-0.15)} = 31$$

Then $p = d(31) = 9.3 - 0.15(31) = \$4.65$ is the price which maximizes the revenue.

119. (A) The revenue function is

$$R(x) = xd(x) = x(3.5 - 0.00007x)$$
$$= 3.5x - 0.00007x^2.$$

The domain is given by $x(3.5 - 0.00007x) \ge 0$ or $0 \le x \le 50{,}000$. The cost function $C(x)$ is given by $C(x)$ = Fixed Cost + Variable Cost = $24{,}500 + 0.35x$. The domain is given by $x \ge 0$ or $[0, \infty)$.
The company will break even when $R(x) = C(x)$. Solve

$$3.5x - 0.00007x^2 = 24{,}5000 + 0.35x$$
$$0 = 0.00007x^2 - 3.15x + 24{,}500$$
$$x = \frac{-(-3.15) \pm \sqrt{(-3.15)^2 - 4(0.00007)(24{,}500)}}{2(0.00007)}$$
$$x = 10{,}000 \text{ or } 35{,}000.$$

(B)

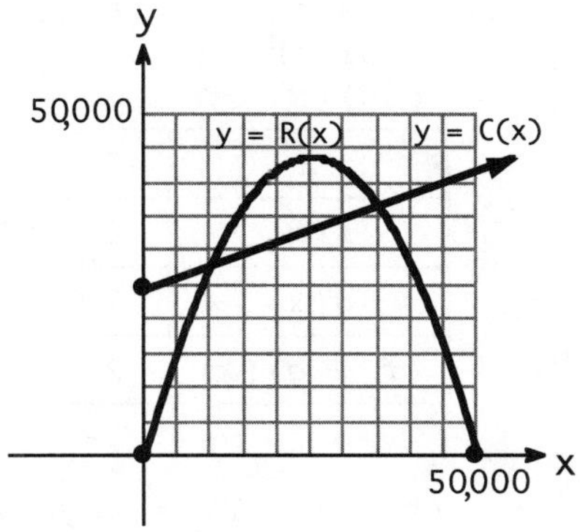

The company will break even for sales of 10,000 or 35,000 gallons.

(C) The company makes a profit for those sales levels for which the graph of the revenue function is above the graph of the cost function, that is, if the sales are between 10,000 and 35,000 gallons. The company suffers a loss for those sales levels for which the graph of the revenue function is below the graph of the cost function, that is, if the sales are between 0 and 10,000 gallons or between 35,000 and 50,000 gallons.

(D) The profit function is given by $P(x) = R(x) - C(x)$. Thus

$$P(x) = (3.5x - 0.00007x^2) - (24{,}500 + 0.35x)$$
$$= -0.00007x^2 + 3.15x - 24{,}500$$

The maximum value of this function occurs at the vertex of its parabola graph. This is given by the formulas (h, k) where

$$h = -\frac{b}{2a} = -\frac{3.15}{2(-0.00007)} = 22{,}500$$

and

$$k = C - \frac{b^2}{4a} = -24{,}500 - \frac{3.15^2}{4(-0.00007)} = 10{,}937.5$$

That is, the maximum profit is \$10,937.50 when 22,500 gallons are sold. Substitute $x = 22{,}500$ to find $p = d(22{,}500) = 3.5 - 0.00007(22{,}500) =$ \$1.92 per gallon.

Section 3-5

1. Construct a table of values of $f(x)$ and $g(x)$ from the graph, then add to obtain $(f + g)(x)$.

x	-3	-2	-1	0	1	2	3
$f(x)$	1	0	-1	-2	-3	-2	-1
$g(x)$	2	3	2	1	0	-1	-2
$(f + g)(x)$	3	3	1	-1	-3	-3	-3

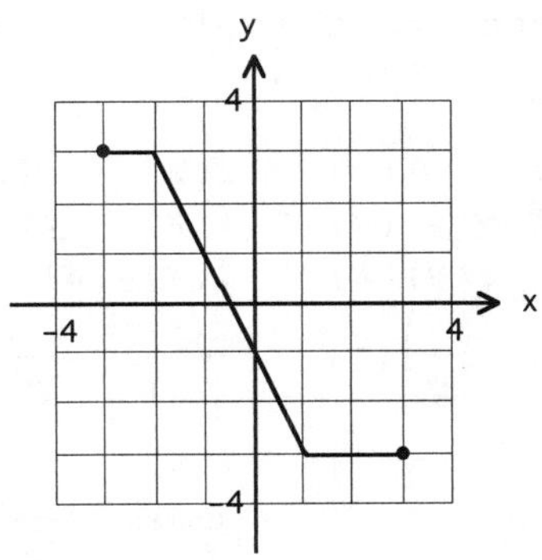

3. Construct a table of values of $f(x)$ and $g(x)$ from the graph, then multiply to obtain $(fg)(x)$.

x	-3	-2	-1	0	1	2	3
$f(x)$	1	0	-1	-2	-3	-2	-1
$g(x)$	2	3	2	1	0	-1	-2
$(f + g)(x)$	2	0	-2	-2	0	2	2

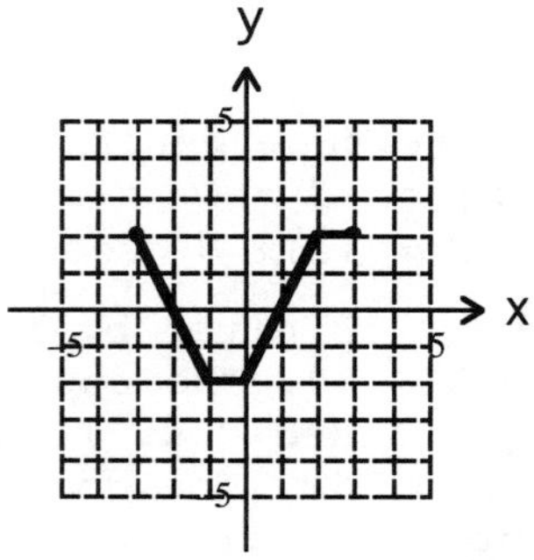

5. $(f \circ g)(-1) = f[g(-1)]$. From the graph of g, $g(-1) = 2$. From the graph of f, $f[g(-1)] = f(2) = -2$.

7. $(g \circ f)(-2) = g[f(-2)]$. From the graph of f, $f(-2) = 0$. From the graph of g, $g[f(-2)] = g(0) = 1$.

9. From the graph of g, $g(1) = 0$. From the graph of f, $f[g(1)] = f(0) = -2$.

11. From the graph of f, $f(2) = -2$. From the graph of g, $g[f(2)] = g(-2) = 3$.

13. $(f + g)(-3) = f(-3) + g(-3) = [2 - (-3)] + \sqrt{3 - (-3)} = 5 + \sqrt{6}$

15. $(g - f)(-5) = g(-5) - f(-5) = \sqrt{3 - (-5)} - [2 - (-5)] = \sqrt{8} - 7 = 2\sqrt{2} - 7$

17. $\left(\frac{f}{g}\right)(4) = \frac{f(4)}{g(4)}$. However, $g(4)$ does not exist, hence neither does $\frac{f}{g}(4)$.

19. $\left(\frac{g}{f}\right)(1) = \frac{g(1)}{f(1)} = \frac{\sqrt{3 - 1}}{2 - 1} = \frac{\sqrt{2}}{1} = \sqrt{2}$

21. $\left(\frac{f}{g}\right)(3) = \frac{f(3)}{g(3)}$. However, $g(3) = \sqrt{3 - 3} = 0$, hence $\left(\frac{f}{g}\right)(3)$ does not exist.

23. $g \circ f(-2) = g(f(-2)) = g(2 - (-2)) = g(4)$. However, $g(4)$ does not exist, hence neither does $g \circ f(-2)$.

25. $f \circ g(1) = f(g(1)) = f(\sqrt{3 - 1}) = f(\sqrt{2}) = 2 - \sqrt{2}$

27. $g \circ g(-7) = g(g(-7)) = g(\sqrt{3 - (-7)}) = g(\sqrt{10})$. However, $g(\sqrt{10})$ does not exist, hence neither does $g \circ g(-7)$.

29. $(f + g)(x) = f(x) + g(x) = 4x + x + 1 = 5x + 1$ Domain: $(-\infty, \infty)$
$(f - g)(x) = f(x) - g(x) = 4x - (x + 1) = 3x - 1$ Domain: $(-\infty, \infty)$
$(fg)(x) = f(x)g(x) = 4x(x + 1) = 4x^2 + 4x$ Domain: $(-\infty, \infty)$
$\left(\frac{f}{g}\right)(x) = \frac{f(x)}{g(x)} = \frac{4x}{x + 1}$ Domain: $\{x \mid x \neq -1\}$, *or* $(-\infty, -1) \cup (-1, \infty)$

Common Error:
$f(x) - g(x) \neq 4x - x + 1$. The parentheses are necessary.

31. $(f + g)(x) = f(x) + g(x) = 2x^2 + x^2 + 1 = 3x^2 + 1$ Domain: $(-\infty, \infty)$
$(f - g)(x) = f(x) - g(x) = 2x^2 - (x^2 + 1) = x^2 - 1$ Domain: $(-\infty, \infty)$
$(fg)(x) = f(x)g(x) = 2x^2(x^2 + 1) = 2x^4 + 2x^2$ Domain: $(-\infty, \infty)$
$\left(\frac{f}{g}\right)(x) = \frac{f(x)}{g(x)} = \frac{2x^2}{x^2 + 1}$ Domain: $(-\infty, \infty)$
(since $g(x)$ is never 0.)

33. $(f + g)(x) = f(x) + g(x) = 3x + 5 + x^2 - 1$
$= x^2 + 3x + 4$ Domain: $(-\infty, \infty)$
$(f - g)(x) = f(x) - g(x) = 3x + 5 - (x^2 - 1)$
$= 3x + 5 - x^2 + 1 = -x^2 + 3x + 6$ Domain: $(-\infty, \infty)$
$(fg)(x) = f(x)g(x) = (3x + 5)(x^2 - 1)$
$= 3x^3 - 3x + 5x^2 - 5 = 3x^3 + 5x^2 - 3x - 5$ Domain: $(-\infty, \infty)$
$\left(\frac{f}{g}\right)(x) = \frac{f(x)}{g(x)} = \frac{3x + 5}{x^2 - 1}$ Domain: $\{x \mid x \neq \pm 1\}$, *or*
$(-\infty, -1) \cup (-1, 1) \cup (1, \infty)$

35. $(f + g)(x) = f(x) + g(x) = \sqrt{2 - x} + \sqrt{x + 3}$
$(f - g)(x) = f(x) - g(x) = \sqrt{2 - x} - \sqrt{x + 3}$
$(fg)(x) = f(x)g(x) = \sqrt{2 - x}\sqrt{x + 3} = \sqrt{(2 - x)(3 + x)} = \sqrt{6 - x - x^2}$
$\left(\frac{f}{g}\right)(x) = \frac{f(x)}{g(x)} = \frac{\sqrt{2 - x}}{\sqrt{x + 3}} = \sqrt{\frac{2 - x}{x + 3}}$
The domains of f and g are:
Domain of $f = \{x \mid 2 - x \geq 0\} = (-\infty, 2]$
Domain of $g = \{x \mid x + 3 \geq 0\} = [-3, \infty)$
The intersection of these domains is $[-3, 2]$.
This is the domain of the functions of $f + g$, $f - g$, and fg.
Since $g(-3) = 0$, $x = -3$ must be excluded from the domain of $\frac{f}{g}$, so its domain is $(-3, 2]$.

37. $(f + g)(x) = f(x) + g(x) = \sqrt{x} + 2 + \sqrt{x} - 4 = 2\sqrt{x} - 2$
$(f - g)(x) = f(x) - g(x) = \sqrt{x} + 2 - (\sqrt{x} - 4) = \sqrt{x} + 2 - \sqrt{x} + 4 = 6$
$(fg)(x) = f(x)g(x) = (\sqrt{x} + 2)(\sqrt{x} - 4) = x - 2\sqrt{x} - 8$
$\left(\frac{f}{g}\right)(x) = \frac{f(x)}{g(x)} = \frac{\sqrt{x} + 2}{\sqrt{x} - 4}$
The domains of f and g are both $\{x \mid x \geq 0\} = [0, \infty)$
This is the domain of $f + g$, $f - g$, and fg. We note that in the donain of $\frac{f}{g}$,
$g(x) \neq 0$. Thus $\sqrt{x} - 4 \neq 0$. To solve this, we solve
$\sqrt{x} - 4 = 0$
$\sqrt{x} = 4$ **Common Error:** $x \neq \sqrt{4}$
$x = 16$
Hence, 16 must be excluded from $\{x \mid x \geq 0\}$ to find the domain of $\frac{f}{g}$.
Domain of $\frac{f}{g} = \{x \mid x \geq 0, x \neq 16\} = [0, 16) \cup (16, \infty)$.

39. $(f + g)(x) = f(x) + g(x) = \sqrt{x^2 + x - 6} + \sqrt{7 + 6x - x^2}$

$(f - g)(x) = f(x) - g(x) = \sqrt{x^2 + x - 6} - \sqrt{7 + 6x - x^2}$

$(fg)(x) = f(x)g(x) = \sqrt{x^2 + x - 6}\sqrt{7 + 6x - x^2} = \sqrt{-x^4 + 5x^3 + 19x^2 - 29x - 42}$

$\left(\frac{f}{g}\right)(x) = \frac{f(x)}{g(x)} = \frac{\sqrt{x^2 + x - 6}}{\sqrt{7 + 6x - x^2}} = \sqrt{\frac{x^2 + x - 6}{7 + 6x - x^2}}$

The domains of f and g are:
Domain of $f = \{x \mid x^2 + x - 6 \ge 0\} = \{x \mid (x + 3)(x - 2) \ge 0\} = (-\infty, -3] \cup [2, \infty)$
Domain of $g = \{x \mid 7 + 6x - x^2 \ge 0\} = \{x \mid (7 - x)(1 + x) \ge 0\} = [-1, 7]$
The intersection of these domains is $[2, 7]$.
This is the domain of the functions $f + g$, $f - g$, and fg.
Since $g(x) = 7 + 6x - x^2 = (7 - x)(1 + x)$, $g(7) = 0$ and $g(-1) = 0$, hence 7 must be excluded from the domain of $\frac{f}{g}$, so its domain is $[2, 7)$.

41. $(f + g)(x) = f(x) + g(x) = x + \frac{1}{x} + x - \frac{1}{x} = 2x$

Common Error:
Domain is not $(-\infty, \infty)$. See below.

$(f - g)(x) = f(x) - g(x) = x + \frac{1}{x} - \left(x - \frac{1}{x}\right) = \frac{2}{x}$

$(fg)(x) = f(x)g(x) = \left(x + \frac{1}{x}\right)\left(x - \frac{1}{x}\right) = x^2 - \frac{1}{x^2}$

$\left(\frac{f}{g}\right)(x) = \frac{f(x)}{g(x)} = \frac{x + \frac{1}{x}}{x - \frac{1}{x}} = \frac{x^2 + 1}{x^2 - 1}$

The domains of f and g are both $\{x \mid x \ne 0\} = (-\infty, 0) \cup (0, \infty)$
This is therefore the domain of $f + g$, $f - g$, and fg. To find the domain of $\frac{f}{g}$, we must exclude from this domain the set of values of x for which $g(x) = 0$.

$$x - \frac{1}{x} = 0$$
$$x^2 - 1 = 0$$
$$x^2 = 1$$
$$x = -1, 1$$

Hence, the domain of $\frac{f}{g}$ is $\{x \mid x \ne 0, -1, \text{ or } 1\}$ or $(-\infty, -1) \cup (-1, 0) \cup (0, 1) \cup (1, \infty)$.

43. $(f \circ g)(x) = f[g(x)] = f(x^2 - x + 1) = (x^2 - x + 1)^3$ Domain: $(-\infty, \infty)$

$(g \circ f)(x) = g[f(x)] = g(x^3) = (x^3)^2 - x^3 + 1 = x^6 - x^3 + 1$ Domain: $(-\infty, \infty)$

45. $(f \circ g)(x) = f[g(x)] = f(2x + 3) = |2x + 3 + 1| = |2x + 4|$ Domain: $(-\infty, \infty)$

$(g \circ f)(x) = g[f(x)] = g(|x + 1|) = 2|x + 1| + 3$ Domain: $(-\infty, \infty)$

47. $(f \circ g)(x) = f[g(x)] = f(2x^3 + 4) = (2x^3 + 4)^{1/3}$ Domain: $(-\infty, \infty)$

$(g \circ f)(x) = g[f(x)] = g(x^{1/3}) = 2(x^{1/3})^3 + 4 = 2x + 4$ Domain: $(-\infty, \infty)$

49. $(f \circ g)(x) = f[g(x)] = f(x - 4) = \sqrt{x - 4}$ Domain: $\{x \mid x \ge 4\}$ or $[4, \infty)$

$(g \circ f)(x) = g[f(x)] = g(\sqrt{x}) = \sqrt{x} - 4$ Domain: $\{x \mid x \ge 0\}$ or $[0, \infty)$

51. $(f \circ g)(x) = f[g(x)] = f\left(\frac{1}{x}\right) = \frac{1}{x} + 2$ Domain: $\{x \mid x \ne 0\}$ or $(-\infty, 0) \cup (0, \infty)$

$(g \circ f)(x) = g[f(x)] = g(x + 2) = \frac{1}{x + 2}$ Domain: $\{x \mid x \ne -2\}$ or $(-\infty, -2) \cup (-2, \infty)$

53. $(f \circ g)(x) = f[g(x)] = f(x^2) = \sqrt{4 - x^2}$
The domain of f is $(-\infty, 4]$. The domain of g is all real numbers. Hence the domain of $f \circ g$ is the set of those real numbers x for which $g(x)$ is in $(-\infty, 4]$, that is, for which $x^2 \le 4$, or $-2 \le x \le 2$.
Domain of $f \circ g = \{x \mid -2 \le x \le 2\} = [-2, 2]$
$(g \circ f)(x) = g[f(x)] = g(\sqrt{4 - x}) = (\sqrt{4 - x})^2 = 4 - x$
The domain of $g \circ f$ is the set of those numbers x in $(-\infty, 4]$ for which $f(x)$ is in $(-\infty, \infty)$, that is, $(-\infty, 4]$.

55. $(f \circ g)(x) = f[g(x)] = f\left(\frac{x}{x - 2}\right) = \frac{\frac{x}{x-2} + 5}{\frac{x}{x-2}} = \frac{x + 5(x - 2)}{x} = \frac{x + 5x - 10}{x} = \frac{6x - 10}{x}$
The domain of f is $\{x \mid x \ne 0\}$. The domain of g is $\{x \mid x \ne 2\}$. Hence the domain of $f \circ g$ is the set of those numbers in $\{x \mid x \ne 2\}$ for which $g(x)$ is in $\{x \mid x \ne 0\}$. Thus we must exclude from $\{x \mid x \ne 2\}$ those numbers x for which $\frac{x}{x - 2} = 0$, or $x = 0$. Hence the domain of $f \circ g$ is $\{x \mid x \ne 0, x \ne 2\}$, or $(-\infty, 0) \cup (0, 2) \cup (2, \infty)$.
$(g \circ f)(x) = g[f(x)] = g\left(\frac{x + 5}{x}\right) = \frac{\frac{x+5}{x}}{\frac{x+5}{x} - 2} = \frac{x + 5}{x + 5 - 2x} = \frac{x + 5}{5 - x}$
The domain of $g \circ f$ is the set of those numbers in $\{x \mid x \ne 0\}$ for which $f(x)$ is in $\{x \mid x \ne 2\}$. Thus we must exclude from $\{x \mid x \ne 0\}$ those numbers x for which $\frac{x + 5}{x} = 2$, or $x + 5 = 2x$, or $x = 5$. Hence the domain of $g \circ f$ is $\{x \mid x \ne 0, x \ne 5\}$ or $(-\infty, 0) \cup (0, 5) \cup (5, \infty)$.

57. $(f \circ g)(x) = f[g(x)] = f\left(\frac{1}{x - 2}\right) = \frac{2\left(\frac{1}{x-2}\right) + 1}{\frac{1}{x-2}} = \frac{2 + x - 2}{1} = x$
The domain of f is $\{x \mid x \ne 0\}$. The domain of g is $\{x \mid x \ne 2\}$. Hence the domain of $f \circ g$ is the set of those numbers in $\{x \mid x \ne 2\}$ for which $g(x)$ is in $\{x \mid x \ne 0\}$. Thus we must exclude from $\{x \mid x \ne 2\}$ those numbers x for which $\frac{1}{x - 2} = 0$; however, there are none. Hence the domain of $f \circ g$ is $\{x \mid x \ne 2\}$ or $(-\infty, 2) \cup (2, \infty)$.
$(g \circ f)(x) = g[f(x)] = g\left(\frac{2x + 1}{x}\right) = \frac{1}{\frac{2x+1}{x} - 2} = \frac{x}{2x + 1 - 2x} = \frac{x}{1} = x$
The domain of $g \circ f$ is the set of those numbers in $\{x \mid x \ne 0\}$ for which $f(x)$ is in $\{x \mid x \ne 2\}$. Thus we must exclude from $\{x \mid x \ne 0\}$ those numbers x for which $\frac{2x + 1}{x} = 2$; however, there are none. Hence the domain of $g \circ f$ is $\{x \mid x \ne 0\}$ or $(-\infty, 0) \cup (0, \infty)$.

59. $(f \circ g)(x) = f[g(x)] = f(\sqrt{9 + x^2}) = \sqrt{25 - (\sqrt{9 + x^2})^2} = \sqrt{25 - (9 + x^2)} = \sqrt{16 - x^2}$
The domain of f is $[-5, 5]$. The domain of g is $(-\infty, \infty)$. Hence the domain of $f \circ g$ is the set of those real numbers x for which $g(x)$ is in $[-5, 5]$, that is, $\sqrt{9 + x^2} \le 5$, or $9 + x^2 \le 25$, or $x^2 \le 16$, or $-4 \le x \le 4$. Hence the domain of $f \circ g$ is $\{x \mid -4 \le x \le 4\}$ or $[-4, 4]$.
$(g \circ f)(x) = g[f(x)] = g(\sqrt{25 - x^2}) = \sqrt{9 + (\sqrt{25 - x^2})^2} = \sqrt{9 + 25 - x^2} = \sqrt{34 - x^2}$
The domain of $g \circ f$ is the set of those numbers x in $[-5, 5]$ for which $g(x)$ is real. Since $g(x)$ is real for all x, the domain of $g \circ f$ is $[-5, 5]$.

Common Error: The domain of $g \circ f$ is not evident from the final form $\sqrt{34 - x^2}$. It is not $[-\sqrt{34}, \sqrt{34}]$.

61. $(f \circ g)(x) = f[g(x)] = f(2x - 2) = \frac{1}{2}(2x - 2) + 1 = x - 1 + 1 = x$

$(g \circ f)(x) = g[f(x)] = g\left(\frac{1}{2}x + 1\right) = 2\left(\frac{1}{2}x + 1\right) - 2 = x + 2 - 2 = x$

Graphing f, g, $f \circ g$, and $g \circ f$, we obtain the graph at the right.

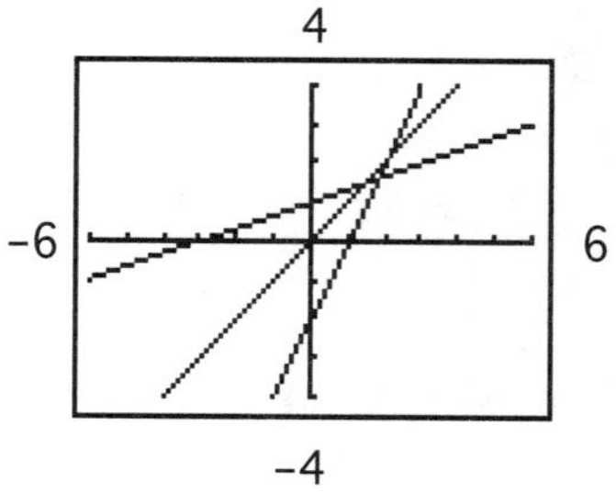

The graphs of f and g are reflections of each other in the line $y = x$, which is the graph of $f \circ g$ and $g \circ f$.

63. $(f \circ g)(x) = f[g(x)] = f\left(-\frac{3}{2}x - \frac{5}{2}\right) = -\frac{2}{3}\left(-\frac{3}{2}x - \frac{5}{2}\right) - \frac{5}{3} = x + \frac{5}{3} - \frac{5}{3} = x$

$(g \circ f)(x) = g[f(x)] = g\left(-\frac{2}{3}x - \frac{5}{3}\right) = -\frac{3}{2}\left(-\frac{2}{3}x - \frac{5}{3}\right) - \frac{5}{2} = x + \frac{5}{2} - \frac{5}{2} = x$

Graphing f, g, $f \circ g$, and $g \circ f$, we obtain the graph at the right.

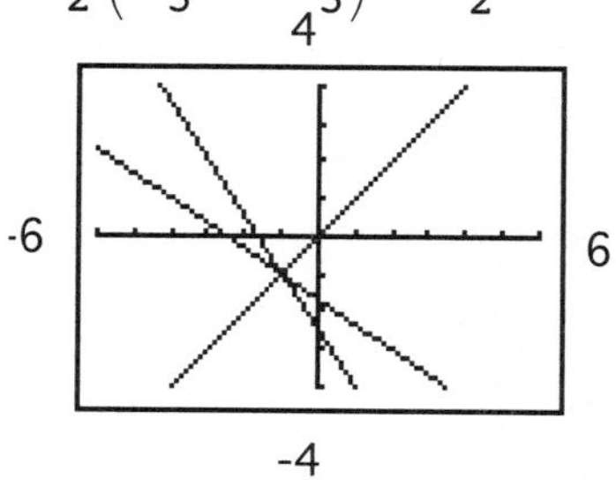

The graphs of f and g are reflections of each other in the line $y = x$, which is the graph of $f \circ g$ and $g \circ f$.

65. $f \circ g(x) = f[g(x)] = f(2\sqrt[3]{x}) = \frac{\left(2\sqrt[3]{x}\right)^3}{8} = \frac{8x}{8} = x$

$g \circ f(x) = g[f(x)] = g\left(\frac{x^3}{8}\right) = 2\sqrt[3]{\frac{x^3}{8}} = 2\left(\frac{x}{2}\right) = x$

Graphing f, g, $f \circ g$, and $g \circ f$, we obtain the graph at the right.

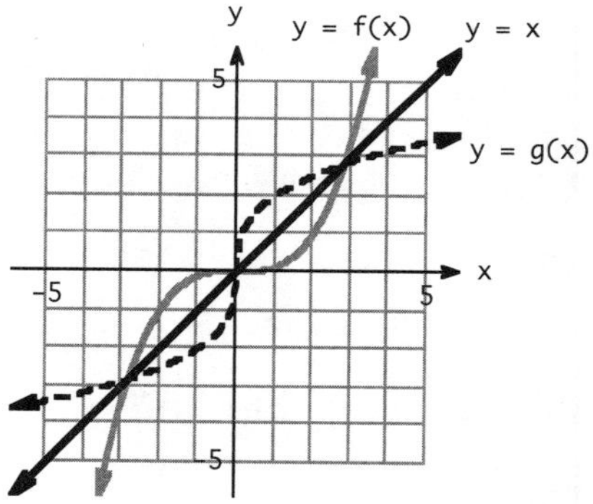

The graphs of f and g are reflections of each other in the line $y = x$, which is the graph of $f \circ g$ and $g \circ f$.

67. $f \circ g(x) = f[g(x)] = f(x^3 + 2) = \sqrt[3]{x^3 + 2 - 2} = \sqrt[3]{x^3} = x$

$g \circ f(x) = g[f(x)] = g(\sqrt[3]{x - 2}) = (\sqrt[3]{x - 2})^3 + 2 = x - 2 + 2 = x$

Graphing f, g, $f \circ g$, and $g \circ f$, we obtain the graph at the right.

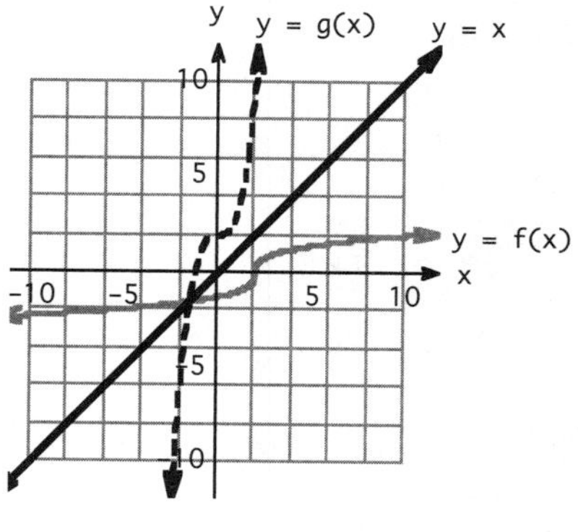

The graphs of $f \circ g$ and $g \circ f$ are reflections of each other in the line $y = x$, which is the graph of $f \circ g$ and $g \circ f$.

69. If we let $g(x) = 2x - 7$, then
$h(x) = [g(x)]^4$
Now if we let $f(x) = x^4$, we have
$h(x) = [g(x)]^4 = f[g(x)] = (f \circ g)(x)$

71. If we let $g(x) = 4 + 2x$, then
$h(x) = \sqrt{g(x)}$
Now if we let $f(x) = x^{1/2}$, we have
$h(x) = \sqrt{g(x)} = [g(x)]^{1/2}$
$= f[g(x)] = (f \circ g)(x)$.

73. If we let $f(x) = x^7$, then
$h(x) = 3f(x) - 5$
Now if we let $g(x) = 3x - 5$, we have
$h(x) = 3f(x) - 5 = g[f(x)] = (g \circ f)(x)$

75. If we let $f(x) = x^{-1/2}$, then
$h(x) = 4f(x) + 3$
Now if we let $g(x) = 4x + 3$, we have
$h(x) = 4f(x) + 3 = g[f(x)] = (g \circ f)(x)$

77. fg and gf are identical, since
$(fg)(x) = f(x)g(x) = g(x)f(x)$ by the commutative law for multiplication of real numbers
$= (gf)(x)$

79. Yes, the function $g(x) = x$ satisfies these conditions.
$(f \circ g)(x) = f(g(x)) = f(x)$, so $f \circ g = f$
$(g \circ f)(x) = g(f(x)) = f(x)$, so $g \circ f = f$

81. Profit is the difference of the amount of money taken in (Revenue) and the amount of money spent (Cost), so $P(x) = R(x) - C(x)$

$$= \left(20x - \frac{1}{200}x^2\right) - (2x + 8{,}000)$$
$$= 20x - \frac{1}{200}x^2 - 2x - 8{,}000 \quad \text{(Distribute!)}$$
$$= 18x - \frac{1}{200}x^2 - 8{,}000$$

we have a profit function, but it's a function of the demand (x), not the price (p). We were given $x = 4{,}000 - 20p$, so we can substitute $4{,}000 - 200p$ in for x to get desired function:

$$P(x) = P(4{,}000 - 200p) = 18(4{,}000 - 200p) - \frac{1}{200}(4{,}000 - 200p)^2 - 8{,}000$$
$$= 72{,}000 - 3{,}600p - \frac{1}{200}(16{,}000{,}000 - 1{,}600{,}000p + 40{,}000p^2) - 8{,}000$$
$$= 72{,}000 - 3{,}600p - 80{,}000 + 8{,}000p - 200p^2 - 8{,}000$$
$$= -16{,}000 + 4{,}400p - 200p^2$$

Now we'll graph the function and use the maximum feature to find the largest profit.

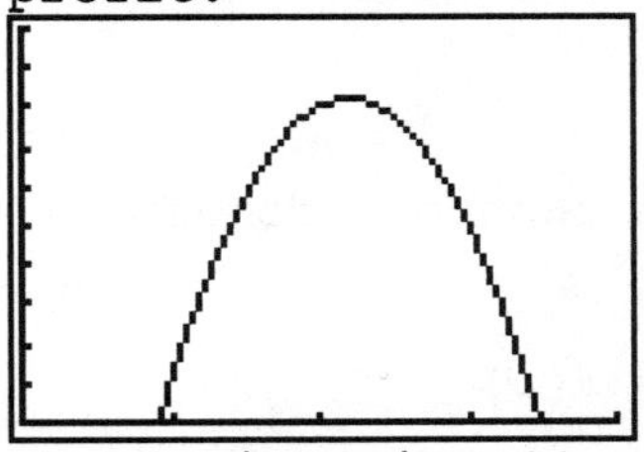
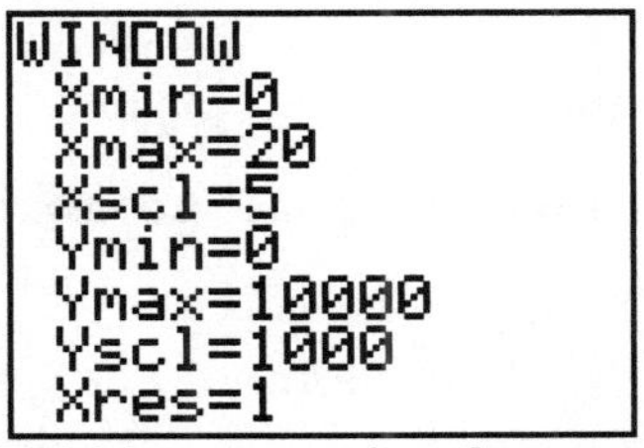

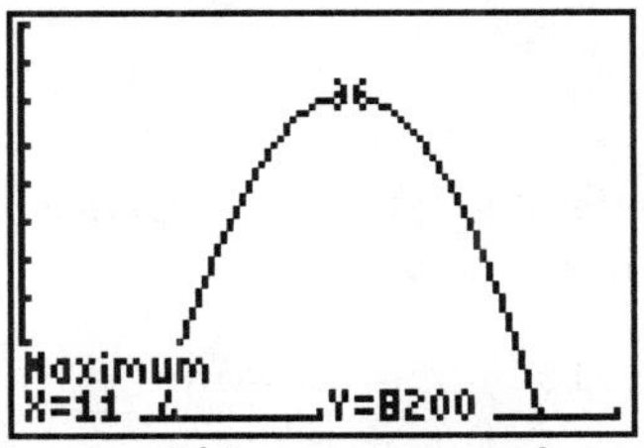

The maximum is (11, 9,200), so the largest profit occurs when the price is \$11.

83. We are given $V(r) = 0.1A(r) = 0.1\pi r^2$ and $r(t) = 0.4t^{1/3}$.
Hence we use composition to express V as a function of the time.

$$(V \circ r)(t) = V[r(t)]$$
$$= 0.1\pi[r(t)]^2$$
$$= 0.1\pi[0.4t^{1/3}]^2$$
$$= 0.1\pi[0.16t^{2/3}]$$
$$= 0.016\pi t^{2/3}$$

We write $V(t) = 0.016\pi t^{2/3}$

85. 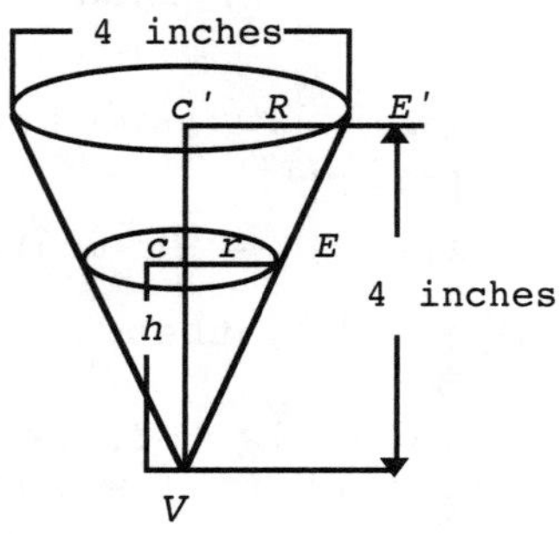

(A) We note: In the figure, triangles VCE and $VC'E'$ are similar. Moreover

R = radius of cup = $\frac{1}{2}$ diameter of cup = $\frac{1}{2}(4) = 2$ inches.

Hence $\frac{r}{2} = \frac{h}{4}$ or $r = \frac{1}{2}h$. We write $r(h) = \frac{1}{2}h$.

(B) Since $V = \frac{1}{3}\pi r^2 h$ and $r = \frac{1}{2}h$, $V = \frac{1}{3}\pi\left(\frac{1}{2}h\right)^2 h = \frac{1}{3}\pi\frac{1}{4}h^2 h = \frac{1}{12}\pi h^3$.

We write $V(h) = \frac{1}{12}\pi h^3$.

(C) Since $V(h) = \frac{1}{12}\pi h^3$ and $h(t) = 0.5\sqrt{t}$, we use composition to express V as a function of t.

$$\begin{aligned}(V \circ h)(t) &= V[h(t)] \\ &= \frac{1}{12}\pi\ [h(t)]^3 \\ &= \frac{1}{12}\pi\ [0.5\sqrt{t}\,]^3 \\ &= \frac{1}{12}\pi(0.125)(t^{1/2})^3 \\ &= \frac{0.125}{12}\pi t^{3/2}\end{aligned}$$

We write $V(t) = \frac{0.125}{12}\pi t^{3/2}$

Section 3-6

1. One-to-one

3. The range element 3 corresponds to more than one domain element. Not one-to-one.

5. One-to-one

7. The range element 7 corresponds to more than one domain element. Not one-to-one.

9. One-to-one

11. Some range elements (0, for example) correspond to more than one domain element. Not one-to-one.

13. One-to-one

15. One-to-one

17. Assume $F(a) = F(b)$

$\frac{1}{2}a + 1 = \frac{1}{2}b + 1$

Then $\frac{1}{2}a = \frac{1}{2}b$

$a = b$

Therefore F is one-to-one.

19. $H(x) = 4 - x^2$

Since $H(1) = 4 - 1^2 = 3$ and $H(-1) = 4 - (-1)^2 = 3$, both $(1, 3)$ and $(-1, 3)$ belong to H.
H is not one-to-one.

21. Assume $M(a) = M(b)$

$\sqrt{a+1} = \sqrt{b+1}$

Then $a + 1 = b + 1$

$a = b$

M is one-to-one.

23. $f(g(x)) = f\left(\frac{1}{3}x - \frac{5}{3}\right) = 3\left(\frac{1}{3}x - \frac{5}{3}\right) + 5 = x - 5 + 5 = x$

$g(f(x)) = g(3x + 5) = \frac{1}{3}(3x + 5) - \frac{5}{3} = x + \frac{5}{3} - \frac{5}{3} = x$

f and g are inverses

25. $f(g(x)) = f(\sqrt[3]{3 - x} - 1) = 2 - ((\sqrt[3]{3 - x} - 1) + 1)^3 = 2 - (\sqrt[3]{3 - x})^3$
$= 2 - (3 - x) = -1 + x$

f and g are not inverses since $(f \circ g)(x)$ is not x.

27. $f(g(x)) = f\left(\frac{3 + 4x}{2 - x}\right) = \dfrac{2\left(\frac{3+4x}{2-x}\right) - 3}{\frac{3+4x}{2-x} + 4} = \dfrac{\frac{6+8x}{2-x} - 3}{\frac{3+4x}{2-x} + 4} = \dfrac{\frac{6+8x}{2-x} - \frac{3(2-x)}{2-x}}{\frac{3+4x}{2-x} + \frac{4(2-x)}{2-x}}$

$= \dfrac{\frac{6+8x-6+3x}{2-x}}{\frac{3+4x+8-4x}{2-x}} = \dfrac{\frac{11x}{2-x}}{\frac{11}{2-x}} = \frac{11x}{2 - x} \cdot \frac{2 - x}{11} = x$

$g(f(x)) = g\left(\frac{2x - 3}{x + 4}\right) = \dfrac{3 + 4\left(\frac{2x-3}{x+4}\right)}{2 - \left(\frac{2x-3}{x+4}\right)} = \dfrac{3 + \frac{8x-12}{x+4}}{2 - \frac{2x-3}{x+4}} = \dfrac{\frac{3(x+4)}{x+4} + \frac{8x-12}{x+4}}{\frac{2(x+4)}{x+4} - \frac{2x-3}{x+4}}$

$= \dfrac{\frac{3x+12+8x-12}{x+4}}{\frac{2x+8-2x+3}{x+4}} = \dfrac{\frac{11x}{x+4}}{\frac{11}{x+4}} = \frac{11x}{x + 4} \cdot \frac{x + 4}{11} = x$

f and g are inverses.

29. $f(g(x)) = f(\sqrt{x - 4}) = 4 + (\sqrt{x - 4})^2 = 4 + x - 4 = x$

$g(f(x)) = g(4 + x^2) = \sqrt{4 + x^2 - 4} = \sqrt{x^2} = x$ as long as $x \geq 0$.

f and g are inverses.

31. $f(g(x)) = f(x^2 - 2) = \sqrt{x^2 - 2 + 2} = \sqrt{x^2} = x$ as long as $x \geq 0$.

$g(f(x)) = g(\sqrt{x + 2}) = (\sqrt{x + 2})^2 - 2 = x + 2 - 2 = x$

f and g are inverses.

33. $f(g(x)) = f(x^2 + 2) = -\sqrt{x^2 + 2 - 2} = -\sqrt{x^2} = -(-x) = x$ as long as $x \leq 0$.

$g(f(x)) = g(-\sqrt{x - 2}) = (-\sqrt{x - 2})^2 + 2 = x - 2 + 2 = x$

f and g are inverses.

35. $f(g(x)) = f(-\sqrt{6 - x}) = 6 - (-\sqrt{6 - x})^2 = 6 - (6 - x) = x$

$g(f(x)) = g(6 - x^2) = -\sqrt{6 - (6 - x^2)} = -\sqrt{x^2} = -(-x) = x$ as long as $x \leq 0$.

f and g are inverses.

37. The function h multiplies the input by 3 and then subtracts 7. The opposite of this is adding 7 to the input and then dividing by 3. The equation that corresponds to these steps is $h^{-1}(x) = \frac{x + 7}{3}$. (Remember that the order of the steps must be reversed.)

39. The function m adds 11 to the input then takes the cube root. The opposite of this is cubing the input, then subtracting 11. The equation that corresponds to these steps is $m^{-1}(x) = x^3 - 11$. (Remember that the order of the steps must be reversed.)

41. The function s multiplies the input by 3, adds 17, than raises the resulting expression to the fifth power. The opposite of this is taking the fifth root of the input, subtracting 17, then dividing by 3. The equation that corresponds to these steps is $s^{-1}(x) = \frac{\sqrt[5]{x} - 17}{3}$. (Remember that the order of the steps must be reversed.)

43. From the graph:
domain of $f = [-4, 4]$
range of $f = [1, 5]$
Therefore:
domain of $f^{-1} = [1, 5]$
range of $f^{-1} = [-4, 4]$

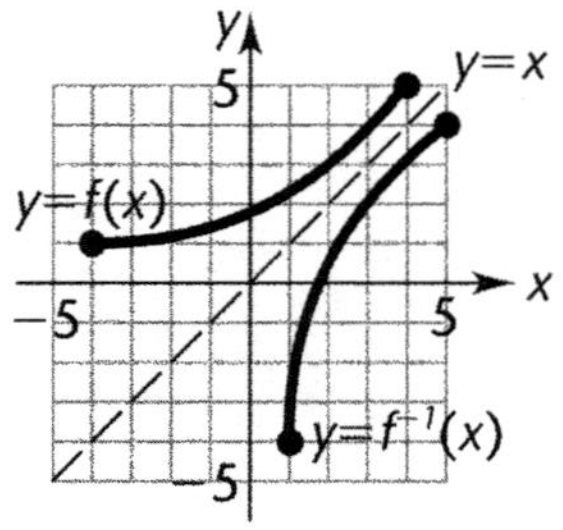

45. From the graph:
domain of $f = [-5, 3]$
range of $f = [-3, 5]$
Therefore:
domain of $f^{-1} = [-3, 5]$
range of $f^{-1} = [-5, 3]$

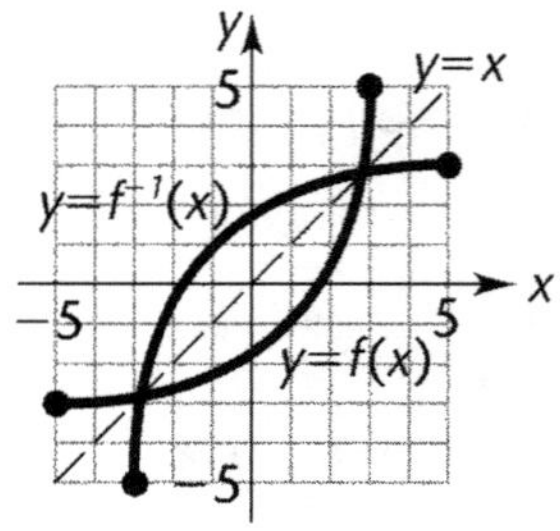

47. The graph of $f(x)$ is a line; f is one-to-one.
The domain of f is therefore $(-\infty, \infty)$ and the range of f is also $(-\infty, \infty)$.

Write $y = f(x)$
$y = 3x$
Solve $y = 3x$ for x.
$\frac{1}{3}y = x$
Interchange x and y:
$\frac{1}{3}x = y$
$f^{-1}(x) = \frac{1}{3}x$

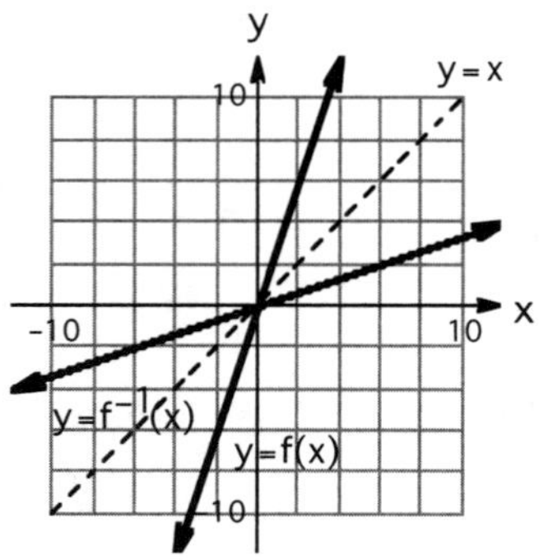

The domain of f^{-1} is $(-\infty, \infty)$ and the range of f^{-1} is $(-\infty, \infty)$.

49. The graph of $f(x)$ is a line; f is one-to-one.
The domain and range of f are therefore $(-\infty, \infty)$.
Write $y = f(x)$
$y = 4x - 3$
Solve $y = 4x - 3$ for x.
$y + 3 = 4x$
$\frac{y + 3}{4} = x$
Interchange x and y:
$\frac{x + 3}{4} = y$
$f^{-1}(x) = \frac{x + 3}{4}$

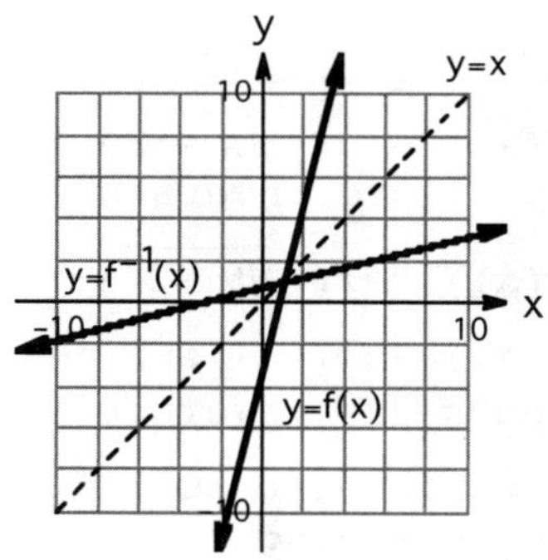

The domain and range of f^{-1} are $(-\infty, \infty)$.

51. The graph of $f(x)$ is a line; f is one-to-one. The domain and range of f are therefore $(-\infty, \infty)$.
Write $y = f(x)$
$y = 0.2x + 0.4$
Solve for x.
$y - 0.4 = 0.2x$
$5y - 2 = x$
Interchange x and y:
$5x - 2 = y$
$f^{-1}(x) = 5x - 2$
The domain and range of f^{-1} are $(-\infty, \infty)$.

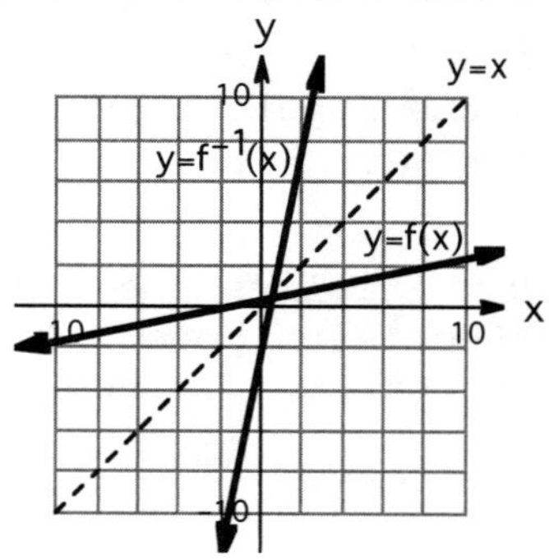

53. From the graph as shown, f is an increasing function with domain $[0, \infty)$ and range $[3, \infty)$; f is one-to-one.

Write $y = f(x)$

$$y = \sqrt{x} + 3$$

Solve for x.

$$y - 3 = \sqrt{x}$$

$$(y - 3)^2 = x$$

Interchange x and y.

$$(x - 3)^2 = y$$

$$f^{-1}(x) = (x - 3)^2 \qquad \text{domain } [3, \infty) \quad \text{range } [0, \infty)$$

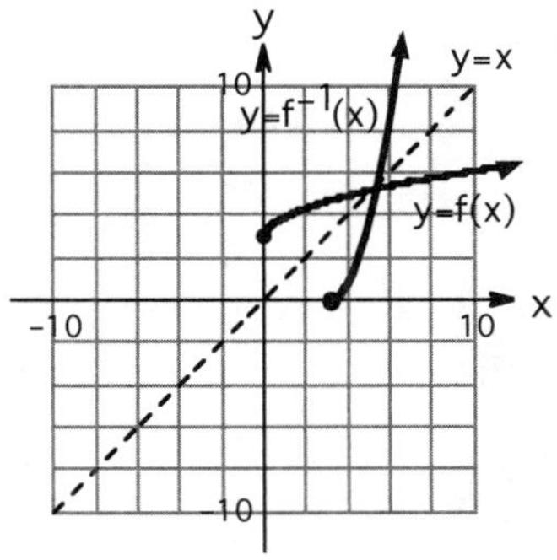

Common error: $f^{-1}(x) \neq (x - 3)^2$ for all x.
The graph is only the right-hand half of the parabola.

55. From the graph as shown, f is a decreasing function with domain $(-\infty, 16]$ and range $[0, \infty)$; f is one-to-one.

Write $y = f(x)$.

$$y = \frac{1}{2}\sqrt{16 - x}$$

Solve for x.

$$2y = \sqrt{16 - x}$$

$$4y^2 = 16 - x$$

$$x = 16 - 4y^2$$

Interchange x and y.

$$y = 16 - 4x^2$$

$$f^{-1}(x) = 16 - 4x^2 \qquad \text{domain } [0, \infty) \quad \text{range } (-\infty, 16]$$

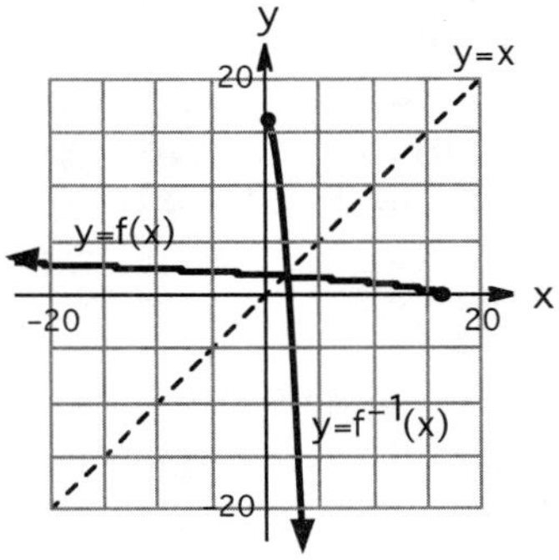

57. From the graph as shown, f is a decreasing function with domain $[1, \infty)$ and range $(-\infty, 3]$; f is one-to-one.

Write $y = f(x)$.

$$y = 3 - \sqrt{x - 1}$$

Solve for x.

$$y - 3 = -\sqrt{x - 1}$$

$$(y - 3)^2 = x - 1$$

$$(y - 3)^2 + 1 = x$$

Interchange x and y.

$$y = (x - 3)^2 + 1$$

$$f^{-1}(x) = (x - 3)^2 + 1 \qquad \text{domain } (-\infty, 3] \quad \text{range } [1, \infty)$$

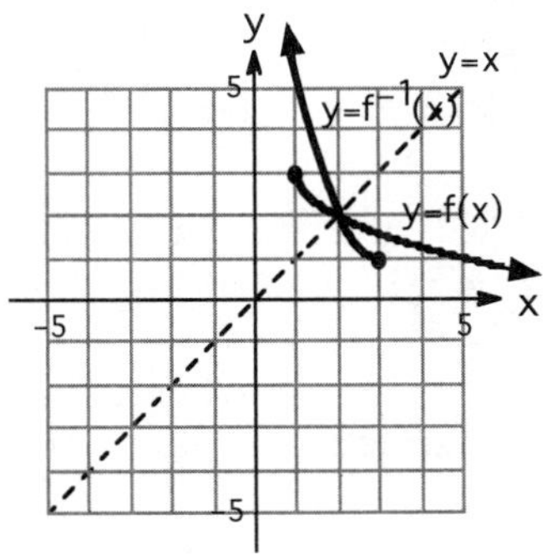

59. The graph of f is the right-hand half of a parabola; f is one-to-one. The domain of f is $[0, \infty)$ and the range is $[5, \infty)$.

Write $y = f(x)$.

$$y = x^2 + 5 \qquad x \geq 0$$

Solve for x.

$$y - 5 = x^2 \qquad x \geq 0$$

$$\sqrt{y - 5} = x \qquad x \geq 0$$

(note: only the positive square root is correct)

Interchange x and y.

$$y = \sqrt{x - 5} \qquad y \geq 0$$

$$f^{-1}(x) = \sqrt{x - 5} \qquad \text{domain } [5, \infty) \quad \text{range } [0, \infty)$$

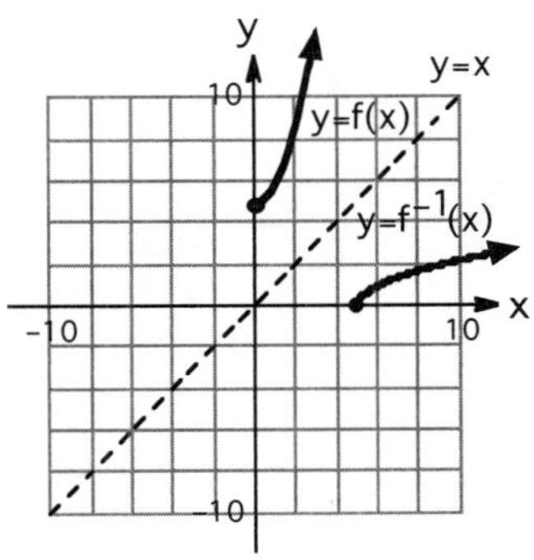

61. The graph of f is the left-hand half of a parabola; f is one-to-one. The domain of f is $(-\infty, 0]$ and the range is $(-\infty, 4]$.
Write $y = f(x)$.

$y = 4 - x^2 \quad x \le 0$

Solve for x.

$y - 4 = -x^2 \quad x \le 0$

$4 - y = x^2 \quad x \le 0$

$-\sqrt{4 - y} = x \quad x \le 0$

(note: only the negative square root is correct)
Interchange x and y.

$y = -\sqrt{4 - x} \quad y \le 0$

$f^{-1}(x) = -\sqrt{4 - x}$ domain $(-\infty, 4]$
range $(-\infty, 0]$

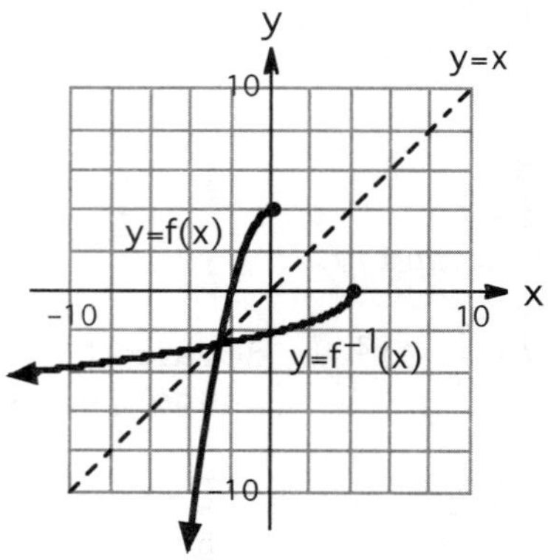

63. The graph of f is the left-hand half of a parabola; f is one-to-one.
Write $6x - x^2 = -(x^2 - 6x) = -(x^2 - 6x + 9) + 9$
$= 9 - (x - 3)^2$
Then the domain of f is given as $(-\infty, 3]$ and the range is seen to be $(-\infty, 9]$.
Write $y = f(x)$.

$y = 9 - (x - 3)^2 \quad x \le 3$

Solve for x.

$y - 9 = -(x - 3)^2 \quad x \le 3$

$9 - y = (x - 3)^2 \quad x \le 3$

$-\sqrt{9 - y} = x - 3 \quad x \le 3$

$3 - \sqrt{9 - y} = x \quad x \le 3$

Interchange x and y.

$y = 3 - \sqrt{9 - x} \quad y \le 3$

$f^{-1}(x) = 3 - \sqrt{9 - x}$ domain $(-\infty, 9]$
range $(-\infty, 3]$

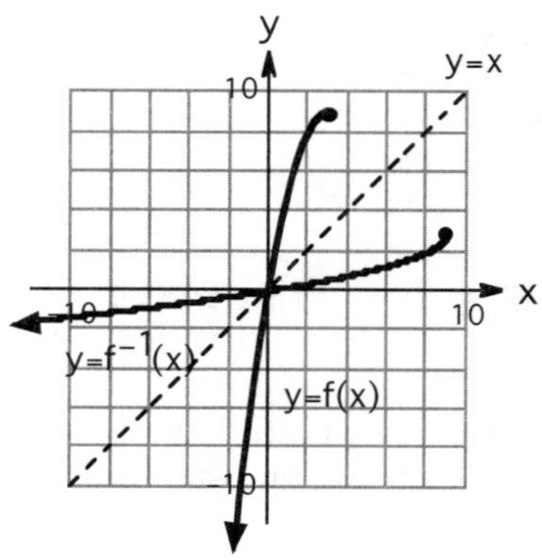

65. The graph of f is the right-hand half of a parabola; f is one-to-one.
Write $x^2 + 8x = (x^2 + 8x + 16) - 16$
$= (x + 4)^2 - 16$
Then the domain of f is given as $[-4, \infty)$ and the range is seen to be $[-16, \infty)$.
Write $y = f(x)$.

$y = (x + 4)^2 - 16 \quad x \ge -4$

Solve for x.

$y + 16 = (x + 4)^2 \quad x \ge -4$

$\sqrt{y + 16} = x + 4 \quad x \ge -4$

$\sqrt{y + 16} - 4 = x \quad x \ge -4$

Interchange x and y.

$y = \sqrt{x + 16} - 4 \quad y \ge -4$

$f^{-1}(x) = \sqrt{x + 16} - 4$ domain $[-16, \infty)$
range $[-4, \infty)$

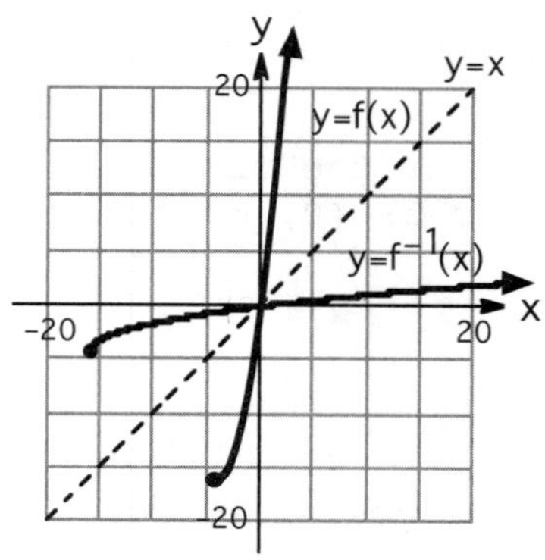

67. The graph of f is the left-hand half of a parabola; f is one-to-one.
The domain of f is $(-\infty, 2]$ and the range is $[0, \infty)$.
Write $y = f(x)$.

$y = (2 - x)^2 \quad x \le 2$

Solve for x.

$\sqrt{y} = 2 - x \quad x \le 2$

$x = 2 - \sqrt{y} \quad x \le 2$

Interchange x and y.

$y = 2 - \sqrt{x} \quad y \le 2$

$f^{-1}(x) = 2 - \sqrt{x}$ domain $[0, \infty)$ range $(-\infty, 2]$

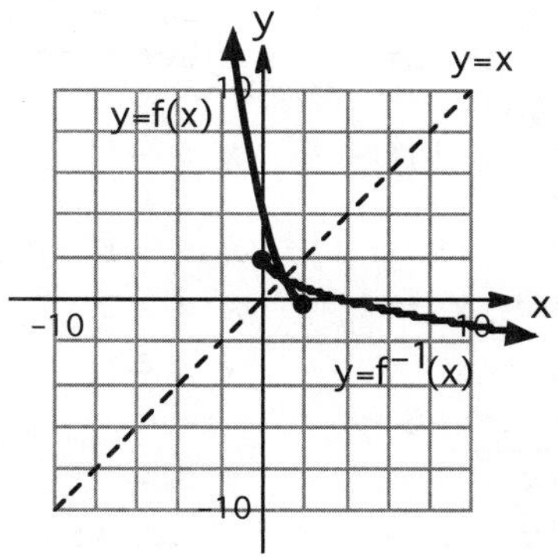

69. The graph of f is the right-hand half of a parabola; f is one-to-one.
The domain of f is $[-1, \infty)$ and the range is $[0, \infty)$.
Write $y = f(x)$.

$y = (1 + x)^2 \quad x \ge -1$

Solve for x.

$\sqrt{y} = 1 + x \quad x \ge -1$

$\sqrt{y} - 1 = x \quad x \ge -1$

Interchange x and y.

$y = \sqrt{x} - 1 \quad y \ge -1$

$f^{-1}(x) = \sqrt{x} - 1$ domain $[0, \infty)$ range $[-1, \infty)$

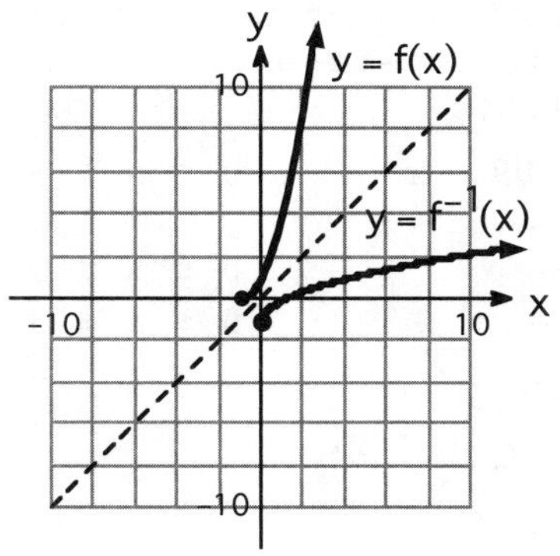

71. The graph of f is the right-hand half of a parabola; f is one-to-one.
The domain of f is $[1, \infty)$ and the range is $[2, \infty)$.
Write $y = f(x)$.

$y = (x - 1)^2 + 2 \quad x \ge 1$

Solve for x.

$y - 2 = (x - 1)^2 \quad x \ge 1$

$\sqrt{y - 2} = x - 1 \quad x \ge 1$

$1 + \sqrt{y - 2} = x \quad x \ge 1$

Interchange x and y.

$y = 1 + \sqrt{x - 2} \quad y \ge 1$

$f^{-1}(x) = 1 + \sqrt{x - 2}$ domain $[2, \infty)$ range $[1, \infty)$

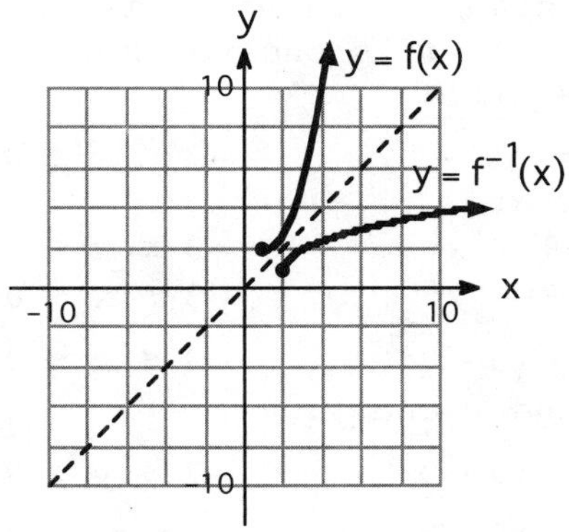

73. The graph of f is the left-hand half of a parabola; f is one-to-one.
Write $x^2 + 2x - 2 = (x^2 + 2x + 1) - 1 - 2$
$= (x + 1)^2 - 3$
Then the domain of f is given as $(-\infty, -1]$ and the range is seen to be $[-3, \infty)$.
Write $y = f(x)$.

$y = (x + 1)^2 - 3 \quad x \le -1$

Solve for x.

$y + 3 = (x + 1)^2 \quad x \le -1$

$-\sqrt{y + 3} = x + 1 \quad x \le -1$

$-\sqrt{y + 3} - 1 = x \quad x \le -1$

Interchange x and y.

$y = -\sqrt{x + 3} - 1 \quad y \le -1$

$f^{-1}(x) = -\sqrt{x + 3} - 1$ domain $[-3, \infty)$ range $(-\infty, -1]$

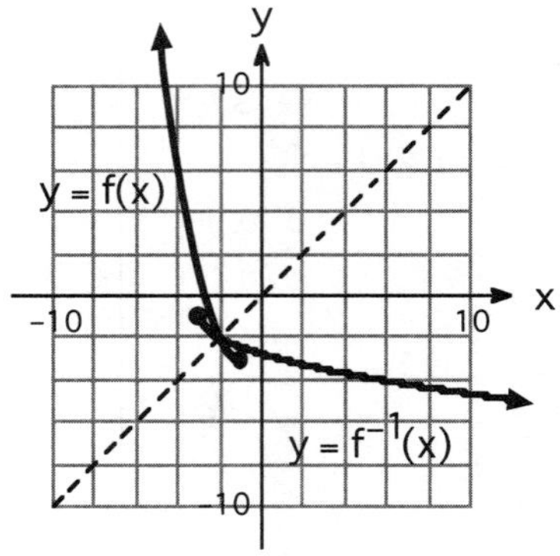

75. From the graph as shown, f is an increasing function with domain $[0, 3]$ and range $[-3, 0]$; f is one-to-one.
Write $y = f(x)$.

$y = -\sqrt{9 - x^2} \quad 0 \le x \le 3$

Solve for x.

$y^2 = 9 - x^2 \quad 0 \le x \le 3$

$x^2 = 9 - y^2 \quad 0 \le x \le 3$

$x = \sqrt{9 - y^2} \quad 0 \le x \le 3$

Interchange x and y.

$y = \sqrt{9 - x^2} \quad 0 \le y \le 3$

$f^{-1}(x) = \sqrt{9 - x^2}$ domain $[-3, 0]$ range $[0, 3]$

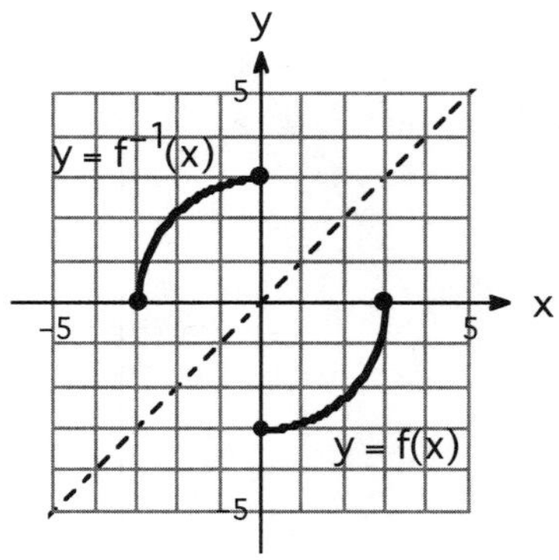

77. From the graph as shown, f is an increasing function with domain $[-3, 0]$ and range $[0, 3]$; f is one-to-one.
Write $y = f(x)$.

$y = \sqrt{9 - x^2} \quad -3 \le x \le 0$

Solve for x.

$y^2 = 9 - x^2 \quad -3 \le x \le 0$

$x^2 = 9 - y^2 \quad -3 \le x \le 0$

$x = -\sqrt{9 - y^2} \quad -3 \le x \le 0$

Interchange x and y.

$y = -\sqrt{9 - x^2} \quad -3 \le y \le 0$

$f^{-1}(x) = -\sqrt{9 - x^2}$ domain $[0, 3]$ range $[-3, 0]$

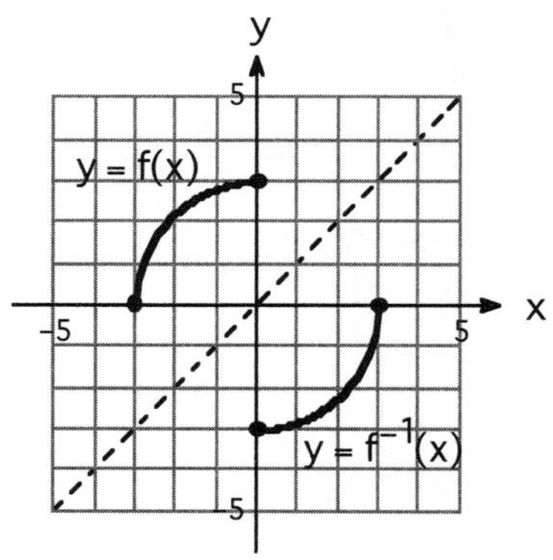

79. From the graph as shown, f is a decreasing function with domain [0, 1] and range [1, 2]; f is one-to-one.

Write $y = f(x)$.

$$y = 1 + \sqrt{1 - x^2} \quad 0 \le x \le 1$$

Solve for x.

$$y - 1 = \sqrt{1 - x^2} \quad 0 \le x \le 1$$
$$(y - 1)^2 = 1 - x^2 \quad 0 \le x \le 1$$
$$y^2 - 2y + 1 = 1 - x^2 \quad 0 \le x \le 1$$
$$x^2 = 2y - y^2 \quad 0 \le x \le 1$$
$$x = \sqrt{2y - y^2} \quad 0 \le x \le 1$$

Interchange x and y.

$$y = \sqrt{2x - x^2} \quad 0 \le y \le 1$$
$$f^{-1}(x) = \sqrt{2x - x^2}$$

domain [1, 2]
range [0, 1]

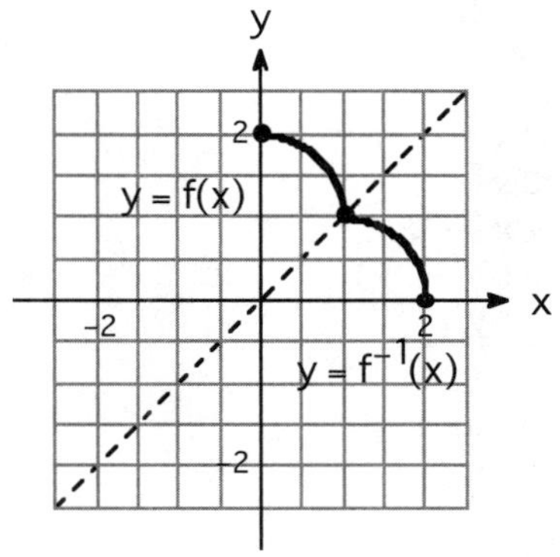

81. From the graph as shown, f is a decreasing function with domain [−1, 0] and range [0, 1]; f is one-to-one.

Write $y = f(x)$.

$$y = 1 - \sqrt{1 - x^2} \quad -1 \le x \le 0$$

Solve for x.

$$y - 1 = -\sqrt{1 - x^2} \quad -1 \le x \le 0$$
$$(y - 1)^2 = 1 - x^2 \quad -1 \le x \le 0$$
$$y^2 - 2y + 1 = 1 - x^2 \quad -1 \le x \le 0$$
$$x^2 = 2y - y^2 \quad -1 \le x \le 0$$
$$x = -\sqrt{2y - y^2} \quad -1 \le x \le 0$$

Interchange x and y.

$$y = -\sqrt{2x - x^2} \quad -1 \le y \le 0$$
$$f^{-1}(x) = -\sqrt{2x - x^2}$$

domain [0, 1]
range [−1, 0]

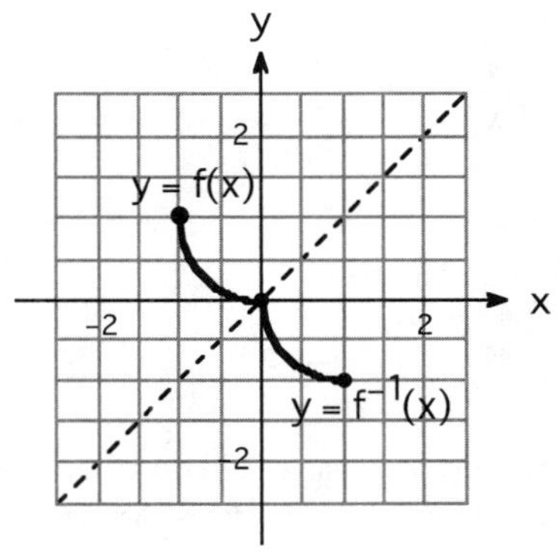

83. Write $y = f(x)$.

$$y = 3 - \frac{2}{x}$$

Solve for x.

$$y - 3 = -\frac{2}{x}$$
$$x(y - 3) = -2$$
$$x = \frac{-2}{y - 3}$$
$$x = \frac{2}{3 - y}$$

Interchange x and y.

$$y = \frac{2}{3 - x}$$
$$f^{-1}(x) = \frac{2}{3 - x}$$

(The student should check that $f \circ f^{-1}(x) = x$ and $f^{-1} \circ f(x) = x$.)

85. Write $y = f(x)$.

$$y = \frac{2}{x - 1}$$

Solve for x.

$$(x - 1)y = 2$$
$$x - 1 = \frac{2}{y}$$
$$x = 1 + \frac{2}{y}$$
$$x = \frac{y}{y} + \frac{2}{y}$$
$$x = \frac{2 + y}{y}$$

Interchange x and y.

$$y = \frac{2 + x}{x}$$
$$f^{-1}(x) = \frac{2 + x}{x}$$

87. Write $y = f(x)$.

$$y = \frac{2x}{x + 1}$$

Solve for x.

$$y(x + 1) = 2x$$
$$xy + y = 2x$$
$$y = 2x - xy$$
$$y = x(2 - y)$$
$$x = \frac{y}{2 - y}$$

Interchange x and y.

$$y = \frac{x}{2 - x}$$
$$f^{-1}(x) = \frac{x}{2 - x}$$

89. Write $y = f(x)$.

$$y = \frac{2x + 5}{3x - 4}$$

Solve for x.

$$y(3x - 4) = 2x + 5$$
$$3xy - 4y = 2x + 5$$
$$3xy - 2x = 4y + 5$$
$$x(3y - 2) = 4y + 5$$
$$x = \frac{4y + 5}{3y - 2}$$

Interchange x and y.

$$y = \frac{4x + 5}{3x - 2}$$
$$f^{-1}(x) = \frac{4x + 5}{3x - 2}$$

91. Write $y = f(x)$.

$$y = 4 - \sqrt[5]{x + 2}$$

Solve for x.

$$y - 4 = -\sqrt[5]{x + 2}$$
$$4 - y = \sqrt[5]{x + 2}$$
$$(4 - y)^5 = x + 2$$
$$x = (4 - y)^5 - 2$$

Interchange x and y.

$$y = (4 - x)^5 - 2$$
$$f^{-1}(x) = (4 - x)^5 - 2$$

93. Since in passing from a function to its inverse, x and y are interchanged, the x intercept of f is the y intercept of f^{-1} and the y intercept of f is the x intercept of f^{-1}.

95. $f(x) = ax + b \quad a \neq 0 \qquad f$ is one-to-one.

Solve $y = f(x)$ for x: $\quad y = ax + b \qquad y - b = ax$

$$x = \frac{y - b}{a} = f^{-1}(y)$$

Proof: Assume $f(u) = f(v)$

$$au + b = av + b$$

Then $au = av$

$$u = v$$

NOTE: We didn't use the graph to check that f is one-to-one since we don't know what numbers a and b are.

Interchange x and y:

$$y = f^{-1}(x) = \frac{x - b}{a} \qquad \text{Domain: } (-\infty, \infty)$$

Check: $f^{-1}[f(x)] = \dfrac{ax + b - b}{a} = \dfrac{ax}{a} = x$

$$f[f^{-1}(x)] = a\frac{x - b}{a} + b = x - b + b = x$$

$$f^{-1}(x) = \frac{x - b}{a}$$

97. For f to be its own inverse, $f(x)$ must equal $f^{-1}(x)$ for all x.

Thus $ax + b = \frac{x - b}{a}$ or

$a^2x + ab = x - b$ or

$a^2x - x = -ab - b$ or

$(a^2 - 1)x = -ab - b$

This can be true for all x only if $a^2 - 1 = 0$ and $-ab - b = 0$.
If $a^2 - 1 = 0$, then $a = -1$ or $a = 1$.
If $a = -1$, then $-ab - b = -(-1)b - b = 0$ for any b.
If $a = 1$, then $-ab - b = -b - b = -2b = 0$ only if $b = 0$.

Summarizing, f will be its own inverse if either $a = -1$ (b arbitrary), or $a = 1$ and $b = 0$.

99. The slope of the line through (a, b) and (b, a) is

$$m_1 = \frac{a - b}{b - a} = \frac{-(b - a)}{b - a} = -1$$

The slope of the line $y = x$ is $1 = m_2$.

Thus, $m_1m_2 = (-1)(1) = -1$. Hence the lines are perpendicular.

101. $p = 100 + 5h$ domain $[0, \infty)$ range $[100, \infty)$.
Solve for h.
$p - 100 = 5h$

$h = \frac{1}{5}(p - 100)$

$h = \frac{1}{5}p - 20$ domain $[100, \infty)$

103. (A) From the graph, d is a decreasing function with domain $[10, 70]$, hence its range is from $d(70) = 200$ to $d(10) = 1{,}000$, $[200, 1000]$

(B) $q = \frac{3{,}000}{0.2p + 1}$

Solve for p.

$q(0.2p + 1) = 3{,}000$

$0.2p + 1 = \frac{3{,}000}{q}$

$0.2p = \frac{3{,}000}{q} - 1$

$p = 5\left(\frac{3{,}000}{q} - 1\right)$

$d^{-1}(q) = p = \frac{15{,}000}{q} - 5$ domain $[200, 1{,}000]$, range $[10, 70]$.

105. (A) If r is linearly related to w, then we are looking for a function whose graph passes through $(w_1, r_1) = (6, 10.50)$ and $(w_2, r_2) = (10, 15.50)$.

We find the slope, then we use the point-slope form to find the equation.

$$M = \frac{r_2 - r_1}{w_2 - w_1} = \frac{15.50 - 10.50}{10 - 6} = 1.25$$

$r - r_1 = M(w - w_1)$

$r - 10.50 = 1.25(w - 6)$

$r - 10.50 = 1.25w - 7.5$

$r = m(w) = 1.25w + 3$ domain $[0, \infty)$, range $[3, \infty)$

(B) Solve $r = 1.25w + 3$ for w.

$r - 3 = 1.25w$

$0.8(r - 3) = w$

$m^{-1}(r) = w = 0.8r - 2.4$ domain $[3, \infty)$, range $[0, \infty)$

107. $L = 0.06s^2 - 1.2s + 26$ domain $[10, \infty)$
Complete the square to find the range.
$L = 0.06(s^2 - 20s + ?) + 26$
$L = 0.06(s^2 - 20s + 100) - 6 + 26$
$L = 0.06(s - 10)^2 + 20$ therefore, range $[20, \infty)$
Solve for s.

$$L - 20 = 0.06(s - 10)^2$$
$$\frac{50}{3}(L - 20) = (s - 10)^2$$
$$\sqrt{\frac{50}{3}(L - 20)} = s - 10 \quad \text{(positive square root only, since } s \geq 10\text{)}$$
$$f^{-1}(2) = s = 10 + \sqrt{\frac{50}{3}(L - 20)} \quad \text{domain } [20, \infty), \text{ range } [10, \infty)$$

CHAPTER 3 REVIEW

1. (A) A function (B) A function
(C) Not a function; two range elements correspond to the domain element 10. *(3-1)*

2. (A) All of the first coordinates are distinct, so this is a function with domain {1, 2, 3}. All of the second coordinates are distinct, so the function is one-to-one. The range is {1, 4, 9}. The inverse function is obtained by reversing the order of the ordered pairs: {(1, 1), (4, 2), (9, 3)}. It has domain {1, 4, 9} and range {1, 2, 3}.

(B) This is not a function: both 1 and 2 are first coordinates that get matched with two different second coordinates.

(C) All of the first coordinates are distinct, so this is a function with domain {-2, -1, 0, 1, 2}. The range is {2}. It's most definitely not one-to-one since every first coordinate gets matched with the same second coordinate.

(D) All of the first coordinates are distinct, so this is a function with domain {-2, -1, 0, 1, 2}. All of the second coordinates are distinct, so the function is one-to-one. The range is {-2, -1, 1, 2, 3}.
The inverse function is obtained by reversing the order of the ordered pairs: {(2, -2), (3, -1), (-1, 0), (-2, 1), (1, 2)}. It has domain {-2, -1, 1, 2, 3} and range {-2, -1, 0, 1, 2}. *(3-1)*

3. (A) Not a function (fails vertical line test)

(B) A function

(C) A function

(D) Not a function (fails vertical line test) *(3-1)*

4. (A) Function

(B) Not a function—two range elements correspond to some domain elements; for example 2 and -2 correspond to 4.

(C) Function

(D) Not a function—two range elements correspond to some domain elements; for example 2 and -2 correspond to 2. *(3-1)*

5. $f(2) = 3(2) + 5 = 11$
$g(-2) = 4 - (-2)^2 = 0$
$k(0) = 5$
Therefore
$f(2) + g(-2) + k(0)$
$= 11 + 0 + 5 = 16$ (3-1)

6. $m(-2) = 2|-2| - 1 = 3$
$g(2) = 4 - (2)^2 = 0$
Therefore $\frac{m(-2) + 1}{g(2) + 4} = \frac{3 + 1}{0 + 4} = 1$ (3-1)

7. $$\frac{f(2 + h) - f(2)}{h} = \frac{[3(2 + h) + 5] - [3(2) + 5]}{h} = \frac{6 + 3h + 5 - 11}{h} = \frac{3h}{h} = 3$$ (3-1)

8. $$\frac{g(a + h) - g(a)}{h} = \frac{[4 - (a + h)^2] - [4 - a^2]}{h} = \frac{4 - a^2 - 2ah - h^2 - 4 + a^2}{h} = \frac{-2ah - h^2}{h} = \frac{h(-2a - h)}{h} = -2a - h$$ (3-1)

9. $(f + g)(x) = f(x) + g(x)$
$= 3x + 5 + 4 - x^2 = 9 + 3x - x^2$ (3-5)

10. $(f - g)(x) = f(x) - g(x) = 3x + 5 - (4 - x^2) = 3x + 5 - 4 + x^2 = x^2 + 3x + 1$ (3-5)

11. $(fg)(x) = f(x)g(x) = (3x + 5)(4 - x^2)$
$= 12x - 3x^3 + 20 - 5x^2 = 20 + 12x - 5x^2 - 3x^3$ (3-5)

12. $\left(\frac{f}{g}\right)(x) = \frac{f(x)}{g(x)} = \frac{3x + 5}{4 - x^2}$
Domain: $\{x \mid 4 - x^2 \neq 0\}$ or $\{x \mid x \neq \pm 2\}$ (3-5)

13. $(f \circ g)(x) = f[g(x)] = f(4 - x^2) = 3(4 - x^2) + 5 = 12 - 3x^2 + 5 = 17 - 3x^2$ (3-5)

14. $(g \circ f)(x) = g[f(x)] = g(3x + 5) = 4 - (3x + 5)^2 = 4 - (9x^2 + 30x + 25)$
$= 4 - 9x^2 - 30x - 25 = -21 - 30x - 9x^2$ (3-5)

15. (A) $f(1) = (1)^2 - 2(1) = -1$

(B) $f(-4) = (-4)^2 - 2(-4) = 24$

(C) $f(2) \cdot f(-1) = [(2)^2 - 2(2)] \cdot [(-1)^2 - 2(-1)] = 0 \cdot 3 = 0$

(D) $\frac{f(0)}{f(3)} = \frac{(0)^2 - 2(0)}{(3)^2 - 2(3)} = \frac{0}{3} = 0$ (3-5)

16. Construct a table of values of $f(x)$ and $g(x)$ from the graph, then subtract to obtain $(f - g)(x)$.

x	-3	-2	-1	0	1	2	3
$f(x)$	3	2	1	0	1	2	3
$g(x)$	-4	-3	-2	-1	0	1	2
$(f - g)(x)$	7	5	3	1	1	1	1

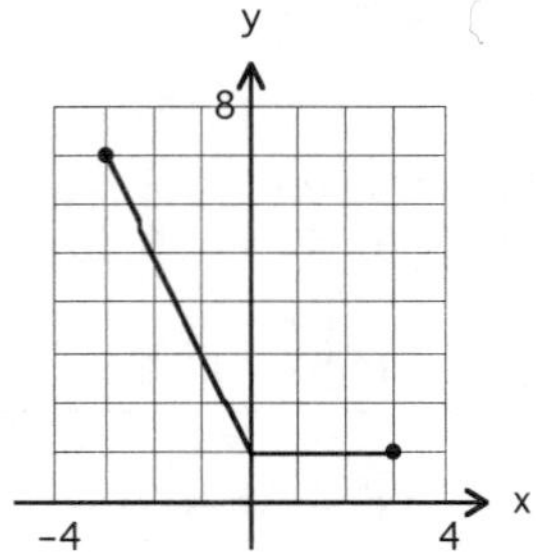

(3-5)

17. Use the top 3 rows of the table above and multiply to get a table of values for $(fg)(x)$. (Note that from the graphs of f and g, we can see that $f\left(\frac{1}{2}\right) = \frac{1}{2}$ and $g\left(\frac{1}{2}\right) = -\frac{1}{2}$, so $(fg)\left(\frac{1}{2}\right) = -\frac{1}{4}$.)

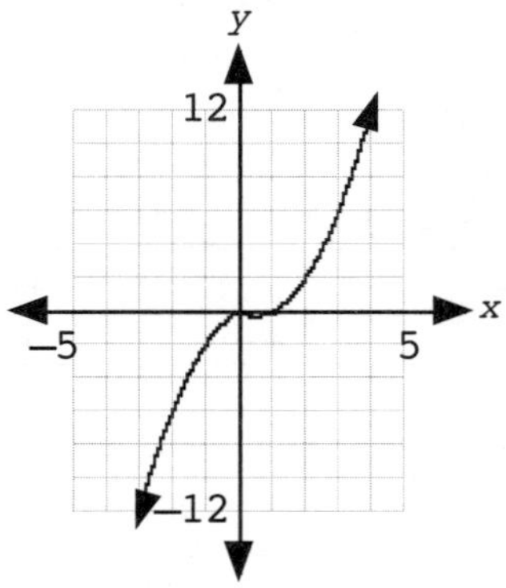

(3-5)

18. $(f \circ g)(-1) = f[g(-1)]$. From the graph of g, $g(-1) = -2$. From the graph of f, $f[g(-1)] = f(-2) = 2$. (3-5)

19. $g \circ f(-2) = g[f(-2)]$. From the graph of f, $f(-2) = 2$. From the graph of g, $g[f(-2)] = g(2) = 1$. (3-5)

20. From the graph of g, $g(1) = 0$. From the graph of f, $f[g(1)] = f(0) = 0$. (3-5)

21. From the graph of f, $f(-3) = 3$. From the graph of g, $g[f(-3)] = g(3) = 2$. (3-5)

22. Some range elements (1 for example) correspond to more than one domain element. Not one-to-one. (3-6)

23. Yes, one-to-one. (3-6)

24. (A) $f(-x) = (-x)^5 + 6(-x) = -x^5 - 6x = -(x^5 + 6x) = -f(x)$. Odd

(B) $g(-t) = (-t)^4 + 3(-t)^2 = t^4 + 3t^2 = g(t)$. Even

(C) $h(-z) = (-z)^5 + 4(-z)^2 = -z^5 + 4z^2$
$-h(z) = -(z^5 + 4z^2) = -z^5 - 4z^2$
Therefore, $h(-z) \neq h(z)$. $h(-z) \neq -h(z)$. h is neither even nor odd. (3-2)

25. When $x = -4$, the corresponding value of $f(x)$ on the graph is 4. $f(-4) = 4$.
When $x = 0$, the corresponding value of $f(x)$ on the graph is -4. $f(0) = -4$.
When $x = 3$, the corresponding value of $f(x)$ on the graph is 0. $f(3) = 0$.
When $x = 5$, there is no corresponding value of $f(x)$ on the graph.
$f(5)$ is not defined. (3-1, 3-2)

26. Two values of x correspond to $f(x) = -2$ on the graph. They are $x = -2$ and $x = 1$. (3-1, 3-2)

27. Domain: [-4, 5). Range: [-4, 4] (3-2)

28. The graph is increasing on [0, 5) and decreasing on [-4, 0]. (3-2)

29. The graph is discontinuous at $x = 0$. (3-2)

30. The graph of $f(x)$ is shifted up 1 unit.

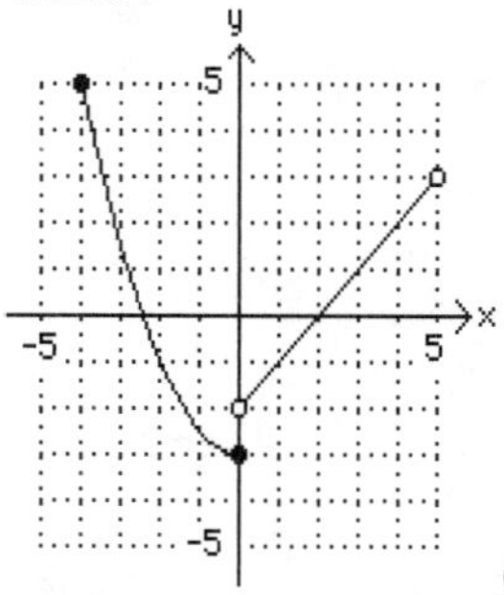

(3-3)

31. The graph of $f(x)$ is shifted left 1 unit.

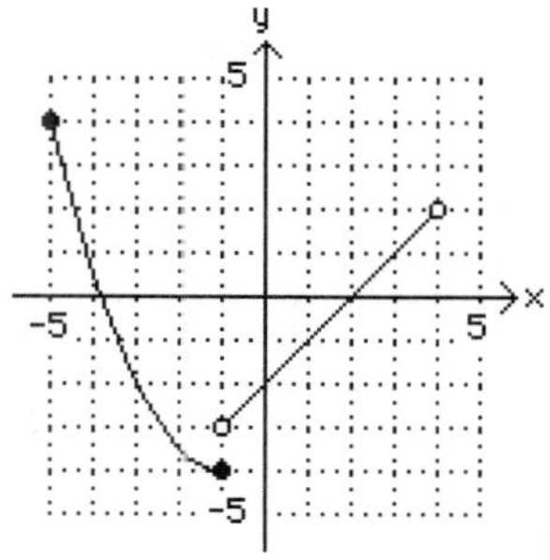

(3-3)

32. The graph of $f(x)$ is reflected in the x axis.

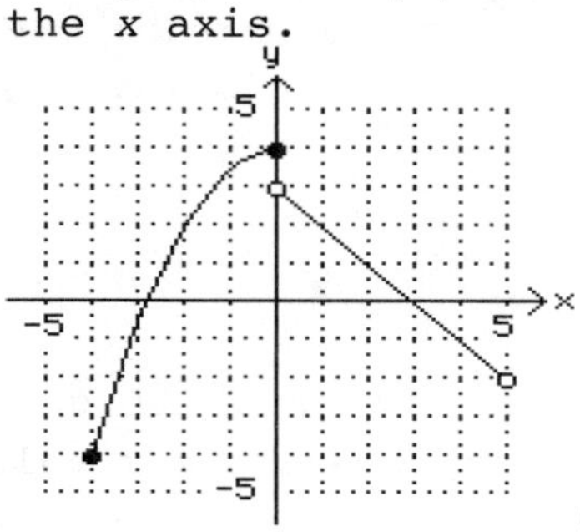

(3-3)

33. The graph of $f(x)$ is vertically shrunk by a factor of 0.5.

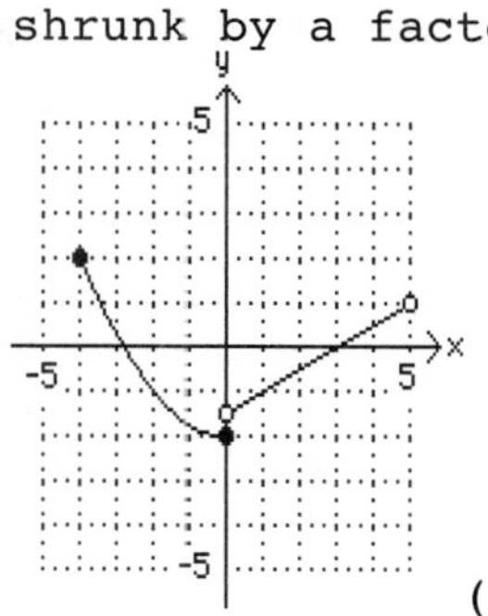

(3-3)

34. The graph is horizontally shrunk by a factor of $\frac{1}{2}$.

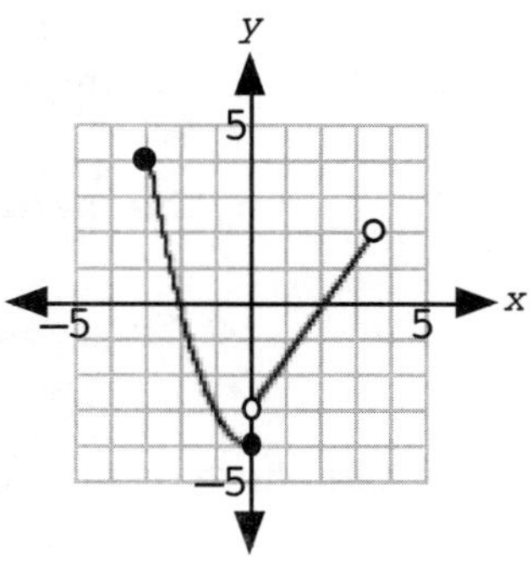

(3-3)

35. The graph is reflected about both axes.

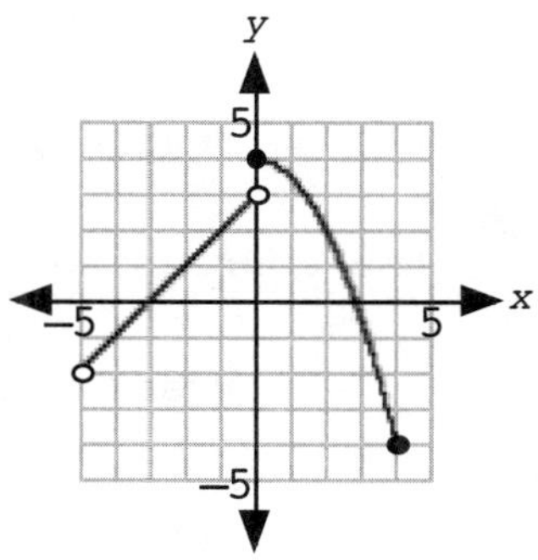

(3-3)

36. (A) The graph that is decreasing, then increasing, and has a minimum at $x = 2$ is g.

(B) The graph that is increasing, then decreasing, and has a maximum at $x = -2$ is m.

(C) The graph that is increasing, then decreasing, and has a maximum at $x = 2$ is n.

(D) The graph that is decreasing, then increasing, and has a minimum at $x = -2$ is f. (3-3)

37. The equation corresponding to graph f is $y = (x + 2)^2 - 4$.

(A) y intercept: Set $x = 0$, then $y = (0 + 2)^2 - 4 = 0$
x intercepts: Set $y = 0$, then $0 = (x + 2)^2 - 4$

$$(x + 2)^2 = 4$$
$$x + 2 = \pm 2$$
$$x = 0, -4$$

(B) $(-2, -4)$

(C) The minimum of -4 occurs at the vertex.

(D) Since y is never less than -4, the range is $[-4, \infty)$.

(E) y is increasing on $[-2, \infty)$.

(F) y is decreasing on $(-\infty, -2]$. (3-4)

38. (A) $\left(\frac{f}{g}\right)(x) = \frac{f(x)}{g(x)} = \frac{x^2 - 4}{x + 3}$ Domain: $\{x \mid x \neq -3\}$, or $(-\infty, -3) \cup (-3, \infty)$

(B) $\left(\frac{g}{f}\right)(x) = \frac{g(x)}{f(x)} = \frac{x + 3}{x^2 - 4}$ Domain: $\{x \mid x^2 - 4 \neq 0\}$, or $\{x \mid x \neq -2, 2\}$ or $(-\infty, -2) \cup (-2, 2) \cup (2, \infty)$

(C) $(f \circ g)(x) = f[g(x)] = f(x + 3) = (x + 3)^2 - 4 = x^2 + 6x + 5$ Domain: $(-\infty, \infty)$

(D) $(g \circ f)(x) = g[f(x)] = g(x^2 - 4) = x^2 - 4 + 3 = x^2 - 1$ Domain: $(-\infty, \infty)$

(3-5)

39. $f(x) = x^2 - 6x + 11$. Complete the square:

$$f(x) = (x^2 - 6x + 9) - 9 + 11$$
$$= (x - 3)^2 + 2$$

Comparing with $f(x) = a(x - h)^2 + k$; $h = 3$ and $k = 2$. Thus, the minimum value is 2 and the vertex is $(3, 2)$. (3-4)

40. (A) Reflected across x axis (B) Shifted down 3 units
(C) Shifted left 3 units (3-5)

41. (A) When $x = 0$, the corresponding y value on the graph is 0, to the nearest integer.

(B) When $x = 1$, the corresponding y value on the graph is 1, to the nearest integer.

(C) When $x = 2$, the corresponding y value on the graph is 2, to the nearest integer.

(D) When $x = -2$, the corresponding y value on the graph is 0, to the nearest integer. (3-2)

42. (A) Two values of x correspond to $y = 0$ on the graph. To the nearest integer, they are -2 and 0.

(B) Two values of x correspond to $y = 1$ on the graph. To the nearest integer, they are -1 and 1.

(C) No value of x corresponds to $y = -3$ on the graph.

(D) To the nearest integer, $x = 3$ corresponds to $y = 3$ on the graph. Also, every value of x such that $x < -2$ corresponds to $y = 3$. (3-2)

43. Domain: $(-\infty, \infty)$. Range: $(-3, \infty)$. (3-2)

44. $[-2, -1]$, $[1, \infty)$. (3-2) **45.** $[-1, 1)$ (3-2) **46.** $(-\infty, -2)$ (3-2)

47. The graph of g is discontinuous at $x = -2$ and $x = 1$. (3-2)

48. $f(x) = 4x^3 - \sqrt{x}$ (3-1)

49. The function f multiplies the square of the domain element by 3, adds 4 times the domain element, and then subtracts 6. (3-1)

50. This equation defines a function. If you solve the equation for y, you get $y = 5 - 0.5x$. This tells us that for any choice of x, we can calculate a unique y that corresponds to it. (3-1)

51. This equation does not define a function since most choices of x will result in two corresponding values of y. For example, the pairs (2, 2) and (2, -2) both make the equation a true statement. (3-1)

52. Since $m(x)$ is a polynomial, the domain is the set of all real numbers R, $(-\infty, \infty)$.
To find the x intercepts, solve $m(x) = 0$:

$$x^2 - 4x + 5 = 0$$

$$x = \frac{-(-4) \pm \sqrt{(-4)^2 - 4(1)(5)}}{2(1)}$$

$$x = \frac{4 \pm \sqrt{-4}}{2} \quad \text{(no real solutions)}$$

There are no x intercepts.
The y intercept is $m(0) = 0^2 - 4(0) + 5 = 5$. (3-1, 3-2)

53. The domain is the set of all real numbers for which $\sqrt{x}$ is defined, that is, $[0, \infty)$. Since $2 + 3\sqrt{x}$ is never less than 2, $r(x)$ is never 0 and there are no x intercepts. The y intercept is $r(0) = 2 + 3\sqrt{0} = 2$. (3-1, 3-2)

54. The rational expression $\frac{1 - x^2}{x^3}$ is defined everywhere except at zeros of the denominator, $x = 0$. The domain is $(-\infty, 0) \cup (0, \infty)$.
A rational expression is 0 if and only if the numerator is zero:

$$1 - x^2 = 0$$
$$1 = x^2$$
$$x = \pm 1$$

The x intercepts of p are -1, 1.
The y intercept is $p(0)$, which is not defined, so there is no y intercept. (3-1, 3-2)

55. The fractional expression $\frac{x}{\sqrt{3 - x}}$ is defined everywhere except at zeros of the denominator, as long as $\sqrt{3 - x}$ is defined, thus

$$3 - x > 0$$
$$x < 3$$

The domain is $(-\infty, 3)$.
A fractional expression is 0 if and only if the numerator is 0. Thus $x = 0$ is the x intercept. The y intercept is $f(0) = \frac{0}{\sqrt{3 - 0}} = 0$. (3-1, 3-2)

56. The rational expression $\frac{2x + 3}{x^2 - 4}$ is defined everywhere except at zeros of the denominator.

$$x^2 - 4 = 0$$
$$x^2 = 4$$
$$x = \pm 2$$

The domain is $(-\infty, -2) \cup (-2, 2) \cup (2, \infty)$.

A rational expression is 0 if and only if the numerator is 0.

$$2x + 3 = 0$$
$$x = -\frac{3}{2}$$

The x intercept is $-\frac{3}{2}$ or -1.5.

The y intercept is $g(0) = \frac{2(0) + 3}{0^2 - 4} = -\frac{3}{4}$ or -0.75. *(3-1, 3-2)*

57. The fractional expression $\frac{1}{4 - \sqrt{x}}$ is defined everywhere except at zeros of the denominator, as long as $\sqrt{x}$ is defined, thus we must restrict $x \geq 0$ as well as $4 - \sqrt{x}$ to being non-zero.

$$4 - \sqrt{x} = 0$$
$$4 = \sqrt{x}$$
$$16 = x$$

The domain is $[0, 16) \cup (16, \infty)$.
Since the numerator is never 0, there are no x intercepts. The y intercept is $h(0) = \frac{1}{4 - \sqrt{0}} = \frac{1}{4}$ or 0.25. *(3-1, 3-2)*

58. (A)
$$\begin{aligned} f(x) &= 0.5x^2 - 4x + 5 \\ &= 0.5(x^2 - 8x) + 5 \\ &= 0.5(x^2 - 8x + 16) + 5 - 0.5(16) \\ &= 0.5(x - 4)^2 - 3 \end{aligned}$$

Therefore the line $x = 4$ is the axis of symmetry of the parabola and $(4, -3)$ is its vertex.

(B) The parabola opens upward, since $0.5 = a > 0$. Thus the parabola is decreasing on $(-\infty, 4]$ and increasing on $[4, \infty)$. The range is $[-3, \infty)$.

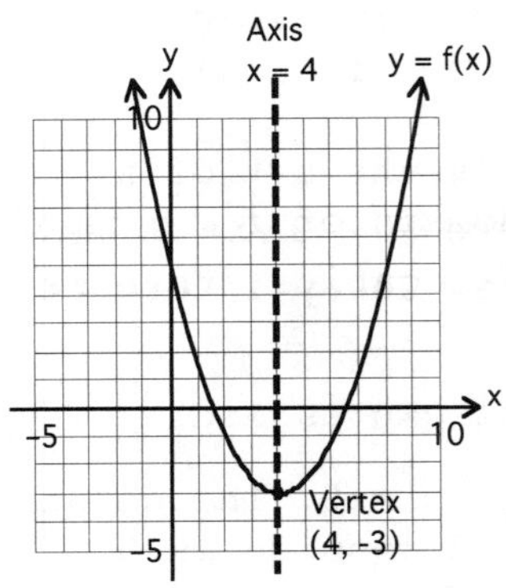

(3-4)

59. g is a linear function. Its graph passes through $(x_1, y_1) = (-1, 0)$ and $(x_2, y_2) = (1, 4)$. Therefore the slope of the line is $m = \frac{y_2 - y_1}{x_2 - x_1} = \frac{4 - 0}{1 - (-1)} = 2$. The equation of the line is obtained from the point-slope form:

$$y - y_1 = m(x - x_1)$$
$$y - 0 = 2[x - (-1)]$$
$$y = 2x + 2$$

The function is given by $g(x) = 2x + 2$.
f is a quadratic function. Its equation must be of the form $f(x) = y = a(x - h)^2 + k$. The vertex of the parabola is at $(1, 2)$.

Therefore the equation must have form

$$y = a(x - 1)^2 + 2$$

Since the parabola passes through $(3, 0)$, these coordinates must satisfy the equation.

$$0 = a(3 - 1)^2 + 2$$
$$0 = 4a + 2$$
$$-2 = 4a$$
$$a = -0.5$$

The equation is

$y = f(x) = -0.5(x - 1)^2 + 2$
$f(x) = -0.5(x^2 - 2x + 1) + 2$
$f(x) = -0.5x^2 + x - 0.5 + 2$
$f(x) = -0.5x^2 + x + 1.5$ (3-4)

60. (A) For $-4 \le x < 0$, $f(x) = -x - 5$, so $f(-4) = -(-4) - 5 = -1$ and $f(-2) = -(-2) - 5 = -3$

For $0 \le x \le 5$, $f(x) = 0.2x^2$, so $f(0) = 0.2(0)^2 = 0$, $f(2) = 0.2(2)^2 = 0.8$, and $f(5) = 0.2(5)^2 = 5$.

(B)

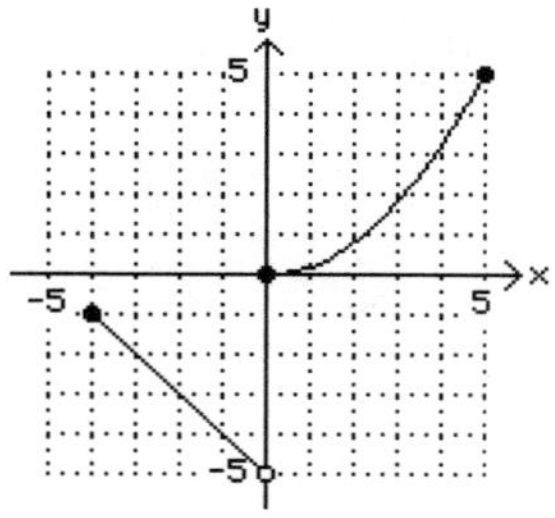

(C) Domain: $[-4, 5]$, range: $(-5, -1] \cup [0, 5]$

(D) The graph is discontinuous at $x = 0$.

(E) Decreasing on $[-4, 0)$, increasing on $[0, 5]$. (3-2, 3-4)

61. $f(x) = \sqrt{x} - 8$ $g(x) = |x|$

(A) $(f \circ g)(x) = f[g(x)] = f(|x|) = \sqrt{|x|} - 8$

$(g \circ f)(x) = g[f(x)] = g(\sqrt{x} - 8) = |\sqrt{x} - 8|$

(B) The domain of f is $\{x \mid x \ge 0\}$. The domain of g is all real numbers. Hence the domain of $f \circ g$ is the set of those real numbers x for which $g(x)$ is non-negative, that is, all real numbers.

The domain of $(g \circ f)$ is the set of all those non-negative numbers x for which $f(x)$ is real, that is all $\{x \mid x \ge 0\}$ or $[0, \infty)$ (3-5)

62. (A) $f(x) = x^3$. The graph passes the horizontal line test, so f is one-to-one. Also, assume

$f(a) = f(b)$
$a^3 = b^3$
$a^3 - b^3 = 0$
$(a - b)(a^2 + ab + b^2) = 0$

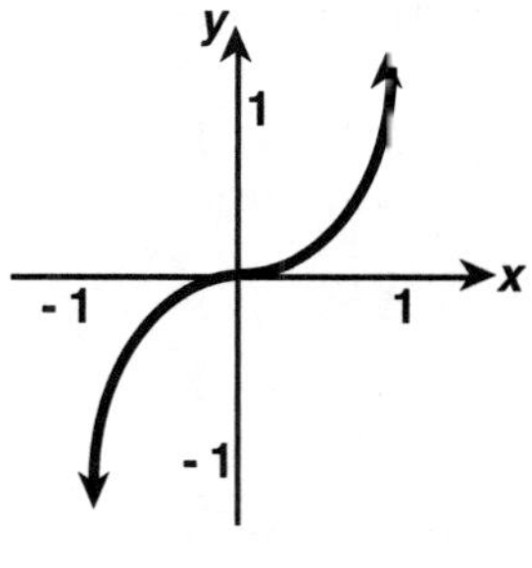

The only real solutions of this equation are those for which $a - b = 0$, hence $a = b$. Thus $f(x)$ is one-to-one.

(B) $g(x) = (x - 2)^2$. Since $g(3) = g(1) = 1$, g is not one-to-one.

(C) $h(x) = 2x - 3$

Assume $h(a) = h(b)$
$2a - 3 = 2b - 3$
Then $2a = 2b$
$a = b$

Thus h is one-to-one.

(D) $F(x) = (x + 3)^2$ $x \ge -3$

The graph passes the horizontal line test, so F is one-to-one.

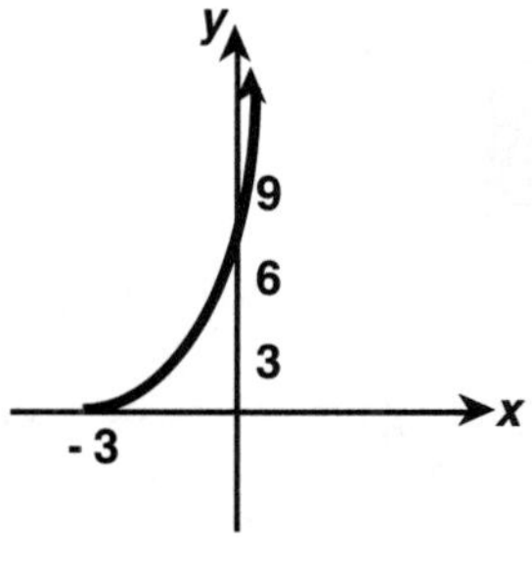

(3-6)

63. Find the composition of the 2 functions in both orders.
$(u \circ v)(x) = u(v(x)) = u(0.25x + 2) = 4(0.25x + 2) - 8 = x + 8 - 8 = x$
$(v \circ u)(x) = v(u(x)) = v(4x - 8) = 0.25(4x - 8) + 2 = x - 2 + 2 = x$
The functions are inverses. (3-6)

64. The function k cubes the input and then adds 5. The opposite of this is subtracting 5 and taking the cube root. (Remember to reverse the order of the steps!) The inverse function is $k^{-1}(x) = \sqrt[3]{x - 5}$. To verify we find the composition of the two functions in both orders.
$(k \circ k^{-1})(x) = k(k^{-1}(x)) = k(\sqrt[3]{x - 5}) = (\sqrt[3]{x - 5})^3 + 5 = x - 5 + x = x$
$(k^{-1} \circ k)(x) = k^{-1}(k(x)) = k^{-1}(x^3 + 5) = \sqrt[3]{(x^3 + 5) - 5} = \sqrt[3]{x^3} = x.$ (3-6)

65. (A) The graph of $f(x)$ is a line; f is one-to-one.
Write $y = f(x)$.
$$y = 3x - 7$$
Solve for x.
$$y + 7 = 3x$$
$$\frac{y + 7}{3} = x$$
Interchange x and y:
$$y = \frac{x + 7}{3}$$
$$f^{-1}(x) = \frac{x + 7}{3}$$

(B) $f^{-1}(5) = \dfrac{5 + 7}{3} = 4$

(C) $f^{-1}[f(x)] = f^{-1}(3x - 7) = \dfrac{3x - 7 + 7}{3} = \dfrac{3x}{3} = x$

(D) Since $a < b$ implies $3a < 3b$, which implies $3a - 7 < 3b - 7$, or $f(a) < f(b)$, f is increasing. (3-6)

66. The graph of $y = x^2$ is vertically stretched by a factor of 2, reflected through the x axis and shifted to the left 3 units. Equation: $y = -2(x + 3)^2$. (3-3)

67. $g(x) = 5 - 3|x - 2|$

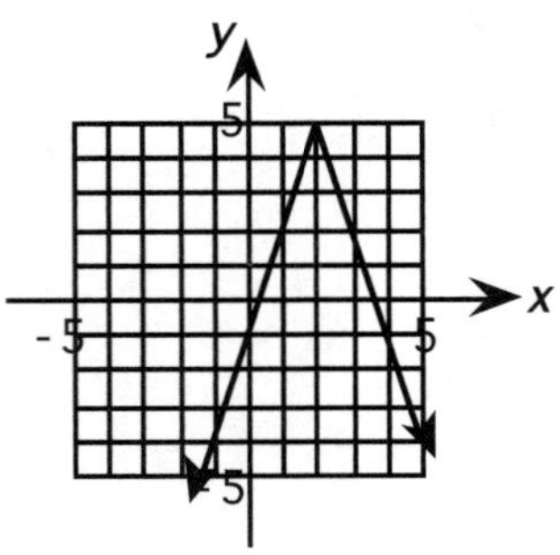

(3-3)

68. The graph of $y = x^2$ has been reflected through the x axis, shifted right 4 units and up 3 units so that the parabola has vertex (4, 3).
Equation: $y = -(x - 4)^2 + 3$. (3-3, 3-4)

69. The graph of $y = \sqrt[3]{x}$ is vertically stretched by a factor of 2, reflected through the x axis, shifted 1 unit left and 1 unit down.
Equation: $y = -2\sqrt[3]{x+1} - 1$. (3-3)

70. It is the same as the graph of g shifted to the right 2 units and down 1 unit, then reflected through the x axis. (3-3)

71. (A) The graph of $y = x^2$ is shifted 2 units to the right, reflected through the x axis, and shifted up 4 units: $y = -(x - 2)^2 + 4$
Check: $y = -(x - 2)^2 + 4$ is graphed on a graphing utility.

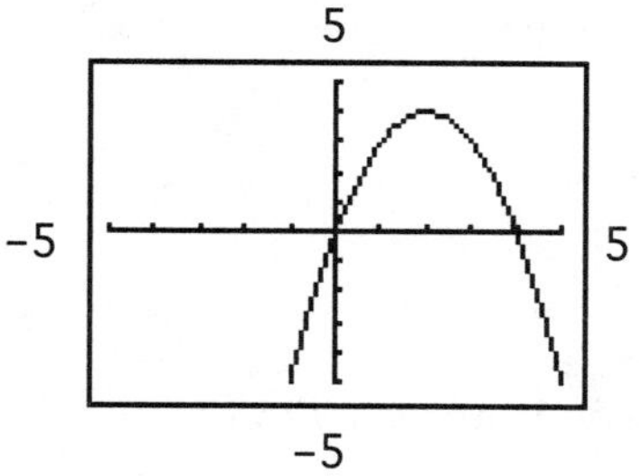

(B) The graph of $y = \sqrt{x}$ is reflected through the x axis, stretched by a factor of 4, and shifted up 4 units: $y = 4 - 4\sqrt{x}$
Check: $y = 4 - 4\sqrt{x}$ is graphed on a graphing utility.

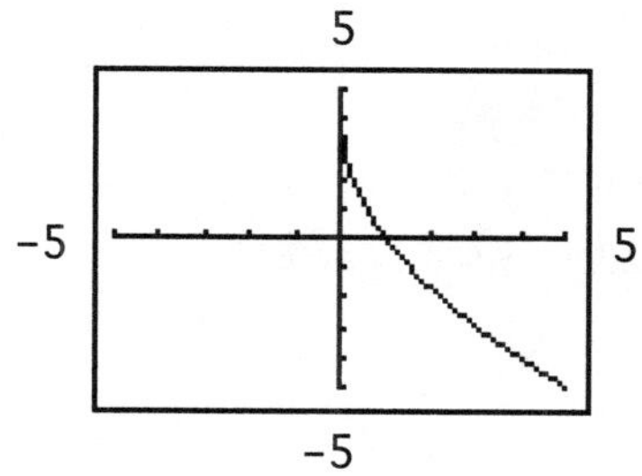

(3-3)

72. $g(x) = 8 - 3|x - 4|$
The -4 shifts 4 units right, the 3 stretches vertically by a factor of 3, and the 8 shifts 8 units up.

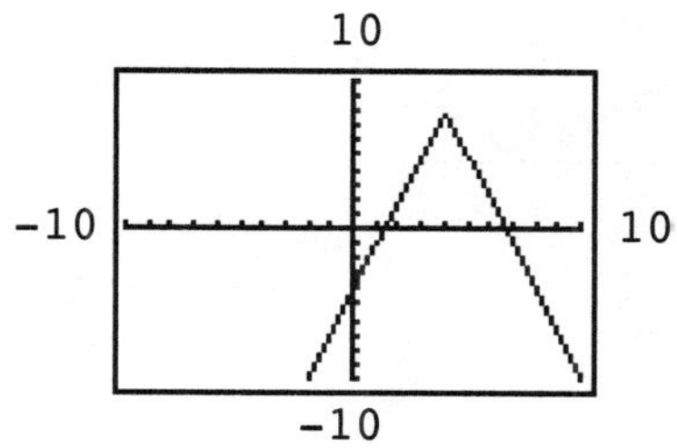

(3-3)

73. The equation is $t(x) = \left(\frac{1}{2}(x + 2)\right)^2 - 4$. The $\frac{1}{2}$ stretches horizontally by 2, the +2 shifts 2 units left, and the -4 shifts 4 units down. This function can be simplified: $\left(\frac{1}{2}(x + 2)\right)^2 - 4 = \left(\frac{1}{2}x + 1\right)^2 - 4 = \frac{1}{4}x^2 + x + 1 - 4 = \frac{1}{4}x^2 + x - 3$.

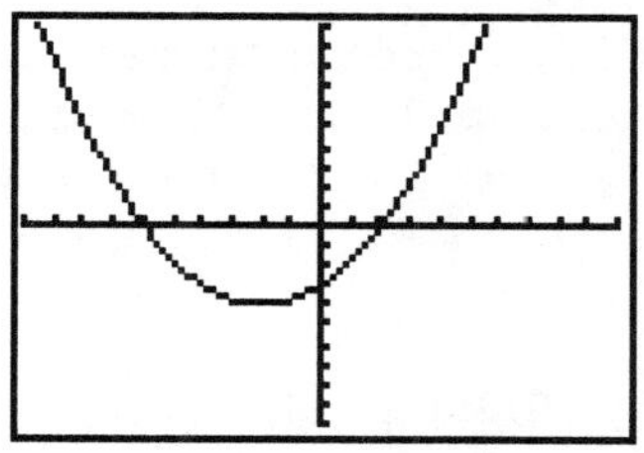

(3-3)

74. The graph of $y = |x|$ is shifted 1 unit left.

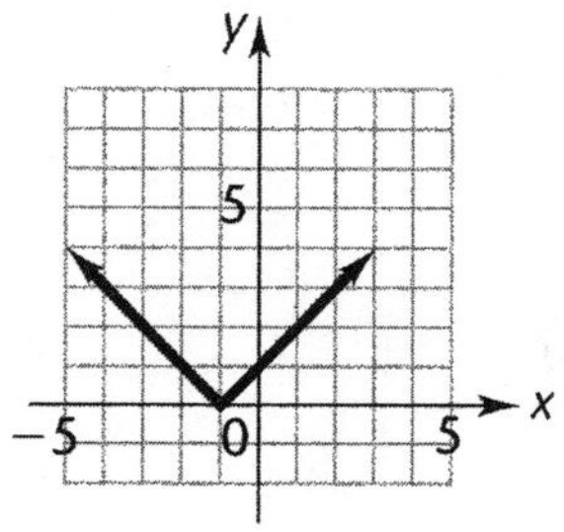

(3-3)

75. The graph of $y = \sqrt[3]{x}$ is shifted 1 unit right, reflected through the x axis, and shifted 1 unit up.

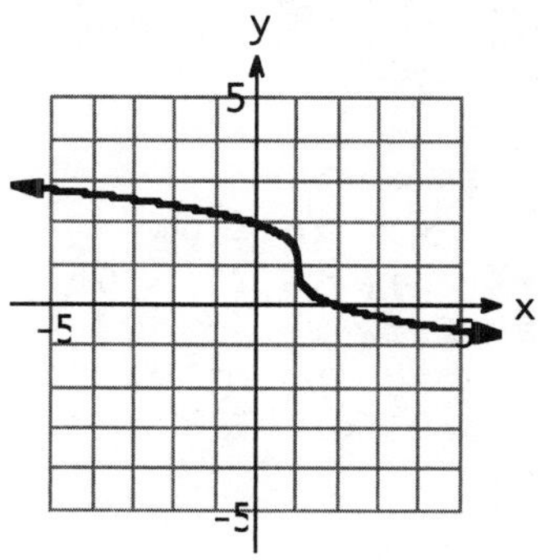

(3-3)

76. The graph of $y = |x|$ is shifted 2 units down.

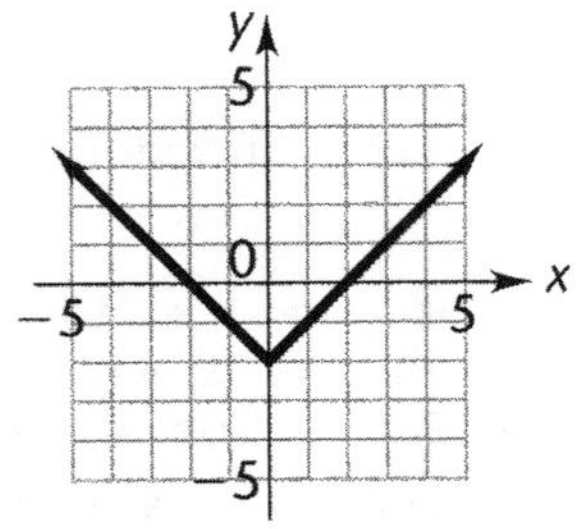

(3-3)

77. The graph of $y = \sqrt{x}$ is reflected through the x axis, stretched vertically by a factor of 3, and shifted up 9 units.

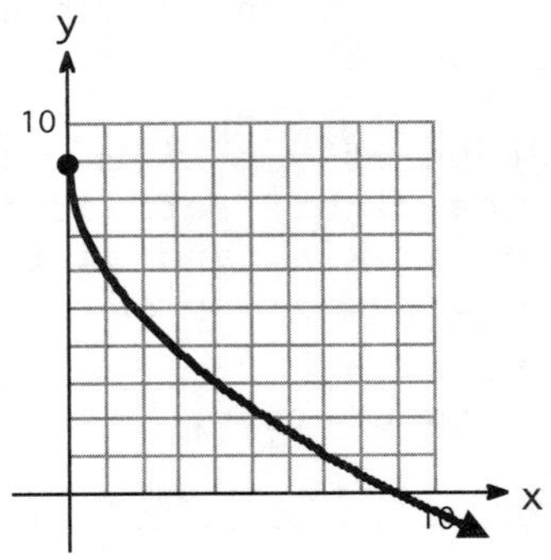

(3-3)

78. The graph of $y = |x|$ is shrunk vertically by a factor of $\frac{1}{2}$.

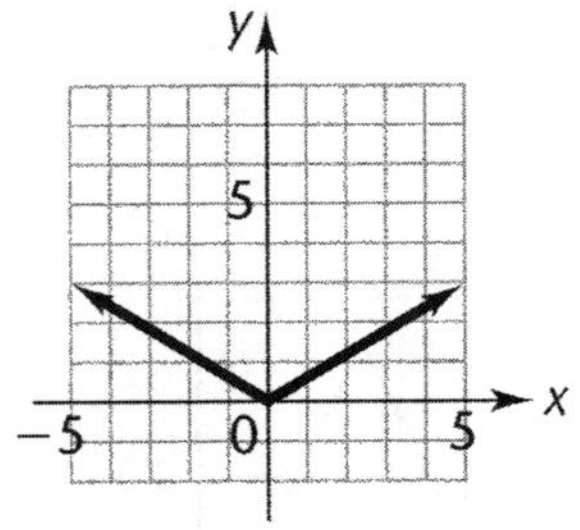

(3-3)

79. $\sqrt[3]{4 - 0.5x} = \sqrt[3]{0.5(8 - x)}$

The graph of $y = \sqrt[3]{x}$ is stretched horizontally by a factor of 2, reflected across the y axis, and shifted 8 units right.

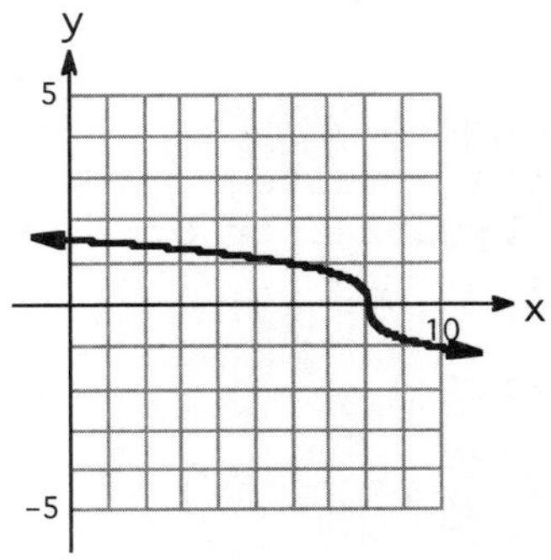

(3-3)

80. The graph of $y = x^3$ is shifted right 1 unit, stretched vertically by a factor of 3, reflected through the x axis, and shifted up 2 units.

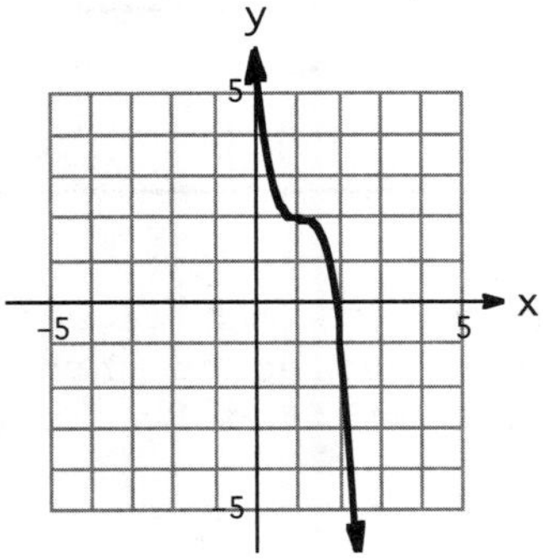

(3-3)

81. The graph of $y = |x|$ is shifted left 1 unit, reflected through the x axis, and shifted down 1 unit.

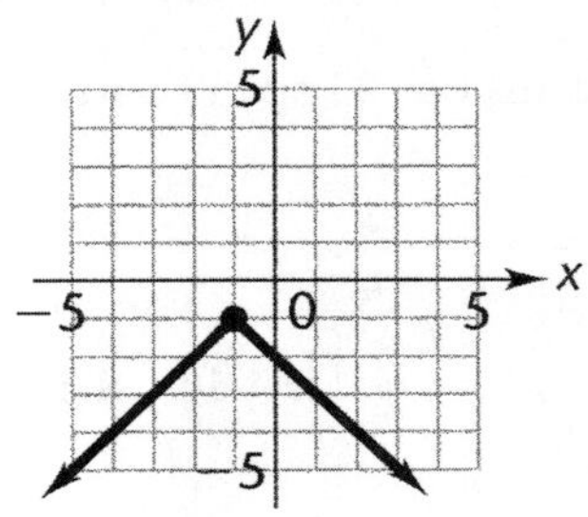

(3-3)

82. $x^2 + x < 20$

$f(x) = x^2 + x - 20 < 0$

$f(x) = (x + 5)(x - 4) < 0$

The zeros of f are -5 and 4. Plotting the graph of $f(x)$, we see that $f(x) < 0$ for $-5 < x < 4$, or $(-5, 4)$.

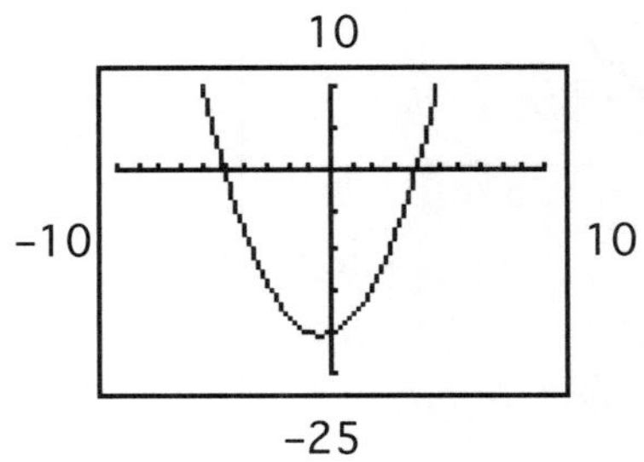

(3-4)

83. $x^2 > 4x + 12$

$f(x) = x^2 - 4x - 12 > 0$

$f(x) = (x + 2)(x - 6) > 0$

The zeros of f are -2 and 6.
Plotting the graph of $f(x)$, we see that $f(x) > 0$ for $x < -2$ and $x > 6$ or $(-\infty, -2) \cup (6, \infty)$.

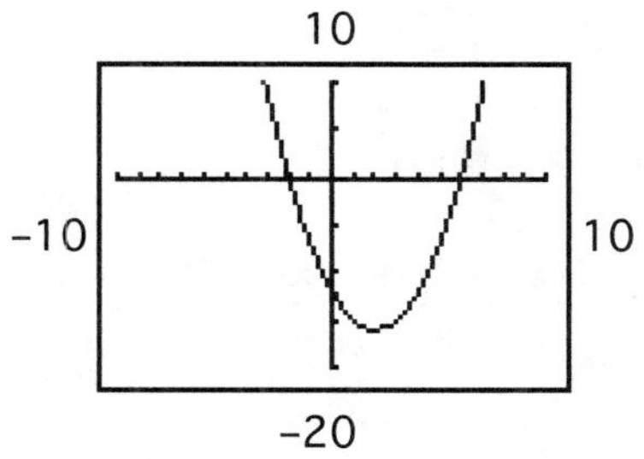

(*3-4*)

84. The domain is the set of all real numbers x such that $\sqrt{25 - x^2}$ is a real number, that is, such that $25 - x^2 \geq 0$ or $-5 \leq x \leq 5$, $[-5, 5]$. (*3-1*)

85. (A) $(fg)(x) = f(x)g(x) = x^2\sqrt{1 - x}$

The domain of f is $(-\infty, \infty)$. The domain of g is $(-\infty, 1]$. Hence the domain of fg is the intersection of these sets, that is, $(-\infty, 1]$.

(B) $\left(\dfrac{f}{g}\right)(x) = \dfrac{f(x)}{g(x)} = \dfrac{x^2}{\sqrt{1 - x}}$

To find the domain of $\dfrac{f}{g}$, we exclude from $(-\infty, 1]$ the set of values of x for which $g(x) = 0$

$\sqrt{1 - x} = 0$

$1 - x = 0$

$x = 1$

Thus the domain of $\dfrac{f}{g}$ is $(-\infty, 1)$

(C) $(f \circ g)(x) = f[g(x)] = f(\sqrt{1 - x}) = [\sqrt{1 - x}]^2 = 1 - x.$

The domain of $f \circ g$ is the set of those numbers in $(-\infty, 1]$ for which $g(x)$ is real, that is $(-\infty, 1]$.

(D) $(g \circ f)(x) = g[f(x)] = g(x^2) = \sqrt{1 - x^2}$

The domain of $g \circ f$ is the set of those real numbers for which $f(x)$ is in $(-\infty, 1]$, that is, $x^2 \leq 1$, or $-1 \leq x \leq 1$. $[-1, 1]$. (*3-5*)

86. (A) Write $y = f(x)$.

$$y = \frac{x + 2}{x - 3}$$

Solve for x.

$$y(x - 3) = x + 2$$
$$xy - 3y = x + 2$$
$$xy - x = 3y + 2$$
$$x(y - 1) = 3y + 2$$
$$x = \frac{3y + 2}{y - 1}$$

Interchange x and y:

$$f^{-1}(x) = y = \frac{3x + 2}{x - 1}$$

(B) $f^{-1}(3) = \dfrac{3(3) + 2}{3 - 1} = \dfrac{11}{2}$

(C) $f^{-1}[f(x)] = f^{-1}\left(\dfrac{x+2}{x-3}\right)$

$$= \frac{3\left(\frac{x+2}{x-3}\right) + 2}{\frac{x+2}{x-3} - 1}$$
$$= \frac{3(x+2) + 2(x-3)}{x + 2 - (x-3)}$$
$$= \frac{3x + 6 + 2x - 6}{x + 2 - x + 3}$$
$$= \frac{5x}{5}$$
$$= x$$

(3-6)

87. (A) Write $y = f(x)$.

$y = \sqrt{x - 1}$

Domain of f: $[1, \infty)$, Range: $[0, \infty)$.

Solve for x.

$y^2 = x - 1$

$y^2 + 1 = x$

Interchange x and y.

$x^2 + 1 = y$

$f^{-1}(x) = x^2 + 1$ Domain of f^{-1}: $[0, \infty)$, Range: $[1, \infty)$

(B) See part (A)

(C)

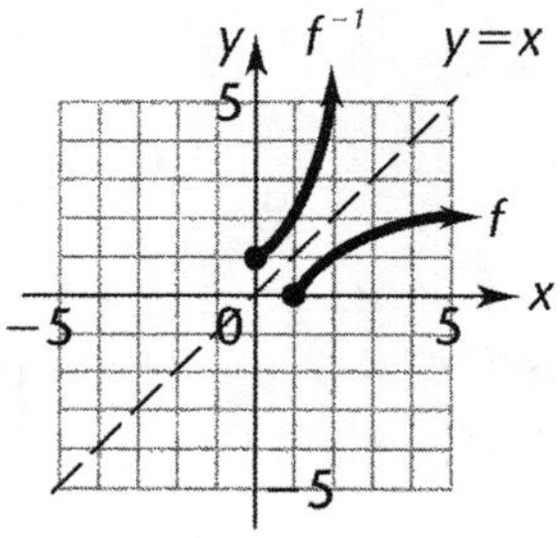

(3-6)

88. (A) Domain of $f = [0, \infty) =$ Range of f^{-1}

Since $x^2 \geq 0$, $x^2 - 1 \geq -1$, so Range of $f = [-1, \infty) =$ Domain of f^{-1}

(B) $f(x) = x^2 - 1 \quad x \geq 0$

f is one-to-one on its domain (steps omitted)

Solve $y = f(x)$ for x:

$y = x^2 - 1 \quad x \geq 0$

$x^2 = y + 1 \quad x \geq 0$

$x = \underbrace{\sqrt{y + 1}} = f^{-1}(y)$

positive square root since $x \geq 0$

Interchange x and y:

$y = f^{-1}(x) = \sqrt{x + 1}$ Domain: $[-1, \infty)$

Check:

$f^{-1}[f(x)] = \sqrt{x^2 - 1 + 1} \quad x \geq 0$

$= \sqrt{x^2} \quad x \geq 0$

$= x$ since $x \geq 0$

$f[f^{-1}(x)] = (\sqrt{x + 1})^2 - 1$

$= x + 1 - 1$

$= x$

(C) $f^{-1}(3) = \sqrt{3 + 1} = 2$ (D) $f^{-1}[f(4)] = 4$ (E) $f^{-1}[f(x)] = x$ (3-6)

89. (A) Reflect the given graph across the y axis:

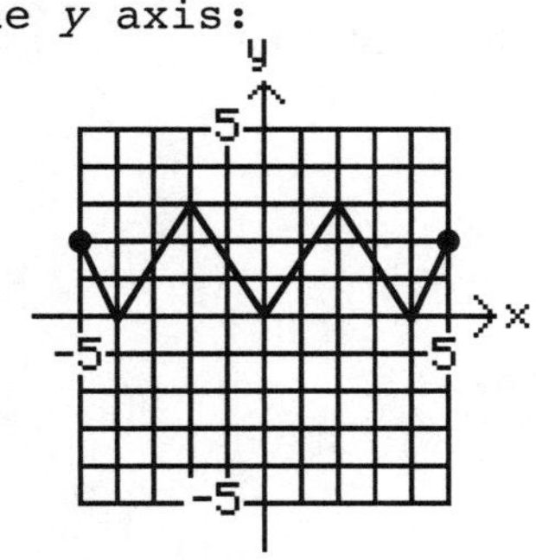

(B) Reflect the given graph across the origin:

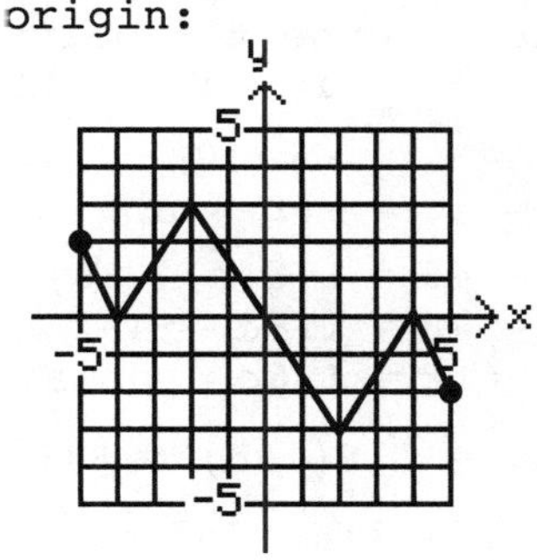

(3-3)

90. (A) The graph must cross the x axis exactly once. Some possible graphs are shown:

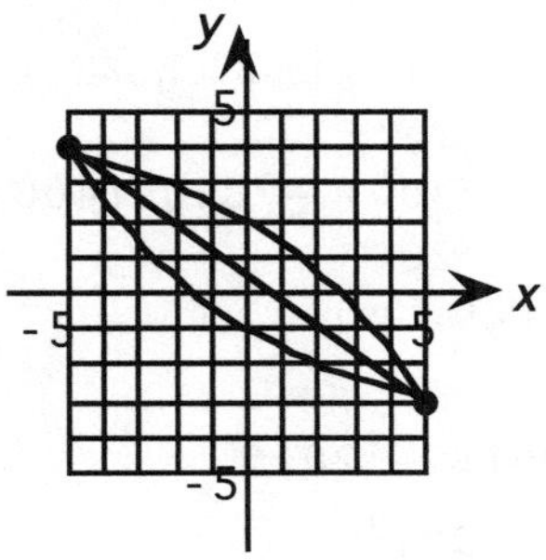

(B) The graph may cross the x axis once, but it may fail to cross the x axis at all. A possible graph of the latter type is shown:

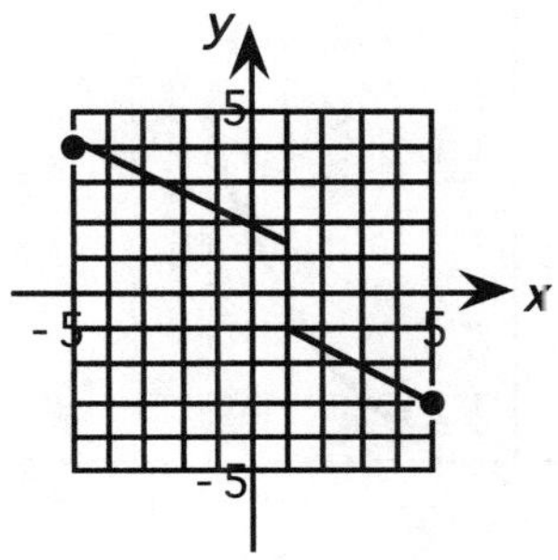

(3-2)

91. If $0 \le x \le 3{,}000$, $E(x) = 200$

If $3{,}000 < x \le 5{,}000$,

$$\begin{aligned} E(x) &= \text{Base salary} + \text{Commission on sales over } \$3{,}000 \\ &= 200 + 0.1(x - 3{,}000) \\ &= 200 + 0.1x - 300 \\ &= 0.1x - 100 \end{aligned}$$

If $x > 5{,}000$,

$$\begin{aligned} E(x) &= \text{Salary} + \text{Bonus} \\ &= (0.1x - 100) + 150 \\ &= 0.1x + 50 \end{aligned}$$

Summarizing,

$$E(x) = \begin{cases} 200 & \text{if } 0 \le x \le 3{,}000 \\ 0.1x - 100 & \text{if } 3{,}000 < x \le 5{,}000 \\ 0.1x + 50 & \text{if } x > 5{,}000 \end{cases}$$

If $0 \le x \le 3{,}000$, $E(x) = 200$, so $E(2{,}000) = \$200$.
If $3{,}000 < x \le 5{,}000$, $E(x) = 0.1x - 100$, so $E(4{,}000) = 0.1(4{,}000) - 100 = \$300$.
If $x > 5{,}000$, $E(x) = 0.1x + 50$, so $E(6{,}000) = 0.1(6{,}000) + 50 = \$650$. (3-2)

92. The function describing the height of the sandbag is $h(t) = 400 - 16t^2$, since the initial height is 400 feet.
Substitute 0 for $h(t)$ and solve.

$$\begin{aligned} 0 &= 400 - 16t^2 \\ 16t^2 &= 400 \\ t^2 &= 25 \\ t &= 5 \text{ seconds after it's dropped} \end{aligned}$$

(3-4)

93. (A) If r is linearly related to c, then we are looking for a function whose graph passes through $(c_1, r_1) = (30, 48)$ and $(c_2, r_2) = (20, 32)$. We find the slope, then we use the point-slope form to find the equation.

$$m = \frac{r_2 - r_1}{c_2 - c_1} = \frac{32 - 48}{20 - 30} = 1.6$$

$$r - r_1 = m(c - c_1)$$
$$r - 48 = 1.6(c - 30)$$
$$r - 48 = 1.6c - 48$$
$$f(c) = r = 1.6c \quad \text{domain: } [10, \infty) \quad \text{range: } [16, \infty)$$

(B) $f(105) = 1.6(105) = \$168$

(C) Solve $r = 1.6c$ for c.

$0.625r = c = f^{-1}(r)$

Domain: $[16, \infty)$, range: $[10, \infty)$

(D) $f^{-1}(39.99) = 0.625(39.99) = \$24.99$ *(3-6)*

94. (A)

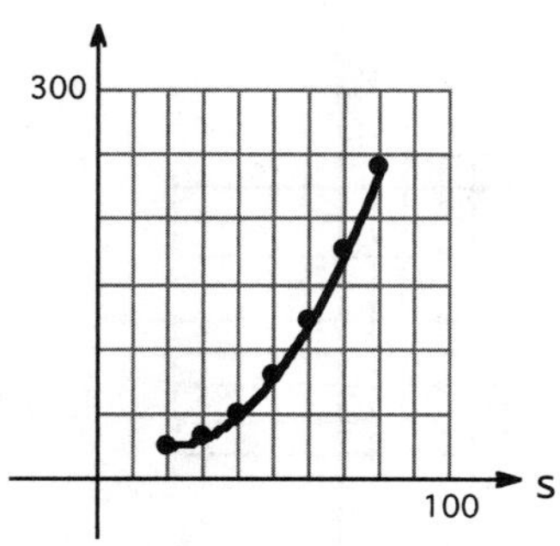

(B) $L = 0.06s^2 - 2.4s + 50$ domain $[20, \infty)$

Complete the square to find the range.

$L = 0.06(s^2 - 40s + ?) + 50$

$L = 0.06(s^2 - 40s + 400) - 0.06(400) + 50$

$L = 0.06(s - 20)^2 + 26$

therefore, range $[26, \infty)$

Solve for s.

$$L - 26 = 0.06(s - 20)^2$$
$$\frac{50}{3}(L - 26) = (s - 20)^2$$
$$\sqrt{\frac{50}{3}(L - 26)} = s - 20 \quad \text{(positive square root only, since } s \ge 20)$$
$$f^{-1}(L) = s = 20 + \sqrt{\frac{50}{3}(L - 26)} \quad \text{domain } [26, \infty), \text{ range } [20, \infty).$$

(C) Substitute $L = 200$ into $f^{-1}(L)$ to obtain

$$S = 20 + \sqrt{\frac{50}{3}(200 - 26)} = 74 \text{ mph}$$ *(3-6)*

95. (A) Examining the graph of p, we obtain

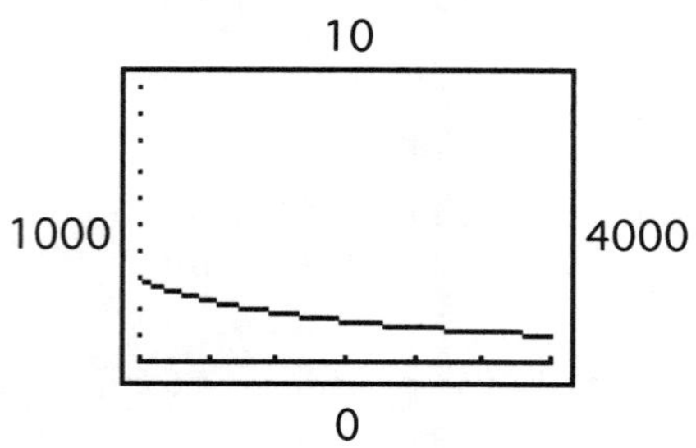

Clearly, d is a decreasing function, passing the horizontal line test, so d is one-to-one.

When $q = 1000$, $p = d(1000) = \dfrac{9}{1 + 0.002(1000)} = 3$

When $q = 4000$, $p = d(4000) = \dfrac{9}{1 + 0.002(4000)} = 1$

Thus the range of d is $1 \le p \le 3$ or $[1, 3]$.

(B) Solve $p = d(q)$ for q.

$$p = \frac{9}{1 + 0.002q}$$

$$p(1 + 0.002q) = 9$$

$$1 + 0.002q = \frac{9}{p}$$

$$0.002q = \frac{9}{p} - 1$$

$$q = \frac{4,500}{p} - 500 = d^{-1}(p)$$

Domain d^{-1} = range d = [1, 3]
Range d^{-1} = domain d = [1000, 4000].

(C) Revenue is price times number sold, so we need to multiply the variable p (price) by the function q (number sold) that we obtained in part B:

$$p \cdot q = p \cdot \left(\frac{4,500}{p} - 500\right) = 4,500 - 500p$$

(D) Revenue is price times number sold, so we need to multiply the function p (price) times the variable q (numbers sold):

$$p \cdot q = \frac{9}{1 + 0.002q} \cdot q = \frac{9q}{1 + 0.002q}$$ *(3-6)*

96. Profit is revenue minus cost so we start by subtracting the revenue and cost functions:

$$P(x) = R(x) - C(x) = (50x - 0.1x^2) - (10x + 1,500)$$
$$= 50x - 0.1x^2 - 10x - 1,500 = -0.1x^2 + 40x - 1,500$$

The variable is x but we're asked to find profit in terms of price (p) so we need to substitute $500 - 10p$ in for x.

$$p(500 - 10p) = -0.1(500 - 10p)^2 + 40(500 - 10p) - 1,500$$
$$= -0.1(250,000 - 10,000p + 100p^2) + 20,000 - 400p - 1,500$$
$$= -25,000 + 1,000p - 10p^2 + 20,000 - 400p - 1,500$$
$$= -10p^2 + 600p - 6,500$$

so $P(p) = -10p^2 + 600p - 6,500$

This is a quadratic function, so its maximum occurs at the vertex of its parabola graph which is given by the formula (h, k) where

$$h = -\frac{b}{2a} = -\frac{600}{2(-10)} = 30$$

$$k = c - \frac{b^2}{4a} = -6,500 - \frac{600^2}{4(-10)} = 2,500$$

Thus the price that produces the largest profit is \$30 and this profit is \$2,500. *(3-4)*

97. (A) From the figure, we see that $A = x(y + y) = 2xy$. Since the fence consists of four pieces of length y and three pieces of length x, we have $3x + 4y = 120$. Hence $4y = 120 - 3x$

$$y = 30 - \frac{3}{4}x$$
$$A = 2x\left(30 - \frac{3}{4}x\right)$$
$$A(x) = 60x - \frac{3}{2}x^2$$

(B) Since both x and y must be positive, we have $x > 0$

$30 - \frac{3}{4}x > 0$ or $-\frac{3}{4}x > -30$ or $x < 40$

Hence $0 < x < 40$ is the domain of A.

(C) The function A is a quadratic function. Completing the square yields:

$$\begin{aligned} A(x) &= -\frac{3}{2}x^2 + 60x \\ &= -\frac{3}{2}(x^2 - 40x) \\ &= -\frac{3}{2}(x^2 - 40x + 400) + \frac{3}{2}\cdot 400 \\ &= -\frac{3}{2}(x - 20)^2 + 600 \end{aligned}$$

Comparing with $f(x) = a(x - h)^2 + k$, the total area will be maximum when $x = 20$. Then

$y = 30 - \frac{3}{4}x = 30 - \frac{3}{4}(20) = 15$ *(3-4)*

98. (A)

$$\begin{aligned} f(1) &= 1 - ([\![\sqrt{1}]\!])^2 = 1 - 1 = 0 \\ f(2) &= 2 - ([\![\sqrt{2}]\!])^2 = 2 - 1 = 1 \\ f(3) &= 3 - ([\![\sqrt{3}]\!])^2 = 3 - 1 = 2 \\ f(4) &= 4 - ([\![\sqrt{4}]\!])^2 = 4 - 4 = 0 \\ f(5) &= 5 - ([\![\sqrt{5}]\!])^2 = 5 - 4 = 1 \end{aligned}$$

(B)
$$\begin{aligned} f(n^2) &= n^2 - ([\![\sqrt{n^2}]\!])^2 \\ &= n^2 - ([\![n]\!])^2 \quad \text{since } \sqrt{n^2} = n \text{ if } n \text{ is positive} \\ &= n^2 - (n)^2 \quad \text{since } [\![n]\!] = n \text{ if } n \text{ is a (positive) integer.} \\ &= 0 \end{aligned}$$

$$\begin{aligned} f(6) &= 6 - ([\![\sqrt{6}]\!])^2 = 6 - 4 = 2 \\ f(7) &= 7 - ([\![7]\!])^2 = 7 - 4 = 3 \\ f(8) &= 8 - ([\![8]\!])^2 = 8 - 4 = 4 \\ f(9) &= 9 - ([\![9]\!])^2 = 9 - 9 = 0 \\ f(10) &= 10 - ([\![10]\!])^2 = 10 - 9 = 1 \\ f(11) &= 11 - ([\![11]\!])^2 = 11 - 9 = 2 \end{aligned}$$

$$\begin{aligned} f(12) &= 12 - ([\![12]\!])^2 = 12 - 9 = 3 \\ f(13) &= 13 - ([\![13]\!])^2 = 13 - 9 = 4 \\ f(14) &= 14 - ([\![14]\!])^2 = 14 - 9 = 5 \\ f(15) &= 15 - ([\![15]\!])^2 = 15 - 9 = 6 \\ f(16) &= 16 - ([\![16]\!])^2 = 16 - 16 = 0 \end{aligned}$$

(C) It determines if an integer is a perfect square integer. If $f(x) = 0$, then x is a perfect square and if $f(x) = 10$, x is not a perfect square integer.

(3-2)

99. Amounts less than \$3,000: Tax is 2% of income (x) $= 0.02x$

Amounts between \$3,000 and \$5,000: Tax is \$60 + 3% of income over 3,000 ($x - 3{,}000$) so tax $= 60 + 0.03(x - 3{,}000) = 0.03x - 30$

Amounts between \$5,000 and \$17,000: Tax is \$120 + 5% of income over 5,000 ($x - 5{,}000$) so tax $= 120 + 0.05(x - 5{,}000) = 0.05x - 130$

Amounts over \$17,000: Tax is \$720 + 5.75% of income over 17,000 ($x - 17{,}000$) so tax $= 720 + 0.0575(x - 17{,}000) = 0.0575x - 257.5$

Combined, we get $$t(x) = \begin{cases} 0.02x & 0 \le x \le 3{,}000 \\ 0.03x - 30 & 3{,}000 < x \le 5{,}000 \\ 0.05x - 130 & 5{,}000 < x \le 17{,}000 \\ 0.0575x - 257.5 & x > 17{,}000 \end{cases}$$

$t(2{,}000) = 0.02(2{,}000) = 40$. The tax on \$2,000 is \$40.

$t(4{,}000) = 0.03(4{,}000) - 30 = 120 - 30 = 90$. The tax on \$4,000 is \$90.

$t(10{,}000) = 0.05(10{,}000) - 130 - 500 - 130 = 370$. The tax on \$10,000 is \$130.

$t(30{,}000) = 0.0575(30{,}000) - 257.5 - 1{,}725 - 257.5 = 1{,}467.5$. The tax on \$30,000 is \$1,467.50. *(3-2)*

CUMULATIVE REVIEW (Chapters 1–3)

1.
$$\frac{7x}{5} - \frac{3 + 2x}{2} = \frac{x - 10}{3} + 2$$
$$30\frac{7x}{5} - 30\frac{(3 + 2x)}{2} = 30\frac{(x - 10)}{3} + 2(30)$$
$$42x - 15(3 + 2x) = 10(x - 10) + 60$$
$$42x - 45 - 30x = 10x - 100 + 60$$
$$12x - 45 = 10x - 40$$
$$2x = 5$$
$$x = \frac{5}{2}$$
(*1-1*)

2.
$$2(3 - y) + 4 \le 5 - y$$
$$6 - 2y + 4 \le 5 - y$$
$$-2y + 10 \le 5 - y$$
$$-y \le -5$$
$$y \ge 5$$
$[5, \infty)$

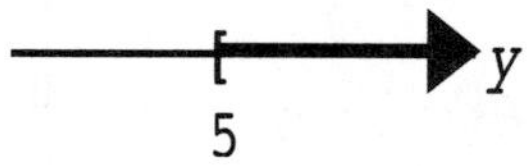

(*1-2*)

3. $|x - 2| < 7$
$$-7 < x - 2 < 7$$
$$-5 < x < 9$$
$(-5, 9)$

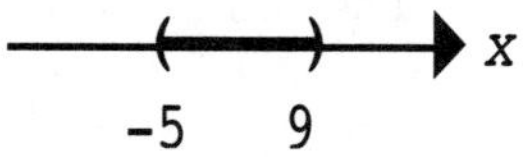

(*1-3*)

4. $x^3 + 3x \ge 10$

$f(x) = x^2 + 3x - 10 \ge 0$

$f(x) = (x + 5)(x - 2) \ge 0$

The zeros of f are -5 and 2.

Plotting the graph of $f(x)$, we see that $f(x) \ge 0$ for $x \le -5$ and $x \ge 2$, or $(-\infty, -5] \cup [2, \infty)$.

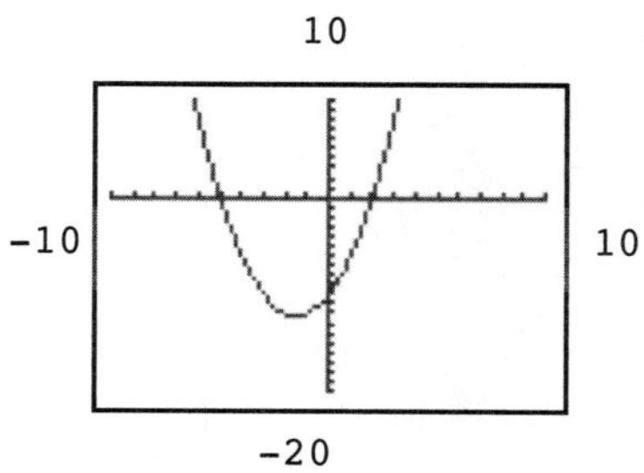

(*3-4*)

5. (A) $(2 - 3i) - (-5 + 7i) = 2 - 3i + 5 - 7i = 7 - 10i$

(B) $(1 + 4i)(3 - 5i) = 3 + 7i - 20i^2 = 3 + 7i + 20 = 23 + 7i$

(C)
$$\frac{5 + i}{2 + 3i} = \frac{(5 + i)}{(2 + 3i)}\frac{(2 - 3i)}{(2 - 3i)} = \frac{10 - 13i - 3i^2}{4 - 9i^2}$$
$$= \frac{10 - 13i + 3}{4 + 9} = \frac{13 - 13i}{13} = 1 - i$$
(*1-4*)

6.
$$3x^2 = -12x$$
$$3x^2 + 12x = 0$$
$$3x(x + 4) = 0$$
$$3x = 0 \quad \text{or} \quad x + 4 = 0$$
$$x = 0 \qquad x = -4$$
(*1-5*)

7.
$$4x^2 - 20 = 0$$
$$4x^2 = 20$$
$$x^2 = 5$$
$$x = \pm\sqrt{5}$$
(*1-5*)

8.
$$x^2 - 6x + 2 = 0$$
$$x^2 - 6x = -2$$
$$x^2 - 6x + 9 = 7$$
$$(x - 3)^2 = 7$$
$$x - 3 = \pm\sqrt{7}$$
$$x = 3 \pm \sqrt{7}$$
(*1-5*)

9.
$$x - \sqrt{12 - x} = 0$$
$$x = \sqrt{12 - x}$$
$$x^2 = 12 - x$$
$$x^2 + x - 12 = 0$$
$$(x + 4)(x - 3) = 0$$
$$x = -4, 3$$

Check:
$$-4 - \sqrt{12 - (-4)} \stackrel{?}{=} 0$$
$$-4 - 4 \stackrel{?}{=} 0$$
$$-8 \neq 0$$
$$3 - \sqrt{12 - 3} \stackrel{?}{=} 0$$
$$3 - 3 \stackrel{?}{=} 0$$
$$0 \stackrel{\surd}{=} 0$$

Solution: 3 (*1-6*)

10. (A) $d(A, B) = \sqrt{(5 - 3)^2 + (6 - 2)^2} = \sqrt{4 + 16} = \sqrt{20} = 2\sqrt{5}$

(B) $m = \dfrac{6 - 2}{5 - 3} = \dfrac{4}{2} = 2$

(C) The slope m_1 of a line perpendicular to AB must satisfy $m_1(2) = -1$.

Therefore $m_1 = -\dfrac{1}{2}$ (*2-2, 3-3*)

11. (A) Center at (0, 0) and radius $\sqrt{2}$
$$x^2 + y^2 = r^2$$
$$x^2 + y^2 = (\sqrt{2})^2$$
$$x^2 + y^2 = 2$$

(B) Center at (-3, 1) and radius $\sqrt{2}$
$$(h, k) = (-3, 1)$$
$$r = \sqrt{2}$$
$$(x - h)^2 + (y - k)^2 = r^2$$
$$[x - (-3)]^2 + (y - 1)^2 = (\sqrt{2})^2$$
$$(x + 3)^2 + (y - 1)^2 = 2$$
(*2-2*)

12.
$$2x - 3y = 6$$
$$-3y = -2x + 6$$
$$y = \frac{2}{3}x - 2$$

slope: $\dfrac{2}{3}$ y intercept: -2 x intercept: 3 (if $y = 0$, $2x = 6$, hence $x = 3$)

x	y
−3	−4
0	−2
3	0

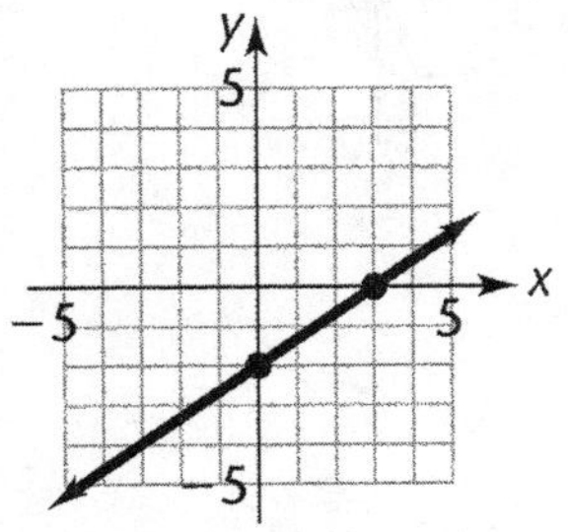

(*2-3*)

13. (A) A function; domain = {1, 2, 3}; range = {1}
(B) Not a function (three range values correspond to the only domain value)
(C) A function; domain = {-2, -1, 0, 1, 2}; range = {-1, 0, 2} (*3-1*)

14. (A) $f(-2) = (-2)^2 - 2(-2) + 5 = 13$
$g(3) = 3 \cdot 3 - 2 = 7$

Therefore

$f(-2) + g(3) = 13 + 7 = 20$

(B)
$$(f + g)(x) = f(x) + g(x)$$
$$= x^2 - 2x + 5 + 3x - 2$$
$$= x^2 + x + 3$$

(C) $(f \circ g)(x) = f[g(x)] = f(3x - 2)$

$$= (3x - 2)^2 - 2(3x - 2) + 5$$
$$= 9x^2 - 12x + 4 - 6x + 4 + 5$$
$$= 9x^2 - 18x + 13$$

(D) $$\frac{f(a + h) - f(a)}{h} = \frac{[(a + h)^2 - 2(a + h) + 5] - (a^2 - 2a + 5)}{h}$$
$$= \frac{a^2 + 2ah + h^2 - 2a - 2h + 5 - a^2 + 2a - 5}{h}$$
$$= \frac{2ah + h^2 - 2h}{h}$$
$$= \frac{h(2a + h - 2)}{h}$$
$$= 2a + h - 2$$ *(3-1, 3-5)*

15. (A) Stretched vertically by a factor of 2. (B) Shifted right 2 units
(C) Shifted down 2 units *(3-3)*

16. Domain: [-2, 3] Range: [-1, 2] *(3-2)*

17. $f(2) = 2$ $f(-2) = -1$.
Therefore it is false to assert either that $f(x) = f(-x)$ for all x in the domain of f or that $-f(x) = f(-x)$ for all x in the domain of f. Thus f is neither even nor odd. *(3-3)*

18. (A) The graph of $f(x)$ is shifted left one unit and reflected across the x axis.

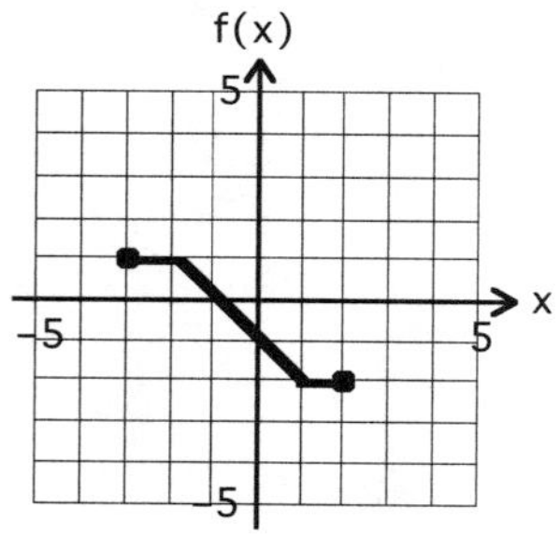

(B) The graph of $f(x)$ is stretched by a factor of two with respect to the y axis and shifted down two units.

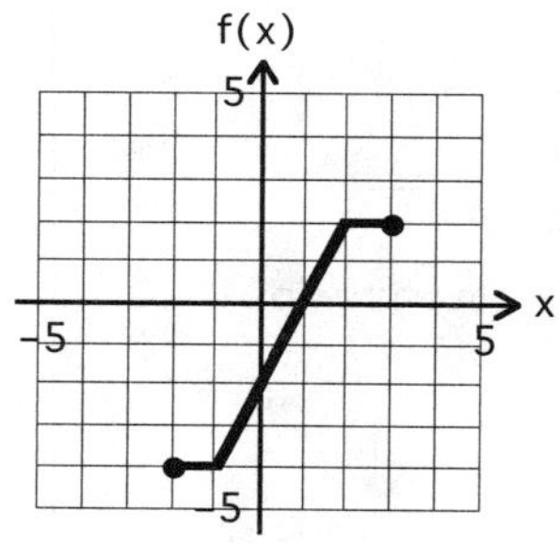

(3-3)

19.
$$\frac{x + 3}{2x + 2} + \frac{5x + 2}{3x + 3} = \frac{5}{6}$$
$$\frac{(x + 3)}{2(x + 1)} + \frac{(5x + 2)}{3(x + 1)} = \frac{5}{6}$$ LCD: $6(x + 1)$ Excluded value: $x \neq -1$
$$6(x + 1)\frac{(x + 3)}{2(x + 1)} + 6(x + 1)\frac{(5x + 2)}{3(x + 1)} = 6(x + 1)\frac{5}{6}$$
$$3(x + 3) + 2(5x + 2) = 5(x + 1)$$
$$3x + 9 + 10x + 4 = 5x + 5$$
$$13x + 13 = 5x + 5$$
$$8x = -8$$
$$x = -1$$
No solution: -1 is excluded *(1-1)*

20. $\frac{3}{x} = \frac{6}{x+1} - \frac{1}{x-1}$ Excluded values: $x \neq 0, 1, -1$

$$\begin{aligned}
x(x+1)(x-1)\frac{3}{x} &= x(x+1)(x-1)\frac{6}{x+1} - x(x+1)(x-1)\frac{1}{x-1} \\
3(x+1)(x-1) &= 6x(x-1) - x(x+1) \\
3(x^2-1) &= 6x^2 - 6x - x^2 - x \\
3x^2 - 3 &= 5x^2 - 7x \\
0 &= 2x^2 - 7x + 3 \\
0 &= (2x-1)(x-3) \\
2x - 1 = 0 \quad &\text{or} \quad x - 3 = 0 \\
2x = 1 \quad & \quad x = 3 \\
x = \frac{1}{2} &
\end{aligned}$$

(2-6)

21.

$$\begin{aligned}
2x + 1 &= 3\sqrt{2x-1} \\
(2x+1)^2 &= 9(2x-1) \\
4x^2 + 4x + 1 &= 18x - 9 \\
4x^2 - 14x + 10 &= 0 \\
2x^2 - 7x + 5 &= 0 \\
(2x-5)(x-1) &= 0 \\
2x - 5 = 0 \quad &\text{or} \quad x - 1 = 0 \\
x = \frac{5}{2} \quad & \quad x = 1
\end{aligned}$$

Check:
$$\begin{aligned}
2\left(\tfrac{5}{2}\right) + 1 &\stackrel{?}{=} 3\sqrt{2\left(\tfrac{5}{2}\right) - 1} \\
5 + 1 &\stackrel{?}{=} 3\sqrt{5-1} \\
6 &\stackrel{\checkmark}{=} 6 \\
2\cdot 1 + 1 &\stackrel{?}{=} 3\sqrt{2\cdot 1 - 1} \\
3 &\stackrel{\checkmark}{=} 3
\end{aligned}$$

Solution: $1, \frac{5}{2}$ (1-6)

22. $|4x - 9| > 3$

$$\begin{aligned}
4x - 9 < -3 \quad &\text{or} \quad 4x - 9 > 3 \\
4x < 6 \quad & \quad 4x > 12 \\
x < \frac{3}{2} \quad &\text{or} \quad x > 3
\end{aligned}$$

$\left(-\infty, \frac{3}{2}\right) \cup (3, \infty)$

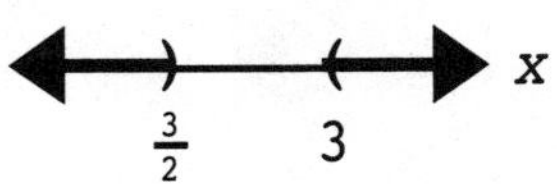

(1-3)

23.

$$\begin{aligned}
\sqrt{(3m-4)^2} &\le 2 \\
|3m - 4| &\le 2 \\
-2 \le 3m - 4 &\le 2 \\
2 \le 3m &\le 6 \\
\frac{2}{3} \le m &\le 2
\end{aligned}$$

$\left[\frac{2}{3}, 2\right]$

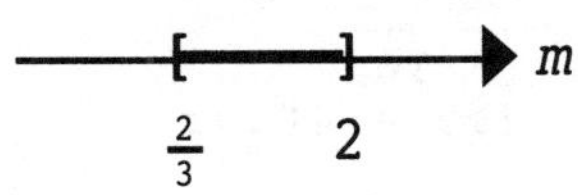

(1-3)

24. $\frac{x+1}{2} > x - 2$

$$\begin{aligned}
x + 1 &> 2x - 4 \\
-x + 1 &> -4 \\
-x &> -5 \\
x &< 5
\end{aligned}$$

$(-\infty, 5)$

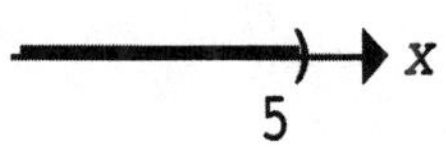

(1-2)

25. $\frac{\sqrt{x-2}}{x-4}$ represents a real number if $x - 2$ is positive or zero, except if $x = 4$. Thus, $x \ge 2$, $x \neq 4$, or $[2, 4) \cup (4, \infty)$.

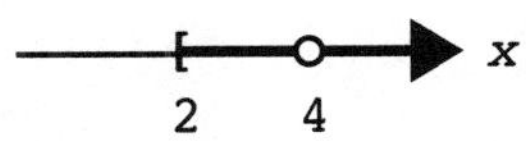

(1-2)

26. (A) $(2 - 3i)^2 - (4 - 5i)(2 - 3i) - (2 + 10i)$

$$= (2)^2 - 2(2)(3i) + (3i)^2 - (8 - 12i - 10i + 15i^2) - 2 - 10i$$
$$= 4 - 12i + 9i^2 - (8 - 22i + 15i^2) - 2 - 10i$$
$$= 4 - 12i + 9i^2 - 8 + 22i - 15i^2 - 2 - 10i$$
$$= 4 - 12i - 9 - 8 + 22i + 15 - 2 - 10i$$
$$= 0 + 0i \quad \text{or} \quad 0$$

(B) $$\frac{3}{5} + \frac{4}{5}i + \frac{1}{\frac{3}{5} + \frac{4}{5}i} = \frac{3}{5} + \frac{4}{5}i + \frac{1}{(\frac{3}{5} + \frac{4}{5}i)} \frac{(\frac{3}{5} - \frac{4}{5}i)}{(\frac{3}{5} - \frac{4}{5}i)}$$
$$= \frac{3}{5} + \frac{4}{5}i + \frac{\frac{3}{5} - \frac{4}{5}i}{\frac{9}{25} - \frac{16}{25}i^2}$$
$$= \frac{3}{5} + \frac{4}{5}i + \frac{\frac{3}{5} - \frac{4}{5}i}{\frac{9}{25} + \frac{16}{25}}$$
$$= \frac{3}{5} + \frac{4}{5}i + \frac{3}{5} - \frac{4}{5}i = \frac{6}{5}$$

(C) $i^{35} = i^{32}i^3 = (i^4)^8(-i) = 1^8(-i) = -i$ *(1-4)*

27. (A) $(5 + 2\sqrt{-9}) - (2 - 3\sqrt{-16}) = (5 + 2i\sqrt{9}) - (2 - 3i\sqrt{16})$

$$= (5 + 6i) - (2 - 12i)$$
$$= 5 + 6i - 2 + 12i$$
$$= 3 + 18i$$

(B) $$\frac{2 + 7\sqrt{-25}}{3 - \sqrt{-1}} = \frac{2 + 7i\sqrt{25}}{3 - i}$$
$$= \frac{2 + 35i}{3 - i}$$
$$= \frac{(2 + 35i)}{(3 - i)} \frac{(3 + i)}{(3 + i)}$$
$$= \frac{6 + 107i + 35i^2}{9 - i^2}$$
$$= \frac{6 + 107i - 35}{9 + 1}$$
$$= \frac{-29 + 107i}{10}$$
$$= -2.9 + 10.7i$$

(C) $$\frac{12 - \sqrt{-64}}{\sqrt{-4}} = \frac{12 - i\sqrt{64}}{i\sqrt{4}} = \frac{12 - 8i}{2i} = \frac{12 - 8i}{2i} \frac{i}{i}$$
$$= \frac{12i - 8i^2}{2i^2} = \frac{12i + 8}{-2} = -4 - 6i$$ *(1-4)*

28. $$1 + \frac{14}{y^2} = \frac{6}{y}$$ Excluded value: $y \neq 0$

$$y^2(1) + y^2 \frac{14}{y^2} = y^2 \frac{6}{y}$$
$$y^2 + 14 = 6y$$
$$y^2 - 6y + 14 = 0$$
$$y^2 - 6y = -14$$
$$y^2 - 6y + 9 = -5$$
$$(y - 3)^2 = -5$$
$$y - 3 = \pm\sqrt{-5}$$
$$y = 3 \pm \sqrt{-5}$$
$$y = 3 \pm i\sqrt{5}$$ *(1-6)*

29. $4x^{2/3} - 4x^{1/3} - 3 = 0$

Let $u = x^{1/3}$, then

$$4u^2 - 4u - 3 = 0$$
$$(2u - 3)(2u + 1) = 0$$
$$u = \frac{3}{2}, -\frac{1}{2}$$
$$x^{1/3} = \frac{3}{2} \qquad x^{1/3} = -\frac{1}{2}$$
$$x = \frac{27}{8} \qquad x = -\frac{1}{8}$$

(*1-6*)

30. $u^4 + u^2 - 12 = 0$

Let $w = u^2$, then

$$w^2 + w - 12 = 0$$
$$(w + 4)(w - 3) = 0$$
$$w = -4, 3$$
$$u^2 = -4 \qquad u^2 = 3$$
$$u = \pm 2i \qquad u = \pm\sqrt{3}$$

(*1-6*)

31.

$$\sqrt{8t - 2} - 2\sqrt{t} = 1$$
$$\sqrt{8t - 2} = 2\sqrt{t} + 1$$
$$8t - 2 = 4t + 4\sqrt{t} + 1$$
$$4t - 3 = 4\sqrt{t}$$
$$16t^2 - 24t + 9 = 16t$$
$$16t^2 - 40t + 9 = 0$$
$$(4t - 1)(4t - 9) = 0$$
$$t = \tfrac{1}{4}, \tfrac{9}{4}$$

Check: $\sqrt{8(\frac{1}{4}) - 2} - 2\sqrt{\frac{1}{4}} \stackrel{?}{=} 1$

$$0 - 2(\tfrac{1}{2}) \stackrel{?}{=} 1$$
$$-1 \neq 1$$
$$\sqrt{8(\tfrac{9}{4}) - 2} - 2\sqrt{\tfrac{9}{4}} \stackrel{?}{=} 1$$
$$\sqrt{16} - 2\sqrt{\tfrac{9}{4}} \stackrel{?}{=} 1$$
$$4 - 2(\tfrac{3}{2}) \stackrel{?}{=} 1$$
$$1 \stackrel{\checkmark}{=} 1$$

Solution: $\frac{9}{4}$ (*1-6*)

32.

$$-3.45 < 1.86 - 0.33x \le 7.92$$
$$-5.31 < -0.33x \le 6.06$$
$$16.09 > x \ge -18.36$$

$-18.36 \le x < 16.09$ or $[-18.36, 16.09)$ (*1-2*)

33. $2.35x^2 + 10.44x - 16.47 = 0$

$$x = \frac{-b \pm \sqrt{b^2 - 4ac}}{2a} \qquad a = 2.35,\ b = 10.44,\ c = -16.47$$
$$x = \frac{-10.44 \pm \sqrt{(10.44)^2 - 4(2.35)(-16.47)}}{2(2.35)}$$
$$= \frac{-10.44 \pm \sqrt{263.8116}}{4.70}$$
$$= \frac{-10.44 \pm 16.2423}{4.70}$$
$$= -5.68, 1.23$$

(*1-5*)

34.

$$\frac{x - 2}{x + 1} = \frac{2y + 1}{y - 2}$$
$$(x + 1)(y - 2)\frac{x - 2}{x + 1} = (x + 1)(y - 2)\frac{2y + 1}{y - 2}$$
$$(y - 2)(x - 2) = (x + 1)(2y + 1)$$
$$xy - 2y - 2x + 4 = 2xy + x + 2y + 1$$
$$3 - 3x = xy + 4y$$
$$3 - 3x = y(x + 4)$$
$$y = \frac{3 - 3x}{x + 4}$$

(*1-1*)

35. (A) All real numbers $(-\infty, \infty)$

(B) From the graph, the possible function values include -2 (only) and all numbers greater than or equal to 1. In set and interval notation: $\{-2\} \cup [1, \infty)$.

(C) $f(-3) = 1$ $f(-2) = 2$ (not -2) $f(2) = -2$ (not 2).
Thus, $f(-3) + f(-2) + f(2) = 1 + 2 + (-2) = 1$

(D) $[-3, -2]$ and $[2, \infty)$

(E) f is discontinuous at $x = -2$ and at $x = 2$. *(3-1, 3-2)*

36. The line $3x + 2y = 12$, or
$2y = -3x + 12$,
or $y = -\frac{3}{2}x + 6$, has slope $-\frac{3}{2}$.

(A) We require a line through (-6, 1) with slope $-\frac{3}{2}$. Applying the point-slope form, we have

$$y - 1 = -\frac{3}{2}(x + 6)$$
$$y - 1 = -\frac{3}{2}x - 9$$
$$y = -\frac{3}{2}x - 8$$

(B) We require a line with slope m satisfying $-\frac{3}{2}m = -1$, or $m = \frac{2}{3}$. Again applying the point-slope form, we have

$$y - 1 = \frac{2}{3}(x + 6)$$
$$y - 1 = \frac{2}{3}x + 4$$
$$y = \frac{2}{3}x + 5$$

(2-3)

37. $x + 4 \ge 0$ is equivalent to $x \ge -4$. Domain of $g(x) = \sqrt{x + 4}$ is $[-4, \infty)$ *(3-1)*

38. $f(x) = x^2 - 2x - 8$. Complete the square.
$f(x) = (x^2 - 2x + 1) - 1 - 8$
$= (x - 1)^2 - 9$

Comparing with $f(x) = a(x - h)^2 + k$, $h = 1$ and $k = -9$. Thus, the vertex is (1, -9), the axis of symmetry is $x = 1$, and the minimum value is -9. $y = f(x)$ can be any number greater than or equal to -9, so the range is $[-9, \infty)$.
y intercept: Set $x = 0$, then $f(0) = -8$ is the y intercept.
x intercepts: Set $f(x) = 0$, then
$0 = x^2 - 2x - 8$
$0 = (x - 4)(x + 2)$
$x = 4$ or $x = -2$ are the x intercepts.

Graph: Locate axis and vertex, then plot several points on either side of the axis.

x	$f(x)$
−3	7
−2	0
−1	−5
0	−8
1	−9
2	−8
3	−5
4	0
5	7

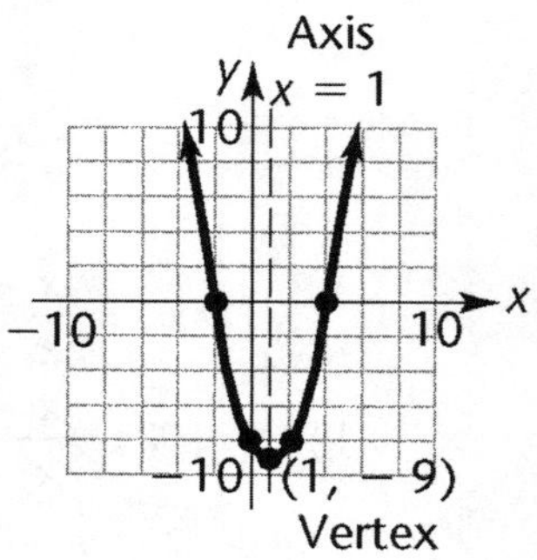

(3-4)

39. $(f \circ g)(x) = f[g(x)] = f\left(\frac{x+3}{x}\right) = \frac{1}{\frac{x+3}{3} - 2} = \frac{x(1)}{x(\frac{x+3}{x} - 2)} = \frac{x}{x + 3 - 2x} = \frac{x}{3 - x}$

The domain of f is all real numbers except 2.
The domain of g is all non-zero real numbers.
The domain of $f \circ g$ is the set of all non-zero real numbers for which $g(x) \neq 2$.

$g(x) = 2$ only if $\frac{x+3}{x} = 2$, that is $x + 3 = 2x$, or $x = 3$.

Thus the domain of $f \circ g$ is the set of all non-zero real numbers except 3, that is, $(-\infty, 0) \cup (0, 3) \cup (3, \infty)$.

Common Error:
The domain of $f \circ g$ cannot be found by looking at the final form $\frac{x}{3 - x}$. It is not $\{x \mid x \neq 3\}$.

(3-5)

40. $f(x) = 2x + 5$
Assume $f(a) = f(b)$

$$2a + 5 = 2b + 5$$
$$2a = 2b$$
$$a = b$$

Thus, f is one-to-one.

Solve $y = f(x)$ for x:

$$y = 2x + 5$$
$$y - 5 = 2x$$
$$x = \frac{y - 5}{2} = f^{-1}(y)$$

Interchange x and y

$y = f^{-1}(x) = \frac{x - 5}{2}$ or $\frac{1}{2}x - \frac{5}{2}$ Domain: $(-\infty, \infty)$ (3-6)

41.

x	$y = x - 1$	x	$y = x^2 + 1$
-1	-2	0	1
-2	-3	1	2
		2	5

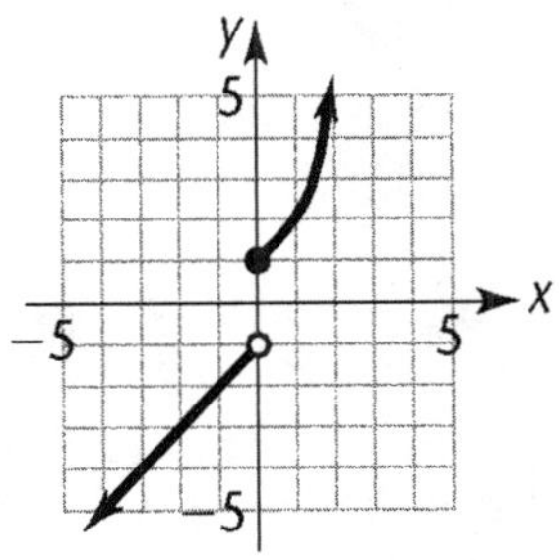

Domain: all real numbers, $(-\infty, \infty)$
Range: $(-\infty, -1) \cup [1, \infty)$
Discontinuous at: $x = 0$

(3-2)

42. (A) This is the same as the graph of $y = \sqrt{x}$ stretched vertically by a factor of 2 and shifted up one unit.

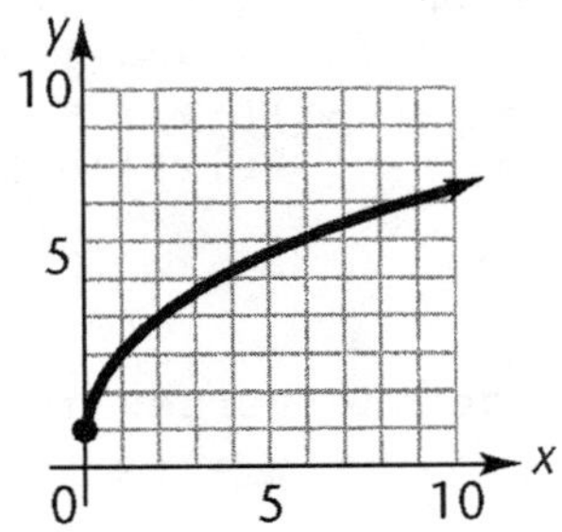

(B) This is the same as the graph of $y = \sqrt{x}$ shifted left 1 unit and reflected through the x axis.

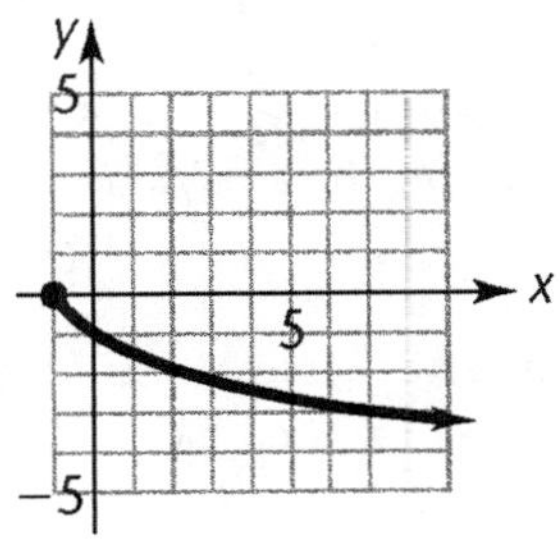

(3-3)

43. The graph of $y = |x|$ is shrunk by a factor of $\frac{1}{2}$, reflected through the x axis, shifted two units to the right and three units up; $y = -\frac{1}{2}|x - 2| + 3$. (3-3)

44. $f(x) = \sqrt{x + 4}$

Assume
$$\begin{aligned} f(a) &= f(b) \\ \sqrt{a+4} &= \sqrt{b+4} \\ a + 4 &= b + 4 \\ a &= b \end{aligned}$$
Thus f is one-to-one.

(A) Solve $y = f(x)$ for x

$y = \sqrt{x + 4}$

$y^2 = x + 4 \quad y \ge 0$

$x = y^2 - 4 \quad y \ge 0 \quad f^{-1}(y) = y^2 - 4$

Interchange x and y:

$y = f^{-1}(x) = x^2 - 4$ Domain: $x \ge 0$

Check: $f^{-1}[f(x)] = (\sqrt{x+4})^2 - 4 = x + 4 - 4 = x$

$f[f^{-1}(x)] = \sqrt{x^2 - 4 + 4} = \sqrt{x^2} = x$ since $x \ge 0$ in the domain of f^{-1}.

(B) Domain of $f = [-4, \infty)$ = Range of f^{-1}

Range of $f = [0, \infty)$ = Domain of f^{-1}

(C)

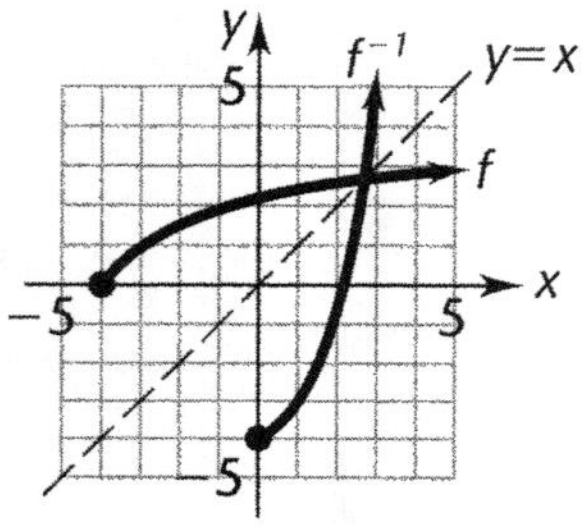

(3-6)

45.
$$\begin{aligned} x^2 - 6x + y^2 + 2y &= 0 \\ (x^2 - 6x + ?) + (y^2 + 2y + ?) &= 0 \\ (x^2 - 6x + 9) + (y^2 + 2y + 1) &= 9 + 1 \\ (x - 3)^2 + (y + 1)^2 &= 10 \\ (x - 3)^2 + [y - (-1)]^2 &= (\sqrt{10})^2 \end{aligned}$$
Center: $C(h, k) = (3, -1)$ Radius $r = \sqrt{10}$

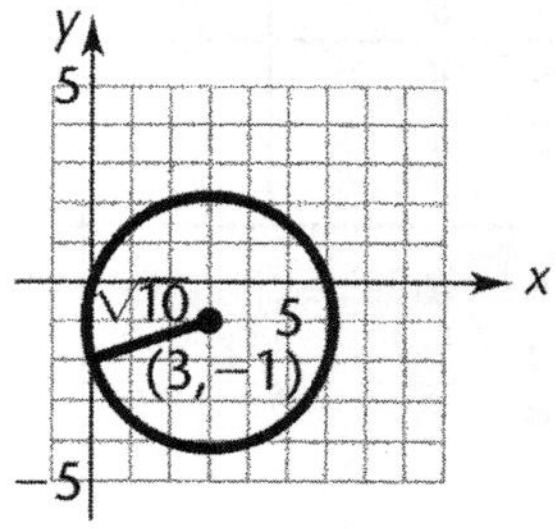

(2-2)

46. $xy + |xy| = 5$

Test y axis
Replace x with $-x$:
$(-x)y + |(-x)y| = 5$
$-xy + |xy| = 5$

Test x axis
Replace y with $-y$:
$x(-y) + |x(-y)| = 5$
$-xy + |xy| = 5$

Test origin
Replace x with $-x$ and y with $-y$:
$(-x)(-y) + |(-x)(-y)| = 5$
$xy + |xy| = 5$

The graph has symmetry with respect to the origin. (2-1)

47. Since the graph is a parabola opening up, $a = +1$. Since the vertex is at $(-2, -3) = (h, k)$, $h = -2$ and $k = -3$. Thus, the equation is $y = (x + 2)^2 - 3$. (3-4)

48.
$$\frac{x + y}{y - \frac{x+y}{x-y}} = 1$$
$$\frac{(x + y)(x - y)}{y(x - y) - (x + y)} = 1$$
$$\frac{x^2 - y^2}{xy - y^2 - x - y} = 1$$
$$x^2 - y^2 = xy - y^2 - x - y$$
$$x^2 + x = xy - y$$
$$x^2 + x = y(x - 1)$$
$$y = \frac{x^2 + x}{x - 1} \quad (1\text{-}1)$$

49.
$$3x^2 = 2\sqrt{2}x - 1$$
$$3x^2 - 2\sqrt{2}x + 1 = 0$$
$$x = \frac{-b \pm \sqrt{b^2 - 4ac}}{2a} \qquad a = 3, \quad b = -2\sqrt{2}, \quad c = 1$$
$$x = \frac{-(-2\sqrt{2}) \pm \sqrt{(-2\sqrt{2})^2 - 4(3)(1)}}{2(3)}$$
$$x = \frac{2\sqrt{2} \pm \sqrt{8 - 12}}{6}$$
$$x = \frac{2\sqrt{2} \pm \sqrt{-4}}{6}$$
$$x = \frac{2\sqrt{2} \pm 2i}{6}$$
$$x = \frac{2(\sqrt{2} \pm i)}{6}$$
$$x = \frac{\sqrt{2} \pm i}{3} \quad (1\text{-}5)$$

50. In this problem, $a = 1$, $b = b$, $c = 1$. Thus, the discriminant $b^2 - 4ac = b^2 - 4 \cdot 1 \cdot 1 = b^2 - 4$. Hence, the number and types of roots depend on the sign of $b^2 - 4 = (b + 2)(b - 2)$. The zeros of this polynomial are 2, -2. Plotting the graph of $y = (b + 2)(b - 2)$, we see $b^2 - 4 > 0$ if $b < -2$ or $b > 2$, and $b^2 - 4 < 0$ if $-2 < b < 2$.

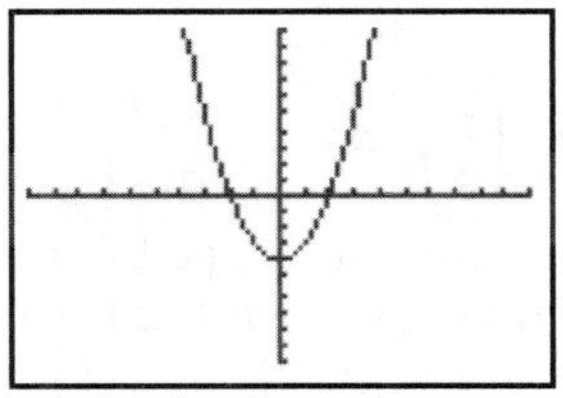

Hence,
if $b^2 - 4 > 0$, there are two distinct real roots. This occurs if $b < -2$ or $b > 2$.
if $b^2 - 4 = 0$, there is one real double root. This occurs if $b = -2$ or $b = 2$.
if $b^2 - 4 < 0$, there are two distinct imaginary roots. This occurs if $-2 < b < 2$. (*1-5*)

51.
$$\sqrt{3 - 2x} - \sqrt{x + 7} = \sqrt{x + 4}$$
$$3 - 2x - 2\sqrt{3 - 2x}\sqrt{x + 7} + x + 7 = x + 4$$
$$-x + 10 - 2\sqrt{3 - 2x}\sqrt{x + 7} = x + 4$$
$$-2\sqrt{3 - 2x}\sqrt{x + 7} = 2x - 6$$
$$-\sqrt{3 - 2x}\sqrt{x + 7} = x - 3$$
$$(3 - 2x)(x + 7) = x^2 - 6x + 9$$
$$3x + 21 - 2x^2 - 14x = x^2 - 6x + 9$$
$$0 = 3x^2 + 5x - 12$$
$$3x^2 + 5x - 12 = 0$$
$$(3x - 4)(x + 3) = 0$$
$$3x - 4 = 0 \qquad x + 3 = 0$$
$$x = \frac{4}{3} \qquad x = -3$$

Check:

$x = \frac{4}{3}$: $\sqrt{3 - 2(\frac{4}{3})} - \sqrt{\frac{4}{3} + 7} \stackrel{?}{=} \sqrt{\frac{4}{3} + 4}$

$\sqrt{\frac{9}{3} - \frac{8}{3}} - \sqrt{\frac{4}{3} + \frac{21}{3}} \stackrel{?}{=} \sqrt{\frac{4}{3} + \frac{12}{3}}$

$\sqrt{\frac{1}{3}} - \sqrt{\frac{25}{3}} \stackrel{?}{=} \sqrt{\frac{16}{3}}$

$\frac{1}{\sqrt{3}} - \frac{5}{\sqrt{3}} \neq \frac{4}{\sqrt{3}}$

Not a solution

$x = -3$:

$\sqrt{3 - 2(-3)} - \sqrt{(-3) + 7} \stackrel{?}{=} \sqrt{(-3) + 4}$

$\sqrt{9} - \sqrt{4} \stackrel{?}{=} \sqrt{1}$

$3 - 2 \stackrel{\checkmark}{=} 1$

A solution

Solution: -3 (*1-6*)

52. $\frac{a + bi}{a - bi} = \frac{(a + bi)(a + bi)}{(a - bi)(a + bi)} = \frac{(a + bi)^2}{a^2 - (bi)^2} = \frac{a^2 + 2abi + b^2i^2}{a^2 - b^2i^2}$

$= \frac{a^2 + 2abi - b^2}{a^2 + b^2}$

$= \frac{a^2 - b^2 + 2abi}{a^2 + b^2}$

$= \frac{a^2 - b^2}{a^2 + b^2} + \frac{2ab}{a^2 + b^2} i$ (*1-4*)

53. (A) $g(x) = \sqrt{4 - x^2}$

The domain of g is those values of x for which $4 - x^2 \ge 0$. Plotting the graph of $y = 4 - x^2$, or $y = (2 - x)(2 + x)$, we see that the zeros of y are -2 and 2, and $4 - x^2 \ge 0$ if $-2 \le x \le 2$. The domain of g is $[-2, 2]$.

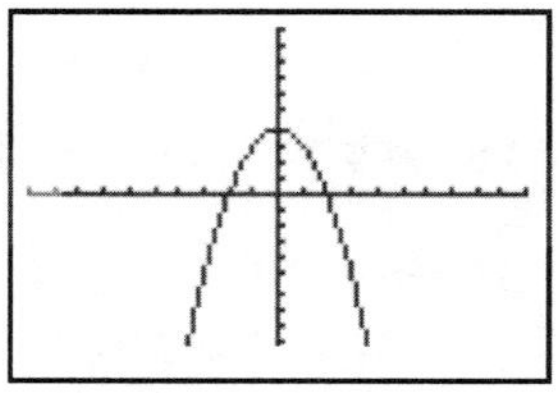

(B) $\left(\frac{f}{g}\right)(x) = \frac{f(x)}{g(x)} = \frac{x^2}{\sqrt{4 - x^2}}$. The domain of $\frac{f}{g}$ is the intersection of the domains of f (all real numbers) and $g([-2, 2])$ with the exclusion of those points (-2 and 2) where $g(x) = 0$. Thus, domain of $f/g = (-2, 2)$.

(C) $(f \circ g)(x) = f[g(x)] = f(\sqrt{4 - x^2}) = (\sqrt{4 - x^2})^2 = 4 - x^2$.

The domain of $f \circ g$ is the set of all real numbers in the domain of $g([-2, 2])$ for which $f(x)$ is real, that is, all numbers in $[-2, 2]$. (*3-5*)

54. (A) $f(x) = x^2 - 2x - 3 \quad x \ge 1$

f passes the horizontal line test (see graph below, in part C), hence f is one-to-one.

Solve $y = f(x)$ for x:

$y = x^2 - 2x - 3 \quad x \ge 1$

$x^2 - 2x = y + 3 \quad x \ge 1$

$x^2 - 2x + 1 = y + 4 \quad x \ge 1$

$(x - 1)^2 = y + 4 \quad x \ge 1$

$x - 1 = \underbrace{\sqrt{y + 4}} \quad x \ge 1$

positive square root only because $x \ge 1$

$x = 1 + \sqrt{y + 4} = f^{-1}(y)$

Interchange x and y:

$y = f^{-1}(x) = 1 + \sqrt{x + 4}$ Domain: $x \ge -4$ Check omitted for lack of space.

(B) Domain of $f^{-1} = [-4, \infty)$
Range of f^{-1} = Domain of $f = [1, \infty)$.

(C)

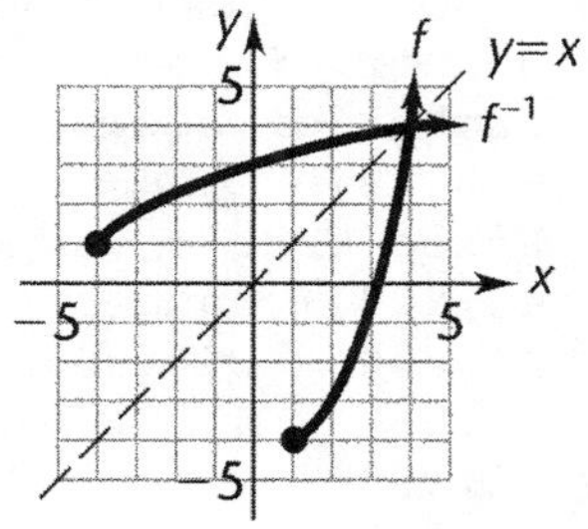

55. Let x = the number

$\frac{1}{x}$ = its reciprocal

Then

$$x - \frac{1}{x} = \frac{3}{2} \quad \text{Excl. value: } x \neq 0$$
$$2x^2 - 2 = 3x$$
$$2x^2 - 3x - 2 = 0$$
$$(2x + 1)(x - 2) = 0$$
$$2x + 1 = 0 \quad \text{or} \quad x - 2 = 0$$
$$x = -\frac{1}{2} \qquad x = 2$$

(*1-5*)

56. Let x = the rate of the current
Then $15 - x$ = the rate of the boat upstream
$15 + x$ = the rate of the boat downstream

Solving $d = rt$ for t, we have $t = \frac{d}{r}$. We use this formula, together with time upstream + time downstream = 4.8.

$$\text{time upstream} = \frac{\text{distance upstream}}{\text{rate upstream}} = \frac{35}{15 - x}$$

$$\text{time downstream} = \frac{\text{distance downstream}}{\text{rate downstream}} = \frac{35}{15 + x}$$

So,

$$\frac{35}{15 - x} + \frac{35}{15 + x} = 4.8 \quad \text{Excluded values: } x \neq 15, -15$$
$$35(15 + x) + 35(15 - x) = 4.8(15 + x)(15 - x)$$
$$525 + 35x + 525 - 35x = 4.8(225 - x^2)$$
$$1050 = 1080 - 4.8x^2$$
$$-30 = -4.8x^2$$
$$6.25 = x^2$$
$$x = 2.5 \text{ miles per hour (discarding the negative answer)}$$

(*1-5*)

57. Let x = amount of distilled water (0% acid)
24 = amount of 90% solution
Then $x + 24$ = amount of 60% solution

acid in distilled water	+	acid in 90% solution	=	acid in 60% solution
0	+	0.9(24)	=	$0.6(x + 24)$

$$21.6 = 0.6x + 14.4$$
$$7.2 = 0.6x$$
$$x = 12 \text{ gallons}$$

(*1-1*)

58. "Break even" means Cost = Revenue

$$\begin{aligned}
\text{Let } x &= \text{number of books sold}\\
\text{Revenue} &= \text{number of books sold} \infty \text{ price per book}\\
&= x(9.65)\\
\text{Cost} &= \text{Fixed Cost + Variable Cost}\\
&= 41{,}800 + \text{number of books} \infty \text{ cost per book}\\
&= 41{,}800 + x(4.90)\\
9.65x &= 41{,}800 + 4.90x\\
4.75x &= 41{,}800\\
x &= \frac{41{,}800}{4.75}\\
x &= 8{,}800 \text{ books}
\end{aligned}$$

(*1-1*)

59. The distance of p from 200 must be no greater than 10.

$|p - 200| \le 10$ (*1-3*)

60. If x is linearly related to p, then we are looking for an equation whose graph passes through $(p_1, x_1) = (3.79, 1{,}160)$ and $(p_2, x_2) = (3.59, 1{,}340)$. We find the slope, and then we use the point-slope form to find the equation.

$$m = \frac{x_2 - x_1}{p_2 - p_1} = \frac{1{,}340 - 1{,}160}{3.59 - 3.79} = \frac{180}{-0.2} = -900$$

$$\begin{aligned}
x - x_1 &= m(p - p_1)\\
x - 1{,}160 &= -900(p - 3.79)\\
x - 1{,}160 &= -900p + 3{,}411\\
x &= -900p + 4{,}571
\end{aligned}$$

If the price is lowered to \$3.29, we are asked for x when $p = 3.29$

$$\begin{aligned}
x &= -900(3.29) + 4{,}571\\
&= 1{,}610 \text{ bottles}
\end{aligned}$$

(*2-4*)

61. If $0 \le x \le 60$, $C(x) = 0.06x$

If $60 < x \le 60 + 90$,

that is,

$$\begin{aligned}
& \qquad\quad \begin{pmatrix}\text{Cost of first}\\ \text{60 calls}\end{pmatrix} + \begin{pmatrix}\text{Cost of next } x - 60\\ \text{calls at 0.05 per call}\end{pmatrix}\\
60 < x \le 150,\ C(x) &= 0.06(60) + 0.05(x - 60)\\
&= 3.60 + 0.05x - 3\\
&= 0.05x + 0.6
\end{aligned}$$

If $60 + 90 < x \le 60 + 90 + 150$,

that is,

$$\begin{aligned}
& \qquad\quad \begin{pmatrix}\text{Cost of first}\\ \text{150 calls}\end{pmatrix} + \begin{pmatrix}\text{Cost of next } x - 150\\ \text{calls at 0.04 per call}\end{pmatrix}\\
150 < x \le 300,\ C(x) &= 0.05(150) + 0.6 + 0.04(x - 150)\\
&= 7.5 + 0.6 + 0.04x - 6\\
&= 0.04x + 2.1
\end{aligned}$$

Finally, if $x > 300$, $\begin{pmatrix}\text{Cost of first}\\ \text{300 calls}\end{pmatrix} + \begin{pmatrix}\text{Cost of next } x - 300\\ \text{calls at 0.03 per call}\end{pmatrix}$

$$\begin{aligned}
C(x) &= 0.04(300) + 2.1 + 0.03(x - 300)\\
&= 12 + 2.1 + 0.03x - 9\\
&= 0.03x + 5.1
\end{aligned}$$

$$\text{Summarizing, } C(x) = \begin{cases} 0.06x & \text{if } 0 \le x \le 60\\ 0.05x + 0.6 & \text{if } 60 < x \le 150\\ 0.04x + 2.1 & \text{if } 150 < x \le 300\\ 0.03x + 5.1 & \text{if } 300 < x \end{cases}$$

x	$C(x) = 0.06x$	x	$C(x) = 0.05x + 0.6$	x	$C(x) = 0.04x + 2.1$	x	$C(x) = 0.03x + 5.1$
0	0	90	5.1	160	8.5	310	14.4
30	1.8	140	7.6	200	10.1	400	17.1
60	3.6						

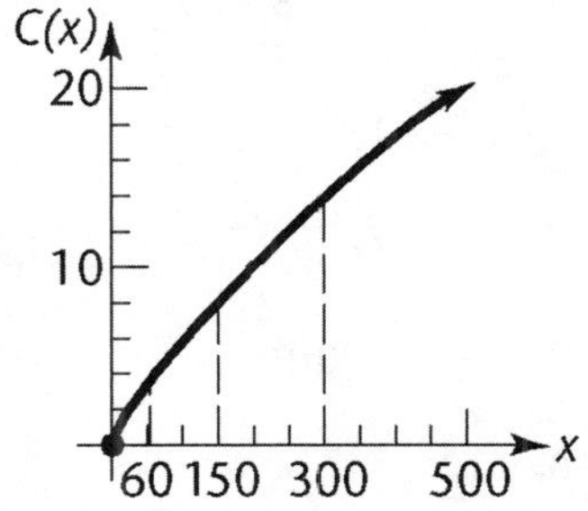

(3-2)

62. (A) Let x = width of pen
Then $2x + \ell = 80$

So $\ell = 80 - 2x$ = length of pen

$A(x) = x\ell = x(80 - 2x) = 80x - 2x^2$

(B) Since all distances must be positive, $x > 0$ and $80 - 2x > 0$, hence $80 > 2x$ or $x < 40$. Therefore $0 < x < 40$ or $(0, 40)$ is the domain of $A(x)$.

(C) Note: 0, 40 were excluded from the domain for geometrical reasons; but can be used to help draw the graph since the *polynomial* $80x - 2x^2$ has domain including these values.

x	$A(x)$
0	0
10	600
20	800
30	600
40	0

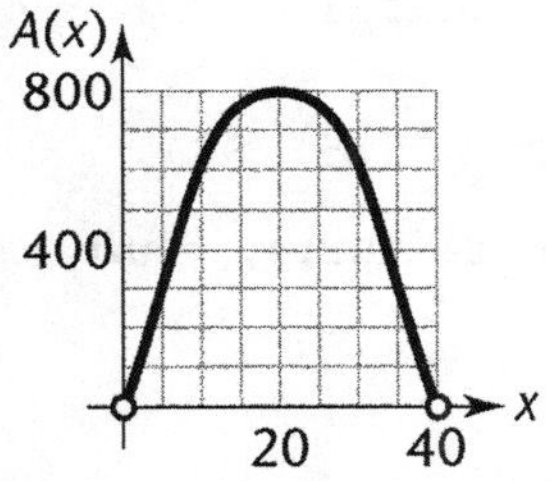

Since the function is quadratic, its maximum value occurs at the vertex of its parabola graph, which is given by the formula (h, k), where

$$h = -\frac{b}{2a} = -\frac{80}{2(-2)} = 20$$

and $$k = c - \frac{b^2}{4a} = 0 - \frac{80^2}{4(-2)} = 800$$

Thus the maximum value of A occurs when $x = 20$ and $\ell = 80 - 2(20) = 40$; the dimensions are 20 feet by 40 feet. (3-4)

63. (A) $f(1) = 1 - 2[\![1/2]\!] = 1 - 2(0) = 1$
$f(2) = 2 - 2[\![2/2]\!] = 2 - 2(1) = 0$
$f(3) = 3 - 2[\![3/2]\!] = 3 - 2(1) = 1$
$f(4) = 4 - 2[\![4/2]\!] = 4 - 2(2) = 0$

(B) If n is an integer, n is either odd or even.
If n is even, it can be written as $2k$, where k is an integer.
Then $f(n) = f(2k) = 2k - 2[\![2k/2]\!] = 2k - 2[\![k]\!] = 2k - 2k = 0$.
Otherwise, n is odd, and n can be written as $2k + 1$, where k is an integer.
Then $f(n) = f(2k + 1) = 2k + 1 - 2[\![(2k + 1)/2]\!] = 2k + 1 - 2[\![k + \tfrac{1}{2}]\!]$
$= 2k + 1 - 2k$
$= 1$

Summarizing, if n is an integer, $f(n) = \begin{cases} 1 & \text{if } n \text{ is an odd integer} \\ 0 & \text{if } n \text{ is an even integer} \end{cases}$ (3-2)

64. (A) If v is linearly related to t, then we are looking for an equation whose graph passes through $(t_1, v_1) = (0, 20000)$ and $(t_2, v_2) = (8, 4000)$. We find the slope, m, and then we use the point-slope form to find the equation.

$$m = \frac{v_2 - v_1}{t_2 - t_1} = \frac{4,000 - 20,000}{8 - 0} = -2,000$$

$$v - v_1 = m(t_2 - t_1)$$
$$v - 20,000 = -2,000(t - 0)$$
$$v - 20,000 = -2,000t$$
$$v = -2,000t + 20,000$$

(B) Solve for t

$$v = -2,000t + 20,000$$
$$v - 20,000 = -2,000t$$
$$t = \frac{v - 20,000}{-2,000}$$
$$t = -0.0005v + 10$$

(3-6)

65. (A) A profit will result if revenue is greater than cost; that is, if

$$R > C$$
$$15p - 2p^2 > 88 - 12p$$
$$-88 + 27p - 2p^2 > 0$$
$$2p^2 - 27p + 88 < 0$$
$$(2p - 11)(p - 8) < 0$$

Zeros: 5.5, 8

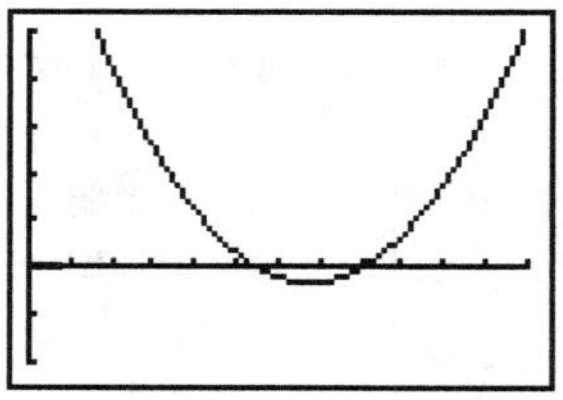

Graphing $f(p) = 2p^2 - 27p + 88$, we see that $2p^2 - 27p + 88 < 0$ and a profit will occur $(R > C)$ for $\$5.5 < p < \$8$ or ($5.5, $8).

(B) A loss will result if cost is greater than revenue, that is if

$$C > R$$
$$88 - 12p > 15p - 2p^2$$
$$2p^2 - 27p + 88 > 0$$

Referring to the graph in part (A), we see that $2p^2 - 27p + 88 > 0$, and a loss will occur $(C > R)$, for $p < \$5.5$ or $p > \$8$. Since a negative price doesn't make sense, we delete any number to the left of 0. Thus, a loss will occur for $\$0 \le p < \$5.5$ or $p > \$8$. [$0, $5.5) ≈ ($8, ∞) (3-4)

66. Let x = the distance from port A to port B

Then $115 - x$ = the distance from port B to port C

Applying the Pythagorean theorem, we have

$$x^2 + (115 - x)^2 = 85^2$$
$$x^2 + 13,225 - 230x + x^2 = 7,225$$
$$2x^2 - 230x + 6,000 = 0$$
$$x^2 - 115x + 3,000 = 0$$
$$(x - 40)(x - 75) = 0$$
$$x = 40, 75$$
$$115 - x = 75, 40$$

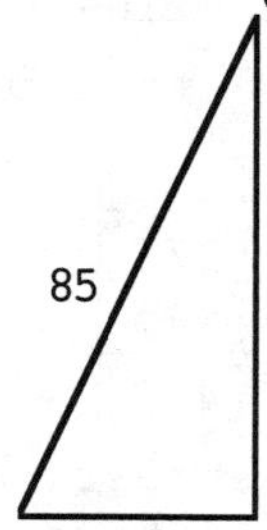

Based on the given information, there are two possible solutions:

40 miles from A to B and 75 miles from B to C *or*

75 miles from A to B and 40 miles from B to C

(1-5)

67. $s = a + bt^2$

(A) We are given: When $t = 5$, $s = 2{,}100$
When $t = 10$, $s = 900$

Substituting these values into the given equation, we have

$$2{,}100 = a + b(5)^2$$
$$900 = a + b(10)^2 \text{ or}$$
$$2{,}100 = a + 25b$$
$$900 = a + 100b$$

Solve the first equation for a in terms of b and substitute into the second equation to eliminate a.

$$a = 2{,}100 - 25b$$
$$900 = 2{,}100 - 25b + 100b$$
$$-1{,}200 = 75b$$
$$b = -16$$
$$a = 2{,}100 - 25b = 2{,}100 - 25(-16) = 2{,}500$$

(B) The height of the balloon is represented by s, the distance of the object above the ground, when $t = 0$. Since we now know

$$s = 2{,}500 - 16t^2$$

from part(A), when $t = 0$, $s = 2{,}500$ feet is the height of the balloon.

(C) The object falls until s, its distance above the ground, is zero. Since

$$s = 2{,}500 - 16t^2$$

we substitute $s = 0$ and solve for t.

$$0 = 2{,}500 - 16t^2$$
$$16t^2 = 2{,}500$$
$$t^2 = \frac{2{,}500}{16}$$
$$t = \frac{50}{4} \text{ (discarding the negative solution)}$$
$$t = 12.5 \text{ seconds}$$

(3-4)

68. (A) 30,000 bushels

(B) The demand decreases by 10,000 bushels to 20,000 bushels.

(C) The demand increases by 10,000 bushels to 40,000 bushels.

(D) As the price varies from \$3.00 to \$3.50 per bushel, the demand varies from 15,000 to 50,000 bushels, increasing with decreasing price and decreasing with increasing price.

(E) From the graph, the following table is constructed:

q	20	25	30	35	40
P	340	332	325	320	315

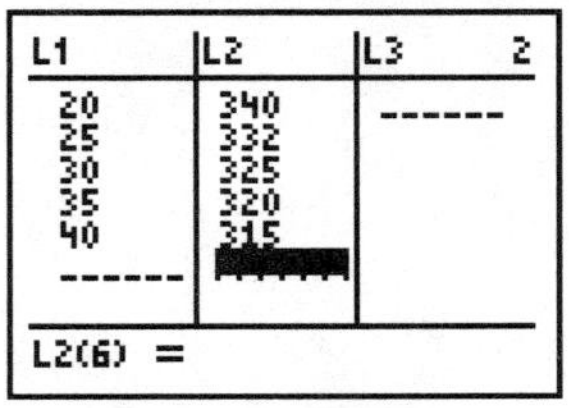

Enter the data

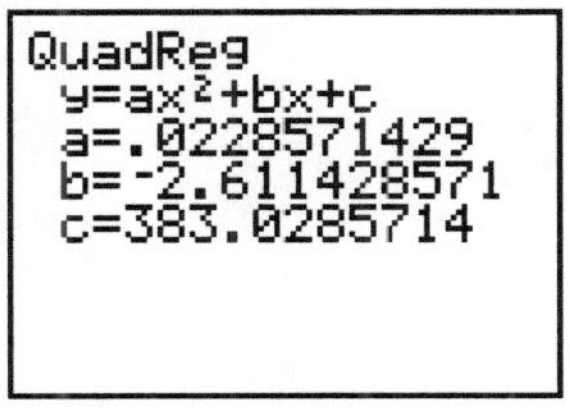

Compute the regression equation.

(3-4)

69. (A)

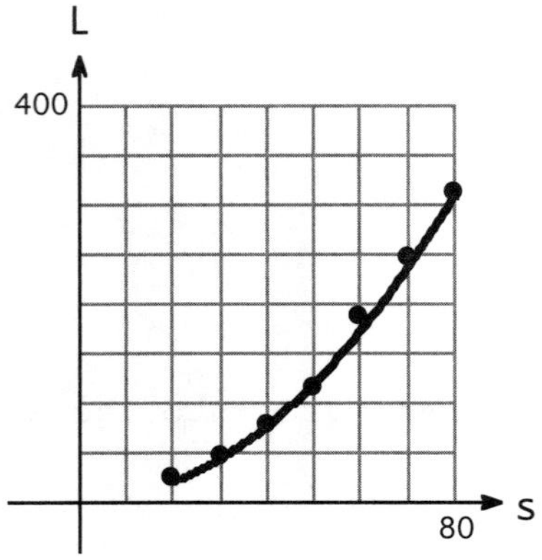

(B) $L = 0.05s^2 - 0.2s + 6.5$ domain $[20, \infty)$
From the graph, L is an increasing function on its domain. Hence the range of L consists of those values of L greater than or equal to
$L(20) = 0.05(20)^2 - 0.2(20) + 6.5 = 22.5$, or $[22.5, \infty)$. Now solve for s.
First, complete the square.

$L = 0.05(s^2 - 4s + ?) + 6.5$

$L = 0.05(s^2 - 4s + 4) - 4(0.05) + 6.5$

$L = 0.05(s - 2)^2 + 6.3$

$L - 6.3 = 0.05(s - 2)^2$

$20L - 126 = (s - 2)^2$

$\sqrt{20L - 126} = s - 2$ (positive square root only, since $s \geq 20$)

$f^{-1}(L) = s = 2 + \sqrt{20L - 126}$,
domain: $[22.5, \infty)$, range: $[20, \infty)$

(C) Substitute $L = 220$ to find

$$s = f^{-1}(220) = 2 + \sqrt{20(220) - 126}$$
$$= 67 \text{ mph}$$

(3-4)

CHAPTER 4 Polynomial and Rational Functions

Section 4-1

1. The degree is odd and the coefficient is positive, so $f(x)$ increases without bound as $x \to \infty$ and decreases without bound as $x \to -\infty$. This matches graph c.

3. The degree is even and the coefficient is positive, so $h(x)$ increases without bound both as $x \to \infty$ and $x \to -\infty$. This matches graph d.

5. The real zeros are the x-intercepts: -1 and 3. The turning point is (1, 4). $P(x) \to -\infty$ as $x \to \infty$ and $P(x) \to -\infty$ as $x \to -\infty$.

7. The real zeros are the x-intercepts: -2 and 1. The turning points are (-1, 4) and (1, 0). $P(x) \to -\infty$ as $x \to -\infty$ and $P(x) \to \infty$ as $x \to \infty$.

9. The graph of a polynomial always increases or decreases without bound as $x \to \infty$ and $x \to -\infty$. This graph does not.

11. The graph of a polynomial always has a finite number of turning points--at most one less than the degree. This graph has infinitely many turning points.

13. Set each factor equal to zero:

$$x = 0 \qquad \begin{aligned} x^2 - 9 &= 0 \\ x^2 &= 9 \\ x &= 3, -3 \end{aligned} \qquad \begin{aligned} x^2 + 4 &= 0 \\ x^2 &= -4 \\ x &= 2i, -2i \end{aligned}$$

The zeros are 0, 3, -3, $2i$, and $-2i$; of these 0, 3, and -3 are x-intercepts.

15. Set each factor equal to zero:

$$\begin{aligned} x + 5 &= 0 \\ x &= -5 \end{aligned} \qquad \begin{aligned} x^2 + 9 &= 0 \\ x^2 &= -9 \\ x &= 3i, -3i \end{aligned} \qquad \begin{aligned} x^2 + 16 &= 0 \\ x^2 &= -16 \\ x &= 4i, -4i \end{aligned}$$

The zeros are -5, $3i$, $-3i$, $4i$, and $4i$. Only -5 is an x intercept.

17.

$$\begin{array}{r|l} & 3x + 2 \\ \hline x + 1 & 3x^2 + 5x + 6 \\ & \underline{3x^2 + 3x} \\ & \qquad 2x + 6 \\ & \qquad \underline{2x + 2} \\ & \qquad\qquad 4 \end{array}$$

$3x + 2$, $R = 4$

19.

$$\begin{array}{r|l} & 2m + 1 \\ \hline 2m - 1 & 4m^2 + 0m - 1 \\ & \underline{4m^2 - 2m} \\ & \qquad 2m - 1 \\ & \qquad \underline{2m - 1} \\ & \qquad\qquad 0 \end{array}$$

$2m + 1$, $R = 0$

21.

$$\begin{array}{r|l} & 4x - 5 \\ \hline 2x + 1 & 8x^2 - 6x + 6 \\ & \underline{8x^2 + 4x} \\ & \quad - 10x + 6 \\ & \quad \underline{- 10x - 5} \\ & \qquad\quad 11 \end{array}$$

$4x - 5$, $R = 11$

23.

$$\begin{array}{r|l} & x^2 + x + 1 \\ \hline x - 1 & x^3 - 0x^2 + 0x - 1 \\ & \underline{x^3 - x^2} \\ & \quad x^2 + 0x \\ & \quad \underline{x^2 - x} \\ & \qquad x - 1 \\ & \qquad \underline{x - 1} \\ & \qquad\quad 0 \end{array}$$

$x^2 + x + 1$, $R = 0$

25.

$$\begin{array}{r} 2y^2 - 5y + 13 \\ y + 2\overline{)2y^3 - y^2 + 3y - 1} \\ \underline{2y^3 + 4y^2} \qquad\qquad \\ -5y^2 + 3y \qquad \\ \underline{-5y^2 - 10y} \qquad \\ 13y - 1 \\ \underline{13y + 26} \\ -27 \end{array}$$

$2y^2 - 5y + 13$, $R = -27$

27.

$$\begin{array}{r|rrr} & 1 & 3 & -7 \\ & & 2 & 10 \\ \hline 2 & 1 & 5 & 3 \end{array}$$

$$\frac{x^2 + 3x - 7}{x - 2} = x + 5 + \frac{3}{x - 2}$$

29.

$$\begin{array}{r|rrr} & 4 & 10 & -9 \\ & & -12 & 6 \\ \hline -3 & 4 & -2 & -3 \end{array}$$

$$\frac{4x^2 + 10x - 9}{x + 3} = 4x - 2 - \frac{3}{x + 3}$$

31.

$$\begin{array}{r|rrrr} & 2 & 0 & -3 & 1 \\ & & 4 & 8 & 10 \\ \hline 2 & 2 & 4 & 5 & 11 \end{array}$$

$$\frac{2x^3 - 3x + 1}{x - 2} = 2x^2 + 4x + 5 + \frac{11}{x - 2}$$

> **Common Error:**
> The first row is *not*
> 2 -3 1
> The 0 must be inserted for the missing power.

33.

$$\begin{array}{r|rrr} & 1 & 4 & -221 \\ & & -17 & 221 \\ \hline -17 & 1 & -13 & 0 \end{array}$$

The remainder is 0, hence -17 is a zero.

35.

$$\begin{array}{r|rrrr} & 2 & 38 & -1 & 19 \\ & & -38 & 0 & 19 \\ \hline -19 & 2 & 0 & -1 & 38 \end{array}$$

The remainder is non-zero, hence -19 is not a zero.

37. $x - 1$ will be a factor of $P(x)$ if $P(1) = 0$. Since $P(x) = x^{18} - 1$, $P(1) = 1^{18} - 1 = 0$. Therefore $x - 1$ is a factor of $x^{18} - 1$.

39. $x + 1$ will be a factor of $P(x)$ if $P(-1) = 0$. Since $P(x) = 3x^3 - 7x^2 - 8x + 2$, $P(-1) = 3(-1)^3 - 7(-1)^2 - 8(-1) + 2 = -3 - 7 + 8 + 2 = 0$. Therefore $x + 1$ is a factor of $3x^3 - 7x^2 - 8x + 2$.

41.

$$\begin{array}{r|rrr} & 3 & -1 & -10 \\ & & -6 & 14 \\ \hline -2 & 3 & -7 & 4 \end{array}$$

$P(-2) = 4$

43.

$$\begin{array}{r|rrrr} & 2 & -5 & 7 & -7 \\ & & 4 & -2 & 10 \\ \hline 2 & 2 & -1 & 5 & 3 \end{array}$$

$P(2) = 3$

45.

$$\begin{array}{r|rrrrr} & 1 & 0 & -10 & 25 & -2 \\ & & -4 & 16 & -24 & -4 \\ \hline -4 & 1 & -4 & 6 & 1 & -6 \end{array}$$

$P(-4) = -6$

47.

$$\begin{array}{r|rrrrr} & 3 & 0 & 0 & -1 & -4 \\ & & -3 & 3 & -3 & 4 \\ \hline -1 & 3 & -3 & 3 & -4 & 0 \end{array}$$

$3x^3 - 3x^2 + 3x - 4$, $R = 0$

49.

$$\begin{array}{r|rrrrrr} & 1 & 0 & 0 & 0 & 0 & 1 \\ & & -1 & 1 & -1 & 1 & -1 \\ \hline -1 & 1 & -1 & 1 & -1 & 1 & 0 \end{array}$$

$x^4 - x^3 + x^2 - x + 1$, $R = 0$

51.

	3	2	0	−4	−1
		−9	21	−63	201
−3	3	−7	21	−67	200

$3x^3 - 7x^2 + 21x - 67$, $R = 200$

53.

	2	−13	0	75	2	0	−50
		10	−15	−75	0	10	50
5	2	−3	−15	0	2	10	0

$2x^5 - 3x^4 - 15x^3 + 2x + 10$, $R = 0$

55.

	4	2	−6	−5	1
		−2	0	3	1
$-\frac{1}{2}$	4	0	−6	−2	2

$4x^3 - 6x - 2$, $R = 2$
(note that the terms must be reordered as shown)

57.

	4	4	−7	−6
		−6	3	6
$-\frac{3}{2}$	4	−2	−4	0

$4x^2 - 2x - 4$, $R = 0$

59.

	3	−2	2	−3	1
		1.2	−0.32	0.672	−0.9312
0.4	3	−0.8	1.68	−2.328	0.0688

$3x^3 - 0.8x^2 + 1.68x - 2.328$, $R = 0.0688$

61.

	3	2	5	0	−7	−3
		−2.4	0.32	−4.256	3.4048	2.87616
−0.8	3	−0.4	5.32	−4.256	−3.5952	−0.12384

$3x^4 - 0.4x^3 + 5.32x^2 - 4.256x - 3.5952$, $R = -0.12384$

63. (A) $P(x) = x^3 - 3x^2$ $n = 3$ $a_n = 1$

Form a synthetic division table; the last line of each division is shown.

	1	−3	0	0		
−2	1	−5	10	−20	=	P(−2)
−1	1	−4	4	−4	=	P(−1)
0	1	−3	0	0	=	P(0)
1	1	−2	−2	−2	=	P(1)
2	1	−1	−2	−4	=	P(2)
3	1	0	0	0	=	P(3)
4	1	1	4	16	=	P(4)

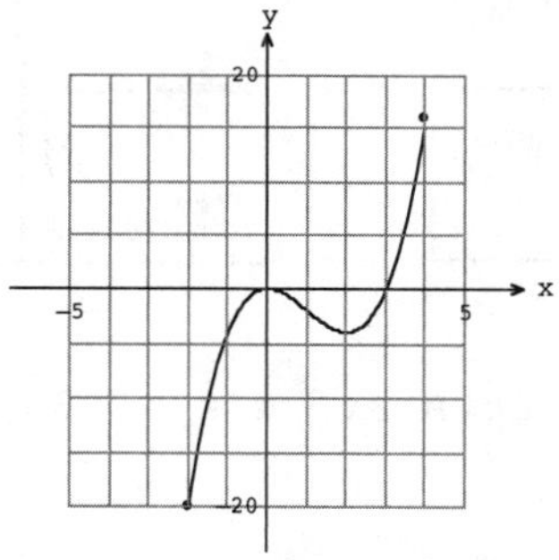

(B) Since $a_n > 0$ and n is odd, $P(x) \to \infty$ as $x \to \infty$ and $P(x) \to -\infty$ as $x \to -\infty$.

(C) From the table and the graph, we conclude that the x intercepts are 0 and 3 and the turning points are (0, 0) and (2, -4).

65. (A) $P(x) = x^4 - 8x^2 + 16$ $n = 4$ $a_n = 1$

Form a synthetic division table; the last line of each synthetic division is shown.

	1	0	−8	0	16		
−3	1	−3	1	−3	25	=	P(−3)
−2	1	−2	−4	8	0	=	P(−2)
−1	1	−1	−7	7	9	=	P(−1)
0	1	0	−8	0	16	=	P(0)
1	1	1	−7	−7	9	=	P(1)
2	1	2	−4	−8	0	=	P(2)
3	1	3	1	3	25	=	P(3)

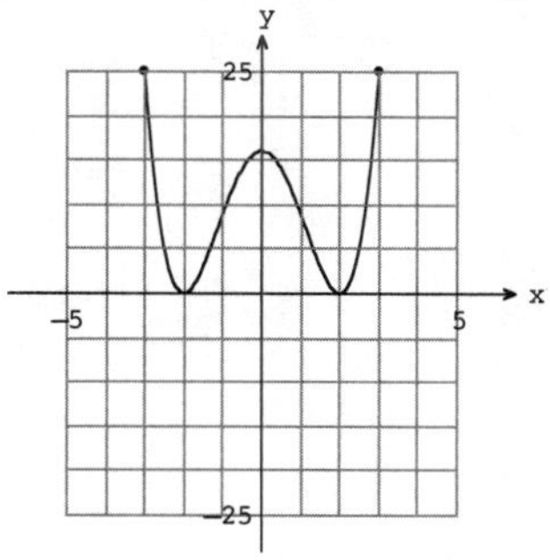

(B) Since $a_n > 0$ and n is even, $P(x) \to \infty$ as $x \to \infty$ and $P(x) \to \infty$ as $x \to -\infty$.

(C) From the table and the graph, we conclude that the x intercepts are -2 and 2 and the turning points are (-2, 0), (0, 16), and (2, 0).

67. (A) $P(x) = -x^5 - 5x^4 + 40x^2 + 80x + 48$ $n = 5$ $a_n = -1$

Form a synthetic division table; the last line of each synthetic division is shown.

	−1	−5	0	40	80	48		
−4	−1	−1	4	24	−16	112	=	$P(-4)$
−3	−1	−2	6	22	14	6	=	$P(-3)$
−2	−1	−3	6	28	24	0	=	$P(-2)$
−1	−1	−4	4	36	44	4	=	$P(-1)$
0	−1	−5	0	40	80	48	=	$P(0)$
1	−1	−6	−6	34	114	162	=	$P(1)$
2	−1	−7	−14	12	104	256	=	$P(2)$
3	−1	−8	−24	−32	−16	0	=	$P(3)$
4	−1	−9	−36	−104	−336	−1296	=	$P(4)$

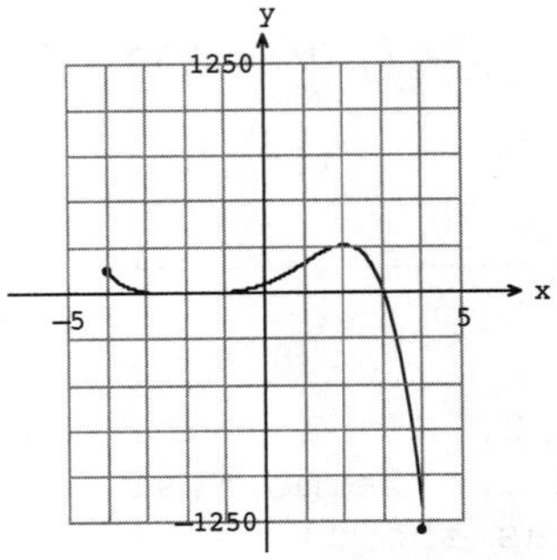

(B) Since $a_n < 0$ and n is odd, $P(x) \to -\infty$ as $x \to \infty$ and $P(x) \to \infty$ as $x \to -\infty$.

(C) From the table and the graph, we conclude that the x intercepts are −2 and 3 and the turning points are (−2, 0) and (2, 256).

69. $P(x) = x^3 - 5x^2 + 2x + 6$ $n = 3$ $a_n = 1$

(A) Since $a_n > 0$ and n is odd $P(x) \to \infty$ as $x \to \infty$ and $P(x) \to -\infty$ as $x \to -\infty$. Since the degree of $P(x)$ is 3, $P(x)$ can have a maximum of three x intercepts and two local extrema.

(B) Graphing and applying built-in routines yields:

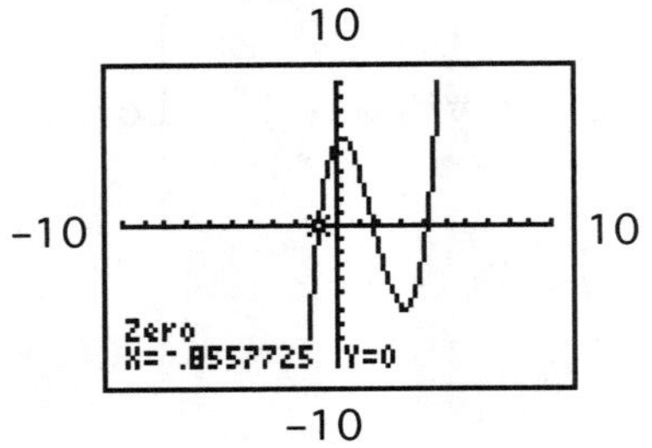

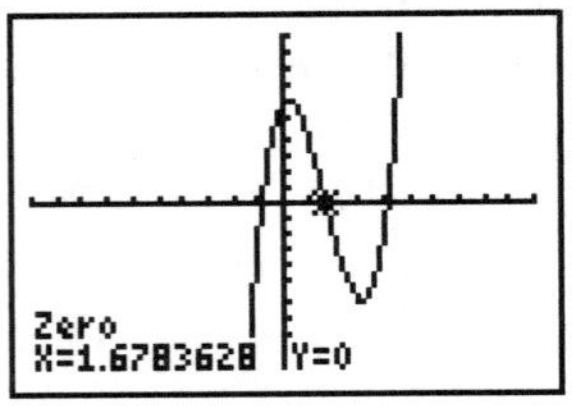

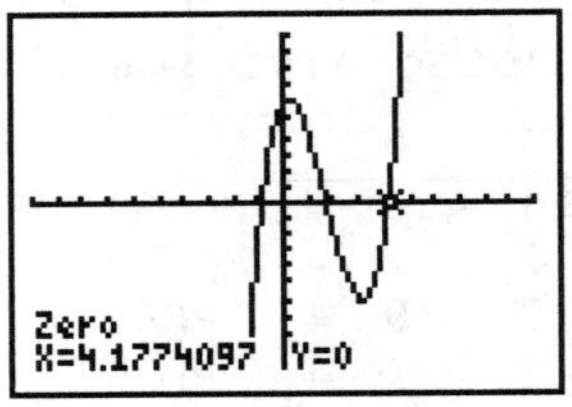

The three x intercepts are $x = -0.86$, 1.68, and 4.18.

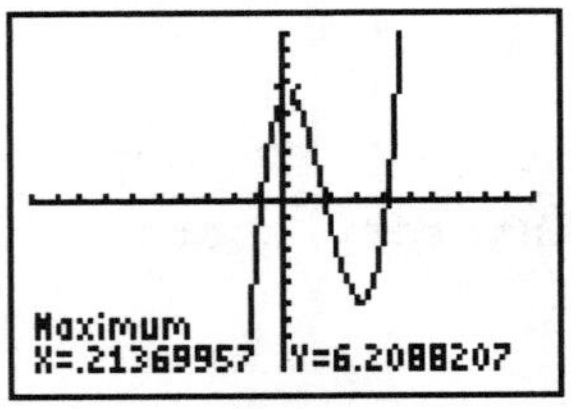

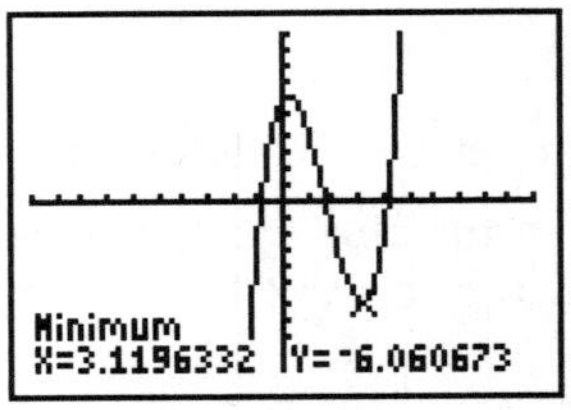

The local maximum is $P(0.21) \approx 6.21$. The local minimum is $P(3.12) \approx -6.06$.

71. $P(x) = -x^3 + 4x^2 + x + 5 \quad n = 3 \quad a_n = -1$

(A) Since $a_n < 0$ and n is odd $P(x) \to -\infty$ as $x \to \infty$ and $P(x) \to \infty$ as $x \to -\infty$. Since the degree of $P(x)$ is 3, $P(x)$ can have a maximum of three x intercepts and two local extrema.

(B) Graphing and applying built-in routines yields the graph at the right.

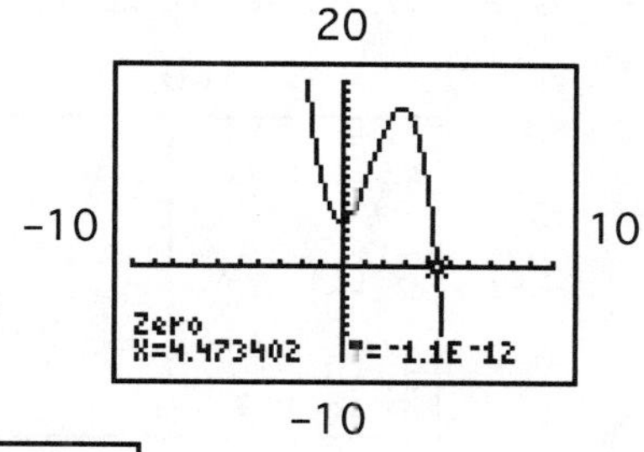

There is actually only one x intercept at $x = 4.47$.

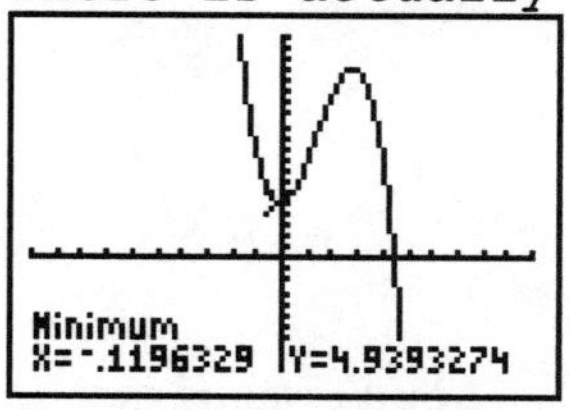

The local minimum is $P(-0.12) \approx 4.94$

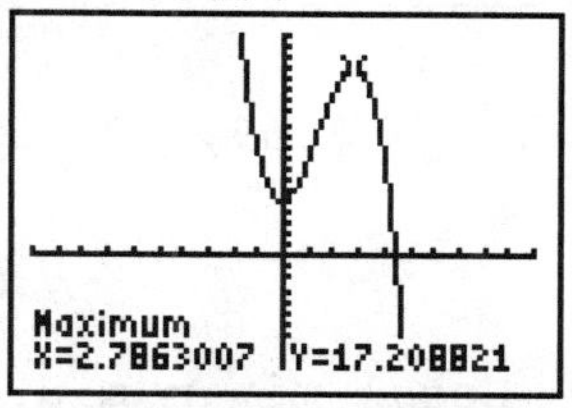

The local maximum is $P(2.79) \approx 17.21$

73. $P(x) = x^4 + x^3 - 5x^2 - 3x + 12 \quad n = 4 \quad a_n = 1$

(A) Since $a_n > 0$ and n is even $P(x) \to \infty$ as $x \to \infty$ and $P(x) \to \infty$ as $x \to -\infty$. Since the degree of $P(x)$ is 4, $P(x)$ can have a maximum of four x intercepts and three local extrema.

(B) There are no x intercepts. Graphing and applying built-in routines yields:

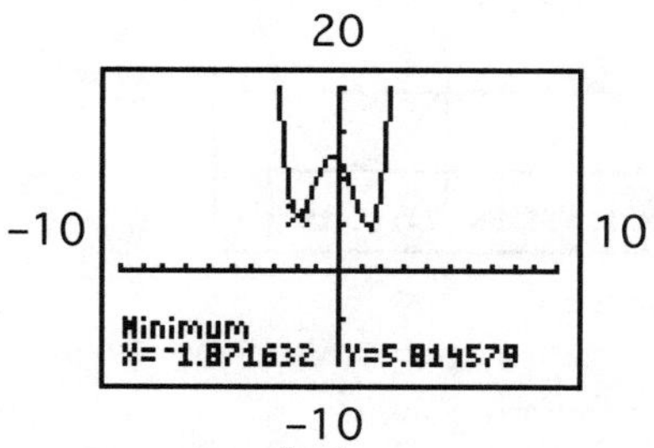

Local minimum at $P(-1.87) \approx 5.81$

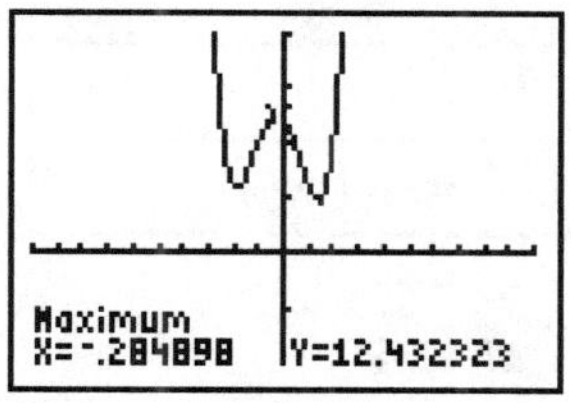

Local maximum at $P(-0.28) \approx 12.43$

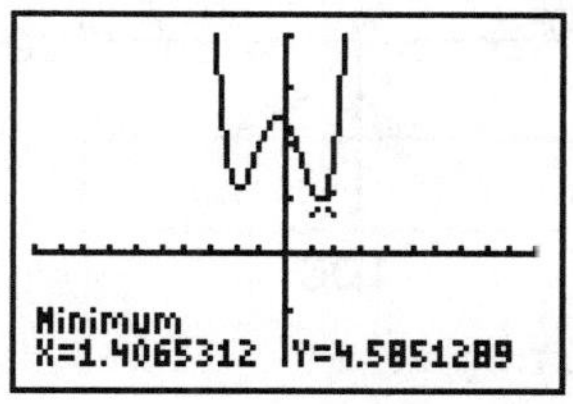

Local minimum at $P(1.41) \approx 4.59$

75. $P(x) = x^3$ is an example of a third-degree polynomial with one x intercept ($x = 0$).

77. No such polynomial exists; the graph of a third-degree polynomial must cross the x axis at least once.

79.

$$\begin{array}{r|cccc} & 1 & -3 & 1 & -3 \\ & & i & -3i - 1 & 3 \\ \hline i & 1 & -3 + i & -3i & 0 \end{array}$$

$x^2 + (-3 + i)x - 3i, \; R = 0$

81. We apply the remainder theorem to evaluate using synthetic division.

(A)
$$\begin{array}{r|ccc} & 1 & 2i & -10 \\ & & 2 - i & 5 \\ \hline 2 - i & 1 & 2 + i & -5 \end{array}$$

$P(2 - i) = -5$

(B)
$$\begin{array}{r|ccc} & 1 & 2i & -10 \\ & & 5 - 5i & 10 - 40i \\ \hline 5 - 5i & 1 & 5 - 3i & -40i \end{array}$$

$P(5 - 5i) = -40i$

(C)
$$\begin{array}{r|ccc} & 1 & 2i & -10 \\ & & 3 - i & 10 \\ \hline 3 - i & 1 & 3 + i & 0 \end{array}$$

$P(3 - i) = 0$

(D)
$$\begin{array}{r|ccc} & 1 & 2i & -10 \\ & & -3 - i & 10 \\ \hline -3 - i & 1 & -3 + i & 0 \end{array}$$

$P(-3 - i) = 0$

83. Graphing and applying built-in routines yields:

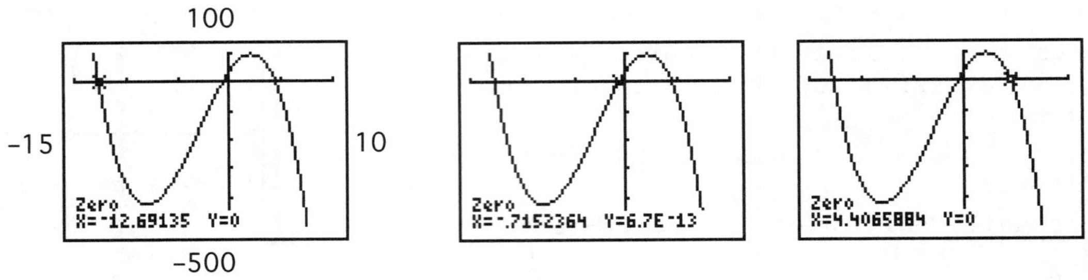

x intercepts at *x* = -12.69, -0.72, 4.41

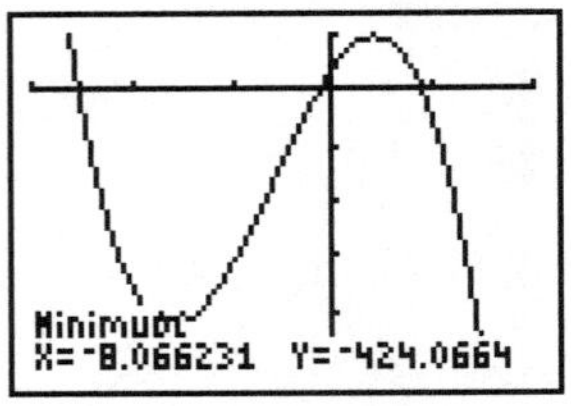

Local minimum at
$P(-8.07) \approx -424.07$

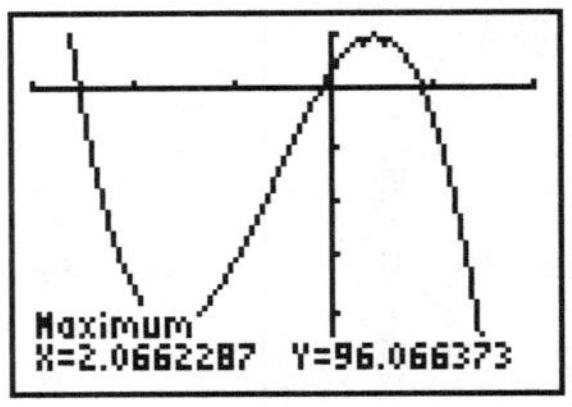

Local maximum at
$P(2.07) \approx 96.07$

85. Graphing and applying built-in routines yields:

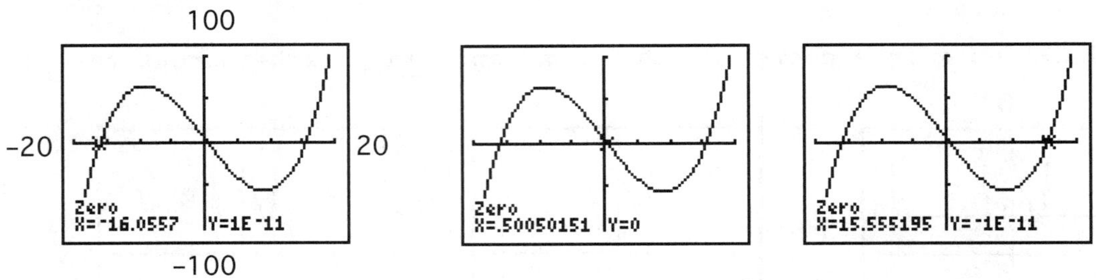

x intercepts at -16.06, 0.50, 15.56

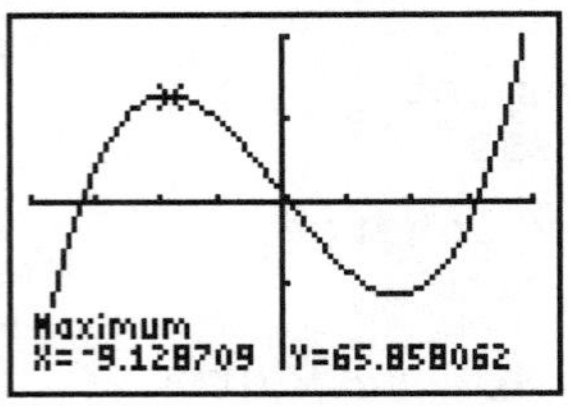

Local maximum at
$P(-9.13) \approx 65.86$

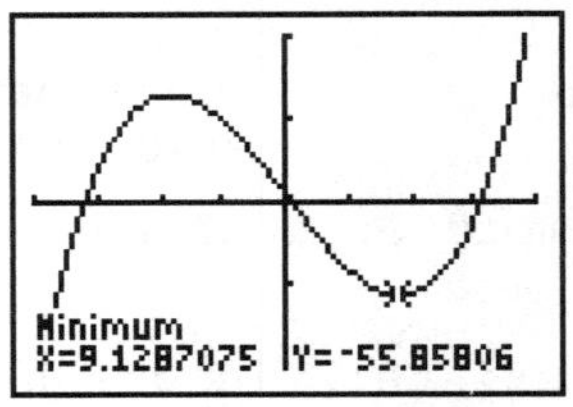

Local minimum at
$P(9.13) \approx -55.86$

87. Graphing and applying built-in routines yields:

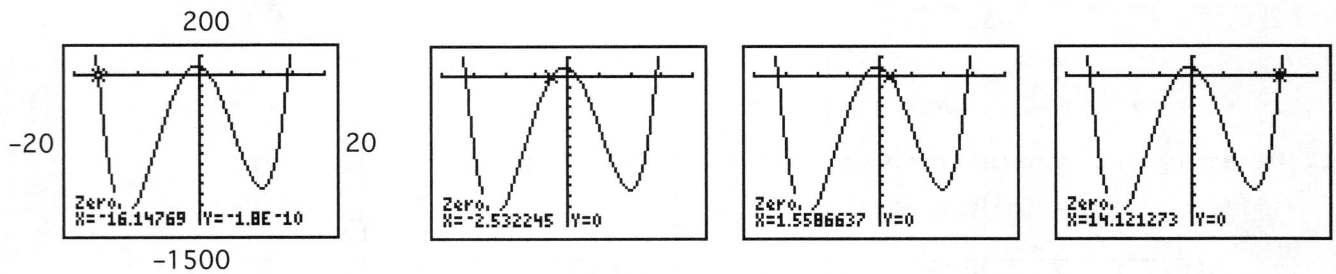

x intercepts at -16.15, -2.53, 1.56, 14.12.

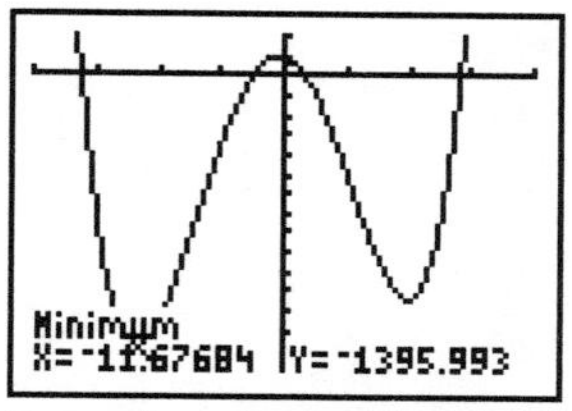

Local minimum at
$P(-11.68) \approx -1,395.99$

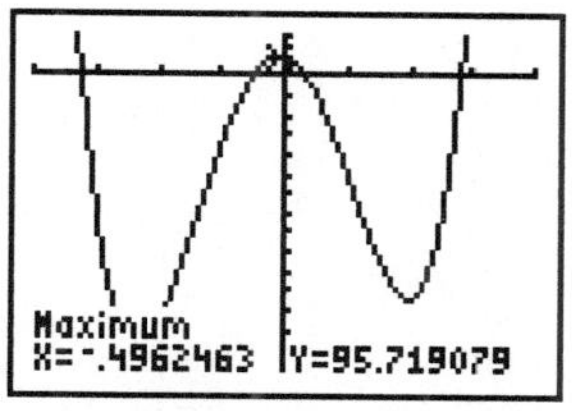

Local maximum at
$P(-0.50) \approx 95.72$

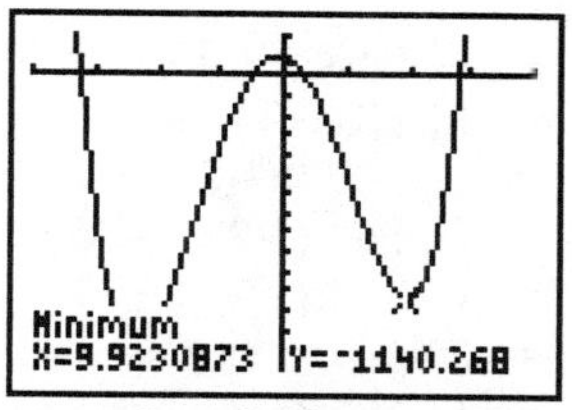

Local minimum at
$P(9.92) \approx -1,140.27$

89. Graphing and applying built-in routines yields:

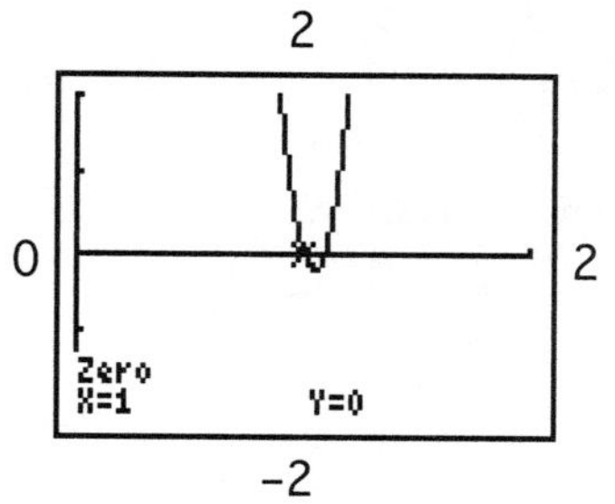

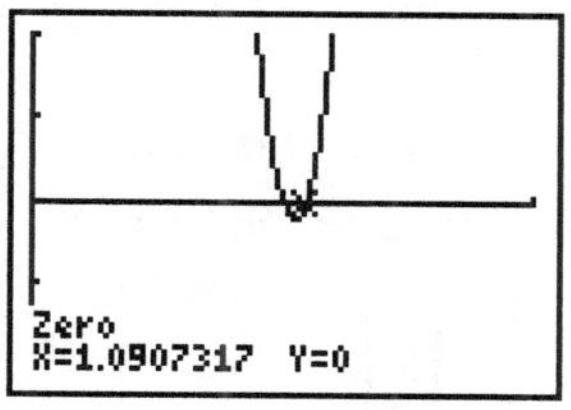

x intercepts at 1, 1.09.

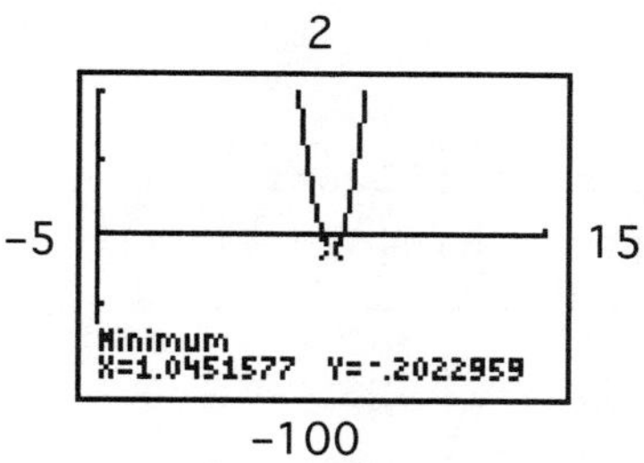

Local minimum at
$P(1.05) \approx -0.20$

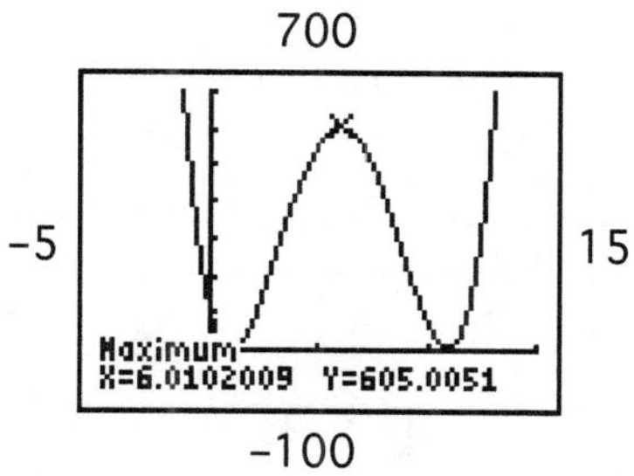

Local maximum at
$P(6.01) \approx 605.01$

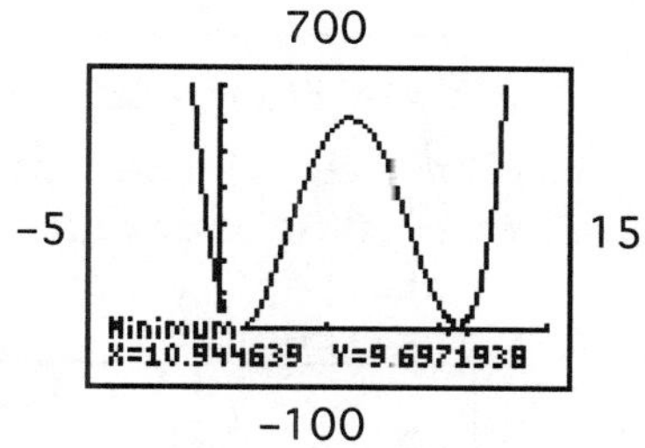

Local minimum at
$P(10.94) \approx 9.70$

91. (A)

$$
\begin{array}{r l}
 & a_2(x) + (a_1 + a_2r) \\
x - r \,\big)\!\! & \overline{a_2x^2 + a_1x \qquad\qquad\qquad + a_0} \\
 & \underline{a_2x^2 - a_2rx} \\
 & \qquad (a_1 + a_2r)x \qquad\qquad + a_0 \\
 & \qquad \underline{(a_1 + a_2r)x - r(a_1 + a_2r)} \\
 & \qquad\qquad\qquad a_0 + r(a_1 + a_2r)
\end{array}
$$

$$
\begin{array}{c|ccc}
 & a_2 & a_1 & a_0 \\
 & & a_2r & (a_1 + a_2r)r \\
\hline
r & a_2 & a_1 + a_2r & a_0 + (a_1 + a_2r)r
\end{array}
$$

In both cases the coefficient of x is a_2, the constant term $a_2r + a_1$, and the remainder is $(a_2r + a_1)r + a_0$.

(B) The remainder expanded is $a_2r^2 + a_1r + a_0 = P(r)$.

93.

$$\begin{aligned}
P(x) &= \{[(2x - 3)x + 2]x - 5\}x + 7 \\
P(-2) &= \{[\{2(-2) - 3\}(-2) + 2](-2) - 5\}(-2) + 7 \\
&= \{[\{-7\}(-2) + 2](-2) - 5\}(-2) + 7 \\
&= \{[16](-2) - 5\}(-2) + 7 \\
&= \{-37\}(-2) + 7 \\
&= 81 \\
P(1.7) &= \{[\{2(1.7) - 3\}(1.7) + 2](1.7) - 5\}(1.7) + 7 \\
&= \{[\{0.4\}(1.7) + 2](1.7) - 5\}(1.7) + 7 \\
&= \{[2.68](1.7) - 5\}(1.7) + 7 \\
&= \{-0.444\}(1.7) + 7 \\
&= 6.2452 \text{ or } 6.2 \text{ to two significant digits.}
\end{aligned}$$

95. (A) If the degree is even the graph has to approach $+\infty$ both to the right and left, so there has to be at least one turning point. $f(x) = x^4$, graphed below, has only 1 turning point.

A degree 4 polynomial can't have any more than 3 turning points. $f(x) = x^4 - 5x^2$, graphed below, has 3 turning points.

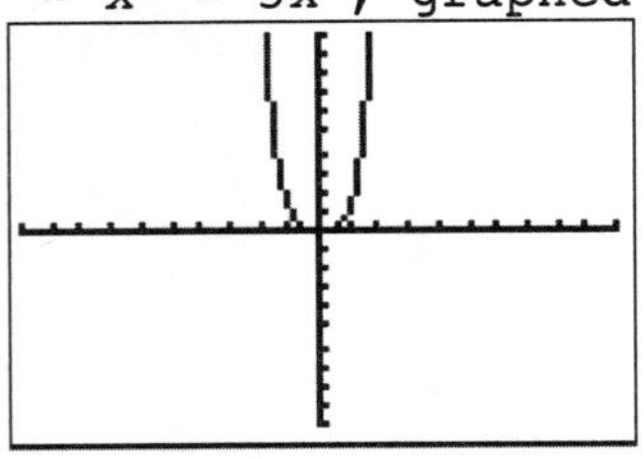 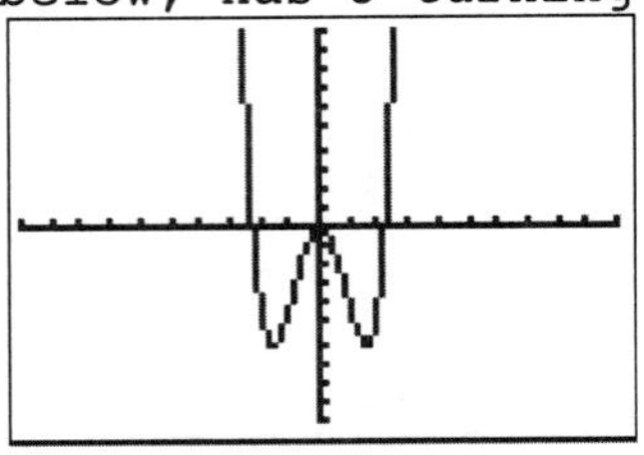

$f(x) = x^4$ $f(x) = x^4 - 5x^2$

(B) An even degree polynomial can have no x-intercepts as illustrated by $f(x) = x^4 + 5$ (graphed below). A polynomial of degree 4 can have at most 4 x intercepts. For example, $f(x) = x^4 - 5x^2 + 5$, graphed below, has 4 x intercepts.

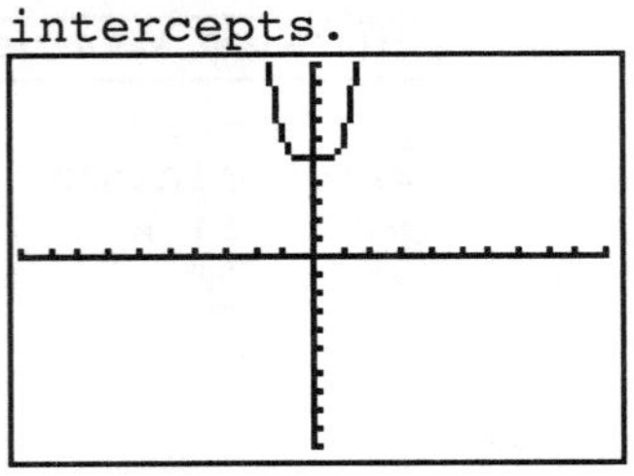 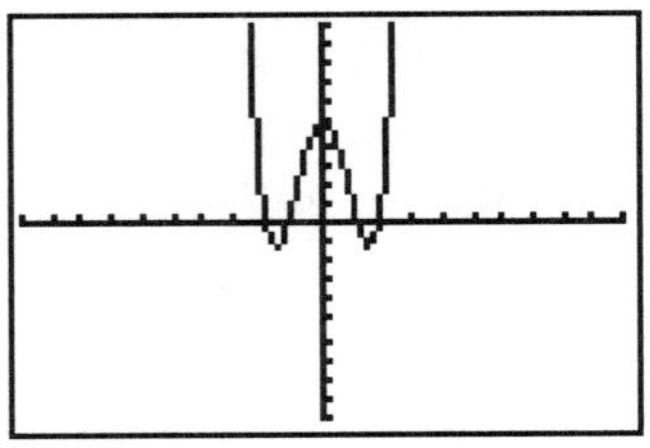

$f(x) = x^4 + 5$ $f(x) = x^4 - 5x^2 + 5$

97. No. If $f(x)$ is an even function, then $f(x) = f(-x)$ for any x. But $f(x) = x^2 + x$ is an even degree polynomial and $f(1) = 2$ while $f(-1) = 0$.

99. (A)

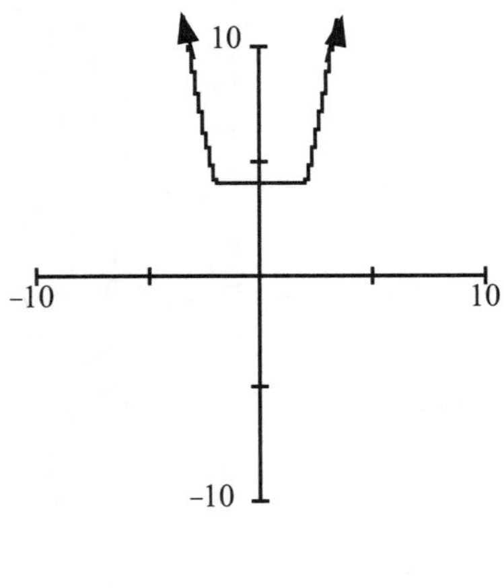

(B) All the points on the horizontal portion of the graph satisfy the definition of local minimum, but none are turning points. The set $\{c \mid -2 \le x \le 2\}$ represents all such numbers c.

101. (A) The revenue $R(x)$ is given by revenue = (number sold)(price per object)

$$\begin{aligned} R(x) &= xp(x) \\ &= x(0.0004x^2 - x + 569) \\ &= 0.0004x^3 - x^2 + 569x \end{aligned}$$

(B) Graphing and applying a built-in routine yields the graph at the right.

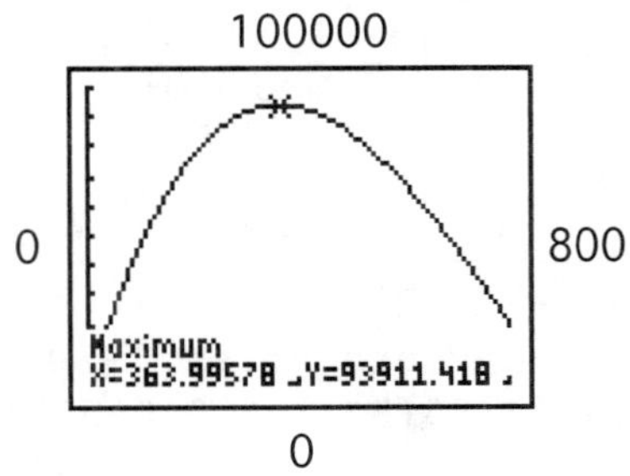

The number of air conditioners that maximizes the revenue = 364. The maximum revenue = $93,911.

The price $p(x) = \frac{R(x)}{x} = \frac{93{,}911}{364} = \$258$

103.

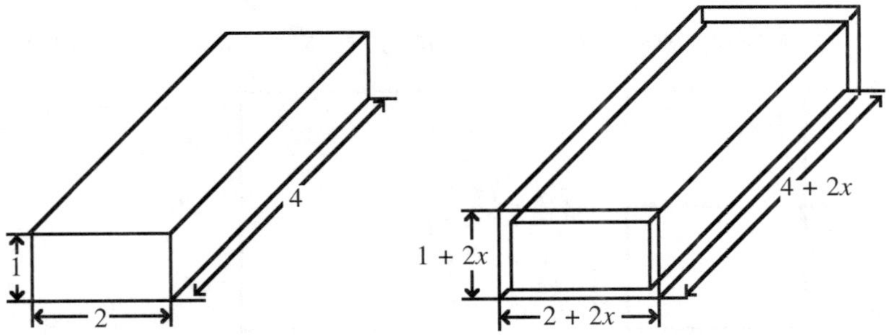

(A) The volume of the shielding is equal to the volume of the shielded box (right hand picture) minus the volume of the unshielded box (left hand picture). Thus, we have

$$\begin{pmatrix}\text{Volume of}\\ \text{shielding}\end{pmatrix} = \begin{pmatrix}\text{Volume of}\\ \text{shielded box}\end{pmatrix} - \begin{pmatrix}\text{Volume of}\\ \text{unshielded box}\end{pmatrix}$$

$$\begin{aligned} V &= (1 + 2x)(2 + 2x)(4 + 2x) - 1 \cdot 2 \cdot 4 \\ &= (2 + 6x + 4x^2)(4 + 2x) - 8 \\ &= 8 + 24x + 16x^2 + 4x + 12x^2 + 8x^3 - 8 \\ &= 28x + 28x^2 + 8x^3 \end{aligned}$$

(B) Substitute $x = 0.05$ to find

$V = 28(0.05) + 28(0.05)^2 + 8(0.05)^3 = 1.471$ cubic feet

105. (A) The independent variable is years since 1960, so enter 0, 10, 20, 30, 40, 42 as L_1. The dependent variable is total expenditures, so enter the total expenditures column as L_2. Then choose the cubic regression from the STAT CALC menu.

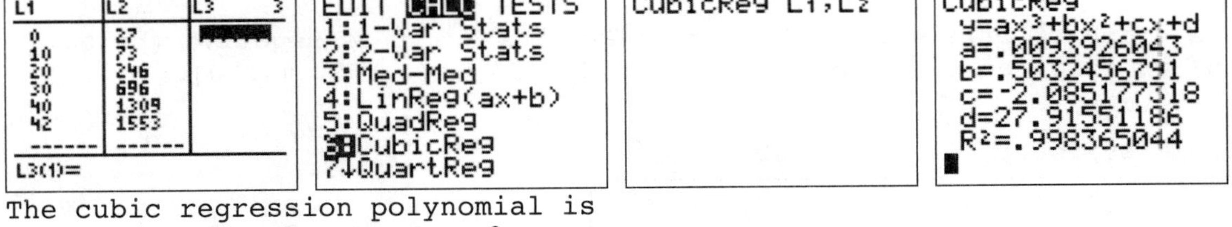

The cubic regression polynomial is

$y = 0.00939x^3 + 0.50325x^2 - 2.08518x + 27.91551$

(B) 2010 is 50 years after 1960, so substitute 50 into the regression model.

$y = 0.00939(50)^3 + 0.50325(50)^2 - 2.08518(50) + 27.91551 = \$2{,}356$ billion.

107. (A) The independent variable is years since 1950, so enter 0, 10, 20, 30, 40, 50 as L_1. The dependent variable is marriage rate, so enter the marriage column as L_2. Then choose the cubic regression from the STAT CALC menu.

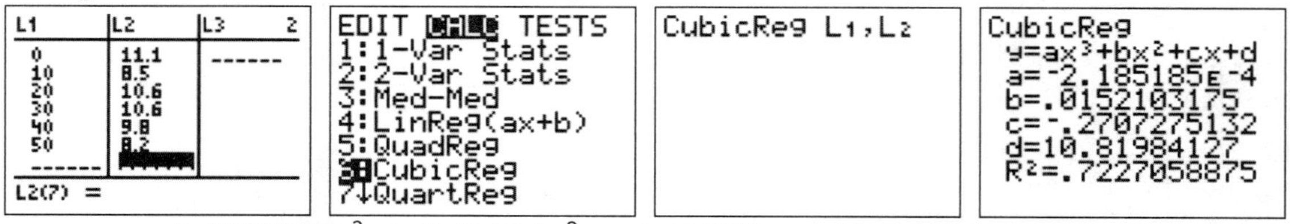

$y = -0.0002185x^3 + 0.01521x^2 - 0.27073x + 10.81984$

(B) 2008 is 58 years after 1950, so substitute 58 into the regression model.
$y = 0.0002185(58)^3 + 0.01521(58)^2 - 0.27073(58) + 10.81984 = 3.6.$

Section 4-2

1. The degree is 2 so there are at most two zeros. Graph $y = x^2 + 5x - 2$ and use the zero command.

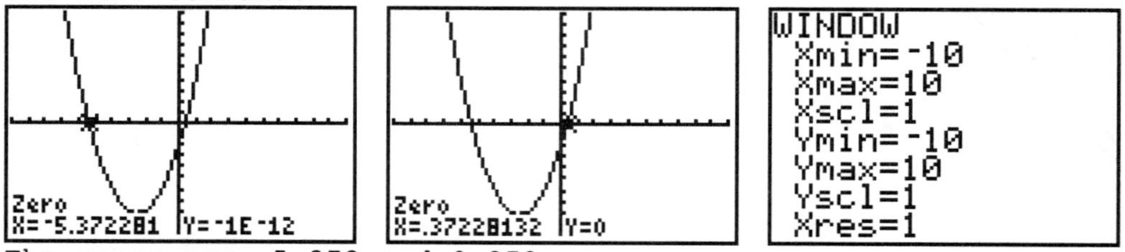

The zeros are -5.372 and 0.372.

3. The degree is 3 so there are at most three zeros. Graph $y = 2x^3 - 5x + 2$ and use the zero command.

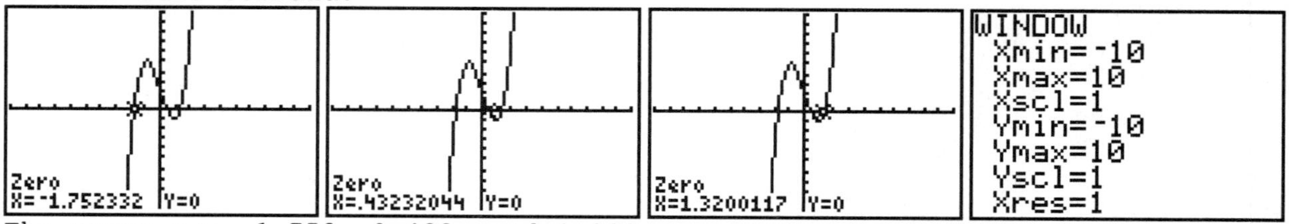

The zeros are -1.752, 0.432, and 1.320.

5. This inequality is true wherever the graph of $P(x)$ is above the x-axis or intersects the x-axis. This is $[-2, -1] \cup \{1\} \cup [3, \infty)$.

7. This inequality is true only when $P(x)$ is above the x-axis but not where it intersects the x-axis. This is $(-2, -1) \cup (3, \infty)$.

9. See the graph of $P(x) = x^2 + 5x - 2$ in Problem 1. It is positive (above the x-axis) on $(-\infty, -5.372) \cup (0.372, \infty)$.

11. See the graph of $P(x) = 2x^3 - 5x + 2$ in Problem 3. It is negative (below the x-axis) or zero (on the x-axis) on $(-\infty, -1.752] \cup [0.432, 1.320]$.

13. We form a synthetic division table.

	1	0	-3	1
0	1	0	-3	1
1	1	1	-2	-1
2	1	2	1	3 an upper bound
-1	1	-1	-2	3
-2	1	-2	1	-1 a lower bound

2 is an upper bound; -2 is a lower bound.

15. We form a synthetic division table.

	1	-3	4	2	-9	
0	1	-3	4	2	-9	
1	1	-2	2	4	-5	
2	1	-1	2	6	3	
3	1	0	4	14	33	an upper bound
-1	1	-4	8	-6	-3	
-2	1	-5	14	-26	43	a lower bound

3 is an upper bound; -2 is a lower bound.

17. We form a synthetic division table.

	1	0	-3	3	2	-2	
0	1	0	-3	3	2	-2	
1	1	1	-2	1	3	1	
2	1	2	1	5	12	22	an upper bound
-1	1	-1	-2	5	-3	1	
-2	1	-2	1	1	0	-2	
-3	1	-3	6	-15	47	-143	a lower bound

2 is an upper bound; -3 is a lower bound.

19. (A) $P(3) = 3^3 - 2(3)^2 - 5(3) + 4 = -2$ Different signs, so there is
$P(4) = 4^3 - 2(4)^2 - 5(4) + 4 = 16$ a zero between 3 and 4.

(B) Midpoint of [3, 4]: 3.5; $P(3.5) = 4.875$; New interval: [3, 3.5]
Midpoint of [3, 3.5]: 3.25; $P(3.25) = 0.953$; New interval: [3, 3.25]
Midpoint of [3, 3.25]: 3.125; $P(3.125) = -0.639$;
New interval: [3.125, 3.25]
Midpoint of [3.125, 3.25]: 3.1875; $P(3.1875) = 0.128$;
New interval: [3.125, 3.1875]
Midpoint of [3.125, 3.1875]: 3.15625; $P(3.15625) = -0.263$

The sign is different for $P(3.15625)$ and $P(3.1875)$; both of these x-values round to 3.2, so 3.2 is the zero to one decimal place accuracy. It took 5 intervals.

> Calculator Note: Here's a convenient way to repeatedly plug different numbers into the same function as we did in Problem 19. Enter the function as y_1 in the equation editor and use the "value" command from the calculate menu. All you have to do is key in the x-value you want to plug in.

21. (A) $P(-2) = -9$ Different signs, so there's a zero between -2 and -1.
$P(-1) = 3$

(B) Midpoint of [-2, -1] = -1.5; $P(-1.5) = -1.375$; New interval: [-1.5, -1]
Midpoint of [-1.5, -1] = -1.25; $P(-1.25) = 1.17$;
New interval: [-1.5, -1.25]
Midpoint of [-1.5, -1.25] = -1.375; $P(-1.375) = -0.006$;
New interval: [-1.375, -1.25]
Midpoint of [-1.375, -1.25] = -1.3125; $P(-1.3125) = 0.606$;
New interval: [-1.375, -1.3125]
Midpoint of [-1.375, -1.3125] = -1.34375; $P(-1.34375) = 0.306$;
New interval: [-1.375, -1.34375]
Midpoint of [-1.375, -1.34375] = -1.359375; $P(-1.359375) = 0.15$

$P(-1.375)$ and $P(-1.359375)$ have opposite signs and both x-values round to -1.4, so -1.4 is the zero to one decimal place accuracy. It took 6 intervals.

23. (A) $P(3) = -2$
$P(4) = 59$ Different signs, so there's a zero between 3 and 4.

(B) Midpoint of [3, 4] = 3.5; $P(3.5) = 17.1$; New interval: [3, 3.5]
Midpoint of [3, 3.5] = 3.25; $P(3.25) = 5.2$; New interval: [3, 3.25]
Midpoint of [3, 3.25] = 3.125; $P(3.125) = 1.1$; New interval: [3, 3.125]
Midpoint of [3, 3.125] = 3.0625; $P(3.0625) = -0.6$

$P(3.0625)$ and $P(3.125)$ have opposite signs and both x-values round to 3.1, so 3.1 is the zero to one decimal place accuracy. It took 4 intervals.

25. (A) $P(-1) = -3$
$P(0) = 3$ Different signs, so there's a zero between -1 and 0.

(B) Midpoint of [-1, 0] = -0.5; $P(-0.5) = .19$; New interval: [-1, -0.5]
Midpoint of [-1, -0.5] = -0.75; $P(-0.75) = -1.5$;
New interval: [-0.75, -0.5]
Midpoint of [-0.75, -0.5] = -0.625; $P(-0.625) = -0.7$;
New interval: [-0.625, -0.5]
Midpoint of [-0.625, -0.5] = -0.5625; $P(-0.5625) = -0.3$;
New interval: [-0.5625, -0.5]
Midpoint of [-0.5625, -0.5] = -0.53125; $P(-0.53125) = -0.02$

$P(-0.5)$ and $P(-0.53125)$ have opposite signs and both x-values round to -0.5, so -0.5 is the zero to one decimal place accuracy. It took 5 intervals.

27. (A) We form a synthetic division table.

	1	-2	3	-8
0	1	-2	3	-8
1	1	-1	2	-6
2	1	0	3	-2
3	1	1	6	10
-1	1	-3	6	-14

From the table, 3 is an upper bound and -1 is a lower bound.

(B) The only interval in which a real zero is indicated is (2, 3). We search for this real zero. We organize our calculations in a table.

Sign Change Interval (a, b)	Midpoint m	Sign of P: $P(a)$	$P(m)$	$P(b)$
(2, 3)	2.5	−	+	+
(2, 2.5)	2.25	−	+	+
(2, 2.25)	2.125	−	−	+
(2.125, 2.25)	2.1875	−	−	+
(2.1875, 2.25)	We stop here	−		+

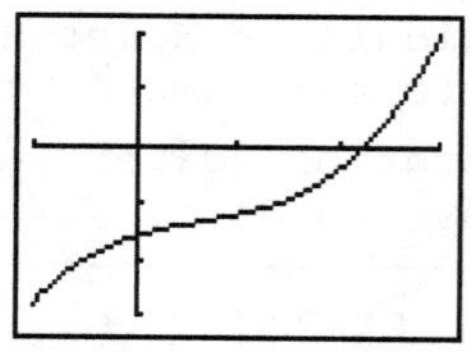

Since each endpoint rounds to 2.2 (using scientific rounding), a real zero lies on this last interval and is given by 2.2 to one decimal place accuracy. A glance at the graph of $P(x)$ on [-1, 3] confirms that this is the only real zero.

29. (A) We form a synthetic division table.

	2	1	2	1
0	2	1	2	1
1	2	3	5	6
-1	2	-1	3	-2

From the table, 1 is an upper bound and -1 is a lower bound.

(B) The only interval in which a real zero is indicated is (-1, 0).
Bisection immediately yields -0.5 as a zero.

31. (A) We form a synthetic division table.

	1	0	1	0	-6
0	1	0	1	0	-6
1	1	1	2	2	-4
2	1	2	5	10	14
-1	1	-1	2	-2	-4
-2	1	-2	5	-10	14

From the table, 2 is an upper bound and -2 is a lower bound.

(B) Real zeros are indicated in the intervals (-2, -1) and (1, 2). Since $P(x)$ is an even function, we need only to search in one of these intervals; if a is a real zero, then $-a$ is also a real zero. We organize our calculations in a table.

Sign Change Interval (a, b)	Midpoint m	Sign of P $P(a)$	$P(m)$	$P(b)$
(1, 2)	1.5	−	+	+
(1, 1.5)	1.25	−	−	+
(1.25, 1.5)	1.375	−	−	+
(1.375, 1.5)	1.4375	−	+	+
(1.375, 1.4375)	We stop here	−		+

Since each endpoint rounds to 1.4, a real zero lies on this last interval and is given by 1.4 to one decimal place accuracy. As noted, -1.4 is also a zero. Algebraically, since $x^4 + x^2 - 6 = (x^2 + 3)(x^2 - 2) = (x^2 + 3)(x - \sqrt{2})(x + \sqrt{2})$, $\sqrt{2} \approx 1.4$ and $-\sqrt{2} \approx -1.4$ are zeros. Thus the approximate answers are confirmed.

33. (A) We form a synthetic division table.

	1	1	3	1	2	-5
0	1	1	3	1	2	-5
1	1	2	5	6	8	3
-1	1	0	3	-2	4	-9

From the table, 1 is an upper bound and -1 is a lower bound.

(B) There is a real zero in the interval (0, 1) indicated in the table. We search for this real zero. We organize our calculations in a table.

Sign Change Interval (a, b)	Midpoint m	Sign of P $P(a)$	$P(m)$	$P(b)$
(0, 1)	0.5	−	−	+
(0.5, 1)	0.75	−	−	+
(0.75, 1)	0.875	−	+	+
(0.75, 0.875)	0.8125	−	−	+
(0.8125, 0.875)	0.84375	−	+	+
(0.8125, 0.84375)	0.828125	−	−	+
(0.828125, 0.84375)	0.8359375	−	+	+
(0.828125, 0.8359375)	0.83203125	−	−	+
(0.83203125, 0.8359375)	0.833984375	−	−	+
(0.833984375, 0.8359375)	0.8349609375	−	+	+
(0.833984375, 0.8349609375)	We stop here	−		+

Since each endpoint rounds to 0.83, a real zero lies on this last interval and is given by 0.83 to two decimal place accuracy. A glance at the graph of $P(x)$ on [-2, 2] confirms that this is the only real zero.

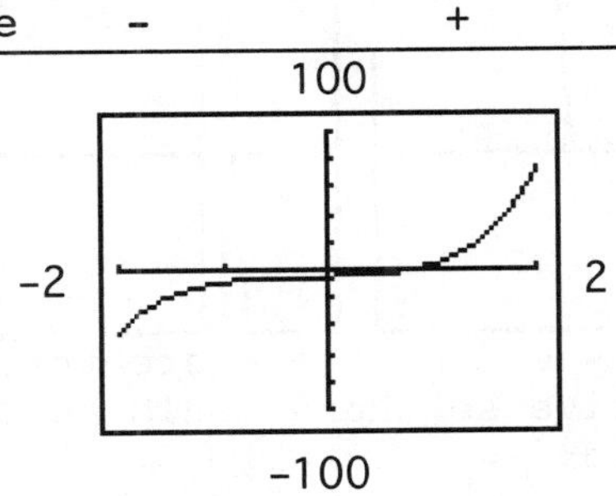

35. Since 1 is a zero of multiplicity 2, $P(x)$ does not change sign at $x = 1$. The bisection method fails. (The TI-83, using a more complicated routine, can find the zero.)

37. Since 3 is a zero of multiplicity 4, $P(x)$ does not change sign at $x = 3$. The bisection method fails. (The TI-83 cannot find this zero.)

39. The graph in a standard window shows two zeros that are turning points. The degree is 4 so there could be more zeros. We'll make a synthetic division table to see if $x = 10$ are bounds for the real zeros.

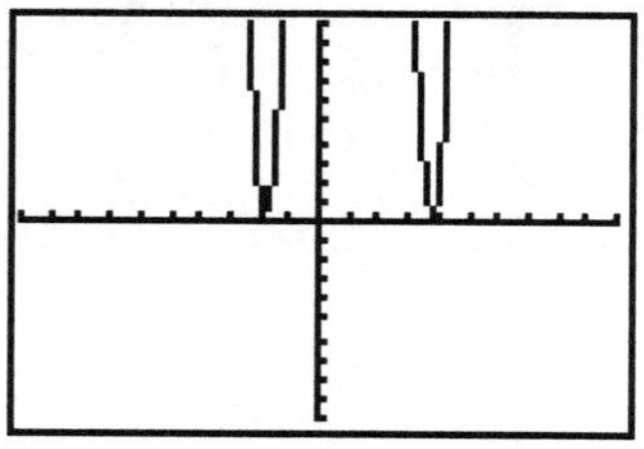

	1	-4	-10	28	49
-10	1	-14	130	-1272	12769
10	1	6	50	528	5329

The signs alternate for -10 so it's a lower bound for the real zeros, and all signs are positive for 10 so it's an upper bound. In other words, the only real zeros are the two we see on the standard screen.

41. The graph in a standard window shows three zeros two of which are turning points. The degree is 5 so there could be more zeros. We'll make a synthetic division table to see if $x = -10$ are bounds for the real zeros.

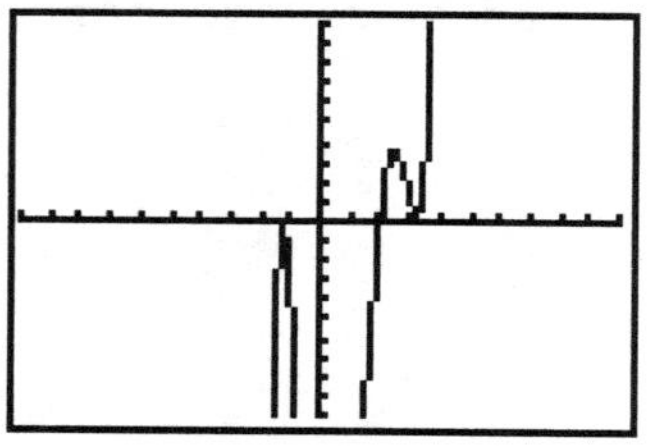

	1	-6	4	24	-16	-32
-10	1	-16	164	-1616	16144	-161472
10	1	4	44	464	4624	46208

The signs alternate for -10 so it's a lower bound for the real zeros, and all signs are positive for 10 so it's an upper bound. In other words, the only real zeros are the two we see on the standard screen.

To find the zero that's not a turning point, we use the zero command; and to find the two that are, we use the maximum and minimum commands:

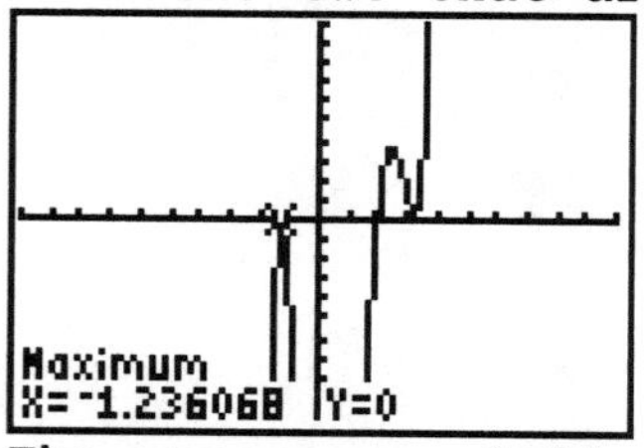

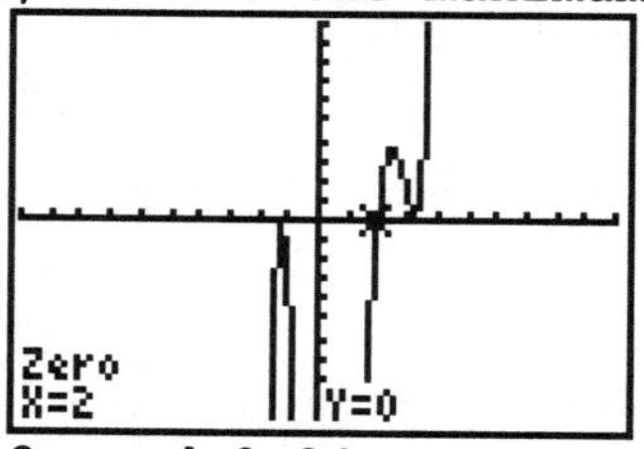

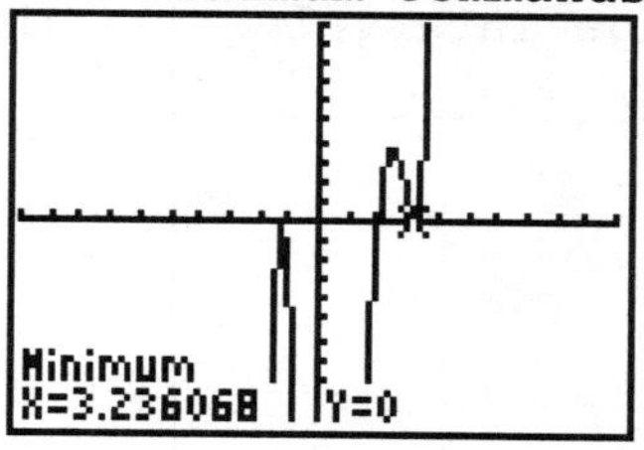

The zeros are -1.24, 2, and 3.24.

43. On a standard viewing window it's difficult to see what's happening close to $x = 2$, so we examine that area closer.

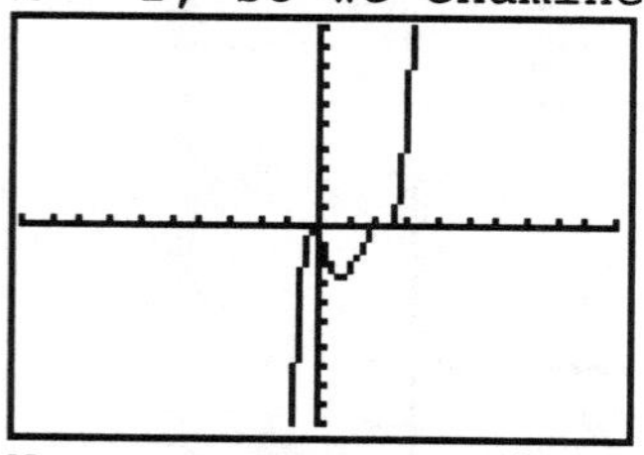
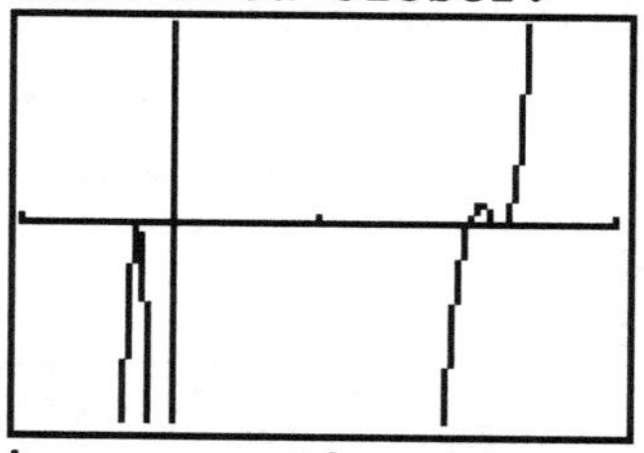
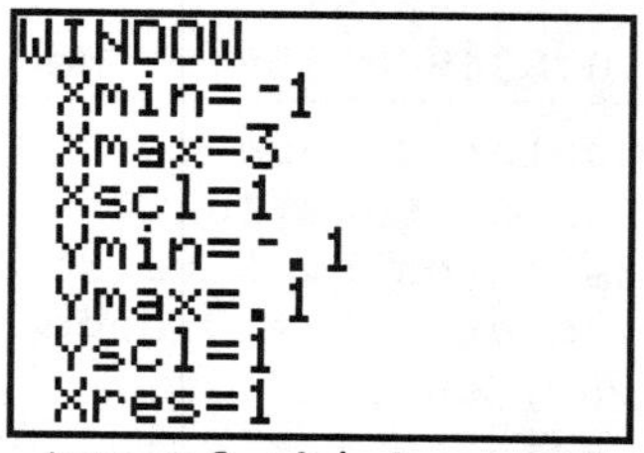

Now we can see that there are three zeros, two of which are turning points. The degree is five so there could be other zeros. A synthetic division table will help us check.

	1	-6	11	-4	-3.75	-0.5
-2	1	-8	27	-58	112.25	-225
10	1	4	51	506	5056.25	50562

The signs alternate for -2 so it's a lower bound for the zeros. The signs are all positive for 10 so it's an upper bound for the zeros and we know that we have seen all of them. To find the zero that's not a turning point we use the zero command; and to find the two that are, we use the maximum and minimum commands.

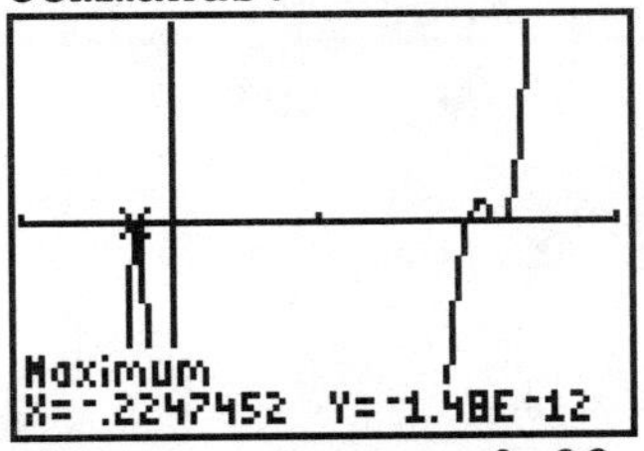

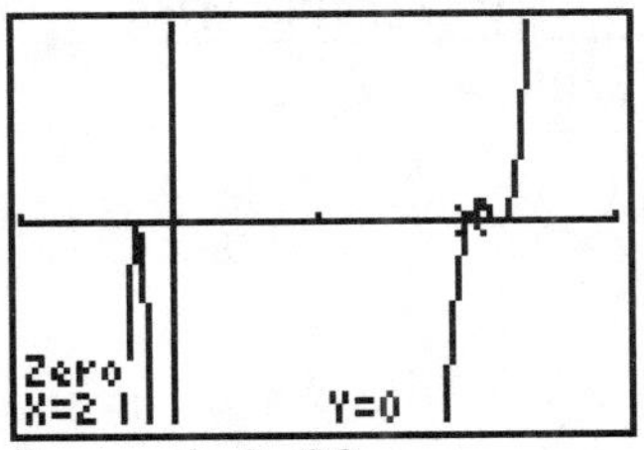

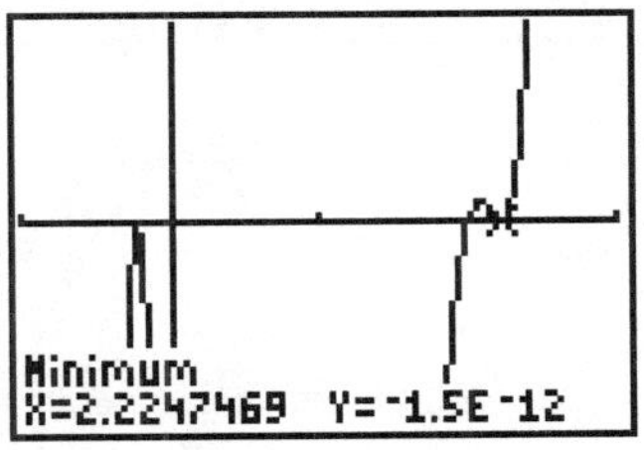

The zeros are -0.22, 2, and 2.22.

45. $x^2 > 9$

$x^2 - 9 > 0$

Let $P(x) = x^2 - 9$, then $P(x) = (x + 3)(x - 3)$. The zeros of P are -3 and 3. They partition the x axis into the three intervals shown in the table. Choose a test number in each interval to determine the sign of P in that interval.

Interval	$(-\infty, -3)$	$(-3, 3)$	$(3, \infty)$
Test number x	-4	0	4
$P(x)$	7	-9	7
Sign of P	+	−	+

We conclude that the solution set of the inequality is $(-\infty, -3) \cup (3, \infty)$.

47. $x^3 \le 16x$

$x^3 - 16x \le 0$

Let $P(x) = x^3 - 16x$, then $P(x) = x(x^2 - 16) = x(x - 4)(x + 4)$. The zeros of P are -4, 0, and 4. They partition the x axis into the four intervals shown in the table. Choose a test number in each interval to determine the sign of P in that interval. The equality holds at the zeros.

Interval	$(-\infty, -4)$	$(-4, 0)$	$(0, 4)$	$(4, \infty)$
Test number x	-5	-2	2	5
$P(x)$	-45	24	-24	45
Sign of P	−	+	−	+

We conclude that the solution set of the inequality is $(-\infty, -4] \cup [0, 4]$.

49. $x^4 + 4 \ge 5x^2$

$x^4 - 5x^2 + 4 \ge 0$

Let $P(x) = x^4 - 5x^2 + 4$, then $P(x) = (x^2 - 4)(x^2 - 1) = (x + 2)(x - 2)(x + 1)(x - 1)$. The zeros of P are -2, -1, 1, 2. They partition the x axis into the five intervals shown in the table. Choose a test number in each interval to determine the sign of P in that interval. The equality holds at the zeros.

Interval	$(-\infty, -2)$	$(-2, -1)$	$(-1, 1)$	$(1, 2)$	$(2, \infty)$
Test number x	-3	-1.5	0	1.5	3
$P(x)$	40	-2.1875	4	-2.1875	40
Sign of P	+	−	+	−	+

We conclude that the solution set of the inequality is $(-\infty, -2] \cup [-1, 1] \cup [2, \infty)$.

51. $x^2 + 7x - 3 \le x^3 + x + 4$

$-x^3 + x^2 + 6x - 7 \le 0$

Find the zeros of $P(x) = -x^3 + x^2 + 6x - 7$ graphically:

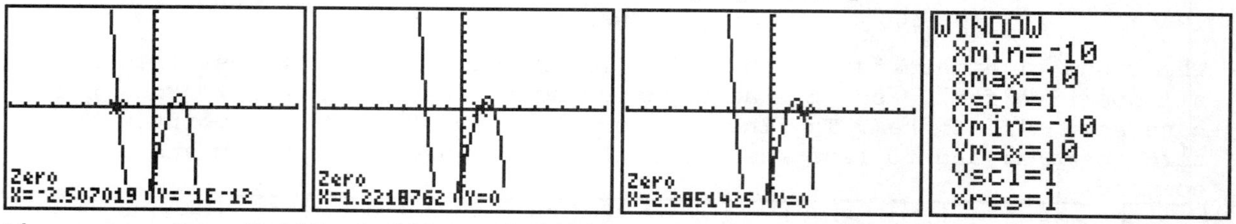

The zeros are -2.507, 1.222, and 2.285. $P(x)$ is negative or zero (below, or on, the x axis) on $[-2.507, 1.222] \cup [2.285, \infty)$.

53. $x^4 < 8x^3 - 17x^2 + 9x - 2$

$x^4 - 8x^3 + 17x^2 - 9x + 2 < 0$

Find the zeros of $P(x) = x^4 - 8x^3 + 17x^2 - 9x + 2$ graphically:

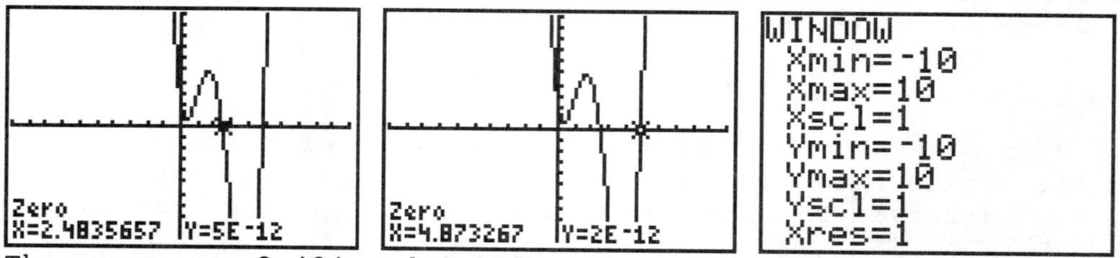

The zeros are 2.484 and 4.873. $P(x)$ is negative (below the x axis) on $(2.484, 4.873)$.

55. $(x^2 + 2x - 2)^2 \ge 2$

$(x^2 + 2x - 2)^2 - 2 \ge 0$

Find the zeros of $P(x) = (x^2 + 2x - 2)^2 - 2$ graphically:

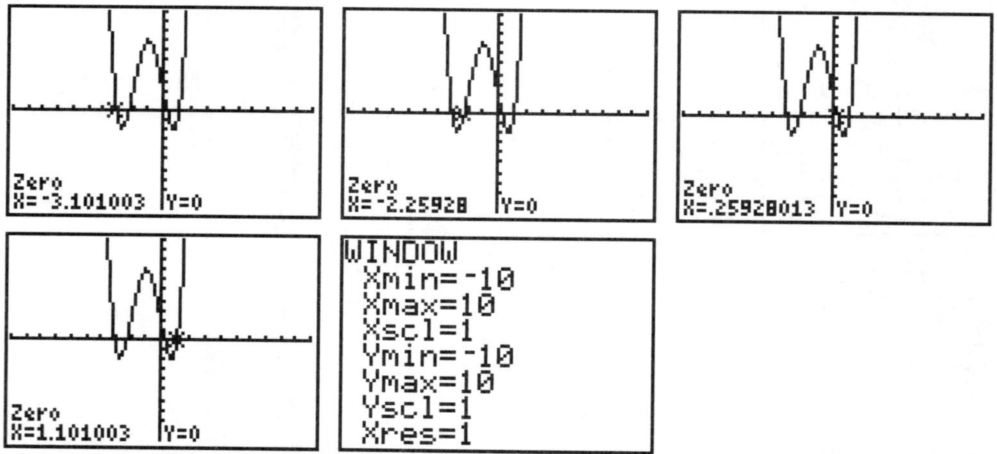

The zeros are -3.101, -2.259, 0.259, and 1.101. $P(x)$ is positive or zero (above, or on, the x axis) on $(-\infty, -3.101] \cup [-2.259, 0.259] \cup [1.101, \infty)$.

57. (A) We form a synthetic division table.

	1	-24	-25	10
0	1	-24	-25	10
10	1	-14	-165	-1640
20	1	-4	-105	-2090
30	1	6	155	4660
-10	1	-34	315	-3140

From the table, 30 is an upper bound and -10 is a lower bound.

(B) Graphing $P(x)$ using the window suggested by part (A), we obtain

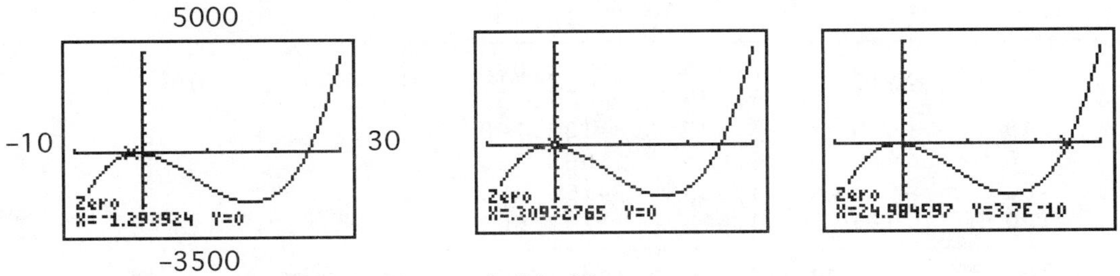

The zeros are -1.29, 0.31, and 24.98.

59. (A) We form a synthetic division table.

	1	12	-900	0	5,000
0	1	12	-900	0	5,000
10	1	22	-680	-680	-63,000
20	1	32	-260	-5200	-99,000
30	1	42	360	10,800	329,000
-10	1	2	-920	9,200	-87,000
-20	1	-8	-740	14,800	-291,000
-30	1	-18	-360	10,800	-319,000
-40	1	-28	220	-8,800	357,000

From the table, 30 is an upper bound and -40 is a lower bound.

(B) Graphing $P(x)$ using the window suggested in part (A), we obtain

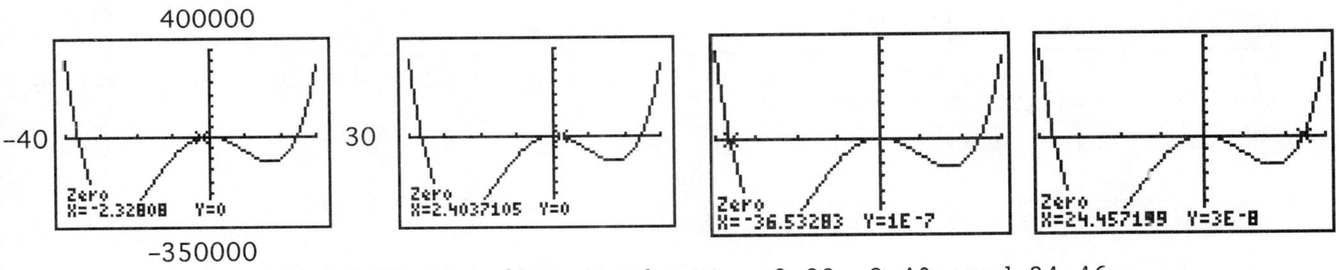

The zeros are -36.53 (details not shown), -2.33, 2.40, and 24.46.

61. (A) We form a synthetic division table.

	1	0	-100	-1,000	-5,000
0	1	0	-100	-1,000	-5,000
10	1	10	0	-1,000	-15,000
20	1	20	300	5,000	95,000
-10	1	-10	0	-1,000	5,000

From the table, 20 is an upper bound and -10 is a lower bound.

(B) Graphing $P(x)$ using the window suggested in part (A), we obtain

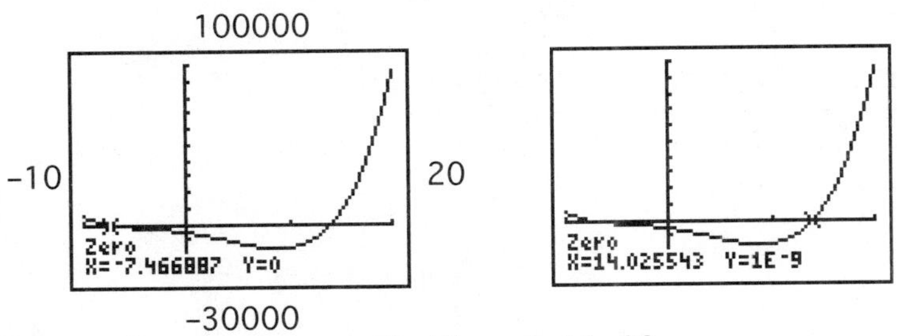

The real zeros are -7.47 and 14.03.

63. (A) We form a synthetic division table.

	4	-40	-1,475	7,875	-10,000
0	4	-40	-1,475	7,875	-10,000
10	4	0	-1,475	-6,875	-78,750
20	4	40	-675	-5,625	-122,500
30	4	80	925	35,625	1,058,750
-10	4	-80	-675	14,625	-156,250
-20	4	-120	925	-10,625	202,500

From the table, 30 is an upper bound and -20 is a lower bound.

(B) Graphing $P(x)$ using the window suggested in part (A), we obtain the graph shown at the right.

It appears that there are zeros between -20 and -10, and between 20 and 30, and there may be a zero of even multiplicity between 0 and 10.

500000

-20 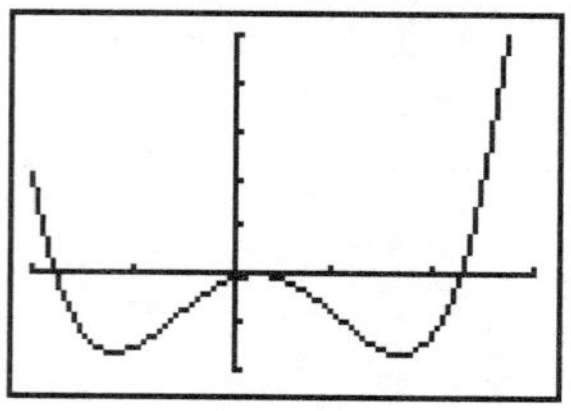 30

-200000

Applying zero and maximum routines, we obtain

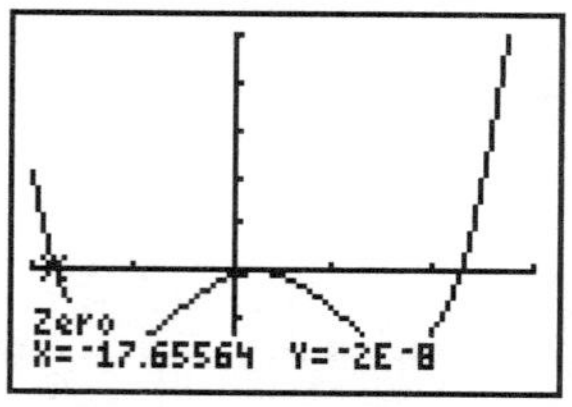

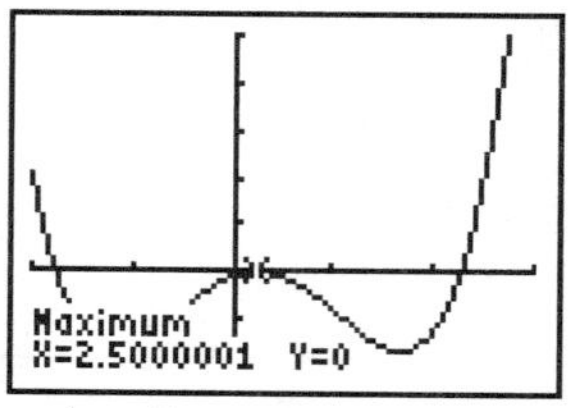

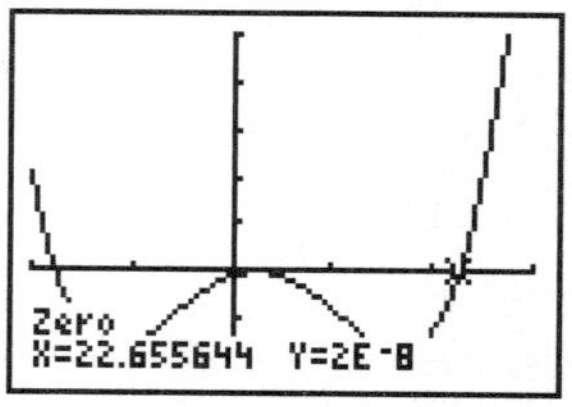

-17.66 and 20.66 are simple zeros and 2.50 is probably a double zero. To check that it is really a zero, we note

$$\begin{array}{r|rrrrr} & 4 & -40 & -1475 & 7875 & -10000 \\ & & 10 & -75 & -3875 & 10000 \\ \hline 2.50 & 4 & -30 & -1550 & 4000 & 0 \end{array}$$

Thus 2.50 is definitely a zero. You can check that it is a double zero using a second synthetic division.

65. (A) We form a synthetic division table.

	0.01	-0.1	-12	0	0	9,000
0	0.01	-0.1	-12	0	0	9,000
10	0.01	0	-12	-120	-1,200	-3,000
20	0.01	0.1	-10	-200	-4,000	-71,000
30	0.01	0.2	-6	-180	-5,400	-153,000
40	0.01	0.3	0	0	0	9,000
-10	0.01	-0.2	-10	100	-1,000	19,000
-20	0.01	-0.3	-6	120	-2,400	57,000
-30	0.01	-0.4	0	0	0	9,000
-40	0.01	-0.5	8	-320	12,800	-503,000

From the table, 40 is an upper bound and -40 is a lower bound.

(B) Graphing $P(x)$ using a window suggested by part (A), we obtain

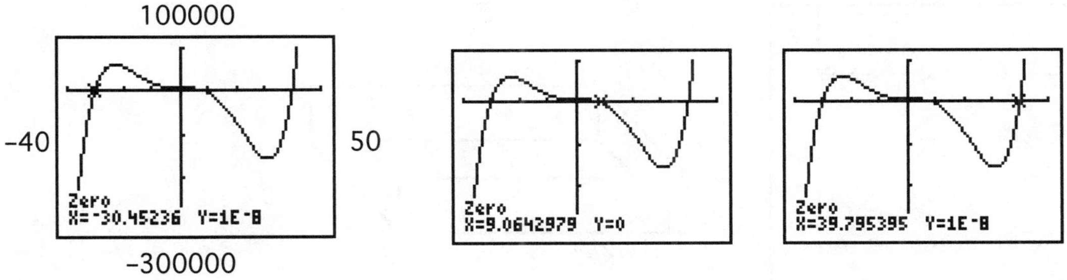

The real zeros are -30.45, 9.06, and 39.80.

A graph in a smaller window confirms that there are no other real zeros between -10 and 10, and therefore most likely no other real zeros.

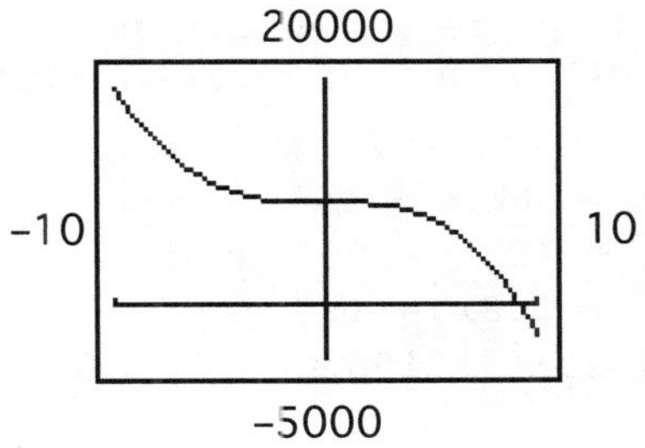

67. Yes. According to the Remainder Theorem, if the remainder is 10 when $P(x)$ is divided by $x + 4$, then $P(-4) = 10$. Similarly, if the remainder is -8 when dividing by $x + 5$, then $P(-5) = -8$. Since $P(-5)$ and $P(-4)$ have opposite signs, there must be a zero between -5 and -4.

69. If all coefficients of $P(x)$ are nonnegative, then no positive replacement of x could possibly make $P(x) = 0$; hence, there are no positive zeros of $P(x)$. All zeros must therefore be less than or equal to 0, and 0 is an upper bound for the zeros.

71. Let (x, x^2) be a point on the graph of $y = x^2$. Then the distance from (1, 2) to (x, x^2) must equal 1 unit. Applying the distance formula, we have,

$$\begin{aligned} \sqrt{(x-1)^2 + (x^2-2)^2} &= 1 \\ (x-1)^2 + (x^2-2)^2 &= 1 \\ x^2 - 2x + 1 + x^4 - 4x^2 + 4 &= 1 \\ x^4 - 3x^2 - 2x + 4 &= 0 \end{aligned}$$

Let $P(x) = x^4 - 3x^2 - 2x + 4$

The only rational zero is 1.

$$\begin{array}{r|rrrrr} & 1 & 0 & -3 & -2 & 4 \\ & & 1 & 1 & -2 & -4 \\ \hline 1 & 1 & 1 & -2 & -4 & 0 \end{array}$$

$P(x) = (x - 1)(x^3 + x^2 - 2x - 4)$

We examine $Q(x) = x^3 + x^2 - 2x - 4$. Graphing $y = Q(x)$ we obtain the graph shown at the right.

$Q(x)$ (and $P(x)$) has a second zero at $x = 1.7$. Therefore, the two real zeros of $P(x)$ are 1 and 1.7. Hence the two required points are (1, 1) and $(1.7, 1.7^2) = (1.7, 2.9)$.

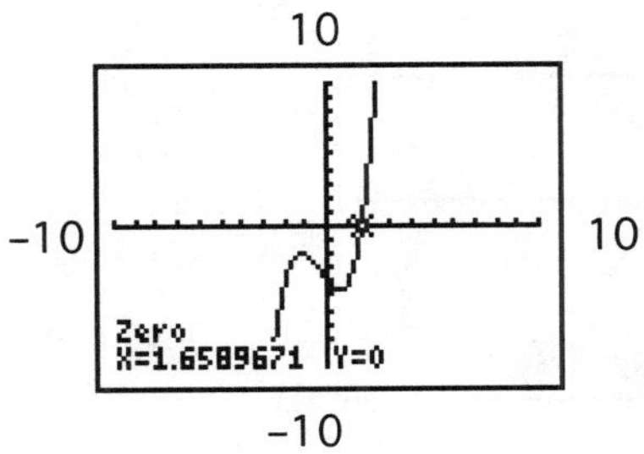

73. 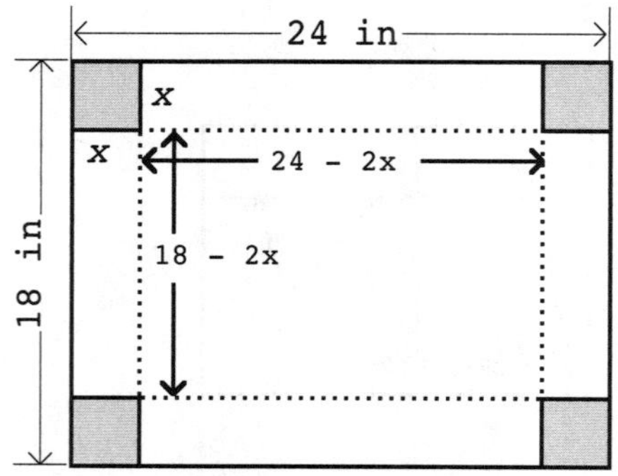

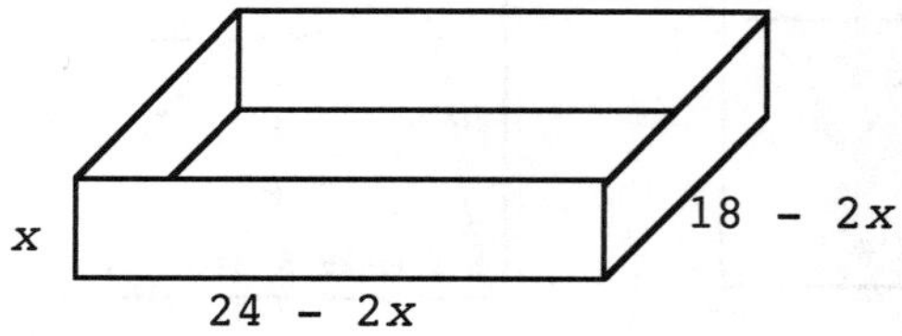

From the above figures it should be clear that
V = length × width × height = $(24 - 2x)(18 - 2x)x \quad 0 < x < 9$ (why?)

We solve $(24 - 2x)(18 - 2x)x = 600$

$$432x - 84x^2 + 4x^3 = 600$$
$$4x^3 - 84x^2 + 432x - 600 = 0$$
$$x^3 - 21x^2 + 108x - 150 = 0$$

Let $P(x) = x^3 - 21x^2 + 108x - 150$. Graphing $y = P(x)$ on the interval $0 < x < 9$, we obtain

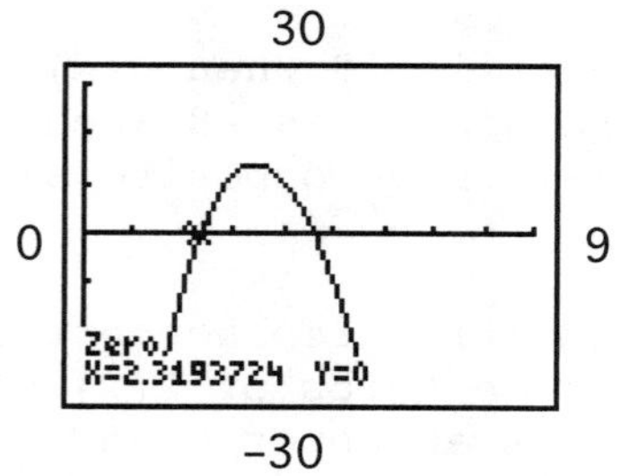

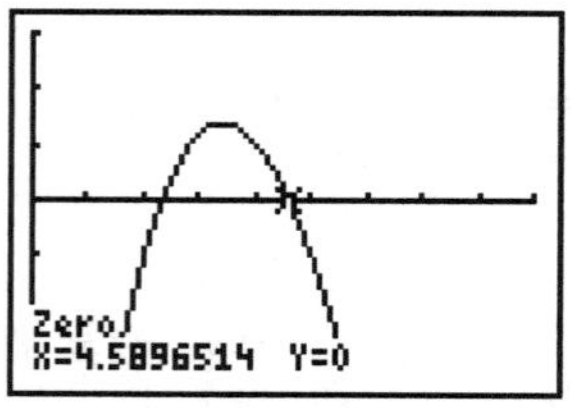

The zeros of $P(x)$ are 2.3 and 4.6, $0 < x < 9$.
x = 2.3 inches or 4.6 inches.

75. We note:

$$\begin{pmatrix}\text{Volume}\\ \text{of}\\ \text{tank}\end{pmatrix} = \begin{pmatrix}\text{Volume of}\\ \text{two hemispheres}\\ \text{of radius } x\end{pmatrix} + \begin{pmatrix}\text{Volume of cylinder}\\ \text{with radius } x,\\ \text{height } 10 - 2x\end{pmatrix}$$

$$20\pi = \frac{4}{3}\pi x^3 + \pi x^2(10 - 2x)$$
$$20 = \frac{4}{3}x^3 + 10x^2 - 2x^3$$
$$60 = 4x^3 + 30x^2 - 6x^3$$
$$2x^3 - 30x^2 + 60 = 0$$
$$x^3 - 15x^2 + 30 = 0$$

From physical considerations, $x > 0$ and $10 - 2x > 0$, hence we are interested only in solutions in (0, 5).

Let $P(x) = x^3 - 15x^2 + 30$
Graphing $y = P(x)$ on the interval $0 < x < 5$, we obtain

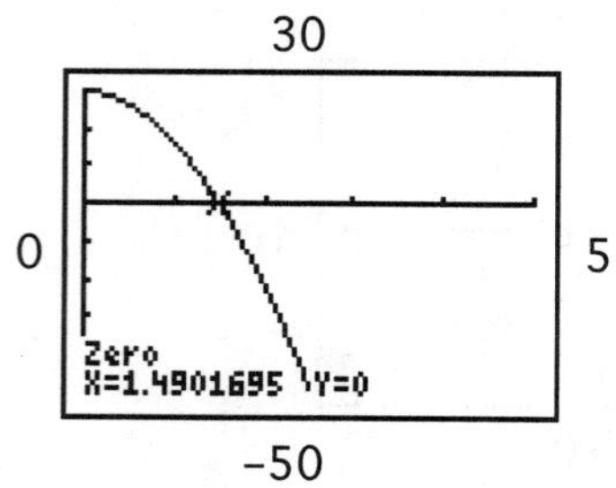

The only zero of $P(x)$, $0 < x < 5$, is 1.5.
x = 1.5 feet.

Section 4-3

1.

Zero	Multiplicity
-8	3
6	2
	5 = degree of $P(x)$

3.

Zero	Multiplicity
-4	3
3	2
-1	1
	6 = degree of $P(x)$

5. $x^3(2x+1)^2 = (x-0)^3\ 2^2\left(x+\frac{1}{2}\right)^2$

Zero	Multiplicity
0	3
-1/2	2
	5 = degree of $P(x)$

7. $(x^2+4)^3(x^2-4)^5(x+2i)$
$= [(x+2i)(x-2i)]^3[(x-2)(x+2)]^5(x+2i)$
$= (x+2i)^4(x-2i)^3(x-2)^5(x+2)^5$

Zero	Multiplicity
$-2i$	4
$2i$	3
2	5
-2	5
	17 = degree of $P(x)$

9. $(x^3-9x)(x^2+9)(x+9)^2 = x(x^2-9)[x^2-(-9)](x+9)^2$
$= x(x-3)(x+3)(x-3i)(x+3i)(x+9)^2$

Zero	Multiplicity
0	1
3	1
-3	1
$3i$	1
$-3i$	1
-9	2
	7 = degree of $P(x)$

11. $P(x) = (x-3)^2(x+4)$; degree 3

13. $P(x) = (x+7)^3[x-(-3+\sqrt{2})][x-(-3-\sqrt{2})]$; degree 5

15. $P(x) = [x-(2-3i)][x-(2+3i)](x+4)^2$; degree 4

17. Since -2, 1, and 3 are zeros, $P(x) = (x+2)(x-1)(x-3)$ is the lowest degree polynomial that has this graph. The degree of $P(x)$ is 3.

19. Since -2 and 1 are zeros, each with multiplicity 2, $P(x) = (x+2)^2(x-1)^2$ is the lowest degree polynomial that has this graph. The degree of $P(x)$ is 4.

21. -3, -2, and 2 are zeros, each with multiplicity 1.
0 is a zero with multiplicity 2.
$P(x) = x^2(x+3)(x+2)(x-2)$ is the lowest degree polynomial that has this graph. The degree of $P(x)$ is 5.

23. (A) Let $u = x^2$; then $x^4 + 5x^2 + 4 = u^2 + 5u + 4 = (u+4)(u+1)$. Thus
$$P(x) = (x^2+4)(x^2+1)$$

(B) The zeros of $x^2 + 4$ are $2i$ and $-2i$, so it factors as $(x-2i)(x+2i)$. The zeros of $x^2 + 1$ are i and $-i$, so it factors as $(x-i)(x+i)$. Thus
$$P(x) = (x-2i)(x+2i)(x-i)(x+i)$$

25. (A) Factor by grouping:

$$P(x) = (x^3 - x^2) + (25x - 25) = x^2(x - 1) + 25(x - 1) = (x^2 + 25)(x - 1)$$

(B) The zeros of $x^2 + 25$ are $5i$ and $-5i$ so it factors as $(x - 5i)(x + 5i)$. Thus

$$P(x) = (x - 5i)(x + 5i)(x - 1)$$

27. Possible factors of 6 are ±1, ±2, ±3, ±6. Possible factors of 1 are ±1. Therefore the possible rational zeros are ±1, ±2, ±3, ±6.

29. Possible factors of 4 are ±1, ±2, ±4. Possible factors of 3 are ±1, ±3. Therefore the possible rational zeros are $\pm 1, \pm 2, \pm 4, \pm\frac{1}{3}, \pm\frac{2}{3}, \pm\frac{4}{3}$.

31. Possible factors of 3 are ±1, ±3. Possible factors of 12 are ±1, ±2, ±3, ±4, ±6, ±12. Therefore the possible rational zeros are $\pm 1, \pm 3, \pm\frac{1}{2}, \pm\frac{3}{2}, \pm\frac{1}{3}, \pm\frac{1}{4}, \pm\frac{3}{4}, \pm\frac{1}{6}, \pm\frac{1}{12}$.

33. $P(x) = x^3 + 9x^2 + 24x + 16$. Since -1 is a zero of $P(x)$, we can write

$$\begin{array}{r|rrrr} & 1 & 9 & 24 & 16 \\ & & -1 & -8 & -16 \\ \hline -1 & 1 & 8 & 16 & 0 \end{array}$$

$P(x) = (x - 1)(x^2 + 8x + 16) = (x - 1)\ Q(x)$
Since $Q(x)$ is a quadratic with obvious factors we can write
$P(x) = (x - 1)(x + 4)(x + 4)$.

35. $P(x) = x^4 + 2x^2 + 1$; Since i is a double zero, we can write

$$\begin{array}{r|rrrrr} & 1 & 0 & 2 & 0 & 1 \\ & & i & -1 & i & -1 \\ \hline i & 1 & i & 1 & i & 0 \end{array}$$

$$\begin{array}{r|rrrr} & & i & -2 & -i \\ \hline i & 1 & 2i & -1 & 0 \end{array} \Rightarrow x^2 + 2ix - 1$$

$$x = \frac{-2i \pm \sqrt{4i^2 - 4(-1)}}{2}$$

$$= \frac{-2i \pm \sqrt{0}}{2} = -i \quad \text{(double zero)}$$

$$P(x) = (x - i)(x - i)(x + i)(x + i)$$

37. $P(x) = 2x^3 - 17x^2 + 90x - 41$. Since $\frac{1}{2}$ is a zero of $P(x)$, we can write

$$\begin{array}{r|rrrr} & 2 & -17 & 90 & -41 \\ & & 1 & -8 & 41 \\ \hline \frac{1}{2} & 2 & -16 & 82 & 0 \end{array}$$

$$P(x) = \left(x - \frac{1}{2}\right)(2x^2 - 16x + 82) = \left(x - \frac{1}{2}\right)2(x^2 - 8x + 41) = (2x - 1)(x^2 - 8x + 41)$$

$$= (2x - 1)\ Q(x)$$

Since $Q(x)$ is a quadratic, we can find its zeros by the quadratic formula

$x^2 - 8x + 41 = 0 \qquad a = 1, \; b = -8, \; c = 41$

$$\begin{aligned} x &= \frac{-b \pm \sqrt{b^2 - 4ac}}{2a} \\ &= \frac{-(-8) \pm \sqrt{(-8)^2 - 4(1)(41)}}{2(1)} \\ &= \frac{8 \pm \sqrt{-100}}{2} \\ &= \frac{8 \pm 10i}{2} \\ &= 4 \pm 5i \end{aligned}$$

Since the zeros of $Q(x)$ are $4 + 5i$ and $4 - 5i$, by the factor theorem

$Q(x) = [x - (4 + 5i)][x - (4 - 5i)]$

and $P(x) = (2x - 1)[x - (4 + 5i)][x - (4 - 5i)]$

39. $P(x) = x^3 - 4x^2 + 6x - 4$. Since $1 + i$ is a zero, we can write

$$\begin{array}{r|rrrr} & 1 & -4 & 6 & -4 \\ & & 1+i & -4-2i & 4 \\ \hline 1+i & 1 & -3+i & 2-2i & 0 \end{array}$$

By Theorem 3, the conjugate of $1 + i$, that is, $1 - i$, is also a zero, so we can write

$$\begin{array}{r|rrr} & 1 & -3+i & 2-2i \\ & & 1-i & -2+2i \\ \hline 1-i & 1 & -2 & 0 \end{array}$$

$P(x) = [x - (1 - i)][x - (1 + i)](x - 2)$

41. This can be solved using synthetic division; however, it is simpler to note:

Let $u = x^2$; then $x^4 + 8x^2 - 9 = u^2 + 8u - 9 = (u - 1)(u + 9)$.

Then

$P(x) = (x^2 - 1)(x^2 + 9) = (x^2 - 1)[x^2 - (-9)] = (x - 1)(x + 1)(x - 3i)(x + 3i)$.

43. Let $P(x) = 2x^3 - 5x^2 + 1$.

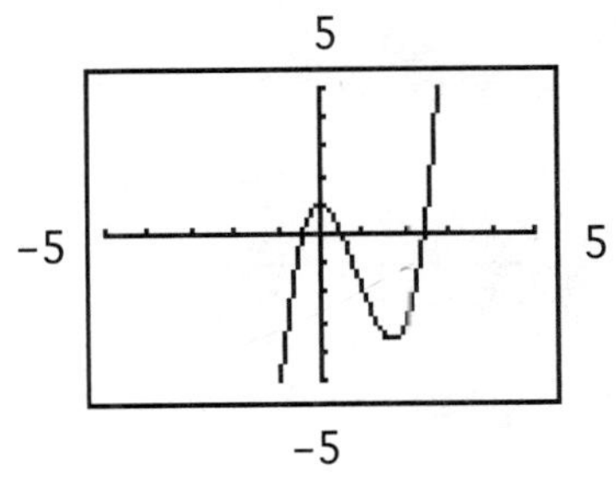

Possible rational zeros: $\pm 1, \pm\frac{1}{2}$.

Examining the graph of $y = P(x)$, we see that there is a zero between -1 and 0, between 0 and 1, and between 1 and 2.

We test the likely candidates, $\frac{1}{2}$ and $-\frac{1}{2}$, and find

$$\begin{array}{r|rrrr} & 2 & -5 & 0 & 1 \\ & & 1 & -2 & -1 \\ \hline \frac{1}{2} & 2 & -4 & -2 & 0 \end{array}$$

$\frac{1}{2}$ is a zero.

So $P(x) = \left(x - \frac{1}{2}\right)(2x^2 - 4x - 2) = (2x - 1)(x^2 - 2x - 1)$

To find the remaining zeros, we solve $x^2 - 2x - 1 = 0$, by completing the square:

$$\begin{aligned} x^2 - 2x &= 1 \\ x^2 - 2x + 1 &= 2 \\ (x - 1)^2 &= 2 \\ x - 1 &= \pm\sqrt{2} \\ x &= 1 \pm \sqrt{2} \end{aligned}$$

Hence the zeros are $\frac{1}{2}$, $1 \pm \sqrt{2}$. These are the roots of the equation.

45. Let $P(x) = x^4 + 4x^3 - x^2 - 20x - 20$. Possible rational zeros: ±1, ±2, ±4, ±5, ±10, ±20.

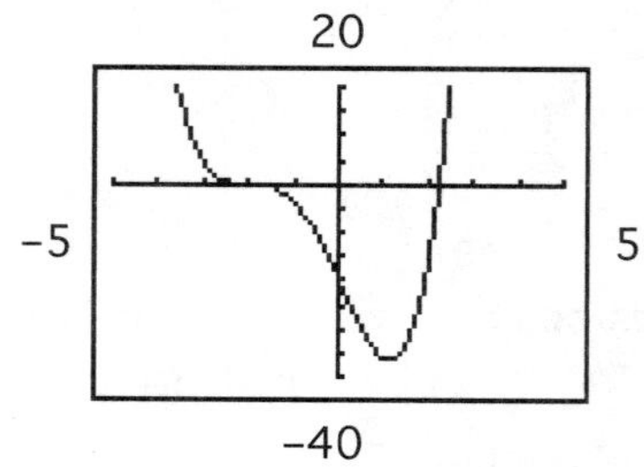

Examining the graph of $y = P(x)$, we see that there is a zero between 2 and 3, and possibly a double zero near -2. We test -2.

$$\begin{array}{r|rrrrr} & 1 & 4 & -1 & -20 & -20 \\ & & -2 & -4 & 10 & 20 \\ \hline -2 & 1 & 2 & -5 & -10 & 0 \end{array}$$

-2 is a zero. Testing whether it is a double zero, we examine the reduced polynomial $x^3 + 2x^2 - 5x - 10$.

$$\begin{array}{r|rrrr} & 1 & 2 & -5 & -10 \\ & & -2 & 0 & 10 \\ \hline -2 & 1 & 0 & -5 & 0 \end{array}$$

-2 is a double zero.

So $P(x) = (x + 2)^2(x^2 - 5) = (x + 2)^2(x - \sqrt{5})(x + \sqrt{5})$. So the zeros of the polynomial are -2 (double), $\pm\sqrt{5}$. These are the roots of the equation.

47. Let $P(x) = x^4 - 2x^3 - 5x^2 + 8x + 4$. Possible rational zeros: ±1, ±2, ±4.

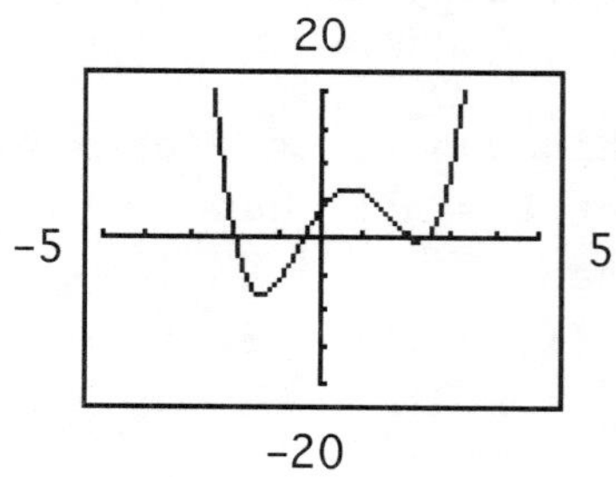

Examining the graph of $y = P(x)$ we see that there is a zero between -1 and 0 and between 2 and 3. The possible integer zeros are -2 and 2, which we test.

$$\begin{array}{r|rrrrr} & 1 & -2 & -5 & 8 & 4 \\ & & 2 & 0 & -10 & -4 \\ \hline 2 & 1 & 0 & -5 & -2 & 0 \end{array}$$

2 is a zero. We examine the reduced polynomial $x^3 - 5x - 2$.

$$\begin{array}{r|rrrr} & 1 & 0 & -5 & -2 \\ & & -2 & 4 & 2 \\ \hline -2 & 1 & -2 & -1 & 0 \end{array}$$

-2 is a zero.

So $P(x) = (x - 2)(x + 2)(x^2 - 2x - 1)$. The zeros of $x^2 - 2x - 1$ are $1 \pm \sqrt{2}$. Hence the zeros of the polynomial are ±2, $1 \pm \sqrt{2}$. These are the roots of the equation.

49. Factoring is the simplest method of solving. Let $u = x^2$.

$$\begin{aligned} x^4 + 10x^2 + 9 &= 0 \\ u^2 + 10u + 9 &= 0 \\ (u + 9)(u + 1) &= 0 \\ u = -9, &\quad u = -1 \\ x^2 = -9, &\quad x^2 = -1 \\ x = 3i, -3i&, i, -i \end{aligned}$$

51. Let $P(x) = 2x^5 - 3x^4 - 2x + 3$. Possible rational zeros: $\pm 1, \pm 3, \pm\frac{1}{2}, \pm\frac{3}{2}$.

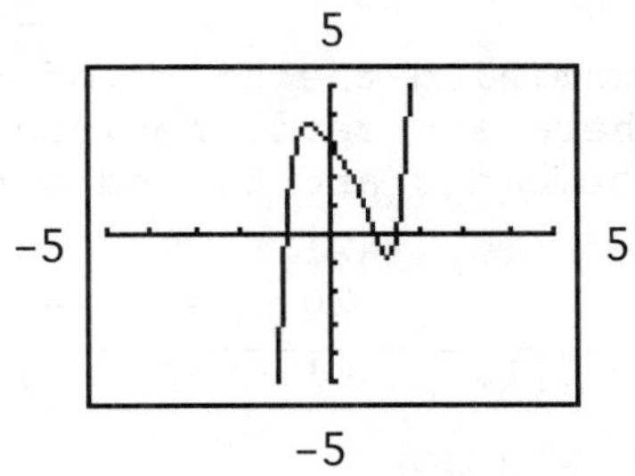

Examining the graph of $y = P(x)$ we see that there is a zero between 1 and 2, and possible zeros at -1 and 1. We test the likely candidates and find

$$\begin{array}{r|rrrrrr} & 2 & -3 & 0 & 0 & -2 & 3 \\ & & 2 & -1 & -1 & -1 & -3 \\ \hline 1 & 2 & -1 & -1 & -1 & -3 & 0 \end{array}$$

1 is a zero. We examine the reduced polynomial and find

$$\begin{array}{r|rrrrr} & 2 & -1 & -1 & -1 & -3 \\ & & -2 & 3 & -2 & 3 \\ \hline -1 & 2 & -3 & 2 & -3 & 0 \end{array}$$

-1 is a zero. We examine the new reduced polynomial and find

$$\begin{array}{r|rrrr} & 2 & -3 & 2 & -3 \\ & & 3 & 0 & 3 \\ \hline \frac{3}{2} & 2 & 0 & 2 & 0 \end{array}$$

$\frac{3}{2}$ is a zero.

So $P(x) = (x - 1)(x + 1)\left(x - \frac{3}{2}\right)(2x^2 + 2)$

$$\begin{aligned} &= (x - 1)(x + 1)\left(x - \frac{3}{2}\right)2(x^2 + 1) \\ &= (x - 1)(x + 1)(2x - 3)(x^2 + 1) \\ &= (x - 1)(x + 1)(2x - 3)(x - i)(x + i) \end{aligned}$$

The zeros are $\pm 1, \frac{3}{2}, \pm i$. These are the roots of the equation.

53. The possible rational zeros are $\pm 1, \pm 2, \pm 3, \pm 5, \pm 6, \pm 10, \pm 15, \pm 30$.

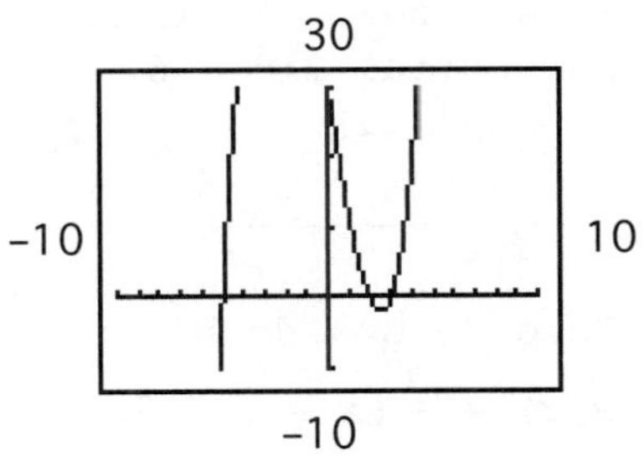

Examining a portion of the graph of $y = P(x)$ we see that there are possible integer zeros at 2, 3, and -5. We test the likely candidates and find

$$\begin{array}{r|rrrr} & 1 & 0 & -19 & 30 \\ & & 2 & 4 & -30 \\ \hline 2 & 1 & 2 & -15 & 0 \end{array}$$

2 is a zero.

So $P(x) = (x - 2)(x^2 + 2x - 15)$

$$= (x - 2)(x - 3)(x + 5)$$

So the zeros of $P(x)$ are 2, 3, -5.

55. $P(x) = x^4 - \frac{21}{10}x^3 + \frac{2}{5}x = \frac{1}{10}(10x^4 - 21x^3 + 4x) = \frac{1}{10}x(10x^3 - 21x^2 + 4)$.

Clearly, 0 is a zero.

We examine $Q(x) = 10x^3 - 21x^2 + 4$.

Possible factors of 4 are ±1, ±2, ±4.

Possible factors of 10 are ±1, ±2, ±5, ±10.

Hence the possible rational zeros of $Q(x)$ are $\pm 1, \pm 2, \pm 4, \pm\frac{1}{2}, \pm\frac{1}{5}, \pm\frac{2}{5}, \pm\frac{4}{5}, \pm\frac{1}{10}$.

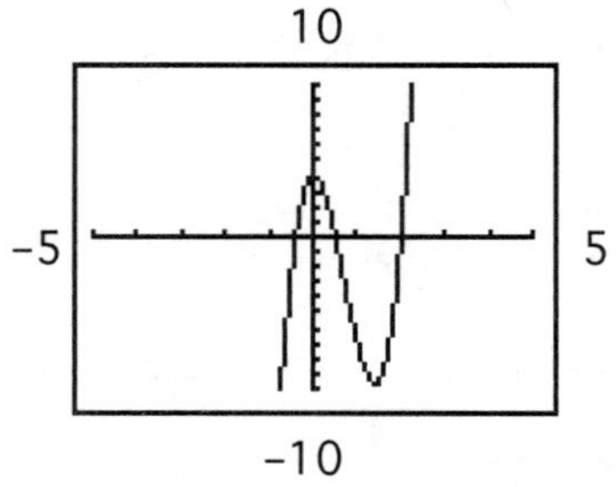

Examining the graph of $y = Q(x)$ we see that there are zeros between −1 and 0, and between 0 and 1, and a possible integer zero at 2. Testing 2, we obtain

	10	−21	0	4
		20	−2	−4
2	10	−1	−2	0

2 is a zero.

So $P(x) = \frac{1}{10}x(x - 2)(10x^2 - x - 2)$

$= \frac{1}{10}x(x - 2)(5x + 2)(2x - 1)$

So the zeros of $P(x)$ are $0, 2, -\frac{2}{5}, \frac{1}{2}$.

57. $P(x) = x^4 - 5x^3 + \frac{15}{2}x^2 - 2x - 2 = \frac{1}{2}(2x^4 - 10x^3 + 15x^2 - 4x - 4)$.

Possible factors of −4 are ±1, ±2, ±4.

Possible factors of 2 are ±1, ±2.

Hence the possible rational zeros are $\pm 1, \pm 2, \pm 4, \pm\frac{1}{2}$.

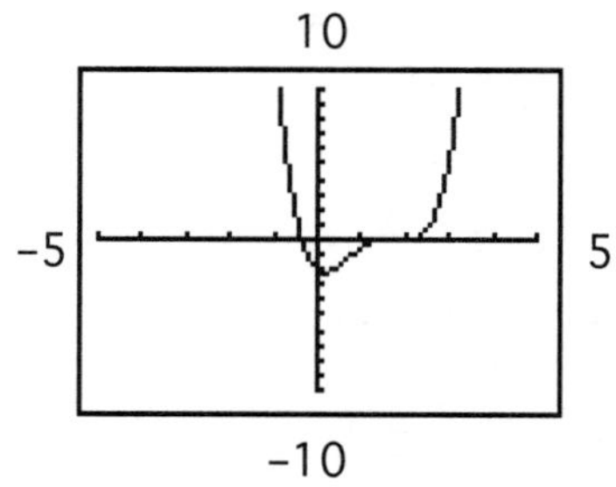

Examining the graph of $y = P(x)$ we see that there are zeros between −1 and 0, and between 1 and 2, and a possible double zero at 2. Testing 2, we obtain

	2	−10	15	−4	−4
		4	−12	6	4
2	2	−6	3	2	0

2 is a zero. Testing whether it is a double zero, we examine the reduced polynomial $2x^3 - 6x^2 + 3x + 2$.

	2	−6	3	2
		4	−4	−2
2	2	−2	−1	0

Thus 2 is a double zero of $P(x)$.

$P(x) = \frac{1}{2}(x - 2)^2(2x^2 - 2x - 1)$

To find the remaining zeros, we solve $2x^2 - 2x - 1 = 0$.

Applying the quadratic formula with $a = 2$, $b = -2$, $c = -1$, we obtain

$$x = \frac{-(-2) \pm \sqrt{(-2)^2 - 4(2)(-1)}}{2(2)} = \frac{2 \pm \sqrt{12}}{4} = \frac{1 \pm \sqrt{3}}{2} \text{ or } \frac{1}{2} \pm \frac{1}{2}\sqrt{3}.$$

So the zeros of $P(x)$ are 2 (double), $\frac{1}{2} \pm \frac{1}{2}\sqrt{3}$.

59. Let $u = x^2$. Then

$$P(x) = x^4 + 11x^2 + 30 = u^2 + 11u + 30 = (u + 6)(u + 5) = (x^2 + 6)(x^2 + 5).$$

The zeros of $x^2 + 6$ are $i\sqrt{6}$ and $-i\sqrt{6}$; the zeros of $x^2 + 5$ are $i\sqrt{5}$ and $-i\sqrt{5}$. So the zeros of $P(x)$ are $\pm i\sqrt{6}$, $\pm i\sqrt{5}$.

61. Possible factors of 5 are ±1, ±5. Possible factors of 3 are ±1, ±3. Hence the possible rational zeros are ±1, ±5, $\pm\frac{1}{3}$, $\pm\frac{5}{3}$.

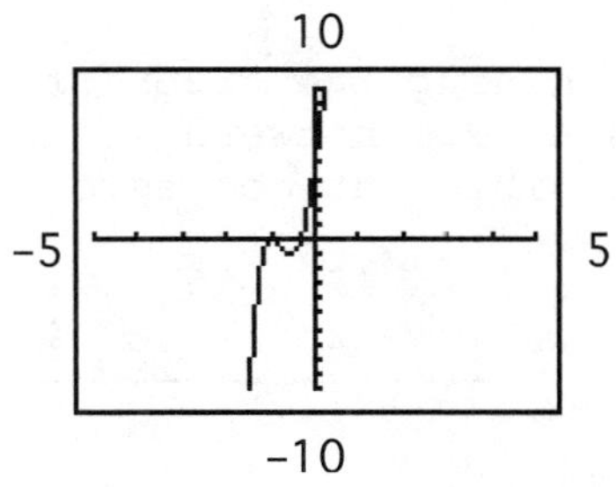

Examining the graph of $y = P(x)$, we see that there is a zero between -1 and 0, and a possible double zero at -1. Testing -1, we obtain

	3	-5	-8	16	21	5
		-3	8	0	-16	-5
-1	3	-8	0	16	5	0

-1 is a zero. Testing whether it is a double zero, we examine the reduced polynomial $3x^4 - 8x^3 + 16x + 5$.

	3	-8	0	16	5
		-3	11	-11	-5
-1	3	-11	11	5	0

Thus -1 is a double zero. The only remaining rational candidate between -1 and 0 is $-\frac{1}{3}$. Testing $-\frac{1}{3}$ in the reduced polynomial $3x^3 - 11x^2 + 11x + 5$, we obtain

	3	-11	11	5
		-1	4	-5
$-\frac{1}{3}$	3	-12	15	0

$-\frac{1}{3}$ is a zero.

So $P(x) = (x + 1)^2\left(x + \frac{1}{3}\right)(3x^2 - 12x + 15) = (x + 1)^2\left(x + \frac{1}{3}\right)3(x^2 - 4x + 5)$

To find the remaining zeros, we solve $x^2 - 4x + 5 = 0$ by completing the square.

$$\begin{aligned} x^2 - 4x &= -5 \\ x^2 - 4x + 4 &= -1 \\ (x - 2)^2 &= -1 \\ x - 2 &= \pm i \\ x &= 2 \pm i \end{aligned}$$

So the zeros of $P(x)$ are -1 (double), $-\frac{1}{3}$, $2 \pm i$.

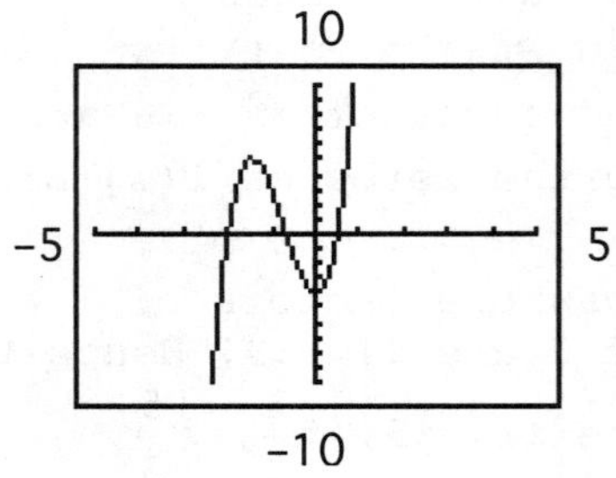

63. The possible rational zeros are ± 1, ± 2, ± 4, $\pm\frac{1}{2}$, $\pm\frac{1}{3}$, $\pm\frac{2}{3}$, $\pm\frac{4}{3}$, $\pm\frac{1}{6}$.

Examining the graph of $y = P(x)$, we see that there is a zero between -1 and 0, between 0 and 1, and a possible integer zero at -2. Testing -2, we obtain

	6	13	0	-4
		-12	-2	4
-2	6	1	-2	0

-2 is a zero.

$P(x) = (x + 2)(6x^2 + x - 2) = (x + 2)(3x + 2)(2x - 1)$

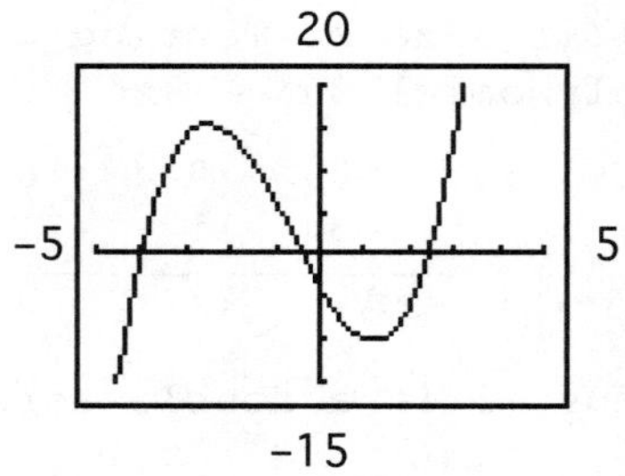

65. The possible rational zeros are ± 1, ± 2, ± 4.

Examining the graph of $y = P(x)$, we see that there is a zero between -1 and 0, between 2 and 3, and a possible integer zero at -4. Testing -4, we obtain

	1	2	-9	-4
		-4	8	4
-4	1	-2	-1	0

-4 is a zero.

So $P(x) = (x + 4)(x^2 - 2x - 1)$. The zeros of $x^2 - 2x - 1$ are $1 \pm \sqrt{2}$.
Hence $P(x) = (x + 4)[x - (1 + \sqrt{2})][x - (1 - \sqrt{2})]$

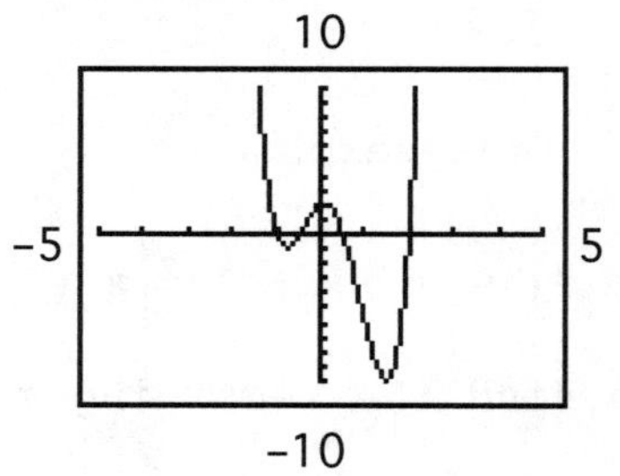

67. The possible rational zeros are ± 1, ± 2, $\pm\frac{1}{2}$, $\pm\frac{1}{4}$.
Examining the graph of $y = P(x)$, we see that there is a zero between -1 and 0, between 0 and 1, and possible integer zeros at -1 and 2. Testing 2, we obtain

	4	-4	-9	1	2
		8	8	-2	-2
2	4	4	-1	-1	0

2 is a zero. Testing -1, we examine the reduced polynomial $4x^3 + 4x^2 - x - 1$.

	4	4	-1	-1
		-4	0	1
-1	4	0	-1	0

-1 is a zero.

So $P(x) = (x - 2)(x + 1)(4x^2 - 1) = (x - 2)(x + 1)(2x - 1)(2x + 1)$.

69. If $2 - 5i$ is a zero, $2 + 5i$ is a zero.

$$\begin{aligned} P(x) &= [x - (2 - 5i)][x - (2 + 5i)] \\ &= [x - 2 + 5i][x - 2 - 5i] \\ &= [(x - 2) + 5i][(x - 2) - 5i] \\ &= (x - 2)^2 - (5i)^2 \\ &= x^2 - 4x + 4 + 25 \\ &= x^2 - 4x + 29 \end{aligned}$$

71. If $6 + i$ is a zero, $6 - i$ is a zero.

$$P(x) = a[x - (6 + i)][x - (6 - i)]$$
$$= a[x - 6 - i][x - 6 + i]$$
$$= a[(x - 6) - i][(x - 6) + i]$$
$$= a[(x - 6)^2 - i^2]$$
$$= a(x^2 - 12x + 36 + 1)$$
$$P(x) = a(x^2 - 12x + 37)$$
$$P(x) = ax^2 - 12ax + 37a.$$

Since $P(0) = 74 = 37a$, $a = 2$. Hence

$$P(x) = 2x^2 - 24x + 74$$

73. If $8i$ is a zero, $-8i$ is a zero.

$$P(x) = (x + 5)(x - 8i)(x + 8i)$$
$$= (x + 5)(x^2 + 64)$$
$$= x^3 + 64x + 5x^2 + 320$$
$$P(x) = x^3 + 5x^2 + 64x + 320$$

75. If i is a zero, $-i$ is a zero. If $1 - i$ is a zero, $1 + i$ is a zero.

$$P(x) = a(x - i)(x + i)[x - (1 - i)][x - (1 + i)]$$
$$= a(x^2 + i)[x - 1 + i][x - 1 - i]$$
$$= a(x^2 + 1)[(x - 1)^2 - i^2]$$
$$= a(x^2 + 1)[x^2 - 2x + 1 + 1]$$
$$= a(x^2 + 1)(x^2 - 2x + 2)$$
$$= a(x^4 - 2x^3 + 2x^2 + x^2 - 2x + 2)$$
$$= a(x^4 - 2x^3 + 3x^2 - 2x + 2)$$

Since $P(1) = 10 = a(1^4 - 2\cdot1^3 + 3\cdot1^2 - 2\cdot1 + 2)$, $10 = 2a$, $5 = a$. Hence

$$P(x) = 5(x^4 - 2x^3 + 3x^2 - 2x + 2)$$
$$= 5x^4 - 10x^3 + 15x^2 - 10x + 10$$

77. $[x - (4 - 5i)][x - (4 + 5i)] = [x - 4 + 5i][x - 4 - 5i]$

$$= [(x - 4) + 5i][(x - 4) - 5i]$$
$$= (x - 4)^2 - 25i^2$$
$$= x^2 - 8x + 16 + 25$$
$$= x^2 - 8x + 41$$

79. $[x - (3 + 4i)][x - (3 - 4i)] = [x - 3 - 4i][x - 3 + 4i]$

$$= [(x - 3) - 4i][(x - 3) + 4i]$$
$$= (x - 3)^2 - 16i^2$$
$$= x^2 - 6x + 9 + 16$$
$$= x^2 - 6x + 25$$

81. $[x - (a + bi)][x - (a - bi)] = [x - a - bi][x - a + bi]$

$$= [(x - a) - bi][(x - a) + bi]$$
$$= (x - a)^2 - b^2i^2$$
$$= (x - a)^2 + b^2$$
$$= x^2 - 2ax + a^2 + b^2$$

83. If $3 - i$ is a zero, then $3 + i$ is a zero. So $Q(x) = [x - (3 - i)][x - (3 + i)]$ divides $P(x)$ evenly. Applying problem 81, $Q(x) = x^2 - 6x + 9 + 1 = x^2 - 6x + 10$. Dividing, we see

$$\begin{array}{r|l} & x \quad + 1 \\ \hline x^2 - 6x + 10 & x^3 - 5x^2 + 4x + 10 \\ & \underline{x^3 - 6x^2 + 10x} \\ & \qquad x^2 - 6x + 10 \\ & \qquad \underline{x^2 - 6x + 10} \\ & \qquad\qquad\qquad 0 \end{array}$$

So $P(x) = (x + 1)Q(x)$ and the two other zeros are -1 and $3 + i$.

85. If $-5i$ is a zero, then so is $5i$. So $Q(x) = (x - 5i)(x + 5i)$ divides $P(x)$ evenly. $Q(x) = x^2 - 25i^2 = x^2 + 25$. Dividing, we see

$$\begin{array}{r|l} & x - 3 \\ \hline x^2 + 25 & x^3 - 3x^2 + 25x - 75 \\ & \underline{x^3 \qquad + 25x} \\ & \quad - 3x^2 \qquad - 75 \\ & \quad \underline{- 3x^2 \qquad - 75} \\ & \qquad\qquad\qquad 0 \end{array}$$

So $P(x) = (x - 3)Q(x)$ and the two other zeros are $5i$ and 3.

87. If $2 + i$ is a zero, then $2 - i$ is a zero. So $Q(x) = [x - (2 + i)][x - (2 - i)]$ divides $P(x)$ evenly. Applying problem 81, $Q(x) = x^2 - 4x + 4 + 1 = x^2 - 4x + 5$. Dividing, we see

$$\begin{array}{r|l} & x^2 - 2 \\ \hline x^2 - 4x + 5 & x^4 - 4x^3 + 3x^2 + 8x - 10 \\ & \underline{x^4 - 4x^3 + 5x^2} \\ & \qquad\quad - 2x^2 + 8x - 10 \\ & \qquad\quad \underline{- 2x^2 + 8x - 10} \\ & \qquad\qquad\qquad\qquad 0 \end{array}$$

So $P(x) = Q(x)(x^2 - 2)$. $x^2 - 2$ has two zeros: $\sqrt{2}$ and $-\sqrt{2}$. Summarizing, $P(x)$ has 4 zeros: $2 + i$, $2 - i$, $\sqrt{2}$, $-\sqrt{2}$.

89. $\sqrt{6}$ is a root of $x^2 = 6$ or $x^2 - 6 = 0$. The possible rational roots of this equation are ± 1, ± 2, ± 3, ± 6. Since none of them satisfies $x^2 = 6$, there are no rational roots, hence $\sqrt{6}$ is not rational.

91. $\sqrt[3]{5}$ is a root of $x^3 = 5$ or $x^3 - 5 = 0$. The possible rational roots of this equation are ± 1, ± 5. Since none of them satisfies $x^3 = 5$, there are no rational roots, hence $\sqrt[3]{5}$ is not rational.

93. The possible rational zeros are ± 1, ± 2, ± 3, ± 6. Examining the graph, we can see that there are two real zeros, one between -3 and -2, and the other between 2 and 3. Since none of the possible rational zeros actually are zeros, there must not be any rational zeros.

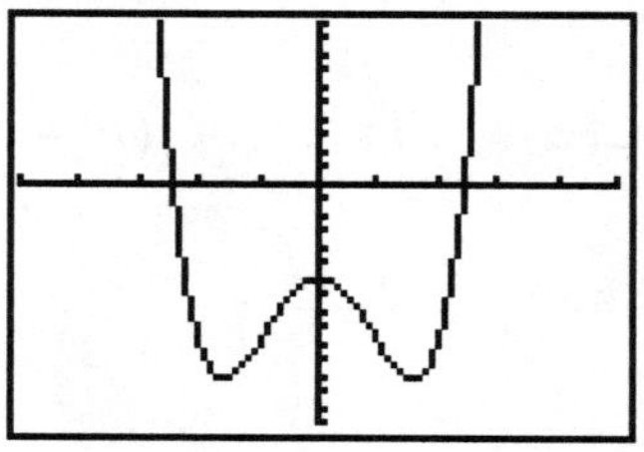

95. The only possible rational zeros are ±1, but $P(1) = -3$ and $P(-1) = 1$, so there are no rational zeros. A look at the graph in a standard window shows 3 real zeros which is the maximum number possible for a third degree polynomial.

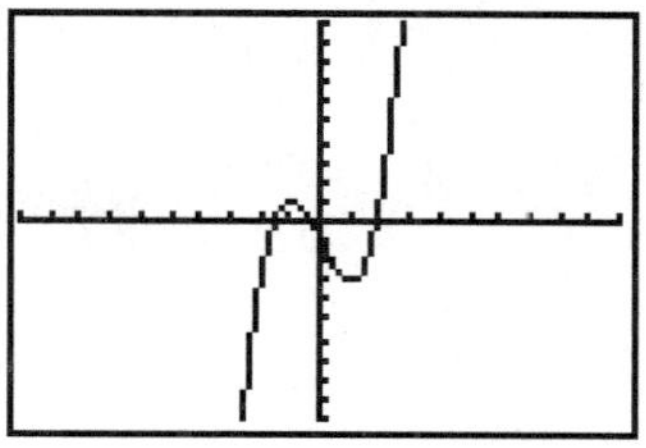

97. Here is a graph of $P(x) = 3x^3 - 37x^2 + 84x - 24$.

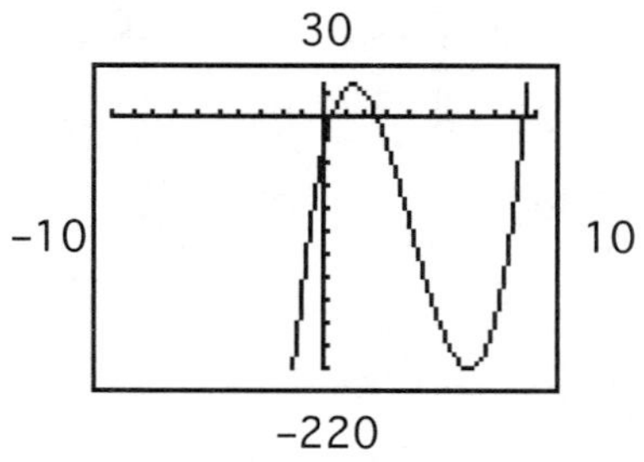

The rational zero theorem gives ±1, ±2, ±3, ±4, ±6, ±8, ±12, ±24, $\pm\frac{1}{3}$, $\pm\frac{2}{3}$, $\pm\frac{4}{3}$, $\pm\frac{8}{3}$ as possible rational zeros. However, the graph suggests that $P(x)$ has no integer zeros; there are no negative zeros, and the positive zeros would appear to lie between 0 and 1, 2 and 3, and 9 and 10. This suggests that we consider only the possible zeros $\frac{1}{3}$, $\frac{2}{3}$, and $\frac{8}{3}$. Testing $\frac{1}{3}$, we obtain

	3	-37	84	-24
		1	-12	24
$\frac{1}{3}$	3	-36	72	0

$\frac{1}{3}$ is a zero.

So $P(x) = \left(x - \frac{1}{3}\right)(3x^2 - 36x + 72) = \left(x - \frac{1}{3}\right)3(x^2 - 12x + 24)$

To find the remaining zeros, we solve $x^2 - 12x + 24 = 0$ by completing the square:

$$\begin{aligned} x^2 - 12x &= -24 \\ x^2 - 12x + 36 &= 12 \\ (x - 6)^2 &= 12 \\ x - 6 &= \pm\sqrt{12} \\ x &= 6 \pm \sqrt{12} \text{ or } 6 \pm 2\sqrt{3} \end{aligned}$$

So the zeros of $P(x)$ are $\frac{1}{3}$, $6 \pm 2\sqrt{3}$.

99. Here is a graph of $P(x) = 4x^4 + 4x^3 + 49x^2 + 64x - 240$.

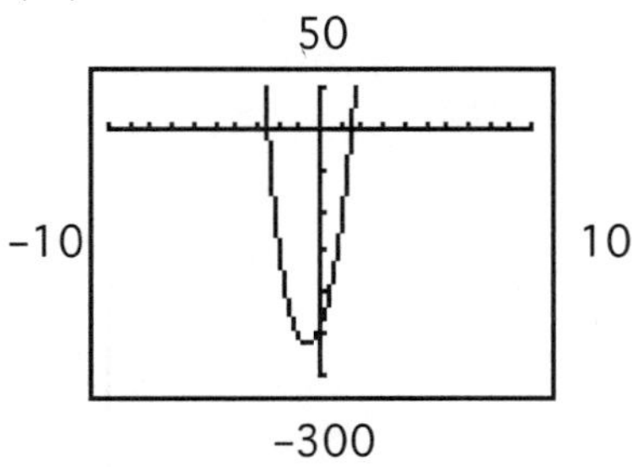

The rational zero theorem gives ±1, ±2, ±3, ±4, ±5, ±6, ±8, ±10, ±12, ±15, ±16, ±20, ±24, ±30, ±40, ±48, ±60, ±80, ±120, ±240, $\pm\frac{1}{2}$, $\pm\frac{3}{2}$, $\pm\frac{5}{2}$, $\pm\frac{15}{2}$, $\pm\frac{1}{4}$, $\pm\frac{3}{4}$, $\pm\frac{5}{4}$, $\pm\frac{15}{4}$ as possible rational zeros. However, the graph suggests that $P(x)$ has no integer zeros, and the zeros appear to lie between -3 and -2, and 1 and 2.

This suggests that we consider only the possible zeros $\frac{3}{2}$, $\frac{5}{4}$, and $-\frac{5}{2}$.

Testing $\frac{3}{2}$, we obtain

	4	4	49	64	-240
		6	15	96	240
$\frac{3}{2}$	4	10	64	160	0

$\frac{3}{2}$ is a zero.

So $P(x) = \left(x - \frac{3}{2}\right)(4x^3 + 10x^2 + 64x + 160) = \left(x - \frac{3}{2}\right)2(2x^3 + 5x^2 + 32x + 80)$

We consider the reduced polynomial $Q(x) = 2x^3 + 5x^2 + 32x + 80$. The graph suggests that the other real zero is negative, so we try $-\frac{5}{2}$ next.

	2	5	32	80
		-5	0	-80
$-\frac{5}{2}$	2	0	32	0

$-\frac{5}{2}$ is a zero.

So $P(x) = \left(x - \frac{3}{2}\right)2\left(x + \frac{5}{2}\right)(2x^2 + 32) = 4\left(x - \frac{3}{2}\right)\left(x + \frac{5}{2}\right)(x^2 + 16)$.

The remaining zeros of $P(x)$ are the zeros of $x^2 + 16$, that is, $\pm 4i$. So the zeros of $P(x)$ are $\frac{3}{2}$, $-\frac{5}{2}$, $\pm 4i$.

101. Here is a graph of $P(x) = 4x^4 - 44x^3 + 145x^2 - 192x + 90$. The rational zero theorem gives ± 1, ± 2, ± 3, ± 5, ± 6, ± 9, ± 10, ± 15, ± 18, ± 30, ± 45, ± 90, $\pm\frac{1}{2}$, $\pm\frac{3}{2}$, $\pm\frac{5}{2}$, $\pm\frac{9}{2}$, $\pm\frac{15}{2}$, $\pm\frac{45}{2}$, $\pm\frac{1}{4}$, $\pm\frac{3}{4}$, $\pm\frac{5}{4}$, $\pm\frac{9}{4}$, $\pm\frac{15}{4}$, $\pm\frac{45}{4}$ as possible rational zeros. However, the graph suggests that the only possible integer zeros are 1 and 2; there are no negative zeros, and the zeros appear to lie between 0 and 3, and between 6 and 7. This suggests that we consider only the possible zeros 1, 2, $\frac{1}{2}$, $\frac{3}{2}$, $\frac{5}{2}$, $\frac{1}{4}$, $\frac{3}{4}$, $\frac{5}{4}$, $\frac{9}{4}$. Testing the possible candidates, we obtain $P(1) = 3$, $P(2) = -2$, and

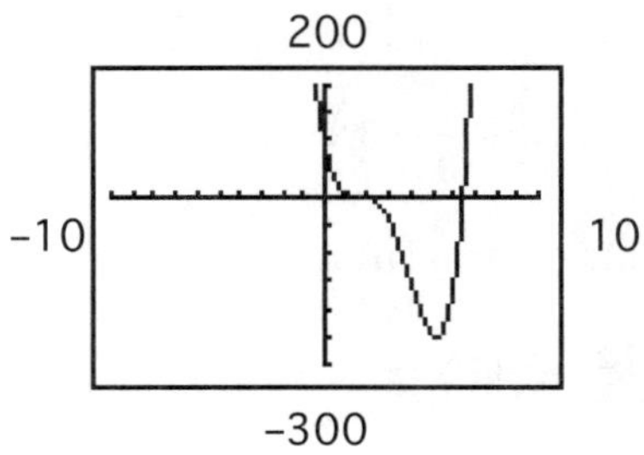

	4	-44	145	-192	90
		6	-57	132	-90
$\frac{3}{2}$	4	-38	88	-60	0

$\frac{3}{2}$ is a zero.

So $P(x) = \left(x - \frac{3}{2}\right)(4x^3 - 38x^2 + 88x - 60) = \left(x - \frac{3}{2}\right)2(2x^3 - 19x^2 + 44x - 30)$

We consider the reduced polynomial $Q(x) = 2x^3 - 19x^2 + 44x - 30$. The remaining possibilities from the reduced list are $\frac{1}{2}$, $\frac{3}{2}$, $\frac{5}{2}$. Testing for a possible double zero at $\frac{3}{2}$, we obtain

$$\begin{array}{r|rrrr} & 2 & -19 & 44 & -30 \\ & & 3 & -24 & 30 \\ \hline \frac{3}{2} & 2 & -16 & 20 & 0 \end{array}$$

$\frac{3}{2}$ is a double zero.

So $P(x) = \left(x - \frac{3}{2}\right)^2 2(2x^2 - 16x + 20) = \left(x - \frac{3}{2}\right)^2 4(x^2 - 8x + 10)$

To find the remaining zeros, we solve $x^2 - 8x + 10 = 0$, by completing the square.

$$\begin{aligned} x^2 - 8x &= -10 \\ x^2 - 8x + 16 &= 6 \\ (x - 4)^2 &= 6 \\ x - 4 &= \pm\sqrt{6} \\ x &= 4 \pm \sqrt{6} \end{aligned}$$

So the zeros of $P(x)$ are $\frac{3}{2}$ (double), $4 \pm \sqrt{6}$.

103. (A) Since 1 is a root, let $P(x) = x^3 - 1$ and write

$$\begin{array}{r|rrrr} & 1 & 0 & 0 & -1 \\ & & 1 & 1 & 1 \\ \hline 1 & 1 & 1 & 1 & 0 \end{array}$$

$P(x) = (x - 1)(x^2 + x + 1)$

The other cube roots of 1 will be solutions to $x^2 + x + 1 = 0$.

Applying the quadratic formula with $a = b = c = 1$, we have

$x = \frac{-1 \pm \sqrt{(1)^2 - 4(1)(1)}}{2(1)} = \frac{-1 \pm \sqrt{-3}}{2} = \frac{-1 \pm i\sqrt{3}}{2}$. Thus, $-\frac{1}{2} + \frac{\sqrt{3}}{2}i$ and $-\frac{1}{2} - \frac{\sqrt{3}}{2}i$ are the other cube roots of 1.

(B) Thus, there are 3 distinct cube roots of 1.

105. $P(x)$ can have at most n and must have at least one real zero. Each zero of $P(x)$ represents a point where $P(x) = y = 0$ so the graph of $P(x)$ will cross the x axis at and only at zeros of $P(x)$. Thus there can be a maximum of n axis crossings and there is a minimum of 1 axis crossing.

107.
$$\begin{aligned} P(2 + i) &= (2 + i)^2 + 2i(2 + i) - 5 \\ &= 4 + 4i + i^2 + 4i + 2i^2 - 5 \\ &= 4 + 4i - 1 + 4i - 2 - 5 \\ &= -4 + 8i \end{aligned}$$

So $P(2 + i) \neq 0$ and $2 + i$ is not a zero of $P(x)$. This does not contradict the theorem, since $P(x)$ is not a polynomial with real coefficients (the coefficient of x is the imaginary number $2i$).

109. Let x = the amount of increase.

Then old volume $= 1 \times 2 \times 3 = 6$

new volume $= (x + 1)(x + 2)(x + 3) = x^3 + 6x^2 + 11x + 6$

Since (new volume) = 10 (old volume), we must solve

$$x^3 + 6x^2 + 11x + 6 = 10(6)$$
$$x^3 + 6x^2 + 11x + 6 = 60$$
$$P(x) = x^3 + 6x^2 + 11x - 54 = 0$$

The possible rational zeros are ±1, ±2, ±3, ±6, ±9, ±18, ±27, ±54. Testing in order, we obtain $P(1) = -36$ $P(2) = 0$

	1	6	11	-54
		2	16	54
2	1	8	27	0

Thus, $x^3 + 6x^2 + 11x - 54 = (x - 2)(x^2 + 8x + 27)$. Since $x^2 + 8x + 27 = 0$ has no positive zeros, 2 is the only positive solution of the equation. The increase must equal 2 feet.

111.

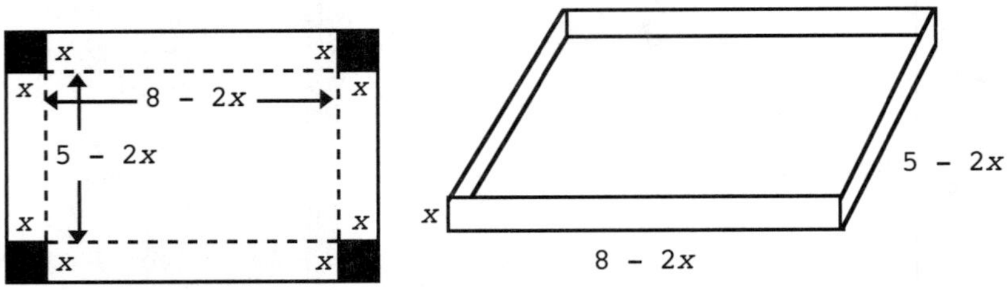

From the figure, it should be clear that
Volume $= x(5 - 2x)(8 - 2x) = 14$

Since x, $5 - 2x$, and $8 - 2x$ must all be positive, the domain of x is $0 < x < \frac{5}{2}$ or (0, 2.5). We solve $x(5 - 2x)(8 - 2x) = 14$, or $4x^3 - 26x^2 + 40x = 14$, for x in this domain.

$$4x^3 - 26x^2 + 40x = 14$$
$$4x^3 - 26x^2 + 40x - 14 = 0$$
$$P(x) = 2x^3 - 13x^2 + 20x - 7 = 0$$

Possible rational zeros: ±1, ±7, $\pm\frac{1}{2}$, $\pm\frac{7}{2}$. Testing the possibilities, we obtain

$P(1) = 2$ $P(7) = 182$ $P\left(\frac{1}{2}\right) = 0$

	2	-13	20	-7
		1	-6	7
$\frac{1}{2}$	2	-12	14	0

$\frac{1}{2}$ is a zero.

So $2x^3 - 13x^2 + 20x - 7 = \left(x - \frac{1}{2}\right)(2x^2 - 12x + 14) = (2x - 1)(x^2 - 6x + 7)$

To find the remaining zeros, we solve $x^2 - 6x + 7 = 0$, by completing the square:

$$\begin{aligned} x^2 - 6x &= -7 \\ x^2 - 6x + 9 &= 2 \\ (x - 3)^2 &= 2 \\ x - 3 &= \pm\sqrt{2} \\ x &= 3 \pm \sqrt{2} \end{aligned}$$

Hence the zeros are $\frac{1}{2}$, $3 - \sqrt{2}$, $3 + \sqrt{2}$, or 0.5, 1.6, 4.4 to two significant digits. We discard $3 + \sqrt{2}$ or 4.4, since it is not in the interval (0, 2.5). The square should be 0.5 × 0.5 inches or 1.6 × 1.6 inches.

Section 4-4

1. This graph has a vertical asymptote $x = 2$, and a horizontal asymptote $y = -2$. This corresponds to $g(x)$.

3. This graph has a vertical asymptote $x = 2$, and a horizontal asymptote $y = 2$. This corresponds to $h(x)$.

5. (A) ∞ (B) $-\infty$ (C) 2 (D) 2

7. (A) $-\infty$ (B) ∞ (C) 2 (D) 2

9. $\frac{2x - 4}{x + 1}$ *Domain:* $d(x) = x + 1$ zero: $x = -1$ domain: $(-\infty, -1) \cup (-1, \infty)$

   x intercepts: $n(x) = 2x - 4$ zero: $x = 2$ x intercept: 2

11. $\frac{x^2 - 1}{x^2 - 16}$ *Domain:* $d(x) = x^2 - 16$ zeros: $x^2 - 16 = 0$

$$\begin{aligned} x^2 &= 16 \\ x &= \pm 4 \end{aligned}$$

   domain: $(-\infty, -4) \cup (-4, 4) \cup (4, \infty)$

   x intercepts: $n(x) = x^2 - 1$ zeros: $x^2 - 1 = 0$

$$\begin{aligned} x^2 &= 1 \\ x &= \pm 1 \end{aligned}$$

   x intercepts: -1, 1

13. $\frac{x^2 - x - 6}{x^2 - x - 12}$ *Domain:* $d(x) = x^2 - x - 12$ zeros: $x^2 - x - 12 = 0$

$$\begin{aligned} (x + 3)(x - 4) &= 0 \\ x &= -3, 4 \end{aligned}$$

   domain: $(-\infty, -3) \cup (-3, 4) \cup (4, \infty)$

   x intercepts: $n(x) = x^2 - x - 6$ zeros: $x^2 - x - 6 = 0$

$$\begin{aligned} (x + 2)(x - 3) &= 0 \\ x &= -2, 3 \end{aligned}$$

   x intercepts: -2, 3

15. $\frac{x}{x^2 + 4}$ *Domain:* $d(x) = x^2 + 4$ no real zeros

   Domain: all real numbers

   x intercepts: $n(x) = x$ zero: $x = 0$

   x intercept: 0

17. $\dfrac{4 - x^2}{x}$ *Domain:* $d(x) = x$ zero: 0

Domain: $(-\infty, 0) \cup (0, \infty)$

x intercepts: $n(x) = 4 - x^2$ zeros: $4 - x^2 = 0$

$(2 + x)(2 - x) = 0$

$x = -2, 2$

x intercepts: $-2, 2$

19. $\dfrac{2x}{x - 4}$ *vertical asymptotes:* $d(x) = x - 4$ zero: $x = 4$

vertical asymptote $x = 4$

> **Common Error:** $x = 0$ is not a vertical asymptote.

horizontal asymptotes: Since $n(x)$ and $d(x)$ have the same degree, the line $y = 2$ is a horizontal asymptote.

21. $\dfrac{2x^2 + 3x}{3x^2 - 48}$ *vertical asymptotes:* $d(x) = 3x^2 - 48$ zeros: $3x^2 - 48 = 0$

$3(x + 4)(x - 4) = 0$

$x = -4, 4$

vertical asymptotes: $x = -4$, $x = 4$

horizontal asymptotes: Since $n(x)$ and $d(x)$ have the same degree, the line $y = \dfrac{2}{3}$ is a horizontal asymptote.

23. $\dfrac{2x}{x^4 + 1}$ *vertical asymptote:* $d(x) = x^4 + 1$ No real zeros:

No vertical asymptotes

horizontal asymptotes: Since the degree of $n(x)$ is less than the degree of $d(x)$, the x axis is a horizontal asymptote
horizontal asymptote: $y = 0$

25. $\dfrac{6x^4}{3x^2 - 2x - 5}$ *vertical asymptotes:* $d(x) = 3x^2 - 2x - 5$ zeros: $3x^2 - 2x - 5 = 0$

$(3x - 5)(x + 1) = 0$

$x = \dfrac{5}{3}, -$

vertical asymptotes: $x = -1$, $x = \dfrac{5}{3}$

horizontal asymptotes: Since the degree of $n(x)$ is greater than the degree of $d(x)$, there are no horizontal asymptotes.

27. $\dfrac{4x^2 + 4x - 24}{x^2 - 2x} = \dfrac{4(x^2 + x - 6)}{x(x - 2)} = \dfrac{4(x + 3)(x - 2)}{x(x - 2)} = \dfrac{4(x + 3)}{x}$, $x \neq 2$

vertical asymptote: $x = 0$

Horizontal asymptotes: Since $n(x)$ and $d(x)$ have the same degree, the horizontal asymptote is $y = 4$.

> **Common error:** $x = 2$ is not an asymptote. The graph has a hole at (2, 10) and the function is not defined at $x = 2$, however, the function does not exhibit asymptotic behavior near $x = 2$.

29. The graph has more than one horizontal asymptote.

31. The graph has a sharp corner at (0, 0).

33. $\dfrac{x^2 + 2x}{x} = \dfrac{x(x + 2)}{x} = x + 2$ if $x \neq 0$. The graph of f is the same as the graph of g except that f has a hole at (0, 2).

35. $\frac{x+2}{x^2+10x+16} = \frac{x+2}{(x+2)(x+8)} = \frac{1}{x+8}$ if $x \neq -2$. The graph of f is the same as the graph of g except that f has a hole at $\left(-2, \frac{1}{6}\right)$.

37. $f(x) = \frac{1}{x-4} = \frac{n(x)}{d(x)}$

Intercepts. There are no real zeros of $n(x) = 1$. No x intercept

$f(0) = -\frac{1}{4}$ $\quad y = -\frac{1}{4}$ $\quad y$ intercept

Vertical asymptotes. $d(x) = x - 4$ zeros: 4 $\quad x = 4$

Horizontal asymptotes. Since the degree of $n(x)$ is less that the degree of $d(x)$, the x axis is a horizontal asymptote.

Complete the sketch. Plot a few points.

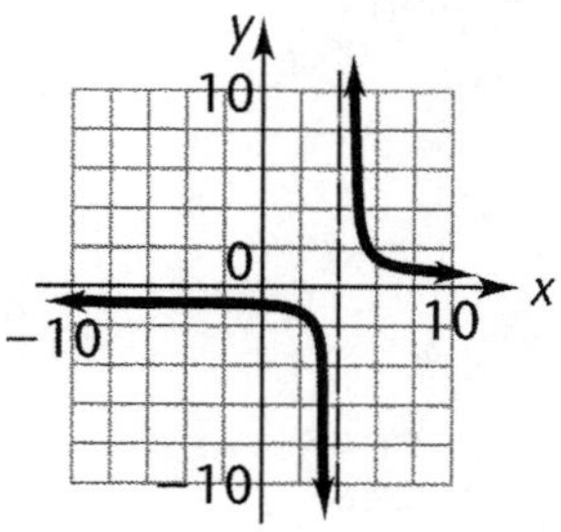

x	$f(x)$
4.5	2
5	1
6	.5
3.5	−2
3	−1
2	−.5

Common Error:
Entering $1/x - 4$ in the calculator.
Parentheses are needed: $1/(x - 4)$.

39. $f(x) = \frac{x}{x+1} = \frac{n(x)}{d(x)}$

Intercepts. Real zeros of $n(x) = x$ $\quad x = 0$ $\quad x$ intercept

$f(0) = 0$ $\quad y = 0$ $\quad y$ intercept

The graph crosses the coordinate axes only at the origin.

Vertical asymptotes. $d(x) = x + 1$ zeros: −1 $\quad x = -1$

Horizontal asymptotes. Since $n(x)$ and $d(x)$ have the same degree, the line $y = 1$ is a horizontal asymptote.

Complete the sketch. Plot a few additional points.

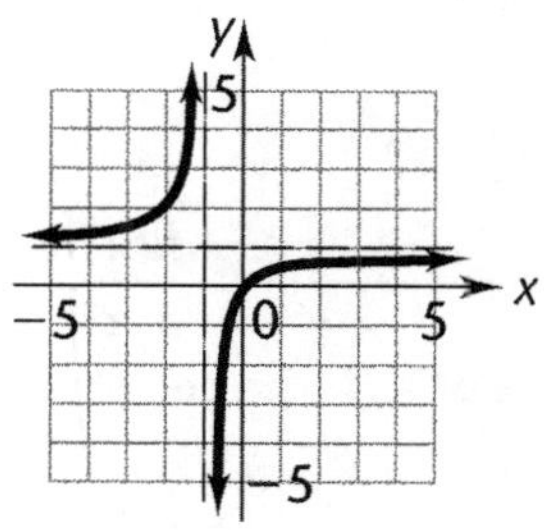

41. $h(x) = \frac{x}{2x-2} = \frac{n(x)}{d(x)}$

Intercepts. Real zeros of $n(x) = x$ $\quad x = 0$ $\quad x$ intercept

$h(0) = 0$ $\quad y = 0$ $\quad y$ intercept

The graph crosses the coordinate axes only at the origin.

Vertical asymptotes. $d(x) = 2x - 2$ zeros: 1 $\quad x = 1$

Horizontal asymptotes. Since $n(x)$ and $d(x)$ have the same degree, the line $y = \frac{1}{2}$ is a horizontal asymptote.

Complete the sketch. Plot a few additional points.

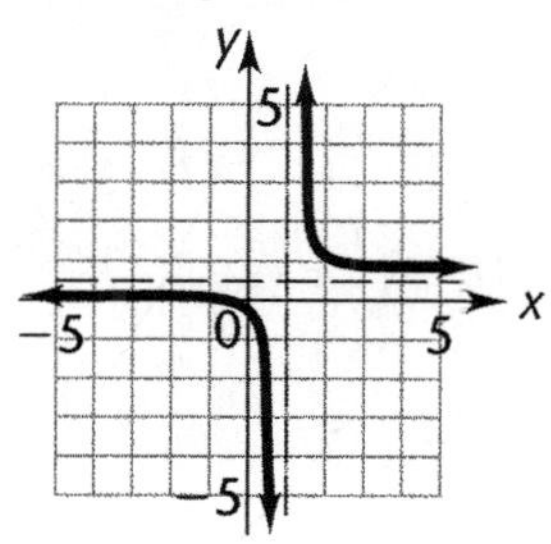

43. $f(x) = \dfrac{2x - 4}{x + 3} = \dfrac{n(x)}{d(x)}$

Intercepts. Real zeros of $n(x) = 2x - 4$ $x = 2$ x intercept

$f(0) = -\frac{4}{3}$ y intercept

Vertical asymptotes. $d(x) = x + 3$ zeros: -3 $x = -3$

Horizontal asymptotes. Since $n(x)$ and $d(x)$ have the same degree, the line $y = 2$ is a horizontal asymptote.

Complete the sketch. Plot a few additional points.

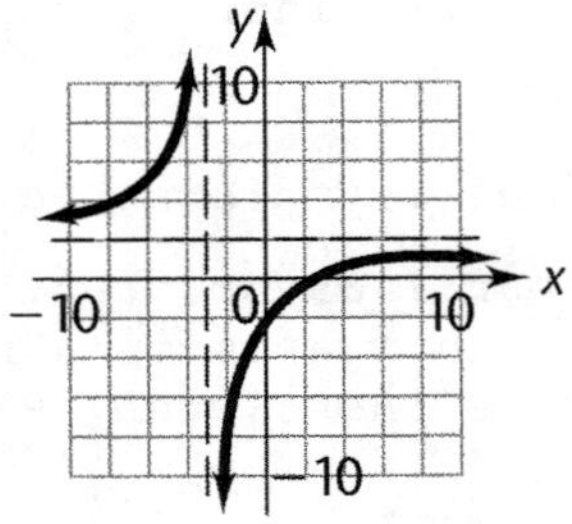

45. $g(x) = \dfrac{1 - x^2}{x^2} = \dfrac{n(x)}{d(x)}$

Intercepts. Real zeros of $n(x) = 1 - x^2$ $1 - x^2 = 0$

$x^2 = 1$

$x = \pm 1$ x intercepts

$g(0)$ is not defined no y intercepts

Vertical asymptotes. $d(x) = x^2$ zeros: 0 $x = 0$

Horizontal asymptotes. Since $n(x)$ and $d(x)$ have the same degree, the line $y = -1$ is a horizontal asymptote.

Complete the sketch. Plot a few additional points.

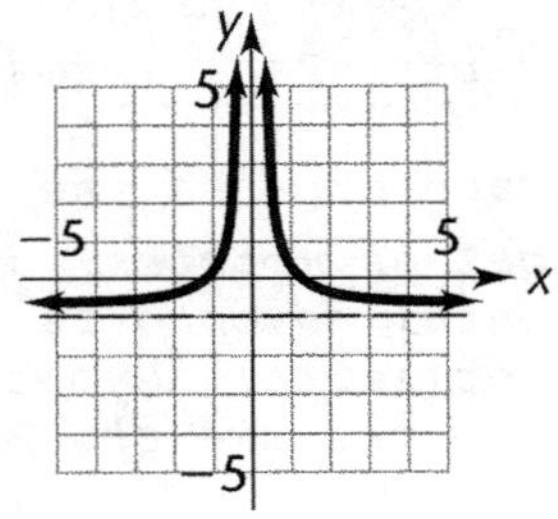

47. $f(x) = \dfrac{9}{x^2 - 9} = \dfrac{n(x)}{d(x)}$

Intercepts. There are no real zeros of $n(x) = 9$. No x intercept

$f(0) = -1$ y intercept

Vertical asymptotes. $d(x) = x^2 - 9$ zeros: $x^2 - 9 = 0$

$x^2 = 9$

$x = \pm 3$

Horizontal asymptotes. Since the degree of $n(x)$ is less than the degree of $d(x)$, the x axis is a horizontal asymptote.

Complete the sketch. Plot a few additional points.

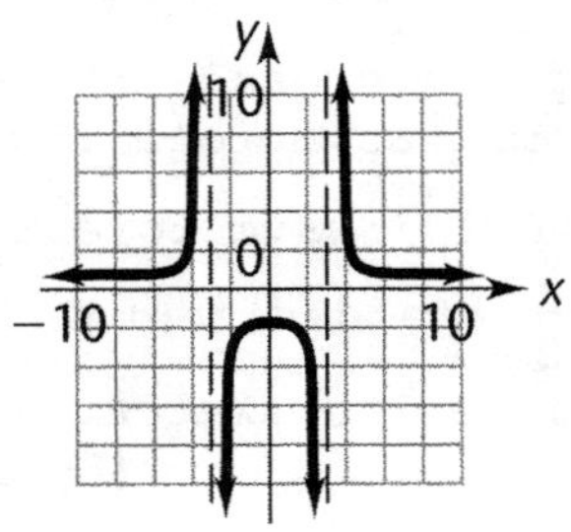

49. $f(x) = \dfrac{x}{x^2 - 1} = \dfrac{n(x)}{d(x)}$

Intercepts. Real zeros of $n(x) = x$ $\quad x = 0$ $\quad x$ intercept

$f(0) = 0$ $\quad y = 0$ $\quad y$ intercept

The graph crosses the coordinate axes only at the origin.

Vertical asymptotes. $d(x) = x^2 - 1$ zeros: $x^2 - 1 = 0$

$$x^2 = 1$$
$$x = \pm 1$$

Horizontal asymptotes. Since the degree of $n(x)$ is less than the degree of $d(x)$, the x axis is a horizontal asymptote.

Complete the sketch. Plot a few additional points.

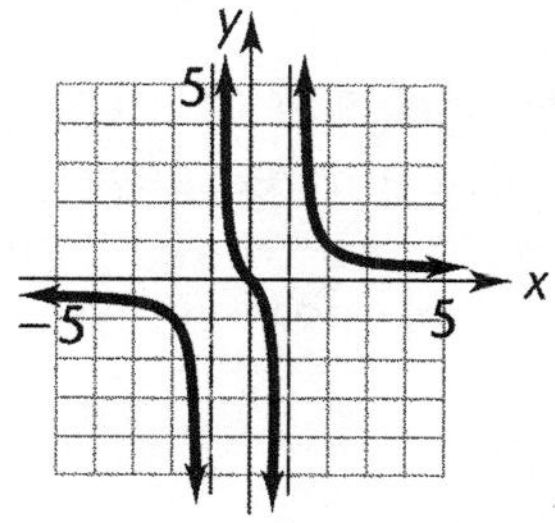

51. $g(x) = \dfrac{2}{x^2 + 1} = \dfrac{n(x)}{d(x)}$

Intercepts. There are no real zeros of $n(x) = 2$. $\quad$ No x intercept

$g(0) = 2$ $\quad y$ intercept

Vertical asymptotes. There are no real zeros of $d(x) = x^2 + 1$

No vertical asymptotes

Horizontal asymptotes. Since the degree of $n(x)$ is less than the degree of $d(x)$, the x axis is a horizontal asymptote.

Complete the sketch. Plot a few additional points.

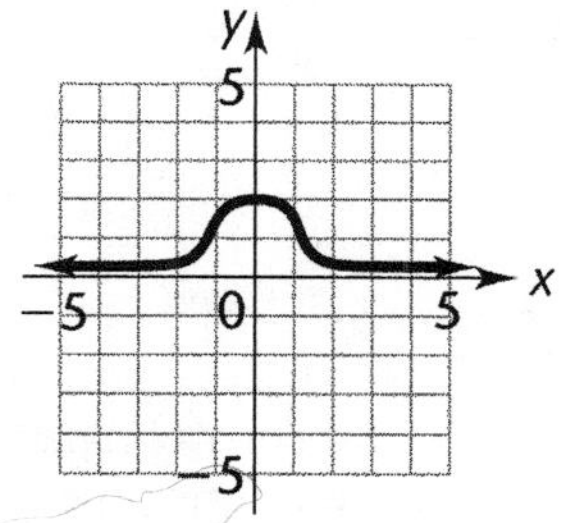

53. $f(x) = \dfrac{12x^2}{(3x + 5)^2} = \dfrac{n(x)}{d(x)}$

Intercepts. Real zeros of $n(x) = 12x^2$ $\quad x = 0$ $\quad x$ intercept

$f(0) = 0$ $\quad y = 0$ $\quad y$ intercept

The graph crosses the coordinate axes only at the origin.

Vertical asymptotes. $d(x) = (3x + 5)^2$ zeros: $(3x + 5)^2 = 0$

$$3x + 5 = 0$$
$$x = -\frac{5}{3}$$

Horizontal asymptotes. Since $n(x)$ and $d(x)$ have the same degree, the line $y = \dfrac{12}{3^2} = \dfrac{4}{3}$ is a horizontal asymptote.

Complete the sketch. Plot a few additional points.

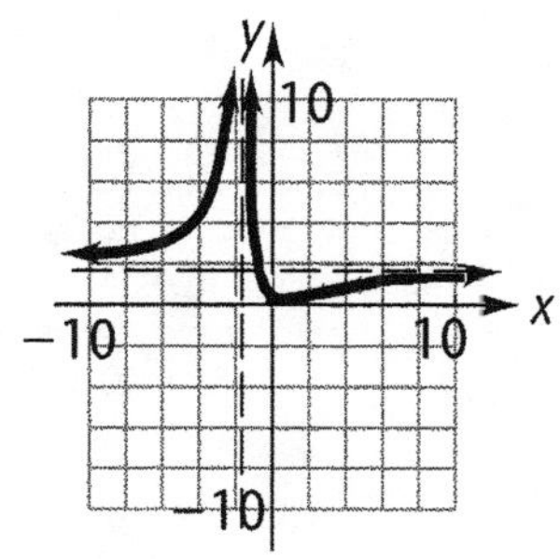

55. $f(x) = \dfrac{x^2 - 1}{x^2 + 7x + 10} = \dfrac{n(x)}{d(x)}$

Intercepts. Real zeros of $n(x) = x^2 - 1$

$$x^2 - 1 = 0$$
$$x^2 = 1$$
$$x = \pm 1 \quad x \text{ intercepts}$$

$$f(0) = -\frac{1}{10} \quad y \text{ intercept}$$

Vertical asymptotes. Real zeros of $d(x) = x^2 + 7x + 10$

$$x^2 + 7x + 10 = 0$$
$$(x + 2)(x + 5) = 0$$
$$x = -2, -5$$

Horizontal asymptotes. Since $n(x)$ and $d(x)$ have the same degree, the line $y = 1$ is a horizontal asymptote.

Complete the sketch. Plot a few additional points.

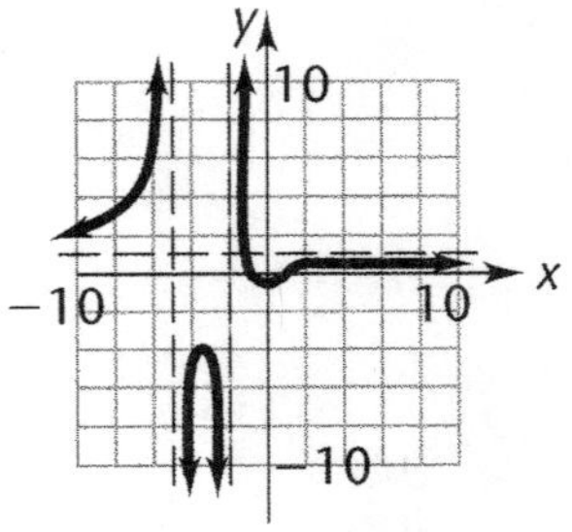

57. $f(x) = \dfrac{x^2 - 3x - 10}{x - 5} = \dfrac{(x - 5)(x + 2)}{x - 5} = x + 2 \quad x \neq 5.$

The graph is the same as the graph of $y = x + 2$ except that f has a hole at $(5, 7)$. There are no asymptotes. The line has x intercept $x = -2$ and y intercept $y = 2$.

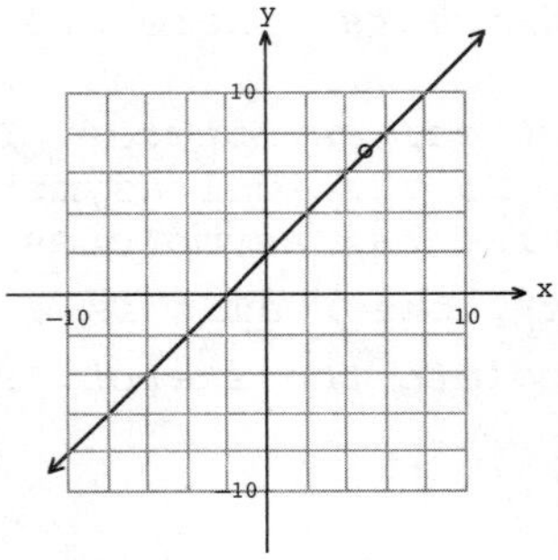

59. To have zeros -2, -1, 1, and 2, the numerator must have factors $(x + 2)$, $(x + 1)$, $(x - 1)$, and $(x - 2)$. To have a horizontal asymptote at $y = 3$, the degree of the denominator should be the same as the numerator, and the leading coefficient of the numerator should be 3 times as large. To have no vertical asymptote, the denominator should have no real zeros.

$$f(x) = \frac{3(x + 2)(x - 2)(x + 1)(x - 1)}{x^4 + 1} \quad \text{or} \quad f(x) = \frac{3(x^2 - 4)(x^2 - 1)}{x^4 + 1} \text{ will work.}$$

61. To get $y = 2x + 5$ as an oblique asymptote, our function should look like $f(x) = 2x + 5 + \dfrac{r(x)}{q(x)}$ where the degree of $q(x)$ is greater than the degree of $r(x)$. If $q(x) = x - 10$, we'll have $x = 10$ as vertical asymptote. In that case $r(x)$ will have to be constant so that it's power is less than $q(x)$; $r(x) = 100$ will do.

$$f(x) = 2x + 5 + \frac{100}{x - 10}$$
$$= \frac{(2x + 5)(x - 10)}{x - 10} + \frac{100}{x - 10}$$
$$= \frac{(2x + 5)(x - 10) + 100}{x - 10}$$

> **Common Error:**
> It is not correct to "multiply both sides" by $x - 2$.

63. $\frac{x}{x-2} \le 0$

Let $f(x) = \frac{p(x)}{q(x)} = \frac{x}{x-2}$

The zero of $p(x) = x$ is 0. The zero of $q(x) = x - 2$ is 2. These two zeros partition the x axis into the three intervals shown in the table. A test number is chosen from each interval to determine the sign of $f(x)$.

Interval	Test number x	$f(x)$	Sign of f
$(-\infty, 0)$	-1	$\frac{1}{3}$	+
$(0, 2)$	1	-1	−
$(2, \infty)$	3	3	+

The equality is satisfied at $x = 0$, but not at $x = 2$. We conclude that the solution set is $[0, 2)$.

65. $\frac{x^2 - 16}{5x - 2} > 0$

Let $f(x) = \frac{p(x)}{q(x)} = \frac{x^2 - 16}{5x - 2}$

The zeros of $p(x) = x^2 - 16 = (x + 4)(x - 4)$ are -4 and 4.

The zero of $q(x) = 5x - 2$ is $\frac{2}{5}$.

These three zeros partition the x axis into the four intervals shown in the table. A test number is chosen from each interval to determine the sign of $f(x)$.

Interval	Test number x	$f(x)$	Sign of f
$(-\infty, -4)$	-5	$-\frac{1}{3}$	−
$\left(-4, \frac{2}{5}\right)$	0	8	+
$\left(\frac{2}{5}, 4\right)$	1	-5	−
$(4, \infty)$	5	$\frac{9}{23}$	+

We conclude that the solution set is $\left(-4, \frac{2}{5}\right) \cup (4, \infty)$.

67. $\frac{x^2 + 4x - 20}{3x} \ge 4$

$$\frac{x^2 + 4x - 20}{3x} - 4 \ge 0$$

$$\frac{x^2 + 4x - 20}{3x} - \frac{12x}{3x} \ge 0$$

$$\frac{x^2 - 8x - 20}{3x} \ge 0$$

Let $f(x) = \frac{p(x)}{q(x)} = \frac{x^2 - 8x - 20}{3x}$.

The zeros of $p(x) = x^2 - 8x - 20 = (x + 2)(x - 10)$ are -2 and 10. The zero of $q(x)$ is 0.

These three zeros partition the x axis into the four intervals shown in the table. A test number is chosen from each interval to determine the sign of $f(x)$.

Interval	Test number x	$f(x)$	Sign of f
$(-\infty, -2)$	-3	$-\frac{13}{9}$	$-$
$(-2, 0)$	-1	$\frac{11}{3}$	$+$
$(0, 10)$	1	-9	$-$
$(10, \infty)$	11	$\frac{13}{33}$	$+$

The equality holds at -2 and 10, but not at 0. We conclude that the solution set is $[-2, 0) \cup [10, \infty)$.

69. $\dfrac{5x}{x^2 - 1} < \dfrac{9}{x}$

$$\frac{5x}{x^2 - 1} - \frac{9}{x} < 0$$

$$\frac{5x^2 - 9(x^2 - 1)}{x(x^2 - 1)} < 0$$

$$\frac{-4x^2 + 9}{x(x^2 - 1)} < 0$$

Let $f(x) = \dfrac{p(x)}{q(x)} = \dfrac{-4x^2 + 9}{x(x^2 - 1)}$.

The zeros of $p(x) = -4x^2 + 9 = 9 - 4x^2 = (3 + 2x)(3 - 2x)$ are $-\frac{3}{2}$ and $\frac{3}{2}$. The zeros of $q(x) = x(x^2 - 1) = x(x + 1)(x - 1)$ are 0, 1, and -1.
These five zeros partition the x axis into the six intervals shown in the table. A test number is chosen from each interval to determine the sign of $f(x)$.

Interval	Test number x	$f(x)$	Sign of f
$\left(-\infty, -\frac{3}{2}\right)$	-2	$\frac{7}{6}$	$+$
$\left(-\frac{3}{2}, -1\right)$	$-\frac{5}{4}$	$-\frac{176}{45}$	$-$
$(-1, 0)$	$-\frac{1}{2}$	$\frac{64}{3}$	$+$
$(0, 1)$	$\frac{1}{2}$	$-\frac{64}{3}$	$-$
$\left(1, \frac{3}{2}\right)$	$\frac{5}{4}$	$\frac{176}{45}$	$+$
$\left(\frac{3}{2}, \infty\right)$	2	$-\frac{7}{6}$	$-$

We conclude that the solution set is $\left(-\frac{3}{2}, -1\right) \cup (0, 1) \cup \left(\frac{3}{2}, \infty\right)$.

71. $\dfrac{x^2 + 7x + 3}{x + 2} > 0$

zeros of the numerator $x^2 + 7x + 3 = 0$ $a = 1,\ b = 7,\ c = 3$

$$x = \frac{-7 \pm \sqrt{7^2 - 4(1)(3)}}{2(1)} = \frac{-7 \pm \sqrt{37}}{2} = \frac{-7 \pm 6.083}{2} = -6.541,\ -0.459$$

zeros of the denominator: $x = -2$

The three zeros partition the x axis into the four intervals shown in the table below. A test number is chosen from each interval to determine the sign of the rational expression.

Interval	$(-\infty, -6.541)$	$(-6.541, -2)$	$(-2, -0.459)$	$(-0.459, \infty)$
Test number x	-10	-5	-1	0
$f(x)$	$-\frac{33}{8}$	$\frac{7}{3}$	-3	$\frac{3}{2}$
Sign of f	$-$	$+$	$-$	$+$

The expressions is positive on $(-6.541, -2) \cup (-0.459, \infty)$. We need to exclude all endpoints, so this is the solution.

Graphical check:

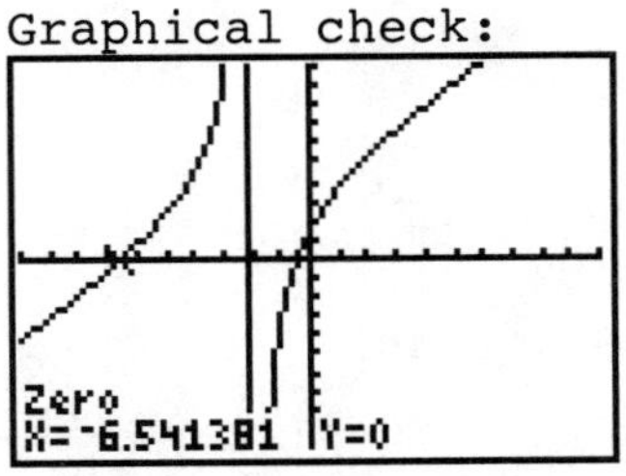

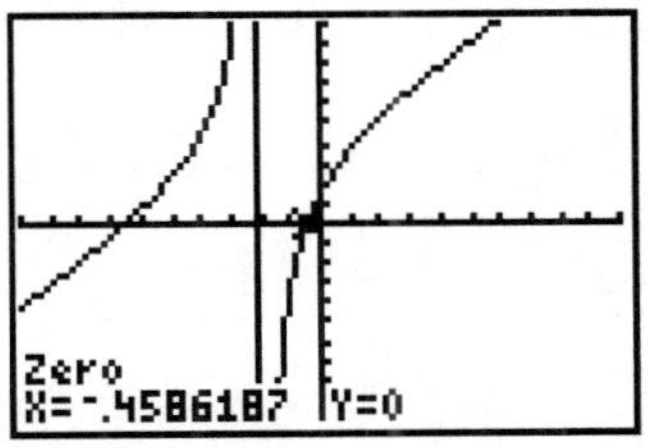

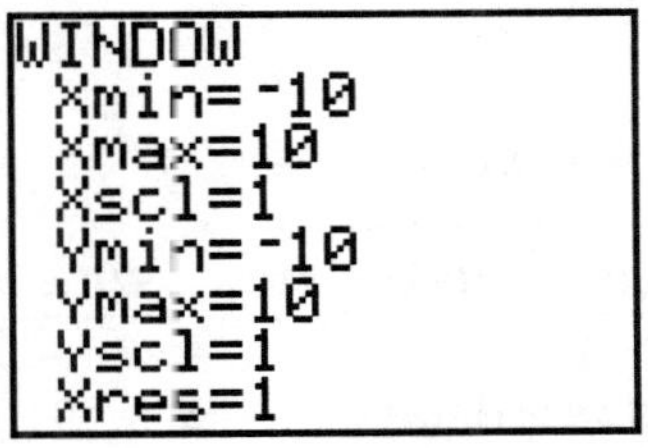

The graph is above the x axis on $(-6.541, -2) \cup (-0.459, \infty)$.

73. $\dfrac{5}{x^2} - \dfrac{1}{x + 3} < 0$ (Simplify the left side)

$$\frac{5(x + 3)}{x^2(x + 3)} - \frac{x^2}{x^2(x + 3)} < 0$$

$$\frac{5x + 15 - x^2}{x^2(x + 3)} < 0$$

$$\frac{-x^2 + 5x + 15}{x^2(x + 3)} < 0$$

zeros of the numerator: $-x^2 + 5x + 15 = 0$ $a = -1,\ b = 5,\ c = 15$

$$x = \frac{-5 \pm \sqrt{5^2 - 4(-1)(15)}}{2(-1)} = \frac{-5 \pm \sqrt{85}}{-2} = \frac{-5 \pm 9.220}{-2} = 7.110,\ -2.110$$

zeros of the denominator: $x^2(x + 3) = 0$

$x = 0,\ -3$

The four zeros partition the x axis into the five intervals shown in the table below. A test number is chosen from each interval to determine the sign of the rational expression.

Interval	$(-\infty, -3)$	$(-3, -2.110)$	$(-2.110, 0)$	$(0, 7.110)$	$(7.110, \infty)$
Test number x	-5	-2.5	-1	1	8
$f(x)$	$\frac{7}{10}$	-1.2	$\frac{9}{2}$	$\frac{19}{4}$	$-\frac{9}{704}$
Sign of f	$+$	$-$	$+$	$+$	$-$

The expression is negative on (-3, -2.110) ∪ (7.110, ∞). We need to exclude all endpoints so this is the solution.

Graphical check:

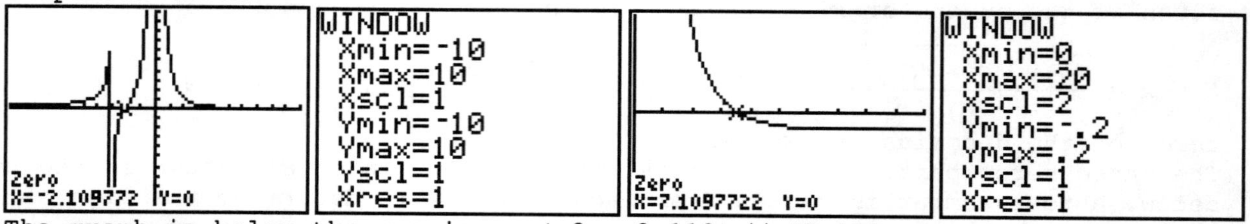

The graph is below the x axis on (-3, -2.110) ∪ (7.110, ∞).

75. $\frac{9}{x} - \frac{5}{x^2} \le 1$ (Simplify)

$$\frac{9}{x} - \frac{5}{x^2} - 1 \le 0$$

$$\frac{9x}{x^2} - \frac{5}{x^2} - \frac{x^2}{x^2} \le 0$$

$$\frac{-x^2 + 9x - 5}{x^2} \le 0$$

zeros of the numerator: $-x^2 + 9x - 5 = 0$ $a = -1$, $b = 9$, $c = -5$

$$x = \frac{-9 \pm \sqrt{9^2 - 4(-1)(-5)}}{2(-1)} = \frac{-9 \pm \sqrt{61}}{-2} = \frac{-9 \pm 7.810}{-2} = 0.595,\ 8.405$$

zeros of the denominator: $x = 0$

These three zeros partition the x axis into the four intervals shown in the table below. A test number is chosen from each interval to determine the sign of the rational expression.

Interval	(-∞, 0)	(0, 0.595)	(0.595, 8.405)	(8.405, ∞)
Test number x	-1	.5	1	10
$f(x)$	-15	-3	3	$-\frac{3}{20}$
Sign of f	−	−	+	−

The expression is negative on (-∞, 0), (0, 0.595), and (8.405, ∞). It's equal to zero when x = 0.595 and 8.405, so we include those endpoints, but we exclude x = 0 since the expression is undefined there. The solution is (-∞, 0) ∪ (0, 0.595] ∪ [8.405, ∞).

Graphical check:

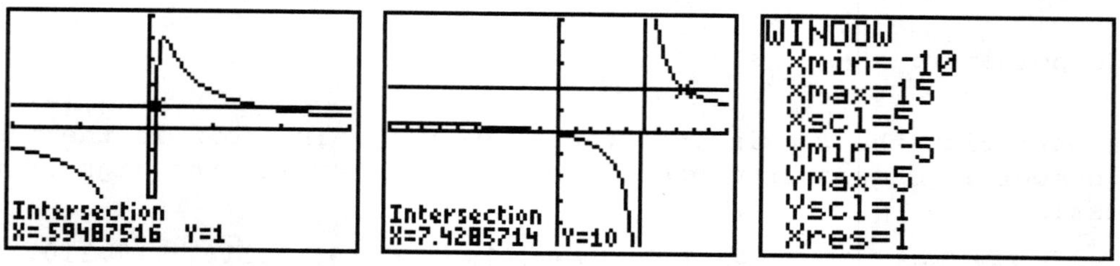

The graph of $y_1 = \frac{9}{x} - \frac{5}{x^2}$ is below the graph of $y_2 = 1$ on (-∞, 0), (0, 0.595) and (8.405, ∞) and they intersect at x = 0.595 and x = 8.405.

77. $\dfrac{3x + 2}{x - 5} > 10$ (Simplify)

$$\frac{3x + 2}{x - 5} - 10 > 0$$

$$\frac{3x + 2}{x - 5} - \frac{10(x - 5)}{x - 5} > 0$$

$$\frac{3x + 2 - 10x + 50}{x - 5} > 0$$

$$\frac{-7x + 52}{x - 5} > 0$$

zeros of the numerator: $-7x + 52 = 0$

$$-7x = -52$$

$$x = 7.429$$

zeros of the denominator: $x = 5$

These two zeros partition the x axis into the 3 intervals shown in the table below. A test number is chosen from each interval to determine the sign of the rational expression.

Interval	$(-\infty, 5)$	$(5, 7.429)$	$(7.429, \infty)$
Test number x	0	6	8
$f(x)$	$-\frac{52}{5}$	10	$-\frac{4}{3}$
Sign of f	$-$	$+$	$-$

The expression is positive on $(5, 7.429)$. We need to exclude the endpoints, so this is the solution.

Graphical check:

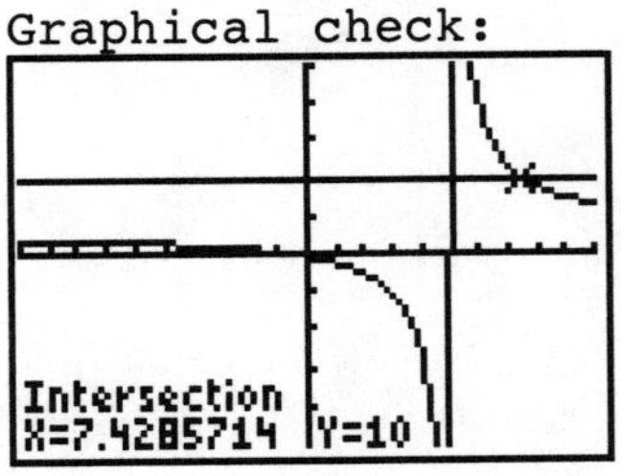

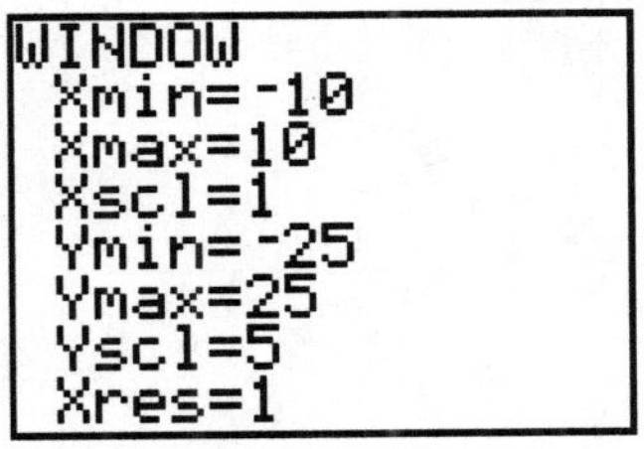

The graph of $y_1 = \dfrac{3x + 2}{x - 5}$ is above the graph of $y_2 = 10$ on $(5, 7.429)$.

79. $\dfrac{4}{x + 1} \geq \dfrac{7}{x}$ (Simplify)

$$\frac{4}{x + 1} - \frac{7}{x} \geq 0$$

$$\frac{4x}{x(x + 1)} - \frac{7(x + 1)}{x(x + 1)} \geq 0$$

$$\frac{4x - 7x - 7}{x(x + 1)} \geq 0$$

$$\frac{-3x - 7}{x(x + 1)} \geq 0$$

zeros of the numerator: $-3x - 7 = 0$

$$-3x = 7$$

$$x = -2.333$$

zeros of the denominator: $x(x + 1) = 0$

$$x = 0, -1$$

These three zeros partition the x axis into the four intervals shown in the table below. A test number is chosen from each interval to determine the sign of the rational expression.

Interval	$(-\infty, -2.333)$	$(-2.333, -1)$	$(-1, 0)$	$(0, \infty)$
Test number x	-3	-2	$-.5$	1
$f(x)$	$\frac{1}{3}$	$-\frac{1}{2}$	22	-5
Sign of f	+	−	+	−

The expression is positive on $(-\infty, -2.333) \cup (-1, 0)$. We include $x = -2.333$ because it makes the expression zero, but exclude 0 and -1 as they make the expression undefined. The solution is $(-\infty, -2.333] \cup (-1, 0)$.

Graphical check:

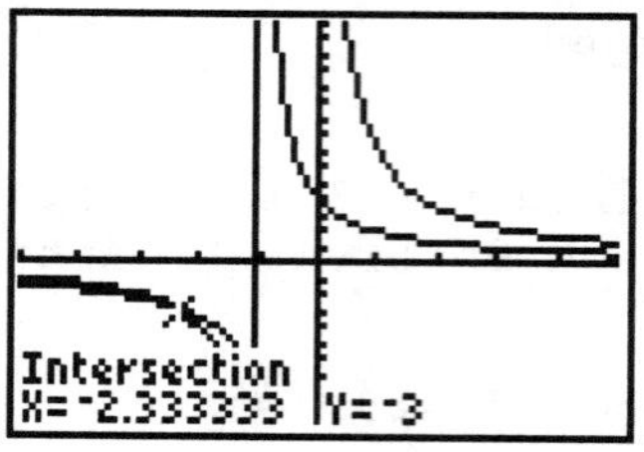

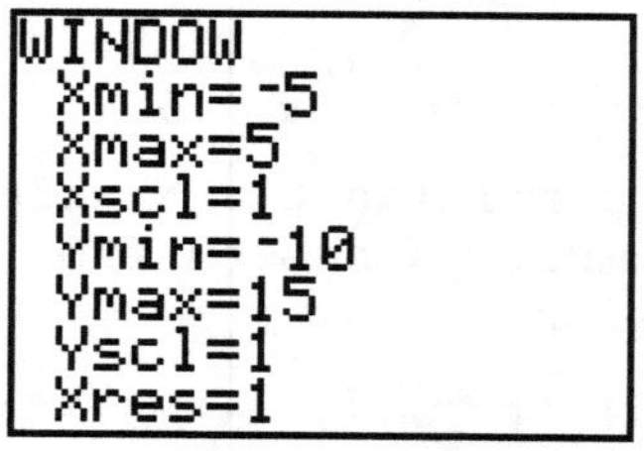

The graph of $y_1 = \frac{4}{x + 1}$ is above the graph of $y_2 = \frac{7}{x}$ on $(-\infty, -2.333] \cup (-1, 0)$.

81. $f(x) = \frac{2x^2}{x - 1} = \frac{n(x)}{d(x)}$

Vertical asymptotes. Real zeroes of $d(x) = x - 1$ $x = 1$
Horizontal asymptote. Since the degree of $n(x)$ is greater than the degree of $d(x)$, there is no horizontal asymptote.
Oblique asymptote.

$$\begin{array}{r|l} & 2x + 2 \\ \hline x - 1 & 2x^2 \\ & \underline{2x^2 - 2x} \\ & \quad 2x \\ & \quad \underline{2x - 2} \\ & \qquad 2 \end{array}$$

Thus, $f(x) = 2x + 2 + \frac{2}{x - 1}$. Hence, the line $y = 2x + 2$ is an oblique asymptote.

83. $p(x) = \frac{x^3}{x^2 + 1} = \frac{n(x)}{d(x)}$

Vertical asymptotes. There are no real zeros of $d(x) = x^2 + 1$.
No vertical asymptotes.
Horizontal asymptotes. Since the degree of $n(x)$ is greater than the degree of $d(x)$, there is no horizontal asymptote.
Oblique asymptote:

$$\begin{array}{r|l} & x \\ \hline x^2 + 1 & x^3 \\ & \underline{x^3 + x} \\ & \quad - x \end{array}$$

Thus, $p(x) = x + \frac{-x}{x^2 + 1}$. Hence, the line $y = x$ is an oblique asymptote.

85. $r(x) = \dfrac{2x^2 - 3x + 5}{x} = \dfrac{n(x)}{d(x)}$

Vertical asymptotes. Real zeros of $d(x) = x$ $x = 0$

Horizontal asymptote. Since the degree of $n(x)$ is greater than the degree of $d(x)$, there is no horizontal asymptote.

Oblique asymptote. $\dfrac{2x^2 - 3x + 5}{x} = \dfrac{2x^2}{x} - \dfrac{3x}{x} + \dfrac{5}{x}$

$$= 2x - 3 + \frac{5}{x}$$

Thus $r(x) = 2x - 3 + \dfrac{5}{x}$. Hence the line $y = 2x - 3$ is an oblique asymptote.

87. Here is a computer-generated graph of $f(x)$.

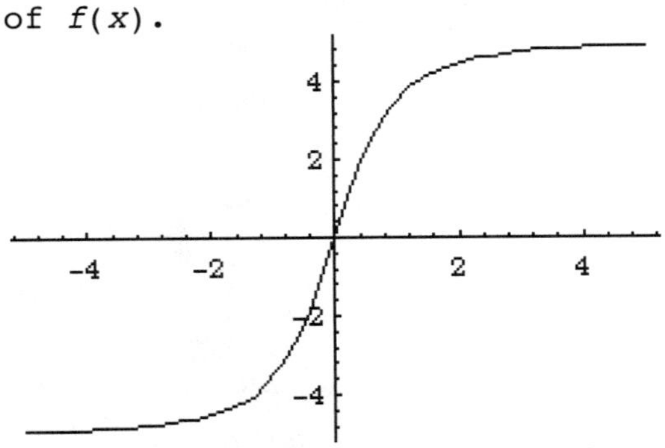

From the graph, we can see that $f(x) \to 5$ as $x \to \infty$ and $f(x) \to -5$ as $x \to -\infty$; the lines $y = 5$ and $y = -5$ are horizontal asymptotes.

89. Here is a computer-generated graph of $f(x)$.

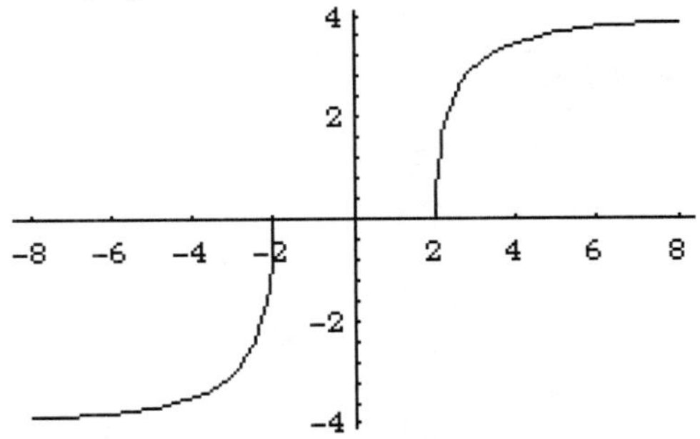

From the graph, we can see that $f(x) \to 4$ as $x \to \infty$ and $f(x) \to 4$ as $x \to -\infty$; the lines $y = 4$ and $y = -4$ are horizontal asymptotes.

91. $f(x) = \dfrac{x^2 + 1}{x} = \dfrac{n(x)}{d(x)}$

Intercepts. There are no real zeros of $n(x) = x^2 + 1$. No x intercept
$f(0)$ is not defined No y intercept

Vertical asymptotes. Real zeros of $d(x) = x$. $x = 0$

Horizontal asymptote. Since the degree of $n(x)$ is greater than the degree of $d(x)$, there is no horizontal asymptote.

Oblique asymptote. $f(x) = \dfrac{x^2 + 1}{x} = \dfrac{x^2}{x} + \dfrac{1}{x} = x + \dfrac{1}{x}$

Hence, the line $y = x$ is an oblique asymptote.

Complete the sketch. Plot a few points.

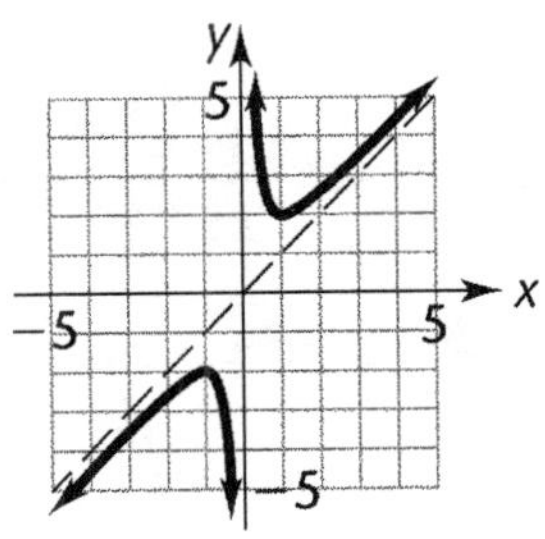

x	$f(x)$
$\frac{1}{2}$	$\frac{5}{2}$
1	2
2	$\frac{5}{2}$
$-\frac{1}{2}$	$-\frac{5}{2}$
-1	-2
-2	$-\frac{5}{2}$

93. $k(x) = \dfrac{x^2 - 4x + 3}{2x - 4} = \dfrac{n(x)}{d(x)}$

Intercepts. Real zeros of $n(x) = x^2 - 4x + 3$

$$x^2 - 4x + 3 = 0$$
$$(x - 1)(x - 3) = 0$$
$$x = 1,\ 3 \quad x \text{ intercepts}$$

$k(0) = -\dfrac{3}{4}$ y intercept

Vertical asymptotes. Real zeros of $d(x) = 2x - 4$

$$2x - 4 = 0$$
$$2x = 4$$
$$x = 2$$

Horizontal asymptote. Since the degree of $n(x)$ is greater than the degree of $d(x)$, there is no horizontal asymptote.

Oblique asymptote:

$$\begin{array}{r} \frac{1}{2}x - 1 \\ 2x - 4\,\overline{)\,x^2 - 4x + 3} \\ \underline{x^2 - 2x} \\ -2x + 3 \\ \underline{-2x + 4} \\ -1 \end{array}$$

Thus, $k(x) = \dfrac{1}{2}x - 1 + \dfrac{-1}{2x - 4}$. Hence, the line $y = \dfrac{1}{2}x - 1$ is an oblique asymptote.

Complete the sketch. Plot a few additional points.

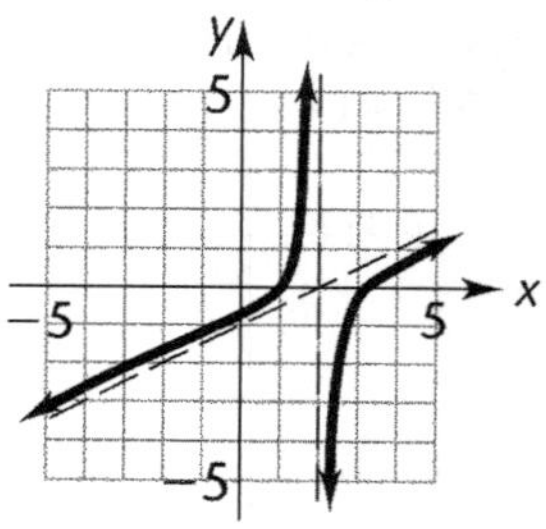

95. $F(x) = \dfrac{8 - x^3}{4x^2} = \dfrac{n(x)}{d(x)}$

Intercepts. Real zeros of $n(x) = 8 - x^3$

$$8 - x^3 = 0$$
$$(2 - x)(4 + 2x + x^2) = 0$$
$$2 - x = 0 \qquad 4 + 2x + x^2 = 0$$
$$x = 2 \qquad \text{No real zeros}$$

$x = 2$ x intercept

$F(0)$ is not defined. No y intercept.

Vertical asymptotes. Real zeros of $d(x) = 4x^2$. $x = 0$

Horizontal asymptote. Since the degree of $n(x)$ is greater than the degree of $d(x)$, there is no horizontal asymptote.

Oblique asymptote. $F(x) = \dfrac{8 - x^3}{4x^2} = \dfrac{8}{4x^2} - \dfrac{x^3}{4x^2} = -\dfrac{1}{4}x + \dfrac{2}{x^2}$. Hence, the line $y = -\dfrac{1}{4}x$ is an oblique asymptote.

Complete the sketch. Plot a few additional points.

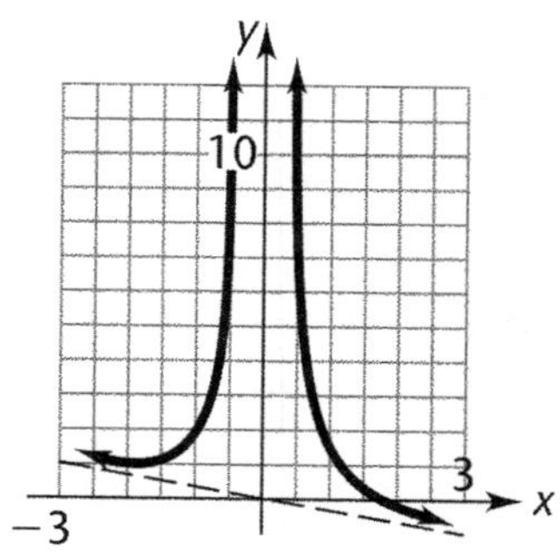

97. $f(x) = \dfrac{x^2 - 4}{x - 2}$. $f(x)$ is not defined if $x - 2 = 0$, that is, $x = 2$

Domain: $(-\infty, 2) \cup (2, \infty)$

$f(x) = \dfrac{(x - 2)(x + 2)}{(x - 2)}$

$f(x) = x + 2$

The graph is a straight line with slope 1 and y intercept 2, except that the point (2, 4) is not on the graph.

There are no asymptotes.

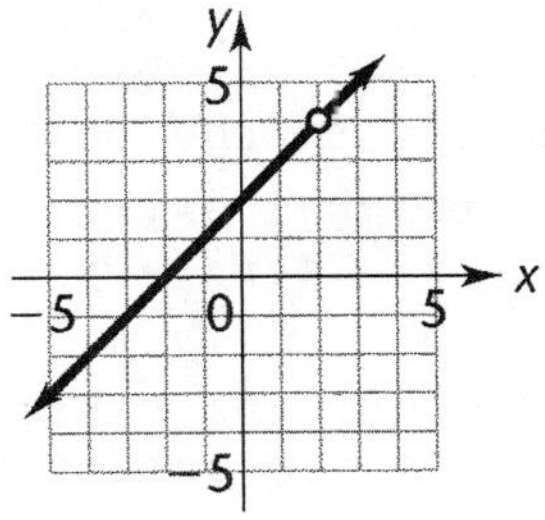

99. $r(x) = \dfrac{x + 2}{x^2 - 4}$. $f(x)$ is not defined if $x^2 - 4 = 0$, that is,

$x^2 = 4$

$x = \pm 2$

Domain: $(-\infty, -2) \cup (-2, 2) \cup (2, \infty)$

$r(x) = \dfrac{x + 2}{(x + 2)(x - 2)}$

$r(x) = \dfrac{1}{x - 2}$

The graph is the same as the graph of the function $\dfrac{1}{x - 2}$, except that the point $\left(-2, -\dfrac{1}{4}\right)$ is not on the graph.

Intercepts: $y = -\dfrac{1}{2}$. No x intercept.

Vertical asymptote: $x = 2$

Horizontal asymptote: $y = 0$

Complete the sketch. Plot a few additional points.

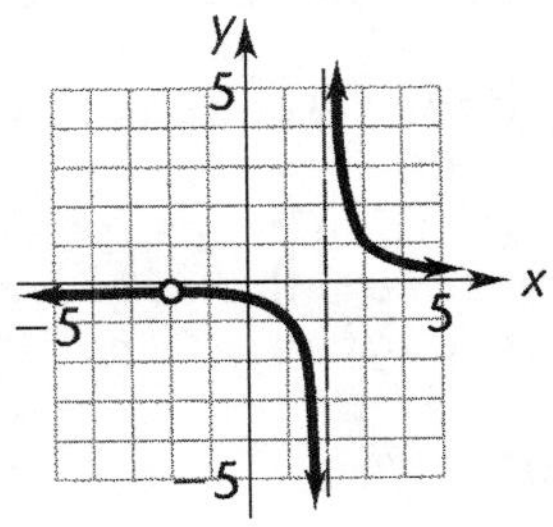

101. $N(t) = \dfrac{50t}{t + 4}$ $t \geq 0$

Intercepts: Real zeros of $50t$: $t = 0$ $N(0) = 0$

Vertical asymptotes: None, since -4, the only zero of $t + 4$, is not in the domain of N.

Horizontal asymptote: $N = 50$. As $t \to \infty$, $N \to 50$

Complete the sketch. Plot a few additional points.

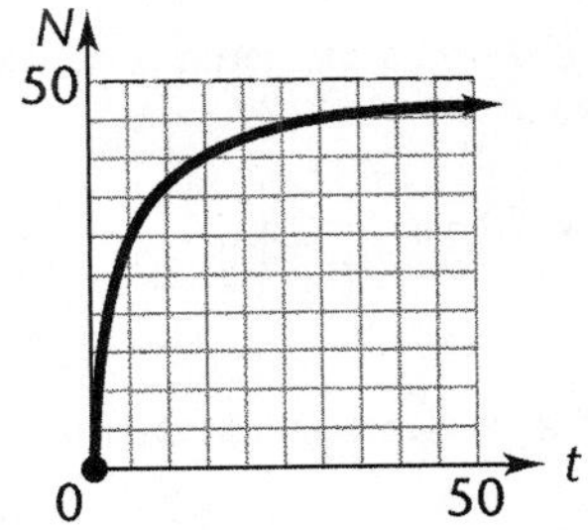

103. $N(t) = \dfrac{5t + 30}{t} \quad t \geq 1$

Intercepts: Real zeros of $5t + 30$, $t \geq 1$. None, since -6, the only zero of $5t + 30$, is not in the domain of N.
Vertical asymptotes: None, since 0, the only zero of t, is not in the domain of N.

Horizontal asymptote: $N = 5$. As $t \to \infty$, $N \to 5$

Complete the sketch. Plot a few points.

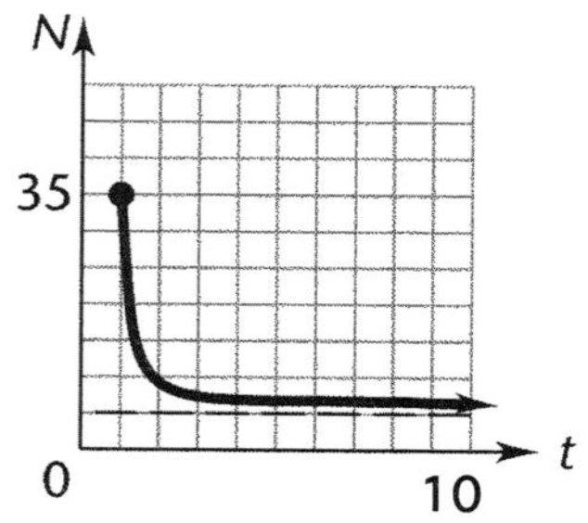

t	$N(t)$
1	35
3	15
5	11
10	8

105. (A) $\bar{C}(n) = \dfrac{C(n)}{n} = \dfrac{2,500 + 175n + 25n^2}{n} = 25n + 175 + \dfrac{2,500}{n}$

(B) The minimum value of the function $\bar{C}(n)$ is $\bar{C}\left(\sqrt{\dfrac{c}{a}}\right)$, where $a = 25$ and $c = 2,500$

$\min \bar{C}(n) = \bar{C}\left(\sqrt{\dfrac{2,500}{25}}\right) = \bar{C}(\sqrt{100}) = \bar{C}(10)$

This minimum occurs when $n = 10$, after 10 years.

(C) *Intercepts:* Real zeros of $2,500 + 175n + 25n^2$.
None. No n intercepts.
0 is not in the domain of n, so there are no $\bar{C}$ intercepts.
Vertical asymptotes: Real zeros of n. The line $n = 0$ is a vertical asymptote.
Sign behavior: $\bar{C}$ is always positive since $n \geq 0$.
Horizontal asymptote: None, since the degree of $C(n)$ is greater than the degree of n.

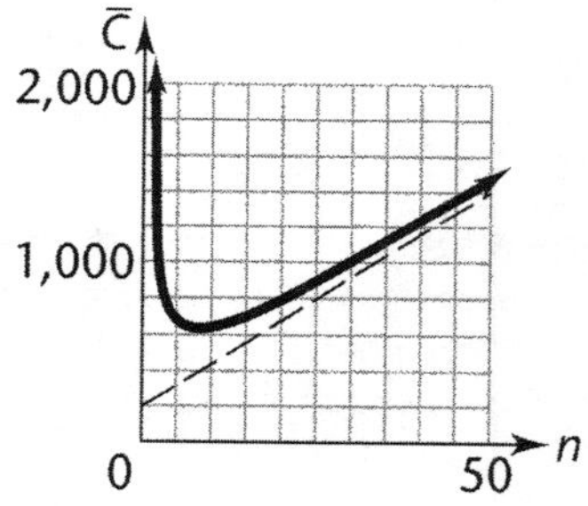

Oblique asymptote: The line $\bar{C} = 25n + 175$ is an oblique asymptote.

107. (A) Since Area = length × width, length $= \dfrac{\text{Area}}{\text{width}} = \dfrac{225}{x}$. Then total length of fence = 2 × width + 2 × length

$L(x) = 2x + \dfrac{450}{x} = \dfrac{2x^2 + 450}{x}$

(B) x can be any positive number, thus, domain $= (0, \infty)$

(C) The minimum value of the function $L(x)$ is $L\left(\sqrt{\dfrac{c}{a}}\right)$ where $a = 2$ and $c = 450$.

$\min L(x) = L\left(\sqrt{\dfrac{450}{2}}\right) = L(\sqrt{225}) = L(15)$

This minimum occurs when $x = 15$.

Width = 15 feet. Length $= \dfrac{225}{15} = 15$ feet.

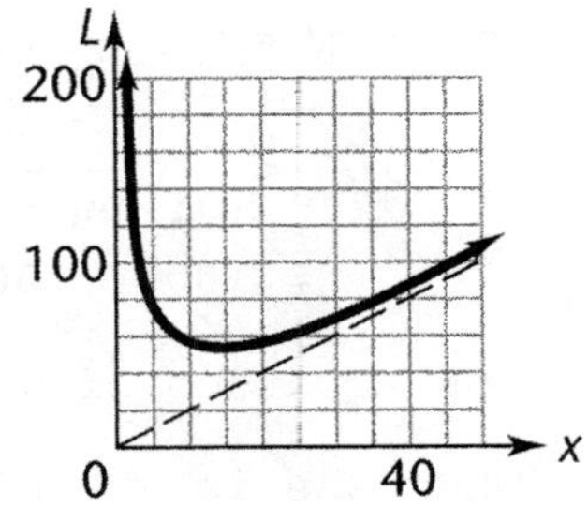

(D) *Intercepts:* Real zeros of $2x^2 + 450$. None, hence, no x intercepts. 0 is not in the domain of L, so there are no L intercepts.
Vertical asymptotes: Real zeros of x. The line $x = 0$ is a vertical asymptote.
Sign behavior: $L(x)$ is always positive since $x > 0$.
Horizontal asymptote: None, since the degree of $2x^2 + 450$ is greater than the degree of x.
Oblique asymptote: The line $L = 2x$ is an oblique asymptote.

Section 4-5

1. $F = \frac{k}{x}$ **3.** $R = kST$ **5.** $L = km^3$ **7.** $A = kc^2d^2$ **9.** $P = kx$ **11.** $h = \frac{k}{\sqrt{s}}$

13. $R = k\frac{m}{d^2}$ **15.** $D = k\frac{xy^2}{z}$

17. Write $u = k\sqrt{v}$. Substitute $u = 3$ and $v = 4$ and solve for k.

$$3 = k\sqrt{4}$$
$$3 = 2k$$
$$k = \frac{3}{2}$$

The equation of variation is $u = \frac{3}{2}\sqrt{v}$. When $v = 10$, $u = \frac{3}{2}\sqrt{10}$.

19. Write $L = \frac{k}{M^2}$. Substitute $L = 9$ and $M = 9$ and solve for k.

$$9 = \frac{k}{9^2}$$
$$k = 9^3 = 729$$

The equation of variation is $L = \frac{729}{M^2}$. When $M = 6$, $L = \frac{729}{6^2} = \frac{81}{4}$.

21. Write $Q = k\frac{mn^2}{P}$. Substitute $Q = 2$, $m = 3$, $n = 6$, and $P = 12$ and solve for k.

$$2 = k\frac{3(6)^2}{12}$$
$$2 = 9k$$
$$k = \frac{2}{9}$$

The equation of variation is $Q = \frac{2}{9}\frac{mn^2}{P}$.

When $m = 4$, $n = 18$, and $P = 2$,

$$Q = \frac{2}{9}\frac{4(18)^2}{2} = 144$$

23. $t = \frac{k}{T}$ **25.** $L = k\frac{wh^2}{x}$ **27.** $N = K\frac{F}{d}$

29. Write $f = kx$. Then $f_1 = kx_1$ and $f_2 = kx_2$. Therefore

$$\frac{kx_1}{kx_2} = \frac{f_1}{f_2}$$
$$\frac{x_1}{x_2} = \frac{f_1}{f_2}$$

31. Write $w = \frac{k}{d^2}$. Substitute $w = 100$ and $d = 4{,}000$ and solve for k.

$$100 = \frac{k}{(4,000)^2}$$

$$k = 100(4{,}000)^2 = 1.6 \times 10^9$$

The equation of variation is $w = \frac{1.6 \times 10^9}{d^2}$.

When she is 400 miles above the earth's surface, $d = 4{,}000 + 400 = 4{,}400$. Substitute to find

$$w = \frac{1.6 \times 10^9}{(4,400)^2} = 83 \text{ lbs to the nearest pound}$$

33. Write $I = k\frac{E}{R}$. Substitute $I = 22$, $E = 110$, and $R = 5$ and solve for k.

$$22 = k\frac{110}{5}$$

$$k = 1$$

The equation of variation is $I = \frac{E}{R}$. When $E = 220$ and $R = 11$,

$$I = \frac{220}{11} = 20 \text{ amperes}$$

35. Write $P = kv^3$. Let $P_1 = kv_1^3$ and $v_2 = 2v_1$, then $P_2 = kv_2^3 = k(2v_1)^3 = 8kv_1^3 = 8P_1$. Hence P must be multiplied by 8 to double the speed of the boat.

37. Write $f = k\frac{\sqrt{T}}{L}$. Let $f_1 = k\frac{\sqrt{T_1}}{L_1}$, $T_2 = 4T_1$, and $L_2 = 2L_1$, then

$$f_2 = k\frac{\sqrt{T_2}}{L_2} = k\frac{\sqrt{4T_1}}{2L_1} = k\frac{\sqrt{T_1}}{L_1} = f_1.$$ There would be no net effect.

39. Write $t = k\frac{r}{v}$. Substitute $t = 1.42$, $r = 4{,}050$, and $v = 18{,}000$ and solve for k.

$$1.42 = k\frac{4,050}{18,000}$$

$$k = 1.42\frac{18,000}{4,050}$$

$$k = 6.311$$

The equation of variation is $t = 6.311\frac{r}{v}$. When $r = 4{,}300$ and $v = 18{,}500$,

$$t = 6.311\frac{4,300}{18,500} = 1.47 \text{ hours}$$

41. Write $d = kh$. Substitute $d = 4$ and $h = 500$ and solve for k.

$$4 = k(500)$$

$$k = \frac{1}{125}$$

The equation of variation is $d = \frac{1}{125}h$. When $h = 2{,}500$,

$$d = \frac{1}{125}2{,}500 = 20 \text{ days.}$$

43. Write $L = kv^2$. Let $L_1 = kv_1^2$ and $v_2 = 2v_1$. Then

$L_2 = kv_2^2 = k(2v_1)^2 = 4kv_1^2 = 4L_1$.

The length would be quadrupled.

45. Write $P = kAv^2$. Substitute $P = 120$, $A = 100$, and $v = 20$ and solve for k.

$$120 = k(100)20^2$$
$$120 = 40{,}000k$$
$$k = \frac{3}{1000}$$

The equation of variation is $P = \frac{3}{1000}Av^2$. When $A = 200$ and $v = 30$,

$$P = \frac{3}{1000}(200)(30)^2 = 540 \text{ lb}$$

47. (A) $\Delta S = kS$

(B) Substitute $\Delta S = 1$ and $S = 50$ and solve for k.

$$1 = k(50)$$
$$k = \frac{1}{50}$$

The equation of variation is $\Delta S = \frac{1}{50}S$. When $S = 500$, $\Delta S = \frac{1}{50}(500) = 10$ oz.

(C) Substitute $\Delta S = 1$ and $S = 60$ and solve for k.

$$1 = k(60)$$
$$k = \frac{1}{60}$$

The equation of variation is $\Delta S = \frac{1}{60}S$. When $S = 480$, $\Delta S = \frac{1}{60}(480) = 8$ candlepower.

49. Write $V = kr^3$. Let $V_1 = kr_1^3$ and $r_2 = 2r_1$, then

$V_2 = kr_2^3 = k(2r_1)^3 = 8kr_1^3 = 8V_1$

The volume would be 8 times the original.

51. Write frequency $= f$, length $= x$, $f = \frac{k}{x}$. Substitute $f = 16$ and $x = 32$ and solve for k.

$$16 = \frac{k}{3^2}$$
$$k = 512$$

The equation of variation is $f = \frac{512}{x}$. When $x = 16$, $f = \frac{512}{16} = 32$ times per second.

CHAPTER 4 REVIEW

1. The zeros are -1 and 3; the turning points are $(-1, 0)$, $(1, 2)$, and $(3, 0)$; $P(x) \to \infty$ as $x \to -\infty$ and $P(x) \to \infty$ as $x \to \infty$. (4-1)

2.

$$\begin{array}{r|rrrr} & 2 & 3 & 0 & -1 \\ & & -4 & 2 & -4 \\ \hline -2 & 2 & -1 & 2 & -5 \end{array}$$

$2x^3 + 3x^2 - 1$
$= (x + 2)(2x^2 - x + 2) - 5$ (4-1)

3.

$$\begin{array}{r|rrrrrr} & 1 & -4 & 0 & 9 & 0 & -8 \\ & & 3 & -3 & -9 & 0 & 0 \\ \hline 3 & 1 & -1 & -3 & 0 & 0 & -8 \end{array}$$

$P(3) = -8$ (4-1)

4. 2, -4, -1 (4-1)

5. Since complex zeros come in conjugate pairs, $1 - i$ is a zero. (4-3)

6. (A) Since the graph has x intercepts -2, 0, and 2, these are zeros of $P(x)$. Therefore, $P(x) = (x + 2)x(x - 2) = x^3 - 4x$.

(B) $P(x) \to \infty$ as $x \to \infty$ and $P(x) \to -\infty$ as $x \to -\infty$. (4-1)

7. We form a synthetic division table:

	1	-4	0	2	
-2	1	-6	12	-22	both are lower bounds, since
-1	1	-5	5	-3	both rows alternate in sign
3	1	-1	-3	-7	
4	1	0	0	2	upper bound

(4-2)

8. We investigate $P(1)$ and $P(2)$ by forming a synthetic division table.

	2	-3	1	-5
1	2	-1	0	-5
2	2	1	3	1

Since $P(1)$ and $P(2)$ have opposite signs, there is at least one real zero between 1 and 2. (4-2)

9. The factors of 6 are ±1, ±2, ±3, ±6. (4-3)

10. Using the possibilities found in problem 9, we form a synthetic division table:

	1	-4	1	6	
1	1	-3	-2	4	
2	1	-2	-3	0	2 is a zero

Thus $x^3 - 4x^2 + x + 6 = (x - 2)(x^2 - 2x - 3) = (x - 2)(x - 3)(x + 1)$.
The rational zeros are 2, 3, -1. (4-3)

11. (A) $f(x) = \dfrac{2x - 3}{x + 4} = \dfrac{n(x)}{d(x)}$

The domain of f is the set of all real numbers x such that $d(x) = x + 4 \neq 0$, that is, $(-\infty, -4) \cup (-4, \infty)$. f has an x intercept where $n(x) = 2x - 3 = 0$, that is, $x = \dfrac{3}{2}$.

(B) $g(x) = \dfrac{3x}{x^2 - x - 6} = \dfrac{n(x)}{d(x)}$

The domain of g is the set of all real numbers x such that $d(x) = x^2 - x - 6 \neq 0$, that is, $(x + 2)(x - 3) \neq 0$, that is, $x \neq -2, 3$, or $(-\infty, -2) \cup (-2, 3) \cup (3, \infty)$. g has an x intercept where $n(x) = 3x = 0$, that is, $x = 0$. (4-4)

12. (A) Horizontal asymptote: since $n(x)$ and $d(x)$ have the same degree, the line $y = 2$ is a horizontal asymptote. Vertical asymptotes: zeros of $d(x)$

$$x + 4 = 0$$
$$x = -4$$

(B) Horizontal asymptote: since the degree of $n(x)$ is less than the degree of $d(x)$, the line $y = 0$ is a horizontal asymptote.
Vertical asymptotes: zeros of $d(x)$

$$x^2 - x - 6 = 0$$
$$(x + 2)(x - 3) = 0$$
$$x = -2,\ x = 3$$

(4-4)

13. The graph of a polynomial has to approach ∞ or $-\infty$ as $x \to \infty$ and $x \to -\infty$. This graph approaches zero. (4-1)

14. $F = k\sqrt{x}$ (4-5) **15.** $G = kxy^2$ (4-5) **16.** $H = \dfrac{k}{z^3}$ (4-5)

17. $R = kx^2y^2$ (4-5) **18.** $S = \frac{k}{u^2}$ (4-5) **19.** $T = k\frac{v}{w}$ (4-5)

20. $P(x) = x^3 - 3x^2 - 3x + 4$

(A)

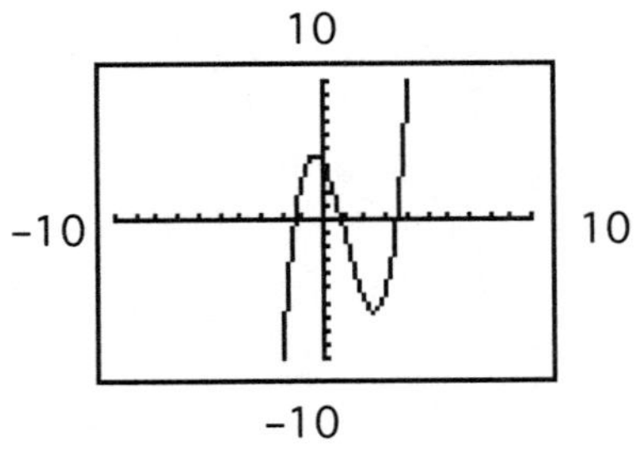

The graph of $P(x)$ has three x intercepts and two turning points. $P(x) \to \infty$ as $x \to \infty$ and $P(x) \to -\infty$ as $x \to -\infty$.

(B) Applying a built-in routine yields:

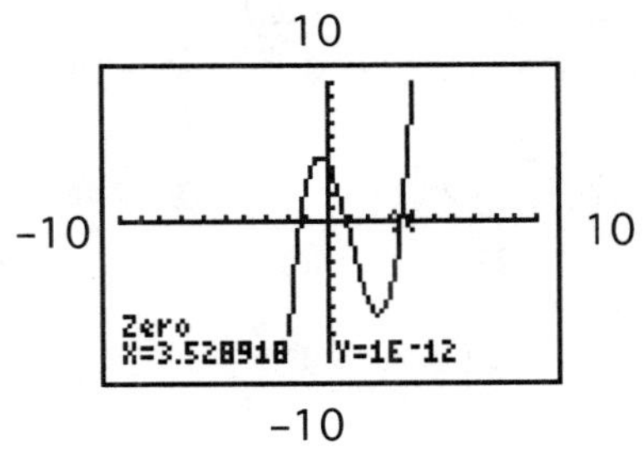

The largest x intercept is 3.53. (4-1)

21. We use synthetic division:

$$\begin{array}{r|rrrrr} & 8 & -14 & -13 & -4 & 7 \\ & & 2 & -3 & -4 & -2 \\ \hline \frac{1}{4} & 8 & -12 & -16 & -8 & 5 \end{array}$$

Thus,

$$P(x) = \left(x - \frac{1}{4}\right)(8x^3 - 12x^2 - 16x - 8) + 5$$

$$P\left(\frac{1}{4}\right) = 5$$ (4-1)

22.

$$\begin{array}{r|rrrr} & 4 & -8 & -3 & -3 \\ & & -2 & 5 & -1 \\ \hline -\frac{1}{2} & 4 & -10 & 2 & -4 \end{array}$$

$$P\left(-\frac{1}{2}\right) = -4$$ (4-1)

23. The quadratic formula tells us that $x^2 - 2x - 1 = 0$ if

$$x = \frac{-b \pm \sqrt{b^2 - 4ac}}{2a} \qquad a = 1,\ b = -2,\ c = -1$$

$$x = \frac{-(-2) \pm \sqrt{(-2)^2 - 4(1)(-1)}}{2(1)}$$

$$= \frac{2 \pm \sqrt{8}}{2}$$

$$= 1 \pm \sqrt{2}$$

Since $1 \pm \sqrt{2}$ are zeros of $x^2 - 2x - 1$, its factors are $x - (1 + \sqrt{2})$ and $x - (1 - \sqrt{2})$, that is, $x^2 - 2x - 1 = [x - (1 + \sqrt{2})][x - (1 - \sqrt{2})]$ (4-1)

24. $x + 1$ will be a factor of $P(x)$ if $P(-1) = 0$.
$P(-1) = 9(-1)^{26} - 11(-1)^{17} + 8(-1)^{11} - 5(-1)^4 - 7 = 9 + 11 - 8 - 5 - 7 = 0$, so the answer is yes, $x + 1$ is a factor. (4-1)

25. The possible rational zeros are ±1, ±2, ±4, ±8, $\pm\frac{1}{2}$. We form a synthetic division table:

	2	−3	−18	−8
1	2	−1	−19	−27
2	2	1	−16	−40
4	2	5	2	0

So $2x^3 - 3x^2 - 18x - 8 = (x - 4)(2x^2 + 5x + 2)$
$= (x - 4)(2x + 1)(x + 2)$

Zeros: 4, $-\frac{1}{2}$, −2 *(4-3)*

26. $(x - 4)(2x + 1)(x + 2)$ *(4-3)*

27. The possible rational zeros are ±1, ±5. We form a synthetic division table:

	1	−3	0	5
1	1	−2	−2	3
5	1	2	10	55
−1	1	−4	4	1
−5	1	−8	40	−195

There are no rational zeros, since all possibilities fail. *(4-3)*

28. $P(x) = 2x^4 - x^3 + 2x - 1$.

Possible rational zeros: ±1, $\pm\frac{1}{2}$

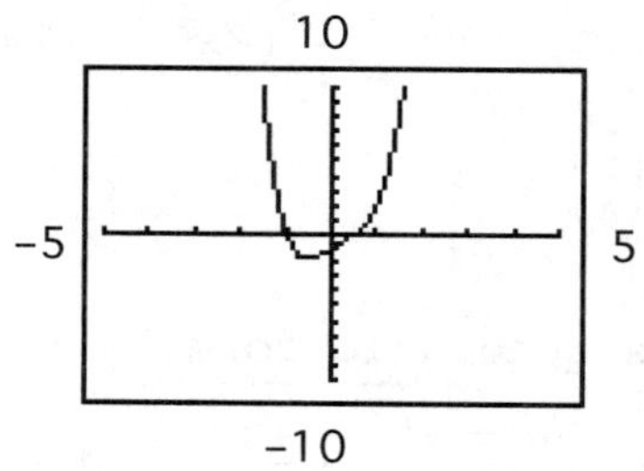

Examining the graph of $y = P(x)$, we see that there is a zero between 0 and 1, and possibly a zero at −1.

We test the likely candidates, and find

$$\begin{array}{r|rrrrr} & 2 & -1 & 0 & 2 & -1 \\ & & -2 & 3 & -3 & 1 \\ \hline -1 & 2 & -3 & 3 & -1 & 0 \end{array}$$

−1 is a zero. So $P(x) = (x + 1)(2x^3 - 3x^2 + 3x - 1) = (x + 1)Q(x)$

Testing $\frac{1}{2}$ in the reduced polynomial $Q(x)$, we obtain

$$\begin{array}{r|rrrr} & 2 & -3 & 3 & -1 \\ & & 1 & -1 & 1 \\ \hline \frac{1}{2} & 2 & -2 & 2 & 0 \end{array}$$

$\frac{1}{2}$ is a zero.

So $P(x) = (x + 1)\left(x - \frac{1}{2}\right)(2x^2 - 2x + 2) = (x + 1)(2x - 1)(x^2 - x + 1)$

To find the remaining zeros, we solve $x^2 - x + 1 = 0$, by the quadratic formula.

$x^2 - x + 1 = 0$

$$x = \frac{-b \pm \sqrt{b^2 - 4ac}}{2a} \qquad a = 1,\ b = -1,\ c = 1$$

$$x = \frac{-(-1) \pm \sqrt{(-1)^2 - 4(1)(1)}}{2(1)}$$

$$x = \frac{1 \pm \sqrt{-3}}{2}$$

$$x = \frac{1 \pm i\sqrt{3}}{2}$$

The four zeros are -1, $\frac{1}{2}$, and $\frac{1 \pm i\sqrt{3}}{2}$ (*4-3*)

29. $(x+1)\left(x - \frac{1}{2}\right)2\left(x - \frac{1 + i\sqrt{3}}{2}\right)\left(x - \frac{1 - i\sqrt{3}}{2}\right) = (x+1)(2x-1)\left(x - \frac{1 + i\sqrt{3}}{2}\right)\left(x - \frac{1 - i\sqrt{3}}{2}\right)$ (*4-3*)

30. The degree is 9. The zeros are 1 (multiplicity 3), -1 (multiplicity 4), i and $-i$. (*4-3*)

31. (A) Let $u = x^2$. Then

$$P(x) = x^4 + 5x^2 - 36$$
$$= u^2 + 5u - 36 = (u+9)(u-4) = (x^2+9)(x^2-4) = (x^2+9)(x+2)(x-2)$$

(B) $x^2 + 9$ has zeros $-3i$ and $3i$, so it factors as $(x + 3i)(x - 3i)$. Thus

$P(x) = (x + 3i)(x - 3i)(x + 2)(x - 2)$ (*4-3*)

32. (A) Examining the graph of $P(x)$, we see that there may be zeros of even multiplicity between -1 and 0, and between 4 and 5, and a possible integer zero at 2.

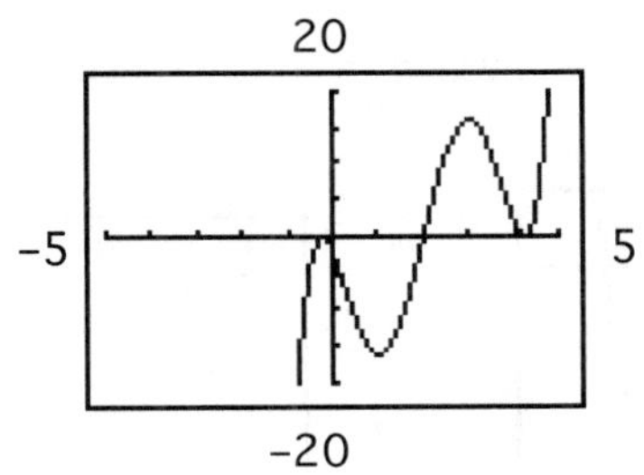

Testing 2, we obtain

	1	-10	30	-20	-15	-2
		2	-16	28	16	2
2	1	-8	14	8	1	0

2 is a zero. Applying built-in routines, we obtain

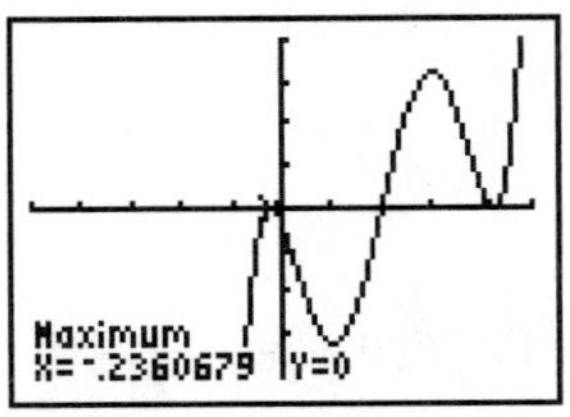

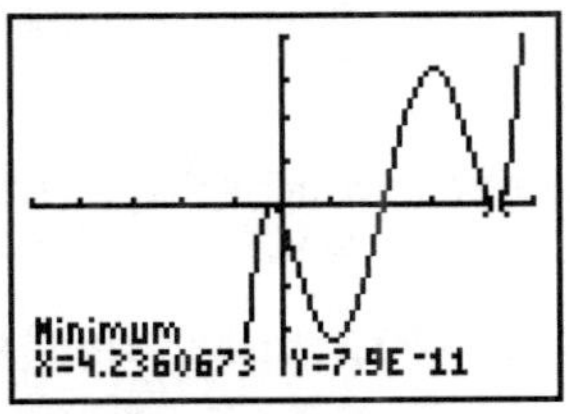

There are zeros of even multiplicity at -0.24 and 4.24. Since $P(x)$ is a fifth-degree polynomial, they must be double zeros and 2 must be a simple zero.

(B) -0.24 can be approximated with a maximum routine; 2 can be approximated by the bisection method; 4.24 can be approximated with a minimum routine. (*4-2*)

33. (A) We form a synthetic division table.

	1	-2	-30	0	-25	
0	1	-2	-30	0	-25	
1	1	-1	-31	-31	-56	
2	1	0	-30	-60	-145	
3	1	1	-27	-81	-268	
4	1	2	-22	-88	-377	
5	1	3	-15	-75	-400	
6	1	4	-6	-36	-241	
7	1	5	5	35	220	7 is an upper bound
-1	1	-3	-27	27	-52	
-2	1	-4	-22	44	-113	
-3	1	-5	-15	45	-160	
-4	1	-6	-6	24	-121	
-5	1	-7	5	-25	100	-5 is a lower bound

(B) We search for the real zero in (6, 7) indicated in the table. We organize our calculations in a table.

Sign Change Interval (a, b)	Midpoint m	Sign of P: $P(a)$	$P(m)$	$P(b)$
(6, 7)	6.5	-	-	+
(6.5, 7)	6.75	-	+	+
(6.5, 6.75)	6.625	-	+	+
(6.5, 6.625)	6.5625	-	-	+
(6.5625, 6.625)	We stop here	-		+

Since each endpoint rounds to 6.6, a real zero lies on this interval and is given by 6.6 to one decimal place accuracy. 4 intervals were required.

(C) Graphing and applying a built-in routine, we obtain:

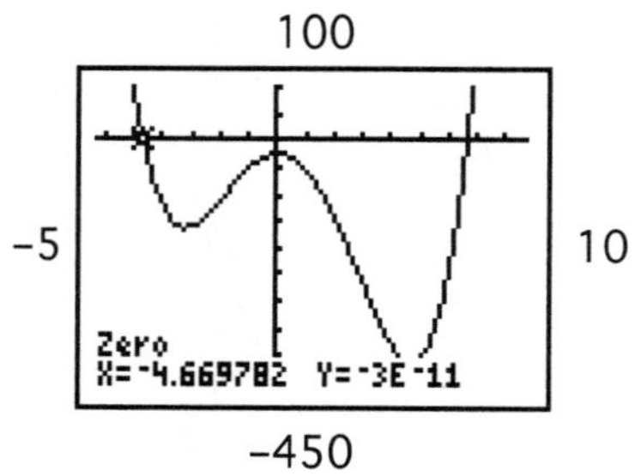

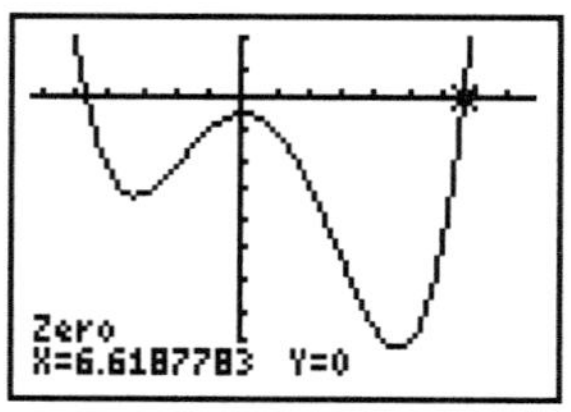

The zeros are -4.67 and 6.62. *(4-2)*

34. $f(x) = \dfrac{x - 1}{2x + 2} = \dfrac{n(x)}{d(x)}$

(A) The domain of f is the set of all real numbers x such that $d(x) = 2x + 2 \neq 0$, that is $(-\infty, -1) \cup (-1, \infty)$.

f has an x intercept where $n(x) = x - 1 = 0$, that is, $x = 1$. $f(0) = -\frac{1}{2}$, hence f has a y intercept at $y = -\frac{1}{2}$

(B) Vertical asymptote: $x = -1$. Horizontal asymptote: since $n(x)$ and $d(x)$ have the same degree, the line $y = \frac{1}{2}$ is a horizontal asymptote.

(C) Complete the sketch. Plot a few additional points.

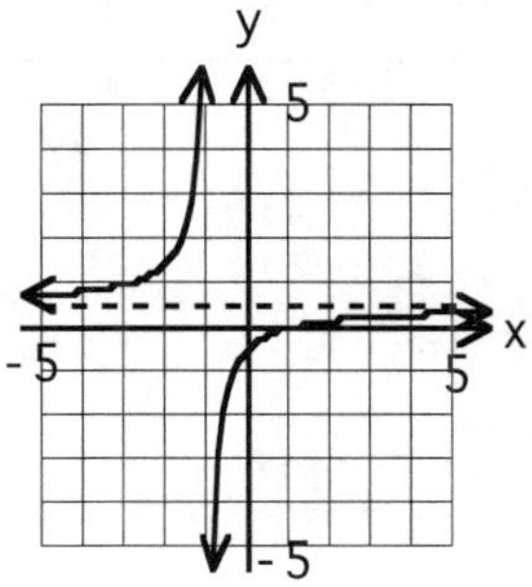

(4-4)

35. (A) Find the zeros of $P(x) = x^3 - 5x + 4$ graphically:

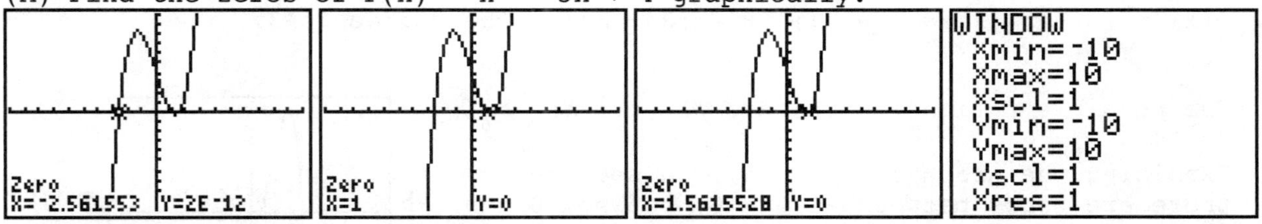

The zeros are -2.562, 1, and 1.562. (Two are very close together, so you may need to look closely at that portion of the graph to see that there are two.) $P(x)$ is negative (below the x axis) on $(-\infty, -2.562) \cup (1, 1.562)$.

(B) Keeping $y_1 = x^3 - 5x + 4$, graph $y_2 = 2$.

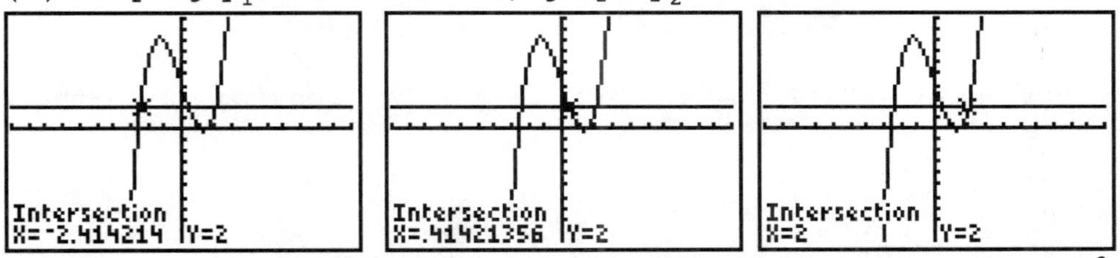

The graphs intersect at -2.414, 0.414, and 2. The graph of $x^3 - 5x + 4$ is below the graph of 2 on $(-\infty, -2.414) \cup (0.414, 2)$. (4-2)

36. The graph of a rational function can only be discontinuous at $x = 0$ because of a break in the graph. (4-4)

37. Write $B = \frac{k}{\sqrt{c}}$. Substitute $B = 5$ and $C = 4$ and solve for k.

$$5 = \frac{k}{\sqrt{4}}$$
$$k = 10$$

The equation of variation is $B = \frac{10}{\sqrt{c}}$. When $c = 25$,

$$B = \frac{10}{\sqrt{25}} = \frac{10}{5} = 2$$ (4-5)

38. Write $D = kxy$. Substitute $D = 10$, $x = 3$, and $y = 2$ and solve for k.

$$10 = k(3)(2)$$
$$k = \frac{5}{3}$$

The equation of variation is $D = \frac{5}{3}xy$. When $x = 9$ and $y = 8$,

$$D = \frac{5}{3}(9)(8) = 120$$ (4-5)

39.

$$\begin{array}{r|lccc} & 1 & 0 & 3 & 2 \\ & & 1+i & 2i & 1+5i \\ \hline 1+i & 1 & 1+i & 3+2i & 3+5i \end{array}$$

$$(1+i)^2 = (1+i)(1+i) = 1 + 2i + i^2 = 1 + 2i - 1 = 2i$$
$$(1+i)(3+2i) = 3 + 5i + 2i^2 = 3 + 5i - 2 = 1 + 5i$$
$$P(x) = [x^2 + (1+i)x + (3+2i)][x - (1+i)] + 3 + 5i$$ (4-1)

40. $P(x) = \left(x + \frac{1}{2}\right)^2 (x+3)(x-1)^3$. The degree is 6. (4-3)

41. $P(x) = (x+5)[x - (2-3i)][x - (2+3i)]$. The degree is 3. (4-3)

42. The possible rational zeros are $\pm 1, \pm 2, \pm 4, \pm\frac{1}{2}$.

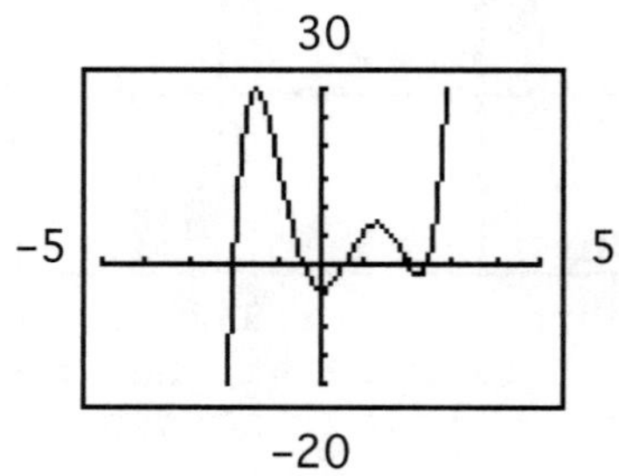

Examining the graph of $y = P(x)$, we see that there are zeros between -1 and 0, between 0 and 1, between 2 and 3, and possible integer zeros at -2 and 2. Testing 2, we obtain

$$\begin{array}{r|rrrrrr} & 2 & -5 & -8 & 21 & 0 & -4 \\ & & 4 & -2 & -20 & 2 & 4 \\ \hline 2 & 2 & -1 & -10 & 1 & 2 & 0 \end{array}$$

2 is a zero. So $P(x) = (x - 2)(2x^4 - x^3 - 10x^2 + x + 2)$. Testing -2 in the reduced polynomial, we obtain

$$\begin{array}{r|rrrrr} & 2 & -1 & -10 & 1 & 2 \\ & & -4 & 10 & 0 & -2 \\ \hline -2 & 2 & -5 & 0 & 1 & 0 \end{array}$$

-2 is a zero.

Hence $P(x) = (x - 2)(x + 2)(2x^3 - 5x^2 + 1)$. $2x^3 - 5x^2 + 1$ has been shown (see Section 4-2, problem 37 for details) to have zeros $\frac{1}{2}$, $1 \pm \sqrt{2}$.

Hence $P(x)$ has zeros $\frac{1}{2}$, ± 2, $1 \pm \sqrt{2}$. (4-3)

43. $(x - 2)(x + 2)\left(x - \frac{1}{2}\right)2[x - (1 - \sqrt{2})][x - (1 + \sqrt{2})]$

$= (x - 2)(x + 2)(2x - 1)[x - (1 - \sqrt{2})][x - (1 + \sqrt{2})]$ (4-3)

44. Graphing $y = P(x)$, we obtain:

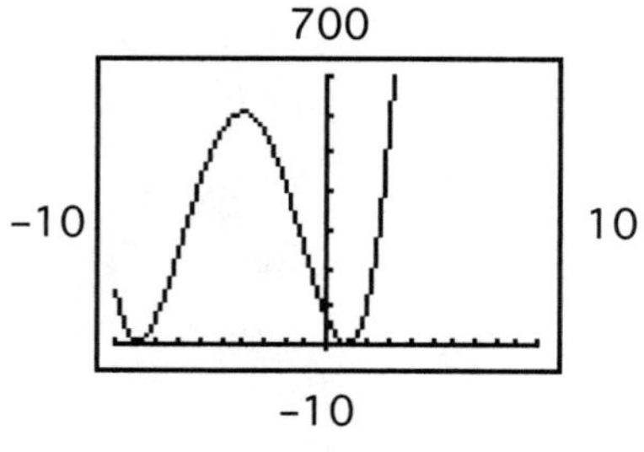

Clearly there is a local maximum near -4. Applying a built-in routine, we obtain

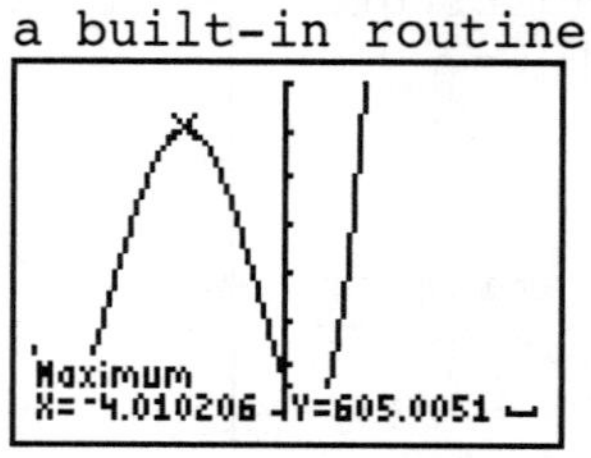

To find the other required points, we need different windows. Redrawing and applying built-in routines, we obtain

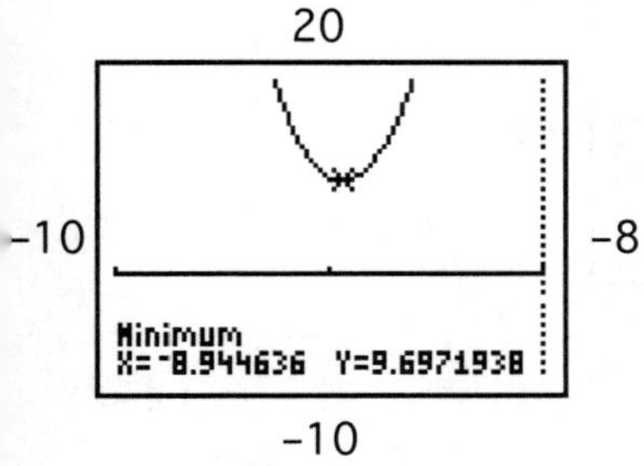

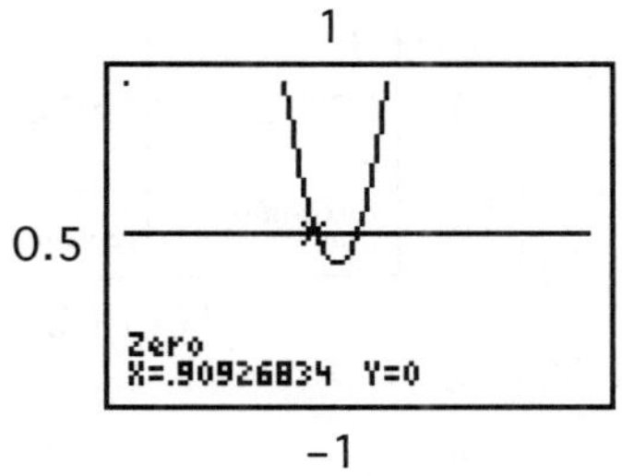

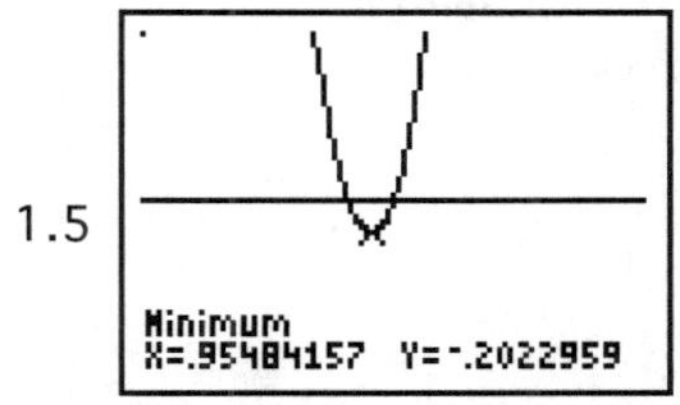

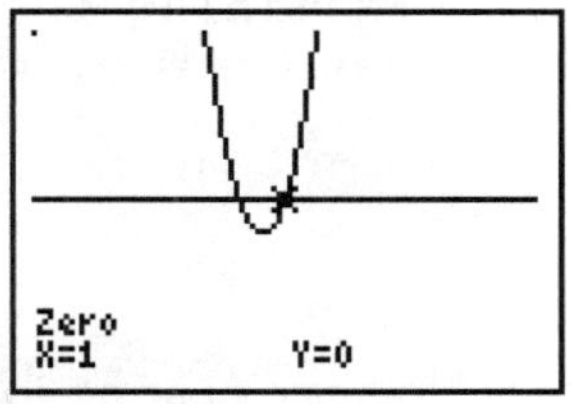

Summarizing, $P(x)$ has zeros at 0.91 and 1. It has a local minimum $P(-8.94) \approx 9.7$. It has a local maximum $P(-4.01) \approx 605.01$. It has a local minimum $P(0.95) \approx -0.20$. (*4-2*)

45. $P(x)$ changes sign three times. Therefore, it has three zeros and its minimal degree is 3. (*4-1*)

46. Since $1 + 2i$ is a zero, $1 - 2i$ is also a zero. Hence $[x - (1 - 2i)][x - (1 + 2i)] = [(x - 1) + 2i][(x - 1) - 2i] = (x - 1)^2 - 4i^2$ $= x^2 - 2x + 5$ is a factor. Since $P(x)$ is a cubic polynomial, it must be of the form $a(x - r)(x^2 - 2x + 5)$. Since the constant term of this polynomial, $-5ar$, must be an integer, r must be a rational number. Thus there can be no irrational zeros. (*4-3*)

47. (A) Since $x^3 - 27$ is a cubic polynomial, it has 3 zeros and there are 3 cube roots of 27.

(B) $x^3 - 27 = (x - 3)(x^2 + 3x + 3^2) = (x - 3)(x^2 + 3x + 9)$. We solve $x^2 + 3x + 9 = 0$ by applying the quadratic formula with $a = 1$, $b = 3$, $c = 9$, to obtain $x = \frac{-3 \pm \sqrt{3^2 - 4(1)(9)}}{2(1)} = \frac{-3 \pm \sqrt{-27}}{2} = \frac{-3 \pm 3i\sqrt{3}}{2}$ or $-\frac{3}{2} \pm \frac{3i}{2}\sqrt{3}$ (*4-3*)

48. (A) We form a synthetic division table.

	1	2	-500	0	-4,000
0	1	2	-500	0	-4,000
10	1	12	-380	-3,800	-42,000
20	1	22	-60	-1,200	-28,000
30	1	32	460	13,800	410,000
-10	1	-8	-420	4,200	-46,000
-20	1	-18	-140	2,800	-60,000
-30	1	-28	340	-10,200	302,000

From the table, 30 is an upper bound and -30 is a lower bound.

(B) Graphing $P(x)$ in the window suggested by part (A), we obtain

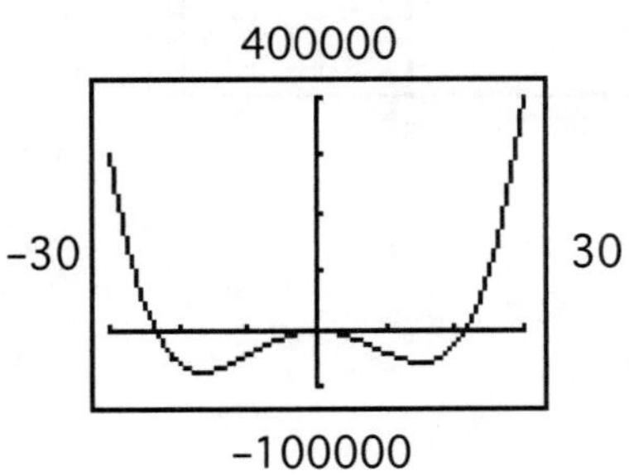

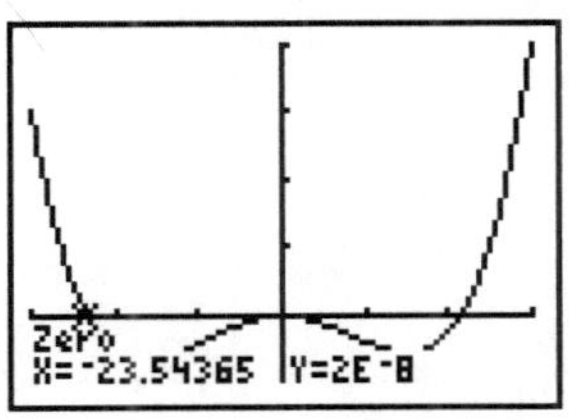

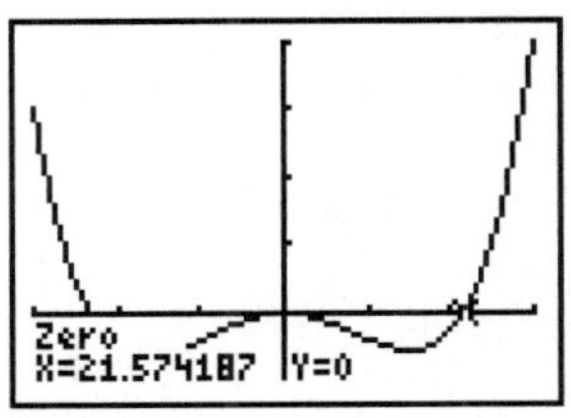

The zeros are -23.54 and 21.57. Examining the graph more closely near $x = 0$, we obtain

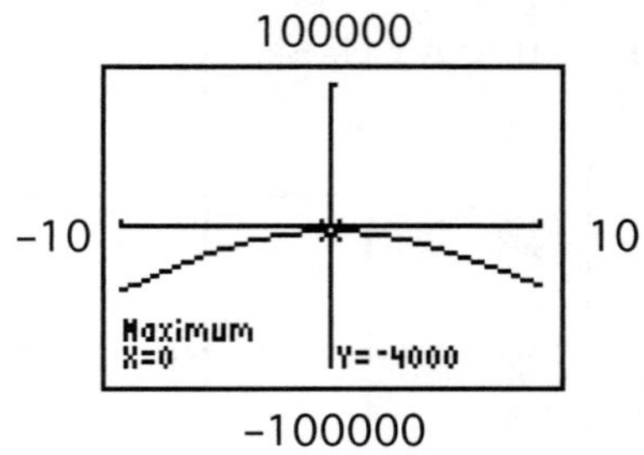

There is no other real zero. (*4-2*)

49. $f(x) = \dfrac{x^2 + 2x + 3}{x + 1} = \dfrac{n(x)}{d(x)}$

Intercepts. There are no real zeros of $n(x) = x^2 + 2x + 3$. No x intercept
$f(0) = 3$ y intercept
Vertical asymptotes. Real zeros of $d(x) = x + 1$ $x = -1$
Horizontal asymptotes. Since the degree of $n(x)$ is greater than the degree of $d(x)$, there is no horizontal asymptote.
Oblique asymptote:

$$\begin{array}{r|l} & x + 1 \\ \hline x + 1 & x^2 + 2x + 3 \\ & \underline{x^2 + x} \\ & \quad x + 3 \\ & \quad \underline{x + 1} \\ & \qquad 2 \end{array}$$

Thus, $f(x) = x + 1 + \dfrac{2}{x + 1}$.
Hence, the line $y = x + 1$ is an oblique asymptote.

Complete the sketch. Plot a few additional points.

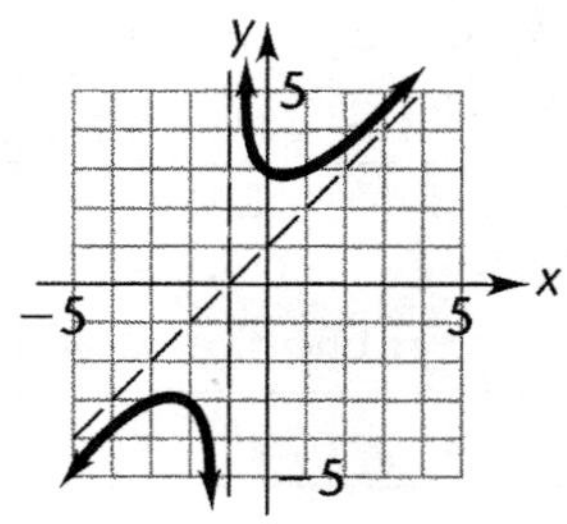

(*4-4*)

50. Graphing $y = f(x)$, we obtain the graph at the right.

From the graph, we can see that $f(x) \to 2$ as $x \to \infty$ and $f(x) \to -2$ as $x \to -\infty$; the lines $y = 2$ and $y = -2$ are horizontal asymptotes.

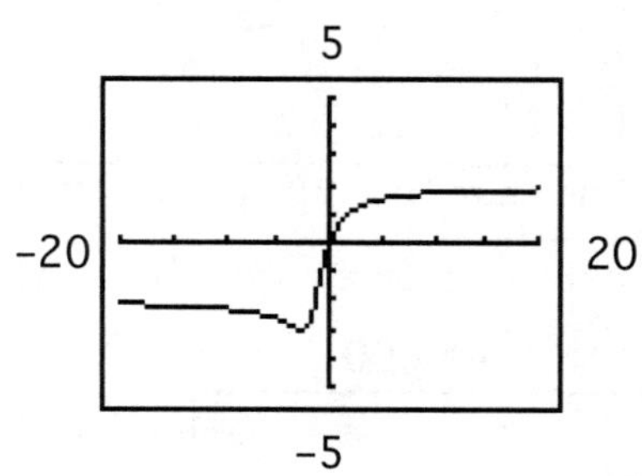

(*4-4*)

51. (A) $\dfrac{x - 2}{5 - x} \le 0$

Common Error:
It is not correct to "multiply both sides" by $5 - x$.

Let $f(x) = \dfrac{p(x)}{q(x)} = \dfrac{x - 2}{5 - x}$

The zero of $p(x) = x - 2$ is 2. The zero of $q(x) = 5 - x$ is 5. These two zeros partition the x axis into the three intervals shown in the table. A test number is chosen from each interval to determine the sign of f.

Interval	Test number x	$f(x)$	Sign of f
$(-\infty, 2)$	1	$-\frac{1}{4}$	−
$(2, 5)$	4	2	+
$(5, \infty)$	6	−4	−

The equality is satisfied at $x = 2$, but not at $x = 5$. We conclude that the solution set is $(-\infty, 2] \cup (5\ \infty)$.

(B) $\dfrac{17}{x + 3} > \dfrac{5}{x}$

$$\frac{17}{x + 3} - \frac{5}{x} > 0$$

$$\frac{17x - 5(x + 3)}{x(x + 3)} > 0$$

$$\frac{12x - 15}{x(x + 3)} > 0$$

Let $f(x) = \dfrac{p(x)}{q(x)} = \dfrac{12x - 15}{x(x + 3)}$

The zero of $p(x) = 12x - 15$ is $\dfrac{15}{12}$ or $\dfrac{5}{4}$. The zeros of $q(x) = x(x + 3)$ are 0 and −3. These three zeros partition the x axis into the four intervals shown in the table. A test number is chosen from each interval to determine the sign of $f(x)$.

Interval	Test number x	$f(x)$	Sign of f
$(-\infty, -3)$	−4	$-\frac{63}{4}$	−
$(-3, 0)$	−1	$\frac{27}{2}$	+
$\left(0, \frac{5}{4}\right)$	1	$-\frac{3}{4}$	−
$\left(\frac{5}{4}, \infty\right)$	2	$\frac{9}{10}$	+

We conclude that the solution set is $(-3, 0) \cup \left(\dfrac{5}{4}, \infty\right)$. (4-4)

52. (A) $\dfrac{x^2 - 3}{x^3 - 3x + 1} \le 0$

zeros of the numerator: $x^2 - 3 = 0$

$$x^2 = 3$$

$$x = \sqrt{3}, -\sqrt{3} \text{ or } 1.732, -1.732$$

zeros of the denominator: Find graphically:

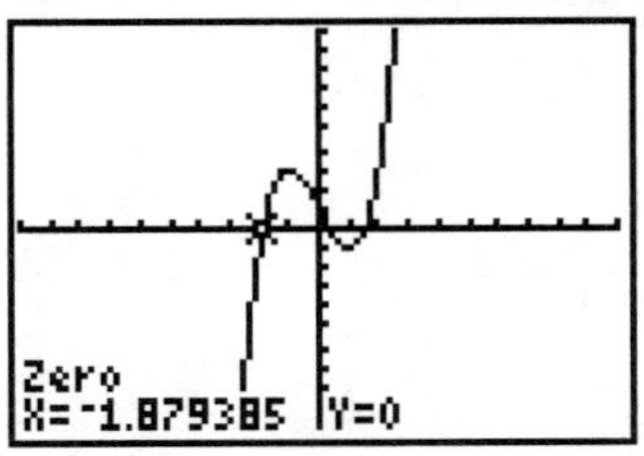

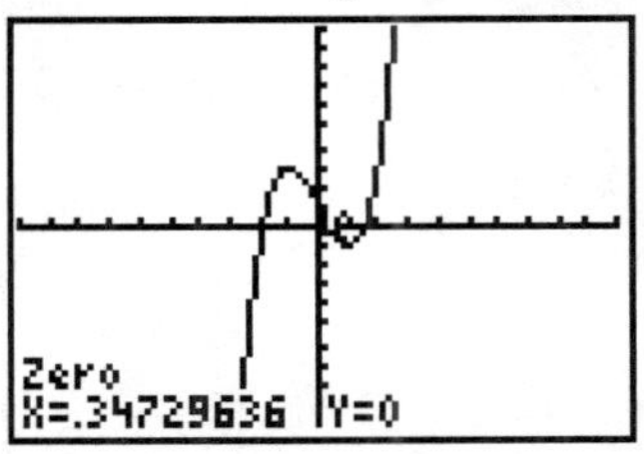

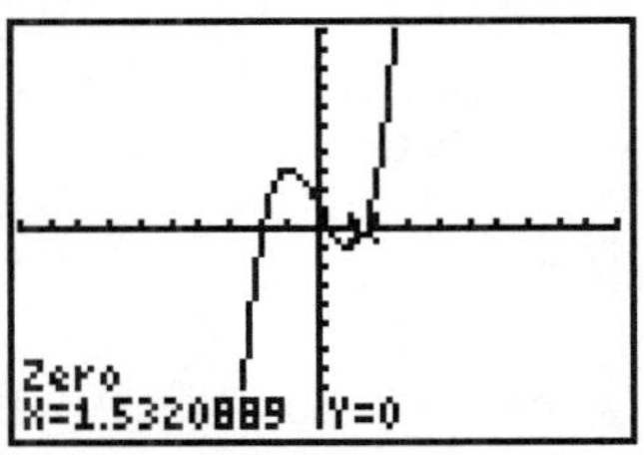

The zeros are -1.879, 0.347, 1.532.

Interval	$(-\infty, -1.879)$	$(-1.879, -1.732)$	$(-1.732, 0.347)$	$(0.347, 1.532)$	$(1.532, 1.732)$	$(1.732, \infty)$
Test number x	-2	-1.8	0	1	1.6	2
$f(x)$	-1	.423	-3	2	-1.49	$\frac{1}{3}$
Sign of f	−	+	−	+	−	+

The expression is negative on $(-\infty, -1.879)$, $(-1.732, 0.347)$, and $(1.532, 1.732)$. We should include -1.732 and 1.732 (since the expression is zero there) and exclude -1.879, 0.347, and 1.532 (since the expression is undefined there.)

The solution is $(-\infty, -1.879) \cup [-1.732, 0.347) \cup (1.532, 1.732]$.

Graphical check:

The graph is below the x axis on $(-\infty, -1.879) \cup (-1.732, 0.347) \cup (1.532, 1.732)$, and intersects the x axis at -1.732 and 1.732.

(B) $\dfrac{x^2 - 3}{x^3 - 3x + 1} > \dfrac{5}{x^2}$

$$\frac{x^2 - 3}{x^3 - 3x + 1} - \frac{5}{x^2} > 0$$

$$\frac{x^2(x^2 - 3)}{x^2(x^3 - 3x + 1)} - \frac{5(x^3 - 3x + 1)}{x^2(x^3 - 3x + 1)} > 0$$

$$\frac{x^4 - 3x^2 - 5x^3 + 15x - 5}{x^2(x^3 - 3x + 1)} > 0$$

zeros of the numerator: Find graphically:

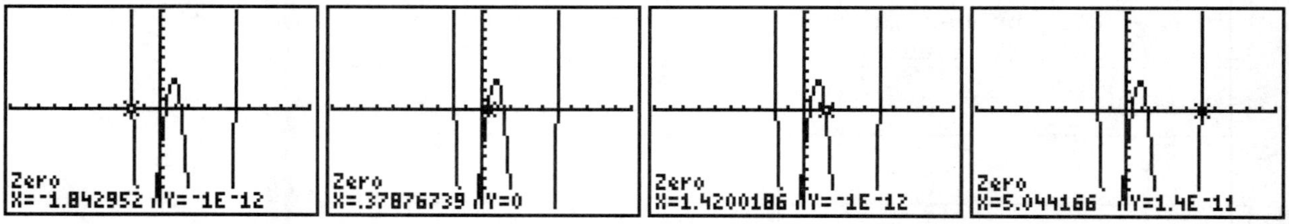

The zeros are -1.843, 0.379, 1.420, and 5.044.

zeros of the denominator:

$x^2(x^3 - 3x + 1 = 0$

$x^2 = 0 \qquad x^3 - 3x + 1 = 0$

$x = 0 \qquad x = -1.879, 0.347, 1.532$ (From part A)

Interval	$(-\infty, -1.879)$	$(-1.879, -1.843)$	$(-1.843, 0)$	$(0, 0.347)$	$(0.347, 0.379)$
Test number x	-2	-1.85	-1	0.1	0.35
$f(x)$	-2.25	0.47	-5.7	-504.3	363
Sign of f	−	+	−	−	+

Interval	$(0.379, 1.420)$	$(1.420, 1.532)$	$(1.532, 5.044)$	$(5.044, \infty)$
Test number x	1	1.5	2	6
$f(x)$	-3	3.8	-0.92	.03
Sign of f	−	+	−	+

The expression is positive on $(-1.879, -1.843) \cup (0.347, 0.379) \cup (1.420, 1.532) \cup (5.044, \infty)$ *(4-4)*

53. The potential rational zeros are ±1, ±2, and ±4. A look at the graph shows 3 zeros, none of which occur at integer values, so none of the candidates for rational zeros are actually zeros.

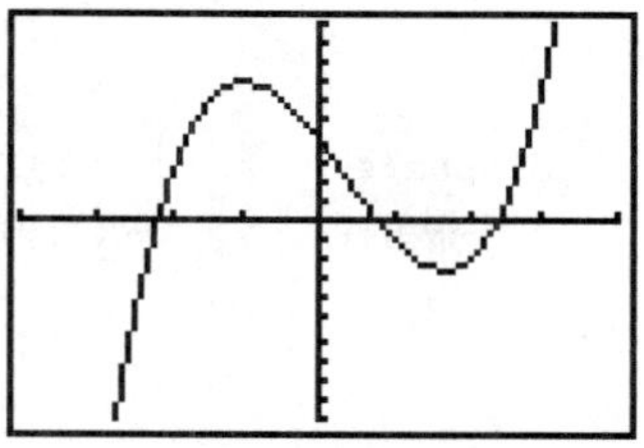

(*4-3*)

54. To have real zeros -3, 0, and 2, we should have factors $(x + 3)$, x, and $(x - 2)$ in the numerator. To have vertical asymptotes $x = -1$ and $x = 4$, we should have factors $(x + 1)$ and $(x - 4)$ in the numerator. At the moment, our function is

$$f(x) = \frac{x(x + 3)(x - 2)}{(x + 1)(x - 4)}$$

But the degree of the numerator is greater than the degree of the denominator, so there will be no horizontal asymptote. We can fix that by squaring one of the factors in the denominator. Multiplying the numerator by 5 will then make $y = 5$ a horizontal asymptote.

$$f(x) = \frac{5x(x + 3)(x - 2)}{(x + 1)^2(x - 4)}$$

(There are other answers.) (*4-4*)

55. In the given figure, let y = height of door
$2x$ = width of door
Then Area of door = 48 = $2xy$

Since (x, y) is a point on the parabola $y = 16 - x^2$, its coordinates satisfy the equation of the parabola.

Hence
$$48 = 2x(16 - x^2)$$
$$48 = 32x - 2x^3$$
$$2x^3 - 32x + 48 = 0$$
$$x^3 - 16x + 24 = 0$$

The possible rational solutions of this equation are ±1, ±2, ±3, ±4, ±6, ±8, ±12, ±24. Testing the likely candidates, we obtain

	1	0	-16	24
		2	4	-24
2	1	2	-12	0

2 is a solution.
Thus the equation can be factored
$(x - 2)(x^2 + 2x - 12) = 0$.
To find the remaining zeros, we solve $x^2 + 2x - 12 = 0$, by completing the square.

$$x^2 + 2x = 12$$
$$x^2 + 2x + 1 = 13$$
$$(x + 1)^2 = 13$$
$$x + 1 = \pm\sqrt{13}$$
$$x = -1 + \sqrt{13} \text{ (discarding the negative solution)} \approx 2.61$$

Thus the positive zeros are $x = 2, 2.61$. Thus the dimensions of the door are either $2x = 4$ feet by $16 - x^2 = 12$ feet, or $2x = 5.2$ feet by $16 - x^2 = 9.2$ feet. (*4-2*)

56. We note:

$$\begin{pmatrix}\text{Volume}\\ \text{of}\\ \text{silo}\end{pmatrix} = \begin{pmatrix}\text{Volume of}\\ \text{hemisphere}\\ \text{of radius } x\end{pmatrix} + \begin{pmatrix}\text{Volume of}\\ \text{cylinder with}\\ \text{radius } x\text{, height 18}\end{pmatrix}$$

$$\begin{aligned} 486\pi &= \frac{2}{3}\pi x^3 + \pi x^2 \cdot 18 \\ 486 &= \frac{2}{3}x^3 + 18x^2 \\ 0 &= x^3 + 27x^2 - 729 \end{aligned}$$

There are no rational zeros of $P(x) = x^3 + 27x^2 - 729$ (details omitted). Examining the graph of $y = P(x)$, we obtain, for positive x: The radius is 4.8 feet.

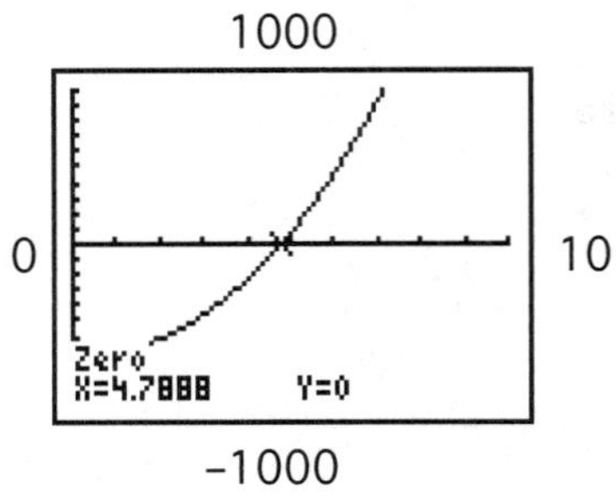

(4-2)

57.

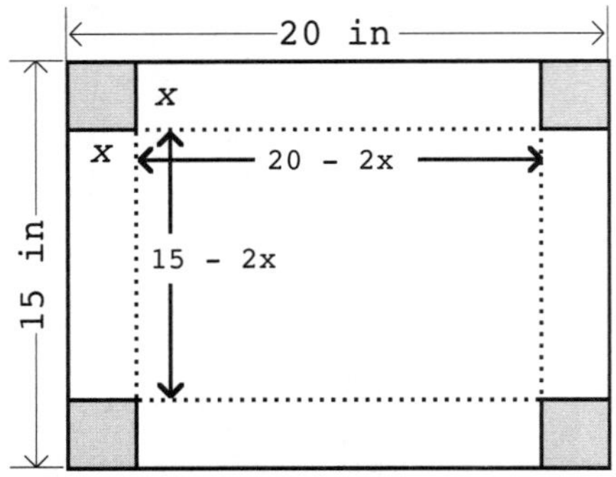

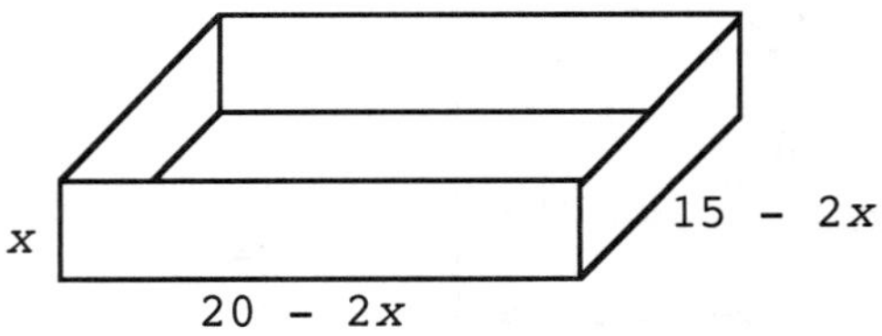

From the above figures it should be clear that

V = length × width × height = $(20 - 2x)(15 - 2x)x \quad 0 < x < 7.5$

We solve

$$\begin{aligned} (20 - 2x)(15 - 2x)x &= 300 \\ 300x - 70x^2 + 4x^3 &= 300 \\ 4x^3 - 70x^2 + 300x - 300 &= 0 \end{aligned}$$

There are no rational zeros of $P(x) = 4x^3 - 70x^2 + 300x - 300$ (details omitted).

Examining the graph of $y = P(x)$ for $0 < x < 7.5$, we obtain

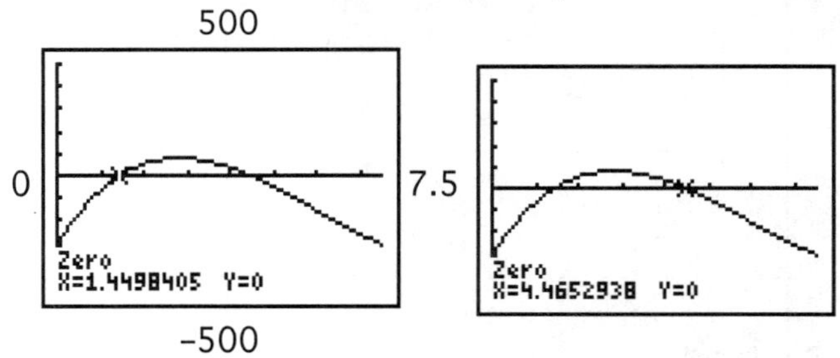

$x = 1.5$ inches or $x = 4.5$ inches.

(4-2)

58. Let $(x, 2x)$ be a point on the graph of $y = 2x$. Then the distance from (1, 4) to $(x, 2x)$ must equal 3 units. Applying the distance formula, we have

$$\begin{aligned} \sqrt{(x - 1)^2 + (2x - 4)^2} &= 3 \\ (x - 1)^2 + (2x - 4)^2 &= 9 \\ x^2 - 2x + 1 + 4x^2 - 16x + 16 &= 9 \\ 5x^2 - 18x + 8 &= 0 \end{aligned}$$

Applying the quadratic formula with $a = 5$, $b = -18$, $c = 8$, we have

$$x = \frac{-(-18) \pm \sqrt{(-18)^2 - 4(5)(8)}}{2(5)}$$

$$x = \frac{18 \pm \sqrt{164}}{10}$$

To one decimal place, this yields $x = 0.5$ and $x = 3.1$. Thus, the two points required are (0.5, 1) and (3.1, 6.2).

59. Write $F = \frac{k}{r}$. Let $F_1 = \frac{k}{r_1}$ and $r_2 = 2r_1$. Then

$$F_2 = \frac{k}{r_2} = \frac{k}{2r_1} = \frac{1}{2}\,\frac{k}{r_1} = \frac{1}{2}F_1.$$

The force would be $\frac{1}{2}$ the original. (4-5)

60. $v = k\frac{\sqrt{T}}{\sqrt{w}}$ (4-5)

61. Write $A = kWt$. Substitute $A = A_0$, $W = 10$ and $t = 8$ and solve for k.

$$A_0 = k(10)(8)$$
$$k = \frac{A_0}{80}$$

The equation of variation for this job is $A = \frac{A_0}{80}Wt$.

Substitute $A = A_0$ (again) and $W = 4$ and solve for t.

$$A_0 = \frac{A_0}{80}(4)t$$
$$A_0 = \frac{A_0}{20}t$$
$$t = 20 \text{ days}$$

(4-5)

62. Write $I = kpr$. Substitute $I = 8$, $p = 100$, and $r = 0.04$ and solve for k.

$$8 = k(100)(0.04)$$
$$k = 2$$

The equation of variation is $I = 2pr$. When $p = 150$ and $r = 0.03$,

$$I = 2(150)(0.03)$$
$$I = \$9.00$$

(4-5)

63. (A)

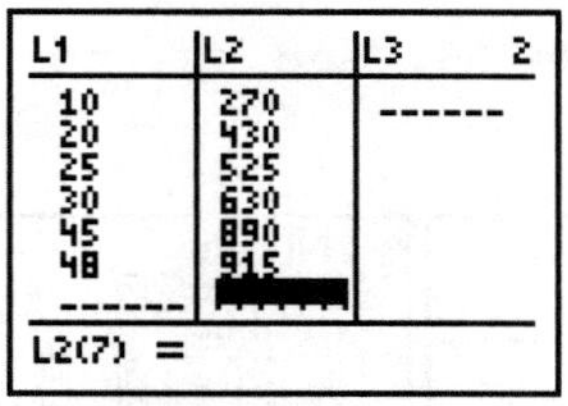

Enter the data.

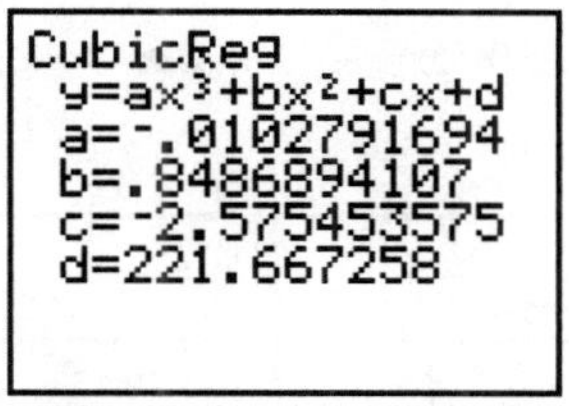

Compute the regression equation.

(B)

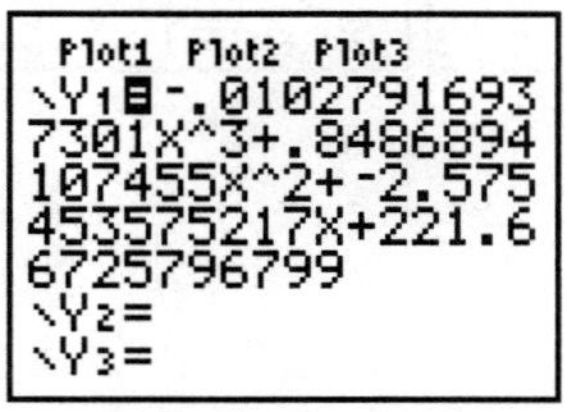

Transfer the regression equation to the equation editor.

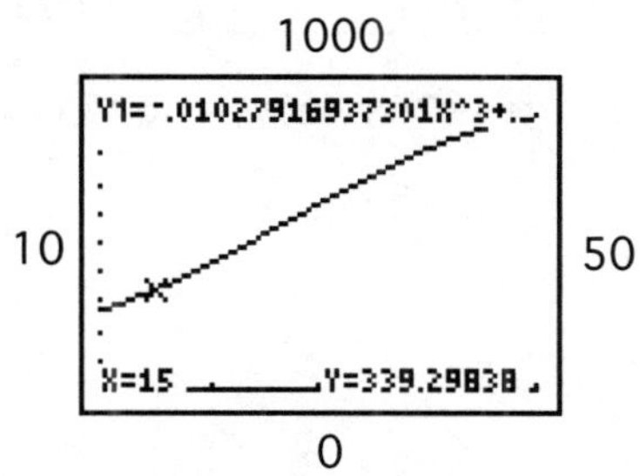

Graph the regression equation and determine the value of y corresponding to $x = 15$.

The model predicts that 339 refrigerators would be sold if 15 ads were placed.

(C) 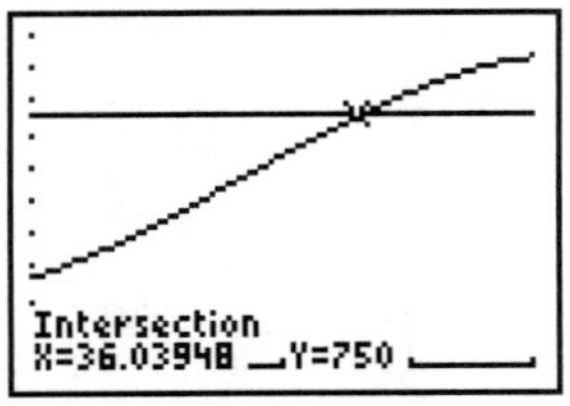

Use a built-in routine to determine the value of x that corresponds to $y = 750$.

The model predicts that 750 refrigerators would be sold if 36 ads were placed.

(4-1)

64. (A) Since the percentage figures in the second column decrease, then increase, then decrease again with increasing age, this might correspond to a function with a local minimum and a local maximum, thus a cubic equation rather than linear or quadratic.

(B) The independent variable is age and the dependent variable is percentage, so we enter the age column as L_1 and the percentage column as L_2. Then choose cubic regression from the STAT CALC menu.

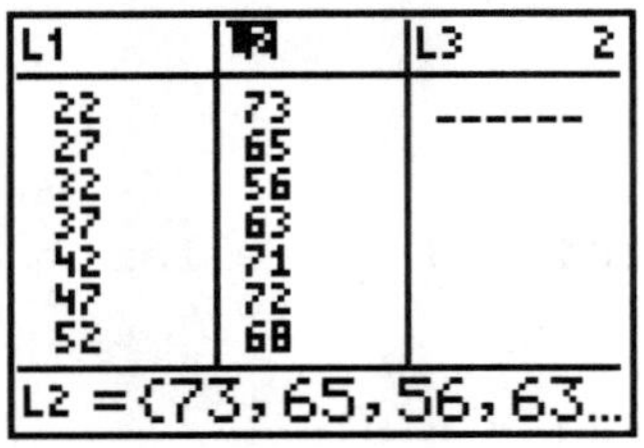

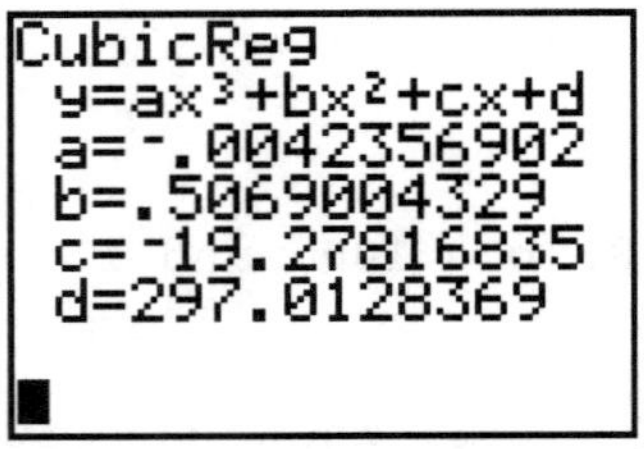

(Note that the entire list is not visible on the graphic.)

The cubic regression model is $y = -0.00424x^3 + 0.507x^2 - 19.3x + 297$.

(C) Graph $y_1 = -0.00424x^3 + 0.507x^2 - 19.3x + 297$, $y_2 = 65$, and use the intersect command.

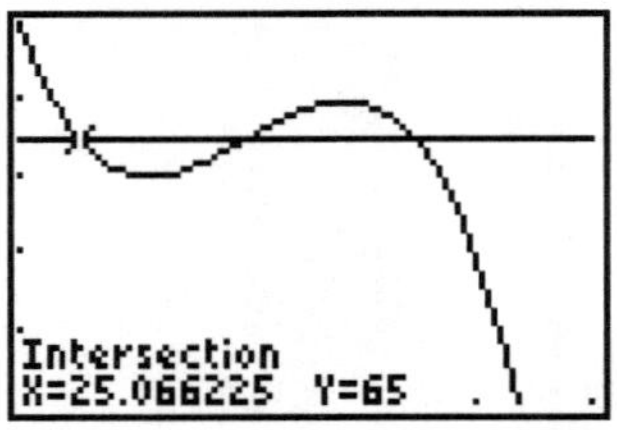

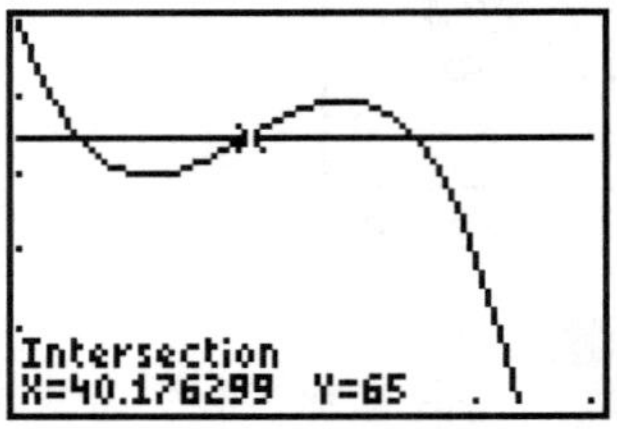

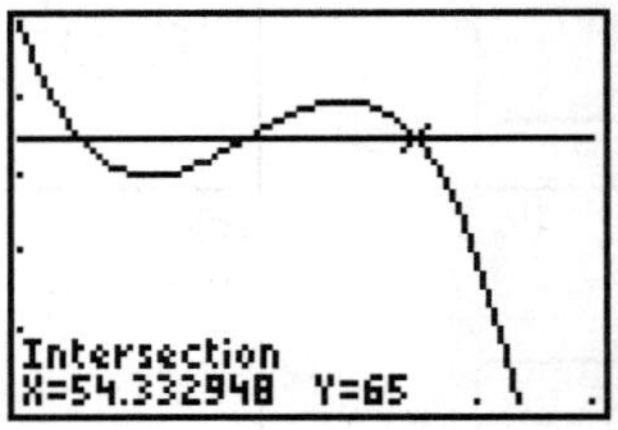

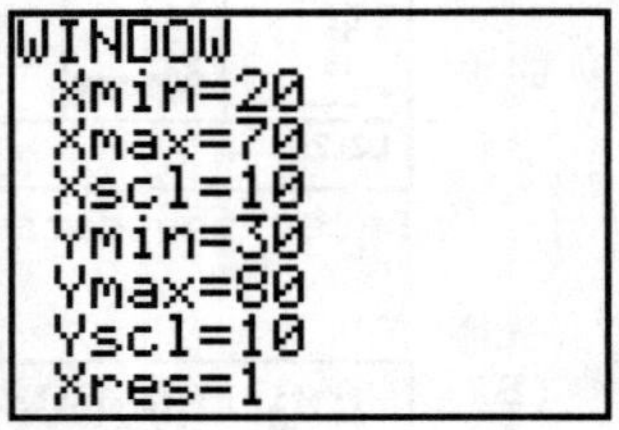

The percentage is 65 for ages 25, 40, and 54. *(4-1)*

CHAPTER 5 Exponential and Logarithmic Functions

Section 5-1

1. (A) The graph of $y = (0.2)^x$ is decreasing and passes through the point $(-1, 0.2^{-1}) = (-1, 5)$. This corresponds tc graph g.

(B) The graph of $y = 2^x$ is increasing and passes through the point (1, 2). This corresponds to graph n.

(C) The graph of $y = \left(\frac{1}{3}\right)^x$ is decreasing and passes through the point (-1, 3). This corresponds to graph f.

(D) The graph of $y = 4^x$ is increasing and passes through the point (1, 4). This corresponds to graph m.

3. 16.24 5. 7.524 7. 1.649 9. 4.469

11. $2^e 5^e = (2\cdot5)^e = 10^e$ 13. $\left(e^{\sqrt{2}}\right)^{\sqrt{2}} = e^{\sqrt{2}\cdot\sqrt{2}} = e^2$

15. $10^{3x-1}10^{4-x} = 10^{3x-1+4-x} = 10^{2x+3}$ 17. $\frac{3x}{3^{1-x}} = 3^{x-(1-x)} = 3^{x-1+x} = 3^{2x-1}$

19. $\left(\frac{4^x}{5^y}\right)^{3z} = \frac{4^{3xz}}{5^{3yz}}$ 21. $\frac{e^{5x}}{e^{2x+1}} = e^{5x-(2x+1)} = e^{5x-2x-1} = e^{3x-1}$

23. This is a simple exponential function. It has horizontal asymptote $y = 0$.

25. The graph is the same as the graph of $y = e^{-x}$ shifted up 2 units. Horizontal asymptote: $y = 2$.

27. The graph is the same as the graph of $y = 2e^{3t}$ shifted down 5 units. Horizontal asymptote: $y = -5$.

29. The graph is the same as the graph of $y = -e^{-x^2}$ shifted up 1 unit. Horizontal asymptote: $y = 1$.

31. No horizontal asymptote.

33. The graph of g is the same as the graph of f stretched vertically by a factor of 3. Therefore g is increasing and the graph has horizontal asymptote $y = 0$.

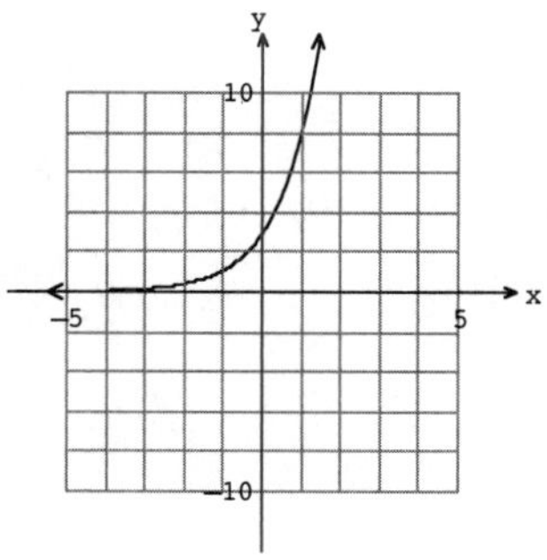

35. The graph of g is the same as the graph of f reflected through the y axis and shrunk vertically by a factor of $\frac{1}{3}$. Therefore g is decreasing and the graph has horizontal asymptote $y = 0$.

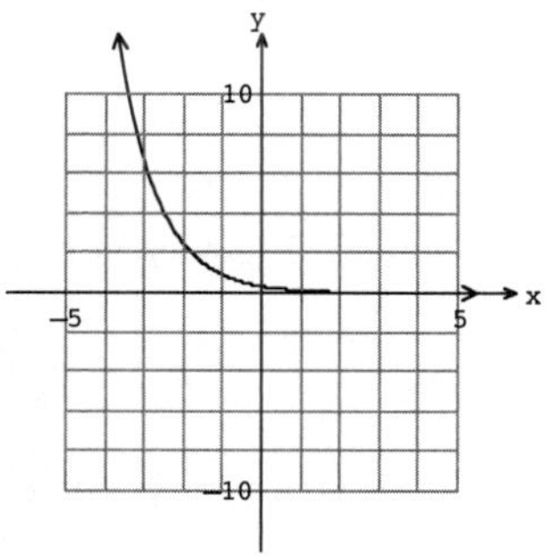

37. The graph of g is the same as the graph of f shifted upward 2 units. Therefore g is increasing and the graph has horizontal asymptote $y = 2$.

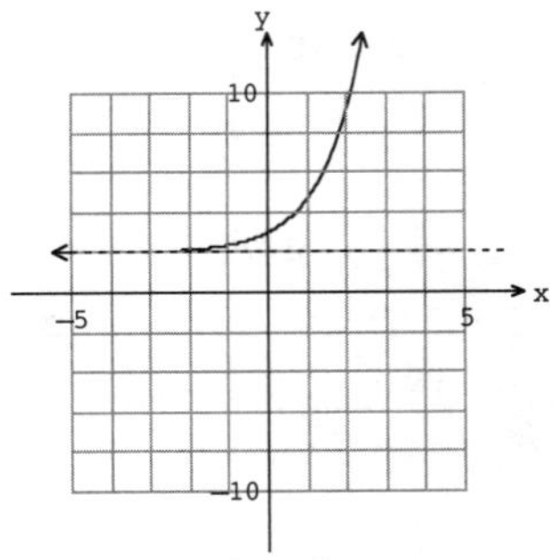

39. The graph of g is the same as the graph of f reflected through the x axis and stretched vertically by a factor of 2. Therefore g is decreasing and the graph has horizontal asymptote $y = 0$.

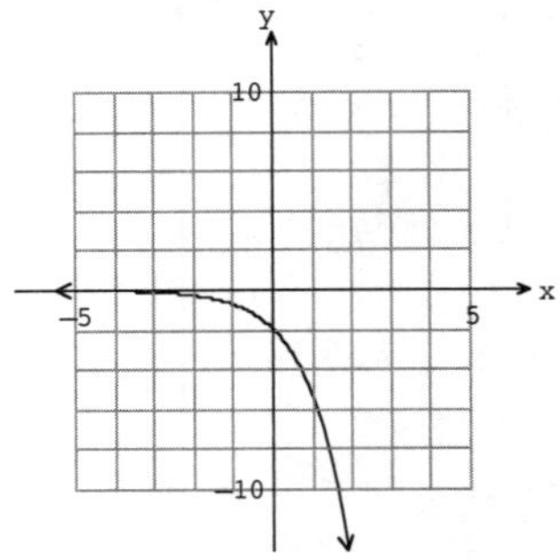

41. The graph of g is the same as the graph of f shifted 2 units to the left. Therefore g is increasing and the graph has horizontal asymptote $y = 0$.

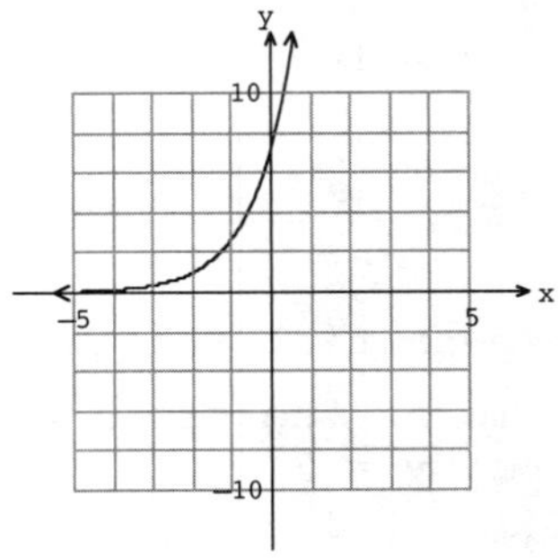

43. $5^{3x} = 5^{4x-2}$ if and only if

$$\begin{aligned} 3x &= 4x - 2 \\ -x &= -2 \\ x &= 2 \end{aligned}$$

45. $7^{x^2} = 7^{2x+3}$ if and only if

$$\begin{aligned} x^2 &= 2x + 3 \\ x^2 - 2x - 3 &= 0 \\ (x - 3)(x + 1) &= 0 \\ x &= -1, 3 \end{aligned}$$

47. $\left(\frac{1}{2}\right)^{x+4} = \left(\frac{1}{2}\right)^{3x-5}$ if and only if

$$\begin{aligned} x + 4 &= 3x - 5 \\ 9 &= 2x \\ x &= \frac{9}{2} \end{aligned}$$

49. $\left(\frac{4}{5}\right)^{6x+1} = \frac{5}{4}$

$\left(\frac{4}{5}\right)^{6x+1} = \left(\frac{4}{5}\right)^{-1}$ if and only if

$$\begin{aligned} 6x + 1 &= -1 \\ 6x &= -2 \\ x &= -\frac{1}{3} \end{aligned}$$

51. $(1 - x)^5 = (2x - 1)^5$ if and only if

$$\begin{aligned} 1 - x &= 2x - 1 \\ -3x &= -2 \\ x &= \frac{2}{3} \end{aligned}$$

53. $2xe^{-x} = 0$ if $2x = 0$ or $e^{-x} = 0$. Since e^{-x} is never 0, the only solution is $x = 0$.

55. $x^2e^x - 5xe^x = 0$
$xe^x(x - 5) = 0$
$x = 0$ or $e^x = 0$ or $x - 5 = 0$
never $\quad x = 5$
$x = 0, 5$

57. $9^{x^2} = 3^{3x-1}$
$(3^2)^{x^2} = 3^{3x-1}$
$3^{2x^2} = 3^{3x-1}$
if and only if
$2x^2 = 3x - 1$
$2x^2 - 3x + 1 = 0$
$(2x - 1)(x - 1) = 0$
$x = \frac{1}{2}, 1$

59. $25^{x+3} = 125^x$
$(5^2)^{x+3} = (5^3)^x$
$5^{2x+6} = 5^{3x}$ if and only if
$2x + 6 = 3x$
$x = 6$

61. $4^{2x+7} = 8^{x+2}$
$(2^2)^{2x+7} = (2^3)^{x+2}$
$2^{4x+14} = 2^{3x+6}$ if and only if
$4x + 14 = 3x + 6$
$x = -8$

63. $100^{x^2} = 1,000^{10}$
$(10^2)^{x^2} = (10^3)^{10}$
$10^{2x^2} = 10^{30}$
$2x^2 = 30$
$x^2 = 15$
$x = \sqrt{15}$ or $x = -\sqrt{15}$

65. $\left(\frac{1}{9}\right)^{5x+1} = 27$
$(3^{-2})^{5x+1} = 3^3$
$3^{-10x-2} = 3^3$ if and only if
$-10x - 2 = 3$
$-10x = 5$
$x = -\frac{1}{2}$

67. $a^2 = a^{-2}$
$a^2 = \frac{1}{a^2}$
$a^4 = 1 \quad (a \neq 0)$
$a^4 - 1 = 0$
$(a - 1)(a + 1)(a^2 + 1) = 0$
$a = 1$ or $a = -1$
This does not violate the exponential property mentioned because $a = 1$ and a negative are excluded from consideration in the statement of the property.

69. The graph of $y = 1^x$ is a horizontal straight line since $1^x = 1$ for all real x. This graph is neither increasing nor decreasing and has no asymptotes. It is not the graph of an exponential function.

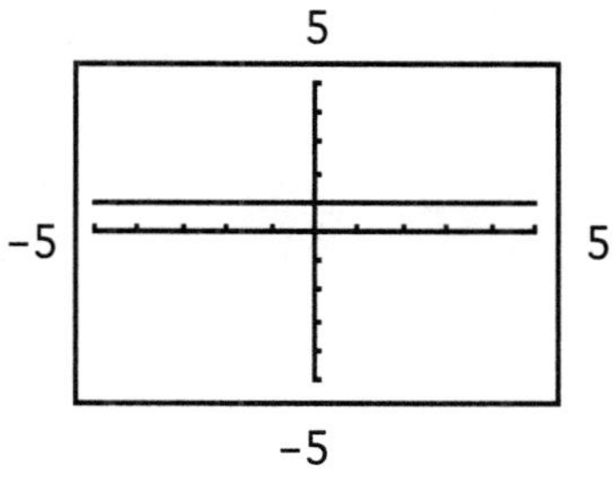

71. The graph of $y = f(x) - 2$ is the graph of f shifted down two units.

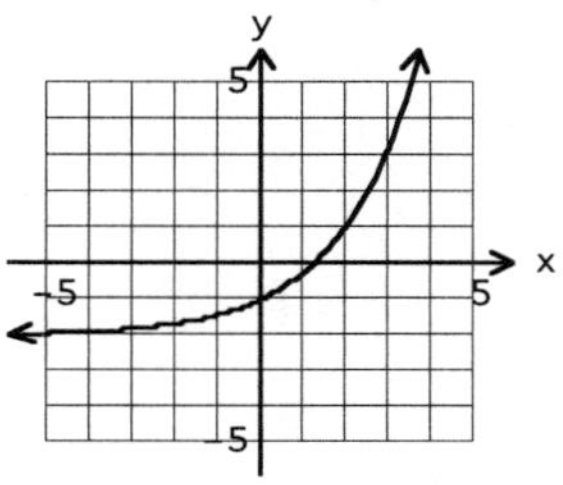

73. The graph of $y = f(x - 2)$ is the graph of f shifted right two units.

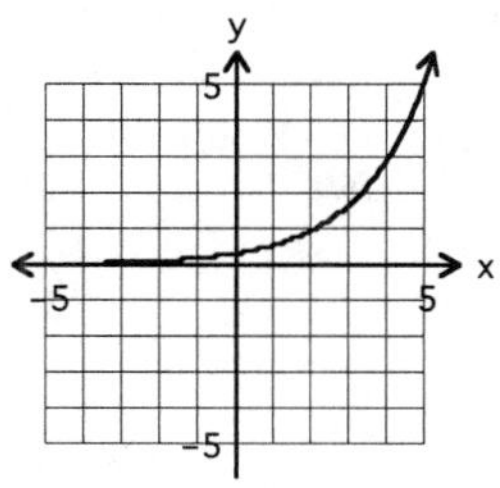

75. The graph of $y = 2f(x) - 4$ is the graph of f stretched vertically by a factor of 2 and shifted down 4 units.

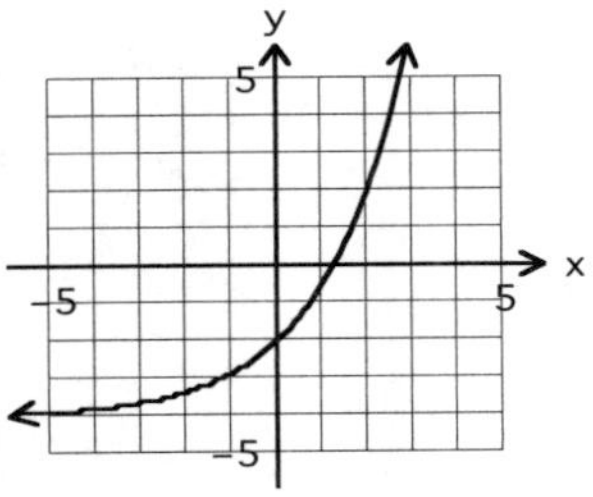

77. The graph of $y = 2 - 3f(x - 4)$ is the graph of f shifted right 4 units, stretched vertically by a factor of 3, reflected in the x axis, and shifted up 2 units.

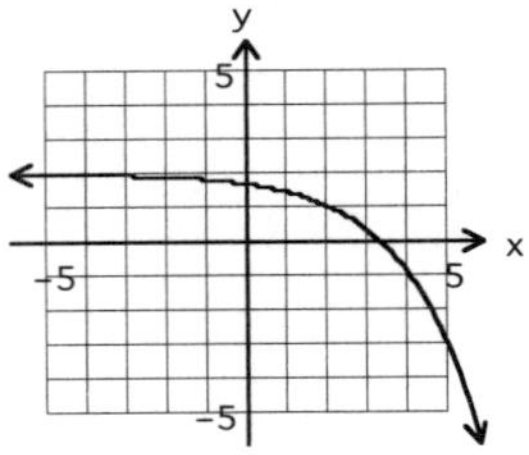

79. The graph of g is the same as the graph of f reflected through the x axis; g is increasing; horizontal asymptote: $y = 0$.

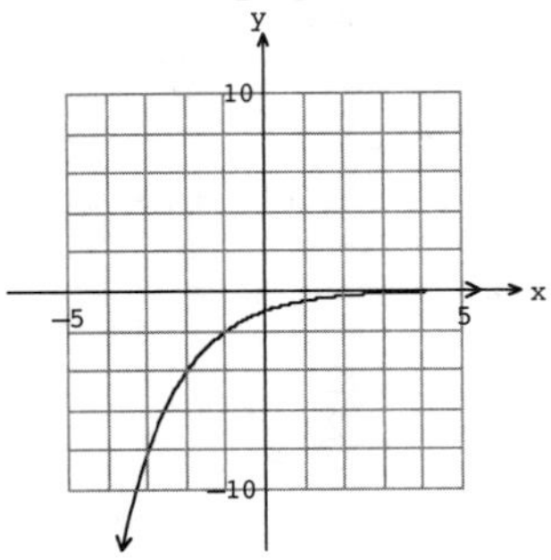

81. The graph of g is the same as the graph of f stretched horizontally by a factor of 2 and shifted upward 3 units; g is decreasing; horizontal asymptote: $y = 3$.

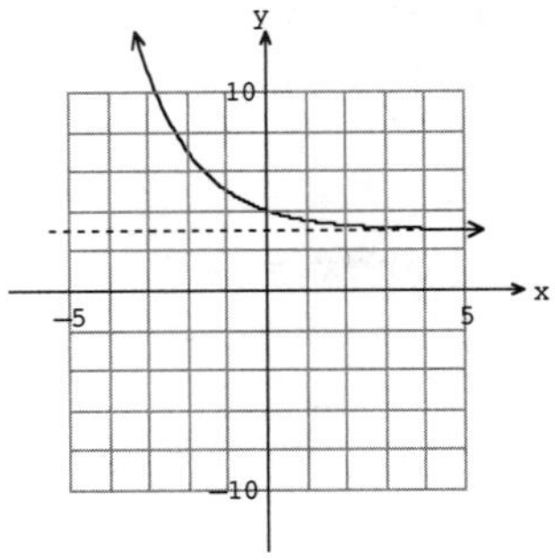

83. The graph of g is the same as the graph of f stretched vertically by a factor of 500; g is increasing; horizontal asymptote: $y = 0$.

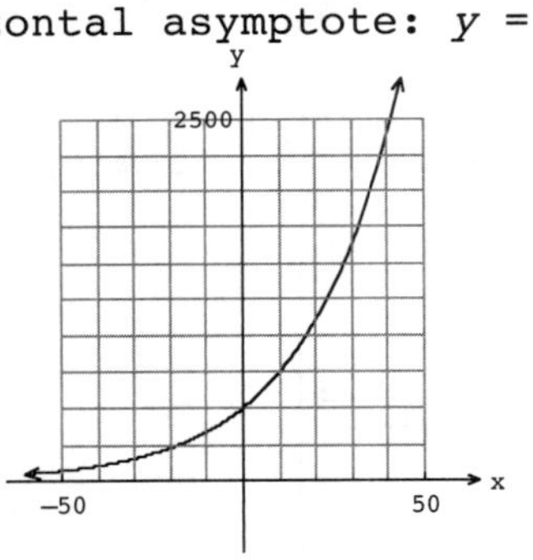

85. The graph of g is the same as the graph of f shifted 3 units to the right, stretched vertically by a factor of 2, and shifted upward 1 unit; g is increasing; horizontal asymptote: $y = 1$.

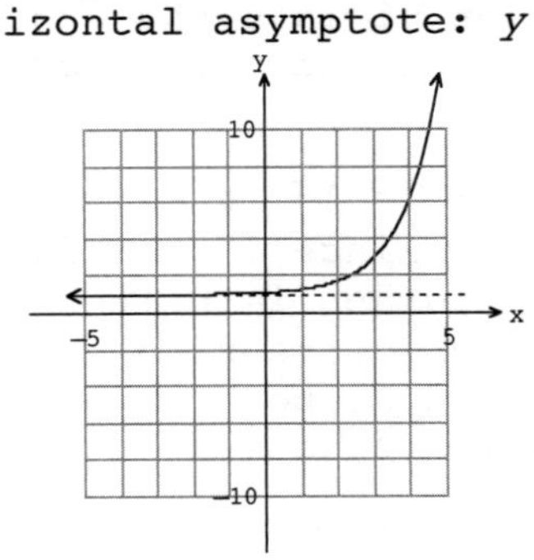

87. The graph of g is the same as the graph of f shifted 2 units to the right, reflected in the origin, stretched vertically by a factor of 4, and shifted upward 3 units; g is increasing; horizontal asymptote: $y = 3$.

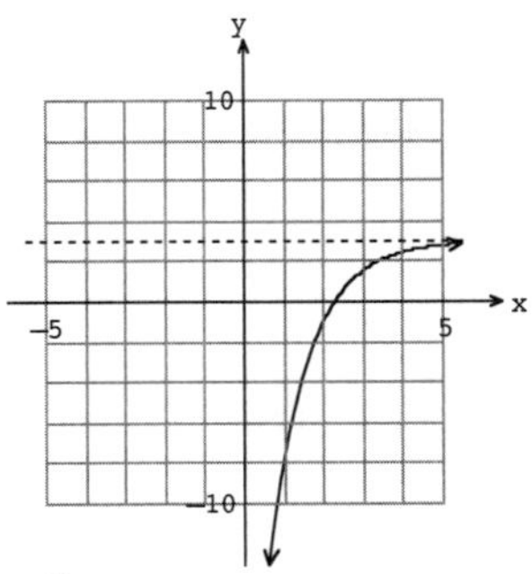

89. $$\frac{-2x^3e^{-2x} - 3x^2e^{-2x}}{x^6} = \frac{x^2e^{-2x}(-2x - 3)}{x^6} = \frac{e^{-2x}(-2x - 3)}{x^4}$$

91. $(e^x + e^{-x})^2 + (e^x - e^{-x})^2 = (e^x)^2 + 2(e^x)(e^{-x}) + (e^{-x})^2 + (e^x)^2 - 2(e^x)(e^{-x}) + (e^{-x})^2$

$$= e^{2x} + 2 + e^{-2x} + e^{2x} - 2 + e^{-2x}$$

$$= 2e^{2x} + 2e^{-2x}$$

Common Errors:
$(e^x)^2 \neq e^{x^2}$
$e^{2x} + e^{2x} \neq e^{4x}$

93. Examining the graph of $y = f(x)$, we obtain

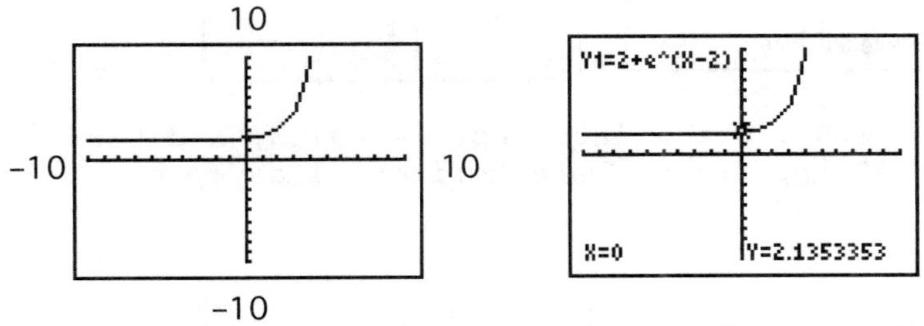

There are no local extrema and no x intercepts. The y intercept is 2.14. As $x \to -\infty$, $y \to 2$, so the line $y = 2$ is a horizontal asymptote.

95. Examining the graph of $y = m(x)$, we obtain

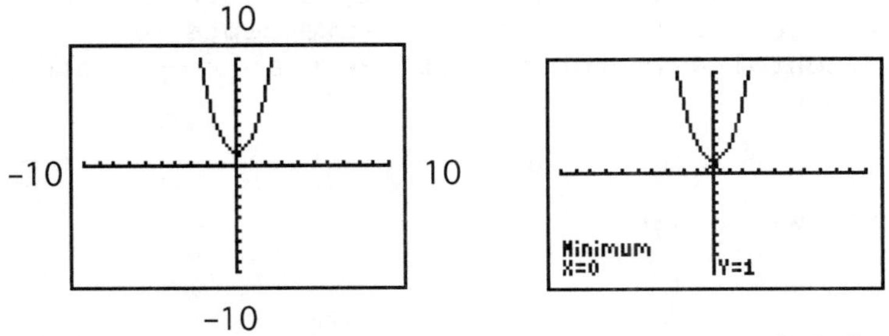

There is a local minimum at $m(0) = 1$, and 1 is the y intercept. There is no x intercept and no horizontal asymptote.

97. Examining the graph of $y = S(x)$, we obtain

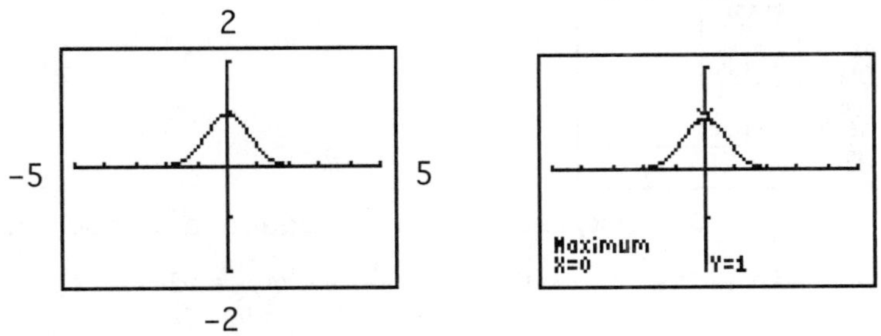

There is a local maximum at $S(0) = 1$, and 1 is the y intercept. There is no x intercept. As $x \to \infty$ or $x \to -\infty$, $y \to 0$, so the line $y = 0$ (the x axis) is a horizontal asymptote.

99. Examining the graph of $y = F(x)$, we obtain

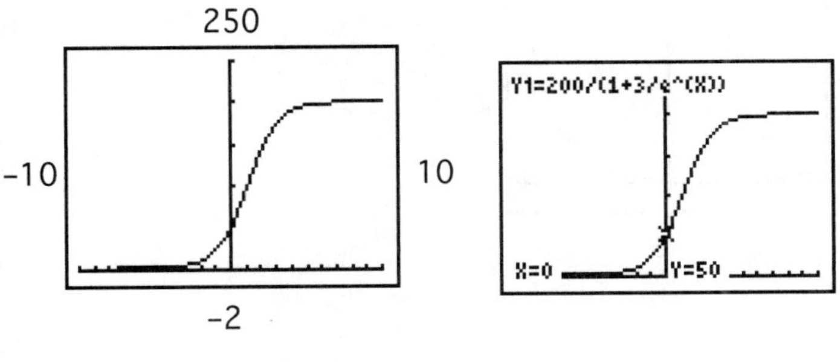

There are no local extrema and no x intercepts.

When $x = 0$, $F(0) = \dfrac{200}{1 + 3e^{-0}} = 50$ is the y intercept. As $x \to -\infty$, $y \to 0$, so the line $y = 0$ (the x axis) is a horizontal asymptote. As $x \to \infty$, $y \to 200$, so the line $y = 200$ is also a horizontal asymptote.

101.

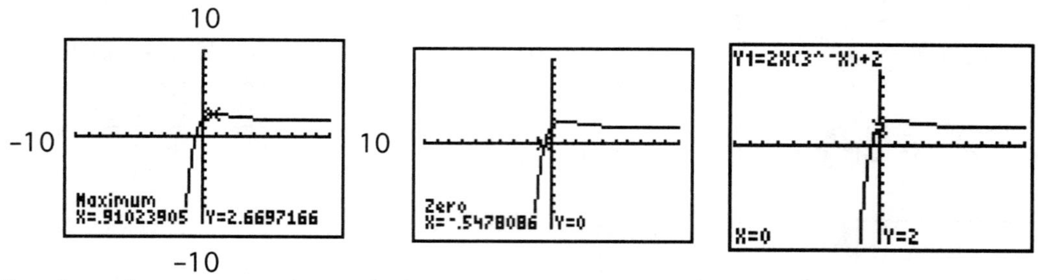

The local maximum is $m(0.91) \approx 2.67$. The x intercept is -0.55, and the y intercept is 2. The line $y = 2$ appears to be a horizontal asymptote. As $x \to -\infty$, $m(x) \to -\infty$.

103.

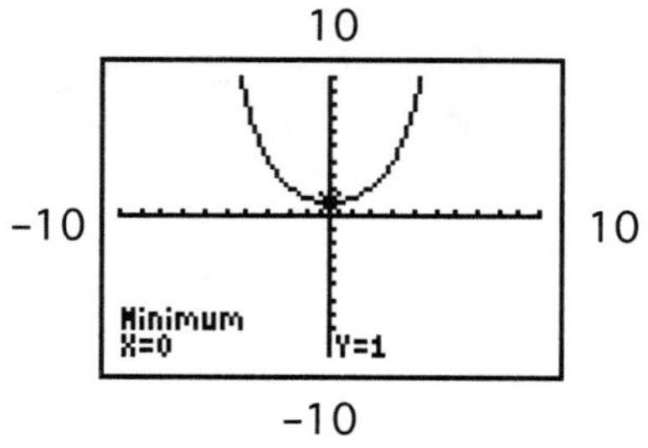

The local minimum is $f(0) = 1$, so zero is the y intercept. There are no x intercepts or horizontal asymptotes; $f(x) \to \infty$ as $x \to \infty$ and $x \to -\infty$.

105. Examining the graph of $y = f(x)$, we obtain

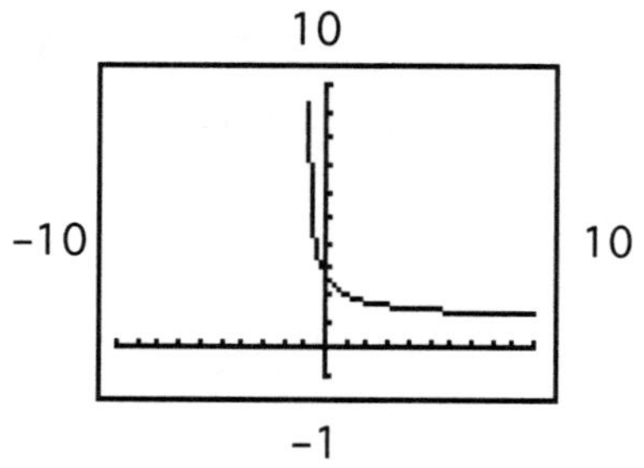

As $x \to 0$, $f(x) = (1 + x)^{1/x}$ seems to approach a value near 3. A table of values near $x = 0$ yields

X	Y1
0	ERROR
.1	2.5937
.01	2.7048
.001	2.7169
1E-4	2.7181
-.001	2.7196
-1E-4	2.7184

Y1=(1+X)^X-1

Although $f(0)$ is not defined, as $x \to 0$, $f(x)$ seems to approach a number near 2.718. In fact, it approaches e, since as $x \to 0$, $u = \dfrac{1}{x} \to \infty$, and

$f(x) = \left(1 + \dfrac{1}{u}\right)^u$ must approach e as $u \to \infty$.

107.

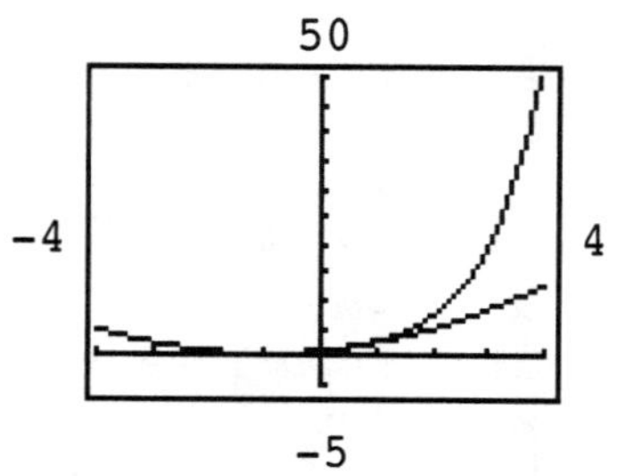

109.

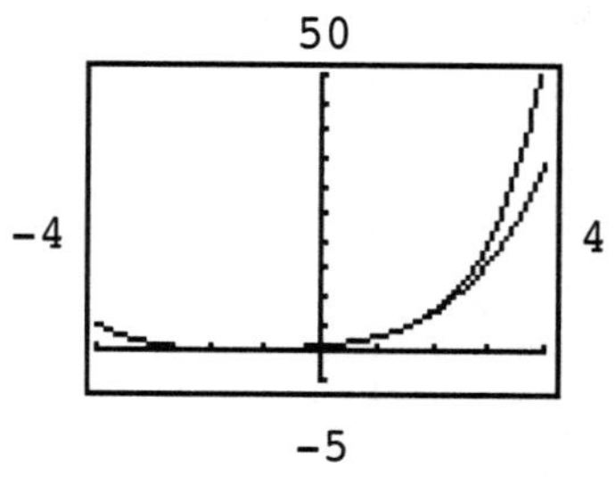

111. Here are graphs of $f_1(x) = \dfrac{x}{e^x}$, $f_2(x) = \dfrac{x^2}{e^x}$, and $f_3(x) = \dfrac{x^3}{e^x}$. In each case as $x \to \infty$, $f_n(x) \to 0$. The line $y = 0$ is a horizontal asymptote.
As $x \to -\infty$, $f_1(x) \to -\infty$ and $f_3(x) \to -\infty$, while $f_2(x) \to \infty$. It appears that as $x \to -\infty$, $f_n(x) \to \infty$ if n is even and $f_n(x) \to -\infty$ if n is odd.

f_1:

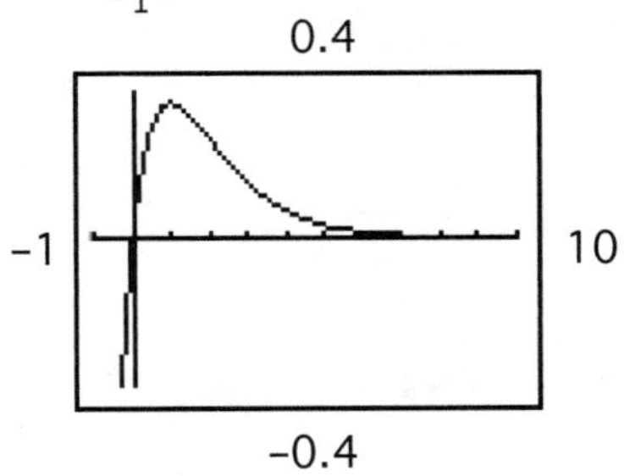

f_2:

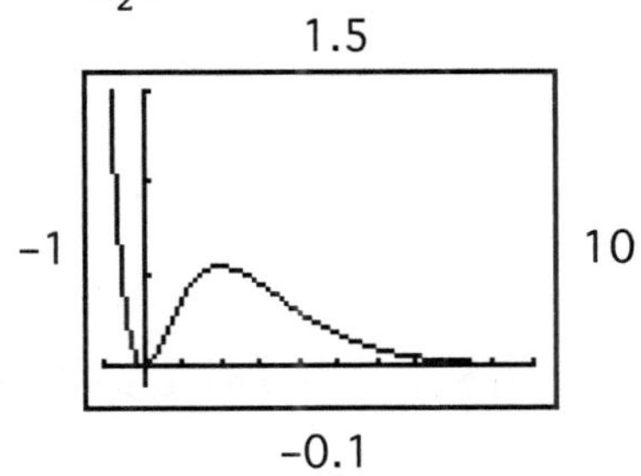

f_3:

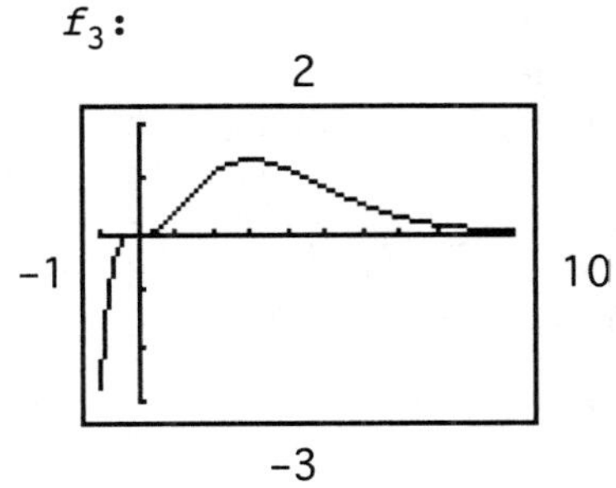

As confirmation of these observations, we show the graph of $f_4 = \dfrac{x^4}{e^x}$ (not required).

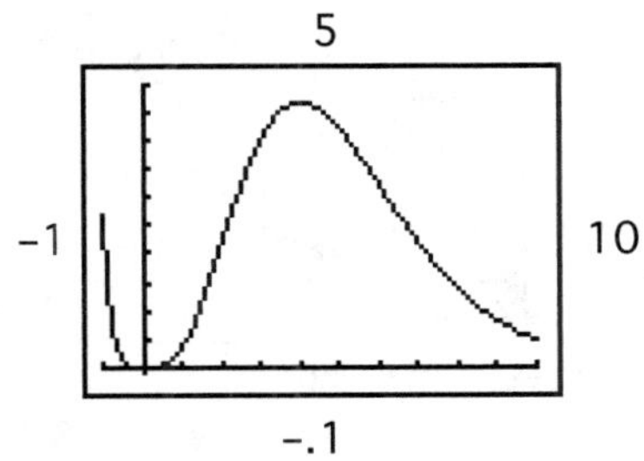

113. The graph of a polynomial other than the constant polynomial $P(x) = C$ for any real number C, doesn't have a horizontal asymptote.

115. We use the compound interest formula

$$A = P\left(1 + \frac{r}{m}\right)^n \text{ to find } P\text{: } P = \frac{A}{\left(1 + \frac{r}{m}\right)^n}$$

$m = 365 \quad r = 0.0825 \quad A = 40{,}000 \quad n = 365 \cdot 17$

$$P = \frac{40{,}000}{\left(1 + \frac{0.0825}{365}\right)^{365 \cdot 17}} = \$9{,}841$$

117. We use the Continuous Compound Interest Formula

$A = Pe^{rt}$

$P = 5{,}250 \quad r = 0.1138 \quad A = 5{,}250e^{0.1138t}$

$$A = P\left(1 + \frac{r}{m}\right)^n$$

For the first account, $P = 3000$, $r = 0.08$, $m = 365$. Let $y1 = A$, then $y1 = 3000(1 + 0.08/365)^x$ where x is the number of compounding periods (days).
For the second account, $P = 5000$, $r = 0.05$, $m = 365$. Let $y2 = A$, then $y2 = 5000(1 + 0.05/365)^x$
Examining the graphs of $y1$ and $y2$, we obtain the graphs at the right.
The graphs intersect at $x = 6216.15$ days.

Comparing the amounts in the accounts, we see that the first account is worth more than the second for $x \geq 6217$ days.

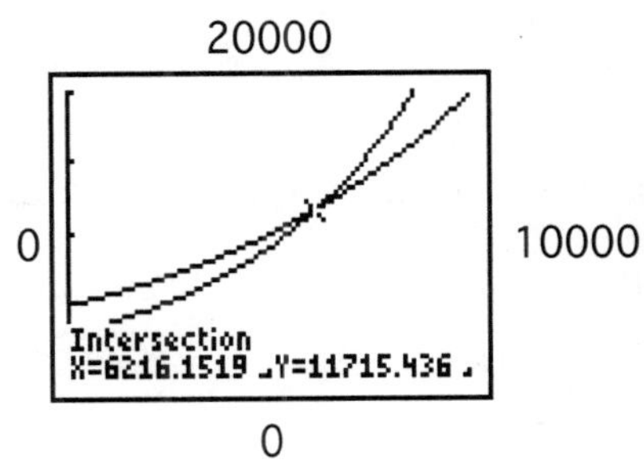

X	Y1	Y2
6215	11712	11714
6216	11715	11715
6217	11718	11717
[illegible]	11720	11718
6219	11723	11720
6220	11725	11722
6221	11728	11723

X=6218

121. We use the compound interest formula

$$A = P\left(1 + \frac{r}{m}\right)^n$$

For the first account, $P = 10,000$, $r = 0.089$, $m = 365$. Let $y1 = A$, then $y1 = 10000(1 + 0.089/365)^x$ where x is the number of compounding periods (days). For the second account, $P = 10,000$, $r = 0.09$, $m = 4$. Let $y2 = A$, then $y2 = 10000(1 + 0.09/4)^{4x/365}$ where x is the number of days. Examining the graphs of $y1$ and $y2$, we obtain the graph at the right.

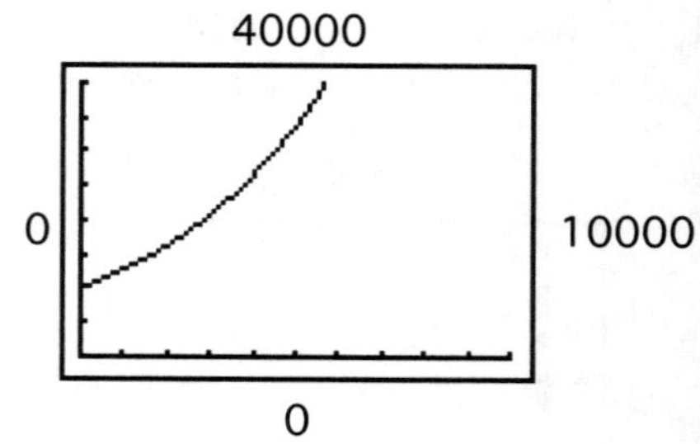

The two graphs are indistinguishable from one another. Examining a table of values, we obtain:

X	Y1	Y2
0	10000	10000
91.25	10225	10225
182.5	10455	10455
273.75	10690	10690
365	10931	10931
456.25	11177	11177
547.5	11428	11428

X=0

X	Y1	Y2
1186.3	13354	13354
1277.5	13654	13655
1368.8	13961	13962
1460	14275	14276
1551.3	14597	14597
1642.5	14925	14926
[illegible]	15261	15262

X=1733.75

The two accounts are extremely close in value, but after 1277.5 days (14 quarters) the second account begins to be observably larger than the first. The first account will never be larger than the second.

123. We use the Continuous Compound Interest Formula

$A = Pe^{rt}$

$P = \dfrac{A}{e^{rt}}$ or $P = Ae^{-rt}$

$A = 30,000 \quad r = 0.09 \quad t = 10$

$P = 30,000e^{(-0.09)(10)}$

$P = \$12,197.09$

125. Gill Savings: Use the Continuous Compound Interest Formula

$A = Pe^{rt} \quad P = 1,000 \quad r = 0.083 \quad t = 2.5$

$A = 1,000e^{(0.083)(2.5)}$

$A = \$1,230.60$

Richardson S & L: Use the Compound Interest Formula

$A = P\left(1 + \dfrac{r}{m}\right)^n \quad P = 1,000 \quad r = 0.084 \quad n = (4)(2.5)$

$A = 1,000\left(1 + \dfrac{0.084}{4}\right)^{(4)(2.5)}$

$A = \$1,231.00$

U.S.A. Savings: Use the Compound Interest Formula

$A = P\left(1 + \dfrac{r}{m}\right)^n \quad P = 1,000 \quad r = 0.0825 \quad n = (365)(2.5)$

$A = 1,000\left(1 + \dfrac{0.0825}{365}\right)^{(365)(2.5)}$

$A = \$1,229.03$

127. We use the compound interest formula

$$A = P\left(1 + \frac{r}{m}\right)^n \quad m = \frac{365}{7}$$

$$P = 4{,}000 \quad r = 0.11$$

$$A = 4{,}000\left(1 + \frac{0.11}{365/7}\right)^n$$

$$= 4{,}000\left(1 + \frac{0.77}{365}\right)^n$$

(A) $n = \left(\frac{365}{7}\right)0.5$, hence

$$A = 4{,}000\left(1 + \frac{0.77}{365}\right)^{365(0.5)/7}$$

$$= \$4{,}225.92$$

(B) $n = \left(\frac{365}{7}\right)10$, hence

$$A = 4{,}000\left(1 + \frac{0.77}{365}\right)^{365(10)/7}$$

$$= \$12{,}002.75$$

Section 5-2

1.

n	L
1	2
2	4
3	8
4	16
5	32
6	64
7	128
8	256
9	512
10	1,024

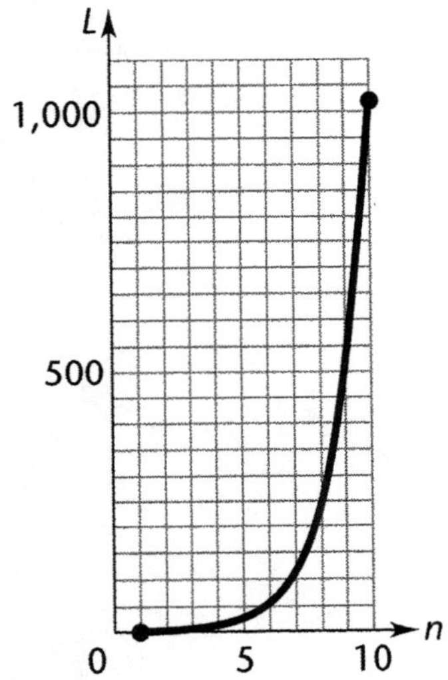

3. We use the Doubling Time Growth Model:

$P = P_0 2^{t/d}$

Substituting $P_0 = 10$ and $d = 2.4$, we have

$P = 10(2^{t/2.4})$

(A) $t = 7$, hence $P = 10(2^{7/2.4})$
$= 75.5$

76 flies

(B) $t = 14$, hence $P = 10(2^{14/2.4})$
$= 570.2$

570 flies

5. We use the Half-Life Decay Model

$$A = A_0\left(\frac{1}{2}\right)^{t/h} = A_0 2^{-t/h}$$

Substituting $A_0 = 25$ and $h = 12$, we have

$A = 25(2^{-t/12})$

(A) $t = 5$, hence $A = 25(2^{-5/12}) = 19$ pounds

(B) $t = 20$, hence $A = 25(2^{-20/12}) = 7.9$ pounds

7. We use the Unlimited Growth model
$y = Ce^{kt}$
$C = 6.5 \times 10^9 \quad r = 0.0114 \quad t = 10$
$y = 6.5 \times 10^9 e^{(0.0114)(10)}$
$y = 7.3 \times 10^9$, 7.3 billion

9. We use the Unlimited Growth model.
$y = Ce^{kt}$
For Russia, $C = 1.43 \times 10^8$, $r = -0.0037$, $y_1 = 1.43 \times 10^8 e^{-0.0037t}$
For Nigeria, $C = 1.29 \times 10^8$, $r = 0.0256$, $y_2 = 1.29 \times 10^8 e^{0.0256t}$
Here is a graph of y_1 and y_2 (vertical axis understood $\times 10^8$)

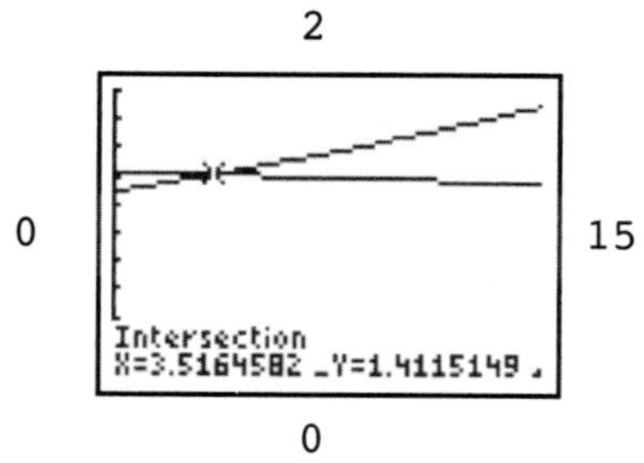

From the graph, assuming $t = 0$ in 2005, it appears that the two populations will be equal when t is approximately 3.5, in 2008. After that the population of Nigeria will be greater than that of Russia.

11. A table of values can be generated by a graphing utility and yields

t	P
0	75
10	72
20	70
30	68
40	65
50	63
60	61
70	59
80	57
90	55
100	53

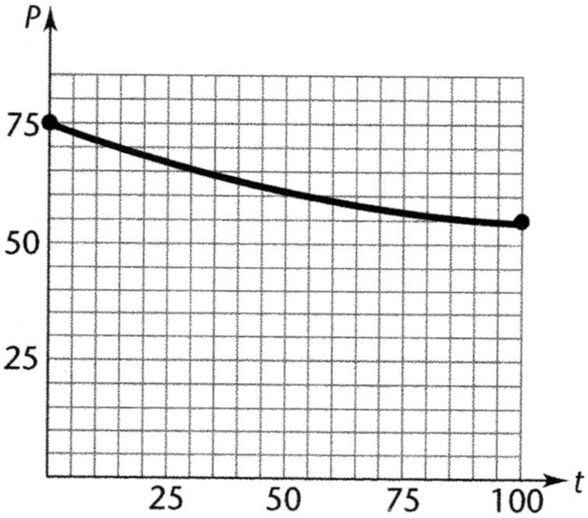

13. $I = I_0 e^{-0.00942d}$

(A) $d = 50 \quad I = I_0 e^{-0.00942(50)} = 0.62I_0 \quad 62\%$

(B) $d = 100 \quad I = I_0 e^{-0.00942(100)} = 0.39I_0 \quad 39\%$

15. (A) We use the Unlimited Growth model $y = Ce^{kt}$ with $C = 39.4$ million, $k = 0.032$, and $t = 6$ since 2010 is 6 years after 2004.
$y = 39.4 \times 10^6 e^{0.032(6)}$
$y = 47.7$ million people living with HIV in 2010.

(B) Here $C = 39.4$ million, $k = 0.032$, and $t = 11$ since 2015 is 11 years after 2004.
$y = 39.4 \times 10^6 e^{0.032(11)}$
$y = 56.0$ million people living with HIV in 2015.

17. $T = T_m + (T_0 - T_m)e^{-kt}$
$T_m = 40° \quad T_0 = 72° \quad k = 0.4 \quad t = 3$
$T = 40 + (72 - 40)e^{-0.4(3)}$
$T = 50°$

19. As t increases without bound, $e^{-0.2t}$ approaches 0, hence $q = 0.0009(1 - e^{-0.2t})$ approaches 0.0009. Hence 0.0009 coulomb is the maximum charge on the capacitor.

21. (A) Examining the graph of $N(t)$, we obtain the graphs below.

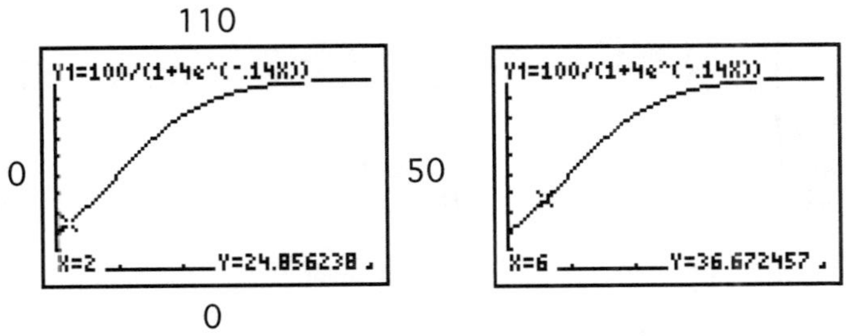

After 2 years, 25 deer will be present. After 6 years, 37 deer will be present.

(B) Applying a built-in routine, we obtain the graph at the right.

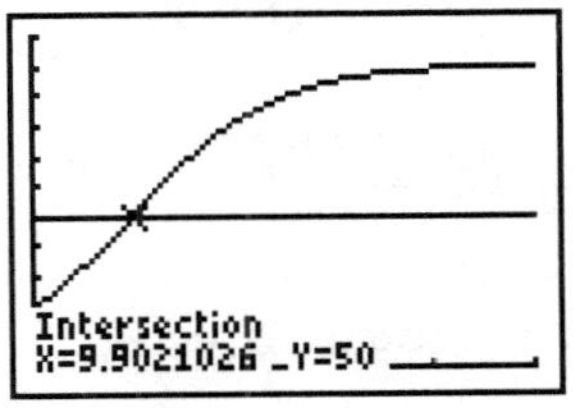

It will take 10 years for the herd to grow to 50 deer.

(C) As t increases without bound, $e^{-0.14t}$ approaches 0, hence $N = \dfrac{100}{1 + 4e^{-0.14t}}$ approaches 100. Hence 100 is the number of deer the island can support.

23.

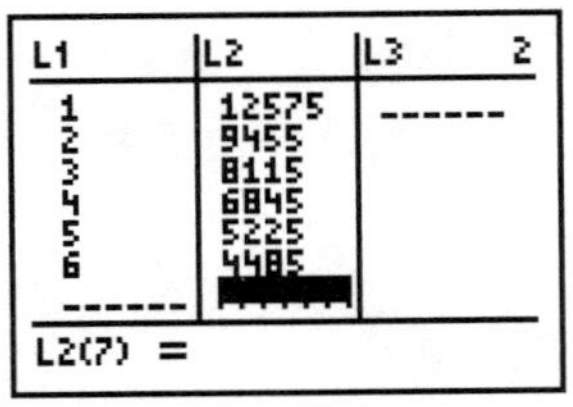

Enter the data.

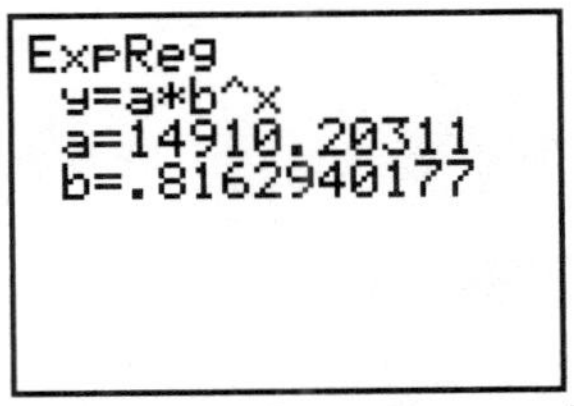

Compute the regression equation.

The model gives $y = 14910.20311*(0.8162940177)^x$. Clearly, when $x = 0$, $y = \$14,910$ is the estimated purchase price. Applying a built-in routine, we obtain the graph at the right.

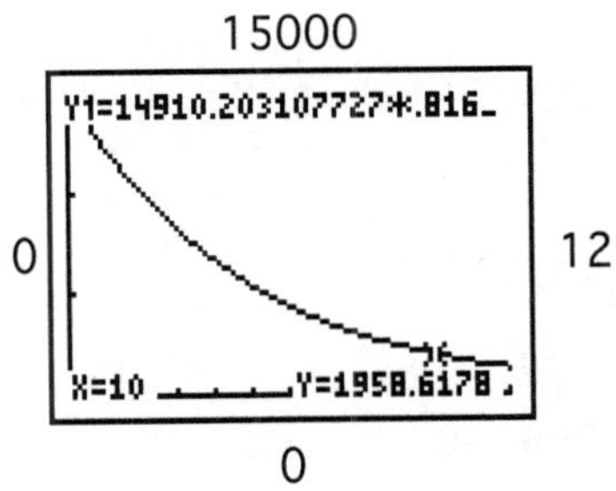

When $x = 10$, the estimated value of the van is \$1,959.

25. (A) The independent variable is years since 1980, so enter 0, 5, 10, 15, 18, and 19 as L_1. The dependent variable is power generation in North America, so enter the North America column as L_2. Then use the logistic regression command from the STAT CALC menu.

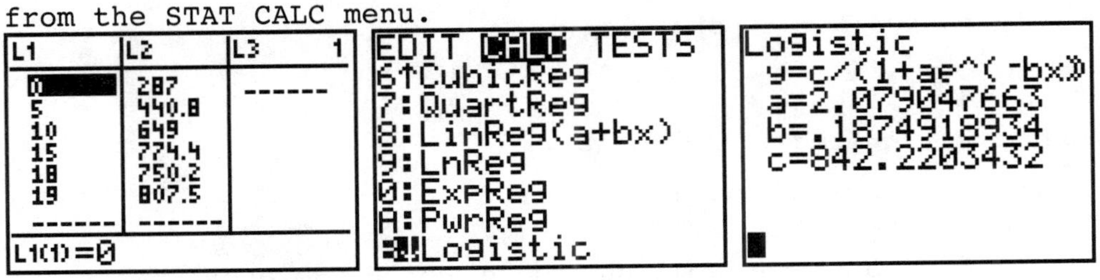

The model is

$$y = \frac{842}{1 + 2.08e^{-0.187x}}$$

(B) 2010 is 30 years after 1980, so we evaluate our model for $x = 30$ using the value command.

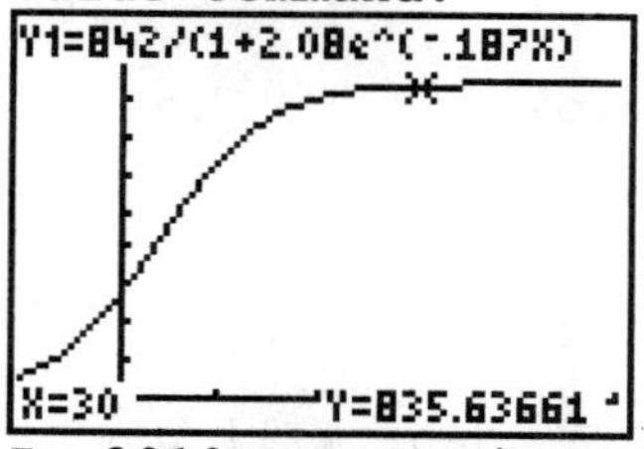

WINDOW
Xmin=-10
Xmax=50
Xscl=10
Ymin=-50
Ymax=1000
Yscl=100
Xres=1

In 2010, approximately 836 billion kilowatt hours will be generated.

Section 5-3

1. $81 = 3^4$ **3.** $0.001 = 10^{-3}$ **5.** $3 = 81^{1/4}$ **7.** $6^{-2} = \frac{1}{36}$

9. $16 = \left(\frac{1}{2}\right)^{-4}$ **11.** $\log_{10} 0.0001 = -4$ **13.** $\log_4 8 = \frac{3}{2}$ **15.** $\log_{32} \frac{1}{2} = -\frac{1}{5}$

17. $\log_{49} 7 = \frac{1}{2}$ **19.** $\log_{2/3} \frac{8}{27} = 3$ **21.** 0 **23.** 1

25. 4 **27.** $\log_{10} 0.01 = \log_{10} 10^{-2} = -2$ **29.** $\log_3 27 = \log_3 3^3 = 3$

31. $\log_{1/2} 2 = \log_{1/2}\left(\frac{1}{2}\right)^{-1} = -1$ **33.** 5

35. $\log_5 \sqrt[3]{5} = \log_5 5^{1/3} = \frac{1}{3}$ **37.** $\sqrt{x}$

39. $e^{2 \log_e x} = e^{\log_e x^2} = x^2$ **Common Error:** $e^{2 \log_e x} \neq 2x$

41. 4.6923 **43.** 3.9905

45. $\log_7 13 = \frac{\ln 13}{\ln 7} = 1.3181$ using the change of base formula

47. $\log_5 120.24 = \frac{\ln 120.24}{\ln 5} = 2.9759$ using the change of base formula

49. 200,800 **51.** $6.648 \times 10^{-4} = 0.0006648$ **53.** 47.73

55. 0.6760

57. Write $\log_2 x = 2$ in equivalent exponential form. $x = 2^2 = 4$

59. $\log_4 16 = \log_4 4^2 = 2$
$y = 2$

61. Write $\log_b 16 = 2$ in equivalent exponential form.
$16 = b^2$
$b^2 = 16$
$b = 4$ since bases are required to be positive

63. Write $\log_b 1 = 0$ in equivalent exponential form.
$1 = b^0$
This statement is true if b is any real number except 0. However, bases are required to be positive and 1 is not allowed, so the original statement is true if b is any positive real number except 1.

65. Write $\log_4 x = \frac{1}{2}$ in equivalent exponential form.
$x = 4^{1/2} = 2$

67. $\log_{1/3} 9 = \log_{1/3} 3^2 = \log_{1/3} \dfrac{1}{\left(\frac{1}{3}\right)^2} = \log_{1/3} \left(\dfrac{1}{3}\right)^{-2} = -2$

69. Write $\log_b 1000 = \dfrac{3}{2}$ in equivalent exponential form

$$\begin{aligned} 1000 &= b^{3/2} \\ 10^3 &= b^{3/2} \\ (10^3)^{2/3} &= (b^{3/2})^{2/3} \\ 10^{3(2/3)} &= b^{3/2(2/3)} \\ 10^2 &= b \\ b &= 100 \end{aligned}$$

(If two numbers are equal the results are equal if they are raised to the same exponent.)

71. Write $\log_8 x = -\dfrac{4}{3}$ in equivalent exponential form.

$$8^{-4/3} = x$$
$$x = (8^{1/3})^{-4} = 2^{-4} = \frac{1}{16}$$

73. Write $\log_{16} 8 = y$ in equivalent exponential form.

$$\begin{aligned} 16^y &= 8 \\ (2^4)^y &= 2^3 \\ 2^{4y} &= 2^3 \text{ if and only if} \\ 4y &= 3 \\ y &= \frac{3}{4} \end{aligned}$$

75. 4.959 **77.** 7.861 **79.** 3.301 **81.** 2.280

83. $\log_b x^2y = \log_b x^2 + \log_b y = 2 \log_b x + \log_b y$

85. $\log_b \dfrac{(x+1)^5}{(x-2)^4} = \log_b(x+1)^5 - \log_b(x-2)^4$
$= 5 \log_b(x+1) - 4 \log_b(x-2)$

Common Error:
$\log_b(x+1)$ cannot be further simplified. It is not equal to $\log_b x + \log_b 1$ or to $\log_b x + 1$.

87. $\log_b x\sqrt{x+1} = \log_b x + \log_b\sqrt{x+1}$
$= \log_b x + \log_b(x+1)^{1/2}$
$= \log_b x + \dfrac{1}{2} \log_b(x+1)$

89. $5 \log_b x + 4 \log_b y = \log_b x^5 + \log_b y^4$
$= \log_b x^5y^4$

91. $-2 \log_b(x+1) + 4 \log_b(x-1) = \log_b(x-1)^4 - \log_b(x+1)^2$
$= \log_b \dfrac{(x-1)^4}{(x+1)^2}$

93. $\dfrac{1}{3} \log_b(x+2) - \dfrac{1}{2} \log_b(x+3) = \log_b(x+2)^{1/3} - \log_b(x+3)^{1/2}$
$= \log_b \dfrac{(x+2)^{1/3}}{(x+3)^{1/2}}$

95. $x = 12.725$ **97.** $x = -25.715$

99. The graph of g is the same as the graph of f shifted upward 3 units; g is increasing.
Domain: $(0, \infty)$
Vertical asymptote: $x = 0$

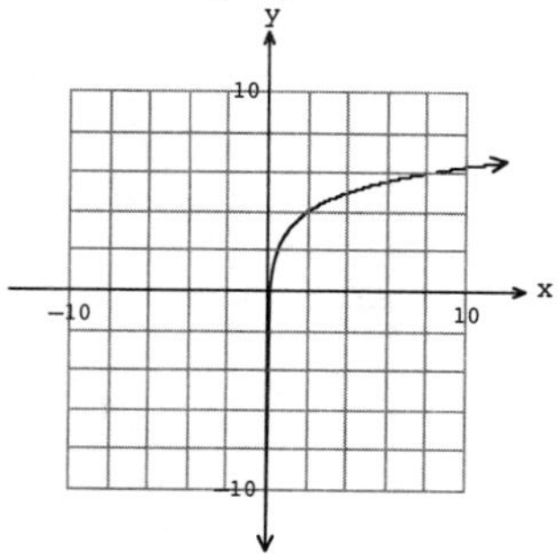

101. The graph of g is the same as the graph of f shifted 2 units to the right; g is decreasing.
Domain: $(2, \infty)$
Vertical asymptote: $x = 2$

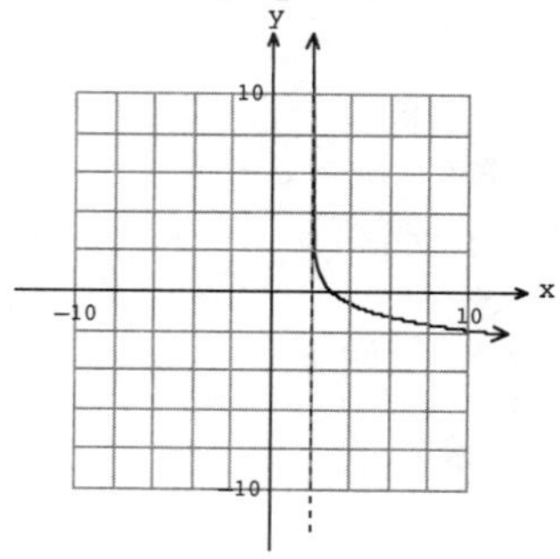

103. The graph of g is the same as the graph of f reflected through the x axis and shifted downward 1 unit; g is decreasing.
Domain: $(0, \infty)$
Vertical asymptote: $x = 0$

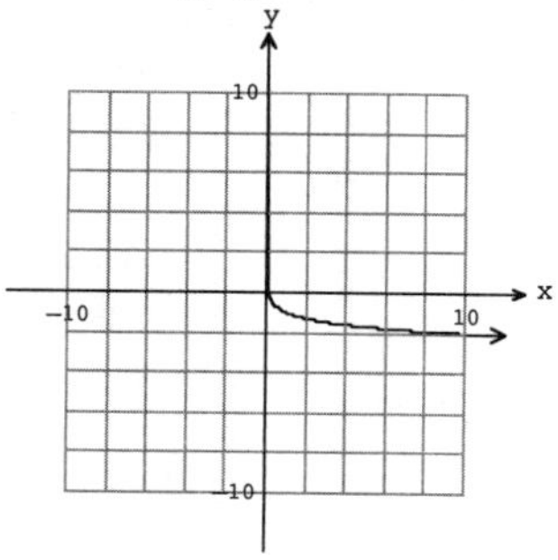

105. The graph of g is the same as the graph of f reflected through the x axis, stretched vertically by a factor of 3, and shifted upward 5 units. g is decreasing.
Domain: $(0, \infty)$
Vertical asymptote: $x = 0$

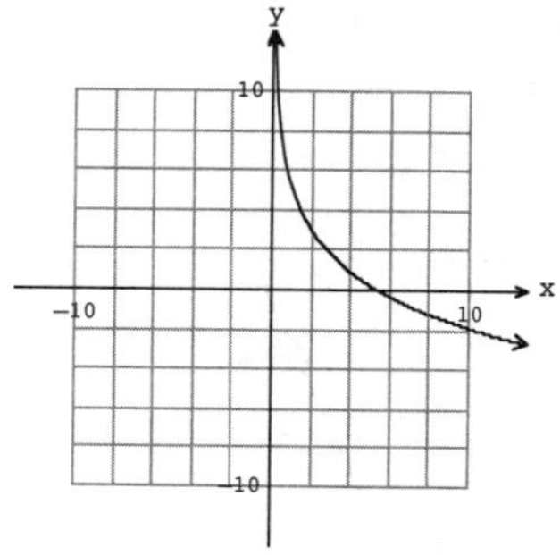

107. Write $y = \log_5 x$
In exponential form:
$5^y = x$
Interchange x and y:
$5^x = y$
Therefore $f^{-1}(x) = 5^x$.

109. Write $y = 4\log_3(x + 3)$
$\frac{y}{4} = \log_3(x + 3)$
In exponential form:
$3^{y/4} = x + 3$
$x = 3^{y/4} - 3$
Interchange x and y:
$y = 3^{x/4} - 3$
Therefore $f^{-1}(x) = 3^{x/4} - 3$

111. Write $y = 4 - 2\log(x + 1)$
$y - 4 = -2\log(x + 1)$
$\frac{4 - y}{2} = \log(x + 1)$
$2 - \frac{y}{2} = \log(x + 1)$
In exponential form:
$10^{2-(y/2)} = x + 1$
$x = 10^{2-(y/2)} - 1$
$x = 10^2 10^{-y/2} - 1$
$x = 100(10)^{-y/2} - 1$
Interchange x and y:
$y = 100(10)^{-x/2} - 1$
Therefore $f^{-1}(x) = 100(10)^{-x/2} - 1$

113. Write $y = -1 + \frac{1}{2}\ln(x - 5)$
$y + 1 = \frac{1}{2}\ln(x - 5)$
$2(y + 1) = \ln(x - 5)$
In exponential form:
$e^{2(y+1)} = x - 5$
$x = e^{2(y+1)} + 5$
Interchange x and y:
$y = e^{2(x+1)} + 5$
Therefore $f^{-1}(x) = e^{2(x+1)} + 5$

115. (A) Write $y = \log_3(2 - x)$

In exponential form:

$$3^y = 2 - x$$
$$x = 2 - 3^y$$

Interchange x and y:

$$y = 2 - 3^x$$

Therefore, $f^{-1}(x) = 2 - 3^x$

(B) The graph is the same as the graph of $y = 3^x$ reflected through the x axis and shifted 2 units upward.

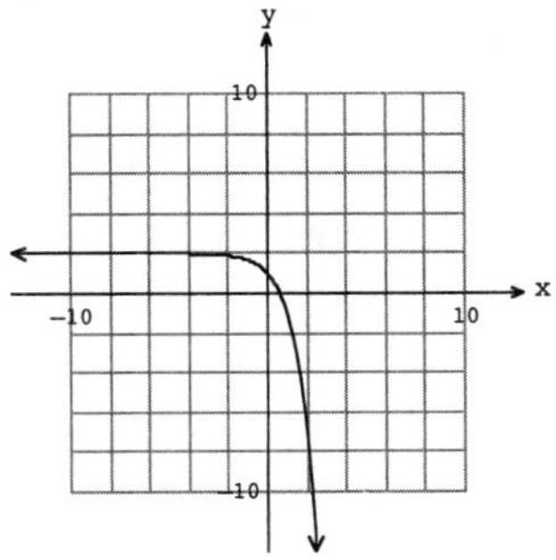

(C)

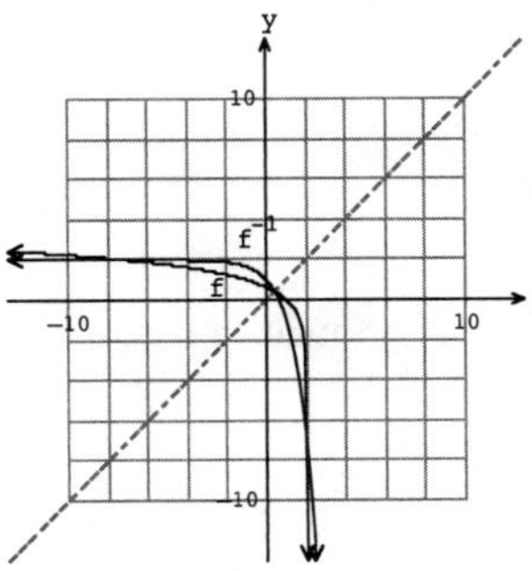

117. The inequality sign in the last step reverses because $\log \frac{1}{3}$ is negative.

119. (A)

x	$\log x$	$\log(\log x)$
10	1	0
100	2	$\log 2 \approx 0.30$
1000	3	$\log 3 \approx 0.48$
10^6	6	$\log 6 \approx 0.78$
10^{100}	100	2
$10^{10^{10}}$	10^{10}	10

(B) Since the range of $\log x$ is all real numbers, the domain of $\log(\log x)$ is those real numbers for which $\log x$ is in the domain of $\log x$, that is, $\log x > 0$, or $x > 1$. The domain is thus $(1, \infty)$. Since if $y = \log(\log x)$, $10^y = \log x$ or $x = 10^{10^y}$, y can take on any real value and the range of g is $(-\infty, \infty)$.

(C) The graph of f has a vertical asymptote at $x = 0$ and no horizontal asymptote. The graph of g has a vertical asymptote at $x = 1$ and no horizontal asymptote. The graph of f has an x intercept at $x = 1$. The graph of g has an x intercept at $x = 10$. Here are graphs of f and g in the same coordinate system.

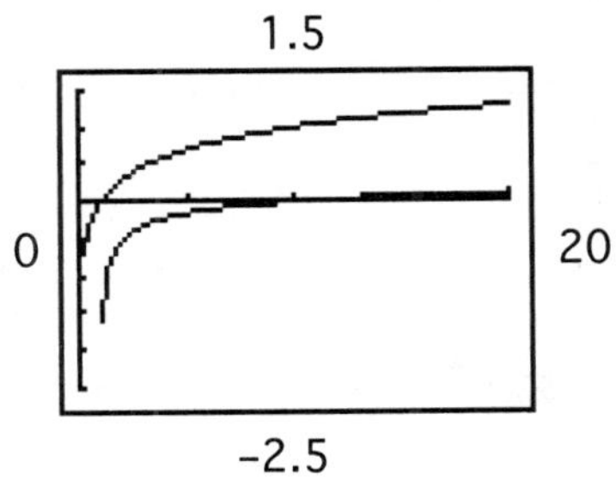

121.

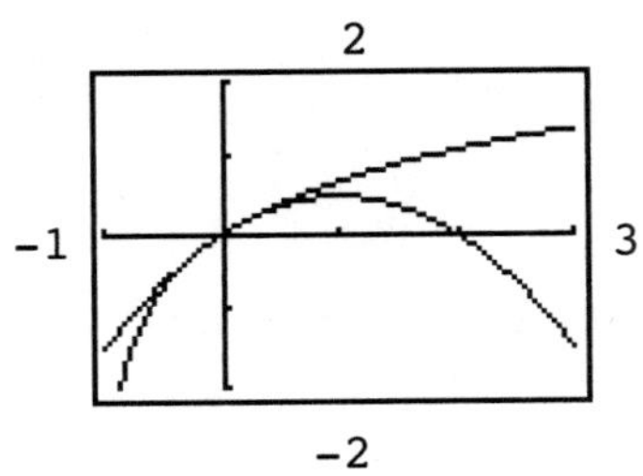

123.

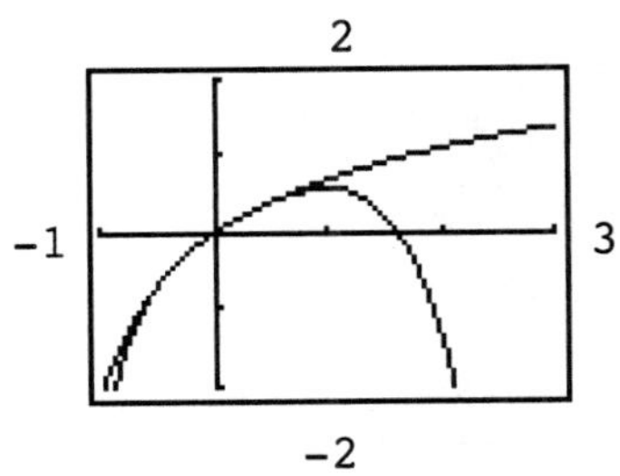

125. $\log_b m^p = \log_b \underbrace{(m \cdot m \cdot m \cdot \cdots \cdot m)}_{p \text{ factors}}$

$= \underbrace{\log_b m + \log_b m + \log_b m + \cdots + \log_b m}_{p \text{ terms}}$

$= p \log_b m$

Section 5-4

1. We use the decibel formula

$D = 10 \log \frac{I}{I_0}$

(A) $I = I_0$

$D = 10 \log \frac{I_0}{I_0}$

$D = 10 \log 1$

$D = 0$ decibels

(B) $I_0 = 1.0 \times 10^{-12}$ $I = 1.0$

$D = 10 \log \frac{1.0}{1.0 \times 10^{-12}}$

$D = 120$ decibels

3. We use the decibel formula

$D = 10 \log \frac{I}{I_0}$

$I_2 = 1000 I_1$

$D_1 = 10 \log \frac{I}{I_0}$ $D_2 = 10 \log \frac{I_2}{I_0}$

$D_2 - D_1 = 10 \log \frac{I_2}{I_0} - 10 \log \frac{I_1}{I_0}$

$= 10 \log\left(\frac{I_2}{I_0} \div \frac{I_1}{I_0}\right)$

$= 10 \log \frac{I_2}{I_1}$

$= 10 \log \frac{1000 I_1}{I_1}$

$= 10 \log 1000$

$= 30$ decibels

5. We use the magnitude formula

$M = \frac{2}{3} \log \frac{E}{E_0}$

$E_0 = 10^{4.40}$ $E = 1.99 \times 10^{17}$

$M = \frac{2}{3} \log \frac{1.99 \times 10^{17}}{10^{4.40}}$

$= 8.6$

7. We use the magnitude formula

$M = \frac{2}{3} \log \frac{E}{E_0}$

For the Long Beach earthquake,

$6.3 = \frac{2}{3} \log \frac{E_1}{E_0}$

$9.45 = \log \frac{E_1}{E_0}$

$\frac{E_1}{E_0} = 10^{9.45}$

$E_1 = E_0 \cdot 10^{9.45}$

For the Anchorage earthquake,

$8.3 = \frac{2}{3} \log \frac{E_2}{E_0}$

$12.45 = \log \frac{E_2}{E_0}$

$\frac{E_2}{E_0} = 10^{12.45}$

$E_2 = E_0 \cdot 10^{12.45}$

Hence, the Anchorage earthquake compares to the Long Beach earthquake as follows:

$\frac{E_2}{E_1} = \frac{E_0 \cdot 10^{12.45}}{E_0 \cdot 10^{9.45}} = 10^3$

$E_2 = 10^3 E_1$, or 1000 times as powerful

9. We use the rocket equation.

$$v = c \ln \frac{W_t}{W_b}$$
$$v = 2.57 \ln (19.8)$$
$$v = 7.67 \text{ km/s}$$

11. (A) $pH = -\log[H^+] = -\log(4.63 \times 10^{-9}) = 8.3.$ Since this is greater than 7, the substance is basic.

(B) $pH = -\log[H^+] = -\log(9.32 \times 10^{-4}) = 3.0$ Since this is less than 7, the substance is acidic.

13. Since $pH = -\log[H^+]$, we have

$$5.2 = -\log[H^+], \text{ or}$$
$$[H^+] = 10^{-5.2} = 6.3 \times 10^{-6} \text{ moles per liter}$$

15. $m = 6 - 2.5 \log \dfrac{L}{L_0}$

(A) We find m when $L = L_0$

$$m = 6 - 2.5 \log \frac{L_0}{L_0}$$
$$m = 6 - 2.5 \log 1$$
$$m = 6$$

(B) We compare L_1 for $m = 1$ with L_2 for $m = 6$

$$1 = 6 - 2.5 \log \frac{L_1}{L_0} \qquad 6 = 6 - 2.5 \log \frac{L_2}{L_0}$$
$$-5 = -2.5 \log \frac{L_1}{L_0} \qquad 0 = -2.5 \log \frac{L_2}{L_0}$$
$$2 = \log \frac{L_1}{L_0} \qquad 0 = \log \frac{L_2}{L_0}$$
$$\frac{L_1}{L_0} = 10^2 \qquad \frac{L_2}{L_0} = 1$$
$$L_1 = 100L_0 \qquad L_2 = L_0$$

Hence $\dfrac{L_1}{L_2} = \dfrac{100L_0}{L_0} = 100$. The star of magnitude 1 is 100 times brighter.

17. (A) Enter the x values shown in the table as L_1. Enter the yield values as L_2. Use the logarithmic regression model from the STAT CALC menu.

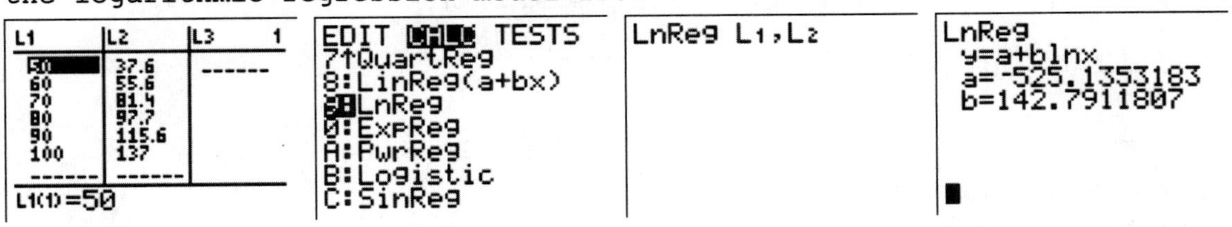

The model is $y = -525.1353183 + 142.7911807 \ln x$. Evaluating this for $x = 103$ (year 2003) yields 136.7 bushels per acre. Evaluating for $x = 110$ (year 2010) yields 146.0 bushels per acre.

(B) The actual yield was higher than predicted by the model. Entering this as further data would tend to raise the curve slightly and increase the predicted estimate for 2010.

Section 5-5

1.
$$10^{-x} = 0.0347$$
$$-x = \log_{10} 0.0347$$
$$x = -\log_{10} 0.0347$$
$$x = 1.46$$

3.
$$10^{3x+1} = 92$$
$$3x + 1 = \log_{10} 92$$
$$3x = \log_{10} 92 - 1$$
$$x = \frac{\log_{10} 92 - 1}{3}$$
$$x = 0.321$$

5.
$$e^x = 3.65$$
$$x = \ln 3.65$$
$$x = 1.29$$

7.
$$\begin{aligned} e^{2x-1} + 68 &= 207 \\ e^{2x-1} &= 139 \\ 2x - 1 &= \ln 139 \\ x &= \frac{1 + \ln 139}{2} \\ x &= 2.97 \end{aligned}$$

9.
$$\begin{aligned} 5^x &= 4^{2x+1} \\ \ln 5^x &= \ln 4^{2x+1} \\ x \ln 5 &= (2x + 1)\ln 4 \\ x \ln 5 &= 2x \ln 4 + \ln 4 \\ x \ln 5 - 2x \ln 4 &= \ln 4 \\ x(\ln 5 - 2 \ln 4) &= \ln 4 \\ x &= \frac{\ln 4}{\ln 5 - 2 \ln 4} \\ x &= -1.19 \end{aligned}$$

11.
$$\begin{aligned} 2^3 2^{-x} &= 0.426 \\ 2^{-x} &= \frac{0.426}{2^3} \\ \ln 2^{-x} &= \ln \frac{0.426}{8} \\ -x \ln 2 &= \ln \frac{0.426}{8} \\ x &= \frac{\ln \frac{0.426}{8}}{-\ln 2} \\ x &= 4.23 \end{aligned}$$

13.
$$\begin{aligned} \log 5 + \log x &= 2 \\ \log(5x) &= 2 \\ 5x &= 10^2 \\ 5x &= 100 \\ x &= 20 \end{aligned}$$

15.
$$\begin{aligned} \log x + \log(x - 3) &= 1 \\ \log[x(x - 3)] &= 1 \\ x(x - 3) &= 10^1 \\ x^2 - 3x &= 10 \\ x^2 - 3x - 10 &= 0 \\ (x - 5)(x + 2) &= 0 \\ x &= 5 \text{ or } -2 \end{aligned}$$

Common Error: $\log(x - 3) \neq \log x - \log 3$

Check:

$\log 5 + \log(5 - 3) \overset{\checkmark}{=} 1$

$\log(-2) + \log(-2 - 3)$ is not defined.

$x = 5$

17.
$$\begin{aligned} \log(x + 1) - \log(x - 1) &= 1 \\ \log \frac{x + 1}{x - 1} &= 1 \\ \frac{x + 1}{x - 1} &= 10^1 \\ \frac{x + 1}{x - 1} &= 10 \\ x + 1 &= 10(x - 1) \\ x + 1 &= 10x - 10 \\ 11 &= 9x \\ x &= \frac{11}{9} \end{aligned}$$

Common Error: $\frac{x + 1}{x - 1} \neq \log 1$

Check:

$$\log\left(\frac{11}{9} + 1\right) - \log\left(\frac{11}{9} - 1\right) \overset{?}{=} 1$$

$$\log \frac{20}{9} - \log \frac{2}{9} \overset{?}{=} 1$$

$$\log 10 \overset{\checkmark}{=} 1$$

19.
$$\begin{aligned} 2 &= 1.05^x \\ \ln 2 &= x \ln 1.05 \\ \frac{\ln 2}{\ln 1.05} &= x \\ x &= 14.2 \end{aligned}$$

21. $e^{-1.4x} + 5 = 0$

No solution.
Both terms on the left side are always positive, so they can never add to 0.

23.
$$\begin{aligned} 123 &= 500e^{-0.12x} \\ \frac{123}{500} &= e^{-0.12x} \\ \ln\left(\frac{123}{500}\right) &= -0.12x \\ \frac{\ln\left(\frac{123}{500}\right)}{-0.12} &= x \\ x &= 11.7 \end{aligned}$$

25.
$$\begin{aligned} e^{-x^2} &= 0.23 \\ -x^2 &= \ln 0.23 \\ x^2 &= -\ln 0.23 \\ x &= \pm\sqrt{-\ln 0.23} \\ x &= \pm 1.21 \end{aligned}$$

27.
$$\begin{aligned} \log(5 - 2x) &= \log(3x + 1) \\ 5 - 2x &= 3x + 1 \\ 4 &= 5x \\ x &= \frac{4}{5} \end{aligned}$$

29. $\log x - \log 5 = \log 2 - \log(x - 3)$

$$\log \frac{x}{5} = \log \frac{2}{x-3}$$

$$\frac{x}{5} = \frac{2}{x-3}$$

Excluded value: $x \neq 3$

$$5(x-3)\frac{x}{5} = 5(x-3)\frac{2}{x-3}$$

$$(x-3)x = 10$$

$$x^2 - 3x = 10$$

$$x^2 - 3x - 10 = 0$$

$$(x-5)(x+2) = 0$$

$$x = 5, -2$$

Check:

$\log 5 - \log 5 \overset{\checkmark}{=} \log 2 - \log 2$

$\log(-2)$ is not defined

Solution: 5

33. $\ln x = \ln(2x - 1) - \ln(x - 2)$

$$\ln x = \ln \frac{2x-1}{x-2}$$

$$x = \frac{2x-1}{x-2}$$

Excluded value: $x \neq 2$

$$x(x-2) = (x-2)\frac{2x-1}{x-2}$$

$$x(x-2) = 2x - 1$$

$$x^2 - 2x = 2x - 1$$

$$x^2 - 4x + 1 = 0$$

$$x = \frac{-b \pm \sqrt{b^2 - 4ac}}{2a}$$

$a = 1,\ b = -4,\ c = 1$

$$x = \frac{-(-4) \pm \sqrt{(-4)^2 - 4(1)(1)}}{2(1)}$$

$$x = \frac{4 \pm \sqrt{12}}{2} = 2 \pm \sqrt{3}$$

Check:

$\ln(2 + \sqrt{3}) \overset{?}{=} \ln[2(2 + \sqrt{3}) - 1] - \ln[(2 + \sqrt{3}) - 2]$

$\ln(2 + \sqrt{3}) \overset{?}{=} \ln(3 + 2\sqrt{3}) - \ln\sqrt{3}$

$\ln(2 + \sqrt{3}) \overset{?}{=} \ln\left(\frac{3 + 2\sqrt{3}}{\sqrt{3}}\right)$

$\ln(2 + \sqrt{3}) \overset{\checkmark}{=} \ln(\sqrt{3} + 2)$

$\ln(x - 2)$ is not defined if $x = 2 - \sqrt{3}$

Solution: $2 + \sqrt{3}$

35. $\ln(x + 1) = \ln(3x + 3)$

$$x + 1 = 3x + 3$$

$$-2x = 2$$

$$x = -1$$

Check: $\ln(-1 + 1)$ is not defined

No solution.

37.

$$(\ln x)^3 = \ln x^4$$

$$(\ln x)^3 = 4 \ln x$$

$$(\ln x)^3 - 4 \ln x = 0$$

$$\ln x[(\ln x)^2 - 4] = 0$$

$$\ln x(\ln x - 2)(\ln x + 2) = 0$$

$\ln x = 0$, $x = 1$; $\ln x - 2 = 0$, $\ln x = 2$, $x = e^2$; $\ln x + 2 = 0$, $\ln x = -2$, $x = e^{-2}$

Check:

$(\ln 1)^3 \overset{?}{=} \ln 1^4$, $0 \overset{\checkmark}{=} 0$

$(\ln e^2)^3 \overset{?}{=} \ln(e^2)^4$, $8 \overset{\checkmark}{=} 8$

$(\ln e^{-2})^3 \overset{?}{=} \ln(e^{-2})^4$, $-8 \overset{\checkmark}{=} -8$

Solution: $1, e^2, e^{-2}$

39. $\ln(\ln x) = 1$

$$\begin{aligned} \ln x &= e^1 \\ \ln x &= e \\ x &= e^e \end{aligned}$$

41. $x^{\log x} = 100x$

We start by taking logarithms of both sides.

$$\begin{aligned} \log(x^{\log x}) &= \log 100x \\ \log x \log x &= \log 100 + \log x \\ (\log x)^2 &= 2 + \log x \\ (\log x)^2 - \log x - 2 &= 0 \\ (\log x - 2)(\log x + 1) &= 0 \end{aligned}$$

$$\begin{aligned} \log x - 2 &= 0 & \log x + 1 &= 0 \\ \log x &= 2 & \log x &= -1 \\ x &= 10^2 & x &= 10^{-1} \\ x &= 100 & x &= 0.1 \end{aligned}$$

Check: $100^{\log 100} \stackrel{?}{=} 100 \cdot 100$ $\quad 10^4 \stackrel{\checkmark}{=} 10^4$

$(0.1)^{\log(0.1)} \stackrel{?}{=} 100(0.1)$ $\quad 0.1^{-1} \stackrel{?}{=} 10$ $\quad 10 \stackrel{\checkmark}{=} 10$

43. x intercept: Solve $f(x) = 0$

$$\begin{aligned} 4 - e^{x/2} &= 0 \\ 4 &= e^{x/2} \\ \ln 4 &= \ln e^{x/2} \\ \ln 4 &= \frac{x}{2} \\ x &= 2 \ln 4 \approx 2.77 \end{aligned}$$

y intercept: $f(0) = 4 - e^{0/2} = 3$

45. x intercept: Solve $g(x) = 0$

$$\begin{aligned} 2 + \ln(x + 1) &= 0 \\ \ln(x + 1) &= -2 \\ x + 1 &= e^{-2} \\ x &= e^{-2} - 1 \approx -0.865 \end{aligned}$$

Check: $2 + \ln(e^{-2} - 1 + 1) = 2 - 2 \stackrel{\checkmark}{=} 0$

y intercept: $g(0) = 2 + \ln(0 + 1) = 2$

47. x intercept: Solve $h(x) = 0$

$2^{x+3} + 1 = 0$

No solution. Both terms on the left side are positive, so they can never add to 0. No x intercept.

y intercept: $h(0) = 2^{0+3} + 1 = 9$

49. x intercept: Solve $k(x) = 0$

$$\begin{aligned} \log_{1/3} x - 5 &= 0 \\ \log_{1/3} x &= 5 \\ \left(\frac{1}{3}\right)^5 &= x \\ x &= \frac{1}{243} \end{aligned}$$

Check: $\log_{1/3}\left(\frac{1}{243}\right) - 5 = 5 - 5 \stackrel{\checkmark}{=} 0$

y intercept: $k(0)$ is not defined. No y intercept.

51. x intercept: Solve $m(x) = 0$

$$\begin{aligned} \log(x + 8) - \log(2x + 1) &= 0 \\ \log \frac{x + 8}{2x + 1} &= 0 \\ \frac{x + 8}{2x + 1} &= 10^0 \\ \frac{x + 8}{2x + 1} &= 1 \\ x + 8 &= 2x + 1 \\ x &= 7 \end{aligned}$$

Check: $\log(7 + 8) - \log(2.7 + 1) = \log 15 - \log 15 \stackrel{\checkmark}{=} 0$

y intercept: $m(0) = \log(0 + 8) - \log(2 \cdot 0 + 1) = \log 8 \approx 0.903.$

53.
$$\begin{aligned} A &= Pe^{rt} \\ \frac{A}{P} &= e^{rt} \\ \ln \frac{A}{P} &= rt \\ \frac{1}{t} \ln \frac{A}{P} &= r \\ r &= \frac{1}{t} \ln \frac{A}{P} \end{aligned}$$

55.
$$\begin{aligned} D &= 10 \log \frac{I}{I_0} \\ \frac{D}{10} &= \log \frac{I}{I_0} \\ \frac{I}{I_0} &= 10^{D/10} \\ I &= I_0(10^{D/10}) \end{aligned}$$

57.
$$\begin{aligned} M &= 6 - 2.5 \log \frac{I}{I_0} \\ 6 - M &= 2.5 \log \frac{I}{I_0} \\ \frac{6 - M}{2.5} &= \log \frac{I}{I_0} \\ \frac{I}{I_0} &= 10^{(6-M)/2.5} \\ I &= I_0[10^{(6-M)/2.5}] \end{aligned}$$

59.
$$\begin{aligned} I &= \frac{E}{R}(1 - e^{-Rt/L}) \\ RI &= E(1 - e^{-Rt/L}) \\ \frac{RI}{E} &= 1 - e^{-Rt/L} \\ \frac{RI}{E} - 1 &= -e^{-Rt/L} \\ -\left(\frac{RI}{E} - 1\right) &= e^{-Rt/L} \\ -\frac{RI}{E} + 1 &= e^{-Rt/L} \\ 1 - \frac{RI}{E} &= e^{-Rt/L} \\ \ln\left(1 - \frac{RI}{E}\right) &= -\frac{Rt}{L} \\ -\frac{L}{R} \ln\left(1 - \frac{RI}{E}\right) &= t \\ t &= -\frac{L}{R} \ln\left(1 - \frac{RI}{E}\right) \end{aligned}$$

61.
$$\begin{aligned} y &= \frac{e^x + e^{-x}}{2} \\ 2y &= e^x + e^{-x} \\ 2y &= e^x + \frac{1}{e^x} \\ 2ye^x &= (e^x)^2 + 1 \\ 0 &= (e^x)^2 - 2ye^x + 1 \end{aligned}$$

This equation is quadratic in e^x

$$e^x = \frac{-b \pm \sqrt{b^2 - 4ac}}{2a} \qquad a = 1,\ b = -2y,\ c = 1$$

$$\begin{aligned} e^x &= \frac{-(-2y) \pm \sqrt{(-2y)^2 - 4(1)(1)}}{2(1)} \\ e^x &= \frac{2y \pm \sqrt{4y^2 - 4}}{2} \\ e^x &= \frac{2(y \pm \sqrt{y^2 - 1})}{2} \\ e^x &= y \pm \sqrt{y^2 - 1} \\ x &= \ln(y \pm \sqrt{y^2 - 1}) \end{aligned}$$

63.
$$\begin{aligned} y &= \frac{e^x - e^{-x}}{e^x + e^{-x}} \\ y &= \frac{e^x - \frac{1}{e^x}}{e^x + \frac{1}{e^x}} \\ y &= \frac{e^x e^x - \frac{1}{e^x} e^x}{e^x e^x + \frac{1}{e^x} e^x} \\ y &= \frac{e^{2x} - 1}{e^{2x} + 1} \\ y(e^{2x} + 1) &= e^{2x} - 1 \\ ye^{2x} + y &= e^{2x} - 1 \\ 1 + y &= e^{2x} - ye^{2x} \\ 1 + y &= (1 - y)e^{2x} \\ e^{2x} &= \frac{1 + y}{1 - y} \\ 2x &= \ln \frac{1 + y}{1 - y} \\ x &= \frac{1}{2} \ln \frac{1 + y}{1 - y} \end{aligned}$$

65. Graphing $y = 2^{-x} - 2x$ and applying a built-in routine, we obtain

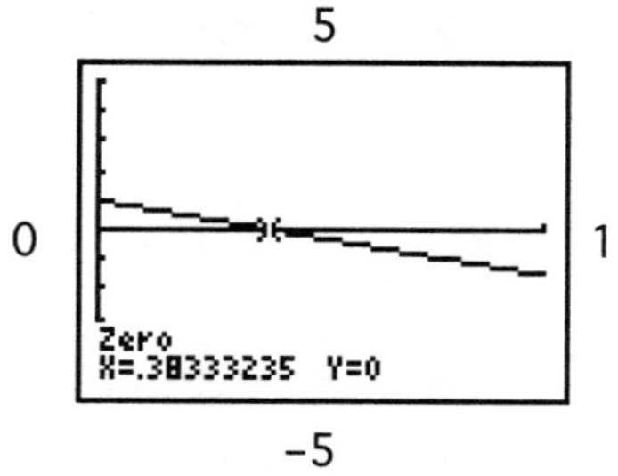

The required solution of $2^{-x} - 2x = 0$, $0 \le x \le 1$, is 0.38.

67. Graphing $y = x3^{x} - 1$ and applying a built-in routine, we obtain

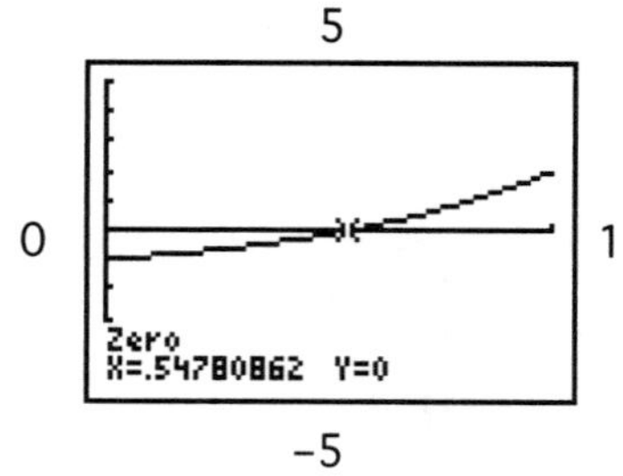

The required solution of $x3^{x} - 1 = 0$, $0 \le x \le 1$, is 0.55.

69. Graphing $y = e^{-x} - x$ and applying a built-in routine, we obtain

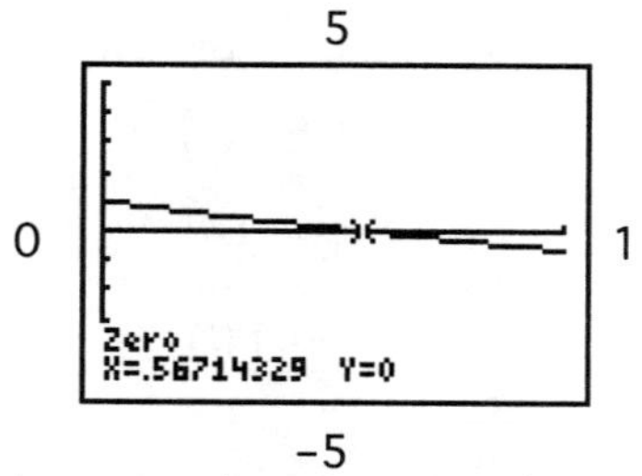

The required solution of $e^{-x} - x = 0$, $0 \le x \le 1$, is 0.57.

71. Graphing $y = xe^{x} - 2$ and applying a built-in routine, we obtain

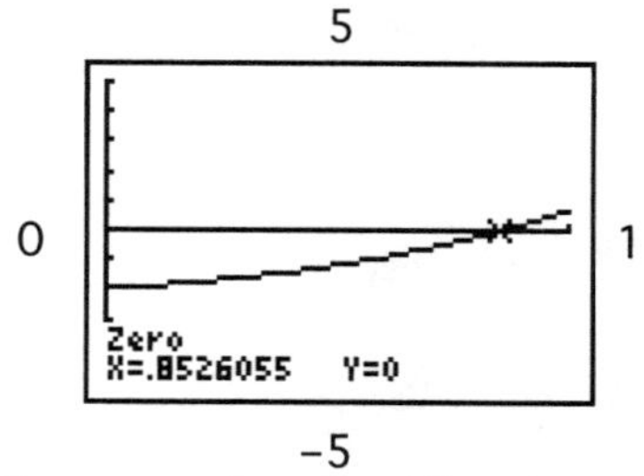

The required solution of $xe^{x} - 2 = 0$, $0 \le x \le 1$, is 0.85.

73. Graphing $y = \ln x + 2x$ and applying a built-in routine, we obtain

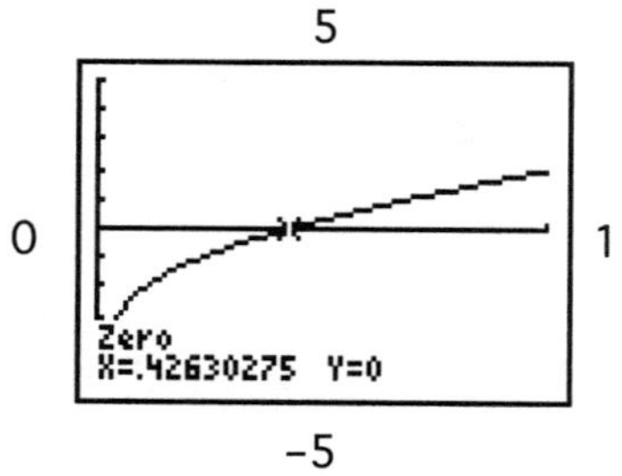

The required solution of $\ln x + 2x = 0$, $0 \le x \le 1$, is 0.43.

75. Graphing $y = \ln x + e^{x}$ and applying a built-in routine, we obtain

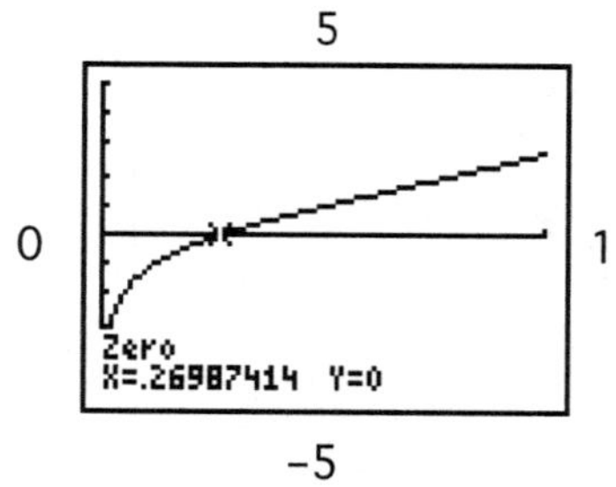

The required solution of $\ln x + e^{x} = 0$, $0 \le x \le 1$, is 0.27.

77. To find the doubling time we replace A in $A = P(1 + 0.07)^n$ with $2P$ and solve for n.

$$2P = P(1.07)^n$$
$$2 = (1.07)^n$$
$$\ln 2 = n \ln 1.07$$
$$n = \frac{\ln 2}{\ln 1.07}$$

$n = 10$ years to the nearest year

79. We solve $A = Pe^{rt}$ for r, with $A = 2{,}500$, $P = 1{,}000$, $t = 10$

$$2{,}500 = 1{,}000e^{r(10)}$$
$$2.5 = e^{10r}$$
$$10r = \ln (2.5)$$
$$r = \frac{1}{10} \ln 2.5$$
$$r = 0.0916 \text{ or } 9.16\%$$

81. We solve $P = P_0e^{rt}$ for t with $P = 2P_0$, $r = 0.0114$.

$$2P_0 = P_0e^{0.0114t}$$
$$2 = e^{0.0114t}$$
$$\ln 2 = 0.0114t$$
$$t = \frac{\ln 2}{0.0114}$$
t = 61 years to the nearest year

83. We solve $A = A_0e^{-0.000124t}$ for t with $A = 0.1A_0$

$$0.1A_0 = A_0e^{-0.000124t}$$
$$0.1 = e^{-0.000124t}$$
$$\ln 0.1 = -0.000124t$$
$$t = \frac{\ln 0.1}{-0.000124}$$
t = 18,600 years old

85. We solve $q = 0.0009(1 - e^{-0.2t})$ for t with $q = 0.0007$

$$0.0007 = 0.0009(1 - e^{-0.2t})$$
$$\frac{0.0007}{0.0009} = 1 - e^{-0.2t}$$
$$\frac{7}{9} = 1 - e^{-0.2t}$$
$$-\frac{2}{9} = -e^{-0.2t}$$
$$\frac{2}{9} = e^{-0.2t}$$
$$\ln \frac{2}{9} = -0.2t$$
$$t = \frac{\ln \frac{2}{9}}{-0.2}$$
t = 7.52 seconds

87. First, we solve $T = T_m + (T_0 - T_m)e^{-kt}$ for k, with $T = 61.5°$, $T_m = 40°$, $T_0 = 72°$, $t = 1$

$$61.5 = 40 + (72 - 40)e^{-k(1)}$$
$$21.5 = 32e^{-k}$$
$$\frac{21.5}{32} = e^{-k}$$
$$\ln \frac{21.5}{32} = -k$$
$$k = -\ln \frac{21.5}{32}$$
$$k = 0.40$$

Now we solve $T = T_m + (T_0 - T_m)e^{-0.40t}$ for t, with $T = 50°$, $T_m = 40°$, $T_0 = 72°$

$$50 = 40 + (72 - 40)e^{-0.40t}$$
$$10 = 32e^{-0.40t}$$
$$\frac{10}{32} = e^{-0.40t}$$
$$\ln \frac{10}{32} = -0.40t$$
$$t = \frac{\ln^{10/32}}{-0.40}$$
t = 2.9 hours

CHAPTER 5 REVIEW

1. (A) The graph of $y = \log_2 x$ passes through (1, 0) and (2, 1). This corresponds to graph m.

(B) The graph of $y = 0.5^x$ passes through (0, 1) and (1, 0.5). This corresponds to graph f.

(C) The graph of $y = \log_{0.5} x$ passes through (1, 0) and (0.5, 1). This corresponds to graph n.

(D) The graph of $y = 2^x$ passes through (0, 1) and (1, 2). This corresponds to graph g. *(5-1, 5-3)*

2. $\log m = n$ *(5-3)* **3.** $\ln x = y$ *(5-3)* **4.** $x = 10^y$ *(5-3)* **5.** $y = e^x$ *(5-3)*

6. $\frac{7^{x+2}}{7^{2-x}} = 7^{(x+2)-(2-x)}$
$= 7^{x+2-2+x}$
$= 7^{2x}$ *(5-1)*

7. $\left(\frac{e^x}{e^{-x}}\right)^x = [e^{x-(-x)}]^x$
$= (e^{2x})^x = e^{2x\cdot x}$
$= e^{2x^2}$ *(5-1)*

8. $\log_2 x = 3$
$x = 2^3$
$x = 8$ *(5-3)*

9. $\log_x 25 = 2$
$25 = x^2$
$x = 5$
since bases are restricted positive (5-3)

10. $\log_3 27 = x$
$\log_3 3^3 = x$
$x = 3$ (5-3)

11. $10^x = 17.5$
$x = \log_{10} 17.5$
$x = 1.24$ (5-5)

12. $e^x = 143{,}000$
$x = \ln 143{,}000$
$x = 11.9$ (5-5)

13. $\ln x = -0.01573$
$x = e^{-0.01573}$
$x = 0.984$ (5-3)

14. $\log x = 2.013$
$x = 10^{2.013}$
$x = 103$ (5-3)

15. 1.145 (5-3)

16. Not defined. ($-e$ is not in the domain of the logarithm function.) (5-3)

17. 2.211 (5-3)

18. 11.59 (5-1)

19. $\ln(2x - 1) = \ln(x + 3)$
$2x - 1 = x + 3$
$x = 4$

Check:
$\ln(2\cdot 4 - 1) \overset{?}{=} \ln(4 + 3)$
$\ln 7 \overset{\checkmark}{=} \ln 7$ (5-5)

20. $\log(x^2 - 3) = 2\log(x - 1)$
$\log(x^2 - 3) = \log(x - 1)^2$
$x^2 - 3 = (x - 1)^2$
$x^2 - 3 = x^2 - 2x + 1$
$-3 = -2x + 1$
$-4 = -2x$
$x = 2$

Check:
$\log(2^2 - 3) \overset{?}{=} 2\log(2 - 1)$
$\log 1 \overset{?}{=} 2\log 1$
$0 \overset{\checkmark}{=} 0$ (5-5)

21. $e^{x^2-3} = e^{2x}$
$x^2 - 3 = 2x$
$x^2 - 2x - 3 = 0$
$(x - 3)(x + 1) = 0$
$x = 3, -1$
(5-5)

22. $4^{x-1} = 2^{1-x}$
$(2^2)^{x-1} = 2^{1-x}$
$2^{2(x-1)} = 2^{1-x}$
$2(x - 1) = 1 - x$
$2x - 2 = 1 - x$
$3x = 3$
$x = 1$ (5-5)

23. $2x^2e^{-x} = 18e^{-x}$
$2x^2e^{-x} - 18e^{-x} = 0$
$2e^{-x}(x^2 - 9) = 0$
$2e^{-x}(x - 3)(x + 3) = 0$
$2e^{-x} = 0$ $\quad x - 3 = 0$ $\quad x + 3 = 0$
never $\quad x = 3$ $\quad x = -3$
Solution: 3, −3 (5-5)

24. $\log_{1/4} 16 = x$
$\log_{1/4} 4^2 = x$
$\log_{1/4} \left(\frac{1}{4}\right)^{-2} = x$
$x = -2$ (5-5)

25. $\log_x 9 = -2$
$x^{-2} = 9$
$\frac{1}{x^2} = 9$
$1 = 9x^2$
$\frac{1}{9} = x^2$
$x = \pm\sqrt{\frac{1}{9}}$
$x = \frac{1}{3}$
since bases are restricted positive
(5-5)

26. $\log_{16} x = \frac{3}{2}$
$16^{3/2} = x$
$64 = x$
$x = 64$ (5-5)

27. $\log_x e^5 = 5$
$e^5 = x^5$
$x = e$ (5-5)

28. $10^{\log_{10} x} = 33$
$\log_{10} x = \log_{10} 33$
$x = 33$ (5-5)

29. $\ln x = 0$
$e^0 = x$
$x = 1$ (5-5)

30. $x = 2(10^{1.32})$
$x = 41.8$ (5-1)

31. $x = \log_5 23$
$x = \dfrac{\log 23}{\log 5}$ or $\dfrac{\ln 23}{\ln 5}$
$x = 1.95$ (5-3)

32. $\ln x = -3.218$
$x = e^{-3.218}$
$x = 0.0400$ (5-3)

33. $x = \log(2.156 \times 10^{-7})$
$x = -6.67$ (5-3)

34. $x = \dfrac{\ln 4}{\ln 2.31}$
$x = 1.66$ (5-3)

35.
$$\begin{aligned} 25 &= 5(2)^x \\ \frac{25}{5} &= 2^x \\ 5 &= 2^x \\ \ln 5 &= x \ln 2 \\ \frac{\ln 5}{\ln 2} &= x \\ x &= 2.32 \end{aligned}$$
(5-5)

36.
$$\begin{aligned} 4{,}000 &= 2{,}500e^{0.12x} \\ \frac{4{,}000}{2{,}500} &= e^{0.12x} \\ 0.12x &= \ln \frac{4{,}000}{2{,}500} \\ x &= \frac{1}{0.12} \ln \frac{4{,}000}{2{,}500} \\ x &= 3.92 \end{aligned}$$
(5-5)

37.
$$\begin{aligned} 0.01 &= e^{-0.05x} \\ -0.05x &= \ln 0.01 \\ x &= \frac{\ln 0.01}{-0.05} \\ x &= 92.1 \end{aligned}$$
(5-5)

38.
$$\begin{aligned} 5^{2x-3} &= 7.08 \\ (2x - 3)\log 5 &= \log 7.08 \\ 2x - 3 &= \frac{\log 7.08}{\log 5} \\ x &= \frac{1}{2}\left[3 + \frac{\log 7.08}{\log 5}\right] \\ x &= 2.11 \end{aligned}$$
(5-5)

39.
$$\begin{aligned} \frac{e^x - e^{-x}}{2} &= 1 \\ e^x - e^{-x} &= 2 \\ e^x - \frac{1}{e^x} &= 2 \\ e^x e^x - e^x\left(\frac{1}{e^x}\right) &= 2e^x \\ (e^x)^2 - 1 &= 2e^x \\ (e^x)^2 - 2e^x - 1 &= 0 \end{aligned}$$

This equation is quadratic in e^x: $e^x = \dfrac{-b \pm \sqrt{b^2 - 4ac}}{2a}$ $a = 1$, $b = -2$, $c = -1$

$$e^x = \frac{-(-2) \pm \sqrt{(-2)^2 - 4(1)(-1)}}{2}$$
$$e^x = \frac{2 \pm \sqrt{8}}{2}$$
$$e^x = 1 \pm \sqrt{2}$$
$x = \ln(1 \pm \sqrt{2})$ $1 - \sqrt{2}$ is negative, hence not in the domain of the logarithm function.

$x = \ln(1 + \sqrt{2})$
$x = 0.881$ (5-5)

40. $\log 3x^2 - \log 9x = 2$

$$\log \frac{3x^2}{9x} = 2$$

$$\frac{3x^2}{9x} = 10^2$$

$$\frac{x}{3} = 100$$

$$x = 300$$

Check:

$$\log(3\cdot 300^2) - \log(9\cdot 300) \stackrel{?}{=} 2$$

$$\log(270{,}000) - \log(2{,}700) \stackrel{?}{=} 2$$

$$\log \frac{270{,}000}{2{,}700} \stackrel{?}{=} 2$$

$$\log 100 \stackrel{\surd}{=} 2$$

(5-5)

41. $\log x - \log 3 = \log 4 - \log(x + 4)$

$$\log \frac{x}{3} = \log \frac{4}{x + 4}$$

$$\frac{x}{3} = \frac{4}{x + 4}$$

excluded value: $x \neq -4$

$$3(x + 4)\frac{x}{3} = 3(x + 4)\frac{4}{x + 4}$$

$$(x + 4)x = 12$$

$$x^2 + 4x = 12$$

$$x^2 + 4x - 12 = 0$$

$$(x + 6)(x - 2) = 0$$

$$x = -6 \quad x = 2$$

Check: $\log(-6)$ is not defined

$$\log 2 - \log 3 \stackrel{?}{=} \log 4 - \log(2 + 4)$$

$$\log \frac{2}{3} \stackrel{?}{=} \log \frac{4}{6}$$

$$\log \frac{2}{3} \stackrel{\surd}{=} \log \frac{2}{3}$$

Solution: 2

(5-5)

42. $\ln(x + 3) - \ln x = 2 \ln 2$

$$\ln \frac{x + 3}{x} = \ln 2^2$$

$$\frac{x + 3}{x} = 2^2$$

$$\frac{x + 3}{x} = 4$$

$$x + 3 = 4x$$

$$3 = 3x$$

$$x = 1$$

Check:

$$\ln(1 + 3) - \ln 1 \stackrel{?}{=} 2 \ln 2$$

$$\ln 4 - 0 \stackrel{?}{=} 2 \ln 2$$

$$\ln 4 \stackrel{\surd}{=} \ln 4$$

(5-5)

43. $\ln(2x + 1) - \ln(x - 1) = \ln x$

$$\ln \frac{2x + 1}{x - 1} = \ln x$$

$$\frac{2x + 1}{x - 1} = x \quad \text{Excluded value: } x \neq 1$$

$$(x - 1)\frac{2x + 1}{x - 1} = x(x - 1)$$

$$2x + 1 = x^2 - x$$

$$0 = x^2 - 3x - 1$$

$$x = \frac{-b \pm \sqrt{b^2 - 4ac}}{2a} \quad a = 1,\ b = -3,\ c = -1$$

$$x = \frac{-(-3) \pm \sqrt{(-3)^2 - 4(1)(-1)}}{2(1)}$$

$$x = \frac{3 \pm \sqrt{13}}{2}$$

Check: $\ln\left(\frac{3 - \sqrt{13}}{2}\right)$ is not defined

$$\ln\left(2 \cdot \frac{3 + \sqrt{13}}{2} + 1\right) - \ln\left(\frac{3 + \sqrt{13}}{2} - 1\right) \stackrel{?}{=} \ln\left(\frac{3 + \sqrt{13}}{2}\right)$$

$$\ln(3 + \sqrt{13} + 1) - \ln\left(\frac{3 + \sqrt{13} - 2}{2}\right) \stackrel{?}{=} \ln\left(\frac{3 + \sqrt{13}}{2}\right)$$

$$\ln(4 + \sqrt{13}) - \ln\left(\frac{1 + \sqrt{13}}{2}\right) \stackrel{?}{=} \ln\left(\frac{3 + \sqrt{13}}{2}\right)$$

$$\ln\left(\frac{4 + \sqrt{13}}{1} \cdot \frac{2}{1 + \sqrt{13}}\right) \stackrel{?}{=} \ln\left(\frac{3 + \sqrt{13}}{2}\right)$$

$$\ln\left(\frac{\left(4 + \sqrt{13}\right)2}{1 + \sqrt{13}}\right) \stackrel{?}{=} \ln\left(\frac{3 + \sqrt{13}}{2}\right)$$

$$\ln\left(\frac{\left(4 + \sqrt{13}\right) 2 \left(1 - \sqrt{13}\right)}{\left(1 + \sqrt{13}\right)\left(1 - \sqrt{13}\right)}\right) \stackrel{?}{=} \ln\left(\frac{3 + \sqrt{13}}{2}\right)$$

$$\ln\left(\frac{2\left(4 - 3\sqrt{13} - 13\right)}{1 - 13}\right) \stackrel{?}{=} \ln\left(\frac{3 + \sqrt{13}}{2}\right)$$

$$\ln\left(\frac{-18 - 6\sqrt{13}}{-12}\right) \stackrel{?}{=} \ln\left(\frac{3 + \sqrt{13}}{2}\right)$$

$$\ln\left(\frac{3 + \sqrt{13}}{2}\right) \stackrel{\checkmark}{=} \ln\left(\frac{3 + \sqrt{13}}{2}\right)$$

Solution: $\frac{3 + \sqrt{13}}{2}$ (5-5)

44.
$$(\log x)^3 = \log x^9$$
$$(\log x)^3 = 9 \log x$$
$$(\log x)^3 - 9 \log x = 0$$
$$\log x[(\log x)^2 - 9] = 0$$
$$\log x(\log x - 3)(\log x + 3) = 0$$
$\log x = 0$ $\quad \log x - 3 = 0$ $\quad \log x + 3 = 0$
$x = 1$ $\quad \log x = 3$ $\quad \log x = -3$
$x = 10^3$ $\quad x = 10^{-3}$

Check:
$$(\log 1)^3 \stackrel{?}{=} \log 1^9$$
$$0 \stackrel{\checkmark}{=} 0$$
$$(\log 10^3)^3 \stackrel{?}{=} \log(10^3)^9$$
$$27 \stackrel{\checkmark}{=} 27$$
$$(\log 10^{-3})^3 \stackrel{?}{=} \log(10^{-3})^9$$
$$-27 \stackrel{\checkmark}{=} -27$$
Solution: $1, 10^3, 10^{-3}$ (5-5)

45. $\ln(\log x) = 1$
$\log x = e$
$x = 10^e$ (5-5)

46. $(e^x + 1)(e^{-x} - 1) - e^x(e^{-x} - 1) = e^xe^{-x} - e^x + e^{-x} - 1 - e^xe^{-x} + e^x$
$= 1 - e^x + e^{-x} - 1 - 1 + e^x = e^{-x} - 1$ (5-1)

47. $(e^x + e^{-x})(e^x - e^{-x}) - (e^x - e^{-x})^2 = (e^x)^2 - (e^{-x})^2 - [(e^x)^2 - 2e^xe^{-x} + (e^{-x})^2]$
$= e^{2x} - e^{-2x} - [e^{2x} - 2 + e^{-2x}]$
$= e^{2x} - e^{-2x} - e^{2x} + 2 - e^{-2x}$
$= 2 - 2e^{-2x}$ (5-1)

48. The graph of g is the same as the graph of f reflected through the x axis, shrunk vertically by a factor of $\frac{1}{3}$, and shifted upward 3 units; g is decreasing.
Domain: all real numbers
Horizontal asymptote: $y = 3$

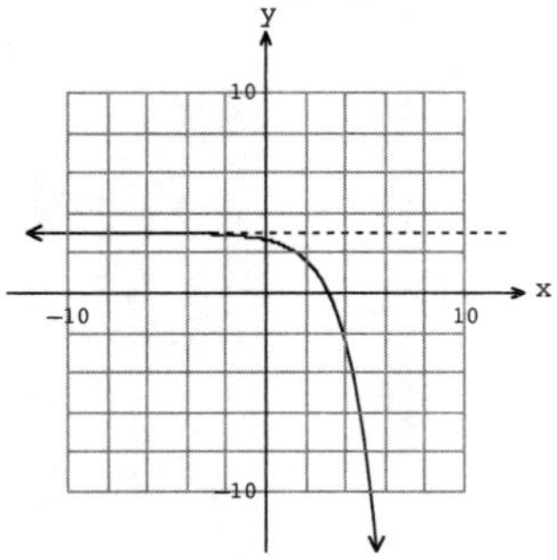

(5-1)

49. The graph of g is the same as the graph of f stretched vertically by a factor of 2 and shifted downward 4 units; g is increasing.
Domain: all real numbers
Horizontal asymptote: $y = -4$

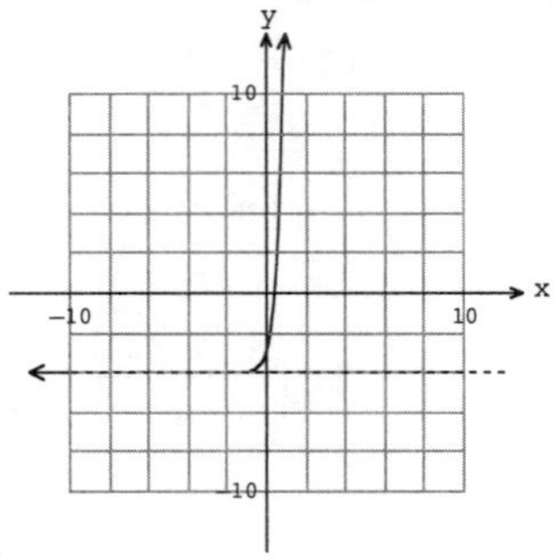

(5-1)

50. The graph of g is the same as the graph of f shifted downward 2 units; g is increasing.
Domain: $(0, \infty)$
Vertical asymptote: $x = 0$

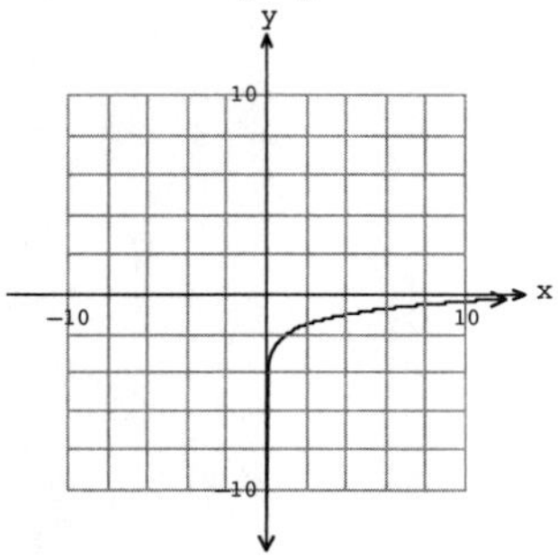

(5-3)

51. The graph of g is the same as the graph of f stretched vertically by a factor of 2 and shifted upward 1 unit; g is decreasing.
Domain: $(0, \infty)$
Vertical asymptote: $x = 0$

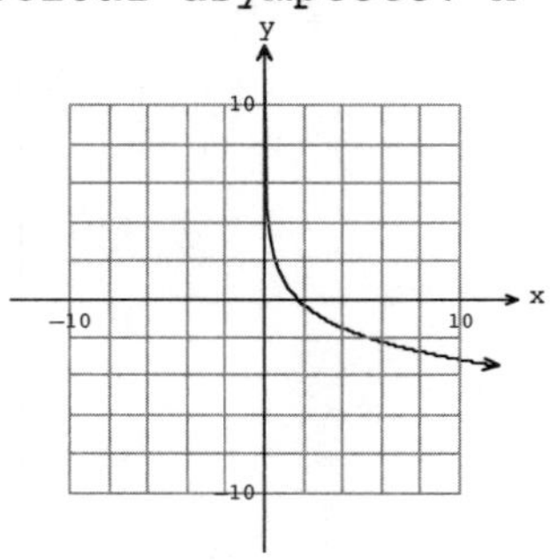

(5-3)

52. If the graph of $y = e^x$ is reflected in the x axis, y is replaced by $-y$ and the graph becomes the graph of $-y = e^x$ or $y = -e^x$.
If the graph of $y = e^x$ is reflected in the y axis, x is replaced by $-x$ and the graph becomes the graph of $y = e^{-x}$ or $y = \frac{1}{e^x}$ or $y = \left(\frac{1}{e}\right)^x$. (5-1)

53. (A) For $x > -1$, $y = e^{-x/3}$ decreases from $e^{1/3}$ to 0 while $\ln(x + 1)$ increases from $-\infty$ to ∞. Consequently, the graphs can intersect at exactly one point.

(B) Graphing $y1 = e^{-x/3}$ and $y2 = 4 \ln(x + 1)$ we obtain

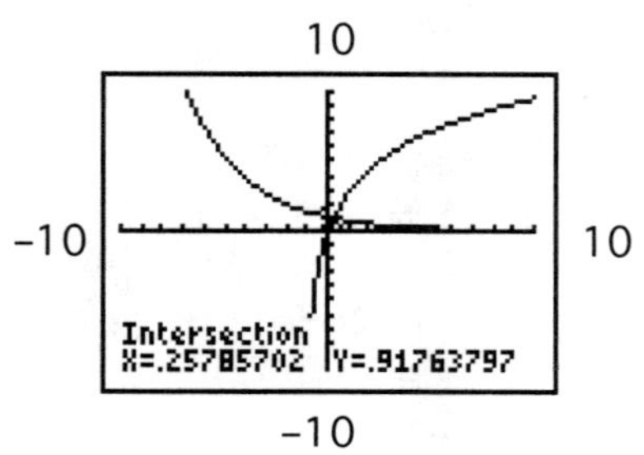

The solution of $e^{-x/3} = 4 \ln(x + 1)$ is $x = 0.258$. (5-5)

54. Examining the graph of $f(x) = 4 - x^2 + \ln x$, we obtain

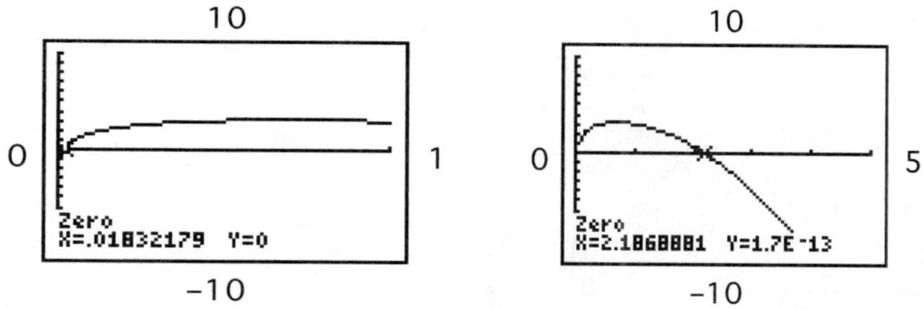

The zeros are at 0.018 and 2.187. (5-5)

55. Graphing $y1 = 10^{x-3}$ and $y2 = 8 \log x$, we obtain

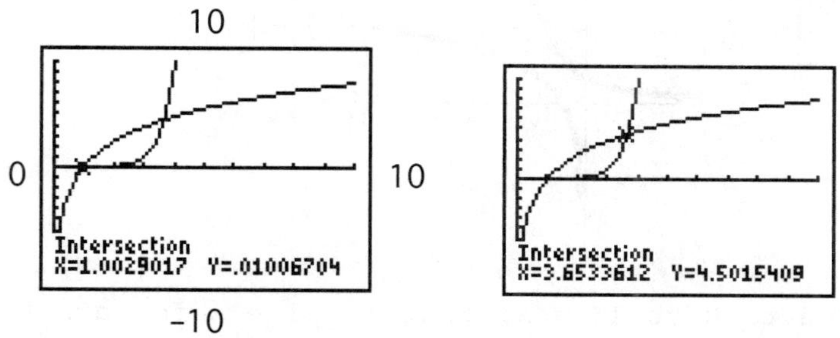

The graphs intersect at (1.003, 0.010) and (3.653, 4.502). (5-5)

56.

$$D = 10 \log \frac{I}{I_0}$$
$$\frac{D}{10} = \log \frac{I}{I_0}$$
$$10^{D/10} = \frac{I}{I_0}$$
$$I_0 10^{D/10} = I$$
$$I = I_0(10^{D/10})$$

(5-5)

57.

$$y = \frac{1}{\sqrt{2\pi}} e^{-x^2/2}$$
$$\sqrt{2\pi}\, y = e^{-x^2/2}$$
$$-\frac{x^2}{2} = \ln(\sqrt{2\pi}\, y)$$
$$x^2 = -2 \ln(\sqrt{2\pi}\, y)$$
$$x = \pm\sqrt{-2 \ln(\sqrt{2\pi}\, y)}$$

(5-5)

58.

$$x = -\frac{1}{k} \ln \frac{I}{I_0}$$
$$-kx = \ln \frac{I}{I_0}$$
$$\frac{I}{I_0} = e^{-kx}$$
$$I = I_0(e^{-kx})$$

(5-5)

59.

$$r = P \frac{i}{1 - (1 + i)^{-n}}$$
$$\frac{r}{P} = \frac{i}{1 - (1 + i)^{-n}}$$
$$\frac{P}{r} = \frac{1 - (1 + i)^{-n}}{i}$$
$$\frac{Pi}{r} = 1 - (1 + i)^{-n}$$
$$\frac{Pi}{r} - 1 = -(1 + i)^{-n}$$
$$1 - \frac{Pi}{r} = (1 + i)^{-n}$$
$$\ln\left(1 - \frac{Pi}{r}\right) = -n \ln(1 + i)$$
$$\frac{\ln\left(1 - \frac{Pi}{r}\right)}{-\ln(1 + i)} = n$$
$$n = -\frac{\ln\left(1 - \frac{Pi}{r}\right)}{-\ln(1 + i)}$$

(5-5)

60.
$$\ln y = -5t + \ln c$$
$$\ln y - \ln c = -5t$$
$$\ln\left(\frac{y}{c}\right) = -5t$$
$$\frac{y}{c} = e^{-5t}$$
$$y = ce^{-5t} \qquad (5\text{-}5)$$

61.

x	$y = \log_2 x$	$x = \log_2 y$	y
1	0	0	1
2	1	1	2
4	2	2	4
8	3	3	8

Domain $f = (0, \infty) =$ Range f^{-1}
Range $f = (-\infty, \infty) =$ Domain f^{-1}

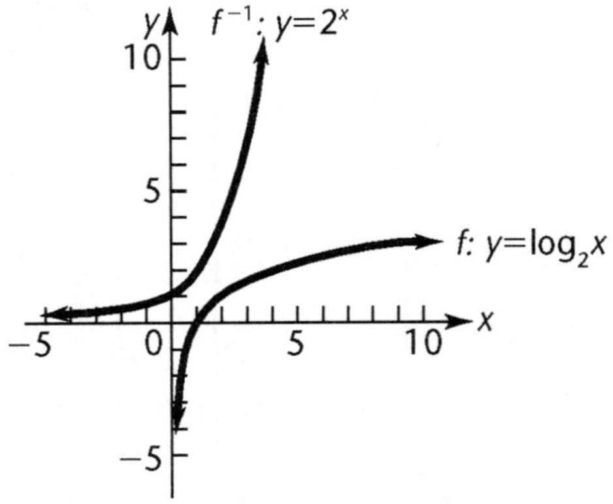

(5-3)

62. If $\log_1 x = y$, then we would have to have $1^y = x$; that is, $1 = x$ for arbitrary positive x, which is impossible. (5-3)

63. Let $u = \log_b M$ and $v = \log_b N$; then $M = b^u$ and $N = b^v$. Thus, $\log(M/N) = \log_b(b^u/b^v) = \log_b b^{u-v} = u - v = \log_b M - \log_b N$. (5-3)

64. We solve $P = P_0(1.03)^t$ for t, using $P = 2P_0$.
$$2P_0 = P_0(1.03)^t$$
$$2 = (1.03)^t$$
$$\ln 2 = t \ln 1.03$$
$$\frac{\ln 2}{\ln 1.03} = t$$
$$t = 23.4 \text{ years} \qquad (5\text{-}2)$$

65. We solve $P = P_0e^{0.03t}$ for t using $P = 2P_0$.
$$2P_0 = P_0e^{0.03t}$$
$$2 = e^{0.03t}$$
$$\ln 2 = 0.03t$$
$$\frac{\ln 2}{0.03} = t$$
$$t = 23.1 \text{ years} \qquad (5\text{-}2)$$

66. A_0 = original amount
$0.01A_0$ = 1 percent of original amount
We solve $A = A_0e^{-0.000124t}$ for t, using $A = 0.01A_0$.
$$0.01A_0 = A_0e^{-0.000124t}$$
$$0.01 = e^{-0.000124t}$$
$$\ln 0.01 = -0.000124t$$
$$\frac{\ln 0.01}{-0.000124} = t$$
$$t = 37{,}100 \text{ years} \qquad (5\text{-}2)$$

67. (A) When $t = 0$, $N = 1$. As t increases by 1/2, N doubles.
Hence $N = 1\cdot(2)^{t \div 1/2}$
$N = 2^{2t}$ (or $N = 4^t$)

(B) We solve $N = 4^t$ for t, using $N = 10^9$
$$10^9 = 4^t$$
$$9 = t \log 4$$
$$t = \frac{9}{\log 4}$$
$$t = 15 \text{ days} \qquad (5\text{-}2)$$

68. We use $A = Pe^{rt}$ with $P = 1$, $r = 0.03$, and $t = 2000$.
$A = 1e^{0.03(2000)}$
$A = 1.1 \times 10^{26}$ dollars (5-1)

69. (A)

t	P
0	1,000
5	670
10	449
15	301
20	202
25	135
30	91

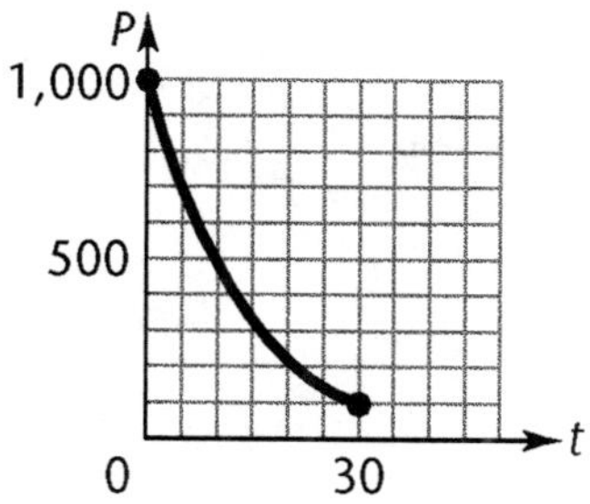

(B) As t tends to infinity, P appears to tend to 0. *(5-1)*

70. $M = \frac{2}{3}\log\frac{E}{E_0}$ $\quad E_0 = 10^{4.40}$

We use $E = 1.99 \times 10^{14}$

$$M = \frac{2}{3}\log\frac{1.99\times 10^{14}}{10^{4.40}}$$
$$M = \frac{2}{3}\log(1.99 \times 10^{9.6})$$
$$M = \frac{2}{3}(\log 1.99 + 9.6)$$
$$M = \frac{2}{3}(0.299 + 9.6)$$
$$M = 6.6$$

(5-4)

71. We solve $M = \frac{2}{3}\log\frac{E}{E_0}$ for E, using $E_0 = 10^{4.40}$, $M = 8.3$

$$8.3 = \frac{2}{3}\log\frac{E}{10^{4.40}}$$
$$\frac{3}{2}(8.3) = \log\frac{E}{10^{4.40}}$$
$$12.45 = \log\frac{E}{10^{4.40}}$$
$$\frac{E}{10^{4.40}} = 10^{12.45}$$
$$E = 10^{4.40}\cdot 10^{12.45}$$
$$E = 10^{16.85} \text{ or } 7.08\times 10^{16} \text{ joules}$$

(5-4)

72. We use the given formula twice, with

$$I_2 = 100{,}000I_1$$
$$D_1 = 10\log\frac{I_1}{I_0} \quad D_2 = 10\log\frac{I_2}{I_0}$$
$$D_2 - D_1 = 10\log\frac{I_2}{I_0} - 10\log\frac{I_1}{I_0} = 10\log\left(\frac{I_2}{I_0} \div \frac{I_1}{I_0}\right) = 10\log\frac{I_2}{I_1}$$
$$= 10\log\frac{100{,}000I_1}{I_1} = 10\log 100{,}000 = 50 \text{ decibels}$$

The level of the louder sound is 50 decibels more. *(5-2)*

73. $I = I_0e^{-kd}$

To find k, we solve for k using

$I = \frac{1}{2}I_0$ and $d = 73.6$

$$\frac{1}{2}I_0 = I_0e^{-k(73.6)}$$
$$\frac{1}{2} = e^{-73.6k}$$
$$-73.6k = \ln\frac{1}{2}$$
$$k = \frac{\ln\frac{1}{2}}{-73.6}$$
$$k = 0.00942$$

We now find the depth at which 1% of the surface light remains. We solve $I = I_0e^{-0.00942d}$ for d with $I = 0.01I_0$

$$0.01I_0 = I_0e^{-0.00942d}$$
$$0.01 = e^{-0.00942d}$$
$$-0.00942d = \ln 0.01$$
$$d = \frac{\ln 0.01}{-0.00942}$$
$$d = 489 \text{ feet}$$

(5-2)

74. We solve $N = \dfrac{30}{1 + 29e^{-1.35t}}$ for t with $N = 20$.

$$20 = \frac{30}{1 + 29e^{-1.35t}}$$
$$\frac{1}{20} = \frac{1 + 29e^{-1.35t}}{30}$$
$$1.5 = 1 + 29e^{-1.35t}$$
$$0.5 = 29e^{-1.35t}$$
$$\frac{0.5}{29} = e^{-1.35t}$$
$$-1.35t = \ln \frac{0.5}{29}$$
$$t = \frac{\ln \frac{0.5}{29}}{-1.35}$$
$$t = 3 \text{ years}$$

(5-2)

75. (A) The independent variable is years since 1975, so enter 5, 10, 15, 20, 25 as L_1. The dependent variable is Medicare expenditures so enter the expenditures column as L_2. Then use the exponential regression from the STAT CALC menu.

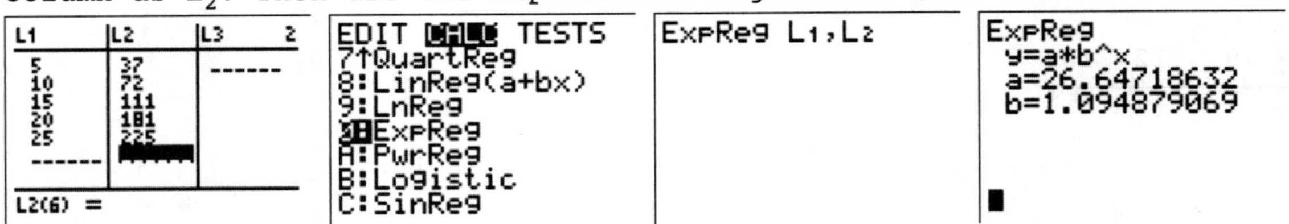

Evaluating this y at $x = 35$ (year 2010) yields $636 billion.

(B) Solving $26.64718632 \cdot (1.094879069^x) = 500$ algebraically yields $x = \dfrac{\ln(500 \div 26.64718632)}{\ln 1.094879069} \approx 32$, which corresponds to year 2007. A graphical solution using the intersection routine yields the same answer.

76. With the same data as in the previous problem, use the logarithmic regression from the STAT CALC menu.

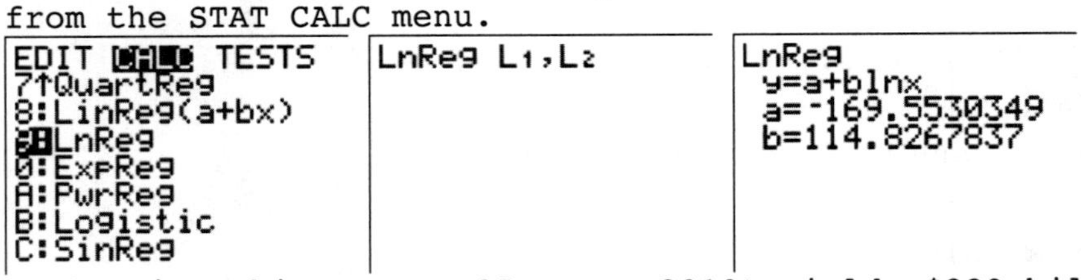

Evaluating this at $x = 35$ (year 2010) yields $239 billion.

(B) Plotting the data points and the two models on the same coordinate system shows the following:

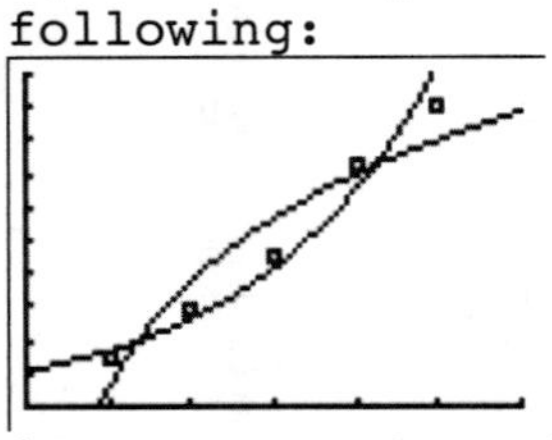

Visually, it appears that the exponential model (unfortunately) is a better fit. (5-4)

Chapters 4 & 5 Cumulative Review

1. (A) Since the graph has x intercepts -1, 1, and 2, and -1 is at least a double zero (the graph is tangent to the x axis at $x = -1$), the lowest degree equation would be $P(x) = (x + 1)^2(x - 1)(x - 2)$.

   (B) $P(x) \to \infty$ as $x \to \infty$ and as $x \to -\infty$. (4-1)

2. (A) The graph of $y = \left(\frac{3}{4}\right)^x$ passes through (0, 1) and $\left(1, \frac{3}{4}\right)$.

   This corresponds to graph m.

   (B) The graph of $y = \left(\frac{4}{3}\right)^x$ passes through (0, 1) and $\left(1, \frac{4}{3}\right)$.

   This corresponds to graph g.

   (C) The graph of $y = \left(\frac{3}{4}\right)^x + \left(\frac{4}{3}\right)^x$ passes through (0, 2).

   This corresponds to graph n.

   (D) The graph of $y = \left(\frac{4}{3}\right)^x - \left(\frac{3}{4}\right)^x$ passes through (0, 0).

   This corresponds to graph f. (5-1)

3.
$$\begin{array}{r|rrrr} & 3 & 5 & -18 & -3 \\ & & -9 & 12 & 18 \\ \hline -3 & 3 & -4 & -6 & 15 \end{array}$$

   $3x^3 + 5x^2 - 18x - 3 = (x + 3)(3x^2 - 4x - 6) + 15$ (4-1)

4. -2, 3, 5 (4-1)

5. We investigate $P(1)$ and $P(2)$ by forming a synthetic division table.

	4	-5	-3	-1
1	4	-1	-4	-5
2	4	3	3	5

   Since $P(1)$ and $P(2)$ have opposite signs, there is at least one real zero between 1 and 2. (4-2)

6. The possible rational zeros are ±1, ±2, ±4, ±8. We form a synthetic division table.

	1	1	-10	8	
1	1	2	-8	0	1 is a zero

   Thus $x^3 + x^2 - 10x + 8 = (x - 1)(x^2 + 2x - 8) = (x - 1)(x - 2)(x + 4)$.
   The rational zeros are 1, 2, -4. (4-3)

7. (A) $x = \log_{10} y$ or $x = \log y$ (B) $x = e^y$ (5-3)

8. (A) $(2e^x)^3 = 2^3(e^x)^3 = 8e^{3x}$ (B) $\frac{e^{3x}}{e^{-2x}} = e^{3x-(-2x)} = e^{5x}$ (5-1)

9. (A)
$$\begin{aligned} \log_3 x &= 2 \\ x &= 3^2 \\ x &= 9 \end{aligned}$$

   (B)
$$\begin{aligned} \log_3 81 &= x \\ \log_3 3^4 &= x \\ x &= 4 \end{aligned}$$

   (C)
$$\begin{aligned} \log_x 4 &= -2 \\ x^{-2} &= 4 \\ \frac{1}{x^2} &= 4 \\ 1 &= 4x^2 \\ x^2 &= \frac{1}{4} \\ x &= \frac{1}{2} \end{aligned}$$

   since bases are restricted positive (5-3)

10. (A) $10^x = 2.35$
$x = \log 2.35$
$x = 0.371$

(B) $e^x = 87{,}500$
$x = \ln 87{,}500$
$x = 11.4$

(C) $\log x = -1.25$
$x = 10^{-1.25}$
$x = 0.0562$

(D) $\ln x = 2.75$
$x = e^{2.75}$
$x = 15.6$ (5-3)

11. $E = k\frac{P}{x^3}$ (4-5) **12.** $F = k\frac{q_1q_2}{r^2}$ (4-5)

13. The graph of a nonconstant polynomial cannot have a horizontal asymptote. (4-1)

14. The graph of a rational function that has a horizontal asymptote has to approach the horizontal asymptote both as $x \to -\infty$ and $x \to \infty$. (4-4)

15. $f(x) = 3 \ln x - \sqrt{x}$ (5-3)

16. The function f multiplies the base e raised to the power of one-half of the domain element by 100 and then subtracts 50. (5-1)

17. $f(x) = \frac{2x + 8}{x + 2} = \frac{n(x)}{d(x)}$

(A) The domain of f is the set of all real numbers x such that $d(x) = x + 2 \neq 0$, that is $(-\infty, -2) \cup (-2, \infty)$ or $x \neq -2$. f has an x intercept where $n(x) = 2x + 8 = 0$, that is, $x = -4$. $f(0) = 4$, hence f has a y intercept at $y = 4$.

(B) *Vertical asymptote:* $x = -2$
Horizontal asymptote: Since $n(x)$ and $d(x)$ have the same degree, the line $y = 2$ is a horizontal asymptote.

Complete the sketch.
Plot a few more points.

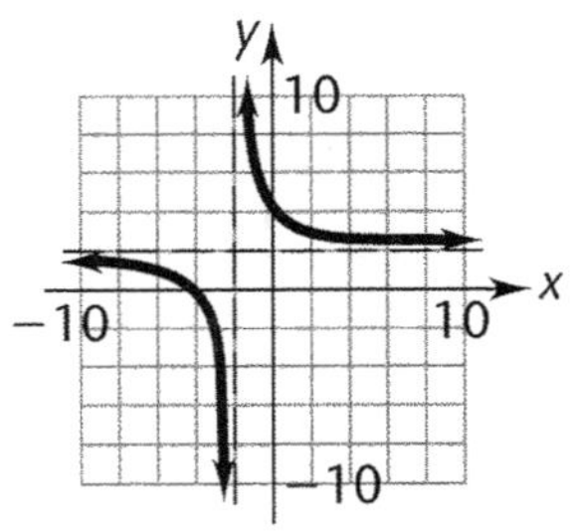

(4-4)

18. $P(x)$ can be factored further:
$P(x) = x(x^2 + 4)(x + 4)$
The zeros are 0, $2i$, $-2i$, and -4. Only 0 and -4 are x intercepts. (4-1)

19. From the previous problem, the real zeros of $P(x) = (x^3 + 4x)(x + 4)$ are 0 and -4. They partition the x axis into the three intervals shown in the table. Choose a test number in each interval to determine the sign of P in that interval. The equality holds at the zeros.

Interval	$(-\infty, -4)$	$(-4, 0)$	$(0, \infty)$
Test number x	-5	-1	1
$P(x)$	145	-15	25
Sign of P	+	−	+

We conclude that the solution set of the inequality is [-4, 0]. (4-2)

20.

$$\begin{array}{r|rrrr} & 2 & -5 & 3 & 2 \\ & & 1 & -2 & \frac{1}{2} \\ \hline \frac{1}{2} & 2 & -4 & 1 & \frac{5}{2} \end{array}$$

$P\left(\frac{1}{2}\right) = \frac{5}{2}$ (*4-1*)

21. $x - 1$ will be a factor of $P(x)$ if $P(1) = 0$

$P(1) = 1^{25} - 1^{20} + 1^{15} + 1^{10} - 1^5 + 1$

$= 1 - 1 + 1 + 1 - 1 + 1 = 2 \neq 0,$

so $x - 1$ is not a factor. $x + 1$ will be a factor of $P(x)$ if $P(-1) = 0$

$P(-1) = (-1)^{25} - (-1)^{20} + (-1)^{15} + (-1)^{10} - (-1)^5 + 1$

$= -1 - 1 - 1 + 1 + 1 + 1 = 0,$

so $x + 1$ is a factor of $P(x)$. (*4-1*)

22. (A) We form a synthetic division table:

	1	0	-8	0	3
-3	1	-3	1	-3	12 = $P(-3)$
-2	1	-2	-4	8	-13 = $P(-2)$
-1	1	-1	-7	7	-4 = $P(-1)$
0	1	0	-8	0	3 = $P(0)$
1	1	1	-7	-7	-4 = $P(1)$
2	1	2	-4	-8	-13 = $P(2)$
3	1	3	1	3	12 = $P(3)$

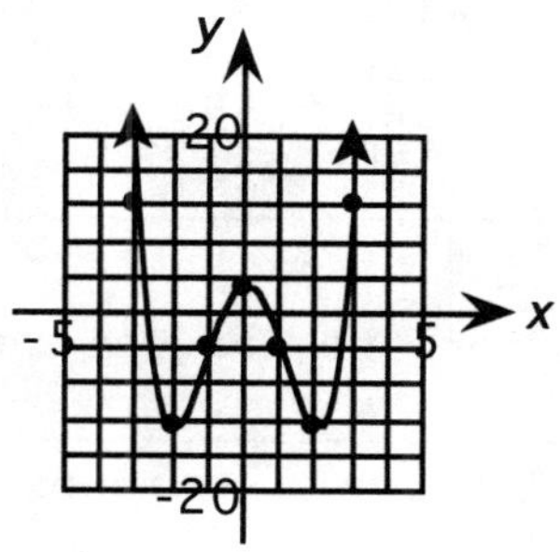

The graph of $P(x)$ has four x intercepts and three turning points; $P(x) \to \infty$ as $x \to \infty$ and as $x \to -\infty$

(B) There are real zeros in the intervals (-3, -2), (-1, 0), (0, 1), and (2, 3) indicated in the table. We search for the real zero in (2, 3). Examining the graph of P in a graphing utility we obtain the graph at the right.

The largest x intercept is 2.76.

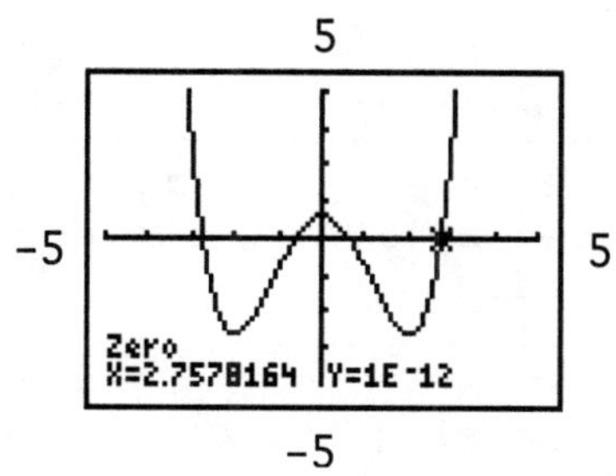

(*4-1*)

23. Examining the graph of $P(x)$, we see that there may be a zero of even multiplicity between -1 and 0, a zero of odd multiplicity near 2, and a zero of even multiplicity between 2 and 3.

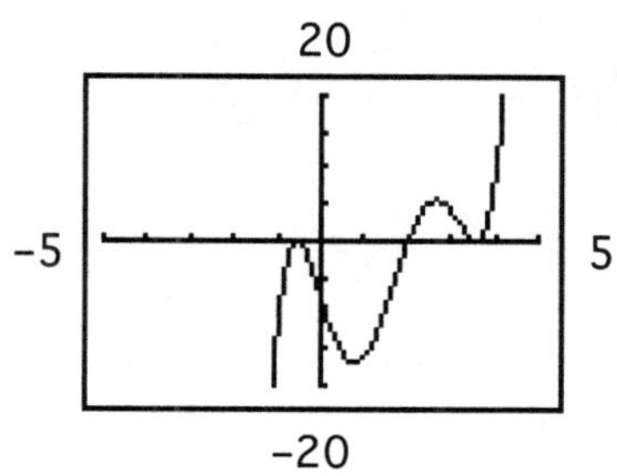

Applying a maximum routine between -1 and 0, a zero routine near 2, and a minimum routine between 2 and 3, we obtain

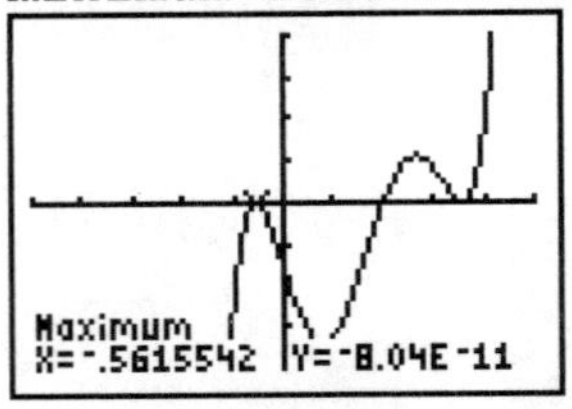

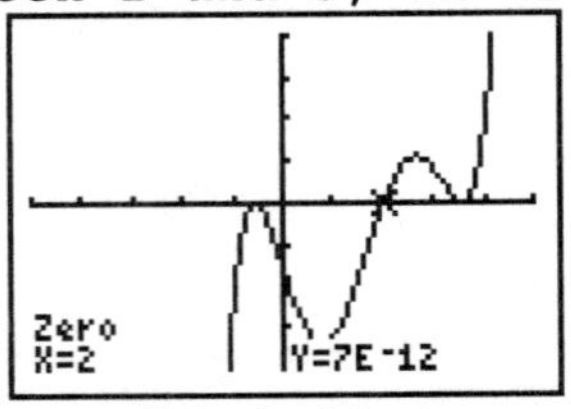

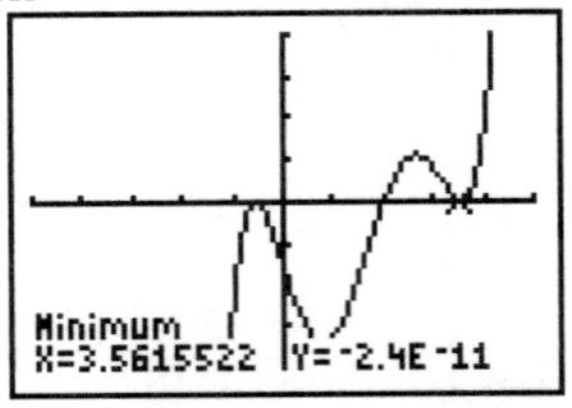

-0.56 is a zero of even multiplicity, 2 is a zero of odd multiplicity, and 3.56 is a zero of even multiplicity. Since $P(x)$ is a fifth-degree polynomial, -0.56 and 3.56 must be double zeros and 2 must be a simple zero. (*4-2*)

24. (A) We form a synthetic division table:

	1	2	-20	0	-30
0	1	2	-20	0	-30
1	1	3	-17	-17	-47
2	1	4	-12	-24	-78
3	1	5	-5	-15	-75
4	1	6	4	16	34
-1	1	1	-21	21	-51
-2	1	0	-20	40	-110
-3	1	-1	-17	51	-183
-4	1	-2	-12	48	-222
-5	1	-3	-5	25	-155
-6	1	-4	4	-24	114

From the table, 4 is an upper bound and -6 is a lower bound.

(B) We search for the largest real zero in (3, 4). We organize our calculations in a table.

Sign Change Interval (a, b)	Midpoint m	Sign of P: $P(a)$	$P(m)$	$P(b)$
(3, 4)	3.5	-	-	+
(3.5, 4)	3.75	-	-	+
(3.75, 4)	3.875	-	+	+
(3.75, 3.875)	3.8125	-	+	+
(3.75, 3.8125)	We stop here	-	-	+

Since each endpoint rounds to 3.8, a real zero lies on this last interval and is given by 3.8 to one decimal place accuracy. 4 further intervals were required.

(C) Examining the graph of $P(x)$, we obtain the graphs at the right.

The real zeros are -5.68 and 3.80.

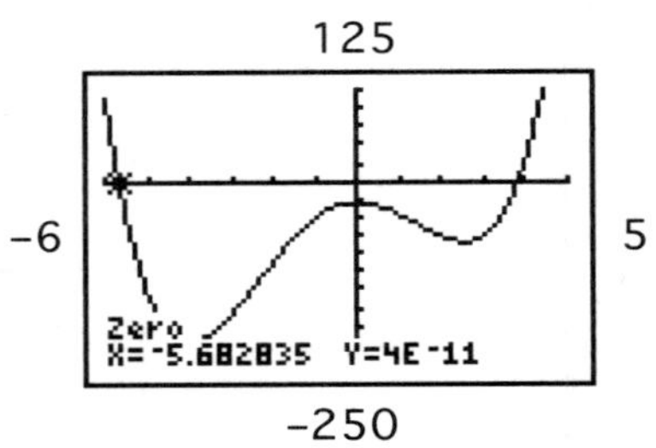

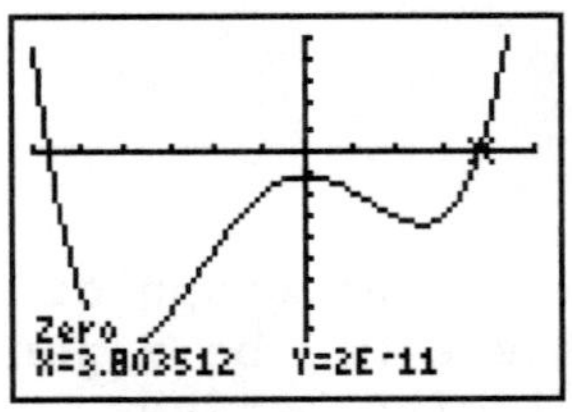

(*4-2*)

25. The possible rational zeros are ±1, ±3, ±5, ±15, $\pm\frac{1}{2}$, $\pm\frac{3}{2}$, $\pm\frac{5}{2}$, $\pm\frac{15}{2}$, $\pm\frac{1}{4}$, $\pm\frac{3}{4}$, $\pm\frac{5}{4}$, $\pm\frac{15}{4}$. We form a synthetic division table.

	4	-20	29	-15	
1	4	-16	13	-2	
3	4	-8	5	0	3 is a zero

So $P(x) = (x - 3)(4x^2 - 8x + 5)$

To find the remaining zeros, we solve $4x^2 - 8x + 5 = 0$ by the quadratic formula.

$4x^2 - 8x + 5 = 0$

$$x = \frac{-b \pm \sqrt{b^2 - 4ac}}{2a} \qquad a = 4,\ b = -8,\ c = 5$$

$$x = \frac{-(-8) \pm \sqrt{(-8)^2 - 4(4)(5)}}{2(4)}$$

$$x = \frac{8 \pm \sqrt{64 - 80}}{8}$$

$$x = \frac{8 \pm \sqrt{-16}}{8}$$

$$x = \frac{8 \pm 4i}{8}$$

$$x = 1 \pm \frac{1}{2}i$$

The zeros are 3, $1 \pm \frac{1}{2}i$ *(4-3)*

26. The possible rational zeros are ±1, ±2, ±3, ±4, ±6, ±12. We form a synthetic division table.

	1	5	1	-15	-12	
1	1	6	7	-8	-20	
2	1	7	15	15	18	a zero between 1 and 2; 2 is an upperbound
-1	1	4	-3	-12	0	-1 is a zero

We now examine $x^3 + 4x^2 - 3x - 12$. This factors by grouping into $(x + 4)(x^2 - 3)$, however, if we don't notice this, we find the remaining possible rational zeros to be -1, -2, -3, -4, -6, -12. We form a synthetic division table.

	1	4	-3	-12	
-1	1	3	-6	-6	not a double zero
-2	1	2	-7	2	a zero between -1 and -2
-3	1	1	-6	6	
-4	1	0	-3	0	-4 is a zero

So $P(x) = (x + 1)(x + 4)(x^2 - 3) = (x + 1)(x + 4)(x - \sqrt{3})(x + \sqrt{3})$.
The four zeros are -1, -4, $\pm\sqrt{3}$. *(4-3)*

27.

$$\begin{aligned} 2^{x^2} &= 4^{x+4} \\ 2^{x^2} &= (2^2)^{x+4} \\ 2^{x^2} &= 2^{2(x+4)} \\ x^2 &= 2(x+4) \\ x^2 &= 2x+8 \\ x^2 - 2x - 8 &= 0 \\ (x-4)(x+2) &= 0 \end{aligned}$$

$x - 4 = 0 \qquad x + 2 = 0$

$x = 4 \qquad\quad x = -2$ (5-5)

28.

$$\begin{aligned} 2x^2e^{-x} + xe^{-x} &= e^{-x} \\ 2x^2e^{-x} + xe^{-x} - e^{-x} &= 0 \\ e^{-x}(2x^2 + x - 1) &= 0 \\ e^{-x}(2x-1)(x+1) &= 0 \end{aligned}$$

$e^{-x} = 0 \qquad 2x - 1 = 0 \qquad x + 1 = 0$

never $\qquad x = \frac{1}{2} \qquad x = -1$

Solutions: $\frac{1}{2}$, -1 (5-5)

29. $e^{\ln x} = 2.5$

$x = 2.5$ (5-5)

30. $\log_x 10^4 = 4$

$x^4 = 10^4$

$x = 10$ (5-5)

31. $\log_9 x = -\frac{3}{2}$

$9^{-3/2} = x$

$x = \frac{1}{27}$ (5-5)

32. $\ln(x+4) - \ln(x-4) = 2 \ln 3$

$$\begin{aligned} \ln \frac{x+4}{x-4} &= \ln 3^2 \\ \frac{x+4}{x-4} &= 3^2 \\ \frac{x+4}{x-4} &= 9 \qquad x \neq 4 \\ x + 4 &= 9(x-4) \\ x + 4 &= 9x - 36 \\ -8x &= -40 \\ x &= 5 \end{aligned}$$

Check: $\ln(5+4) - \ln(5-4) \stackrel{?}{=} 2 \ln 3$

$\ln 9 - \ln 1 \stackrel{?}{=} 2 \ln 3$

$\ln 9 - 0 \stackrel{\checkmark}{=} \ln 9$

Solution: $x = 5$ (5-5)

33.

$$\begin{aligned} \ln(2x^2 + 2) &= 2 \ln(2x - 4) \\ \ln(2x^2 + 2) &= \ln(2x-4)^2 \\ 2x^2 + 2 &= (2x-4)^2 \\ 2x^2 + 2 &= 4x^2 - 16x + 16 \\ 0 &= 2x^2 - 16x + 14 \\ 0 &= 2(x-7)(x-1) \end{aligned}$$

$x = 7, 1$

Check: $x = 7$

$\ln(2\cdot7^2 + 2) \stackrel{?}{=} 2 \ln(2\cdot7 - 4)$

$\ln 100 \stackrel{?}{=} 2 \ln 10$

$\ln 100 \stackrel{\checkmark}{=} \ln 100$

$x = 1$

$\ln(2\cdot1^2 + 2) \stackrel{?}{=} 2 \ln(2\cdot1 - 4)$

$\ln(4) \neq 2 \ln(-2)$

Solution: $x = 7$ (5-5)

34.

$$\begin{aligned} \log x + \log(x+15) &= 2 \\ \log[x(x+15)] &= 2 \\ x(x+15) &= 10^2 \\ x^2 + 15x &= 100 \\ x^2 + 15x - 100 &= 0 \\ (x-5)(x+20) &= 0 \end{aligned}$$

$x = 5, -20$

Check: $x = 5$

$\log 5 + \log(5 + 15) \stackrel{?}{=} 2$

$\log 5 + \log 20 \stackrel{?}{=} 2$

$\log 100 \stackrel{\checkmark}{=} 2$

$x = -20$

$\log(-20) + \log(-20 + 15) \neq 2$

Solution: $x = 5$ (5-5)

35. $\log(\ln x) = -1$
$\ln x = 10^{-1}$
$\ln x = 0.1$
$x = e^{0.1}$ (5-5)

36.
$$4(\ln x)^2 = \ln x^2$$
$$4(\ln x)^2 = 2\ln x$$
$$4(\ln x)^2 - 2\ln x = 0$$
$$2\ln x(2\ln x - 1) = 0$$
$2\ln x = 0$ $\quad$ $2\ln x - 1 = 0$
$\ln x = 0$ $\quad$ $\ln x = \frac{1}{2}$
$x = 1$ $\quad$ $x = e^{0.5}$

Check: $x = 1$
$$4(\ln 1)^2 \stackrel{?}{=} \ln 1^2$$
$$0 \stackrel{\checkmark}{=} 0$$
$x = e^{0.5}$
$$4(\ln e^{0.5})^2 \stackrel{?}{=} \ln(e^{0.5})^2$$
$$4(0.5)^2 \stackrel{?}{=} \ln(e^{2(0.5)})$$
$$4(0.25) \stackrel{?}{=} \ln e$$
$$1 \stackrel{\checkmark}{=} 1$$
Solution: 1, $e^{0.5}$ (5-5)

37. $x = \log_3 41$
We use the change of base formula
$x = \dfrac{\log 41}{\log 3}$
$x = 3.38$ (5-5)

38. $\ln x = 1.45$
$x = e^{1.45}$
$x = 4.26$ (5-5)

39. $4(2^x) = 20$
$2^x = 5$
$x \log 2 = \log 5$
$x = \dfrac{\log 5}{\log 2}$
$x = 2.32$ (5-5)

40. $10e^{-0.5x} = 1.6$
$e^{-0.5x} = 0.16$
$-0.5x = \ln 0.16$
$x = \dfrac{\ln 0.16}{-0.5}$
$x = 3.67$ (5-5)

41.
$$\frac{e^x - e^{-x}}{e^x + e^{-x}} = \frac{1}{2}$$
$$\frac{e^x - \frac{1}{e^x}}{e^x + \frac{1}{e^x}} = \frac{1}{2}$$
$$\frac{e^x\left(e^x - \frac{1}{e^x}\right)}{e^x\left(e^x + \frac{1}{e^x}\right)} = \frac{1}{2}$$
$$\frac{(e^x)^2 - 1}{(e^x)^2 + 1} = \frac{1}{2}$$
$$\frac{e^{2x} - 1}{e^{2x} + 1} = \frac{1}{2}$$
$$2(e^{2x} + 1)\frac{e^{2x} - 1}{e^{2x} + 1} = 2(e^{2x} + 1)\frac{1}{2}$$
$$2(e^{2x} - 1) = e^{2x} + 1$$
$$2e^{2x} - 2 = e^{2x} + 1$$
$$e^{2x} = 3$$
$$2x = \ln 3$$
$$x = \frac{1}{2}\ln 3$$
$$x = 0.549$$
(5-5)

42. Write $G = kx^2$. Substitute $G = 10$ and $x = 5$ and solve for k.

$$10 = k(5)^2$$
$$10 = 25k$$
$$k = 0.4$$

The equation of variation is $G = 0.4x^2$.
When $x = 7$

$$G = 0.4(7)^2 = 19.6 \quad (4\text{-}5)$$

43. Write $H = \dfrac{k}{r^3}$. Substitute $H = 162$ and $r = 2$ and solve for k.

$$162 = \frac{k}{2^3}$$
$$k = 8(162)$$
$$k = 1296$$

The equation of variation is $H = \dfrac{1296}{r^3}$.

When $r = 3$

$$H = \frac{1296}{3^3} = 48 \quad (4\text{-}5)$$

44. The graph of f is the same as the graph of the exponential function $g(x) = 2^x$ shifted upward 3 units. Therefore, f has domain all real numbers, range $(3, \infty)$, and horizontal asymptote $y = 3$. (5-1)

45. The domain of f is the set of real numbers for which $x - 1 > 0$, that is, $x > 1$, or $(1, \infty)$. The graph of f is the same as the graph of the logarithmic function $g(x) = \log_3 x$ shifted right 1 unit, reflected through the x axis, and shifted up 2 units. Therefore, f has range all real numbers and vertical asymptote $x = 1$. (5-3)

46. f is a polynomial function. Therefore, it has domain all real numbers and its graph has no asymptotes. Since $f(x) \to -\infty$ as $x \to \infty$ and $f(x) \to \infty$ as $x \to -\infty$, it has range all real numbers. (4-1)

47. f is a polynomial function. Therefore it has domain all real numbers and its graph has no asymptotes. Since $x^4 \ge 0$ for all x, we can write

$$2x^4 \ge 0 \text{ all } x$$
$$3 + 2x^4 \ge 0 \quad \text{all } x$$

Hence the range of f is $[3, \infty)$ (4-1)

48. $f(x) = \dfrac{5}{x + 3} = \dfrac{n(x)}{d(x)}$

Domain: $d(x) = x + 3$ zero: -3
domain: $(-\infty, -3) \cup (-3, \infty)$

Range: If we set $y = \dfrac{5}{x + 3}$ then $x + 3 = \dfrac{5}{y}$, $x = 3 - \dfrac{5}{y}$. Hence, y can take on any value except 0.
range: $(-\infty, 0) \cup (0, \infty)$

Vertical asymptote: $x = -3$
Horizontal asymptote: Since the degree of $n(x) <$ the degree of $d(x)$ the x axis, $y = 0$ is a horizontal asymptote. (4-4)

49. The graph of f is the same as the graph of the exponential function $g(x) = e^x$ reflected through the y axis, stretched vertically by a factor of 20, and shifted downward 15 units. Therefore, f has domain all real numbers, range $(-15, \infty)$, and horizontal asymptote $y = -15$. (5-1)

50. The domain of f is the set of real numbers for which $x + 2 > 0$, that is, $x > -2$ or $(-2, \infty)$. The graph of f is the same as the graph of the logarithmic function $g(x) = \ln x$ shifted left 2 units and upward 8 units. Therefore, f has range all real numbers and vertical asymptote $x = -2$. (5-3)

51. If the graph of $y = \ln x$ is reflected in the x axis, y is replaced by $-y$ and the graph becomes the graph of $-y = \ln x$ or $y = -\ln x$.
If the graph of $y = \ln x$ is reflected in the y axis, x is replaced by $-x$ and the graph becomes the graph of $y = \ln(-x)$. (5-3)

52. (A) For $x > 0$, $y = e^{-x}$ decreases from 1 to 0 while $\ln x$ increases from $-\infty$ to ∞. Consequently, the graphs can intersect at exactly one point.

(B) Graphing $y1 = e^{-x}$ and $y2 = \ln x$, we obtain

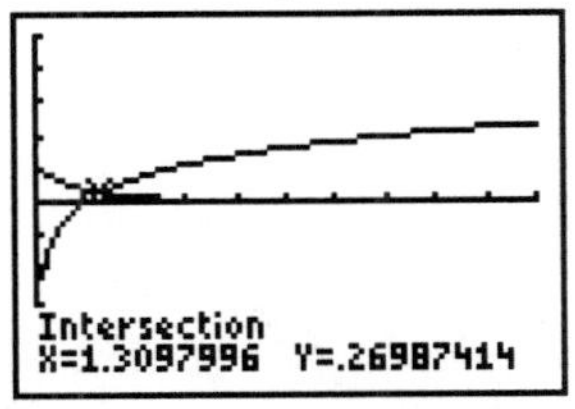

The solution is 1.31. (5-3)

53. (A) Let $u = x^2$. Then

$$P(x) = x^4 + 9x^2 + 18$$
$$P(x) = u^2 + 9u + 18$$
$$P(x) = (u + 6)(u + 3)$$
$$P(x) = (x^2 + 6)(x^2 + 3)$$

(B) $x^2 + 6$ has $i\sqrt{6}$ and $-i\sqrt{6}$ as zeros, so it factors as $x^2 + 6 = (x + i\sqrt{6})(x - i\sqrt{6})$. $x^2 + 3$ has $i\sqrt{3}$ and $-i\sqrt{3}$ as zeros, so it factors as

$$x^2 + 3 = (x + i\sqrt{3})(x - i\sqrt{3}).$$
$$P(x) = (x + i\sqrt{6})(x - i\sqrt{6})(x + i\sqrt{3})(x - i\sqrt{3})$$ (4-3)

54. (A) Let $u = x^2$. Then

$$P(x) = x^4 - 23x^2 - 50$$
$$P(x) = u^2 - 23u - 50$$
$$P(x) = (u - 25)(u + 2)$$
$$P(x) = (x^2 - 25)(x^2 + 2)$$
$$P(x) = (x + 5)(x - 5)(x^2 + 2)$$

(B) $x^2 + 2$ has zeros $i\sqrt{2}$ and $-i\sqrt{2}$ so it factors as

$$x^2 + 2 = (x + i\sqrt{2})(x - i\sqrt{2}).$$
$$P(x) = (x + 5)(x - 5)(x + i\sqrt{2})(x - i\sqrt{2})$$ (4-3)

55. $f(x) = \dfrac{x^2 + 4x + 8}{x + 2} = \dfrac{n(x)}{d(x)}$

Intercepts. There are no real zeros of $n(x) = x^2 + 4x + 8$. No x intercept
$f(0) = 4$ y intercept

Vertical asymptotes. Real zeros of $d(x) = x + 2$. $x = -2$

Horizontal asymptote. Since the degree of $n(x)$ is greater than the degree of $d(x)$, there is no horizontal asymptote.

Oblique asymptote.

$$\begin{array}{r} x + 2 \\ x + 2\,\overline{)\,x^2 + 4x + 8} \\ \underline{x^2 + 2x} \\ 2x + 8 \\ \underline{2x + 4} \\ 4 \end{array}$$

Thus, $f(x) = x + 2 + \dfrac{4}{x + 2}$

Hence, the line $y = x + 2$ is an oblique asymptote.

Complete the sketch.
Graphing utility:

10

-10 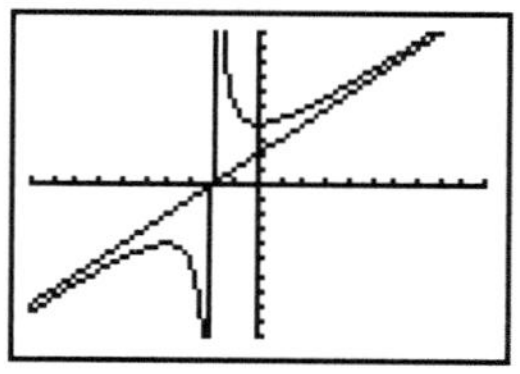 10

-10

Hand sketch:

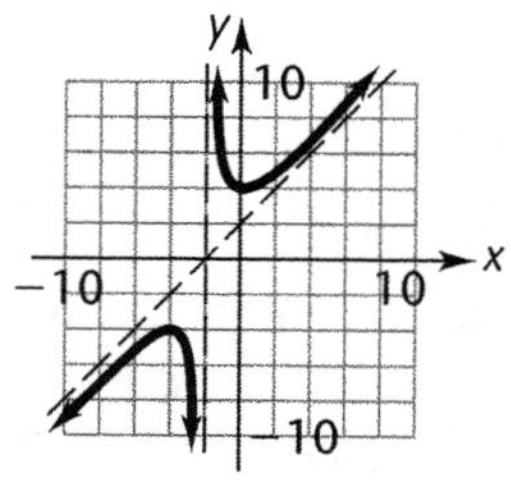

56. A preliminary graph of $y = P(x)$ is shown at the right.

1000

-1 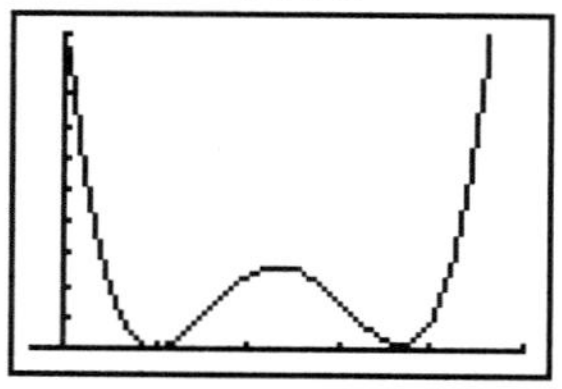 15

0

Examining the behavior near the maxima and minima, we obtain

.5

2.9 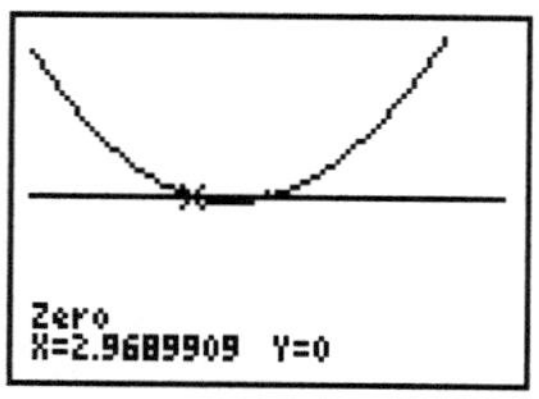 3.1

-.5

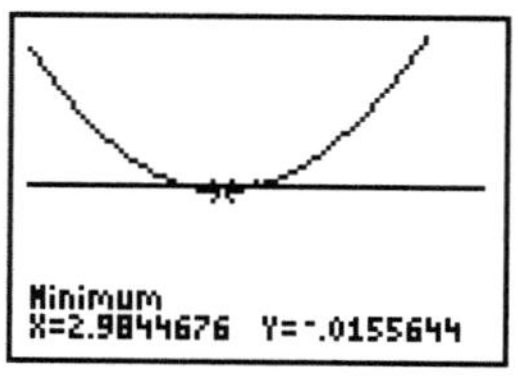

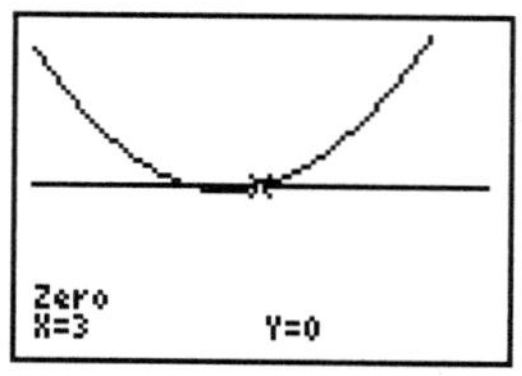

There are zeros at $x = 2.97$ and $x = 3$ and a local minimum at $P(2.98) \approx -0.02$.

1000

-1 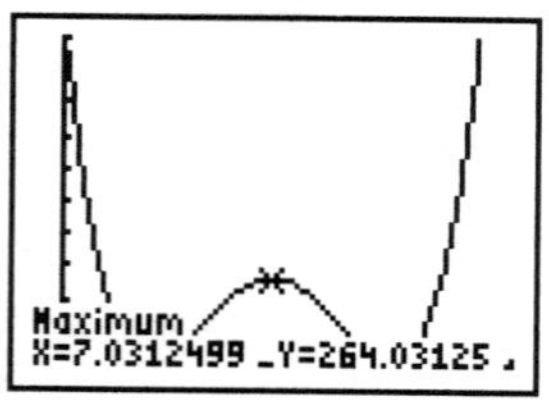 15

0

100

10 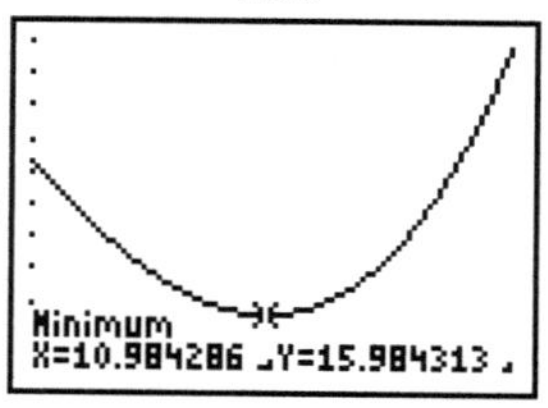 12

0

There is also a local maximum at $P(7.03) \approx 264.03$ and a local minimum at $P(10.98) \approx 15.98$.

(4-2)

57. $[x - (-1)]^2(x - 0)^3[x - (3 + 5i)][x - (3 - 5i)]$
$= (x + 1)^2x^3(x - 3 - 5i)(x - 3 + 5i)$

degree 7 *(4-3)*

58. Yes, for example:
$P(x) = (x + i)(x - i)(x + \sqrt{2})(x - \sqrt{2}) = x^4 - x^2 - 2$ has irrational zeros $\sqrt{2}$ and $-\sqrt{2}$. *(4-3)*

59. (A) We form a synthetic division table:

	1	9	-500	0	20,000
0	1	9	-500	0	20,000
10	1	19	-310	-3,100	-11,000
20	1	29	80	1,600	52,000
-10	1	-1	-490	4,900	-29,000
-20	1	-11	-280	5,600	-92,000
-30	1	-21	130	-3,900	137,000

From the table, 20 is an upper bound and -30 is a lower bound.

(B) Examining the graph of $y = P(x)$, we obtain

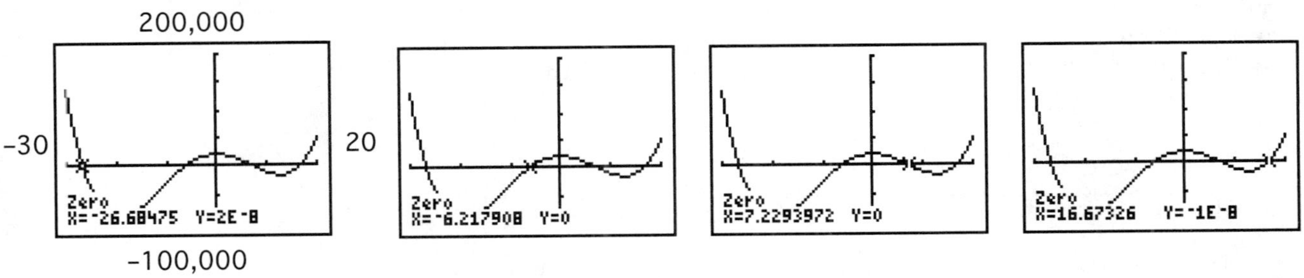

The zeros of $P(x)$ are -26.68, -6.22, 7.23, and 16.67. *(4-2)*

60. The possible rational zeros of $P(x)$ are ±1, ±2, ±3, ±4, ±6, ±12. Examining the graph of $y = P(x)$, we obtain the graph at the right.

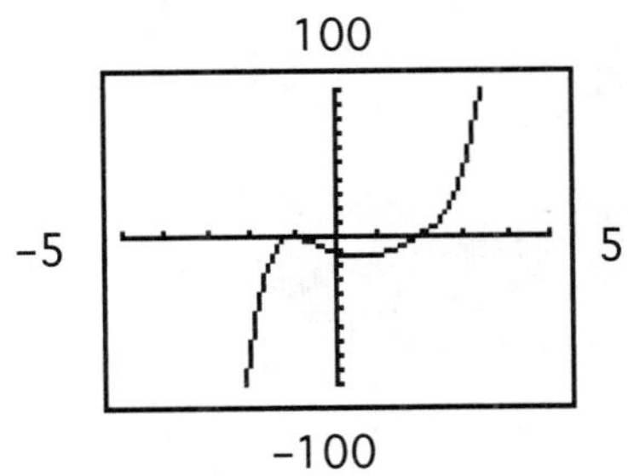

It appears that there may be a zero of even multiplicity near -1 and a zero of odd multiplicity near 2. We test the likely candidates and find

	1	-4	3	10	-10	-12
		2	-4	-2	16	12
2	1	-2	-1	8	6	0

2 is a zero. $P(x) = (x - 2)(x^4 - 2x^3 - x^2 + 8x + 6) = (x - 2)\ Q(x)$. Testing -1 in $Q(x)$ we obtain

	1	-2	-1	8	6
		-1	3	-2	-6
-1	1	-3	2	6	0

-1 is a zero. $Q(x) = (x + 1)(x^3 - 3x^2 + 2x + 6) = (x + 1)\ R(x)$. Testing -1 again in $R(x)$ we obtain

	1	-3	2	6
		-1	4	-6
-1	1	-4	6	0

-1 is a double zero of $Q(x)$. $P(x) = (x - 2)(x + 1)^2(x^2 - 4x + 6)$.

We complete the solution by solving $x^2 - 4x + 6 = 0$ by completing the square.

$$x^2 - 4x = -6$$
$$x^2 - 4x + 4 = -2$$
$$(x - 2)^2 = -2$$
$$x - 2 = \pm i\sqrt{2}$$
$$x = 2 \pm i\sqrt{2}$$

Thus the zeros of $P(x)$ are 2, -1 (double), and $2 \pm i\sqrt{2}$.

$$P(x) = (x - 2)(x + 1)^2[x - (2 + i\sqrt{2})][x - (2 - i\sqrt{2}\sqrt{2})]$$
$$= (x - 2)(x + 1)^2(x - 2 - i\sqrt{2})(x - 2 + i\sqrt{2})$$

(4-3)

61. The possible rational zeros of $P(x)$ are ±1, ±2, ±4. Examining the graph of $y = P(x)$, we obtain the graph at the right.

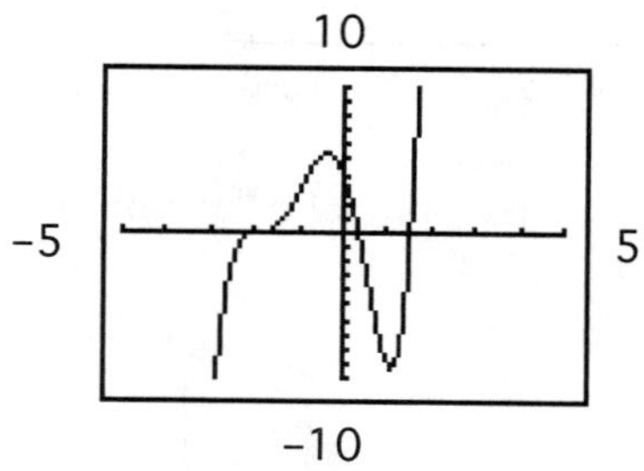

It appears that there may be zeros of unclear multiplicity near -2, as well as zeros between 1 and 2. The only possible rational zero allowed by the graph from the above list is -2. Testing it, we obtain

	1	4	1	-11	-8	4
		-2	-4	6	10	-4
-2	1	2	-3	-5	2	0

-2 is a zero. Testing whether it is a double zero, we obtain

	1	2	-3	-5	2
		-2	0	6	-2
-2	1	0	-3	1	0

Thus $P(x) = (x + 2)^2(x^3 - 3x + 1) = (x + 2)^2Q(x)$. Examining the graph of $Q(x)$, we obtain

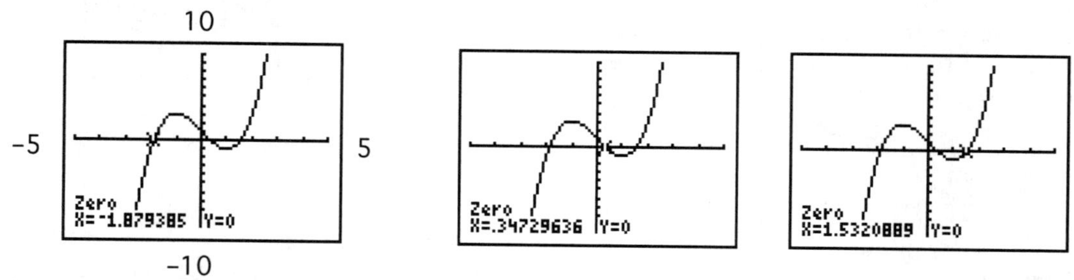

Thus, the zeros of $P(x)$ are -2 (double), -1.88, 0.35, and 1.53. (4-3)

62. Since the real zeros are 5 and 8, we should have factors of $x - 5$ and $x - 8$ in the numerator. A factor of $x - 1$ in the denominator provides vertical asymptote 1, but we need the degree of the numerator and denominator to be equal in order to have a non-zero horizontal asymptote. So we'll make the denominator $(x - 1)^2$. Finally, a factor of 3 in the numerator makes $y = 3$ the horizontal asymptote: $f(x) = \dfrac{3(x - 8)(x - 5)}{(x - 1)^2}$. (4-4)

63.
$$A = P\frac{(1+i)^n - 1}{i}$$
$$Ai = P[(1+i)^n - 1]$$
$$\frac{Ai}{P} = (1+i)^n - 1$$
$$1 + \frac{Ai}{P} = (1+i)^n$$
$$\ln\left(1 + \frac{Ai}{P}\right) = n\ln(1+i)$$
$$n = \frac{\ln\left(1 + \frac{Ai}{P}\right)}{\ln(1+i)}$$
(5-5)

64.
$$\ln y = 5x + \ln A$$
$$\ln y - \ln A = 5x$$
$$\ln\left(\frac{y}{A}\right) = 5x$$
$$\frac{y}{A} = e^{5x}$$
$$y = Ae^{5x}$$
(5-5)

65.
$$y = \frac{e^x - 2e^{-x}}{2}$$
$$2y = e^x - 2e^{-x}$$
$$2y = e^x - \frac{2}{e^x}$$
$$2ye^x = e^xe^x - e^x\left(\frac{2}{e^x}\right)$$
$$2ye^x = (e^x)^2 - 2$$
$$0 = (e^x)^2 - 2ye^x - 2$$

This equation is quadratic in e^x

$$e^x = \frac{-b \pm \sqrt{b^2 - 4ac}}{2a} \qquad a = 1,\ b = -2y,\ c = -2$$
$$e^x = \frac{-(-2y) \pm \sqrt{(-2y)^2 - 4(1)(-2)}}{2(1)}$$
$$e^x = \frac{2y \pm \sqrt{4y^2 + 8}}{2}$$
$$e^x = y \pm \sqrt{y^2 + 2}$$

Note: Since $0 < 2$, $y^2 < y^2 + 2$, $\sqrt{y^2} < \sqrt{y^2 + 2}$ and $y < \sqrt{y^2 + 2}$ for all real y.
Hence $y - \sqrt{y^2 + 2}$ is always negative. Also, $y + \sqrt{y^2 + 2}$ is always positive.

$x = \ln(y + \sqrt{y^2 + 2})$ (5-5)

66. $\dfrac{x^3 - x}{x^3 - 8} \geq 0$

Let $f(x) = \dfrac{p(x)}{q(x)} = \dfrac{x^3 - x}{x^3 - 8}$

The zeros of $p(x) = x^3 - x = x(x - 1)(x + 1)$ are -1, 0, and 1. The real zero of $q(x) = x^3 - 8 = (x - 2)(x^2 + 2x + 4)$ is 2. These four zeros partition the x axis into the five intervals shown in the table. A test number is chosen from each interval to determine the sign of $f(x)$.

Interval	Test number x	$f(x)$	Sign of f
$(-\infty, -1)$	-2	0.375	+
$(-1, 0)$	$-\frac{1}{2}$	-0.05	−
$(0, 1)$	$\frac{1}{2}$	0.05	+
$(1, 2)$	$\frac{3}{2}$	-0.4	−
$(2, \infty)$	3	1.3	+

The equality holds at −1, 0, and 1, but not at 2. We conclude that the solution set is $(-\infty, -1] \cup [0, 1] \cup (2, \infty)$. (4-4)

67. Algebraic solution:

$$\frac{4x}{x^2 - 1} - 3 < 0$$

$$\frac{4x}{x^2 - 1} - \frac{3(x^2 - 1)}{x^2 - 1} < 0$$

$$\frac{4x - 3x^2 + 3}{x^2 - 1} < 0$$

$$\frac{-3x^2 + 4x + 3}{x^2 - 1} < 0$$

zeros of the numerator: $a = -3$, $b = 4$, $c = 3$

$$x = \frac{-4 \pm \sqrt{16 - 4(-3)(3)}}{2(-3)}$$

$$= \frac{-4 \pm \sqrt{52}}{-6}$$

$$= -0.535, 1.869$$

zeros of the denominator:

$$x^2 - 1 = 0$$

$$x^2 = 1$$

$$x = \pm 1$$

These four zeros divide the real line into the five intervals shown below.

Interval	$(-\infty, -1)$	$(-1, -0.535)$	$(-0.535, 1)$	$(1, 1.869)$	$(1.869, \infty)$
Test number x	-2	-0.8	0.8	$\frac{3}{2}$	3
$f(x)$	-5.9	5.9	-11.9	1.8	-1.5
Sign of f	−	+	−	+	−

The expression is negative on $(-\infty, -1) \cup (-0.535, 1) \cup (1.869, \infty)$.

Graphical solution:

Graph $y_1 = \frac{4x}{x^2 - 1}$, $y_2 = 3$ and find where the graph of y_1 is below the graph of y_2.

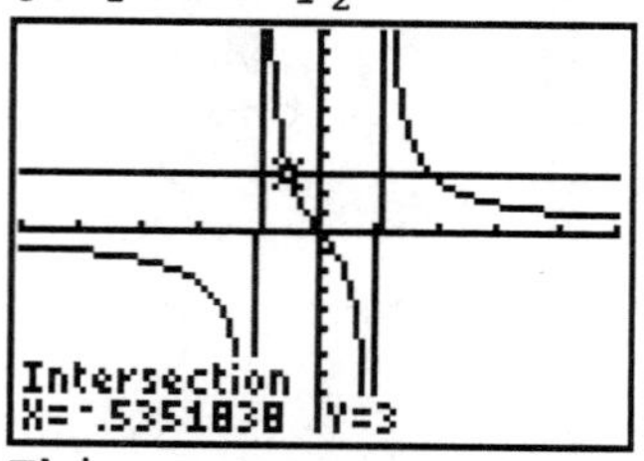

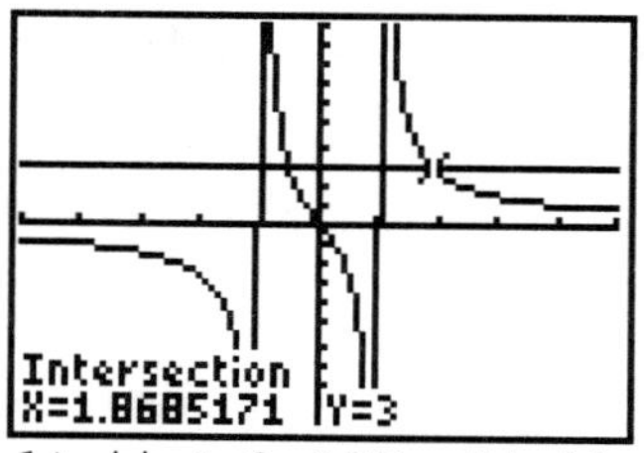

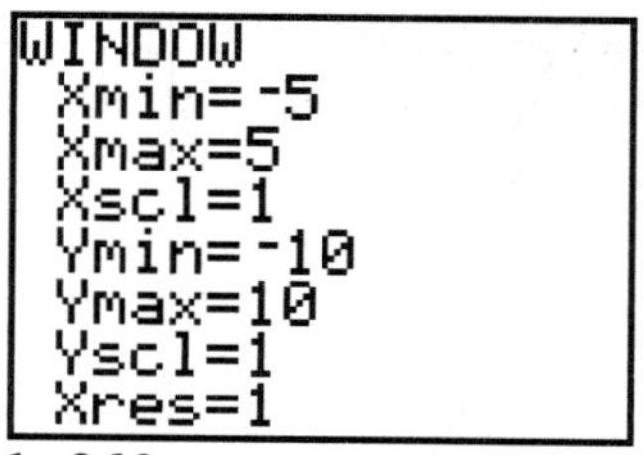

This occurs on $(-\infty, -1) \cup (-0.535, 1) \cup (1.869, \infty)$. (4-4)

68. We are given y = length of container
x = width of one end
Hence $4x$ = girth of container

Length + girth $= y + 4x = 10$
So $y = 10 - 4x$
Since Volume $= 8 = x^2y$, we have

$$8 = x^2(10 - 4x)$$
$$8 = 10x^2 - 4x^3$$
$$4x^3 - 10x^2 + 8 = 0$$
$$2x^3 - 5x^2 + 4 = 0$$

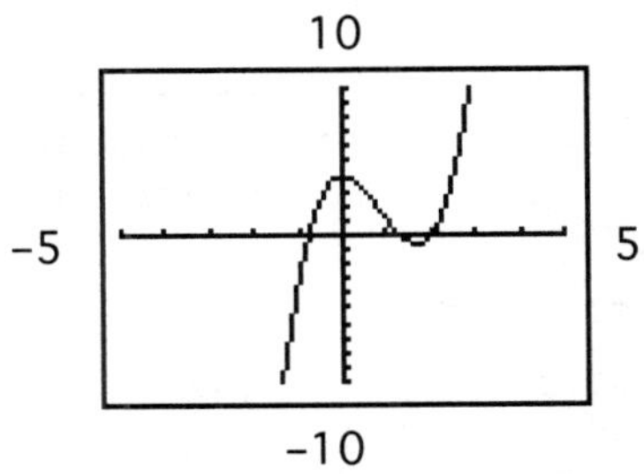

The possible rational solutions of this equation are $\pm 1, \pm 2, \pm 4, \pm \frac{1}{2}$.

Examining the graph of $y = 2x^3 - 5x^2 + 4$, we obtain the graph at the right.

It appears that there may be an integer zero at 2, a zero between -1 and 0, and a zero between 1 and 2. Testing 2, we obtain

	2	-5	0	4
		4	-2	-4
2	2	-1	-2	0

2 is a zero.

Thus, the equation can be factored
$$(x - 2)(2x^2 - x - 2) = 0$$

To find the remaining zeros, we solve $2x^2 - x - 2 = 0$ by the quadratic formula.

$$2x^2 - x - 2 = 0$$

$$x = \frac{-b \pm \sqrt{b^2 - 4ac}}{2a} \qquad a = 2,\ b = -1,\ c = -2$$

$$x = \frac{-(-1) \pm \sqrt{(-1)^2 - 4(2)(-2)}}{2(2)}$$

$$x = \frac{1 + \sqrt{17}}{4} \quad \text{(discarding the negative solution)}$$

Thus, the possible side lengths of the end are $x = 2$ feet and

$x = \frac{1 + \sqrt{17}}{4} \approx 1.28$ feet. *(4-2)*

69.

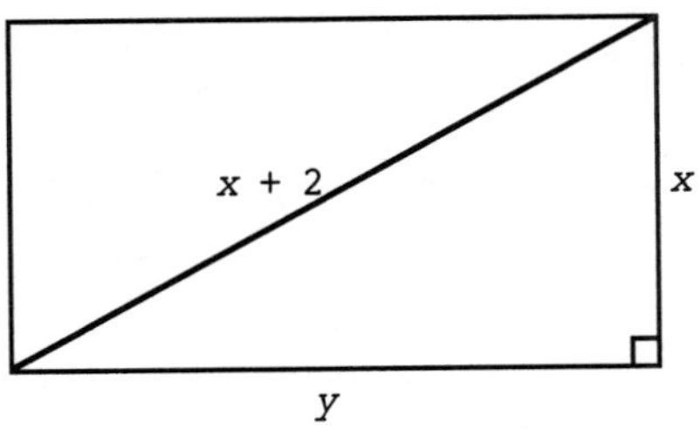

Labeling the rectangle as in the diagram, we have

$$\text{Area} = xy = 6$$

Hence $y = \frac{6}{x}$

Applying the Pythagorean theorem, we have

$$(x + 2)^2 = x^2 + y^2$$
$$(x + 2)^2 = x^2 + \left(\frac{6}{x}\right)^2$$
$$x^2 + 4x + 4 = x^2 + \frac{36}{x^2}$$

$$4x + 4 = \frac{36}{x^2}$$
$$4x^3 + 4x^2 = 36$$
$$4x^3 + 4x^2 - 36 = 0$$
$$x^3 + x^2 - 9 = 0$$

There are no rational zeros of $P(x) = x^3 + x^2 - 9$. Examining the graph of $y = P(x)$, we obtain the graph at the right.

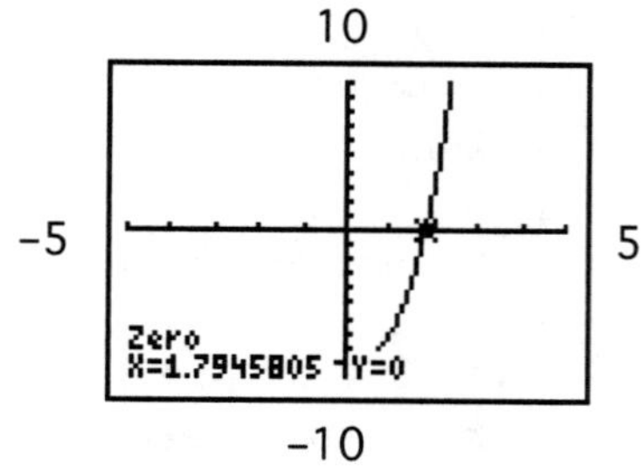

The only real zero is $x = 1.79$.

Then $y = \frac{6}{1.79} = 3.35$. The dimensions of the rectangle are 1.79 feet by 3.35 feet. (4-2)

70. We use the Doubling Time Growth Model:

$$P = P_0 2^{t/d}$$

Substituting $P_0 = 60$ million and $d = 23$, we have

$$P = 60(2^{t/23}) \text{ million}$$

(A) $t = 5$, hence $P = 60(2^{5/23})$ million $= 69.8$ million

(B) $t = 30$, hence $P = 60(2^{30/23})$ million $= 148$ million (5-2)

71. We solve $P = P_0(1.07)^t$ for t, using $P = 2P_0$

$$2P_0 = P_0(1.07)^t$$
$$2 = (1.07)^t$$
$$\ln 2 = t \ln 1.07$$
$$\frac{\ln 2}{\ln 1.07} = t$$
$$t = 10.2 \text{ years}$$

(5-1)

72. We solve $P = P_0e^{0.07t}$ for t, using $P = 2P_0$

$$2P_0 = P_0e^{0.07t}$$
$$2 = e^{0.07t}$$
$$\ln 2 = 0.07t$$
$$t = \frac{\ln 2}{0.07}$$
$$t = 9.90 \text{ years}$$

(5-1)

73. First, we solve $M = \frac{2}{3}\log\left(\frac{E}{E_0}\right)$ for E.

$$M = \frac{2}{3}\log\left(\frac{E}{E_0}\right)$$
$$\frac{3M}{2} = \log\frac{E}{E_0}$$
$$\frac{E}{E_0} = 10^{3m/2}$$
$$E = E_0(10^{3m/2})$$

We now compare E_1 for $M = 8.3$ with E_2 for $M = 7.1$.

$$E_1 = E_0(10^{3\cdot 8.3/2}) \qquad E_2 = E_0(10^{3\cdot 7.1/2})$$
$$E_1 = E_0(10^{12.45}) \qquad E_2 = E_0(10^{10.65})$$

Hence $\dfrac{E_1}{E_2} = \dfrac{E_0(10^{12.45})}{E_0(10^{10.65})} = 10^{12.45-10.65} = 10^{1.8}$

$E_1 = 10^{1.8}E_2$ or $63.1E_2$.

The 1906 earthquake was 63.1 times as powerful. (5-4)

74. We solve $D = 10 \log \dfrac{I}{I_0}$ for I, with $D = 88$, $I_0 = 10^{-12}$

$$88 = 10 \log \frac{I}{10^{-12}}$$
$$8.8 = \log \frac{I}{10^{-12}}$$
$$10^{8.8} = \frac{I}{10^{-12}}$$
$$I = 10^{8.8}\cdot 10^{-12}$$
$$I = 10^{-3.2}$$
$$I = 6.31 \times 10^{-4} \text{ } w/m^2$$

(5-4)

75. $t^2 = kd^3$ (4-5)

76. Write $v = \dfrac{k}{\sqrt{w}}$. Substitute $v = 0.3$ and $w = w_1$ and solve for k.

$$0.3 = \frac{k}{\sqrt{w_1}}$$

$k = 0.3\sqrt{w_1}$ where w_1 = weight of oxygen molecule.

The equation of variation is

$$v = \frac{0.3\sqrt{w_1}}{\sqrt{w_2}}$$

If $w_2 = \dfrac{w_1}{16}$ (weight of hydrogen molecule) then

$$v = \frac{0.3\sqrt{w_1}}{\sqrt{w_1/16}} = \frac{0.3\sqrt{w_1}}{\sqrt{w_1}/4} = 4(0.3) = 1.2 \text{ mile/second}$$

(4-5)

77. The independent variable is years since 1970, so enter 0, 5, 10, 15, 20, 25, 30 as L_1. The dependent variable is life expectancy, so enter the figures from this column as L_2.

L1	L2	L3 3
0	70.8	
5	72.6	
10	73.7	
15	74.7	
20	75.4	
25	75.9	
30	77	

L3(1)=

(A) The linear regression routine from the STAT CALC menu yields

```
LinReg
 y=ax+b
 a=.1921428571
 b=71.41785714
```

Evaluating this at x = 40 (year 2010) yields 79.1 years.

(B) The quadratic regression routine yields

```
QuadReg
 y=ax²+bx+c
 a=-.0033809524
 b=.2935714286
 c=70.9952381
```

Evaluating this at x = 40 yields 77.3 years.

(C) The cubic regression routine yields

```
CubicReg
 y=ax³+bx²+cx+d
 a=2.6666667E-4
 b=-.0153809524
 c=.4269047619
 d=70.7952381
```

Evaluating this at x = 40 yields 80.3 years.

(D) The exponential regression routine yields

```
ExpReg
 y=a*b^x
 a=71.43663414
 b=1.002600141
```

Evaluating this at x = 40 yields 79.3 years. *(4-1, 5-2)*

78. Clearly, the quadratic regression routine answer of 77.3 years is closest to the Census Bureau projection. *(4-1, 5-2)*

CHAPTER 6 Trigonometric Functions

Section 6-1

1. Since 1 rotation corresponds to 360°, $\frac{1}{9}$ rotation corresponds to $\frac{1}{9}(360°) = 40°$

3. Since 1 rotation corresponds to 360°, $\frac{3}{4}$ rotation corresponds to $\frac{3}{4}(360°) = 270°$

5. Since 1 rotation corresponds to 360°, $\frac{9}{8}$ rotation corresponds to $\frac{9}{8}(360°) = 405°$

7. $\theta = \frac{s}{r} = \frac{24 \text{ centimeters}}{4 \text{ centimeters}} = 6$ radians

9. $\theta = \frac{s}{r} = \frac{30 \text{ feet}}{12 \text{ feet}} = 2.5$ radians

11. Since 1 rotation corresponds to 2π radians, $\frac{1}{8}$ rotation corresponds to $\frac{1}{8}(2\pi) = \frac{\pi}{4}$ radians

13. As in problem 11, $\frac{3}{4}$ rotation corresponds to $\frac{3}{4}(2\pi) = \frac{3\pi}{2}$ radians

15. As in problem 11, $\frac{13}{12}$ rotation corresponds to $\frac{13}{12}(2\pi) = \frac{13\pi}{6}$ radians

17. Using the relation $\theta_R = \frac{\pi \text{ rad}}{180°}\theta_D$,

we have:

$$\frac{\pi \text{ rad}}{180°}30° = \frac{\pi}{6}\text{rad}$$
$$\frac{\pi \text{ rad}}{180°}60° = \frac{\pi}{3}\text{rad}$$
$$\frac{\pi \text{ rad}}{180°}90° = \frac{\pi}{2}\text{rad}$$
$$\frac{\pi \text{ rad}}{180°}120° = \frac{2\pi}{3}\text{rad}$$
$$\frac{\pi \text{ rad}}{180°}150° = \frac{5\pi}{6}\text{rad}$$
$$\frac{\pi \text{ rad}}{180°}180° = \pi \text{ rad}$$

19. Using the relation $\theta_R = \frac{\pi \text{ rad}}{180°}\theta_D$,

we have:

$$\frac{\pi \text{ rad}}{180°}(-45°) = -\frac{\pi}{4}\text{rad}$$
$$\frac{\pi \text{ rad}}{180°}(-90°) = -\frac{\pi}{2}\text{rad}$$
$$\frac{\pi \text{ rad}}{180°}(-135°) = -\frac{3\pi}{4}\text{rad}$$
$$\frac{\pi \text{ rad}}{180°}(-180°) = -\pi \text{ rad}$$

21. Using the relation $\theta_D = \frac{180°}{\pi \text{ rad}}\theta_R$,

we have:

$$\frac{180°}{\pi}\frac{\pi}{3} = 60°$$
$$\frac{180°}{\pi}\frac{2\pi}{3} = 120°$$
$$\frac{180°}{\pi}\pi = 180°$$
$$\frac{180°}{\pi}\frac{4\pi}{3} = 240°$$
$$\frac{180°}{\pi}\frac{5\pi}{3} = 300°$$
$$\frac{180°}{\pi}2\pi = 360°$$

23. Using the relation $\theta_D = \frac{180°}{\pi \text{ rad}}\theta_R$,

we have:

$$\frac{180°}{\pi}\left(-\frac{\pi}{2}\right) = -90°$$
$$\frac{180°}{\pi}(-\pi) = -180°$$
$$\frac{180°}{\pi}\left(-\frac{3\pi}{2}\right) = -270°$$
$$\frac{180°}{\pi}(-2\pi) = -360°$$

25. True. "Standard position" means that both angles have the positive x-axis as their initial sides. If they have the same initial side and the same measure, they must have the same terminal side as well.

27. True. Angles that are complementary have measures that add up to 90°, so if both are positive and add to 90° the measure of each has to be between 0° and 90°. That's the definition of acute.

29. False. The terminal side of the angle -315° (or $-\frac{7\pi}{4}$) is in quadrant I.

31. $5°51'33'' = \left(5 + \frac{51}{60} + \frac{33}{3600}\right)^\circ = 5.859°$

33. $354°8'29'' = \left(5 + \frac{8}{60} + \frac{29}{3600}\right)^\circ = 354.141°$

35. $3.042° = 3°(0.042 \cdot 60)' = 3°2.52' = 3°2'(0.52 \cdot 60)'' = 3°2'31''$

37. $403.223° = 403°(0.223 \cdot 60)' = 403°13.38' = 403°13'(0.38 \cdot 60)'' = 403°13'23''$

39. Using the relation $\theta_R = \frac{\pi \text{ rad}}{180°}\theta_D$, we have $\frac{\pi \text{ rad}}{180°}64° = 1.117$ rad

41. Using the relation $\theta_R = \frac{\pi \text{ rad}}{180°}\theta_D$, we have $\frac{\pi \text{ rad}}{180°}108.413° = 1.892$ rad

43. First we convert 13°25'14" to decimal degrees:

$$13°25'14'' = \left(13 + \frac{25}{60} + \frac{14}{3600}\right)^\circ = 13.421°$$

Then we use the relation $\theta_R = \frac{\pi \text{ rad}}{180°}\theta_D$ to find $\theta_R = \frac{\pi \text{ rad}}{180°}(13.421°) = 0.234$ radians.

45. Using the relation $\theta_D = \frac{180°}{\pi \text{ rad}}\theta_R$, we have $\frac{180°}{\pi}(0.93) = 53.29°$

47. Using the relation $\theta_D = \frac{180°}{\pi \text{ rad}}\theta_R$, we have $\frac{180°}{\pi}(1.13) = 64.74°$

49. Using the relation $\theta_D = \frac{180°}{\pi \text{ rad}}\theta_R$, we have $\frac{180°}{\pi}(-2.35) = -134.65°$

For Problems 51– 70 the following sketch is useful:

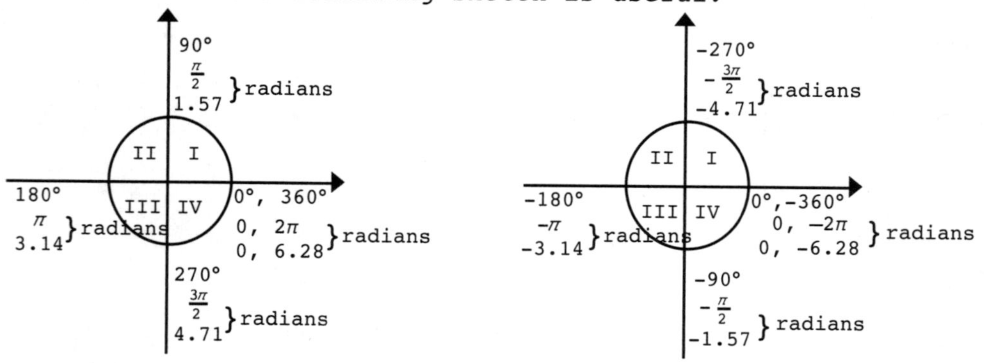

51. From the sketch, we find, since 180° < 187° < 270°, 187° is a III quadrant angle.

53. Since -270° < -200° < -180°, -200° is a II quadrant angle.

55. Since 3.14 < 4 < 4.71, 4 is a III quadrant angle.

57. 270° is a quadrantal angle.

59. Since -1.57 < -1 < 0, -1 is a IV quadrant angle.

61. Since $\frac{3\pi}{2} < \frac{5\pi}{3} < 2\pi$, $\frac{5\pi}{3}$ is a IV quadrant angle.

63. Since $-\frac{3\pi}{2} < \frac{-7\pi}{6} < -\pi$, $\frac{-7\pi}{6}$ is a II quadrant angle.

65. $-\pi$ is a quadrantal angle.

67. 820° is coterminal with 820° − 2(360°) = 100°. Since 90° < 100° < 180°, 100° (and therefore 820°) is a II quadrant angle.

69. $\frac{13\pi}{4}$ is coterminal with $\frac{13\pi}{4} - 2\pi = \frac{13\pi - 8\pi}{4} = \frac{5\pi}{4}$. Since $\pi < \frac{5\pi}{4} < \frac{3\pi}{2}$, $\frac{5\pi}{4}$ (and therefore $\frac{13\pi}{4}$) is a III quadrant angle.

71. A central angle of radian measure 1 is an angle subtended by an arc of the same length as the radius of the circle.

73. Of all the coterminal angles $150^\circ + n360^\circ$, $150^\circ + 360^\circ = 510^\circ$ is in the correct interval.

75. Of all the coterminal angles $-80^\circ + n360^\circ$, $-80^\circ + 360^\circ = 280^\circ$ is in the correct interval.

77. Of all the coterminal angles $210^\circ + n360^\circ$, $210^\circ - 720^\circ = -510^\circ$ and $210^\circ - 1080^\circ = -870^\circ$ are in the correct interval.

79. Of all the coterminal angles $\frac{\pi}{4} + n2\pi$, $\frac{\pi}{4} + 2\pi = \frac{9\pi}{4}$ and $\frac{\pi}{4} + 4\pi = \frac{17\pi}{4}$ are in the correct interval.

81. Of all the coterminal angles $-\frac{7\pi}{6} + n2\pi$, $-\frac{7\pi}{6} + 2\pi = \frac{5\pi}{6}$, $-\frac{7\pi}{6} + 4\pi = \frac{17\pi}{6}$, and $-\frac{7\pi}{6} + 6\pi = \frac{29\pi}{6}$ are in the correct interval.

83. Of all the coterminal angles $\frac{\pi}{2} + n2\pi$, $\frac{\pi}{2} - 2\pi = -\frac{3\pi}{2}$ and $\frac{\pi}{2}$ itself are in the correct interval.

85. We apply $\frac{s}{c} = \frac{\theta^\circ}{360^\circ}$ with $s = 500$ mi, and $\theta^\circ = 7.5^\circ$

$$\text{Then} \quad \frac{500}{c} = \frac{7.5^\circ}{360^\circ} \quad \text{or} \quad \frac{500}{c} = \frac{7.5}{360}$$

$$360c \cdot \frac{500}{c} = 360c \cdot \frac{7.5}{360}$$

$$180{,}000 = 7.5c$$

$$c = 24{,}000 \text{ mi}$$

87. The 7.5° angle and θ have a common side. (An extended vertical pole in Alexandria will pass through the center of the earth.) The sun's rays are essentially parallel when they arrive at the earth. Thus, the other two sides of the angles are parallel, since a sun ray to the bottom of the well, when extended, will pass through the center of the earth. From geometry we know that the alternate interior angles made by a line intersecting two parallel lines are equal. Therefore, $\theta = 7.5^\circ$.

89. To calculate angular speed, we need radians and seconds.

200 revolutions × 2π radians = 400π radians

1 minute = 60 sec.

$$\text{angular speed} = \frac{400\pi \text{ radians}}{60 \text{ sec}} = 20.94 \frac{\text{rad}}{\text{sec}}$$

To calculate linear speed, we need the circumference of the wheel.

$c = \pi d = 6\pi$

In 200 revolutions, the linear distance is $200 \times 6\pi = 1200\pi$ feet.

$$\text{linear speed} = \frac{1200\pi \text{ feet}}{60 \text{ sec}} = 62.83 \text{ ft/sec}$$

91. 

From the figure, it should be clear that the minute (larger) hand is displaced π radians from noon, while the hour (smaller) hand is displaced $\theta = \frac{4\frac{1}{2}}{12}$ of a full revolution, or $\frac{4\frac{1}{2}}{12} \cdot 2\pi$ radians from noon. Hence the larger angle between the hands has measure $\theta + \pi$, or

$$\frac{4\frac{1}{2}}{12} \cdot 2\pi + \pi = \frac{3}{4}\pi + \pi = \frac{7}{4}\pi \text{ radians.}$$

93. We use $\theta = \frac{s}{r}$ with $r = \frac{1}{2}(10) = 5$ centimeters and $s = 10$ meters $\times\ 100 \frac{\text{centimeters}}{\text{meter}}$ $= 1000$ centimeters. Then $\theta = \frac{1000}{5} = 200$ radians.

95. In one year the line sweeps out one full revolution, or 2π radians. In one week the line sweeps out $\frac{1}{52}$ of one full revolution, or $\frac{1}{52} \cdot 2\pi = \frac{\pi}{26}$ radians.

$\frac{\pi}{26} = 0.12$ radian to two decimal places.

97. Following example 4, we reason that points on the circumference of each wheel travel the same distance. Then $s_1 = r_1\theta_1 = r_2\theta_2 = s_2$. The radius of the front wheel is $\frac{1}{2}(40)$ or 20 centimeters. The radius of the back wheel is $\frac{1}{2}(60)$ or 30 centimeters. So

$$20\theta_1 = 8(30)$$
$$\theta_1 = \frac{240}{20}$$
$$\theta_1 = 12 \text{ radians}$$

99. Front wheel: $d = 40$ cm; $c = \pi d = 40\pi$ cm. One complete revolution is 2π radians. So 2π rad $= 40\pi$ cm or 1 cm $= \frac{1}{20}$ radian. In one hour, the linear distance is 10 km which is 1,000,000 cm. 1,000,000 cm $= \frac{1}{20} \cdot 1{,}000{,}000$ radian $= 50{,}000$ radians

$$\frac{50{,}000 \text{ radians}}{\text{hr}} \times \frac{1 \text{ hr}}{3600 \text{ sec}} = 13.9 \text{ rad/sec}$$

Back wheel: $d = 60$ cm; $c = 60\pi$ cm

2π rad $= 60\pi$ cm or 1 cm $= \frac{1}{30}$ radian

1,000,000 cm $= \frac{1}{30} \cdot 1{,}000{,}000$ radian $= 33{,}333.3$ radians

$$\frac{33{,}333.3 \text{ radians}}{\text{hr}} \times \frac{1 \text{ hr}}{3600 \text{ sec}} = 9.26 \text{ rad/sec}$$

101. We use $c \approx s = r\theta$ with $r = 9.3 \times 10^7$ mi and $\theta = 9.3 \times 10^{-3}$ rad.
Then $c \approx (9.3 \times 10^7)(9.3 \times 10^{-3}) = 865{,}000$ mi

103. We use $c \approx s = r\theta$. $r = 750$ ft. θ must be converted to radians to use the formula, thus, using $\theta_R = \frac{\pi \text{ rad}}{180°}2.5°$ we have $\theta_R = \frac{\pi}{72}$ rad.
Then $c \approx 750 \cdot \frac{\pi}{72} = 33$ ft.

Section 6-2

For Problems 1-15, the following sketch is useful:

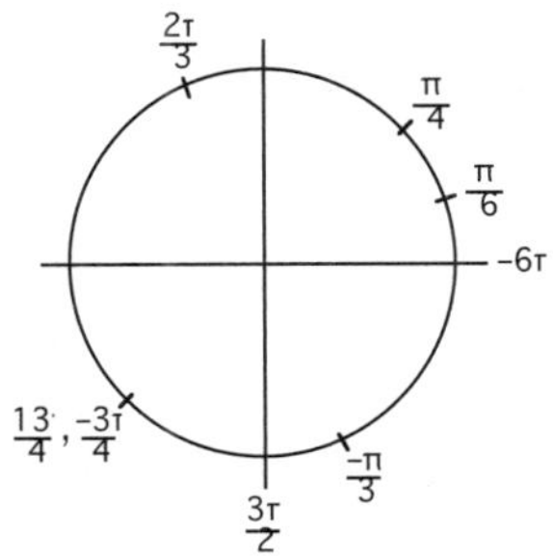

1. $w\left(\frac{3\pi}{2}\right) = (0, -1)$ **3.** $w(-6\pi) = (1, 0)$ **5.** $w\left(\frac{\pi}{4}\right) = \left(\frac{1}{\sqrt{2}}, \frac{1}{\sqrt{2}}\right)$

7. $w\left(\frac{\pi}{6}\right) = \left(\frac{\sqrt{3}}{2}, \frac{1}{2}\right)$ **9.** $w\left(-\frac{\pi}{3}\right) = \left(\frac{1}{2}, -\frac{\sqrt{3}}{2}\right)$ **11.** $w\left(\frac{2\pi}{3}\right) = \left(-\frac{1}{2}, \frac{\sqrt{3}}{2}\right)$

13. $w\left(-\frac{3\pi}{4}\right) = \left(\frac{-1}{\sqrt{2}}, \frac{-1}{\sqrt{2}}\right)$ **15.** $w\left(\frac{13\pi}{4}\right) = \left(\frac{-1}{\sqrt{2}}, \frac{-1}{\sqrt{2}}\right)$

It's helpful to remember these rules for finding the coordinates of the circular points when x is a multiple of π:

1) If x has a denominator of 6, like $\frac{5\pi}{6}$, the coordinates are $\left(\pm\frac{\sqrt{3}}{2}, \pm\frac{1}{2}\right)$.
Locate the quadrant to decide on the appropriate signs.

2) If x has a denominator of 4, like $-\frac{3\pi}{4}$, the coordinates are $\left(\pm\frac{1}{\sqrt{2}}, \pm\frac{1}{\sqrt{2}}\right)$.
Locate the quadrant to decide on the appropriate signs.

3) If x has a denominator of 3, like $\frac{7\pi}{3}$, the coordinates are $\left(\pm\frac{1}{2}, \pm\frac{\sqrt{3}}{2}\right)$.
Locate the quadrant to decide on the appropriate signs.

4) If the denominator is 2, or there is no denominator, like $\frac{3\pi}{2}$ or -6π, x is on one of the coordinate axes and you should be able to find the coordinates by just locating x on the unit circle.

IMPORTANT: These rules are only valid if x is a <u>reduced</u> <u>fraction</u>! For example, $\frac{6\pi}{4}$ has denominator 4 but can be reduced to $\frac{3\pi}{2}$.

17. $w\left(\frac{3\pi}{2}\right) = (0, -1)$ so $\sin\left(\frac{3\pi}{2}\right) = -1$

19. $w(-6\pi) = (1, 0)$ so $\cos(-6\pi) = 1$

21. $w\left(\frac{\pi}{4}\right) = \left(\frac{1}{\sqrt{2}}, \frac{1}{\sqrt{2}}\right)$ so $\sec\left(\frac{\pi}{4}\right) = \frac{1}{\frac{1}{\sqrt{2}}} = \sqrt{2}$

23. $w\left(\frac{\pi}{6}\right) = \left(\frac{\sqrt{3}}{2}, \frac{1}{2}\right)$ so $\tan\left(\frac{\pi}{6}\right) = \frac{\frac{1}{2}}{\frac{\sqrt{3}}{2}} = \frac{1}{\sqrt{3}}$

25. $w\left(-\frac{\pi}{3}\right) = \left(\frac{1}{2}, -\frac{\sqrt{3}}{2}\right)$ so $\sin\left(-\frac{\pi}{3}\right) = -\frac{\sqrt{3}}{2}$

27. $w\left(\frac{2\pi}{3}\right) = \left(-\frac{1}{2}, \frac{\sqrt{3}}{2}\right)$ so $\csc\left(\frac{2\pi}{3}\right) = \frac{1}{\frac{\sqrt{3}}{2}} = \frac{2}{\sqrt{3}}$

29. $w\left(-\frac{3\pi}{4}\right) = \left(\frac{-1}{\sqrt{2}}, \frac{-1}{\sqrt{2}}\right)$ so $\cos\left(-\frac{3\pi}{4}\right) = \frac{-1}{\sqrt{2}}$

31. $w\left(\frac{13\pi}{4}\right) = \left(\frac{-1}{\sqrt{2}}, \frac{-1}{\sqrt{2}}\right)$ so $\cot\left(\frac{13\pi}{4}\right) = \frac{\frac{-1}{\sqrt{2}}}{\frac{-1}{\sqrt{2}}} = 1$

33. Since $\cos x = a$, it is negative in quadrants II and III.

35. Since $\sin x = b$, it is positive in quadrants I and II.

37. Since $\cot x = \frac{a}{b}$, it is negative if a and b have opposite signs. This occurs in quadrants II and IV.

39. -0.6573 | **Common Error:** $\cos 2.288 \neq 0.9992$; calculator must be in *radian* mode. |

41. -14.60 **43.** 1.000 **45.** 0.8138 **47.** 0.5290

49. 0.4226 (calculator in degree mode) **51.** -1.573 (calculator in radian mode)

53. 0.8439 **55.** -0.3363

57. $\sin(113°27'13'') = \sin\left(113 + \frac{27}{60} + \frac{13}{3600}\right)^\circ = 0.9174$

59. False. The domain of the wrapping function is the set of all real numbers.

61. False. The wrapping function is not one-to-one. $W(0) = W(2\pi)$ but $0 \neq 2\pi$.

63. True. $a^2 + b^2 = 1$ describes all points (a, b) on the unit circle.

65. False. The reciprocal of $\sin x$ is $\csc x$, unless $\sin x = 0$.

67. False. The secant function is not one-to-one. $\sec 0 = \sec 2\pi$, but $0 \neq 2\pi$.

69. False. The domain of $\sin x$ is all real numbers; the domain of $\csc x$ excludes all integer multiples of π.

71. True. All integer multiples of π are turning points.

73. True. $\frac{\pi}{2} + k\pi$, k any integer, represents all zeros.

75.

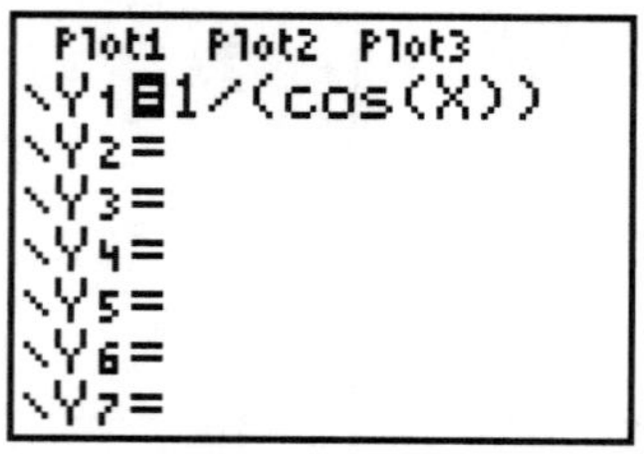

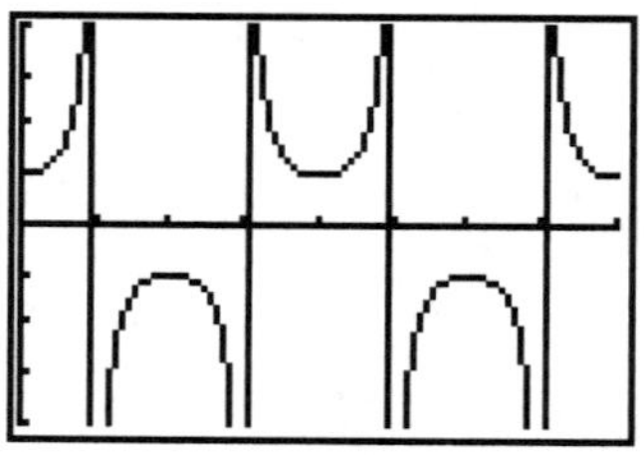

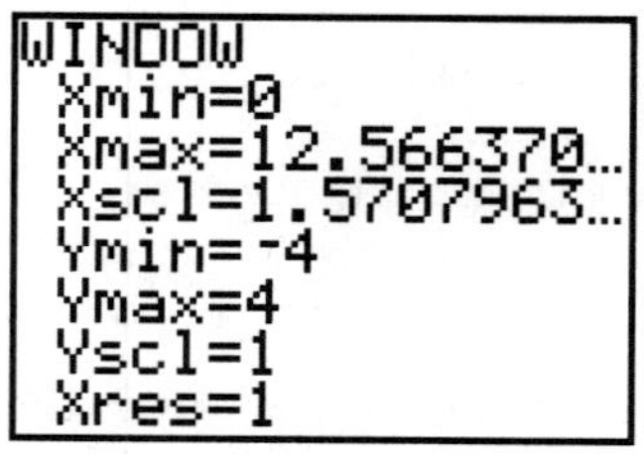

Since there is no secant button, we enter $y = \sec x$ as $y = 1/\cos x$. There are no zeros and 3 turning points. They occur when the graph is at height 1 or -1. Since $\sec x = \frac{1}{\cos x}$, $\sec x = \pm 1$ when $\cos x = \pm 1$. This occurs for $x = \pi$, 2π, and 3π, so the turning points are $(\pi, -1)$, $(2\pi, 1)$, and $(3\pi, -1)$.

77. 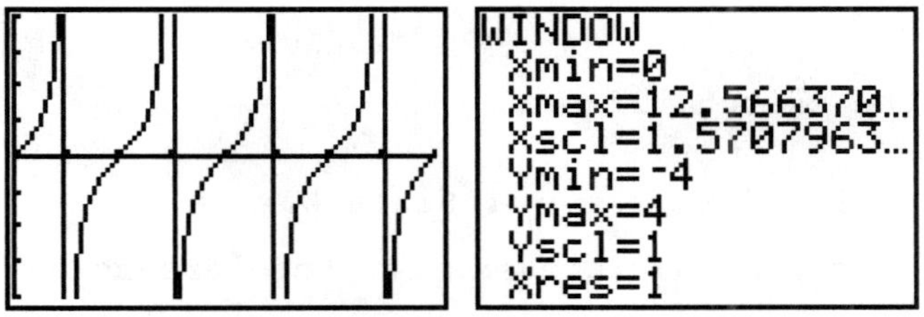

There are 5 zeros. Since $\tan x = \frac{\sin x}{\cos x}$, the zeros occur when $\sin x = 0$. The x values are 0, π, 2π, 3π, and 4π. There are no turning points.

For Problems 79—88, the following sketches are useful.

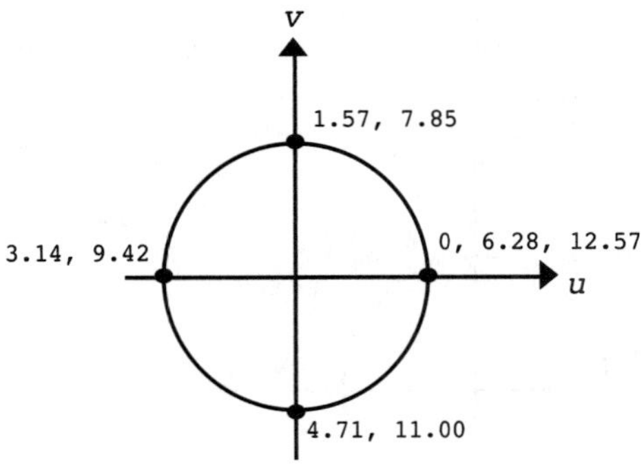

Sketch 1: counterclockwise wrapping

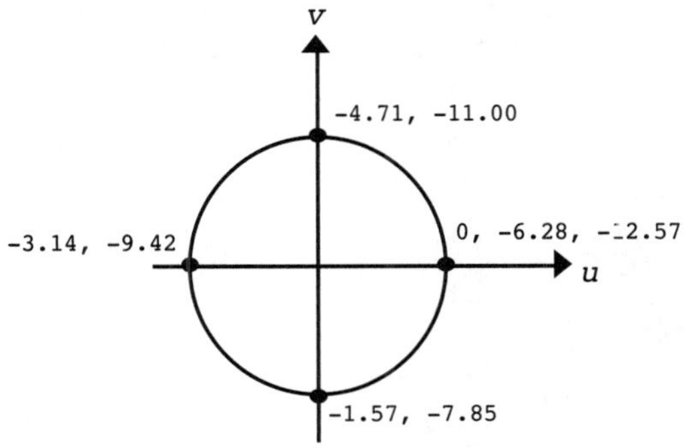

Sketch 2: clockwise wrapping

79. Using sketch 1 and $1.57 < 2 < 3.14$, we see that $W(2)$ is in the second quadrant. Hence, if $W(2) = (a, b)$, then $a < 0$ and $b > 0$, that is, a is negative and b is positive.

81. Using sketch 1 and $1.57 < 3 < 3.14$, we see that $W(3)$ is in the second quadrant. Hence, if $W(3) = (a, b)$, then $a < 0$ and $b > 0$, that is, a is negative and b is positive.

83. Using sketch 1 and $4.71 < 5 < 6.28$, we see that $W(5)$ is in the fourth quadrant. Hence, if $W(5) = (a, b)$, then $a > 0$ and $b < 0$, that is, a is positive and b is negative.

85. Using sketch 2 and $-3.14 < -2.5 < -1.57$, we see that $W(-2.5)$ is in the third quadrant. Hence, if $W(-2.5) = (a, b)$, then $a < 0$ and $b < 0$, that is, a and b are negative.

87. Using sketch 2 and $-6.28 < -6.1 < -4.71$, we see that $W(-6.1)$ is in the first quadrant. Hence, if $W(-6.1) = (a, b)$, then $a > 0$ and $b > 0$, that is, a and b are positive.

89. 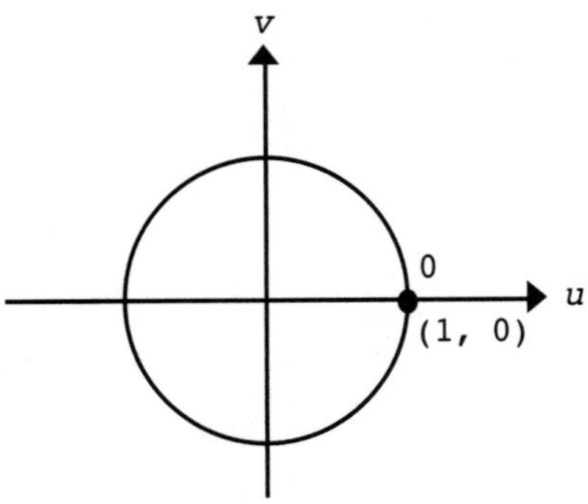

From the sketch, if $W(x) = (1, 0)$, $0 \le x < 2\pi$, then $x = 0$. Since $W(x) = W(x + 2k\pi)$, k any integer, if there are no restrictions on x, then $x = 0 + 2k\pi$ or $x = 2k\pi$, k any integer.

91.

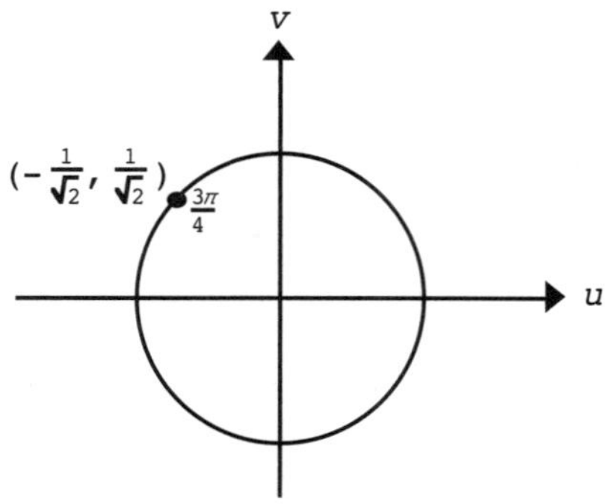

From the sketch, if $W(x) = \left(-\frac{1}{\sqrt{2}}, \frac{1}{\sqrt{2}}\right)$, $0 \le x < 2\pi$, then $x = \frac{3\pi}{4}$. Since $W(x) = W(x + 2k\pi)$, k any integer, if there are no restrictions on x, then $x = \frac{3\pi}{4} + 2k\pi$, k any integer.

93. $W(x)$ is the coordinates of a point on a unit circle that is $|x|$ units from (1, 0), in a counterclockwise direction if x is positive and in a clockwise direction if x is negative. $W(x + 4\pi)$ has the same coordinates as $W(x)$, since we return to the same point every time we go around the unit circle any integer multiple of 2π units (the circumference of the circle) in either direction.

95.

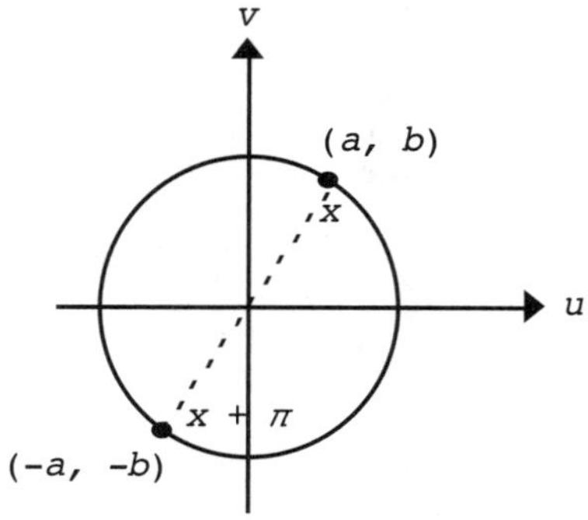

Since (see sketch) $W(x)$ and $W(x + \pi)$ are symmetrically placed with respect to the origin, $W(x) = (a, b)$ requires $W(x + \pi) = (-a, -b)$. T

97.

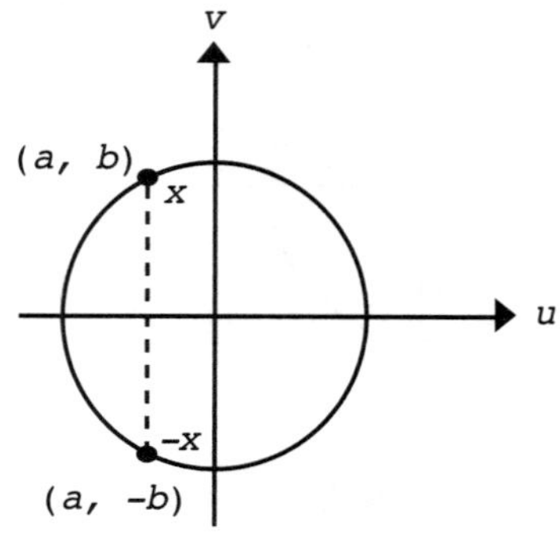

Since (see sketch) $W(x)$ and $W(-x)$ are symmetrically placed with respect to the horizontal axis, $W(x) = (a, b)$ requires $W(-x) = (a, -b)$. F

99. Since for all real numbers x, $W(x) = W(x + 2\pi)$, $W(x) = (a, b)$ requires $W(x + 2\pi) = (a, b)$. T

101. Locate $W(0.4) = (a, b) = (0.9, 0.4)$ on the figure.

(A) $\sin 0.4 = b = 0.4$ (B) $\cos 0.4 = a = 0.9$ (C) $\tan 0.4 = \frac{b}{a} = \frac{0.4}{0.9} = 0.4$

103. (A) Locate $W(2.2) = (a, b) = (-0.6, 0.8)$ on the figure.

$\sec 2.2 = \frac{1}{a} = \frac{1}{-0.6} = -2$ (to one significant digit)

(B) Locate $W(5.9) = (a, b) = (0.9, -0.4)$ on the figure.

$\tan 5.9 = \frac{b}{a} = \frac{-0.4}{0.9} = -0.4$

(C) Locate $W(3.8) = (a, b) = (-0.8, -0.6)$ on the figure.

$\cot 3.8 = \frac{a}{b} = \frac{-0.8}{-0.6} = 1$(to one significant digit)

105. $\sin x < 0$ in quadrants III and IV; $\cot x < 0$ in quadrants II and IV; therefore, both are true in quadrant IV.

107. $\cos x < 0$ in quadrants II and III; $\sec x > 0$ in quadrants I and IV; therefore, it is not possible to have both true for the same value of x.

109. $\cos x = a$ is always defined. There are no values for which it is undefined.

111. $\tan x = \frac{b}{a}$ is undefined if and only if $a = 0$. This occurs at points on the vertical axis. The only values of x between 0 and 2π for which $W(x)$ is on the vertical axis are $\frac{\pi}{2}$ and $\frac{3\pi}{2}$.

113. $\sec x = \frac{1}{a}$ is undefined if and only if $a = 0$. This occurs at points on the vertical axis. The only values of x between 0 and 2π for which $W(x)$ is on the vertical axis are $\frac{\pi}{2}$ and $\frac{3\pi}{2}$.

115. Given $n = 12$, $r = 5$, we use the given formula $A = \frac{1}{2}nr^2 \sin \frac{2\pi}{n}$ to obtain:

$$A = \frac{1}{2}(12)(5)^2 \sin \frac{2\pi}{12}$$
$$= 150 \sin \frac{\pi}{6}$$
$$= 150\left(\frac{1}{2}\right) = 75 \text{ square meters}$$

117. Given $n = 3$, $r = 4$, we use the given formula $A = \frac{1}{2}nr^2 \sin \frac{2\pi}{n}$ to obtain:

$$A = \frac{1}{2}(3)(4)^2 \sin \frac{2\pi}{3}$$
$$= 24 \sin \frac{2\pi}{3}$$
$$= 24\left(\frac{\sqrt{3}}{2}\right) = 12\sqrt{3} \approx 20.78 \text{ square inches}$$

119.
$$a_1 = 0.5$$
$$a_2 = a_1 + \cos a_1 = 0.5 + \cos 0.5 = 1.377583$$
$$a_3 = a_2 + \cos a_2 = 1.377583 + \cos 1.377583 = 1.569596$$
$$a_4 = a_3 + \cos a_3 = 1.569596 + \cos 1.569596 = 1.570796$$
$$a_5 = a_4 + \cos a_4 = 1.570796 + \cos 1.570796 = 1.570796$$

$\frac{\pi}{2} = 1.570796$ to six decimal places.

Section 6-3

1. b/c **3.** c/b **5.** b/a **7.** $\cos\theta$ **9.** $\sec\theta$ **11.** $\cot\theta$

13. 60.55° **15.** 82.90° **17.** 37.09°

19.

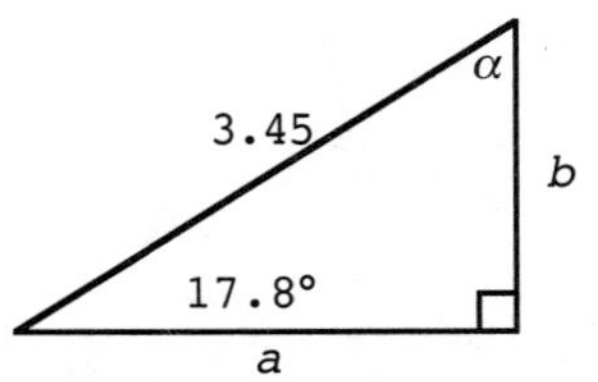

Solve for α:
$\alpha = 90° - 17.8° = 72.2°$

Solve for b: $\sin\beta = \frac{b}{c}$

$$\sin 17.8° = \frac{b}{3.45}$$
$$b = 3.45 \sin 17.8° = 1.05$$

Solve for a: $\cos\beta = \frac{a}{c}$

$$\cos 17.8° = \frac{a}{3.45}$$
$$a = 3.45 \cos 17.8° = 3.28$$

21.

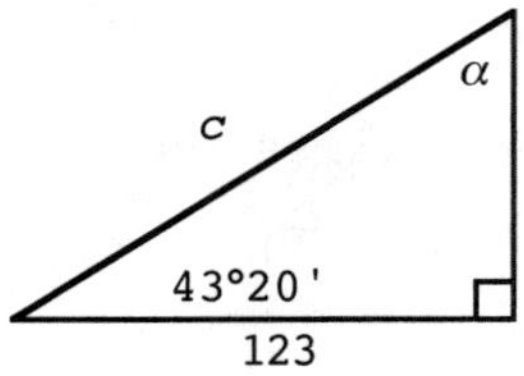

Solve for α:
$\alpha = 90° - 43°20' = 46°40'$

Solve for b: $\tan \beta = \dfrac{b}{a}$

$\tan 43°20' = \dfrac{b}{123}$

$b = 123 \tan 43°20'$
$= 116$

Solve for c: $\sec \beta = \dfrac{c}{a}$

$\sec 43°20' = \dfrac{c}{123}$

$c = 123 \sec 43°20'$
$= 169$

23.

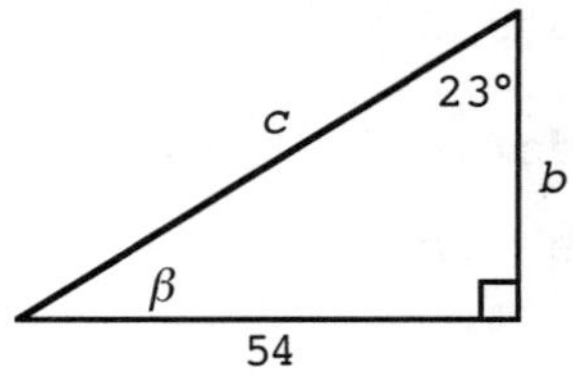

Solve for β:
$\beta = 90° - 23°0' = 67°0'$

Solve for b: $\tan \beta = \dfrac{b}{a}$

$\tan 67°0' = \dfrac{b}{54}$

$b = 54 \tan 67°0'$
$= 127$

Solve for c: $\csc \alpha = \dfrac{c}{a}$

$\csc 23°0' = \dfrac{c}{54}$

$c = 54 \csc 23°0'$
$= 138$

25.

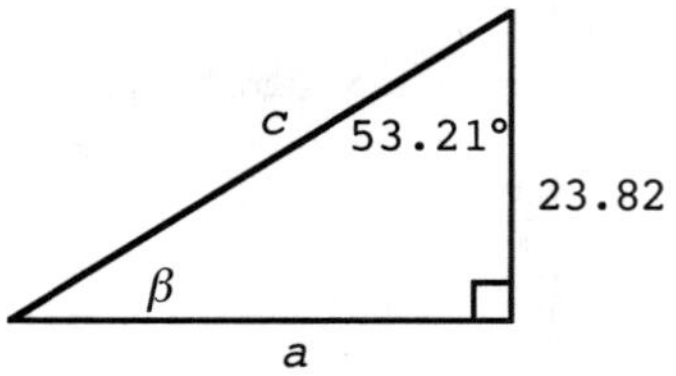

Solve for β:
$\beta = 90° - 53.21° = 36.79°$

Solve for a: $\tan \alpha = \dfrac{a}{b}$

$\tan 53.21° = \dfrac{a}{23.82}$

$a = 23.82 \tan 53.21°$
$= 31.85$

Solve for c: $\sec \alpha = \dfrac{c}{b}$

$\sec 53.21° = \dfrac{c}{23.82}$

$c = 23.82 \sec 53.21°$
$= 39.77$

27.

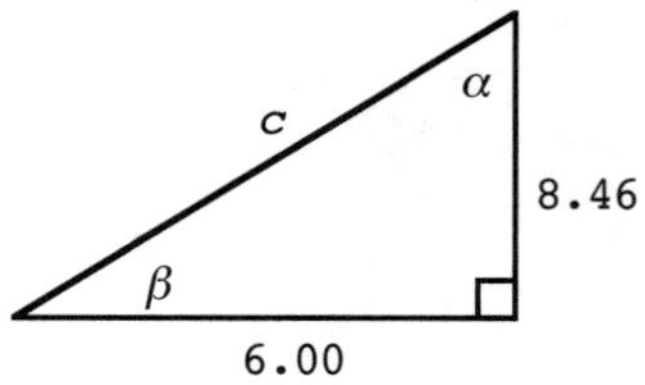

Solve for β: $\tan \beta = \dfrac{b}{a}$

$\tan \beta = \dfrac{8.46}{6.00}$

$\beta = \tan^{-1} \dfrac{8.46}{6.00}$
$= 54.7°$ or $54°40'$

Solve for α:
$\alpha = 90° - 54°40' = 35°20'$

Solve for c: $\csc \beta = \dfrac{c}{b}$

$\csc 54°40' = \dfrac{c}{8.46}$

$c = 8.46 \csc 54°40'$
$= 10.4$

29.

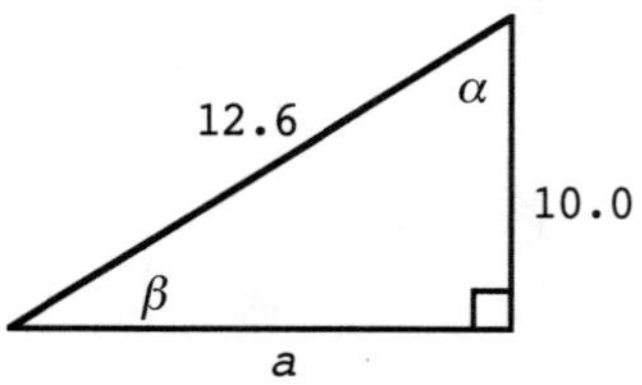

Solve for β: $\sin \beta = \dfrac{b}{c}$

$\sin \beta = \dfrac{10.0}{12.6}$

$\beta = \sin^{-1} \dfrac{10.0}{12.6}$
$= 52.5°$ or $52°30'$

Solve for α:
$\alpha = 90° - 52°30' = 37°30'$

Solve for a: $\cot \beta = \dfrac{a}{b}$

$\cot 52°30' = \dfrac{a}{10.0}$

$a = 10.0 \cot 52°30'$
$= 7.67$

31. False. Knowing only the three angles can't tell us anything about the lengths of the sides. The two triangles at the right have the exact same angles but totally different side lengths.

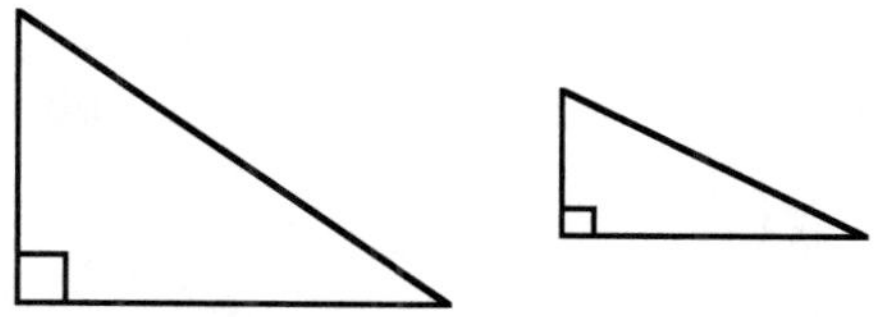

33. True.

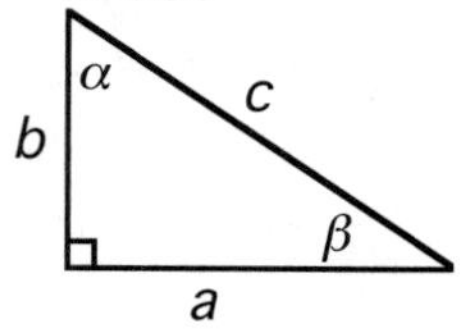

According to the diagram,

$\sin \alpha = \dfrac{\text{Opp}}{\text{Hyp}} = \dfrac{a}{c}$ and $\cos \beta = \dfrac{\text{Adj}}{\text{Hyp}} = \dfrac{a}{c}$.

35. False.

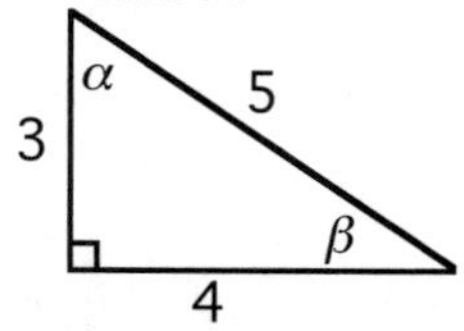

According to the diagram,

$\sec \alpha = \dfrac{\text{Hyp}}{\text{Adj}} = \dfrac{5}{3}$ and $\cos \beta = \dfrac{\text{Adj}}{\text{Hyp}} = \dfrac{4}{5}$.

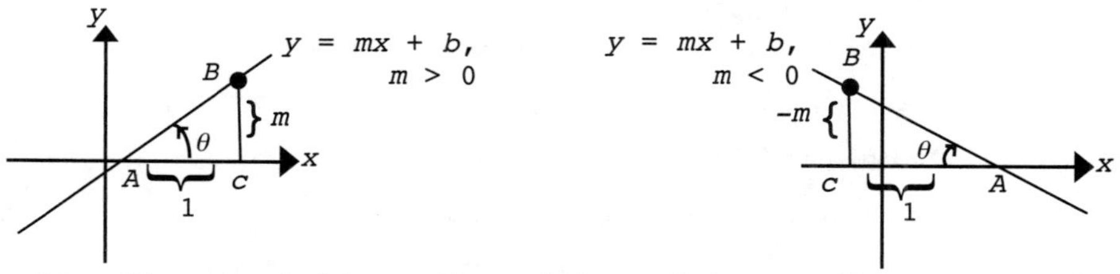

37. See the figure. A line of positive slope m will form a right triangle ABC as shown. $\tan \theta = \dfrac{m}{1}$.

Here $\tan \theta = \dfrac{1}{2}$

$$\theta = \tan^{-1} \frac{1}{2} = 26.6°$$

39. See the figure. A line of positive slope m will form a right triangle ABC as shown. $\tan \theta = \dfrac{m}{1}$.

Here $\tan \theta = 5$

$$\theta = \tan^{-1} 5 = 78.7°$$

41. See the figure. A line of negative slope m will form a right triangle ABC as shown. $\tan \theta = \dfrac{-m}{1}$.

Here $\tan \theta = -\dfrac{-2}{1} = 2$

$$\theta = \tan^{-1} 2 = 63.4°$$

43. See the figure. An angle θ can be formed by either of two lines, one with positive slope m and the other with negative slope m.
Thus $m = \pm\tan \theta$
Here $m = \pm\tan 20° = \pm 0.36$

45. See the figure. An angle θ can be formed by either of two lines, one with positive slope m and the other with negative slope m.
Thus $m = \pm\tan \theta$
Here $m = \pm\tan 80° = \pm 5.67$

47. See the figure. An angle θ can be formed by either of two lines, one with positive slope m and the other with negative slope m.
Thus $m = \pm\tan\theta$

Here $m = \pm\tan\left(\frac{\pi}{30}\right) = \pm 0.11$

49. (A) In triangle OAD, $\cos\theta = \frac{\text{Adj}}{\text{Hyp}} = \frac{OA}{1} = OA$

(B) In triangle OED, angle $EOD = 90° - \theta$, angle $OED = 90° - (90° - \theta) = \theta$.
Thus $\cot OED = \frac{\text{Adj}}{\text{Opp}} = \frac{DE}{1} = DE = \cot\theta$

(C) In triangle ODC, $\sec\theta = \frac{\text{Hyp}}{\text{Adj}} = \frac{OC}{1} = OC$

51. (A) As θ approaches 90°, $OA = \cos\theta$ approaches 0.
(B) As θ approaches 90°, $DE = \cot\theta$ approaches 0.
(C) As θ approaches 90°, $OC = \sec\theta$ increases without bound.

53. (A) As θ approaches 0°, $AD = \sin\theta$ approaches 0.
(B) As θ approaches 0°, $CD = \tan\theta$ approaches 0.
(C) As θ approaches 0°, $OE = \csc\theta$ increases without bound.

55.

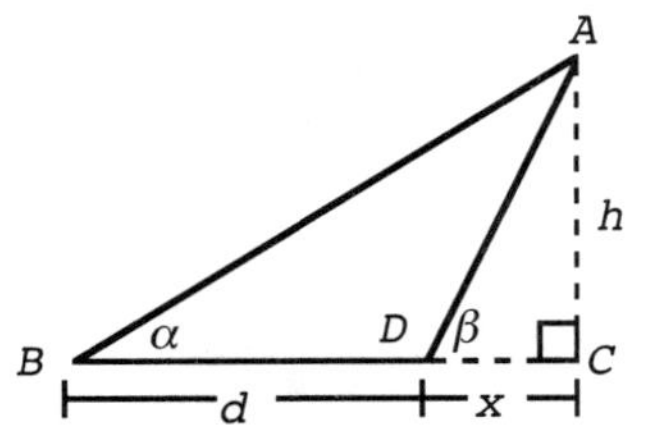

Label as shown at the left.
In right triangle ADC, $\cot\beta = \frac{x}{h}$
In right triangle ABC, $\cot\alpha = \frac{d + x}{h}$

Hence $x = h\cot\beta$ and $d + x = h\cot\alpha$

$$d = h\cot\alpha - x$$
$$d = h\cot\alpha - h\cot\beta$$
$$d = h(\cot\alpha - \cot\beta)$$
$$h = \frac{d}{\cot\alpha - \cot\beta}$$

57. Sketch a figure:

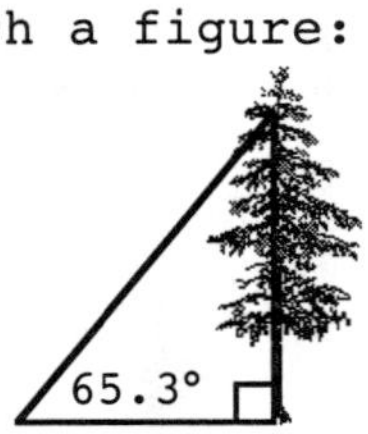

Let h = height of tree
From the figure, it should be clear that $\tan 65.3° = \frac{h}{105}$

$$h = 105\tan 65.3°$$
$$= 228 \text{ feet}$$

59. Sketch a figure:

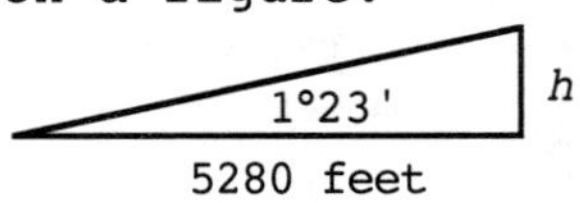

Let h = how far train climbs.

$\tan 1°23' = \frac{h}{5280}$

$$h = 5280\tan 1°23'$$
$$= 127.5 \text{ feet}$$

61. Sketch a figure:

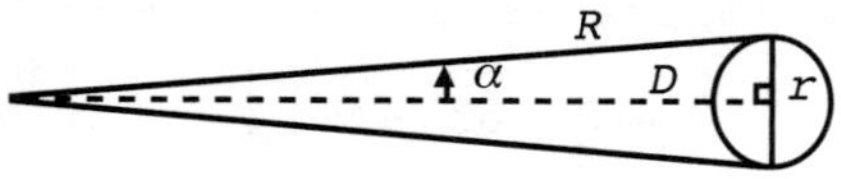

Note $\alpha = \frac{1}{2}(32') = 16'$

diameter $= d = 2r$

$$\tan \alpha = \frac{r}{D}$$

$$\tan 16' = \frac{r}{239{,}000}$$

$$r = 239{,}000 \tan 16'$$

$$d = 2(239{,}000)\tan 16'$$

$$= 2225 \text{ miles}$$

Alternatively, we can write

$$\sin \alpha = \frac{r}{R}$$

$$\sin 16' = \frac{r}{239{,}000}$$

$$r = 239{,}000 \sin 16'$$

$$d = 2(239{,}000)\sin 16'$$

$$= 2225 \text{ miles}$$

Although it is not clear whether 239,000 miles is to be interpreted as D, R, or $D - r$, at this accuracy it does not matter.

63. Sketch a figure:

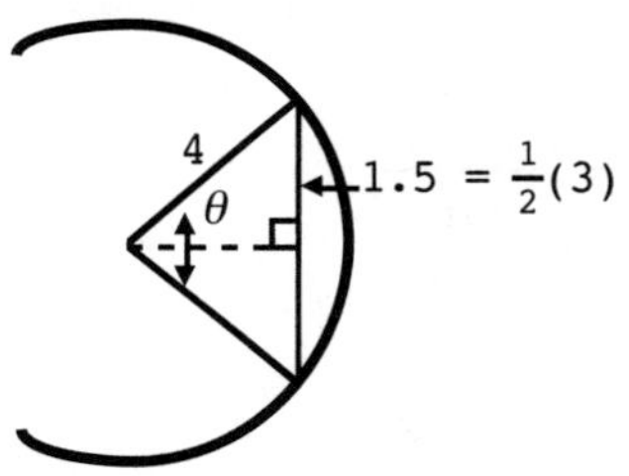

We will find $\frac{1}{2}\theta$ and double it.

$$\sin \frac{1}{2}\theta = \frac{1.5}{4}$$

$$\frac{1}{2}\theta = \sin^{-1} \frac{1.5}{4}$$

$$\theta = 2 \sin^{-1} \frac{1.5}{4}$$

$$= 44°$$

65. We use $g = \dfrac{v}{t \sin \theta}$ with $v = 4.1$,

$t = 3.0$, $\theta = 8.0°$

$$g = \frac{4.1}{3.0 \sin 8.0°}$$

$g = 9.8$ meters/second^2

67.

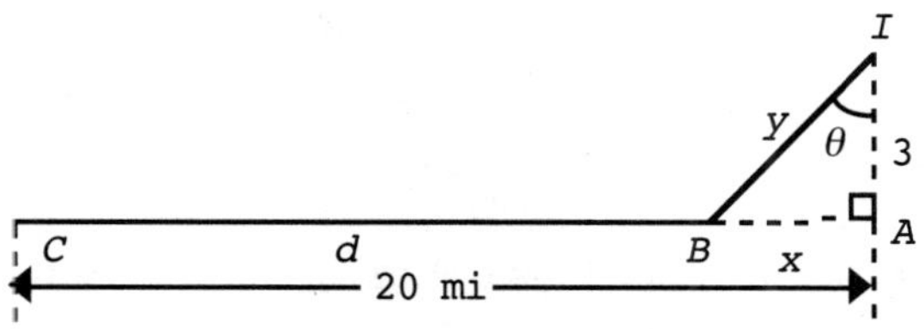

(A) We note that the cable consists of water section IB, and shore section $CB = 20$ mi $- AB$.

Let $y = IB$ and $d = CB = 20 - x$

In right triangle ABI

$$\sec \theta = \frac{y}{3 \text{ mi}} \quad \text{and} \quad \tan \theta = \frac{x}{3 \text{ mi}}$$

$$y = 3 \sec \theta \text{ mi} \qquad x = 3 \tan \theta \text{ mi}$$

Thus the cost of the cable =

$$\begin{pmatrix}\text{Cost of Water}\\\text{Section Per Mile}\end{pmatrix}\begin{pmatrix}\text{Number of}\\\text{Water miles} = y\end{pmatrix} + \begin{pmatrix}\text{Cost of Shore}\\\text{Section Per Mile}\end{pmatrix}\begin{pmatrix}\text{Number of}\\\text{Shore miles} = d\end{pmatrix}$$

$$C(\theta) = (25{,}000 \frac{\text{dollars}}{\text{mi}})(3 \sec \theta \text{ mi}) + (15{,}000 \frac{\text{dollars}}{\text{mi}})(20 - 3 \tan \theta \text{ mi})$$

$$C(\theta) = 75{,}000 \sec \theta + 300{,}000 - 45{,}000 \tan \theta$$

(B)

θ	$C(\theta)$
10°	$368,222
20°	$363,435
30°	$360,622
40°	$360,146
50°	$363,050

69. Sketch a figure:

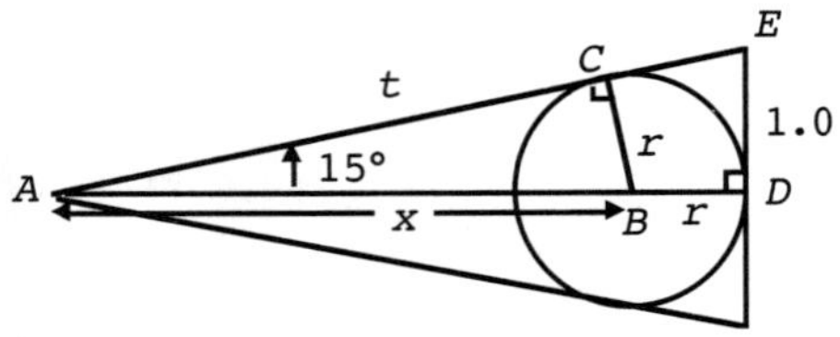

In triangle *ABC*, we have $\frac{r}{x} = \sin 15°$

In triangle *ADE*, we have $\frac{1.0}{r + x} = \tan 15°$

Eliminating x, we see that

$$\frac{r + x}{1.0} = \frac{1}{\tan 15°} = \cot 15° \text{ (reciprocal identity)}$$

$$x = \cot 15° - r$$

Substituting, we have

$$\frac{r}{\cot 15° - r} = \sin 15°$$

$$r = \sin 15°(\cot 15° - r)$$

$$r = \sin 15° \cot 15° - \sin 15° r$$

$$r(1 + \sin 15°) = \sin 15° \cot 15°$$

$$r = \frac{\sin 15° \cot 15°}{1 + \sin 15°}$$

$$r = 0.77 \text{ meters}$$

Section 6-4

1. sine: 2π; cotangent: π; cosecant: 2π

3. (A) Since the range of the cosine function is [-1, 1], the largest and smallest y values on its graph are, respectively, 1 and -1. The largest deviation of the function from the x axis is therefore 1 unit.

   (B) Since the range of the tangent function is all real numbers, the graph deviates indefinitely far from the x axis.

   (C) Since the range of the cosecant function is all real numbers $y \ge 1$ or $y \le -1$, the graph deviates indefinitely far from the x axis.

5. (A) $-2\pi, -\pi, 0, \pi, 2\pi$ (B) $-\frac{3\pi}{2}, -\frac{\pi}{2}, \frac{\pi}{2}, \frac{3\pi}{2}$ (C) No x intercepts

7. (A) Defined for all real x (B) $-\frac{3\pi}{2}, -\frac{\pi}{2}, \frac{\pi}{2}, \frac{3\pi}{2}$ (C) $-2\pi, -\pi, 0, \pi, 2\pi$

9. (A) There are no vertical asymptotes (B) $-\frac{3\pi}{2}, -\frac{\pi}{2}, \frac{\pi}{2}, \frac{3\pi}{2}$

(C) $-2\pi, -\pi, 0, \pi, 2\pi$

11. (A) A shift of $\pi/2$ to the left will transform the cosecant graph into the secant graph. [The answer is not unique--see part (B).]

(B) The graph of $y = -\csc(x - \pi/2)$ is a $\pi/2$ shift to the right and a reflection in the x axis of the graph of $y = \csc x$. The result is the graph of $y = \sec x$. The graph of $y = -\csc\left(x + \frac{\pi}{2}\right)$ is a $\pi/2$ shift to the left and a reflection in the x axis of the graph of $y = \csc x$. The result is not the graph of $y = \sec x$.

13. Let $f(x) = \frac{\tan x}{x}$

$$f(-x) = \frac{\tan(-x)}{(-x)} = \frac{-\tan x}{-x} = \frac{\tan x}{x} = f(x)$$

$y = \frac{\tan x}{x}$ is even.

15. Let $f(x) = \frac{\csc x}{x}$

$$f(-x) = \frac{\csc(-x)}{-x} = \frac{\frac{1}{\sin(-x)}}{-x} = \frac{\frac{1}{-\sin x}}{-x} = \frac{-\csc x}{-x} = \frac{\csc x}{x} = f(x)$$

$y = \frac{\csc x}{x}$ is even.

17. Let $f(x) = \sin x \cos x$

$$f(-x) = \sin(-x)\cos(-x) = -\sin x \cos x = -f(x)$$

$y = \sin x \cos x$ is odd.

19. Let $f(x) = x^2 \sin x$

$$f(-x) = (-x)^2 \sin(-x) = x^2(-\sin x) = -x^2 \sin x = -f(x)$$

$y = x^2 \sin x$ is odd.

21. $(a, b) = (6, 8)$

$r = \sqrt{a^2 + b^2} = \sqrt{6^2 + 8^2} = \sqrt{100} = 10$

$\sin\theta = \frac{b}{r} = \frac{8}{10} = \frac{4}{5}$ $\csc\theta = \frac{r}{b} = \frac{10}{8} = \frac{5}{4}$

$\cos\theta = \frac{a}{r} = \frac{6}{10} = \frac{3}{5}$ $\sec\theta = \frac{r}{a} = \frac{10}{6} = \frac{5}{3}$

$\tan\theta = \frac{b}{a} = \frac{8}{6} = \frac{4}{3}$ $\cot\theta = \frac{a}{b} = \frac{6}{8} = \frac{3}{4}$

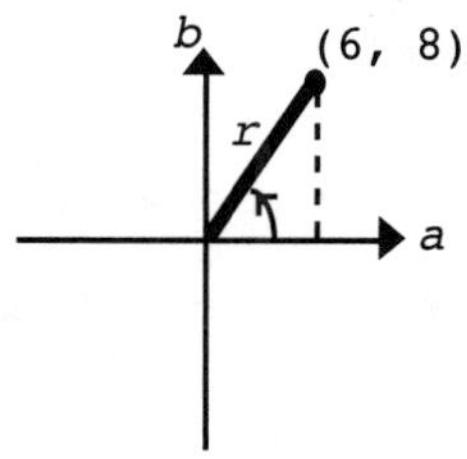

23. $(a, b) = (-1, \sqrt{3})$

$r = \sqrt{a^2 + b^2} = \sqrt{(-1)^2 + (\sqrt{3})^2} = \sqrt{4} = 2$

$\sin\theta = \frac{b}{r} = \frac{\sqrt{3}}{2}$ $\csc\theta = \frac{r}{b} = \frac{2}{\sqrt{3}}$

$\cos\theta = \frac{a}{r} = \frac{-1}{2} = -\frac{1}{2}$ $\sec\theta = \frac{r}{a} = \frac{2}{-1} = -2$

$\tan\theta = \frac{b}{a} = \frac{\sqrt{3}}{-1} = -\sqrt{3}$ $\cot\theta = \frac{a}{b} = \frac{-1}{\sqrt{3}} = -\frac{1}{\sqrt{3}}$

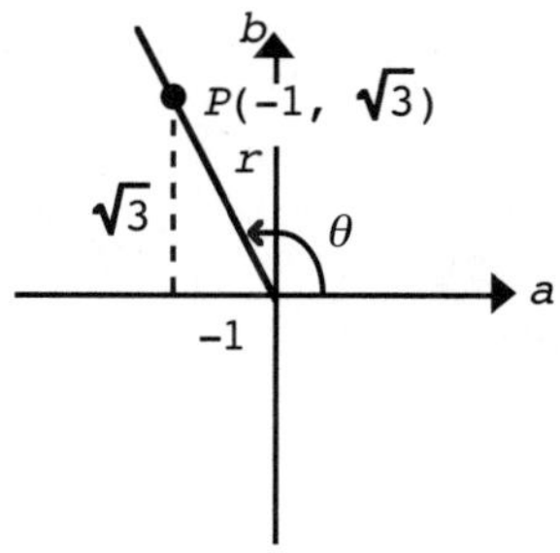

25.

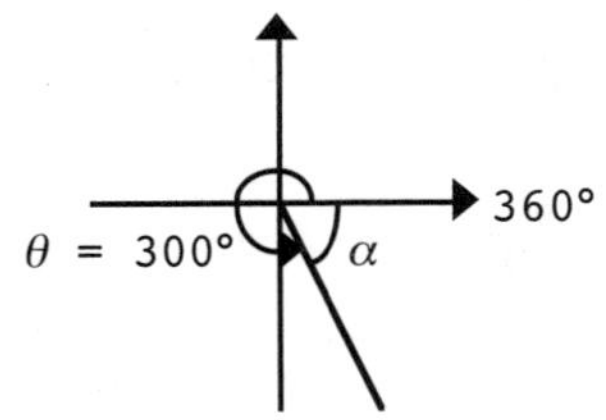

$\alpha = 360° - 300° = 60°$

27.

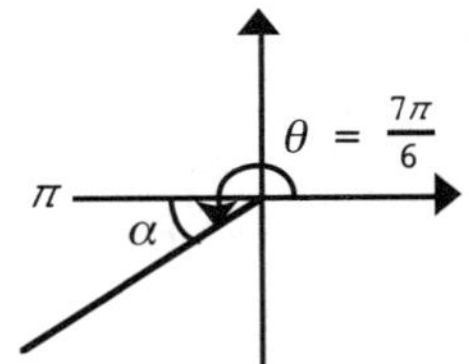

$\alpha = \dfrac{7\pi}{6} - \pi = \dfrac{\pi}{6}$

29.

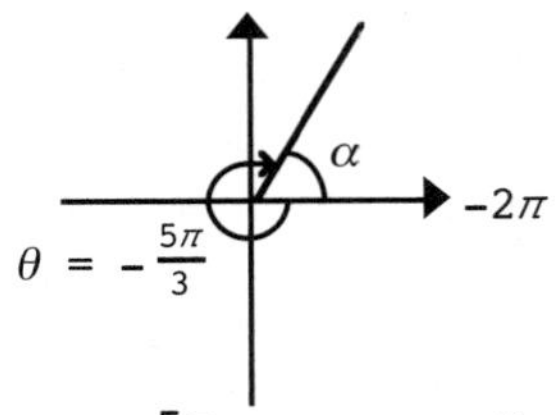

$\alpha = -\dfrac{5\pi}{3} + 2\pi = \dfrac{\pi}{3}$

31. $\cos\theta = \dfrac{a}{r} = \dfrac{-1}{2}$ because $r > 0$.
a is negative in the II and III quadrants. The smallest positive θ is associated with a 60° reference triangle in the II quadrant as shown in the figure below.

$\theta = 120°$ or $\dfrac{2\pi}{3}$ radians

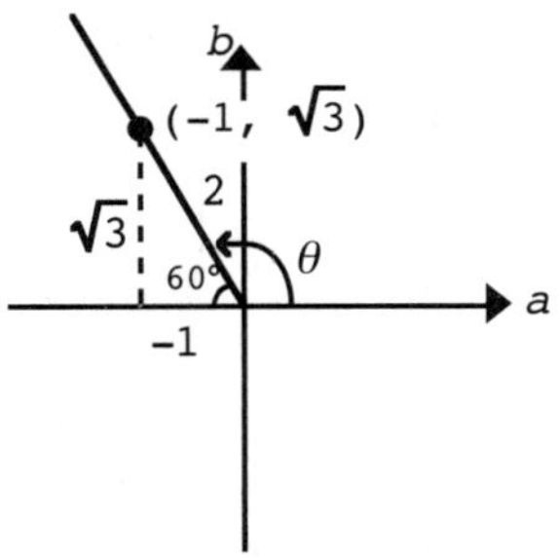

33. $\sin\theta = \dfrac{b}{r} = \dfrac{-1}{2}$ because $r > 0$.
b is negative in the III and IV quadrants. The smallest positive θ is associated with a 30° reference triangle in the III quadrant as drawn.

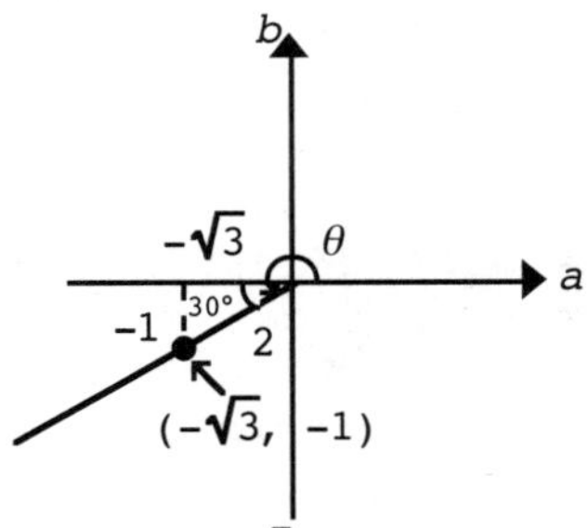

$\theta = 210°$ or $\dfrac{7\pi}{6}$ radians

35. $\csc\theta = \dfrac{r}{b} = \dfrac{2}{-\sqrt{3}}$ because $r > 0$.
b is negative in the III and IV quadrants. The smallest positive θ is associated with a 60° reference triangle in the III quadrant as drawn.

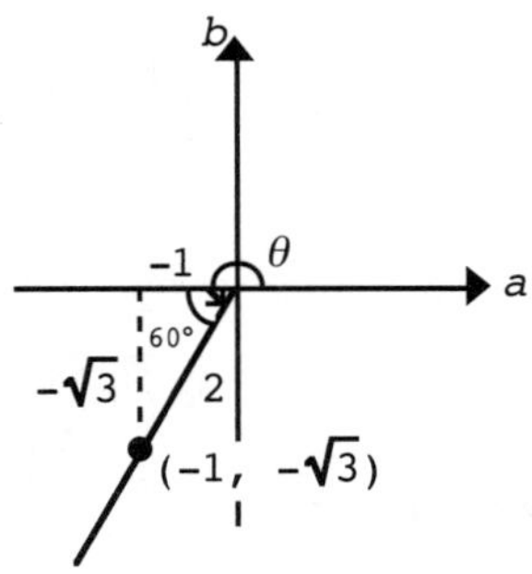

$\theta = 240°$ or $\dfrac{4\pi}{3}$ radians

37. Since $\sin\theta = \frac{3}{5} > 0$ and $\cos\theta < 0$, θ is a II quadrant angle. We sketch a reference triangle. Since $\sin\theta = \frac{b}{r} = \frac{3}{5}$, we know that $b = 3$ and $r = 5$. Use the Pythagorean theorem to find a.

$$a^2 + 3^2 = 5^2$$
$$a^2 = 16$$
$$a = -4$$

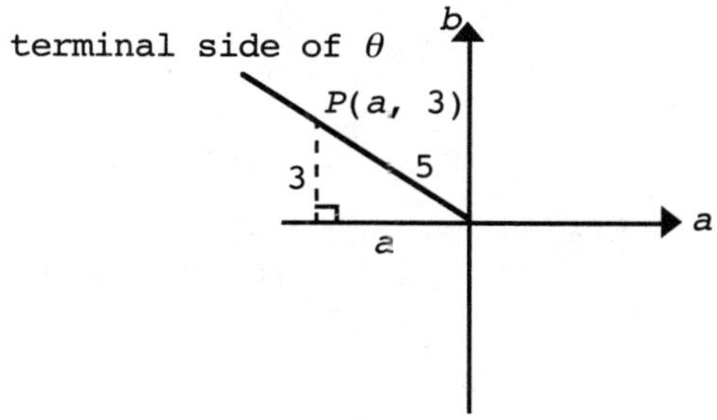

(a must be negative because θ is a II quadrant angle)

Using $(a, b) = (-4, 3)$ and $r = 5$, we have

$\cos\theta = \frac{a}{r} = \frac{-4}{5} = -\frac{4}{5}$ $\quad\tan\theta = \frac{b}{a} = \frac{3}{-4} = -\frac{3}{4}$

$\sec\theta = \frac{r}{a} = \frac{5}{-4} = -\frac{5}{4}$ $\quad\cot\theta = \frac{a}{b} = \frac{-4}{3} = -\frac{4}{3}$

$\csc\theta = \frac{r}{b} = \frac{5}{3}$

39. Since $\cos\theta = -\frac{\sqrt{5}}{3} < 0$ and $\cot\theta > 0$, θ is a III quadrant angle. We sketch a reference triangle. Since $\cos\theta = \frac{a}{r} = -\frac{\sqrt{5}}{3}$ we know that $a = -\sqrt{5}$ and $r = 3$. Use the Pythagorean theorem to find b.

$$(-\sqrt{5})^2 + b^2 = 3^2$$
$$b^2 = 4$$
$$b = -2$$

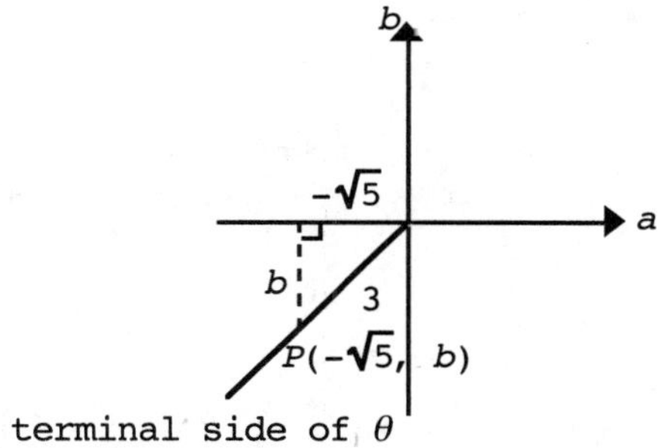

(b must be negative because θ is a III quadrant angle)

Using $(a, b) = (-\sqrt{5}, -2)$ and $r = 3$, we have

$\sin\theta = \frac{b}{r} = -\frac{2}{3}$ $\quad\sec\theta = \frac{r}{a} = -\frac{3}{\sqrt{5}}$

$\tan\theta = \frac{b}{a} = \frac{-2}{-\sqrt{5}} = \frac{2}{\sqrt{5}}$ $\quad\cot\theta = \frac{a}{b} = \frac{-\sqrt{5}}{-2} = \frac{\sqrt{5}}{2}$

$\csc\theta = \frac{r}{b} = \frac{3}{-2} = -\frac{3}{2}$

41. In these situations $P(a, b)$ is restricted so that $a = 0$. In this case, functions for which a is in the denominator are not defined. These functions are tangent and secant.

43. $\cos\theta = \frac{a}{r} = -\frac{\sqrt{3}}{2} = \frac{-\sqrt{3}}{2}$

Thus $(a, b) = (-\sqrt{3}, 1)$ or $(-\sqrt{3}, -1)$

θ is associated with a 30° reference triangle in the II quadrant or the III quadrant as drawn:

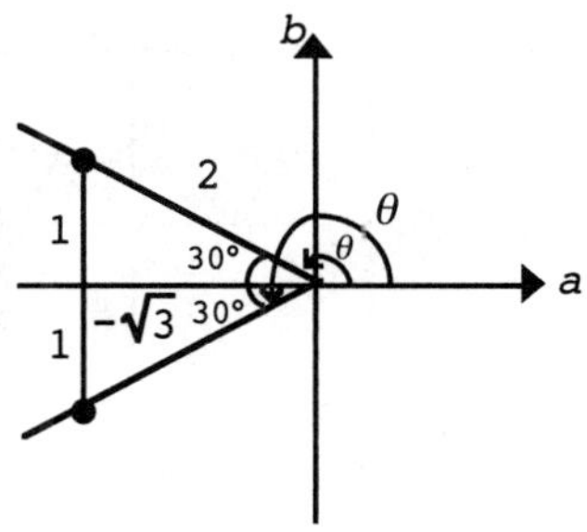

$\theta = 150°$ or $210°$

45. $\tan \theta = \frac{b}{a} = \frac{1}{1} = \frac{-1}{-1}$

Thus $(a, b) = (1, 1)$ or $(-1, -1)$

θ is associated with a $\frac{\pi}{4}$ reference triangle in the I quadrant or the III quadrant as drawn:

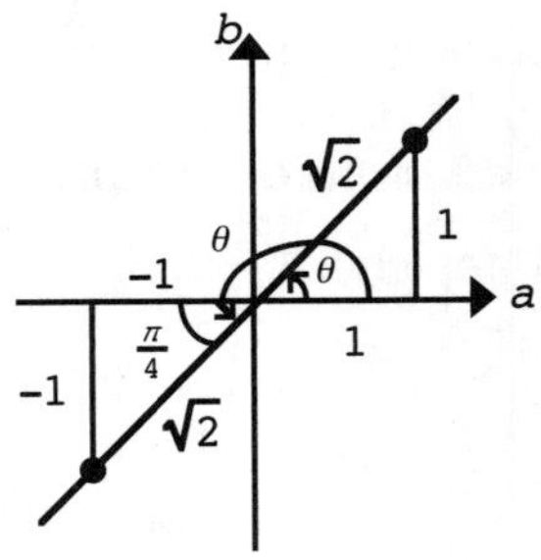

$\theta = \frac{\pi}{4}$ or $\frac{5\pi}{4}$

47. False. sec x and csc x are never 0.

49. True. The definition of periodicity implies

$f(x + 2p) = f(x + p + p) = f(x + p) = f(x)$

for all x.

51. False. $f(x) = x + 1$ is neither even nor odd.

53. False. sin x and cos x have period 2π; however, $\frac{\sin x}{\cos x} = \tan x$ has period π.

55. True. If $f(-x) = -f(x)$ and $g(-x) = -g(x)$, then $(fg)(-x) = f(-x)g(-x) = [-f(x)][-g(x)] = f(x)g(x) = (fg)(x)$ for all x in the domain of $fg(x)$.

57. If $f(x + p) = f(x)$ for all x, then

$$a(x + p) + b = ax + b$$
$$ap + b = b$$
$$a = 0$$

Hence $f(x) = b$, b any real number.

59. If $f(-x) = f(x)$ for all x, then

$$a(-x) + b = ax + b$$
$$0 = 2ax$$

Hence a must be 0,

$f(x) = b$, b any real number

61. (A)

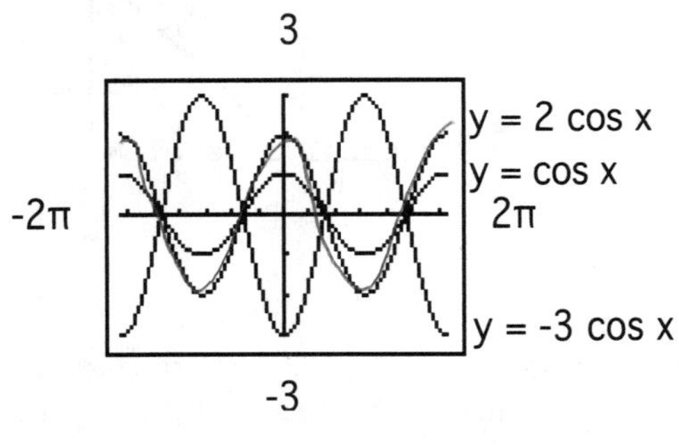

(B) The x intercepts do not change.

(C) The deviation of $y = \cos x$ from the x axis is 1 unit; the deviation of $y = 2 \cos x$ from the x axis is 2 units; the deviation of $y = -3 \cos x$ from the x axis is 3 units.

(D) The deviation of the graph from the x axis is changed by changing A. The deviation appears to be $|A|$.

63. (A)

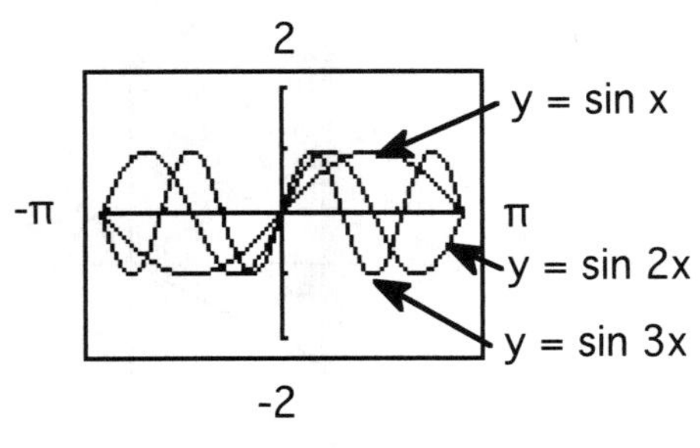

(B) 1 period of $y = \sin x$ appears. 2 periods of $y = \sin 2x$ appear. 3 periods of $y = \sin 3x$ appear.

(C) n periods of $y = \sin nx$ would appear.

65. (A)

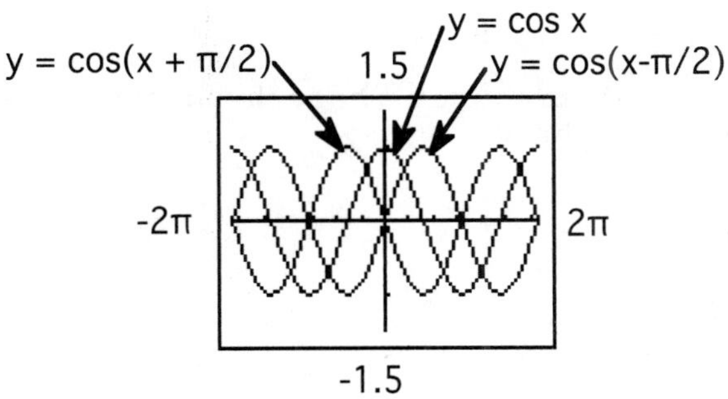

(B) The graph of $y = \cos x$ is shifted $|C|$ units to the right if $C < 0$ and $|C|$ units to the left if $C > 0$.

67. For each case, the number is not in the domain of the function and an error message of some type will appear.

69. Here are graphs of $f(x) = \sin x$ and $g(x) = x$, $-1 \le x \le 1$.

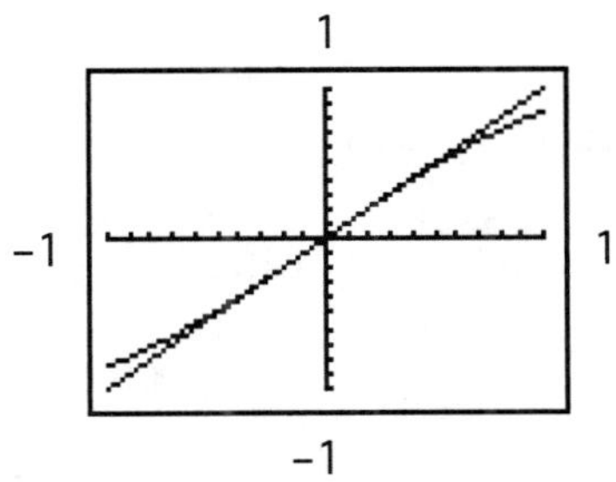

(A) The graphs become more indistinguishable the closer x is to the origin.

(B)

x	-0.3	-0.2	-0.1	0.0	0.1	0.2	0.3
$\sin x$	-0.296	-0.199	-0.100	0.000	0.100	0.199	0.296

71. (A) Since $\theta_R = \dfrac{s}{r}$ and r = radius of circle = 4, we have $\theta_R = \dfrac{7}{4}$ or 1.75 radians

(B) Since $\sin\theta = \dfrac{b}{r}$ and $\cos\theta = \dfrac{a}{r}$, we can write

$$a = r\cos\theta = 4\cos\frac{7}{4} = -0.713$$

$$b = r\sin\theta = 4\sin\frac{7}{4} = 3.936$$

$(a, b) = (-0.713, 3.936)$

73. We know that $s = r\theta$. $(a, b) = (6\sqrt{3}, 6)$.
From the Pythagorean theorem,

$$r = \sqrt{a^2 + b^2} = \sqrt{(6\sqrt{3})^2 + 6^2} = 12$$

Since $\tan\theta = \dfrac{b}{a} = \dfrac{6}{6\sqrt{3}} = \dfrac{1}{\sqrt{3}}$, $\theta = \dfrac{\pi}{6}$.

Hence $s = r\theta = 12\left(\dfrac{\pi}{6}\right) = 2\pi$ units.

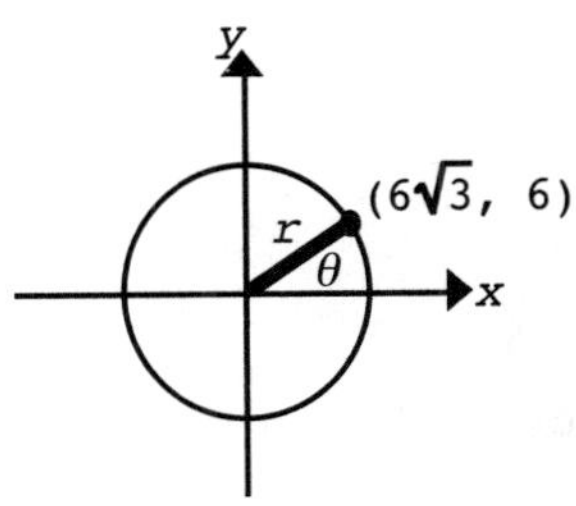

75. $I = k\cos\theta$
For $\theta = 0°$, $I = k\cos 0° = k$
For $\theta = 30°$, $I = k\cos 30° = 0.866k$
For $\theta = 60°$, $I = k\cos 60° = 0.5k$

77. From the figure, the following relations are clear:

$a^2 + b^2 = r^2 \quad r = 1 \quad \frac{a}{1} = \cos\theta \quad \frac{b}{1} = \sin\theta$

Using the Pythagorean theorem in the right triangle whose hypotenuse is the rod connecting the piston to the wheel, we have

$$(y - b)^2 + a^2 = 4^2$$
$$(y - b)^2 = 4^2 - a^2$$
$$y - b = \sqrt{4^2 - a^2}$$
$$y = b + \sqrt{4^2 - a^2}$$

Since $a = \cos\theta$ and $b = \sin\theta$ and $\theta = 6\pi t$, we have

$$y = \sin 6\pi t + \sqrt{16 - (\cos 6\pi t)^2}$$

79. $A = n\tan\left(\frac{180°}{n}\right)$

(A) For $n = 8$, $A = 8\tan\frac{180°}{8} = 3.31371$

For $n = 100$, $A = 100\tan\frac{180°}{100} = 3.14263$

For $n = 1000$, $A = 1000\tan\frac{180°}{1000} = 3.14160$

For $n = 10{,}000$, $A = 10{,}000\tan\frac{180°}{10{,}000} = 3.14159$

(B) as $n \to \infty$, A seems to approach $\pi(= 3.1415926\ldots)$, the area of the circle.

81. (A) Using the formula given:

For $\theta = 88.7°$, $m = \tan\theta = \tan 88.7° = 44.07$

For $\theta = 162.3°$, $m = \tan\theta = \tan 162.3° = -0.32$

(B) Using the formula for inclination, the slope m is given by $m = \tan 137°$ $= -0.93$. We now use the point-slope form of the equation of a line.

$$y - y_0 = m(x - x_0)$$
$$y - 5 = -0.93[x - (-4)]$$
$$y - 5 = -0.93x - 3.72$$
$$y = -0.93x + 1.28$$

Section 6-5

1. Amplitude $= |A| = |3| = 3$ Period $= \frac{2\pi}{B} = \frac{2\pi}{1} = 2\pi$

3. Amplitude $= |A| = \left|-\frac{1}{2}\right| = \frac{1}{2}$ Period $= \frac{2\pi}{B} = \frac{2\pi}{1} = 2\pi$

5. Amplitude $= |A| = |1| = 1$ Period $= \frac{2\pi}{B} = \frac{2\pi}{3}$

7. Amplitude is not defined for the cotangent function. Period $= \frac{\pi}{B} = \frac{\pi}{4}$

9. Amplitude is not defined for the tangent function. Period $= \frac{\pi}{B} = \frac{\pi}{8\pi} = \frac{1}{8}$

11. Amplitude is not defined for the cosecant function. Period $= \frac{2\pi}{B} = \frac{2\pi}{1/2} = 4\pi$

13. Amplitude $= |A| = |1| = 1$. Period $= \frac{2\pi}{B} = \frac{2\pi}{\pi} = 2$

The basic sine function $\sin t$ has zeros when $t = k\pi$, k an integer.

We examine $\pi x = k\pi$, $-2 \le x \le 2$ and find $x = k$ falls in this interval when $x = -2, -1, 0, 1, 2$.

15. Amplitude is not defined for the cotangent function. Period $= \frac{\pi}{1/2} = 2\pi$.

The basic cotangent function $\cot t$ has zeros when $t = \frac{\pi}{2} + k\pi$, k an integer.

We examine $\frac{x}{2} = \frac{\pi}{2} + k\pi$, $0 < x < 4\pi$, and find $x = \pi + 2k\pi$ falls in this interval when $x = \pi$ and $x = 3\pi$.

17. Amplitude $= |A| = |3| = 3$. Period $= \frac{2\pi}{B} = \frac{2\pi}{2} = 1$

The basic cosine function $\cos t$ has turning points when $t = k\pi$, k an integer.

We examine $2x = k\pi$, $-\pi \le x \le \pi$, and find that $x = \frac{k\pi}{2}$ falls in this interval when $x = -\pi, -\frac{\pi}{2}, 0, \frac{\pi}{2}, \pi$. The turning points are therefore $(-\pi, 3)$, $\left(-\frac{\pi}{2}, -3\right)$, $(0, 3)$, $\left(\frac{\pi}{2}, -3\right)$, $(\pi, 3)$.

19. Amplitude is not defined for the secant function. Period $= \frac{2\pi}{B} = \frac{2\pi}{\pi} = 2$.
The basic secant function $\sec t$ has turning points when $t = k\pi$, k an integer. We examine $\pi x = k\pi$, $-1 \le x \le 3$, and find that $x = k$ falls in this interval when $x = -1, 0, 1, 2, 3$. The turning points are therefore $(-1, -2)$, $(0, 2)$, $(1, -2)$, $(2, 2)$, $(3, -2)$.

21. $A = 3$ $P = \frac{\pi}{2} = \frac{2\pi}{4}$. Hence $B = 4$,

$y = 3 \sin 4x$, $-\frac{\pi}{4} \le x \le \frac{\pi}{2}$.

23. $|A| = 10$ $P = 2 = \frac{2\pi}{\pi}$. Hence $B = \pi$, $A = -10$, since the graph has the form of the standard sine curve turned upside down. $y = -10 \sin \pi x$ $-1 \le x \le 2$.

25. $A = 5$ $P = 8\pi = 2\pi \cdot 4 = 2\pi \div \frac{1}{4}$.

Hence $B = \frac{1}{4}$,

$y = 5 \cos \frac{1}{4}x$ $-4\pi \le x \le 8\pi$.

27. $|A| = 0.5$, $P = 8 = 2\pi \cdot \frac{4}{\pi} = 2\pi \div \frac{\pi}{4}$.

Hence $B = \frac{\pi}{4}$, $A = -0.5$, since the graph has the form of the standard cosine curve turned upside down.

$y = -0.5 \cos \frac{\pi x}{4}$ $-4 \le x \le 8$

29. $y = 4 \cos x$

Amplitude $= |A| = 4$ Period $= 2\pi$

Phase Shift $= 0$

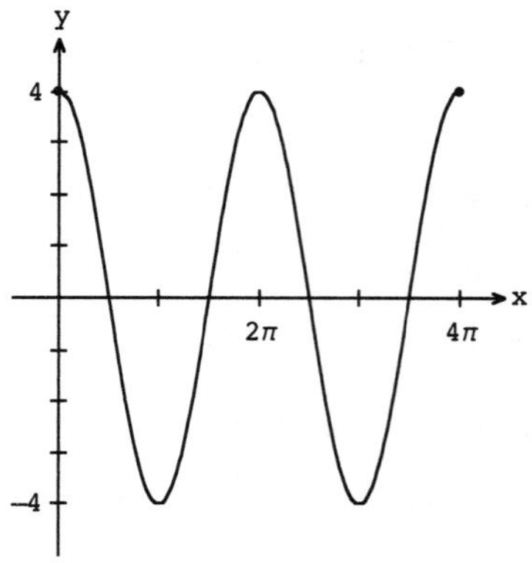

31. $y = \frac{1}{2} \sin\left(x + \frac{\pi}{4}\right)$ completes one cycle as $x + \frac{\pi}{4}$ varies from $x + \frac{\pi}{4} = 0$ to $x + \frac{\pi}{4} = 2\pi$, that is as x varies from $-\frac{\pi}{4}$ to $-\frac{\pi}{4} + 2\pi$.

Amplitude: $|A| = \frac{1}{2}$ Period: 2π Phase shift: $-\frac{\pi}{4}$

Divide the interval $\left[-\frac{\pi}{4}, -\frac{\pi}{4} + 2\pi\right]$ into four equal parts and sketch one cycle of $y = \frac{1}{2} \sin\left(x + \frac{\pi}{4}\right)$. Then extend the graph to cover $[-2\pi, 2\pi]$, deleting the small portion beyond $x = 2\pi$.

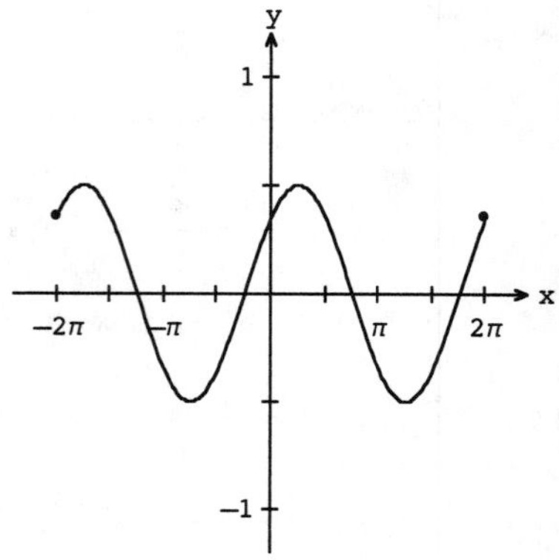

33. $y = \cot\left(x - \frac{\pi}{6}\right)$ completes one period as $x - \frac{\pi}{6}$ varies from $x - \frac{\pi}{6} = 0$ to $x - \frac{\pi}{6} = \pi$, that is as x varies from $\frac{\pi}{6}$ to $\frac{\pi}{6} + \pi$.

Period: π Phase shift: $\frac{\pi}{6}$

Sketch one cycle of $y = \cot\left(x - \frac{\pi}{6}\right)$, the graph of $y = \cot x$ shifted $\frac{\pi}{6}$ units to the right, on the interval $\left(\frac{\pi}{6}, \frac{\pi}{6} + \pi\right)$. Then extend the graph to cover $[-\pi, \pi]$, deleting the portion beyond π.

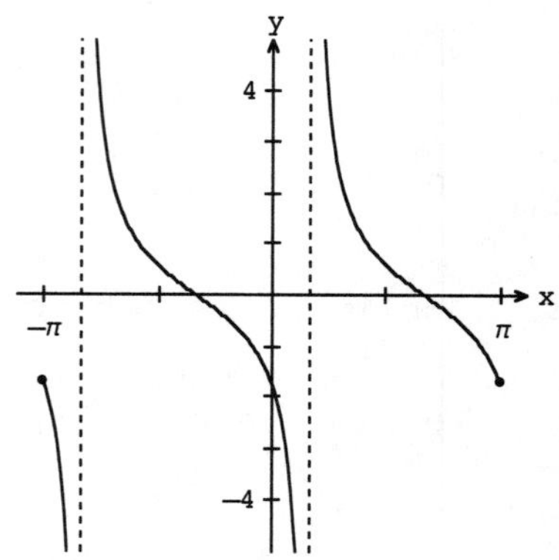

35. $y = 3 \tan 2x$ completes one period as $2x$ varies from $2x = -\frac{\pi}{2}$ to $2x = \frac{\pi}{2}$ that is, as x varies from $-\frac{\pi}{4}$ to $\frac{\pi}{4}$.

Period: $\frac{\pi}{2}$ Phase shift: 0

Sketch one cycle of $y = 3 \tan 2x$, the graph of $y = \tan x$ stretched vertically by a factor of 3 and shrunk horizontally by a factor of 2, on the interval $\left(-\frac{\pi}{4}, \frac{\pi}{4}\right)$. Then extend the graph to cover $[0, 2\pi]$.

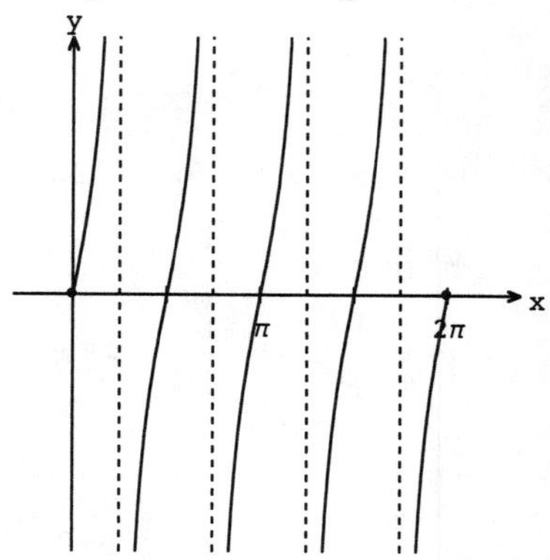

37. $y = 2\pi \sin \frac{\pi x}{2}$ completes one cycle as $\frac{\pi x}{2}$ varies from

$\frac{\pi x}{2} = 0$ to $\frac{\pi x}{2} = 2\pi$

that is, as x varies from 0 to 4.
Amplitude = $|A| = 2\pi$ Period: 4   Phase shift: 0
Divide the interval [0, 4] into four equal parts and sketch one cycle of $y = 2\pi \sin \frac{\pi x}{2}$, then extend the graph to cover [0, 12].

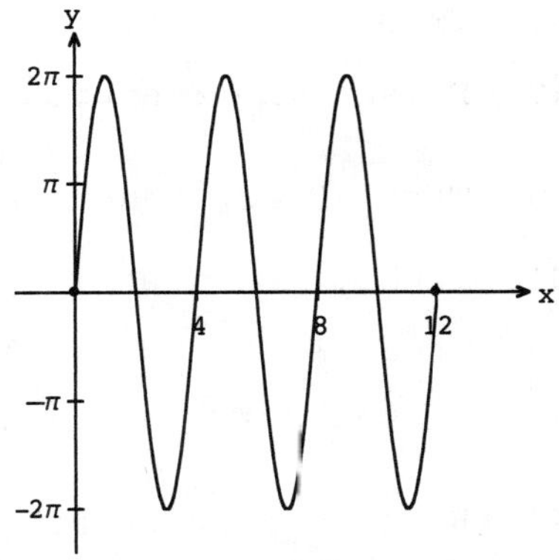

39. $y = -3 \sin\left[2\pi\left(x + \frac{1}{2}\right)\right]$ completes one cycle as $2\pi\left(x + \frac{1}{2}\right)$ varies from

$$2\pi\left(x + \frac{1}{2}\right) = 0 \quad \text{to} \quad 2\pi\left(x + \frac{1}{2}\right) = 2\pi$$
$$x + \frac{1}{2} = 0 \qquad\qquad x + \frac{1}{2} = 1$$
$$x = -\frac{1}{2} \qquad\qquad x = \frac{1}{2}$$

Amplitude = $|A| = |-3| = 3$ Period: $\frac{1}{2} - \left(-\frac{1}{2}\right) = 1$

Phase shift: $-\frac{1}{2}$

Divide the interval $\left[-\frac{1}{2}, \frac{1}{2}\right]$ into four equal parts and sketch one cycle of $y = -3 \sin\left[2\pi\left(x + \frac{1}{2}\right)\right]$—an upside-down sine curve. Then extend the graph to [-1, 2].

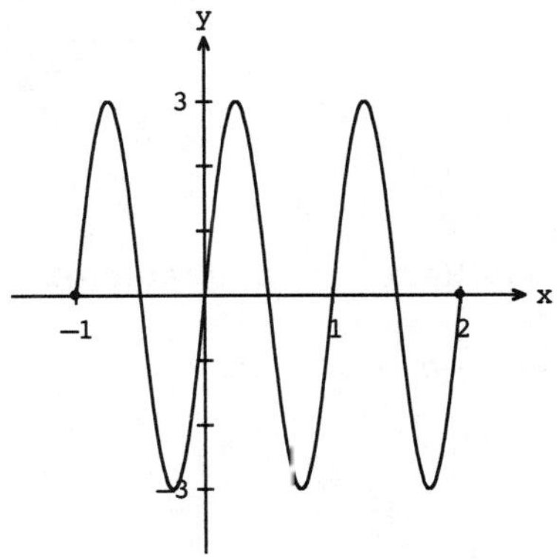

41. $y = \sec(x + \pi)$ has period 2π and completes one period as $x + \pi$ varies from $x + \pi = 0$ to $x + \pi = 2\pi$, that is, as x varies from $-\pi$ to π.
Phase shift: $-\pi$
The required graph is this one period, the graph of $y = \sec x$ shifted π units to the left.

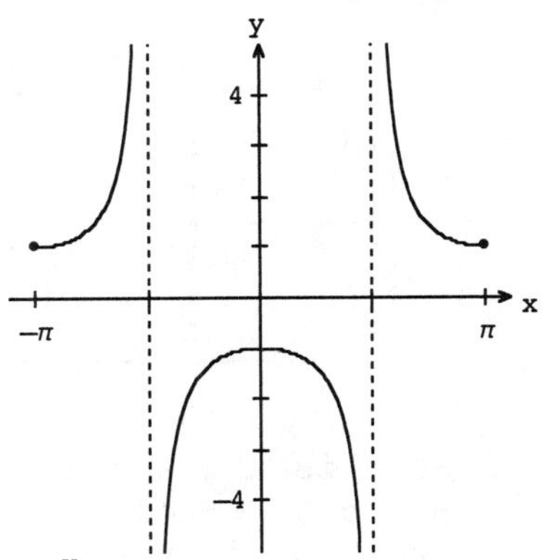

43. $y = 10 \csc \pi x$ completes one period as πx varies from $\pi x = 0$ to $\pi x = 2\pi$, that is, as x varies from 0 to 2.
Period: 2   Phase shift: 0
Sketch one period of $y = 10 \csc \pi x$, the graph of $y = \csc x$ stretched vertically by a factor of 10 and shrunk horizontally by a factor of $\frac{1}{\pi}$, on this interval. Then extend the graph to [0, 3].

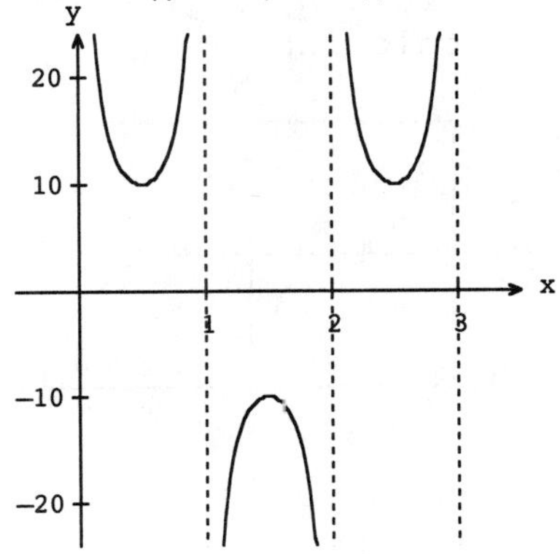

45. True. If $x = 0$, $y = A \sin B(0) = 0$. Thus the point $(0, 0)$ is on the graph.

47. False. The function $y = \sin\left(x - \frac{\pi}{4}\right)$ is neither even nor odd.

49. True. Every function of form $A \sin(Bx + C)$ or $A \cos(Bx + C)$ has period $\frac{2\pi}{B}$.

51. True. $|A \sin(Bx + C)| \leq |A|$ and $|A \cos(Bx + C)| \leq |A|$ for all x.

53. True. $A \sin(3Bx + C)$ and $A \cos(3Bx + C)$ define simple harmonics.

55. False. $y = 3 - \sin x$ cannot be represented as $y = A \sin(Bx + C)$.

57. Here is a graph of $y = \cos^2 x - \sin^2 x$, $-\pi \leq x \leq \pi$.

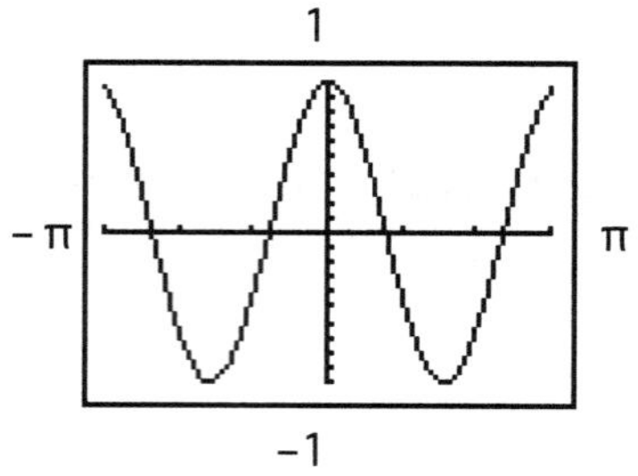

The graph has amplitude 1 and period π. It appears to be the graph of $y = A \cos Bx$ with $A = 1$, and $B = 2\pi \div P = 2\pi \div \pi = 2$, that is, $y = \cos 2x$.

59. Here is a graph of $y = 2 \sin^2 x$, $-\pi \leq x \leq \pi$.

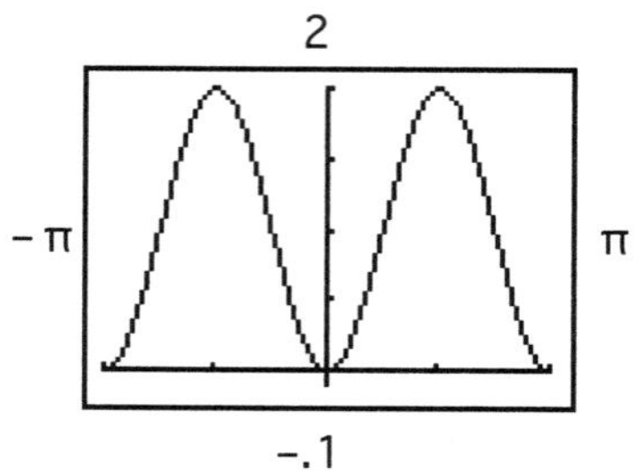

The graph has amplitude $\frac{2 - 0}{2} = 1$ and period π. It appears to be the graph of $y = \cos 2x$ turned upside down and shifted up one unit, that is, $y = -\cos 2x + 1$ or $y = 1 - \cos 2x$.

61. Here is a graph of $y = \cot x - \tan x$ drawn by a graphing calculator.

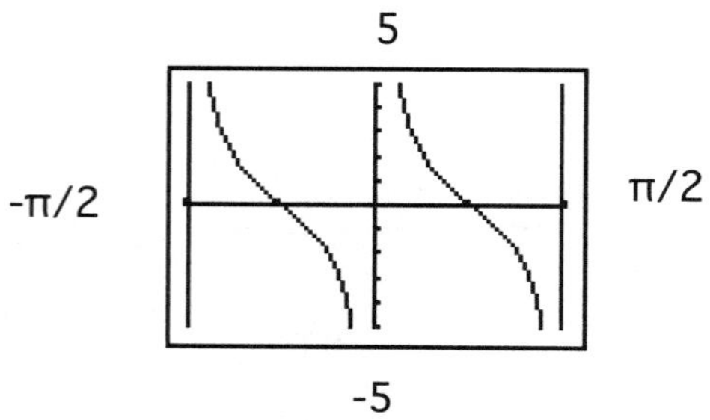

The graph appears to have the form $y = A \cot Bx$. Since the period is $\frac{\pi}{2}$, set $\frac{\pi}{2} = \frac{\pi}{B}$ to obtain $B = 2$. The graph of $y = A \cot 2x$ shown appears to pass through $\left(\frac{\pi}{8}, 2\right)$, thus $2 = A \cot 2\left(\frac{\pi}{8}\right)$

$$2 = A \cot \frac{\pi}{4}$$

$$2 = A$$

The equation of the graph can be written $y = 2 \cot 2x$.

63. Here is a graph of $y = \csc x + \cot x$ drawn by a graphing calculator.

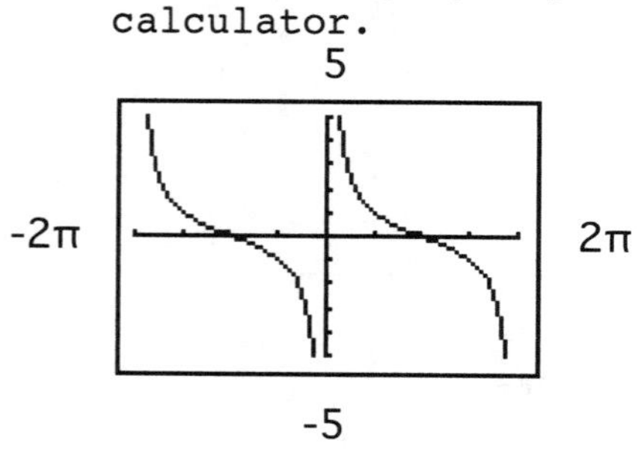

The graph appears to have the form $y = A \cot Bx$. Since the period is 2π, set $2\pi = \frac{\pi}{B}$ to obtain $B = \frac{1}{2}$. The graph of $y = A \cot \frac{1}{2}x$ shown appears to pass through $\left(\frac{\pi}{2}, 1\right)$, thus

$$1 = A \cot \frac{1}{2}\left(\frac{\pi}{2}\right)$$

$$1 = A \cot \frac{\pi}{4}$$

$$1 = A$$

The equation of the graph can be written $y = \cot \frac{1}{2}x$.

65. Here is a graph of $y = \sin 3x + \cos 3x \cot 3x$ drawn by a graphing calculator.

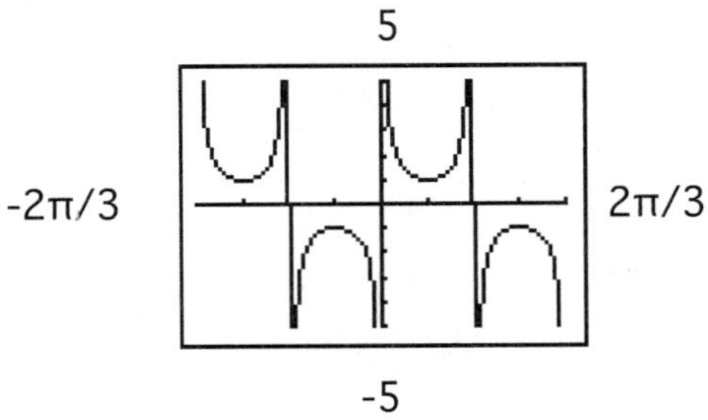

The graph appears to have the form $y = A \csc Bx$. Since the period is $\frac{2\pi}{3}$, set $\frac{2\pi}{3} = \frac{2\pi}{B}$ to obtain $B = 3$. The graph of $y = A \csc 3x$ shown appears to pass through $\left(\frac{\pi}{6}, 1\right)$, thus

$$1 = A \csc 3\left(\frac{\pi}{6}\right)$$

$$1 = A \csc \frac{\pi}{2}$$

$$1 = A$$

The equation of the graph can be written $y = \csc 3x$.

67. Here is a graph of $y = \dfrac{\sin 4x}{1 + \cos 4x}$ drawn by a graphing calculator.

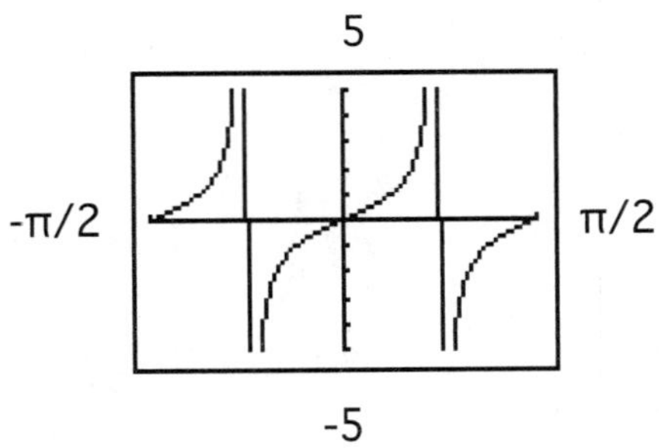

The graph appears to have the form $y = A \tan Bx$. Since the period is $\frac{\pi}{2}$, set $\frac{\pi}{B} = \frac{\pi}{2}$ to obtain $B = 2$. The graph of $y = A \tan 2x$ shown appears to pass through $\left(\frac{\pi}{8}, 1\right)$, thus

$$1 = A \tan 2\left(\frac{\pi}{8}\right)$$

$$1 = A \tan \frac{\pi}{4}$$

$$1 = A$$

The equation of the graph can be written $y = \tan 2x$.

69. $|A| = 4$. Hence $A = 4$ or -4. The graph completes one full cycle as x varies over the interval $[-1, 3]$. Since $-\frac{C}{B}$ is required between 0 and 2, we cannot simply set $-\frac{C}{B} = -1$. We must (mentally) extend the curve so that the phase shift is positive. Then the (extended) graph is that of an upside down sine curve that completes one full cycle as x varies over the interval $[1, 5]$. Hence

$$A = -4$$

$$-\frac{C}{B} = 1 \qquad -\frac{C}{B} + \frac{2\pi}{B} = 5$$

$$C = -B \qquad \frac{2\pi}{B} = 4 \qquad B = \frac{2\pi}{4} = \frac{\pi}{2} \qquad C = -\frac{\pi}{2}$$

The equation is then

$$y = A \sin(Bx + C)$$

$$y = -4 \sin\left(\frac{\pi}{2}x - \frac{\pi}{2}\right)$$

71. $|A| = \frac{1}{2}$. Hence $A = \frac{1}{2}$ or $-\frac{1}{2}$. The graph completes one full cycle of the cosine function as x varies over the (mentally extended) intervals $[-\pi, 7\pi]$ or $[3\pi, 11\pi]$. Since the phase shift is required between 0 and 4π, we must set $-\frac{C}{B} = 3\pi$. Then the (extended) graph has the form of a standard cosine curve. Hence

$$A = \frac{1}{2} \qquad -\frac{C}{B} + \frac{2\pi}{B} = 11\pi$$

$$-\frac{C}{B} = 3\pi \qquad \frac{2\pi}{B} = 8\pi \qquad B = \frac{2\pi}{8\pi} = \frac{1}{4} \qquad C = -3\pi B = -\frac{3\pi}{4}$$

$$C = -3\pi B$$

The equation is then

$$y = A\cos(Bx + C)$$

$$y = \frac{1}{2}\cos\left(\frac{1}{4}x - \frac{3\pi}{4}\right)$$

73. $y = 3.5\sin\left[\frac{\pi}{2}(t + 0.5)\right]$

$A = 3.5$

Solve $\frac{\pi}{2}(t + 0.5) = 0 \qquad \frac{\pi}{2}(t + 0.5) = 2\pi$

$t + 0.5 = 0 \qquad t + 0.5 = 4$

$t = -0.5$ (Phase shift) $\qquad t = -0.5 + 4 = 3.5$ (Period $P = 4$)

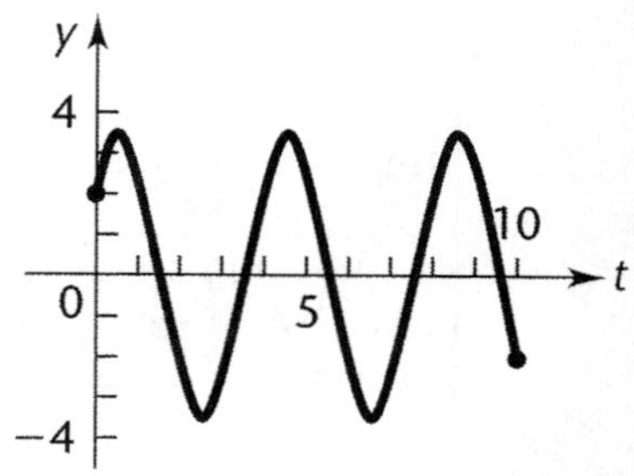

The graph completes one full cycle as t varies over the interval $[-0.5, 3.5]$.

75. $y = 50\cos[2\pi(t - 0.25)]$

$A = 50$

Solve $2\pi(t - 0.25) = 0 \qquad 2\pi(t - 0.25) = 2\pi$

$t - 0.25 = 0 \qquad t - 0.25 = 1$

$t = 0.25$ (Phase shift) $\qquad t = 0.25 + 1 = 1.25$ (Period $P = 1$)

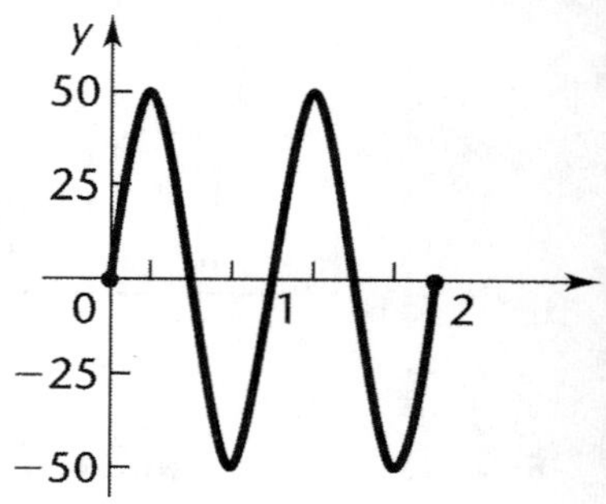

The graph completes one full cycle as t varies over the interval $[0.25, 1.25]$.

77. Here is a graph of $y = \sqrt{2}\sin x + \sqrt{2}\cos x$, $-2\pi \le x \le 2\pi$. It appears that this is a sine curve shifted to the left, with $A = 2$ and, since $P = \frac{2\pi}{B}$ and P appears to be 2π, $B = \frac{2\pi}{P} = \frac{2\pi}{2\pi} = 1$.

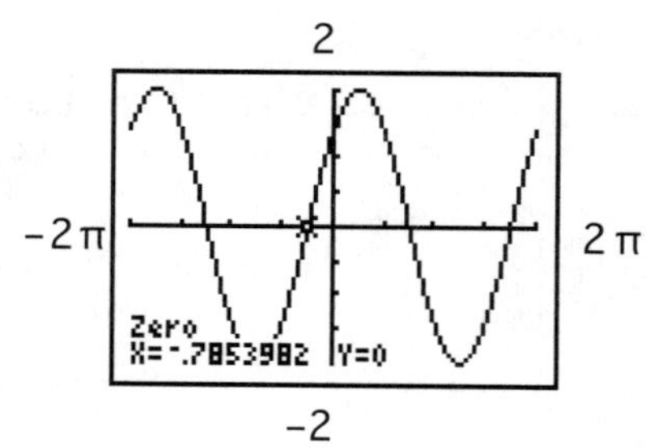

The x intercept closest to the origin, to three decimal places, is -0.785. To find C, substitute $B = 1$ and $x = -0.785$ into the phase-shift formula $x = -\frac{C}{B}$ and solve for C:

$$x = -\frac{C}{B}$$

$$-0.785 = -\frac{C}{1}$$

$$C = 0.785$$

The equation is thus $y = 2\sin(x + 0.785)$.

79. Here is a graph of $y = \sqrt{3}\sin x - \cos x$, $-2\pi \le x \le 2\pi$. It appears that this is a sine curve shifted to the right, with $A = 2$ and, since $P = \frac{2\pi}{B}$ and P appears to be 2π, $B = \frac{2\pi}{P} = \frac{2\pi}{2\pi} = 1$.

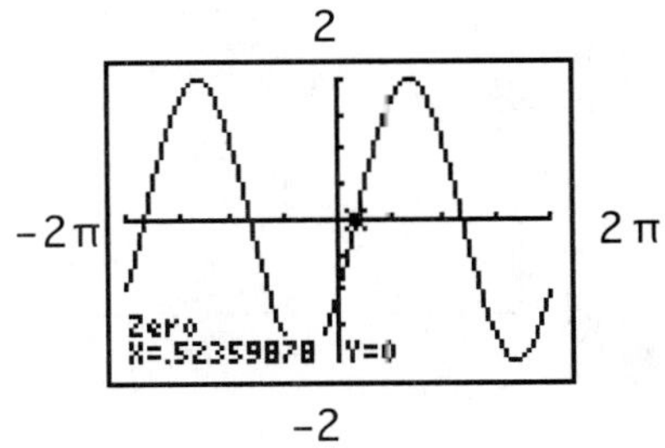

The x intercept closest to the origin, to three decimal places, is 0.524. To find C, substitute $B = 1$ and $x = 0.524$ into the phase-shift formula $x = -\frac{C}{B}$ and solve for C:

$$x = -\frac{C}{B}$$
$$0.524 = -\frac{C}{1}$$
$$C = -0.524$$

The equation is thus $y = 2\sin(x - 0.524)$.

81. Here is a graph of $y = 4.8\sin 2x - 1.4\cos 2x$, $-\pi \le x \le \pi$. It appears that this is a sine curve shifted to the right, with $A = 5$ and, since $P = \frac{2\pi}{B}$ and P appears to be π, $B = \frac{2\pi}{P} = \frac{2\pi}{\pi} = 2$.

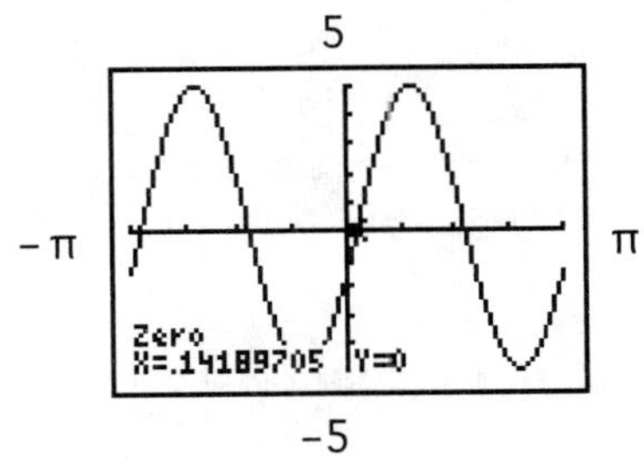

The x intercept closest to the origin, to three decimal places, is 0.142. To find C, substitute $B = 2$ and $x = 0.142$ into the phase-shift formula $x = -\frac{C}{B}$ and solve for C:

$$x = -\frac{C}{B}$$
$$0.142 = -\frac{C}{2}$$
$$C = -0.284$$

The equation is thus $y = 5\sin(2x - 0.284)$.

83. 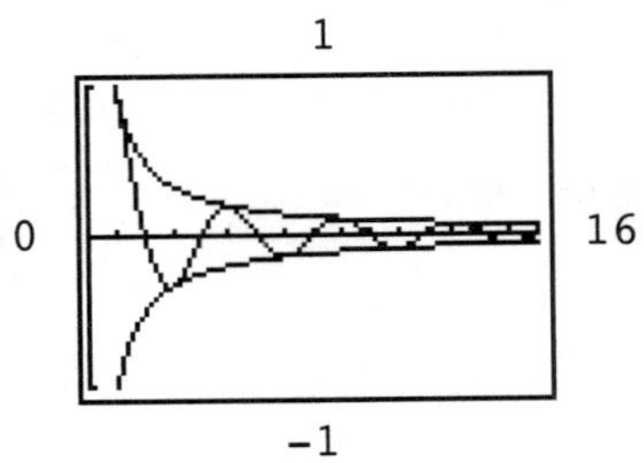

The amplitude is decreasing with time. This is often referred to as a *damped sine wave*. Examples are the vertical motion of a car after going over a bump (which is damped by the suspension system) and the slowing down of a pendulum that is released away from the vertical line of suspension (air resistance and friction).

85. 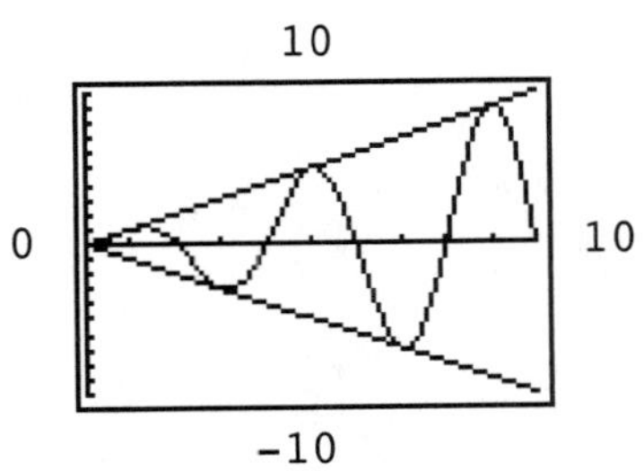

The amplitude is increasing with time. In physical and electrical systems this is referred to as *resonance*. Some examples are the swinging of a bridge during high winds and the movement of tall buildings during an earthquake. Some bridges and buildings are destroyed when the resonance reaches the elastic limits of the structure.

87.

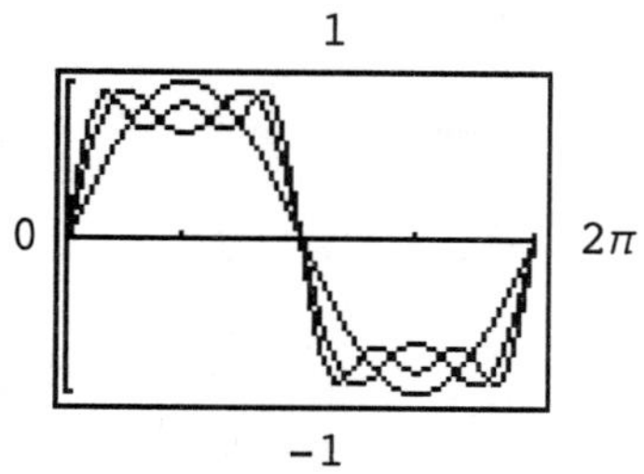

89. $A = \dfrac{1}{3}$

$P = \dfrac{2\pi}{8} = \dfrac{\pi}{4}$

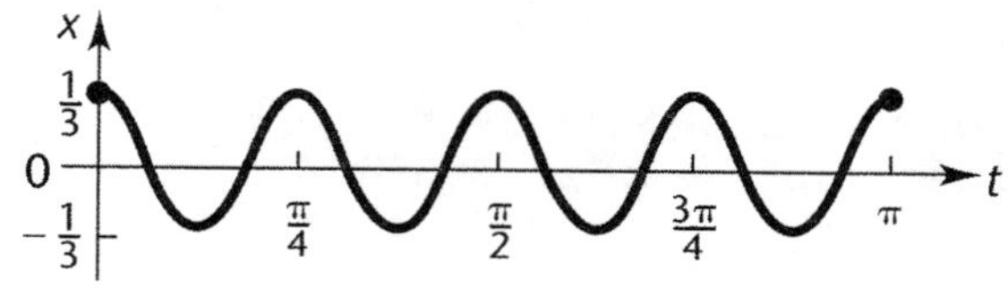

91. When $t = 0$ $y = -8$. Hence $-8 = A \cos B(0)$, that is, $A = -8$. Since the period is 0.5 seconds, $\dfrac{2\pi}{B} = 0.5$, $B = \dfrac{2\pi}{0.5} = 4\pi$. Hence the equation is $y = -8 \cos 4\pi t$.

93. The graph is the same as the graph of $y = \cos \dfrac{n\pi}{26}$, shifted 1.5 units up.

$A = 1$

$P = 2\pi \div \dfrac{\pi}{26} = 52$

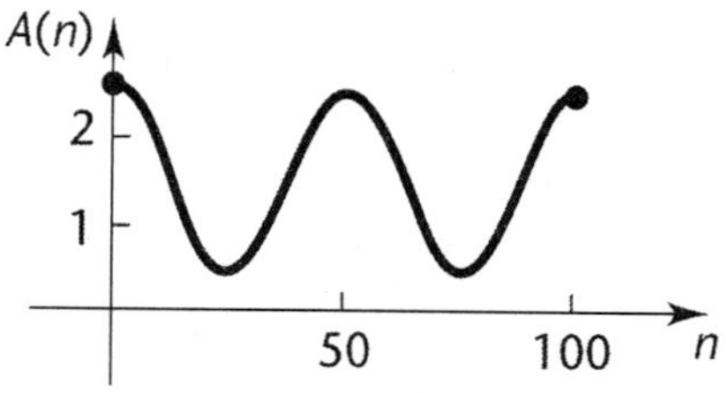

The graph shows the seasonal changes of sulfur dioxide pollutant in the atmosphere; more is produced during winter months because of increased heating.

95. $I = 15 \cos\left(120\pi t + \dfrac{\pi}{2}\right)$

$A = 15$

Solve $120\pi t + \dfrac{\pi}{2} = 0$ $\qquad$ $120\pi t + \dfrac{\pi}{2} = 2\pi$

$120\pi t = -\dfrac{\pi}{2}$ $\qquad$ $120\pi t = -\dfrac{\pi}{2} + 2\pi$

$t = -\dfrac{\pi}{2} \div 120\pi$ $\qquad$ $t = \left(-\dfrac{\pi}{2} + 2\pi\right) \div 120\pi$

$t = -\dfrac{1}{240}$ $\qquad$ $t = -\dfrac{1}{240} + \dfrac{1}{60} = \dfrac{3}{240}$ or $\dfrac{1}{80}$

Phase Shift $= -\dfrac{1}{240}$ $\qquad$ Period $P = \dfrac{1}{60}$

The graph completes one full cycle as t varies over the interval $\left[-\dfrac{1}{240}, \dfrac{3}{240}\right]$.

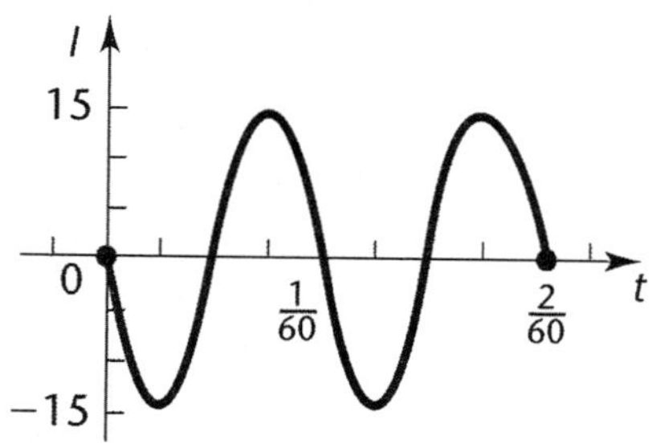

97. If the disk rotates through an angle θ in t seconds, we see

$$\theta = 3\frac{\text{revolutions}}{\text{second}} \cdot 2\pi\frac{\text{radians}}{\text{revolution}} \cdot t \text{ seconds}$$
$$= 6\pi t \text{ radians}$$

Then $\frac{y}{R} = \sin\theta = \sin 6\pi t$

$y = R \sin 6\pi t$

Since the disk has radius 3, $y = 3 \sin 6\pi t$.

This function has $A = 3$, $P = \frac{2\pi}{6\pi} = \frac{1}{3}$.

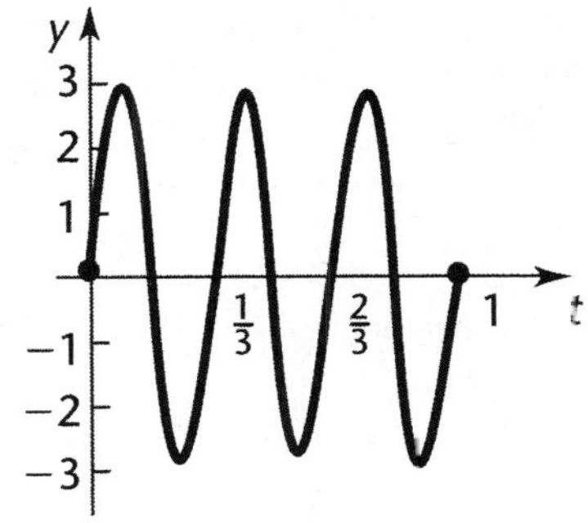

99. (A) In triangle MNP, we can write

$$\cos\theta = \frac{20}{c}$$
$$c\cos\theta = 20$$
$$c = \frac{20}{\cos\theta}$$
$$c = 20\sec\theta$$

Since $\theta = \frac{\pi t}{2}$, the equation for c in terms of t is $c = 20\sec\frac{\pi t}{2}$. The equation is valid for $0 \le t < 1$, since $\sec\frac{\pi t}{2}$ is undefined when $t = 1$.

(B) The graph has an asymptote at $t = 1$. Sketch the portion of $y = \sec\frac{\pi t}{2}$ on $[0, 1)$.

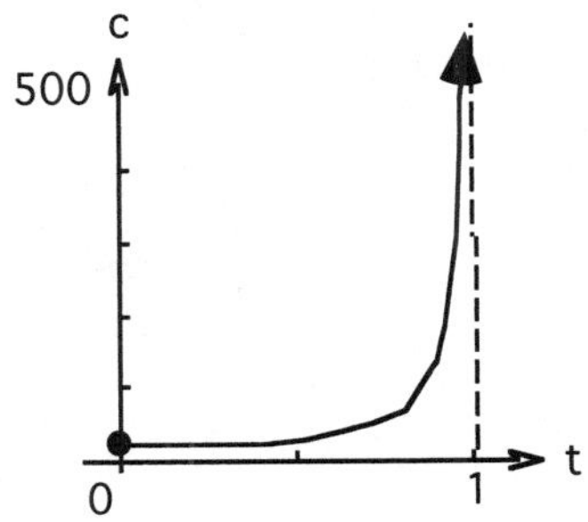

(C) The length of the light beam starts at 20 ft and increases slowly at first, then increases rapidly without end.

101. (A)

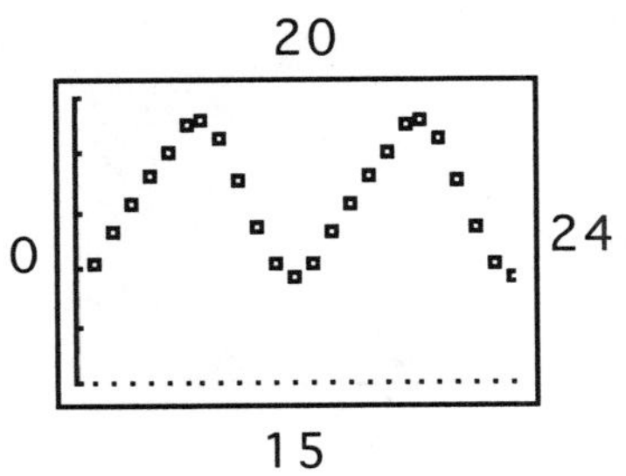

(B) From the table, Max y = 19:35 and Min y = 16:51.

Hence $A = \frac{\text{Max } y - \text{Min } y}{2} = \frac{19\frac{35}{60} - 16\frac{51}{60}}{2} = 1.37$

$$B = \frac{2\pi}{\text{Period}} = \frac{2\pi}{12} = \frac{\pi}{6}$$

$C = -1.75$ from the graph

$k = \text{Min } y + A = 16\frac{51}{60} + 1.37 = 18.22$

Thus, $y = 18.22 + 1.37\sin(\frac{\pi x}{6} - 1.75)$

(C)

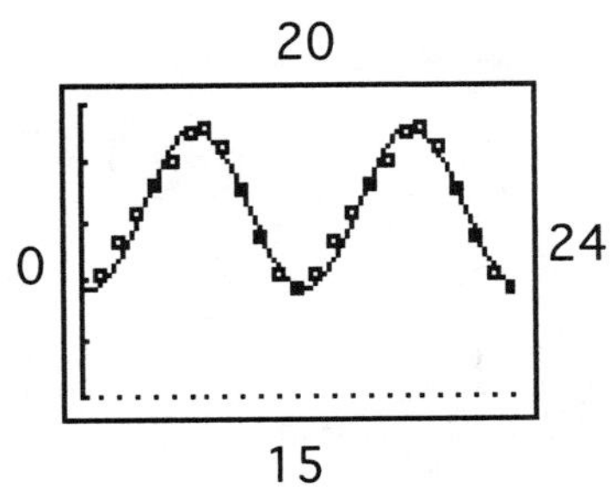

(D)

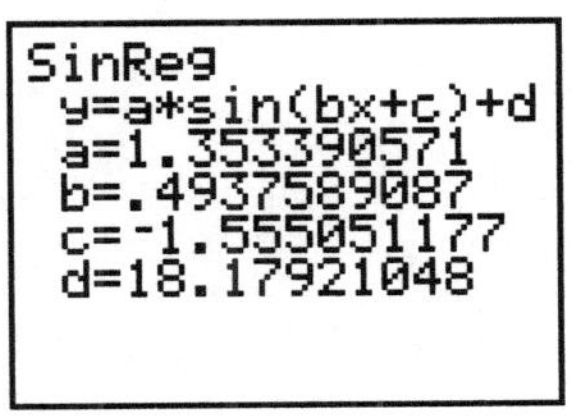

Compute the regression equation.

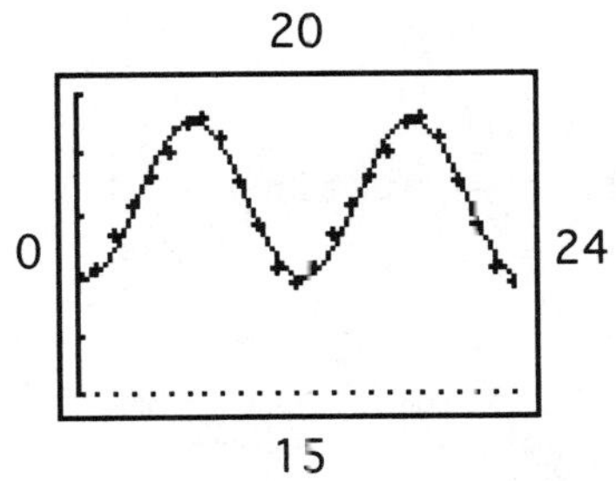

Graph the data and the regression equation.

Section 6-6

1. $y = \cos^{-1} 0$ is equivalent to
$\cos y = 0 \quad 0 \le y \le \pi$
$y = \frac{\pi}{2}$

3. $y = \arcsin \frac{\sqrt{3}}{2}$ is equivalent to
$\sin y = \frac{\sqrt{3}}{2} \quad -\frac{\pi}{2} \le y \le \frac{\pi}{2}$
$y = \frac{\pi}{3}$

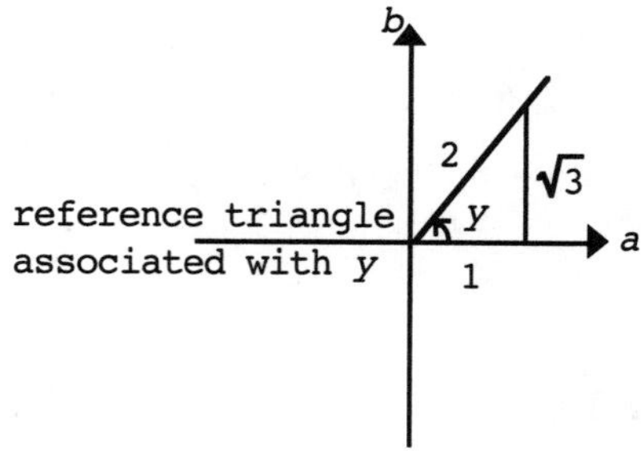

5. $y = \arctan \sqrt{3}$ is equivalent to
$\tan y = \sqrt{3} \quad -\frac{\pi}{2} < y < \frac{\pi}{2}$
$y = \frac{\pi}{3}$
(see sketch for problem 3)

7. $y = \sin^{-1} \frac{\sqrt{2}}{2}$ is equivalent to
$\sin y = \frac{\sqrt{2}}{2} \quad -\frac{\pi}{2} \le y \le \frac{\pi}{2}$
$y = \frac{\pi}{4}$

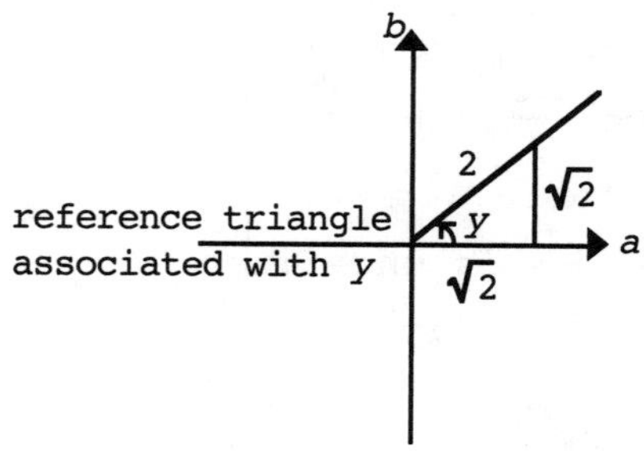

9. $y = \arccos 1$ is equivalent to
$\cos y = 1 \quad 0 \le y \le \pi$
$y = 0$

11. $y = \sin^{-1} \frac{1}{2}$ is equivalent to
$\sin y = \frac{1}{2} \quad -\frac{\pi}{2} \le y \le \frac{\pi}{2}$
$y = \frac{\pi}{6}$

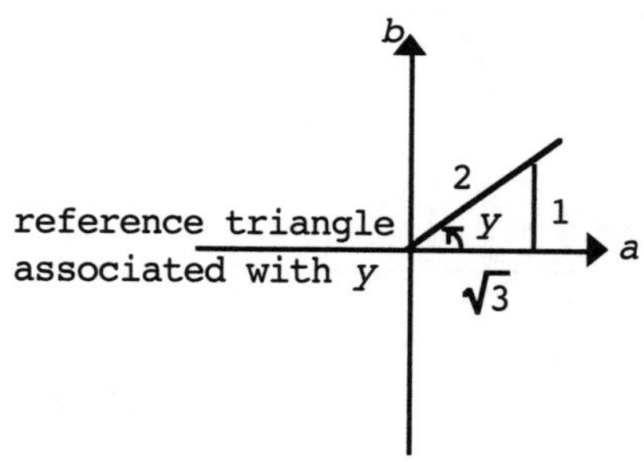

13. 1.144 **15.** 1.561 **17.** $\arccos x$ is not defined if $x > 1$.

19. $y = \arcsin\left(-\frac{\sqrt{2}}{2}\right)$ is equivalent to
$\sin y = -\frac{\sqrt{2}}{2} \quad -\frac{\pi}{2} \le y \le \frac{\pi}{2}$
$y = -\frac{\pi}{4}$

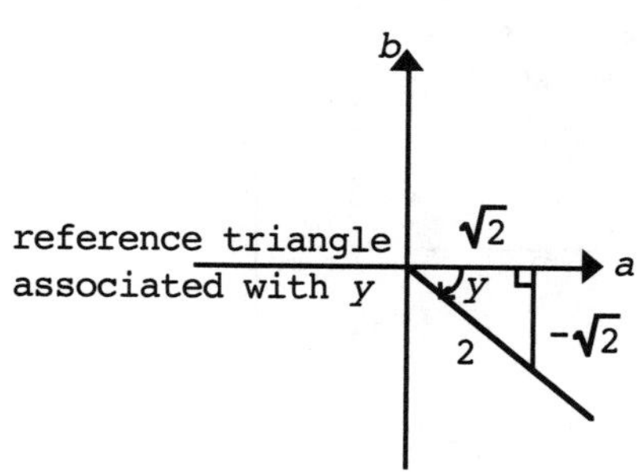

21. $y = \tan^{-1}(-\sqrt{3})$ is equivalent to

$\tan y = -\sqrt{3} \quad -\frac{\pi}{2} < y < \frac{\pi}{2}$

$y = -\frac{\pi}{3}$

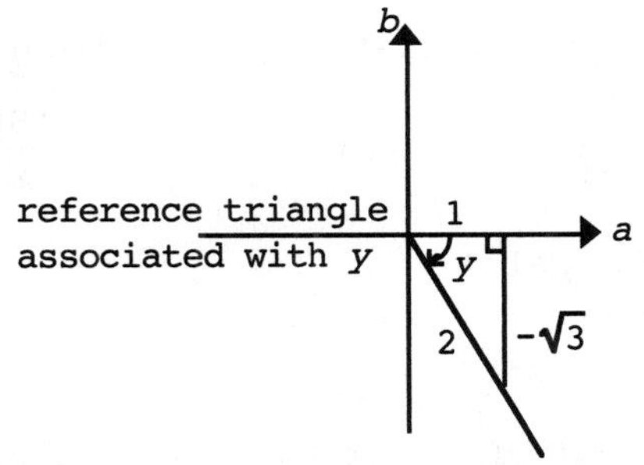

23. $\tan(\tan^{-1} 25) = 25$ by the tangent-inverse tangent identity.

25. $\cos^{-1}(\cos 2.3) = 2.3$ by the cosine-inverse cosine identity.

27. Let $y = \cos^{-1}\frac{\sqrt{3}}{2}$; then $\cos y = \frac{\sqrt{3}}{2}$, $0 \le y \le \pi$. Draw the reference triangle associated with y; then $\sin y = \sin\left[\cos^{-1}\frac{\sqrt{3}}{2}\right]$ can be determined directly from the triangle.

$\cos y = \frac{a}{r} = \frac{\sqrt{3}}{2} \quad a = \sqrt{3} \quad r = 2$

$(\sqrt{3})^2 + b^2 = 2^2$

$b^2 = 1$

$b = 1$

(positive since y is a quadrant I or II angle)

Therefore

$\sin y = \sin = \left[\cos^{-1}\frac{\sqrt{3}}{2}\right]\frac{1}{2}.$

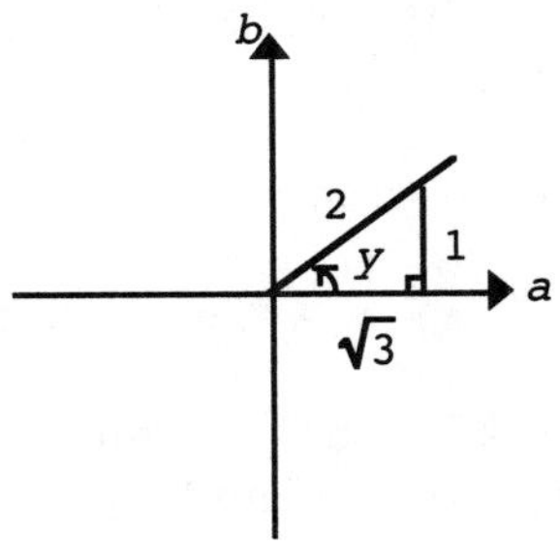

29. Let $y = \tan(-1)$; then $\tan y = -1$, $-\frac{\pi}{2} < y < \frac{\pi}{2}$. Draw the reference triangle associated with y; then $\csc y = \csc[\tan^{-1}(-1)]$ can be determined directly from the triangle.

$\tan y = \frac{b}{a} = \frac{-1}{1} \quad b = -1 \quad a = 1$

$(-1)^2 + 1^2 = r^2$

$r^2 = 2$

$r = \sqrt{2}$

$\csc y = \frac{r}{b} = \frac{\sqrt{2}}{-1} = -\sqrt{2}$

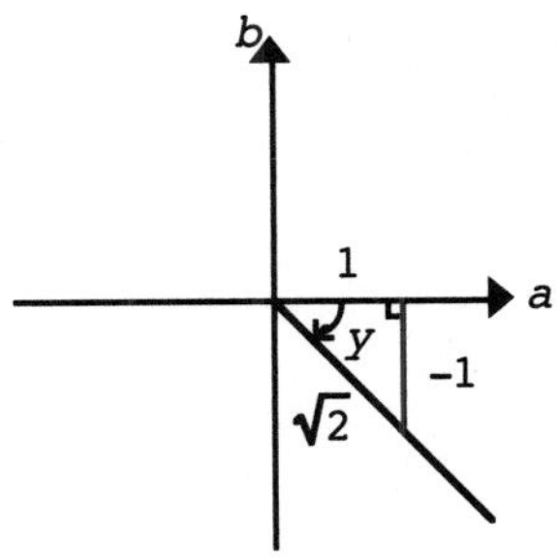

31. $\sin^{-1}(\sin \pi) = \sin^{-1}(0) = 0$

33. $\cos^{-1}\left(\cos\left(\frac{4\pi}{3}\right)\right) = \cos^{-1}\left(-\frac{1}{2}\right) = \frac{2\pi}{3}$

35. -1.472 37. $\cot[\cos^{-1}(-0.7003)] = \frac{1}{\tan[\cos^{-1}(-0.7003)]} = -0.9810$ 39. 2.645

41. In problem 19 we found $\sin^{-1}\left(-\frac{\sqrt{2}}{2}\right) = \arcsin\left(-\frac{\sqrt{2}}{2}\right) = -\frac{\pi}{4}$. Using the relation $\theta_D = \frac{180°}{\pi \text{ rad}}\theta_R$, we have $\sin^{-1}\left(-\frac{\sqrt{2}}{2}\right) = \frac{180°}{\pi}\left(-\frac{\pi}{4}\right) = -45°$.

43. In problem 21 we found $\arctan(-\sqrt{3}) = \tan^{-1}(-\sqrt{3}) = -\frac{\pi}{3}$. Using the relation $\theta_D = \frac{180°}{\pi \text{ rad}}\theta_R$, we have $\arctan(-\sqrt{3}) = \frac{180°}{\pi}\left(-\frac{\pi}{3}\right) = -60°$.

45. In problem 23 we found $\cos^{-1}(-1) = \pi$. Since π rad = 180°, we have $\cos^{-1}(-1) = 180°$.

Calculator in degree mode for problems 47 — 52.

47. 43.51° **49.** -21.48° **51.** -89.93°

53. $\sin^{-1}(\sin 2) = 1.1416 \neq 2$. For the identity $\sin^{-1}(\sin x) = x$ to hold, x must be in the restricted domain of the sine function; that is, $-\frac{\pi}{2} \leq x \leq \frac{\pi}{2}$. The number 2 is not in the restricted domain.

55. True. A periodic function cannot be one-to-one, since $f(x + p) = f(x)$ but $x + p \neq x$.

57. False. None of them are periodic.

59. True. $\sin^{-1}(-x) = -\sin^{-1}(x)$ for all x, $-1 \leq x \leq 1$

61. True. $\sin^{-1} x$, $\tan^{-1} x$, and $\sec^{-1} x$ are increasing functions on any interval on which they are defined, $\cos^{-1} x$, $\cot^{-1} x$, and $\csc^{-1} x$ similarly decreasing.

63.

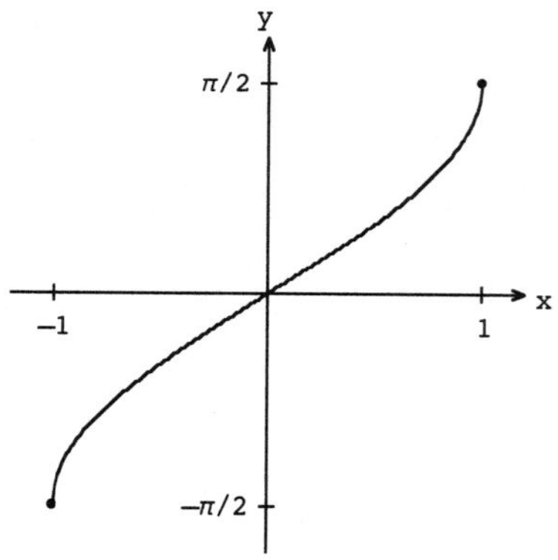

65.

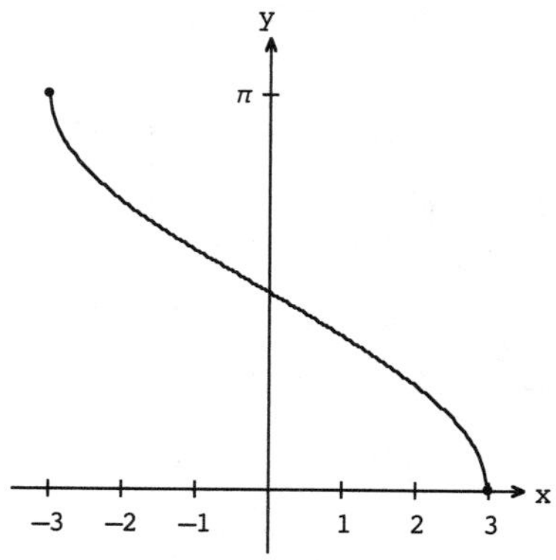

67.

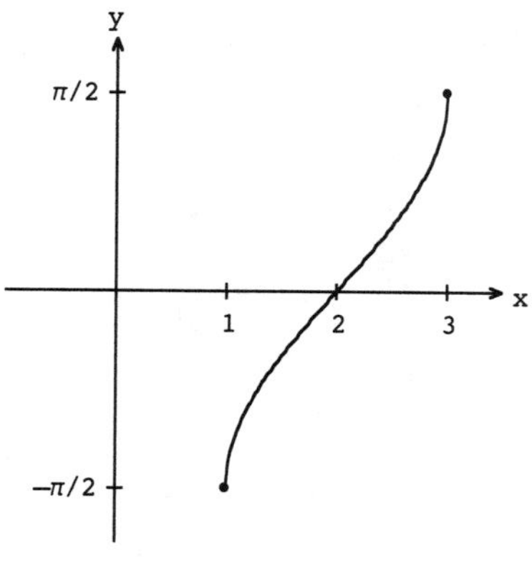

69.

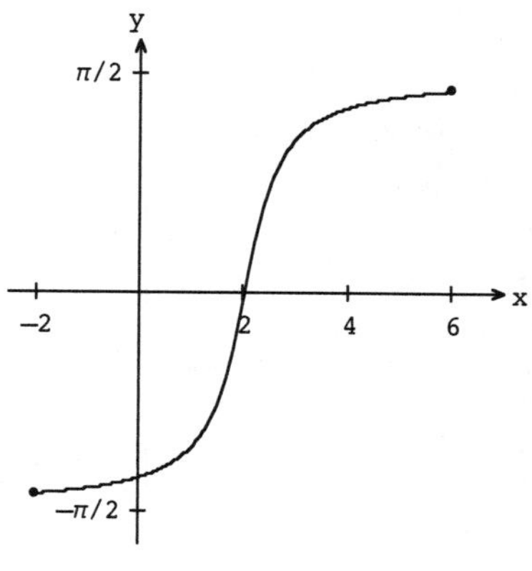

71. (A)

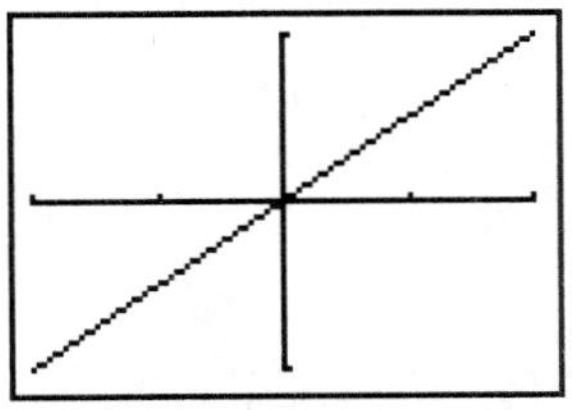

(B) The domain of $\cos^{-1}$ is restricted to $-1 \le x \le 1$; hence no graph will appear for other x.

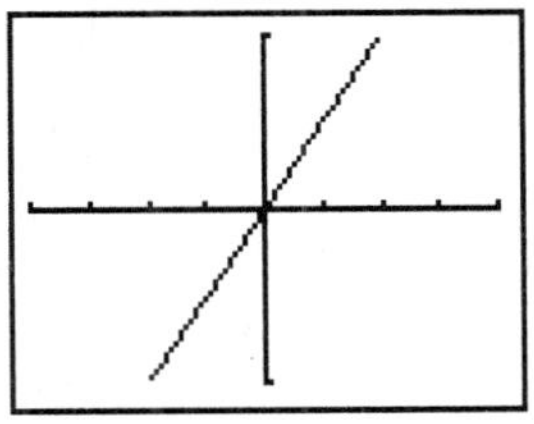

73. Let $y = \sin^{-1} x$. Then $x = \sin y$, $-\frac{\pi}{2} \le y \le \frac{\pi}{2}$. If $\sin y = \frac{b}{r} = x = \frac{x}{1}$, then let $b = x$, $r = 1$.

$$a^2 + b^2 = r^2$$
$$a^2 + x^2 = 1^2$$
$$a^2 = 1 - x^2$$
$$a = \sqrt{1 - x^2}$$ (a must be positive since y is a I or IV quadrant angle)

$$\cos y = \cos(\sin^{-1} x) = \frac{a}{r} = \frac{\sqrt{1 - x^2}}{1} = \sqrt{1 - x^2}$$

75. Let $y = \arctan x$. Then $x = \tan y$, $-\frac{\pi}{2} < y < \frac{\pi}{2}$. If $\tan y = \frac{b}{a} = x = \frac{x}{1}$, then let $b = x$, $a = 1$.

$$a^2 + b^2 = r^2$$
$$1^2 + x^2 = r^2$$
$$r = \sqrt{1 + x^2}$$ (r is always taken positive)

$$\cos y = \cos(\tan^{-1} x) = \frac{a}{r} = \frac{1}{\sqrt{1 + x^2}}$$

77. $f(x) = 4 + 2\cos(x - 3)$. $3 \le x \le 3 + \pi$. With this restriction, f is one-to-one (proof omitted). Solve $y = f(x)$ for x:

$$y = 4 + 2\cos(x - 3) \qquad 3 \le x \le 3 + \pi$$
$$y - 4 = 2\cos(x - 3) \qquad 0 \le x - 3 \le \pi$$
$$\frac{y - 4}{2} = \cos(x - 3) \qquad -1 \le \cos(x - 3) \le 1$$
$$\cos^{-1}\frac{y - 4}{2} = x - 3 \qquad -2 \le 2\cos(x - 3) \le 2$$
$$x = 3 + \cos^{-1}\frac{y - 4}{2} = f^{-1}(y) \qquad 2 \le 4 + 2\cos(x - 3) \le 6$$

range of f: $2 \le y \le 6$

Interchange x and y

$f^{-1}(x) = 3 + \cos^{-1}\frac{x - 4}{2}$ $\quad 2 \le 4 + 2\cos(y - 3) \le 6$. Thus range of f = domain of f^{-1}: $2 \le x \le 6$.

79. (A)

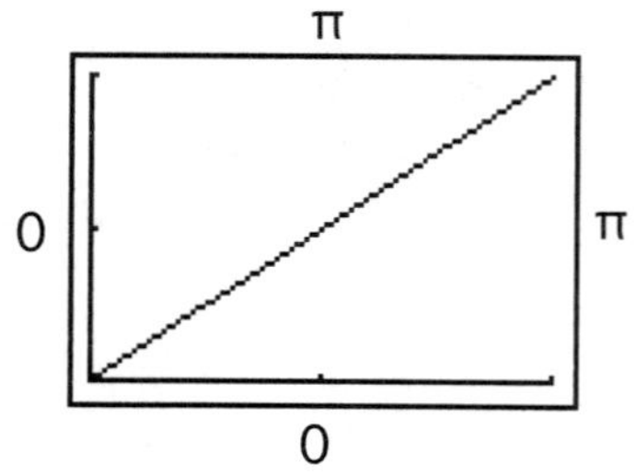

(B) The domain for cos x is $(-\infty, \infty)$ and the range is $[-1, 1]$, which is the domain for $\cos^{-1} x$. Thus, $y = \cos^{-1}(\cos x)$ has a graph over the interval $(-\infty, \infty)$, but $\cos^{-1}(\cos x) = x$ only on the restricted domain of cos x, $[0, \pi]$.

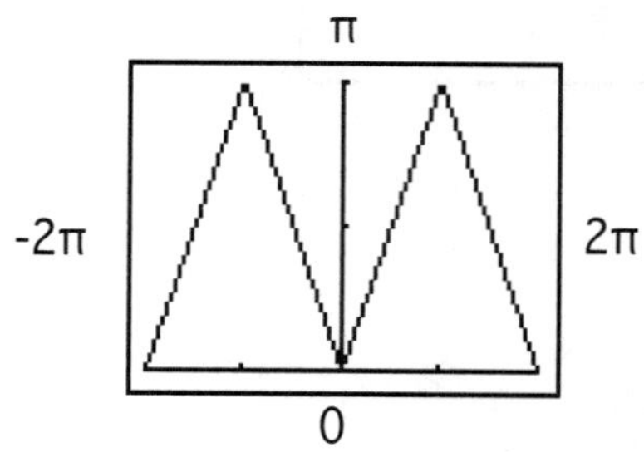

81. For a 28mm lens, $x = 28$, thus $\theta = 2\tan^{-1}\dfrac{21.634}{28} = 1.31567$ radians. In decimal degrees, $\theta_D = \dfrac{180°}{\pi}(1.31567 \text{ rad}) = 75.38°$.

For a 100 mm lens, $x = 100$, thus $\theta = 2\tan^{-1}\dfrac{21.634}{100} = 0.42611$ radians. In decimal degrees, $\theta_D = \dfrac{180°}{\pi}(0.42611 \text{ rad}) = 24.41°$.

83. (A)

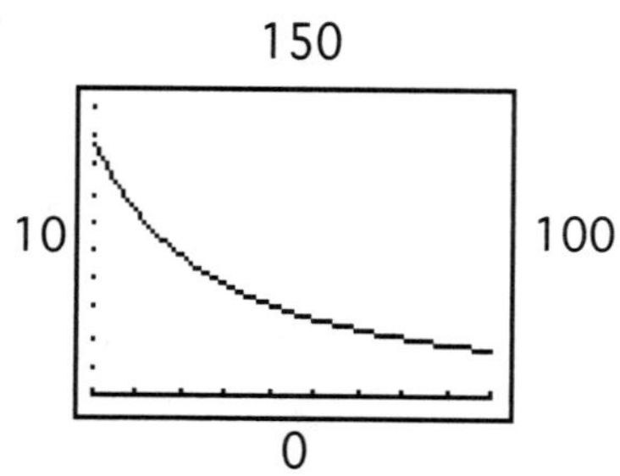

(B) 59.44 mm

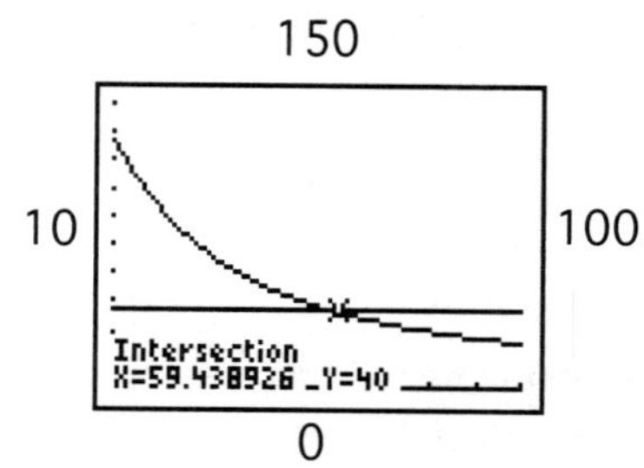

85.

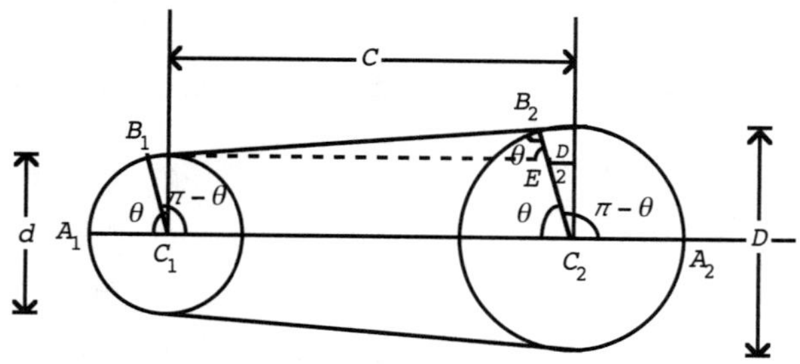

From the above figure, the following should be clear:

Length of belt $= 2[arc\ \widehat{A_1B_1} + B_1B_2 + arc\ \widehat{B_2A_2}]$

$arc\ A_1B_1 = \dfrac{d}{2}(\theta)$

$arc\ B_2A_2 = \dfrac{D}{2}(\pi - \theta)$

To find B_1B_2 we note:

C_1C_2 has length C

B_1E is constructed parallel to C_1C_2. EC_2 is parallel to B_1C_1. Hence $EB_1C_1C_2$ is a parallelogram. EB_1 has length C. EB_1B_2 is a right triangle. Thus

(1) $\cos\theta = \frac{B_2E}{EB_1} = \frac{\frac{D}{2} - \frac{d}{2}}{C} = \frac{D - d}{2C}$

(2) $\sin\theta = \frac{B_1B_2}{EB_1}$, so $B_1B_2 = EB_1 \sin\theta = C\sin\theta$

Finally,
Length of belt

$$L = 2[\text{arc } A_1B_1 + B_1B_2 + \text{arc } B_2A_2]$$
$$= 2\left[\frac{d}{2}\theta + C\sin\theta + \frac{D}{2}(\pi - \theta)\right]$$
$$= d\theta + 2C\sin\theta + D(\pi - \theta)$$
$$L = \pi D + (d - D)\theta + 2C\sin\theta$$

and (from (1) above)

$$\theta = \cos^{-1}\frac{D - d}{2C}$$

Substituting the given values, we have
$D = 4$, $d = 2$, $C = 6$ (calculator in radian mode)

$$\theta = \cos^{-1}\frac{4 - 2}{2\cdot 6} = \cos^{-1}\frac{1}{6}$$
$$L = 4\pi + (2 - 4)\cos^{-1}\frac{1}{6} + 2\cdot 6\sin\left(\cos^{-1}\frac{1}{6}\right)$$
$$\approx 21.59 \text{ inches}$$

87. (A)

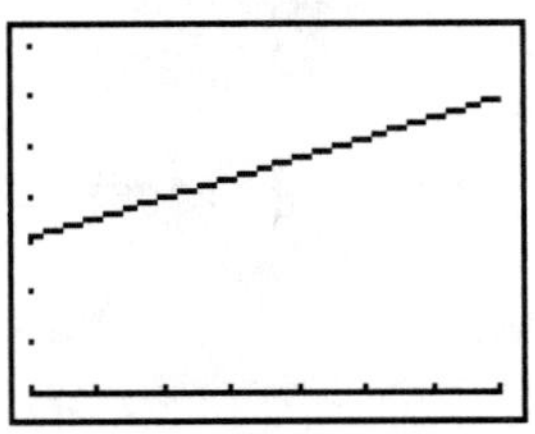

(B) 7.22 inches

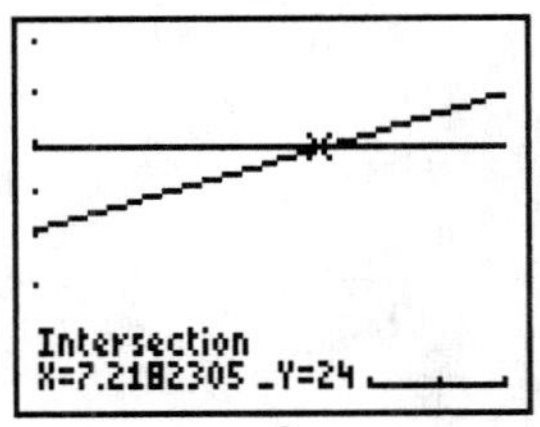

89. (A) Following the hint, we draw AC. Then, since the central angle in a circle subtended by an arc is twice any inscribed angle subtended by the same arc, angle ACB has measure 2θ. Thus, $d = r\cdot 2\theta = 2r\theta$

In triangle ECP, $\tan\theta = \frac{x}{r}$, hence $\theta = \tan^{-1}\frac{x}{r}$

$$d = 2r\tan^{-1}\frac{x}{r}$$

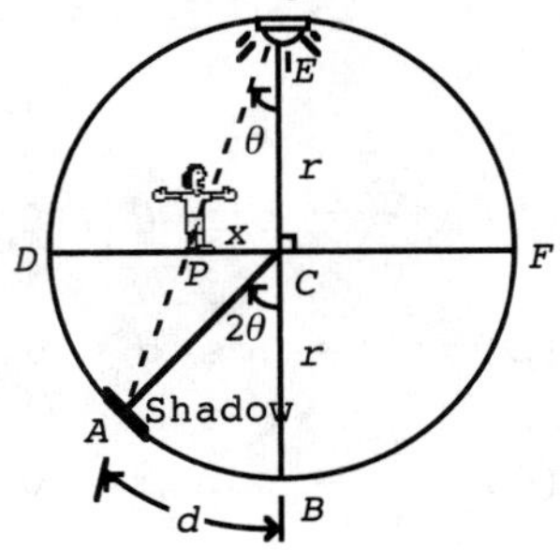

(B) Substituting the given values, we have
$r = 100 \quad x = 40$

$$d = 2\cdot 100\tan^{-1}\frac{40}{100}$$
$$= 200\tan^{-1}\frac{40}{100}$$
$$= 200\tan^{-1}(0.4) \text{ (calculator in radian mode)}$$
$$= 76.10 \text{ feet}$$

CHAPTER 6 REVIEW

1. $\theta = \dfrac{s}{r} = \dfrac{15 \text{ centimeters}}{6 \text{ centimeters}} = \dfrac{15}{6} = 2.5$ radians (6-3)

2. $s = r\theta = (3 \text{ centimeters})(2.5 \text{ radians}) = 7.5$ centimeters (6-3)

3. Solve for α:

$\alpha = 90° - 35.2° = 54.8°$

Solve for a:

$$\cos \beta = \frac{a}{c}$$

$$\cos 35.2° = \frac{a}{20.2}$$

$$a = 20.2 \cos 35.2° = 16.5 \text{ ft}$$

Solve for b:

$$\sin \beta = \frac{b}{c}$$

$$\sin 35.2° = \frac{b}{20.2}$$

$$b = 20.2 \sin 35.2° = 11.6 \text{ ft}$$

(6-3)

4. (A)

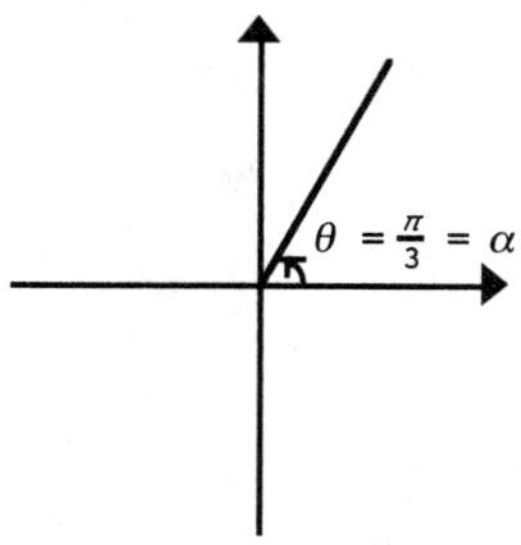

$\alpha = \dfrac{\pi}{3}$

(B)

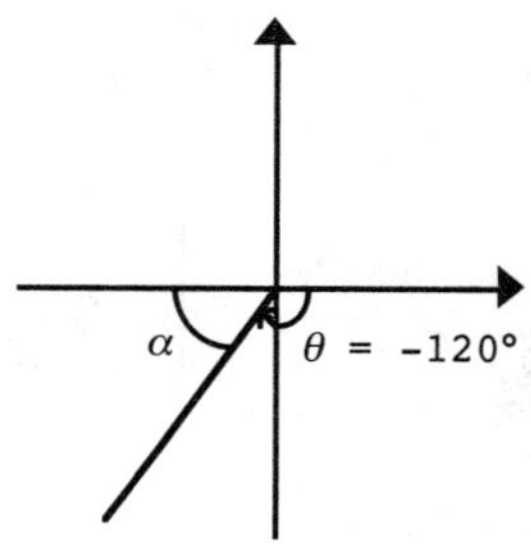

$\alpha = 180° + (-120°) = 60°$

(C)

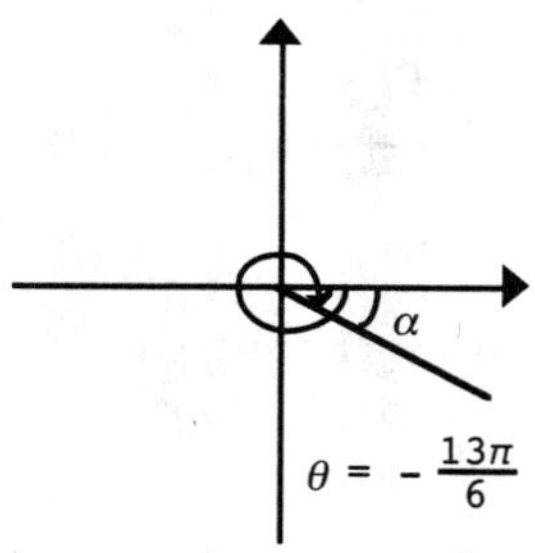

$\alpha = \left|2\pi - \dfrac{13\pi}{6}\right| = \dfrac{\pi}{6}$

(D)

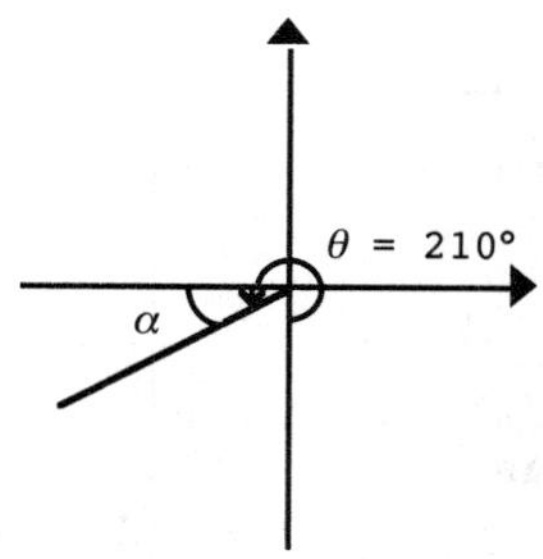

$a = 210° - 180° = 30°$ (6-4)

5. (A) $\dfrac{b}{r} = \sin \theta < 0$ if $b < 0$. This occurs in quadrants III, IV.

(B) $\dfrac{a}{r} = \cos \theta < 0$ if $a < 0$. This occurs in quadrants II, III.

(C) $\dfrac{b}{a} = \tan \theta < 0$ if a and b have opposite signs. This occurs in quadrants II, IV.

(6-2)

6.

$$a^2 + b^2 = r^2$$
$$4^2 + (-3)^2 = r^2$$
$$25 = r^2$$
$$r = 5$$
$$\sin\theta = \frac{b}{r} = \frac{-3}{5} = -\frac{3}{5}$$
$$\sec\theta = \frac{r}{a} = \frac{5}{4}$$
$$\cot\theta = \frac{a}{b} = \frac{4}{-3} = -\frac{4}{3}$$

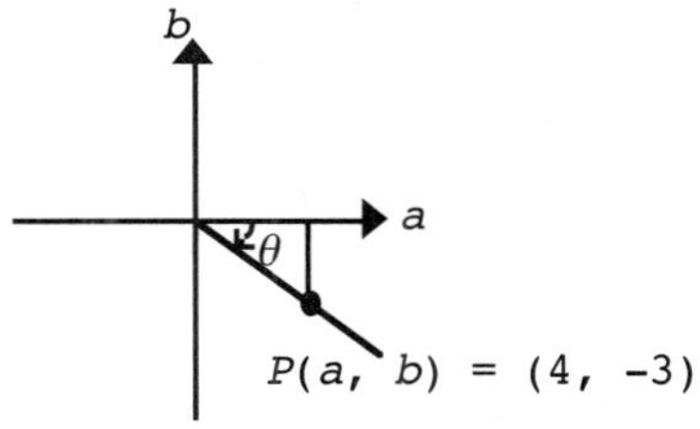

(6-4)

7.

$\theta°$	θ rad	$\sin\theta$	$\cos\theta$	$\tan\theta$	$\csc\theta$	$\sec\theta$	$\cot\theta$
0°	0	0	1	0	ND *	1	ND
30°	$\pi/6$	$1/2$	$\sqrt{3}/2$	$1/\sqrt{3}$	2	$2/\sqrt{3}$	$\sqrt{3}$
45°	$\pi/4$	$1/\sqrt{2}$	$1/\sqrt{2}$	1	$\sqrt{2}$	$\sqrt{2}$	1
60°	$\pi/3$	$\sqrt{3}/2$	$1/2$	$\sqrt{3}$	$2/\sqrt{3}$	2	$1/\sqrt{3}$
90°	$\pi/2$	1	0	ND	1	ND	0
180°	π	0	−1	0	ND	−1	ND
270°	$3\pi/2$	−1	0	ND	−1	ND	0
360°	2π	0	1	0	ND	1	ND

*ND = not defined

(6-1, 6-2)

8. (A) 2π (B) 2π (C) π (6-4)

9. (A) Domain = all real numbers, Range = [−1, 1]

(B) Domain is set of all real numbers except $x = \frac{2k+1}{2}\pi$, k an integer,

Range = all real numbers (6-4)

10.

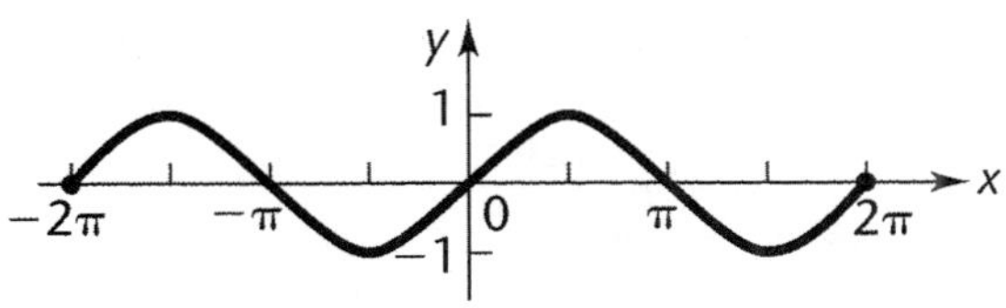

(6-4)

11.

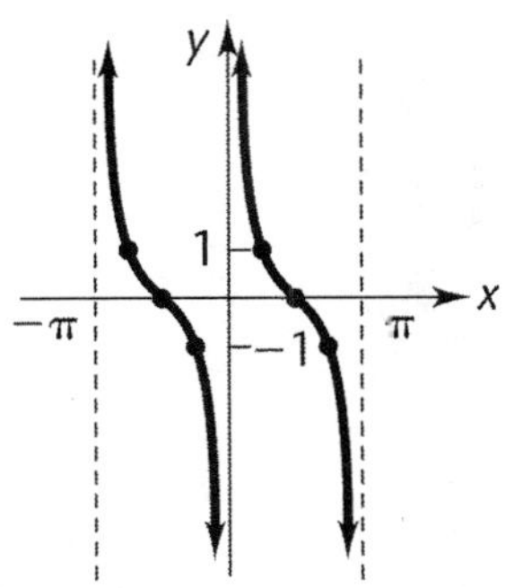

(6-4)

12. The central angle in a circle subtended by an arc of half the length of the radius. (6-1)

13. If the graph of $y = \sin x$ is shifted $\frac{\pi}{2}$ units to the left, the result will be the graph of $y = \cos x$. (6-4)

14. $\theta_D = \frac{180°}{\pi}\theta_R = \frac{180°}{\pi}(1.37) = 78.50°$ (6-1)

15. Solve for β:

$\tan \beta = \frac{b}{a}$

$\tan \beta = \frac{13.3}{15.7}$

$\beta = \tan^{-1} \frac{13.3}{15.7}$

$= 40.3°$

Solve for α:

$\alpha = 90° - 40.3° = 49.7°$

Solve for c:

$\sec \beta = \frac{c}{a}$

$\sec 40.3° = \frac{c}{15.7}$

$c = 15.7 \sec 40.3°$

$= 20.6$ cm (6-3)

16. (A) Since $-270° < -210° < -180°$, this is a II quadrant angle.

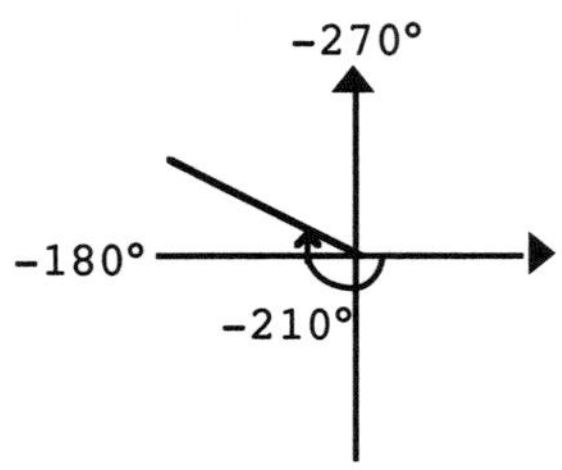

(C) Since $3.14 < 4.2 < 4.71$, this is a III quadrant angle.

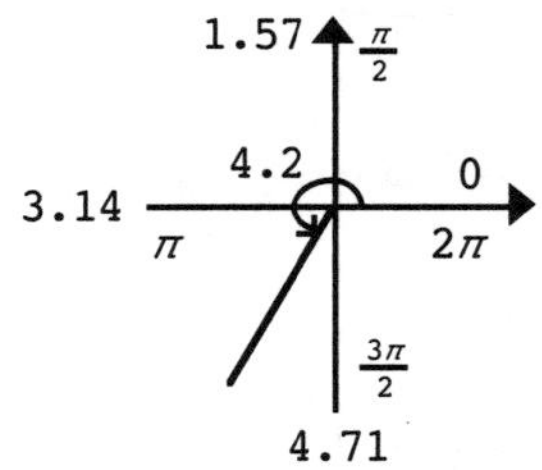

(B) Since $\frac{5\pi}{2}$ is coterminal with $\frac{\pi}{2}$, this is a quadrantal angle. (6-1)

17. (A) Since $-240° + 360° = 120°$, this is coterminal with 120°.
(B) Since $-\frac{7\pi}{6} + 2\pi = \frac{5\pi}{6}$, which is equivalent to 150°, this is not coterminal with 120°.
(C) Since $840° - 2(360°) = 120°$, this is coterminal with 120°. (6-1)

18. (B) and (C), since 3 radians is equivalent to the real number 3, and cosine is periodic with period 2π. (A) is not the same as cos 3, since 3° is equivalent to $\frac{\pi}{180°}3°$, not 3. (6-2)

19. (A) $\frac{b}{a} = \tan x$ is not defined if $a = 0$. This occurs if $\theta = \frac{\pi}{2}, \frac{3\pi}{2}$.
(B) $\frac{a}{b} = \cot x$ is not defined if $b = 0$. This occurs if $\theta = 0, \pi$.
(C) $\frac{r}{b} = \csc x$ is not defined if $b = 0$. This occurs if $\theta = 0, \pi$. (6-4)

20. Since the coordinates of a point on a unit circle are given by $P(a, b) = P(\cos x, \sin x)$, we evaluate $P(\cos(-8.305), \sin(-8.305))$--using a calculator set in radian mode--to obtain $P(-0.436, -0.900)$. Note that $x = -8.305$, since P is moving clockwise. The quadrant in which $P(a, b)$ lies can be determined by the signs of a and b. In this case P is in the third quadrant, since a is negative and b is negative. (6-1, 6-2)

21. $(a, b) = (1, 0) \quad r = 1$

$\tan 0 = \frac{b}{a} = \frac{0}{1} = 0$

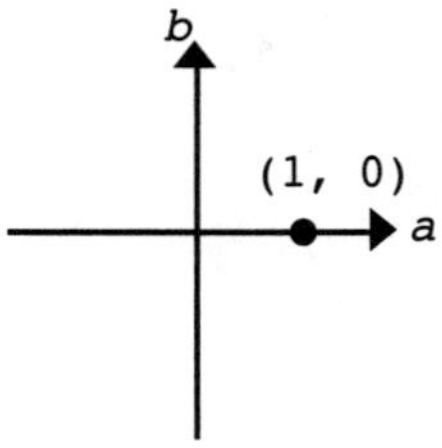

(6-2)

22. $(a, b) = (0, 1) \quad r = 1$

$\sec 90° = \frac{r}{a} = \frac{1}{0}$ Not defined

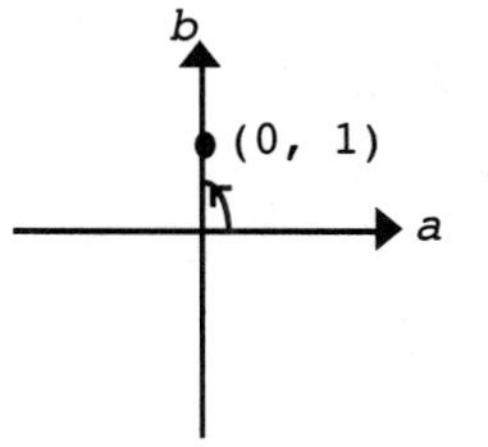

(6-2)

23. $y = \cos^{-1} 1$ is equivalent to

$\cos y = 1 \quad 0 \le y \le \pi$

$y = 0$ (6-6)

24.

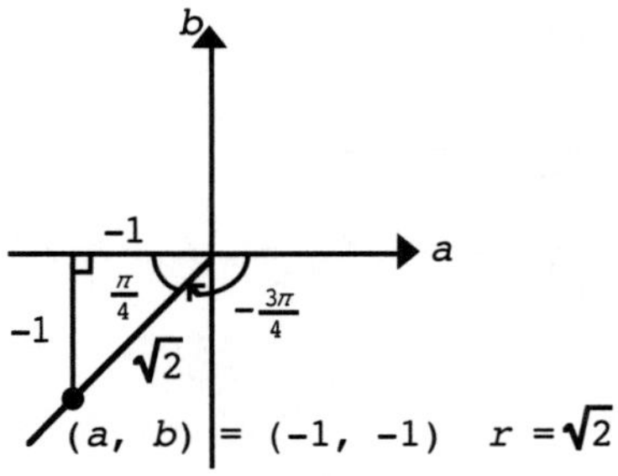

$\cos\left(-\frac{3\pi}{4}\right) = \frac{a}{r} = \frac{-1}{\sqrt{2}} = -\frac{1}{\sqrt{2}} = -\frac{\sqrt{2}}{2}$

(6-2)

25. $y = \sin^{-1} \frac{\sqrt{2}}{2}$ is equivalent to

$\sin y = \frac{\sqrt{2}}{2} \qquad -\frac{\pi}{2} \le y \le \frac{\pi}{2}$

$\sin y = \frac{b}{r} = \frac{\sqrt{2}}{2} \qquad b = \sqrt{2} \quad r = 2$

$a^2 + (\sqrt{2})^2 = 2^2$

$a^2 = 2$

$a = \sqrt{2}$

(positive since y is a quadrant I or IV angle)

$y = \frac{\pi}{4}$

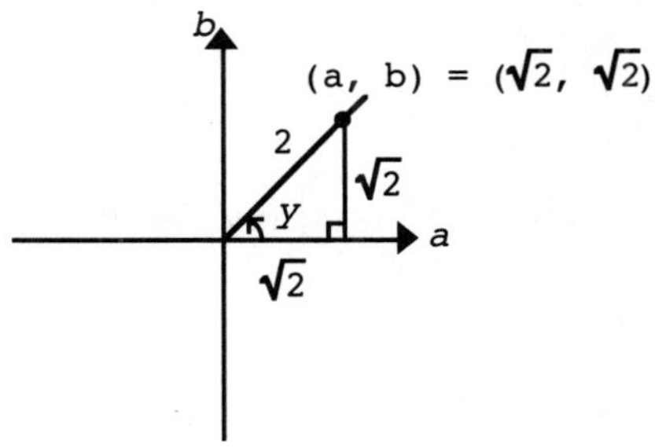

(6-6)

26.

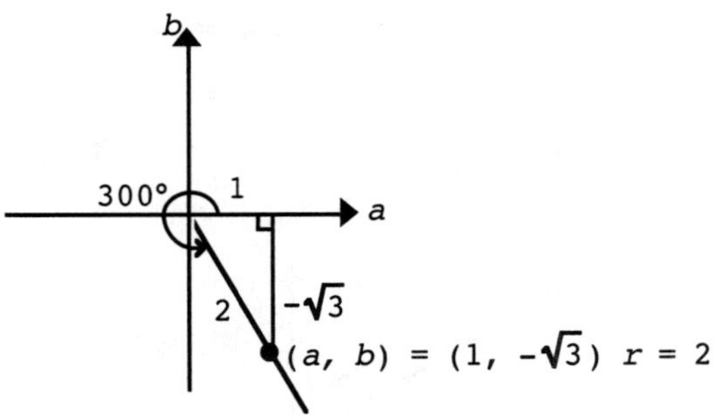

$\csc 300° = \frac{r}{b} = \frac{2}{-\sqrt{3}}$

$= -\frac{2}{\sqrt{3}}$ or $\frac{-2\sqrt{3}}{3}$ (6-2)

27. $y = \arctan \sqrt{3}$ is equivalent to

$\tan y = \sqrt{3} \qquad -\frac{\pi}{2} < y < \frac{\pi}{2}$

$\tan y = \frac{b}{a} = \frac{\sqrt{3}}{1} \qquad b = \sqrt{3} \quad a = 1$

$r^2 = 1^2 + (\sqrt{3})^2$

$r^2 = 4$

$r = 2$

$y = \frac{\pi}{3}$

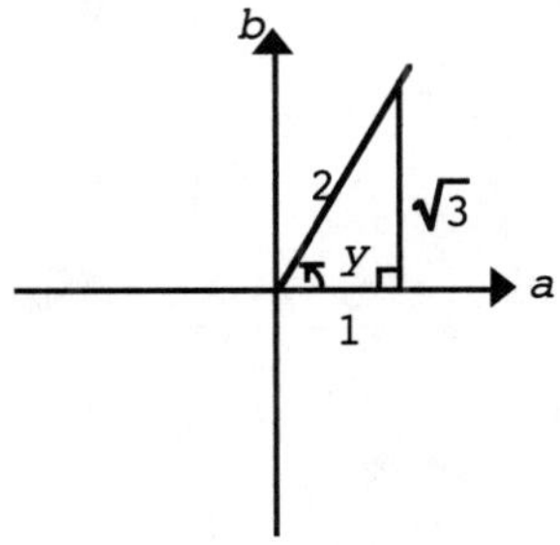

(6-6)

28.

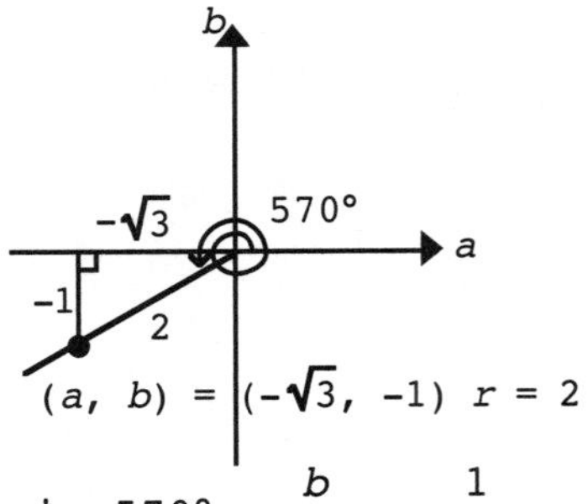

$$\sin 570° = \frac{b}{r} = -\frac{1}{2} \qquad (6\text{-}2)$$

29. $y = \tan^{-1}(-1)$ is equivalent to

$\tan y = -1 \qquad -\frac{\pi}{2} < y < \frac{\pi}{2}$

$\tan y = \frac{b}{a} = \frac{-1}{1} \qquad b = -1 \quad a = 1$

$r^2 = (-1)^2 + 1^2$

$r^2 = 2$

$r = \sqrt{2}$

$y = -\frac{\pi}{4}$

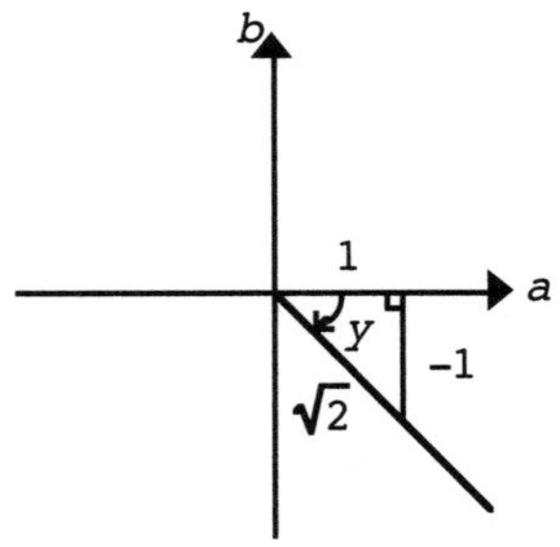

(6-6)

30.

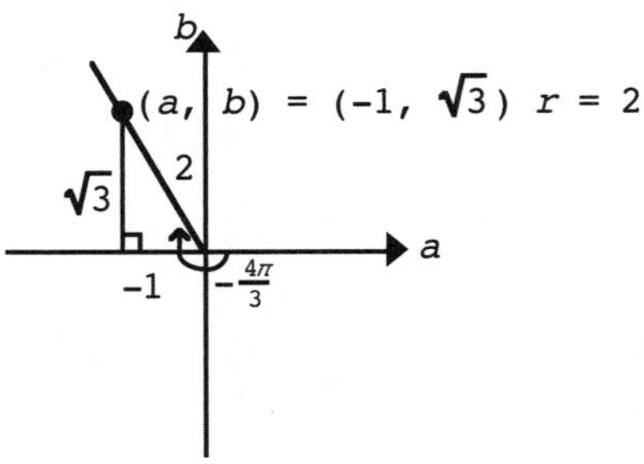

$$\cot\left(-\frac{4\pi}{3}\right) = \frac{a}{b} = \frac{-1}{\sqrt{3}} = -\frac{1}{\sqrt{3}} \text{ or } -\frac{\sqrt{3}}{3}$$

(6-2)

31. $y = \arcsin\left(-\frac{1}{2}\right)$ is equivalent to

$\sin y = -\frac{1}{2} \qquad -\frac{\pi}{2} \le y \le \frac{\pi}{2}$

$\sin y = \frac{b}{r} = \frac{-1}{2} \qquad b = -1 \quad r = 2$

$a = \sqrt{3}$

$y = -\frac{\pi}{6}$

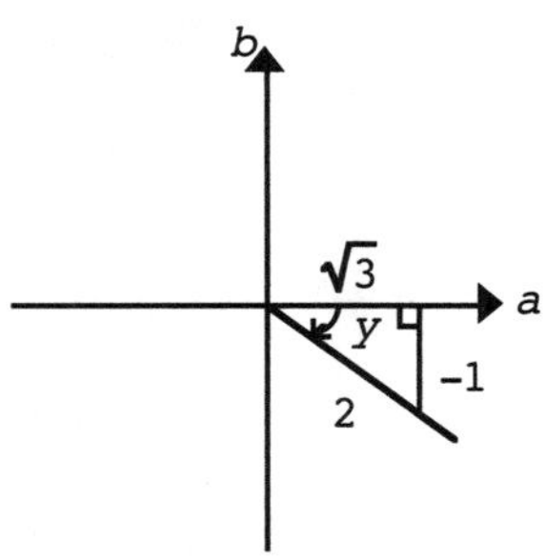

(6-6)

32. $y = \cos^{-1}\left(-\frac{\sqrt{3}}{2}\right)$ is equivalent to

$\cos y = -\frac{\sqrt{3}}{2} \qquad 0 \le y \le \pi$

$\cos y = \frac{a}{r} = \frac{-\sqrt{3}}{2} \qquad a = -\sqrt{3} \quad r = 2$

$b = 1$

$y = \frac{5\pi}{6}$

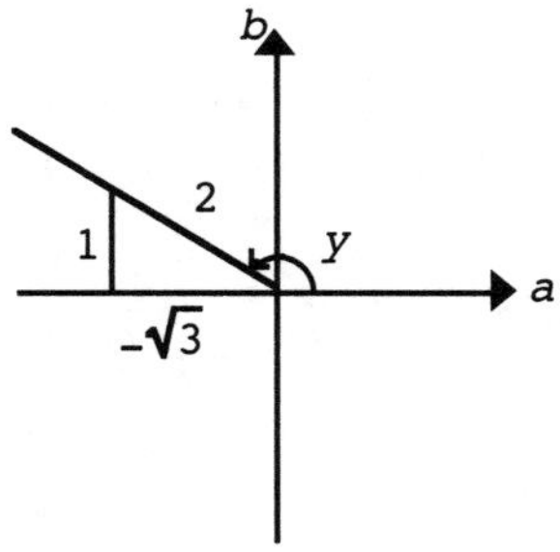

(6-6)

33. $\cos(\cos^{-1} 0.33) = 0.33$ by the cosine-inverse cosine identity. (6-6)

34. Let $y = \tan^{-1}(-1)$, then $\tan y = -1$, $-\frac{\pi}{2} < y < \frac{\pi}{2}$. Using the drawing in problem 29, we have $(a, b) = (1, -1)$, $r = \sqrt{2}$.

$\csc y = \csc[\tan^{-1}(-1)] = \frac{r}{b} = \frac{\sqrt{2}}{-1} = -\sqrt{2}$ (6-6)

35. Let $y = \arccos\left(-\frac{1}{2}\right)$, then $\cos y = -\frac{1}{2}$, $0 \le y \le \pi$. Draw the reference triangle associated with y, then $\sin y = \sin\left[\arccos\left(-\frac{1}{2}\right)\right]$ can be determined directly from the triangle.

$\cos y = \frac{a}{r} = -\frac{1}{2} = \frac{-1}{2}$ $a = -1$ $r = 2$

$b = \sqrt{3}$

$\sin y = \frac{b}{r} = \frac{\sqrt{3}}{2}$

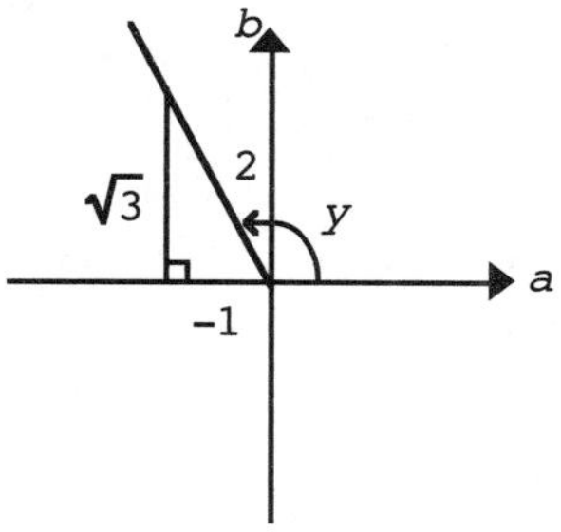

(6-6)

36. Let $y = \sin^{-1}\frac{-4}{5}$, then $\sin y = \frac{-4}{5}$, $-\frac{\pi}{2} \le y \le \frac{\pi}{2}$. Draw the reference triangle associated with y, then $\tan y = \tan\left[\sin^{-1}\left(-\frac{4}{5}\right)\right]$ can be determined directly from the triangle.

$\sin y = \frac{b}{r} = \frac{-4}{5}$ $b = -4$ $r = 5$ $a = 3$

$\tan y = \frac{b}{a} = \frac{-4}{3}$

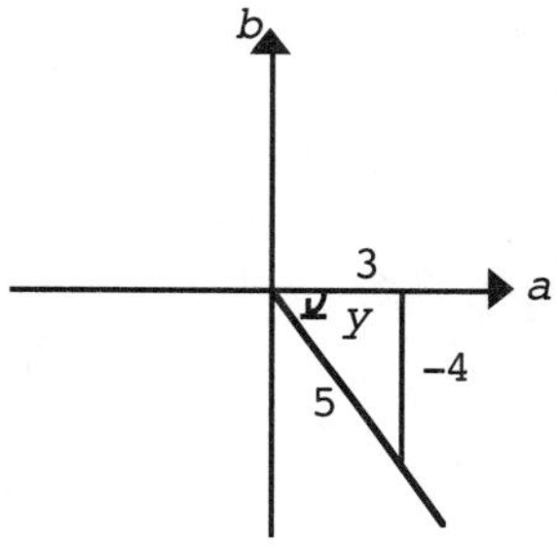

(6-6)

37. 0.4431 (6-2)

38. $\tan\left(93 + \frac{46}{60} + \frac{17}{3600}\right)^\circ = -15.17$ (6-2)

39. $\sec(-2.073) = \frac{1}{\cos(-2.073)} = -2.077$ (6-2)

40. -0.9750 (6-6)

41. arccos x is not defined if $x < -1$ (6-6)

42. 1.557 (6-6)

43. 1.095 (6-6)

44. Since $\tan 1.345 = 4.353 > 1$, $\sin^{-1}(\tan 1.345)$ is not defined. (6-6)

45. (A) $\theta = \arcsin^{-1}\left(-\frac{1}{2}\right)$ is equivalent to

$\sin\theta = -\frac{1}{2}$ $-\frac{\pi}{2} \le \theta \le \frac{\pi}{2}$

$\theta = -\frac{\pi}{6}$

Using the relation $\theta_D = \frac{180^\circ}{\pi \text{ rad}}\theta_R$, we have

$\sin^{-1}\left(-\frac{1}{2}\right) = \frac{180^\circ}{\pi}\cdot\left(-\frac{\pi}{6}\right) = -30^\circ$

reference triangle associated with θ

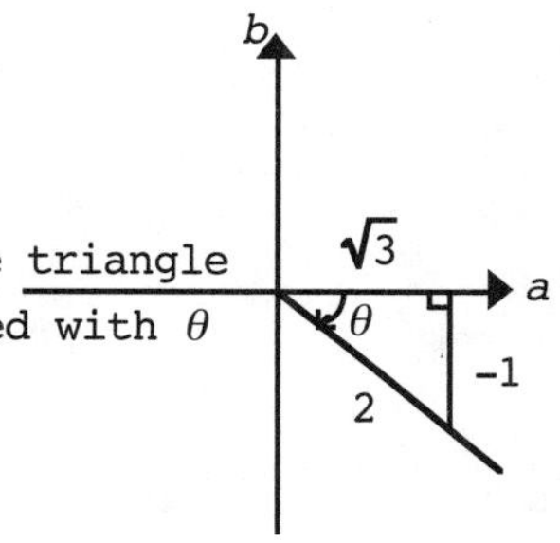

(B) $\theta = \arccos\left(-\frac{1}{2}\right)$ is equivalent to

$$\cos\theta = -\frac{1}{2} \qquad 0 \le \theta \le \pi$$
$$\theta = \frac{2\pi}{3}$$

Using the relation $\theta_D = \frac{180°}{\pi \text{ rad}}\theta_R$, we have

$$\arccos\left(-\frac{1}{2}\right) = \frac{180°}{\pi}\cdot\frac{2\pi}{3} = 120°.$$

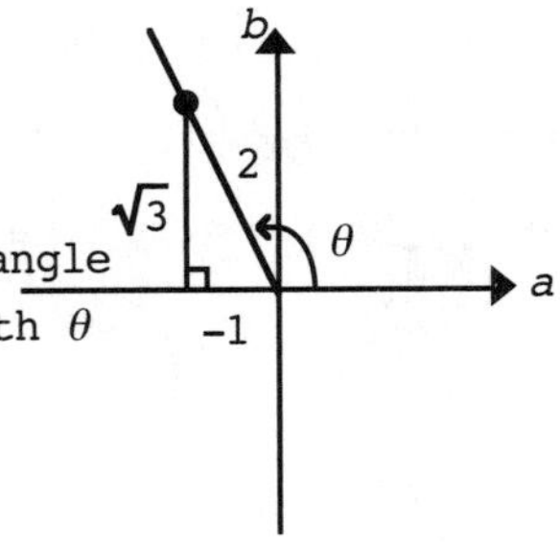

(6-6)

46. Calculator in degree mode: (A) $\Theta = 151.20°$ (B) $\Theta = 82.28°$ (6-6)

47. $\cos^{-1}[\cos(-2)] = 2$ For the identity $\cos^{-1}(\cos x) = x$ to hold, x must be in the restricted domain of the cosine function; that is, $0 \le x \le \pi$. The number -2 is not in the restricted domain. (6-6)

48. Amplitude = $|-2| = 2$

Period = $\frac{2\pi}{\pi} = 2$

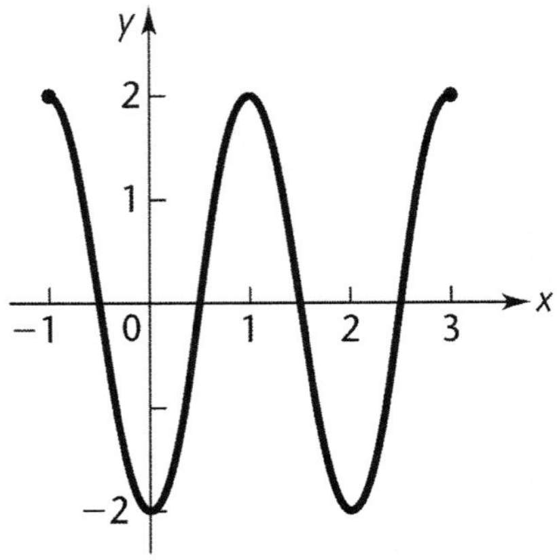

(6-5)

49. $y = -2 + 3\sin\left(\frac{x}{2}\right)$

For the graph of $y = 3\sin\frac{x}{2}$, we note: $A = 3$, $P = 2\pi \div \frac{1}{2} = 4\pi$, phase shift = 0. We graph $y = 3\sin\frac{x}{2}$, then vertically translate the graph down 2 units.

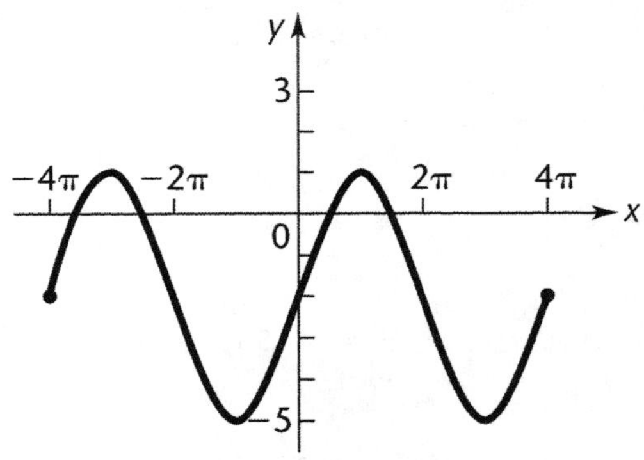

(6-5)

50. $A = 6 \quad P = \pi = 2\pi \div 2$. Hence $B = 2$

$y = 6\cos 2x; \quad -\frac{\pi}{2} \le x \le \pi$ (6-5)

51. $|A| = 0.5 \quad P = 2 = 2\pi \div \pi$. Hence $B = \pi \quad A = -0.5$, since the graph has the form of the standard sine curve turned upside down.
$y = -0.5\sin\pi x; \ -1 \le x \le 2$ (6-5)

52. If the graph of $y = \tan x$ is shifted $\frac{\pi}{2}$ units to the right and reflected in the x axis, the result will be the graph of $y = \cot x$. (6-4)

53. (A) $\sin(-x)\cot(-x) = \sin(-x)\dfrac{\cos(-x)}{\sin(-x)}$ Quotient Identity

$= \cos(-x)$ Algebra

$= \cos x$ Identities for negatives

(B) $\dfrac{\sin^2 x}{1-\sin^2 x} = \dfrac{\sin^2 x}{\cos^2 x}$ Pythagorean Identity

$= \left(\dfrac{\sin x}{\cos x}\right)^2$ Algebra

$= \tan^2 x$ Quotient Identity (6-4)

54. $y = 3\sin\left(\dfrac{x}{2} + \dfrac{\pi}{2}\right)$

$A = 3$

Solve $\dfrac{x}{2} + \dfrac{\pi}{2} = 0$ $\dfrac{x}{2} + \dfrac{\pi}{2} = 2\pi$

$\dfrac{x}{2} = -\dfrac{\pi}{2}$ $\dfrac{x}{2} = -\dfrac{\pi}{2} + 2\pi$

$x = -\pi$ $x = -\pi + 4\pi$

Phase shift   Period

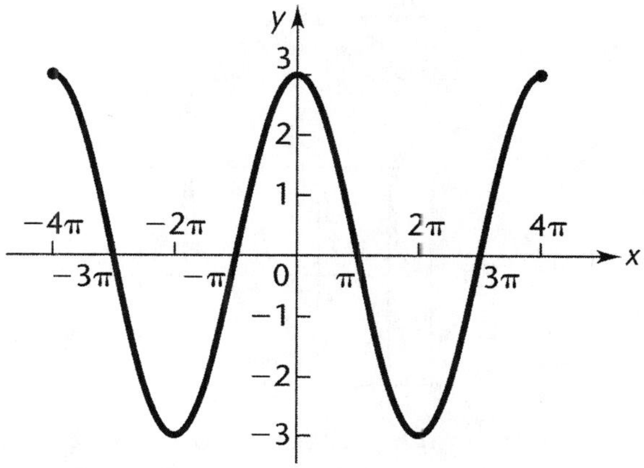

The graph completes one full cycle as x varies over the interval $[-\pi, 3\pi]$.

(6-5)

55. $y = -2\cos\left(\dfrac{\pi}{2}x - \dfrac{\pi}{4}\right)$

amplitude $= |A| = |-2| = 2$

Solve $\dfrac{\pi}{2}x - \dfrac{\pi}{4} = 0$ $\dfrac{\pi}{2}x - \dfrac{\pi}{4} = 2\pi$

$\dfrac{\pi}{2}x = \dfrac{\pi}{4}$ $\dfrac{\pi}{2}x = \dfrac{\pi}{4} + 2\pi$

$x = \dfrac{\pi}{4} \div \dfrac{\pi}{2}$ $x = \left(\dfrac{\pi}{4} + 2\pi\right) \div \dfrac{\pi}{2}$

$x = \dfrac{1}{2}$ $x = \dfrac{1}{2} + 4$

Phase shift   Period = 4   (6-5)

56. Domain = $[-1, 1]$

Range = $[0, \pi]$

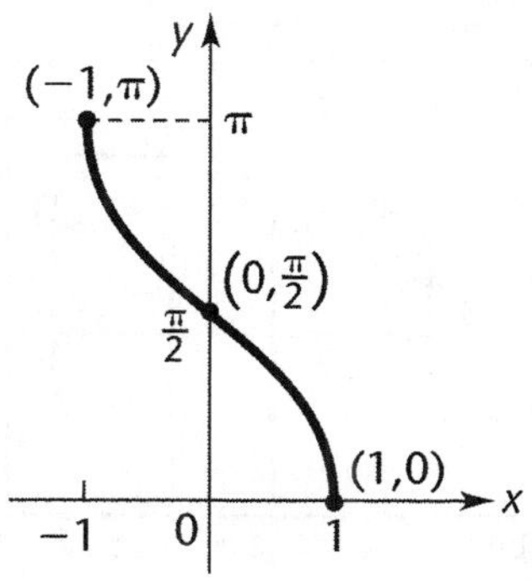

(6-6)

57. Here is the graph of $y = \dfrac{1}{1 + \tan^2 x}$ in a graphing calculator.

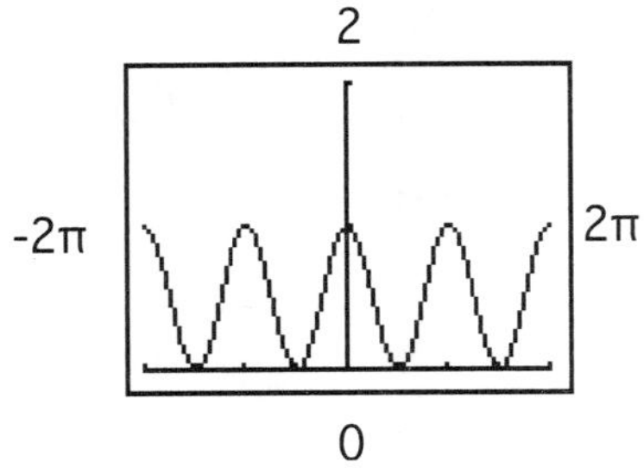

The graph has amplitude $\dfrac{1 - 0}{2} = \dfrac{1}{2}$ and period π. It appears to be the graph of $y = \dfrac{1}{2}\cos 2x$ shifted up $\dfrac{1}{2}$ unit, that is, $y = \dfrac{1}{2}\cos 2x + \dfrac{1}{2}$. (6-5)

58. (A) Here is the graph of $y = \dfrac{2\sin^2 x}{\sin 2x}$ in a graphing calculator.

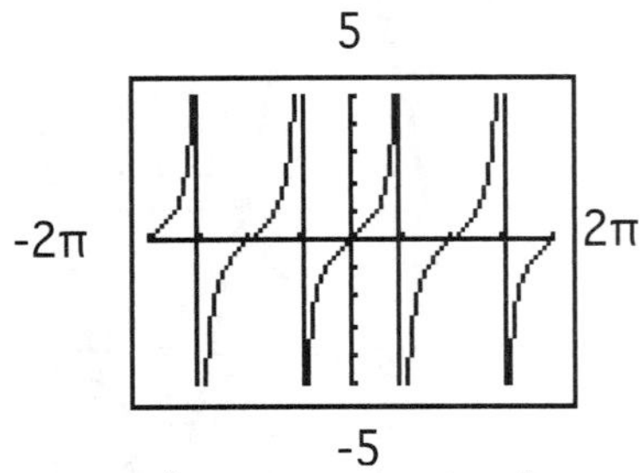

The graph appears to have the form $y = A \tan Bx$. Since the period is π, $B = 1$. The graph of $y = A \tan x$ shown appears to pass through $\left(\dfrac{\pi}{4}, 1\right)$, thus

$$1 = A \tan \frac{\pi}{4}$$
$$1 = A$$

The equation of the graph can be written $y = \tan x$.

(B) Here is the graph of $y = \dfrac{2\cos^2 x}{\sin 2x}$ in a graphing calculator.

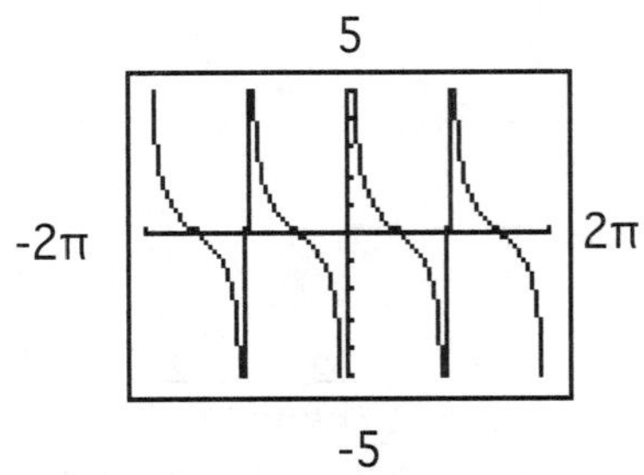

The graph appears to have the form $y = A \cot Bx$. Since the period is π, $B = 1$. The graph of $y = A \cot x$ shown appears to pass through $\left(\dfrac{\pi}{4}, 1\right)$, thus

$$1 = A \cot \frac{\pi}{4}$$
$$1 = A$$

The equation of the graph can be written $y = \cot x$. (6-5)

59. (A) $f(-x) = \dfrac{1}{1 + \tan^2(-x)}$
$= \dfrac{1}{1 + (-\tan x)^2}$
$= \dfrac{1}{1 + \tan^2 x}$
$= f(x)$

$f(x)$ is even.

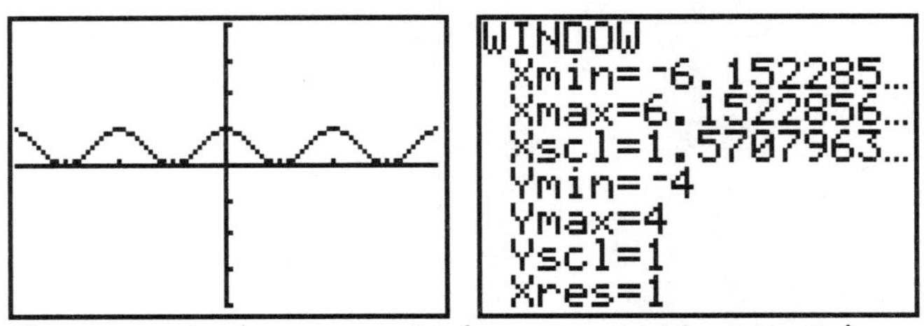

The graph is symmetric about the y-axis so $f(x)$ is even.

(B) $g(-x) = \dfrac{1}{1 + \tan(-x)}$
$= \dfrac{1}{1 - \tan x}$

This is not equal to $g(x)$ nor is it equal to $-g(x)$ which would be $\dfrac{1}{-1 - \tan x}$. So $g(x)$ is neither even nor odd.

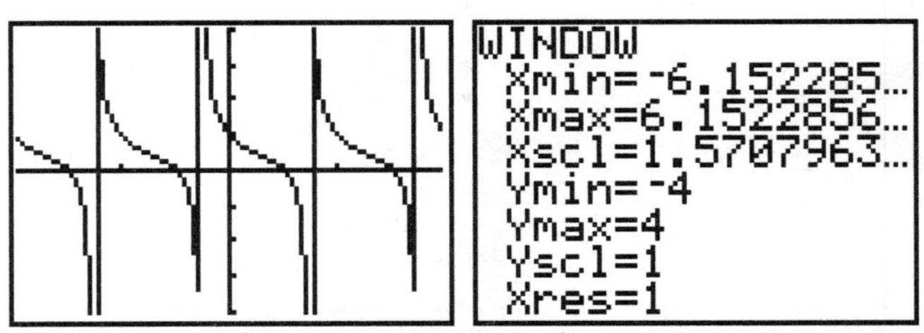

The graph is not symmetric about the y-axis nor is is symmetric about the origin. (6-4)

60. False.

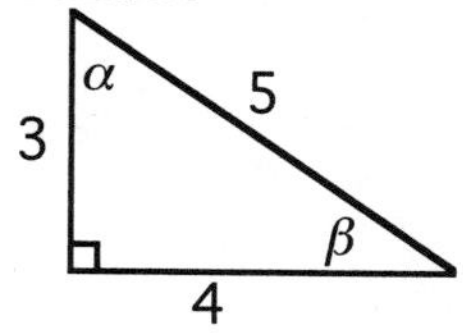

According to the diagram,

$\sin \alpha = \frac{\text{Opp}}{\text{Hyp}} = \frac{4}{5}$ while $\csc \beta = \frac{\text{Hyp}}{\text{Opp}} = \frac{5}{4}$.

(6-3)

61. True. The two acute angles in a right triangle add up to 90° so if they're equal, both are 45°.

$\sin 45° = \frac{1}{\sqrt{2}} \approx 0.71$ $\qquad \cos 45° = \frac{1}{\sqrt{2}} \approx 0.71$

$\tan 45° = 1$ $\qquad \cot 45° = 1$

$\sec 45° = \sqrt{2} \approx 1.4$ $\qquad \csc 45° = \sqrt{2} \approx 1.4$

(6-3)

62. (A) Since $\theta_R = \frac{s}{r}$ and r = radius of circle = distance of A from center = 8,

we have $\theta_R = \frac{20}{8} = 2.5$ radians.

(B) Since $\sin \theta = \frac{b}{r}$ and $\cos \theta = \frac{a}{r}$, we can write

$a = r \cos \theta = 8 \cos 2.5 = -6.41$ (calculator in radian mode)

$b = r \sin \theta = 8 \sin 2.5 = 4.79$

$(a, b) = (-6.41, 4.79)$

(6-1, 6-2)

63. (A) $\cos x = \frac{a}{r} = -\frac{1}{2} = \frac{-1}{2}$. a is negative in the II and III quadrants. The least positive x is associated with a $\frac{\pi}{3}$ reference triangle in the II quadrant as drawn: $x = \frac{2\pi}{3}$

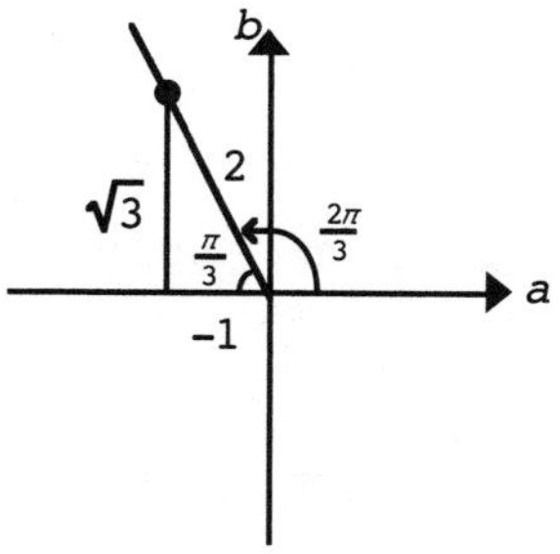

(B) $\csc x = -\sqrt{2} = \frac{\sqrt{2}}{-1} = \frac{r}{b}$, b is negative in the III and IV quadrants. The least positive x is associated with a $\frac{\pi}{4}$ reference triangle in the III quadrant as drawn: $x = \frac{5\pi}{4}$

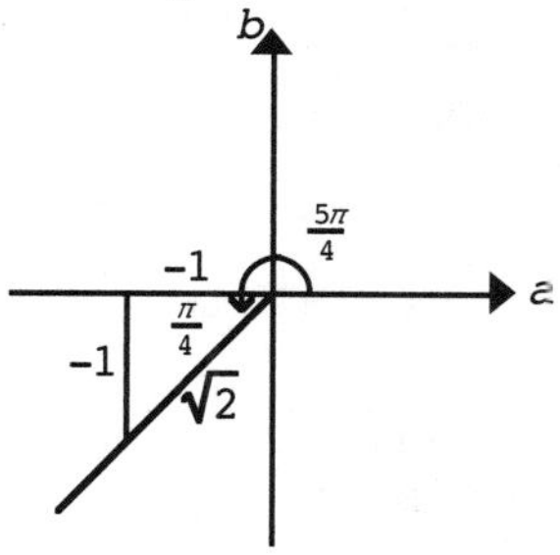

(6-2)

64. The dashed curve is the graph of $y = \cos x$. The solid curve is the required graph of $y = \sec x$, $-\frac{\pi}{2} < x < \frac{3\pi}{2}$

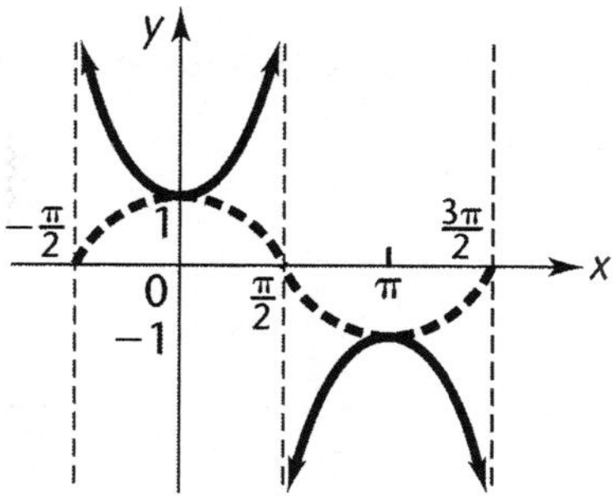

(6-4)

65.

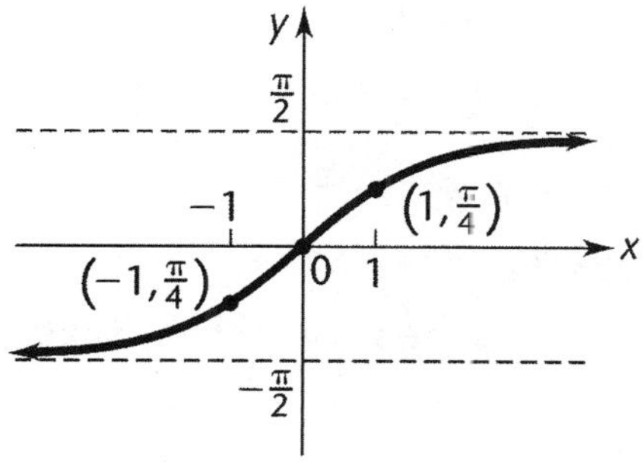

Domain = all real numbers

Range = $\left(-\frac{\pi}{2}, \frac{\pi}{2}\right)$

(6-6)

66. $y = -5\tan\left(\pi x + \frac{\pi}{2}\right)$

Solve $\pi x + \frac{\pi}{2} = 0$ $\qquad$ $\pi x + \frac{\pi}{2} = \pi$

$\pi x = -\frac{\pi}{2}$ $\qquad$ $\pi x = -\frac{\pi}{2} + \pi$

$x = -\frac{1}{2}x$ $\qquad$ $= -\frac{1}{2} + 1$

Phase Shift $= -\frac{1}{2}$ $\qquad$ Period $P = 1$

(6-5)

67. One cycle of $y = 3\csc\left(\frac{x}{2} - \frac{\pi}{4}\right)$ is completed as $\frac{x}{2} - \frac{\pi}{4}$ varies from 0 to 2π. Solve each equation for x:

$\frac{x}{2} - \frac{\pi}{4} = 0$ $\qquad$ $\frac{x}{2} - \frac{\pi}{4} = 2\pi$

$\frac{x}{2} = \frac{\pi}{4}$ $\qquad$ $\frac{x}{2} = \frac{\pi}{4} + 2\pi$

$x = \frac{\pi}{2}$ $\qquad$ $x = \frac{\pi}{2} + 4\pi$

Phase shift $= \frac{\pi}{2}$ $\qquad$ Period $= 4\pi$ (6-5)

68. From the figure, it should be clear that $\cos(-x) = \cos x$ (P and P_1 have the same x-coordinate)
$\sin(-x) = -\sin x$ (P and P_1 have opposite y-coordinates)

Therefore $\tan(-x) = \frac{\sin(-x)}{\cos(-x)} = \frac{-\sin x}{\cos x}$
$= -\tan x$

It follows that the graph of
(A) sine has origin symmetry
(B) cosine has y axis symmetry
(C) tangent has origin symmetry

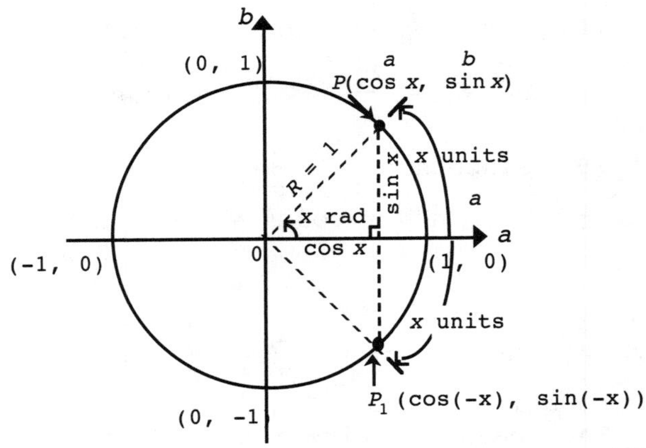

(6-4)

69. Let $y = \sin^{-1} x$. Then $x = \sin y$, $-\frac{\pi}{2} \le y \le \frac{\pi}{2}$. If $\sin y = \frac{b}{r} = \frac{x}{1}$, then let

$b = x$, $r = 1$
$a^2 + x^2 = 1^2$
$a^2 = 1 - x^2$
$a = \sqrt{1 - x^2}$

We choose the positive sign because y is a I or IV quadrant angle.

$\sec y = \sec(\sin^{-1} x) = \frac{r}{a} = \frac{1}{\sqrt{1 - x^2}}$ (6-6)

70. For each case, the number is not in the domain of the function and an error message of some type will appear. (6-2, 6-6)

71. $|A| = 2$. Hence $A = 2$ or -2. The graph completes one full cycle as x varies over the interval $\left[-\frac{5}{4}, \frac{3}{4}\right]$. Since $-\frac{C}{B}$ is required between -1 and 0, we cannot simply set $-\frac{C}{B} = -\frac{5}{4}$. We must (mentally) extend the curve so that the (extended) graph is that of a standard sine curve that completes one full cycle as x varies over the interval $\left[-\frac{1}{4}, \frac{7}{4}\right]$. Hence

$A = 2$

$-\frac{C}{B} = -\frac{1}{4}$ $\qquad$ $-\frac{C}{B} + \frac{2\pi}{B} = \frac{7}{4}$

$C = \frac{B}{4}$ $\qquad$ $\frac{2\pi}{B} = 2$ $\qquad$ $B = \frac{2\pi}{2} = \pi$ $\qquad$ $C = \frac{\pi}{4}$

The equation is then

$$y = A\sin(Bx + C)$$

$$y = 2\sin\left(\pi x + \frac{\pi}{4}\right) \qquad (6\text{-}5)$$

72. Here is the graph of $y = 1.2\sin 2x + 1.6\cos 2x$ in a graphing calculator.

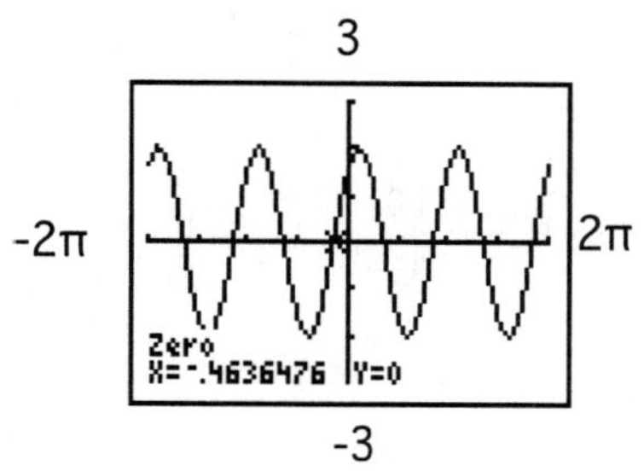

It appears that this is a sine curve shifted to the left, with $A = 2$ and, since $P = \frac{2\pi}{B}$ and P appears to be π, $B = \frac{2\pi}{P} = \frac{2\pi}{\pi} = 2$. From the graphing calculator, we find that the x intercept closest to the origin, to three decimal places, is -0.464. To find C, substitute $B = 2$ and $x = -0.464$ into the phase-shift formula $x = -\frac{C}{B}$ and solve for C.

$$x = -\frac{C}{B}$$

$$-0.464 = -\frac{C}{2}$$

$$C = 0.928$$

The equation is thus $y = 2\sin(2x + 0.928)$. (6-5)

73. (A)

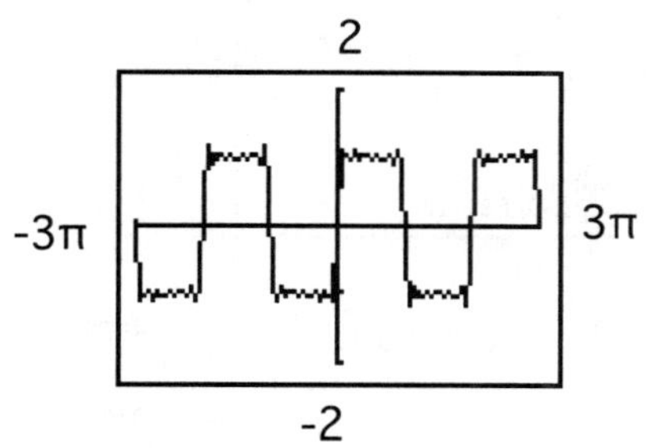

(B)

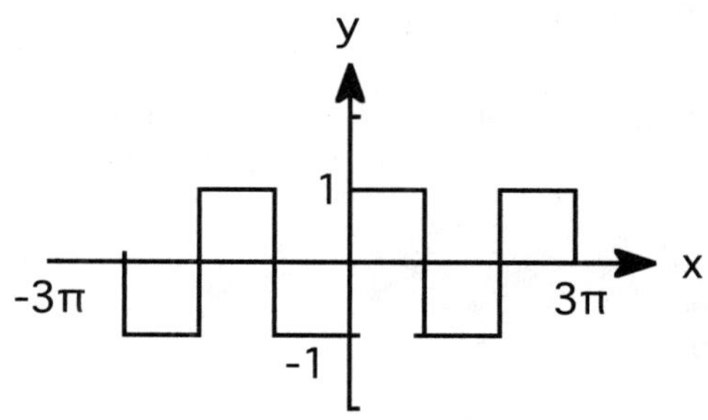

(6-5)

74. In one year the line sweeps out one full revolution, or 2π radians in 365 days. In 73 days the line sweeps out $\frac{73}{365}$ of one full revolution, or $\frac{73}{365} \cdot 2\pi = \frac{2\pi}{5}$ radians. (6-1)

75. Sketch a figure.

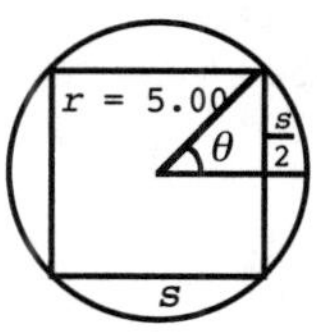

From geometry we know that

$$\theta = \frac{1}{8}(360°) = 45°$$

$$\frac{s}{2} = r\sin 45° = r\frac{1}{\sqrt{2}}$$

Hence $P = 4s = 8\left(\frac{s}{2}\right) = 8r\left(\frac{1}{\sqrt{2}}\right) = 8(5.00)\left(\frac{1}{\sqrt{2}}\right) = 28.3$ cm (6-2)

76. $d = 40$ feet, so $c = \pi d = 40\pi$ feet
In 80 revolutions, the wheel turns a total of $80 \times 2\pi$ radians $= 160\pi$ radians.
$$\frac{160\pi \text{ rad}}{1 \text{ min}} \times \frac{1 \text{ min}}{60 \text{ sec}} = 8.38\frac{\text{rad}}{\text{sec}}$$
In 80 revolutions, the linear distance traveled is $80 \times 40\pi$ feet $= 3200\pi$ feet.
$$\frac{3200\pi \text{ feet}}{1 \text{ min}} \times \frac{1 \text{ min}}{60 \text{ sec}} = 167.55\frac{\text{ft}}{\text{sec}}$$
(*6-1*)

77. When $t = 0$, $I = 30$. Hence $30 = A \cos B(0)$, that is, $A = 30$. Since the period is $\frac{1}{60}$ second, $\frac{2\pi}{B} = \frac{1}{60}$, $B = 2\pi(60) = 120\pi$. Hence the equation is $I = 30 \cos 120\pi t$.
(*6-5*)

78.

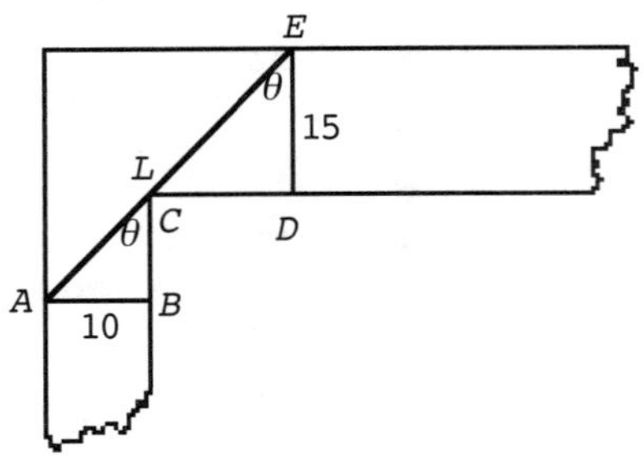

(A) In the figure as labeled, $L = AC + CE$.
In triangle ABC, since $\sin\theta = \frac{10}{AC}$, $AC \sin\theta = 10$,
$$AC = \frac{10}{\sin\theta} = 10 \csc\theta$$
In triangle CDE, since $\cos\theta = \frac{15}{CE}$, $CE \cos\theta = 15$,
$$CE = \frac{15}{\cos\theta} = 15 \sec\theta$$
Then $L = AC + CE = 10 \csc\theta + 15 \sec\theta \qquad 0 < \theta < \frac{\pi}{2}$

(B)

θ radians	0.4	0.5	0.6	0.7	0.8	0.9	1.0
L feet	42.0	38.0	35.9	35.1	35.5	36.9	39.6

From the table, the shortest distance L to the nearest foot, is 35 feet. This is the length of the longest log that can make the corner.

(C) Here is a graph of
$L = 10 \csc\theta + 15 \sec\theta$
from a graphing calculator.

The minimum value of L is shown as 35.1 feet, to one decimal place.

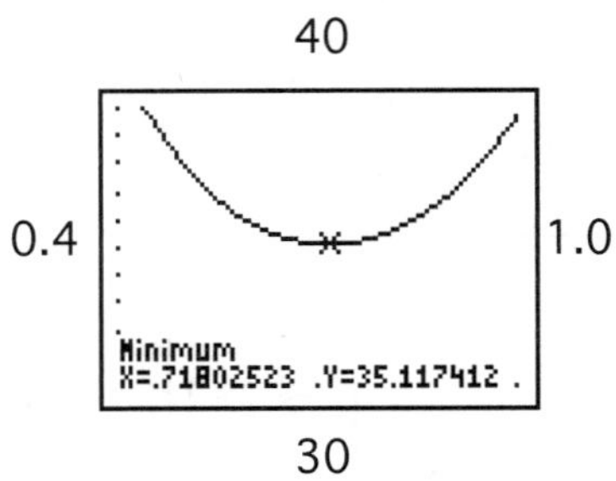

(D) As $\theta \to 0$, $L = 10 \csc\theta + 15 \sec\theta$ approaches an asymptote of $\csc\theta$;
L increases without bound.
As $\theta \to \frac{\pi}{2}$, $L = 10 \csc\theta + 15 \sec\theta$ approaches an asymptote of $\sec\theta$;
L increases without bound.
(*6-2, 6-3*)

79. (A) $|A| = \frac{R_{\max} - R_{\min}}{2} = \frac{7 - 1}{2} = 3$. Hence $A = 3$ or -3. The graph appears to be shifted up from a graph of an upside down cosine curve that completes one full cycle as t varies over the interval $[0, 12]$. Thus $A = -3$.

Then $P = 12 = 2\pi \cdot \frac{6}{\pi} = 2\pi \div \frac{\pi}{6}$. Hence $B = \frac{\pi}{6}$.

$$k = \frac{R_{\max} + R_{\min}}{2} = \frac{7 + 1}{2} = 4.$$

Thus $R(t) = 4 - 3 \cos \frac{\pi}{6}t$.

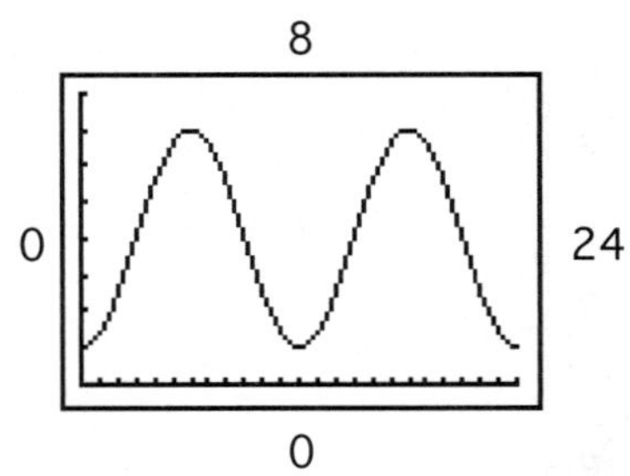

(B) The graph shows the seasonal changes in soft drink consumption. Most is consumed in August and the least in February.
(*6-5*)

80. (A)

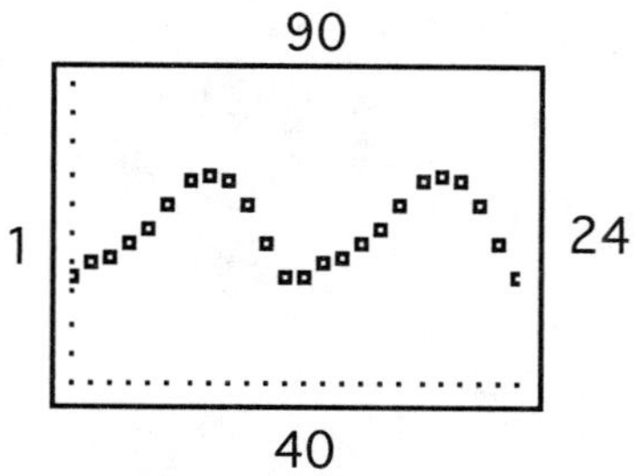

(B) $|A| = \dfrac{y_{max} - y_{min}}{2} = \dfrac{75 - 58}{2} = 8.5$. Hence $A = 8.5$ or -8.5.

$k = \dfrac{y_{max} + y_{min}}{2} = \dfrac{75 + 58}{2} = 66.5$. $P = 12 = 2\pi \cdot \dfrac{6}{\pi} = 2\pi \div \dfrac{\pi}{6}$. Hence $B = \dfrac{\pi}{6}$.

The x intercept closest to the origin is estimated from the graph as 4.5.

To find C, substitute $B = \dfrac{\pi}{6}$ and $x = 4.5$ into the phase shift formula $x = -\dfrac{C}{B}$ and solve for C.

$$x = -\frac{C}{B}$$

$$4.5 = -C \div \frac{\pi}{6}$$

$$C = -4.5\left(\frac{\pi}{6}\right)$$

$$C = -2.4$$

With this value of C, the graph is seen to be shifted up from the graph of a standard sine curve, thus $A = 8.5$. The equation is thus

$y = 66.5 + 8.5 \sin\left(\dfrac{\pi}{6}x - 2.4\right)$.

(C)

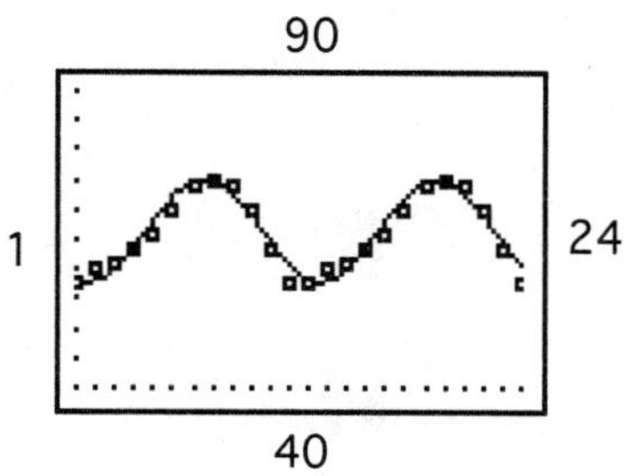

(D)

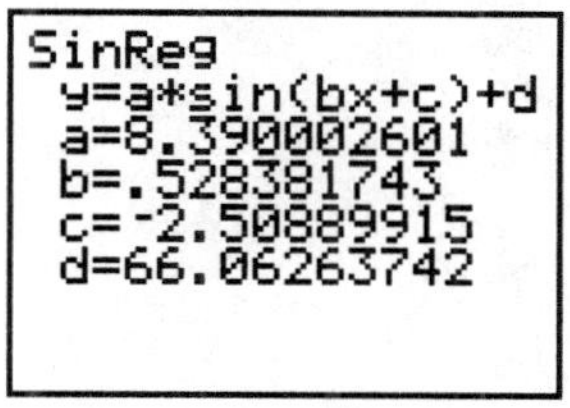

Compute the regression equation.

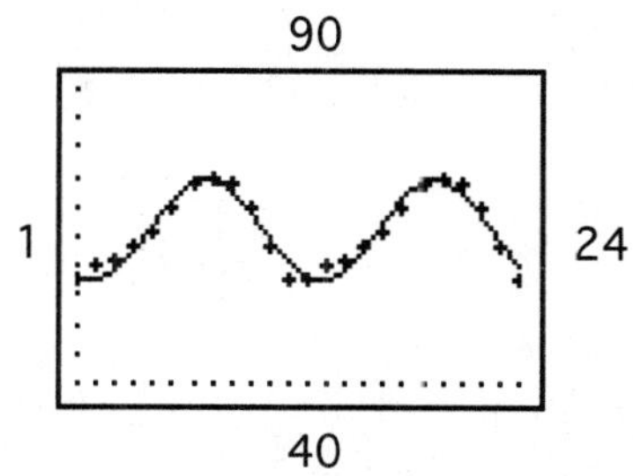

Graph the data and the regression equation.

(6-5)

CHAPTER 7 Trigonometric Identities and Conditional Equations

Section 7-1

1. $\sin\theta\sec\theta = \sin\theta\,\dfrac{1}{\cos\theta}$ Reciprocal Identity

$= \dfrac{\sin\theta}{\cos\theta}$ Algebra

$= \tan\theta$ Quotient Identity

3. $\cot u\sec u\sin u = \dfrac{\cos u}{\sin u}\,\dfrac{1}{\cos u}\sin u$ Reciprocal and Quotient Identities

$= 1$ Algebra

5. $\dfrac{\sin(-x)}{\cos(-x)} = \dfrac{-\sin x}{\cos x}$ Identities for Negatives

$= -\tan x$ Quotient Identity

7. $\dfrac{\tan\alpha\cot\alpha}{\csc\alpha} = \dfrac{\tan\alpha\,\frac{1}{\tan\alpha}}{\csc\alpha}$ Reciprocal Identity

$= \dfrac{1}{\csc\alpha}$ Algebra

$= \dfrac{1}{\frac{1}{\sin\alpha}}$ Reciprocal Identity

$= \sin\alpha$ Algebra

9. $\csc u(\cos u + \sin u) = \dfrac{1}{\sin u}(\cos u + \sin u)$ Reciprocal Identity

$= \dfrac{\cos u}{\sin u} + \dfrac{\sin u}{\sin u}$ Algebra

$= \dfrac{\cos u}{\sin u} + 1$ Algebra

$= \cot u + 1$ Quotient Identity

Key Algebraic Steps:

$$\frac{1}{b}(a + b) = \frac{a}{b} + \frac{b}{b} = \frac{a}{b} + 1$$

11. $\dfrac{\cos x - \sin x}{\sin x\cos x} = \dfrac{\cos x}{\sin x\cos x} - \dfrac{\sin x}{\sin x\cos x}$ Algebra

$= \dfrac{1}{\sin x} - \dfrac{1}{\cos x}$ Algebra

$= \csc x - \sec x$ Reciprocal Identities

13. $\dfrac{\sin^2 t}{\cos t} + \cos t = \dfrac{\sin^2 t}{\cos t} + \dfrac{\cos t}{1}$ Algebra

$= \dfrac{\sin^2 t}{\cos t} + \dfrac{\cos^2 t}{\cos t}$ Algebra

$= \dfrac{\sin^2 t + \cos^2 t}{\cos t}$ Algebra

$= \dfrac{1}{\cos t}$ Pythagorean Identity

$= \sec t$ Reciprocal Identity

Key Algebraic Steps:

$$\frac{a^2}{b} + b = \frac{a^2}{b} + \frac{b}{1} = \frac{a^2}{b} + \frac{b^2}{b} = \frac{a^2 + b^2}{b}$$

15. $\dfrac{\cos x}{1 - \sin^2 x} = \dfrac{\cos x}{\sin^2 x + \cos^2 x - \sin^2 x}$ Pythagorean Identity

$= \dfrac{\cos x}{\cos^2 x}$ Algebra

$= \dfrac{1}{\cos x}$ Algebra

$= \sec x$ Reciprocal Identity

17. $(1 - \cos u)(1 + \cos u) = 1 - \cos^2 u$ Algebra

$= \sin^2 u + \cos^2 u - \cos^2 u$ Pythagorean Identity

$= \sin^2 u$ Algebra

19. $1 - 2\sin^2 x = \cos^2 x + \sin^2 x - 2\sin^2 x$ Pythagorean Identity

$= \cos^2 x - \sin^2 x$ Algebra

21. $(\sec t + 1)(\sec t - 1) = \sec^2 t - 1$ Algebra

$= \tan^2 t + 1 - 1$ Pythagorean Identity

$= \tan^2 t$ Algebra

23. $\csc^2 x - \cot^2 x = 1 + \cot^2 x - \cot^2 x$ Pythagorean Identity

$= 1$ Algebra

25. $\dfrac{\cos x + \tan x}{\sin x} = \dfrac{\cos x}{\sin x} + \dfrac{\tan x}{\sin x}$ Algebra

$= \cot x + \dfrac{\frac{\sin x}{\cos x}}{\sin x}$ Quotient Identities

$= \cot x + \dfrac{\sin x}{\cos x} \div \sin x$ Algebra

$= \cot x + \dfrac{\sin x}{\cos x} \cdot \dfrac{1}{\sin x}$ Algebra

$= \cot x + \dfrac{1}{\cos x}$ Algebra

$= \cot x + \sec x$ Reciprocal Identities

Key Algebraic Steps:

$$a + \frac{\frac{b}{c}}{b} = a + \frac{b}{c} \div b = a + \frac{b}{c} \cdot \frac{1}{b} = a + \frac{1}{c}$$

27. Plug in $x = -1$.
Left side: $\sqrt{(-1)^2} = \sqrt{1} = 1$
Right side: -1
The equation is not an identity.

29. Plug in $x = 2$.
Left side: $2^4 + 2^3 - 2^2 - 2 = 18$
Right side: $2^5 + 2^4 - 2^3 - 2^2 = 36$
The equation is not an identity.

31. Plug in $x = \frac{\pi}{4}$.
Left side: $\sin \frac{\pi}{4} \cos \frac{\pi}{4} = \frac{1}{\sqrt{2}} \cdot \frac{1}{\sqrt{2}} = \frac{1}{2}$
Right side: $\tan \frac{\pi}{4} = 1$
The equation is not an identity.

33. Plug in $x = \frac{\pi}{6}$.
Left side: $\cos \frac{\pi}{6} = \frac{\sqrt{3}}{2}$
Right side: $1 - \sin \frac{\pi}{6} = 1 - \frac{1}{2} = \frac{1}{2}$
The equation is not an identity.

35.

2

-π 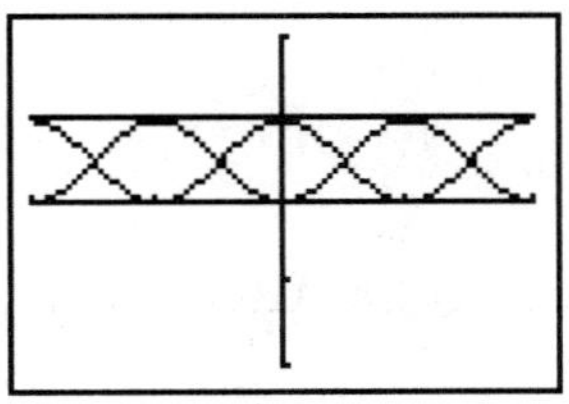 π

-2

The straight line is the graph of $y = \sin^2 x + \cos^2 x$, illustrating the identity $\sin^2 x + \cos^2 x = 1$.

37.

2

-π 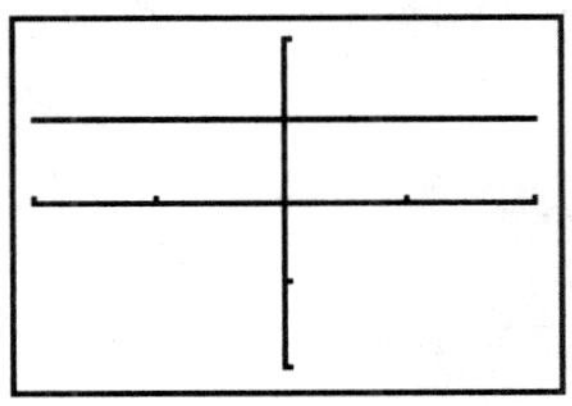 π

-2

Both graphs are the same horizontal line, illustrating the identity
$$\frac{\cos x}{\cot x \sin x} = 1$$

39. Simplify the left side:
$$\frac{x^2 - 9}{x + 3} = \frac{\cancel{(x + 3)}(x - 3)}{\cancel{x + 3}} = x - 3$$
The equation is an identity.

41. Simplify the left side:
$$\sqrt{x^2 + 4x + 4} = \sqrt{(x + 2)(x + 2)} = \sqrt{(x + 2)^2} = |x + 2|$$
The equation is not an identity: the left side must be positive, but the right can be negative. For example, if we plug in $x = -4$, we get
$$\sqrt{(-4)^2 + 4(-4) + 4} \stackrel{?}{=} -4 + 2$$
$$\sqrt{4} \stackrel{?}{=} -2$$
$$2 \stackrel{?}{=} -2$$

43. Plug in $x = \frac{\pi}{4}$: $\sin\frac{\pi}{4} - \cos\frac{\pi}{4} \overset{?}{=} 1$

$$\frac{1}{\sqrt{2}} - \frac{1}{\sqrt{2}} \overset{?}{=} 1$$

obviously this is false so the equation is not an identity.

45. Since $\sin^2 x + \cos^2 x = 1$, we suspect that this equation is not an identity.

Plug in $x = 0$: $\sin^2(0) - \cos^2(0) \overset{?}{=} 1$

$$0 - 1 \overset{?}{=} 1$$

This is false so the equation is not an identity.

> Remember: If you can find even one x-value that 1) makes the equation false, and 2) is in the domain of each side, then the equation is NOT an identity.

47.

$$\frac{1-(\sin x - \cos x)^2}{\sin x} = \frac{1-(\sin^2 x - 2\sin x\cos x + \cos^2 x)}{\sin x} \qquad \text{Algebra}$$

$$= \frac{1-(-2\sin x\cos x + \sin^2 x + \cos^2 x)}{\sin x} \qquad \text{Algebra}$$

$$= \frac{1-(-2\sin x\cos x + 1)}{\sin x} \qquad \text{Pythagorean Identity}$$

$$= \frac{2\sin x\cos x}{\sin x} \qquad \text{Algebra}$$

$$= 2\cos x \qquad \text{Algebra}$$

Key Algebraic Steps:

$$(a-b)^2 = a^2 - 2ab + b^2 = -2ab + a^2 + b^2$$

49.

$$\frac{\cot\theta + 1}{\csc\theta} = \frac{\frac{\cos\theta}{\sin\theta} + 1}{\frac{1}{\sin\theta}} \qquad \text{Reciprocal and Quotient Identities}$$

$$= \frac{\sin\theta\,\frac{\cos\theta}{\sin\theta} + 1\cdot\sin\theta}{\sin\theta\,\frac{1}{\sin\theta}} \qquad \text{Algebra}$$

$$= \cos\theta + \sin\theta \qquad \text{Algebra}$$

Key Algebraic Steps:

$$\frac{\frac{b}{a}+1}{\frac{1}{a}} = \frac{a\frac{b}{a} + 1a}{a\frac{1}{a}} = b + a$$

51.

$$\frac{1+\cos y}{1-\cos y} = \frac{(1+\cos y)}{(1-\cos y)}\frac{(1-\cos y)}{(1-\cos y)} \qquad \text{Algebra}$$

$$= \frac{1-\cos^2 y}{(1-\cos y)^2} \qquad \text{Algebra}$$

$$= \frac{\sin^2 y + \cos^2 y - \cos^2 y}{(1-\cos y)^2} \qquad \text{Pythagorean Identity}$$

$$= \frac{\sin^2 y}{(1-\cos y)^2} \qquad \text{Algebra}$$

53. $\tan^2 x - \sin^2 x = \dfrac{\sin^2 x}{\cos^2 x} - \sin^2 x$ Quotient Identity

$= \sin^2 x \dfrac{1}{\cos^2 x} - \sin^2 x \cdot 1$ Algebra

$= \sin^2 x\left(\dfrac{1}{\cos^2 x} - 1\right)$ Algebra

$= \sin^2 x(\sec^2 x - 1)$ Reciprocal Identity

$= \sin^2 x \tan^2 x$ Pythagorean Identity

$= \tan^2 x \sin^2 x$ Algebra

55. $\dfrac{\csc\theta}{\cot\theta + \tan\theta} = \dfrac{\frac{1}{\sin\theta}}{\frac{\cos\theta}{\sin\theta} + \frac{\sin\theta}{\cos\theta}}$ Reciprocal and Quotient Identities

$= \dfrac{\cos\theta\sin\theta\frac{1}{\sin\theta}}{\cos\theta\sin\theta\frac{\cos\theta}{\sin\theta} + \cos\theta\sin\theta\frac{\sin\theta}{\cos\theta}}$ Algebra

$= \dfrac{\cos\theta}{\cos^2\theta + \sin^2\theta}$ Algebra

$= \dfrac{\cos\theta}{1}$ Pythagorean Identity

$= \cos\theta$

Key Algebraic Steps:

$$\frac{\frac{1}{a}}{\frac{b}{a} + \frac{a}{b}} = \frac{ab \cdot \frac{1}{a}}{ab \cdot \frac{b}{a} + ab \cdot \frac{a}{b}} = \frac{b}{b^2 + a^2}$$

57. $\ln\tan x = \ln\left(\dfrac{\sin x}{\cos x}\right)$ Quotient Identity

$= \ln\sin x - \ln\cos x$ Algebra

59. $\ln\cot x = \ln\left(\dfrac{1}{\tan x}\right)$ Reciprocal Identity

$= \ln 1 - \ln\tan x$ Algebra

$= 0 - \ln\tan x$ Algebra

$= -\ln\tan x$ Algebra

61. $\dfrac{\sec A - 1}{\sec A + 1} = \dfrac{\frac{1}{\cos A} - 1}{\frac{1}{\cos A} + 1}$ Reciprocal Identity

$= \dfrac{\cos A\frac{1}{\cos A} - 1 \cdot \cos A}{\cos A\frac{1}{\cos A} + 1 \cdot \cos A}$ Algebra

$= \dfrac{1 - \cos A}{1 + \cos A}$ Algebra

Key Algebraic Steps:

$$\frac{\frac{1}{a} - 1}{\frac{1}{a} + 1} = \frac{a \cdot \frac{1}{a} - 1 \cdot a}{a \cdot \frac{1}{a} + 1 \cdot a} = \frac{1 - a}{1 + a}$$

63.

$\sin^4 w - \cos^4 w = (\sin^2 w)^2 - (\cos^2 w)^2$	Algebra
$= (\sin^2 w + \cos^2 w)(\sin^2 w - \cos^2 w)$	Algebra
$= 1(\sin^2 w - \cos^2 w)$	Pythagorean Identity
$= \sin^2 w - \cos^2 w$	Algebra
$= 1 - \cos^2 w - \cos^2 w$	Pythagorean Identity
$= 1 - 2\cos^2 w$	Algebra

Key Algebraic Steps:

$$a^4 - b^4 = (a^2)^2 - (b^2)^2 = (a^2 + b^2)(a^2 - b^2)$$

65.

$\sec x - \dfrac{\cos x}{1 + \sin x} = \dfrac{1}{\cos x} - \dfrac{\cos x}{1 + \sin x}$	Reciprocal Identity
$= \dfrac{1 + \sin x - \cos^2 x}{(1 + \sin x)\cos x}$	Algebra
$= \dfrac{\sin^2 x + \cos^2 x + \sin x - \cos^2 x}{(1 + \sin x)\cos x}$	Pythagorean Identity
$= \dfrac{\sin^2 x + \sin x}{(1 + \sin x)\cos x}$	Algebra
$= \dfrac{(1 + \sin x)\sin x}{(1 + \sin x)\cos x}$	Algebra
$= \dfrac{\sin x}{\cos x}$	Algebra
$= \tan x$	Quotient Identity

Key Algebraic Steps:

$$\frac{1}{a} - \frac{a}{1 + b} = \frac{1 + b - a^2}{(1 + b)a}$$

$$\frac{b^2 + b}{(1 + b)a} = \frac{(1 + b)b}{(1 + b)a} = \frac{b}{a}$$

67.

$\dfrac{\cos^2 z - 3\cos z + 2}{\sin^2 z} = \dfrac{\cos^2 z - 3\cos z + 2}{\sin^2 z + \cos^2 z - \cos^2 z}$	Algebra
$= \dfrac{\cos^2 z - 3\cos z + 2}{1 - \cos^2 z}$	Pythagorean Identity
$= \dfrac{(\cos z - 1)(\cos z - 2)}{(1 - \cos z)(1 + \cos z)}$	Algebra
$= \dfrac{-(\cos z - 2)}{1 + \cos z}$	Algebra
$= \dfrac{2 - \cos z}{1 + \cos z}$	Algebra

Key Algebraic Steps:

$$\frac{a^2 - 3a + 2}{1 - a^2} = \frac{(a - 1)(a - 2)}{(1 - a)(1 + a)} = \frac{-(a - 2)}{1 + a} = \frac{2 - a}{1 + a}$$

69. $\dfrac{\cos^3\theta - \sin^3\theta}{\cos\theta - \sin\theta} = \dfrac{(\cos\theta - \sin\theta)(\cos^2\theta + \cos\theta\sin\theta + \sin^2\theta)}{\cos\theta - \sin\theta}$ Algebra

$= \cos^2\theta + \cos\theta\sin\theta + \sin^2\theta$ Algebra

$= 1 + \cos\theta\sin\theta$ Pythagorean Identity

$= 1 + \sin\theta\cos\theta$ Algebra

Key Algebraic Steps:

$$\frac{a^3 - b^3}{a - b} = \frac{(a - b)(a^2 + ab + b^2)}{a - b}$$
$$= a^2 + ab + b^2$$

Common Error: $\dfrac{a^3 - b^3}{a - b} \neq a^2 + b^2$

71. $(\sec x - \tan x)^2 = \left(\dfrac{1}{\cos x} - \dfrac{\sin x}{\cos x}\right)^2$ Quotient and Reciprocal Identities

$= \left(\dfrac{1 - \sin x}{\cos x}\right)^2$ Algebra

$= \dfrac{(1 - \sin x)^2}{\cos^2 x}$ Algebra

$= \dfrac{(1 - \sin x)^2}{1 - \sin^2 x}$ Pythagorean Identity

$= \dfrac{(1 - \sin x)(1 - \sin x)}{(1 - \sin x)(1 + \sin x)}$ Algebra

$= \dfrac{1 - \sin x}{1 + \sin x}$ Algebra

73. $\dfrac{\csc^4 x - 1}{\cot^2 x} = \dfrac{(\csc^2 x)^2 - (1)^2}{\cot^2 x}$ Algebra

$= \dfrac{(\csc^2 x - 1)(\csc^2 x + 1)}{\cot^2 x}$ Algebra

$= \dfrac{(1 + \cot^2 x - 1)(\csc^2 x + 1)}{\cot^2 x}$ Pythagorean Identity

$= \dfrac{\cot^2 x(\csc^2 x + 1)}{\cot^2 x}$ Algebra

$= \csc^2 x + 1$ Algebra

$= 1 + \cot^2 x + 1$ Pythagorean Identity

$= 2 + \cot^2 x$ Algebra

75. $\dfrac{1 + \sin v}{\cos v} = \dfrac{(1 + \sin v)(1 - \sin v)}{\cos v(1 - \sin v)}$ Algebra

$= \dfrac{1 - \sin^2 v}{\cos v(1 - \sin v)}$ Algebra

$= \dfrac{\cos^2 v}{\cos v(1 - \sin v)}$ Pythagorean Identity

$= \dfrac{\cos v}{1 - \sin v}$ Algebra

77. Graph both sides of the equation in the same viewing window.

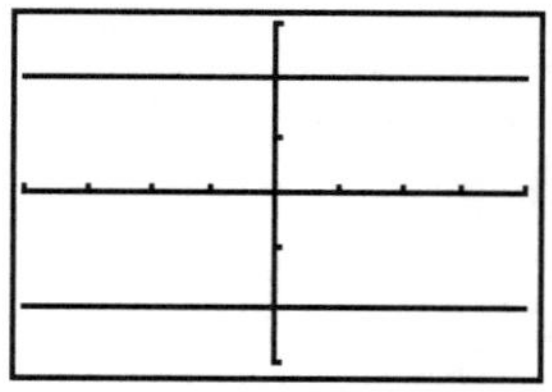

$\dfrac{\sin(-x)}{\cos(-x)\tan(-x)} = -1$ is not an identity, since the graphs do not match.

Try $x = -\dfrac{\pi}{4}$

Left side: $\dfrac{\sin\left[-\left(-\frac{\pi}{4}\right)\right]}{\cos\left[-\left(-\frac{\pi}{4}\right)\right]\tan\left[-\left(-\frac{\pi}{4}\right)\right]} = \dfrac{\frac{1}{\sqrt{2}}}{\frac{1}{\sqrt{2}}\cdot 1} = 1$

Right side: $= -1$

This verifies that the equation is not an identity.

79. Graph both sides of the equation in the same viewing window.

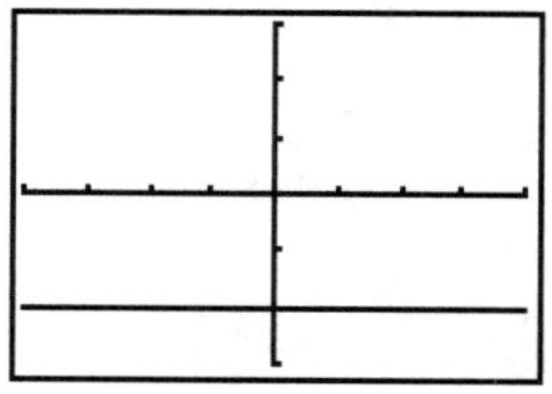

$\dfrac{\sin x}{\cos x\tan(-x)} = -1$ appears to be an identity, which we now verify:

$\dfrac{\sin x}{\cos x\tan(-x)} = \dfrac{\sin x}{\cos x(-\tan x)}$	Identities for Negatives
$= -\dfrac{\sin x}{\cos x\tan x}$	Algebra
$= -\dfrac{\sin x}{\cos x\cdot\frac{\sin x}{\cos x}}$	Quotient Identity
$= -\dfrac{\sin x}{\sin x}$	Algebra
$= -1$	Algebra

81. Graph both sides of the equation in the same viewing window.

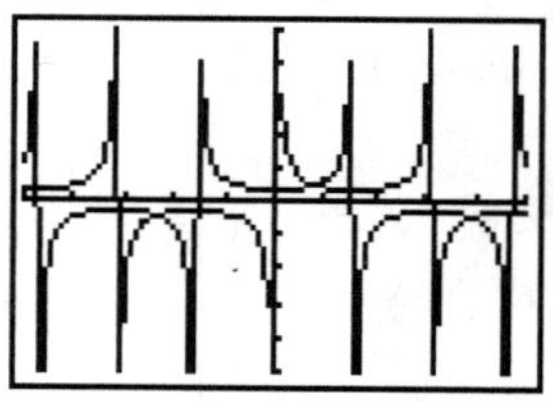

$\sin x + \dfrac{\cos^2 x}{\sin x} = \sec x$ is not an identity, since the graphs do not match.

Try $x = -\dfrac{\pi}{4}$

Left side: $\sin\left(-\dfrac{\pi}{4}\right) + \dfrac{\cos^2\left(-\frac{\pi}{4}\right)}{\sin\left(-\frac{\pi}{4}\right)} = -\dfrac{1}{\sqrt{2}} + \dfrac{\left(\frac{1}{\sqrt{2}}\right)^2}{\left(-\frac{1}{\sqrt{2}}\right)} = -\dfrac{1}{\sqrt{2}} - \dfrac{1}{\sqrt{2}} = -\sqrt{2}$

Right side: $\sec\left(-\dfrac{\pi}{4}\right) = \sqrt{2}$

This verifies that the equation is not an identity.

83. Graph both sides of the equation in the same viewing window.

$\sin x + \dfrac{\cos^2 x}{\sin x} = \csc x$ appears to be an identity, which we now verify:

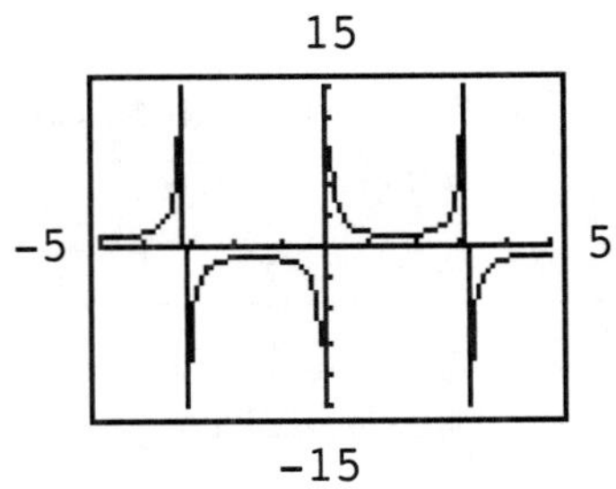

$$\sin x + \frac{\cos^2 x}{\sin x} = \frac{\sin x}{1} + \frac{\cos^2 x}{\sin x} \qquad \text{Algebra}$$

$$= \frac{\sin^2 x}{\sin x} + \frac{\cos^2 x}{\sin x} \qquad \text{Algebra}$$

$$= \frac{\sin^2 x + \cos^2 x}{\sin x} \qquad \text{Algebra}$$

$$= \frac{1}{\sin x} \qquad \text{Pythagorean Identity}$$

$$= \csc x \qquad \text{Reciprocal Identity}$$

85. Graph both sides of the equation in the same viewing window.

$\dfrac{\tan x}{\sin x - 2\tan x} = \dfrac{1}{\cos x - 2}$ appears to be an identity, which we now verify:

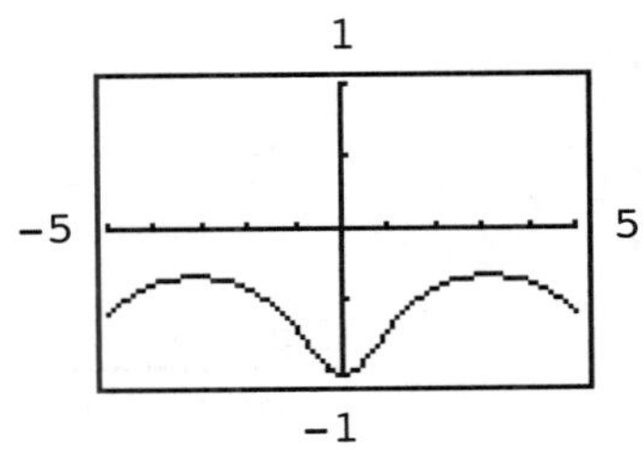

$$\frac{\tan x}{\sin x - 2\tan x} = \frac{\frac{\sin x}{\cos x}}{\sin x - 2\frac{\sin x}{\cos x}} \qquad \text{Quotient Identity}$$

$$= \frac{\cos x \cdot \frac{\sin x}{\cos x}}{\sin x \cos x - 2\cos x \cdot \frac{\sin x}{\cos x}} \qquad \text{Algebra}$$

$$= \frac{\sin x}{\sin x \cos x - 2\sin x} \qquad \text{Algebra}$$

$$= \frac{\sin x}{\sin x(\cos x - 2)} \qquad \text{Algebra}$$

$$= \frac{1}{\cos x - 2} \qquad \text{Algebra}$$

87. Graph both sides of the equation in the same viewing window.

$\dfrac{\tan x}{\sin x + 2\tan x} = \dfrac{1}{\cos x - 2}$ is not an identity, since the graphs do not match.

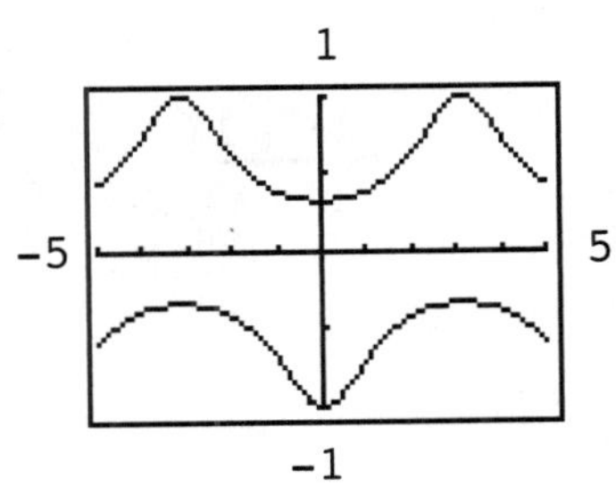

Try $x = \frac{\pi}{3}$

Left side: $\dfrac{\tan\frac{\pi}{3}}{\sin\frac{\pi}{3} + 2\tan\frac{\pi}{3}} = \dfrac{\sqrt{3}}{\frac{\sqrt{3}}{2} + 2\sqrt{3}} = \dfrac{2\sqrt{3}}{\sqrt{3} + 4\sqrt{3}} = \dfrac{2\sqrt{3}}{5\sqrt{3}} = \dfrac{2}{5}$

Right side: $\dfrac{1}{\cos\frac{\pi}{3} - 2} = \dfrac{1}{\frac{1}{2} - 2} = \dfrac{1}{-\frac{3}{2}} = -\dfrac{2}{3}$

This verifies that the equation is not an identity.

89.

$$\frac{2\sin^2 x + 3\cos x - 3}{\sin^2 x} = \frac{2(1 - \cos^2 x) + 3\cos x - 3}{\sin^2 x} \quad \text{Pythagorean Identity}$$

$$= \frac{2 - 2\cos^2 x + 3\cos x - 3}{\sin^2 x} \quad \text{Algebra}$$

$$= \frac{-2\cos^2 x + 3\cos x - 1}{\sin^2 x} \quad \text{Algebra}$$

$$= \frac{-2\cos^2 x + 3\cos x - 1}{1 - \cos^2 x} \quad \text{Pythagorean Identity}$$

$$= \frac{-(2\cos^2 x - 3\cos x + 1)}{-(\cos^2 x - 1)} \quad \text{Algebra}$$

$$= \frac{-(2\cos x - 1)(\cos x - 1)}{-(\cos x + 1)(\cos x - 1)} \quad \text{Algebra}$$

$$= \frac{2\cos x - 1}{\cos x + 1} \quad \text{Algebra}$$

$$= \frac{2\cos x - 1}{1 + \cos x} \quad \text{Algebra}$$

Key Algebraic Steps:

$$\frac{-2a^2 + 3a - 1}{1 - a^2} = \frac{-(2a^2 - 3a + 1)}{-(a^2 - 1)}$$

$$= \frac{-(2a - 1)(a - 1)}{-(a + 1)(a - 1)}$$

$$= \frac{2a - 1}{a + 1}$$

$$= \frac{2a - 1}{1 + a}$$

91. $$\frac{\tan u + \sin u}{\tan u - \sin u} - \frac{\sec u + 1}{\sec u - 1} = \frac{\frac{\sin u}{\cos u} + \sin u}{\frac{\sin u}{\cos u} - \sin u} - \frac{\sec u + 1}{\sec u - 1}$$ Quotient Identity

$$= \frac{\frac{1}{\cos u}\sin u + 1 \cdot \sin u}{\frac{1}{\cos u}\sin u - 1 \cdot \sin u} - \frac{\sec u + 1}{\sec u - 1}$$ Algebra

$$= \frac{\sin u\left(\frac{1}{\cos u} + 1\right)}{\sin u\left(\frac{1}{\cos u} - 1\right)} - \frac{\sec u + 1}{\sec u - 1}$$ Algebra

$$= \frac{\frac{1}{\cos u} + 1}{\frac{1}{\cos u} - 1} - \frac{\sec u + 1}{\sec u - 1}$$ Algebra

$$= \frac{\sec u + 1}{\sec u - 1} - \frac{\sec u + 1}{\sec u - 1}$$ Reciprocal Identity

$$= 0$$ Algebra

Key Algebraic Steps:

$$\frac{\frac{b}{a} + b}{\frac{b}{a} - b} = \frac{b \cdot \frac{1}{a} + b}{b \cdot \frac{1}{a} - b} = \frac{b\left(\frac{1}{a} + 1\right)}{b\left(\frac{1}{a} - 1\right)} = \frac{\frac{1}{a} + 1}{\frac{1}{a} - 1}$$

93. $$\frac{\tan\beta + \cot\alpha}{\tan\beta\cot\alpha} = \frac{\tan\beta}{\tan\beta\cot\alpha} + \frac{\cot\alpha}{\tan\beta\cot\alpha}$$ Algebra

$$= \frac{1}{\cot\alpha} + \frac{1}{\tan\beta}$$ Algebra

$$= \tan\alpha + \cot\beta$$ Reciprocal Identities

95.

Statement	Reason
$\cot^2 x + 1 = \left(\frac{\cos x}{\sin x}\right)^2 + 1$	$\cot x = \frac{\cos x}{\sin x}$
$= \frac{\cos^2 x}{\sin^2 x} + 1$	Algebra
$= \frac{\cos^2 x + \sin^2 x}{\sin^2 x}$	Algebra
$= \frac{1}{\sin^2 x}$	$\cos^2 x + \sin^2 x = 1$
$= \left(\frac{1}{\sin x}\right)^2$	Algebra
$= \csc^2 x$	$\frac{1}{\cos x} = \sec x$

97. Here is a graph of $f(x)$ from a graphing calculator.

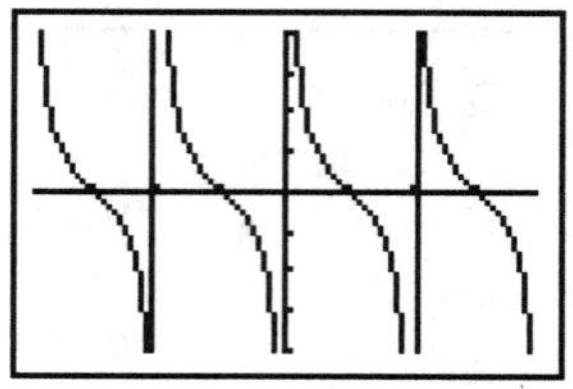

The graph appears identical with the graph of $g(x) = \cot x$. We attempt to verify that $\frac{1 - \sin^2 x}{\tan x} + \sin x \cos x = \cot x$ is an identity.

$$\frac{1 - \sin^2 x}{\tan x} + \sin x \cos x = \frac{1}{\tan x} - \frac{\sin^2 x}{\tan x} + \sin x \cos x \qquad \text{Algebra}$$

$$= \cot x - \frac{\sin^2 x}{\frac{\sin x}{\cos x}} + \sin x \cos x \qquad \text{Quotient Identity}$$

$$= \cot x - \frac{\sin^2 x \cos x}{\sin x} + \sin x \cos x \qquad \text{Algebra}$$

$$= \cot x - \sin x \cos x + \sin x \cos x \qquad \text{Algebra}$$

$$= \cot x \qquad \text{Algebra}$$

Thus $f(x) = g(x)$ is an identity for $g(x) = \cot x$.

99. Here is a graph of $f(x)$ from a graphing calculator.

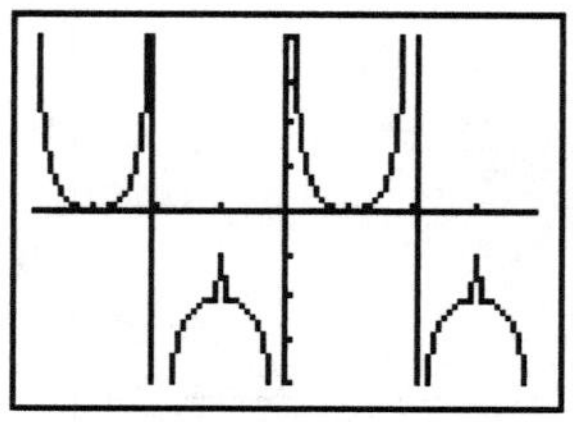

The graph appears similar to the graph of the cosecant function, but shifted down 1 unit, and with peculiar behavior at $-\frac{\pi}{2}$ and $\frac{3\pi}{2}$ where $f(x)$ is not defined, although $\csc x$ is defined. We set $g(x) = \csc x - 1$ and attempt to verify that

$$\frac{\cos^2 x}{1 + \sin x - \cos^2 x} = \csc x - 1 \text{ is an identity.}$$

$$\frac{\cos^2 x}{1 + \sin x - \cos^2 x} = \frac{1 - \sin^2 x}{1 + \sin x - (1 - \sin^2 x)} \qquad \text{Pythagorean Identity}$$

$$= \frac{1 - \sin^2 x}{\sin x + \sin^2 x} \qquad \text{Algebra}$$

$$= \frac{(1 - \sin x)(1 + \sin x)}{\sin x(1 + \sin x)} \qquad \text{Algebra}$$

$$= \frac{1 - \sin x}{\sin x} \qquad \text{Algebra}$$

$$= \frac{1}{\sin x} - \frac{\sin x}{\sin x} \qquad \text{Algebra}$$

$$= \csc x - 1 \qquad \text{Reciprocal Identity; Algebra}$$

Thus $f(x) = g(x)$ is an identity for $g(x) = \csc x - 1$.

101. Here is a graph of $f(x)$ from a graphing calculator.

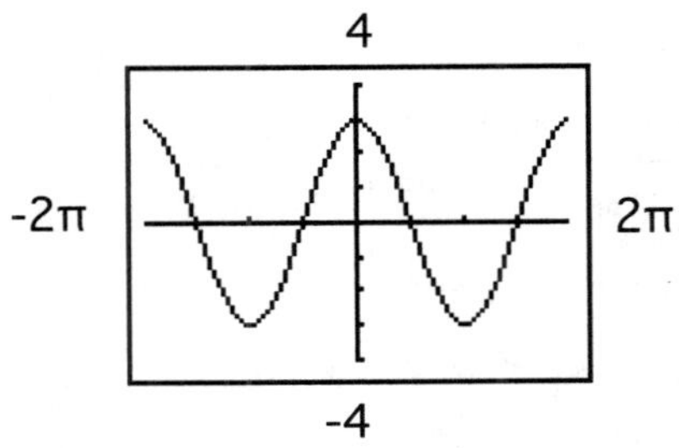

The graph appears identical with the graph of $g(x) = 3\cos x$. We attempt to verify that

$$\frac{1 + \cos x - 2\cos^2 x}{1 - \cos x} + \frac{\sin^2 x}{1 + \cos x}$$

$= 3\cos x$ is an identity.

$$\frac{1 + \cos x - 2\cos^2 x}{1 - \cos x} + \frac{\sin^2 x}{1 + \cos x}$$

$$= \frac{1 + \cos x - 2\cos^2 x}{1 - \cos x} - \frac{1 - \cos^2 x}{1 + \cos x}$$

$$= \frac{(1 - \cos x)(1 + 2\cos x)}{1 - \cos x} - \frac{(1 - \cos x)(1 + \cos x)}{1 + \cos x}$$

$$= 1 + 2\cos x - (1 - \cos x)$$

$$= 1 + 2\cos x - 1 + \cos x$$

$$= 3\cos x$$

Thus, $f(x) = g(x)$ is an identity for $g(x) = 3\cos x$.

103. Since $\sqrt{1 - \cos^2 x} = \sqrt{\sin^2 x}$, the equation will be true when $\sqrt{\sin^2 x} = -\sin x$, that is, when $\sin x$ is negative. This occurs in Quadrants III, IV.

105. Since $\sqrt{1 - \cos^2 x} = \sqrt{\sin^2 x}$, the equation will be true when $\sqrt{\sin^2 x} = \sin x$, that is, when $\sin x$ is positive. This occurs in Quadrants I, II.

107. Since $\sqrt{1 - \sin^2 x} = \sqrt{\cos^2 x} = |\cos x|$ is an identity, this will hold in all quadrants.

109. Since $\sqrt{1 - \sin^2 x} = \sqrt{\cos^2 x} = \cos x$ if $\cos x$ is positive, this will hold when $\dfrac{\sin x}{\sqrt{1 - \sin^2 x}} = \dfrac{\sin x}{\sqrt{\cos^2 x}} = \dfrac{\sin x}{\cos x} = \tan x$, that is, when $\cos x$ is positive. This occurs in Quadrants I, IV.

111.
$$\sqrt{a^2 - u^2} = \sqrt{a^2 - (a\sin x)^2}$$
$$= \sqrt{a^2 - a^2\sin^2 x}$$
$$= \sqrt{a^2(1 - \sin^2 x}$$
$$= \sqrt{a^2}\sqrt{1 - \sin^2 x}$$
$$= \sqrt{a^2}\sqrt{\cos^2 x}$$

Common Error: $\sqrt{1 - \sin^2 x} \neq 1 - \sin x$

Since $a > 0$, $\sqrt{a^2} = a$.
$\cos x$ will be ≥ 0 if x is a quadrant I or IV angle. But the restriction $-\dfrac{\pi}{2} < x < \dfrac{\pi}{2}$ requires that x is such an angle. Therefore $\sqrt{\cos^2 x} = \cos x$

$$\sqrt{a^2 - u^2} = a\cos x$$

113. $\sqrt{a^2 + u^2} = \sqrt{a^2 + (a \tan x)^2}$

$= \sqrt{a^2 + a^2 \tan^2 x}$

$= \sqrt{a^2(1 + \tan^2 x}$

$= \sqrt{a^2}\sqrt{1 + \tan^2 x}$

$= \sqrt{a^2}\sqrt{\sec^2 x}$

Since $a > 0$, $\sqrt{a^2} = a$.

$\sec x$ will be ≥ 0 if x is a quadrant I or IV angle. But the restriction $0 < x < \frac{\pi}{2}$ requires that x is a quadrant I angle. Therefore $\sqrt{\sec^2 x} = \sec x$

$\sqrt{a^2 + u^2} = a \sec x$

Exercise 7-2

1. Plug in $x = 1$, $y = 1$.
Left side: $(1 + 1)^2 = 2^2 = 4$
Right side: $1^2 + 1^2 = 2$
The equation is not an identity.

3. Plug in $x = 2$, $y = \frac{\pi}{6}$.

Left side: $2 \sin \frac{\pi}{6} = 1$

Right side: $\sin\left(2 \cdot \frac{\pi}{6}\right) = \frac{\sqrt{3}}{2}$

The equation is not an identity.

5. Plug in $x = \frac{\pi}{6}$, $y = \frac{\pi}{6}$.

Left side: $\cos\left(\frac{\pi}{6} + \frac{\pi}{6}\right) = \cos \frac{\pi}{3} = \frac{1}{2}$

Right side: $\cos \frac{\pi}{6} + \cos \frac{\pi}{6} = \frac{\sqrt{3}}{2} + \frac{\sqrt{3}}{2} = \sqrt{3}$

The equation is not an identity.

7. Plug in $x = \frac{\pi}{3}$, $y = \frac{\pi}{6}$.

Left side: $\tan\left(\frac{\pi}{3} - \frac{\pi}{6}\right) = \tan \frac{\pi}{6} = \frac{1}{\sqrt{3}}$

Right side: $\tan \frac{\pi}{3} - \tan \frac{\pi}{6} = \sqrt{3} - \frac{1}{\sqrt{3}} = \frac{2}{\sqrt{3}}$

The equation is not an identity.

9. Expand and simplify the left side:

$\tan(x - \pi) = \frac{\tan x - \tan \pi}{1 + \tan x \tan \pi}$ Difference Identity

$= \frac{\tan x - 0}{1 + \tan x \cdot 0}$ $\tan \pi = \frac{\sin \pi}{\cos \pi} = \frac{0}{-1} = 0$

$= \frac{\tan x}{1}$ Algebra

$= \tan x$ Algebra

The equation is an identity.

11. Expand and simplify the left side:

$\sin(x - \pi) = \sin x \cos \pi - \cos x \sin \pi$ — Difference Identity

$= \sin x(-1) - \cos x(0)$ — $\cos \pi = -1;\ \sin \pi = 0$

$= -\sin x$ — Algebra

The equation is not an identity.

13. Expand and simplify the left side:

$\csc(2\pi - x) = \dfrac{1}{\sin(2\pi - x)}$ — $\csc x = \dfrac{1}{\sin x}$

$= \dfrac{1}{\sin 2\pi \cos x - \cos 2\pi \sin x}$ — Difference Identity

$= \dfrac{1}{0 \cdot \cos x - 1 \cdot \sin x}$ — $\sin 2\pi = 0;\ \cos 2\pi = 1$

$= \dfrac{1}{-\sin x}$ — Algebra

$= -\csc x$ — $\csc x = \dfrac{1}{\sin x}$

The equation is not an identity.

15. Expand and simplify the left side:

$\sin\left(x - \dfrac{\pi}{2}\right) = \sin x \cos \dfrac{\pi}{2} - \cos x \sin \dfrac{\pi}{2}$ — Difference Identity

$= \sin x \cdot 0 - \cos x \cdot 1$ — $\cos \dfrac{\pi}{2} = 0;\ \sin \dfrac{\pi}{2} = 1$

$= -\cos x$ — Algebra

The equation is an identity.

17. $\cot\left(\dfrac{\pi}{2} - x\right) = \dfrac{\cos\left(\frac{\pi}{2} - x\right)}{\sin\left(\frac{\pi}{2} - x\right)}$ — Quotient Identity

$= \dfrac{\cos \frac{\pi}{2} \cos x + \sin \frac{\pi}{2} \sin x}{\sin \frac{\pi}{2} \cos x - \cos \frac{\pi}{2} \sin x}$ — Difference Identities

$= \dfrac{0 \cos x + 1 \sin x}{1 \cos x - 0 \sin x}$ — Known Values

$= \dfrac{\sin x}{\cos x}$ — Algebra

$= \tan x$ — Quotient Identity

19. $\csc\left(\dfrac{\pi}{2} - x\right) = \dfrac{1}{\sin\left(\frac{\pi}{2} - x\right)}$ — Reciprocal Identity

$= \dfrac{1}{\sin \frac{\pi}{2} \cos x - \cos \frac{\pi}{2} \sin x}$ — Difference Identity

$= \dfrac{1}{1 \cos x - 0 \sin x}$ — Known Values

$= \dfrac{1}{\cos x}$ — Algebra

$= \sec x$ — Reciprocal Identity

21. $\sin(30° - x) = \sin 30° \cos x - \cos 30° \sin x$ Difference Identity

$= \frac{1}{2}\cos x - \frac{\sqrt{3}}{2}\sin x$ Known Values

$= \frac{1}{2}(\cos x - \sqrt{3}\sin x)$ Algebra

23. $\sin(180° - x) = \sin 180° \cos x - \cos 180° \sin x$ Difference Identity

$= 0 \cos x - (-1) \sin x$ Known Values

$= \sin x$ Algebra

25. $\tan\left(x + \frac{\pi}{3}\right) = \dfrac{\tan x + \tan \frac{\pi}{3}}{1 - \tan x \tan \frac{\pi}{3}}$ Sum Identity

$= \dfrac{\tan x + \sqrt{3}}{1 - \sqrt{3}\tan x}$ Known Values

27. $\sec 75° = \sec(30° + 45°)$

$$= \frac{1}{\cos(30° + 45°)}$$

$$= \frac{1}{\cos 30° \cos 45° - \sin 30° \sin 45°}$$

$$= \frac{1}{\frac{\sqrt{3}}{2}\frac{1}{\sqrt{2}} - \frac{1}{2}\frac{1}{\sqrt{2}}}$$

$$= \frac{2\sqrt{2} \cdot 1}{2\sqrt{2}\frac{\sqrt{3}}{2\sqrt{2}} - 2\sqrt{2}\frac{1}{2\sqrt{2}}}$$

$$= \frac{2\sqrt{2}}{\sqrt{3} - 1}$$

29. $\sin\frac{7\pi}{12} = \sin\left(\frac{\pi}{3} + \frac{\pi}{4}\right)$

$$= \sin\frac{\pi}{3}\cos\frac{\pi}{4} + \cos\frac{\pi}{3}\sin\frac{\pi}{4}$$

$$= \frac{\sqrt{3}}{2}\cdot\frac{1}{\sqrt{2}} + \frac{1}{2}\cdot\frac{1}{\sqrt{2}}$$

$$= \frac{\sqrt{3}}{2\sqrt{2}} + \frac{1}{2\sqrt{2}}$$

$$= \frac{\sqrt{3} + 1}{2\sqrt{2}}$$

31. $\cos 74° \cos 44° + \sin 74° \sin 44° = \cos(74° - 44°) = \cos 30° = \frac{\sqrt{3}}{2}$

33. $\dfrac{\tan 27° + \tan 18°}{1 - \tan 27° \tan 18°} = \tan(27° + 18°) = \tan 45° = 1$

35.

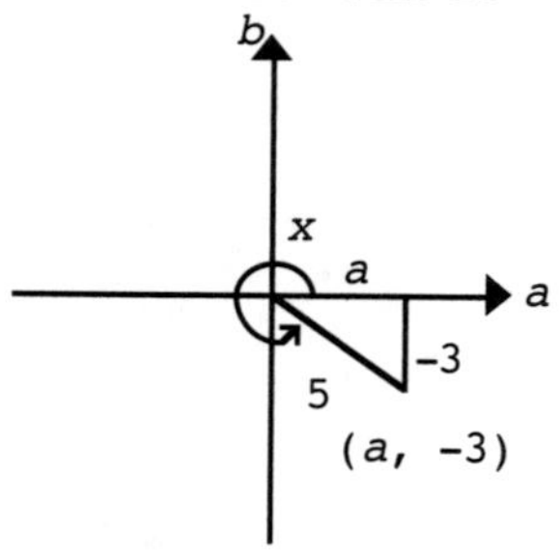

$a = \sqrt{5^2 - (-3)^2} = 4$

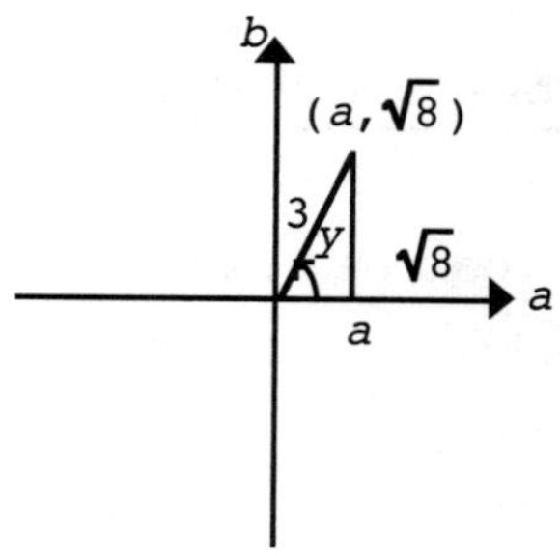

$a = \sqrt{3^2 - (\sqrt{8})^2} = 1$

$\cos x = \frac{4}{5}$

$\tan x = -\frac{3}{4}$

$\sin(x - y) = \sin x \cos y - \cos x \sin y$

$$= \left(-\frac{3}{5}\right)\frac{1}{3} - \frac{4}{5}\left(\frac{\sqrt{8}}{3}\right)$$

$$= \frac{-3 - 4\sqrt{8}}{15}$$

$\cos y = \frac{1}{3}$

$\tan y = \frac{\sqrt{8}}{1} = \sqrt{8}$

$\tan(x + y) = \dfrac{\tan x + \tan y}{1 - \tan x \tan y}$

$$= \frac{-\frac{3}{4} + \sqrt{8}}{1 - \left(-\frac{3}{4}\right)\sqrt{8}}$$

$$= \frac{-3 + 4\sqrt{8}}{4 + 3\sqrt{8}} = \frac{4\sqrt{8} - 3}{4 + 3\sqrt{8}}$$

37.

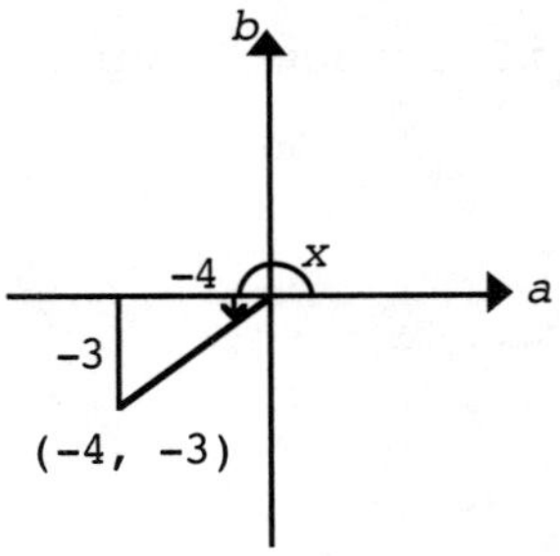

$r = \sqrt{(-4)^2 + (-3)^2} = 5$

$\sin x = -\frac{3}{5}$

$\cos x = -\frac{4}{5}$

$$\begin{aligned}\sin(x - y) &= \sin x \cos y - \cos x \sin y\\ &= \left(-\frac{3}{5}\right)\left(\frac{2}{\sqrt{5}}\right) - \left(-\frac{4}{5}\right)\left(-\frac{1}{\sqrt{5}}\right)\\ &= -\frac{6}{5\sqrt{5}} - \frac{4}{5\sqrt{5}}\\ &= \frac{-10}{5\sqrt{5}}\\ &= \frac{-2}{\sqrt{5}}\end{aligned}$$

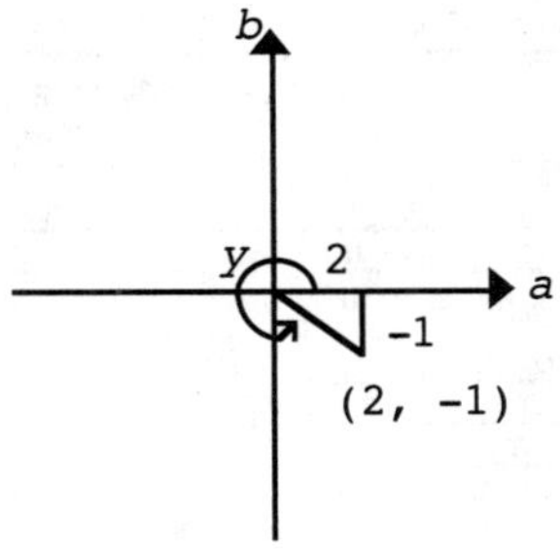

$r = \sqrt{2^2 + (-1)^2} = \sqrt{5}$

$\sin y = \frac{-1}{\sqrt{5}}$

$\cos y = \frac{2}{\sqrt{5}}$

$$\begin{aligned}\tan(x + y) &= \frac{\tan x + \tan y}{1 - \tan x \tan y}\\ &= \frac{\frac{3}{4} + \left(-\frac{1}{2}\right)}{1 - \left(\frac{3}{4}\right)\left(-\frac{1}{2}\right)}\\ &= \frac{6 - 4}{8 + 3}\\ &= \frac{2}{11}\end{aligned}$$

39. $\cos 2x = \cos(x + x)$ Algebra

$= \cos x \cos x - \sin x \sin x$ Sum Identity

$= \cos^2 x - \sin^2 x$ Algebra

41. $\cot(x + y) = \dfrac{\cos(x + y)}{\sin(x + y)}$ Quotient Identity

$= \dfrac{\cos x \cos y - \sin x \sin y}{\sin x \cos y + \cos x \sin y}$ Sum Identities

$= \dfrac{\frac{\cos x \cos y}{\sin x \sin y} - \frac{\sin x \sin y}{\sin x \sin y}}{\frac{\sin x \cos y}{\sin x \sin y} + \frac{\cos x \sin y}{\sin x \sin y}}$ Algebra

$= \dfrac{\frac{\cos x}{\sin x} \cdot \frac{\cos y}{\sin y} - 1}{\frac{\cos y}{\sin y} + \frac{\cos x}{\sin x}}$ Algebra

$= \dfrac{\cot x \cot y - 1}{\cot y + \cot x}$ Quotient Identity

43. $\tan 2x = \tan(x + x)$ Algebra

$= \dfrac{\tan x + \tan x}{1 - \tan x \tan x}$ Sum Identity

$= \dfrac{2 \tan x}{1 - \tan^2 x}$ Algebra

45. $\dfrac{\sin(v+u)}{\sin(v-u)} = \dfrac{\sin v\cos u + \cos v\sin u}{\sin v\cos u - \cos v\sin u}$ Sum and Difference Identities

$= \dfrac{\dfrac{\sin v\cos u}{\sin v\sin u} + \dfrac{\cos v\sin u}{\sin v\sin u}}{\dfrac{\sin v\cos u}{\sin v\sin u} - \dfrac{\cos v\sin u}{\sin v\sin u}}$ Algebra

$= \dfrac{\dfrac{\cos u}{\sin u} + \dfrac{\cos v}{\sin v}}{\dfrac{\cos u}{\sin u} - \dfrac{\cos v}{\sin v}}$ Algebra

$= \dfrac{\cot u + \cot v}{\cot u - \cot v}$ Quotient Identity

47. $\cot x - \tan y = \dfrac{\cos x}{\sin x} - \dfrac{\sin y}{\cos y}$ Quotient Identities

$= \dfrac{\cos x\cos y}{\sin x\cos y} - \dfrac{\sin x\sin y}{\sin x\cos y}$ Algebra

$= \dfrac{\cos x\cos y - \sin x\sin y}{\sin x\cos y}$ Algebra

$= \dfrac{\cos(x+y)}{\sin x\cos y}$ Sum Identity

49. $\tan(x-y) = \dfrac{\tan x - \tan y}{1 + \tan x\tan y}$ Difference Identity

$= \dfrac{\dfrac{1}{\cot x} - \dfrac{1}{\cot y}}{1 + \dfrac{1}{\cot x}\dfrac{1}{\cot y}}$ Reciprocal Identity

$= \dfrac{\cot x\cot y\left(\dfrac{1}{\cot x} - \dfrac{1}{\cot y}\right)}{\cot x\cot y\left(1 + \dfrac{1}{\cot x}\dfrac{1}{\cot y}\right)}$ Algebra

$= \dfrac{\cot y - \cot x}{\cot x\cot y + 1}$ Algebra

51. $\dfrac{\cos(x+h) - \cos x}{h} = \dfrac{\cos x\ \cos h - \sin x\sin h - \cos x}{h}$ Sum Identity

$= \dfrac{\cos x\cos h - \cos x - \sin x\sin h}{h}$ Algebra

$= \dfrac{\cos x(\cos h - 1) - \sin x\sin h}{h}$ Algebra

$= \cos x\left(\dfrac{\cos h - 1}{h}\right) - \sin x\ \dfrac{\sin h}{h}$ Algebra

53. $\sin(x-y) = \sin(5.288 - 1.769) = -0.3685$

$\sin x\cos y - \cos x\sin y = \sin 5.288\cos 1.769 - \cos 5.288\sin 1.769 = -0.3685$

$\tan(x+y) = \tan(5.288 + 1.769) = 0.9771$

$\dfrac{\tan x + \tan y}{1 - \tan x\tan y} = \dfrac{\tan 5.288 + \tan 1.769}{1 - \tan 5.288\tan 1.769} = 0.9771$

55. $\sin(x - y) = \sin(42.08° - 68.37°) = -0.4429$

$\sin x \cos y - \cos x \sin y = \sin 42.08° \cos 68.37° - \cos 42.08° \sin 68.37° = -0.4429$

$\tan(x + y) = \tan(42.08° + 68.37°) = -2.682$

$$\frac{\tan x + \tan y}{1 - \tan x \tan y} = \frac{\tan 42.08° + \tan 68.37°}{1 - \tan 42.08° \tan 68.37°} = -2.682$$

57. Evaluate each side for a particular set of values of x and y for which each side is defined. If the left side is not equal to the right side, then the equation is not an identity. For example, for $x = 2$ and $y = 1$, both sides are defined, but are not equal.

59. Let $y1 = \sin\left(x + \frac{\pi}{6}\right)$

Then $y2 = \sin x \cos \frac{\pi}{6} + \cos x \sin \frac{\pi}{6}$

$= \sin x \cdot \frac{\sqrt{3}}{2} + \cos x \cdot \frac{1}{2} = \frac{\sqrt{3}}{2}\sin x + \frac{1}{2}\cos x.$

The graphs coincide as shown at the right.

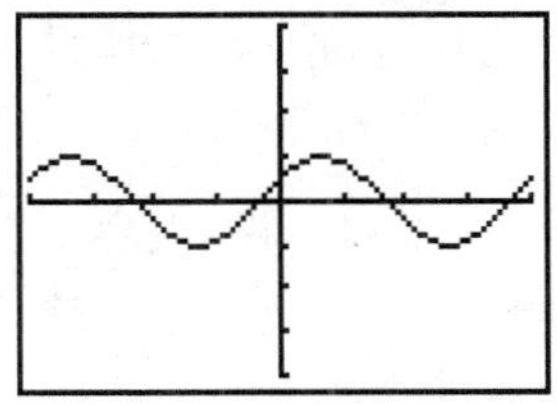

61. Let $y1 = \cos\left(x - \frac{3\pi}{4}\right)$

Then $y2 = \cos x \cos \frac{3\pi}{4} + \sin x \sin \frac{3\pi}{4}$

$= \cos x\left(-\frac{\sqrt{2}}{2}\right) + \sin x\left(\frac{\sqrt{2}}{2}\right)$

$= -\frac{\sqrt{2}}{2} \cos x + \frac{\sqrt{2}}{2} \sin x$

The graphs coincide as shown at the right.

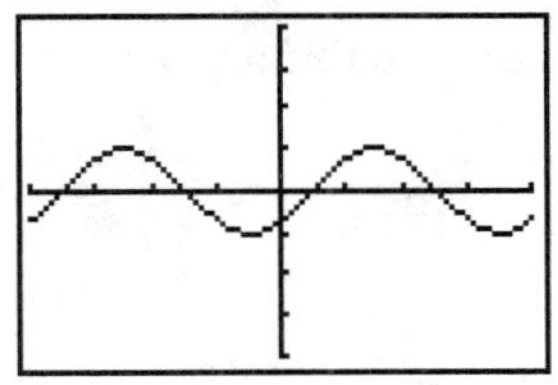

63. Let $y1 = \tan\left(x + \frac{2\pi}{3}\right)$

Then $y2 = \dfrac{\tan x + \tan \frac{2\pi}{3}}{1 - \tan x \tan \frac{2\pi}{3}}$

$= \dfrac{\tan x + (-\sqrt{3})}{1 - \tan x(-\sqrt{3})} = \dfrac{\tan x - \sqrt{3}}{1 + \sqrt{3} \tan x}$

The graphs coincide as shown at the right.

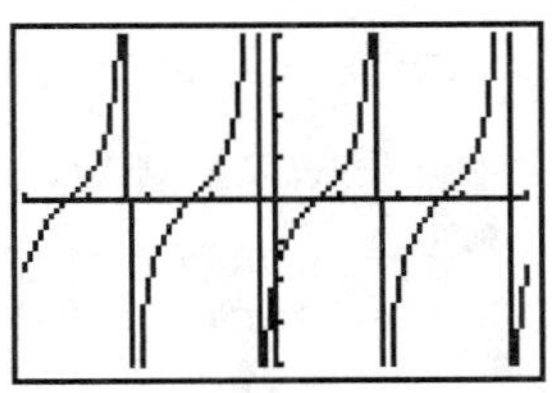

65. Let $u = \cos^{-1}\left(-\frac{4}{5}\right)$, $v = \sin^{-1}\left(-\frac{3}{5}\right)$

Then we are asked to evaluate $\sin(u + v)$ which is $\sin u \cos v + \cos u \sin v$ from the sum identity. We know $\sin v = \sin\left[\sin^{-1}\left(-\frac{3}{5}\right)\right] = -\frac{3}{5}$ and

$\cos u = \cos\left[\cos^{-1}\left(-\frac{4}{5}\right)\right] = -\frac{4}{5}$ from the function-inverse function identities.

It remains to find $\cos v$ and $\sin u$. Note: $0 \le u \le \pi$ and $-\frac{\pi}{2} \le v \le \frac{\pi}{2}$

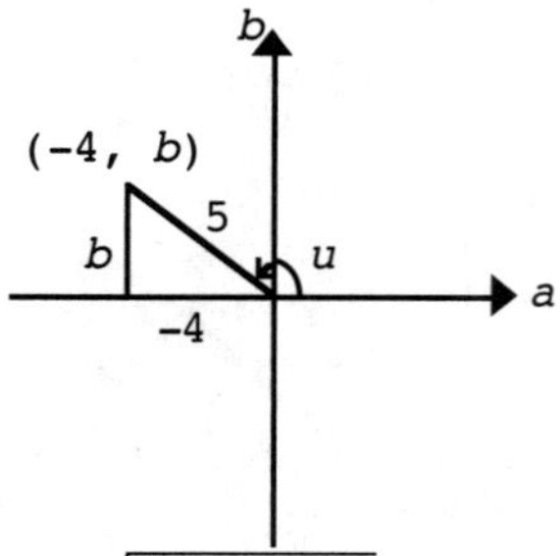

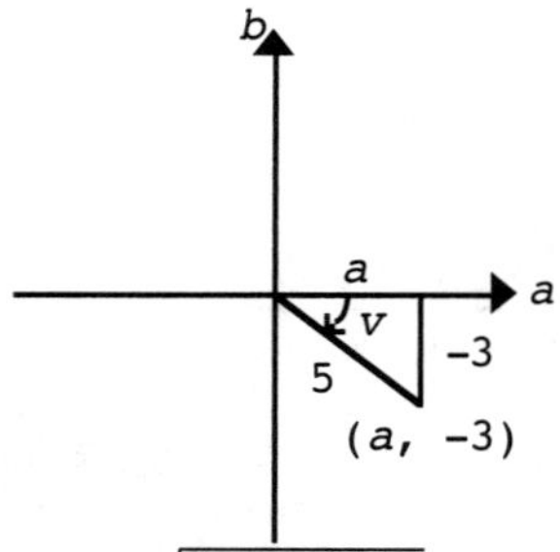

$b = \sqrt{5^2 - (-4)^2} = 3$ $\quad$ $a = \sqrt{5^2 - (-3)^2} = 4$

$\sin u = \frac{3}{5}$ $\quad$ $\cos v = \frac{4}{5}$

Then $\sin\left[\cos^{-1}\left(-\frac{4}{5}\right) + \sin^{-1}\left(-\frac{3}{5}\right)\right] = \sin(u + v)$

$= \sin u \cos v + \cos u \sin v$

$= \left(\frac{3}{5}\right)\left(\frac{4}{5}\right) + \left(-\frac{4}{5}\right)\left(-\frac{3}{5}\right)$

$= \frac{12}{25} + \frac{12}{25}$

$= \frac{24}{25}$

67. We could proceed as in problem 65. Alternatively, we can shorten the process by recognizing $\arccos \frac{1}{2} = \frac{\pi}{3}$ and $\arcsin(-1) = -\frac{\pi}{2}$. Then $\sin[\arccos \frac{1}{2} + \arcsin(-1)]$ $= \sin\left(\frac{\pi}{3} + -\frac{\pi}{2}\right) = \sin \frac{\pi}{3} \cos\left(-\frac{\pi}{2}\right) + \cos \frac{\pi}{3} \sin\left(-\frac{\pi}{2}\right) = \frac{\sqrt{3}}{2}(0) + \left(\frac{1}{2}\right)(-1) = -\frac{1}{2}$.

69. Let $u = \sin^{-1} x$, $v = \cos^{-1} y$. Then $x = \sin u$, $-\frac{\pi}{2} \le u \le \frac{\pi}{2}$, $y = \cos v$, $0 \le y \le \pi$.

Then $\cos u = \sqrt{1 - \sin^2 u}$ (in Quadrants I, IV) $= \sqrt{1 - x^2}$

$\sin v = \sqrt{1 - \cos^2 v}$ (in Quadrants I, II) $= \sqrt{1 - y^2}$

Hence $\sin(\sin^{-1} x + \cos^{-1} y) = \sin(u + v) = \sin u \cos v + \cos u \sin v$

$= xy + \sqrt{1 - x^2}\sqrt{1 - y^2}$

71. $\cos(x + y + z) = \cos[(x + y) + z]$

$= \cos(x + y)\cos z - \sin(x + y)\sin z$

$= (\cos x \cos y - \sin x \sin y)\cos z - (\sin x \cos y + \cos x \sin y)\sin z$

$= \cos x \cos y \cos z - \sin x \sin y \cos z - \sin x \cos y \sin z - \cos x \sin y \sin z$

73. Let $y1 = \cos 1.2x \cos 0.8x - \sin 1.2x \sin 0.8x$
Then $y2 = \cos(1.2x + 0.8x) = \cos 2x$
The graphs coincide as shown at the right.

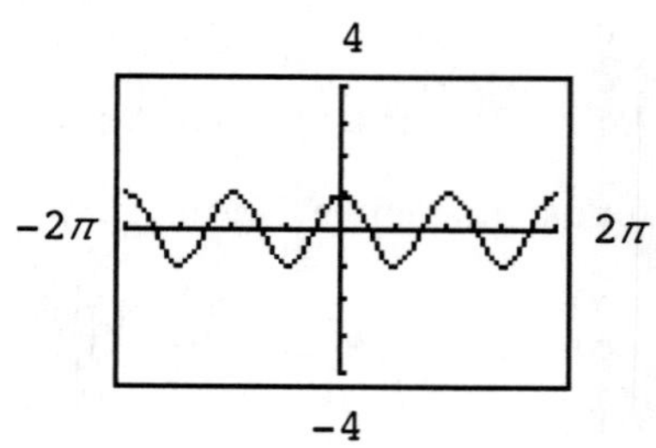

75. $\tan(\theta_2 - \theta_1) = \dfrac{\tan\theta_2 - \tan\theta_1}{1 + \tan\theta_2 \tan\theta_1}$ Difference Identity

$= \dfrac{m_2 - m_1}{1 + m_2 m_1}$ Given

$= \dfrac{m_2 - m_1}{1 + m_1 m_2}$ Algebra

77.

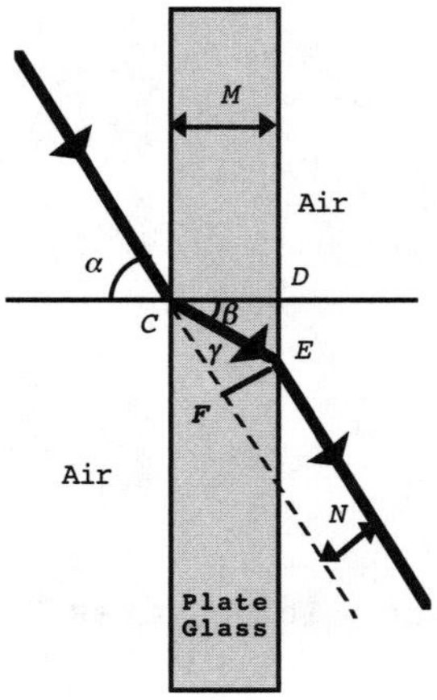

Note: In the text figure we have drawn EF perpendicular to CF, the track of the incident ray. ΔCDE and ΔCEF are right triangles.
Angle $DCF = \beta + \gamma = \alpha$. Hence, $\gamma = \alpha - \beta$.
Denote EC by x. Then,

in ΔCDE $\dfrac{M}{x} = \cos\beta = \sin(90° - \beta)$

in ΔCEF $\dfrac{N}{x} = \sin\gamma = \sin(\alpha - \beta)$

Therefore $x = \dfrac{M}{\sin(90° - \beta)} = \dfrac{N}{\sin(\alpha - \beta)}$

$M\sin(\alpha - \beta) = N\sin(90° - \beta)$ Algebra

$M(\sin\alpha\cos\beta - \cos\alpha\sin\beta) = N(\sin 90°\cos\beta - \cos 90°\sin\beta)$ Difference Identities

$M(\sin\alpha\cos\beta - \cos\alpha\sin\beta) = N(1\cos\beta - 0\sin\beta)$ Known Values

$M(\sin\alpha\cos\beta - \cos\alpha\sin\beta) = N\cos\beta$ Algebra

$M\sin\alpha\cos\beta - M\cos\alpha\sin\beta = N\cos\beta$ Algebra

$M\sin\alpha\cos\beta - N\cos\beta = M\cos\alpha\sin\beta$ Algebra

$\cos\beta(M\sin\alpha - N) = M\cos\alpha\sin\beta$ Algebra

$\dfrac{M\sin\alpha - N}{M\cos\alpha} = \dfrac{\sin\beta}{\cos\beta}$ Algebra

$\tan\beta = \dfrac{M\sin\alpha - N}{M\cos\alpha}$ Quotient Identity

$\tan\beta = \dfrac{M\sin\alpha}{M\cos\alpha} - \dfrac{N}{M\cos\alpha}$ Algebra

$\tan\beta = \dfrac{\sin\alpha}{\cos\alpha} - \dfrac{N}{M}\dfrac{1}{\cos\alpha}$ Algebra

$\tan\beta = \tan\alpha - \dfrac{N}{M}\sec\alpha$ Quotient and Reciprocal Identities

79. (A) From the text figure:

In right triangle ABE, we have (1) $\cot\alpha = \dfrac{AB}{AE} = \dfrac{AB}{h}$

In right triangle BCD, we have (2) $\cot\alpha = \dfrac{BC}{CD} = \dfrac{BC}{H}$

In right triangle $EE'D$, we have (3) $\tan\beta = \dfrac{E'D}{EE'} = \dfrac{H - h}{AC} = \dfrac{H - h}{AB + BC}$

From (3), $H - h = (AB + BC)\tan\beta$

From (1) and (2), $AB = h\cot\alpha$ and $BC = H\cot\alpha$

Hence, substituting, we have (4) $H - h = (h\cot\alpha + H\cot\alpha)\tan\beta$, or

$= (h + H)\cot\alpha\tan\beta$

Solving for H in terms of h yields:

$$H - h = h \cot \alpha \tan \beta + H \cot \alpha \tan \beta$$

$$H - H \cot \alpha \tan \beta = h \cot \alpha \tan \beta + h$$

$$H(1 - \cot \alpha \tan \beta) = h(\cot \alpha \tan \beta + 1)$$

$$H = h \frac{1 + \cot \alpha \tan \beta}{1 - \cot \alpha \tan \beta}$$

(B) $H = h \dfrac{1 + \dfrac{\cos \alpha}{\sin \alpha} \dfrac{\sin \beta}{\cos \beta}}{1 - \dfrac{\cos \alpha}{\sin \alpha} \dfrac{\sin \beta}{\cos \beta}}$ Quotient Identities

$$H = h \frac{\left(1 + \frac{\cos \alpha}{\sin \alpha} \frac{\sin \beta}{\cos \beta}\right) \cdot \sin \alpha \cos \beta}{\left(1 - \frac{\cos \alpha}{\sin \alpha} \frac{\sin \beta}{\cos \beta}\right) \cdot \sin \alpha \cos \beta} \qquad \text{Algebra}$$

$$H = h \frac{\sin \alpha \cos \beta + \cos \alpha \sin \beta}{\sin \alpha \cos \beta - \cos \alpha \sin \beta} \qquad \text{Algebra}$$

$$H = h \frac{\sin(\alpha + \beta)}{\sin(\alpha - \beta)} \qquad \text{Sum and Difference Identities}$$

(C) Substitute the given values to obtain

$$H = 4.90 \frac{\sin(46.23° + 46.15°)}{\sin(46.23° - 46.15°)}$$

$$= 4.90 \frac{\sin(92.38°)}{\sin(0.08°)}$$

$H = 3510$ ft (to three significant digits)

Section 7-3

1. $\cos 2(30°) = \cos 60° = \frac{1}{2}$

$$\cos^2 30° - \sin^2 30° = \left(\frac{\sqrt{3}}{2}\right)^2 - \left(\frac{1}{2}\right)^2 = \frac{3}{4} - \frac{1}{4} = \frac{2}{4} = \frac{1}{2}$$

3. $\tan 2\left(\frac{\pi}{3}\right) = \tan \frac{2\pi}{3} = -\sqrt{3}$

$$\frac{2}{\cot \frac{\pi}{3} - \tan \frac{\pi}{3}} = \frac{2}{\frac{1}{\sqrt{3}} - \sqrt{3}} = \frac{2\sqrt{3}}{\sqrt{3}\left(\frac{1}{\sqrt{3}} - \sqrt{3}\right)} = \frac{2\sqrt{3}}{1 - 3} = \frac{2\sqrt{3}}{-2} = -\sqrt{3}$$

5. $\sin \frac{\pi}{2} = 1 \qquad \sqrt{\dfrac{1 - \cos \pi}{2}} = \sqrt{\dfrac{1 - (-1)}{2}} = \sqrt{\dfrac{2}{2}} = \sqrt{1} = 1$

7. $\sin 22.5° = \sin\left(\frac{1}{2}\cdot 45°\right)$

$= +\sqrt{\frac{1-\cos 45°}{2}}$

(We use + since 22.5° is a I quadrant angle)

$= \sqrt{\frac{1-\frac{1}{\sqrt{2}}}{2}}$

$= \sqrt{\frac{\sqrt{2}-1}{2\sqrt{2}}}$

$= \sqrt{\frac{\sqrt{2}(\sqrt{2}-1)}{2\cdot 2}}$

$= \sqrt{\frac{2-\sqrt{2}}{4}}$

$= \frac{\sqrt{2-\sqrt{2}}}{2}$

9. $\cos 67.5° = \cos\left(\frac{1}{2}\cdot 135°\right)$

$= +\sqrt{\frac{1+\cos 135°}{2}}$

(We use + since 67.5° is a I quadrant angle)

$= \sqrt{\frac{1+\left(-\frac{\sqrt{2}}{2}\right)}{2}}$

$= \sqrt{\frac{1-\frac{\sqrt{2}}{2}}{2}}$

$= \sqrt{\frac{2-\sqrt{2}}{4}}$

$= \frac{\sqrt{2-\sqrt{2}}}{2}$

11.

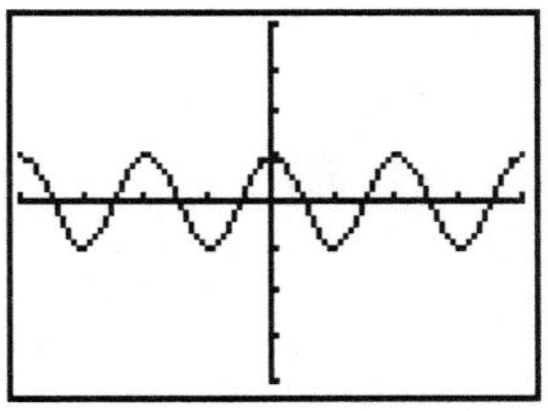

13.

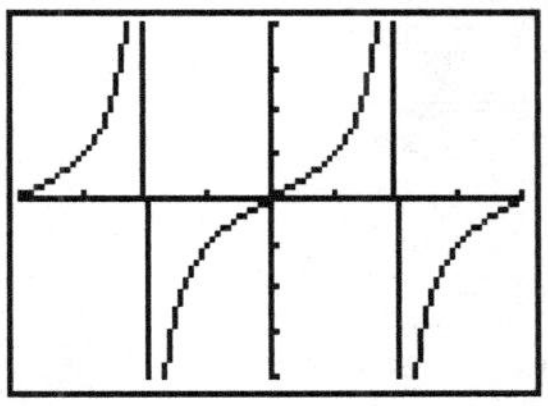

15.

$(\sin x + \cos x)^2$	$= \sin^2 x + 2\sin x\cos x + \cos^2 x$	Algebra
	$= 1 + 2\sin x\cos x$	Pythagorean Identity
	$= 1 + \sin 2x$	Double-angle Identity

17.

$\frac{1}{2}(1-\cos 2x)$	$= \frac{1}{2}[1-(1-2\sin^2 x)]$	Double-angle Identity
	$= \frac{1}{2}[1-1+2\sin^2 x)]$	Algebra
	$= \frac{1}{2}(2\sin^2 x)$	Algebra
	$= \sin^2 x$	Algebra

19.

$\tan x \sin 2x$	$= \frac{\sin x}{\cos x} 2\sin x\cos x$	Quotient and Double-angle Identities
	$= 2\sin^2 x$	Algebra
	$= 2\sin^2 x - 1 + 1$	Algebra
	$= 1 - (1-2\sin^2 x)$	Algebra
	$= 1 - \cos 2x$	Double-angle Identity

21. $\sin^2 \frac{x}{2} = \left(\pm\sqrt{\frac{1 - \cos x}{2}}\right)^2$ Half-angle Identity

$= \left(\sqrt{\frac{1 - \cos x}{2}}\right)^2$ Algebra

$= \frac{1 - \cos x}{2}$ Algebra

23. $\cot 2x = \frac{1}{\tan 2x}$ Reciprocal Identity

$= \frac{1}{\frac{2\tan x}{1 - \tan^2 x}}$ Double-angle Identity

$= \frac{1 - \tan^2 x}{2\tan x}$ Algebra

25. $\cot \frac{\theta}{2} = \frac{\cos \frac{\theta}{2}}{\sin \frac{\theta}{2}}$ Quotient Identity

$= \frac{\pm\sqrt{\frac{1+\cos\theta}{2}}}{\pm\sqrt{\frac{1-\cos\theta}{2}}}$ Half-angle Identities

$= \pm\sqrt{\frac{1 + \cos\theta}{1 - \cos\theta}}$ Algebra

$\left|\cot \frac{\theta}{2}\right| = \sqrt{\frac{1 + \cos\theta}{1 - \cos\theta}}$ Algebra

$= \sqrt{\frac{1 + \cos\theta}{1 - \cos\theta} \cdot \frac{1 - \cos\theta}{1 - \cos\theta}}$ Algebra

$= \sqrt{\frac{1 - \cos^2\theta}{(1 - \cos\theta)^2}}$ Algebra

$= \sqrt{\frac{\sin^2\theta}{(1 - \cos\theta)^2}}$ Pythagorean Identity

$= \left|\frac{\sin\theta}{1 - \cos\theta}\right|$ Algebra

Since $1 - \cos\theta \geq 0$ and $\sin\theta$ has the same sign as $\cot \frac{\theta}{2}$, we may drop the absolute value signs to obtain

$$\cot \frac{\theta}{2} = \frac{\sin\theta}{1 - \cos\theta}$$

[To show that $\sin\theta$ has the same sign as $\cot \frac{\theta}{2}$, we note the following cases:

If $0 < \theta < \pi$, $\sin\theta > 0$ then $0 < \frac{\theta}{2} < \frac{\pi}{2}$, $\cot \frac{\theta}{2} > 0$.

If $\pi = \theta$, $\cot \frac{\theta}{2} = \cot \frac{\pi}{2} = 0$, $\frac{\sin \theta}{1 - \cos \theta} = \frac{\sin \pi}{1 - \cos \pi} = \frac{0}{1 - (-1)} = 0$

If $\pi < \theta < 2\pi$, $\sin \theta < 0$, then $\frac{\pi}{2} < \frac{\theta}{2} < \pi$, $\cot \frac{\theta}{2} < 0$.

The truth of the statement for other values of θ follows since $\sin(\theta + 2k\pi) = \sin \theta$ and $\cot \frac{\theta + 2k\pi}{2} = \cot \frac{\theta}{2}$.]

Key Algebraic Steps:

If $y = \pm\sqrt{\frac{1 + a}{1 - a}}$, then

$$|y| = \sqrt{\frac{1 + a}{1 - a}} = \sqrt{\frac{(1 + a)(1 - a)}{(1 - a)(1 - a)}} = \sqrt{\frac{1 - a^2}{(1 - a)^2}}$$

27.

$\cos 2u = \cos^2 u - \sin^2 u$	Double-angle Identity
$= \dfrac{\cos^2 u - \sin^2 u}{1}$	Algebra
$= \dfrac{\cos^2 u - \sin^2 u}{\cos^2 u + \sin^2 u}$	Pythagorean Identity
$= \dfrac{\frac{\cos^2 u}{\cos^2 u} - \frac{\sin^2 u}{\cos^2 u}}{\frac{\cos^2 u}{\cos^2 u} + \frac{\sin^2 u}{\cos^2 u}}$	Algebra
$= \dfrac{1 - \frac{\sin^2 u}{\cos^2 u}}{1 + \frac{\sin^2 u}{\cos^2 u}}$	Algebra
$= \dfrac{1 - \tan^2 u}{1 + \tan^2 u}$	Quotient Identity

29.

$2 \csc 2x = 2 \cdot \dfrac{1}{\sin 2x}$	Reciprocal Identity
$= \dfrac{2}{\sin 2x}$	Algebra
$= \dfrac{2}{2 \sin x \cos x}$	Double-angle Identity
$= \dfrac{1}{\sin x \cos x}$	Algebra
$= \dfrac{\cos^2 x + \sin^2 x}{\sin x \cos x}$	Pythagorean Identity
$= \dfrac{\frac{\cos^2 x}{\cos^2 x} + \frac{\sin^2 x}{\cos^2 x}}{\frac{\sin x \cos x}{\cos^2 x}}$	Algebra
$= \dfrac{1 + \frac{\sin^2 x}{\cos^2 x}}{\frac{\sin x}{\cos x}}$	Algebra

$$= \frac{1 + \tan^2 x}{\tan x}$$ Quotient Identity

31. $$\frac{1 - \tan^2 \frac{\alpha}{2}}{1 + \tan^2 \frac{\alpha}{2}} = \frac{1 - \left(\pm\sqrt{\frac{1-\cos\alpha}{1+\cos\alpha}}\right)^2}{1 + \left(\pm\sqrt{\frac{1-\cos\alpha}{1+\cos\alpha}}\right)^2}$$ Half-angle Identity

$$= \frac{1 - \frac{1-\cos\alpha}{1+\cos\alpha}}{1 + \frac{1-\cos\alpha}{1+\cos\alpha}}$$ Algebra

$$= \frac{(1 + \cos\alpha)\left(1 - \frac{1-\cos\alpha}{1+\cos\alpha}\right)}{(1 + \cos\alpha)\left(1 + \frac{1-\cos\alpha}{1+\cos\alpha}\right)}$$ Algebra

$$= \frac{1 + \cos\alpha - (1 - \cos\alpha)}{1 + \cos\alpha + 1 - \cos\alpha}$$ Algebra

$$= \frac{1 + \cos\alpha - 1 + \cos\alpha}{1 + \cos\alpha + 1 - \cos\alpha}$$ Algebra

$$= \frac{2\cos\alpha}{2}$$ Algebra

$$= \cos\alpha$$

Key Algebraic Steps:

$$\frac{1 - \left(\pm\sqrt{\frac{1-x}{1+x}}\right)^2}{1 + \left(\pm\sqrt{\frac{1-x}{1+x}}\right)^2} = \frac{1 - \frac{1-x}{1+x}}{1 + \frac{1-x}{1+x}} = \frac{(1 + x)\left(1 - \frac{1-x}{1+x}\right)}{(1 + x)\left(1 + \frac{1-x}{1+x}\right)} = \frac{1 + x - (1 - x)}{1 + x + 1 - x}$$

$$= \frac{1 + x - 1 + x}{1 + x + 1 - x} = \frac{2x}{2} = x$$

33. Plug in $x = \frac{\pi}{6}$.

Left side: $\tan\left(2 \cdot \frac{\pi}{6}\right) = \tan\frac{\pi}{3} = \sqrt{3}$

Right side: $2\tan\frac{\pi}{6} = \frac{2}{\sqrt{3}}$

The equation is not an identity.

35. Plug in $x = \frac{\pi}{3}$.

Left side: $\sin\frac{\pi/3}{2} = \sin\frac{\pi}{6} = \frac{1}{2}$

Right side: $\frac{1}{2}\sin\frac{\pi}{3} = \frac{\sqrt{3}}{4}$

The equation is not an identity.

37. Plug in $x = \frac{4\pi}{3}$.

Left side: $\cos\frac{4\pi/3}{2} = \cos\frac{2\pi}{3} = -\frac{1}{2}$

Right side: $\sqrt{\frac{1 + \cos 4\pi/3}{2}} = \sqrt{\frac{1 - \frac{1}{2}}{2}} = \frac{1}{2}$

The equation is not an identity.

39. Plug in $x = \dfrac{4\pi}{3}$.

Left side: $\tan \dfrac{4\pi/3}{2} = \tan \dfrac{2\pi}{3} = -\sqrt{3}$

Right side: $\sqrt{\dfrac{1 - \cos \frac{4\pi}{3}}{1 + \cos \frac{4\pi}{3}}} = \sqrt{\dfrac{1 - \left(-\frac{1}{2}\right)}{1 + \left(-\frac{1}{2}\right)}} = \sqrt{3}$

The equation is not an identity.

41. The equation is not an identity. Plug in $x = \dfrac{\pi}{4}$:

$$\sin\left(4 \cdot \frac{\pi}{4}\right) \stackrel{?}{=} 4 \sin \frac{\pi}{4} \cos \frac{\pi}{4}$$

$$\sin \pi \stackrel{?}{=} 4 \cdot \frac{1}{\sqrt{2}} \cdot \frac{1}{\sqrt{2}}$$

$$0 \stackrel{?}{=} 4 \cdot \frac{1}{2}$$

This is false.

43. Expand the left side using a double-angle formula:

$\cot 2x = \dfrac{1}{\tan(2x)}$	$\cot x = \dfrac{1}{\tan x}$
$= \dfrac{1}{\frac{2 \tan x}{1 \tan^2 x}}$	Double-angle formula
$= \dfrac{1 \tan^2 x}{2 \tan x}$	Algebra

Now simplify the left side:

$\dfrac{\tan x(\cot^2 x - 1)}{2} = \dfrac{\tan x\left(\frac{1}{\tan^2 x} - 1\right)}{2}$	$\cot^2 x = \dfrac{1}{\tan^2 x}$
$= \dfrac{\frac{1}{\tan x} - \tan x}{2}$	Algebra
$= \dfrac{\tan x}{\tan x} \cdot \dfrac{\left(\frac{1}{\tan x} - \tan x\right)}{2}$	Algebra
$= \dfrac{1 - \tan^2 x}{2 \tan x}$	Algebra

Both sides can be simplified to the same expression, so the equation is an identity.

45. The equation is not an identity. Plug in $x = \pi$.

$$\cos(2\pi) \stackrel{?}{=} 1 - 2 \cos^2 \pi$$

$$1 \stackrel{?}{=} 1 - 2(-1)^2$$

$$1 \stackrel{?}{=} 1 - 2$$

This is false.

47. $a = -\sqrt{5^2 - 3^2} = -4$

$\cos x = -\dfrac{4}{5} \qquad \tan x = -\dfrac{3}{4}$

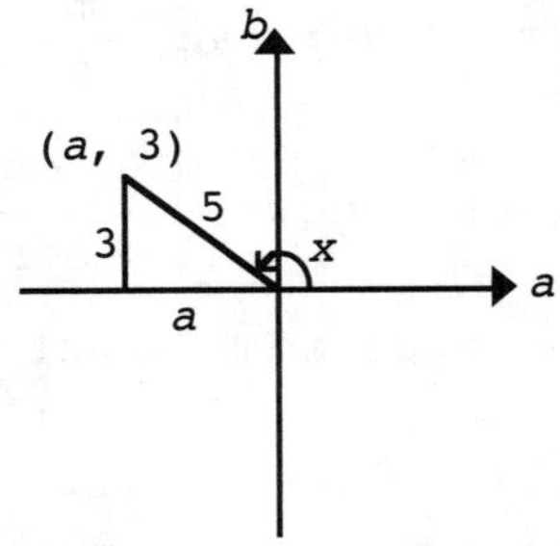

$\sin 2x = 2 \sin x \cos x = 2\left(\dfrac{3}{5}\right)\left(-\dfrac{4}{5}\right) = -\dfrac{24}{25}$

$\cos 2x = 1 - 2 \sin^2 x = 1 - 2\left(\dfrac{3}{5}\right)^2 = 1 - \dfrac{18}{25} = \dfrac{7}{25}$

$\tan 2x = \dfrac{\sin 2x}{\cos 2x} = \left(-\dfrac{24}{25}\right) \div \left(\dfrac{7}{25}\right) = -\dfrac{24}{25} \cdot \dfrac{25}{7} = -\dfrac{24}{7}$

49. $r = \sqrt{12^2 + (-5)^2} = 13$

$\sin x = -\dfrac{5}{13} \qquad \cos x = \dfrac{12}{13}$

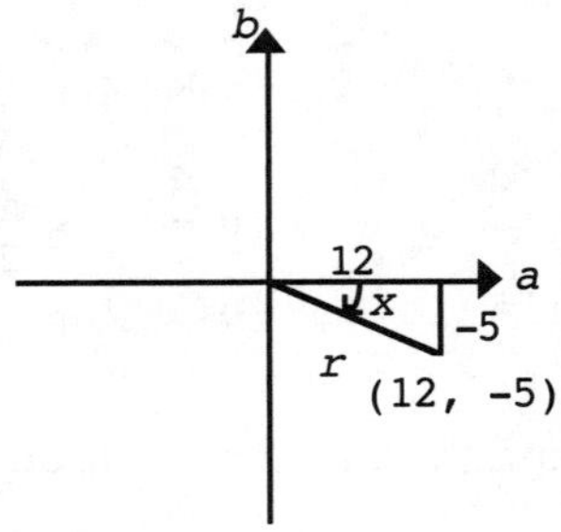

$\sin 2x = 2 \sin x \cos x = 2\left(-\dfrac{5}{13}\right)\left(\dfrac{12}{13}\right) = -\dfrac{120}{169}$

$\cos 2x = 2 \cos^2 x - 1 = 2\left(\dfrac{12}{13}\right)^2 - 1 = \dfrac{288}{169} - 1 = \dfrac{119}{169}$

$\tan 2x = \dfrac{\sin 2x}{\cos 2x} = \left(-\dfrac{120}{169}\right) \div \left(\dfrac{119}{169}\right) = -\dfrac{120}{169} \cdot \dfrac{169}{119} = -\dfrac{120}{119}$

51. Since $\pi < x < \dfrac{3\pi}{2}$, $\dfrac{\pi}{2} < \dfrac{x}{2} < \dfrac{3\pi}{4}$, $\sin \dfrac{x}{2}$ will be positive, $\cos \dfrac{x}{2}$, $\tan \dfrac{x}{2}$ will be negative.

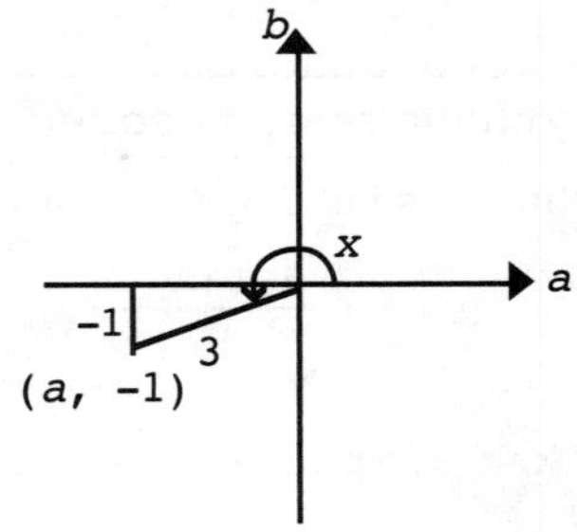

$a = -\sqrt{3^2 - (-1)^2} = -\sqrt{8}$

$\cos x = \dfrac{-\sqrt{8}}{3}$

$$\sin \frac{1}{2}x = \sqrt{\frac{1 - \cos x}{2}} = \sqrt{\frac{1 - \left(-\frac{\sqrt{8}}{3}\right)}{2}} = \sqrt{\frac{1 + \frac{\sqrt{8}}{3}}{2}} = \sqrt{\frac{3 + \sqrt{8}}{6}} = \sqrt{\frac{3 + 2\sqrt{2}}{6}}$$

$$\cos \frac{1}{2}x = -\sqrt{\frac{1 + \cos x}{2}} = -\sqrt{\frac{1 + \left(-\frac{\sqrt{8}}{3}\right)}{2}} = -\sqrt{\frac{1 - \frac{\sqrt{8}}{3}}{2}} = -\sqrt{\frac{3 - \sqrt{8}}{6}} = -\sqrt{\frac{3 - 2\sqrt{2}}{6}}$$

$$\tan \frac{1}{2}x = -\sqrt{\frac{1 - \cos x}{1 + \cos x}} = -\sqrt{\frac{1 - \left(-\frac{\sqrt{8}}{3}\right)}{1 + \left(-\frac{\sqrt{8}}{3}\right)}} = -\sqrt{\frac{3 + \sqrt{8}}{3 - \sqrt{8}}} = -\sqrt{\frac{(3 + \sqrt{8})(3 + \sqrt{8})}{(3 - \sqrt{8})(3 + \sqrt{8})}}$$

$$= -\sqrt{(3 + \sqrt{8})^2} = -(3 + \sqrt{8}) = -3 - \sqrt{8} = -3 - 2\sqrt{2}$$

53. Since $-\pi < x < -\frac{\pi}{2}$, $-\frac{\pi}{2} < \frac{x}{2} < -\frac{\pi}{4}$, $\cos\frac{x}{2}$ will be positive, $\sin\frac{x}{2}$, $\tan\frac{x}{2}$ will be negative.

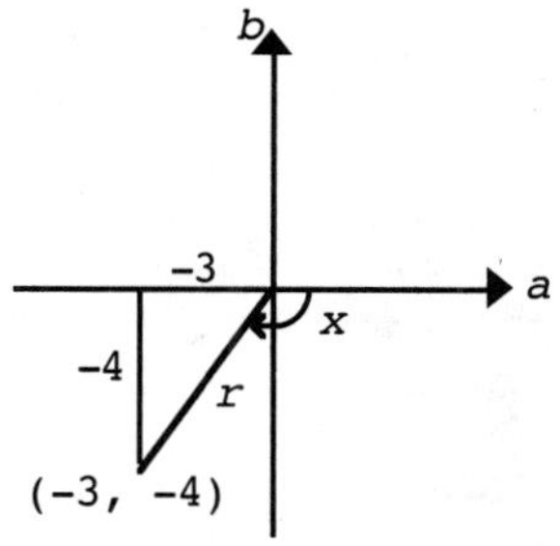

$r = \sqrt{(-3)^2 + (-4)^2} = 5$

$\sin x = -\frac{4}{5} \qquad \cos x = -\frac{3}{5}$

$$\sin\frac{1}{2}x = -\sqrt{\frac{1 - \cos x}{2}} = -\sqrt{\frac{1 - \left(-\frac{3}{5}\right)}{2}}$$
$$= -\sqrt{\frac{1 + \frac{3}{5}}{2}} = -\sqrt{\frac{\frac{8}{5}}{2}} = -\sqrt{\frac{4}{5}} = -\frac{2\sqrt{5}}{5}$$

$$\cos\frac{1}{2}x = \sqrt{\frac{1 + \cos x}{2}} = \sqrt{\frac{1 + \left(-\frac{3}{5}\right)}{2}} = \sqrt{\frac{\frac{2}{5}}{2}} = \sqrt{\frac{1}{5}} = \frac{\sqrt{5}}{5}$$

$$\tan\frac{1}{2}x = \frac{\sin\frac{1}{2}x}{\cos\frac{1}{2}x} = \frac{-2\sqrt{5}}{5} \div \frac{\sqrt{5}}{5} = -\frac{2\sqrt{5}}{5} \cdot \frac{5}{\sqrt{5}} = -2$$

55. (A) 2θ is a second quadrant angle, since θ is a first quadrant angle and $\tan 2\theta$ is negative for 2θ in the second quadrant and not for 2θ in the first.

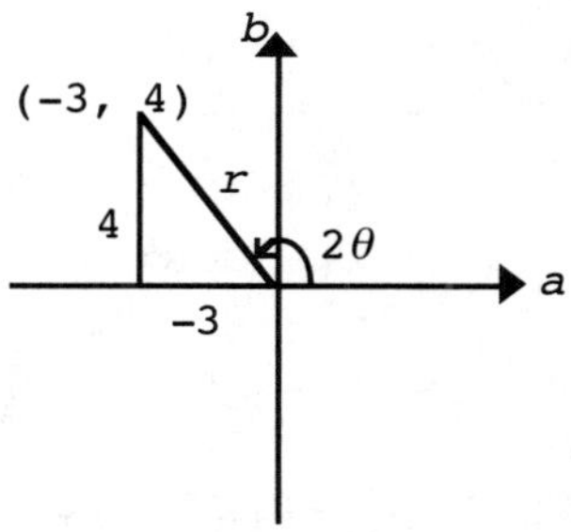

(B) Construct a reference triangle for 2θ in the second quadrant with $(a, b) = (-3, 4)$. Use the Pythagorean theorem to find $r = 5$.
Thus, $\sin 2\theta = \frac{4}{5}$ and $\cos 2\theta = -\frac{3}{5}$.

(C) The double angle identities $\cos 2\theta = 1 - 2\sin^2\theta$ and $\cos 2\theta = 2\cos^2\theta - 1$.

(D) Use the identities in part (C) in the form

$$\sin\theta = \sqrt{\frac{1 - \cos 2\theta}{2}} \quad \text{and} \quad \cos\theta = \sqrt{\frac{1 + \cos 2\theta}{2}}$$

The positive radicals are used because θ is in quadrant one.

(E) $$\sin\theta = \sqrt{\frac{1 - \left(-\frac{3}{5}\right)}{2}} = \sqrt{\frac{5 + 3}{10}} = \sqrt{\frac{8}{10}} = \sqrt{\frac{4}{5}} = \frac{2}{\sqrt{5}} = \frac{2\sqrt{5}}{5}$$

$$\cos\theta = \sqrt{\frac{1 + \left(-\frac{3}{5}\right)}{2}} = \sqrt{\frac{5 - 3}{10}} = \sqrt{\frac{2}{10}} = \sqrt{\frac{1}{5}} = \frac{1}{\sqrt{5}} = \frac{\sqrt{5}}{5}$$

57. (A) $\tan[2(252.06°)] = -0.72335$

$$\frac{2\tan x}{1 - \tan^2 x} = \frac{2\tan(252.06°)}{1 - \tan^2(252.06°)} = -0.72335$$

(B) $\cos\frac{252.06°}{2} = -0.58821$

$$-\sqrt{\frac{1 + \cos 252.06°}{2}} = -0.58821$$

59. (A) $\tan[2(0.93457)] = -3.2518$

$$\frac{2\tan(0.93457)}{1 - \tan^2(0.93457)} = -3.2518$$

(B) $\cos\frac{0.93457}{2} = 0.89279$

$$\sqrt{\frac{1 + \cos 0.93457}{2}} = 0.89279$$

61.

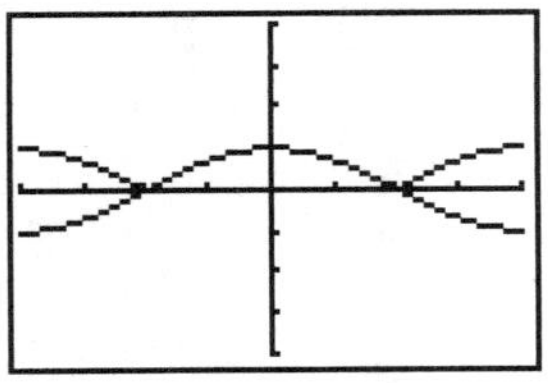

The graphs appear to coincide on the interval $[-\pi, \pi]$.

63.

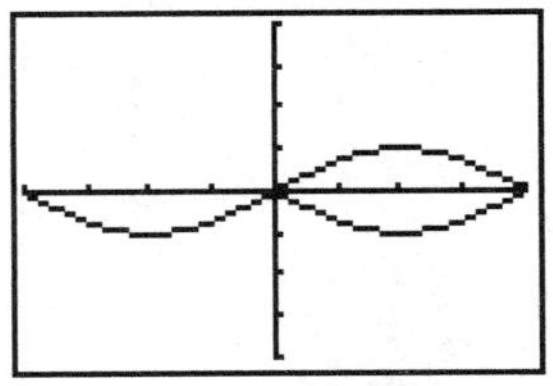

The graphs appear to coincide on the interval $[-2\pi, 0]$.

65.

$$\begin{aligned}
\cos 3x &= \cos(2x + x) && \text{Algebra}\\
&= \cos 2x \cos x - \sin 2x \sin x && \text{Sum Identity}\\
&= (2\cos^2 x - 1)\cos x - 2 \sin x \cos x \sin x && \text{Double-angle Identities}\\
&= 2\cos^3 x - \cos x - 2\sin^2 x \cos x && \text{Algebra}\\
&= 2\cos^3 x - \cos x - 2(1 - \cos^2 x)\cos x && \text{Pythagorean Identity}\\
&= 2\cos^3 x - \cos x - 2\cos x + 2\cos^3 x && \text{Algebra}\\
&= 4\cos^3 x - 3\cos x && \text{Algebra}
\end{aligned}$$

67.

$$\begin{aligned}
\cos 4x &= \cos 2(2x) && \text{Algebra}\\
&= 2\cos^2 2x - 1 && \text{Double-angle Identity}\\
&= 2(2\cos^2 x - 1)^2 - 1 && \text{Double-angle Identity}\\
&= 2(4\cos^4 x - 4\cos^2 x + 1) - 1 && \text{Algebra}\\
&= 8\cos^4 x - 8\cos^2 x + 2 - 1 && \text{Algebra}\\
&= 8\cos^4 x - 8\cos^2 x + 1 && \text{Algebra}
\end{aligned}$$

69. Let $u = \cos^{-1} \frac{3}{5}$. Then $\cos u = \frac{3}{5}$, $0 < u < \pi$. $\cos\left(2\cos^{-1}\frac{3}{5}\right) = \cos 2u$

$$= 2\cos^2 u - 1 = 2\left(\frac{3}{5}\right)^2 - 1 = -\frac{7}{25}$$

71. Let $u = \cos^{-1}\left(-\frac{4}{5}\right)$. Then $\cos u = -\frac{4}{5}$, $0 < u < \pi$.

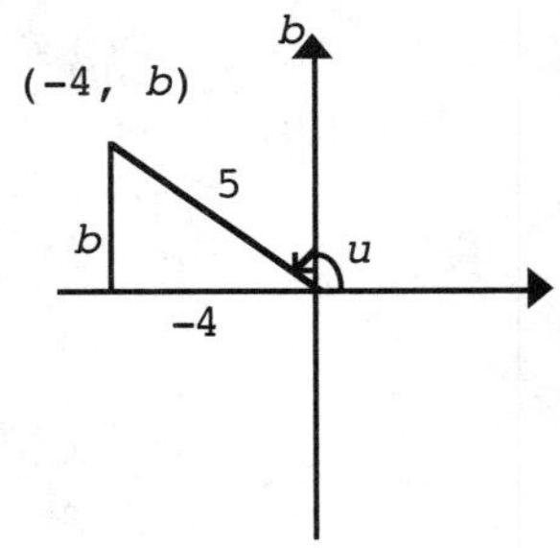

$$b = \sqrt{5^2 - (-4)^2} = 3$$

$$\tan u = -\frac{3}{4}$$

$$\tan\left[2\cos^{-1}\left(-\frac{4}{5}\right)\right] = \tan 2u = \frac{2\tan u}{1 - \tan^2 u} = \frac{2\left(-\frac{3}{4}\right)}{1 - \left(-\frac{3}{4}\right)^2}$$

$$= \frac{\frac{-3}{2}}{1 - \frac{9}{16}} = \frac{\frac{-3}{2}}{\frac{7}{16}} = \left(-\frac{3}{2}\right) \div \frac{7}{16}$$

$$= -\frac{3}{2} \cdot \frac{16}{7} = -\frac{24}{7}$$

73. Let $u = \cos^{-1}\left(-\frac{3}{5}\right)$. Then $\cos u = -\frac{3}{5}$, $0 < u < \pi$. Since $0 < \frac{1}{2}u < \frac{\pi}{2}$, $\cos \frac{1}{2}u$ is positive. So

$$\cos\left[\frac{1}{2}\cos^{-1}\left(-\frac{3}{5}\right)\right] = \cos\frac{1}{2}u = \sqrt{\frac{1 + \cos u}{2}} = \sqrt{\frac{1 + -\frac{3}{5}}{2}} = \sqrt{\frac{\frac{2}{5}}{2}} = \sqrt{\frac{1}{5}} = \frac{\sqrt{5}}{5}$$

75. Here is the graph of $f(x) = \csc x - \cot x$ in a graphing calculator.

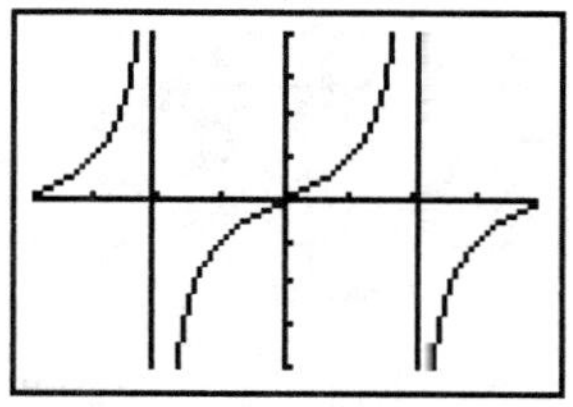

The graph appears to have the form $y = A \tan Bx$.

Since the period is 2π, $\frac{\pi}{B} = 2\pi$, thus $B = \frac{1}{2}$.

The graph of $A \tan \frac{1}{2}x$ shown appears to pass through $\left(\frac{\pi}{2}, 1\right)$, thus $1 = A \tan \frac{1}{2} \cdot \frac{\pi}{2}$

$$1 = A \tan \frac{\pi}{4}$$
$$1 = A$$

The equation of the graph appears to be $y = \tan \frac{1}{2}x$. We verify that $\csc x - \cot x = \tan \frac{1}{2}x$ is an identity:

$\csc x - \cot x = \frac{1}{\sin x} - \frac{\cos x}{\sin x}$ Quotient and Reciprocal Identities

$= \frac{1 - \cos x}{\sin x}$ Algebra

$= \tan \frac{1}{2}x$ Half-angle Identity

77. Here is the graph of $f(x) = \frac{1 - 2\cos 2x}{2\sin x - 1}$ in a graphing calculator.

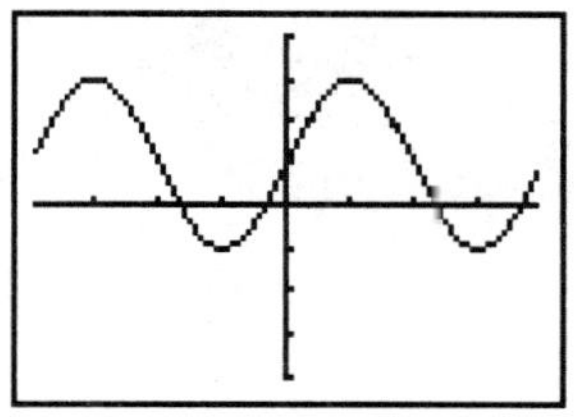

The graph appears to be a standard sine curve shifted up 1 unit. It has amplitude $\frac{3 - (-1)}{2} = 2$ and period 2π, thus, it appears to be the graph of $y = 1 + 2 \sin x$. We verify that $\frac{1 - 2\cos 2x}{2\sin x - 1} = 1 + 2 \sin x$ is an identity:

$\frac{1 - 2\cos 2x}{2\sin x - 1} = \frac{1 - 2(1 - 2\sin^2 x)}{2\sin x - 1}$ Double-angle Identity

$= \frac{1 - 2 + 4\sin^2 x}{2\sin x - 1}$ Algebra

$= \frac{4\sin^2 x - 1}{2\sin x - 1}$ Algebra

$= \frac{(2\sin x - 1)(2\sin x + 1)}{2\sin x - 1}$ Algebra

$= 2 \sin x + 1$ Algebra

79. Here is the graph of $f(x) = \frac{1}{\cot x \sin 2x - 1}$ in a graphing calculator.

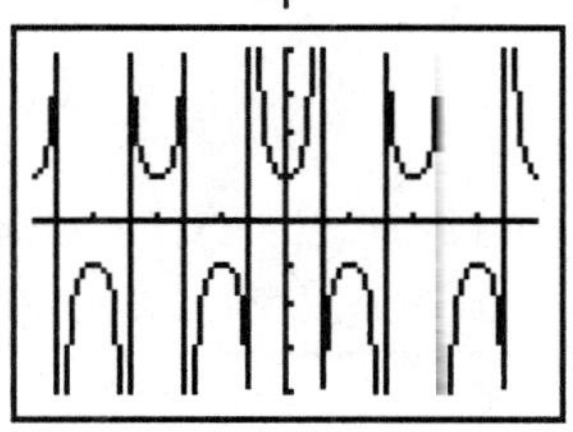

The graph appears to have the form $y = A \sec Bx$.

Since the period is π, $\frac{2\pi}{B} = \pi$, thus $B = 2$.

The graph of $y = A \sec 2x$ shown appears to pass through $(0, 1)$, thus

$$1 = A \sec 2 \cdot 0$$
$$1 = A \sec 0$$
$$1 = A$$

The equation of the graph appears to be $y = \sec 2x$. We verify that

$\dfrac{1}{\cot x \sin 2x - 1} = \sec 2x$ is an identity.

$$\begin{aligned}
\frac{1}{\cot x \sin 2x - 1} &= \frac{1}{\frac{\cos x}{\sin x}\sin 2x - 1} && \text{Quotient Identity}\\
&= \frac{1}{\frac{\cos x}{\sin x}2\sin x\cos x - 1} && \text{Double-angle Identity}\\
&= \frac{1}{2\cos^2 x - 1} && \text{Algebra}\\
&= \frac{1}{\cos 2x} && \text{Double-angle Identity}\\
&= \sec 2x && \text{Reciprocal Identity}
\end{aligned}$$

81. From the figure we see:

ACD is a right triangle, hence $\cos\theta = \frac{7}{8}$, $\theta = \cos^{-1}\frac{7}{8}$

ABC is a right triangle, with angle $BAC = 2\theta$. Hence $\cos 2\theta = \frac{7}{x}$. Using the hint,

$$\begin{aligned}
\cos 2\theta &= 2\cos^2\theta - 1\\
&= 2\left(\frac{7}{8}\right)^2 - 1\\
&= 2\cdot\frac{49}{64} - 1\\
&= \frac{34}{64}\\
&= \frac{17}{32}
\end{aligned}$$

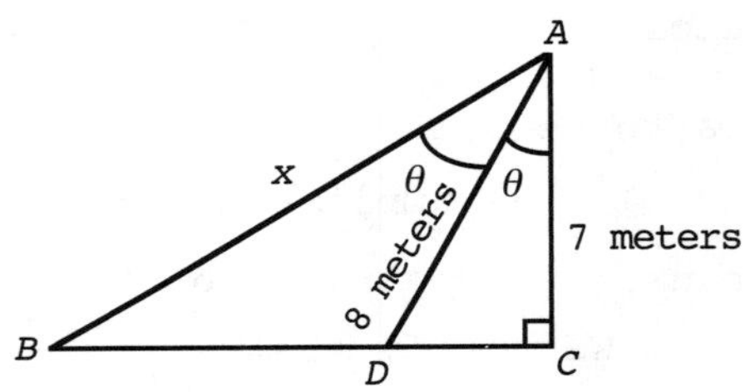

Thus

$$\begin{aligned}
\frac{17}{32} &= \frac{7}{x}\\
17x &= 32(7)\\
x &= \frac{32(7)}{17} = \frac{224}{17}
\end{aligned}$$

To three decimal places

$$x = \frac{224}{17} = 13.176 \text{ meters}$$

$$\theta = \cos^{-1}\frac{7}{8} = 28.955° \quad \text{(calculator in degree mode)}$$

83. (A) Since $2\sin\theta\cos\theta = \sin 2\theta$ by the double-angle identity, we can write

$$d = \frac{2v_0^2\sin\theta\cos\theta}{32\ \text{ft/sec}^2}$$

$$d = \frac{2v_0^2(2\sin\theta\cos\theta)}{32\ \text{ft/sec}^2}$$

$$d = \frac{v_0^2\sin 2\theta}{32\ \text{ft/sec}^2}$$

(B) For fixed v_0, the quantity $\dfrac{v_0^2 \sin 2\theta}{32}$ will be maximum when $\sin 2\theta$ achieves its maximum value, namely 1. Then

$$\sin 2\theta = 1$$
$$2\theta = 90°$$
$$\theta = 45°$$

its maximum value, namely 1. Then
sin 2θ = 1
2θ = 90°
θ = 45°

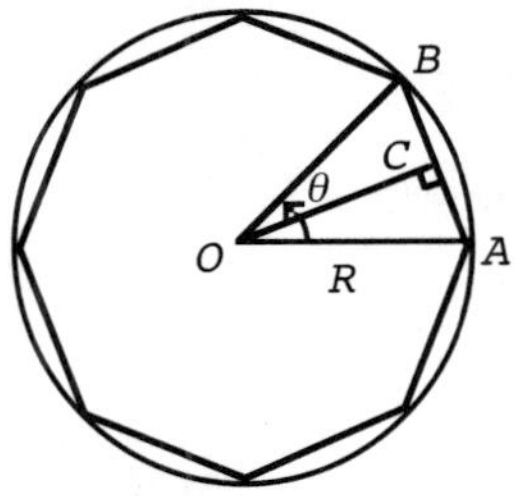

85. (A) Here is a drawing of a typical situation. (The case $n = 8$ is shown, but the reasoning is independent of n):

We note: Area of polygon $= n(\text{Area of triangle } OAB)$

$$= 2n(\text{Area of triangle } OAC)$$
$$= 2n\left(\frac{1}{2} \cdot OC \cdot AC\right)$$
$$= n \cdot OC \cdot AC$$

In triangle OAC, angle $AOC = \dfrac{1}{2}\theta = \dfrac{1}{2} \cdot \dfrac{2\pi}{n} = \dfrac{\pi}{n}$

$$\frac{OC}{R} = \cos AOC = \cos \frac{\pi}{n} \qquad \frac{AC}{R} = \sin AOC = \sin \frac{\pi}{n}$$

Thus, $OC = R \cos \dfrac{\pi}{n}$, $AC = R \sin \dfrac{\pi}{n}$

Hence,

Area of polygon $= n\, R \cos \dfrac{\pi}{n}\; R \sin \dfrac{\pi}{n}$

$$= nR^2 \sin \frac{\pi}{n} \cos \frac{\pi}{n}$$
$$= \frac{1}{2} nR^2 \cdot 2 \sin \frac{\pi}{n} \cos \frac{\pi}{n}$$
$$= \frac{1}{2} nR^2 \sin \frac{2\pi}{n} \text{ by the double-angle identity for sine}$$

(B) TABLE 1

n	10	100	1,000	10,000
A_n	2.93893	3.13953	3.14157	3.14159

(C) A_n appears to approach π, the area of the circle with radius 1.

(D) A_n will not exactly equal the area of the circumscribing circle for any n no matter how large n is chosen; however, A_n can be made as close to the area of the circumscribing circle as we like by making n sufficiently large.

Section 7-4

1. $\sin x \cos y = \dfrac{1}{2}[\sin(x + y) + \sin(x - y)]$

$$\sin 3m \cos m = \frac{1}{2}[\sin(3m + m) + \sin(3m - m)]$$
$$= \frac{1}{2}(\sin 4m + \sin 2m)$$
$$= \frac{1}{2}\sin 4m + \frac{1}{2}\sin 2m$$

3. $$\begin{aligned}\sin x \sin y &= \frac{1}{2}[\cos(x - y) - \cos(x + y)]\\ \sin u \sin 3u &= \frac{1}{2}[\cos(u - 3u) - \cos(u + 3u)]\\ &= \frac{1}{2}(\cos(-2u) - \cos 4u)\\ &= \frac{1}{2}(\cos 2u - \cos 4u)\\ &= \frac{1}{2}\cos 2u - \frac{1}{2}\cos 4u\end{aligned}$$

5. $$\begin{aligned}\sin x + \sin y &= 2 \sin \frac{x + y}{2} \cos \frac{x - y}{2}\\ \sin 3t + \sin t &= 2 \sin \frac{3t + t}{2} \cos \frac{3t - t}{2}\\ &= 2 \sin 2t \cos t\end{aligned}$$

7. $$\begin{aligned}\cos x - \cos y &= -2 \sin \frac{x + y}{2} \sin \frac{x - y}{2}\\ \cos 5w - \cos 9w &= -2 \sin \frac{5w + 9w}{2} \sin \frac{5w - 9w}{2}\\ &= -2 \sin 7w \sin(-2w)\\ &= 2 \sin 7w \sin 2w\end{aligned}$$

9. $$\begin{aligned}\sin x \cos y &= \frac{1}{2}[\sin(x + y) + \sin(x - y)]\\ \sin 195° \cos 75° &= \frac{1}{2}[\sin(195° + 75°) + \sin(195° - 75°)]\\ &= \frac{1}{2}[\sin 270° + \sin 120°]\\ &= \frac{1}{2}\left(-1 + \frac{\sqrt{3}}{2}\right)\\ &= \frac{1}{2}\left(\frac{\sqrt{3} - 2}{2}\right)\\ &= \frac{\sqrt{3} - 2}{4}\end{aligned}$$

11. $$\begin{aligned}\cos x \cos y &= \frac{1}{2}[\cos(x + y) + \cos(x - y)]\\ \cos 15° \cos 75° &= \frac{1}{2}[\cos(15° + 75°) + \cos(15° - 75°)]\\ &= \frac{1}{2}[\cos 90° + \cos(-60°)]\\ &= \frac{1}{2}\left[0 + \frac{1}{2}\right]\\ &= \frac{1}{4}\end{aligned}$$

13.

$$\cos x + \cos y = 2\cos\frac{x+y}{2}\cos\frac{x-y}{2}$$

$$\begin{aligned}\cos 285° + \cos 195° &= 2\cos\frac{285° + 195°}{2}\cos\frac{285° - 195°}{2}\\ &= 2\cos 240° \cos 45°\\ &= 2\left(-\frac{1}{2}\right)\left(\frac{\sqrt{2}}{2}\right)\\ &= -\frac{\sqrt{2}}{2}\end{aligned}$$

15.

$$\cos x - \cos y = -2\sin\frac{x+y}{2}\sin\frac{x-y}{2}$$

$$\begin{aligned}\cos 15° - \cos 105° &= -2\sin\frac{15° + 105°}{2}\sin\frac{15° - 105°}{2}\\ &= -2\sin 60° \sin(-45°)\\ &= -2\,\frac{\sqrt{3}}{2}\left(-\frac{\sqrt{2}}{2}\right)\\ &= \frac{\sqrt{6}}{2}\end{aligned}$$

17.

$$\begin{aligned}\cos(x+y) &= \cos x\cos y - \sin x\sin y\\ \cos(x-y) &= \cos x\cos y + \sin x\sin y\\ \hline \cos(x+y) + \cos(x-y) &= 2\cos x\cos y \quad \text{(adding the above)}\\ \cos x\cos y &= \frac{1}{2}[\cos(x+y) + \cos(x-y)]\end{aligned}$$

19. Let $x = u + v$ and $y = u - v$ and solve for u and v in terms of x and y:

$$\begin{aligned}x &= u + v\\ y &= u - v\end{aligned}$$

$$x + y = 2u \qquad x - y = 2v$$

$$u = \frac{x+y}{2} \qquad v = \frac{x-y}{2}$$

Substitute these results into $\sin u \sin v = \frac{1}{2}[\cos(u - v) - \cos(u + v)]$ to obtain:

$$\sin\frac{x+y}{2}\sin\left(\frac{x-y}{2}\right) = \frac{1}{2}[\cos y - \cos x]$$

or

$$-\sin\frac{x+y}{2}\sin\frac{x-y}{2} = \frac{1}{2}[\cos x - \cos y]$$

or

$$\cos x - \cos y = -2\sin\frac{x+y}{2}\sin\frac{x-y}{2}$$

21.

$\dfrac{\sin 2t + \sin 4t}{\cos 2t - \cos 4t} = \dfrac{2\sin\frac{2t+4t}{2}\cos\frac{2t-4t}{2}}{-2\sin\frac{2t+4t}{2}\sin\frac{2t-4t}{2}}$	Sum-product Identities
$= \dfrac{2\sin 3t\cos(-t)}{-2\sin 3t\sin(-t)}$	Algebra
$= \dfrac{2\sin 3t\cos t}{2\sin 3t\sin t}$	Identities for Negatives
$= \dfrac{\cos t}{\sin t}$	Algebra
$= \cot t$	Quotient Identity

23. $\dfrac{\sin x - \sin y}{\cos x - \cos y} = \dfrac{2\cos\frac{x+y}{2}\sin\frac{x-y}{2}}{-2\sin\frac{x+y}{2}\sin\frac{x-y}{2}}$ Sum-product Identities

$= -\dfrac{\cos\frac{x+y}{2}}{\sin\frac{x+y}{2}}$ Algebra

$= -\cot\dfrac{x+y}{2}$ Quotient Identity

25. $\dfrac{\cos x + \cos y}{\sin x - \sin y} = \dfrac{2\cos\frac{x+y}{2}\cos\frac{x-y}{2}}{2\cos\frac{x+y}{2}\sin\frac{x-y}{2}}$ Sum-product Identities

$= \dfrac{\cos\frac{x-y}{2}}{\sin\frac{x-y}{2}}$ Algebra

$= \cot\dfrac{x-y}{2}$ Quotient Identity

27. $\dfrac{\cos x + \cos y}{\cos x - \cos y} = \dfrac{2\cos\frac{x+y}{2}\cos\frac{x-y}{2}}{-2\sin\frac{x+y}{2}\sin\frac{x-y}{2}}$ Sum-product Identities

$= -\dfrac{\cos\frac{x+y}{2}\cos\frac{x-y}{2}}{\sin\frac{x+y}{2}\sin\frac{x-y}{2}}$ Algebra

$= -\cot\dfrac{x+y}{2}\cot\dfrac{x-y}{2}$ Quotient Identity

29. Let $x = y = 0$

Left side: $\sin 0 \cos 0 = 0$

Right side: $\sin 0 + \cos 0 = 1$

The equation is not an identity.

31. Let $x = 0$ and $y = \dfrac{\pi}{2}$.

Left side: $\sin 0 \sin\dfrac{\pi}{2} = 0$

Right side: $\sin\left(0 + \dfrac{\pi}{2}\right) = 1$

The equation is not an identity.

33. Let $x = \dfrac{\pi}{3}$ and $y = \dfrac{\pi}{3}$.

Left side: $\cos\dfrac{\pi}{3} + \cos\dfrac{\pi}{3} = 1$

Right side: $\cos\dfrac{\pi}{3}\cos\dfrac{\pi}{3} = \dfrac{1}{4}$

The equation is not an identity.

35. Let $x = \frac{\pi}{2}$ and $y = 0$.

Left side: $\sin \frac{\pi}{2} - \sin 0 = 1$

Right side: $\cos \frac{\frac{\pi}{2} + 0}{2} \sin \frac{\frac{\pi}{2} - 0}{2} = \cos \frac{\pi}{4} \sin \frac{\pi}{4} = \frac{1}{\sqrt{2}} \cdot \frac{1}{\sqrt{2}} = \frac{1}{2}$

The equation is not an identity.

37. $\sin x - \sin y = 2 \cos \frac{x + y}{2} \sin \frac{x - y}{2}$ Sum-product Identity

$\sin 3x - \sin x = 2 \cos \frac{3x + x}{2} \sin \frac{3x - x}{2}$ Sum-product Identity

$= 2 \cos \frac{4x}{2} \sin \frac{2x}{2}$ Algebra

$= 2 \cos 2x \sin x$ Algebra

The equation is an identity.

39. Try using a sum-product identity:

$\cos x - \cos y = -2 \sin \frac{x + y}{2} \sin \frac{x - y}{2}$

$\cos 3x - \cos x = -2 \sin \frac{3x + x}{2} \sin \frac{3x - x}{2}$

$= -2 \sin 2x \sin x$

This is not an identity--the right side lacks the negative sign.

41. $\cos x + \cos y = 2 \cos \frac{x + y}{2} \cos \frac{x - y}{2}$ Sum-product Identity

$\cos x + \cos 5x = 2 \cos \frac{x + 5x}{2} \cos \frac{x - 5x}{2}$ Sum-product Identity

$= 2 \cos \frac{6x}{2} \cos -\frac{4x}{2}$ Algebra

$= 2 \cos 3x \cos(-2x)$ Algebra

$= 2 \cos 3x \cos 2x$ $\cos(-2x) = \cos(2x)$

The equation is an identity.

43. (A) $\cos 172.63° \sin 20.177° = -0.34207$

$\frac{1}{2}[\sin(172.63° + 20.177°) - \sin(172.63° - 20.177°)] = -0.34207$

(B) $\cos 172.63° + \cos 20.177° = -0.05311$

$2 \cos \frac{172.63° + 20.177°}{2} \cos \frac{172.63° - 20.177°}{2} = -0.05311$

(calculator in degree mode)

45. (A) $\cos 1.1255 \sin 3.6014 = -0.19115$

$\frac{1}{2}[\sin(1.1255 + 3.6014) - \sin(1.1255 - 3.6014)] = -0.19115$

(B) $\cos 1.1255 + \cos 3.6014 = -0.46541$

$2 \cos \frac{1.1255 + 3.6014}{2} \cos \frac{1.1255 - 3.6014}{2} = -0.46541$

(calculator in radian mode)

47. $\sin 2x + \sin x = 2 \sin \frac{2x + x}{2} \cos \frac{2x - x}{2}$

$= 2 \sin \frac{3x}{2} \cos \frac{x}{2}$

Graph $y1 = \sin 2x + \sin x$ and

$y2 = 2 \sin \frac{3x}{2} \cos \frac{x}{2}$

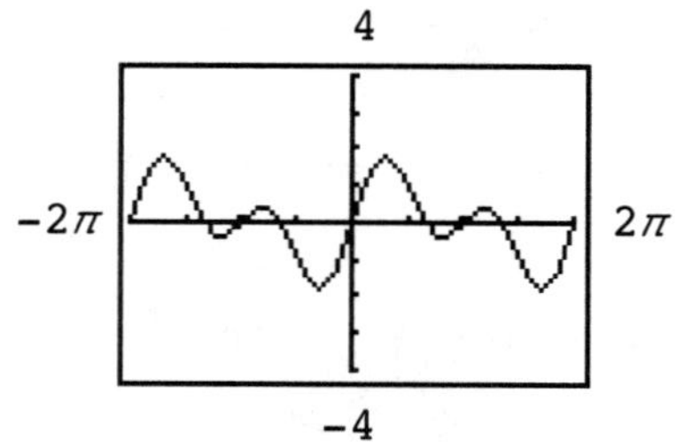

49. $\cos 1.7x - \cos 0.3x$

$= -2 \sin \frac{1.7x + 0.3x}{2} \sin \frac{1.7x - 0.3x}{2}$

$= -2 \sin x \sin 0.7x$

Graph $y1 = \cos 1.7x - \cos 0.3x$

$y2 = -2 \sin x \sin 0.7x$

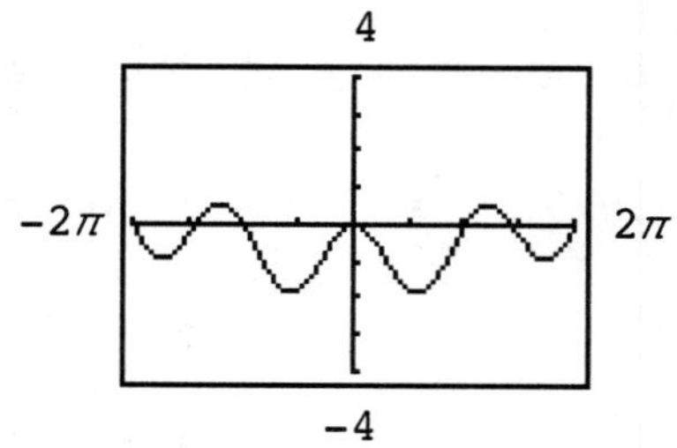

51. $\sin 3x \cos x = \frac{1}{2}[\sin(3x + x) + \sin(3x - x)]$

$= \frac{1}{2}(\sin 4x + \sin 2x)$

Graph $y1 = \sin 3x \cos x$ and

$y2 = \frac{1}{2}(\sin 4x + \sin 2x)$

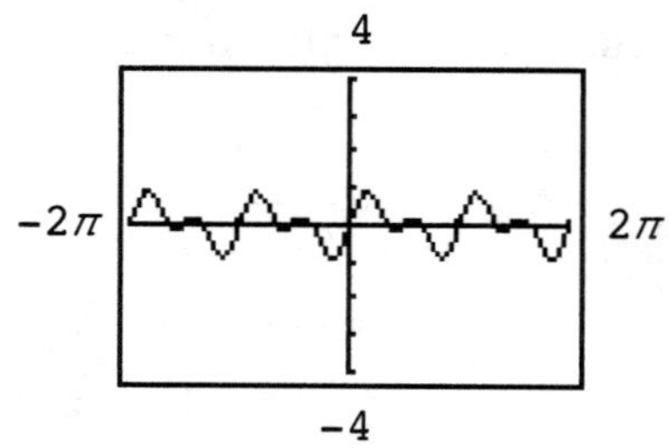

53. $\sin 2.3x \sin 0.7x$

$= \frac{1}{2}[\cos(2.3x - 0.7x) - \cos(2.3x + 0.7x)]$

$= \frac{1}{2}(\cos 1.6x - \cos 3x)$

Graph $y1 = \sin 2.3x \sin 0.7x$

$y2 = \frac{1}{2}(\cos 1.6x - \cos 3x)$

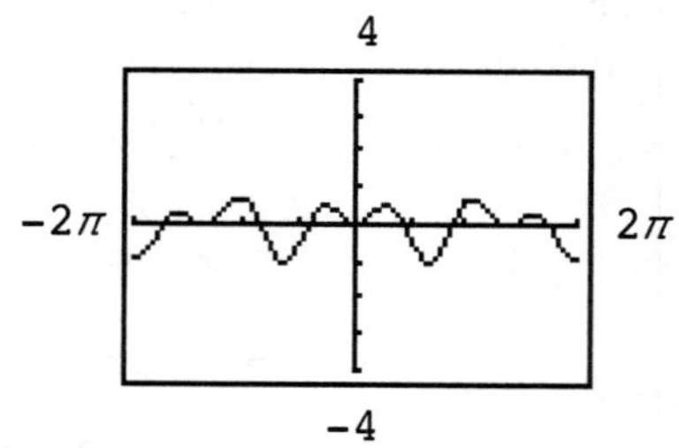

55. $\cos x \cos y \cos z = \cos x \, \frac{1}{2}[\cos(y + z) + \cos(y - z)]$ Product-sum Identity

$= \frac{1}{2} \cos x \cos(y + z) + \frac{1}{2} \cos x \cos(y - z)$ Algebra

$= \frac{1}{2}\left\{\frac{1}{2}[\cos(x + y + z) + \cos(x - \{y + z\})]\right\}$

$+ \frac{1}{2}\left\{\frac{1}{2}[\cos(x + y - z) + \cos(x - \{y - z\})]\right\}$ Product-sum Identity

$= \frac{1}{4} \cos(x + y + z) + \frac{1}{4} \cos(x - y - z)$

$+ \frac{1}{4} \cos(x + y - z) + \frac{1}{4} \cos(x - y + z)$ Algebra

$= \frac{1}{4}[\cos(x + y - z) + \cos(x - y - z)$

$+ \cos(z + x - y) + \cos(x + y + z)]$ Algebra

$$= \frac{1}{4}[\cos(x + y - z) + \cos\{-(y + z - x)\} + \cos(z + x - y) + \cos(x + y + z)] \quad \text{Algebra}$$

$$= \frac{1}{4}[\cos(x + y - z) + \cos(y + z - x) + \cos(z + x - y) + \cos(x + y + z)] \quad \text{Identity for Negatives}$$

57. (A)

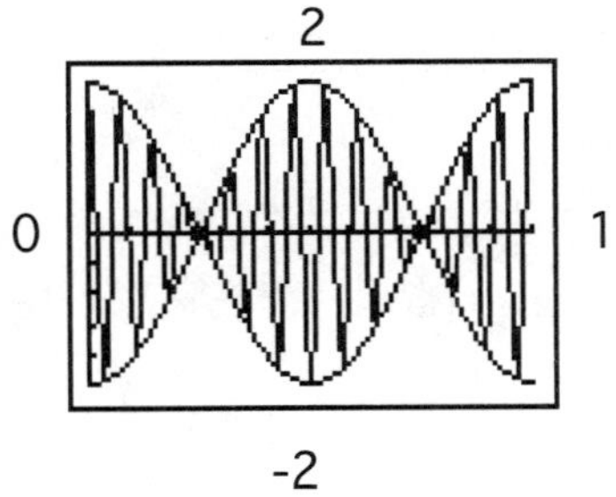

(B) $2\cos(28\pi x)\cos(2\pi x)$

$$= 2 \cdot \frac{1}{2}[\cos(28\pi x + 2\pi x) + \cos(28\pi x - 2\pi x)]$$

$$= \cos 30\pi x + \cos 26\pi x$$

Graphing $y1 = \cos 30\pi x + \cos 26\pi x$ will yield the same result as before.

59. (A)

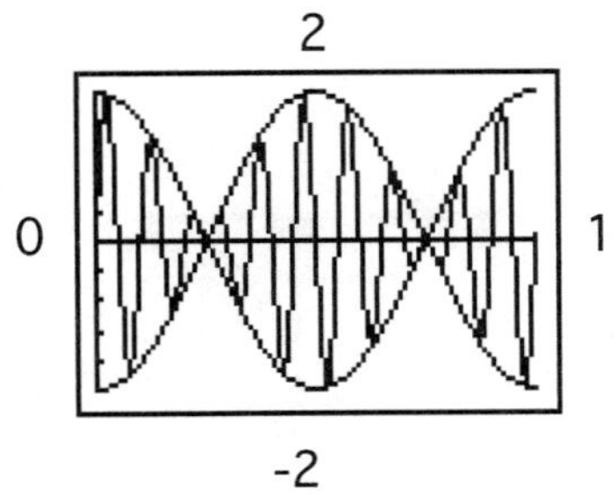

(B) $2\sin(20\pi x)\cos(2\pi x)$

$$= 2 \cdot \frac{1}{2}[\sin(20\pi x + 2\pi x) + \sin(20\pi x - 2\pi x)]$$

$$= \sin 22\pi x + \sin 18\pi x$$

Graphing $y1 = \sin 22\pi x + \sin 18\pi x$ will yield the same result as before.

61. (A) $\cos x - \cos y = -2\sin\frac{x + y}{2}\sin\frac{x - y}{2}$

In this case

$$\cos 128\pi t - \cos 144\pi t = -2\sin\frac{128\pi t + 144\pi t}{2}\sin\frac{128\pi t - 144\pi t}{2}$$

$$\cos 128\pi t - \cos 144\pi t = -2\sin 136\pi t\sin(-8\pi t)$$

$$\cos 128\pi t - \cos 144\pi t = 2\sin 136\pi t\sin 8\pi t$$

Multiplying both sides by 0.5, we have

$$0.5\cos 128\pi t - 0.5\cos 144\pi t = \sin 136\pi t\sin 8\pi t$$

(B) $y = 0.5\cos 128\pi t$

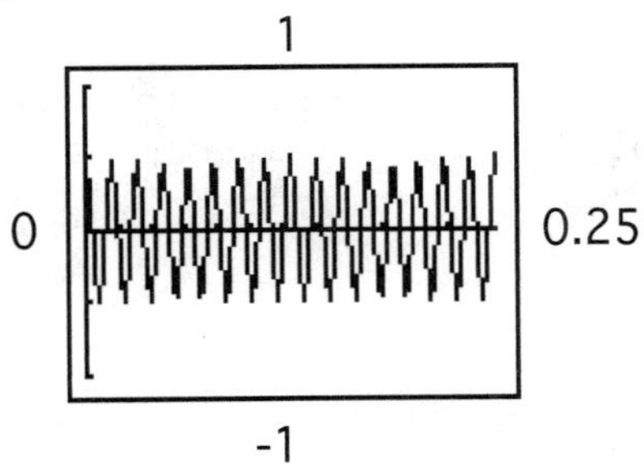

$y = -0.5\cos 144\pi t$

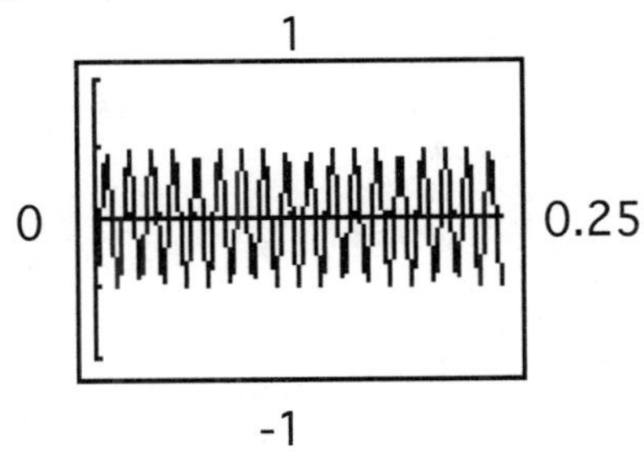

$y = 0.5\cos 128\pi t - 0.5\cos 144\pi t$

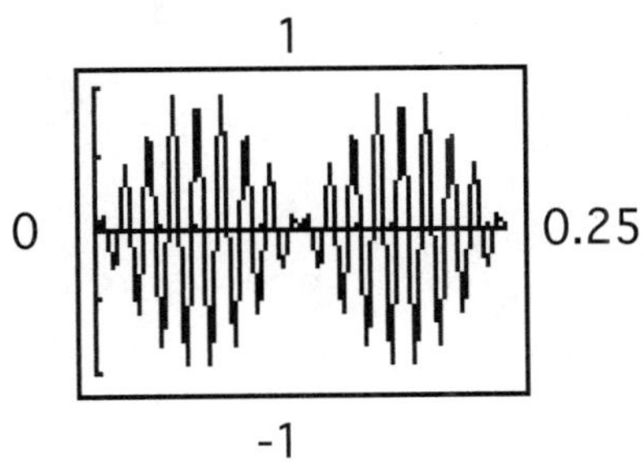

$y = \sin 8\pi t\sin 136\pi t$

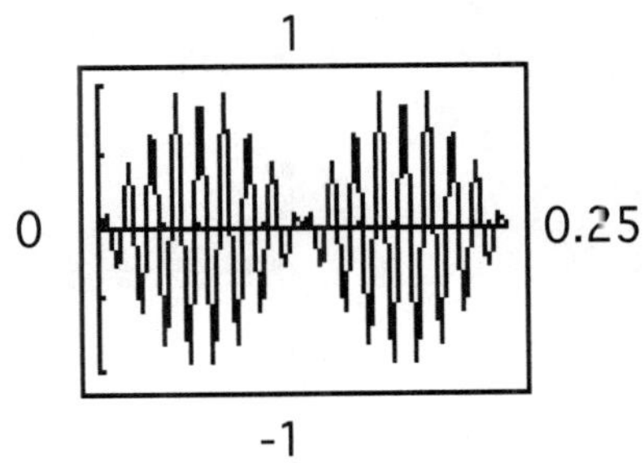

The latter two graphs are the same, illustrating the identity proved in part (A).

Section 7-5

1. $\sin x + 1 = 0$
$\sin x = -1$
Sketch a graph of $y = \sin x$ and $y = -1$ on the interval $[0, 2\pi)$.
$x = \frac{3\pi}{2}$
The checking steps in this and subsequent problems are omitted for lack of space.

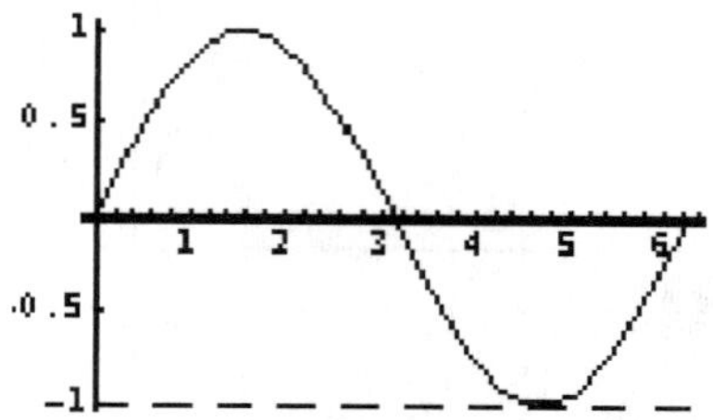

3. We have found all solutions of $\sin x + 1 = 0$ over one period in problem 1. Since the sine function is periodic with period 2π, all solutions are given by
$x = \frac{3\pi}{2} + 2k\pi$, k any integer

5. $\tan \theta - 1 = 0$
$\tan \theta = 1$
Sketch a graph of $y = \tan \theta$ and $y = 1$ on the interval $[0, 360°)$
$\theta = 45°, 225°$

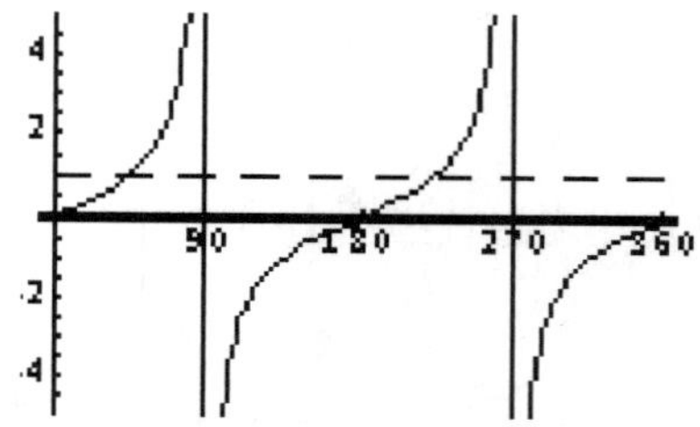

7. We have found all solutions of $\tan \theta - 1 = 0$ over two periods in problem 5. Since the tangent function is periodic with period 180°, all solutions are given by
$\theta = 45° + k180°$ k any integer

9. $2 \sin x - 1 = 0$
$\sin x = \frac{1}{2}$
Sketch a graph of $y = \sin x$ and $y = \frac{1}{2}$ on the interval $[0, 2\pi)$
$x = \frac{\pi}{6}, \frac{5\pi}{6}$

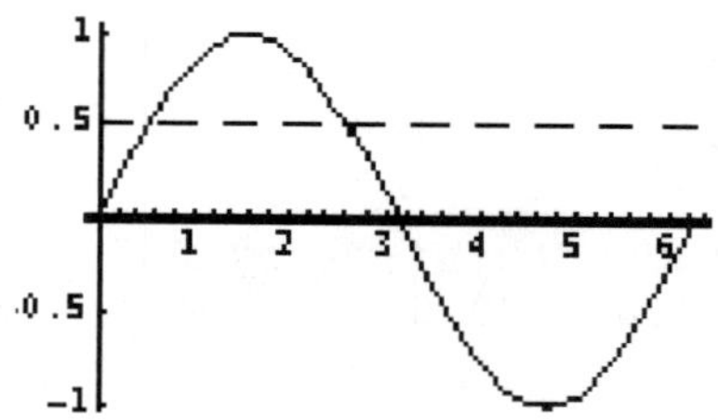

11. We have found all solutions of $2 \sin x - 1 = 0$ over one period in problem 9. Since the sine function is periodic with period 2π, all solutions are given by
$$\left.\begin{array}{l} x = \frac{\pi}{6} + 2k\pi \\ x = \frac{5\pi}{6} + 2k\pi \end{array}\right\} k \text{ any integer}$$

13. $2 \sin \theta + \sqrt{3} = 0$
$\sin \theta = -\frac{\sqrt{3}}{2}$
Sketch a graph of $y = \sin \theta$ and $y = -\frac{\sqrt{3}}{2}$ on the interval $[0, 360°)$.
$\theta = 240°, 300°$

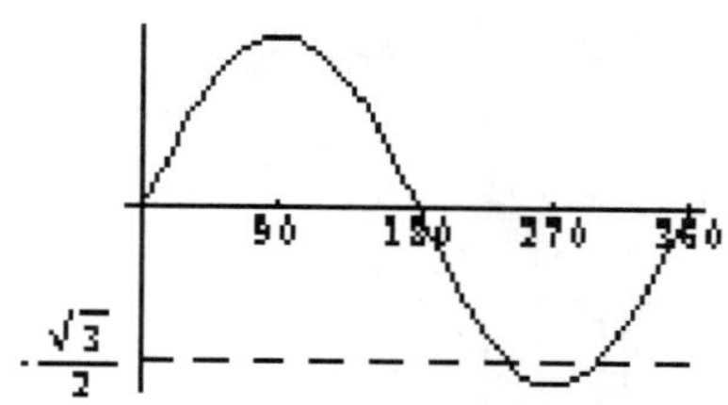

15. We have found all solutions of $2 \sin \theta + \sqrt{3} = 0$ over one period in problem 13. Since the sine function is periodic with period 360°, all solutions are given by
$$\left.\begin{array}{l} \theta = 240° + k360° \\ \theta = 300° + k360° \end{array}\right\} k \text{ any integer}$$

17. $\tan x - \sqrt{3} = 0$

$\tan x = \sqrt{3}$

Sketch a graph of $y = \tan x$ and $y = \sqrt{3}$ on the interval $[0, 2\pi)$.

$x = \dfrac{\pi}{3}, \dfrac{4\pi}{3}$

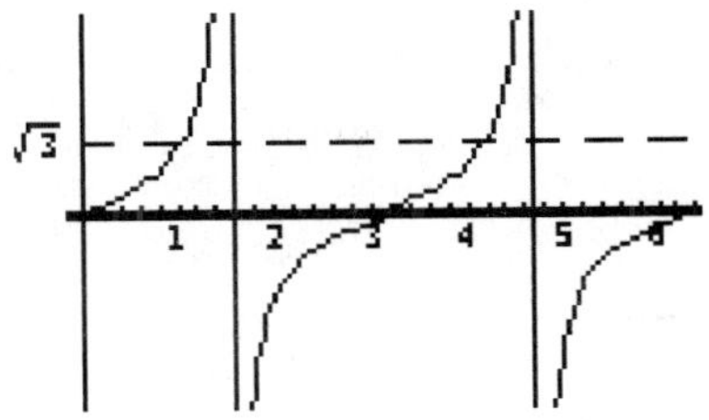

19. We have found all solutions of $\tan x - \sqrt{3} = 0$ over two periods in problem 17. Since the tangent function is periodic with period π, all solutions are given by $x = \dfrac{\pi}{3} + k\pi$, k any integer.

21. $7 \cos x - 3 = 0 \quad 0 \le x < 2\pi$

$\cos x = \dfrac{3}{7}$

Sketch a graph of $y = \cos x$ and $y = \dfrac{3}{7}$, x on the interval $[0, 2\pi)$.

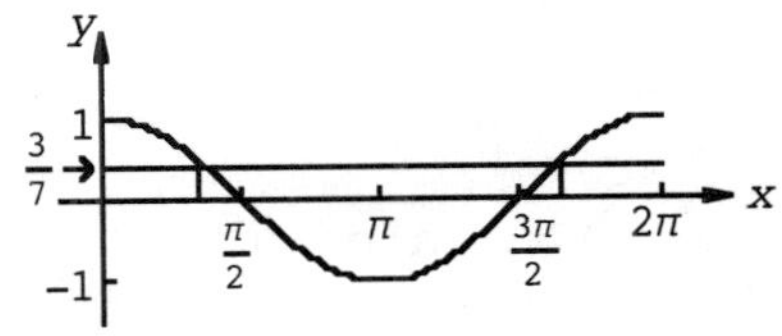

$$x = \begin{cases} \cos^{-1}\dfrac{3}{7} = 1.1279 & \text{First quadrant solution} \\ 2\pi \cos^{-1}\dfrac{3}{7} = 5.1553 & \text{Fourth quadrant solution} \end{cases}$$

23. $2 \tan \theta - 7 = 0 \quad 0° \le \theta < 180°$

$\tan \theta = \dfrac{7}{2}$

Sketch a graph of $y = \tan \theta$ and $y = \dfrac{7}{2}$ on the interval $[0°, 180°)$

$\theta = \tan^{-1}\dfrac{7}{2} = 74.0546°$

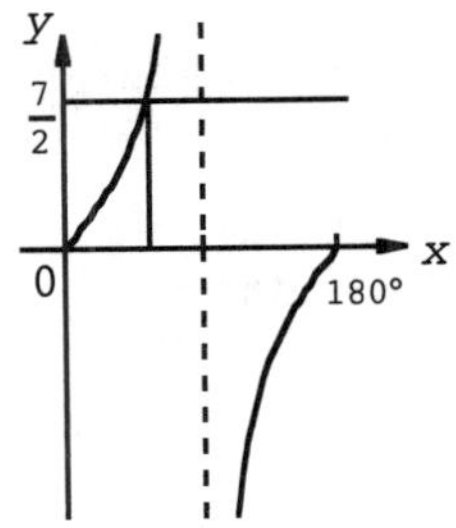

25. $1.3224 \sin x + 0.4732 = 0$

$\sin x = -\dfrac{0.4732}{1.3224}$

Solve over one period $[0, 2\pi)$:

$\sin x = -0.3578$

Sketch a graph of $y = \sin x$ and $y = -0.3578$, on the interval $[0, 2\pi)$

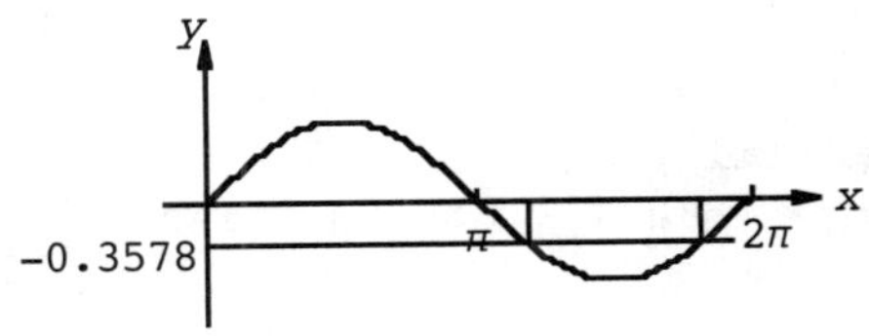

$$x = \begin{cases} \pi - \sin^{-1}(-0.3578) = 3.5075 & \text{Third quadrant solution} \\ 2\pi + \sin^{-1}(-0.3578) = 5.9172 & \text{Fourth quadrant solution} \end{cases}$$

Since the same function is periodic with period 2π, all solutions are given by

$$x = \begin{cases} 3.5075 + 2k\pi \\ 5.9172 + 2k\pi \end{cases} \quad k \text{ any integer}$$

27. Here is a graph of $y1 = 1 - x$ and $y2 = 2 \sin x$, on the interval $0 \le x \le 2\pi$.

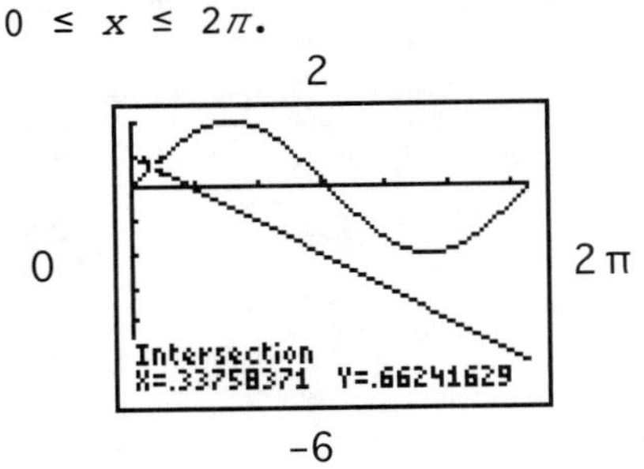

To four decimal places, the only solution is 0.3376.

Check: $1 - x = 1 - 0.3376 = 0.6624$

$2 \sin x = 2 \sin(0.3376) = 0.6624$

29. Here is a graph of $y1 = \tan \frac{x}{2}$ and $y2 = 8 - x$, on the interval $0 \le x \le \pi$.

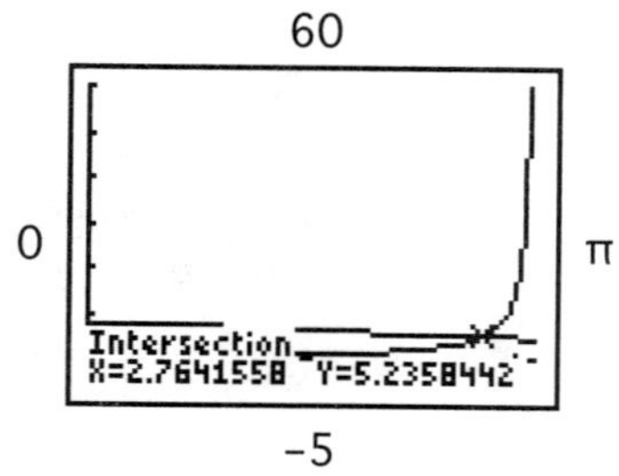

To four decimal places, the only solution is 2.7642.

Check: $\tan \frac{x}{2} = \tan \frac{2.7642}{2} = 5.2358$

$8 - x = 8 - 2.7642 = 5.2358$

31.

$$2 \sin^2 \theta + \sin 2\theta = 0 \quad \text{all } \theta$$
$$2 \sin^2 \theta + 2 \sin \theta \cos \theta = 0$$
$$2 \sin \theta(\sin \theta + \cos \theta) = 0$$

Either $2 \sin \theta = 0$

$\sin \theta = 0$

Solutions over $0° \le \theta < 360°$ are $\theta = 0°, 180°$

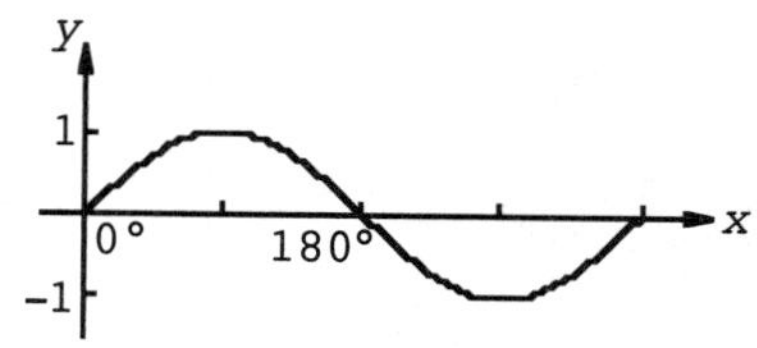

Thus, solutions, if θ is allowed to range over all possible values, are $\theta = 0° + k\ 180°$ or $\theta = k\ 180°$ k any integer.

Or $\sin \theta + \cos \theta = 0$

$$\sin \theta = -\cos \theta$$
$$\frac{\sin \theta}{\cos \theta} = -1$$
$$\tan \theta = -1$$

Solutions over $0° < \theta < 180°$ are $\theta = 135°$

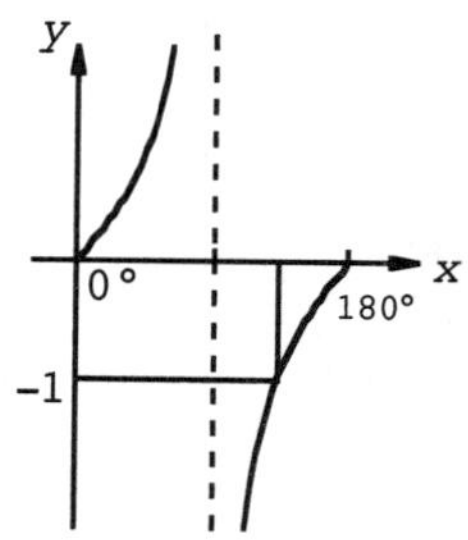

Thus, solutions, if θ is allowed to range over all possible values, are $\theta = 135° + k\ 180°$ k any integer

Solutions: $k\ 180°$, $135° + k\ 180°$ k any integer

33.

$$\tan x = -2 \sin x \qquad 0 \le x < 2\pi$$
$$\frac{\sin x}{\cos x} = -2 \sin x$$
$$\sin x = -2 \sin x \cos x \qquad \cos x \ne 0$$
$$2 \sin x \cos x + \sin x = 0$$
$$\sin x(2 \cos x + 1) = 0$$

$\sin x = 0$ $\qquad$ $2 \cos x + 1 = 0$

$x = 0, \pi$ $\qquad$ $2 \cos x = -1$

$$\cos x = -\frac{1}{2}$$
$$x = \frac{2\pi}{3}, \frac{4\pi}{3}$$

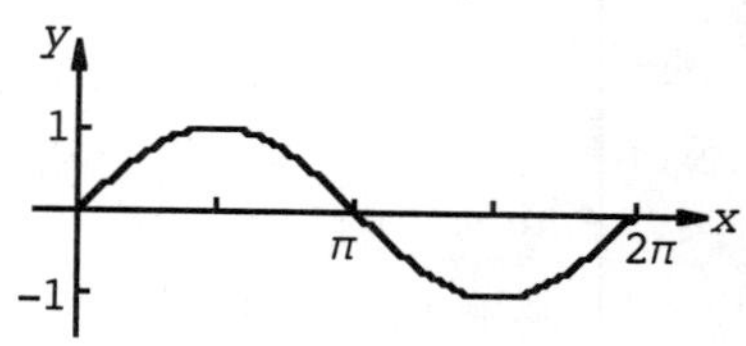

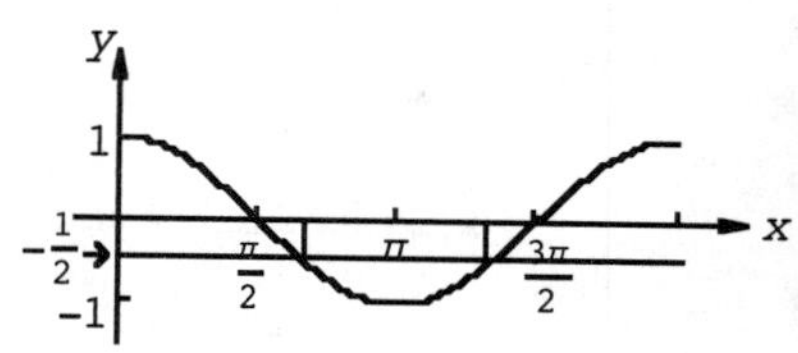

Solutions: $0, \frac{2\pi}{3}, \pi, \frac{4\pi}{3}$

35. $\sec \frac{x}{2} + 2 = 0 \quad 0 \le x \le 2\pi$ is equivalent to

$$\sec \frac{x}{2} + 2 = 0 \quad 0 \le \frac{x}{2} \le \pi$$

$$\sec \frac{x}{2} = -2$$

$$\cos \frac{x}{2} = -\frac{1}{2}$$

$$\frac{x}{2} = \frac{2\pi}{3}$$

$$x = \frac{4\pi}{3}$$

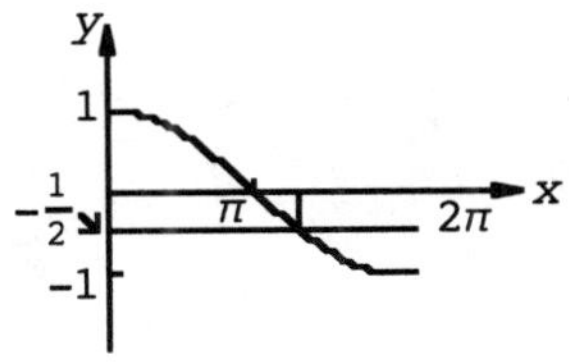

37.

$$2\cos^2\theta + 3\sin\theta = 0 \quad 0° \le \theta < 360°$$

$$2(1 - \sin^2\theta) + 3\sin\theta = 0$$

$$2 - 2\sin^2\theta + 3\sin\theta = 0$$

$$-2\sin^2\theta + 3\sin\theta + 2 = 0$$

$$2\sin^2\theta - 3\sin\theta - 2 = 0$$

$$(2\sin\theta + 1)(\sin\theta - 2) = 0$$

$$2\sin\theta + 1 = 0 \qquad \sin\theta - 2 = 0$$

$$2\sin\theta = -1 \qquad \sin\theta = 2$$

$$\sin\theta = -\frac{1}{2} \qquad \text{No solution}$$

$$\theta = 210°, 330°$$

Solutions: $\theta = 210°, 330°$

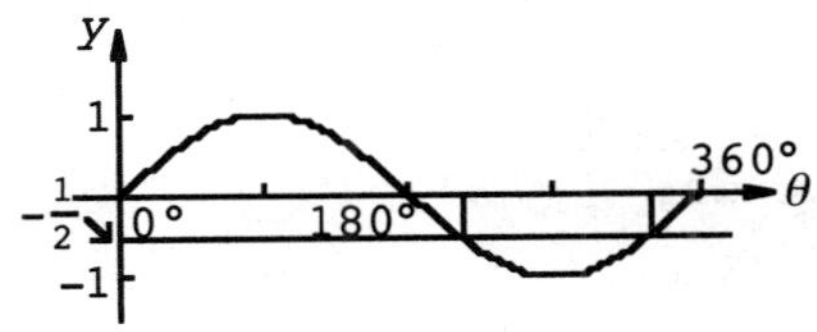

39.

$$\cos 2\theta + \cos\theta = 0 \quad 0° \le \theta < 360°$$

$$2\cos^2\theta - 1 + \cos\theta = 0$$

$$2\cos^2\theta + \cos\theta - 1 = 0$$

$$(2\cos\theta - 1)(\cos\theta + 1) = 0$$

$$2\cos\theta - 1 = 0 \qquad \cos\theta + 1 = 0$$

$$2\cos\theta = 1 \qquad \cos\theta = -1$$

$$\theta = 180°$$

$$\cos\theta = \frac{1}{2}$$

$$\theta = 60°, 300°$$

Solutions: $\theta = 60°, 180°, 300°$

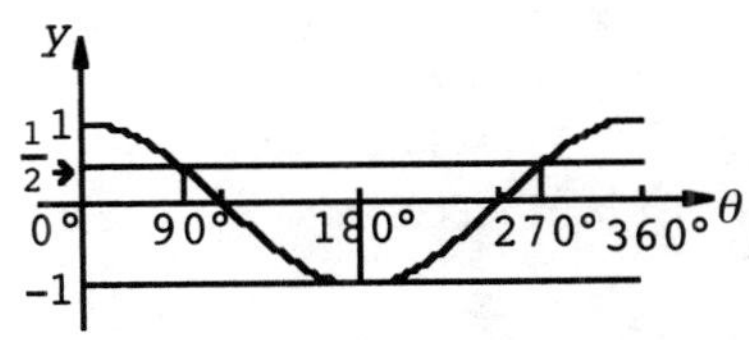

41. $2\sin^2\frac{x}{2} - 3\sin\frac{x}{2} + 1 = 0 \quad 0 \le x \le 2\pi$ is equivalent to

$$2\sin^2\frac{x}{2} - 3\sin\frac{x}{2} + 1 = 0 \quad 0 \le \frac{x}{2} \le \pi$$

$$\left(2\sin\frac{x}{2} - 1\right)\left(\sin\frac{x}{2} - 1\right) = 0$$

$$2\sin\frac{x}{2} - 1 = 0 \qquad \sin\frac{x}{2} - 1 = 0$$

Common Error: $2\sin\frac{x}{2} \ne \sin x$

$$2\sin\frac{x}{2} = 1 \qquad \sin\frac{x}{2} = 1$$

$$\sin\frac{x}{2} = \frac{1}{2} \qquad \frac{x}{2} = \frac{\pi}{2}$$

$$\frac{x}{2} = \frac{\pi}{6}, \frac{5\pi}{6} \qquad x = \pi$$

$$x = \frac{\pi}{3}, \frac{5\pi}{3}$$

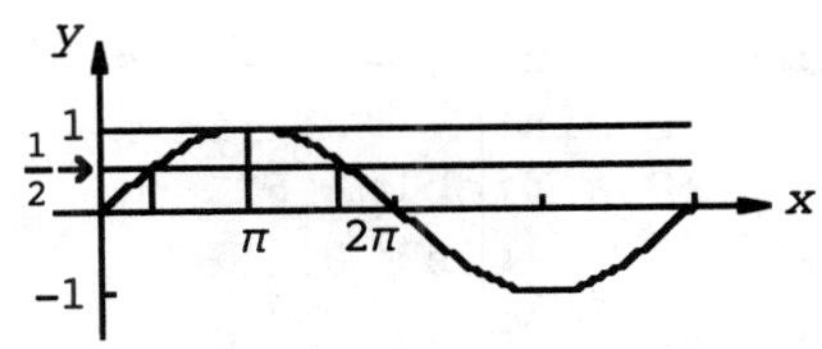

Solutions: $x = \frac{\pi}{3}, \frac{5\pi}{3}, \pi$

43. Since $\cos^2 x + \sin^2 x = 1$ is an identity, the solutions of this equation are all x, $0 \le x < 2\pi$.

45. Since $|2 \sin \theta| \le 2$ and $\cos \theta \ge -1$ for all θ, the left side of this equation cannot be greater than 2 and the right side cannot be less than 4. The equation has no solutions.

47.
$$6 \sin^2 \theta + 5 \sin \theta = 6 \quad 0° \le \theta \le 90°$$
$$6 \sin^2 \theta + 5 \sin \theta - 6 = 0$$
$$(3 \sin \theta - 2)(2 \sin \theta + 3) = 0$$

$3 \sin \theta - 2 = 0 \quad 0° \le \theta \le 90°$ | $2 \sin \theta + 3 = 0 \quad 0° \le \theta \le 90°$

$3 \sin \theta = 2$ | $2 \sin \theta = -3$

$\sin \theta = \dfrac{2}{3}$ | $\sin \theta = -\dfrac{3}{2}$

$\theta = \sin^{-1} \dfrac{2}{3}$ (calculator in degree mode) | No solution

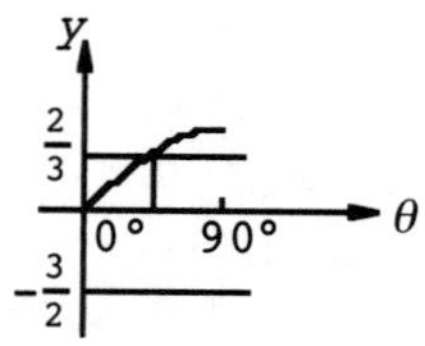

$\theta = 41.81°$

Solution: $\theta = 41.81°$

49.
$$3 \cos^2 x - 8 \cos x = 3 \quad 0 \le x \le \pi$$
$$3 \cos^2 x - 8 \cos x - 3 = 0$$
$$(3 \cos x + 1)(\cos x - 3) = 0$$

$3 \cos x + 1 = 0 \quad 0 \le x \le \pi$ | $\cos x - 3 = 0 \quad 0 \le x \le \pi$

$3 \cos x = -1$ | $\cos x = 3$

$\cos x = -\dfrac{1}{3}$ | No solution

$x = \cos^{-1}\left(-\dfrac{1}{3}\right)$ (calculator in radian mode)

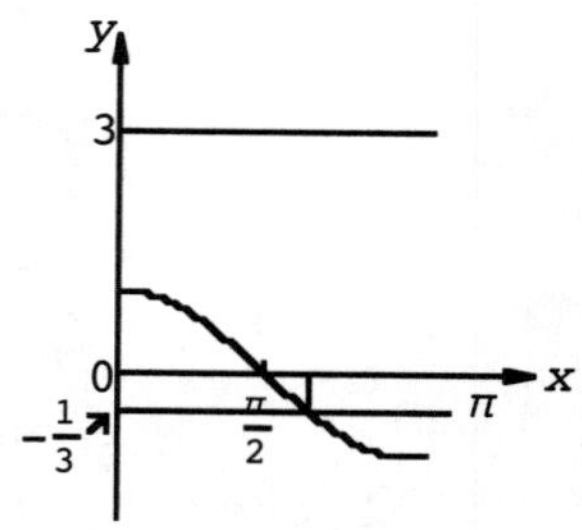

$x = 1.911$

Solution: $x = 1.911$

51.
$$2 \sin x = \cos 2x \quad 0 \le x < 2\pi$$
$$2 \sin x = 1 - 2 \sin^2 x$$
$$2 \sin^2 x + 2 \sin x - 1 = 0 \quad \text{Quadratic in } \sin x$$
$$\sin x = \frac{-b \pm \sqrt{b^2 - 4ac}}{2a} \quad a = 2,\ b = 2,\ c = -1$$
$$\sin x = \frac{-2 \pm \sqrt{2^2 - 4(2)(-1)}}{2(2)} = \frac{-2 \pm \sqrt{12}}{4}$$

$\sin x = 0.3660$ | $\sin x = -1.366$ No solution

$$x = \begin{cases} \sin^{-1} 0.3660 \\ \pi - \sin^{-1} 0.3660 \end{cases} = \begin{cases} 0.3747 \\ 2.767 \end{cases}$$

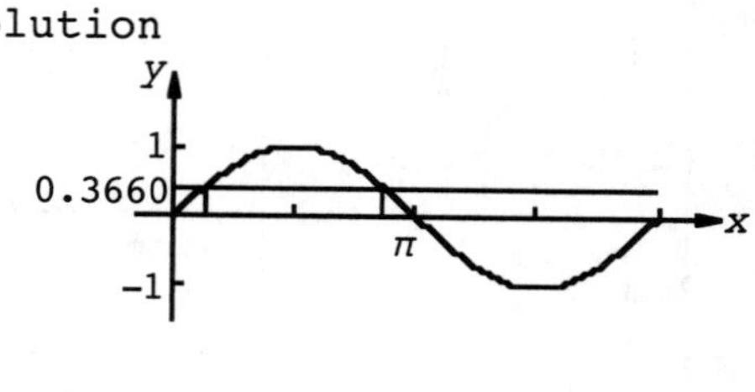

Solutions: $x = 0.3747, 2.767$

53. $2 \sin^2 x = 1 - 2 \sin x$

$2 \sin^2 x + 2 \sin x - 1 = 0 \quad x$ any real number

See Problem 51.

$$x = \begin{cases} \sin^{-1} 0.3660 \\ \pi - \sin^{-1} 0.3660 \end{cases} = \begin{cases} 0.3747 \\ 2.767 \end{cases}$$ are the solutions over one period $0 \le x < 2\pi$

If x can range over all real numbers,

$$x = \begin{cases} 0.3747 + 2k\pi \\ 2.767 + 2k\pi \end{cases} \quad k \text{ any integer}$$

55. Examining the graphs of $y1 = 2 \sin x$ and $y2 = \cos 2x$, $0 \le x < 2\pi$, we obtain

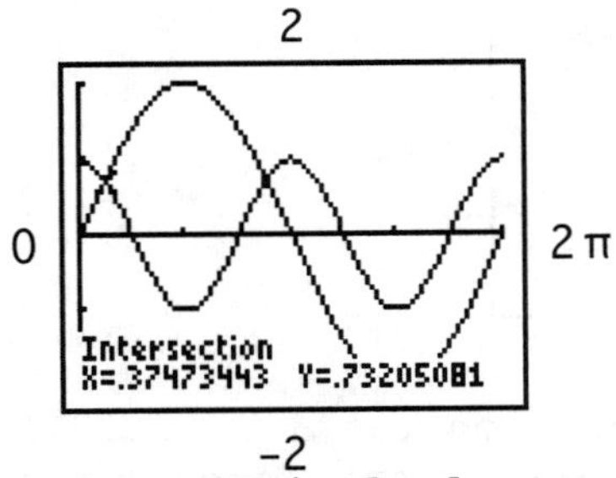

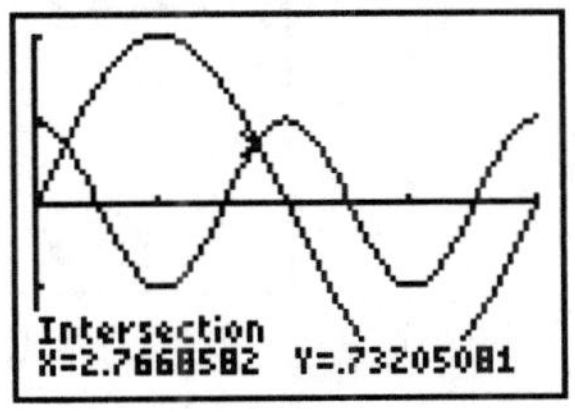

To four decimal places, the solutions are 0.3747 and 2.7669.

Check: 0.3747: $2 \sin x = 0.7321$ 2.7669: $2 \sin x = 2 \sin 2.7669 = 0.7321$
$\cos 2x = 0.7321$ $\cos 2x = \cos 2.7669 = 0.7321$

In later problems, the checking is left to the student.

57. Examining the graphs of $y1 = 2 \sin^2 x$ and $y2 = 1 - 2 \sin x$, $0 \le x \le 2\pi$, we obtain the graphs at the right.

To four decimal places, the solutions are 0.3747 and 2.7669. Thus, the solutions over all real x are given by

$$x = \begin{cases} 0.3747 + 2k\pi \\ 2.7669 + 2k\pi \end{cases} \quad k \text{ any integer}$$

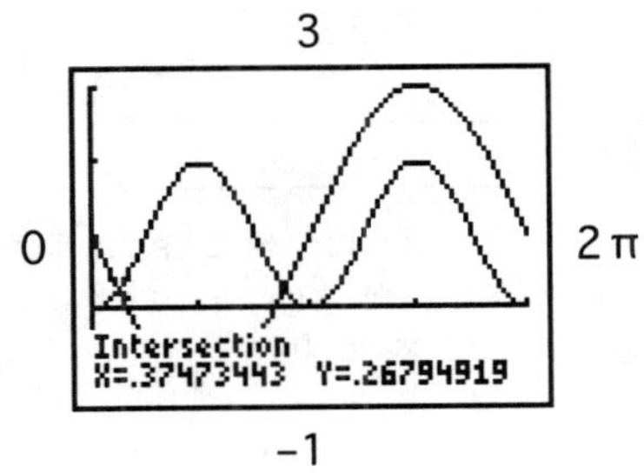

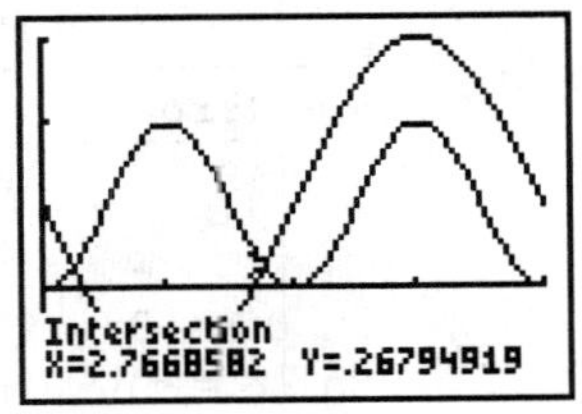

59. Examining the graphs of $y1 = \cos 2x$ and $y2 = x^2 - 2$, we obtain the graphs at the right.

The graphs intersect at $x = -1.1530$ and $x = 1.1530$. The cosine graph is above the parabola, and $\cos 2x > x^2 - 2$, on the interval $(-1.1530, 1.1530)$.

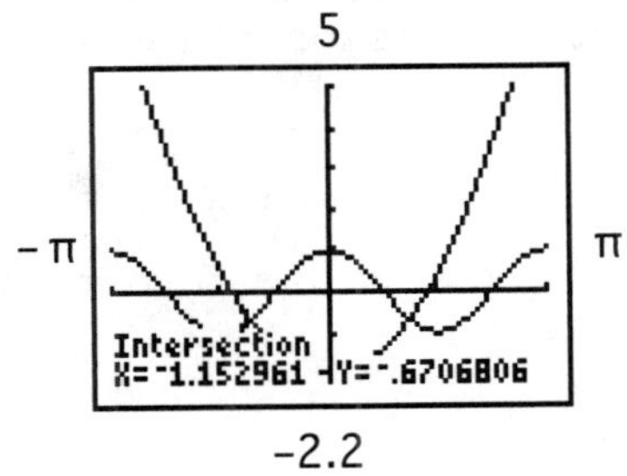

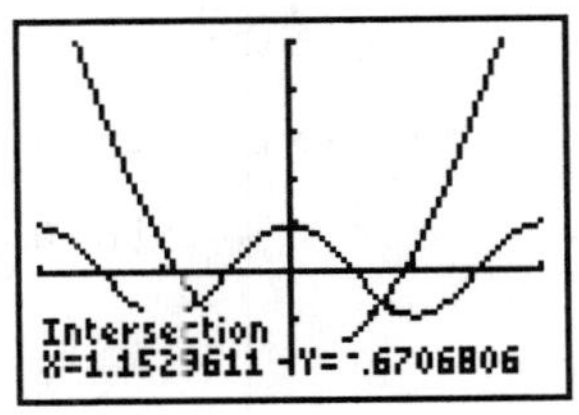

61. Examining the graphs of $y1 = \cos(2x + 1)$ and $y2 = 0.5x - 2$, we obtain

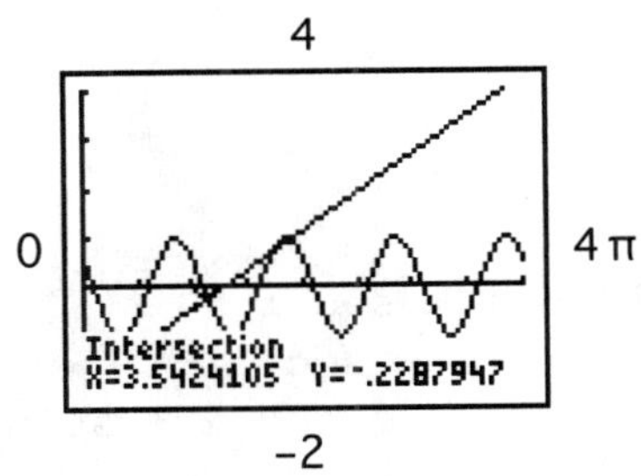

There is clearly one point of intersection at $x = 3.5424$. Examining the graphs again on a smaller interval, we obtain

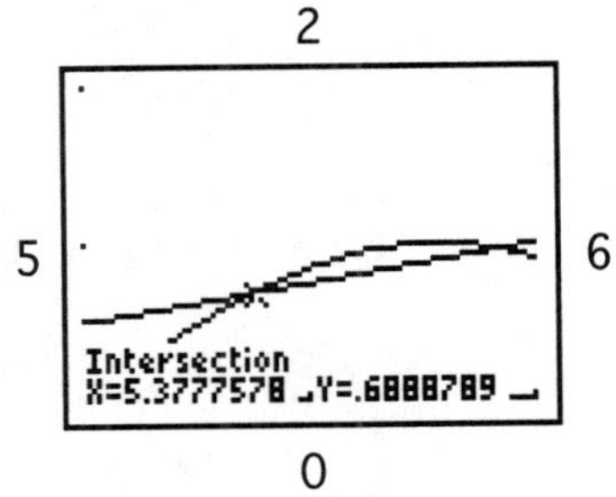

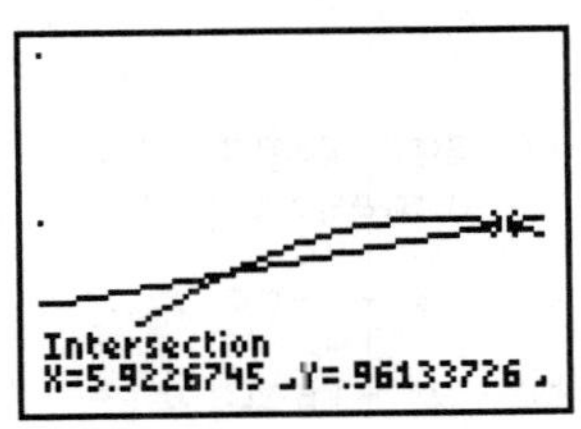

Combining information from all three graphs, the graph of the cosine function is below or intersects the straight line, $\cos(2x + 1) \le 0.5x - 2$, on the two intervals $[3.5424, 5.3778]$ and $[5.9227, \infty)$.

63. Examining the graphs of $y1 = e^{\sin x}$ and $y2 = 2x - 1$, we obtain the graph at the right.

The solution is 1.8183.

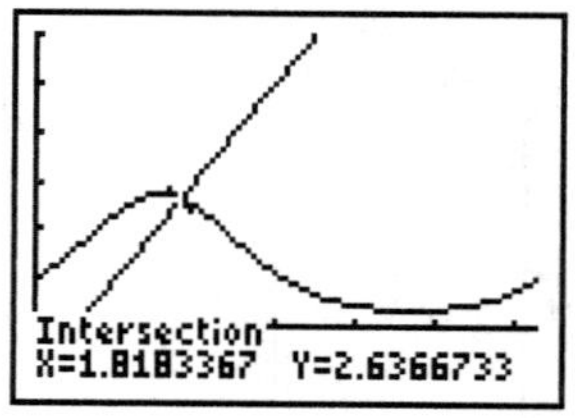

65. $\tan^{-1}(-5.377)$ has exactly one value, -1.387; the equation $\tan x = -5.377$ has infinitely many solutions, which are found by adding $k\pi$, k any integer, to each solution in one period of $\tan x$.

67. $\cos x - \sin x = 1 \qquad 0 \le x < 2\pi$

$$\pm\sqrt{1-\sin^2 x} - \sin x = 1$$
$$\pm\sqrt{1-\sin^2 x} = 1 + \sin x$$
$$1 - \sin^2 x = (1+\sin x)^2 \qquad \text{Squaring both sides}$$
$$1 - \sin^2 x = 1 + 2\sin x + \sin^2 x$$

Common Error: $(1 + \sin x)^2 \ne 1 + \sin^2 x$

$$0 = 2\sin x + 2\sin^2 x$$
$$0 = 2\sin x(1 + \sin x)$$

$2\sin x = 0$ $1 + \sin x = 0$

$\sin x = 0$ $\sin x = -1$

$x = 0, \pi$ $x = \frac{3\pi}{2}$

In squaring both sides we may have introduced extraneous solutions; hence, it is necessary to check solutions of these equations in the original equation.

$x = 0$

$\cos x - \sin x = 1$

$\cos 0 - \sin 0 \stackrel{?}{=} 1$

$1 - 0 \stackrel{\checkmark}{=} 1$

A solution

$x = \pi$

$\cos x - \sin x = 1$

$\cos \pi - \sin \pi \stackrel{?}{=} 1$

$-1 - 0 \ne 1$

Not a solution

$x = \frac{3\pi}{2}$

$\cos x - \sin x = 1$

$\cos \frac{3\pi}{2} - \sin \frac{3\pi}{2} \stackrel{?}{=} 1$

$0 - (-1) \stackrel{\checkmark}{=} 1$

A solution

Solutions: $x = 0, \frac{3\pi}{2}$

69. $\tan x - \sec x = 1 \qquad 0 \le x < 2\pi$

$$\pm\sqrt{\sec^2 x - 1} - \sec x = 1$$
$$\pm\sqrt{\sec^2 x - 1} = 1 + \sec x$$
$$\sec^2 x - 1 = (1 + \sec x)^2 \qquad \text{Squaring both sides}$$
$$\sec^2 x - 1 = 1 + 2\sec x + \sec^2 x$$
$$0 = 2 + 2\sec x$$
$$-2\sec x = 2$$
$$\sec x = -1$$
$$x = \pi$$

In squaring both sides we may have introduced extraneous solutions; hence, it is necessary to check this apparent solution in the original equation.

$\tan x - \sec x = 1$

$\tan \pi - \sec \pi \stackrel{?}{=} 1$

$0 - (-1) \stackrel{\checkmark}{=} 1$

A solution

Solution: $x = \pi$

71. Examining the graphs of
$y1 = \sin \frac{1}{x}$ and
$y2 = 1.5 - 5x$,
$0.04 \le x \le 0.2$,
we obtain the graphs at the right.

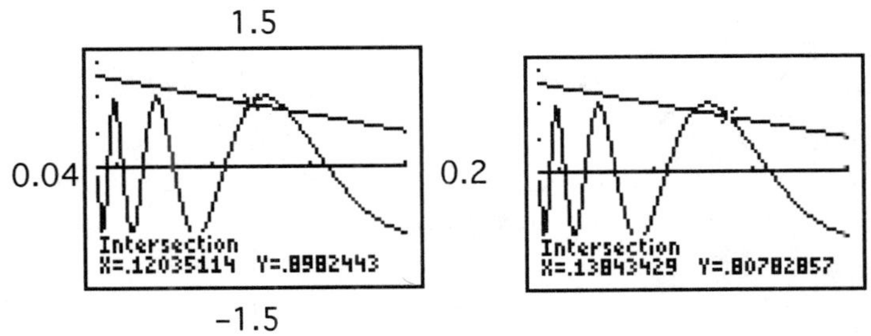

The graphs intersect, and $\sin \frac{1}{x} = 1.5 - 5x$, at $x = 0.1204$ and $x = 0.1384$.

73. (A) Here are graphs of $f(x) = \sin \frac{1}{x}$ on [0.1, 0.5] and [0.2, 5].

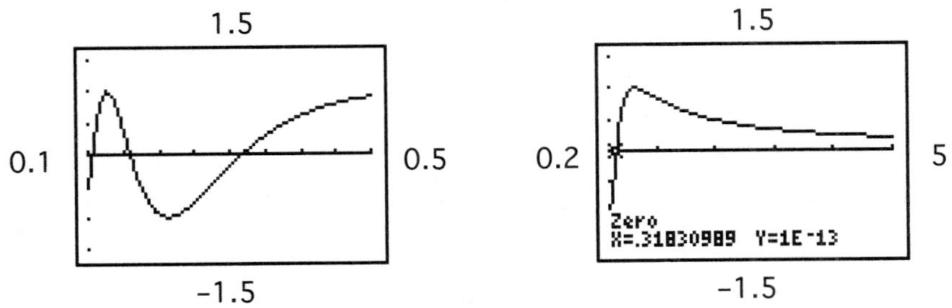

The largest zero for f is 0.3183. As x increases without bound, $\frac{1}{x}$ tends to 0 through positive numbers, and $\sin \frac{1}{x}$ tends to 0 through positive numbers. $y = 0$ is a horizontal asymptote for the graph of f.

(B) Here is a computer-generated graph of $f(x) = \sin \frac{1}{x}$ on [0.015, 0.1].

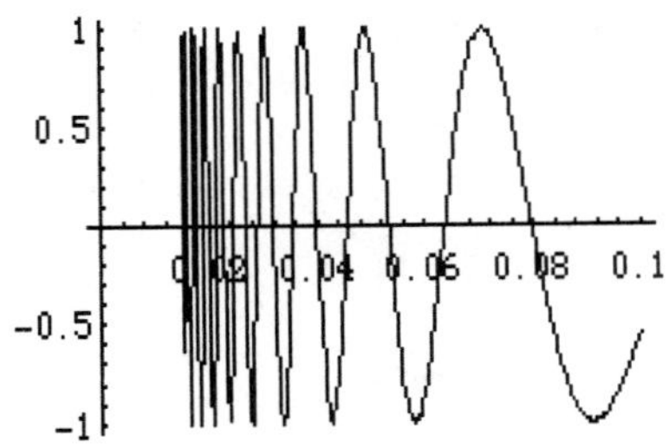

The student may provide other exploratory graphs. Infinitely many zeros exist between 0 and b, for any b, however small. The exploration graphs suggest this conclusion, which is reinforced by the following reasoning: Note that for each interval $(0, b]$, however small, as x tends to zero through positive numbers, $\frac{1}{x}$ increases without bound, and as $\frac{1}{x}$ increases without bound, $\sin \frac{1}{x}$ will cross the x axis an unlimited number of times. The function f does not have a smallest zero, because, between 0 and b, no matter how small b is, there is always an unlimited number of zeros.

75. $I = 30 \sin 120\pi t \quad I = -10$

$-10 = 30 \sin 120\pi t$

$\sin 120\pi t = -\frac{10}{30}$

$120\pi t = \pi + \sin^{-1} \frac{10}{30}$ (third quadrant) will yield the least positive solution of the equation.

$t = \frac{1}{120\pi}\left(\pi + \sin^{-1} \frac{10}{30}\right)$

$= 0.009235$ sec (calculator in radian mode)

77. We want the least positive θ so that $I\cos^2\theta$ is 40% of I, that is

$$I\cos^2\theta = 0.40I \qquad 0° \le \theta \le 180°$$
$$\cos^2\theta = 0.40$$
$$\cos\theta = \pm\sqrt{0.40}$$
$$\theta = \cos^{-1}\sqrt{0.40}$$ will yield the least positive solution of the equation
$$\theta = 50.77°$$ (calculator in degree mode)

79. We are to solve $3.09 \times 10^7 = \dfrac{3.44 \times 10^7}{1 - 0.206\cos\theta}$

For convenience, we can divide both sides of this equation by 10^7

$$3.09 = \frac{3.44}{1 - 0.206\cos\theta}$$
$$3.09(1 - 0.206\cos\theta) = 3.44$$
$$3.09 - (3.09)(0.206)\cos\theta = 3.44$$
$$-(3.09)(0.206)\cos\theta = 3.44 - 3.09$$
$$\cos\theta = \frac{3.44 - 3.09}{-(3.09)(0.206)}$$
$$\cos\theta = -0.5498$$

$\theta = 180° - \cos^{-1}(0.5498)$ (second quadrant) will yield the least positive solution of this equation.

or

$\theta = \cos^{-1}(-0.5498)$

$\theta = 123°$

81. Use $A = \dfrac{1}{2}R^2(\theta - \sin\theta)$ with $R = 8$ and $A = 48$.

$$48 = \frac{1}{2}\cdot 8^2(\theta - \sin\theta)$$
$$48 = 32(\theta - \sin\theta)$$
$$1.5 = \theta - \sin\theta$$

Examining the graph of $y1 = 1.5$ and $y2 = \theta - \sin\theta$, $0 \le \theta \le \pi$, we obtain

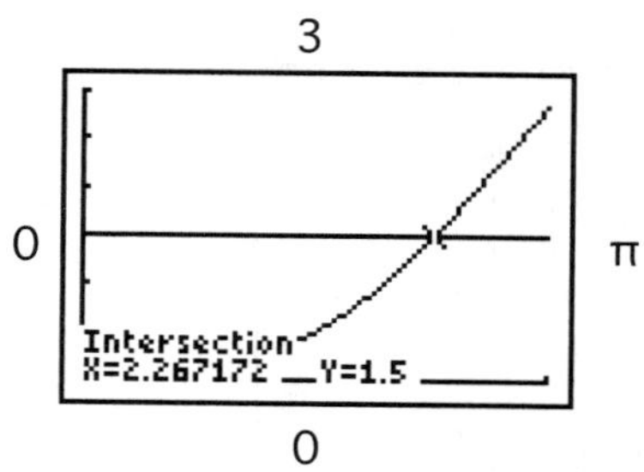

To three decimal places, $\theta = 2.267$ radians.

83. Using the figure in the text, we can write:

$$\sin\theta = \frac{a}{R} \qquad (1)$$
$$a^2 + (R - b)^2 = R^2 \qquad (2)$$ applying the Pythagorean theorem to the right triangle with sides R, a, and $R - b$.
$$L = R\cdot 2\theta \qquad (3)$$

Thus, given a and b, we can solve equation (2) for R:

$$a^2 + R^2 - 2Rb + b^2 = R^2$$
$$-2Rb = -a^2 - b^2$$
$$R = \frac{a^2 + b^2}{2b}$$

We can solve equation (1) for θ to obtain $\theta = \sin^{-1}\dfrac{a}{R}$ (since θ is an acute angle) Hence, from equation (3),

$$L = R\cdot 2\theta = \frac{a^2 + b^2}{2b}\cdot 2\sin^{-1}\frac{a}{\frac{a^2+b^2}{2b}}$$
$$L = \frac{a^2 + b^2}{b}\sin^{-1}\frac{2ab}{a^2 + b^2} \qquad (4)$$

(A) Substituting the given values $a = 5.5$ and $b = 2.5$,

$$L = \frac{(5.5)^2 + (2.5)^2}{2.5} \sin^{-1} \frac{2(5.5)(2.5)}{(5.5)^2 + (2.5)^2}$$

$L = 12.4575$ mm.

(B) Substituting $L = 12.4575$ and $a = 5.4$ mm yields

$$12.4575 = \frac{(5.4)^2 + b^2}{b} \sin^{-1} \frac{2(5.4)b}{(5.4)^2 + b^2}$$

$$12.4575 = \frac{29.16 + b^2}{b} \sin^{-1} \frac{10.8b}{29.16 + b^2} \qquad (5)$$

Examining the graph of $y1 = 12.4575$ and

$$y2 = \frac{29.16 + x^2}{x} \sin^{-1} \frac{10.8x}{29.16 + x^2},$$

$2 \le x \le 3$, we obtain the graph at the right.

The solution is $b = 2.6496$ mm.

13

2 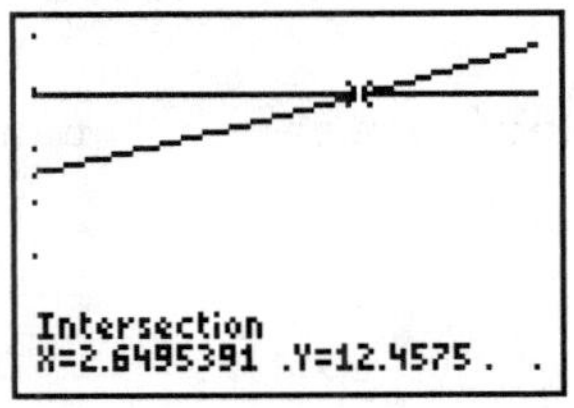 3

10

85. $r = 2 \sin \theta \quad 0° \le \theta \le 360°$

$r = \sin 2\theta$

We solve this system of equations by equating the right sides:

$2 \sin \theta = \sin 2\theta$

$2 \sin \theta = 2 \sin \theta \cos \theta$

$0 = 2 \sin \theta \cos \theta - 2 \sin \theta$

$0 = 2 \sin \theta(\cos \theta - 1)$

$2 \sin \theta = 0$	$\cos \theta - 1 = 0$
$\sin \theta = 0$	$\cos \theta = 1$
$\theta = 0°, 180°, 360°$	$\theta = 0°, 360°$

If we substitute these values of θ in either of the original equations, we obtain

$r = 2 \sin 0°$	$r = 2 \sin 180°$	$r = 2 \sin 360°$
$r = 0$	$r = 0$	$r = 0$

Thus, the solutions of the system of equations are

$(r, \theta) = (0, 0°)$, $(r, \theta) = (0, 180°)$, and $(r, \theta) = (0, 360°)$

87. $2xy = 1$

$2(u \cos \theta - v \sin \theta)(u \sin \theta + v \cos \theta) = 1$ (substitution)

$2(u \cos \theta u \sin \theta + u \cos \theta v \cos \theta - v \sin \theta u \sin \theta - v \sin \theta v \cos \theta) = 1$ (multiplication)

$2u^2 \cos \theta \sin \theta + uv(2 \cos^2 \theta - 2 \sin^2 \theta) - 2v^2 \sin \theta \cos \theta = 1$

We are to find the least positive θ so that the coefficient of the uv term will be zero.

$2 \cos^2 \theta - 2 \sin^2 \theta = 0$

$2(\cos^2 \theta - \sin^2 \theta) = 0$

$2 \cos 2\theta = 0$

$\cos 2\theta = 0$

$2\theta = \cos^{-1} 0$ yields the least positive θ

$2\theta = 90°$

$\theta = 45°$

CHAPTER 7 REVIEW

1. $\tan x + \cot x = \dfrac{\sin x}{\cos x} + \dfrac{\cos x}{\sin x}$ Quotient Identities

$= \dfrac{\sin^2 x + \cos^2 x}{\cos x \sin x}$ Algebra

$= \dfrac{1}{\cos x \sin x}$ Pythagorean Identity

$= \dfrac{1}{\cos x}\,\dfrac{1}{\sin x}$ Algebra

$= \sec x \csc x$ Reciprocal Identity (7-1)

2. $\sec^4 x - 2\sec^2 x \tan^2 x + \tan^4 x = (\sec^2 x - \tan^2 x)^2$ Algebra

$= (1 + \tan^2 x - \tan^2 x)^2$ Pythagorean Identity

$= 1^2$ Algebra

$= 1$ (7-1)

3. $\dfrac{1}{1 - \sin x} + \dfrac{1}{1 + \sin x} = \dfrac{1 + \sin x + 1 - \sin x}{(1 - \sin x)(1 + \sin x)}$ Algebra

$= \dfrac{2}{1 - \sin^2 x}$ Algebra

$= \dfrac{2}{\sin^2 x + \cos^2 x - \sin^2 x}$ Pythagorean Identity

$= \dfrac{2}{\cos^2 x}$ Algebra

$= 2\sec^2 x$ Reciprocal Identity (7-1)

4. $\cos\left(x - \dfrac{3\pi}{2}\right) = \cos x \cos \dfrac{3\pi}{2} + \sin x \sin \dfrac{3\pi}{2}$ Difference Identity

$= \cos x(0) + \sin x(-1)$ Known Values

$= -\sin x$ Algebra (7-2)

5. $\sin x \cos y = \dfrac{1}{2}[\sin(x + y) + \sin(x - y)]$

$\sin 5\alpha \cos 3\alpha = \dfrac{1}{2}[\sin(5\alpha + 3\alpha) + \sin(5\alpha - 3\alpha)]$

$= \dfrac{1}{2}[\sin 8\alpha + \sin 2\alpha]$

$= \dfrac{1}{2}\sin 8\alpha + \dfrac{1}{2}\sin 2\alpha$ (7-4)

6. $\cos x - \cos y = -2\sin \dfrac{x + y}{2} \sin \dfrac{x - y}{2}$

$\cos 7x - \cos 5x = -2\sin \dfrac{7x + 5x}{2} \sin \dfrac{7x - 5x}{2}$

$= -2\sin 6x \sin x$ (7-4)

7. $\sin\left(x + \frac{9\pi}{2}\right) = \sin x \cos \frac{9\pi}{2} + \cos x \sin \frac{9\pi}{2}$ Sum Identity

$= \sin x \cos \frac{\pi}{2} + \cos x \sin \frac{\pi}{2}$ Periodicity of sine and cosine

$= \sin x(0) + \cos x(1)$ Known Values

$= \cos x$ (7-2)

8. $\sqrt{2} \cos \theta + 1 = 0$ all θ
Solve over one period [0°, 360°):

$$\cos \theta = -\frac{1}{\sqrt{2}}$$

Sketch a graph of $y = \cos \theta$ and $y = -\frac{1}{\sqrt{2}}$, θ in [0°, 360°).

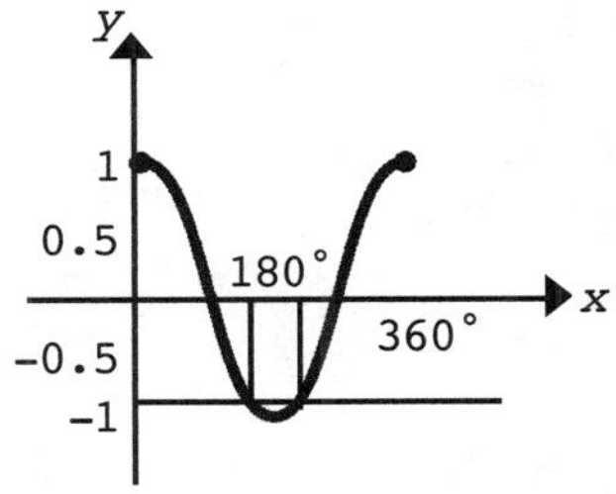

$\theta = 135°, 225°$

Since the cosine function is periodic with period 360°, all solutions are given by $\theta = 135° + k360°$, $\theta = 225° + k360°$, k any integer. (7-5)

9. $\sin x \tan x - \sin x = 0$ all x
Solve over one period, $[0, 2\pi)$ for $\sin x$, $[0, \pi)$ for $\tan x$:

$\sin x(\tan x - 1) = 0$

$\sin x = 0$ $\quad$ $\tan x - 1 = 0$

$\tan x = 1$

$x = \frac{\pi}{4}$

$x = 0, \pi$

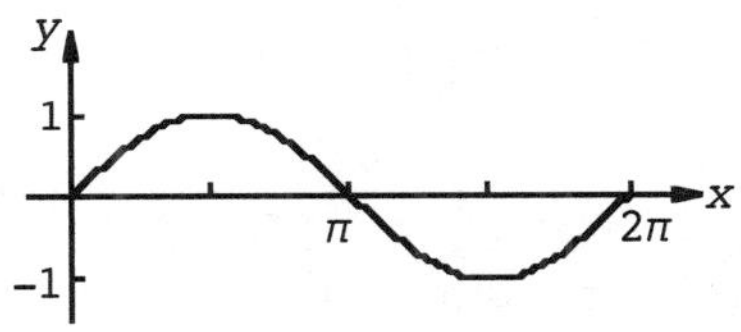

Thus, solutions if x is allowed to range over all values, are $x = 0 + 2k\pi$, $\pi + 2k\pi$, k any integer, and $x = \frac{\pi}{4} + k\pi$, k any integer. These solutions can also be written as $x = k\pi$ or $x = \frac{\pi}{4} + k\pi$, k any integer.

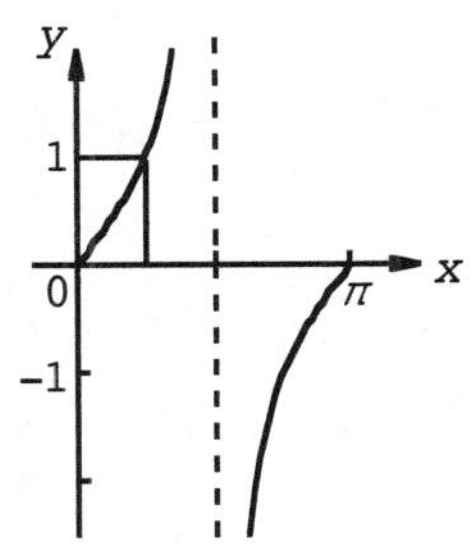

(7-5)

10. $\sin x = 0.7088$ all real x
Solve over one period $[0, 2\pi)$:
Sketch a graph of $y = \sin x$ and $y = 0.7088$, x in $[0, 2\pi)$.

$$x = \begin{cases} \sin^{-1} 0.7088 = 0.7878 & \text{First quadrant solution} \\ \pi - \sin^{-1} 0.7088 = 2.3538 & \text{Second quadrant solution} \end{cases}$$
(calculator in radian mode)

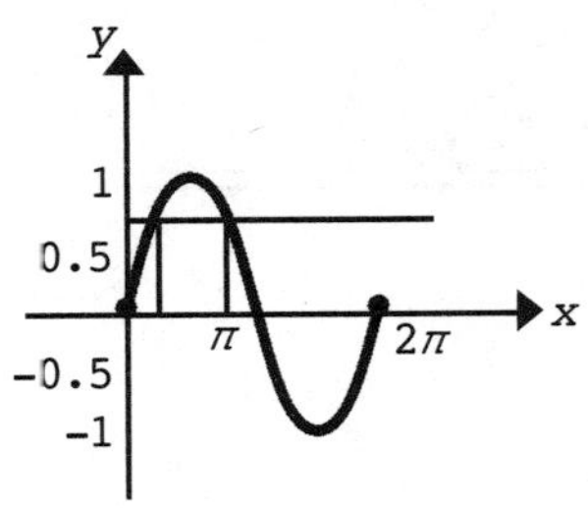

Since the sine function is periodic with period 2π, all solutions are given by

$$x = \begin{cases} 0.7878 + 2k\pi \\ 2.3538 + 2k\pi \end{cases} \quad k \text{ any integer}$$

(7-5)

11. $\cos\theta = 0.2557$ all θ

Solve over one period, $[0°, 360°)$:

Sketch a graph of $y = \cos\theta$ and $y = 0.2557$, θ in $[0°, 360°)$.

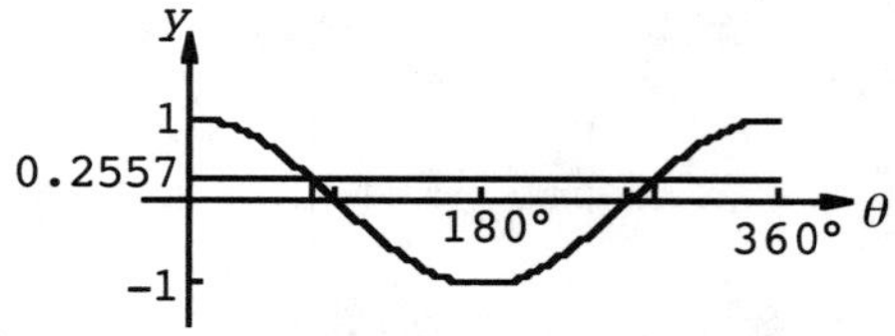

$$x = \begin{cases} \cos^{-1} 0.2557 = 75.1849° & \text{First quadrant solution} \\ 360° - \cos^{-1} 0.2557 = 284.8151° & \text{Fourth quadrant solution} \end{cases}$$

(calculator in degree mode)

Since the cosine function is periodic with period 360°, all solutions are given by

$$x = \begin{cases} 75.1849° + k360° \\ 284.8151° + k360° \end{cases} \quad k \text{ any integer}$$

(7-5)

12. $\cot x = -0.1692 \quad -\frac{\pi}{2} < x < \frac{\pi}{2}$

Sketch a graph of $y = \cot x$ and $y = -0.1692$, x in $\left(-\frac{\pi}{2}, \frac{\pi}{2}\right)$

$x = -\cot^{-1}(0.1692)$ Fourth quadrant solution, (calculator in radian mode)

$= -1.4032$

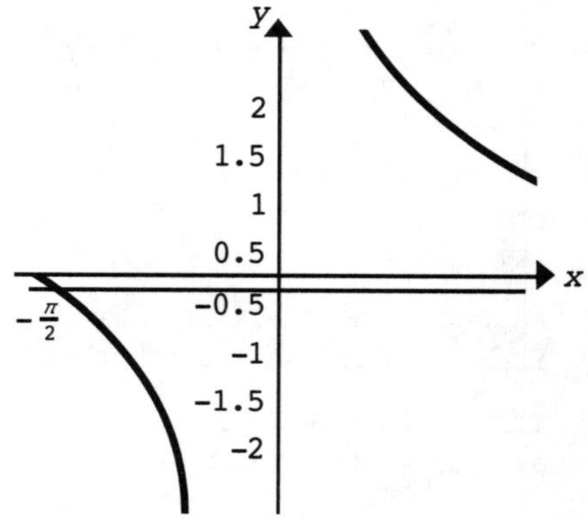

(7-5)

13. $3\tan(11 - 3x) = 23.46 \quad -\frac{\pi}{2} < x < \frac{\pi}{2}$

$$\tan(11 - 3x) = \frac{23.46}{3}$$

$$11 - 3x = \tan^{-1}\left(\frac{23.46}{3}\right)$$

$$-3x = \tan^{-1}\left(\frac{23.46}{3}\right) - 11$$

$$x = \frac{\tan^{-1}\left(\frac{23.46}{3}\right) - 11}{-3}$$

$$x = 3.1855$$

(7-5)

14. (A) Graph both sides of the equation in the same viewing window.

$(\sin x + \cos x)^2 = 1 - 2\sin x \cos x$ is not an identity, since the graphs do not match.

Try $x = \frac{\pi}{4}$.

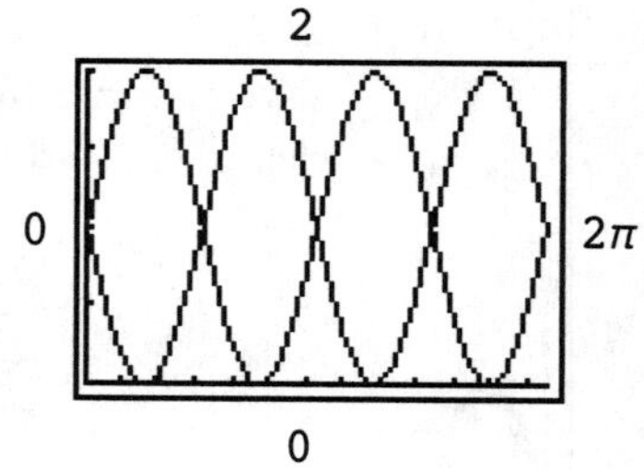

Left side: $\left(\sin\frac{\pi}{4} + \cos\frac{\pi}{4}\right)^2 = \left(\frac{1}{\sqrt{2}} + \frac{1}{\sqrt{2}}\right)^2 = \left(\frac{2}{\sqrt{2}}\right)^2 = 2$

Right side: $1 - 2\sin\frac{\pi}{4}\cos\frac{\pi}{4} = 1 - 2\left(\frac{1}{\sqrt{2}}\right)\left(\frac{1}{\sqrt{2}}\right) = 0$

This verifies that the equation is not an identity.

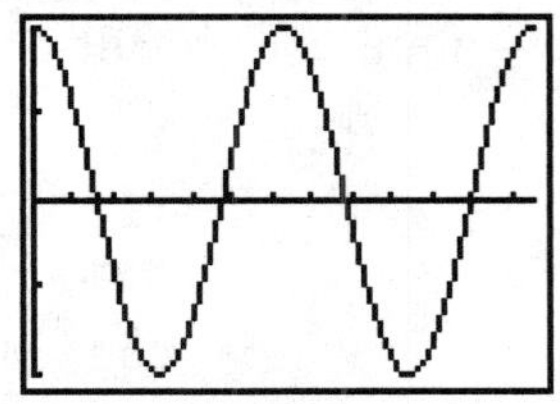

(B) Graph both sides of the equation in the same viewing window.

$\cos^2 x - \sin^2 x = 1 - 2\sin^2 x$ appears to be an identity, which we now verify:

$\cos^2 x - \sin^2 x = 1 - \sin^2 x - \sin^2 x$ Pythagorean Identity
$= 1 - 2\sin^2 x$ Algebra (7-1)

15. $\dfrac{1 - 2\cos x - 3\cos^2 x}{\sin^2 x} = \dfrac{1 - 2\cos x - 3\cos^2 x}{1 - \cos^2 x}$ Pythagorean Identity

$= \dfrac{(1 - 3\cos x)(1 + \cos x)}{(1 + \cos x)(1 - \cos x)}$ Algebra

$= \dfrac{1 - 3\cos x}{1 - \cos x}$ Algebra (7-1)

16. $(1 - \cos x)(\csc x + \cot x) = (1 - \cos x)\left(\dfrac{1}{\sin x} + \dfrac{\cos x}{\sin x}\right)$ Reciprocal and Quotient Identities

$= (1 - \cos x)\dfrac{(1 + \cos x)}{\sin x}$ Algebra

$= \dfrac{(1 - \cos x)(1 + \cos x)}{\sin x}$ Algebra

$= \dfrac{1 - \cos^2 x}{\sin x}$ Algebra

$= \dfrac{\sin^2 x}{\sin x}$ Pythagorean Identity

$= \sin x$ Algebra

Key Algebraic Steps: $(1 - a)\left(\dfrac{1}{b} + \dfrac{a}{b}\right) = (1 - a)\left(\dfrac{1 + a}{b}\right)$

$= \dfrac{(1 - a)(1 + a)}{b}$

$= \dfrac{1 - a^2}{b}$ (7-1)

17. $\dfrac{1 + \sin x}{\cos x} = \dfrac{(1 + \sin x)(1 - \sin x)}{\cos x(1 - \sin x)}$ Algebra

$= \dfrac{1 - \sin^2 x}{\cos x(1 - \sin x)}$ Algebra

$= \dfrac{\cos^2 x}{\cos x(1 - \sin x)}$ Pythagorean Identity

$= \dfrac{\cos x}{1 - \sin x}$ Algebra (7-1)

18. $\dfrac{1 - \tan^2 x}{1 + \tan^2 x} = \dfrac{1 - \tan^2 x}{\sec^2 x}$ Pythagorean Identity

$= \dfrac{1 - \frac{\sin^2 x}{\cos^2 x}}{\frac{1}{\cos^2 x}}$ Reciprocal and Quotient Identities

$= \dfrac{\cos^2 x\left[1 - \frac{\sin^2 x}{\cos^2 x}\right]}{\cos^2 x \frac{1}{\cos^2 x}}$ Algebra

$= \dfrac{\cos^2 x - \sin^2 x}{1}$ Algebra

$= \cos^2 x - \sin^2 x$ Algebra

$= \cos 2x$ Double-angle Identity (7-3)

19. $\cot \dfrac{x}{2} = \dfrac{1}{\tan \frac{x}{2}}$ Reciprocal Identity

$= \dfrac{1}{\frac{1 - \cos x}{\sin x}}$ Half-angle Identity

$= \dfrac{\sin x}{1 - \cos x}$ Algebra (7-3)

20. $\cot x - \tan x = \dfrac{\cos x}{\sin x} - \dfrac{\sin x}{\cos x}$ Quotient Identities

$= \dfrac{\cos^2 x - \sin^2 x}{\sin x \cos x}$ Algebra

$= \dfrac{2(\cos^2 x - \sin^2 x)}{2 \sin x \cos x}$ Algebra

$= \dfrac{2[\cos^2 x - (1 - \cos^2 x)]}{2 \sin x \cos x}$ Pythagorean Identity

$= \dfrac{2[2\cos^2 x - 1]}{2 \sin x \cos x}$ Algebra

$= \dfrac{4\cos^2 x - 2}{2 \sin x \cos x}$ Algebra

$= \dfrac{4\cos^2 x - 2}{\sin 2x}$ Double-angle Identity (7-3)

21. $\left(\dfrac{1 - \cot x}{\csc x}\right)^2 = \dfrac{(1 - \cot x)^2}{\csc^2 x}$ Algebra

$= \dfrac{1 - 2\cot x + \cot^2 x}{\csc^2 x}$ Algebra

$= \dfrac{1 + \cot^2 x - 2\cot x}{\csc^2 x}$ Algebra

$= \dfrac{\csc^2 x - 2\cot x}{\csc^2 x}$ Pythagorean Identity

$= \dfrac{\csc^2 x}{\csc^2 x} - \dfrac{2\cot x}{\csc^2 x}$ Algebra

$= 1 - \dfrac{2\cot x}{\csc^2 x}$ Algebra

$= 1 - \dfrac{2\frac{\cos x}{\sin x}}{\frac{1}{\sin^2 x}}$ Quotient and Reciprocal Identities

$= 1 - 2 \sin x \cos x$ Algebra

$= 1 - \sin 2x$ Double-angle Identity (7-3)

22. $\tan m + \tan n = \dfrac{\sin m}{\cos m} + \dfrac{\sin n}{\cos n}$ Quotient Identity

$= \dfrac{\sin m \cos n + \cos m \sin n}{\cos m \cos n}$ Algebra

$= \dfrac{\sin(m + n)}{\cos m \cos n}$ Sum Identity (7-2)

23. $\tan(x + y) = \dfrac{\tan x + \tan y}{1 - \tan x \tan y}$ Sum Identity

$= \dfrac{\frac{1}{\cot x} + \frac{1}{\cot y}}{1 - \frac{1}{\cot x}\frac{1}{\cot y}}$ Reciprocal Identity

$= \dfrac{\cot y + \cot x}{\cot x \cot y - 1}$ Algebra

$= \dfrac{\cot x + \cot y}{\cot x \cot y - 1}$ Algebra (7-2)

24.

$$\cos x \sin y = \frac{1}{2}[\sin(x + y) - \sin(x - y)]$$

$$\begin{aligned}
\cos 195° \sin 75° &= \frac{1}{2}[\sin(195° + 75°) - \sin(195° - 75°)] \\
&= \frac{1}{2}[\sin 270° - \sin 120°] \\
&= \frac{1}{2}\left(-1 - \frac{\sqrt{3}}{2}\right) \\
&= \frac{1}{2}\,\frac{-2 - \sqrt{3}}{2} \\
&= \frac{-2 - \sqrt{3}}{4}
\end{aligned}$$

(7-4, 6-2)

25.

$$\cos x + \cos y = 2 \cos \frac{x + y}{2} \cos \frac{x - y}{2}$$

$$\begin{aligned}
\cos 195° + \cos 105° &= 2 \cos \frac{195° + 105°}{2} \cos \frac{195° - 105°}{2} \\
&= 2 \cos 150° \cos 45° \\
&= 2\left(-\frac{\sqrt{3}}{2}\right)\left(\frac{\sqrt{2}}{2}\right) \\
&= -\frac{\sqrt{6}}{2}
\end{aligned}$$

(7-4, 6-2)

26. We suspect that this is not an identity since $\cot^2 x = \csc^2 x - 1$ is a Pythagorean Identity. (Note the difference in sign.) To verify, plug in $x = \dfrac{\pi}{2}$.

$$\cot^2\left(\frac{\pi}{2}\right) \stackrel{?}{=} \csc^2\left(\frac{\pi}{2}\right) + 1$$

$$\frac{\cos^2\left(\frac{\pi}{2}\right)}{\sin^2\left(\frac{\pi}{2}\right)} \stackrel{?}{=} \frac{1}{\sin^2\left(\frac{\pi}{2}\right)} + 1$$

$$\frac{0}{1} \stackrel{?}{=} 1 + 1$$

Not true, so the equation is not an identity. (7-1)

27. Graphing $y_1 = \cos 3x$, $y_2 = \cos x(\cos 2x - 2\sin^2 x)$, we see that the graphs match so we suspect that this is an identity.

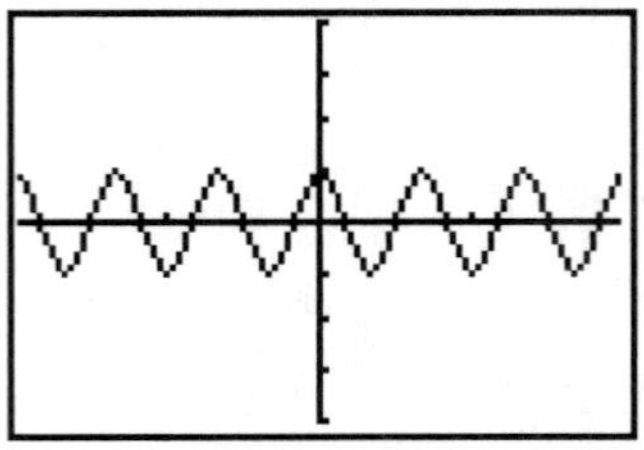

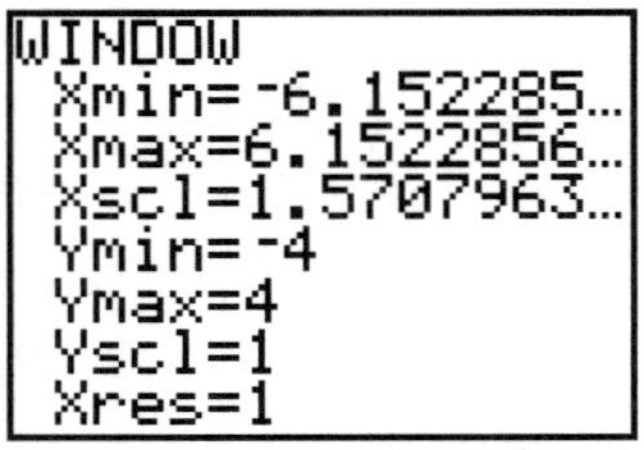

$\cos 3x = \cos(2x + x)$ Algebra

$= \cos 2x \cos x - \sin 2x \sin x$ Sum Identity

$= \cos 2x \cos x - 2 \sin x \cos x \sin x$ Double-angle Identity for sine

$= \cos 2x \cos x - 2 \sin^2 x \cos x$ Algebra

$= \cos x(\cos 2x - 2 \sin^2 x)$ Factoring

The identity is verified. (7-2)

28. $\sin\left(x + \frac{3\pi}{2}\right) = \sin x \cos \frac{3\pi}{2} + \cos x \sin \frac{3\pi}{2}$ Sum Identity

$= \sin x(0) + \cos x(-1)$ Evaluation

$= -\cos x$ Algebra

The equation is not an identity due to an incorrect sign. (7-2)

29. $\cos\left(x - \frac{3\pi}{2}\right) = \cos x \cos \frac{3\pi}{2} + \sin x \sin \frac{3\pi}{2}$ Difference Identity

$= \cos x(0) + \sin x(-1)$ Evaluation

$= -\sin x$ Algebra

The equation is not an identity due to an incorrect sign. (7-2)

30. $4 \sin^2 x - 3 = 0 \quad 0 \le x < 2\pi$

$$4 \sin^2 x = 3$$

$$\sin^2 x = \frac{3}{4}$$

$$\sin x = \pm\frac{\sqrt{3}}{2}$$

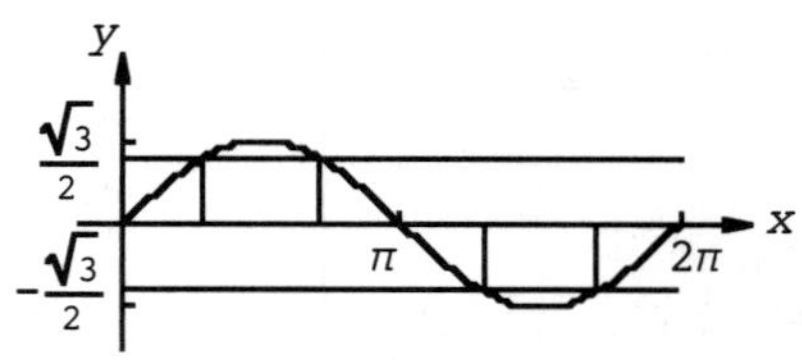

$\sin x = \frac{\sqrt{3}}{2}$ $\qquad$ $\sin x = -\frac{\sqrt{3}}{2}$

$x = \frac{\pi}{3}, \frac{2\pi}{3}$ $\qquad$ $x = \frac{4\pi}{3}, \frac{5\pi}{3}$

Solutions: $x = \frac{\pi}{3}, \frac{2\pi}{3}, \frac{4\pi}{3}, \frac{5\pi}{3}$ (7-5)

31. $2\sin^2\theta + \cos\theta = 1 \quad 0° \le \theta \le 180°$

$$\begin{aligned} 2\sin^2\theta + \cos\theta &= 1 \\ 2(1 - \cos^2\theta) + \cos\theta &= 1 \\ 2 - 2\cos^2\theta + \cos\theta &= 1 \\ 1 - 2\cos^2\theta + \cos\theta &= 0 \\ -2\cos^2\theta + \cos\theta + 1 &= 0 \\ 2\cos^2\theta - \cos\theta - 1 &= 0 \\ (2\cos\theta + 1)(\cos\theta - 1) &= 0 \end{aligned}$$

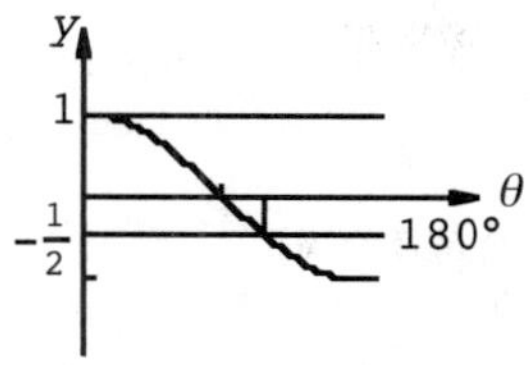

$2\cos\theta + 1 = 0 \qquad \cos\theta - 1 = 0$

$\cos\theta = -\frac{1}{2} \qquad \cos\theta = 1$

$\theta = 120° \qquad \theta = 0°$ (7-5)

Solutions: $\theta = 0°, 120°$

32. $2\sin^2 x - \sin x = 0$ all real x

Solve over one period $[0, 2\pi)$:

$\sin x(2\sin x - 1) = 0$

$\sin x = 0 \qquad 2\sin x - 1 = 0$

$x = 0, \pi \qquad \sin x = \frac{1}{2}$

$x = \frac{\pi}{6}, \frac{5\pi}{6}$

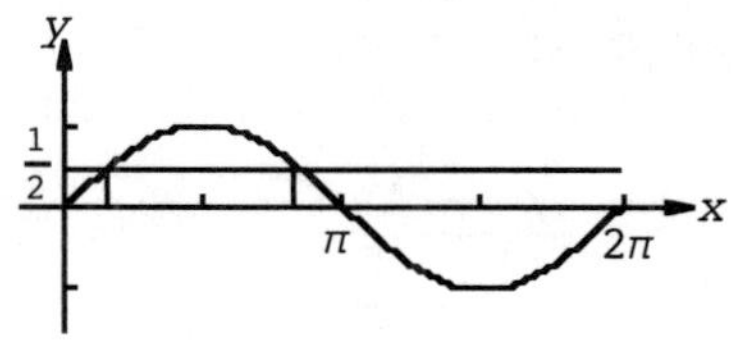

Since the sine function is periodic with period 2π, all solutions are given by

$x = 0 + 2k\pi,\ x = \pi + 2k\pi,\ x = \frac{\pi}{6} + 2k\pi,\ x = \frac{5\pi}{6} + 2k\pi$, k any integer.

The first two can also be written together as $x = k\pi$, k any integer. (7-5)

33. $\sin 2x = \sqrt{3}\sin x$ all real x

Solve over one period $[0, 2\pi)$:

$$\begin{aligned} \sin 2x - \sqrt{3}\sin x &= 0 \\ 2\sin x\cos x - \sqrt{3}\sin x &= 0 \\ \sin x(2\cos x - \sqrt{3}) &= 0 \end{aligned}$$

$\sin x = 0 \qquad 2\cos x - \sqrt{3} = 0$

$x = 0, \pi \qquad \cos x = \frac{\sqrt{3}}{2}$

$x = \frac{\pi}{6}, \frac{11\pi}{6}$

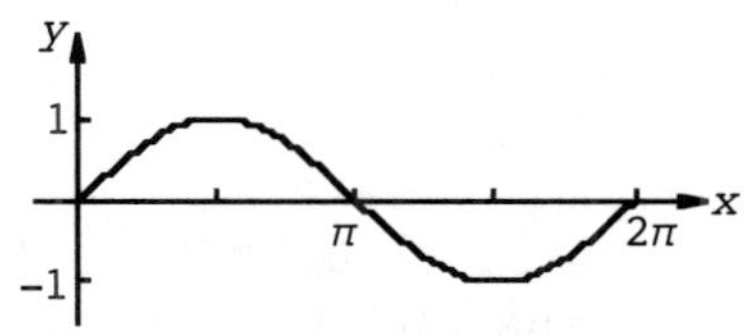

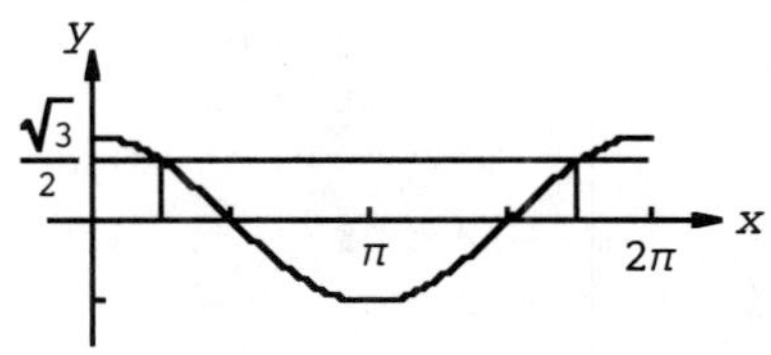

Since the sine and cosine functions are periodic with period 2π, all solutions are given by

$x = 0 + 2k\pi,\ x = \pi + 2k\pi,\ x = \frac{\pi}{6} + 2k\pi,\ x = \frac{11\pi}{6} + 2k\pi$, k any integer.

The first two can also be written together as $x = k\pi$, k any integer. (7-5)

34. $2\sin^2\theta + 5\cos\theta + 1 = 0$ all θ

Solve over one period $[0°, 360°)$

$$\begin{aligned} 2(1 - \cos^2\theta) + 5\cos\theta + 1 &= 0 \\ 2 - 2\cos^2\theta + 5\cos\theta + 1 &= 0 \\ -2\cos^2\theta + 5\cos\theta + 3 &= 0 \\ 2\cos^2\theta - 5\cos\theta - 3 &= 0 \\ (2\cos\theta + 1)(\cos\theta - 3) &= 0 \end{aligned}$$

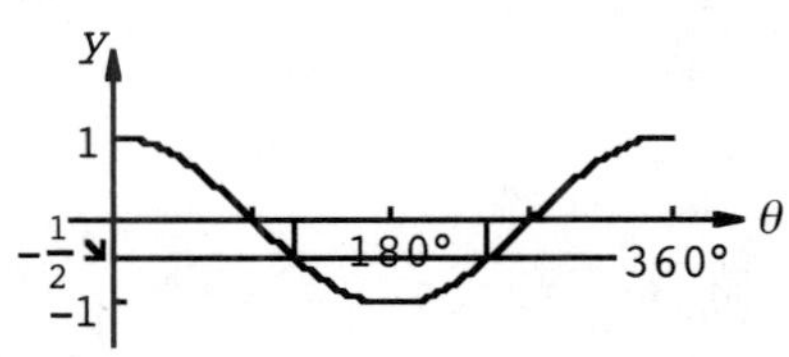

$2\cos\theta + 1 = 0 \qquad \cos\theta - 3 = 0$

$\cos\theta = -\frac{1}{2} \qquad \cos\theta = 3$

$\theta = 120°, 240° \qquad$ No solution

Since the cosine function is periodic with period 360°, all solutions are given by $\theta = 120° + k360°$, $\theta = 240° + k360°$, k any integer. (7-5)

35. $\tan \theta = 0.2557$ all θ
Solve over one period $[0°, 180°)$
Sketch a graph of $y = \tan \theta$ and $y = 0.2557$, on the interval $[0°, 180°)$

$\theta = \tan^{-1} 0.2557 = 14.34°$ Since the tangent function is periodic with period 180°, all solutions are given by

$\theta = 14.34° + k180°$

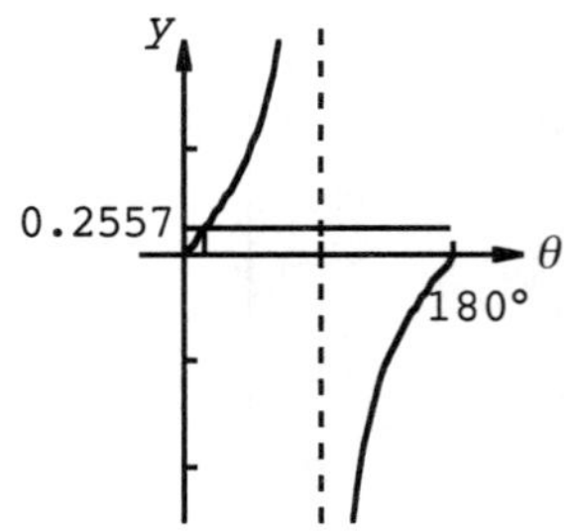

(7-5)

36. $\sin^2 x + 2 = 4 \sin x$ all real x
Solve over on period, $[0, 2\pi)$:
$\sin^2 x - 4 \sin x + 2 = 0$ Quadratic in $\sin x$

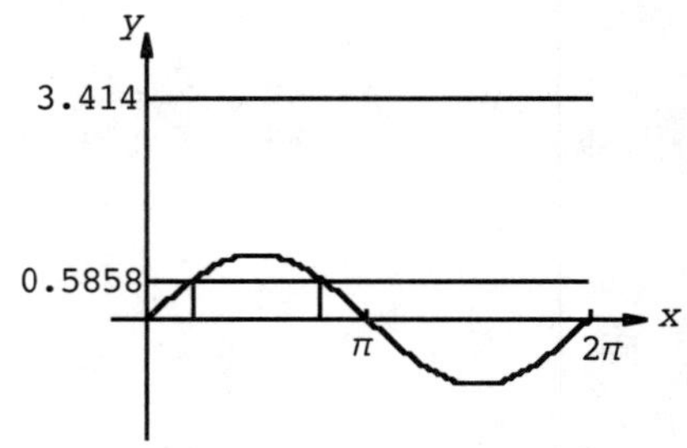

$$\sin x = \frac{-b \pm \sqrt{b^2 - 4ac}}{2a} \quad a = 1,\ b = -4,\ c = 2$$

$$\sin x = \frac{-(-4) \pm \sqrt{(-4)^2 - 4(1)(2)}}{2(1)}$$

$$\sin x = \frac{4 \pm \sqrt{8}}{2}$$

$$\sin x = \frac{4 + \sqrt{8}}{2} \qquad \sin x = \frac{4 - \sqrt{8}}{2}$$

$$\sin x = 3.414 \qquad \sin x = 0.5858$$

No solution

$$x = \begin{cases} \sin^{-1} 0.5858 \\ \pi - \sin^{-1} 0.5858 \end{cases} = \begin{cases} 0.6259 & \text{First quadrant solution} \\ 2.516 & \text{Second quadrant solution} \end{cases}$$

Since the sine function is periodic with period 2π, all solutions are given by

$$x = \begin{cases} 0.6259 + 2k\pi \\ 2.516 + 2k\pi \end{cases} \quad k \text{ any integer}$$

(7-5)

37. $\tan^2 x = 2 \tan x + 1 \qquad 0 \le x < \pi$
$\tan^2 x - 2 \tan x - 1 = 0$ Quadratic in $\tan x$

$$\tan x = \frac{-b \pm \sqrt{b^2 - 4ac}}{2a} \quad a = 1,\ b = -2,\ c = -1$$

$$\tan x = \frac{-(-2) \pm \sqrt{(-2)^2 - 4(1)(-1)}}{2(1)} \qquad \tan x = \frac{2 + \sqrt{8}}{2} \qquad \tan x = \frac{2 - \sqrt{8}}{2}$$

$$\tan x = \frac{(2) \pm \sqrt{8}}{2} \qquad \tan x = 2.414 \qquad \tan x = -0.4142$$

$$x = \tan^{-1} 2.414 \qquad x = \pi + \tan^{-1}(-0.4142)$$

$$x = 1.178 \qquad x = 2.749$$

Solutions: 1.178, 2.749

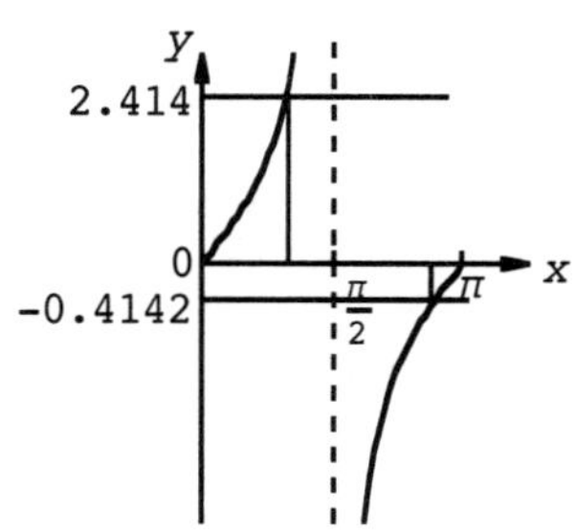

(7-5)

38. Examining the graph of $y1 = 3 \sin 2x$ and $y2 = 2x - 2.5$, we obtain the graph at the right.

To four decimal places, the solution in $[0, 2\pi]$ is given by $x = 1.4903$, and there is no other solution.

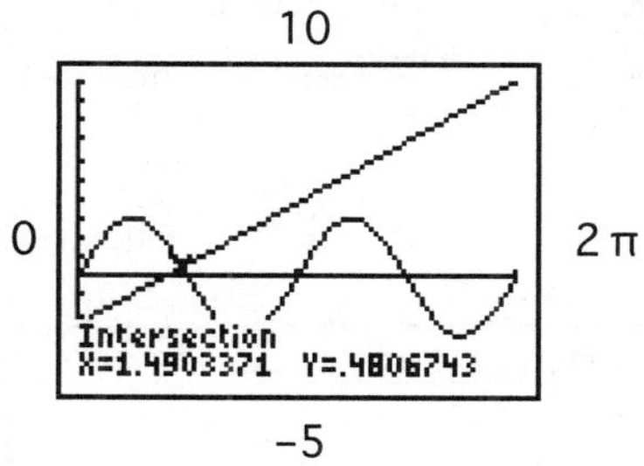

(7-5)

39. From the graph in the previous problem, the graph of $y = 3 \sin 2x$ is above the graph of $y = 2x - 2.5$, that is, $3 \sin 2x > 2x - 2.5$, for x in the interval $(-\infty, 1.4903)$. (7-5)

40. Examining the graph of $y1 = 2 \sin^2 x - \cos 2x$ and $y2 = 1 - x^2$, $-\pi \le x \le \pi$, we obtain

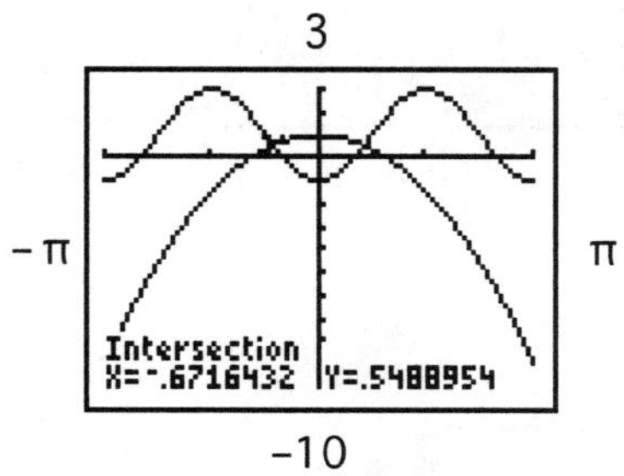

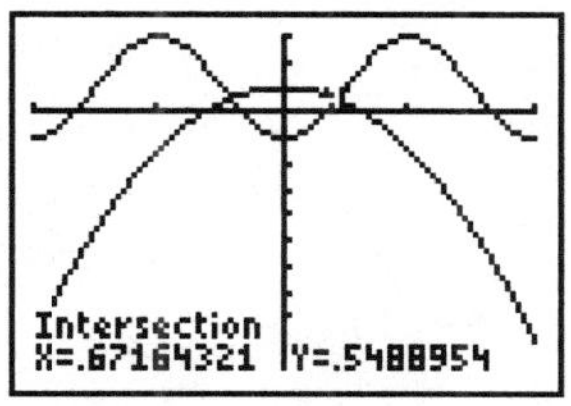

To four decimal places, the solutions in $[-\pi, \pi]$ are given by -0.6716 and 0.6716, and there are no other solutions. (7-5)

41. From the graph in the previous problem, the graph of $y = 2 \sin^2 x - \cos 2x$ is below or intersects the graph of $y = 1 - x^2$, that is, $2 \sin^2 x - \cos 2x \le 1 - x^2$, for x in the interval $[-0.6716, 0.6716]$. (7-5)

42. (A) Testing $x = 0$ and $y = \frac{\pi}{4}$ yields

$$\tan\left(0 + \frac{\pi}{4}\right) \stackrel{?}{=} \tan 0 + \tan \frac{\pi}{4}$$

$$\tan \frac{\pi}{4} \stackrel{?}{=} 0 + \tan \frac{\pi}{4}$$

$$1 \stackrel{\checkmark}{=} 0 + 1$$

Yes, $x = 0$ and $y = \frac{\pi}{4}$ is a solution.

(B) Consider $x = \frac{\pi}{3}$ and $y = \frac{\pi}{3}$. Then the left side $= \tan\left(\frac{\pi}{3} + \frac{\pi}{3}\right) = \tan \frac{2\pi}{3} = -\sqrt{3}$. The right side $= \tan \frac{\pi}{3} + \tan \frac{\pi}{3} = \sqrt{3} + \sqrt{3} = 2\sqrt{3}$. Since the left side is not equal to the right side for at least one set of values for which both are defined, the equation is a conditional equation. (7-1)

43. $\sin^{-1} 0.3351$ has exactly one value, while the equation $\sin x = 0.3351$ has infinitely many solutions. (6-6, 7-5)

44. (A) Graph both sides of the equation in the same viewing window.

$\dfrac{\tan x}{\sin x + 2\tan x} = \dfrac{1}{\cos x - 2}$ is not an identity, since the graphs do not match.

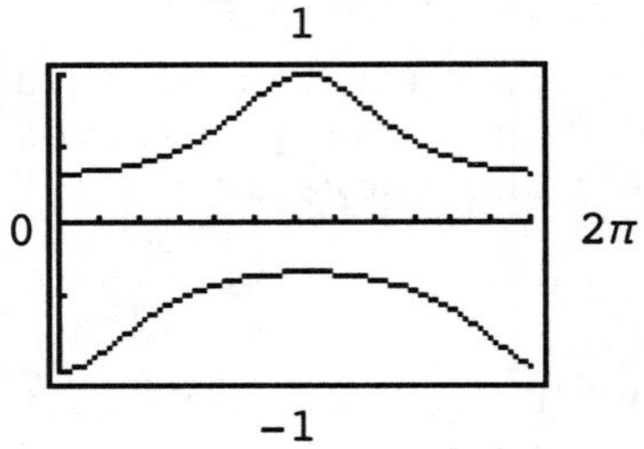

Try $x = \dfrac{\pi}{4}$.

Left side: $\dfrac{\tan\frac{\pi}{4}}{\sin\frac{\pi}{4} + 2\tan\frac{\pi}{4}} = \dfrac{1}{\frac{1}{\sqrt{2}} + 2\cdot 1} = \dfrac{\sqrt{2}}{1 + 2\sqrt{2}}$

Right side: $\dfrac{1}{\cos\frac{\pi}{4} - 2} = \dfrac{1}{\frac{1}{\sqrt{2}} - 2} = \dfrac{\sqrt{2}}{1 - 2\sqrt{2}}$

This verifies that the equation is not an identity.

(B) Graph both sides of the equation in the same viewing window.

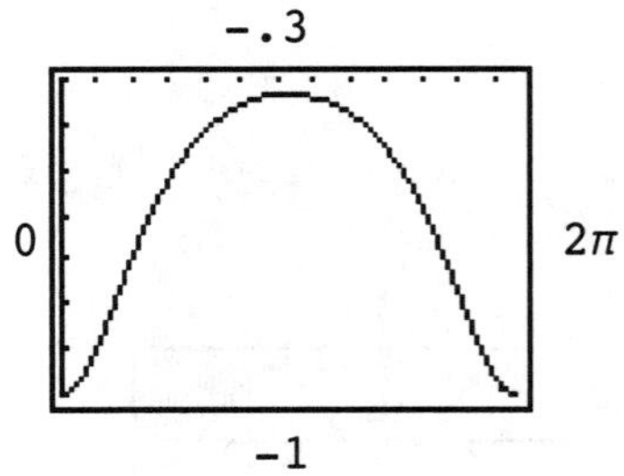

$\dfrac{\tan x}{\sin x + 2\tan x} = \dfrac{1}{\cos x - 2}$ appears to be an identity, which we now verify.

$$\frac{1}{\cos x - 2} = \frac{\tan x}{\tan x(\cos x - 2)} \quad \text{Algebra}$$

$$= \frac{\tan x}{\tan x\cos x - 2\tan x} \quad \text{Algebra}$$

$$= \frac{\tan x}{\frac{\sin x}{\cos x}\cos x - 2\tan x} \quad \text{Quotient Identity}$$

$$= \frac{\tan x}{\sin x - 2\tan x} \quad \text{Algebra}$$

(7-1)

45. Let $y1 = \cos\left(x - \dfrac{\pi}{3}\right)$

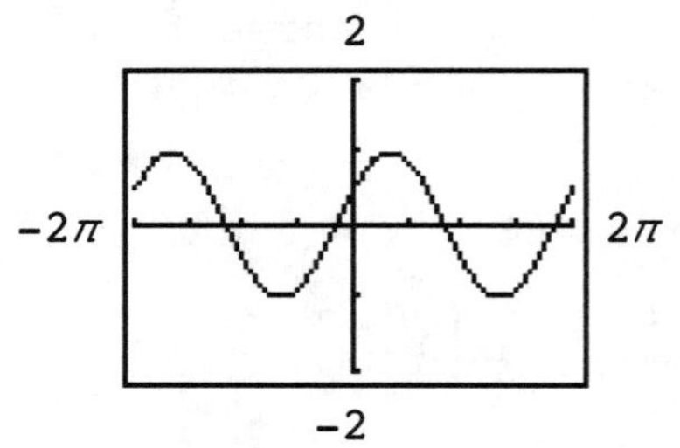

Then $y2 = \cos x \cos\dfrac{\pi}{3} + \sin x \sin\dfrac{\pi}{3}$

$= \cos x\left(\dfrac{1}{2}\right) + \sin x\left(\dfrac{\sqrt{3}}{2}\right)$

$= \dfrac{1}{2}\cos x + \dfrac{\sqrt{3}}{2}\sin x.$

The graphs coincide as shown at the right.

(7-2)

46. (A) $\tan\dfrac{x}{2} = 2\sin x \qquad 0 \le x < 2\pi$

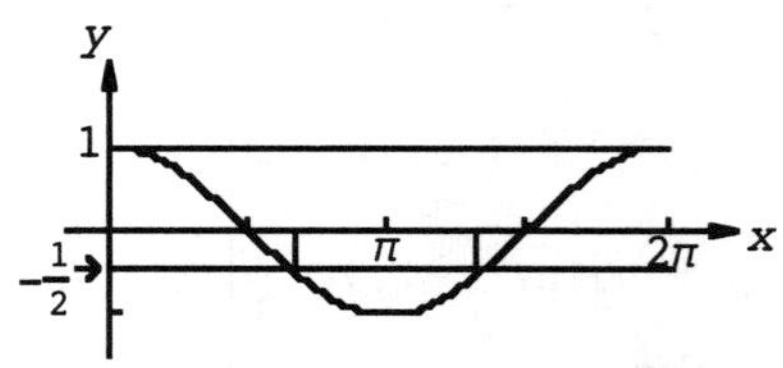

$\dfrac{1 - \cos x}{\sin x} = 2\sin x$

$1 - \cos x = 2\sin^2 x \qquad \sin x \ne 0$

$1 - \cos x = 2(1 - \cos^2 x)$

$1 - \cos x = 2 - 2\cos^2 x$

$2\cos^2 x - \cos x - 1 = 0$

$(2\cos x + 1)(\cos x - 1) = 0$

$2\cos x + 1 = 0 \qquad \cos x - 1 = 0$

$\cos x = -\dfrac{1}{2} \qquad \cos x = 1$

$x = \dfrac{2\pi}{3}, \dfrac{4\pi}{3} \qquad x = 0$

Note that even though we multiplied both sides of the original equation by $\sin x$, which is 0 when $x = 0$, $x = 0$ is still a solution of the original equation, as is shown by checking:

Left side: $\tan \frac{0}{2} = \tan 0 = 0$

Right side: $2 \sin 0 = 2 \cdot 0 = 0$

(B) Examining the graphs of $y1 = \tan \frac{x}{2}$ and $y2 = 2 \sin x$, $0 \le x \le 2\pi$ (drawn in dot mode), we obtain

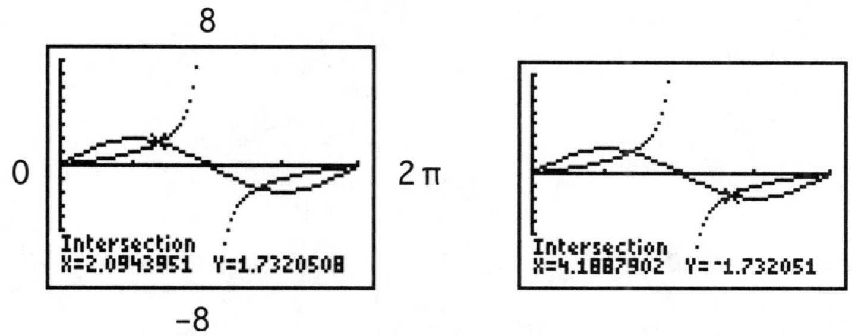

The graph confirms the solution 0 (checked in the previous part of the problem). The other intersections are found at $x = 2.0944$ and $x = 4.1888$. (*7-5*)

47. Examining the graph of $y1 = 3 \cos(x - 1)$ and $y2 = 2 - x^2$, $-\pi \le x \le \pi$, we obtain

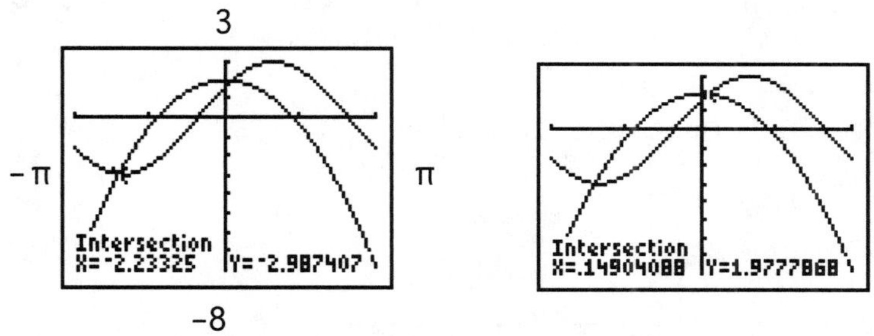

To three decimal places, the solutions on $[-\pi, \pi]$ are 0.149 and -2.233, and there are no other solutions. (*7-5*)

48. Since $\frac{\pi}{2} \le x \le \pi$, $\frac{\pi}{4} \le \frac{x}{2} \le \frac{\pi}{2}$, hence $\sin \frac{x}{2}$ is positive.

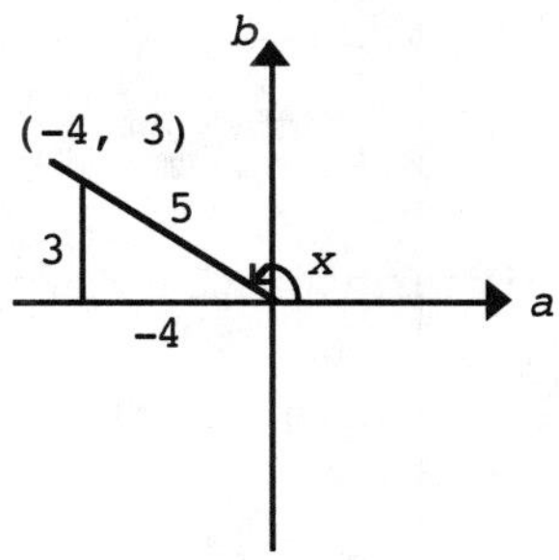

$\sin x = \frac{3}{5}$ $\quad \cos x = -\frac{4}{5}$

(A) $$\sin \frac{x}{2} = \sqrt{\frac{1 - \cos x}{2}} = \sqrt{\frac{1 - \left(-\frac{4}{5}\right)}{2}} = \sqrt{\frac{\frac{9}{5}}{2}} = \sqrt{\frac{9}{10}}$$

$$= \frac{3}{\sqrt{10}} \text{ or } \frac{3\sqrt{10}}{10}$$

(B) $$\cos 2x = \cos^2 x - \sin^2 x = \left(-\frac{4}{5}\right)^2 - \left(\frac{3}{5}\right)^2 = \frac{16}{25} - \frac{9}{25} = \frac{7}{25}$$ (*7-3*)

49. Let $u = \tan^{-1}\left(-\frac{3}{4}\right)$.

Then $-\frac{3}{4} = \tan u \quad -\frac{\pi}{2} < u < \frac{\pi}{2}$.

$\sin u = -\frac{3}{5} \quad \cos u = \frac{4}{5}$

$\sin\left[2\tan^{-1}\left(-\frac{3}{4}\right)\right] = \sin 2u = 2 \sin u \cos u$

$= 2\left(-\frac{3}{5}\right)\left(\frac{4}{5}\right) = -\frac{24}{25}$

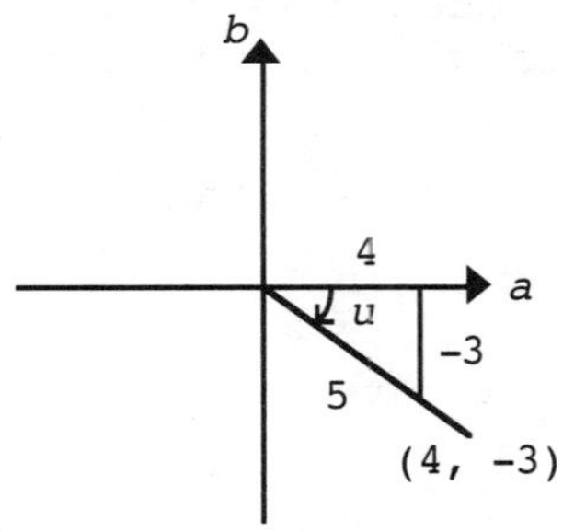

(7-3)

50. Let $u = \sin^{-1} \frac{3}{5}$

Then $\frac{3}{5} = \sin u, \ -\frac{\pi}{2} \le u \le \frac{\pi}{2}$.

Ordinarily we would also set $v = \cos^{-1} \frac{4}{5}$, but a glance at the reference triangle indicates that

$\cos^{-1} \frac{4}{5} = \sin^{-1} \frac{3}{5} = u$

$\sin\left(\sin^{-1} \frac{3}{5} + \cos^{-1} \frac{4}{5}\right) = \sin(u + u) = \sin 2u$

$= 2 \sin u \cos u$

$= 2\left(\frac{3}{5}\right)\left(\frac{4}{5}\right) = \frac{24}{25}$

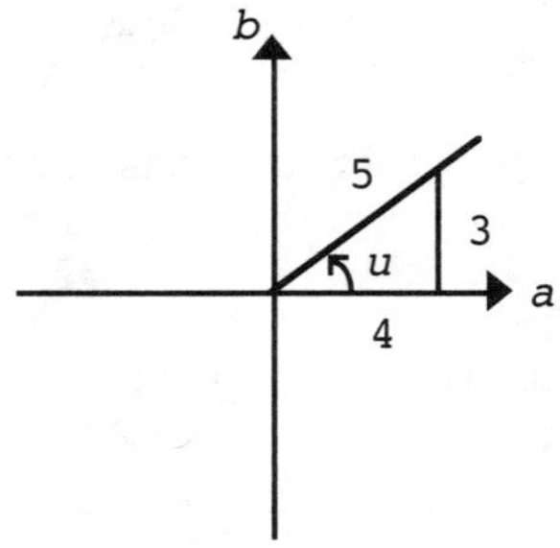

(7-3)

51. (A) $\cos^2 2x = \cos 2x + \sin^2 2x \quad 0 \le x < \pi$ is equivalent to

$\cos^2 2x = \cos 2x + \sin^2 2x \quad 0 \le 2x < 2\pi$

$\cos^2 2x = \cos 2x + 1 - \cos^2 2x$

$2\cos^2 2x - \cos 2x - 1 = 0$

$(2\cos 2x + 1)(\cos 2x - 1) = 0$

$2\cos 2x + 1 = 0$ or $\cos 2x - 1 = 0$

$\cos 2x = -\frac{1}{2} \qquad \cos 2x = 1$

$2x = \frac{2\pi}{3}, \frac{4\pi}{3} \qquad 2x = 0$

$x = \frac{\pi}{3}, \frac{2\pi}{3} \qquad x = 0$

Solutions: $x = 0, \frac{\pi}{3}, \frac{2\pi}{3}$

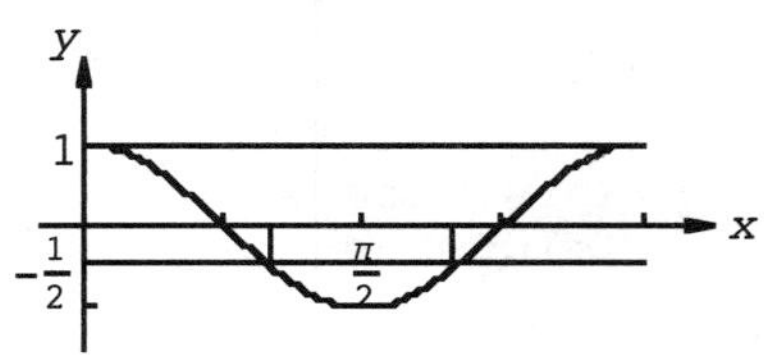

(B) Examining the graph of $y1 = \cos^2 2x$ and $y2 = \cos 2x + \sin^2 2x$, $0 \le x < \pi$, we obtain

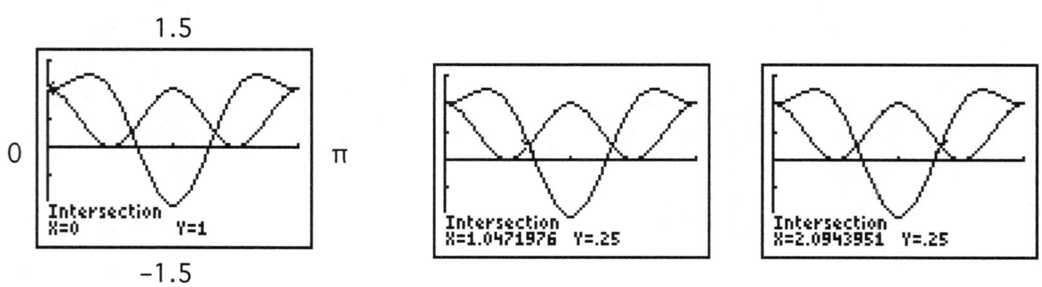

The solutions on $[0, \pi]$ are 0, and, to four decimal places, 1.0472 and 2.0944.

(7-5)

52. (A) Here is a computer-generated graph of $f(x) = \sin \dfrac{1}{x - 1}$, $0 \le x \le 2$.

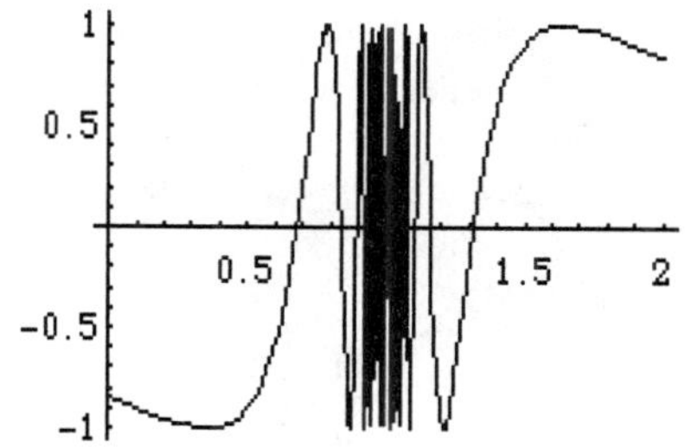

The function appears to have a smallest zero in the interval [0.6, 0.7] and a largest zero in the interval [1.3, 1.4]. Zooming in on these zeros, we obtain the following graphs.

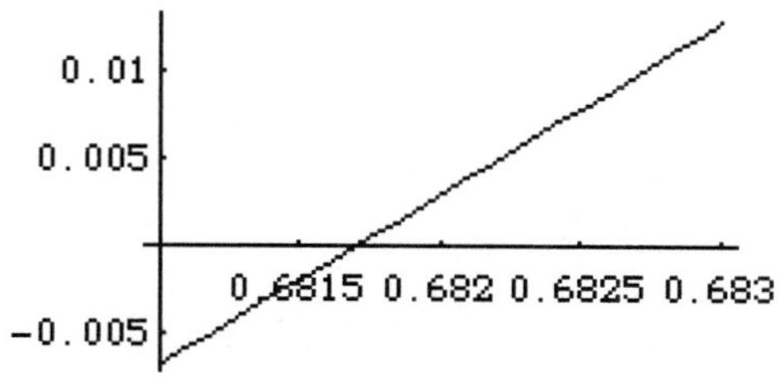

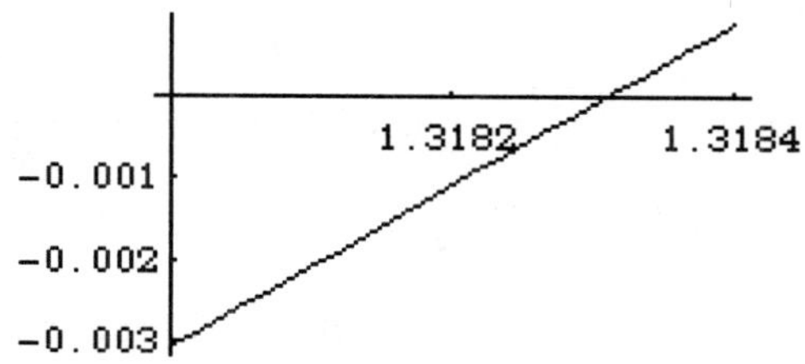

To four decimal places, the smallest zero is 0.6817 and the largest is 1.3183.

(B) As x increases without bound, $\dfrac{1}{x - 1}$ tends to 0 through positive numbers and $\sin \dfrac{1}{x - 1}$ tends to 0 through positive numbers. $y = 0$ is a horizontal asymptote for the graph of f.

(C) The exploratory graphs are left to the student. There are infinitely many zeros in any interval containing $x = 1$. The number $x = 1$ is not a zero because $\sin \dfrac{1}{x - 1}$ is not defined at $x = 1$. *(7-5)*

53. We note that $\tan \theta = \dfrac{3}{x}$ and $\tan 2\theta = \dfrac{3 + 6}{x} = \dfrac{9}{x}$ (see figure).

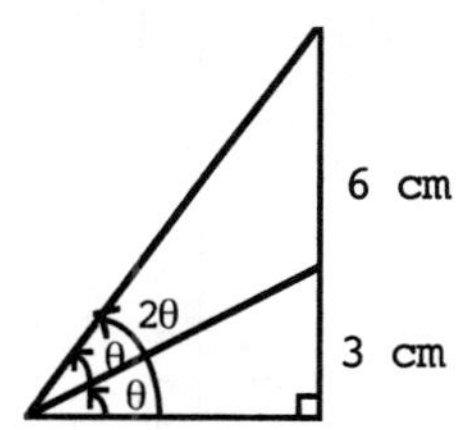

Then, $\tan 2\theta = \dfrac{2 \tan \theta}{1 - \tan^2 \theta}$

$$\frac{9}{x} = \frac{2\left(\frac{3}{x}\right)}{1 - \left(\frac{3}{x}\right)^2} = \frac{\frac{6}{x}}{1 - \frac{9}{x^2}} = \frac{x^2 \cdot \frac{6}{x}}{x^2 \cdot 1 - x^2 \cdot \frac{9}{x^2}} = \frac{6x}{x^2 - 9}$$

$$x(x^2 - 9)\frac{9}{x} = x(x^2 - 9)\frac{6x}{x^2 - 9} \qquad x \ne 0, 3, -3$$

$$9(x^2 - 9) = 6x \cdot x$$
$$9x^2 - 81 = 6x^2$$
$$3x^2 - 81 = 0$$
$$x^2 = 27$$
$$x = \sqrt{27} \quad \text{(we discard the negative solution)}$$

To 3 decimal places, $x = 5.196$ cm.

Since $\tan \theta = \dfrac{3}{x} = \dfrac{3}{\sqrt{27}} = \dfrac{1}{\sqrt{3}}$, $\theta = 30.000°$. *(7-3)*

54. $I = 50 \sin 120\pi(t - 0.001) \qquad I = 40$

$$40 = 50 \sin 120\pi(t - 0.001)$$

$$\sin 120\pi(t - 0.001) = \frac{40}{50}$$

$$120\pi(t - 0.001) = \sin^{-1}\frac{40}{50}$$ will yield the least positive solution of the equation

$$t - 0.001 = \frac{1}{120\pi}\sin^{-1}\frac{40}{50}$$

$$t = 0.001 + \frac{1}{120\pi}\sin^{-1}\frac{40}{50}$$

$$= 0.00346 \text{ sec}$$ (calculator in radian mode) (7-5)

55. (A) $0.6 \cos 184\pi t - 0.6 \cos 208\pi t = 0.6(\cos 184\pi t - \cos 208\pi t)$ Algebra

$$= 0.6\left[(-2)\sin\frac{184\pi t + 208\pi t}{2}\sin\frac{184\pi t + 208\pi t}{2}\right]$$ Sum-Product Identity

$$= -1.2 \sin 196\pi t \sin(-12\pi t)$$ Algebra

$$= -1.2 \sin 196\pi t(-\sin 12\pi t)$$ Identities for Negatives

$$= 1.2 \sin 12\pi t \sin 196\pi t$$ Algebra

(B) The required graphs are as shown.

$y = 0.6 \cos 184\pi t$

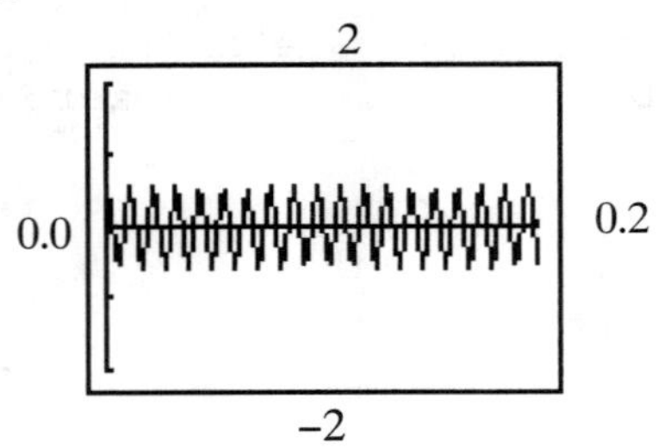

$y = -0.6 \cos 208\pi t$

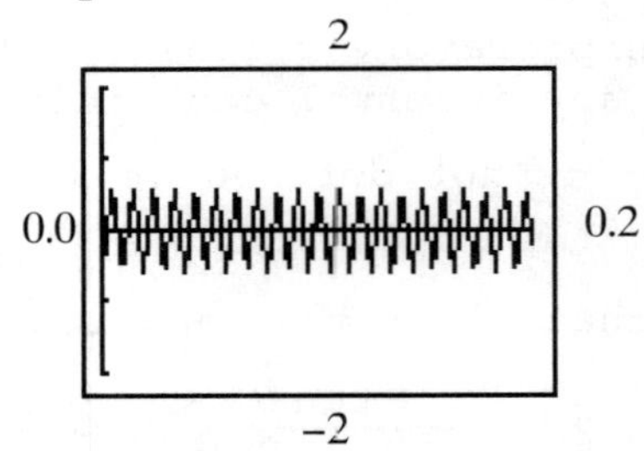

$y = 0.6 \cos 184\pi t - 0.6 \cos 208\pi t$

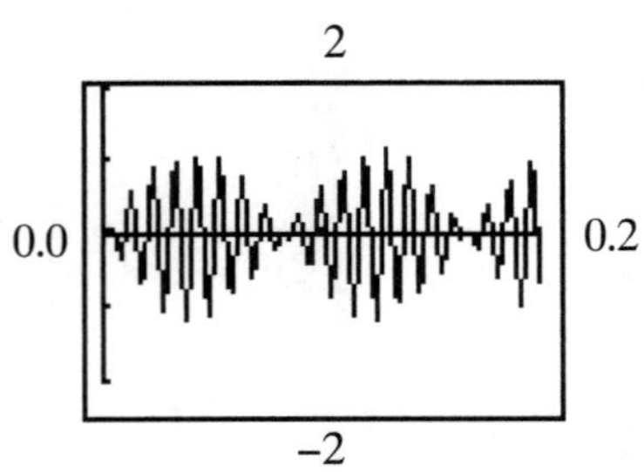

$y = 1.2 \sin 12\pi t \sin 196\pi t$

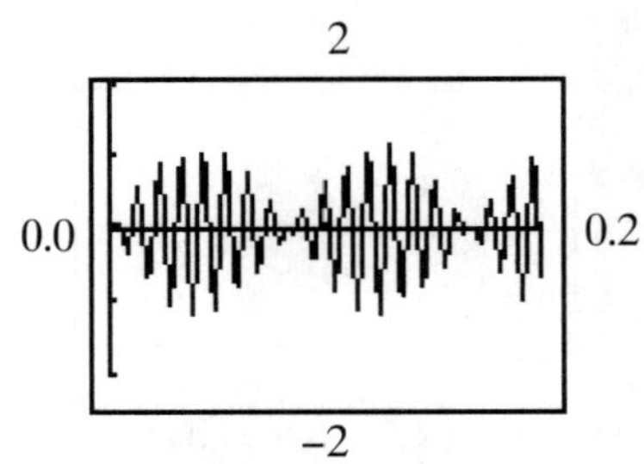

(7-4)

56.

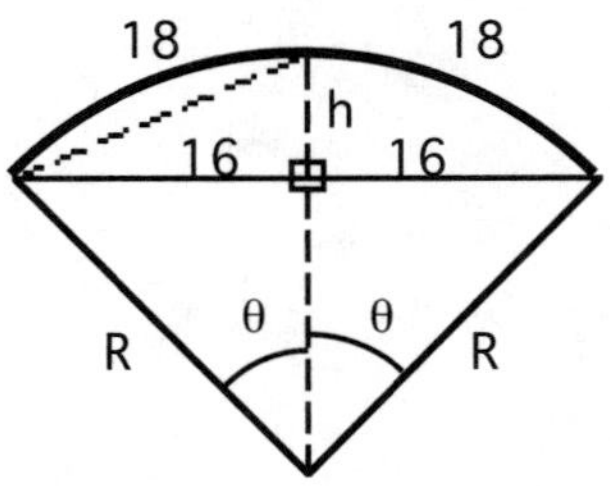

From the figure, $R\theta = 18$ and $\sin\theta = \frac{16}{R}$. The definition of radian measure is $\theta = \frac{S}{R}$ where S is the arc length and R the radius. From these two equations, solving each for R in terms of θ and setting the results equal to each other, we obtain

$$R = \frac{18}{\theta} \qquad R = \frac{16}{\sin\theta}$$

$$\frac{18}{\theta} = \frac{16}{\sin\theta}$$

$$18\sin\theta = 16\theta$$

$$\sin\theta = \frac{8}{9}\theta \quad \text{as required.}$$

Examining the graph of $y1 = \sin\theta$ and $y2 = \frac{8}{9}\theta$, $0 \le \theta \le \frac{\pi}{2}$, we obtain

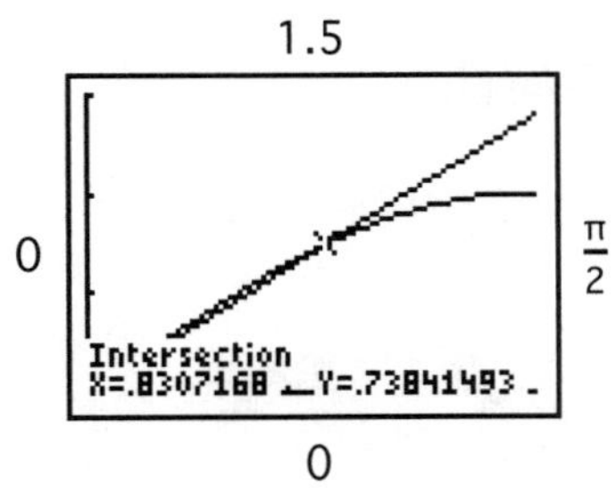

To four decimal places, $\theta = 0.8307$.

Then $R = \frac{18}{\theta} = \frac{18}{0.8307} = 21.668$ ft.

Since $\frac{R-h}{16} = \cot\theta$, $h = R - 16\cot\theta$
$= 7.057$ ft.

(7-5)

CHAPTER 8 Additional Topics in Trigonometry

Section 8-1

1. 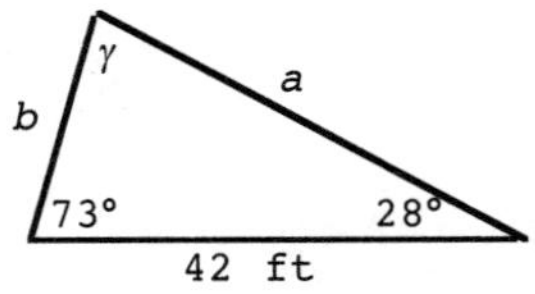

Solve for γ:

$$\begin{aligned}\alpha + \beta + \gamma &= 180° \\ \gamma &= 180° - (\alpha + \beta) \\ &= 180° - (73° + 28°) \\ &= 79°\end{aligned}$$

Solve for b:

$$\begin{aligned}\frac{\sin\beta}{b} &= \frac{\sin\gamma}{c} \\ b &= \frac{c\sin\beta}{\sin\gamma} \\ &= \frac{42\sin 28°}{\sin 79°} \\ &= 20 \text{ ft}\end{aligned}$$

Solve for a:

$$\begin{aligned}\frac{\sin\alpha}{a} &= \frac{\sin\gamma}{c} \\ a &= \frac{c\sin\alpha}{\sin\gamma} \\ &= \frac{42\sin 73°}{\sin 79°} \\ &= 41 \text{ ft}\end{aligned}$$

3. 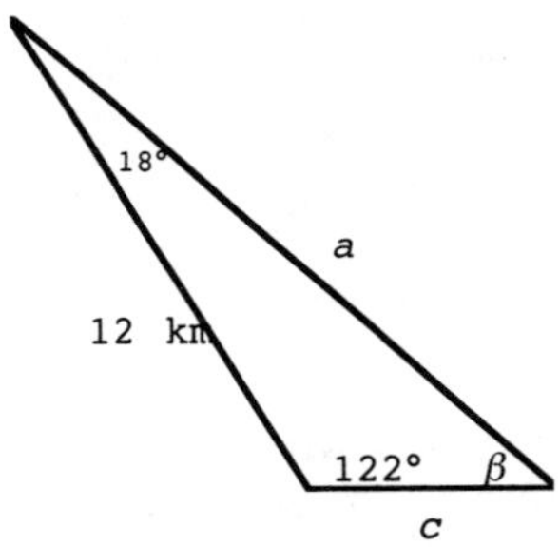

Solve for β:

$$\begin{aligned}\alpha + \beta + \gamma &= 180° \\ \beta &= 180° - (\alpha + \gamma) \\ &= 180° - (122° + 18°) \\ &= 40°\end{aligned}$$

Solve for a:

$$\begin{aligned}\frac{\sin\alpha}{a} &= \frac{\sin\beta}{b} \\ a &= \frac{b\sin\alpha}{\sin\beta} \\ &= \frac{12\sin 122°}{\sin 40°} \\ &= 16 \text{ km}\end{aligned}$$

Solve for c:

$$\begin{aligned}\frac{\sin\beta}{b} &= \frac{\sin\gamma}{c} \\ c &= \frac{b\sin\gamma}{\sin\beta} \\ &= \frac{12\sin 18°}{\sin 40°} \\ &= 5.8 \text{ km}\end{aligned}$$

5. 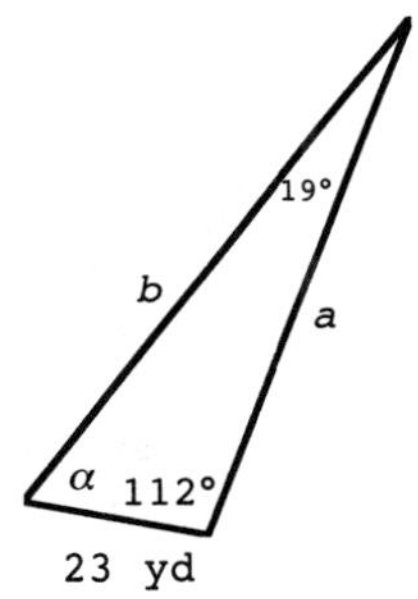

Solve for α:

$$\begin{aligned}\alpha + \beta + \gamma &= 180° \\ \alpha &= 180° - (\beta + \gamma) \\ &= 180° - (112° + 19°) \\ &= 49°\end{aligned}$$

Solve for a:

$$\begin{aligned}\frac{\sin\alpha}{a} &= \frac{\sin\gamma}{c} \\ a &= \frac{c\sin\alpha}{\sin\gamma} \\ &= \frac{23\sin 49°}{\sin 19°} \\ &= 53 \text{ yd}\end{aligned}$$

Solve for b:

$$\begin{aligned}\frac{\sin\beta}{b} &= \frac{\sin\gamma}{c} \\ b &= \frac{c\sin\beta}{\sin\gamma} \\ &= \frac{23\sin 112°}{\sin 19°} \\ &= 66 \text{ yd}\end{aligned}$$

7. 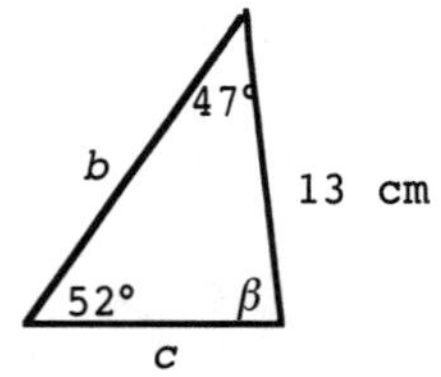

Solve for β:

$$\alpha + \beta + \gamma = 180°$$
$$\beta = 180° - (\alpha + \gamma) = 180° - (52° + 47°) = 81°$$

Solve for c:

$$\frac{\sin\alpha}{a} = \frac{\sin\gamma}{c}$$
$$c = \frac{a\sin\gamma}{\sin\alpha} = \frac{13\sin 47°}{\sin 52°} = 12 \text{ cm}$$

Solve for b:

$$\frac{\sin\alpha}{a} = \frac{\sin\beta}{b}$$
$$b = \frac{a\sin\beta}{\sin\alpha} = \frac{13\sin 81°}{\sin 52°} = 16 \text{ cm}$$

9. 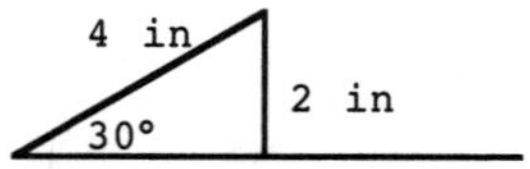

We are given two sides and a non-included angle (SSA). α is acute
$h = b \sin \alpha = 4 \sin 30° = 2$
$a = 2 = h$
One triangle can be constructed.

11. 

We are given two sides and a non-included angle (SSA). α is acute
$6 = a \geq b = 4$
One triangle can be constructed.

13. 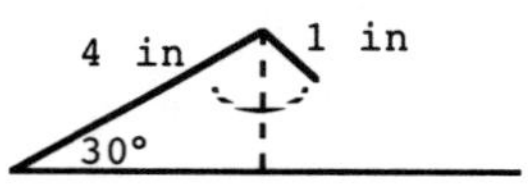

We are given two sides and a non-included angle (SSA). α is acute
$h = b \sin \alpha = 4 \sin 30° = 2$
$0 < a = 1 < 2 = h$
No triangle can be constructed.

15. 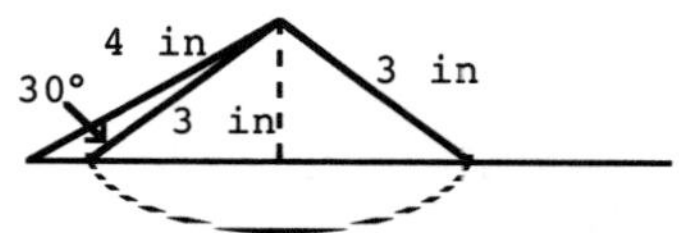

We are given two sides and a non-included angle (SSA). α is acute
$h = b \sin \alpha = 4 \sin 30° = 2$
$h < a < b$
Two triangles can be constructed.

17. 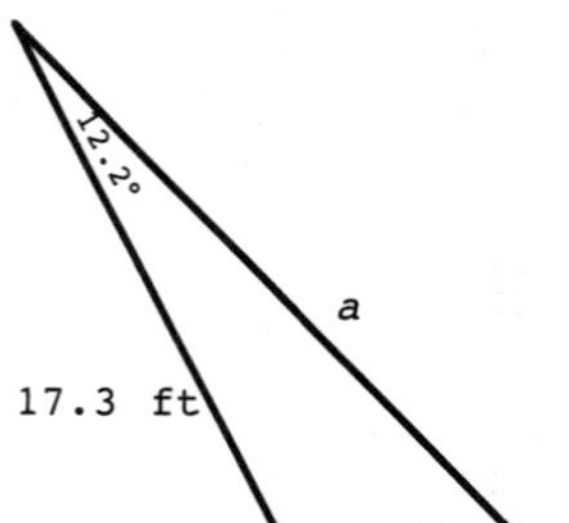

Solve for β:

$$\alpha + \beta + \gamma = 180°$$
$$\beta = 180° - (\alpha + \gamma) = 180° - (118.3° + 12.2°) = 49.5°$$

Solve for a:

$$\frac{\sin\alpha}{a} = \frac{\sin\beta}{b}$$
$$a = \frac{b\sin\alpha}{\sin\beta} = \frac{17.3\sin 118.3°}{\sin 49.5°} = 20.0 \text{ ft}$$

Solve for c:

$$\frac{\sin\beta}{b} = \frac{\sin\gamma}{c}$$
$$c = \frac{b\sin\gamma}{\sin\beta} = \frac{17.3\sin 12.2°}{\sin 49.5°} = 4.81 \text{ ft}$$

19.

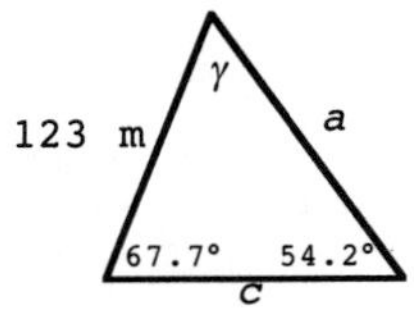

Solve for γ:

$$\begin{aligned}\alpha + \beta + \gamma &= 180^\circ\\ \gamma &= 180^\circ - (\alpha + \beta)\\ &= 180^\circ - (67.7^\circ + 54.2^\circ)\\ &= 58.1^\circ\end{aligned}$$

Solve for c:

$$\begin{aligned}\frac{\sin\beta}{b} &= \frac{\sin\gamma}{c}\\ c &= \frac{b\sin\gamma}{\sin\beta}\\ &= \frac{123\sin 58.1^\circ}{\sin 54.2^\circ}\\ &= 129\text{ m}\end{aligned}$$

Solve for a:

$$\begin{aligned}\frac{\sin\alpha}{a} &= \frac{\sin\beta}{b}\\ a &= \frac{b\sin\alpha}{\sin\beta}\\ &= \frac{123\sin 67.7^\circ}{\sin 54.2^\circ}\\ &= 140\text{ m}\end{aligned}$$

21.

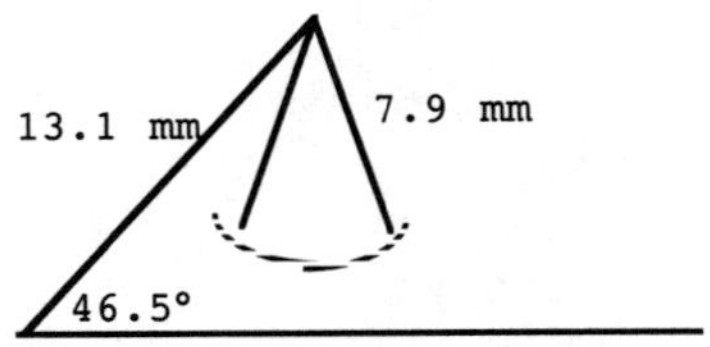

α is acute

$h = b\sin\alpha = 13.1\sin 46.5^\circ = 9.50$

$a = 7.9 < 9.50 = h$

No triangle is possible.

If we try to draw a triangle with these values, we will be unsuccessful. No solution.

23.

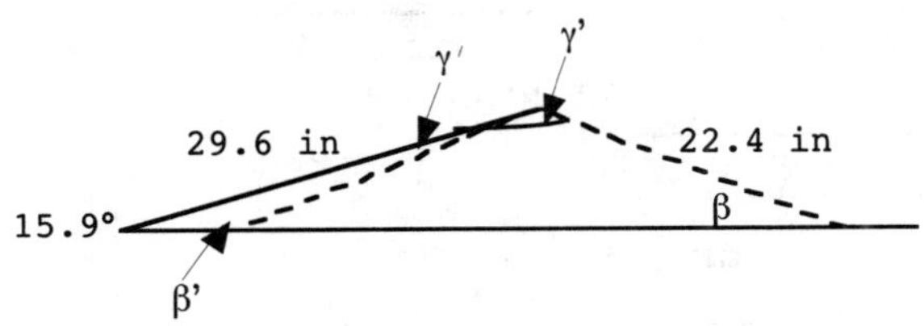

We're given two sides and a non-included angle (SSA).

α is acute

$h = b\sin\alpha = 29.6\sin 15.9^\circ = 8.11$

$a = 22.4$; we have $h < a < b$, so two triangles are possible.

Solve for β:

$$\begin{aligned}\frac{\sin\alpha}{a} &= \frac{\sin\beta}{b}\\ \sin\beta &= \frac{b\sin\alpha}{a}\\ &= \frac{29.6\sin 15.9^\circ}{22.4}\\ &= 0.362\end{aligned}$$

$\beta = \sin^{-1}(0.362) = 21.2^\circ$

$\beta' = 180^\circ - 21.2^\circ = 158.8^\circ$

Triangle I $\quad \beta = 21.2^\circ$

Solve for γ:

$$\begin{aligned}\alpha + \beta + \gamma &= 180^\circ\\ &= 180^\circ - (\alpha + \beta)\\ &= 180^\circ - (15.9^\circ + 21.2^\circ)\\ &= 142.9^\circ\end{aligned}$$

Solve for c:

$$\begin{aligned}\frac{\sin\alpha}{a} &= \frac{\sin\gamma}{c}\\ c\sin\alpha &= a\sin\gamma\\ c &= \frac{a\sin\gamma}{\sin\alpha}\\ &= \frac{22.4\sin 142.9^\circ}{\sin 15.9^\circ}\\ &= 49.3\text{ in}\end{aligned}$$

Triangle II $\quad \beta' = 158.8^\circ$

Solve for γ':

$$\begin{aligned}\alpha + \beta' + \gamma' &= 180^\circ\\ \gamma' &= 180^\circ - (\alpha + \beta')\\ &= 180^\circ - (15.9^\circ + 158.8^\circ)\\ &= 5.3^\circ\end{aligned}$$

Solve for c':

$$\frac{\sin\alpha}{a} = \frac{\sin\gamma'}{c'}$$
$$c'\sin\alpha = a\sin\gamma'$$
$$c' = \frac{a\sin\gamma'}{\sin\alpha} = \frac{22.4\sin 5.3°}{\sin 15.9°} = 7.55 \text{ in}$$

25. 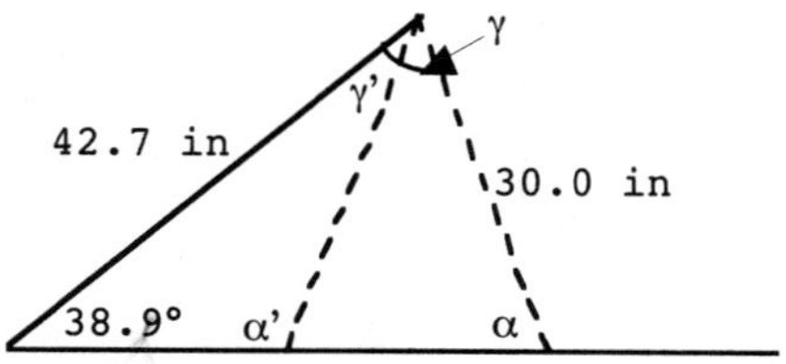

We're given two sides and a non-included angle (SSA).
β is acute.

$h = a\sin\beta = 42.7\sin 38.9° = 26.8$

$h < b < a$, so there are 2 triangles.

Solve for α:

$$\frac{\sin\alpha}{a} = \frac{\sin\beta}{b}$$
$$\sin\alpha = \frac{a\sin\beta}{b} = \frac{42.7\sin 38.9°}{30.0} = 0.894$$

$\alpha = \sin^{-1}(0.894) = 63.4°$

$\alpha' = 180 - 63.4° = 116.6°$

Triangle I $\alpha = 63.4°$

Solve for γ:

$$\alpha + \beta + \gamma = 180°$$
$$\gamma = 180° - (\alpha + \beta) = 180° - (63.4° + 38.9°) = 77.7°$$

Solve for c:

$$\frac{\sin\beta}{b} = \frac{\sin\gamma}{c}$$
$$c\sin\beta = b\sin\gamma$$
$$c = \frac{b\sin\gamma}{\sin\beta} = \frac{30.0\sin 77.7°}{\sin 38.9°} = 46.7 \text{ in}$$

Triangle II $\alpha' = 116.6°$

Solve for γ':

$$\alpha' + \beta + \gamma' = 180°$$
$$\gamma' = 180° - (\alpha + \beta) = 180° - (116.6° + 38.9°) = 24.5°$$

Solve for c':

$$\frac{\sin\beta}{b} = \frac{\sin\gamma'}{c'}$$
$$c'\sin\beta = b\sin\gamma'$$
$$c' = \frac{b\sin\gamma'}{\sin\beta} = \frac{30.0\sin 24.5°}{\sin 38.9°} = 19.8 \text{ in}$$

27. 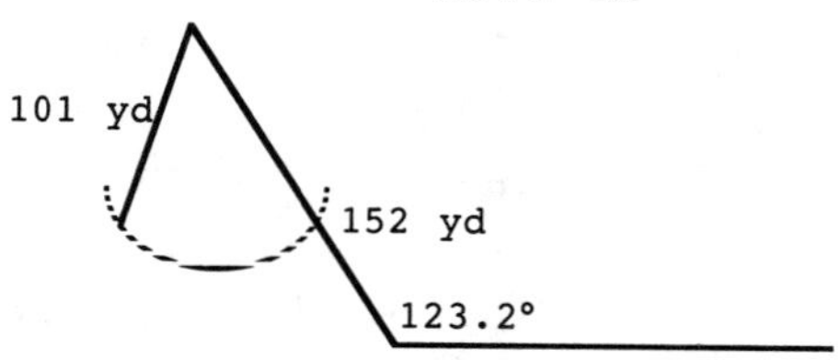

α is obtuse

$101 = a < b = 152$

No triangle is possible.

If we try to draw a triangle with these values, we will be unsuccessful.
No solution.

29. From a rough sketch, we see that there is only one triangle possible:

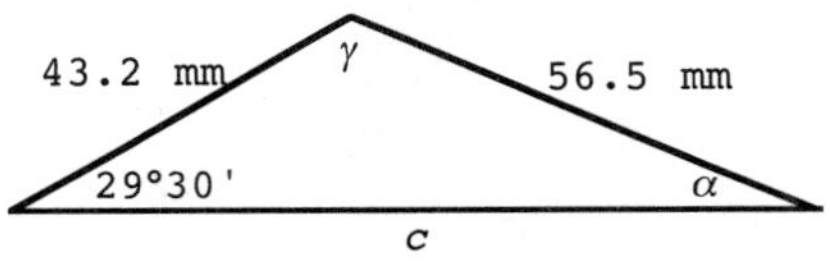

Solve for α:

$$\frac{\sin\alpha}{a} = \frac{\sin\beta}{b}$$

$$\sin\alpha = \frac{a\sin\beta}{b} = \frac{43.2\sin 29°30'}{56.5}$$

$$\alpha = \sin^{-1}\left(\frac{43.2\sin 29°30'}{56.5}\right) = 22°10'$$

Solve for γ:

$$\alpha + \beta + \gamma = 180°$$

$$\gamma = 180° - (\alpha + \beta) = 180° - (22°10' + 29°30') = 128°20'$$

Solve for c:

$$\frac{\sin\beta}{b} = \frac{\sin\gamma}{c}$$

$$c = \frac{b\sin\gamma}{\sin\beta} = \frac{56.5\sin 128°20'}{\sin 29°30'} = 89.9 \text{ mm}$$

31.

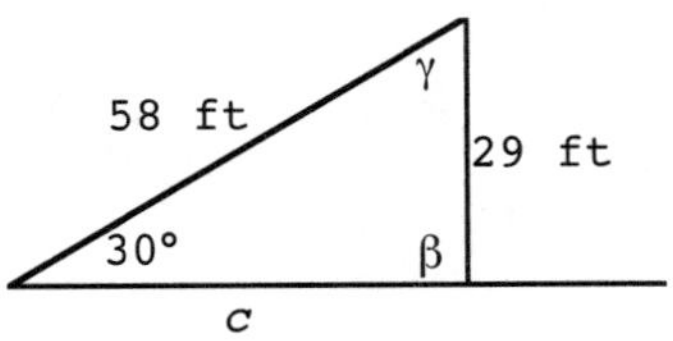

We're given two sides and a non-included angle (SSA). α is acute.

$h = b\sin\alpha = 58\sin 30° = 29$

$h = a$, so there is 1 triangle.

Solve for β:

$$\frac{\sin\alpha}{a} = \frac{\sin\beta}{b}$$

$$\sin\beta = \frac{b\sin\alpha}{a} = \frac{58\sin 30°}{29} = 1$$

so $\beta = 90°$

Solve for c:

$$\frac{\sin\alpha}{a} = \frac{\sin\gamma}{c}$$

$$c\sin\alpha = a\sin\gamma$$

$$c = \frac{a\sin\gamma}{\sin\alpha} = \frac{29\sin 60°}{\sin 30°} = 50 \text{ ft}$$

Solve for γ:

$$\alpha + \beta + \gamma = 180°$$

$$\gamma = 180° - (\alpha + \beta) = 180° - (30° + 90°) = 60°$$

33.

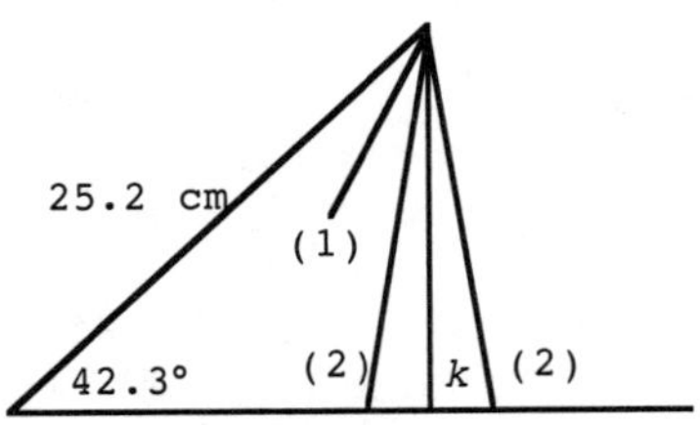

From the diagram, we can see that k = altitude of any possible triangle. Thus,

$\sin\alpha = \frac{k}{b}$ $\quad k = b\sin\alpha$

$k = (25.2 \text{ cm})\sin 42.3° = 17.0 \text{ cm}$

If $0 < a < k$, there is no solution: (1) in diagram.
If $a = k$, there is one solution.
If $k < a < b$, there are two solutions: (2) in diagram.

35. Using the given and the calculated data, we have:

$$(a - b)\cos\frac{\gamma}{2} = c\sin\frac{\alpha - \beta}{2}$$

$$(41 - 20)\cos\frac{79°}{2} = 42\sin\frac{73° - 28°}{2}$$

$$16.204 \approx 16.073$$

37. Sketch a figure:

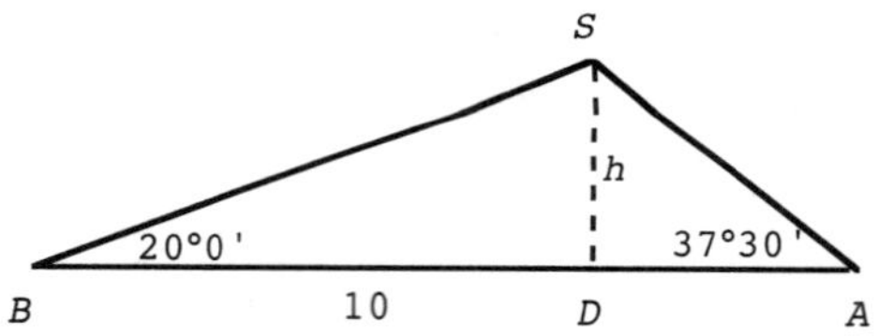

Using the law of sines, we have, in triangle ABS:

$$\frac{AS}{\sin ABS} = \frac{AB}{\sin ASB}$$

We are given angle $ABS = 20°0'$ and $AB = 10$. To find angle ASB, we use $ABS + ASB + BAS = 180°$, hence

$$\begin{aligned} ASB &= 180° - (ABS + BAS) \\ &= 180° - (20°0' + 37°30') \\ &= 180° - 57°30' \\ &= 122°30' \end{aligned}$$

Hence

$$\begin{aligned} \frac{AS}{\sin 20°0'} &= \frac{10}{\sin 122°30'} \\ AS &= \frac{10 \sin 20°0'}{\sin 122°30'} \\ &= 4.06 \text{ miles} = \text{distance of ship from point } A. \end{aligned}$$

To find

$h = SD$ = distance of ship from shore, we note in right triangle ADS

$$\begin{aligned} \frac{h}{AS} &= \sin DAS \\ h &= AS \sin DAS \\ &= 4.06 \sin 37°30' \\ &= 2.47 \text{ miles} \end{aligned}$$

39. Redrawing the figure and labeling the sides, we have:

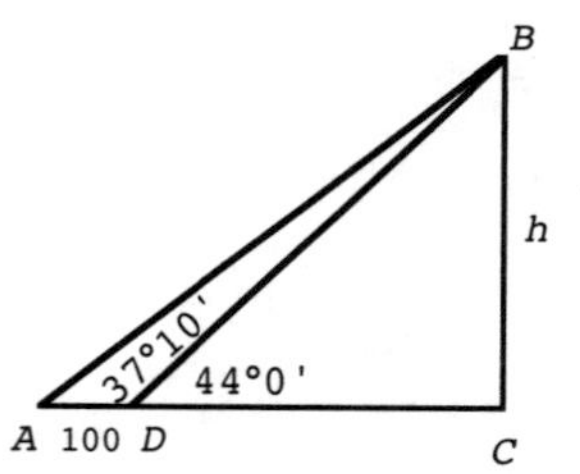

In triangle ABC, we have $\frac{AC}{h} = \cot 37°10'$.

Hence $AC = h \cot 37°10'$

In triangle DBC, we have $\frac{DC}{h} = \cot 44°$.

Hence $DC = h \cot 44°$.

Hence $100 = AC - DC = h \cot 37°10' - h \cot 44°$

Solving for h, we have

$$\begin{aligned} 100 &= h(\cot 37°10' - \cot 44°) \\ \text{So } h &= \frac{100}{\cot 37°10 - \cot 44°} \\ &= 353 \text{ feet} \end{aligned}$$

Alternatively, we can use the law of sines in triangle ADB to find BD, after noting that angle $ADB = 180°$ - angle $BDC = 180° - 44° = 136°$, and therefore angle $ABD = 180° - (136° + 37°10') = 6°50'$. Hence

$$\begin{aligned} \frac{BD}{\sin DAB} &= \frac{AD}{\sin ABD} \\ BD &= \frac{AD \sin DAB}{\sin ABD} \\ &= \frac{100 \sin 37°10'}{\sin 6°50'} \\ &= 508 \text{ feet} \end{aligned}$$

Hence, in right triangle *BCD*, we have

$$\frac{h}{BD} = \sin BDC$$

$$\begin{aligned} h &= BD \sin BDC \\ &= 508 \sin 44° \\ &= 353 \text{ feet} \end{aligned}$$

41. 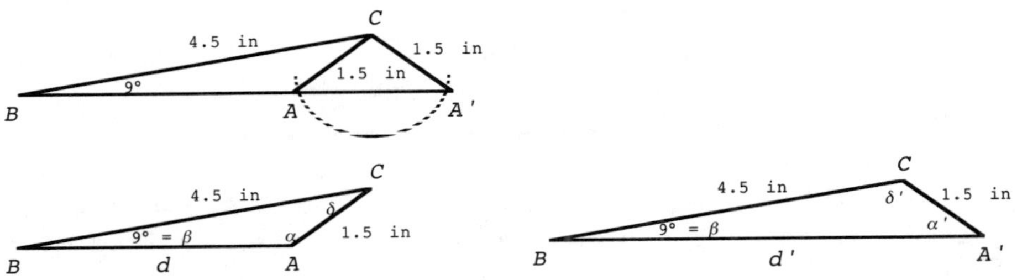

There are two possible triangles, hence two possible values of the required distance. Call them *d* and *d'*.

The law of sines gives two possible values for the angle that the crankshaft makes with the center line; we denote them α and α' in the figures.

We start by calculating angle *BAC*, or α, and angle *BA'C*, or α', from the law of sines.

$$\frac{BC}{\sin \alpha} = \frac{AC}{\sin CBA}$$

$$\begin{aligned} \sin \alpha &= \frac{BC \sin CBA}{AC} \\ &= \frac{4.5 \sin 9°}{1.5} \\ &= 0.4693 \end{aligned}$$

Hence the two possibilities are

α obtuse

$$\begin{aligned} \alpha &= 180° - \sin^{-1} 0.4693 \\ &= 152° \end{aligned}$$

α acute

$$\begin{aligned} \alpha' &= \sin^{-1} 0.4693 \\ &= 28° \end{aligned}$$

Hence the two possibilities for angle *BCA*, or δ, become

$$\begin{aligned} \delta &= 180° - (\alpha + \beta) \\ &= 180° - (152° + 9°) \\ &= 19° \end{aligned}$$

$$\begin{aligned} \delta' &= 180° - (\alpha' + \beta) \\ &= 180° - (28° + 9°) \\ &= 143° \end{aligned}$$

Applying the law of sines again to calculate *d* and *d'* from these two values of δ, we have

$$\frac{d}{\sin \delta} = \frac{AC}{\sin \beta}$$

$$\begin{aligned} d &= \frac{AC \sin \delta}{\sin \beta} \\ &= \frac{1.5 \sin 19°}{\sin 9°} \\ &= 3.1 \text{ in} \end{aligned}$$

$$\frac{d'}{\sin \delta'} = \frac{AC}{\sin \beta}$$

$$\begin{aligned} d' &= \frac{AC \sin \delta}{\sin \beta} \\ &= \frac{1.5 \sin 143°}{\sin 9°} \\ &= 5.8 \text{ in} \end{aligned}$$

43. Sketch a figure.

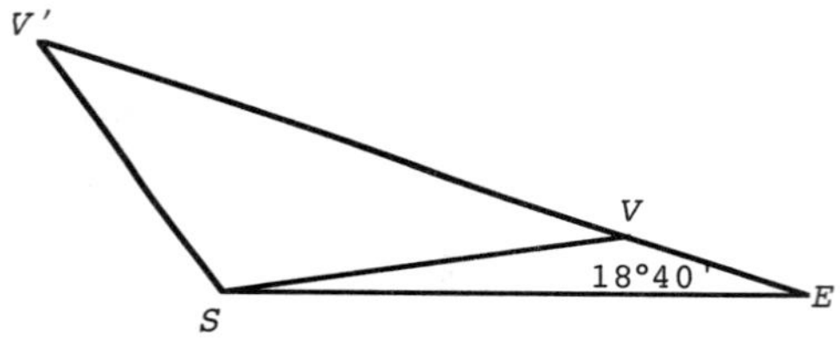

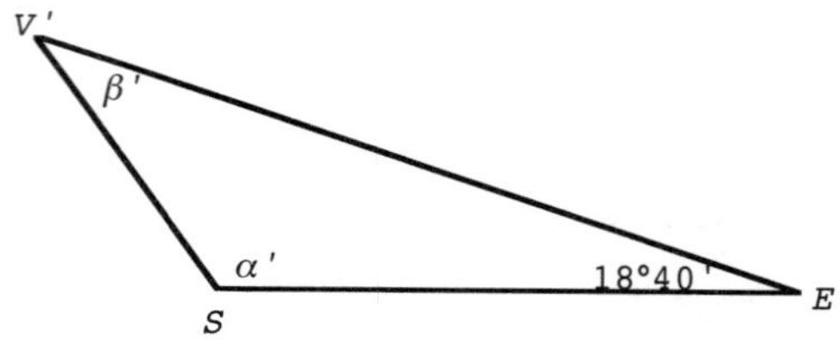

We are given $SE = 1.495 \times 10^8$, $SV = SV' = 1.085 \times 10^8$, angle $SEV = 18°40'$. There are two possible triangles, hence two possible values of the required distance. Call them EV and EV'.

The law of sines gives two possible values for angle VSE; we denote them α and α' in the figures.

We start by calculating angle EVS, or β, and angle $EV'S$, or β', from the law of sines.

$$\frac{SE}{\sin \beta} = \frac{SV}{\sin SEV}$$

$$\sin \beta = \frac{SE \sin SEV}{SV}$$

$$= \frac{1.495 \times 10^8 \sin 18°40'}{1.085 \times 10^8}$$

$$= 0.4410$$

Hence the two possibilities are

β obtuse

$$\beta = 180° - \sin^{-1} 0.4410 = 153°50'$$

β acute

$$\beta' = \sin^{-1} 0.4410 = 26.2° = 26°10'$$

Hence the two possibilities for angle VSE, or α, become

$$\alpha = 180° - (\beta + \gamma) = 180° - (153°50' + 18°40') = 7°30'$$

$$\alpha' = 180° - (\beta' + \gamma') = 180° - (26°10' + 18°40') = 135°10'$$

Applying the law of sines again to calculate EV and EV' from these two values of α, we have

$$\frac{EV}{\sin \alpha} = \frac{SV}{\sin SEV}$$

$$EV = \frac{SV \sin \alpha}{\sin SEV} = \frac{1.085 \times 10^8 \sin(7°30')}{\sin(18°40')} = 4.42 \times 10^7 \text{ kilometers}$$

$$\frac{EV'}{\sin \alpha'} = \frac{SV'}{\sin SEV}$$

$$EV' = \frac{SV' \sin \alpha'}{\sin SEV} = \frac{1.085 \times 10^8 \sin(135°10')}{\sin(18°40')} = 2.39 \times 10^8 \text{ kilometers}$$

45. Redrawing the figure and labeling, we have:

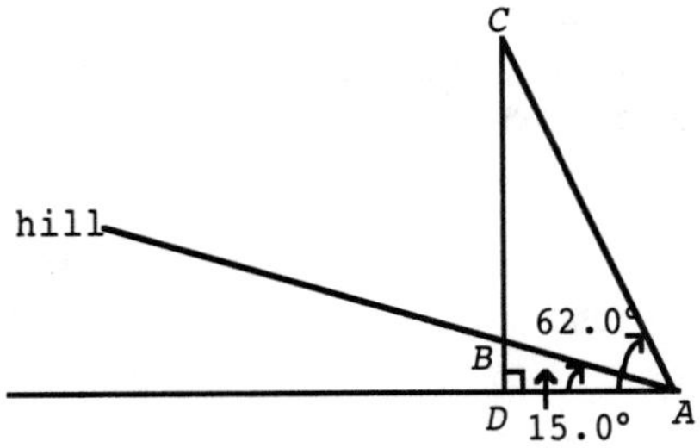

In the figure, note: Triangle *ADC* is a right triangle, hence $\angle ACB = 90° - 62.0° = 28.0°$. Triangle *ABC* is not a right triangle; but, from the law of sines,

$$\frac{\sin CAB}{BC} = \frac{\sin ACB}{AB}$$

$$\angle CAB = \angle CAD - \angle BAD = 62.0° - 15.0° = 47.0°$$

$$BC = \frac{AB \sin CAB}{\sin ACB} = \frac{(102 \text{ ft}) \sin 47.0°}{\sin 28.0°} = 159 \text{ ft.}$$

47. Labeling the diagram as shown, we note: triangle *OAB* is isosceles, hence $\alpha = \beta$. Thus, $\alpha + \beta + \theta = \alpha + \alpha + 98.9° = 180°$

$$2\alpha = 81.1°$$

$$\alpha = 40.55°$$

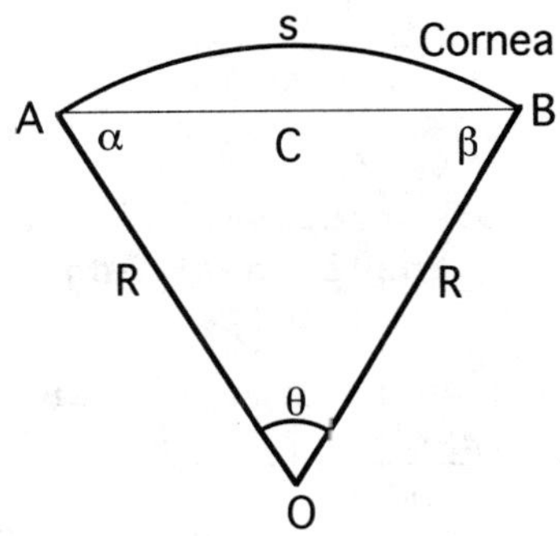

Now apply the law of sines to find *R*.

$$\frac{\sin \alpha}{R} = \frac{\sin \theta}{AB}$$

$$R = \frac{AB \sin \alpha}{\sin \theta} = \frac{(11.8 \text{ mm}) \sin 40.55°}{\sin 98.9°} = 7.76 \text{ mm}$$

To find *s*, we use the formula $s = R\theta_{\text{rad}} = R\frac{\pi}{180°}\theta_{\text{deg}}$,

thus

$$s = (7.76 \text{ mm})\frac{\pi}{180°}(98.9°) = 13.4 \text{ mm}$$

49. Redrawing the figure and labeling, we have:

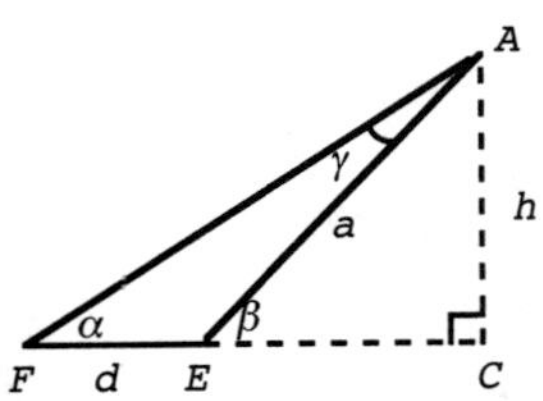

Triangle *ACE* is a right triangle, hence $\frac{h}{a} = \sin \beta$ and $h = a \sin \beta$. Triangle *AEF* is not a right triangle, but from the law of sines,

$$\frac{\sin \gamma}{d} = \frac{\sin \alpha}{a}$$

Hence, $a = \dfrac{d \sin \alpha}{\sin \gamma}$

We can find γ since the exterior angle of a triangle has measure equal to the sum of the two nonadjacent interior angles. Hence

$$\alpha + \gamma = \beta$$

$$\gamma = \beta - \alpha$$

Hence

$$h = a \sin \beta$$

$$= \frac{d \sin \alpha}{\sin \gamma} \sin \beta$$

$$= d\frac{\sin \alpha \sin \beta}{\sin \gamma}$$

$$h = d\frac{\sin \alpha \sin \beta}{\sin(\beta - \alpha)}$$

Section 8-2

1. Angle γ is acute. A triangle can have at most one obtuse angle. Since α is acute, then, if the triangle has an obtuse angle it must be the angle opposite the longer of the two sides, b and c. Thus, γ, the angle opposite the shorter of the two sides, c, must be acute.

3. 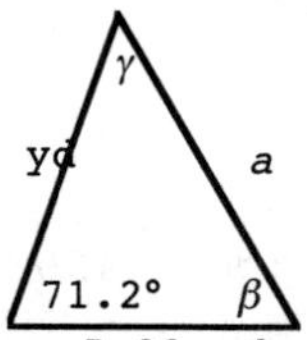

We are given two sides and the included angle (SAS).

Solve for a:

We use the law of cosines

$$\begin{aligned} a^2 &= b^2 + c^2 - 2bc\cos\alpha \\ &= (5.32)^2 + (5.03)^2 - 2(5.32)(5.03)\cos 71.2° \\ &= 36.355898\ldots \\ a &= 6.03 \text{ yd.} \end{aligned}$$

> **Common Error:**
> "$a^2 + b^2 = c^2$"
> The Pythagorean relationship
> $a^2 + b^2 = c^2$
> applies only in right triangles. It is not applicable to oblique triangles.

Solve for β:

Note that since a is the longest side, the angle opposite a, that is, α, must be the largest angle. Hence both β and γ must be less than 71.2°, hence, β and γ must be acute.

We solve for β using the law of sines.

$$\frac{\sin\alpha}{a} = \frac{\sin\beta}{b}$$

$$\sin\beta = \frac{b\sin\alpha}{a} = \frac{5.32\sin 71.2°}{6.03}$$

$$\begin{aligned} \beta &= \sin^{-1}\left(\frac{5.32\sin 71.2°}{6.03}\right) \\ &= 56.6° \end{aligned}$$

Solve for γ:

$$\begin{aligned} \alpha + \beta + \gamma &= 180° \\ \gamma &= 180° - (\alpha + \beta) \\ &= 180° - (71.2° + 56.6°) \\ &= 52.2° \end{aligned}$$

5. 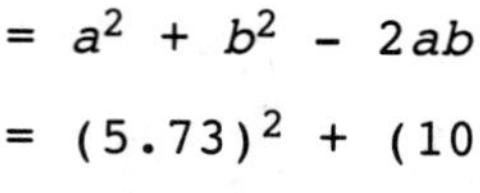

We are given two sides and the included angle (SAS).

Solve for c:

We use the law of cosines

$$\begin{aligned} c^2 &= a^2 + b^2 - 2ab\cos\gamma \\ &= (5.73)^2 + (10.2)^2 - 2(5.73)(10.2)\cos 120°20' \\ &= 195.90685\ldots \\ c &= 14.0 \text{ mm} \end{aligned}$$

Solve for β:
Note that since the sum of β and α is 180° - 120°20' = 59°40', each must be less than 90°. That is, both β and α are acute. We use the law of sines.

$$\frac{\sin\beta}{b} = \frac{\sin\gamma}{c}$$

$$\sin\beta = \frac{b\sin\gamma}{c} = \frac{10.2\sin 120°20'}{14.0}$$

$$\beta = \sin^{-1}\left(\frac{10.2\sin 120°20©}{14.0}\right)$$

$$= 39°0'$$

Solve for α:

$$\alpha + \beta + \gamma = 180°$$

$$\alpha = 180° - (\beta + \gamma)$$

$$= 180° - (39°0' + 120°20')$$

$$= 20°40'$$

7. If the triangle has an obtuse angle, then it must be the angle opposite the longest side; in this case, β.

9. 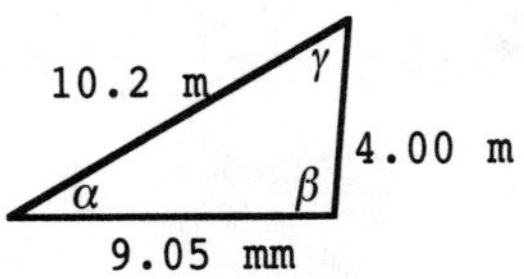

We are given three sides of the triangle (SSS). We solve for the largest angle first, using the law of cosines.

Solve for β:

$$b^2 = a^2 + c^2 - 2ac\cos\beta$$

$$\cos\beta = \frac{a^2 + c^2 - b^2}{2ac}$$

$$\beta = \cos^{-1}\left(\frac{a^2 + c^2 - b^2}{2ac}\right)$$

$$= \cos^{-1}\left(\frac{(4.00)^2 + (9.05)^2 - (10.02)^2}{2(4.00)(9.05)}\right)$$

$$= 94.9°$$

Solve for α:
We use the law of sines. Since the sum of α and γ is 180° - 94.9° = 85.1°, both α and γ must be acute.

$$\frac{\sin\alpha}{a} = \frac{\sin\beta}{b}$$

$$\sin\alpha = \frac{a\sin\beta}{b}$$

$$= \frac{4.00\sin 94.9°}{10.2}$$

$$\alpha = \sin^{-1}\left(\frac{4.00\sin 94.9°}{10.2}\right)$$

$$= 23.0°$$

Solve for γ:

$$\alpha + \beta + \gamma = 180°$$

$$\gamma = 180° - (\alpha + \beta)$$

$$= 180° - (23.0° + 94.9°)$$

$$= 62.1°$$

11. 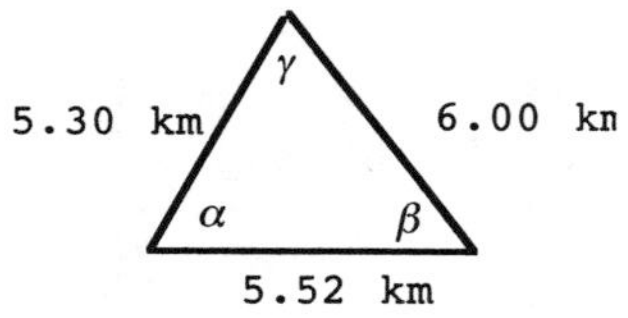

We are given three sides of the triangle (SSS). We solve for the largest angle first, using the law of cosines.

Solve for α:

$$a^2 = b^2 + c^2 - 2bc\cos\alpha$$

$$\cos\alpha = \frac{b^2 + c^2 - a^2}{2bc}$$

$$\alpha = \cos^{-1}\left(\frac{b^2 + c^2 - a^2}{2bc}\right)$$

$$= \cos^{-1}\left(\frac{(5.30)^2 + (5.52)^2 - (6.00)^2}{2(5.30)(5.52)}\right)$$

$$= 67.3°$$

Solve for β:
We use the law of sines. Both β and γ must be less than 67.3°, hence acute.

$$\frac{\sin\alpha}{a} = \frac{\sin\beta}{b}$$

$$\sin\beta = \frac{b\sin\alpha}{a}$$

$$= \frac{5.30\sin 67.3°}{6.00}$$

$$\beta = \sin^{-1}\left(\frac{5.30\sin 67.3°}{6.00}\right)$$

$$= 54.6°$$

Solve for γ:

$$\alpha + \beta + \gamma = 180°$$

$$\gamma = 180° - (\alpha + \beta)$$

$$= 180° - (67.3° + 54.6°)$$

$$= 58.1°$$

13. 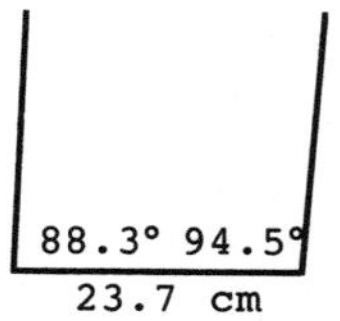

It is impossible to form a triangle with this data, since angles α and γ together add up to more than 180°. No solution.

15. 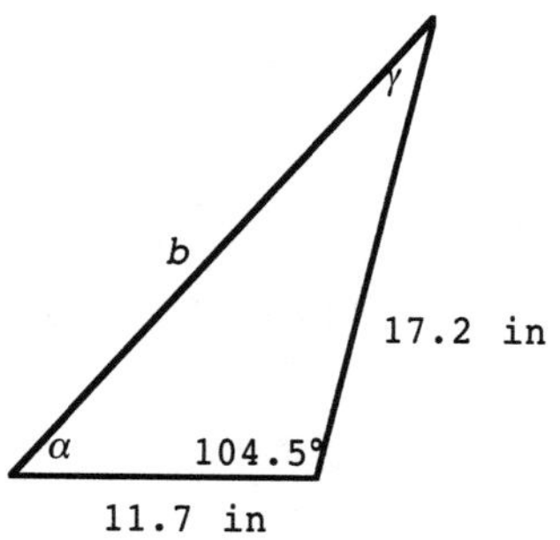

We are given two sides and the included angle (SAS). We use the law of cosines to find the third side, then the law of sines to find a second angle.

Solve for b:

$$b^2 = a^2 + c^2 - 2ac\cos\beta$$

$$= (17.2)^2 + (11.7)^2 - 2(17.2)(11.7)\cos 104.5°$$

$$= 533.50294\ldots$$

$$b = 23.1 \text{ in}$$

Solve for α:
Note that since the sum of α and γ is $180° - 104.5° = 75.5°$, both must be acute.

$$\frac{\sin\alpha}{a} = \frac{\sin\beta}{b}$$
$$\sin\alpha = \frac{a\sin\beta}{b} = \frac{17.2\sin 104.5°}{23.1}$$
$$\alpha = \sin^{-1}\left(\frac{17.2\sin 104.5°}{23.1}\right) = 46.1°$$

Solve for γ:

$$\alpha + \beta + \gamma = 180°$$
$$\gamma = 180° - (\alpha + \beta) = 180° - (46.1° + 104.5°) = 29.4°$$

17. 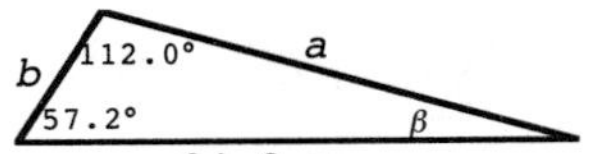

We are given two angles and a non-included side (AAS). We use the law of sines.

Solve for β:

$$\alpha + \beta + \gamma = 180°$$
$$\beta = 180° - (\alpha + \gamma) = 180° - (57.2° + 112.0°) = 10.8°$$

Solve for b:

$$\frac{\sin\beta}{b} = \frac{\sin\gamma}{c}$$
$$b = \frac{c\sin\beta}{\sin\gamma} = \frac{24.8\sin 10.8°}{\sin 112.0°} = 5.01 \text{ m}$$

Solve for a:

$$\frac{\sin\alpha}{a} = \frac{\sin\gamma}{c}$$
$$a = \frac{c\sin\alpha}{\sin\gamma} = \frac{24.8\sin 57.2°}{\sin 112.0°} = 22.5 \text{ m}$$

19. We are given two sides and a non-included angle (SSA). From a rough sketch, we see that there is only one triangle possible.

We use the law of sines.

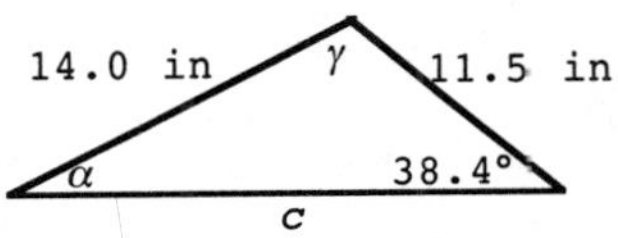

Solve for α:

$$\frac{\sin\alpha}{a} = \frac{\sin\beta}{b}$$
$$\sin\alpha = \frac{a\sin\beta}{b} = \frac{11.5\sin 38.4°}{14.0}$$
$$\alpha = \sin^{-1}\left(\frac{11.5\sin 38.4°}{14.0}\right) \quad \alpha \text{ is acute}$$
$$= 30.7°$$

Solve for γ:

$$\alpha + \beta + \gamma = 180°$$
$$\gamma = 180° - (\alpha + \beta) = 180° - (30.7° + 38.4°) = 110.9°$$

Solve for c:

$$\frac{\sin\alpha}{a} = \frac{\sin\gamma}{c}$$
$$c = \frac{a\sin\gamma}{\sin\alpha} = \frac{11.5\sin 110.9°}{\sin 30.7°} = 21.0 \text{ in}$$

21.

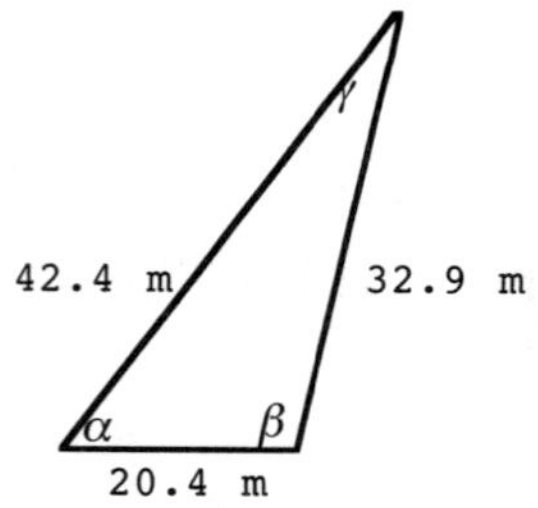

We are given three sides (SSS). We solve for the largest angle, β (largest because it is opposite the largest side, b) using the law of cosines. We then solve for a second angle using the law of sines.

Solve for β:

$$b^2 = a^2 + c^2 - 2ac\cos\beta$$

$$\cos\beta = \frac{a^2 + c^2 - b^2}{2ac}$$

$$\beta = \cos^{-1}\left(\frac{a^2 + c^2 - b^2}{2ac}\right)$$

$$= \cos^{-1}\left(\frac{(32.9)^2 + (20.4)^2 - (42.4)^2}{2(32.9)(20.4)}\right)$$

$$= 102.9°$$

Solve for α:

We use the law of sines. Since the sum of α and γ is $180° - 102.9° = 77.1°$, both α and γ must be acute.

$$\frac{\sin\alpha}{a} = \frac{\sin\beta}{b}$$

$$\sin\alpha = \frac{a\sin\beta}{b}$$

$$= \frac{32.9\sin 102.9°}{42.4}$$

$$\alpha = \sin^{-1}\left(\frac{32.9\sin 102.9°}{42.4}\right)$$

$$= 49.1°$$

Solve for γ:

$$\alpha + \beta + \gamma = 180°$$

$$\gamma = 180° - (\alpha + \beta)$$

$$= 180° - (49.1° + 102.9°)$$

$$= 28.0°$$

23.

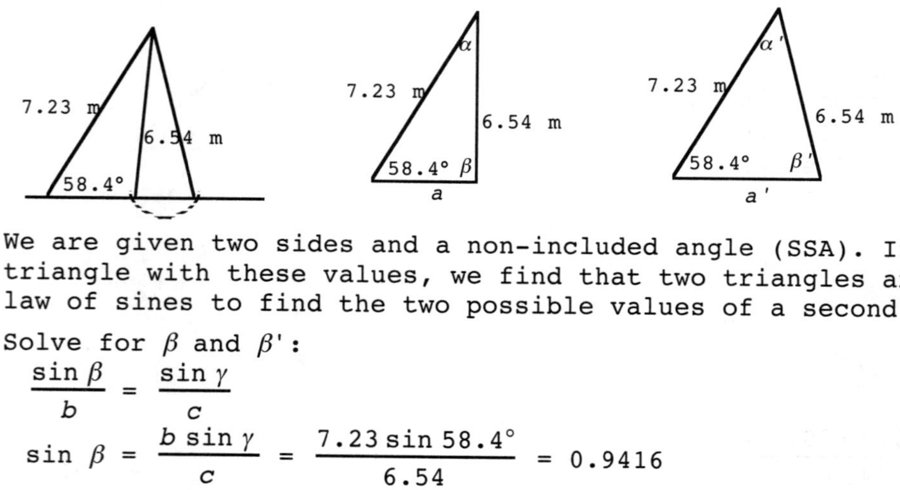

We are given two sides and a non-included angle (SSA). If we try to draw a triangle with these values, we find that two triangles are possible. We use the law of sines to find the two possible values of a second angle.

Solve for β and β':

$$\frac{\sin\beta}{b} = \frac{\sin\gamma}{c}$$

$$\sin\beta = \frac{b\sin\gamma}{c} = \frac{7.23\sin 58.4°}{6.54} = 0.9416$$

Angle β can be either obtuse or acute:

$$\beta = 180° - \sin^{-1} 0.9416$$

$$= 180° - 70.3°$$

$$= 109.7°$$

$$\beta' = \sin^{-1} 0.9416$$

$$= 70.3°$$

Solve for α and α':

$$\alpha = 180° - (\beta + \gamma)$$

$$= 180° - (109.7° + 58.4)$$

$$= 11.9°$$

$$\alpha' = 180° - (\beta' + \gamma)$$

$$= 180° - (70.3° + 58.4°)$$

$$= 51.3°$$

Solve for a and a':

$$\frac{\sin\alpha}{a} = \frac{\sin\gamma}{c}$$
$$a = \frac{c\sin\alpha}{\sin\gamma} = \frac{6.54\sin 11.9°}{\sin 58.4°} = 1.58 \text{ m}$$

$$\frac{\sin\alpha'}{a'} = \frac{\sin\gamma}{c}$$
$$a' = \frac{c\sin\alpha'}{\sin\gamma} = \frac{6.54\sin 51.3°}{\sin 58.4°} = 5.99 \text{ m}$$

In summary:

Triangle I: $\beta = 109.7°$, $\alpha = 11.9°$, $a = 1.58$ m

Triangle II: $\beta' = 70.3°$, $\alpha' = 51.3°$, $a' = 5.99$ m

25.

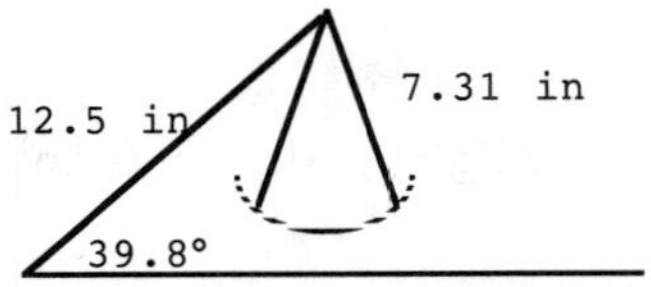

β is acute

$h = a \sin\beta = 12.5 \sin 39.8° = 8.00$

$b = 7.31 < 8.00 = h$

We are given two sides and a non-included angle (SSA). If we try to draw a triangle with these values, we will be unsuccessful. No solution.

27.

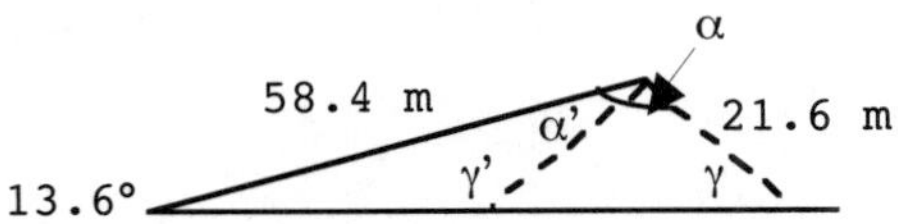

We're given two sides and a non-included angle (SSA).

β is acute

$h = c \sin\beta = 58.4 \sin 13.6° = 13.7$

$h < b < c$; This in combination with the diagram indicates that there are 2 triangles.

Find γ:

$$\frac{\sin\beta}{b} = \frac{\sin\gamma}{c}$$
$$\sin\gamma = \frac{c\sin\beta}{b} = \frac{58.4\sin 13.6°}{21.6} = 0.63575$$

$\gamma = \sin^{-1}(0.63575) = 39.5°$

$\gamma' = 180° - 39.5° = 140.5°$

Triangle I $\quad \gamma = 39.5°$

Find α:

$\alpha + \beta + \gamma = 180°$

$\alpha = 180° - (\beta + \gamma)$

$= 180° - (13.6° + 39.5°)$

$= 126.9°$

Triangle II $\quad \gamma' = 140.5°$

Find α':

$\alpha' + \beta + \gamma' = 180°$

$\alpha' = 180° - (\beta + \gamma)$

$= 180° - (13.6° + 140.5°)$

$= 25.9°$

Find a:

$$\frac{\sin\beta}{b} = \frac{\sin\alpha}{a}$$
$$a \sin\beta = b \sin\alpha$$
$$a = \frac{b\sin\alpha}{\sin\beta} = \frac{21.6\sin 126.9°}{\sin 13.6°} = 73.5 \text{ m}$$

Find a':

$$\frac{\sin\beta}{b} = \frac{\sin\alpha'}{a'}$$
$$a' \sin\beta = b \sin\alpha'$$
$$a' = \frac{b\sin\alpha'}{\sin\beta} = \frac{21.6\sin 25.9°}{\sin 13.6°} = 40.1 \text{ m}$$

29. The law of cosines states that $c^2 = a^2 + b^2 - 2ab \cos\gamma$ for any triangle. If $\gamma = 90°$, $\cos\gamma = 0$; hence,

$c^2 = a^2 + b^2 - 2ac \cos 90°$

$c^2 = a^2 + b^2 - 0$

$c^2 = a^2 + b^2$

31. We write the law of cosines in its three equivalent forms:

$$\begin{aligned} c^2 &= a^2 + b^2 - 2ab\cos\gamma \\ a^2 &= b^2 + c^2 - 2bc\cos\alpha \\ b^2 &= a^2 + c^2 - 2ac\cos\beta \end{aligned}$$

$$a^2 + b^2 + c^2 = 2a^2 + 2b^2 + 2c^2 - 2ab\cos\gamma - 2bc\cos\alpha - 2ac\cos\beta \quad \text{(Adding)}$$

Subtracting $2a^2 + 2b^2 + 2c^2$ from both sides, we obtain:

$-a^2 - b^2 - c^2 = -2ab\cos\gamma - 2bc\cos\alpha - 2ac\cos\beta$

Dividing both sides by $-2abc$, we obtain

$$\frac{-a^2 - b^2 - c^2}{-2abc} = \frac{-2ab\cos\gamma}{-2abc} + \frac{-2bc\cos\alpha}{-2abc} + \frac{-2ac\cos\beta}{-2abc}$$

Removing common factors and rearranging the terms on the right, we obtain

$$\frac{a^2 + b^2 + c^2}{2abc} = \frac{\cos\alpha}{a} + \frac{\cos\beta}{b} + \frac{\cos\gamma}{c}$$

as required.

33. From the law of cosines, we have

$$\begin{aligned} (AB)^2 &= (AC)^2 + (BC)^2 - 2(AC)(BC)\cos ACB \\ &= 91^2 + 71^2 - 2(91)(71)\cos 96° \\ &= 14{,}672.7168\ldots \\ AB &= 120 \text{ yds} \end{aligned}$$

35. Sketch a figure:

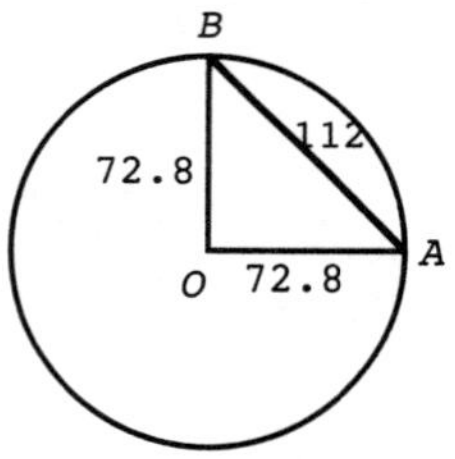

To find $\angle AOB$, we apply the law of cosines:

$$AB^2 = OA^2 + OB^2 - 2(OA)(OB)\cos AOB$$

$$\cos AOB = \frac{OA^2 + OB^2 - AB^2}{2(OA)(OB)} = \frac{72.8^2 + 72.8^2 - 112^2}{2(72.8)(72.8)}$$

$$\angle AOB = \cos^{-1}\left(\frac{72.8^2 + 72.8^2 - 112^2}{2(72.8)(72.8)}\right) = 100.6°$$

37. Sketch a figure:

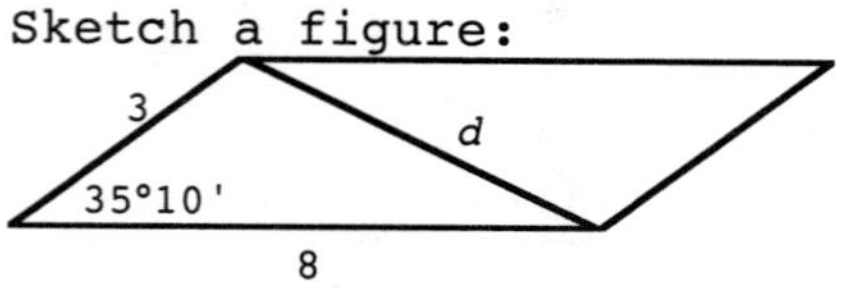

From the law of cosines, we have

$$\begin{aligned} d^2 &= 3^2 + 8^2 - 2(3)(8)\cos 35°10' \\ &= 33.760960\ldots \\ d &= 5.81 \text{ feet} \end{aligned}$$

39. Sketch a figure:

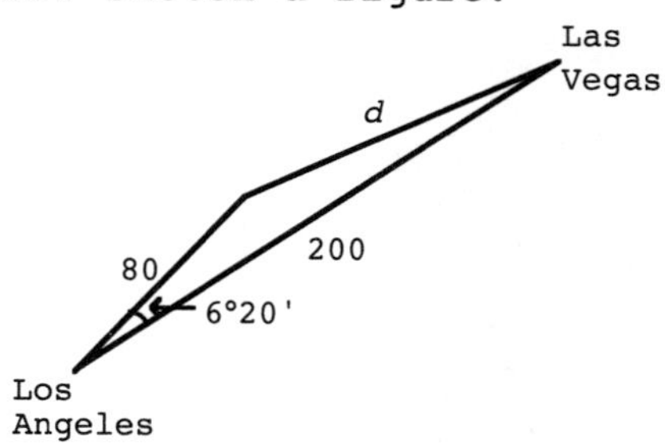

From the law of cosines, we have

$$\begin{aligned} d^2 &= 80^2 + 200^2 - 2(80)(200)\cos(6°20') \\ &= 14595.296\ldots \\ d &= 121 \text{ miles} \end{aligned}$$

41. Sketch a figure:

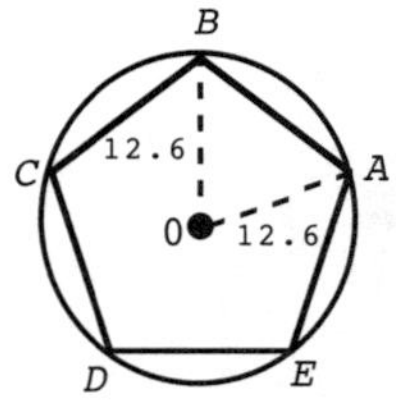

The perimeter *ABCDE* is 5 times the length of side *AB*.

Angle $AOB = \frac{1}{5}(360°) = 72°$

We use the law of cosines in triangle *AOB* to determine side *AB*.

$$(AB)^2 = (12.6)^2 + (12.6)^2 - 2(12.6)(12.6)\cos 72°$$
$$= 219.40092\ldots$$
$$AB = 14.812$$

Perimeter = 5(*AB*) = 74.1 meters

43.

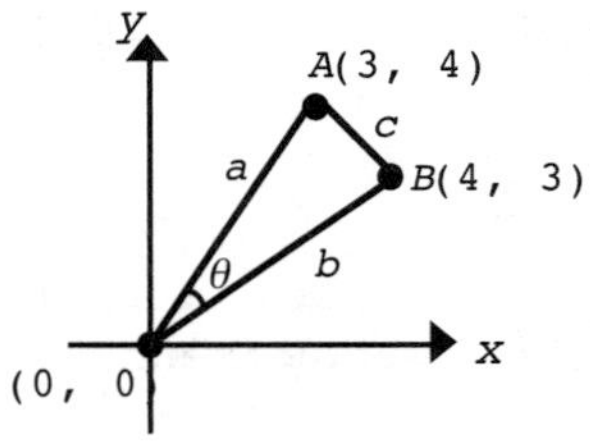

From the given information and the distance formula (Section 2-1), we can determine the lengths of the three sides of the triangle.

$$d(P_1, P_2) = \sqrt{(x_2 - x_1)^2 + (y_2 - y_1)^2}$$
$$c = d(A, B) = \sqrt{(4 - 3)^2 + (3 - 4)^2} = \sqrt{1 + 1} = \sqrt{2}$$
$$b = d(O, B) = \sqrt{(4 - 0)^2 + (3 - 0)^2} = \sqrt{16 + 9} = 5$$
$$a = d(O, A) = \sqrt{(3 - 0)^2 + (4 - 0)^2} = \sqrt{9 + 16} = 5$$

From the law of cosines we can now determine θ.

$$c^2 = a^2 + b^2 - 2ab\cos\theta$$
$$\cos\theta = \frac{a^2 + b^2 - c^2}{2ab}$$
$$\theta = \cos^{-1}\frac{a^2 + b^2 - c^2}{2ab}$$
$$= \cos^{-1}\left(\frac{5^2 + 5^2 - (\sqrt{2})^2}{2(5)(5)}\right)$$
$$= \cos^{-1} 0.96$$
$$= 0.284 \text{ rad}$$

45. From the figure it should be clear that the sides of the triangle are:

$a = 2.03 + 5.00 = 7.03$
$b = 2.03 + 8.20 = 10.23$
$c = 8.20 + 5.00 = 13.20$

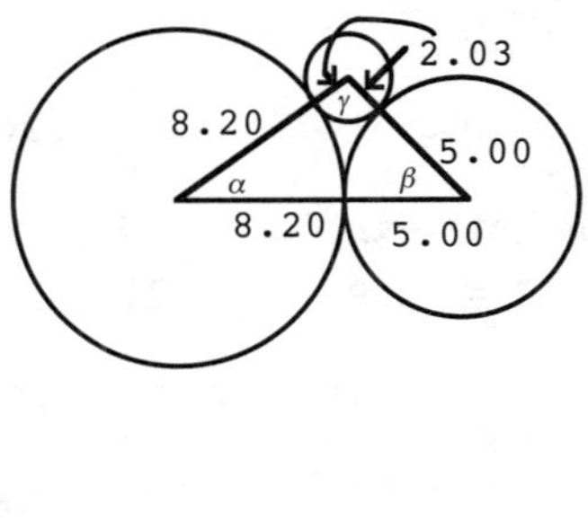

Solve for α:

We use the law of cosines.

$$a^2 = b^2 + c^2 - 2bc\cos\alpha$$
$$\cos\alpha = \frac{b^2 + c^2 - a^2}{2bc}$$
$$\alpha = \cos^{-1}\left(\frac{b^2 + c^2 - a^2}{2bc}\right)$$
$$\alpha = \cos^{-1}\left(\frac{(10.23)^2 + (13.20)^2 - (7.03)^2}{2(10.23)(13.20)}\right)$$
$$= \cos^{-1} 0.8497$$
$$= 31°50'$$

Solve for β:
We use the law of sines.

$$\frac{\sin \alpha}{a} = \frac{\sin \beta}{b}$$

$$\sin \beta = \frac{b \sin \alpha}{a}$$

$$\beta = \sin^{-1}\left(\frac{b \sin \alpha}{a}\right) \quad \beta \text{ is acute (see figure.)}$$

$$= \sin^{-1}\left(\frac{(10.23)(\sin 31°50©)}{7.03}\right)$$

$$= \sin^{-1} 0.7673$$

$$= 50°10'$$

Solve for γ:

$$\gamma = 180° - (\alpha + \beta)$$
$$= 180° - (31°50' + 50°10')$$
$$= 98°0'$$

47. The three sides of triangle ABC are each in turn the hypotenuse of a right triangle formed with two edges of the solid. Hence, by the Pythagorean theorem,

$AB^2 = 4.3^2 + 8.1^2 = 84.10$ $\quad AB = 9.17$ cm
$AC^2 = 2.8^2 + 8.1^2 = 73.45$ $\quad AC = 8.57$ cm
$BC^2 = 2.8^2 + 4.3^2 = 26.33$ $\quad BC = 5.13$ cm

To find $\angle CAB$, we apply the law of cosines.

$$BC^2 = AC^2 + AB^2 - 2(AC)(AB)\cos CAB$$

$$\cos CAB = \frac{AC^2 + AB^2 - BC^2}{2(AC)(AB)} = \frac{73.45 + 84.10 - 26.33}{2(8.57)(9.17)}$$

$$\angle CAB = \cos^{-1}\left(\frac{73.45 + 84.10 - 26.33}{2(8.57)(9.17)}\right)$$

$$= 33°$$

49. Redrawing the figure and re-labeling somewhat:

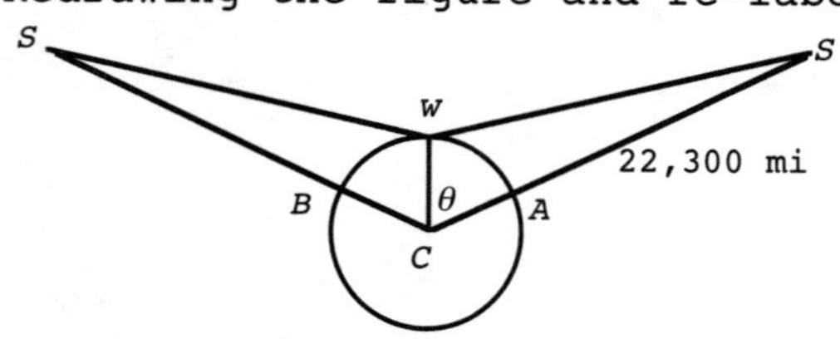

$$\text{Angle } \theta = \frac{1}{2} \text{ angle } ACB = \frac{1}{2}(130°) = 65°$$

CW = radius of earth = 3,964 mi

$CS = CA + AS = 3,964 + 22,300 = 26,264$ mi

From the law of cosines applied to triangle SWC, we have

$$SW^2 = CS^2 + CW^2 - 2(CS)(CW)\cos \theta$$
$$= (26,264)^2 + (3,964)^2 - 2(26,264)(3,964)\cos 65°$$
$$\approx 617,513,000$$
$$SW = 24,800 \text{ mi}$$

Exercise 8-3

1. The coordinates of P are given by
$$\begin{aligned}(x_p, y_p) &= (x_b - x_a, y_b - y_a)\\ &= (10 - 4, 11 - 6)\\ &= (6, 5)\end{aligned}$$
Hence $\mathbf{OP} = \langle 6, 5\rangle$

3. The coordinates of P are given by
$$\begin{aligned}(x_p, y_p) &= (x_b - x_a, y_b - y_a)\\ &= ((-4) - 3, 5 - (-9))\\ &= (-7, 14)\end{aligned}$$
Hence $\mathbf{OP} = \langle -7, 14\rangle$

5. $\mathbf{OP} = \mathbf{AB} = \langle -6, 7\rangle$

7. The coordinates of P are given by
$$\begin{aligned}(x_p, y_p) &= (x_b - x_a, y_b - y_a)\\ &= (0 - 5, 0 - 8)\\ &= (-5, -8)\end{aligned}$$
Hence $\mathbf{OP} = \langle -5, -8\rangle$

9. $|\mathbf{v}| = \sqrt{(-10)^2 + 0^2} = \sqrt{100} = 10$

11. $|\mathbf{v}| = \sqrt{5^2 + (-12)^2} = \sqrt{169} = 13$

13. $|\mathbf{v}| = \sqrt{(-24)^2 + 7^2} = \sqrt{625} = 25$

15. $|\mathbf{v}| = \sqrt{1^2 + 1^2} = \sqrt{2}$

Figures for problems 17, 19, 21, 23:

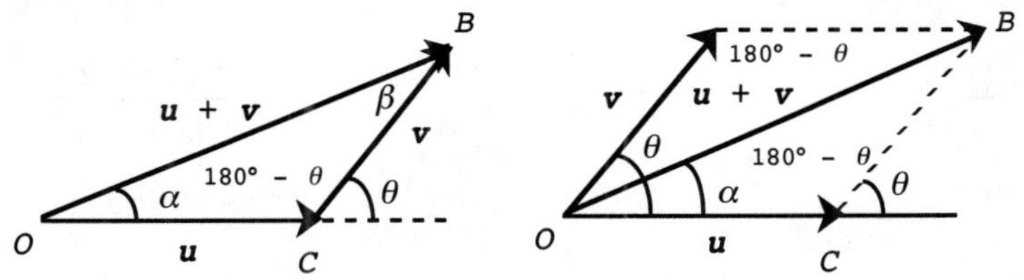

17. $\theta = 68°$. Hence, $\angle OCB = 180° - \theta = 180° - 68° = 112°$
We can find $|\mathbf{u} + \mathbf{v}|$ using the law of cosines:
$$\begin{aligned}|\mathbf{u} + \mathbf{v}|^2 &= |\mathbf{u}|^2 + |\mathbf{v}|^2 - 2|\mathbf{u}||\mathbf{v}|\cos(OCB)\\ &= 66^2 + 22^2 - 2(66)(22)\cos 112°\\ &= 5927.8575\ldots\end{aligned}$$
$$\begin{aligned}|\mathbf{u} + \mathbf{v}| &= \sqrt{5927.8575\ldots}\\ &= 77\text{g}\end{aligned}$$

To find α, we use the law of sines:
$$\frac{\sin\alpha}{|\mathbf{v}|} = \frac{\sin(OCB)}{|\mathbf{u} + \mathbf{v}|} \qquad \frac{\sin\alpha}{22} = \frac{\sin 112°}{77}$$
$$\sin\alpha = \frac{22}{77}\sin 112°$$
$$\alpha = \sin^{-1}\left(\frac{22}{77}\sin 112°\right) = 15°$$

19. $\theta = 53°$. Hence, $\angle OCB = 180° - \theta = 180° - 53° = 127°$

We can find $|\mathbf{u} + \mathbf{v}|$ using the law of cosines:

$$\begin{aligned} |\mathbf{u} + \mathbf{v}|^2 &= |\mathbf{u}|^2 + |\mathbf{v}|^2 - 2|\mathbf{u}||\mathbf{v}|\cos(OCB) \\ &= 21^2 + 3.2^2 - 2(21)(3.2)\cos 127° \\ &= 532.12394\ldots \\ |\mathbf{u} + \mathbf{v}| &= \sqrt{532.12394\ldots} \\ &= 23 \text{ knots} \end{aligned}$$

To find α, we use the law of sines:

$$\begin{aligned} \frac{\sin\alpha}{|\mathbf{v}|} &= \frac{\sin(OCB)}{|\mathbf{u} + \mathbf{v}|} \\ \frac{\sin\alpha}{3.2} &= \frac{\sin 127°}{23} \\ \sin\alpha &= \frac{3.2}{23}\sin 127° \\ \alpha &= \sin^{-1}\left(\frac{3.2}{23}\sin 127°\right) \\ &= 6° \end{aligned}$$

21. $\theta = 79°$. Hence, $\angle OCB = 180° - \theta = 180° - 79° = 101°$. Also, $\angle OBC = \beta = \theta - \alpha = 79° - 25° = 54°$.

We can now find $|\mathbf{u}|$ and $|\mathbf{v}|$ from the law of sines.

$$\begin{aligned} \frac{\sin\beta}{|\mathbf{u}|} &= \frac{\sin(OCB)}{|\mathbf{u} + \mathbf{v}|} \\ |\mathbf{u}| &= \frac{|\mathbf{u} + \mathbf{v}|\sin\beta}{\sin(OCB)} \\ &= \frac{14\sin 54°}{\sin 101°} \\ &= 12 \text{ kg} \end{aligned}$$

$$\begin{aligned} \frac{\sin\alpha}{|\mathbf{v}|} &= \frac{\sin(OCB)}{|\mathbf{u} + \mathbf{v}|} \\ |\mathbf{v}| &= \frac{|\mathbf{u} + \mathbf{v}|\sin\alpha}{\sin(OCB)} \\ &= \frac{14\sin 25°}{\sin 101°} \\ &= 6.0 \text{ kg} \end{aligned}$$

23. $\theta = 69.4°$. Hence, $\angle OCB = 180° - \theta = 180° - 69.4° = 110.6°$. Also, $\angle OBC = \beta = \theta - \alpha = 69.4° - 42.3° = 27.1°$.

We can now find $|\mathbf{u}|$ and $|\mathbf{v}|$ from the law of sines:

$$\begin{aligned} \frac{\sin\beta}{|\mathbf{u}|} &= \frac{\sin(OCB)}{|\mathbf{u} + \mathbf{v}|} \\ |\mathbf{u}| &= \frac{|\mathbf{u} + \mathbf{v}|\sin\beta}{\sin(OCB)} \\ &= \frac{223\sin 27.1°}{\sin 110.6°} \\ &= 109 \text{ mi/hr} \end{aligned}$$

$$\begin{aligned} \frac{\sin\alpha}{|\mathbf{v}|} &= \frac{\sin(OCB)}{|\mathbf{u} + \mathbf{v}|} \\ |\mathbf{v}| &= \frac{|\mathbf{u} + \mathbf{v}|\sin\alpha}{\sin(OCB)} \\ &= \frac{223\sin 42.3°}{\sin 110.6°} \\ &= 160 \text{ mi/hr} \end{aligned}$$

25. (A) $\boldsymbol{u} + \boldsymbol{v} = \langle 2, 1\rangle + \langle -1, 3\rangle = \langle 1, 4\rangle$

(B) $\boldsymbol{u} - \boldsymbol{v} = \langle 2, 1\rangle - \langle -1, 3\rangle = \langle 2, 1\rangle + \langle 1, -3\rangle = \langle 3, -2\rangle$

(C) $$\begin{aligned} 2\boldsymbol{u} - \boldsymbol{v} + 3\boldsymbol{w} &= 2\langle 2, 1\rangle - \langle -1, 3\rangle + 3\langle 3, 0\rangle \\ &= \langle 4, 2\rangle + \langle 1, -3\rangle + \langle 9, 0\rangle \\ &= \langle 14, -1\rangle \end{aligned}$$

27. (A) $\boldsymbol{u} + \boldsymbol{v} = \langle -4, -1\rangle + \langle 2, 2\rangle = \langle -2, 1\rangle$

(B) $\boldsymbol{u} - \boldsymbol{v} = \langle -4, -1\rangle - \langle 2, 2\rangle = \langle -4, -1\rangle + \langle -2, -2\rangle = \langle -6, -3\rangle$

(C) $$\begin{aligned} 2\boldsymbol{u} - \boldsymbol{v} + 3\boldsymbol{w} &= 2\langle -4, -1\rangle - \langle 2, 2\rangle + 3\langle 0, 1\rangle \\ &= \langle -8, -2\rangle + \langle -2, -2\rangle + \langle 0, 3\rangle \\ &= \langle -10, -1\rangle \end{aligned}$$

29. $$\begin{aligned} \boldsymbol{v} = \langle -3, 4\rangle &= \langle -3, 0\rangle + \langle 0, 4\rangle \\ &= -3\langle 1, 0\rangle + 4\langle 0, 1\rangle \\ &= -3\boldsymbol{i} + 4\boldsymbol{j} \end{aligned}$$

31. $\boldsymbol{v} = \langle 3, 0\rangle = 3\langle 1, 0\rangle = 3\boldsymbol{i}$

33. $$\begin{aligned} \boldsymbol{v} = \overrightarrow{AB} &= \langle -3 - 2, 1 - 3\rangle \\ &= \langle -5, -2\rangle \\ &= \langle -5, 0\rangle + \langle 0, -2\rangle \\ &= -5\langle 1, 0\rangle - 2\langle 0, 1\rangle \\ &= -5\boldsymbol{i} - 2\boldsymbol{j} \end{aligned}$$

35. $\boldsymbol{u} + \boldsymbol{v} = 3\boldsymbol{i} - 2\boldsymbol{j} + 2\boldsymbol{i} + 4\boldsymbol{j} = 5\boldsymbol{i} + 2\boldsymbol{j}$

37. $$\begin{aligned} 2\boldsymbol{u} - 3\boldsymbol{v} &= 2(3\boldsymbol{i} - 2\boldsymbol{j}) - 3(2\boldsymbol{i} + 4\boldsymbol{j}) \\ &= 6\boldsymbol{i} - 4\boldsymbol{j} - 6\boldsymbol{i} - 12\boldsymbol{j} \\ &= -16\boldsymbol{j} \end{aligned}$$

39. $$\begin{aligned} 2\boldsymbol{u} - \boldsymbol{v} - 2\boldsymbol{w} &= 2(3\boldsymbol{i} - 2\boldsymbol{j}) - (2\boldsymbol{i} + 4\boldsymbol{j}) - 2(2\boldsymbol{i}) \\ &= 6\boldsymbol{i} - 4\boldsymbol{j} - 2\boldsymbol{i} - 4\boldsymbol{j} - 4\boldsymbol{i} \\ &= -8\boldsymbol{j} \end{aligned}$$

41. $|\boldsymbol{v}| = \sqrt{4^2 + 3^2} = \sqrt{25} = 5$

$\boldsymbol{u} = \frac{1}{|\boldsymbol{v}|}\boldsymbol{v} = \frac{1}{5}\langle 4, 3\rangle = \left\langle \frac{4}{5}, \frac{3}{5}\right\rangle$

43. $|\boldsymbol{v}| = \sqrt{(-1)^2 + 1^2} = \sqrt{2}$

$\boldsymbol{u} = \frac{1}{|\boldsymbol{v}|}\boldsymbol{v} = \frac{1}{\sqrt{2}}\langle -1, 1\rangle = \left\langle -\frac{1}{\sqrt{2}}, \frac{1}{\sqrt{2}}\right\rangle$

45. $|\boldsymbol{v}| = \sqrt{(-8)^2 + 0^2} = \sqrt{64} = 8$

$\boldsymbol{u} = \frac{1}{|\boldsymbol{v}|}\boldsymbol{v} = \frac{1}{8}\langle -8, 0\rangle = \langle -1, 0\rangle$

47. $|\boldsymbol{v}| = \sqrt{5^2 + (\sqrt{11})^2} = \sqrt{36} = 6$

$\boldsymbol{u} = \frac{1}{|\boldsymbol{v}|}\boldsymbol{v} = \frac{1}{6}(5\boldsymbol{i} + \sqrt{11}\boldsymbol{j}) = \frac{5}{6}\boldsymbol{i} + \frac{\sqrt{11}}{6}\boldsymbol{j} = \left\langle \frac{5}{6}, \frac{\sqrt{11}}{6}\right\rangle$

49. False. $\langle 0, 1\rangle$ and $\langle 1, 0\rangle$ both have magnitude 1, but are not equal.

51. True. If $A(x_a, y_a) = B(x_a, y_a)$, then $\langle x_a - x_a, y_a - y_a\rangle = \langle 0, 0\rangle$.

53. False. $|\boldsymbol{i}| = 1$; $|\boldsymbol{i} + \boldsymbol{i}| = |2\boldsymbol{i}| = 2$

55. False. $|\mathbf{0}| = 0$

57. $\mathbf{u} + (\mathbf{v} + \mathbf{w}) = \langle a, b\rangle + (\langle c, d\rangle + \langle e, f\rangle)$

$= \langle a, b\rangle + \langle c + e, d + f\rangle$ Definition of vector addition

$= \langle a + (c + e), b + (d + f)\rangle$ Definition of vector addition

$= \langle (a + c) + e, (b + d) + f\rangle$ Associative property for addition of real numbers

$= \langle a + c, b + d\rangle + \langle e, f\rangle$ Definition of vector addition

$= (\langle a, b\rangle + \langle c, d\rangle) + \langle e, f\rangle$ Definition of vector addition

$= (\mathbf{u} + \mathbf{v}) + \mathbf{w}$

59. $\mathbf{u} + \mathbf{O} = \langle a, b\rangle + \langle 0, 0\rangle$

$= \langle a + 0, b + 0\rangle$ Definition of vector addition

$= \langle a, b\rangle$ Additive identity property for real numbers

$= \mathbf{u}$

61. $(m + n)\mathbf{u} = (m + n)\langle a, b\rangle$

$= \langle (m + n)a, (m + n)b\rangle$ Definition of scalar multiplication

$= \langle ma + na, mb + nb\rangle$ Distributive property for real numbers

$= \langle ma, mb\rangle + \langle na, nb\rangle$ Definition of vector addition

$= m\langle a, b\rangle + n\langle a, b\rangle$ Definition of scalar multiplication

$= m\mathbf{u} + n\mathbf{u}$

63. $m(n\mathbf{u}) = m(n\langle a, b\rangle)$

$= m\langle na, nb\rangle$ Definition of scalar multiplication

$= \langle m(na), m(nb)\rangle$ Definition of scalar multiplication

$= \langle (mn)a, (mn)b\rangle$ Associative property for multiplication of real numbers

$= (mn)\langle a, b\rangle$ Definition of scalar multiplication

$= (mn)\mathbf{u}$

65.

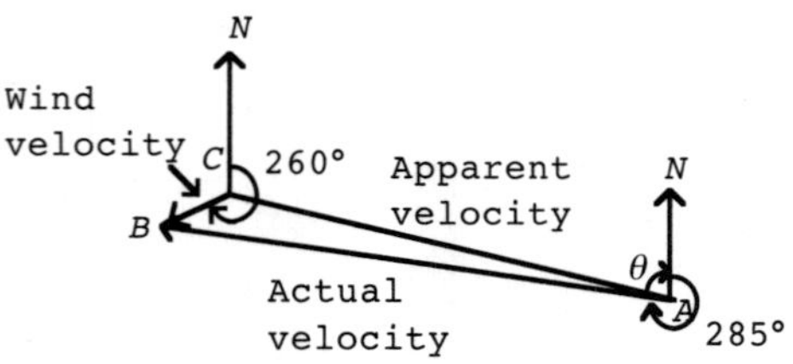

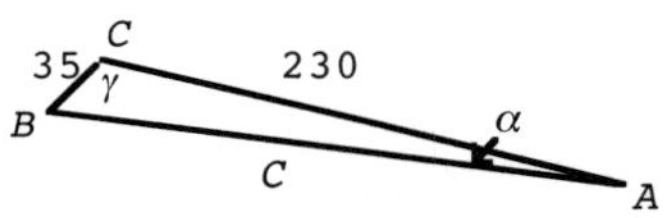

In the (left-hand) vector diagram above, we use the tail-to-tip method of vector addition to obtain the resultant (actual) velocity. Note:

angle $\theta = 360° - 285° = 75°$
angle $ACN = 180° - 75° = 105°$
angle $\gamma = 260° - 105° = 155°$

Then in triangle ABC, we solve for c and α.

Solve for c:
Use the law of cosines.

$$c^2 = a^2 + b^2 - 2ab\cos\gamma$$

$$= 35^2 + 230^2 - 2(35)(230)\cos 155°$$

$$= 68{,}716.5\ldots$$

$$c = \sqrt{68{,}716.5\ldots}$$

$$= 260 \text{ mph}$$

Solve for α:
Use the law of sines

$$\frac{\sin\alpha}{a} = \frac{\sin\gamma}{c}$$

$$\sin\alpha = \frac{a\sin\gamma}{c}$$

$$\alpha = \sin^{-1}\left(\frac{a\sin\gamma}{c}\right) \quad \alpha \text{ is acute since } \gamma \text{ is obtuse}$$

$$= \sin^{-1}\left(\frac{35\sin 155°}{260}\right)$$

$$= 3°$$

Actual heading = 285° - α = 285° - 3° = 282°.

Thus, the plane, relative to the ground, is traveling at 260 mph in a direction of 282°.

67. 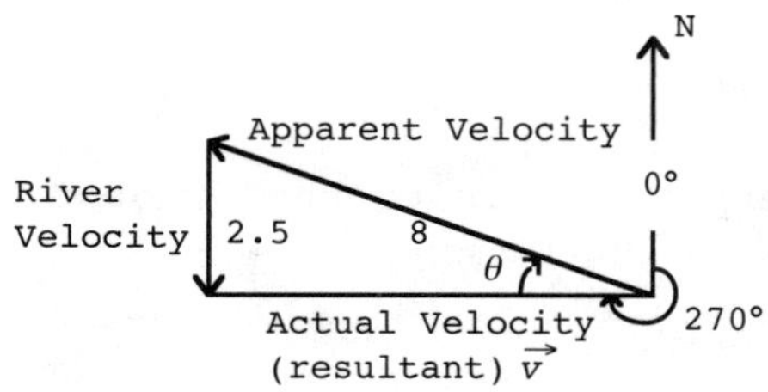

Using the Pythagorean theorem, we find the magnitude of the resultant vector to be

$$|\mathbf{v}| = \sqrt{8^2 - 2.5^2} = 7.6 \text{ knots}$$

To find θ, we see that

$$\sin\theta = \frac{2.5}{8}$$

$$\theta = \sin^{-1}\left(\frac{2.5}{8}\right) = 18°$$

Compass heading = 270° + θ = 270° + 18° = 288°

Actual speed: 7.6 knots.

69. 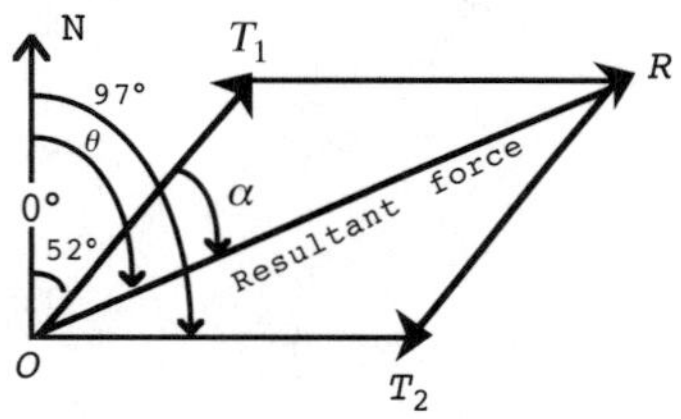

In the figure, we are given OT_1 = 2,300, $T_1R = OT_2$ = 1,900. $\angle T_1OT_2$ = 97° - 52° = 45°, hence $\angle OT_1R$ = 180° - 45° = 135°. Since we know two sides and the included angle in triangle OT_1R, we can use the law of cosines to find OR, the magnitude of the resultant force.

$$OR^2 = OT_1^{\,2} + T_1R^2 - 2(OT_1)(T_1R)\cos(\angle OT_1R)$$

$$= (2{,}300)^2 + (1{,}900)^2 - 2(2{,}300)(1{,}900)\cos 135°$$

$$= 15{,}080{,}113\ldots$$

$$OR = \sqrt{15{,}080{,}113\ldots}$$

$$= 3{,}900 \text{ pounds}$$

We use the law of sines to find α

$$\frac{\sin\alpha}{T_1R} = \frac{\sin(OT_1R)}{OR}$$

$$\frac{\sin\alpha}{1{,}900} = \frac{\sin 135°}{3{,}900}$$

$$\sin\alpha = \frac{1{,}900}{3{,}900}\sin 135°$$

$$\alpha = \sin^{-1}\left(\frac{1{,}900}{3{,}900}\sin 135°\right) = 20°$$

Then the compass direction θ = 52° + α = 72°.

71. 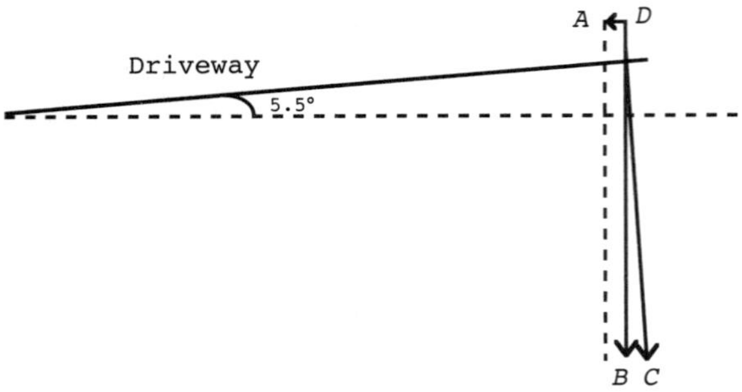

(A) To keep the car at D from rolling down the hill, we need a force with the magnitude of $\mathbf{DA}$ but oppositely directed. To find $|\mathbf{DA}|$, we observe that DAB is a right triangle with $\angle ABD = 5.5°$. Hence

$$\sin 5.5° = \frac{|\mathbf{DA}|}{4,050}$$
$$|\mathbf{DA}| = 4,050 \sin 5.5°$$
$$= 388 \text{ lb}$$

(B) To find $|\mathbf{DC}|$, the magnitude of the force perpendicular to the driveway, we note that DCB is a right triangle with $\angle BDC = 5.5°$. Hence

$$\cos 5.5° = \frac{|\mathbf{DC}|}{4,050}$$
$$|\mathbf{DC}| = 4,050 \cos 5.5°$$
$$= 4,030 \text{ lb}$$

73. 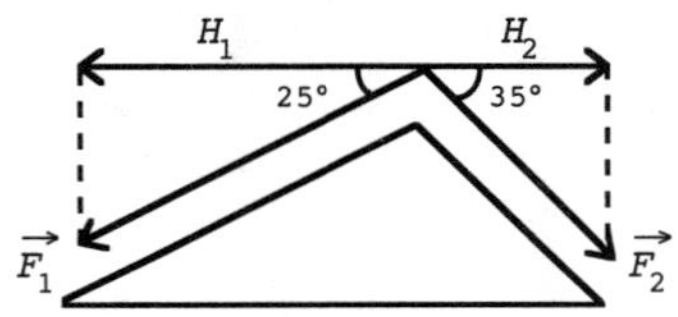

The weights will slide to the left if the horizontal component H_1 pointing left is greater than the horizontal component H_2 pointing right. They will slide to the right if H_2 is greater than H_1.

$$H_1 = |\mathbf{F}_1|\cos 25° = 110 \cos 25° \approx 100 \text{ lb}$$

$$H_2 = |\mathbf{F}_2|\cos 35° = 85 \cos 35° \approx 70 \text{ lb}$$

Since H_1 is greater than H_2, they will slide to the left.

75. First, form a force diagram with all force vectors in standard position at the origin.

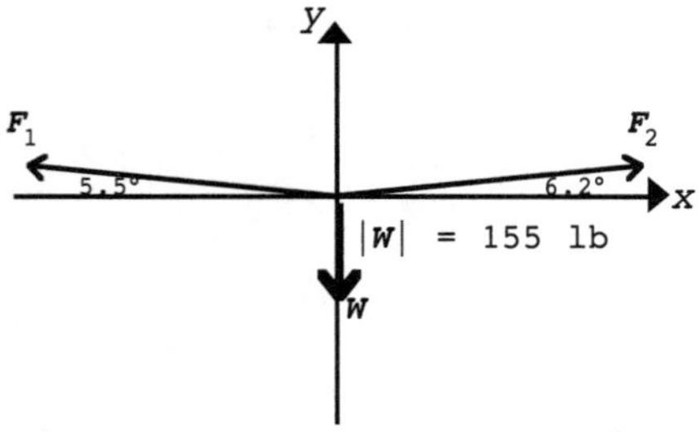

Let $\mathbf{F}_1$ = the tension in left rope
$\mathbf{F}_2$ = the tension in right rope

Write each force vector in terms of $\boldsymbol{i}$ and $\boldsymbol{j}$ unit vectors.

$\mathbf{F}_1 = |\mathbf{F}_1|(-\cos 5.5°)\boldsymbol{i} + |\mathbf{F}_1|(\sin 5.5°)\boldsymbol{j}$

$\mathbf{F}_2 = |\mathbf{F}_2|(\cos 6.2°)\boldsymbol{i} + |\mathbf{F}_2|(\sin 6.2°)\boldsymbol{j}$

$\mathbf{W} = -155\boldsymbol{j}$

For the system to be in static equilibrium, we must have
$\mathbf{F}_1 + \mathbf{F}_2 + \mathbf{W} = \mathbf{O}$

which becomes, on addition,

$$[-|\boldsymbol{F}_1|(\cos 5.5°) + |\boldsymbol{F}_2|(\cos 6.2°)]\boldsymbol{i} + [|\boldsymbol{F}_1|(\sin 5.5°) + |\boldsymbol{F}_2|(\sin 6.2°) - 155]\boldsymbol{j} = 0\boldsymbol{i} + 0\boldsymbol{j}$$

Since two vectors are equal if and only if their corresponding components are equal, we are led to the following system of equations in $|\boldsymbol{F}_1|$ and $|\boldsymbol{F}_2|$:

$$-|\boldsymbol{F}_1|\cos 5.5° + |\boldsymbol{F}_2|\cos 6.2° = 0$$
$$|\boldsymbol{F}_1|\sin 5.5° + |\boldsymbol{F}_2|\sin 6.2° - 155 = 0$$

Solving by standard methods, we obtain

$$|\boldsymbol{F}_1| = \frac{155}{\sin 5.5° + \cos 5.5° \tan 6.2°} = 760 \text{ lb to the left}$$

$$|\boldsymbol{F}_2| = 760 \frac{\cos 5.5°}{\cos 6.2°} = 761 \text{ lb to the right.}$$

77. First, form a force diagram with all force vectors in standard position at the origin.

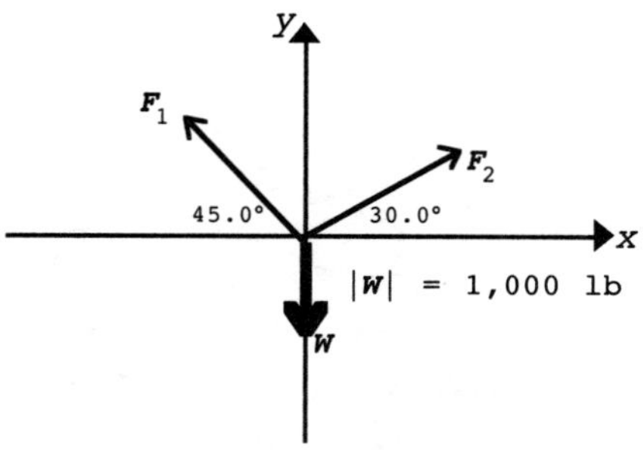

Let $\boldsymbol{F}_1$ = the tension in left cable
$\boldsymbol{F}_2$ = the tension in right cable

Write each force vector in terms of $\boldsymbol{i}$ and $\boldsymbol{j}$ unit vectors.

$$\boldsymbol{F}_1 = |\boldsymbol{F}_1|(-\cos 45.0°)\boldsymbol{i} + |\boldsymbol{F}_1|(\sin 45.0°)\boldsymbol{j}$$
$$\boldsymbol{F}_2 = |\boldsymbol{F}_2|(\cos 30.0°)\boldsymbol{i} + |\boldsymbol{F}_2|(\sin 30.0°)\boldsymbol{j}$$
$$\boldsymbol{W} = -1{,}000\boldsymbol{j}$$

For the system to be in static equilibrium, we must have
$\boldsymbol{F}_1 + \boldsymbol{F}_2 + \boldsymbol{W} = \boldsymbol{0}$

which becomes, on addition,

$$[-|\boldsymbol{F}_1|(\cos 45.0°) + |\boldsymbol{F}_2|(\cos 30.0°)]\boldsymbol{i} + [|\boldsymbol{F}_1|(\sin 45.0°) + |\boldsymbol{F}_2|(\sin 30.0°) - 1{,}000]\boldsymbol{j} = 0\boldsymbol{i} + 0\boldsymbol{j}$$

Since two vectors are equal if and only if their corresponding components are equal, we are led to the following system of equations in $|\boldsymbol{F}_1|$ and $|\boldsymbol{F}_2|$:

$$-|\boldsymbol{F}_1|\cos 45.0° + |\boldsymbol{F}_2|\cos 30.0° = 0$$
$$|\boldsymbol{F}_1|\sin 45.0° + |\boldsymbol{F}_2|\sin 30.0° - 1{,}000 = 0$$

Solving by standard methods, we obtain

$$|\boldsymbol{F}_1| = \frac{1{,}000}{\sin 45.0° + \cos 45.0° \tan 30.0°} = 897 \text{ lb to the left}$$

$$|\boldsymbol{F}_2| = 897 \frac{\cos 45.0°}{\cos 30.0°} = 732 \text{ lb to the right.}$$

79. In the force diagram, figure (b), we are given $|\boldsymbol{c}| = 400$ lb. In triangle *ABC*, figure (a), we note: $\cos ABC = \frac{1}{2}$, $ABC = 60°$. Then write each force in terms of $\boldsymbol{i}$ and $\boldsymbol{j}$ unit vectors.

$1\boldsymbol{a} = |\boldsymbol{a}|\boldsymbol{i}$

$\boldsymbol{b} = |\boldsymbol{b}|(-\cos 60°)\boldsymbol{i} + |\boldsymbol{b}|\sin 60°\ \boldsymbol{j} = -\frac{1}{2}|\boldsymbol{b}|\boldsymbol{i} + \frac{\sqrt{3}}{2}|\boldsymbol{b}|\boldsymbol{j}$

$\boldsymbol{c} = -400\boldsymbol{j}$

For the system to be in static equilibrium, we must have

$$\boldsymbol{a} + \boldsymbol{b} + \boldsymbol{c} = \boldsymbol{0}$$

which becomes, on addition

$$\left[|\boldsymbol{a}| - \frac{1}{2}|\boldsymbol{b}|\right]\boldsymbol{i} + \left[\frac{\sqrt{3}}{2}|\boldsymbol{b}| - 400\right]\boldsymbol{j} = 0\boldsymbol{i} + 0\boldsymbol{j}$$

Since two vectors are equal if and only if their corresponding components are equal, we are led to the following system of equations in $|\boldsymbol{a}|$ and $|\boldsymbol{b}|$:

$$|\boldsymbol{a}| - \frac{1}{2}|\boldsymbol{b}| = 0$$

$$\frac{\sqrt{3}}{2}|\boldsymbol{b}| - 400 = 0$$

Solving, we obtain

$|\boldsymbol{b}| = \frac{2}{\sqrt{3}}(400) = 462$ lb. This corresponds to a tension force of 462 lb in member *CB*.

$|\boldsymbol{a}| = \frac{1}{2}|\boldsymbol{b}| = 231$ lb This corresponds to a compression force of 231 lb in member *AB*.

81. First, form a force diagram with all force vectors in standard position at the origin.

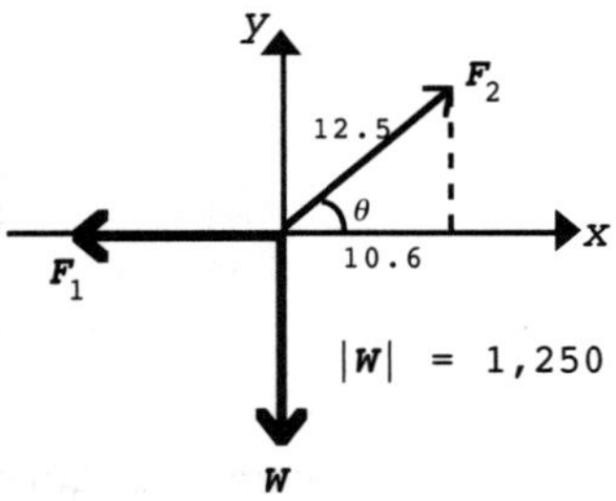

Let $\boldsymbol{F}_1$ = the force on the horizontal member *BC*
$\boldsymbol{F}_2$ = the force on the supporting member *AB*
$\boldsymbol{W}$ = the downward force (1,250 lb)

We note: $\cos\theta = \frac{10.6}{12.5}$ $\theta = \cos^{-1}\left(\frac{10.6}{12.5}\right) = 32.0°$

Then write each force vector in terms of $\boldsymbol{i}$ and $\boldsymbol{j}$ unit vectors.

$\boldsymbol{F}_1 = -|\boldsymbol{F}_1|\boldsymbol{i}$
$\boldsymbol{F}_2 = |\boldsymbol{F}_2|(\cos 32.0°)\boldsymbol{i} + |\boldsymbol{F}_2|(\sin 32.0°)\boldsymbol{j}$
$\boldsymbol{W} = -1{,}250\boldsymbol{j}$

For the system to be in static equilibrium, we must have
$\boldsymbol{F}_1 + \boldsymbol{F}_2 + \boldsymbol{W} = \boldsymbol{0}$

which becomes, on addition,
$[-|\boldsymbol{F}_1| + |\boldsymbol{F}_2|(\cos 32.0°)]\boldsymbol{i} + [|\boldsymbol{F}_2|(\sin 32.0°) - 1{,}250]\boldsymbol{j} = 0\boldsymbol{i} + 0\boldsymbol{j}$

Since two vectors are equal if and only if their corresponding components are equal, we are led to the following system of equations in $|\mathbf{F}_1|$ and $|\mathbf{F}_2|$:

$$-|\mathbf{F}_1| + |\mathbf{F}_2|(\cos 32.0°) = 0$$
$$|\mathbf{F}_2|(\sin 32.0°) - 1{,}250 = 0$$

Solving,

$$|\mathbf{F}_2| = \frac{1{,}250}{\sin 32.0°} = 2{,}360 \text{ lb}$$

$$|\mathbf{F}_1| = |\mathbf{F}_2|\cos 32.0° = 2{,}000 \text{ lb}$$

The force in the member *AB* is directed oppositely to the diagram— a compression of 2,360 lb. The force in the member *BC* is also directed oppositely to the diagram— a tension of 2,000 lb.

Section 8-4

1.

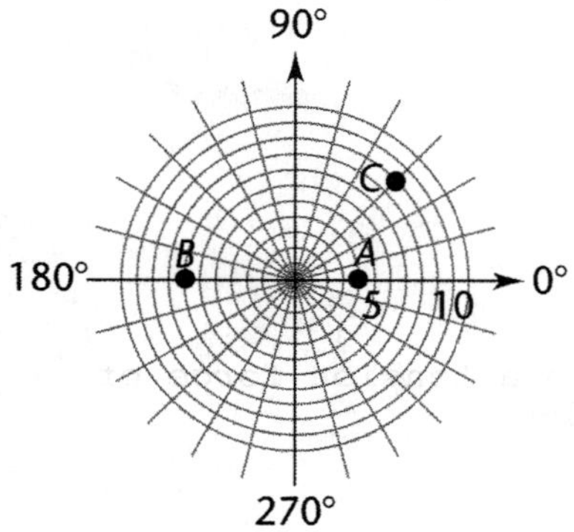

3.

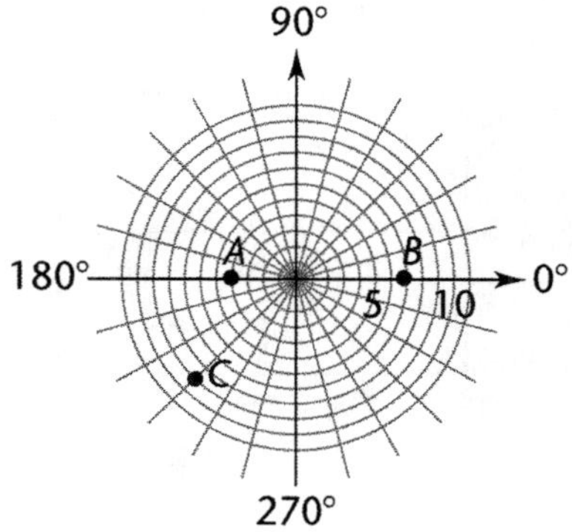

5.

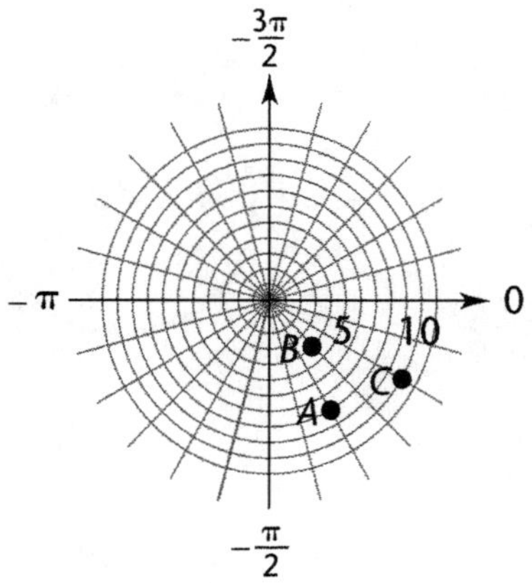

7.

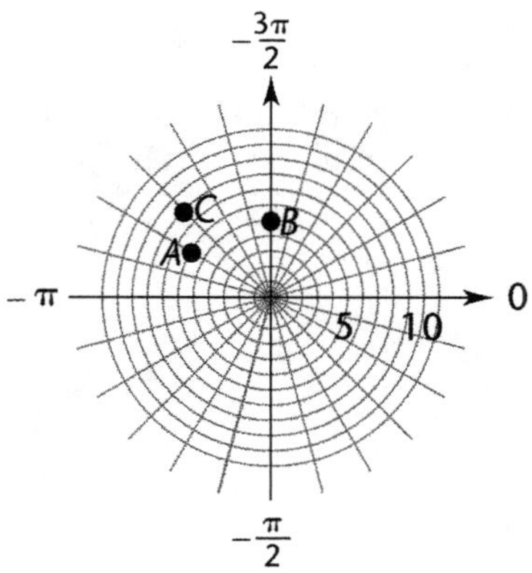

9. See figure.

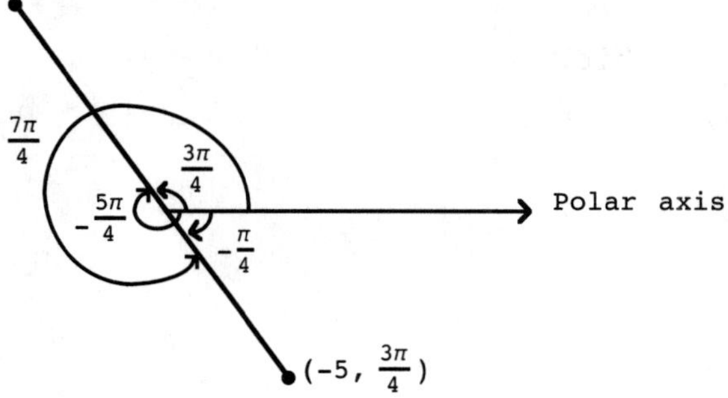

The point with coordinates $\left(-5, \frac{3\pi}{4}\right)$ can equally well be described as $\left(5, -\frac{\pi}{4}\right)$ or $\left(5, \frac{7\pi}{4}\right)$ or $\left(-5, -\frac{5\pi}{4}\right)$.

Thus,

$\left(5, -\frac{\pi}{4}\right)$: The polar axis is rotated $\frac{\pi}{4}$ radians clockwise (negative direction) and the point is located 5 units from the pole along the positive polar axis;

$\left(5, \frac{7\pi}{4}\right)$: The polar axis is rotated $\frac{7\pi}{4}$ radians counterclockwise (positive direction) and the point is located 5 units from the pole along the positive polar axis;

$\left(-5, -\frac{5\pi}{4}\right)$: The polar axis is rotated $\frac{5\pi}{4}$ radians clockwise (negative direction) and the point is located 5 units from the pole along the negative polar axis.

11.

θ	0	$\frac{\pi}{6}$	$\frac{\pi}{4}$	$\frac{\pi}{3}$	$\frac{\pi}{2}$	$\frac{2\pi}{3}$	$\frac{3\pi}{4}$	$\frac{5\pi}{6}$	π
r	0	5	$5\sqrt{2}$	$5\sqrt{3}$	10	$5\sqrt{3}$	$5\sqrt{2}$	5	0
			≈ 7.1	≈ 8.7		≈ 8.7	≈ 7.1		

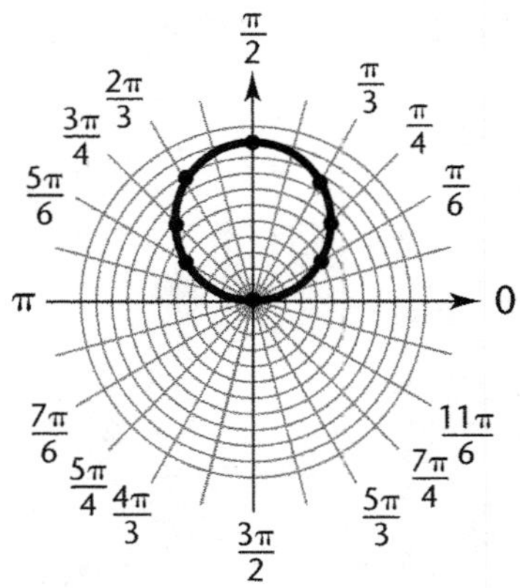

Check:

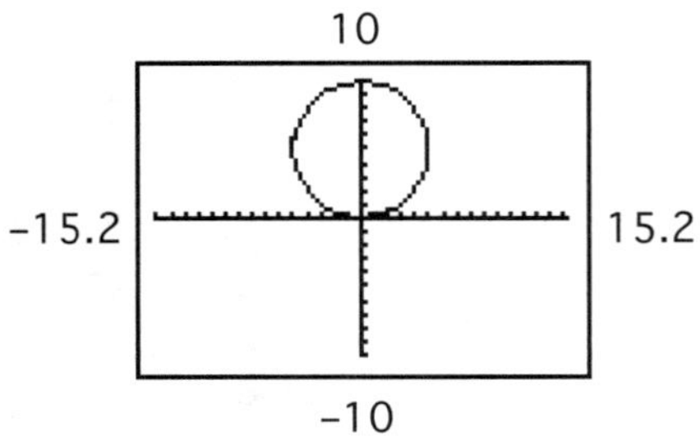

13. The graph consists of all points whose distance from the pole is 8: a circle with center at the pole, and radius 8.

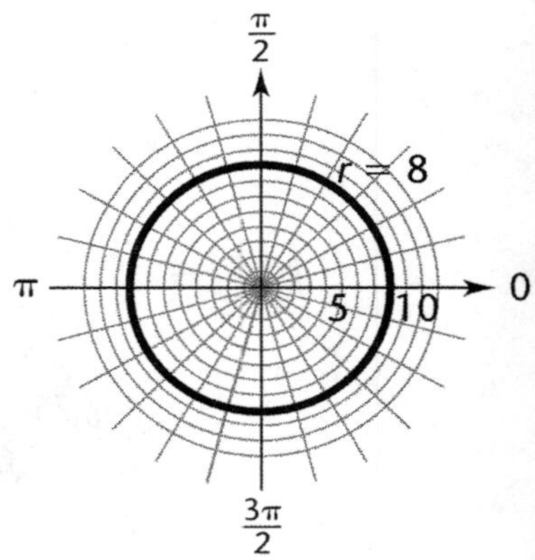

15. The graph consists of all points on a line which forms an angle of $\frac{\pi}{3}$ with the polar axis, and passes through the pole.

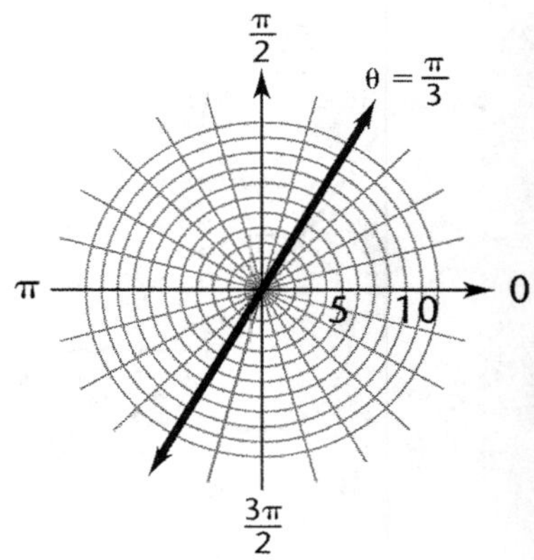

In problems 17— 22, use a calculator set in radian mode.

17. $(r, \theta) = \left(6, \frac{\pi}{6}\right)$

$x = r \cos \theta = 6 \cos \frac{\pi}{6} = 5.196$

$y = r \sin \theta = 6 \sin \frac{\pi}{6} = 3.000$

$(x, y) = (5.196, 3.000)$

19. $(r, \theta) = \left(-2, \frac{7\pi}{8}\right)$

$x = r \cos \theta = -2 \cos \frac{7\pi}{8} = 1.848$

$y = r \sin \theta = -2 \sin \frac{7\pi}{8} = -0.765$

$(x, y) = (1.848, -0.765)$

21. $(r, \theta) = (-4.233, -2.084)$
$x = r \cos \theta = -4.233 \cos(-2.084) = 2.078$
$y = r \sin \theta = -4.233 \sin(-2.084) = 3.688$
$(x, y) = (2.078, 3.688)$

In problems 23—28, use a calculator set in degree mode.

23. $(x, y) = (3.5, 7.1)$

$r = \sqrt{x^2 + y^2} = \sqrt{3.5^2 + 7.1^2} = 7.9$

$\tan \theta = \frac{y}{x} = \frac{7.1}{3.5}$

θ is a first quadrant angle and is to be chosen as that $-180° < \theta \le 180°$.

$\theta = \tan^{-1}\left(\frac{7.1}{3.5}\right) = 64°$

$(r, \theta) = (7.9, 64°)$

25. $(x, y) = (22, -14)$

$r = \sqrt{x^2 + y^2} = \sqrt{22^2 + (-14)^2} = 26$

$\tan \theta = \frac{y}{x} = \frac{-14}{22}$

θ is a fourth quadrant angle and is to be chosen so that $-180° < \theta \le 180°$

$\theta = \tan^{-1}\left(\frac{-14}{22}\right) = -32°$

$(r, \theta) = (26, -32°)$

27. $(x, y) = (-7.33, -2.04)$

$r = \sqrt{x^2 + y^2} = \sqrt{(-7.33)^2 + (-2.04)^2} = 7.61$

$\tan \theta = \frac{y}{x} = \frac{-2.04}{-7.33}$

θ is a third quadrant angle and must be chosen so that $-180° < \theta \le 180°$.

$\theta = -180° + \tan^{-1}\frac{-2.04}{-7.33} = -164.4°$.

$(r, \theta) = (7.61, -164.4°)$

29.

θ	$\sin\theta$	$4\sin\theta$	
0 to $\frac{\pi}{2}$	0 to 1	0 to 4	
$\frac{\pi}{2}$ to 0	1 to 0	4 to 0	
π to $\frac{3\pi}{2}$	0 to −1	0 to −4	Curve is traced out a second time in this region although coordinates seem different.
$\frac{3\pi}{2}$ to 2π	−1 to 0	−4 to 0	

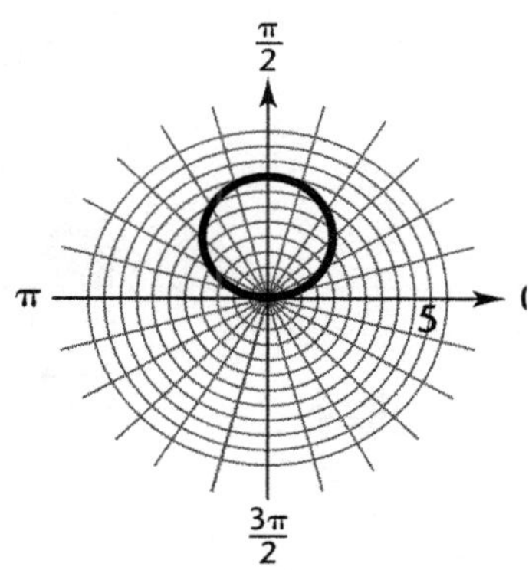

Check:

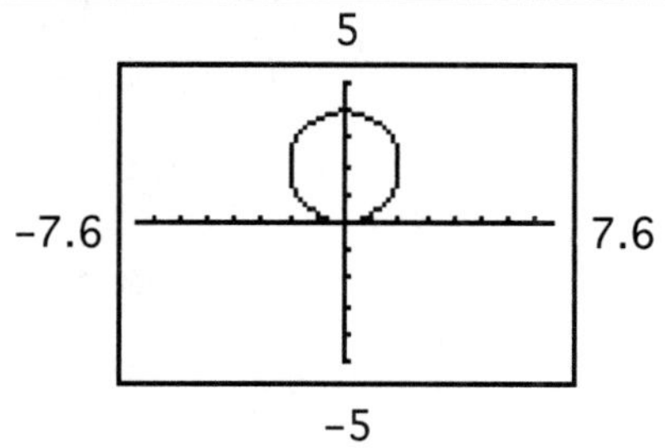

31.

θ	2θ	$\sin 2\theta$	$10 \sin 2\theta$
0 to $\frac{\pi}{4}$	0 to $\frac{\pi}{2}$	0 to 1	0 to 10
$\frac{\pi}{4}$ to $\frac{\pi}{2}$	$\frac{\pi}{2}$ to π	1 to 0	10 to 0
$\frac{\pi}{2}$ to $\frac{3\pi}{4}$	π to $\frac{3\pi}{2}$	0 to -1	0 to -10
$\frac{3\pi}{4}$ to π	$\frac{3\pi}{2}$ to 2π	-1 to 0	-10 to 0
π to $\frac{5\pi}{4}$	2π to $\frac{5\pi}{2}$	0 to 1	0 to 10
$\frac{5\pi}{4}$ to $\frac{3\pi}{2}$	$\frac{5\pi}{2}$ to 3π	1 to 0	10 to 0
$\frac{3\pi}{2}$ to $\frac{7\pi}{4}$	3π to $\frac{7\pi}{2}$	0 to -1	0 to -10
$\frac{7\pi}{4}$ to 2π	$\frac{7\pi}{2}$ to 4π	-1 to 0	-10 to 0

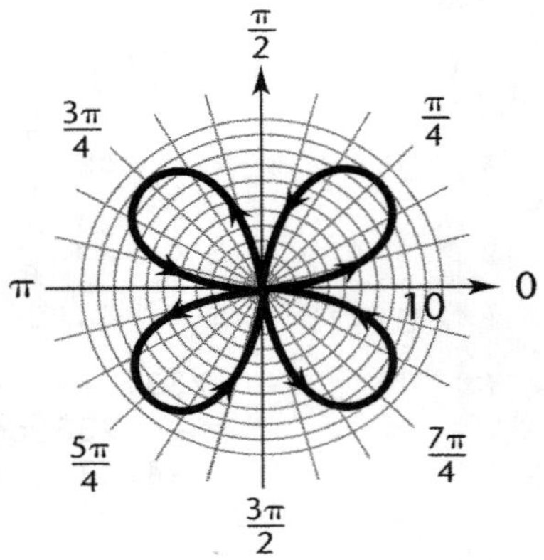

Check:

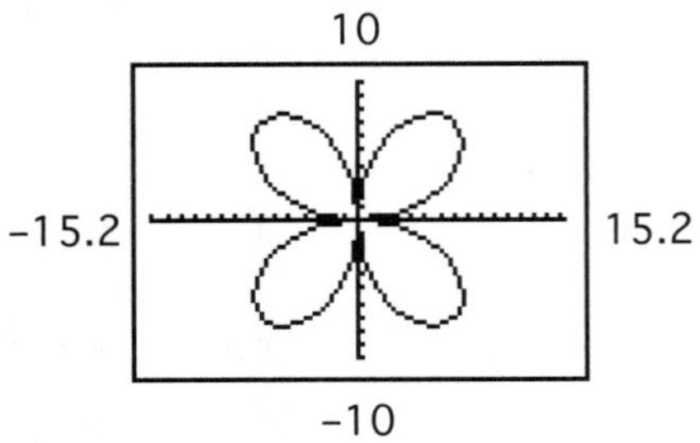

33.

θ	3θ	$\cos 3\theta$	$5 \cos 3\theta$
0 to $\frac{\pi}{6}$	0 to $\frac{\pi}{2}$	1 to 0	5 to 0
$\frac{\pi}{6}$ to $\frac{\pi}{3}$	$\frac{\pi}{2}$ to π	0 to -1	0 to -5
$\frac{\pi}{3}$ to $\frac{\pi}{2}$	π to $\frac{3\pi}{2}$	-1 to 0	-5 to 0
$\frac{\pi}{2}$ to $\frac{2\pi}{3}$	$\frac{3\pi}{2}$ to 2π	0 to 1	0 to 5
$\frac{2\pi}{3}$ to $\frac{5\pi}{6}$	2π to $\frac{5\pi}{2}$	1 to 0	5 to 0
$\frac{5\pi}{6}$ to π	$\frac{5\pi}{2}$ to 3π	0 to -1	0 to -5

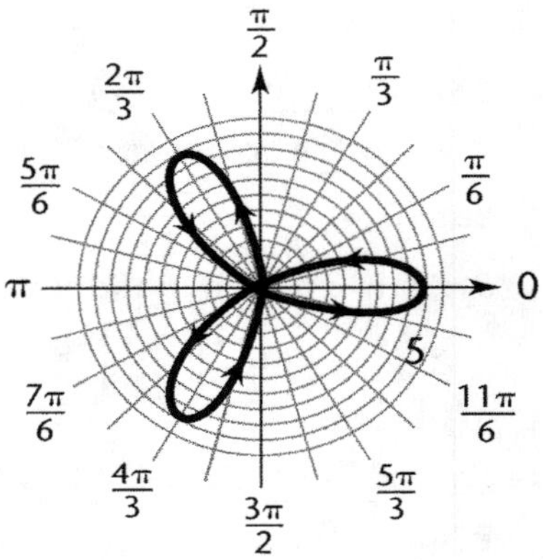

Check:

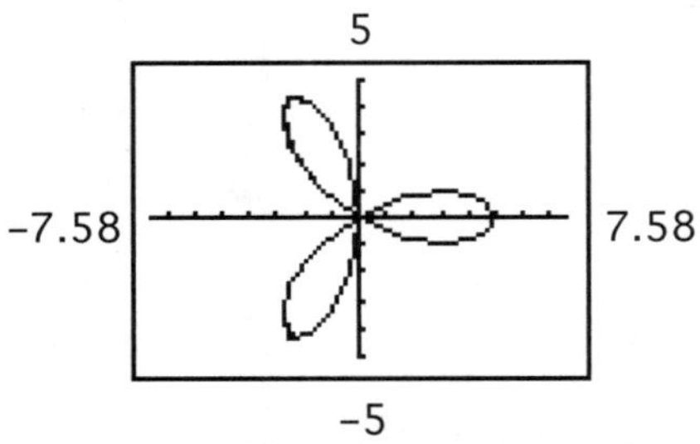

35.

θ	$\sin \theta$	$2 \sin \theta$	$2 + 2 \sin \theta$
0 to $\frac{\pi}{2}$	0 to 1	0 to 2	2 to 4
$\frac{\pi}{2}$ to π	1 to 0	2 to 0	4 to 2
π to $\frac{3\pi}{2}$	0 to -1	0 to -2	2 to 0
$\frac{3\pi}{2}$ to π	-1 to 0	-2 to 0	0 to 2

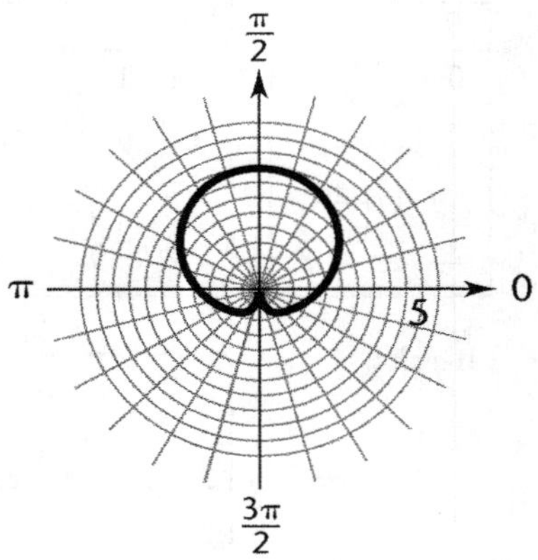

Check:

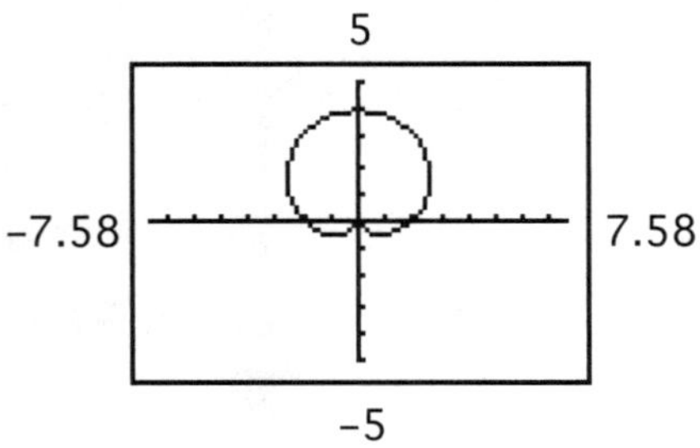

37.

θ	$\sin\theta$	$4\sin\theta$	$2 + 4\sin\theta$
0 to $\frac{\pi}{2}$	0 to 1	0 to 4	2 to 6
$\frac{\pi}{2}$ to π	1 to 0	4 to 0	6 to 2
π to $\frac{3\pi}{2}$	0 to −1	0 to −4	2 to −2
$\frac{3\pi}{2}$ to π	−1 to 0	−4 to 0	−2 to 2

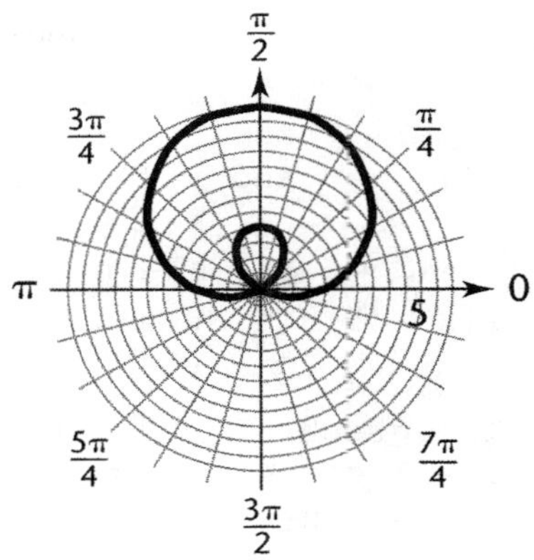

Check:

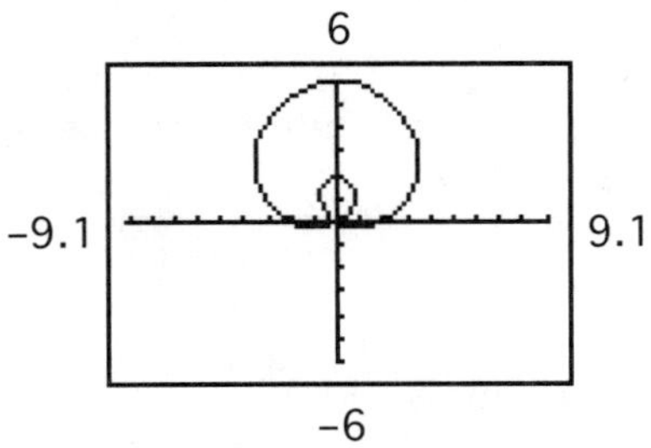

39. $r = 2 + 2\sin\theta$

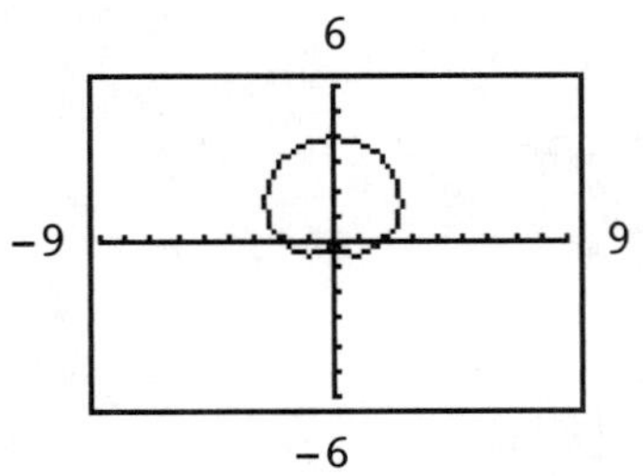

$r = 4 + 2\sin\theta$

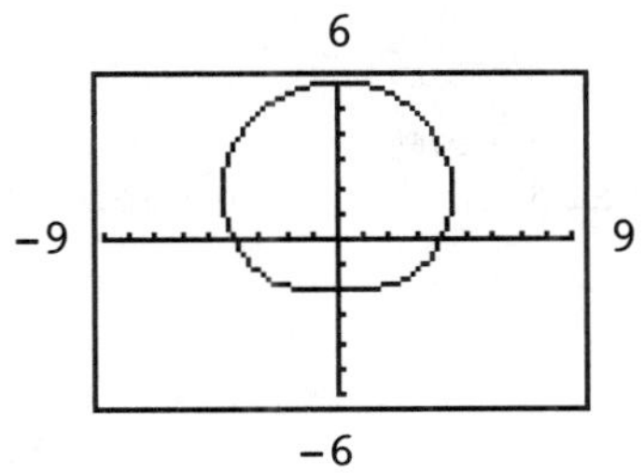

$r = 2 + 4\sin\theta$

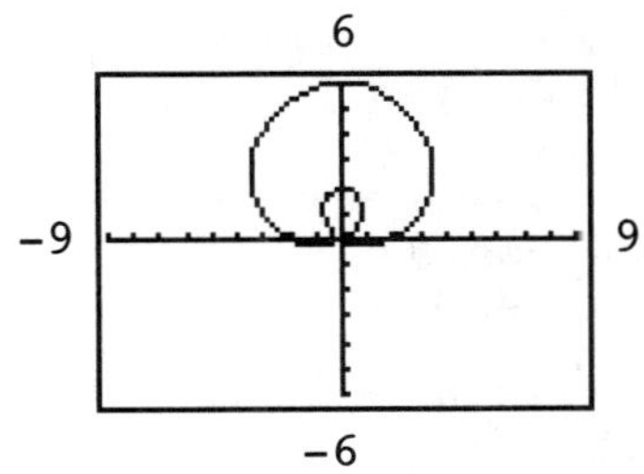

41. (A) $r = 4\sin\theta$

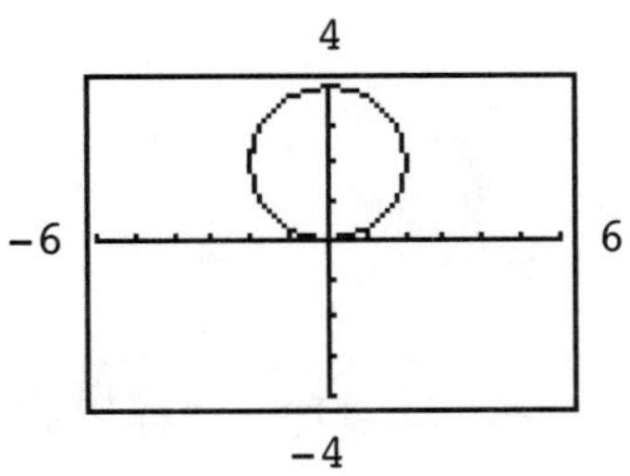

$r = 4\sin 3\theta$

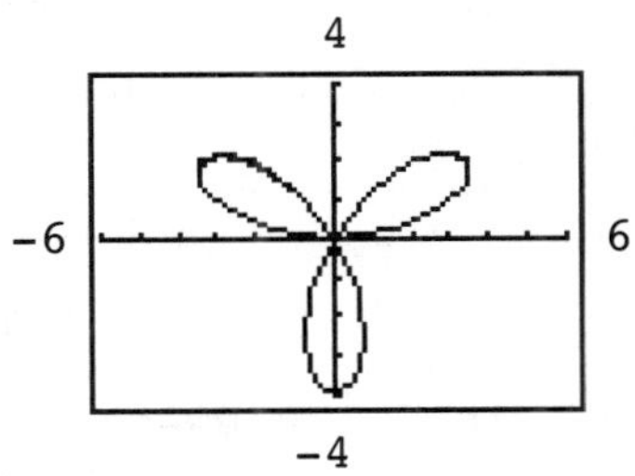

$r = 4\sin 5\theta$

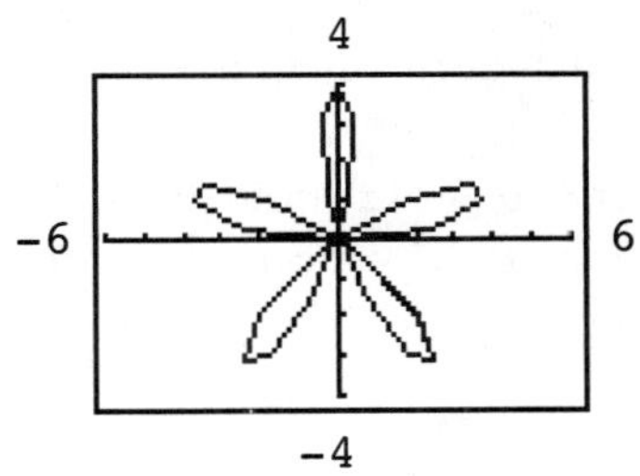

(B) and (C) Since there is one "leaf" for $r = 4\sin\theta$, three leaves for $r = 4\sin 3\theta$, and 5 leaves for $r = 4\sin 5\theta$, reasonable guesses would be seven leaves for $r = 4\sin 7\theta$ and n leaves for $r = a\sin n\theta$, n odd, $a > 0$.

43. (A) $r = 4\sin 2\theta r = 4\sin 4\theta r = 4\sin 6\theta$

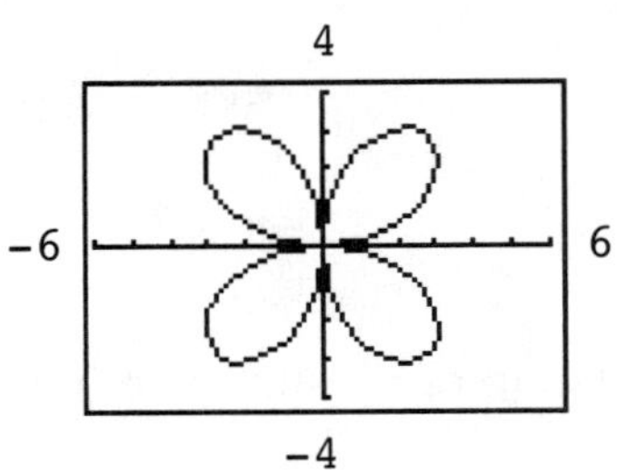

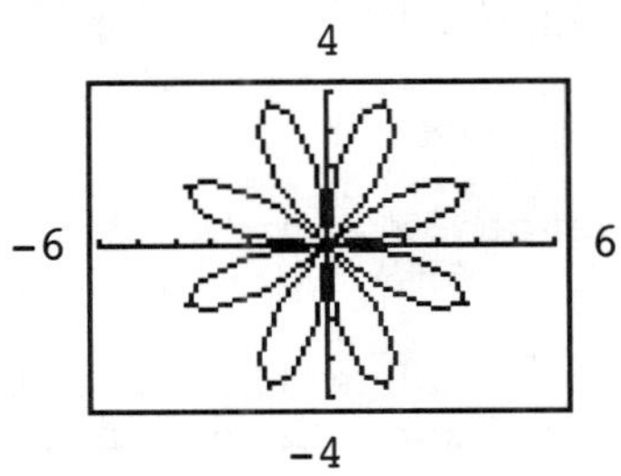

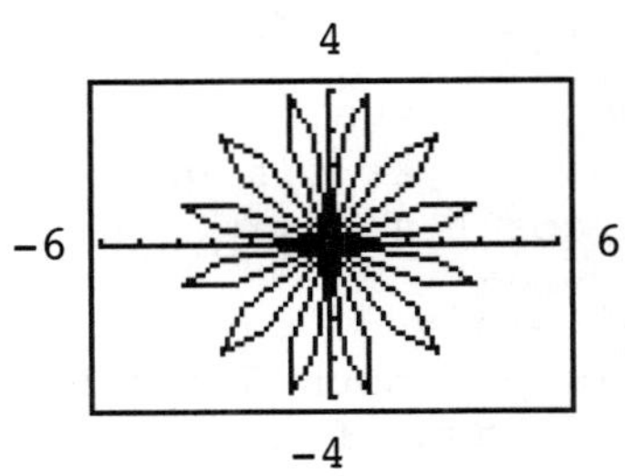

(B) and (C) Since there are four leaves for $r = 4\sin 2\theta$, eight leaves for $r = 4\sin 4\theta$, and twelve leaves for $r = 4\sin 6\theta$, reasonable guesses would be 16 leaves for $r = 4\sin 8\theta$ and $2n$ leaves for $r = a\sin n\theta$, n even, $a > 0$.

45.
$$
\begin{aligned}
y^2 &= 5y - x^2 \\
x^2 + y^2 &= 5y \\
\text{Use } r^2 &= x^2 + y^2 \\
\text{and } y &= r \sin \theta \\
r^2 &= 5r \sin \theta \\
r^2 - 5r \sin \theta &= 0 \\
r(r - 5 \sin \theta) &= 0 \\
r &= 0 \text{ or } r - 5 \sin \theta = 0
\end{aligned}
$$

The graph of $r = 0$ is the pole. Since the pole is included in the graph of $r - 5 \sin \theta = 0$, we can discard $r = 0$ and keep only
$r - 5 \sin \theta = 0$ or $r = 5 \sin \theta$

47.
$$
\begin{aligned}
y &= x \\
y &= r \sin \theta \text{ and } x = r \cos \theta, \text{ hence} \\
r \sin \theta &= r \cos \theta \\
r \sin \theta - r \cos \theta &= 0 \\
r(\sin \theta - \cos \theta) &= 0 \\
r &= 0 \text{ or } \sin \theta - \cos \theta = 0
\end{aligned}
$$

The graph of $r = 0$ is the pole. Since the pole is included in the graph of $\sin \theta - \cos \theta = 0$, we can discard $r = 0$ and keep only
$\sin \theta - \cos \theta = 0$
$$
\begin{aligned}
\sin \theta &= \cos \theta \\
\text{or } \tan \theta &= 1 \quad \text{or} \quad \theta = \frac{\pi}{4}
\end{aligned}
$$

49.
$$
\begin{aligned}
y^2 &= 4x \\
y &= r \sin \theta \text{ and } x = r \cos \theta, \text{ hence} \\
(r \sin \theta)^2 &= 4r \cos \theta \\
r^2 \sin^2 \theta &= 4r \cos \theta \\
r^2 \sin^2 \theta - 4r \cos \theta &= 0 \\
r(r \sin^2 \theta - 4 \cos \theta) &= 0 \\
r &= 0 \text{ or } r \sin^2 \theta - 4 \cos \theta = 0
\end{aligned}
$$

The graph of $r = 0$ is the pole. Since the pole is included in the graph of $r \sin^2 \theta - 4 \cos \theta = 0$, we can discard $r = 0$ and keep only
$r \sin^2 \theta - 4 \cos \theta = 0$
$$
\begin{aligned}
r \sin^2 \theta &= 4 \cos \theta \\
r &= \frac{4 \cos \theta}{\sin^2 \theta} \\
\text{or } r &= 4 \frac{\cos \theta}{\sin \theta} \frac{1}{\sin \theta} \\
\text{or } r &= 4 \cot \theta \csc \theta
\end{aligned}
$$

51. $r(3 \cos \theta - 4 \sin \theta) = -1$
$3r \cos \theta - 4r \sin \theta = -1$
$x = r \cos \theta$ and $y = r \sin \theta$, hence $3x - 4y = -1$

53. $r = -2 \sin \theta$
We multiply both sides by r, which adds the pole to the graph. But the pole is already part of the graph, so we have changed nothing.
$r^2 = -2r \sin \theta$
But $r^2 = x^2 + y^2$ and $r \sin \theta = y$. Hence, $x^2 + y^2 = -2y$.

55.

$$\theta = \frac{\pi}{4}$$

$$\tan\theta = \tan\frac{\pi}{4}$$

$$\tan\theta = 1$$

Use $\tan\theta = \frac{y}{x}$, then

$$\frac{y}{x} = 1$$

$$y = x$$

(Multiplying by x adds the origin to the graph, but the origin is already part of the graph.)

57. Here are computer-generated graphs of $r = 1 + 2\sin(n\theta)$ for $n = 1$, $n = 2$, $n = 3$, and $n = 4$.

$n = 1$

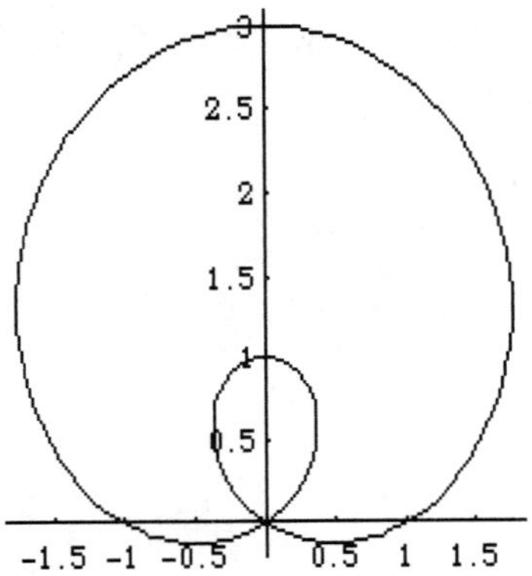

$n = 2$

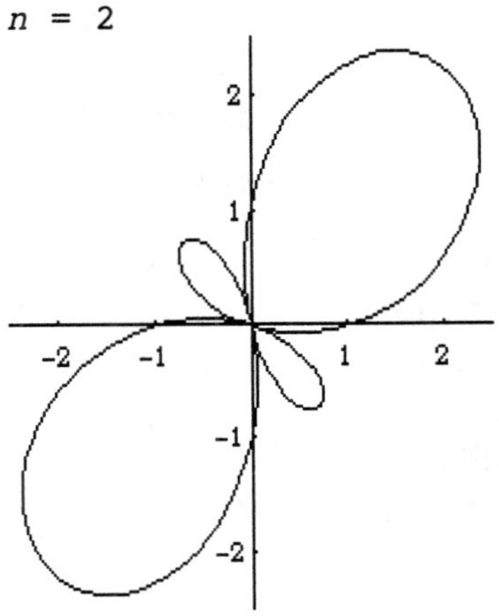

$n = 3$

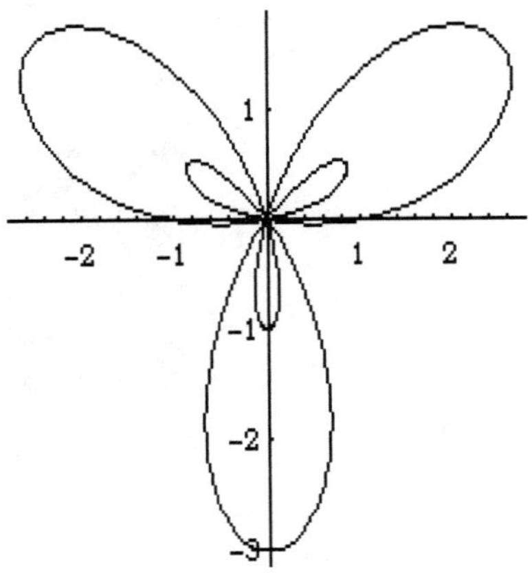

$n = 4$

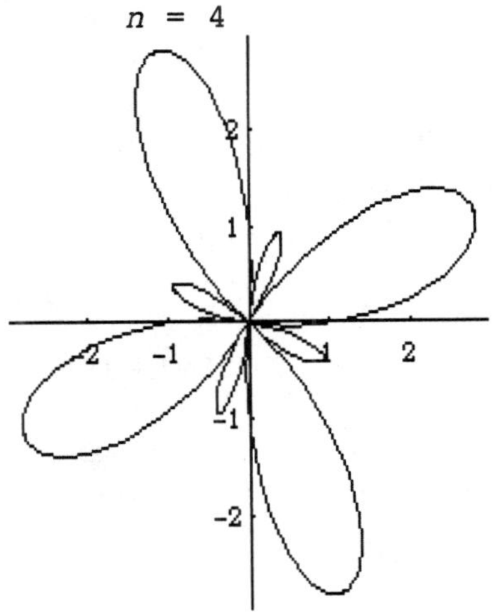

Generalizing from these, we can say that for each value of n, there are n large petals and n small petals. For n odd, the small petals are within the large petals; for n even, the small petals are between the large petals.

59. We solve the system

$r = 4\cos\theta$

$r = -4\sin\theta$

by equating the right sides:

$$4\cos\theta = -4\sin\theta$$
$$\cos\theta = -\sin\theta$$
$$-1 = \tan\theta$$

The only solution of this equation, $0 \le \theta \le \pi$, is $\theta = \frac{3\pi}{4}$. If we substitute this in either of the original equations, we get

$$r = 4\cos\frac{3\pi}{4} = -4\sin\frac{3\pi}{4} = 4\left(\frac{-\sqrt{2}}{2}\right) = -2\sqrt{2}$$

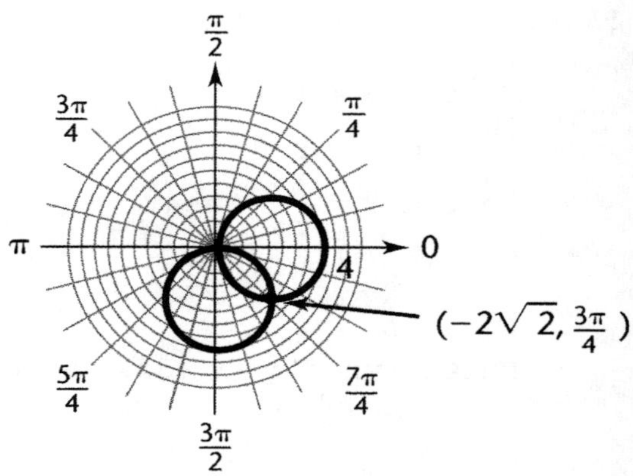

A sketch of the two graphs shows that the pole is on both graphs; however, the pole has no ordered pairs of coordinates that simultaneously satisfy both equations. As $\left(0, \frac{\pi}{2}\right)$ it satisfies the first; as (0,0) it satisfies the second; it is not a solution of the system.

Solution: $\left(-2\sqrt{2}, \frac{3\pi}{4}\right)$

61. We solve the system

$r = 6\cos\theta \quad 0° \le \theta \le 360°$

$r = 6\sin 2\theta$

by equating the right sides:

$$6\cos\theta = 6\sin 2\theta$$
$$\cos\theta = \sin 2\theta$$
$$\cos\theta = 2\sin\theta\cos\theta$$
$$0 = 2\sin\theta\cos\theta - \cos\theta$$
$$0 = \cos\theta(2\sin\theta - 1)$$

$\cos\theta = 0 \qquad\qquad 2\sin\theta - 1 = 0$

$\theta = 90°, 270° \qquad \sin\theta = \frac{1}{2}$

$\theta = 30°, 150°$

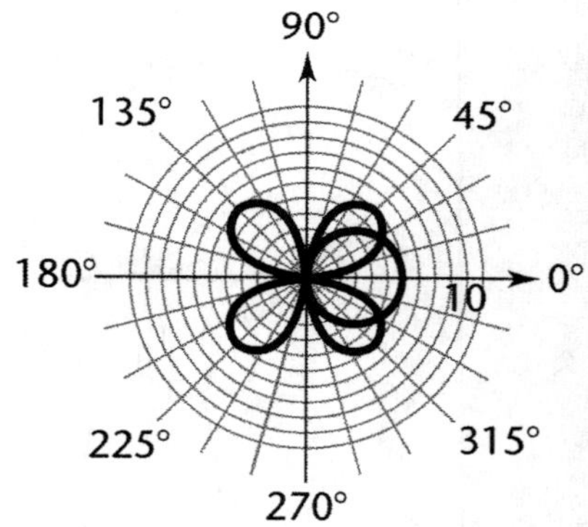

If we substitute these values of θ in either of the original equations, we get the corresponding values of r:

$\theta = 90° \quad r = 6\cos 90° = 0$

$\theta = 270° \quad r = 6\cos 270° = 0$

$\theta = 30° \quad r = 6\cos 30° = 3\sqrt{3}$

$\theta = 150° \quad r = 6\cos 150° = -3\sqrt{3}$

The four solutions of the system are (0, 90°), (0, 270°), $(3\sqrt{3}, 30°)$, and $(-3\sqrt{3}, 150°)$. Note that two of these (0, 90°) and (0, 270°) name the same point (the pole). A sketch of the two graphs shows three points of intersection.

63. $$d = \sqrt{r_1^2 + r_2^2 - 2r_1r_2\cos(\theta_2 - \theta_1)}$$

$$(r_1, \theta_1) = \left(4, \frac{\pi}{4}\right) \quad (r_2, \theta_2) = \left(1, \frac{\pi}{2}\right)$$

$$d = \sqrt{4^2 + 1^2 - 2(4)(1)\cos\left(\frac{\pi}{2} - \frac{\pi}{4}\right)}$$

$$d = \sqrt{17 - 8\cos\frac{\pi}{4}}$$

$d = 3.368$ units

65. 6 knots at 30°, 13 knots at 75°, 12 knots at 135°, 9 knots at 180°

67. (A) $e = 0.4$ $\quad r = \dfrac{8}{1 - 0.4\cos\theta}$

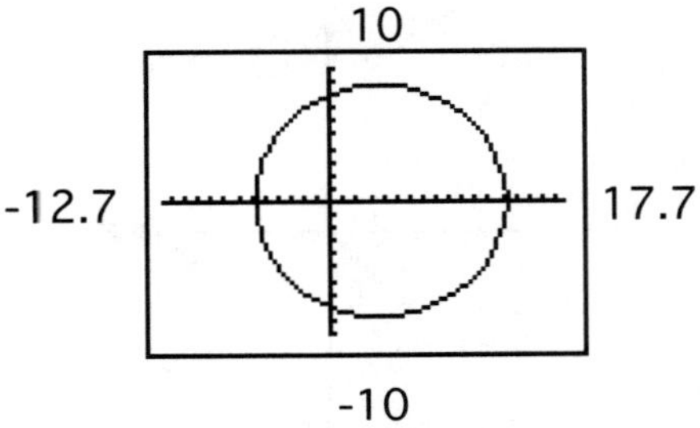

An ellipse

(B) $e = 1$ $\quad r = \dfrac{8}{1 - \cos\theta}$

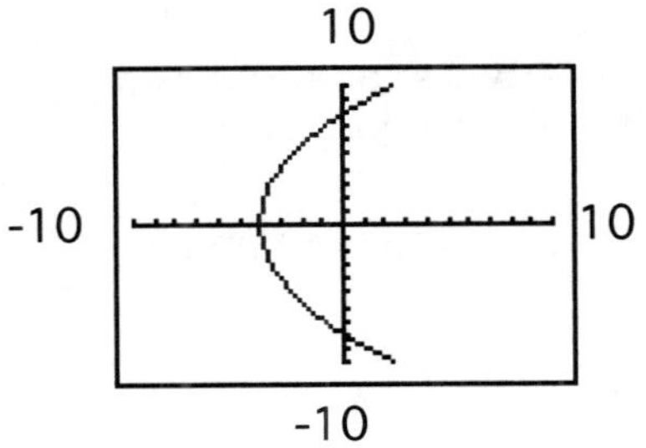

A parabola

(C) $e = 1.6$ $\quad r = \dfrac{8}{1 - 1.6\cos\theta}$

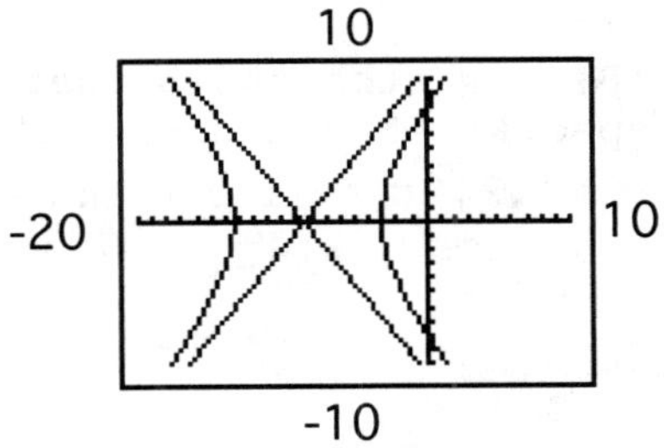

A hyperbola

69. (A)

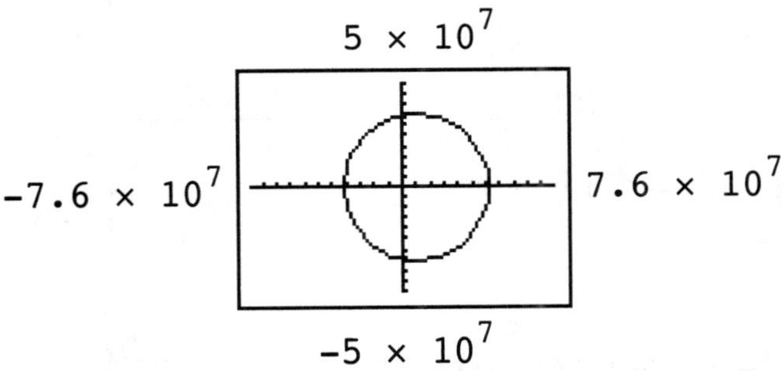

At aphelion, the distance is 4.34×10^7 mi; at perihelion it is 2.85×10^7 mi.

(B) Faster at perihelion. Since the distance from the sun to Mercury is less at perihelion than at aphelion, the planet must move faster near perihelion in order for the line joining Mercury to the sun to sweep out equal areas in equal intervals of time.

Section 8-5

1.

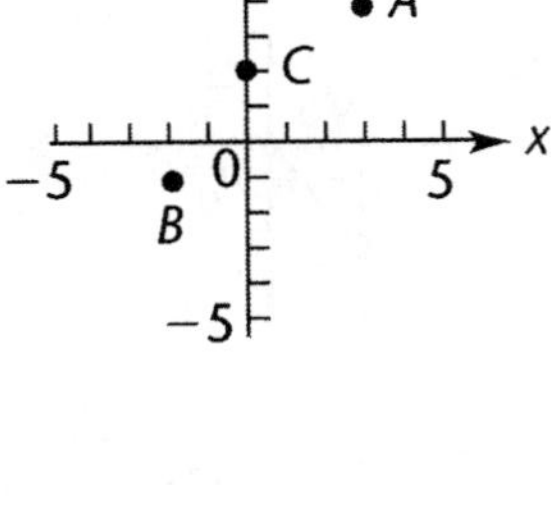

3.

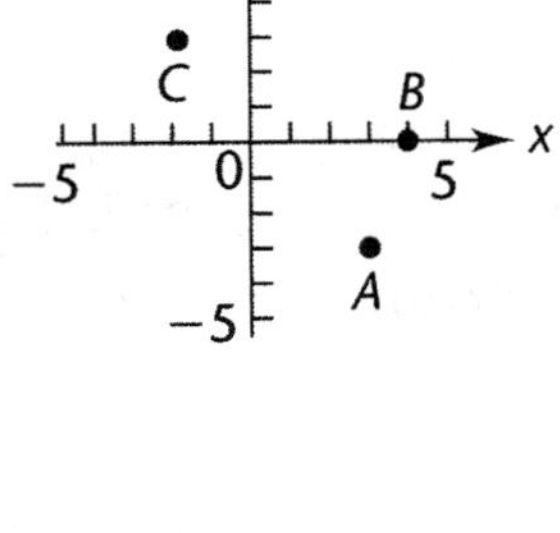

5.

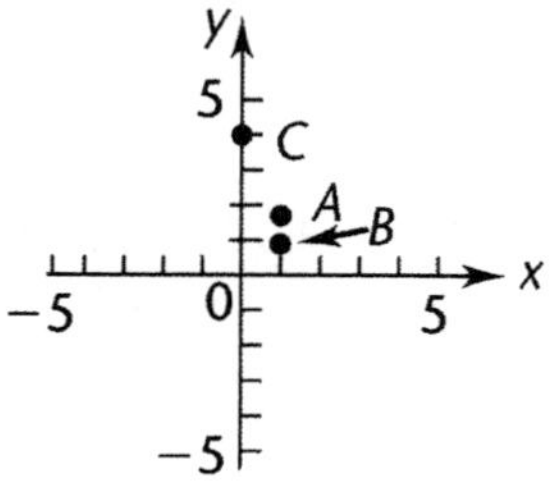

7.

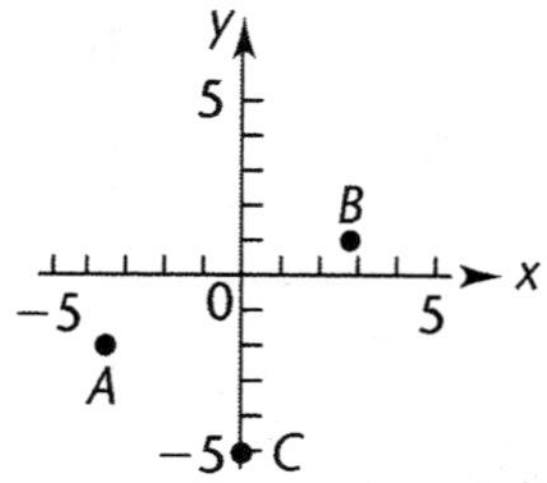

9. (A) A sketch shows that $\sqrt{3} + i$ is associated with a special 30°-60° triangle. Thus, by inspection, $r = 2$, $\theta = 30°$, and

$$\sqrt{3} + i = 2(\cos 30° + i \sin 30°)$$
$$= 2e^{30° i}$$

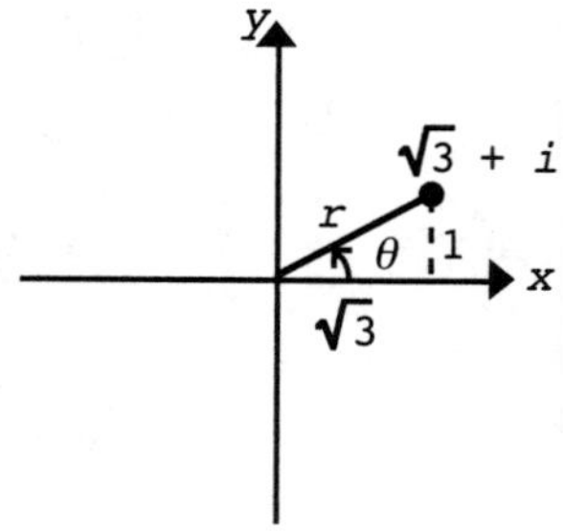

(B) A sketch shows that $-1 - i$ is associated with a special 45° triangle. Thus, by inspection, $r = \sqrt{2}$, $\theta = -135°$, and

$$-1 - i = \sqrt{2}[\cos(-135°) + i \sin(-135°)]$$
$$= \sqrt{2}\, e^{-135° i}$$

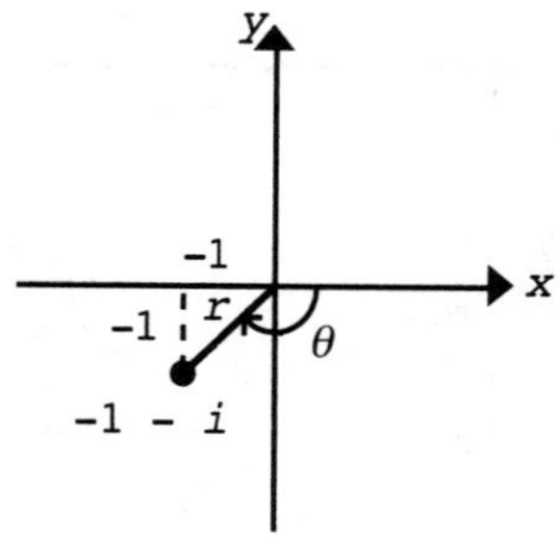

(C) A sketch shows that $5 - 6i$ is not associated with a special triangle.

$$r = \sqrt{5^2 + (-6)^2} = 7.81$$
$$\theta = \tan^{-1} \frac{-6}{5} = -50.19°$$
$$5 - 6i = 7.81[\cos(-50.19°) + i \sin(-50.19°)]$$
$$= 7.81e^{-50.19° i}$$

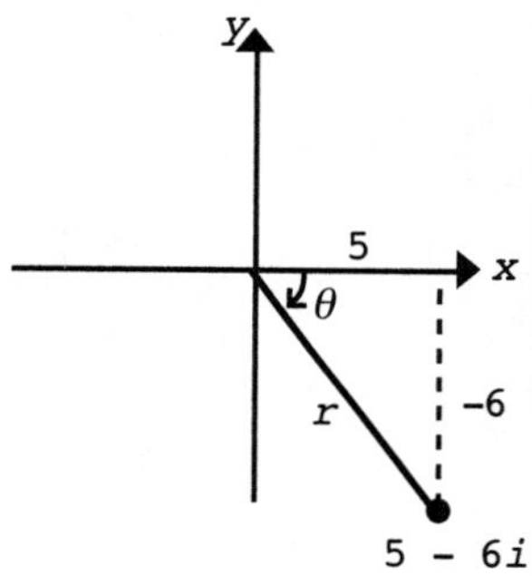

11. (A) A sketch shows that $-i\sqrt{3}$ is associated with the special quadrantal angle $-\frac{\pi}{2}$. Thus, by inspection, $r = \sqrt{3}$, $\theta = -\frac{\pi}{2}$, and

$$-i\sqrt{3} = \sqrt{3}\left[\cos\left(-\frac{\pi}{2}\right) + i \sin\left(-\frac{\pi}{2}\right)\right]$$
$$= \sqrt{3}\, e^{(-\pi/2) i}$$

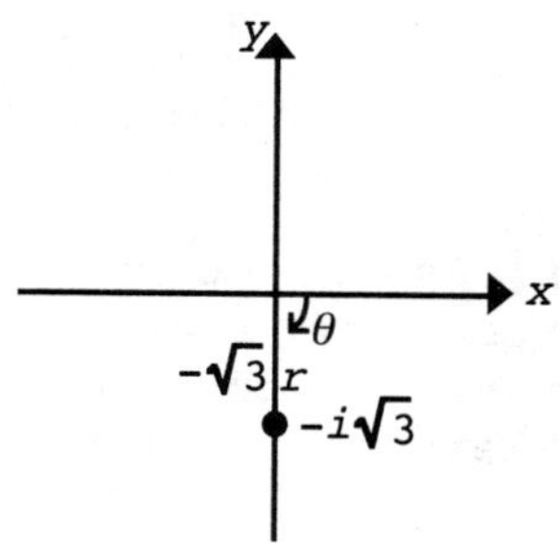

(B) A sketch shows that $-\sqrt{3} - i$ is associated with a special 30°-60° triangle. Thus, by inspection, $r = 2$, $\theta = -\frac{5\pi}{6}$, and

$$-\sqrt{3} - i = 2\left[\cos\left(-\frac{5\pi}{6}\right) + i \sin\left(-\frac{5\pi}{6}\right)\right]$$
$$= 2e^{(-5\pi/6) i}$$

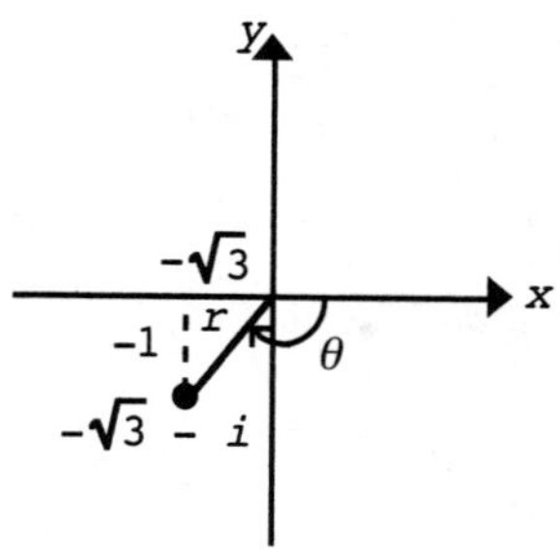

(C) A sketch shows that $-8 + 5i$ is not associated with a special triangle.

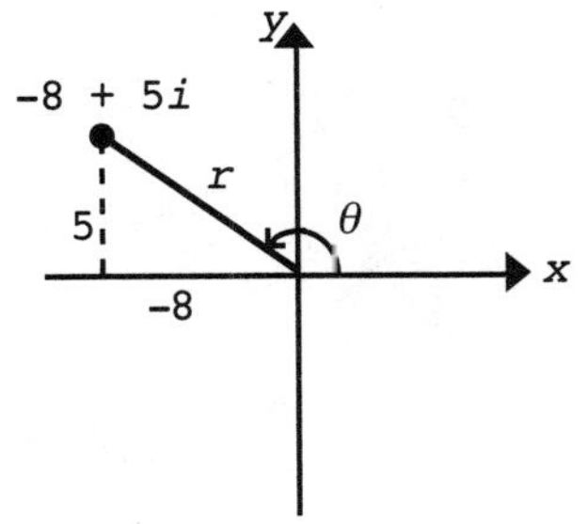

$r = \sqrt{(-8)^2 + 5^2} = 9.43$

$\theta = \pi + \tan^{-1} \dfrac{5}{(-8)} = 2.58$

$$-8 + 5i = 9.43(\cos 2.58 + i \sin 2.58) = 9.43e^{2.58i}$$

13. (A) $x + iy = 2e^{(\pi/3)i} = 2\left(\cos\frac{\pi}{3} + i\sin\frac{\pi}{3}\right) = 2\left(\frac{1}{2} + i\frac{\sqrt{3}}{2}\right) = 1 + i\sqrt{3}$

(B) $x + iy = \sqrt{2}e^{(-45°)i} = \sqrt{2}[\cos(-45°) + i\sin(-45°)] = \sqrt{2}\left[\frac{1}{\sqrt{2}} + i\left(-\frac{1}{\sqrt{2}}\right)\right] = 1 - i$

(C) $x + iy = 3.08e^{2.44i} = 3.08(\cos 2.44 + i \sin 2.44) = -2.35 + 1.99i$

15. (A) $x + iy = 6e^{(\pi/6)i} = 6\left(\cos\frac{\pi}{6} + i\sin\frac{\pi}{6}\right) = 6\left(\frac{\sqrt{3}}{2} + i\frac{1}{2}\right) = 3\sqrt{3} + 3i$

(B) $x + iy = \sqrt{7}e^{(-90°)i} = \sqrt{7}[\cos(-90°) + i\sin(-90°)] = \sqrt{7}(0 - 1i) = -i\sqrt{7}$

(C) $4.09e^{(-122.88°)i} = 4.09[\cos(-122.88°) + i\sin(-122.88°)] = -2.22 - 3.43i$

17. $z_1z_2 = 7e^{82°i} \cdot 2e^{31°i} = 7 \cdot 2e^{i(82°+31°)} = 14e^{113°i}$

$\dfrac{z_1}{z_2} = \dfrac{7e^{82°i}}{2e^{31°i}} = \dfrac{7}{2}e^{i(82°-31°)} = 3.5e^{51°i}$

19. $z_1z_2 = 5e^{52°i} \cdot 2e^{83°i} = 5 \cdot 2e^{i(52°+83°)} = 10e^{135°i}$

$\dfrac{z_1}{z_2} = \dfrac{5e^{52°i}}{2e^{83°i}} = \dfrac{5}{2}e^{i(52°-83°)} = 2.5e^{(-31°)i}$

21. $z_1z_2 = 3.05e^{1.76i} \cdot 11.94e^{2.59i} = 3.05 \cdot 11.94e^{(1.76+2.59)i} = 36.42e^{4.35i}$

$\dfrac{z_1}{z_2} = \dfrac{3.05e^{1.76i}}{11.94e^{2.59i}} = \dfrac{3.05}{11.94}e^{(1.76-2.59)i} = 0.26e^{-0.83i}$

23. $(2e^{30°i})^3 = 2^3e^{3\cdot 30°i} = 8e^{90°i}$

25. $(\sqrt{2}e^{10°i})^6 = (\sqrt{2})^6e^{6\cdot 10°i} = (2^{1/2})^6e^{60°i} = 2^3e^{60°i} = 8e^{60°i}$

27. $(1 + i\sqrt{3})^3 = (2e^{60°i})^3 = 2^3e^{3\cdot 60°i} = 8e^{180°i}$

29. $(-\sqrt{3} - i)^4 = (2e^{-150°i})^4 = 2^4e^{4(-150°i)} = 16e^{-600°i}$

$$= 16[\cos(-600°) + i\sin(-600°)] = 16\left(-\frac{1}{2} + i\frac{\sqrt{3}}{2}\right) = -8 + 8i\sqrt{3}$$

31. $(1 - i)^8 = (\sqrt{2}e^{-45°i})^8 = (\sqrt{2})^8e^{8(-45°i)} = 16e^{-360°i}$

$$= 16[\cos(-360°) + i\sin(-360°)] = 16(1 + 0i) = 16$$

33. $\left(-\frac{1}{2} + \frac{\sqrt{3}}{2}i\right) = (e^{120°i})^3 = e^{360°i} = \cos 360° + i\sin 360° = 1 + 0i = 1$

35. Using the nth-root theorem, all three cube roots of $8e^{30°i}$ are given by

$$8^{1/3}e^{(30°/3+k360°/3)i} = 8^{1/3}e^{(10°+k120°)i} \qquad k = 0, 1, 2$$

Thus,

$$w_1 = 8^{1/3}e^{(10°+0\cdot 120°)i} = 2e^{10°i}$$
$$w_2 = 8^{1/3}e^{(10°+1\cdot 120°)i} = 2e^{130°i}$$
$$w_3 = 8^{1/3}e^{(10°+2\cdot 120°)i} = 2e^{250°i}$$

37. Using the nth-root theorem, all four fourth roots of $81e^{60°i}$ are given by

$$81^{1/4}e^{(60°/4+k360°/4)i} = 81^{1/4}e^{(15°+k90°)i} \qquad k = 0, 1, 2, 3$$

Thus,

$$w_1 = 81^{1/4}e^{(15°+0\cdot 90°)i} = 3e^{15°i}$$
$$w_2 = 81^{1/4}e^{(15°+1\cdot 90°)i} = 3e^{105°i}$$
$$w_3 = 81^{1/4}e^{(15°+2\cdot 90°)i} = 3e^{195°i}$$
$$w_4 = 81^{1/4}e^{(15°+3\cdot 90°)i} = 3e^{285°i}$$

39. First write $1 - i$ in polar form.

$$1 - i = \sqrt{2}e^{(-45°)i}$$

Using the nth-root theorem, all five fifth roots of $\sqrt{2}e^{(-45°)i}$ are given by

$$(\sqrt{2})^{1/5}e^{(-45°/5+k360°/5)i} = (2^{1/2})^{1/5}e^{(-9°+k72°)i} \qquad k = 0, 1, 2, 3, 4$$

Thus,

$$w_1 = 2^{1/10}e^{(-9°+0\cdot 72°)i} = 2^{1/10}e^{(-9°)i}$$
$$w_2 = 2^{1/10}e^{(-9°+1\cdot 72°)i} = 2^{1/10}e^{63°i}$$
$$w_3 = 2^{1/10}e^{(-9°+2\cdot 72°)i} = 2^{1/10}e^{135°i}$$
$$w_4 = 2^{1/10}e^{(-9°+3\cdot 72°)i} = 2^{1/10}e^{207°i}$$
$$w_5 = 2^{1/10}e^{(-9°+4\cdot 72°)i} = 2^{1/10}e^{279°i}$$

41. First write 8 in polar form.

$$8 = 8 + 0i = 8e^{0°i}$$

Using the nth-root theorem, all three cube roots of $8e^{0°i}$ are given by

$$8^{1/3}e^{(0°/3+k360°/3)i} = 8^{1/3}e^{k120°i} \qquad k = 0, 1, 2$$

Thus,

$$w_1 = 2e^{0\cdot 120°i} = 2e^{0°i}$$
$$w_2 = 2e^{1\cdot 120°i} = 2e^{120°i}$$
$$w_3 = 2e^{2\cdot 120°i} = 2e^{240°i}$$

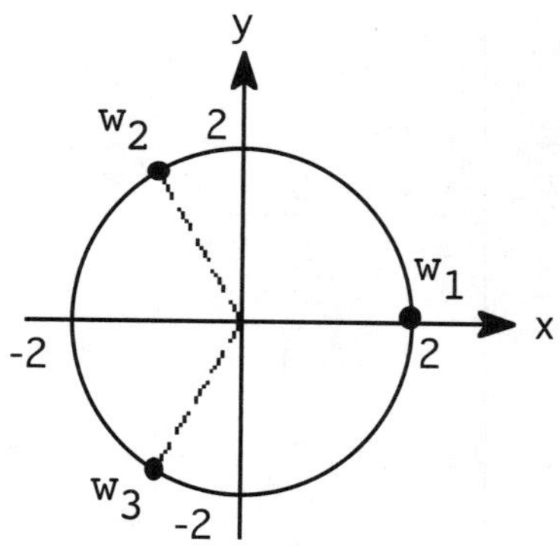

w_1, w_2, w_3 lie on a circle of radius 2, equally spaced so that the angle between successive roots is 120°.

43. First write -16 in polar form.

$$-16 = -16 + 0i = 16e^{180°i}$$

Using the nth-root theorem, all four fourth roots of $16e^{180°i}$ are given by

$$16^{1/4}e^{(180°/4 + k \cdot 360°/4)i} = 16^{1/4}e^{(45° + k \cdot 90°)i} \quad k = 0, 1, 2, 3$$

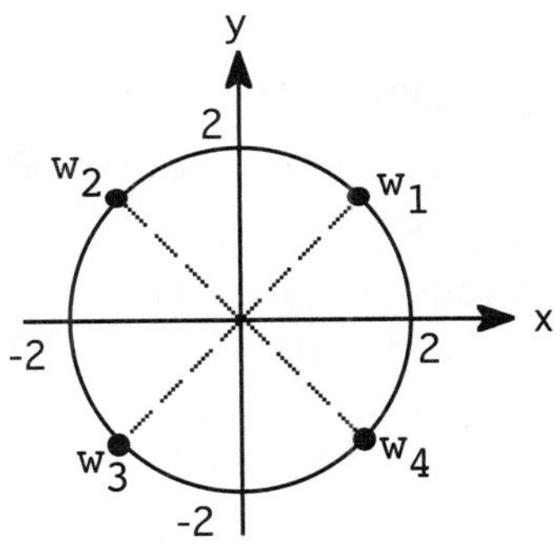

Thus,
$w_1 = 2e^{(45° + 0 \cdot 90°)i} = 2e^{45°i}$
$w_2 = 2e^{(45° + 1 \cdot 90°)i} = 2e^{135°i}$
$w_3 = 2e^{(45° + 2 \cdot 90°)i} = 2e^{225°i}$
$w_4 = 2e^{(45° + 3 \cdot 90°)i} = 2e^{315°i}$

w_1, w_2, w_3, w_4 lie on a circle of radius 2, equally spaced so that the angle between successive roots is 90°.

45. First write i in polar form.

$$i = 0 + 1i = 1e^{90°i}$$

Using the nth-root theorem, all sixth roots of $1e^{90°i}$ are given by

$$1^{1/6}e^{(90°/6 + k \cdot 360°/6)i} = 1e^{(15° + k \cdot 60°)i} \quad k = 0, 1, 2, 3, 4, 5$$

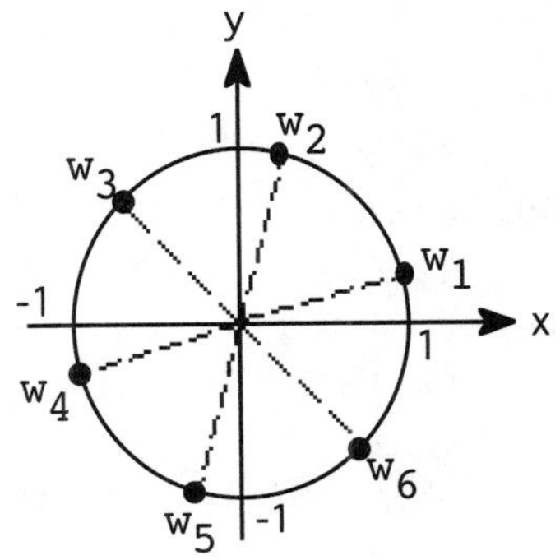

Thus,
$w_1 = 1e^{(15° + 0 \cdot 60°)i} = 1e^{15°i}$
$w_2 = 1e^{(15° + 1 \cdot 60°)i} = 1e^{75°i}$
$w_3 = 1e^{(15° + 2 \cdot 60°)i} = 1e^{135°i}$
$w_4 = 1e^{(15° + 3 \cdot 60°)i} = 1e^{195°i}$
$w_5 = 1e^{(15° + 4 \cdot 60°)i} = 1e^{255°i}$
$w_6 = 1e^{(15° + 5 \cdot 60°)i} = 1e^{315°i}$

These roots lie on a circle of radius 1, equally spaced so that the angle between successive roots is 60°.

47. (A) Substituting $1 + i$ for x in $x^4 + 4 = 0$ yields

$$(1 + i)^4 + 4 \stackrel{?}{=} 0$$
$$((1 + i)^2)^2 + 4 \stackrel{?}{=} 0$$
$$(1 + 2i + i^2)^2 + 4 \stackrel{?}{=} 0$$
$$(2i)^2 + 4 \stackrel{?}{=} 0$$
$$-4 + 4 \stackrel{\checkmark}{=} 0$$

Thus, $1 + i$ is a root of $x^4 + 4 = 0$.

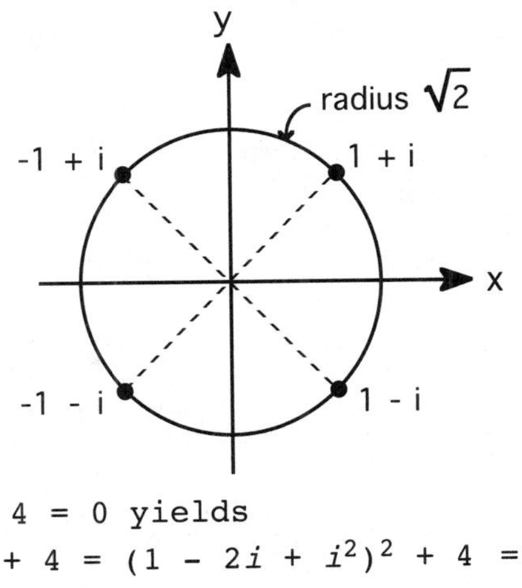

(B) The four roots are equally spaced around the circle. Since there are four roots, the angle between successive roots on the circle is $\frac{360°}{4} = 90°$.

(C) Substituting each root for x in $x^4 + 4 = 0$ yields

$-1 + i$: $(-1 + i)^4 + 4 = ((-1 + i)^2)^2 + 4 = (1 - 2i + i^2)^2 + 4 = (-2i)^2 + 4 = -4 + 4 = 0$

$-1 - i$: $(-1 - i)^4 + 4 = ((-1 - i)^2)^2 + 4 = (1 + 2i + i^2)^2 + 4 = (2i)^2 + 4 = -4 + 4 = 0$

$1 - i$: $(1 - i)^4 + 4 = ((1 - i)^2)^2 + 4 = (1 - 2i + i^2)^2 + 4 = (-2i)^2 + 4 = -4 + 4 = 0$

49. $x^3 + 64 = 0$

$x^3 = -64$

Thus we require the three cube roots of -64.

First write -64 in polar form:

$-64 = 64e^{180°i}$

Using the nth-root theorem, all three cube roots of -64 are given by

$64^{1/3}e^{(180°/3 + k \cdot 360°/3)i} = 4e^{(60° + k \cdot 120°)i} \quad k = 0, 1, 2$

Thus,

$w_1 = 4e^{60°i} = 4(\cos 60° + i \sin 60°) = 2 + 2i\sqrt{3}$

$w_2 = 4e^{180°i} = 4(\cos 180° + i \sin 180°) = -4$

$w_3 = 4e^{300°i} = 4(\cos 300° + i \sin 300°) = 2 - 2i\sqrt{3}$

51. $x^3 - 27 = 0$

$x^3 = 27$

Thus we require the three cube roots of 27.

First write 27 in polar form:

$27 = 27e^{0°i}$

Using the nth-root theorem, all three cube roots of 27 are given by

$27^{1/3}e^{(0°/3 + k \cdot 360°/3)i} = 3e^{k \cdot 120°i} \quad k = 0, 1, 2$

Thus,

$w_1 = 3e^{0°i} = 3(\cos 0° + i \sin 0°) = 3$

$w_2 = 3e^{120°i} = 3(\cos 120° + i \sin 120°) = -\frac{3}{2} + \frac{3\sqrt{3}}{2}i$

$w_3 = 3e^{240°i} = 3(\cos 240° + i \sin 240°) = -\frac{3}{2} - \frac{3\sqrt{3}}{2}i$

53. True. If $z_1 = r_1e^{i0°}$ and $z_2 = r_2e^{i0°}$, then $z_1z_2 = r_1r_2e^{i0°}$.

55. False. 16 has fourth roots 2, -2, $2i$, $-2i$.

57. True. If $z^2 = 1$, then $(z^2)^3 = 1^3$ and $z^6 = 1$.

59. True. If $w^3 = w^4$, then $w^4 - w^3 = 0$, hence either $w^3 = 0$ (not allowed here) or $w = 1$.

61. False. If $w^3 = w^6$, then $w^6 - w^3 = 0$, hence either $w^3 = 0$ (not allowed here) or $w^3 - 1 = 0$, hence $w^3 = 1$ and w is a cube root of 1. However, w need not equal 1. $\frac{-1 + \sqrt{3}i}{2}$ is also a cube root of 1.

63. If w is real and $w^n = z$ then $w = re^{i0°}$ and $w^n = r^ne^{i0°}$; w^n must be real.

65. To show $(r^{1/n}e^{(\theta/n + k \cdot 360°/n)i})^n = re^{i\theta}$ we apply DeMoivre's Theorem:

$$\begin{aligned}(r^{1/n}e^{(\theta/n + k \cdot 360°/n)i})^n &= (r^{1/n})^n e^{n(\theta/n + k \cdot 360°/n)i} \\ &= re^{(\theta + k360°)i} \\ &= r[\cos(\theta + k360°) + i \sin(\theta + k \cdot 360°)] \\ &= r(\cos \theta + i \sin \theta) \quad \text{since } k \text{ is an integer and sine and cosine are periodic with period } 360° \\ &= re^{i\theta}\end{aligned}$$

67. The complex zeros of $P(x) = x^5 - 32$ are the values of x for which $x^5 - 32 = 0$, or $x^5 = 32$. Thus we require the five fifth roots of 32. First write 32 in polar form:

$$32 = 32e^{0°i}$$

Using the nth-root theorem, all five fifth roots of 32 are given by

$$32^{1/5}e^{(0°/5 + k \cdot 360°/5)i} = 2e^{k \cdot 72°i} \qquad k = 0, 1, 2, 3, 4$$

Thus,

$$w_1 = 2e^{0°i}$$
$$w_2 = 2e^{72°i}$$
$$w_3 = 2e^{144°i}$$
$$w_4 = 2e^{216°i}$$
$$w_5 = 2e^{288°i}$$

69. $x^5 + 1 = 0$

$$x^5 = -1$$

Thus we require the five fifth roots of -1. First write -1 in polar form:

$$-1 = 1e^{180°i}$$

Using the nth-root theorem, all five fifth roots of -1 are given by

$$1^{1/5}e^{(180°/5+k\cdot 360°/5)i} = 1e^{(36°+k\cdot 72°)i} \qquad k = 0, 1, 2, 3, 4$$

Thus,

$$w_1 = 1e^{(36° + 0 \cdot 72°)i} = e^{36°i}$$
$$w_2 = 1e^{(36° + 1 \cdot 72°)i} = e^{108°i}$$
$$w_3 = 1e^{(36° + 2 \cdot 72°)i} = e^{180°i}$$
$$w_4 = 1e^{(36° + 3 \cdot 72°)i} = e^{252°i}$$
$$w_5 = 1e^{(36° + 4 \cdot 72°)i} = e^{324°i}$$

71. The linear factors of $x^6 + 64$ will, by the factor theorem, be $x - x_i$, where x_i are zeros of $x^6 + 64$, that is $x^6 + 64 = 0$ or $x^6 = -64$. Hence the x_i are the six sixth roots of -64, or $64e^{180°i}$.
Using the nth-root theorem, all six sixth roots of -64 are given by

$$64^{1/6}e^{(180°/6 + k \cdot 360°/6)i} = 2e^{(30° + k \cdot 60°)i} \qquad k = 0, 1, 2, 3, 4, 5$$

Thus,

$$w_1 = 2e^{30°i} = 2(\cos 30° + i \sin 30°) = 2\left(\frac{\sqrt{3}}{2} + i\frac{1}{2}\right) = \sqrt{3} + i$$
$$w_2 = 2e^{90°i} = 2(\cos 90° + i \sin 90°) = 2(0 + i1) = 2i$$
$$w_3 = 2e^{150°i} = 2(\cos 150° + i \sin 150°) = 2\left(-\frac{\sqrt{3}}{2} + i\frac{1}{2}\right) = -\sqrt{3} + i$$
$$w_4 = 2e^{210°i} = 2(\cos 210° + i \sin 210°) = 2\left(-\frac{\sqrt{3}}{2} - i\frac{1}{2}\right) = -\sqrt{3} - i$$
$$w_5 = 2e^{270°i} = 2(\cos 270° + i \sin 270°) = 2(0 - i1) = -2i$$
$$w_6 = 2e^{330°i} = 2(\cos 330° + i \sin 330°) = 2\left(\frac{\sqrt{3}}{2} - i\frac{1}{2}\right) = \sqrt{3} - i$$

Hence,

$$x^6 + 64 = (x - 2i)(x + 2i)[x - (-\sqrt{3} + i)][x - (-\sqrt{3} - i)]$$
$$[x - (\sqrt{3} + i)][x - (\sqrt{3} - i)]$$

CHAPTER 8 REVIEW

1. 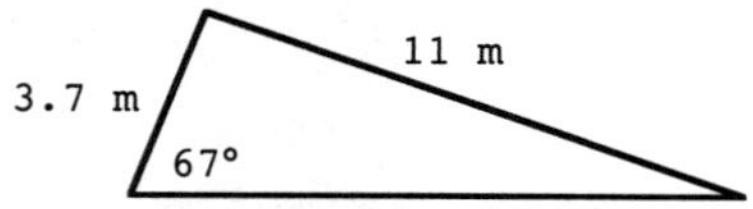

We are given two sides and a non-included angle (SSA). α is acute

$11 = a \geq b = 3.7$

One triangle can be constructed.

(*8-1*)

2. It is impossible to draw or form a triangle with this data, since angles α and β together add up to more than 180°. No triangle can be constructed.

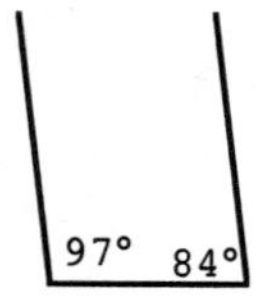

(*8-1*)

3. 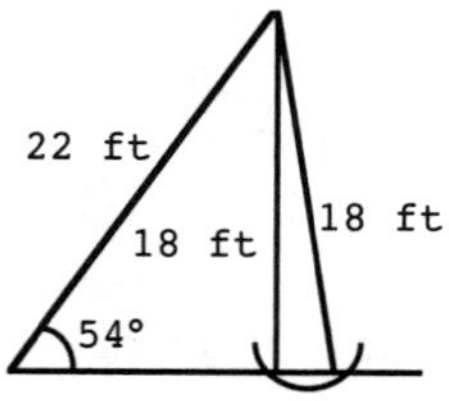

We are given two sides and a non-included angle (SSA). α is acute.

$h = b \sin \alpha = 22 \sin 54° = 17.8$

$h < a < b$

Two triangles can be constructed. (*8-1*)

4. Angle β is acute. A triangle can have at most one obtuse angle. Since α is acute, then, if the triangle has an obtuse angle it must be the angle opposite the longer of the two sides, b and c. Thus, β, the angle opposite the shorter of the two sides, b, must be acute. (*8-2*)

5. We are given two angles and the included side (ASA). We use the law of sines.

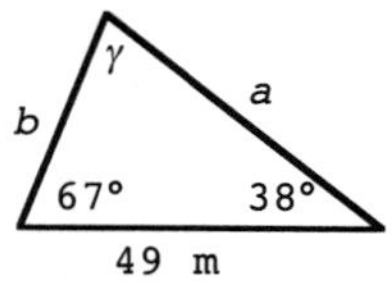

Solve for γ:

$$\begin{aligned}\alpha + \beta + \gamma &= 180° \\ \gamma &= 180° - (\alpha + \beta) \\ &= 180° - (67° + 38°) \\ &= 75°\end{aligned}$$

Solve for b:

$$\begin{aligned}\frac{\sin \beta}{b} &= \frac{\sin \gamma}{c} \\ b &= \frac{c \sin \beta}{\sin \gamma} \\ &= \frac{49 \sin 38°}{\sin 75°} \\ &= 31 \text{ m}\end{aligned}$$

Solve for a:

$$\begin{aligned}\frac{\sin \alpha}{a} &= \frac{\sin \gamma}{c} \\ a &= \frac{c \sin \alpha}{\sin \gamma} \\ &= \frac{49 \sin 67°}{\sin 75°} \\ &= 47 \text{ m}\end{aligned}$$

(*8-1*)

6. 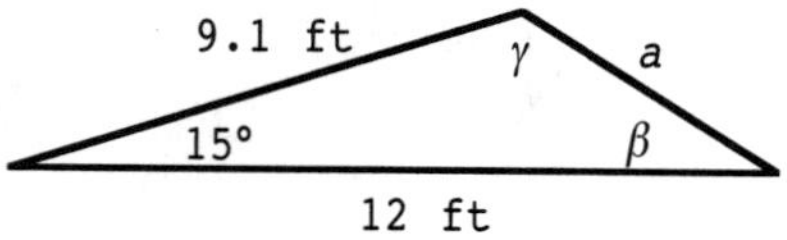

We are given two sides and the included angle (SAS). We use the law of cosines to find the third side, then the law of sines to find a second angle.

Solve for a:

$$\begin{aligned}a^2 &= b^2 + c^2 - 2bc \cos \alpha \\ &= (9.1)^2 + (12)^2 - 2(9.1)(12)\cos 15° \\ &= 15.8518 \\ a &= 4.00 \text{ ft}\end{aligned}$$

Solve for β:

$$\begin{aligned}\frac{\sin \alpha}{a} &= \frac{\sin \beta}{b} \\ \sin \beta &= \frac{b \sin \alpha}{a} \\ &= \frac{9.1 \sin 15°}{4.00} \\ &= 0.5888 \\ \beta &= \sin^{-1}(0.5888) = 36°\end{aligned}$$

There is another solution of $\sin \beta = 0.5915$ which deserves brief consideration, that is, $\beta = 180° - \sin^{-1}(0.5915) = 144°$. This would lead to a contradiction, however, since the largest side of the triangle, c, must be opposite the largest angle, γ, and this would lead to two obtuse angles in the triangle.

Solve for γ:

$$\begin{aligned} \alpha + \beta + \gamma &= 180° \\ \gamma &= 180° - (\alpha + \beta) \\ &= 180° - (15° + 36°) \\ &= 129° \end{aligned}$$

(8-1, 8-2)

7. We are given two sides and a non-included angle (SSA). From a rough sketch we see that there is only one triangle possible.

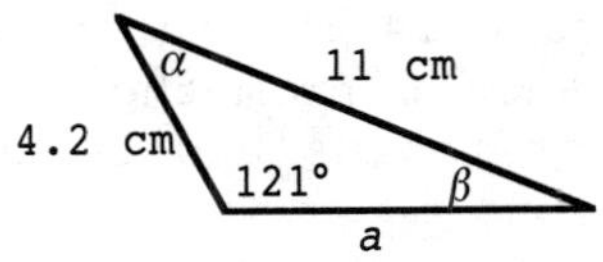

We use the law of sines.

Solve for β:

$$\begin{aligned} \frac{\sin \beta}{b} &= \frac{\sin \gamma}{c} \\ \sin \beta &= \frac{b \sin \gamma}{c} \\ &= \frac{4.2 \sin 121°}{11} \\ \beta &= \sin^{-1} \frac{4.2 \sin 121°}{11} \quad \beta \text{ is acute} \\ &= 19° \end{aligned}$$

Solve for α:

$$\begin{aligned} \alpha + \beta + \gamma &= 180° \\ \alpha &= 180° - (\beta + \gamma) \\ &= 180° - (19° + 121°) \\ &= 40° \end{aligned}$$

Solve for a:

$$\begin{aligned} \frac{\sin \alpha}{a} &= \frac{\sin \gamma}{c} \\ a &= \frac{c \sin \alpha}{\sin \gamma} \\ &= \frac{11 \sin 40°}{\sin 121°} \\ &= 8.2 \text{ cm} \end{aligned}$$

(8-1)

8.

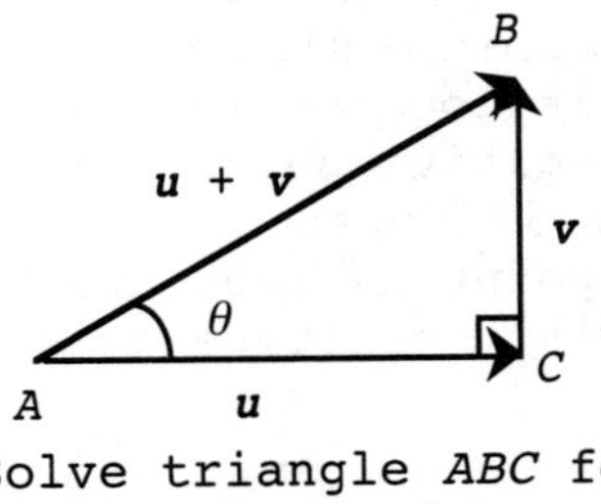

To find $|\boldsymbol{u} + \boldsymbol{v}|$: Apply the Pythagorean theorem to triangle ABC.

$$\begin{aligned} |\boldsymbol{u} + \boldsymbol{v}|^2 = AB^2 &= AC^2 + BC^2 \\ &= |\boldsymbol{u}|^2 + |\boldsymbol{v}|^2 \\ &= 160^2 + 55^2 \\ &= 28{,}625 \\ |\boldsymbol{u} + \boldsymbol{v}| &= \sqrt{28{,}625} = 170 \text{ mi/hr} \end{aligned}$$

Solve triangle ABC for θ

$$\begin{aligned} \tan \theta &= \frac{BC}{AC} = \frac{|\boldsymbol{v}|}{|\boldsymbol{u}|} \\ \theta &= \tan^{-1} \frac{|\boldsymbol{v}|}{|\boldsymbol{u}|} \quad \theta \text{ is acute} \\ &= \tan^{-1} \frac{55}{160} = 19° \end{aligned}$$

(8-3)

9. The algebraic vector $\langle a, b \rangle$ corresponds to the standard geometric vector with terminal point $P(a, b)$ and initial point $O(0, 0)$. The coordinates of the point $P(a, b)$ are given by

$$\begin{aligned} a &= x_B - x_A = 5 - 2 = 3 \\ b &= y_B - y_A = -1 - 6 = -7 \end{aligned}$$

Thus, $\langle a, b \rangle = \langle 3, -7 \rangle$

(8-3)

10. $|\mathbf{v}| = \sqrt{(-3)^2 + (-5)^2} = \sqrt{34}$ (8-3)

11. The graph is the set of all points on a line forming an angle of $\frac{\pi}{6}$ with the polar axis.

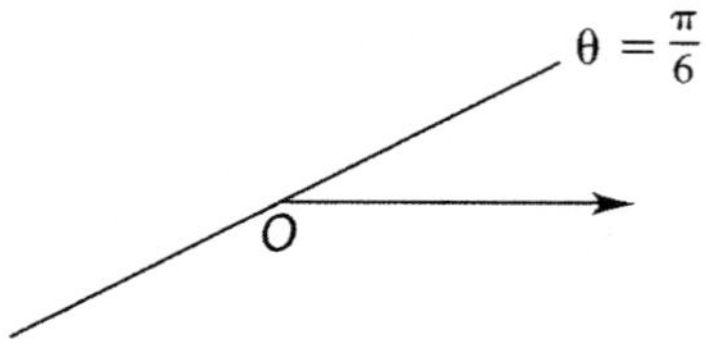

(8-4)

12. The graph is the set of all points at distance 6 from the pole—a circle of radius 6 with center at the pole.

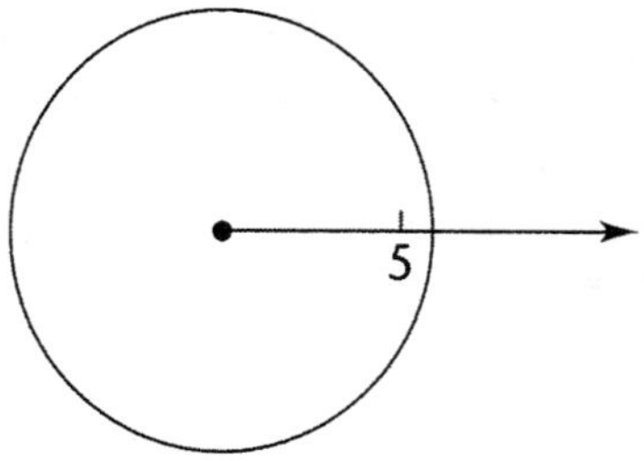

(8-4)

13.

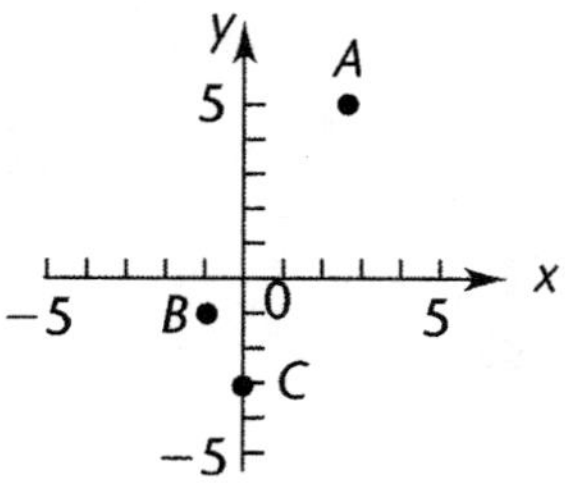

(8-5)

14. See figure.

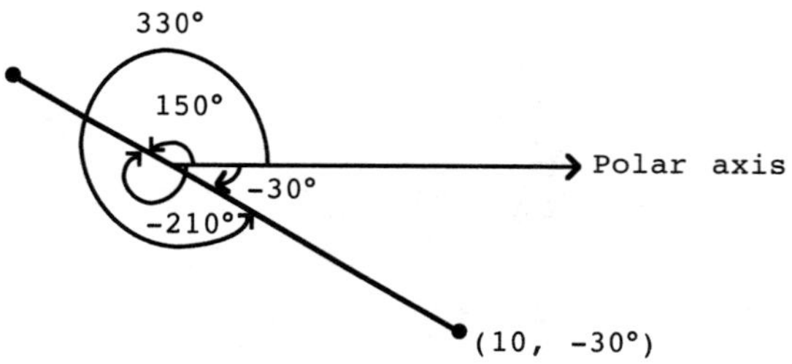

The point with coordinates (10, -30°) can equally well be described as (-10°, -210°) or (-10, 150°) or (10, 330°). Thus, (-10, -210°): The polar axis is rotated 210° clockwise (negative direction) and the point is located 10 units from the pole along the negative polar axis. (-10, 150°): The polar axis is rotated 150° counterclockwise (positive direction) and the point is located 10 units from the pole along the negative polar axis. (10, 330°): The polar axis is rotated 330° counterclockwise and the point is located 10 units from the pole along the positive polar axis. (8-4)

15.

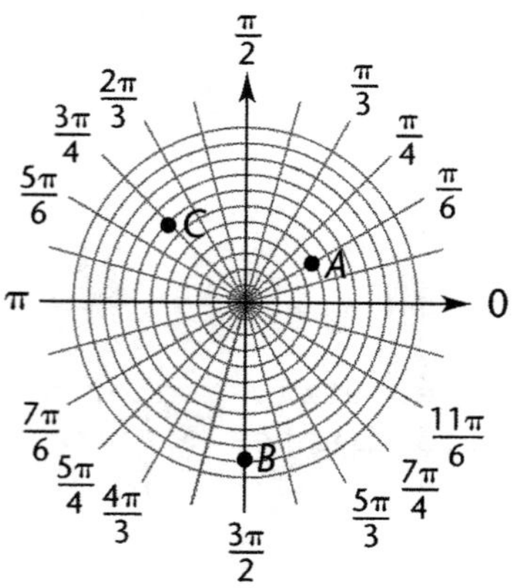

(8-5)

16. (A) A sketch shows that $1 - i\sqrt{3}$ is associated with a special 30°–60° triangle. Thus, by inspection, $r = 2$, $\theta = -60°$, and

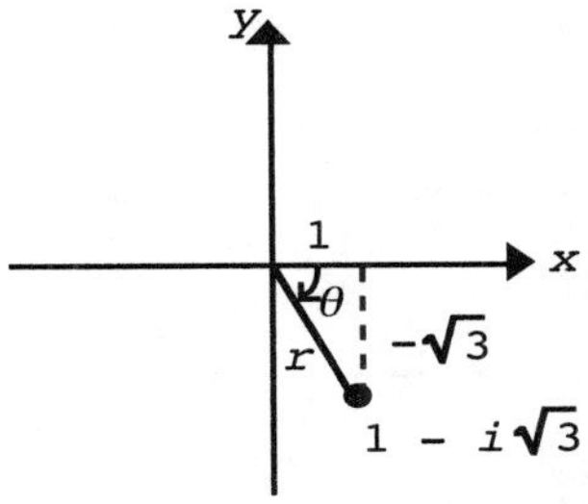

$$1 - i\sqrt{3} = 2[\cos(-60°) + i\sin(-60°)] = 2e^{(-60°)i}$$

(B) $$4e^{(-30°)i} = 4[\cos(-30°) + i\sin(-30°)] = 4\left[\frac{\sqrt{3}}{2} + i\left(-\frac{1}{2}\right)\right] = 2\sqrt{3} - 2i$$ (8-5)

17. (A) $\left[-\frac{1}{2} - \frac{\sqrt{3}}{2}i\right]^3 = (e^{-120°i})^3 = e^{-360°i} = \cos(-360°) + i\sin(-360°) = 1$

(B) is left to the student (8-5)

18. $$(2e^{15°i})^4 = 2^4e^{4(15°i)} = 16e^{60°i} = 16(\cos 60° + i\sin 60°) = 16\left(\frac{1}{2} + i\frac{\sqrt{3}}{2}\right) = 8 + 8i\sqrt{3}$$ (8-5)

19. If the triangle has an obtuse angle, then it must be the angle opposite the longest side; in this case, α. (8-2)

20.

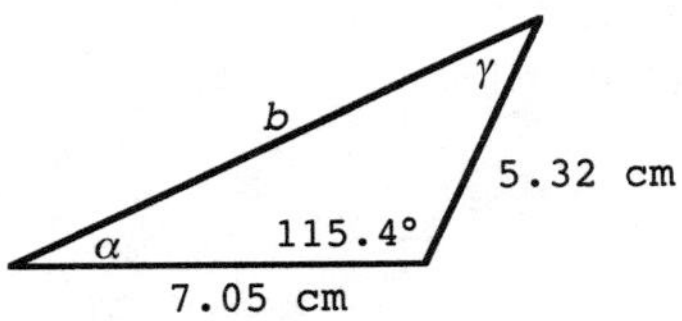

We are given two sides and the included angle (SAS). We use the law of cosines to find the third side, then the law of sines to find a second angle.

Solve for b:

$$\begin{aligned} b^2 &= a^2 + c^2 - 2ac\cos\beta \\ &= (5.32)^2 + (7.05)^2 - 2(5.32)(7.05)\cos 115.4° \\ &= 110.18018\ldots \\ b &= 10.5 \text{ cm} \end{aligned}$$

Solve for α:

Note that since the sum of α and γ is $180° - 115.4° = 64.6°$, both must be acute.

$$\begin{aligned} \frac{\sin\alpha}{a} &= \frac{\sin\beta}{b} \\ \sin\alpha &= \frac{a\sin\beta}{b} \\ &= \frac{5.32\sin 115.4°}{10.5} \\ \alpha &= \sin^{-1}\left(\frac{5.32\sin 115.4°}{10.5}\right) \\ &= 27.2° \end{aligned}$$

Solve for γ:

$$\begin{aligned} \alpha + \beta + \gamma &= 180° \\ \gamma &= 180° - (\alpha + \beta) \\ &= 180° - (27.2° + 115.4°) \\ &= 37.4° \end{aligned}$$ (8-2)

21. 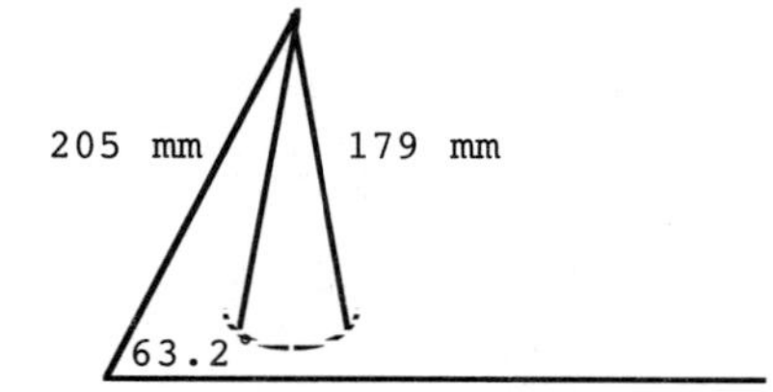

α is acute
$h = b \sin \alpha = 205 \sin 63.2° = 183$
$a = 179 < 183 = h$

We are given two sides and a non-included angle (SSA). If we try to draw a triangle with these values, we will be unsuccessful.
No solution. *(8-1)*

22. 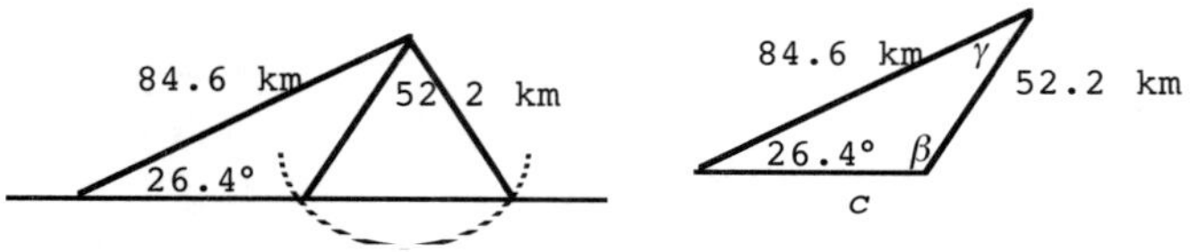

We are given two sides and a non-included angle (SSA). If we try to draw a triangle with these values, we find that two triangles are possible. Following the instructions, we choose the solution for which β is obtuse.

Solve for β:

$$\frac{\sin \beta}{b} = \frac{\sin \alpha}{a}$$

$$\sin \beta = \frac{b \sin \alpha}{a} = \frac{84.6 \sin 26.4°}{52.2} = 0.7206$$

β is chosen obtuse

$$\begin{aligned} \beta &= 180° - \sin^{-1} 0.7206 \\ &= 180° - 46.1° \\ &= 133.9° \end{aligned}$$

Solve for γ:

$$\begin{aligned} \gamma &= 180° - (\alpha + \beta) \\ &= 180° - (26.4° + 133.9°) \\ &= 19.7° \end{aligned}$$

Solve for c:

$$\frac{\sin \alpha}{a} = \frac{\sin \gamma}{c}$$

$$\begin{aligned} c &= \frac{a \sin \gamma}{\sin \alpha} \\ &= \frac{52.2 \sin 19.7°}{\sin 26.4°} \\ &= 39.6 \text{ km} \end{aligned}$$

(8-1)

23. 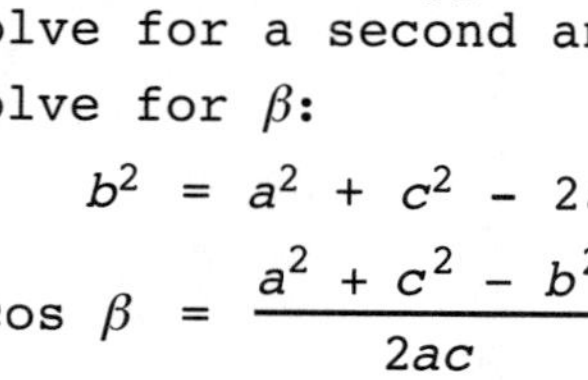

We are given three sides (SSS). We solve for the largest angle, β (largest because it is opposite the largest side, b) using the law of cosines. We then solve for a second angle using the law of sines.

Solve for β:

$$b^2 = a^2 + c^2 - 2ac \cos \beta$$

$$\cos \beta = \frac{a^2 + c^2 - b^2}{2ac}$$

$$\beta = \cos^{-1}\left(\frac{a^2 + c^2 - b^2}{2ac}\right)$$

$$= \cos^{-1}\left(\frac{(19.0)^2 + (26.1)^2 - (27.8)^2}{2(19.0)(26.1)}\right) = 74.2°$$

Solve for α:
We use the law of sines. Since α must be smaller than β, α is acute.

$$\frac{\sin\alpha}{a} = \frac{\sin\beta}{b}$$
$$\sin\alpha = \frac{a\sin\beta}{b} = \frac{19.0\sin 74.2^\circ}{27.8}$$
$$\alpha = \sin^{-1}\left(\frac{19.0\sin 74.2^\circ}{27.8}\right) = 41.1^\circ$$

Solve for γ:

$$\alpha + \beta + \gamma = 180^\circ$$
$$\gamma = 180^\circ - (\alpha + \beta) = 180^\circ - (41.1^\circ + 74.2^\circ) = 64.7^\circ$$

Note: these answers differ slightly from those given in the text. This results from having solved for the angles in a different order. *(8-1, 8-2)*

24. The sum of all of the force vectors must be the zero vector for the object to remain at rest. *(8-3)*

25.

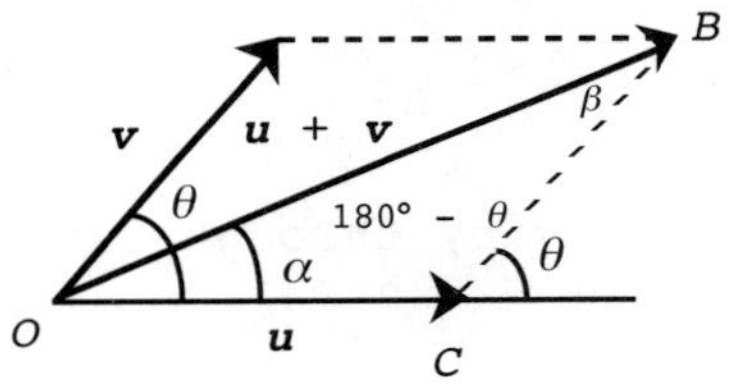

$\theta = 57.2^\circ$.
Hence, $\angle OCB = 180^\circ - \theta = 180^\circ - 57.2^\circ = 122.8^\circ$
We can find $|\boldsymbol{u} + \boldsymbol{v}|$ using the law of cosines:

$$|\boldsymbol{u} + \boldsymbol{v}|^2 = |\boldsymbol{u}|^2 + |\boldsymbol{v}|^2 - 2|\boldsymbol{u}||\boldsymbol{v}|\cos(OCB)$$
$$= (75.2)^2 + (34.2)^2 - 2(75.2)(34.2)\cos 122.8^\circ$$
$$= 9611.0537\ldots$$
$$|\boldsymbol{u} + \boldsymbol{v}| = \sqrt{9611.0537\ldots} = 98.0 \text{ kg}$$

To find α, we use the law of sines:

$$\frac{\sin\alpha}{|\boldsymbol{v}|} = \frac{\sin(OCB)}{|\boldsymbol{u} + \boldsymbol{v}|}$$
$$\frac{\sin\alpha}{34.2} = \frac{\sin 122.8^\circ}{98.0}$$
$$\sin\alpha = \frac{34.2}{98.0}\sin 122.8^\circ$$
$$\alpha = \sin^{-1}\left(\frac{34.2}{98.0}\sin 122.8^\circ\right) = 17.1^\circ$$

(8-3)

26. (A) $\boldsymbol{u} = \langle -3, 9\rangle = \langle -3, 0\rangle + \langle 0, 9\rangle = -3\langle 1, 0\rangle + 9\langle 0, 1\rangle = -3\boldsymbol{i} + 9\boldsymbol{j}$

(B) $\boldsymbol{v} = \langle 0, -2\rangle = -2\langle 0, 1\rangle = -2\boldsymbol{j}$ *(8-3)*

27. (A) $\boldsymbol{u} - \boldsymbol{v} = \langle -2, 3\rangle - \langle 2, -4\rangle = \langle -2, 3\rangle + \langle -2, 4\rangle = \langle -4, 7\rangle$

(B) $3\boldsymbol{u} - \boldsymbol{v} + 2\boldsymbol{w} = 3\langle -2, 3\rangle - \langle 2, -4\rangle + 2\langle -3, 0\rangle = \langle -6, 9\rangle + \langle -2, 4\rangle + \langle -6, 0\rangle = \langle -14, 13\rangle$ *(8-3)*

28. (A) $\boldsymbol{u} - \boldsymbol{v} = (\boldsymbol{i} - 2\boldsymbol{j}) - (3\boldsymbol{i} + 2\boldsymbol{j}) = \boldsymbol{i} - 2\boldsymbol{j} - 3\boldsymbol{i} - 2\boldsymbol{j} = -2\boldsymbol{i} - 4\boldsymbol{j}$

(B) $3\boldsymbol{u} - \boldsymbol{v} + 2\boldsymbol{w} = 3(\boldsymbol{i} - 2\boldsymbol{j}) - (3\boldsymbol{i} + 2\boldsymbol{j}) + 2(-\boldsymbol{j}) = 3\boldsymbol{i} - 6\boldsymbol{j} - 3\boldsymbol{i} - 2\boldsymbol{j} - 2\boldsymbol{j} = -10\boldsymbol{j}$ *(8-3)*

29. $|\mathbf{v}| = \sqrt{(-1)^2 + (-3)^2} = \sqrt{10}$

Then $\mathbf{u} = \frac{1}{|\mathbf{v}|}\mathbf{v} = \frac{1}{\sqrt{10}}\langle -1, -3\rangle = \left\langle \frac{-1}{\sqrt{10}}, \frac{-3}{\sqrt{10}} \right\rangle$ (8-3)

30. We set up a table that indicates how r varies as we let θ vary through each set of quadrant values.

θ	$\cos\theta$	$4\cos\theta$	$6 + 4\cos\theta$
0 to $\frac{\pi}{2}$	1 to 0	4 to 0	10 to 6
$\frac{\pi}{2}$ to π	0 to -1	0 to -4	6 to 2
π to $\frac{3\pi}{2}$	-1 to 0	-4 to 0	2 to 6
$\frac{3\pi}{2}$ to 2π	0 to 1	0 to 4	6 to 10

Check:

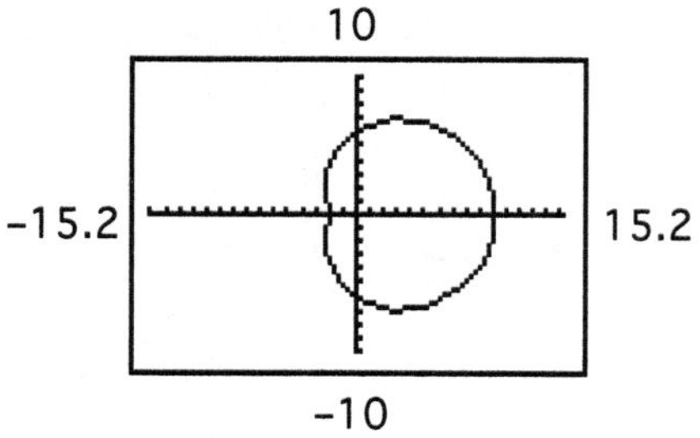

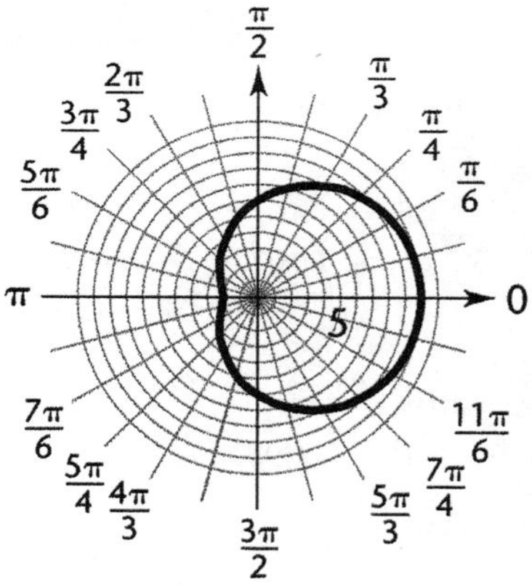

(8-4)

31. We set up a table that indicates how r varies as we let θ vary through each set of quadrant values.

θ	$\sin\theta$	$8\sin\theta$	$8 + 8\sin\theta$
0 to $\frac{\pi}{2}$	0 to 1	0 to 8	8 to 16
$\frac{\pi}{2}$ to π	1 to 0	8 to 0	16 to 8
π to $\frac{3\pi}{2}$	0 to -1	0 to -8	8 to 0
$\frac{3\pi}{2}$ to 2π	-1 to 0	-8 to 0	0 to 8

Check:

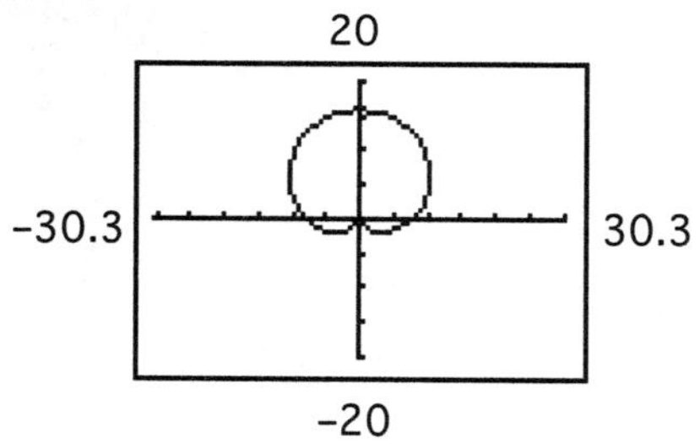

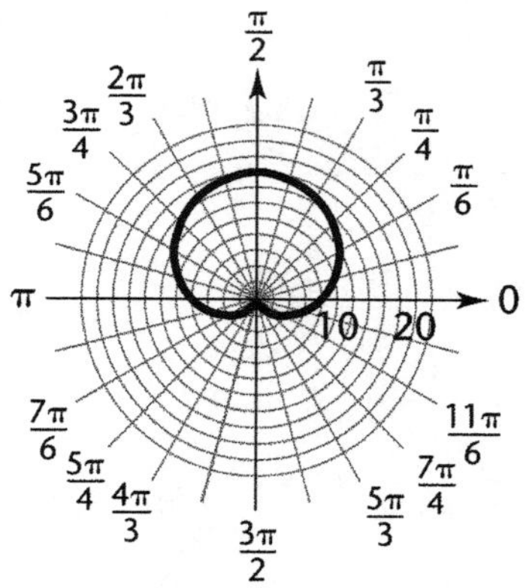

(8-4)

32. We set up a table that indicates how r varies as we let 2θ vary through each set of quadrant values.

θ	2θ	$\cos 2\theta$	$10 \cos 2\theta$
0 to $\frac{\pi}{4}$	0 to $\frac{\pi}{2}$	1 to 0	10 to 0
$\frac{\pi}{4}$ to $\frac{\pi}{2}$	$\frac{\pi}{2}$ to π	0 to −1	0 to −10
$\frac{\pi}{2}$ to $\frac{3\pi}{4}$	π to $\frac{3\pi}{2}$	−1 to 0	−10 to 0
$\frac{3\pi}{4}$ to π	$\frac{3\pi}{2}$ to 2π	0 to 1	0 to 10
π to $\frac{5\pi}{4}$	2π to $\frac{5\pi}{2}$	1 to 0	10 to 0
$\frac{5\pi}{4}$ to $\frac{3\pi}{2}$	$\frac{5\pi}{2}$ to 3π	0 to −1	0 to −10
$\frac{3\pi}{2}$ to $\frac{7\pi}{4}$	3π to $\frac{7\pi}{2}$	−1 to 0	−10 to 0
$\frac{7\pi}{4}$ to 2π	$\frac{7\pi}{2}$ to 4π	0 to 1	0 to 10

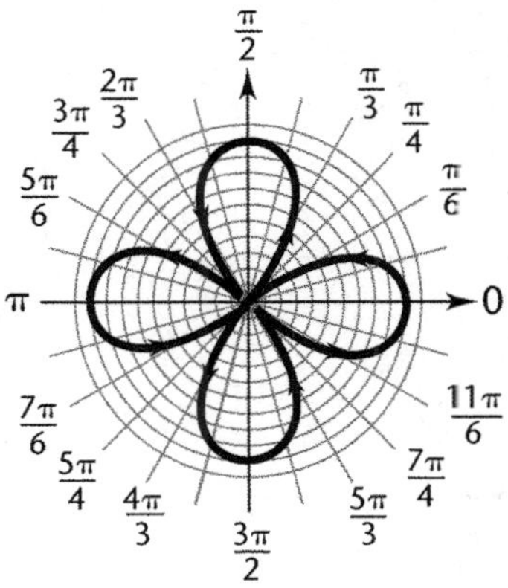

Check:

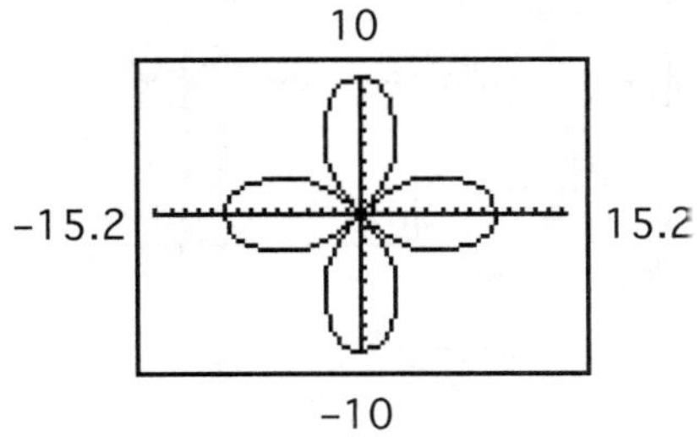

(8-4)

33. We set up a table that indicates how r varies as we let 3θ vary through each set of quadrant values.

θ	3θ	$\sin 3\theta$	$8 \sin 3\theta$
0 to $\frac{\pi}{6}$	0 to $\frac{\pi}{2}$	0 to 1	0 to 8
$\frac{\pi}{6}$ to $\frac{\pi}{3}$	$\frac{\pi}{2}$ to π	1 to 0	8 to 0
$\frac{\pi}{3}$ to $\frac{\pi}{2}$	π to $\frac{3\pi}{2}$	0 to −1	0 to −8
$\frac{\pi}{2}$ to $\frac{2\pi}{3}$	$\frac{3\pi}{2}$ to 2π	−1 to 0	−8 to 0
$\frac{2\pi}{3}$ to $\frac{5\pi}{6}$	2π to $\frac{5\pi}{2}$	0 to 1	0 to 8
$\frac{5\pi}{6}$ to π	$\frac{5\pi}{2}$ to 3π	1 to 0	8 to 0
π to 2π	repeats curve already drawn		

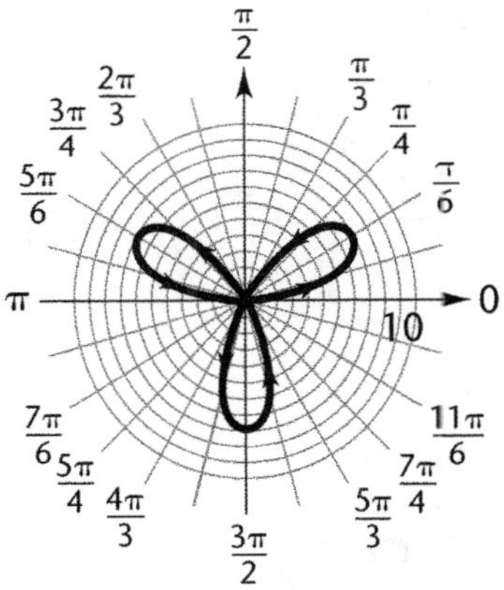

Check:

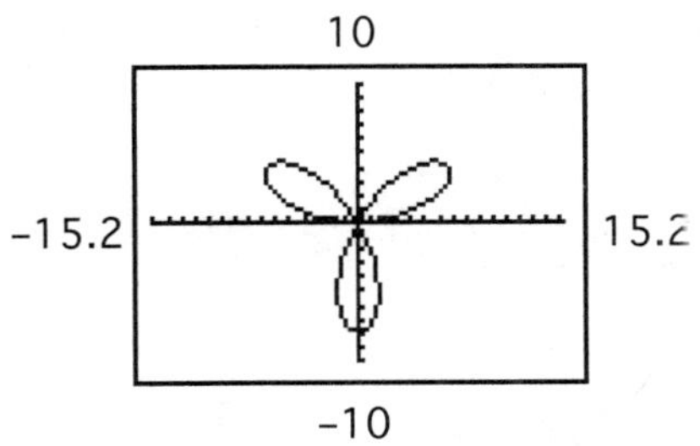

(8-4)

34.

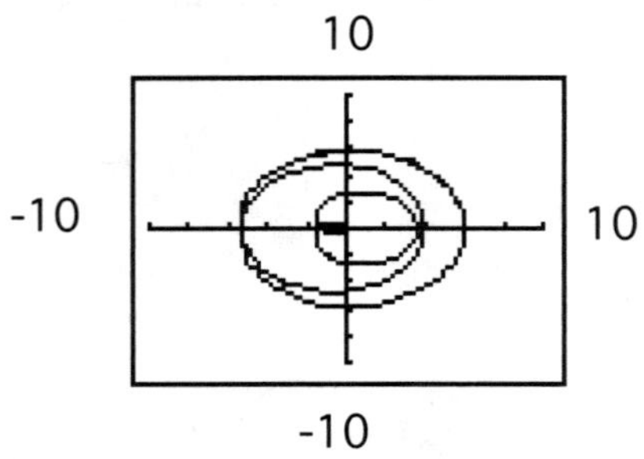

(8-4)

35.

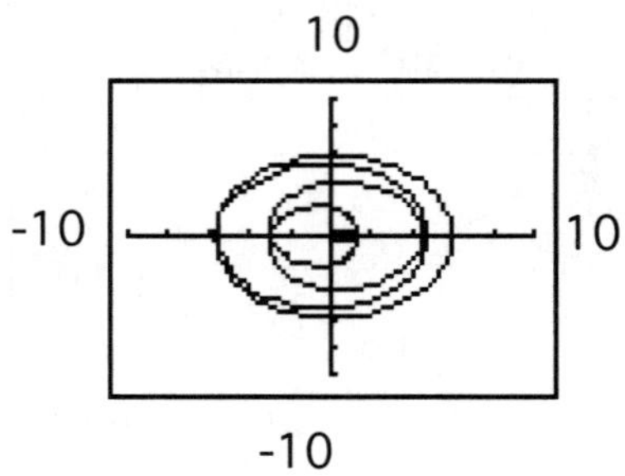

(8-4)

36. $n = 1$

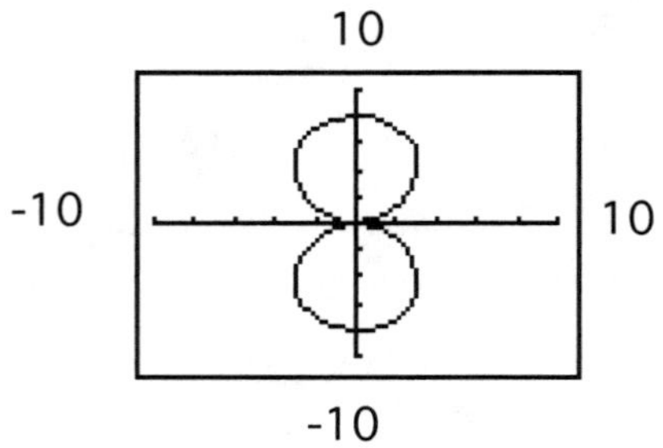

$n = 2$

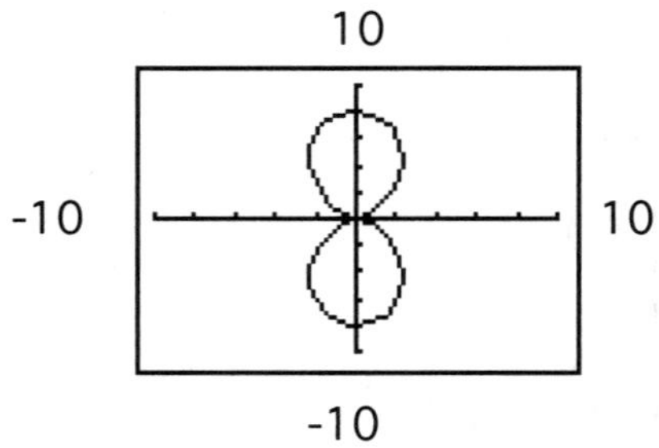

$n = 3$

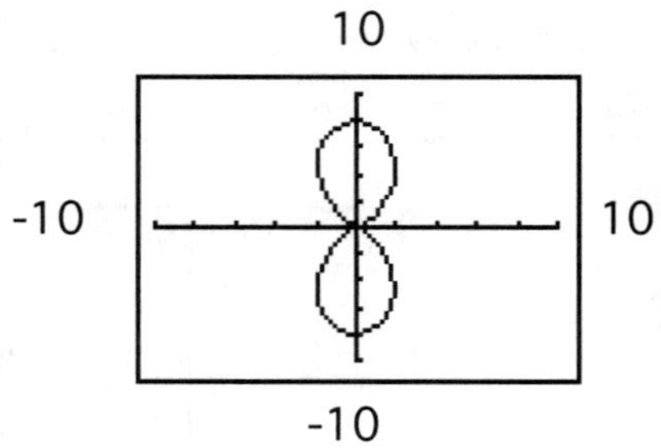

Generalizing from these graphs, we expect 2 leaves for all n. (8-4)

37. (A) $e = 0.55$

$r = \dfrac{3}{1 - 0.55\cos\theta}$

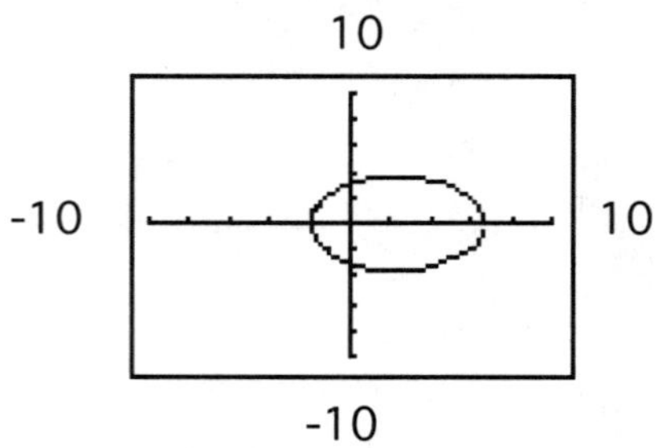

An ellipse

(B) $e = 1$

$r = \dfrac{3}{1 - \cos\theta}$

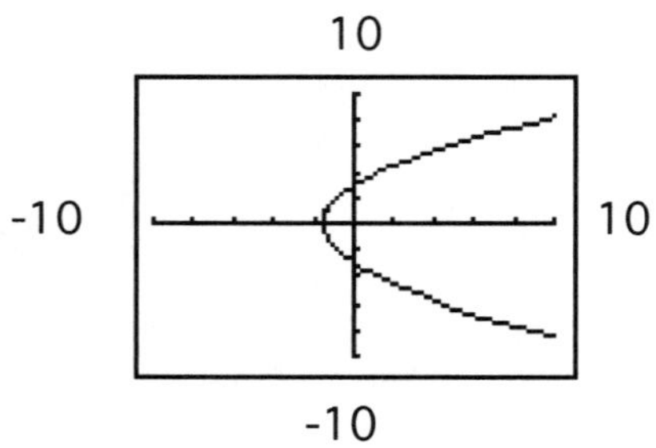

A parabola

(C) $e = 1.7$

$r = \dfrac{3}{1 - 1.7\cos\theta}$

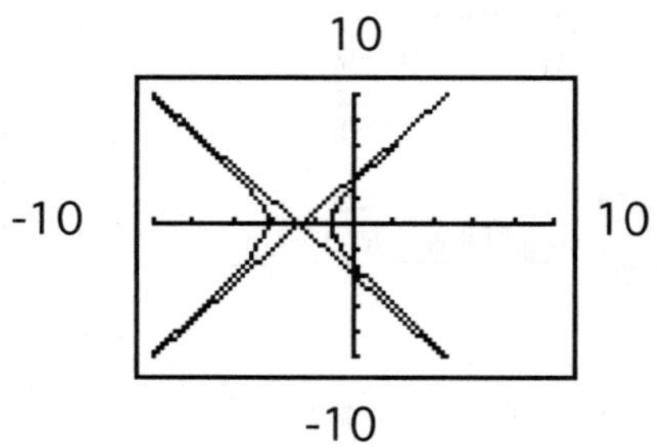

A hyperbola (8-4)

38. $x^2 + y^2 = 6x$

Use $r^2 = x^2 + y^2$ and $x = r\cos\theta$.

$r^2 = 6r\cos\theta$

$r^2 - 6r\cos\theta = 0$

$r(r - 6\cos\theta) = 0$

$r = 0$ or $r - 6\cos\theta = 0$

The graph of $r = 0$ is the pole. Since the pole is included in the graph of $r - 6\cos\theta = 0$, we can discard $r = 0$ and keep only

$r - 6\cos\theta = 0$ or $r = 6\cos\theta$ (8-4)

39. $r = 5\cos\theta$

We multiply both sides by r, which adds the pole to the graph. But the pole is already part of the graph, so we have changed nothing.

$r^2 = 5r\cos\theta$

But $r^2 = x^2 + y^2$ and $r\cos\theta = x$. Hence, $x^2 + y^2 = 5x$. (8-5)

40. A sketch shows that $-1 + i$ is associated with a special 45° triangle. Thus, by inspection, $r = \sqrt{2}$, $\theta = 135°$, and

$$-1 + i = \sqrt{2}(\cos 135° + i \sin 135°) = \sqrt{2}\, e^{135° i}$$

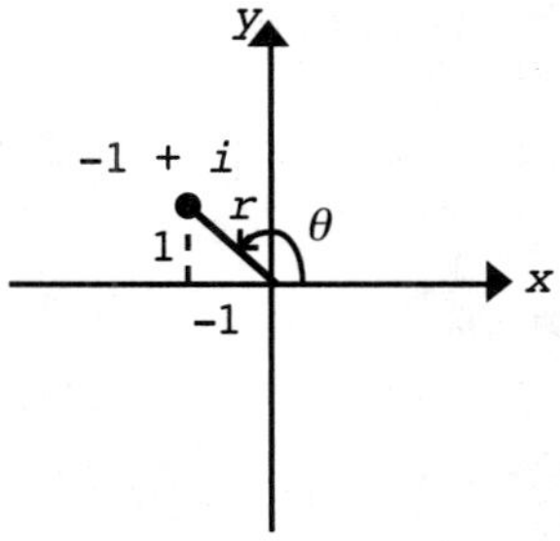

A sketch shows that $-1 - i\sqrt{3}$ is associated with a special 30°-60° triangle. Thus, by inspection, $r = 2$, $\theta = -120°$, and

$$-1 - i\sqrt{3} = 2[\cos(-120°) + i \sin(-120°)] = 2e^{(-120°)i}$$

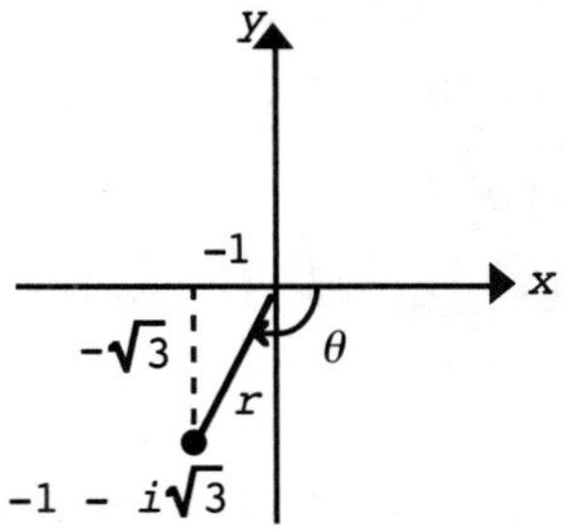

A sketch shows that 5 is associated with the special quadrantal angle 0°. Thus, by inspection, $r = 5$, $\theta = 0°$, and

$$5 = 5 + 0i = 5(\cos 0° + i \sin 0°) = 5e^{0° i}$$

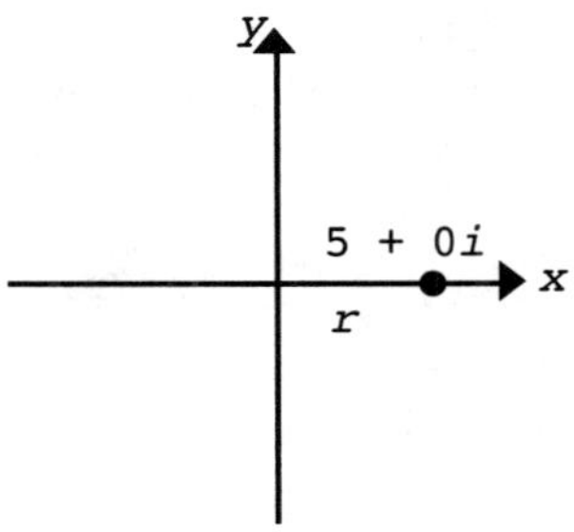

(8-5)

41.
$$z_1 = x + iy = \sqrt{2}\, e^{(\pi/4)i} = \sqrt{2}\left(\cos\frac{\pi}{4} + i \sin\frac{\pi}{4}\right) = \sqrt{2}\left(\frac{1}{\sqrt{2}} + i\,\frac{1}{\sqrt{2}}\right) = 1 + i$$

$$z_2 = x + iy = 3e^{210° i} = 3(\cos 210° + i \sin 210°) = 3\left[-\frac{\sqrt{3}}{2} + i\left(-\frac{1}{2}\right)\right] = \frac{-3\sqrt{3}}{2} - \frac{3}{2}i$$

$$z_3 = x + iy = 2e^{(-2\pi/3)i} = 2\left[\cos\left(-\frac{2\pi}{3}\right) + i \sin\left(-\frac{2\pi}{3}\right)\right] = 2\left[-\frac{1}{2} + i\left(-\frac{\sqrt{3}}{2}\right)\right] = -1 - i\sqrt{3}$$

(8-5)

42. (A) $z_1 z_2 = 8e^{25° i} \cdot 4e^{19° i} = 8 \cdot 4e^{i(25° + 19°)} = 32e^{44° i}$

(B) $\dfrac{z_1}{z_2} = \dfrac{8e^{25° i}}{4e^{19° i}} = \dfrac{8}{4}e^{(25° - 19°)i} = 2e^{6° i}$ (8-5)

43. (A) $(1 + i\sqrt{3})^4 = (2e^{60° i})^4 = 2^4 e^{4 \cdot 60° i} = 16e^{240° i}$

$$= 16[\cos 240° + i \sin 240°] = 16\left[-\frac{1}{2} + i\left(-\frac{\sqrt{3}}{2}\right)\right] = -8 - 8i\sqrt{3}$$

(B) The calculator steps are not shown here. $-8 - 8i\sqrt{3}$ is approximated on a calculator as $-8 - 13.86i$. (8-5)

44. First write i in polar form.

$i = 0 + 1i = 1e^{90°i}$

Using the nth-root theorem, all three cube roots of $1e^{90°i}$ are given by

$1^{1/3}e^{(90°/3 + k360°/3)i} = 1^{1/3}e^{(30° + k120°)i} \quad k = 0, 1, 2$

Thus,

$$w_1 = 1^{1/3}e^{(30° + 0 \cdot 120°)i} = e^{30°i}$$
$$w_2 = 1^{1/3}e^{(30° + 1 \cdot 120°)i} = e^{150°i}$$
$$w_3 = 1^{1/3}e^{(30° + 2 \cdot 120°)i} = e^{270°i}$$

Now convert to rectangular form:

$$e^{30°i} = 1(\cos 30° + i \sin 30°) = \frac{\sqrt{3}}{2} + \frac{1}{2}i$$
$$e^{150°i} = 1(\cos 150° + i \sin 150°) = -\frac{\sqrt{3}}{2} + \frac{1}{2}i$$
$$e^{270°i} = 1(\cos 270° + i \sin 270°) = -i$$

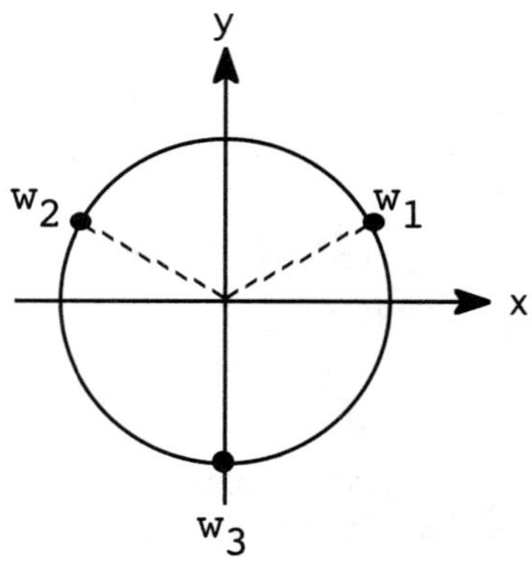

These roots lie on a circle of radius 1, equally spaced so that the angle between successive roots is 120°. (8-5)

45. First write $-4\sqrt{3} + 4i$ in polar form.

$-4\sqrt{3} + 4i = 8e^{150°i}$

Using the nth-root theorem, all three cube roots of $8e^{150°i}$ are given by

$8^{1/3}e^{(150°/3 + k360°/3)i} = 8^{1/3}e^{(50° + k \cdot 120°)i} \quad k = 0, 1, 2$

Thus,

$$w_1 = 8^{1/3}e^{(50° + 0 \cdot 120°)i} = 2e^{50°i}$$
$$w_2 = 8^{1/3}e^{(50° + 1 \cdot 120°)i} = 2e^{170°i}$$
$$w_3 = 8^{1/3}e^{(50° + 2 \cdot 120°)i} = 2e^{290°i}$$

(8-5)

46. $(4e^{15°i})^2 = 4^2e^{2 \cdot 15°i} = 16e^{30°i} = 16(\cos 30° + i \sin 30°) = 16\left(\frac{\sqrt{3}}{2} + i \cdot \frac{1}{2}\right)$

$= 8\sqrt{3} + 8i$

Thus, $4e^{15°i}$ is a square root of $8\sqrt{3} + 8i$. (8-5)

47. $(x, y) = (5.17, -2.53)$

$r = \sqrt{x^2 + y^2} = \sqrt{5.17^2 + (-2.53)^2} = 5.76$

$\tan \theta = \frac{y}{x} = \frac{-2.53}{5.17}$

θ is a fourth quadrant angle and is to be chosen so that $-180° < \theta \le 180°$.

$\theta = \tan^{-1}\frac{-2.53}{5.17} = -26.08°$

$(r, \theta) = (5.76, -26.08°)$ (8-4)

48. $(r, \theta) = (5.81, -2.72)$

$x = r \cos \theta = 5.81 \cos(-2.72) = -5.30$

$y = r \sin \theta = 5.81 \sin(-2.72) = -2.38$

$(x, y) = (-5.30, -2.38)$ (8-4)

49. $r = \sqrt{(-3.18)^2 + (4.19)^2} = 5.26$

$\theta = 180° + \tan^{-1}\dfrac{4.19}{-3.18} = 127.20°$ since $(-3.18, 4.19)$ is in the second quadrant.

$-3.18 + 4.19i = 5.26(\cos 127.20° + i \sin 127.20°) = 5.26e^{127.20° i}$ (8-5)

50. $x + iy = 7.63e^{(-162.27°)i}$

$= 7.63[\cos(-162.27°) + i \sin(-162.27°)]$

$= -7.27 - 2.32i$ (8-5)

51. (A) There are a total of three cube roots and they are spaced equally around a circle of radius 2, so that the angle between successive roots is 120°.

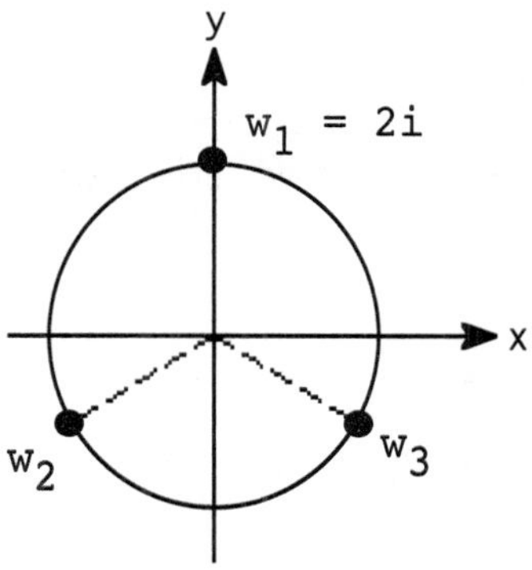

(B) Since $w_1 = 2e^{90° i}$, $w_2 = 2e^{(90° + 120°)i} = 2e^{210° i}$, and $w_3 = 2e^{(210° + 120°)i} = 2e^{330° i}$

Thus, $w_2 = 2(\cos 210° + i \sin 210°) = 2\left[-\dfrac{\sqrt{3}}{2} + i\left(-\dfrac{1}{2}\right)\right] = -\sqrt{3} - i$

$w_3 = 2(\cos 330° + i \sin 330°) = 2\left[\dfrac{\sqrt{3}}{2} + i\left(-\dfrac{1}{2}\right)\right] = \sqrt{3} - i$

(C) $(2i)^3 = 8i^3 = 8(-i) = -8i$

$$\begin{aligned}(-\sqrt{3} - i)^3 &= (-\sqrt{3} - i)(-\sqrt{3} - i)(-\sqrt{3} - i)\\ &= (3 + 2i\sqrt{3} + i^2)(-\sqrt{3} - i)\\ &= (2 + 2i\sqrt{3})(-\sqrt{3} - i)\\ &= -2\sqrt{3} - 2i - 6i + 2\sqrt{3}\\ &= -8i\end{aligned}$$

$$\begin{aligned}(\sqrt{3} - i)^3 &= (\sqrt{3} - i)(\sqrt{3} - i)(\sqrt{3} - i)\\ &= (3 - 2i\sqrt{3} + i^2)(\sqrt{3} - i)\\ &= (2 - 2i\sqrt{3})(\sqrt{3} - i)\\ &= 2\sqrt{3} - 2i - 6i + 2\sqrt{3}i^2\\ &= -8i\end{aligned}$$

(8-5)

52.

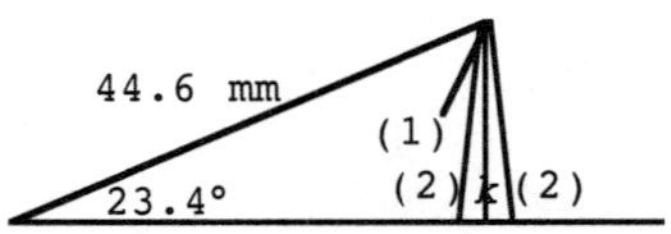

From the diagram, we can see that k = altitude of any possible triangle. Thus,

$\sin \alpha = \dfrac{k}{b}$ $\quad k = b \sin \alpha$

$k = (44.6 \text{ mm})\sin 23.4° = 17.7 \text{ mm}$

If $0 < a < k$, there is no solution: (1) in above diagram.
If $a = k$, there is one solution.
If $k < a < b$, there are two solutions: (2) in above diagram. (8-1)

53. See Exercise 8-2, Problem 29, for a solution. (8-2)

54. (A) $\mathbf{u} + \mathbf{v} = \langle a, b\rangle + \langle c, d\rangle$
$= \langle a + c, b + d\rangle$ Definition of vector addition
$= \langle c + a, d + b\rangle$ Commutative property for addition of real numbers
$= \langle c, d\rangle + \langle a, b\rangle$ Definition of vector addition
$= \mathbf{v} + \mathbf{u}$

(B) $m(\mathbf{u} + \mathbf{v}) = m(\langle a, b\rangle + \langle c, d\rangle)$
$= m\langle a + c, b + d\rangle$ Definition of vector addition
$= \langle m(a + c), m(b + d)\rangle$ Definition of scalar multiplication
$= \langle ma + mc, mb + md\rangle$ Distributive property for real numbers
$= \langle ma, mb\rangle + \langle mc, md\rangle$ Definition of vector addition
$= m\langle a, b\rangle + m\langle c, d\rangle$ Definition of scalar multiplication
$= m\mathbf{u} + m\mathbf{v}$ (8-3)

55. We set up a table that indicates how r varies as we let $\frac{\theta}{2}$ vary through each set of quadrant values.

θ	$\frac{\theta}{2}$	$\cos\frac{\theta}{2}$	$4\cos\frac{\theta}{2}$	$4 + 4\cos\frac{\theta}{2}$
0 to $\frac{\pi}{2}$	0 to $\frac{\pi}{4}$	1 to $\frac{\sqrt{2}}{2}$	4 to $2\sqrt{2}$	8 to $4 + 2\sqrt{2}$
$\frac{\pi}{2}$ to π	$\frac{\pi}{4}$ to $\frac{\pi}{2}$	$\frac{\sqrt{2}}{2}$ to 0	$2\sqrt{2}$ to 0	$4 + 2\sqrt{2}$ to 4
π to $\frac{3\pi}{2}$	$\frac{\pi}{2}$ to $\frac{3\pi}{4}$	0 to $-\frac{\sqrt{2}}{2}$	0 to $-2\sqrt{2}$	4 to $4 - 2\sqrt{2}$
$\frac{3\pi}{2}$ to 2π	$\frac{3\pi}{4}$ to π	$-\frac{\sqrt{2}}{2}$ to -1	$-2\sqrt{2}$ to -4	$4 - 2\sqrt{2}$ to 0
2π to $\frac{5\pi}{2}$	π to $\frac{5\pi}{4}$	-1 to $-\frac{\sqrt{2}}{2}$	-4 to $-2\sqrt{2}$	0 to $4 - 2\sqrt{2}$
$\frac{5\pi}{2}$ to 3π	$\frac{5\pi}{4}$ to $\frac{3\pi}{2}$	$-\frac{\sqrt{2}}{2}$ to 0	$-2\sqrt{2}$ to 0	$4 - 2\sqrt{2}$ to 4
3π to $\frac{7\pi}{2}$	$\frac{3\pi}{2}$ to $\frac{7\pi}{4}$	0 to $\frac{\sqrt{2}}{2}$	0 to $2\sqrt{2}$	4 to $4 + 2\sqrt{2}$
$\frac{7\pi}{2}$ to 4π	$\frac{7\pi}{4}$ to 2π	$\frac{\sqrt{2}}{2}$ to 1	$2\sqrt{2}$ to 4	$4 + 2\sqrt{2}$ to 8

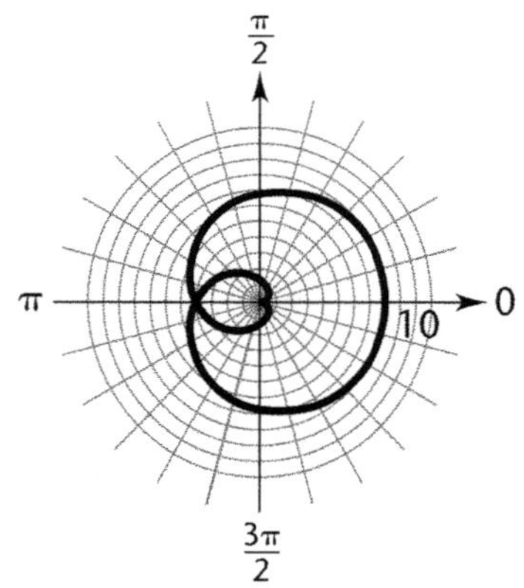

(B)

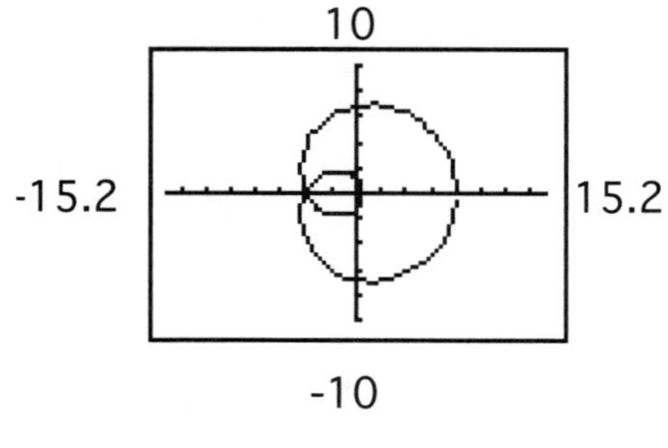

(8-4)

56. (A)

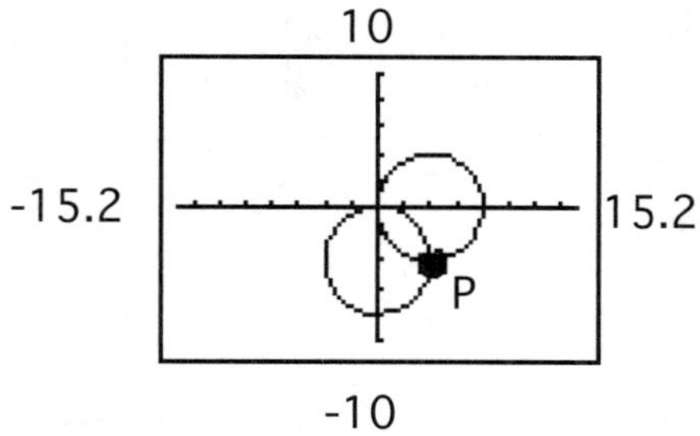

The coordinates of P represent a simultaneous solution.

(B) We solve the system

$r = -8 \sin \theta \qquad 0 \le \theta \le \pi$

$r = 8 \cos \theta$

by equating the right sides:

$-8 \sin \theta = 8 \cos \theta$

$-\sin \theta = \cos \theta$

$-1 = \tan \theta$

The only solution of this equation, $0 \le \theta \le \pi$, is $\theta = \frac{3\pi}{4}$. If we substitute this in either of the original equations, we get

$$r = -8 \sin \frac{3\pi}{4} = 8 \cos \frac{3\pi}{4} = 8\left(-\frac{\sqrt{2}}{2}\right) = -4\sqrt{2}$$

The coordinates of P are $\left(-4\sqrt{2}, \frac{3\pi}{4}\right)$.

(C) The pole is not a simultaneous solution, because the two graphs go through the pole at different values of θ. As (0, 0) the pole satisfies the first equation; as $\left(0, \frac{\pi}{2}\right)$ it satisfies the second; it is not a solution of the system.

(8-4)

57. The solution to $x^8 - 1 = 0$ or $x^8 = 1$ will be the eight eighth roots of 1. First write 1 in polar form:

$1 = 1e^{0°i}$

Using the nth-root theorem, all eight eighth roots of 1 are given by

$1^{1/8}e^{(0°/8 + k \cdot 360°/8)i} = e^{k \cdot 45°i} \qquad k = 0, 1, 2, 3, 4, 5, 6, 7$

Thus,

$w_1 = e^{0 \cdot 45°i} = e^{0°i} = \cos 0° - i \sin 0° = 1$

$w_2 = e^{1 \cdot 45°i} = e^{45°i} = \cos 45° + i \sin 45° = \frac{\sqrt{2}}{2} + i\frac{\sqrt{2}}{2}$

$w_3 = e^{2 \cdot 45°i} = e^{90°i} = \cos 90° + i \sin 90° = i$

$w_4 = e^{3 \cdot 45°i} = e^{135°i} = \cos 135° + i \sin 135° = -\frac{\sqrt{2}}{2} + i\frac{\sqrt{2}}{2}$

$w_5 = e^{4 \cdot 45°i} = e^{180°i} = \cos 180° + i \sin 180° = -1$

$w_6 = e^{5 \cdot 45°i} = e^{225°i} = \cos 225° + i \sin 225° = -\frac{\sqrt{2}}{2} - i\frac{\sqrt{2}}{2}$

$w_7 = e^{6 \cdot 45°i} = e^{270°i} = \cos 270° + i \sin 270° = -i$

$w_8 = e^{7 \cdot 45°i} = e^{315°i} = \cos 315° + i \sin 315° = \frac{\sqrt{2}}{2} - i\frac{\sqrt{2}}{2}$ (8-5)

58. The linear factors of $x^3 - 8i$ will, by the factor theorem, be $x - x_i$, where x_i are zeros of $x^3 - 8i$, that is, $x^3 - 8i = 0$ or $x^3 = 8i$. Hence the x_i are the three cube roots of $8i$, or $8e^{90°i}$.

Using the nth-root theorem, all three cube roots of $8e^{90°i}$ are given by

$8^{1/3}e^{(90°/3 + k \cdot 360°/3)i} = 2e^{(30° + k \cdot 120°)i} \qquad k = 0, 1, 2$

Thus,

$w_1 = 2e^{(30° + 0 \cdot 120°)i} = 2e^{30°i} = 2(\cos 30° + i \sin 30°) = 2\left(\frac{\sqrt{3}}{2} + i\frac{1}{2}\right) = \sqrt{3} + i$

$$w_2 = 2e^{(30° + 1 \cdot 120°)i} = 2e^{150°i} = 2(\cos 150° + i \sin 150°) = 2\left(-\frac{\sqrt{3}}{2} + i\frac{1}{2}\right) = -\sqrt{3} + i$$

$$w_3 = 2e^{(30° + 2 \cdot 120°)i} = 2e^{270°i} = 2(\cos 270° + i \sin 270°) = 2(0 - 1i) = -2i$$

Hence, $x^3 - 8i = [x - (\sqrt{3} + i)][x - (\sqrt{3} + i)](x + 2i)$ (8-5)

59. Sketch a figure.

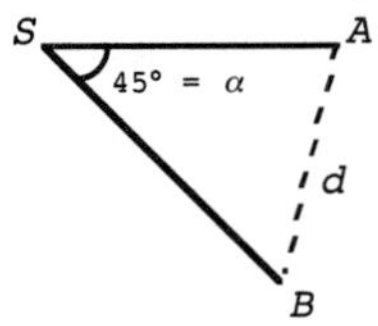

Using $D = rt$, we have

Plane 1: r = 256 miles per hour, t = 2 hours,
D = 256(2) = 512 miles = SA

Plane 2: r = 304 miles per hour, t = 2 hours,
D = 304(2) = 608 miles = SB

The angle α between east and south-east is 45°

Given two sides and the included angle, we find d from the law of cosines:

$$d^2 = (SA)^2 + (SB)^2 - 2(SA)(SB)\cos\alpha$$
$$= (512)^2 + (608)^2 - 2(512)(608)\cos 45°.$$
$$= 191{,}568.97\ldots$$
$$d = 438 \text{ miles}$$

(8-3)

60. Sketch a figure.

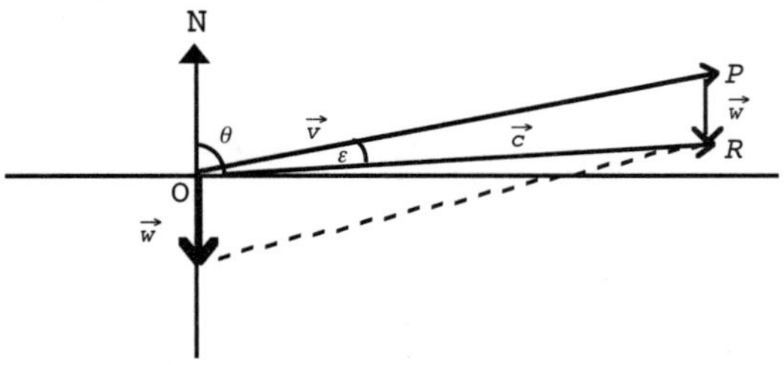

The actual course $\vec{c}$ will be the resultant of $\vec{v}$, the plane's heading, and $\vec{w}$, the wind velocity. In triangle OPR, we know $OP = |\vec{v}| = 450$, $PR = |\vec{w}| = 65$, and angle OPR, which is congruent to θ, the heading of 75°.

Given two sides and the included angle, we use the law of cosines to find $|\vec{c}| = OR$

$$|\vec{c}|^2 = OP^2 + PR^2 - 2(OP)(PR)\cos OPR$$
$$= (450)^2 + (65)^2 - 2(450)(65)\cos 75°$$
$$= 191{,}584.09\ldots$$
$$|\vec{c}| = 438 \text{ miles per hour.}$$

To find the actual direction of the plane, we must find angle POR, or ε. Then the heading will be $\theta + \varepsilon$. We use the law of sines.

$$\frac{\sin\varepsilon}{PR} = \frac{\sin OPR}{OR}$$
$$\frac{\sin\varepsilon}{65} = \frac{\sin 75°}{438}$$
$$\sin\varepsilon = \frac{65\sin 75°}{438}$$
$$\sin\varepsilon = 0.1434$$
$$\varepsilon = \sin^{-1} 0.1434 \text{ since } \varepsilon \text{ is acute}$$
$$\varepsilon = 8°$$

Thus the heading of the plane's actual direction = $\theta + \varepsilon$ = 75° + 8° = 83°. (8-3)

61. 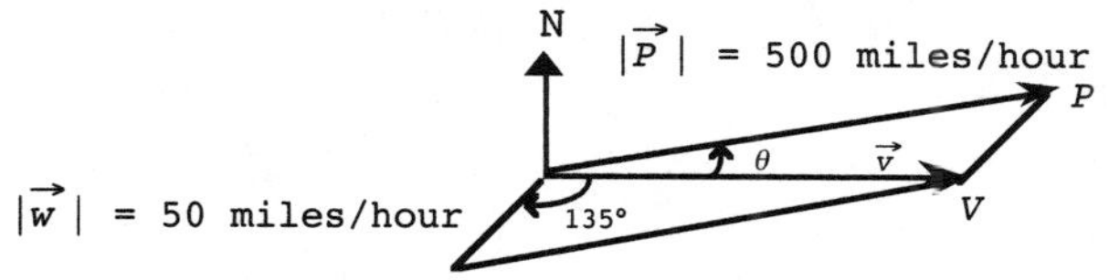

We wish to find $|\vec{v}|$ and θ such that $\vec{v}$, the resultant of $\vec{P}$, the plane's velocity, and $\vec{w}$, the wind's velocity, is in the direction due east. Applying the law of sines to triangle *OPV* in which we have given two sides, *OP* and *PV*, and the angle *OVP* opposite one of them, we obtain

$$\frac{\sin\theta}{PV} = \frac{\sin OVP}{OP}$$

$$\sin\theta = \frac{PV \sin OVP}{OP} = \frac{50 \sin 135^\circ}{500} = 0.0707$$

Since θ must be acute, there is only one possibility:

$$\theta = \sin^{-1} 0.0707 = 4^\circ$$

The heading of the plane will therefore be 90° - 4°, or 86°. We apply the law of sines again to find $|\vec{v}|$, noting that

$$\angle OPV = 180^\circ - (\angle OVP + \theta) = 180^\circ - (135^\circ + 4^\circ) = 41^\circ$$

$$\frac{|\vec{v}|}{\sin OPV} = \frac{OP}{\sin OVP}$$

$$|\vec{v}| = \frac{OP \sin OPV}{\sin OVP}$$

$$|\vec{v}| = \frac{500 \sin 41^\circ}{\sin 135^\circ}$$

$|\vec{v}|$ = 464 miles per hour (8-3)

62. 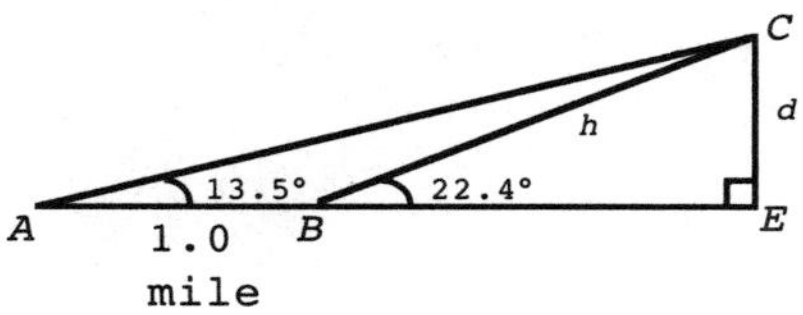

We note: In triangle *ABC*, angle *ABC* is the supplement of given angle *CBE*. Angle *ABC* has measure 180° - 22.4° = 157.6°. In triangle *ABC* we know two angles and the included side, hence we can find angle *ACB* and side *BC*. Side *BC* is then the hypotenuse of right triangle *BCE*, so that we can find *d*.

Solve for angle *ACB*: $13.5^\circ + 157.6^\circ + \angle ACB = 180^\circ$

$$\angle ACB = 180^\circ - 13.5^\circ - 157.6^\circ = 8.9^\circ$$

Solve for *CB* = *h*: $\dfrac{\sin A}{h} = \dfrac{\sin ACB}{AB}$

$$h = \frac{AB \sin A}{\sin ACB} = \frac{1.0 \sin 13.5^\circ}{\sin 8.9^\circ}$$

= 1.51 miles

Solve for *d*: $\sin CBE = \dfrac{d}{h}$

$$d = h \sin CBE$$

$$d = 1.51 \sin 22.4^\circ$$

d = 0.6 miles

(8-1)

63. 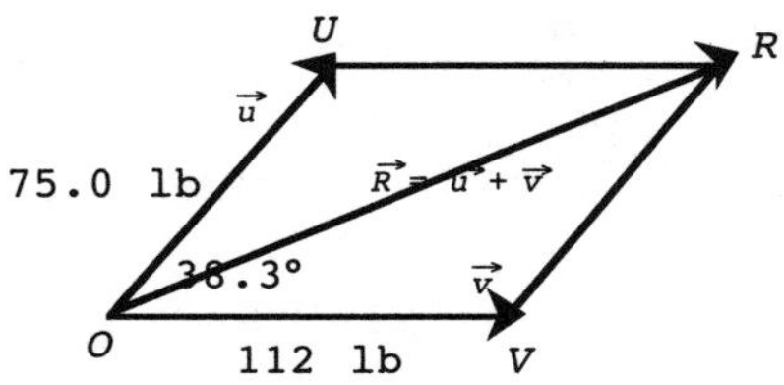

In triangle *OVR*, angle *OVR* has measure 180° - 38.3° = 141.7°. We have *OV* = 112, *VR* = 75.0, hence we can use the law of cosines to find $OR = |\vec{R}|$

$$|\vec{R}|^2 = (OV)^2 + (VR)^2 - 2(OV)(VR)\cos\alpha$$

$$= (112)^2 + (75.0)^2 - 2(112)(75.0)\cos 141.7^\circ$$

$$= 31{,}353.243\ldots$$

$|\vec{R}|$ = 177 pounds

To find the direction of $\vec{R}$ relative to *v*, we find angle *VOR* from the law of sines.

$$\frac{\sin \alpha}{|\vec{R}|} = \frac{\sin VOR}{VR}$$

$$\sin VOR = \frac{VR \sin \alpha}{|\vec{R}|}$$

$$\sin VOR = \frac{75.0 \sin 141.7^\circ}{177}$$

$$\sin VOR = 0.2626$$

$\angle VOR = \sin^{-1} 0.2626$ since this angle must be acute

$\angle VOR = 15.2^\circ$ (8-3)

64. 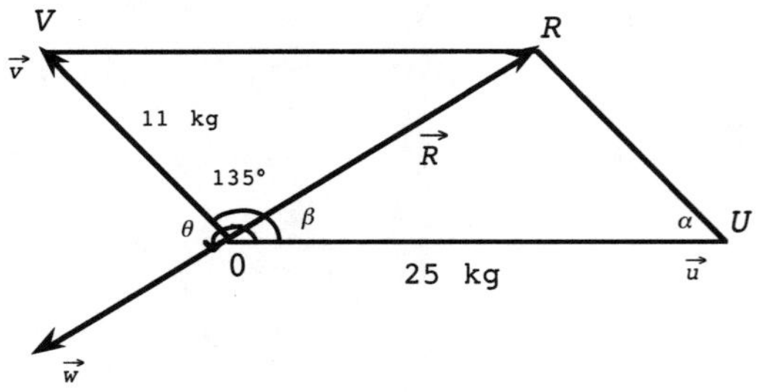

From the figure it should be clear that:
The two forces **u** and **v** act together as their resultant **R**. Needed to balance them, then, is a force **w** in precisely the opposite direction to **R** with the same magnitude as **R**. We calculate $|\mathbf{R}|$, the magnitude of **R**, then its direction β relative to **u** from triangle *OUR*. Then the direction of **w** relative to **u** is
$\theta = 180^\circ + \beta$.

Solve for $|\mathbf{R}|$: We use the law of cosines, with $\alpha = 180^\circ - 135^\circ = 45^\circ$.

$$|\mathbf{R}|^2 = (OU)^2 + (UR)^2 - 2(OU)(UR)\cos \alpha$$
$$= 25^2 + 11^2 - 2(25)(11)\cos 45^\circ$$
$$= 357.09127\ldots$$
$$|\mathbf{R}| = 19 \text{ kg}$$

Solve for β: We use the law of sines.

$$\frac{\sin \beta}{UR} = \frac{\sin \alpha}{|\mathbf{R}|}$$

$$\sin \beta = \frac{UR \sin \alpha}{|\mathbf{R}|}$$

$$= \frac{11 \sin 45^\circ}{19} = 0.4094$$

$\beta = \sin^{-1} 0.4094$ since β must be acute

$\beta = 24^\circ$

Thus the direction of **w** is $\theta = 180^\circ + 24^\circ = 204^\circ$ relative to **u**. (8-3)

65. First, form a force diagram with all forces in standard position at the origin.

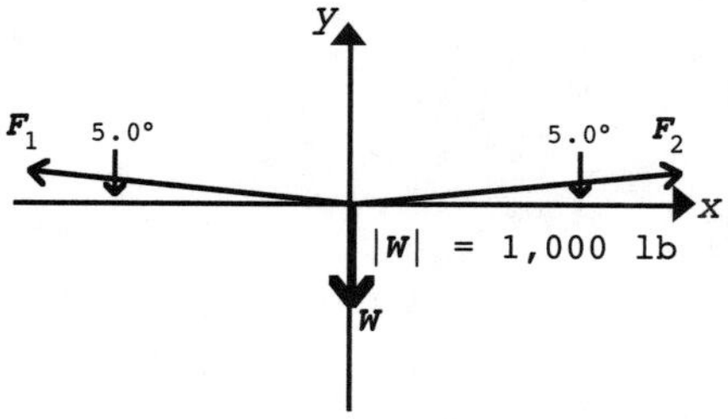

Let F_1 = the tension in left cable
F_2 = the tension in right cable

Write each force vector in terms of $\boldsymbol{i}$ and $\boldsymbol{j}$ unit vectors.

$\boldsymbol{F}_1 = |\boldsymbol{F}_1|(-\cos 5.0°)\boldsymbol{i} + |\boldsymbol{F}_1|(\sin 5.0°)\boldsymbol{j}$

$\boldsymbol{F}_2 = |\boldsymbol{F}_2|(\cos 5.0°)\boldsymbol{i} + |\boldsymbol{F}_2|(\sin 5.0°)\boldsymbol{j}$

$\boldsymbol{W} = -1{,}000\boldsymbol{j}$

For the system to be in static equilibrium, we must have

$\boldsymbol{F}_1 + \boldsymbol{F}_2 + \boldsymbol{W} = \boldsymbol{0}$

which becomes, on addition,

$[-|\boldsymbol{F}_1|(\cos 5.0°) + |\boldsymbol{F}_2|(\cos 5.0°)]\boldsymbol{i} + [|\boldsymbol{F}_1|(\sin 5.0°) + |\boldsymbol{F}_2|(\sin 5.0°) - 1{,}000]\boldsymbol{j} = 0\boldsymbol{i} + 0\boldsymbol{j}$

Since two vectors are equal if and only if their corresponding components are equal, we are led to the following system of equations in $|\boldsymbol{F}_1|$ and $|\boldsymbol{F}_2|$:

$$-|\boldsymbol{F}_1|\cos 5.0° + |\boldsymbol{F}_2|\cos 5.0° = 0$$
$$|\boldsymbol{F}_1|\sin 5.0° + |\boldsymbol{F}_2|\sin 5.0° - 1{,}000 = 0$$

Solving, we obtain

$|\boldsymbol{F}_1| = |\boldsymbol{F}_2|$ (as we expect from the symmetry of the situation)

$|\boldsymbol{F}_1| = \dfrac{1{,}000}{2\sin 5.0°} = 5{,}740$ lb = tension in each half of the cable. *(8-3)*

66. (A)

2×10^8

-2×10^8 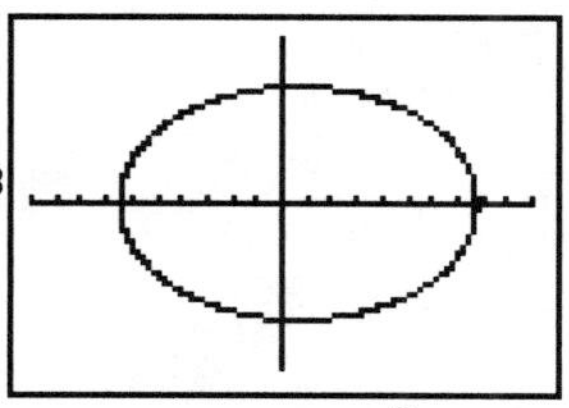 2×10^8

-2×10^8

From the trace function of the utility as applied to the graph, the maximum distance (aphelion) is 1.56×10^8 miles, and the minimum distance (perihelion) is 1.29×10^8 miles.

(B) r is maximum when the denominator $1 - 0.0934\cos\theta$ is minimum. This occurs when $\cos\theta$ is as large as possible, that is, $\cos\theta = 1$.

Then $r = \dfrac{1.41 \times 10^8}{1 - 0.0934} = 1.56 \times 10^8$ miles (aphelion)

r is minimum when the denominator $1 - 0.0934\cos\theta$ is maximum. This occurs when $\cos\theta = -1$.

Then $r = \dfrac{1.41 \times 10^8}{1 + 0.0934} = 1.29 \times 10^8$ miles (perihelion) *(8-4)*

Chapters 6, 7, & 8 Cumulative Review

1. $s = r\theta = (6 \text{ meters})(0.31 \text{ radians}) = 1.86 \text{ meters}$ (6-1)

2. Solve for θ:

$\theta = 90° - 32.7° = 57.3°$

Solve for a:

$$\tan\theta = \frac{b}{a}$$
$$\tan 57.3° = \frac{12.2}{a}$$
$$a = \frac{12.2}{\tan 57.3°} = 7.83 \text{ cm}$$

Solve for c:

$$\sin\theta = \frac{b}{c}$$
$$\sin 57.3° = \frac{12.2}{c}$$
$$c = \frac{12.2}{\sin 57.3°}$$
$$c = 14.5 \text{ cm}$$

(6-3)

3. (A) $\frac{b}{r} = \sin\theta > 0$ if $b > 0$. This occurs in quadrants I, II.

(B) $\frac{a}{r} = \cos\theta > 0$ if $a > 0$. This occurs in quadrants I, IV.

(C) $\frac{b}{a} = \tan\theta > 0$ if a and b have the same sign. This occurs in quadrants I, III. (6-2)

4.
$$a^2 + b^2 = r^2$$
$$(-3)^2 + 4^2 = r^2$$
$$25 = r^2$$
$$r = 5$$

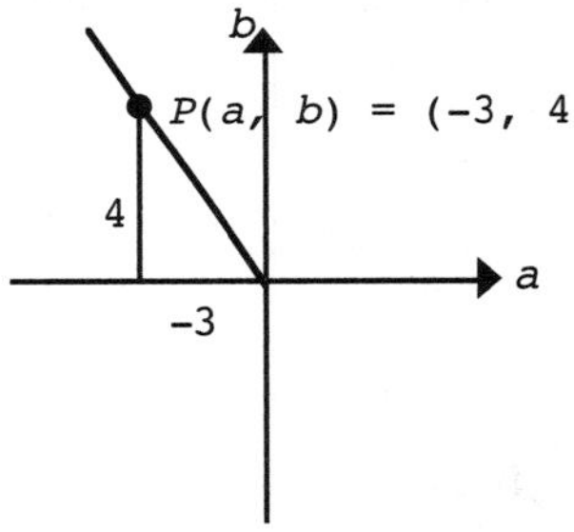

(A) $\cos\theta = \frac{a}{r} = \frac{-3}{5} = -\frac{3}{5}$

(B) $\csc\theta = \frac{r}{b} = \frac{5}{4}$

(C) $\tan\theta = \frac{b}{a} = \frac{4}{-3} = -\frac{4}{3}$ (6-4)

5. (A)

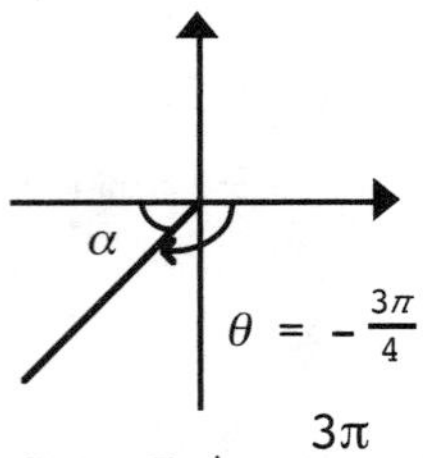

$\alpha = \pi + -\frac{3\pi}{4} = \frac{\pi}{4}$

(B)

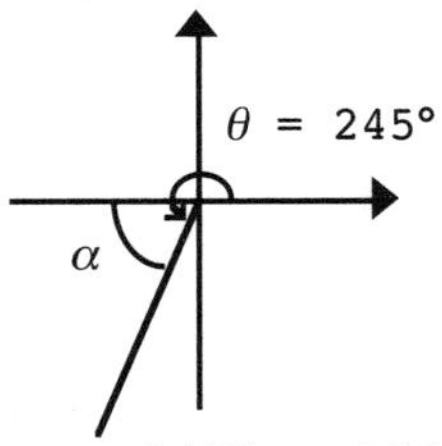

$\alpha = 245° - 180° = 65°$

(C)

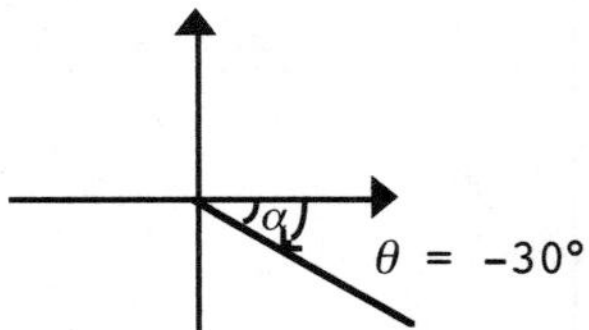

$\alpha = |-30°| = 30°$ (6-4)

6. (A) Domain: all real numbers; Range: $-1 \le y \le 1$; Period: 2π

(B) Domain: all real numbers; Range: $-1 \le y \le 1$; Period: 2π

(C) Domain: all real numbers except $x = \frac{\pi}{2} + k\pi$, k an integer; Range: all real numbers; Period: π (6-4)

7.

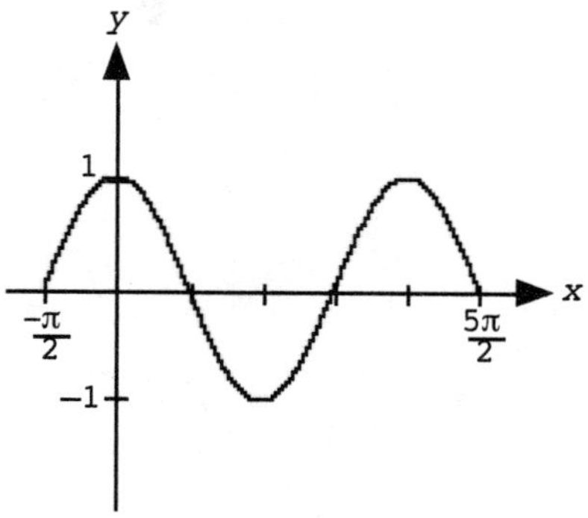

(6-4)

8.

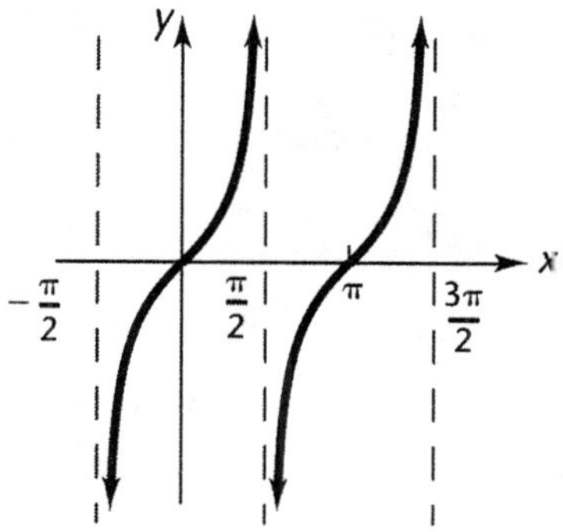

(6-4)

9. See figure.

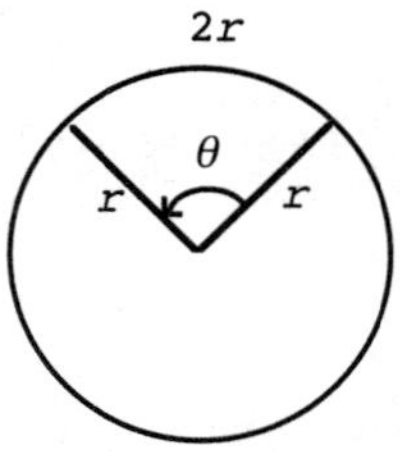

A central angle in a circle with radian measure 2 is the angle subtended by an arc of length twice the measure of the radius.

(6-1)

10. See figure for problem 7 above. If the graph of $y = \cos x$ is shifted $\frac{\pi}{2}$ units to the right, the result will be the graph of $y = \sin x$. (6-4)

11.

$\cot\theta \sec\theta = \dfrac{\cos\theta}{\sin\theta}\dfrac{1}{\cos\theta}$	Quotient and Reciprocal Identities
$= \dfrac{1}{\sin\theta}$	Algebra
$= \csc\theta$	Reciprocal Identity

(7-1)

12.

$\sec x - \cos x = \dfrac{1}{\cos x} - \cos x$	Reciprocal Identity
$= \dfrac{1}{\cos x} - \dfrac{\cos^2 x}{\cos x}$	Algebra
$= \dfrac{1 - \cos^2 x}{\cos x}$	Algebra
$= \dfrac{\sin^2 x + \cos^2 x - \cos^2 x}{\cos x}$	Pythagorean Identity
$= \dfrac{\sin^2 x}{\cos x}$	Algebra
$= \dfrac{\sin x}{\cos x}\sin x$	Algebra
$= \tan x \sin x$	Quotient Identity

(7-1)

13.

$\sin\left(x - \dfrac{\pi}{2}\right) = \sin x \cos\dfrac{\pi}{2} - \cos x \sin\dfrac{\pi}{2}$	Sum Identity
$= \sin x(0) - \cos x(1)$	Known Values
$= -\cos x$	Algebra

(7-2)

14. $\csc 2x = \dfrac{1}{\sin 2x}$ Reciprocal Identity

$= \dfrac{1}{2 \sin x \cos x}$ Double-angle Identity

$= \dfrac{1}{2} \dfrac{1}{\sin x} \dfrac{1}{\cos x}$ Algebra

$= \dfrac{1}{2} \csc x \sec x$ Reciprocal Identities (7-3)

15. (A) Graph both sides of the equation in the same viewing window.

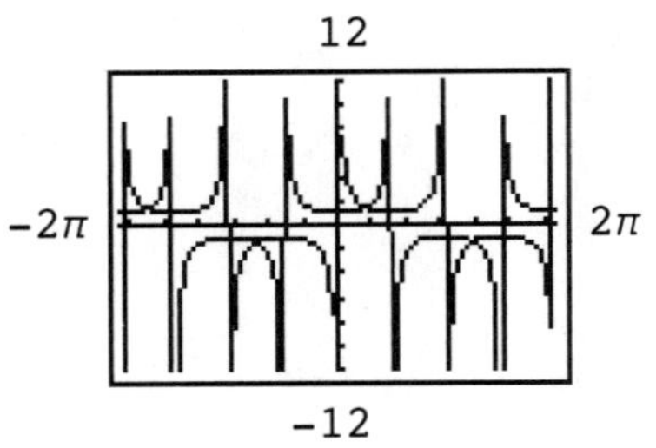

$\dfrac{\sin^2 x}{\cos x} + \cos x = \csc x$ is not an identity, since the graphs do not match.

Try $x = \dfrac{3\pi}{4}$.

Left side: $\dfrac{\sin^2\left(\frac{3\pi}{4}\right)}{\cos\left(\frac{3\pi}{4}\right)} + \cos\left(\dfrac{3\pi}{4}\right) = \dfrac{\frac{1}{2}}{-\frac{1}{\sqrt{2}}} + \left(-\dfrac{1}{\sqrt{2}}\right) = -\sqrt{2}$

Right side: $\csc\left(\dfrac{3\pi}{4}\right) = \sqrt{2}$

This verifies that the equation is not an identity.

(B) Graph both sides of the equation in the same viewing window.

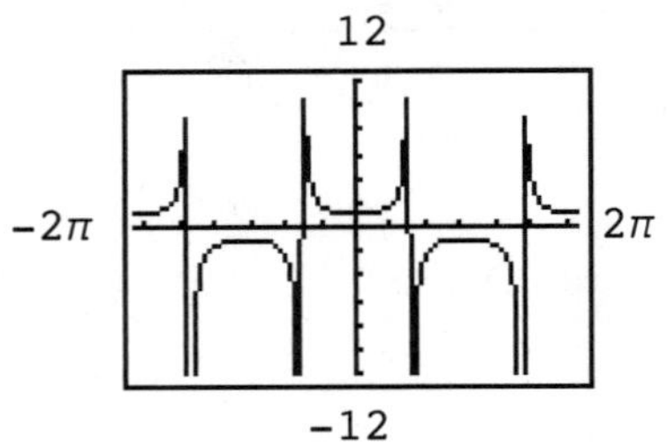

$\dfrac{\sin^2 x}{\cos x} + \cos x = \sec x$ appears to be an identity, which we now verify:

$\dfrac{\sin^2 x}{\cos x} + \cos x = \dfrac{\sin^2 x}{\cos x} + \dfrac{\cos^2 x}{\cos x}$ Algebra

$= \dfrac{\sin^2 x + \cos^2 x}{\cos x}$ Algebra

$= \dfrac{1}{\cos x}$ Pythagorean Theorem

$= \sec x$ Reciprocal Identity

(7-

16. Angle α is acute. A triangle can have at most one obtuse angle. Since β is acute, then, if the triangle has an obtuse angle it must be the angle opposite the longer of the two sides, a and c. Thus, α, the angle opposite the shorter of the two sides, a, must be acute. (8-2)

17. $\sin x = 0.3188 \quad 0 \le x < 2\pi$

Sketch a graph of $y = \sin x$ and $y = 0.3188$

$x = [0, 2\pi)$.

$x = \begin{cases} \sin^{-1} 0.3188 = 0.3245 & \text{First quadrant solution} \\ \pi - \sin^{-1} 0.3188 = 2.8171 & \text{Second quadrant solution} \end{cases}$

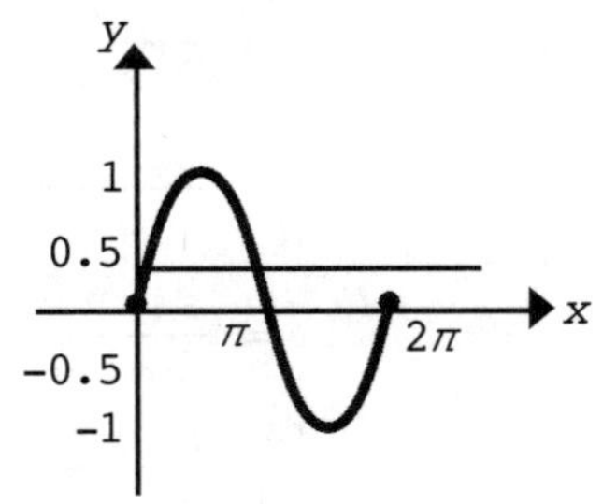

(7-5)

18. $\tan \theta = -4.076 \quad -90° < \theta < 90°$

Sketch a graph of $y = \tan \theta$ and $y = -4.076$

$\theta = (-90°, 90°)$

$\theta = \tan^{-1}(-4.076) = -76.2154°$

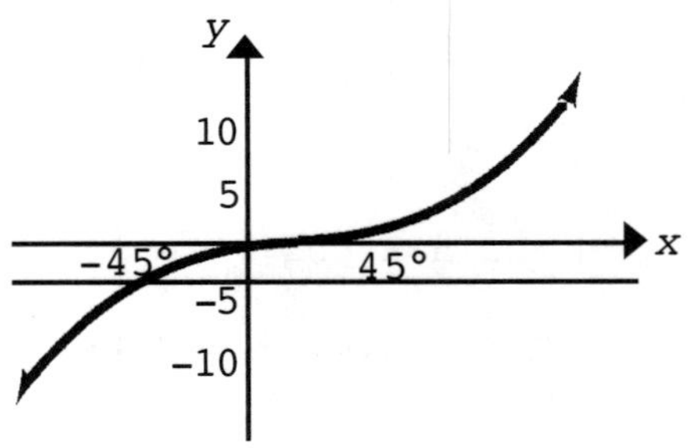

(7-5)

19. We are given two sides and the included angle (SAS). We use the law of cosines to find the third side, then the law of sines to find a second angle.

Solve for b:

$$b^2 = a^2 + c^2 - 2ac\cos\beta$$
$$= 12^2 + 13^2 - 2(12)(13)\cos 121°$$
$$= 473.69188\ldots$$
$$b = 22 \text{ ft}$$

Solve for α:
Note that since the sum of α and γ is $180° - 121° = 59°$, both must be acute.

$$\frac{\sin\alpha}{a} = \frac{\sin\beta}{b}$$
$$\sin\alpha = \frac{a\sin\beta}{b}$$
$$= \frac{12\sin 121°}{22}$$
$$\alpha = \sin^{-1}\left(\frac{12\sin 121°}{22}\right)$$
$$= 28°$$

Solve for γ:

$$\alpha + \beta + \gamma = 180°$$
$$\gamma = 180° - (\alpha + \beta)$$
$$= 180° - (28° + 121°)$$
$$= 31°$$

(8-1, 8-2)

20. The algebraic vector $\langle a, b\rangle$ corresponds to the standard geometric vector with terminal point $P(a, b)$ and initial point $O(0, 0)$. The coordinates of the point $P(a, b)$ are given by

$$a = x_B - x_A = 3 - (-3) = 6$$
$$b = y_B - y_A = -1 - 2 = -3$$

Thus, $\langle a, b\rangle = \langle 6, -3\rangle$ (8-3)

21. See figure.

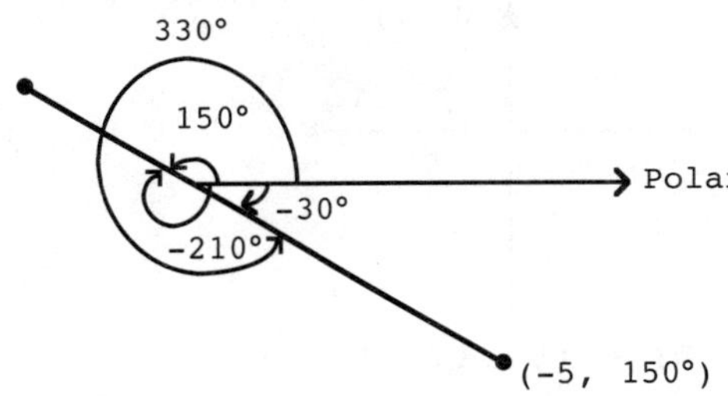

The point with coordinates (-5, 150°) can equally well be described as (5, -30°) or (-5, -210°) or (5, 330°). Thus, (5, -30°): The polar axis is rotated 30° clockwise (negative direction) and the point is located 5 units from the pole along the positive polar axis. (-5, -210°): The polar axis is rotated 210° clockwise (negative direction) and the point is located 5 units from the pole along the negative polar axis. (5, 330°): The polar axis is rotated 330° counterclockwise (positive direction) and the point is located 5 units from the pole along the positive polar axis. (8-4)

22. We set up a table that indicates how r varies as we let θ vary through each set of quadrant values.

θ	$\cos\theta$	$6\cos\theta$	
0 to $\frac{\pi}{2}$	1 to 0	6 to 0	
$\frac{\pi}{2}$ to π	0 to −1	0 to −6	
π to $\frac{3\pi}{2}$	−1 to 0	−6 to 0	Curve is traced out a second time in this region though coordinates seem different.
$\frac{3\pi}{2}$ to 2π	0 to 1	0 to 6	

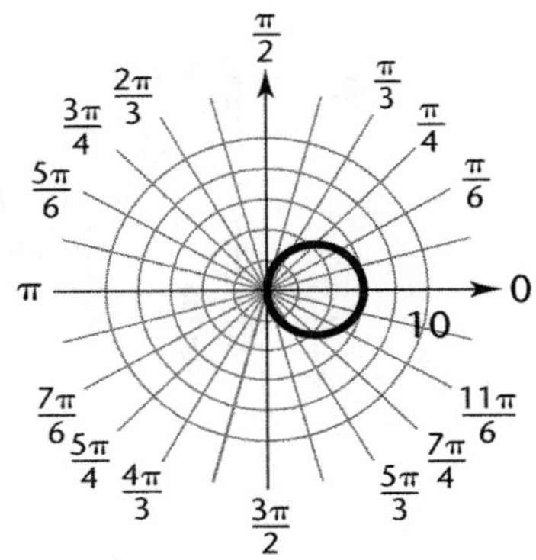

(8-4)

23.

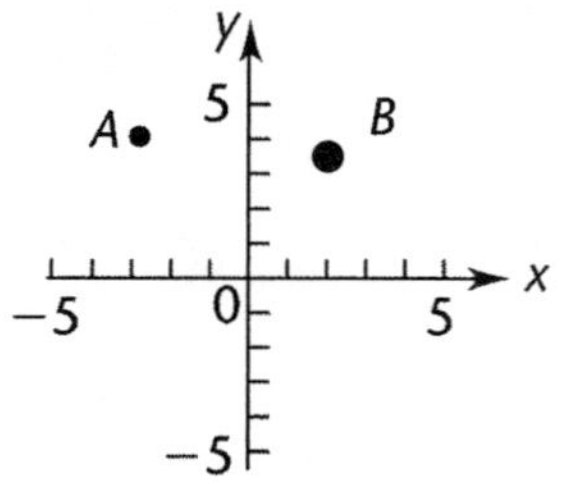

(8-5)

24. $(2e^{10°i})^3 = 2^3e^{3 \cdot 10°i} = 8e^{30°i}$
$= 8(\cos 30° + i \sin 30°)$
$= 8\left(\frac{\sqrt{3}}{2} + i\frac{1}{2}\right)$
$= 4\sqrt{3} + 4i$ (8-5)

25. 30° is not coterminal with 150°; 30° is a I quadrant angle, but 150° is a II quadrant angle.

Since $-\frac{7\pi}{6} + 2\pi = \frac{5\pi}{6}$, which is equivalent to 150°, $-\frac{7\pi}{6}$ is coterminal with 150°.

Since 870° − 2(360°) = 150°, 870° is coterminal with 150°. (6-2)

26. $\theta_D = \frac{180°}{\pi}\theta_R = \frac{180°}{\pi}(1.31) = 75.06°$ (6-1)

27. (A) and (C), since 8 radians is equivalent to the real number 8, and cosine is periodic with period 2π. (B) is not the same as cos 8, since 8° is equivalent to $\frac{\pi}{180°}8°$, not 8. (6-2)

28.

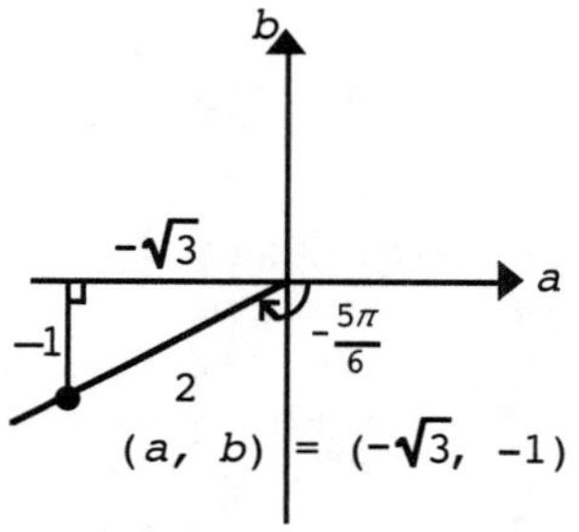

$\sin\left(-\frac{5\pi}{6}\right) = \frac{b}{r} = -\frac{1}{2}$ (6-2)

29.

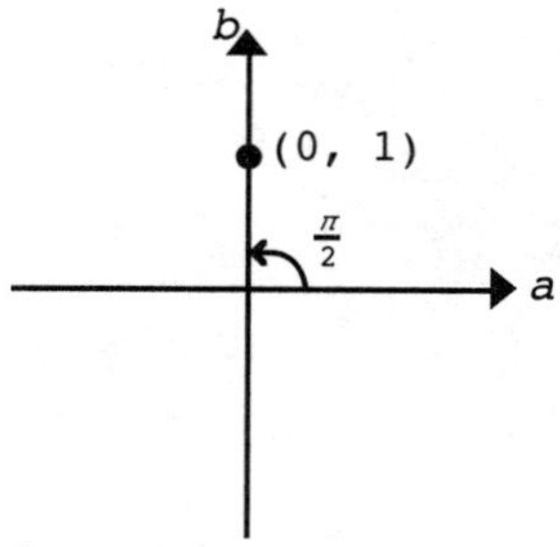

$(a, b) = (0, 1) \quad r = 1$

$\tan\frac{\pi}{2} = \frac{b}{a} = \frac{1}{0}$ Not defined (6-2)

30.

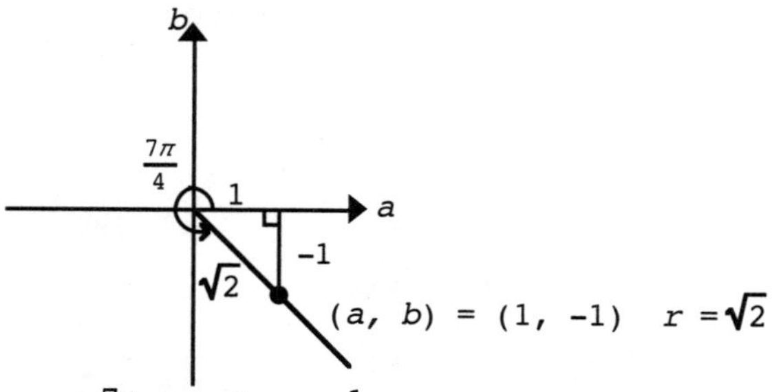

$\cot\frac{7\pi}{4} = \frac{a}{b} = \frac{1}{-1} = -1$ (6-2)

31.

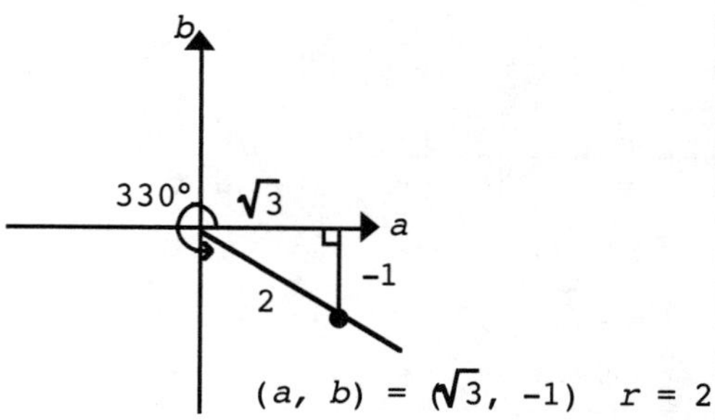

$\sec 330° = \frac{r}{a} = \frac{2}{\sqrt{3}}$ or $\frac{2\sqrt{3}}{3}$ (6-2)

32. $y = \cos^{-1}(-1)$ is equivalent to
$\cos y = -1 \quad 0 \le y \le \pi$
$y = \pi$ (6-6)

33. $\sin^{-1} x$ is not defined if $x > 1$. (6-6)

34. $y = \arccos\left(-\frac{1}{2}\right)$ is equivalent to

$\cos y = -\frac{1}{2} \qquad 0 \le y \le \pi$

$\cos y = \frac{a}{r} = -\frac{1}{2} \qquad a = -1 \quad r = 2$

$(-1)^2 + b^2 = 2^2$

$b^2 = 3$

$b = \sqrt{3}$

(positive since y is a quadrant I or II angle.)

$y = \frac{2\pi}{3}$

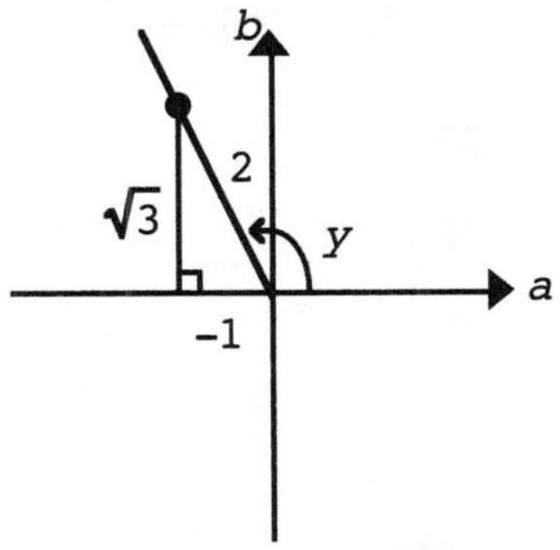

(6-6)

35. $\sin(\sin^{-1} 0.55) = 0.55$ by the sine-inverse sine identity. (6-6)

36. Let $y = \sin^{-1}\left(-\frac{4}{5}\right)$

then $\sin y = -\frac{4}{5} \qquad -\frac{\pi}{2} \le y \le \frac{\pi}{2}$

Draw the reference triangle associated with y, then $\cos y = \cos\left[\sin^{-1}\left(-\frac{4}{5}\right)\right]$ can be determined directly from the triangle.

$\sin y = \frac{b}{r} = \frac{-4}{5} \qquad b = -4 \quad r = 5 \quad a = 3$

$\cos y = \frac{a}{r} = \frac{3}{5}$

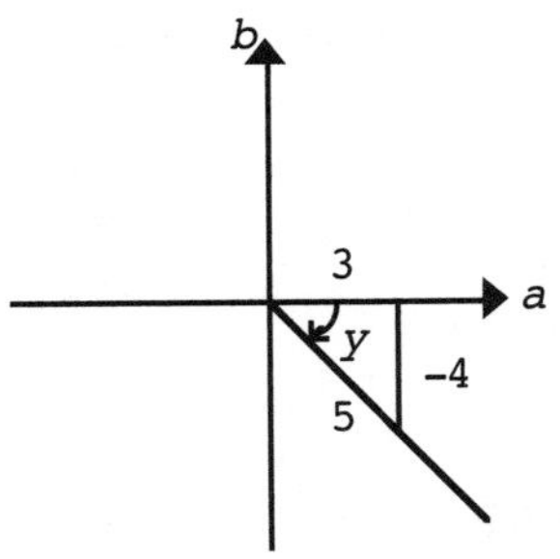

(6-6)

37. Let $y = \tan^{-1}(-2)$

then $\tan y = -2 \qquad -\frac{\pi}{2} < y < \frac{\pi}{2}$

Draw the reference triangle associated with y, then $\cos y = \cos[\tan^{-1}(-2)]$ can be determined directly from the triangle.

$\tan y = \frac{b}{a} = -2 = \frac{-2}{1}$

$b = -2 \quad a = 1 \quad r = \sqrt{5}$

$\cos y = \frac{a}{r} = \frac{1}{\sqrt{5}}$ or $\frac{\sqrt{5}}{5}$

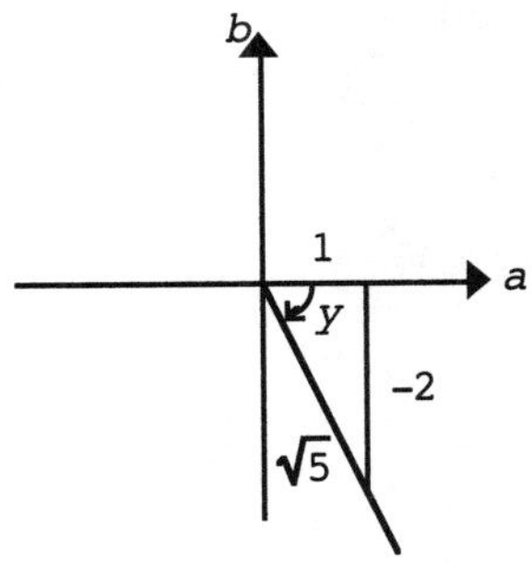

(6-6)

38. (A) 9.871 (B) -3.748 (C) -1.559
(D) Since $\tan 2.314 = -1.088 < -1$, $\cos^{-1}(\tan 2.314)$ is not defined. (6-2, 6-6)

39. $y = 2 - 2 \cos \frac{\pi x}{2}$

For the graph of $y = 2 \cos \frac{\pi x}{2}$, we note: $A = 2$

$P = 2\pi \div \frac{\pi}{2} = 4$, Phase shift = 0. We graph

$y = 2 \cos \frac{\pi x}{2}$ turned upside down, then vertically translate the graph up 2 units.

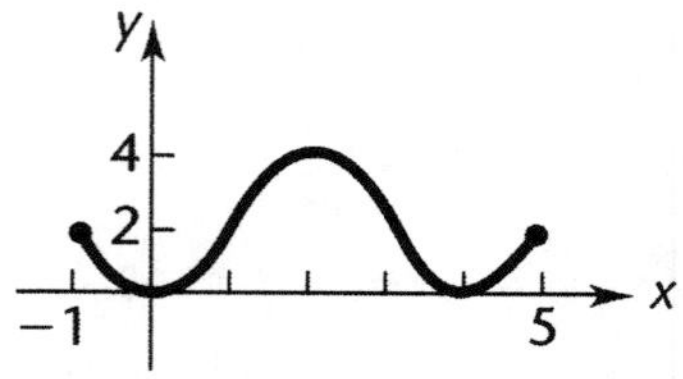

(6-5)

40. (A) $y = \cos^{-1}\left(-\frac{\sqrt{3}}{2}\right)$ is equivalent to

$$\cos y = -\frac{\sqrt{3}}{2} \qquad 0 \le y \le \pi$$

$$y = \frac{5\pi}{6}$$

Using the relation $\theta_D = \frac{180°}{\pi \text{ rad}} \theta_R$,

we have

$$\cos^{-1}\left(-\frac{\sqrt{3}}{2}\right) = \frac{180°}{\pi}\left(\frac{5\pi}{6}\right) = 150°$$

(B) -19.755°

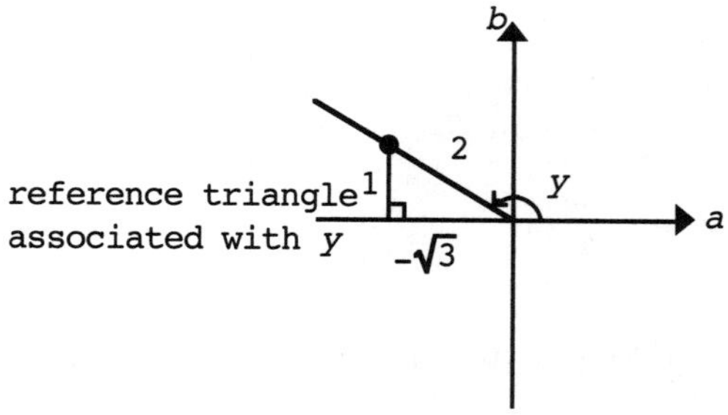

(6-6)

41. $\sin^{-1}(\sin 3) = 0.142$. For the identity $\sin^{-1}(\sin x) = x$ to hold, x must be in the restricted domain of the sine function; that is, $-\frac{\pi}{2} \le x \le \frac{\pi}{2}$. The number 3 is not in the restricted domain. (6-6)

42. Since the coordinates of a point on a unit circle are given by $P(a, b) = P(\cos x, \sin x)$, we evaluate $P(\cos(11.205), \sin(11.205))$—using a calculator set in radian mode--to obtain $P(0.208, -0.978)$. The quadrant in which $P(a, b)$ lies can be determined by the signs of a and b. In this case P is in the fourth quadrant, since a is positive and b is negative. (6-1, 6-2)

43. The equation has infinitely many solutions [$x = \tan^{-1}(-24.5) + k\pi$, k any integer]; $\tan^{-1}(-24.5)$ has a unique value (-1.530 to three decimal places). (6-6)

44. $k = 3$. $A = \frac{1}{2}(5 - 1) = 2$ $P = 2 = \frac{2\pi}{\pi}$. Hence $B = \pi$, $y = 3 + 2 \sin \pi x$ (6-5)

45. $y = 3 \sin(2x - \pi)$

$A = 3$

Solve $2x - \pi = 0$ $2x - \pi = 2\pi$

$$2x = \pi \qquad 2x = \pi + 2\pi$$

$$x = \frac{\pi}{2} \qquad x = \frac{\pi}{2} + \pi$$

Phase shift   Period $P = \pi$

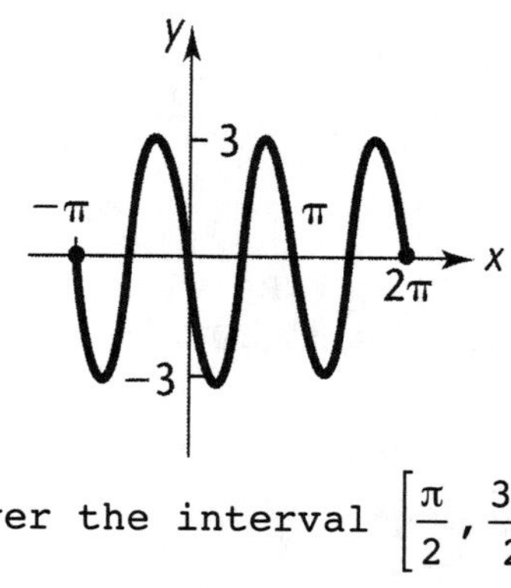

The graph completes one full cycle as x varies over the interval $\left[\frac{\pi}{2}, \frac{3\pi}{2}\right]$ (6-5)

46. One cycle of $y = 2\tan\left(\frac{\pi x}{2} - \frac{\pi}{2}\right)$ is completed as $\frac{\pi x}{2} - \frac{\pi}{2}$ varies from 0 to π.

Solve each equation for x:

$$\frac{\pi x}{2} - \frac{\pi}{2} = 0 \qquad\qquad \frac{\pi x}{2} - \frac{\pi}{2} = \pi$$

$$\frac{\pi x}{2} = \frac{\pi}{2} \qquad\qquad \frac{\pi x}{2} = \frac{\pi}{2} + \pi$$

$$x = 1 \qquad\qquad x = 1 + 2$$

Phase shift = 1   Period = 2

Sketch the graph for one period [1, 3], then extend over the interval (0, 4).

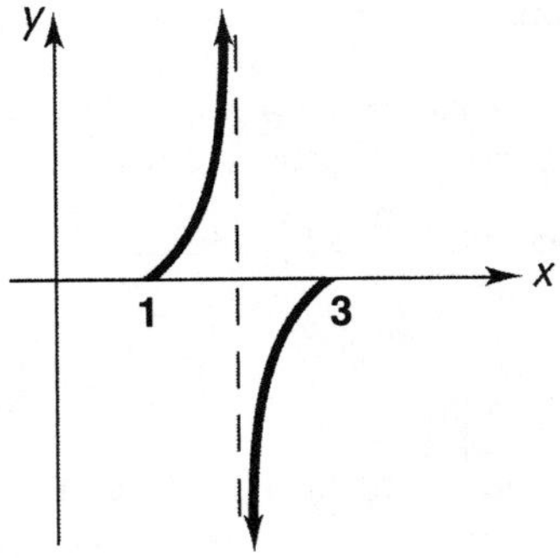

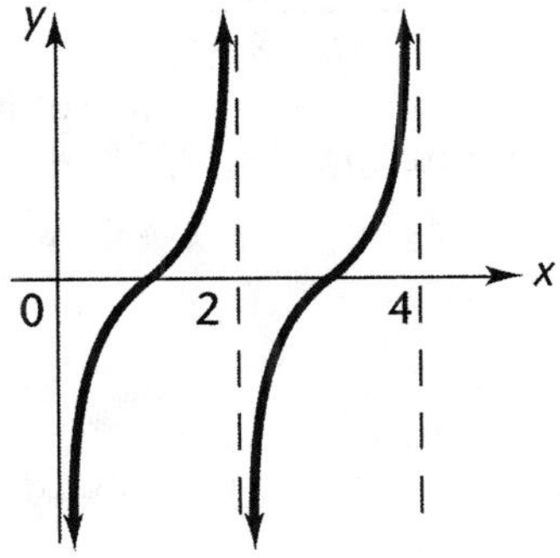

(6-5)

47. The dashed curve is $y = \sin x$. The solid curve is $y = \csc x$.

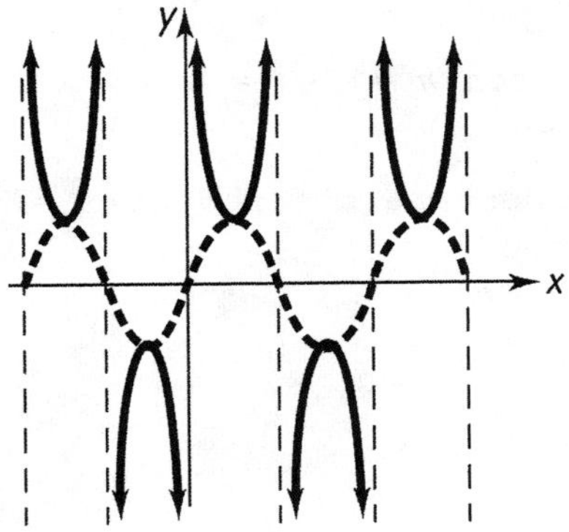

(6-4)

48. See figure.

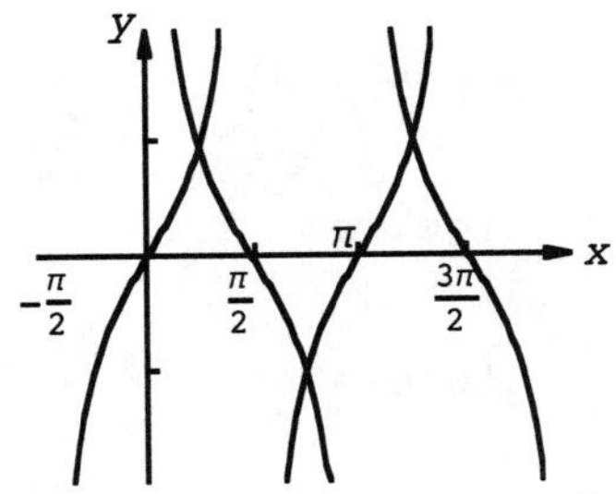

If the graph of $y = \cot x$ is shifted to the left $\frac{\pi}{2}$ units and reflected in the x axis, the result will be the graph of $y = \tan x$.

(6-4)

49. Here is the graph of $y = \dfrac{1}{\cot^2 x + 1}$ in a graphing utility.

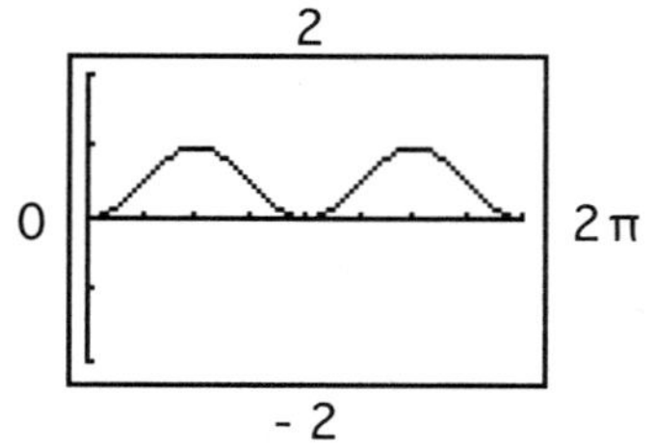

The graph has amplitude $\dfrac{1 - 0}{2} = \dfrac{1}{2}$ and period π. It appears to be the graph of $y = -\dfrac{1}{2}\cos 2x$ shifted up $\dfrac{1}{2}$ unit, that is, $y = \dfrac{1}{2} - \dfrac{1}{2}\cos 2x$. (6-5)

50. Here is the graph of $y = \dfrac{2 - 2\sin^2 x}{\sin 2x}$ in a graphing utility.

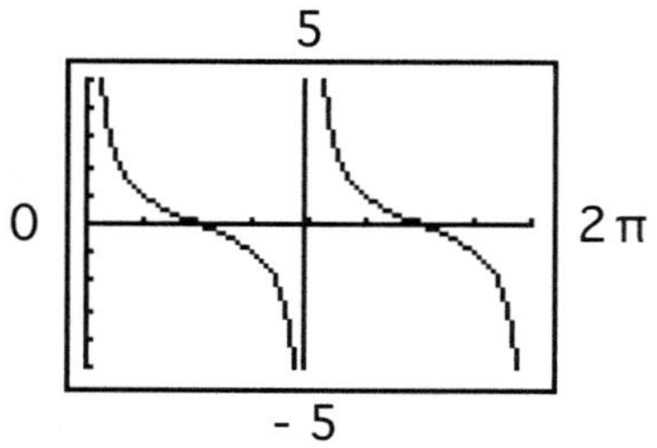

The graph appears to have the form $y = A\cot Bx$. Since the period is π, $B = 1$. The graph of $y = A\cot x$ shown appears to pass through the point $\left(\dfrac{\pi}{4}, 1\right)$, thus

$$1 = A\cot\frac{\pi}{4}$$

$$1 = A$$

The equation of the graph can be written $y = \cot x$. (6-5)

51. (A) Check $x = 0$
Left side:
$\sin 2x = \sin 2\cdot 0 = \sin 0 = 0$
Right side:
$2\sin x = 2\sin 0 = 2\cdot 0 = 0$
0 is a solution

Check $x = \pi$
Left side:
$\sin 2x = \sin 2\pi = 0$
Right side:
$2\sin x = 2\sin \pi = 2\cdot 0 = 0$
π is a solution

(B) The equation is not an identity because both sides are defined at $x = \dfrac{\pi}{2}$; for example, but $\dfrac{\pi}{2}$ is not a solution.

Left side: $\sin 2x = \sin 2\left(\dfrac{\pi}{2}\right) = \sin \pi = 0$

Right side: $2\sin x = 2\sin\dfrac{\pi}{2} = 2\cdot 1 = 2$

The equation is conditional. (7-1)

52.

$\dfrac{\sin u}{1 + \cos u} + \cot u = \dfrac{\sin u}{1 + \cos u} + \dfrac{\cos u}{\sin u}$	Quotient Identity
$= \dfrac{\sin^2 u}{\sin u(1 + \cos u)} + \dfrac{\cos u(1 + \cos u)}{\sin u(1 + \cos u)}$	Algebra
$= \dfrac{\sin^2 u + \cos u(1 + \cos u)}{\sin u(1 + \cos u)}$	Algebra
$= \dfrac{\sin^2 u + \cos u + \cos^2 u}{\sin u(1 + \cos u)}$	Algebra
$= \dfrac{\sin^2 u + \cos^2 u + \cos u}{\sin u(1 + \cos u)}$	Algebra
$= \dfrac{1 + \cos u}{\sin u(1 + \cos u)}$	Pythagorean Identity
$= \dfrac{1}{\sin u}$	Algebra
$= \csc u$	Reciprocal Identity

Key Algebraic Steps:

$$\frac{a}{1+b} + \frac{b}{a} = \frac{a^2}{a(1+b)} + \frac{b(1+b)}{a(1+b)}$$
$$= \frac{a^2 + b(1+b)}{a(1+b)}$$
$$= \frac{a^2 + b + b^2}{a(1+b)}$$
$$= \frac{a^2 + b^2 + b}{a(1+b)}$$

(7-1)

53. $\sec x + \tan x = \frac{1}{\cos x} + \frac{\sin x}{\cos x}$ Reciprocal and Quotient Identities

$= \frac{1 + \sin x}{\cos x}$ Algebra

$= \frac{(1 + \sin x)(1 - \sin x)}{\cos x(1 - \sin x)}$ Algebra

$= \frac{1 - \sin^2 x}{\cos x(1 - \sin x)}$ Algebra

$= \frac{\cos^2 x + \sin^2 x - \sin^2 x}{\cos x(1 - \sin x)}$ Pythagorean Identity

$= \frac{\cos^2 x}{\cos x(1 - \sin x)}$ Algebra

$= \frac{\cos x}{1 - \sin x}$ Algebra (7-1)

54. $\tan \frac{x}{2} = \frac{1 - \cos x}{\sin x}$ Half-angle Identity

$= \frac{1}{\sin x} - \frac{\cos x}{\sin x}$ Algebra

$= \csc x - \cot x$ Reciprocal and Quotient Identities (7-3)

55. $\csc^2 \frac{x}{2} = \frac{1}{\sin^2 \frac{x}{2}}$ Reciprocal Identity

$= \frac{1}{\left(\pm\sqrt{\frac{1-\cos x}{2}}\right)^2}$ Half-angle Identity

$= \frac{1}{\frac{1-\cos x}{2}}$ Algebra

$= 1 \div \frac{1 - \cos x}{2}$ Algebra

$= 1 \cdot \frac{2}{1 - \cos x}$ Algebra

$= \frac{2}{1 - \cos x}$ Algebra

$= \dfrac{2}{1 - \cos x} \dfrac{1 + \cos x}{1 + \cos x}$ Algebra

$= \dfrac{2(1 + \cos x)}{1 - \cos^2 x}$ Algebra

$= \dfrac{2(1 + \cos x)}{\sin^2 x + \cos^2 x - \cos^2 x}$ Pythagorean Identity

$= \dfrac{2(1 + \cos x)}{\sin^2 x}$ Algebra

$= 2 \cdot \dfrac{1}{\sin x} \dfrac{1 + \cos x}{\sin x}$ Algebra

$= 2 \cdot \dfrac{1}{\sin x}\left(\dfrac{1}{\sin x} + \dfrac{\cos x}{\sin x}\right)$ Algebra

$= 2 \csc x(\csc x + \cot x)$ Reciprocal and Quotient Identities

Key Algebraic Steps:

$$\frac{1}{\frac{1-a}{2}} = 1 \div \frac{1 - a}{2} = 1 \cdot \frac{2}{1 - a} = \frac{2}{1 - a} = \frac{2}{1 - a}\frac{1 + a}{1 + a} = \frac{2(1 + a)}{1 - a^2}$$ (7-3)

56. $\dfrac{2}{1 + \cos 2x} = \dfrac{2}{1 + 2\cos^2 x - 1}$ Double-angle Identity

$= \dfrac{2}{2\cos^2 x}$ Algebra

$= \dfrac{1}{\cos^2 x}$ Algebra

$= \sec^2 x$ Reciprocal Identity (7-3)

57. $\dfrac{\cos x + \cos y}{\sin x - \sin y} = \dfrac{2\cos \frac{x+y}{2} \cos \frac{x-y}{2}}{2\cos \frac{x+y}{2} \sin \frac{x-y}{2}}$ Sum-product Identity

$= \dfrac{\cos \frac{x-y}{2}}{\sin \frac{x-y}{2}}$ Algebra

$= \cot \dfrac{x - y}{2}$ Quotient Identity (7-4)

58. (A) Graph both sides of the equation in the same viewing window.

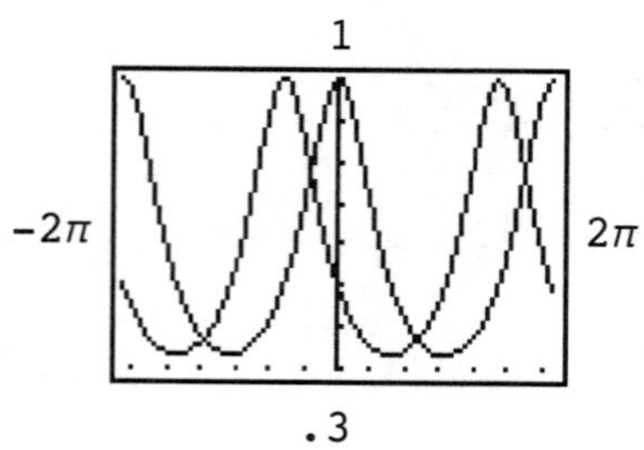

$\dfrac{\tan x}{2\tan x - \sin x} = \dfrac{1}{2 + \sin x}$ is not an identity, since the graphs do not match.

Try $x = \dfrac{\pi}{4}$

Left side: $\dfrac{\tan \frac{\pi}{4}}{2\tan \frac{\pi}{4} - \sin \frac{\pi}{4}} = \dfrac{1}{2 \cdot 1 - \frac{1}{\sqrt{2}}} = \dfrac{\sqrt{2}}{2\sqrt{2} - 1}$

Right side: $\dfrac{1}{2 + \sin \frac{\pi}{4}} = \dfrac{1}{2 + \frac{1}{\sqrt{2}}} = \dfrac{\sqrt{2}}{2\sqrt{2} + 1}$

This verifies that the equation is not an identity.

(B) Graph both sides of the equation in the same viewing window.

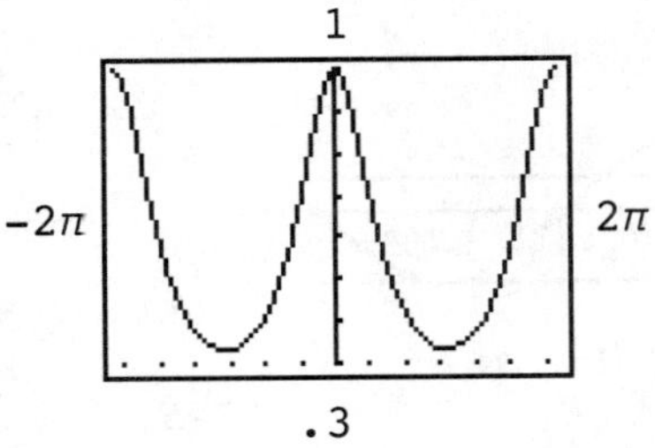

$\dfrac{\tan x}{2\tan x - \sin x} = \dfrac{1}{2 - \cos x}$ appears to be an identity, which we now verify.

$$\frac{1}{2-\cos x} = \frac{\tan x}{\tan x(2-\cos x)} \quad \text{Algebra}$$
$$= \frac{\tan x}{2\tan x - \cos x \tan x} \quad \text{Algebra}$$
$$= \frac{\tan x}{2\tan x - \cos x \cdot \frac{\sin x}{\cos x}} \quad \text{Quotient Identity}$$
$$= \frac{\tan x}{2\tan x - \sin x} \quad \text{Algebra}$$

(7-1)

59.

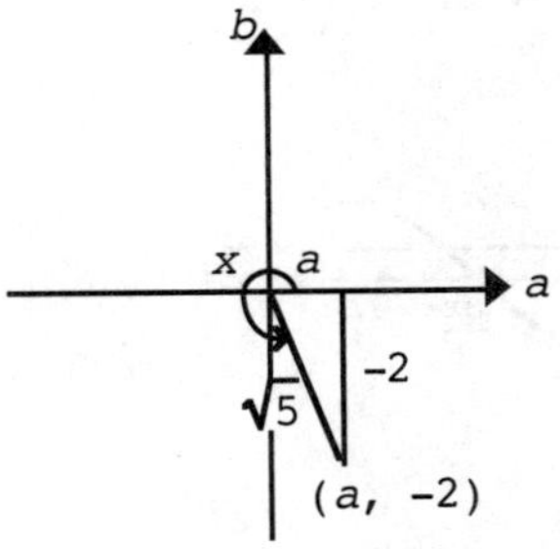

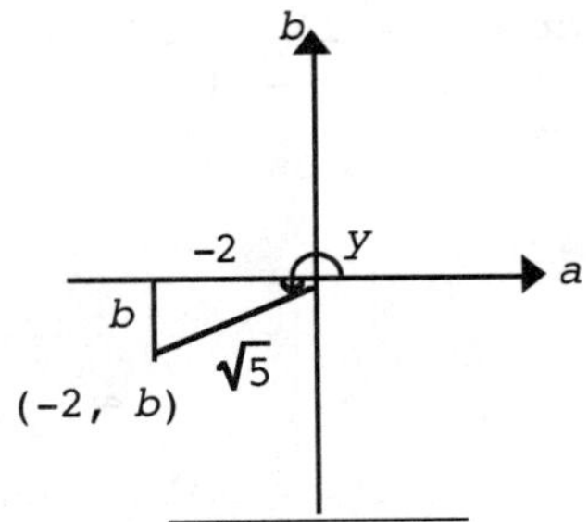

$a = \sqrt{(\sqrt{5})^2 - (-2)^2} = 1$

$\cos x = \dfrac{1}{\sqrt{5}}$

$b = -\sqrt{(\sqrt{5})^2 - (-2)^2} = -1$

$\sin y = \dfrac{-1}{\sqrt{5}}$

$$\cos(x - y) = \cos x \cos y + \sin x \sin y$$
$$= \frac{1}{\sqrt{5}}\left(-\frac{2}{\sqrt{5}}\right) + \left(\frac{-2}{\sqrt{5}}\right)\left(-\frac{1}{\sqrt{5}}\right)$$
$$= \frac{-2}{5} + \frac{2}{5}$$
$$= 0$$

(7-2)

60.

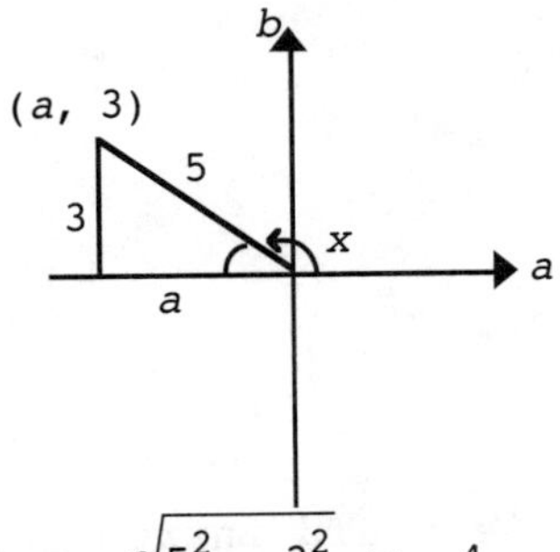

$a = -\sqrt{5^2 - 3^2} = -4$

$\cos x = -\dfrac{4}{5}$

$$\sin 2x = 2\sin x \cos x$$
$$= 2\left(\frac{3}{5}\right)\left(-\frac{4}{5}\right)$$
$$= \frac{-24}{25}$$

Since $\dfrac{\pi}{2} \le x \le \pi$, $\dfrac{\pi}{4} \le \dfrac{x}{2} \le \dfrac{\pi}{2}$,

$\cos \dfrac{x}{2}$ will be positive

$$\cos\frac{x}{2} = \sqrt{\frac{1 + \cos x}{2}}$$
$$= \sqrt{\frac{1 + \left(-\frac{4}{5}\right)}{2}} = \sqrt{\frac{\frac{1}{5}}{2}} = \sqrt{\frac{1}{10}} \text{ or } \frac{\sqrt{10}}{10}$$

(7-3)

61.

$$2\sin^2\theta + \sin\theta = 1 \qquad 0° \le \theta < 360°$$
$$2\sin^2\theta + \sin\theta - 1 = 0$$
$$(2\sin\theta - 1)(\sin\theta + 1) = 0$$
$$2\sin\theta - 1 = 0 \qquad \sin\theta + 1 = 0$$
$$\sin\theta = \frac{1}{2} \qquad \sin\theta = -1$$
$$\theta = 30°,\ 150° \qquad \theta = 270°$$

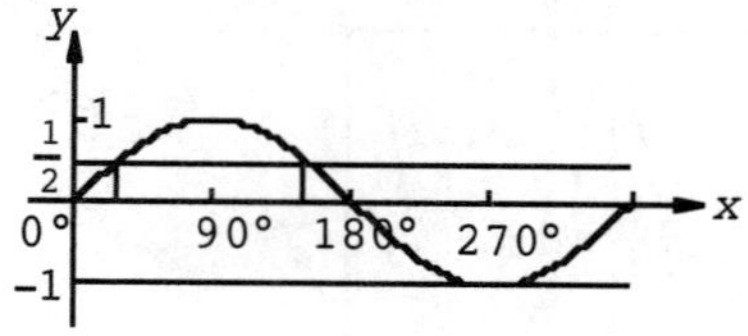

Solutions: $\theta = 30°,\ 150°,\ 270°$ (7-5)

62.

$$\sin 2x = \sin x$$
$$\sin 2x - \sin x = 0$$
$$2\sin x\cos x - \sin x = 0$$
$$\sin x(2\cos x - 1) = 0$$
$$\sin x = 0 \qquad 2\cos x - 1 = 0$$
$$x = k\pi \qquad \cos x = \frac{1}{2}$$

k any integer

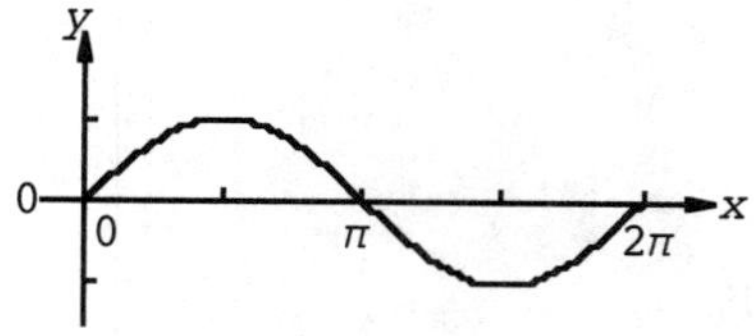

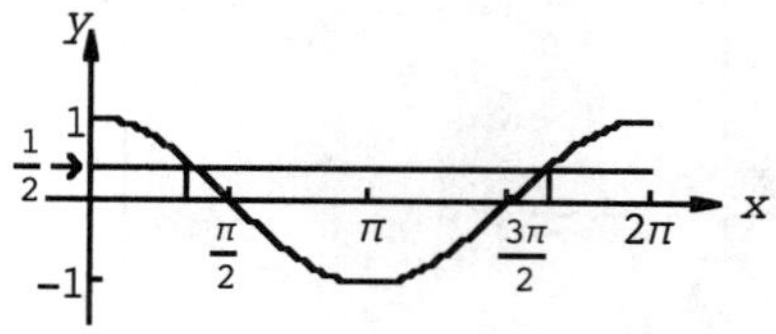

Solutions over one period

$$x = \frac{\pi}{3},\ \frac{5\pi}{3}$$

Solutions over all real x:

$$x = \frac{\pi}{3} + 2k\pi,\quad \frac{5\pi}{3} + 2k\pi = -\frac{\pi}{3} + 2\pi + 2k\pi$$
$$= -\frac{\pi}{3} + (2k+2)\pi$$
$$\text{or } -\frac{\pi}{3} + 2k\pi$$

k any integer

Solutions: $x = k\pi,\ \frac{\pi}{3} + 2k\pi,\ -\frac{\pi}{3} + 2k\pi$, k any integer. (7-5)

63. (A) $\cot x = -2\cos x$

$$\frac{\cos x}{\sin x} = -2\cos x$$
$$\frac{\cos x}{\sin x} + 2\cos x = 0$$
$$\frac{\cos x}{\sin x} + \frac{2\sin x\cos x}{\sin x} = 0$$
$$\frac{\cos x + 2\sin x\cos x}{\sin x} = 0$$

We multiply both sides by $\sin x$, keeping in mind that this could introduce extraneous solutions, hence it is necessary to check the solutions in the original equation.

$$\cos x + 2\sin x\cos x = 0$$
$$\cos x(1 + 2\sin x) = 0$$
$$\cos x = 0 \qquad 1 + 2\sin x = 0$$
$$x = \frac{\pi}{2},\ \frac{3\pi}{2} \qquad \sin x = -\frac{1}{2}$$
$$x = \frac{7\pi}{6},\ \frac{11\pi}{6}$$

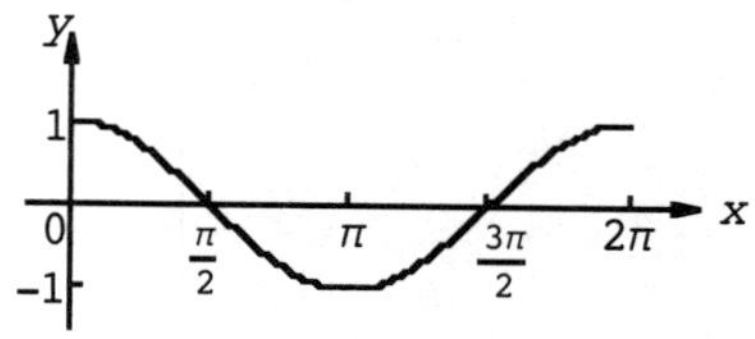

Check:

$x = \frac{\pi}{2}$	$x = \frac{3\pi}{2}$	$x = \frac{7\pi}{6}$	$x = \frac{11\pi}{6}$
$\cot \frac{\pi}{2} \overset{?}{=} -2\cos \frac{\pi}{2}$	$\cot \frac{3\pi}{2} \overset{?}{=} -2\cos \frac{3\pi}{2}$	$\cot \frac{7\pi}{6} \overset{?}{=} -2\cos \frac{7\pi}{6}$	$\cot \frac{11\pi}{6} \overset{?}{=} -2\cos \frac{11\pi}{6}$
$0 \overset{\checkmark}{=} 0$	$0 \overset{\checkmark}{=} 0$	$\sqrt{3} \overset{\checkmark}{=} -2\left(-\frac{\sqrt{3}}{2}\right)$	$-\sqrt{3} \overset{\checkmark}{=} -2\left(\frac{\sqrt{3}}{2}\right)$
A solution	A solution	A solution	A solution

Solutions: $\frac{\pi}{2}, \frac{3\pi}{2}, \frac{7\pi}{6}, \frac{11\pi}{6}$

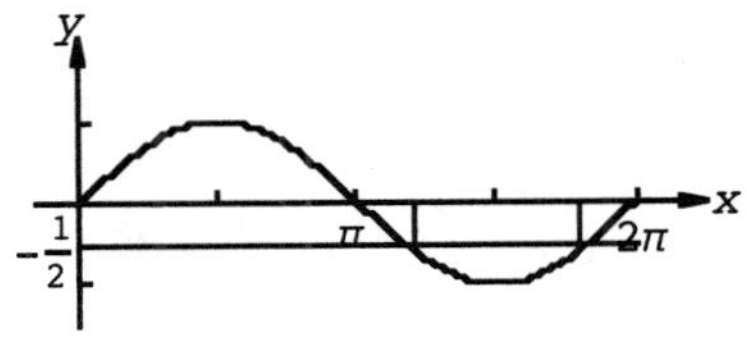

(B) Examining the graphs of $y1 = \cot x$ (drawn as $\cos x/\sin x$ in dot mode) and $y2 = -2 \cos x$, we obtain

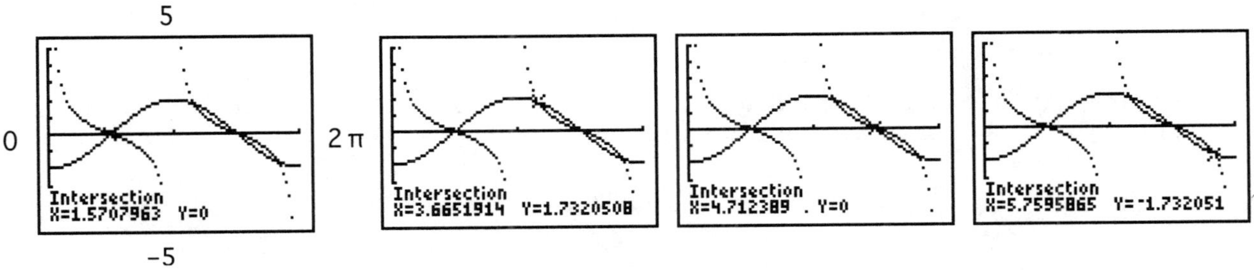

To three decimal places, the graphs intersect and $\cot x = -2 \cos x$, for $x = 1.571$, 3.665, 4.712, and 5.760. (7-5)

64. Examining the graph of $y1 = 2 \cos x$ and $y2 = x - \cos 2x$, $-2\pi \le x \le 2\pi$, we obtain the graph at the right. To three decimal places, the graphs intersect and $2 \cos x = x - \cos 2x$, for $x = 0.926$ and there are no other solutions.

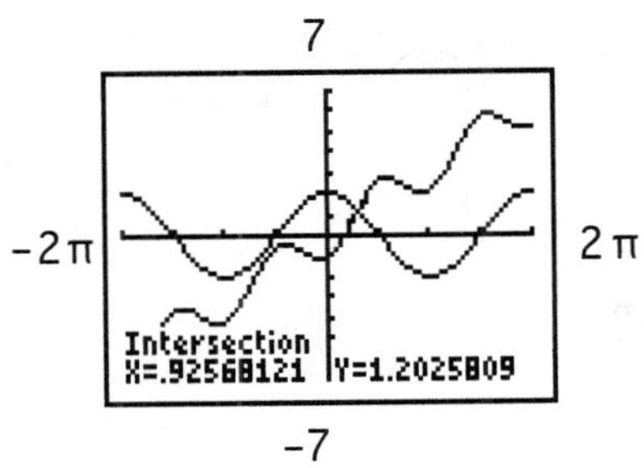

(7-5)

65.

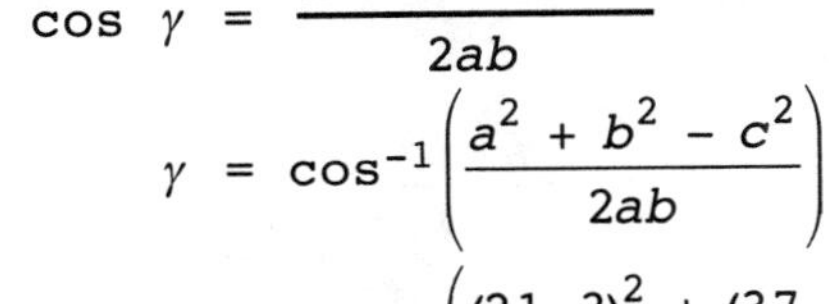

We are given three sides (SSS). We solve for the largest angle, γ (largest because it is opposite the longest side, c) using the law of cosines. We then solve for a second angle using the law of sines.

Solve for γ:

$$c^2 = a^2 + b^2 - 2ab\cos\gamma$$

$$\cos\gamma = \frac{a^2 + b^2 - c^2}{2ab}$$

$$\gamma = \cos^{-1}\left(\frac{a^2 + b^2 - c^2}{2ab}\right)$$

$$= \cos^{-1}\left(\frac{(21.3)^2 + (37.4)^2 - (48.2)^2}{2(21.3)(37.4)}\right)$$

$$= 107.2°$$

Solve for α:
We use the law of sines. Since the sum of α and β is $180° - 107.2° = 72.8°$, both must be acute.

$$\frac{\sin\alpha}{a} = \frac{\sin\gamma}{c}$$
$$\sin\alpha = \frac{a\sin\gamma}{c} = \frac{21.3\sin 107.2°}{48.2}$$
$$\alpha = \sin^{-1}\left(\frac{21.3\sin 107.2°}{48.2}\right) = 25.0°$$

Solve for β:

$$\alpha + \beta + \gamma = 180°$$
$$\beta = 180° - (\alpha + \gamma) = 180° - (25.0° + 107.2°) = 47.8°$$

(8-1, 8-2)

66. 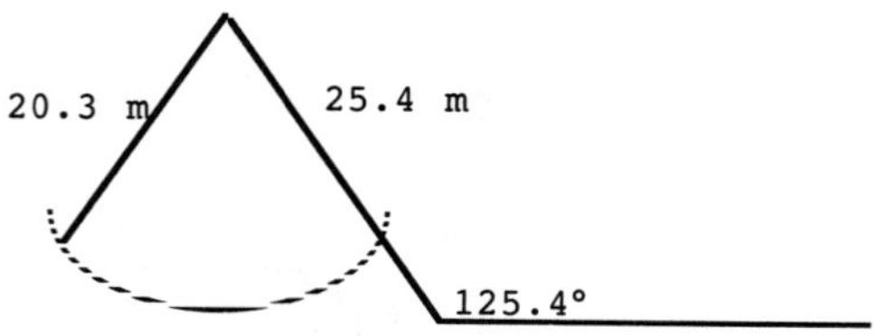

α is obtuse

$20.3 = a < b = 25.4$

We are given two sides and a non-included angle (SSA). If we try to draw a triangle with these values, we will be unsuccessful.
No solution.

(8-1)

67. 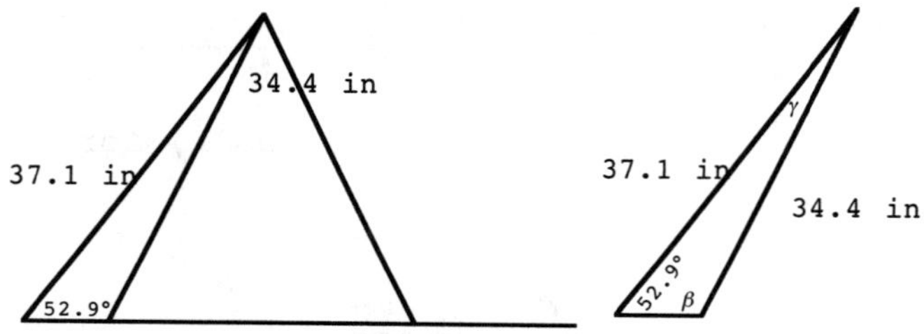

We are given two sides and a non-included angle (SSA). If we try to draw a triangle with these values, we find that two triangles are possible. Following the instructions, we choose the solution for which β is obtuse.

Solve for β:

$$\frac{\sin\beta}{b} = \frac{\sin\alpha}{a}$$
$$\sin\beta = \frac{b\sin\alpha}{a} = \frac{37.1\sin 52.9°}{34.4} = 0.8602 \quad \beta \text{ is chosen obtuse}$$
$$\beta = 180° - \sin^{-1} 0.8602 = 180° - 59.3° = 120.7°$$

Solve for γ:

$$\alpha + \beta + \gamma = 180°$$
$$\gamma = 180° - (\alpha + \beta) = 180° - (52.9° + 120.7°) = 6.4°$$

Solve for c:

$$\frac{\sin\alpha}{a} = \frac{\sin\gamma}{c}$$
$$c = \frac{a\sin\gamma}{\sin\alpha} = \frac{34.4\sin 6.4°}{\sin 52.9°} = 4.81 \text{ in}$$

(8-1)

68. β must be acute. A triangle can have at most one obtuse angle, and since γ is acute, the obtuse angle, if present, must be opposite the longer of the two sides a and b. (*8-2*)

69.

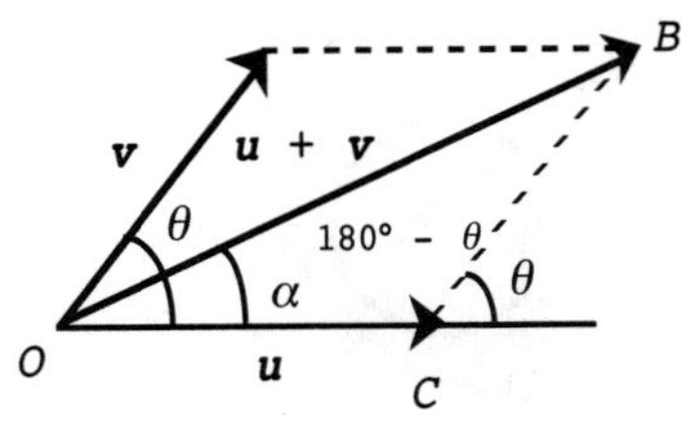

$\theta = 48.3°$. Hence, $\angle OCB = 180° - \theta = 180° - 48.3° = 131.7°$. We can find $|\mathbf{u} + \mathbf{v}|$ using the law of cosines:

$$\begin{aligned} |\mathbf{u} + \mathbf{v}|^2 &= |\mathbf{u}|^2 + |\mathbf{v}|^2 - 2|\mathbf{u}||\mathbf{v}|\cos(OCB) \\ &= (25.3)^2 + (13.4)^2 - 2(25.3)(13.4)\cos 131.7° \\ &= 1270.7028\ldots \\ |\mathbf{u} + \mathbf{v}| &= 35.6 \text{ lb} \end{aligned}$$

To find α, we use the law of sines:

$$\frac{\sin\alpha}{|\mathbf{v}|} = \frac{\sin(OCB)}{|\mathbf{u} + \mathbf{v}|}$$

$$\frac{\sin\alpha}{13.4} = \frac{\sin 131.7°}{35.6}$$

$$\sin\alpha = \frac{13.4}{35.6}\sin 131.7°$$

$$\begin{aligned} \alpha &= \sin^{-1}\left(\frac{13.4}{35.6}\sin 131.7°\right) \\ &= 16.3° \end{aligned}$$

(*8-1, 8-2, 8-3*)

70. (A)

$$\begin{aligned} 2\mathbf{u} - \mathbf{v} + 3\mathbf{w} &= 2\langle -1, 2\rangle - \langle 0, -2\rangle + 3\langle 1, -1\rangle \\ &= \langle -2, 4\rangle + \langle 0, 2\rangle + \langle 3, -3\rangle \\ &= \langle 1, 3\rangle \end{aligned}$$

(B)

$$\begin{aligned} 2\mathbf{u} - \mathbf{v} + 3\mathbf{w} &= 2(2\mathbf{i} - \mathbf{j}) - (\mathbf{i} + 3\mathbf{j}) + 3(2\mathbf{j}) \\ &= 4\mathbf{i} - 2\mathbf{j} - \mathbf{i} - 3\mathbf{j} + 6\mathbf{j} \\ &= 3\mathbf{i} + \mathbf{j} \end{aligned}$$

(*8-3*)

71. $x^2 + y^2 = 8y$

Use $r^2 = x^2 + y^2$ and $y = r\sin\theta$

$$\begin{aligned} r^2 &= 8r\sin\theta \\ r^2 - 8r\sin\theta &= 0 \\ r(r - 8\sin\theta) &= 0 \end{aligned}$$

$r = 0$ or $r - 8\sin\theta = 0$

The graph of $r = 0$ is the pole. Since the pole is included in the graph of $r - 8\sin\theta = 0$, we can discard $r = 0$ and keep only

$r - 8\sin\theta = 0$ or

$r = 8\sin\theta$ (*8-4*)

72. $r = -4\cos\theta$

We multiply both sides by r, which adds the pole to the graph. But the pole is already part of the graph, so we have changed nothing.

$r^2 = -4r\cos\theta$

But $r^2 = x^2 + y^2$ and $r\cos\theta = x$. Hence, $x^2 + y^2 = -4x$. (*8-4*)

73. We set up a table that indicates how r varies as we let θ vary through each set of quadrant values.

θ	$\cos\theta$	$4\cos\theta$	$4 + 4\cos\theta$
0 to $\frac{\pi}{2}$	1 to 0	4 to 0	8 to 4
$\frac{\pi}{2}$ to π	0 to -1	0 to -4	4 to 0
π to $\frac{3\pi}{2}$	-1 to 0	-4 to 0	0 to 4
$\frac{3\pi}{2}$ to 2π	0 to 1	0 to 4	4 to 8

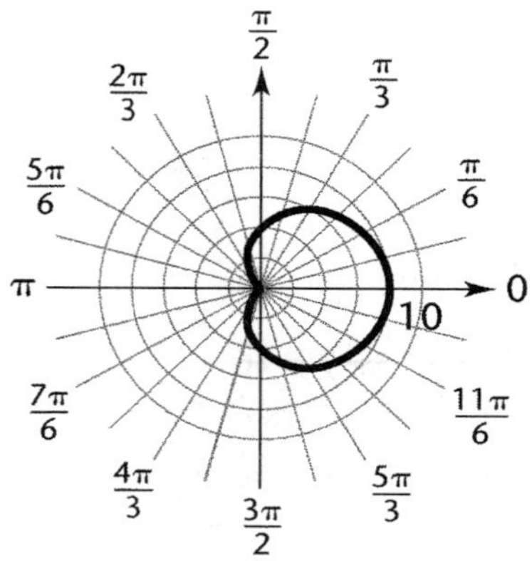

Check:

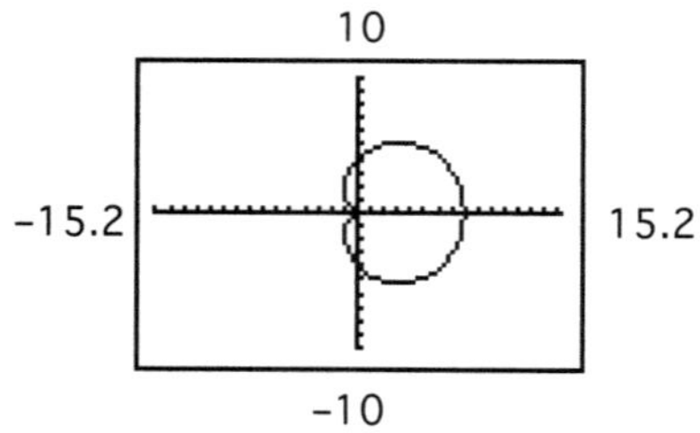

(8-4)

74. We set up a table that indicates how r varies as we let 3θ vary through each set of quadrant values.

θ	3θ	$\sin 3\theta$	$6\sin 3\theta$
0 to $\frac{\pi}{6}$	0 to $\frac{\pi}{2}$	0 to 1	0 to 6
$\frac{\pi}{6}$ to $\frac{\pi}{3}$	$\frac{\pi}{2}$ to π	1 to 0	6 to 0
$\frac{\pi}{3}$ to $\frac{\pi}{2}$	π to $\frac{3\pi}{2}$	0 to -1	0 to -6
$\frac{\pi}{2}$ to $\frac{2\pi}{3}$	$\frac{3\pi}{2}$ to 2π	-1 to 0	-6 to 0
$\frac{2\pi}{3}$ to $\frac{5\pi}{6}$	2π to $\frac{5\pi}{2}$	0 to 1	0 to 6
$\frac{5\pi}{6}$ to π	$\frac{5\pi}{2}$ to 3π	1 to 0	6 to 0
π to 2π	repeats curve already drawn		

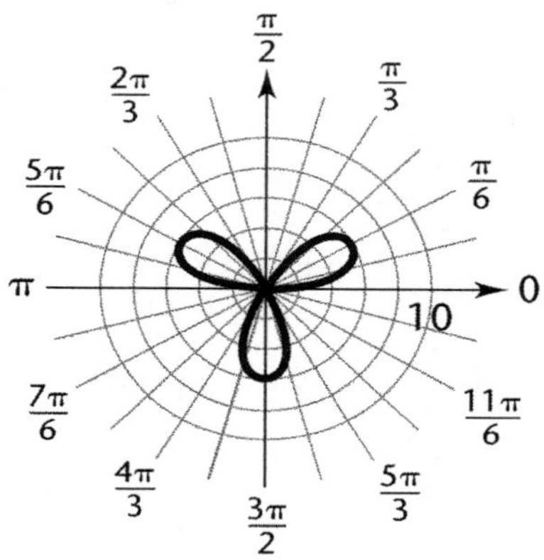

Check:

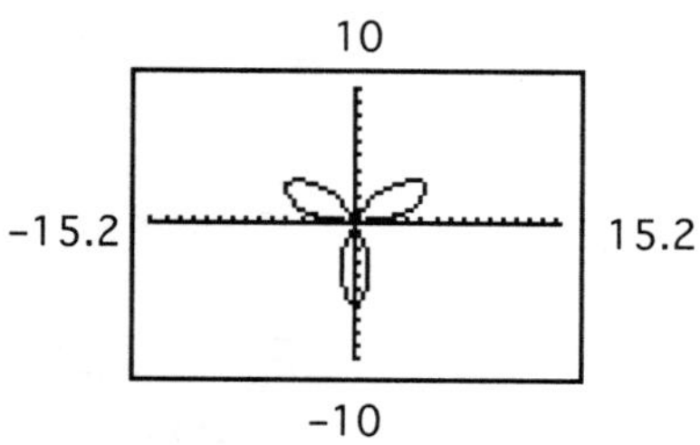

(8-4)

75. $n = 1$

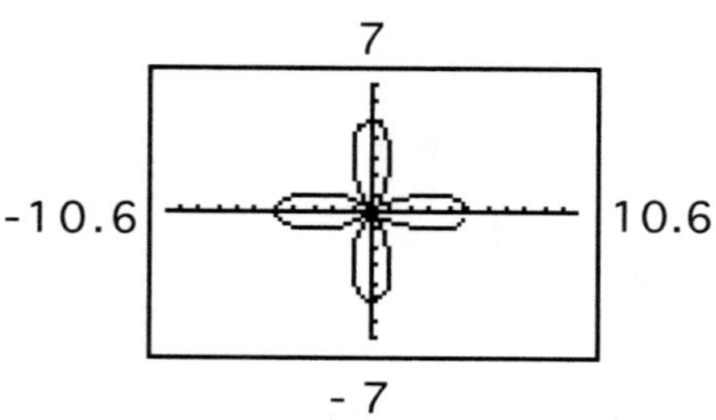

$n = 2$

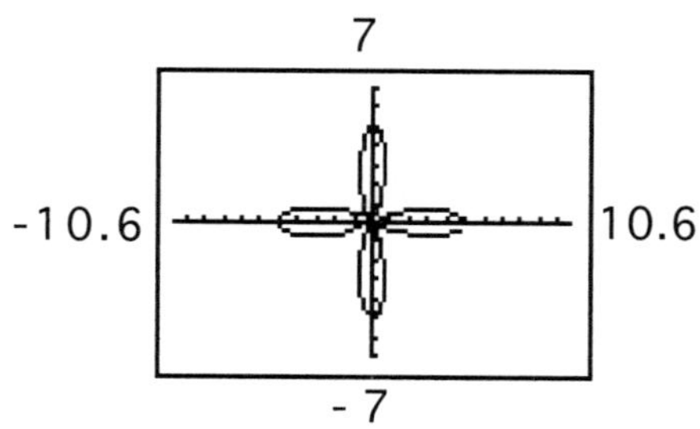

$n = 3$

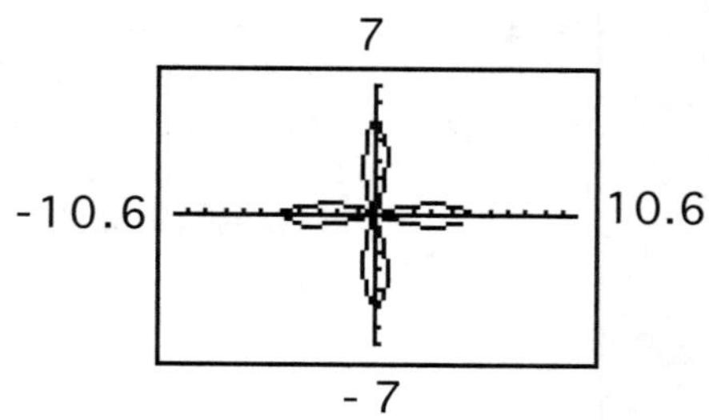

Generalizing from these graphs, we expect 4 leaves for all n. (8-4)

76.

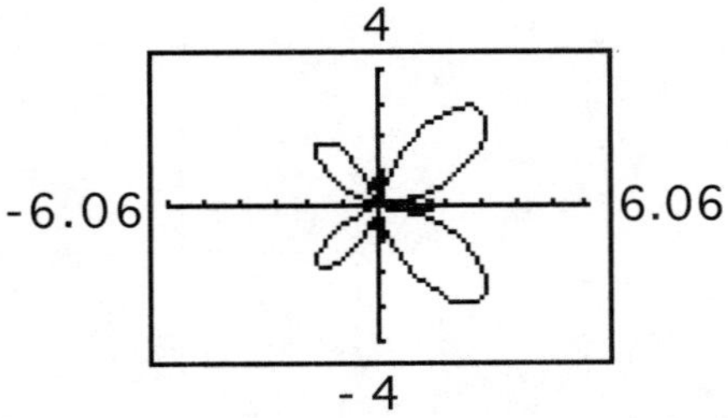

(8-4)

77. $(x, y) = (-2.78, -3.19)$

$r = \sqrt{x^2 + y^2} = \sqrt{(-2.78)^2 + (-3.19)^2} = 4.23$

$\tan\theta = \dfrac{y}{x} = \dfrac{-3.19}{-2.78}$

θ is a third quadrant angle and is to be chosen so that $-180° < \theta \le 180°$.

$\theta = \tan^{-1}\dfrac{-3.19}{-2.78} - 180° = -131.07°$

$(r, \theta) = (4.23, -131.07°)$ (8-4)

78. $(r, \theta) = (6.22, -4.08)$

$x = r\cos\theta = 6.22\cos(-4.08) = -3.68$

$y = r\sin\theta = 6.22\sin(-4.08) = 5.02$

$(x, y) = (-3.68, 5.02)$ (8-4)

79. $2e^{(-\pi/6)i} = 2\left[\cos\left(-\dfrac{\pi}{6}\right) + i\sin\left(-\dfrac{\pi}{6}\right)\right]$

$= 2\left[\dfrac{\sqrt{3}}{2} + i\left(-\dfrac{1}{2}\right)\right]$

$= \sqrt{3} - i$ (8-5)

80. A sketch shows that $-1 + i\sqrt{3}$ is associated with a special 30°-60° triangle. Thus, by inspection, $r = 2$, $\theta = 120°$, and

$-1 + i\sqrt{3} = 2(\cos 120° + i\sin 120°)$
$= 2e^{120°i}$

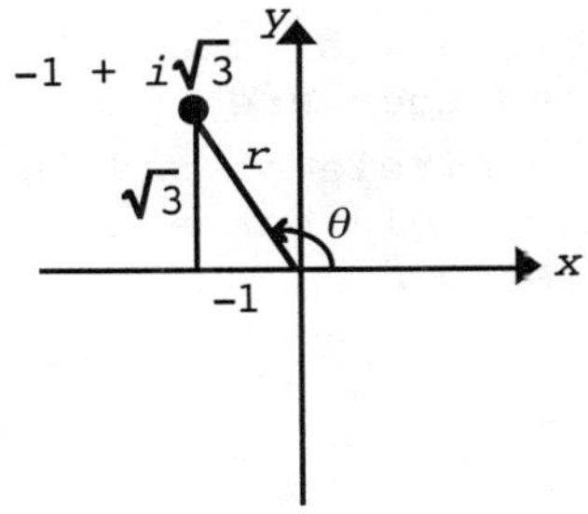

(8-5)

81. $(1 - i\sqrt{3})^6 = (2e^{-60°i})^6 = 2^6 e^{6(-60°i)} = 64e^{-360°i} = 64[\cos(-360°) + i\sin(-360°)]$
$= 64(1) = 64$ or $64 + 0i$ (8-5)

82. First write $-i$ in polar form

$-i = 0 - i = 1e^{-90°i}$

Using the nth-root theorem, all three cube roots of $1e^{-90°i}$ are given by

$1^{1/3}e^{(-90°/3 + k\cdot 360°/3)i} = 1^{1/3}e^{(-30° + k120°)i} \quad k = 0, 1, 2$

Thus,

$w_1 = 1^{1/3}e^{(-30° + 0\cdot 120°)i} = e^{-30°i} = \cos(-30°) + i\sin(-30°) = \dfrac{\sqrt{3}}{2} - i\dfrac{1}{2}$

$w_2 = 1^{1/3}e^{(-30° + 1\cdot 120°)i} = e^{90°i} = \cos 90° + i\sin 90° = i$

$w_3 = 1^{1/3}e^{(-30° + 2\cdot 120°)i} = e^{210°i} = \cos 210° + i\sin 210° = -\dfrac{\sqrt{3}}{2} - i\dfrac{1}{2}$

These roots lie on a circle of radius 1, equally spaced so that the angle between successive roots is 120°.

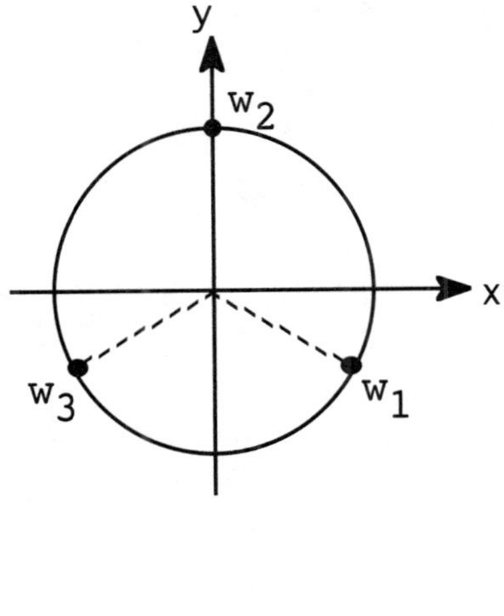

(8-5)

83. $-4.88 - 3.17i$ is not associated with a special triangle.

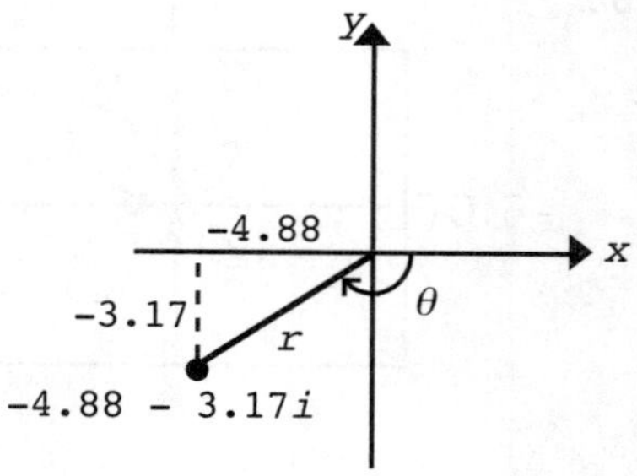

$$r = \sqrt{(-4.88)^2 + (-3.17)^2} = 5.82$$

$$\theta = \tan^{-1}\frac{-3.17}{-4.88} - 180° = -146.99°$$

$$-4.88 - 3.17i = 5.82[\cos(-146.99°) + i\sin(-146.99°)]$$
$$= 5.82e^{(-146.99°)i}$$

(8-5)

84. $x + iy = 6.97e^{163.87°i}$
$= 6.97(\cos 163.87° + i \sin 163.87°)$
$= -6.70 + 1.94i$ (8-5)

85. (A) There are a total of four fourth roots and they are spaced equally around a circle of radius $\sqrt{2}$, so that the angle between successive roots is 90°.

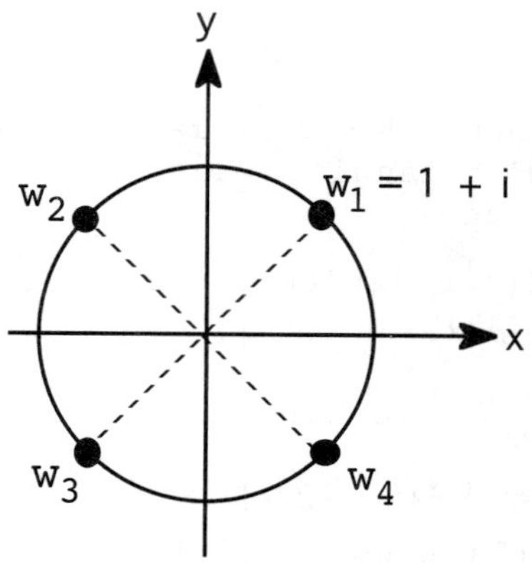

(B) Since $w_1 = \sqrt{2}e^{45°i}$, $w_2 = \sqrt{2}e^{(45°+90°)i} = \sqrt{2}e^{135°i}$,
$w_3 = \sqrt{2}e^{(135°+90°)i} = \sqrt{2}e^{225°i}$, and $w_4 = \sqrt{2}e^{(225°+90°)i} = \sqrt{2}e^{315°i}$

Thus, $w_2 = \sqrt{2}(\cos 135° + i \sin 135°) = -1 + i$
$w_3 = \sqrt{2}(\cos 225° + i \sin 225°) = -1 - i$
$w_4 = \sqrt{2}(\cos 315° + i \sin 315°) = 1 - i$

(C) $(1 + i)^4 = [(1 + i)^2]^2 = (1 + 2i + i^2)^2 = (2i)^2 = -4$
$(-1 + i)^4 = [(-1 + i)^2]^2 = (1 - 2i + i^2)^2 = (-2i)^2 = -4$
$(-1 - i)^4 = [(-1 - i)^2]^2 = (1 + 2i + i^2)^2 = (2i)^2 = -4$
$(1 - i)^4 = [(1 - i)^2]^2 = (1 - 2i + i^2)^2 = (-2i)^2 = -4$ (8-5)

86. On this unit circle,
$P(a, b) = P(\cos\theta, \sin\theta)$
$= P(\cos(s \text{ rad}), \sin(s \text{ rad}))$
$= P(\cos 1.2, \sin 1.2)$
$= (0.362, 0.932)$

(6-2)

87. We graph $y = \sec x$, then translate the graph up one unit.

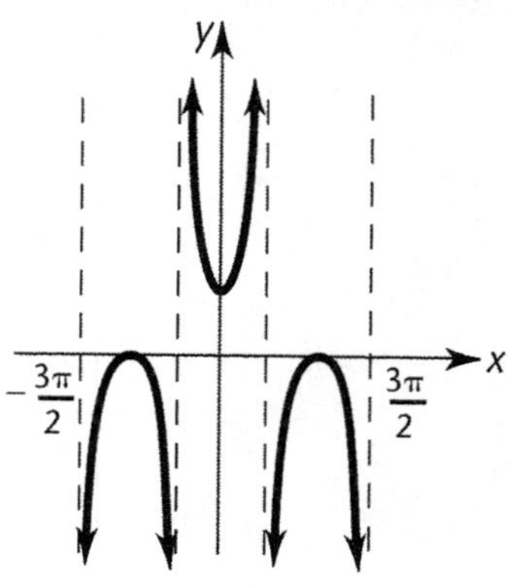

(6-4)

88. $|A| = 3$. The graph completes one full cycle as x varies over the (mentally extended) intervals $\left[-\frac{7}{8}, \frac{1}{8}\right]$ or $\left[\frac{1}{8}, \frac{9}{8}\right]$. Since the phase shift is required between 0 and 1, we must set $-\frac{C}{B} = \frac{1}{8}$. Then the graph has the form of a standard cosine curve. Hence

$$A = 3$$

$$-\frac{C}{B} = \frac{1}{8} \qquad -\frac{C}{B} + \frac{2\pi}{B} = \frac{9}{8}$$

$$\frac{2\pi}{B} = 1 \qquad B = 2\pi \qquad C = -\frac{1}{8}B = -\frac{\pi}{4}$$

The equation is then $y = A\cos(Bx + C)$

$$y = 3\cos\left(2\pi x - \frac{\pi}{4}\right)$$

The amplitude is 3, the period is 1, and the phase shift is $\frac{1}{8}$. (6-5)

89. Here is a graph of $y = 1.6 \sin 2x - 1.2 \cos 2x$ in a graphing utility.

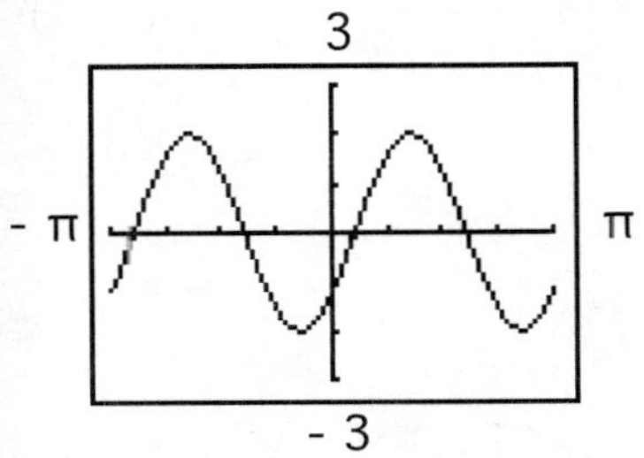

It appears that this is a sine curve shifted to the right, with $A = 2$ and, since $P = \frac{2\pi}{B}$ and P appears to be π, $B = \frac{2\pi}{P} = \frac{2\pi}{\pi} = 2$.

From the graphing utility, we find that the x intercept closest to the origin, to three decimal places, is 0.322. To find C, substitute $B = 2$ and $x = 0.322$ into the phase shift formula $x = -\frac{C}{B}$ and solve for C:

$$x = -\frac{C}{B}$$

$$0.322 = -\frac{C}{2}$$

$$C = -0.644$$

The equation is thus $y = 2\sin(2x - 0.644)$. (6-5)

90. Let $y = \cos^{-1} x$, then $x = \cos y \quad 0 \le y \le \pi$

Then $\csc y = \frac{1}{\sin y} = \frac{1}{\sqrt{1 - \cos^2 y}}$ from the reciprocal and Pythagorean identities.

(The positive sign is chosen for the square root since $\sin y$ is positive in Quadrants I and II.)

Hence $\csc y = \csc(\cos^{-1} x) = \frac{1}{\sqrt{1 - \cos^2 y}} = \frac{1}{\sqrt{1 - x^2}}$. (6-6)

91. Let $u = \cot^{-1} \frac{3}{4}$. Then $\cot u = \frac{3}{4}$, $0 < u < \pi$.

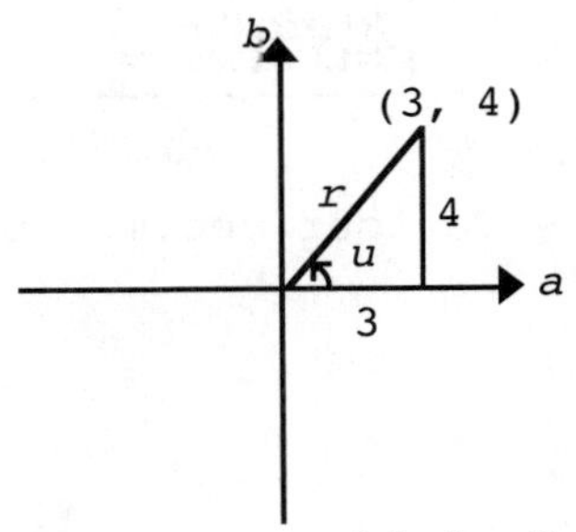

$$r = \sqrt{3^2 + 4^2} = 5$$

$$\sin u = \frac{4}{5} \qquad \cos u = \frac{3}{5}$$

$$\sin\left[2\cot^{-1}\left(\frac{3}{4}\right)\right] = \sin 2u = 2\sin u \cos u = 2\left(\frac{4}{5}\right)\left(\frac{3}{5}\right) = \frac{24}{25}$$

(6-6, 7-3)

92. $b = \sqrt{5^2 - (-3)^2} = 4$

$\cos x = -\frac{3}{5}$

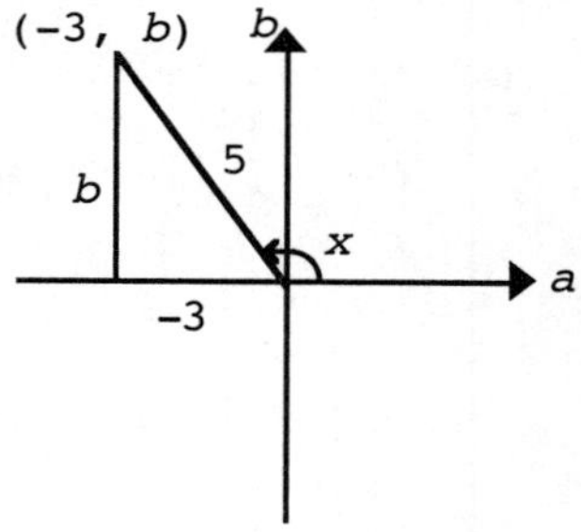

(A) Since $\frac{\pi}{2} \le x \le \pi$, $\frac{\pi}{4} \le \frac{x}{2} \le \frac{\pi}{2}$, $\sin \frac{x}{2}$ will be positive.

$$\sin \frac{x}{2} = \sqrt{\frac{1 - \cos x}{2}} = \sqrt{\frac{1 - \left(-\frac{3}{5}\right)}{2}}$$
$$= \sqrt{\frac{\frac{8}{5}}{2}}$$
$$= \sqrt{\frac{4}{5}}$$
$$= \frac{2}{\sqrt{5}} \text{ or } \frac{2\sqrt{5}}{5}$$

(B) $\cos 2x = 2\cos^2 x - 1 = 2\left(-\frac{3}{5}\right)^2 - 1 = 2\left(\frac{9}{25}\right) - 1 = \frac{18}{25} - 1 = -\frac{7}{25}$ (7-3)

93. (A) $2\sin^2 x = 3\cos x \quad 0 \le x < 2\pi$

$$2(1 - \cos^2 x) = 3\cos x$$
$$2 - 2\cos^2 x = 3\cos x$$
$$0 = 2\cos^2 x + 3\cos x - 2$$
$$0 = (2\cos x - 1)(\cos x + 2)$$

$2\cos x - 1 = 0 \qquad \cos x + 2 = 0$

$\cos x = \frac{1}{2} \qquad \cos x = -2$

$x = \frac{\pi}{3}, \frac{5\pi}{3} \qquad$ No solution

Solutions: $\frac{\pi}{3}, \frac{5\pi}{3}$

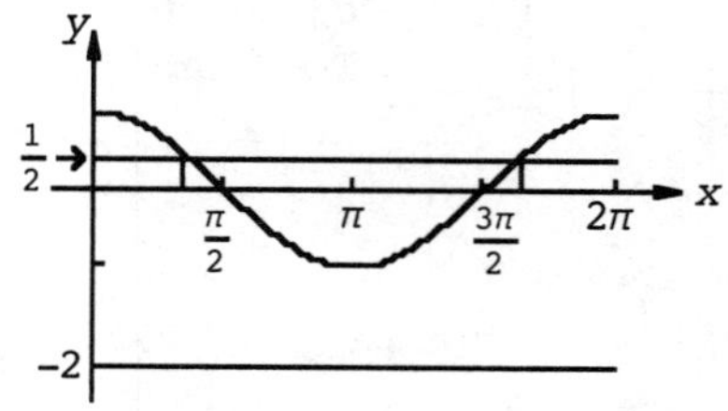

(B) Examining the graph of $y1 = 2\sin^2 x$ and $y2 = 3\cos x$, $0 \le x \le 2\pi$, we obtain

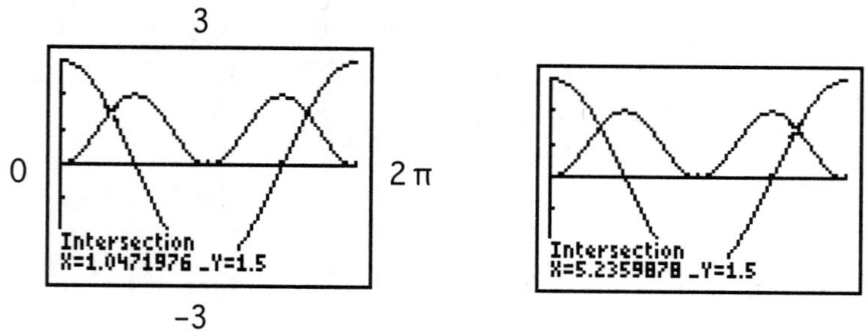

To four decimal places, the solutions are 1.0472 and 5.2360. (7-5)

94. (A) If $r^2 = 36\cos 2\theta$, we can write $r = \pm\sqrt{36\cos 2\theta}$. We then set up a table that indicates how r varies as we let 2θ vary through each set of quadrant values.

θ	2θ	$\cos 2\theta$	$36\cos 2\theta$	$\pm\sqrt{36\cos 2\theta}$
0 to $\frac{\pi}{4}$	0 to $\frac{\pi}{2}$	1 to 0	36 to 0	{6 to 0, −6 to 0
$\frac{\pi}{4}$ to $\frac{\pi}{2}$	$\frac{\pi}{2}$ to π	0 to −1	0 to −36	Not defined
$\frac{\pi}{2}$ to $\frac{3\pi}{4}$	π to $\frac{3\pi}{2}$	−1 to 0	−36 to 0	Not defined
$\frac{3\pi}{4}$ to π	$\frac{3\pi}{2}$ to 2π	0 to 1	0 to 36	{0 to 6, 0 to −6

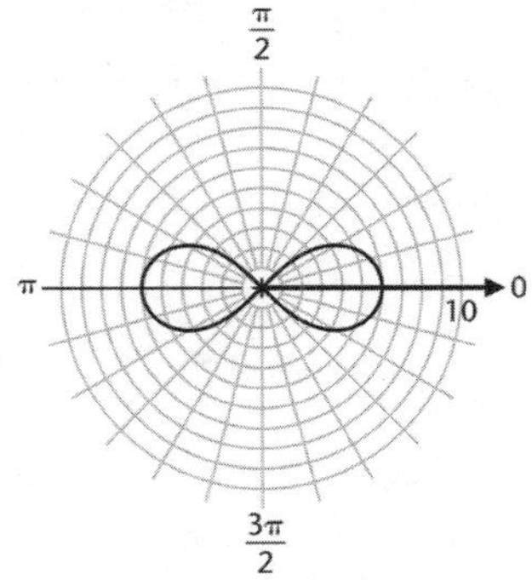

(B)

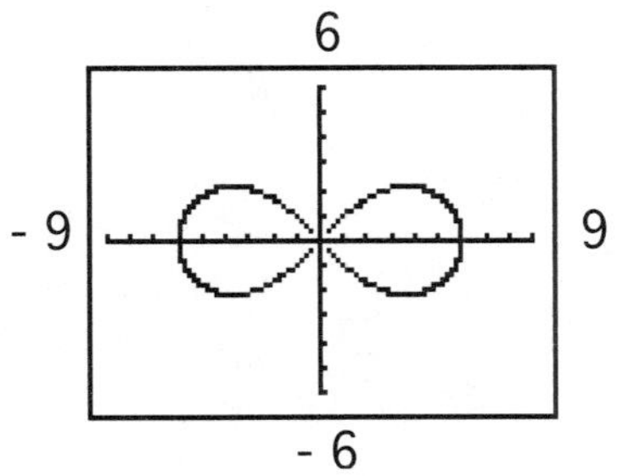

(8-4)

95. (A)

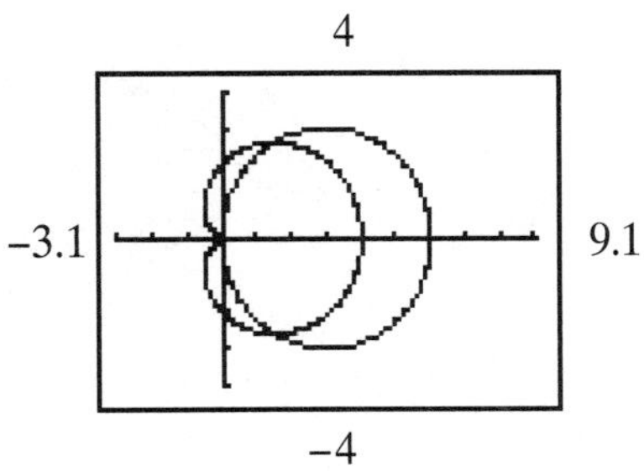

(B) 6

(C) We solve the system

$r = 2 + 2\cos\theta \qquad 0 \le \theta \le 2\pi$

$r = 6\cos\theta$

by equating the right sides:

$6\cos\theta = 2 + 2\cos\theta$

$4\cos\theta = 2$

$\cos\theta = \frac{1}{2}$

$\theta = \frac{\pi}{3}, \frac{5\pi}{3}$

If we substitute these values of θ in either of the original equations, we get the corresponding values of r.

$\theta = \frac{\pi}{3} \qquad r = 6\cos\frac{\pi}{3} = 3$

$\theta = \frac{5\pi}{3} \qquad r = 6\cos\frac{5\pi}{3} = 3$

The two solutions of the system are $\left(3, \frac{\pi}{3}\right)$ and $\left(3, \frac{5\pi}{3}\right)$.

(D) The points on $r2$ and $r1$ arrive at the intersection points for different values of θ, except for the two found in part (C). (8-4)

96. The linear factors of $x^3 + i$ will, by the factor theorem, be $x - x_i$, where x_i are zeros of $x^3 + i$, that is, $x^3 + i = 0$ or $x^3 = -i$. Hence the x_i are the three cube roots of $-i$, or $1e^{(-90°)i}$.

Using the nth-root theorem, all three cube roots of $-i$ are given by

$$1^{1/3}e^{(-90°/3 + k \cdot 360°/3)i} = 1e^{(-30° + k \cdot 120°)i} \quad k = 0, 1, 2$$

Thus,

$$w_1 = 1e^{(-30° + 0 \cdot 120°)i} = e^{-30°i} = \cos(-30°) + i\sin(-30°) = \frac{\sqrt{3}}{2} - \frac{i}{2}$$

$$w_2 = 1e^{(-30° + 1 \cdot 120°)i} = e^{90°i} = \cos 90° + i\sin 90° = i$$

$$w_3 = 1e^{(-30° + 2 \cdot 120°)i} = e^{210°i} = \cos 210° + i\sin 210° = -\frac{\sqrt{3}}{2} - \frac{i}{2}$$

Hence, $x^3 + i = \left[x - \left(\frac{\sqrt{3}}{2} - \frac{i}{2}\right)\right](x - i)\left[x - \left(-\frac{\sqrt{3}}{2} - \frac{i}{2}\right)\right]$ *(8-5)*

97. In one year the line sweeps out one full revolution, or 2π radians. In 5 days the line sweeps out $\frac{5}{365}$ of a full revolution, or $\frac{5}{365} \cdot 2\pi = \frac{2\pi}{73}$ radians. *(6-1)*

98. In the diagram, let E be the point from which the balloon was released. Then BDE and BCE are right triangles.

In Triangle BDE, $\cot 37° = \frac{DE}{h}$

In triangle BCE, $\cot 24° = \frac{1{,}000 + DE}{h}$

Thus, $\cot 24° - \cot 37° = \frac{1{,}000 + DE}{h} - \frac{DE}{h}$

$$\cot 24° - \cot 37° = \frac{1{,}000}{h}$$

$$h(\cot 24° - \cot 37°) = 1{,}000$$

$$h = \frac{1{,}000}{\cot 24° - \cot 37°}$$

$$h = 1{,}088 \text{ m}$$

(6-3)

99. Sketch a figure:

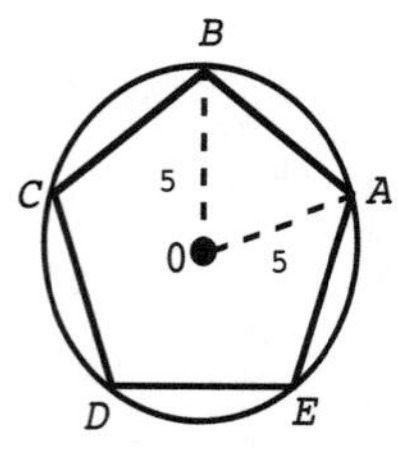

Angle $AOB = \frac{1}{5}(360°) = 72°$

We use the law of cosines in triangle AOB to determine side AB.

$$(AB)^2 = 5^2 + 5^2 - 2(5)(5)\cos 72°$$
$$= 34.54915\ldots$$
$$AB = 5.88 \text{ in}$$

(6-3, 8-2)

100. The three sides of triangle ABC are each in turn the hypotenuse of a right triangle formed with two edges of the solid. Hence, by the Pythagorean theorem,

$AB^2 = 12^2 + 14^2 = 340$ $\quad AB = 18.44$ cm

$AC^2 = 12^2 + 42^2 = 1{,}908$ $\quad AC = 43.68$ cm

$BC^2 = 14^2 + 42^2 = 1{,}960$ $\quad BC = 44.27$ cm

To find $\angle ABC$, we apply the law of cosines.

$$AC^2 = AB^2 + BC^2 - 2(AB)(BC)\cos ABC$$

$$\cos ABC = \frac{AB^2 + BC^2 - AC^2}{2(AB)(BC)} = \frac{340 + 1{,}960 - 1{,}908}{2(18.44)(44.27)}$$

$$\angle ABC = \cos^{-1}\left(\frac{340 + 1{,}960 - 1{,}908}{2(18.44)(44.27)}\right) = 76° \qquad (8\text{-}2)$$

101. When $t = 0$, $I = 50$. Hence $50 = A \cos B(0)$, that is, $A = 50$. Since the period is $\frac{1}{110}$ seconds, $\frac{2\pi}{B} = \frac{1}{110}$, $B = 220\pi$. Hence the equation is

$$I = 50 \cos 220\pi t \qquad (6\text{-}5)$$

102. Sketch a figure.

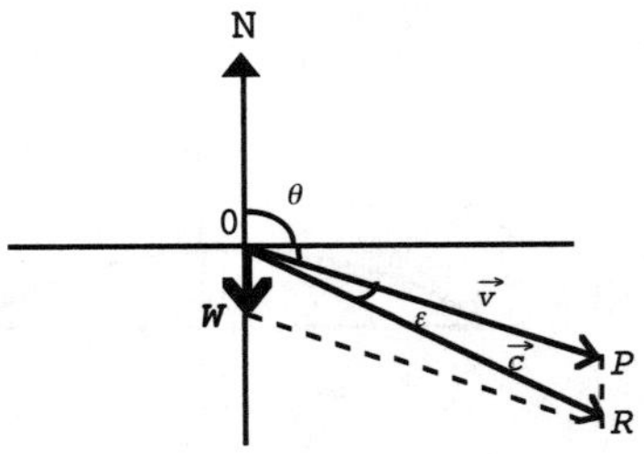

The actual course $\vec{c}$ will be the resultant of $\vec{v}$, the plane's heading, and $\vec{w}$, the wind velocity. In triangle OPR, we know $OP = |\vec{v}| = 260$, $PR = |\vec{w}| = 36$, and angle OPR, which is congruent to θ, the heading of 110°.

Given two sides and the included angle, we use the law of cosines to find

$$|\vec{c}| = OR$$
$$|\vec{c}|^2 = OP^2 + PR^2 - 2(OP)(PR)\cos OPR$$
$$= (260)^2 + (36)^2 - 2(260)(36)\cos 110°$$
$$= 75{,}298.617\ldots$$
$$|\vec{c}| = 274 \text{ miles per hour.}$$

To find the actual direction of the plane, we must find angle POR, or ε. Then the heading will be $\theta + \varepsilon$. We use the law of sines.

$$\frac{\sin \varepsilon}{PR} = \frac{\sin OPR}{OR}$$
$$\frac{\sin \varepsilon}{36} = \frac{\sin 110°}{274}$$
$$\sin \varepsilon = \frac{36 \sin 110°}{274}$$
$$\sin \varepsilon = 0.1233$$
$$\varepsilon = \sin^{-1} 0.1233 \text{ since } \varepsilon \text{ is acute}$$
$$\varepsilon = 7°$$

Thus the heading of the plane's actual direction $= \theta + \varepsilon = 110° + 7° = 117°$.

(8-3)

103. First, form a force diagram with all forces in standard position at the origin.

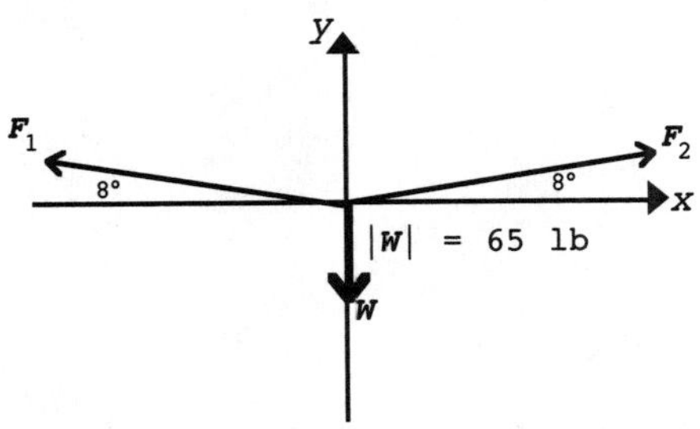

Let $\boldsymbol{F}_1$ = the tension in left cable
$\boldsymbol{F}_2$ = the tension in right cable

Write each force vector in terms of $\boldsymbol{i}$ and $\boldsymbol{j}$ unit vectors.

$$\boldsymbol{F}_1 = |\boldsymbol{F}_1|(-\cos 8°)\boldsymbol{i} + |\boldsymbol{F}_1|(\sin 8°)\boldsymbol{j}$$
$$\boldsymbol{F}_2 = |\boldsymbol{F}_2|(\cos 8°)\boldsymbol{i} + |\boldsymbol{F}_2|(\sin 8°)\boldsymbol{j}$$
$$\boldsymbol{W} = -65\boldsymbol{j}$$

For the system to be in static equilibrium, we must have
$\mathbf{F}_1 + \mathbf{F}_2 + \mathbf{W} = \mathbf{O}$
which becomes, on addition,
$[-|\mathbf{F}_1|\cos 8° + |\mathbf{F}_2|\cos 8°]\mathbf{i} + [|\mathbf{F}_1|\sin 8° + |\mathbf{F}_2|\sin 8° - 65]\mathbf{j} = 0\mathbf{i} + 0\mathbf{j}$

Since two vectors are equal if and only if their corresponding components are equal, we are led to the following system of equations in $|\mathbf{F}_1|$ and $|\mathbf{F}_2|$:

$$-|\mathbf{F}_1|\cos 8° + |\mathbf{F}_2|\cos 8° = 0$$
$$|\mathbf{F}_1|\sin 8° + |\mathbf{F}_2|\sin 8° - 65 = 0$$

Solving, we obtain
$|\mathbf{F}_1| = |\mathbf{F}_2|$ (as we expect from the symmetry of the situation)
$|\mathbf{F}_1| = \dfrac{65}{2\sin 8°} = 234$ lb = tension in each half of the cable. (8-4)

104. (A) Add the perpendicular bisector of the chord as shown in the figure. Then, $\sin\theta = \dfrac{4}{R}$ and $\theta = \dfrac{5}{R}$.
Substituting the second into the first, we obtain
$\sin\dfrac{5}{R} = \dfrac{4}{R}$.

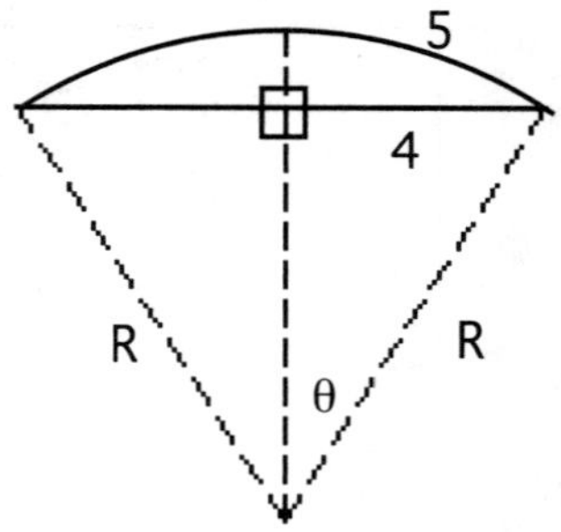

(B) R cannot be isolated on one side of the equation.

(C) Plot $y1 = \sin\dfrac{5}{R}$ and $y2 = \dfrac{4}{R}$ in the same viewing window and solve for R at the point of intersection using a built-in routine (see figure). To three decimal places, $R = 4.420$ cm.

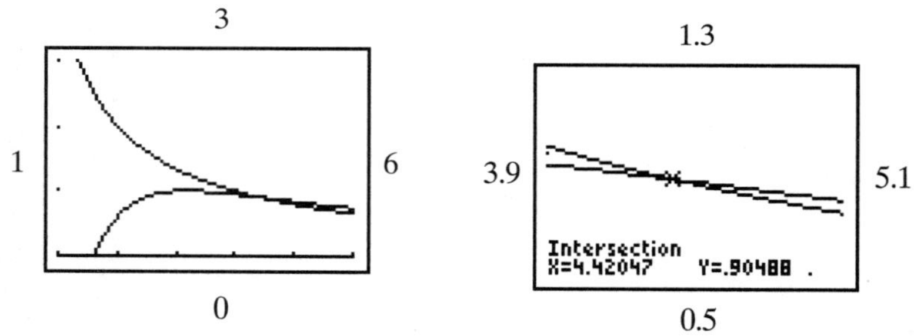

(7-5)

105. (A)

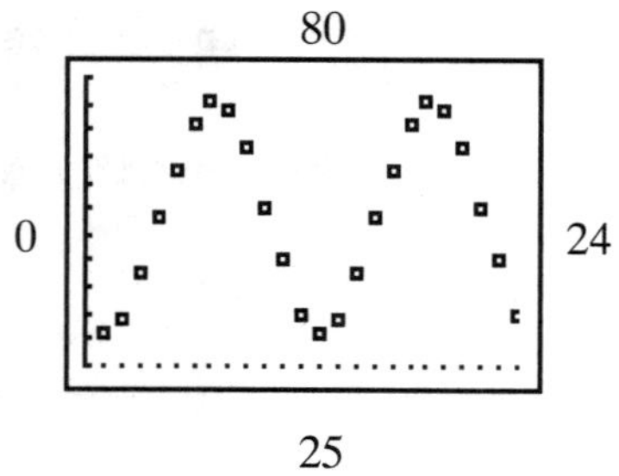

(B) $|A| = \dfrac{y_{max} - y_{min}}{2} = \dfrac{76 - 31}{2} = 22.5$. Hence $A = 22.5$ or -22.5.

$k = \dfrac{y_{max} + y_{min}}{2} = \dfrac{76 + 31}{2} = 53.5 \quad P = 12 = 2\pi \cdot \dfrac{6}{\pi} = 2\pi \div \dfrac{\pi}{6}$. Hence $B = \dfrac{\pi}{6}$.

The x intercept closest to the origin is estimated from the graph as 4.0. To find C, substitute $B = \frac{\pi}{6}$ and $x = 4.0$ into the phase shift formula $x = -\frac{C}{B}$ and solve for C.

$$x = -\frac{C}{B}$$

$$4.0 = -C \div \frac{\pi}{6}$$

$$C = -4.0\left(\frac{\pi}{6}\right)$$

$$C = -2.1$$

With this value of C, the graph is seen to be shifted up from the graph of a standard sine curve, thus $A = 22.5$. The equation is thus

$$y = 53.5 + 22.5 \sin\left(\frac{\pi}{6}x - 2.1\right).$$

(C)

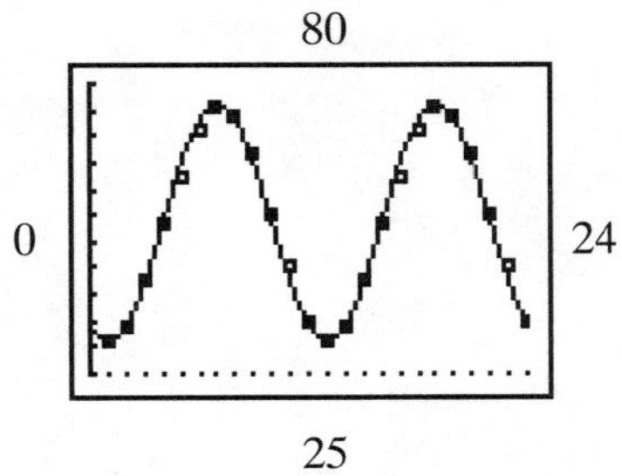

(D)

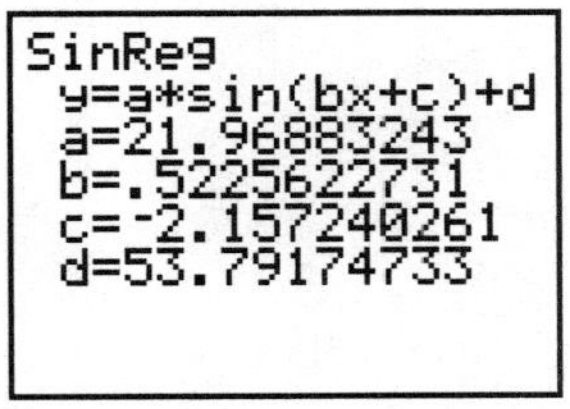

Compute the regression equation.

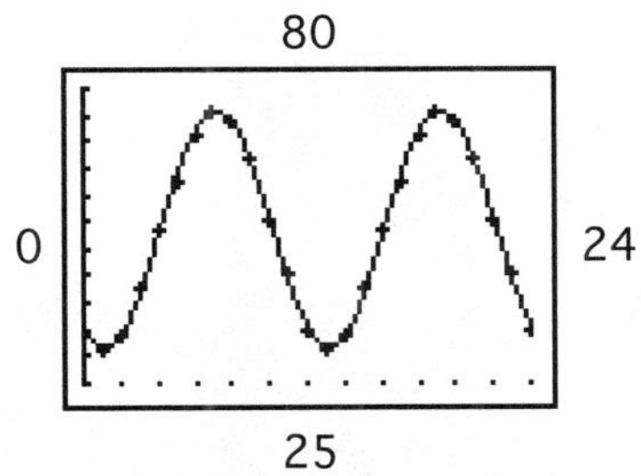

Graph the data and the regression equation.

(6-5)

CHAPTER 9 Additional Topics in Analytic Geometry

Section 9-1

1. To graph $y^2 = 4x$, assign x values that make the right side a perfect square (x must be non-negative for y to be real) and solve for y. Since the coefficient of x is positive, a must be positive, and the parabola opens right.

x	0	1	4
y	0	±2	±4

To find the focus and directrix, solve
$4a = 4$
$a = 1$
Focus: (1, 0) Directrix: $x = -1$

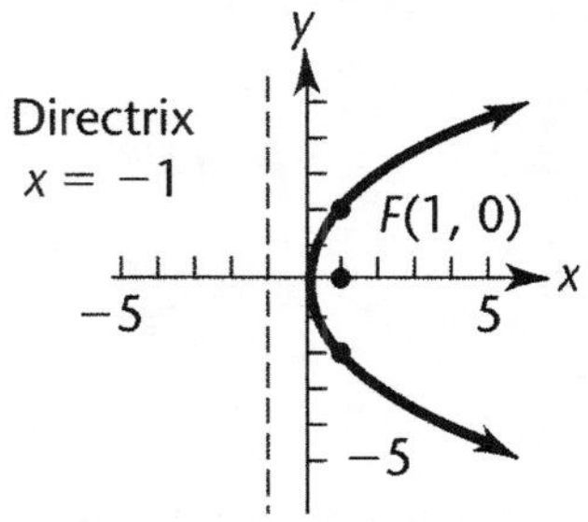

3. To graph $x^2 = 8y$, assign y values that make the right side a perfect square (y must be non-negative for x to be real) and solve for x. Since the coefficient of y is positive, a must be positive, and the parabola opens up.

x	0	±4	±2
y	0	2	$\frac{1}{2}$

To find the focus and directrix, solve
$4a = 8$
$a = 2$
Focus: (0, 2) Directrix: $y = -2$

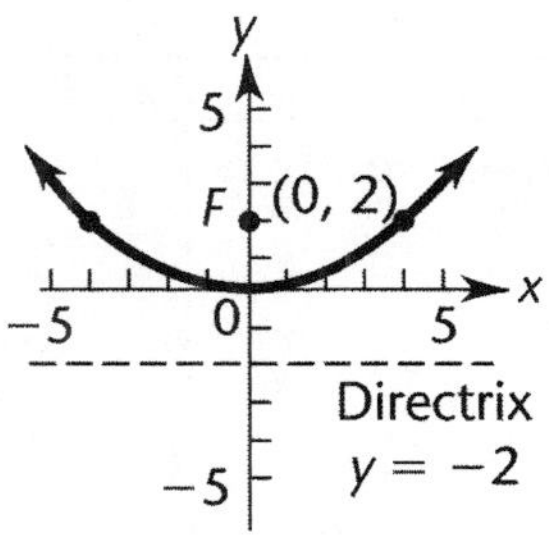

5. To graph $y^2 = -12x$, assign x values that make the right side a perfect square (x must be zero or negative for y to be real) and solve for y. Since the coefficient of x is negative, a must be negative, and the parabola opens left.

x	0	−3	$-\frac{1}{3}$
y	0	±6	±2

To find the focus and directrix, solve
$4a = -12$
$a = -3$
Focus: (−3, 0)
Directrix: $x = -(-3) = 3$

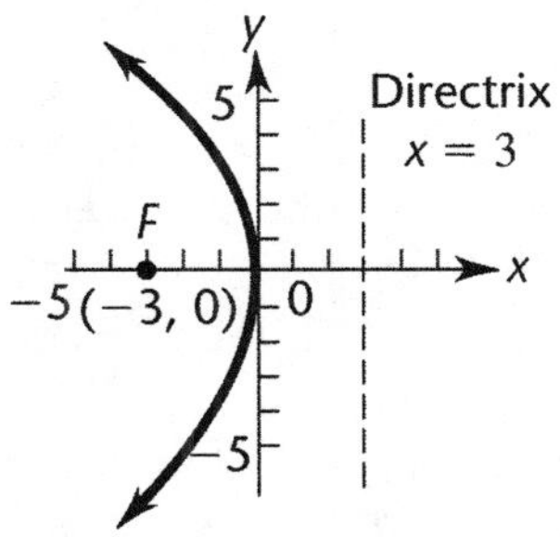

7. To graph $x^2 = -4y$, assign y values that make the right side a perfect square (y must be zero or negative for x to be real) and solve for x. Since the coefficient of y is negative, a must be negative, and the parabola opens down.

x	0	±2	±4
y	0	−1	−4

To find the focus and directrix, solve
$4a = -4$
$a = -1$
Focus: (0, −1)
Directrix: $y = -(-1) = 1$

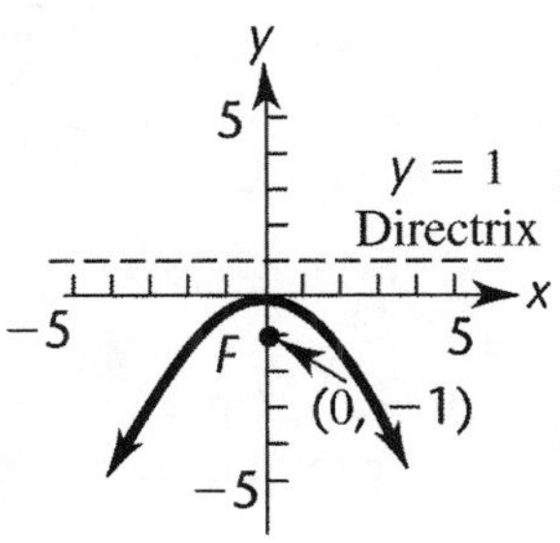

9. To graph $y^2 = -20x$, we may proceed as in problem 5. Alternatively, after noting that x must be zero or negative for y to be real, we may pick convenient values for x and solve for y using a calculator. Since the coefficient of x is negative, a must be negative, and the parabola opens left.

x	0	-1	-2
y	0	$\pm\sqrt{20} \approx \pm 4.5$	$\pm\sqrt{40} \approx \pm 6.3$

To find the focus and directrix, solve

$4a = -20$

$a = -5$

Focus: $(-5, 0)$

Directrix: $x = -(-5) = 5$

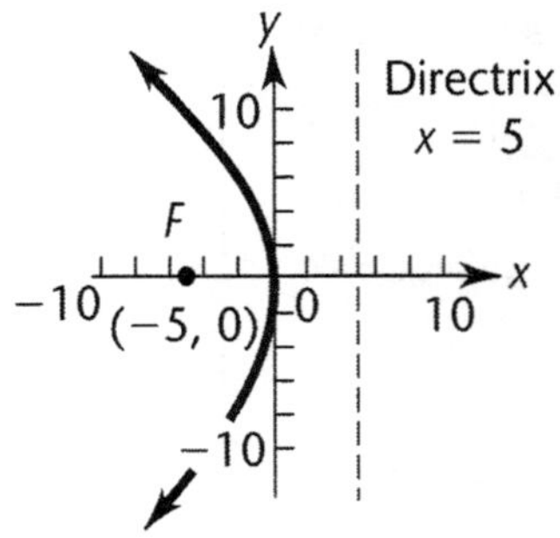

11. To graph $x^2 = 10y$, we may proceed as in Problem 3. Alternatively, after noting that y must be non-negative for x to be real, we may pick convenient values for y and solve for x using a calculator. Since the coefficient of y is positive, a must be positive, and the parabola opens up.

x	0	$\pm\sqrt{10} \approx \pm 3.2$	$\pm\sqrt{20} \approx \pm 4.5$
y	0	1	2

To find the focus and directrix, solve

$4a = 10$

$a = 2.5$

Focus: $(0, 2.5)$

Directrix: $y = -2.5$

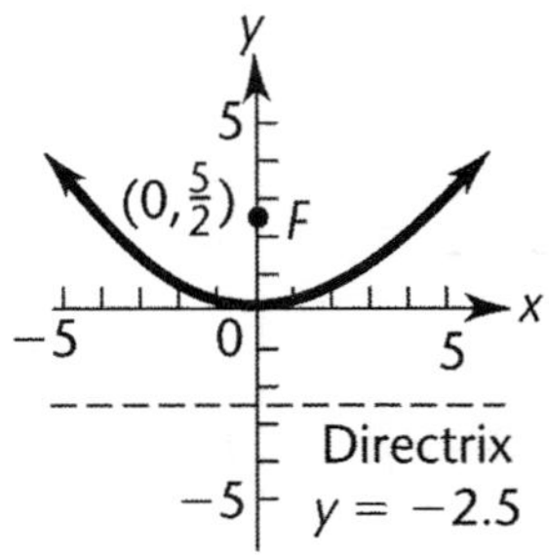

13. Comparing $y^2 = 39x$ with $y^2 = 4ax$, the standard equation of a parabola symmetric with respect to the x axis, we have

$4a = 39$ Focus on x axis

$a = 9.75$

Focus: $(9.75, 0)$

15. Comparing $x^2 = -105y$ with $x^2 = 4ay$, the standard equation of a parabola symmetric with respect to the y axis, we have

$4a = -105$ Focus on y axis

$a = -26.25$

Focus: $(0, -26.25)$

17. Comparing $y^2 = -77x$ with $y^2 = 4ax$, the standard equation of a parabola symmetric with respect to the x axis, we have

$4a = -77$ Focus on x axis

$a = -19.25$

Focus: $(-19.25, 0)$

19. Comparing directrix $y = -3$ with the information in the chart of standard equations for a parabola, we see:

$a = 3$. Axis: the y axis. Equation: $x^2 = 4ay$.

Thus the equation of the parabola must be $x^2 = 4 \cdot 3y$, or $x^2 = 12y$.

21. Comparing focus $(0, -7)$ with the information in the chart of standard equations for a parabola, we see:

$a = -7$ Axis: the y-axis. Equation: $x^2 = 4ay$.

Thus the equation of the parabola must be $x^2 = 4(-7)y$, or $x^2 = -28y$.

23. Comparing directrix $x = 6$ with the information in the chart of standard equations for a parabola, we see:

$6 = -a$. $a = -6$. Axis: the x-axis. Equation: $y^2 = 4ax$.

Thus the equation of the parabola must be $y^2 = 4(-6)x$, or $y^2 = -24x$.

25. Comparing focus (2, 0) with the information in the chart of standard equations for a parabola, we see:

$a = 2$ Axis: the x-axis. Equation: $y^2 = 4ax$.

Thus the equation of the parabola must be $y^2 = 4(2)x$, or $y^2 = 8x$.

27. The parabola is opening up and has an equation of the form $x^2 = 4ay$. Since (4, 2) is on the graph, we have:

$$\begin{aligned} x^2 &= 4ay \\ (4)^2 &= 4a(2) \\ 16 &= 8a \\ 2 &= a \end{aligned}$$

Thus, the equation of the parabola is

$$\begin{aligned} x^2 &= 4(2)y \\ x^2 &= 8y \end{aligned}$$

29. The parabola is opening left and has an equation of the form $y^2 = 4ax$. Since (-3, 6) is on the graph, we have:

$$\begin{aligned} y^2 &= 4ax \\ (6)^2 &= 4a(-3) \\ -3 &= a \end{aligned}$$

Thus, the equation of the parabola is

$$\begin{aligned} y^2 &= 4(-3)x \\ y^2 &= -12x \end{aligned}$$

31. The parabola is opening down and has an equation of the form $x^2 = 4ay$. Since (-6, -9) is on the graph, we have:

$$\begin{aligned} x^2 &= 4ay \\ (-6)^2 &= 4a(-9) \\ 36 &= -36a \\ -1 &= a \end{aligned}$$

Thus, the equation of the parabola is

$$\begin{aligned} x^2 &= 4(-1)y \\ x^2 &= -4y \end{aligned}$$

33. $x^2 = 4y$

$y^2 = 4x$

Solve for y in the first equation, then substitute into the second equation.

$$\begin{aligned} y &= \frac{x^2}{4} \\ \left(\frac{x^2}{4}\right)^2 &= 4x \\ \frac{x^4}{16} &= 4x \\ x^4 &= 64x \\ x^4 - 64x &= 0 \\ x(x^3 - 64) &= 0 \\ x(x - 4)(x^2 + 4x + 8) &= 0 \end{aligned}$$

$x = 0$ $\quad x - 4 = 0$ $\quad x^2 + 4x + 8 = 0$

$\quad\quad x = 4$ $\quad$ No real solutions

For $x = 0$ $\quad$ For $x = 4$

$y = 0$ $\quad y = \frac{4^2}{4}$

$y = 4$ $\quad$ Solutions: (0, 0), (4, 4)

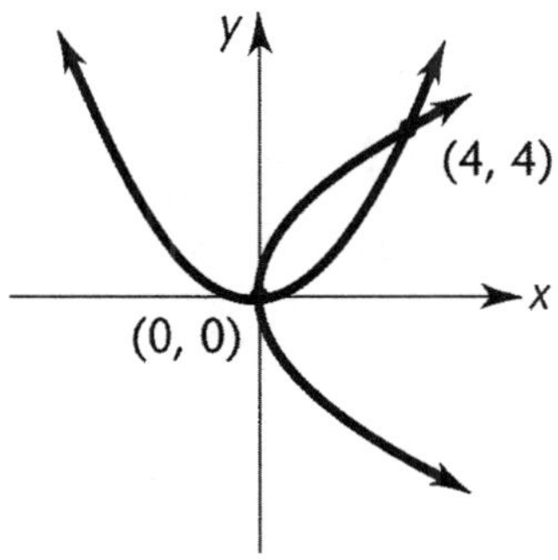

These exact solutions yield more than 3-digit accuracy.

35. $y^2 = 6x$
$x^2 = 5y$

Solve for x in the first equation, then substitute into the second equation.

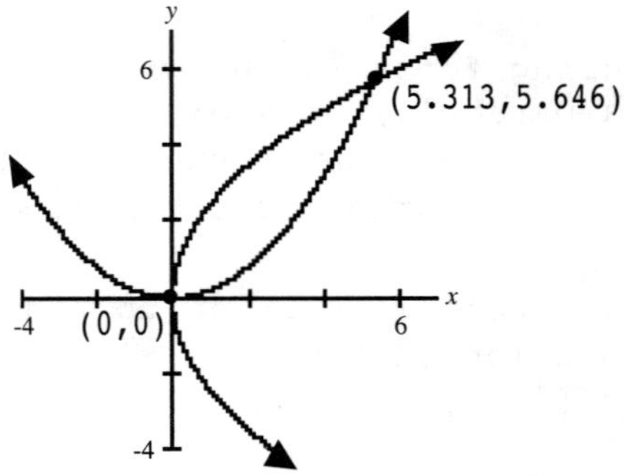

$$x = \frac{y^2}{6}$$
$$\left(\frac{y^2}{6}\right)^2 = 5y$$
$$\frac{y^4}{36} = 5y$$
$$y^4 = 180y$$
$$y^4 - 180y = 0$$
$$y(y^3 - 180) = 0$$
$$y(y - \sqrt[3]{180})(y^2 + y\sqrt[3]{180} + (\sqrt[3]{180})^2) = 0$$

$y = 0$ $\quad y - \sqrt[3]{180} = 0$ $\quad y^2 + y\sqrt[3]{180} + (\sqrt[3]{180})^2 = 0$

$y = \sqrt[3]{180} \approx 5.646$ $\quad$ No real solutions

For $y = 0$ $\quad x = 0$

For $y = \sqrt[3]{180}$ $\quad x = \frac{(\sqrt[3]{180})^2}{6} \approx 5.313$

Solutions: (0, 0), (5.313, 5.646)

37. (A) The line $x = 0$ intersects the parabola $x^2 = 4ay$ only at (0, 0).
The line $y = 0$ intersects the parabola $x^2 = 4ay$ only at (0, 0).
Only these 2 lines intersect the parabola at exactly one point (see part (B)).

(B) A line through (0, 0) with slope $m \neq 0$ has equation $y = mx$.
Solve the system:

$$y = mx$$
$$x^2 = 4ay$$

by substituting y from the first equation into the second equation.

$$x^2 = 4amx$$
$$x^2 - 4amx = 0$$
$$x(x - 4am) = 0$$

$x = 0 \quad x = 4am$

For $x = 0$: $y = 0$ $\quad$ For $x = 4am$: $y = m(4am) = 4am^2$

Solutions: (0, 0), $(4am, 4am^2)$ are the required coordinates.

39. Since A and B lie on the curve $x^2 = 4ay$, their coordinates must satisfy the equation of the curve. Clearly, the y coordinate of A, F, and B is a. Substituting a for y, we have

$$x^2 = 4aa$$
$$x^2 = 4a^2$$
$$x = \pm 2a$$

Therefore, A has coordinates $(-2a, a)$ and B has coordinates $(2a, a)$.

41. According to Explore/Discuss 2, to obtain the magnification by a factor of 2, we replace y and x by $y/2$ and $x/2$.

$$\left(\frac{y}{2}\right)^2 = 4\left(\frac{x}{2}\right)$$
$$\frac{y^2}{4} = 2x$$
$$y^2 = 8x$$

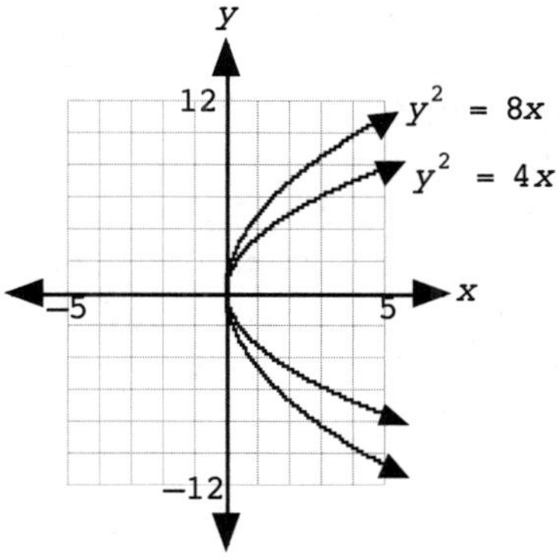

43. According to Explore/Discuss 2, to obtain the magnification by a factor of 0.5, we replace y and x by $\frac{y}{0.5}$ and $\frac{x}{0.5}$.

$$\left(\frac{x}{0.5}\right)^2 = -4\left(\frac{y}{0.5}\right)$$
$$(2x)^2 = -4(2y)$$
$$4x^2 = -8y$$
$$x^2 = -2y$$

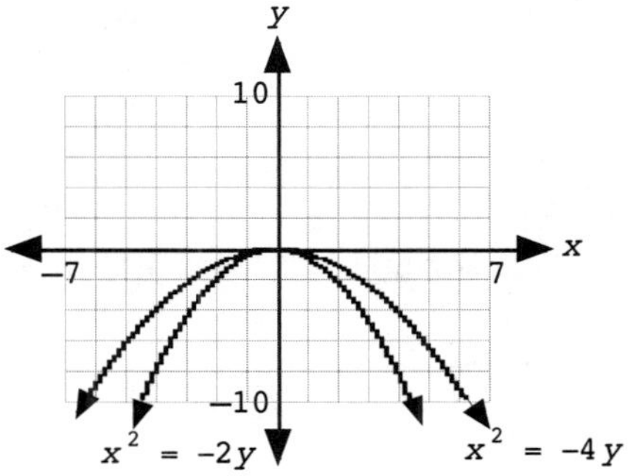

45. Let $P = (x, y)$ be a point on the parabola. Then, by the definition of the parabola, the distance from $P = (x, y)$ to the focus $F = (2, 2)$ must equal the perpendicular distance from P to the directrix at $D = (x, -4)$. Applying the distance formula, we have

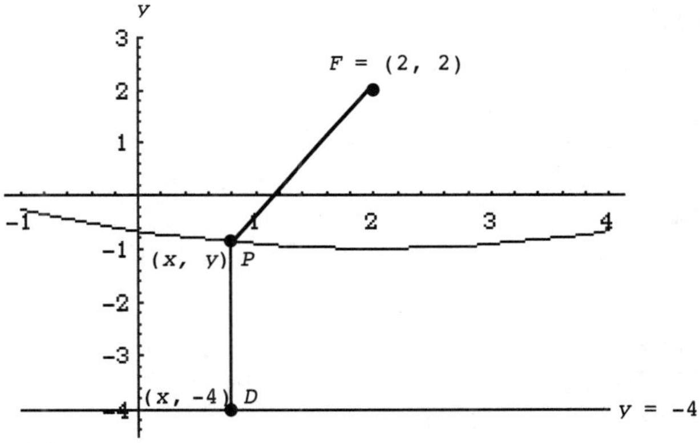

$$d(P, F) = d(P, D)$$
$$\sqrt{(x-2)^2 + (y-2)^2} = \sqrt{(x-x)^2 + [y-(-4)]^2}$$
$$(x-2)^2 + (y-2)^2 = (x-x)^2 + (y+4)^2$$
$$x^2 - 4x + 4 + y^2 - 4y + 4 = 0 + y^2 + 8y + 16$$
$$x^2 - 4x - 12y - 8 = 0$$

47. Let $P = (x, y)$ be a point on the parabola.

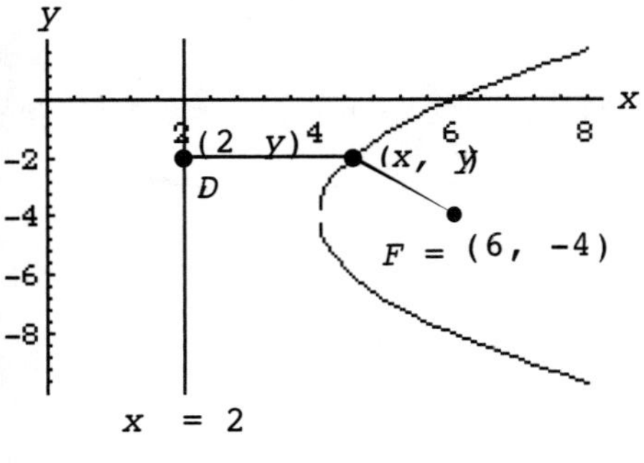

Then, by the definition of the parabola, the distance from $P = (x, y)$ to the focus $F = (6, -4)$ must equal the perpendicular distance from P to the directrix at $D = (2, y)$.

Applying the distance formula, we have

$$\begin{aligned} d(P, F) &= d(P, D) \\ \sqrt{(x-6)^2 + [y-(-4)]^2} &= \sqrt{(x-2)^2 + (y-y)^2} \\ (x-6)^2 + (y+4)^2 &= (x-2)^2 + (y-y)^2 \\ x^2 - 12x + 36 + y^2 + 8y + 16 &= x^2 - 4x + 4 + 0 \\ y^2 + 8y - 8x + 48 &= 0 \end{aligned}$$

49. Using Figure 9-5, we note:
The point $P = (x, y)$ is a point on the parabola if and only if

$$\begin{aligned} d_1 &= d_2 \\ d(P, N) &= d(P, F) \\ \sqrt{(x-x)^2 + (y+a)^2} &= \sqrt{(x-0)^2 + (y-a)^2} \\ (y+a)^2 &= x^2 + (y-a)^2 \\ y^2 + 2ay + a^2 &= x^2 + y^2 - 2ay + a^2 \\ x^2 &= 4ay \end{aligned}$$

51. From the figure, we see that the coordinates of P must be (-100, -50). The parabola is opening down with axis the y axis, hence it has an equation of the form $x^2 = 4ay$. Since (-100, -50) is on the graph, we have

$$\begin{aligned} (-100)^2 &= 4a(-50) \\ 10{,}000 &= -200a \\ a &= -50 \end{aligned}$$

Thus, the equation of the parabola is

$$x^2 = 4(-50)y$$
$$x^2 = -200y$$

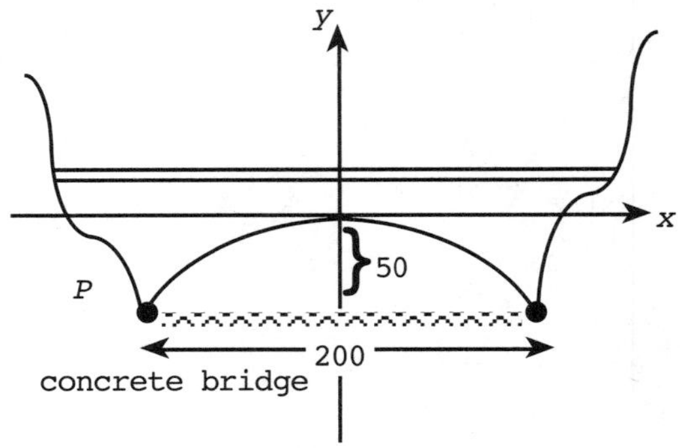

53. (A) From the figure, we see that the parabola is opening up with axis the y axis, hence it has an equation of the form $x^2 = 4ay$. Since the focus is at $(0, a) = (0, 100)$, $a = 100$ and the equation of the parabola is $x^2 = 400y$ or $y = 0.0025x^2$

(B) Since the depth represents the y coordinate y_1 of a point on the parabola with $x = 100$, we have

$$y_1 = 0.0025(100)^2$$
$$\text{depth} = 25 \text{ feet}$$

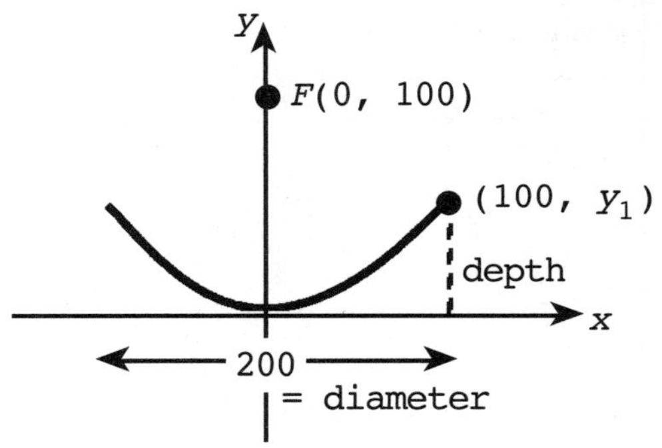

Section 9-2

Except as otherwise noted, the graphing calculator drawings in this and the following sections of this chapter have been plotted in a squared window (zoom menu 5 on the TI-83) to avoid distorting the shapes of the conic sections.

1. When $y = 0$, $\frac{x^2}{25} = 1$. x intercepts: ± 5

When $x = 0$, $\frac{y^2}{4} = 1$. y intercepts: ± 2

Thus, $a = 5$, $b = 2$, and the major axis is on the x axis.

$$\begin{aligned} \text{Foci: } c^2 &= a^2 - b^2 \\ c^2 &= 25 - 4 \\ c^2 &= 21 \\ c &= \sqrt{21} \end{aligned}$$

Foci: $F' = (-\sqrt{21}, 0)$, $F = (\sqrt{21}, 0)$
Major axis length = 2(5) = 10; Minor axis length = 2(2) = 4

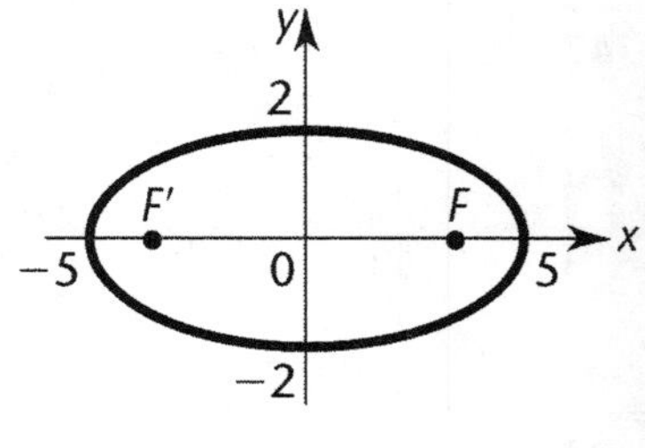

3. When $y = 0$, $\frac{x^2}{4} = 1$. x intercepts: ±2

When $x = 0$, $\frac{y^2}{25} = 1$. y intercepts: ±5

Thus, $a = 5$, $b = 2$, and the major axis is on the y axis.

Foci: $c^2 = a^2 - b^2$

$c^2 = 25 - 4$

$c^2 = 21$

$c = \sqrt{21}$ Foci: $F' = (0, -\sqrt{21})$, $F = (0, \sqrt{21})$

Major axis length = 2(5) = 10 Minor axis length = 2(2) = 4

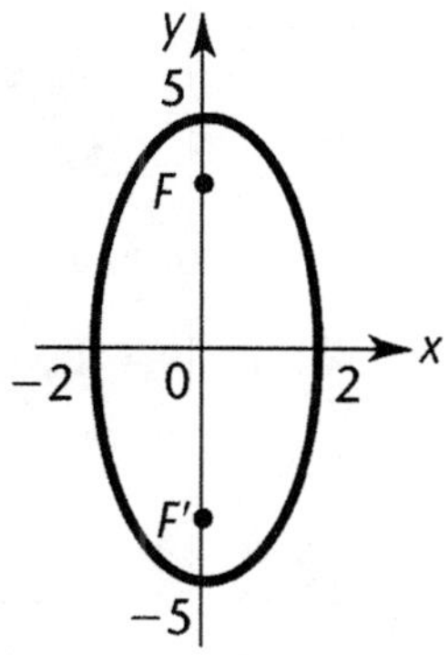

5. First, write the equation in standard form by dividing both sides by 9.

$x^2 + 9y^2 = 9$

$\frac{x^2}{9} + \frac{y^2}{1} = 1$

Locate the intercepts.

When $y = 0$, $\frac{x^2}{9} = 1$. x intercepts: ±3

When $x = 0$, $\frac{y^2}{1} = 1$. y intercepts: ±1

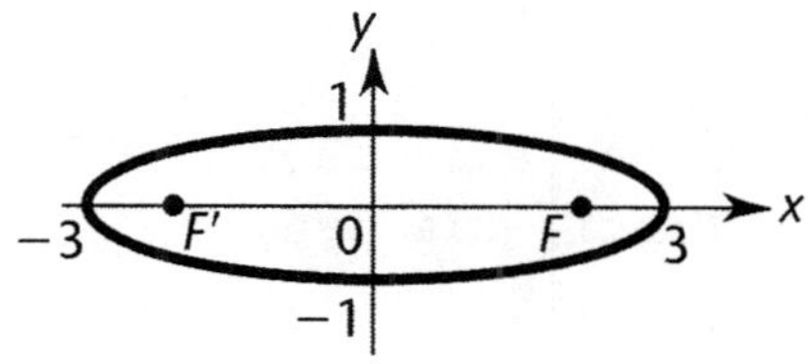

Thus $a = 3$, $b = 1$, and the major axis is on the x axis.

Foci: $c^2 = a^2 - b^2$

$c^2 = 9 - 1$

$c^2 = 8$

$c = \sqrt{8}$ Foci: $F' = (-\sqrt{8}, 0)$, $F = (\sqrt{8}, 0)$

Major axis length = 2(3) = 6

Minor axis length = 2(1) = 2

7. When $y = 0$, $9x^2 = 144$. x intercepts: ±4

When $x = 0$, $16y^2 = 144$. y intercepts: ±3

This corresponds to graph (b).

9. When $y = 0$, $4x^2 = 16$. x intercepts: ±2

When $x = 0$, $y^2 = 16$. y intercepts: ±4

This corresponds to graph (a).

11. First, write the equation in standard form by dividing both sides by 225.

$25x^2 + 9y^2 = 225$

$\frac{x^2}{9} + \frac{y^2}{25} = 1$

Locate the intercepts.

When $y = 0$, $\frac{x^2}{9} = 1$. x intercepts: ±3

When $x = 0$, $\frac{y^2}{25} = 1$. y intercepts: ±5

Thus $a = 5$, $b = 3$, and the major axis is on the y axis.

Foci: $c^2 = a^2 - b^2$

$c^2 = 25 - 9$

$c^2 = 16$

$c = 4$ Foci: $F' = (0, -4)$, $F = (0, 4)$

Major axis length = 2(5) = 10

Minor axis length = 2(3) = 6

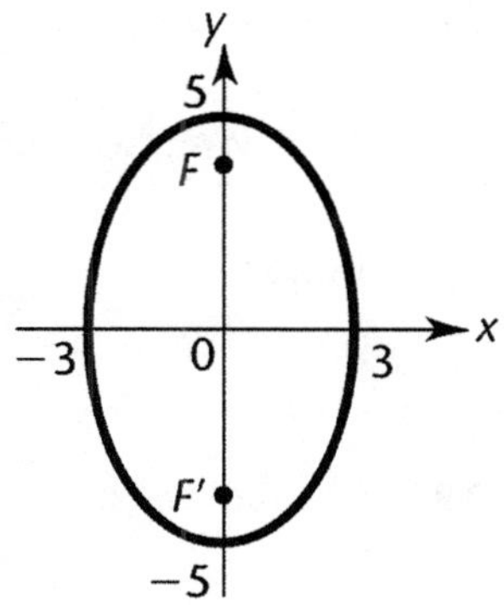

13. First, write the equation in standard form by dividing both sides by 12.

$2x^2 + y^2 = 12$

$\frac{x^2}{6} + \frac{y^2}{12} = 1$

Locate the intercepts.

When $y = 0$, $\frac{x^2}{6} = 1$. x intercepts: $\pm\sqrt{6}$

When $x = 0$, $\frac{y^2}{12} = 1$. y intercepts: $\pm\sqrt{12}$

Thus $a = \sqrt{12}$, $b = \sqrt{6}$, and the major axis is on the y axis.

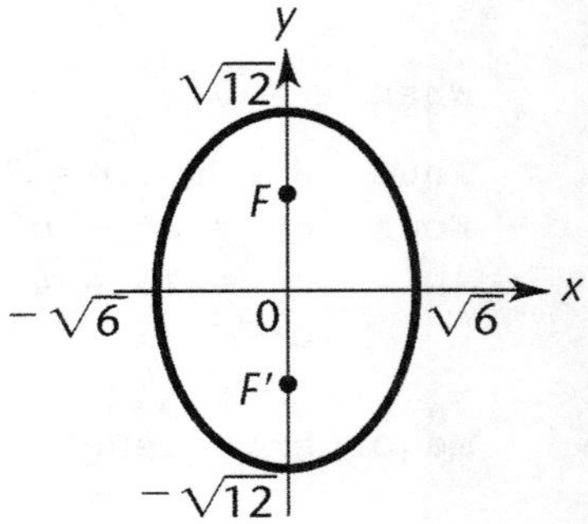

Foci: $c^2 = a^2 - b^2$

$c^2 = 12 - 6$

$c^2 = 6$

$c = \sqrt{6}$ Foci: $F' = (0, -\sqrt{6})$, $F = (0, \sqrt{6})$

Major axis length $= 2\sqrt{12} \approx 6.93$

Minor axis length $= 2\sqrt{6} \approx 4.90$

15. First, write the equation in standard form by dividing both sides by 28.

$4x^2 + 7y^2 = 28$

$\frac{x^2}{7} + \frac{y^2}{4} = 1$

Locate the intercepts.

When $y = 0$, $\frac{x^2}{7} = 1$. x intercepts: $\pm\sqrt{7}$

When $x = 0$, $\frac{y^2}{4} = 1$. y intercepts: ± 2

Thus $a = \sqrt{7}$, $b = 2$, and the major axis is on the x axis.

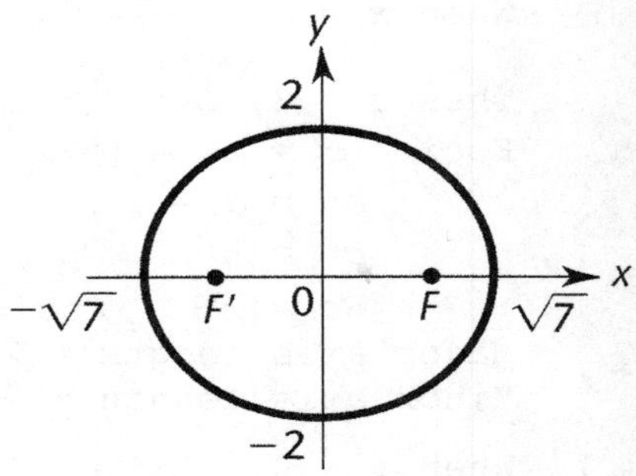

Foci: $c^2 = a^2 - b^2$

$c^2 = 7 - 4$

$c^2 = 3$

$c = \sqrt{3}$ Foci: $F' = (-\sqrt{3}, 0)$, $F = (\sqrt{3}, 0)$

Major axis length $= 2\sqrt{7} \approx 5.29$

Minor axis length $= 2(2) = 4$

17. The x intercepts are ± 5. Thus, if $y = 0$, $\frac{(\pm 5)^2}{M} = 1$, so $M = 25$.

The y intercepts are ± 4. Thus, if $x = 0$, $\frac{(\pm 4)^2}{N} = 1$, so $N = 16$.

Equation: $\frac{x^2}{25} + \frac{y^2}{16} = 1$

19. The x intercepts are ± 3. Thus, if $y = 0$, $\frac{(\pm 3)^2}{M} = 1$, so $M = 9$.

The y intercepts are ± 6. Thus, if $x = 0$, $\frac{(\pm 6)^2}{N} = 1$, so $N = 36$.

Equation: $\frac{x^2}{9} + \frac{y^2}{36} = 1$

21. Make a rough sketch of the ellipse and compute x and y intercepts.

$$\frac{x^2}{a^2} + \frac{y^2}{b^2} = 1$$

$$a = \frac{10}{2} = 5, \quad b = \frac{6}{2} = 3$$

$$\frac{x^2}{25} + \frac{y^2}{9} = 1$$

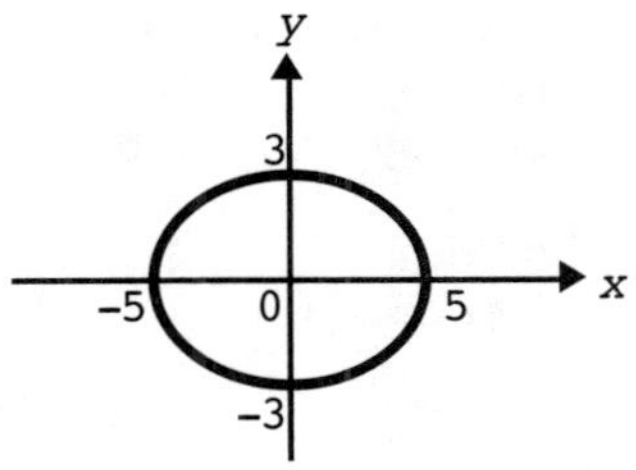

23. Make a rough sketch of the ellipse and compute x and y intercepts.

$$\frac{x^2}{b^2} + \frac{y^2}{a^2} = 1$$

$$a = \frac{22}{2} = 11, \quad b = \frac{16}{2} = 8$$

$$\frac{x^2}{64} + \frac{y^2}{121} = 1$$

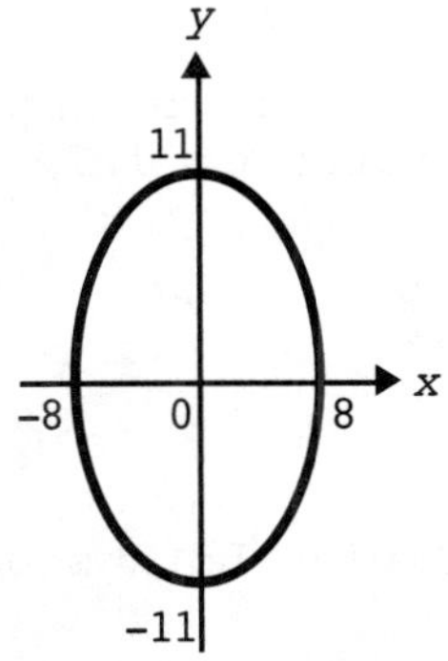

25. Make a rough sketch of the ellipse, locate focus and x intercepts, then determine y intercepts using the special triangle relationship.

$$\frac{x^2}{a^2} + \frac{y^2}{b^2} = 1$$

$$a = \frac{16}{2} = 8$$

$$b^2 = 8^2 - 6^2 = 64 - 36 = 28$$

$$b = \sqrt{28}$$

$$\frac{x^2}{64} + \frac{y^2}{28} = 1$$

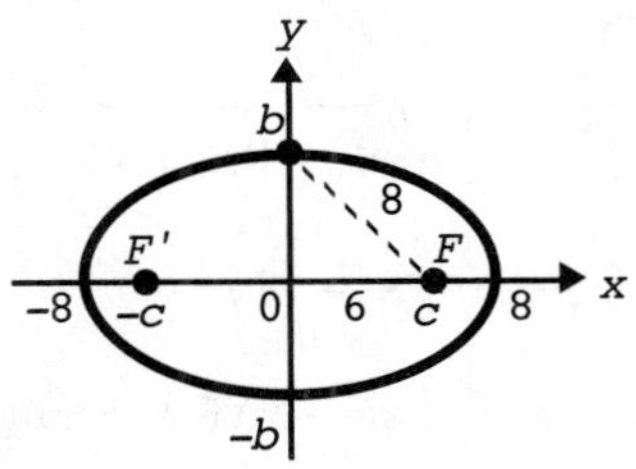

27. Make a rough sketch of the ellipse, locate focus and x intercepts, then determine y intercepts using the special triangle relationship.

$$\frac{x^2}{b^2} + \frac{y^2}{a^2} = 1$$

$$b = \frac{20}{2} = 10$$

$$a^2 = 10^2 + (\sqrt{70})^2 = 100 + 70 = 170$$

$$a = \sqrt{170}$$

$$\frac{x^2}{100} + \frac{y^2}{170} = 1$$

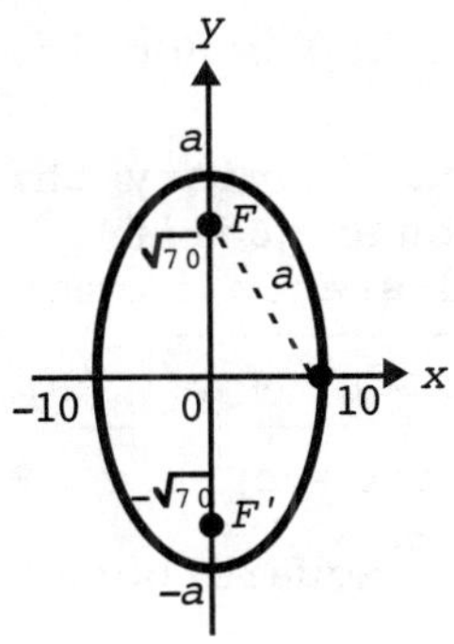

29. The graph does not pass the vertical line test; most vertical lines that intersect an ellipse do so in two places; hence, the equation does not define a function.

31. According to Explore/Discuss 2, to obtain the magnification by a factor of 2, we replace x and y by $x/2$ and $y/2$.

$$16\left(\frac{x}{2}\right)^2 + 9\left(\frac{y}{2}\right)^2 = 144$$
$$16 \cdot \frac{x^2}{4} + 9 \cdot \frac{y^2}{4} = 144$$
$$16x^2 + 9y^2 = 576$$

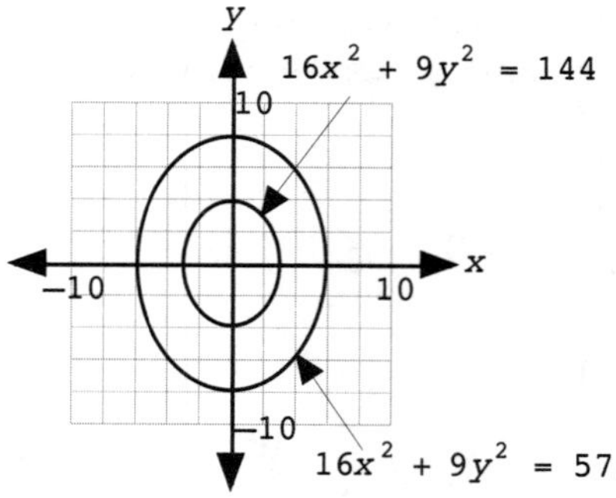

33. According to Explore/Discuss 2, to obtain the magnification by a factor of 0.5, we replace x and y by $\frac{x}{0.5}$ and $\frac{y}{0.5}$.

$$\left(\frac{x}{0.5}\right)^2 + 4\left(\frac{y}{0.5}\right)^2 = 16$$
$$(2x)^2 + 4(2y)^2 = 16$$
$$4x^2 + 16y^2 = 16 \quad \text{(Divide both sides by 4)}$$
$$x^2 + 4y^2 = 4$$

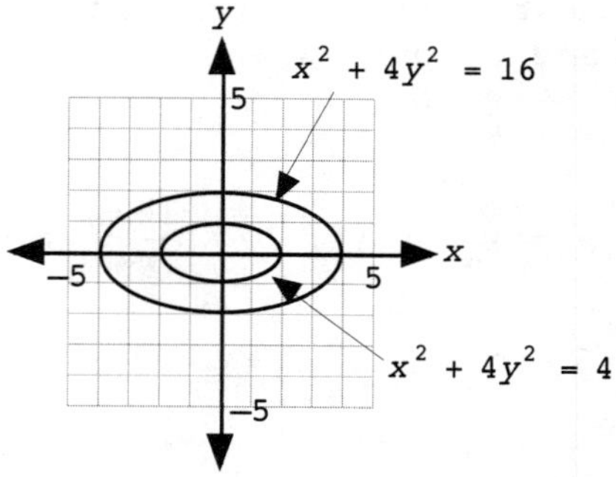

35. From the figure, we see that the point $P = (x, y)$ is a point on the curve if and only if $d_1 = \frac{1}{2}d_2$

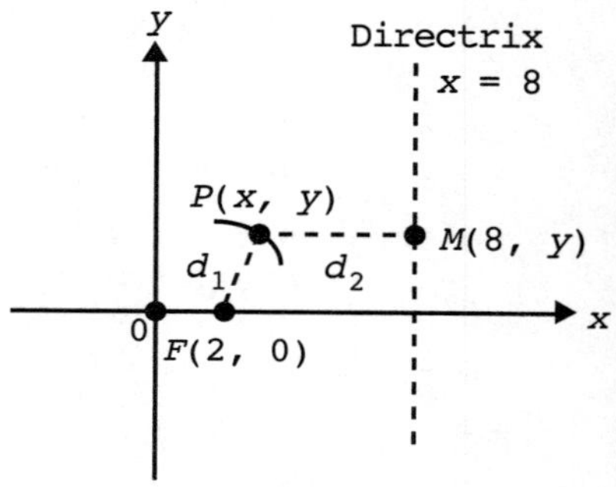

$$d(P, F) = \frac{1}{2}d(P, M)$$
$$\sqrt{(x-2)^2 + (y-0)^2} = \frac{1}{2}\sqrt{(x-8)^2 + (y-y)^2}$$
$$(x-2)^2 + y^2 = \frac{1}{4}(x-8)^2$$
$$x^2 - 4x + 4 + y^2 = \frac{1}{4}(x^2 - 16x + 64)$$
$$4x^2 - 16x + 16 + 4y^2 = x^2 - 16x + 64$$
$$3x^2 + 4y^2 = 48$$
$$\frac{x^2}{16} + \frac{y^2}{12} = 1$$

The curve must be an ellipse since its equation can be written in standard form for an ellipse.

37. The only points that could satisfy this condition are those on the line segment $F'F$, and all of those would qualify.

39. $d_1 + d_2 = 2a$ Use the distance formula:

$\sqrt{(x+c)^2 + y^2} = 2a - \sqrt{(x-c)^2 + y^2}$ Square both sides:

$(x+c)^2 + y^2 = 4a^2 - 4a\sqrt{(x-c)^2 + y^2} + (x-c)^2 + y^2$ Isolate the square root:

$\sqrt{(x-c)^2 + y^2} = a - \frac{cx}{a}$ Square both sides:

$(x-c)^2 + y^2 = a^2 - 2cx + \frac{c^2x^2}{a^2}$ Simplify and collect terms:

$\left(1 - \frac{c^2}{a^2}\right)x^2 + y^2 = a^2 - c^2$ Use $a^2 - c^2 = b^2$, $1 - \frac{c^2}{a^2} = \frac{b^2}{a^2}$, and divide both sides by b^2:

$$\frac{x^2}{a^2} + \frac{y^2}{b^2} = 1$$

41. From the figure we see that the x and y intercepts of the ellipse must be 20 and 12 respectively. Hence the equation of the ellipse must be

$$\frac{x^2}{(20)^2} + \frac{y^2}{(12)^2} = 1$$

or

$$\frac{x^2}{400} + \frac{y^2}{144} = 1$$

To find the clearance above the water 5 feet from the bank, we need the y coordinate y_1 of the point P whose x coordinate is $a - 5 = 15$. Since P is on the ellipse, we have

$$\frac{15^2}{400} + \frac{y_1^2}{144} = 1$$

$$\frac{225}{400} + \frac{y_1^2}{144} = 1$$

$$0.5625 + \frac{y_1^2}{144} = 1$$

$$\frac{y_1^2}{144} = 0.4375$$

$$y_1^2 = 144(0.4375)$$

$$y_1^2 = 63$$

$$y_1 \approx \pm 7.94$$

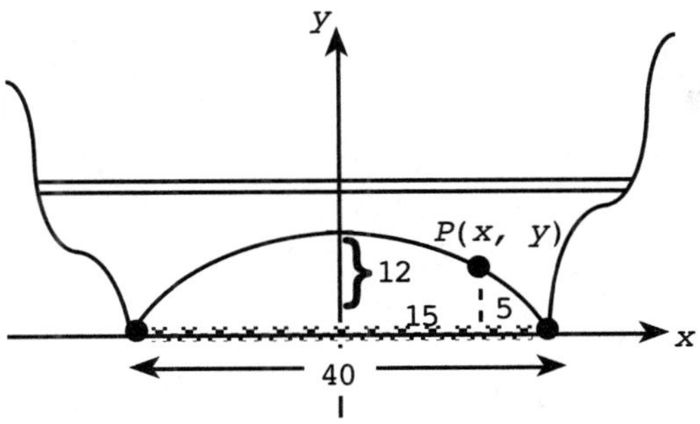

Therefore, the clearance is 7.94 feet, approximately.

43. (A) From the figure we see that the x intercept of the ellipse must be 24.0. Hence the equation of the ellipse must have the form

$$\frac{x^2}{(24.0)^2} + \frac{y^2}{b^2} = 1$$

Since the point (23.0, 1.14) is on the ellipse, its coordinates must satisfy the equation of the ellipse. Hence

$$\frac{(23.0)^2}{(24.0)^2} + \frac{(1.14)^2}{b^2} = 1$$

$$\frac{(1.14)^2}{b^2} = 1 - \frac{(23.0)^2}{(24.0)^2}$$

$$\frac{(1.14)^2}{b^2} = \frac{47}{576}$$

$$b^2 = \frac{576(1.14)^2}{47}$$

$$b^2 = 15.9$$

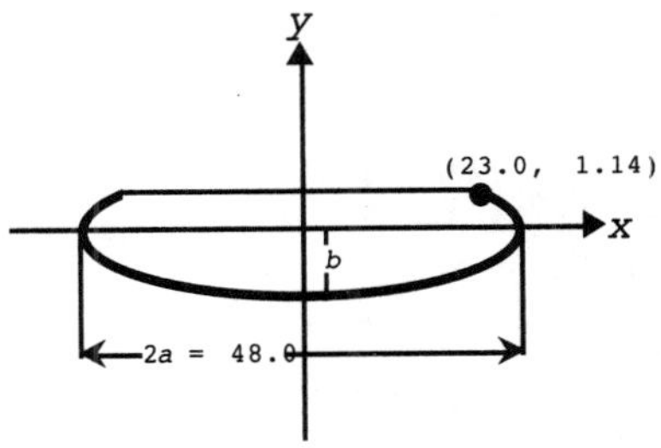

Thus, the equation of the ellipse must be $\frac{x^2}{576} + \frac{y^2}{15.9} = 1$

(B) From the figure, we can see that the width of the wing must equal $1.14 + b = 1.14 + \sqrt{15.9} = 5.13$ feet.

Section 9-3

1. When $y = 0$, $x^2 = 1$ x intercepts: ± 1
When $x = 0$, $-y^2 = 1$. There are no y intercepts.
This corresponds to graph (d).

3. When $y = 0$, $-x^2 = 4$. There are no x intercepts.
When $x = 0$, $y^2 = 4$. y intercepts: ± 2
This corresponds to graph (c).

5. When $y = 0$, $\frac{x^2}{9} = 1$. x intercepts: ± 3 $a = 3$

When $x = 0$, $-\frac{y^2}{4} = 1$. There are no y intercepts, but $b = 2$.

Sketch the asymptotes using the asymptote rectangle, then sketch in the hyperbola.

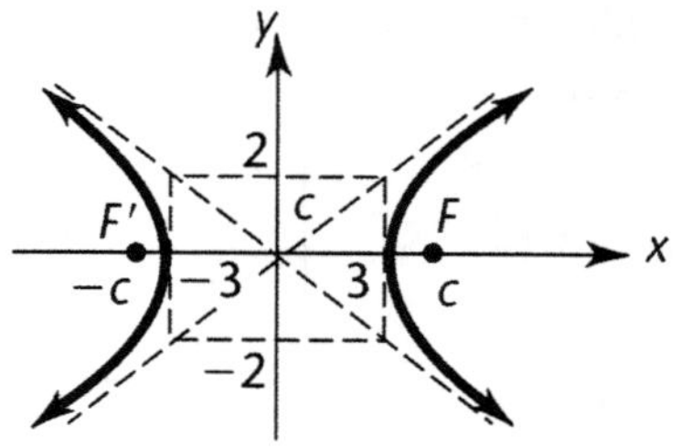

Foci: $c^2 = 3^2 + 2^2$
$c^2 = 13$
$c = \sqrt{13}$
$F' = (-\sqrt{13}, 0)$, $F = (\sqrt{13}, 0)$

Transverse axis length = 2(3) = 6
Conjugate axis length = 2(2) = 4

7. When $y = 0$, $-\frac{x^2}{9} = 1$. There are no x intercepts, but $b = 3$.

When $x = 0$, $\frac{y^2}{4} = 1$. y intercepts: ± 2, $a = 2$

Sketch the asymptotes using the asymptote rectangle, then sketch in the hyperbola.

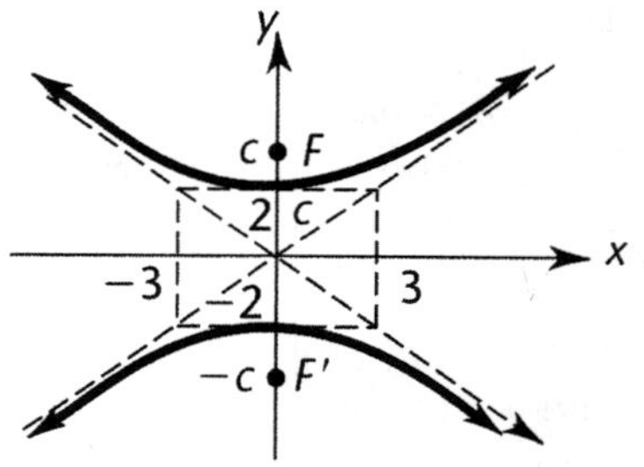

Foci: $c^2 = 2^2 + 3^2$
$c^2 = 13$
$c = \sqrt{13}$
$F' = (0, -\sqrt{13})$, $F = (0, \sqrt{13})$
Transverse axis length = 2(2) = 4
Conjugate axis length = 2(3) = 6

9. First, write the equation in standard form by dividing both sides by 16.

$4x^2 - y^2 = 16$

$\frac{x^2}{4} - \frac{y^2}{16} = 1$

Locate intercepts:
When $y = 0$, $x = \pm 2$. x intercepts: ± 2 $a = 2$

When $x = 0$, $-\frac{y^2}{16} = 1$. There are no y intercepts, but $b = 4$.

Sketch the asymptotes using the asymptote rectangle, then sketch in the hyperbola.

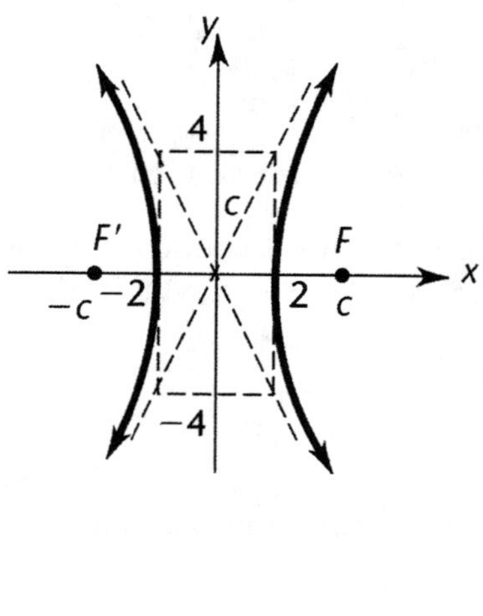

Foci: $c^2 = 2^2 + 4^2$
$c^2 = 20$
$c = \sqrt{20}$
$F' = (-\sqrt{20}, 0)$, $F = (\sqrt{20}, 0)$
Transverse axis length = 2(2) = 4
Conjugate axis length = 2(4) = 8

11. First, write the equation in standard form by dividing both sides by 144.

$9y^2 - 16x^2 = 144$

$\frac{y^2}{16} - \frac{x^2}{9} = 1$

Locate intercepts:

When $y = 0$, $-\frac{x^2}{9} = 1$. There are no x intercepts, but $b = 3$.

When $x = 0$, $\frac{y^2}{16} = 1$, $y = \pm 4$.

y intercepts: ± 4 $a = 4$

Sketch the asymptotes using the asymptote rectangle, then sketch in the hyperbola.

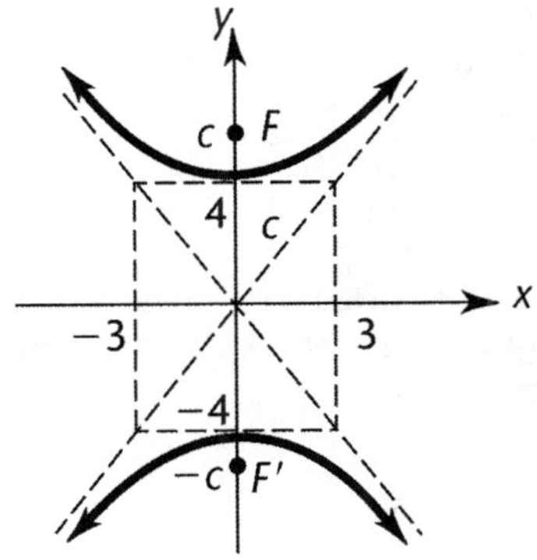

Foci: $c^2 = 4^2 + 3^2$

$c^2 = 25$

$c = 5$

$F' = (0, -5)$, $F = (0, 5)$

Transverse axis length = 2(4) = 8. Conjugate axis length = 2(3) = 6.

13. First, write the equation in standard form by dividing both sides by 12.

$3x^2 - 2y^2 = 12$

$\frac{x^2}{4} - \frac{y^2}{6} = 1$

Locate intercepts: When $y = 0$, $\frac{x^2}{4} = 1$. $x = \pm 2$.

x intercepts: ± 2 $a = 2$

When $x = 0$, $-\frac{y^2}{6} = 1$. There are no y intercepts, but $b = \sqrt{6}$.

Sketch the asymptotes using the asymptote rectangle, then sketch in the hyperbola.

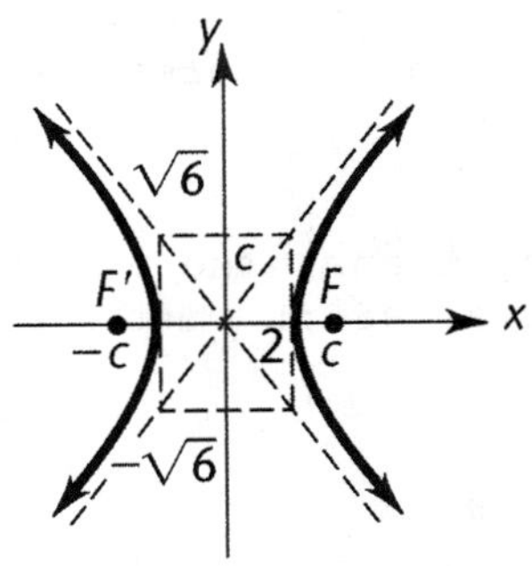

Foci: $c^2 = 2^2 + (\sqrt{6})^2$

$c^2 = 10$

$c = \sqrt{10}$

$F' = (-\sqrt{10}, 0)$, $F = (\sqrt{10}, 0)$

Transverse axis length = 2(2) = 4

Conjugate axis length = $2\sqrt{6} \approx 4.90$

15. First, write the equation in standard form by dividing both sides by 28.

$7y^2 - 4x^2 = 28$

$\frac{y^2}{4} - \frac{x^2}{7} = 1$

Locate intercepts:

When $y = 0$, $-\frac{x^2}{7} = 1$. There are no x intercepts, but $b = \sqrt{7}$.

When $x = 0$, $\frac{y^2}{4} = 1$, $y = \pm 2$. y intercepts: ± 2 $a = 2$

Sketch the asymptotes using the asymptote rectangle, then sketch in the hyperbola.

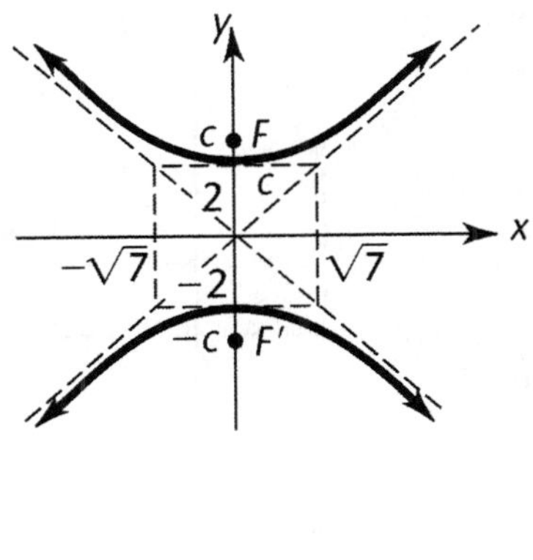

Foci: $c^2 = 2^2 + (\sqrt{7})^2$

$c^2 = 11$

$c = \sqrt{11}$

$F' = (0, -\sqrt{11})$, $F = (0, \sqrt{11})$

Transverse axis length = 2(2) = 4

Conjugate axis length = $2\sqrt{7} \approx 5.29$

17. Since the graph has x intercepts ($x = \pm 3$) but no y intercepts, the equation must be in the form

$$\frac{x^2}{M} - \frac{y^2}{N} = 1$$

and, when $y = 0$, $x = \pm 3$, hence

$$\frac{9}{M} - \frac{0}{N} = 1$$
$$M = 9$$

Since the point (5, 4) is on the graph, its coordinates must satisfy the equation, which is known to be of form

$$\frac{x^2}{9} - \frac{y^2}{N} = 1$$

Hence

$$\frac{5^2}{9} - \frac{4^2}{N} = 1$$
$$\frac{25}{9} - \frac{16}{N} = 1$$
$$-\frac{16}{N} = -\frac{16}{9}$$
$$N = 9$$

The equation is therefore

$$\frac{x^2}{9} - \frac{y^2}{9} = 1$$

19. Since the graph has y intercepts ($y = \pm 4$) but no x intercepts, the equation must be in the form

$$\frac{y^2}{N} - \frac{x^2}{M} = 1$$

and, when $x = 0$, $y = \pm 4$, hence

$$\frac{16}{N} - \frac{0}{M} = 1$$
$$N = 16$$

Since the point (3, 5) is on the graph, its coordinates must satisfy the equation, which is known to be of form

$$\frac{y^2}{16} - \frac{x^2}{M} = 1$$

Hence

$$\frac{5^2}{16} - \frac{3^2}{M} = 1$$
$$\frac{25}{16} - \frac{9}{M} = 1$$
$$-\frac{9}{M} = -\frac{9}{16}$$
$$M = 16$$

The equation is therefore

$$\frac{y^2}{16} - \frac{x^2}{16} = 1$$

21. Since the transverse axis is on the x axis, start with

$$\frac{x^2}{a^2} - \frac{y^2}{b^2} = 1$$

and find a and b

$$a = \frac{14}{2} = 7 \text{ and } b = \frac{10}{2} = 5$$

Thus, the equation is

$$\frac{x^2}{49} - \frac{y^2}{25} = 1$$

23. Since the transverse axis is on the y axis, start with

$$\frac{y^2}{a^2} - \frac{x^2}{b^2} = 1$$

and find a and b

$$a = \frac{24}{2} = 12 \text{ and } b = \frac{18}{2} = 9$$

Thus, the equation is

$$\frac{y^2}{144} - \frac{x^2}{81} = 1$$

25. Since the transverse axis is on the x axis, start with

$$\frac{x^2}{a^2} - \frac{y^2}{b^2} = 1$$

and find a and b

$$a = \frac{18}{2} = 9$$

To find b, sketch the asymptote rectangle, label known parts, and use the Pythagorean Theorem.

$$b^2 = 11^2 - 9^2$$
$$b^2 = 40$$
$$b = \sqrt{40}$$

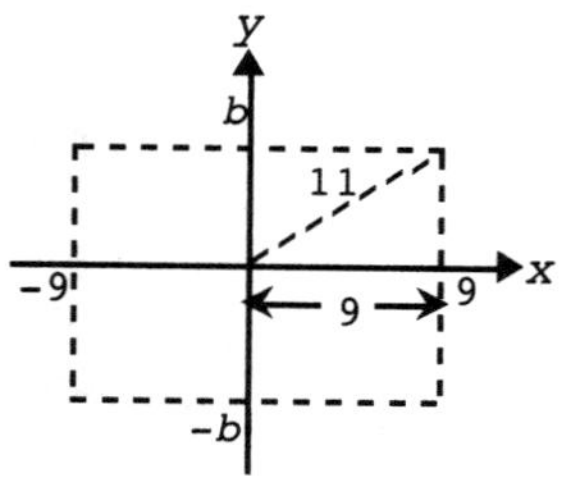

Thus, the equation is

$$\frac{x^2}{81} - \frac{y^2}{40} = 1$$

27. Since the conjugate axis is on the x axis, start with

$$\frac{y^2}{a^2} - \frac{x^2}{b^2} = 1$$

and find a and b

$$b = \frac{14}{2} = 7$$

To find a, sketch the asymptote rectangle, label known parts, and use the Pythagorean Theorem.

$$a^2 = (\sqrt{200})^2 - 7^2$$
$$a^2 = 151$$
$$a = \sqrt{151}$$

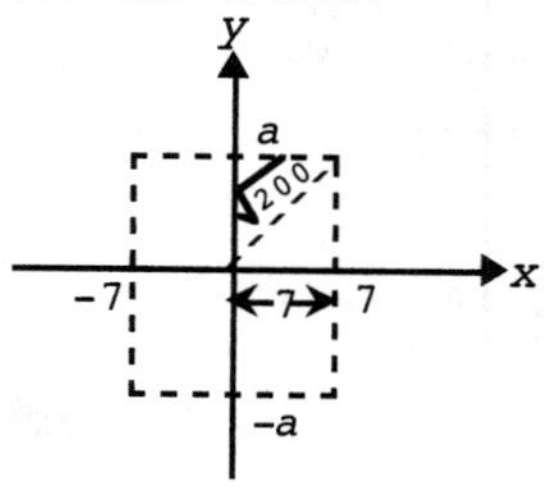

Thus, the equation is

$$\frac{y^2}{151} - \frac{x^2}{49} = 1$$

29. The equation $\frac{x^2}{25} - \frac{y^2}{4} = 1$ is in standard form with $a^2 = 25$, $b^2 = 4$. Since the intercepts are on the x axis, the asymptotes have equations $y = \pm\frac{b}{a}x = \pm\frac{2}{5}x$.

31. The equation $\frac{y^2}{4} - \frac{x^2}{16} = 1$ is in standard form with $a^2 = 4$, $b^2 = 16$. Since the intercepts are on the y axis, the asymptotes have equations $y = \pm\frac{a}{b}x = \pm\frac{2}{4}x$ or $y = \pm\frac{1}{2}x$.

33. $9x^2 - y^2 = 9$. In standard form, this becomes

$$\frac{x^2}{1} - \frac{y^2}{9} = 1$$

Then $a^2 = 1$, $b^2 = 9$. Since the intercepts are on the x axis, the asymptotes have equations $y = \pm\frac{b}{a}x = \pm\frac{3}{1}x$ or $y = \pm 3x$.

35. $2y^2 - 3x^2 = 1$. In standard form, this becomes

$$\frac{y^2}{1/2} - \frac{x^2}{1/3} = 1$$

Then $a^2 = \frac{1}{2}$, $b^2 = \frac{1}{3}$. Since the intercepts are on the y axis, the asymptotes have equations $y = \pm\frac{a}{b}x$. Write $a = \sqrt{\frac{1}{2}}$, $b = \sqrt{\frac{1}{3}}$, $\frac{a}{b} = \sqrt{\frac{1}{2}} \div \sqrt{\frac{1}{3}} = \sqrt{\frac{3}{2}}$ or $\frac{\sqrt{3}}{\sqrt{2}}$.

Asymptotes: $y = \pm\frac{\sqrt{3}}{\sqrt{2}}x$.

37. (A) If a hyperbola has center at (0, 0) and a focus at (1, 0), its equation must be of form

$$\frac{x^2}{a^2} - \frac{y^2}{b^2} = 1$$

with $c = 1$, thus $a^2 + b^2 = 1$ or $b^2 = 1 - a^2$

Therefore there are an infinite number of such hyperbolas. Each has an equation of form

$$\frac{x^2}{a^2} - \frac{y^2}{1 - a^2} = 1$$

Note that since $0 < a < c$, we require $0 < a < 1$.

(B) If an ellipse has center at (0, 0) and a focus at (1, 0), its equation must be of form

$$\frac{x^2}{a^2} + \frac{y^2}{b^2} = 1$$

with $c = 1$, thus $a^2 - 1 = b^2$

Therefore there are an infinite number of such ellipses. Each has an equation of form

$$\frac{x^2}{a^2} + \frac{y^2}{a^2 - 1} = 1$$

Note that since $a > c > 0$, we require $a > 1$.

(C) If a parabola has vertex at (0, 0) and focus at (1, 0), its equation must be of form

$$y^2 = 4ax$$

with $a = 1$.

Therefore there is one such parabola; its equation is $y^2 = 4x$.

39. (A) The points of intersection are the solutions of the system

$$x^2 - y^2 = 1$$
$$y = 0.5x = \frac{1}{2}x$$

We solve using substitution:

$$x^2 - \left(\frac{1}{2}x\right)^2 = 1$$
$$x^2 - \frac{1}{4}x^2 = 1$$
$$\frac{3}{4}x^2 = 1$$
$$x^2 = \frac{4}{3}$$
$$x = \pm\frac{2}{\sqrt{3}}$$

$y = \frac{1}{2}x$, so when $x = \frac{2}{\sqrt{3}}$, $y = \frac{1}{2} \cdot \frac{2}{\sqrt{3}} = \frac{1}{\sqrt{3}}$

when $x = \frac{-2}{\sqrt{3}}$, $y = \frac{1}{2} \cdot \frac{-2}{\sqrt{3}} = \frac{-1}{\sqrt{3}}$

The points of intersection are $\left(\frac{2}{\sqrt{3}}, \frac{1}{\sqrt{3}}\right)$ and $\left(\frac{-2}{\sqrt{3}}, \frac{-1}{\sqrt{3}}\right)$.

(B)

$$x^2 - y^2 = 1$$
$$y = 2x$$
$$x^2 - (2x)^2 = 1$$
$$x^2 - 4x^2 = 1$$
$$-3x^2 = 1$$
$$x^2 = -\frac{1}{3}$$

No solution.
The graphs do not intersect.

Repeat the above calculations using $y = mx$ instead of $y = 2x$.

$$x^2 - y^2 = 1$$
$$y = mx$$
$$x^2 - (mx)^2 = 1$$
$$x^2 - m^2x^2 = 1$$
$$x^2(1 - m^2) = 1$$
$$x^2 = \frac{1}{1 - m^2}$$

If the right side is negative the equation will have no solution, so the graphs intersect only if $1 - m^2 > 0$. This occurs only if m^2 is between zero and 1; that is, for $-1 < m < 1$. In this case, $x = \pm\frac{1}{\sqrt{1 - m^2}}$; $y = mx = \pm\frac{m}{\sqrt{1 - m^2}}$.

So if $-1 < m < 1$, the graphs intersect at

$$\left(\frac{1}{\sqrt{1 - m^2}}, \frac{m}{\sqrt{1 - m^2}}\right) \text{ and } \left(\frac{-1}{\sqrt{1 - m^2}}, \frac{-m}{\sqrt{1 - m^2}}\right).$$

41. (A) The points of intersection are the solutions of the system

$$y^2 - 4x^2 = 1$$
$$y = x$$
$$x^2 - 4x^2 = 1$$
$$-3x^2 = 1$$
$$x^2 = -\frac{1}{3}$$

No solution. The graphs do not intersect.

(B)
$$y^2 - 4x^2 = 1$$
$$y = 3x$$
$$(3x)^2 - 4x^2 = 1$$
$$9x^2 - 4x^2 = 1$$
$$5x^2 = 1$$
$$x^2 = \frac{1}{5}$$
$$x = \pm\frac{1}{\sqrt{5}}$$

$y = 3x$, so when $x = \frac{1}{\sqrt{5}}$, $y = \frac{3}{\sqrt{5}}$ and when $x = \frac{-1}{\sqrt{5}}$, $y = \frac{-3}{\sqrt{5}}$. The points of intersection are $\left(\frac{1}{\sqrt{5}}, \frac{3}{\sqrt{5}}\right)$ and $\left(\frac{-1}{\sqrt{5}}, \frac{-3}{\sqrt{5}}\right)$.

Repeat the above calculations with $y = mx$ in place of $y = 3x$.

$$y^2 - 4x^2 = 1$$
$$y = mx$$
$$(mx)^2 - 4x^2 = 1$$
$$m^2x^2 - 4x^2 = 1$$
$$x^2(m^2 - 4) = 1$$
$$x^2 = \frac{1}{m^2 - 4}$$

This equation will have solutions only if $m^2 - 4$ is positive; this occurs when $m > 2$ or $m < -2$. In this case,

$$x = \pm\frac{1}{\sqrt{m^2 - 4}}; \quad y = mx = \pm\frac{m}{\sqrt{m^2 - 4}}.$$

So if $m > 2$ or $m < -2$ the points of intersection are

$$\left(\frac{1}{\sqrt{m^2 - 4}}, \frac{m}{\sqrt{m^2 - 4}}\right) \text{ and } \left(\frac{-1}{\sqrt{m^2 - 4}}, \frac{-m}{\sqrt{m^2 - 4}}\right).$$

43. (A) $\frac{x^2}{a^2} - \frac{y^2}{b^2} = 1$ Solve for y.

$$-\frac{y^2}{b^2} = 1 - \frac{x^2}{a^2}$$
$$\frac{y^2}{b^2} = \frac{x^2}{a^2} - 1$$
$$y^2 = \frac{b^2x^2}{a^2} - b^2$$
$$y^2 = \frac{b^2}{a^2}x^2\left(1 - \frac{a^2}{x^2}\right)$$
$$y = \pm\frac{b}{a}x\sqrt{1 - \frac{a^2}{x^2}}$$

(B) As $|x| \to \infty$, $\frac{a}{|x|} \to 0$, $\frac{a^2}{x^2} \to 0$, $1 - \frac{a^2}{x^2} \to 1$, and $\sqrt{1 - \frac{a^2}{x^2}} \to 1$.

Hence as $|x| \to \infty$, $y \to \pm\frac{b}{a}x$. Therefore these lines form asymptotes for the hyperbola.

(C) Since $\sqrt{1 - \frac{a^2}{x^2}} < 1$ for $|x| > a$, $\frac{b}{a}x\sqrt{1 - \frac{a^2}{x^2}} < \frac{b}{a}x$, for $x > 0$. Thus in quadrant I, the hyperbola is below its asymptote $y = \frac{b}{a}x$.

However, $-\frac{b}{a}x < -\frac{b}{a}x\sqrt{1 - \frac{a^2}{x^2}}$, for $x > 0$, hence, in quadrant IV, the hyperbola is above its asymptote, $y = -\frac{b}{a}x$. Similar arguments applied for $x < 0$ show that in quadrant II the hyperbola is below its asymptote $y = -\frac{b}{a}x$, and in quadrant III, it is above its asymptote $y = \frac{b}{a}x$.

45. By the triangle inequality, no points can satisfy this condition. The set is empty.

47.
$$|d_1 - d_2| = 2a$$
$$d_1 - d_2 = \pm 2a$$
$$d_1 = \pm 2a + d_2 \quad \text{Use the distance formula}$$
$$\sqrt{(x + c)^2 + y^2} = \pm 2a + \sqrt{(x - c)^2 + y^2} \quad \text{Square both sides}$$
$$(x + c)^2 + y^2 = 4a^2 \pm 4a\sqrt{(x - c)^2 + y^2} + (x - c)^2 + y^2 \quad \text{Isolate the square root.}$$
$$\pm\sqrt{(x - c)^2 + y^2} = a - \frac{cx}{a} \quad \text{Square both sides}$$
$$(x - c)^2 + y^2 = a^2 - 2cx + \frac{c^2x^2}{a^2} \quad \text{Simplify and collect terms}$$
$$\left(1 - \frac{c^2}{a^2}\right)x^2 + y^2 = a^2 - c^2 \quad \text{Use } a^2 - c^2 = b^2,\ 1 - \frac{c^2}{a^2} = \frac{b^2}{a^2} \text{ and divide both sides by } b^2.$$
$$\frac{x^2}{a^2} - \frac{y^2}{b^2} = 1$$

49. From the figure, we see that the point $P(x, y)$ is a point on the curve if and only if
$$d_1 = \frac{3}{2}d_2$$
$$d(P, F) = \frac{3}{2}d(P, M)$$
$$\sqrt{(x - 3)^2 + (y - 0)^2} = \frac{3}{2}\sqrt{\left(x - \frac{4}{3}\right)^2 + (y - y)^2}$$
$$(x - 3)^2 + y^2 = \frac{9}{4}\left(x - \frac{4}{3}\right)^2$$
$$x^2 - 6x + 9 + y^2 = \frac{9}{4}\left(x^2 - \frac{8}{3}x + \frac{16}{9}\right)$$
$$4x^2 - 24x + 36 + 4y^2 = 9x^2 - 24x + 16$$
$$-5x^2 + 4y^2 = -20$$
$$\frac{x^2}{4} - \frac{y^2}{5} = 1$$

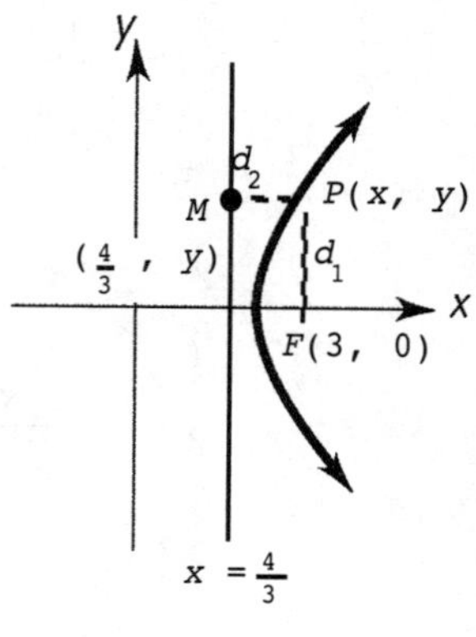

The curve must be a hyperbola since its equation can be written in standard form for a hyperbola.

51. From the figure below, we see that the transverse axis of the hyperbola must be on the y axis and $a = 4$. Hence, the equation of the hyperbola must have the form

$$\frac{y^2}{4^2} - \frac{x^2}{b^2} = 1$$

To find b, we note that the point (8, 12) is on the hyperbola, hence its coordinates satisfy the equation. Substituting, we have

$$\frac{12^2}{4^2} - \frac{8^2}{b^2} = 1$$
$$9 - \frac{64}{b^2} = 1$$
$$-\frac{64}{b^2} = -8$$
$$-64 = -8b^2$$
$$b^2 = 8$$

The equation required is

$$\frac{y^2}{16} - \frac{x^2}{8} = 1$$

Using this equation, we can compute y when $x = 6$ to answer the question asked (see figure).

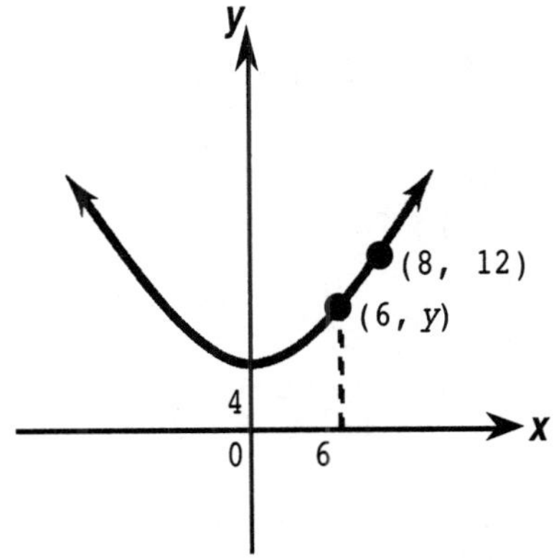

Hyperbola part of dome

$$\frac{y^2}{16} - \frac{6^2}{8} = 1$$
$$\frac{y^2}{16} - \frac{36}{8} = 1$$
$$y^2 - 72 = 16$$
$$y^2 = 88$$
$$y = 9.38 \text{ to two decimal places}$$

The height above the vertex
$= y$ - height of vertex
$= 9.38 - 4$
$= 5.38$ feet

53. From the figure below, we can see:
$FF' = 2c = 120 - 20 = 100$
Thus $c = 50$
$FV = c - a = 120 - 110 = 10$
Thus $a = c - 10 = 50 - 10 = 40$
Since $c^2 = a^2 + b^2$
$50^2 = 40^2 + b^2$
$b = 30$

Thus the equation of the hyperbola, in standard form, is

$$\frac{y^2}{40^2} - \frac{x^2}{30^2} = 1$$

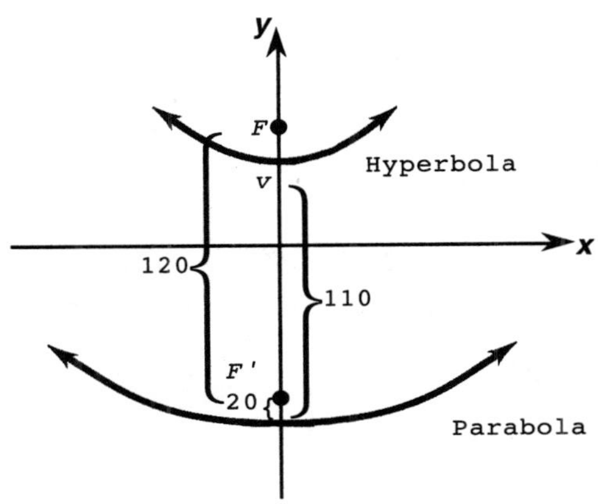

Expressing y in terms of x, we have

$$\frac{y^2}{40^2} = 1 + \frac{x^2}{30^2}$$
$$\frac{y^2}{40^2} = \frac{1}{30^2}(30^2 + x^2)$$
$$y^2 = \frac{40^2}{30^2}(30^2 + x^2)$$
$$y = \frac{4}{3}\sqrt{x^2 + 30^2}$$

discarding the negative solution, since the reflecting hyperbola is above the x axis.

Section 9-4

1. (A) Since $(h, k) = (3, 5)$, use translation formulas
$x' = x - h = x - 3$
$y' = y - k = y - 5$

(B) $x'^2 + y'^2 = 81$

(C) Circle

3. (A) Since $(h, k) = (-7, 4)$, use translation formulas
$x' = x - h = x + 7$
$y' = y - k = y - 4$

(B) $\frac{x'^2}{9} + \frac{y'^2}{16} = 1$

(C) Ellipse

5. (A) Since $(h, k) = (4, -9)$, use translation formulas
$x' = x - h = x - 4$
$y' = y - k = y + 9$

(B) $y'^2 = 16x'$

(C) Parabola

7. (A) Since $(h, k) = (-8, -3)$, use translation formulas
$x' = x - h = x + 8$
$y' = y - k = y + 3$

(B) $\frac{x'^2}{12} + \frac{y'^2}{8} = 1$

(C) Ellipse

9. (A) Divide both sides by 144
$$\frac{(x - 3)^2}{9} - \frac{(y + 2)^2}{16} = 1$$
(B) This is the equation of a hyperbola.

11. (A) Divide both sides by 30.
$$\frac{(x + 5)^2}{5} + \frac{(y + 7)^2}{6} = 1$$
(B) This is the equation of an ellipse.

13. (A) Subtract $24(y - 4)$ from both sides.
$(x + 6)^2 = -24(y - 4)$

(B) This is the equation of a parabola.

For question 15 and 17, plug the given x and y coordinates into the formulas $x' = x \cos \theta + y \sin \theta$, $y' = -x \sin \theta + y \cos \theta$.

15. $(1, 0)$: $x' = 1 \cos 30° + 0 \sin 30°$, $y' = -1 \sin 30° + 0 \cos 30°$
$x' = \frac{\sqrt{3}}{2}$ $\quad y' = -\frac{1}{2}$
$\left(\frac{\sqrt{3}}{2}, -\frac{1}{2}\right)$

$(0, 1)$: $x' = 0 \cos 30° + 1 \sin 30°$, $y' = -0 \sin 30° + 1 \cos 30°$
$x' = \frac{1}{2}$ $\quad y' = \frac{\sqrt{3}}{2}$
$\left(\frac{1}{2}, \frac{\sqrt{3}}{2}\right)$

$(1, -1)$: $x' = 1 \cos 30° + (-1) \sin 30°$, $y' = -1 \sin 30° + (-1) \cos 30°$
$x' = \frac{\sqrt{3}}{2} - \frac{1}{2} = \frac{\sqrt{3} - 1}{2}$ $\quad y' = -\frac{1}{2} - \frac{\sqrt{3}}{2} = \frac{-1 - \sqrt{3}}{2}$
$\left(\frac{\sqrt{3} - 1}{2}, \frac{-1 - \sqrt{3}}{2}\right)$

$(-3, 4)$: $x' = -3 \cos 30° + 4 \sin 30°$, $y' = -(-3) \sin 30° + (-4) \cos 30°$

$$x' = -3 \cdot \frac{\sqrt{3}}{2} + 4 \cdot \frac{1}{2} \qquad y' = 3 \cdot \frac{1}{2} - 4 \cdot \frac{\sqrt{3}}{2}$$

$$x' = \frac{-3\sqrt{3} + 4}{2} \qquad y' = \frac{3 - 4\sqrt{3}}{2}$$

$$\left(\frac{-3\sqrt{3} + 4}{2}, \frac{3 - 4\sqrt{3}}{2}\right)$$

17. $(1, 0)$: $x' = 1 \cos 45° + 0 \sin 45°$, $y' = -1 \sin 45° + 0 \cos 45°$

$$x' = \frac{\sqrt{2}}{2} \qquad y' = -\frac{\sqrt{2}}{2}$$

$$\left(\frac{\sqrt{2}}{2}, -\frac{\sqrt{2}}{2}\right)$$

$(0, 1)$: $x' = 0 \cos 45° + 1 \sin 45°$, $y' = -0 \sin 45° + 1 \cos 45°$

$$x' = \frac{\sqrt{2}}{2} \qquad y' = \frac{\sqrt{2}}{2}$$

$$\left(\frac{\sqrt{2}}{2}, \frac{\sqrt{2}}{2}\right)$$

$(-1, -2)$: $x' = -1 \cos 45° + (-2) \sin 45°$, $y' = -(-1) \sin 45° + (-2) \cos 45°$

$$x' = -\frac{\sqrt{2}}{2} - \frac{2\sqrt{2}}{2} \qquad y' = \frac{\sqrt{2}}{2} - \frac{2\sqrt{2}}{2}$$

$$x' = \frac{3\sqrt{2}}{2} \qquad y' = -\frac{\sqrt{2}}{2}$$

$$\left(\frac{-3\sqrt{2}}{2}, -\frac{\sqrt{2}}{2}\right)$$

$(1, -3)$: $x' = 1 \cos 45° + (-3) \sin 45°$, $y' = -1 \sin 45° + (-3) \cos 45°$

$$x' = \frac{\sqrt{2}}{2} - \frac{3\sqrt{2}}{2} \qquad y' = -\frac{\sqrt{2}}{2} - \frac{3\sqrt{2}}{2}$$

$$x' = -\frac{2\sqrt{2}}{2} = -\sqrt{2} \qquad y' = -\frac{4\sqrt{2}}{2} = -2\sqrt{2}$$

$(-\sqrt{2}, -2\sqrt{2})$

19. We need to use the rotation identities to rewrite the equations of the x axis ($y' = 0$) and the y' axis ($x' = 0$).

First, plug in $\theta = 30°$:

$$x' = x \cos 30° + y \sin 30° \qquad y' = -x \sin 30° + y \cos 30°$$

$$x' = \frac{\sqrt{3}}{2}x + \frac{1}{2}y \qquad y' = -\frac{1}{2}x + \frac{\sqrt{3}}{2}y$$

The x' axis corresponds to $y' = 0$ so we set $-\frac{1}{2}x + \frac{\sqrt{3}}{2}y$ equal to zero and solve for y:

$$-\frac{1}{2}x + \frac{\sqrt{3}}{2}y = 0$$

$$\frac{\sqrt{3}}{2}y = \frac{1}{2}x$$

$$y = \frac{1}{\sqrt{3}}x$$

To find the equation of the y' axis set $\frac{\sqrt{3}}{2}x + \frac{1}{2}y$ equal to zero:

$$\frac{\sqrt{3}}{2}x + \frac{1}{2}y = 0$$
$$\frac{1}{2}y = -\frac{\sqrt{3}}{2}x$$
$$y = -\sqrt{3}x$$

21. We need to use the rotation identities to rewrite the equations of the x axis ($y' = 0$) and the y' axis ($x' = 0$).

First plug in $\theta = 45°$:

$$x' = x\cos 45° + y\sin 45° \qquad y' = -x\sin 45° + y\cos 45°$$
$$x' = \frac{\sqrt{2}}{2}x + \frac{\sqrt{2}}{2}y \qquad y' = -\frac{\sqrt{2}}{2}x + \frac{\sqrt{2}}{2}y$$

The x' axis corresponds to $y' = 0$ so we set $-\frac{\sqrt{2}}{2}x + \frac{\sqrt{2}}{2}y$ equal to zero and solve for y:

$$-\frac{\sqrt{2}}{2}x + \frac{\sqrt{2}}{2}y = 0$$
$$\frac{\sqrt{2}}{2}y = \frac{\sqrt{2}}{2}x$$
$$y = x$$

To find the equation of the y' axis set $\frac{\sqrt{2}}{2}x + \frac{\sqrt{2}}{2}y$ equal to zero:

$$\frac{\sqrt{2}}{2}x + \frac{\sqrt{2}}{2}y = 0$$
$$\frac{\sqrt{2}}{2}y = -\frac{\sqrt{2}}{2}x$$
$$y = -x$$

23.

$$4x^2 + 9y^2 - 16x - 36y + 16 = 0$$
$$4x^2 - 16x + 9y^2 - 36y = -16$$
$$4(x^2 - 4x + ?) + 9(y^2 - 4y + ?) = -16 \qquad \left(\frac{-4}{2}\right)^2 = 4$$
$$4(x^2 - 4x + 4) + 9(y^2 - 4y + 4) = -16 + 16 + 36$$
$$4(x - 2)^2 + 9(y - 2)^2 = 36$$
$$\frac{(x-2)^2}{9} + \frac{(y-2)^2}{4} = 1$$

This is the equation of an ellipse with center at (2, 2). The equations of translation are $x' = x - 2$, $y' = y - 2$. Making these substitutions, we obtain

$$\frac{x'^2}{9} + \frac{y'^2}{4} = 1$$

We graph this in the $x'y'$ system, following the process discussed in Section 9-2.

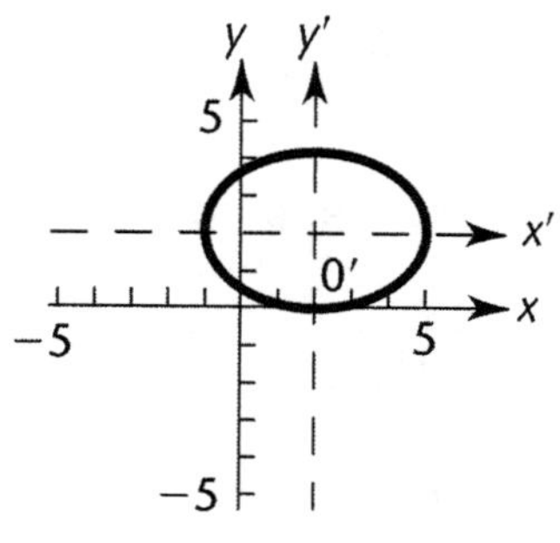

25. $x^2 + 8x + 8y = 0$

$$x^2 + 8x = -8y$$
$$x^2 + 8x + 16 = -8y + 16$$
$$(x + 4)^2 = -8(y - 2)$$

This is the equation of a parabola opening down with vertex at $(h, k) = (-4, 2)$. The equations of translation are $x' = x + 4$, $y' = y - 2$. Making these substitutions, we obtain

$$x'^2 = -8y'$$

We graph this in the $x'y'$ system, following the process discussed in Section 9-1.

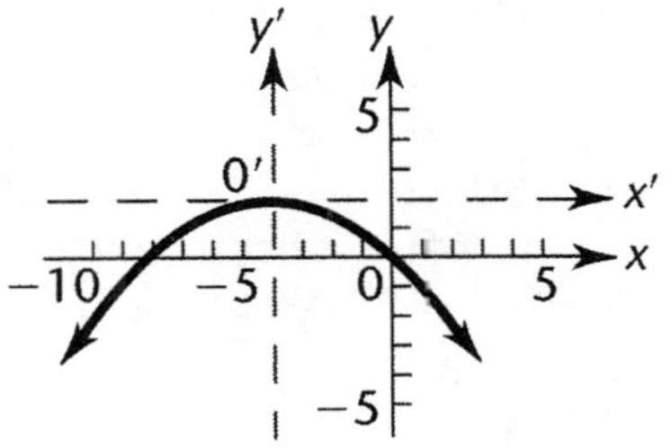

27.

$$x^2 + y^2 + 12x + 10y + 45 = 0$$
$$x^2 + 12x + y^2 + 10y = -45$$
$$x^2 + 12x + 36 + y^2 + 10y + 25 = -45 + 36 + 25$$
$$(x + 6)^2 + (y + 5)^2 = 16$$

This is the equation of a circle with center at $(-6, -5)$ and radius 4. The equations of translation are $x' = x + 6$, $y' = y + 5$. Making these substitutions, we obtain

$$x'^2 + y'^2 = 16$$

We graph this in the $x'y'$ system.

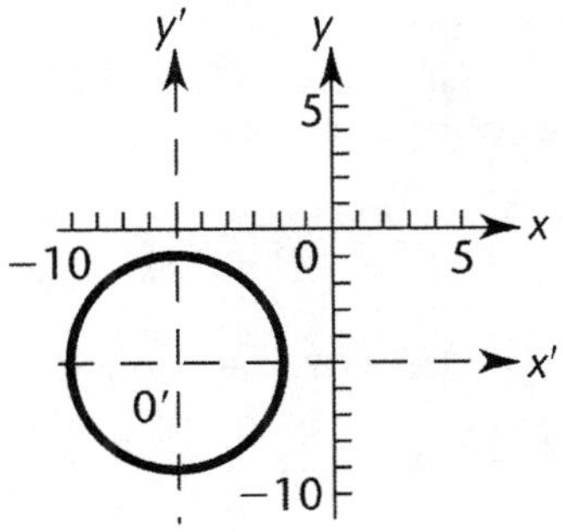

29.

$$-9x^2 + 16y^2 - 72x - 96y - 144 = 0$$
$$-9x^2 - 72x + 16y^2 - 96y = 144$$
$$-9(x^2 + 8x + ?) + 16(y^2 - 6y + ?) = 144$$
$$-9(x^2 + 8x + 16) + 16(y^2 - 6y + 9) = 144 - 144 + 144$$
$$-9(x + 4)^2 + 16(y - 3)^2 = 144$$
$$\frac{(y - 3)^2}{9} - \frac{(x + 4)^2}{16} = 1$$

This is the equation of a hyperbola with center at $(-4, 3)$. The equations of translation are $x' = x + 4$, $y' = y - 3$. Making these substitutions, we obtain

$$\frac{y'^2}{9} - \frac{x'^2}{16} = 1$$

We graph this in the $x'y'$ system, following the process discussed in Section 9-3.

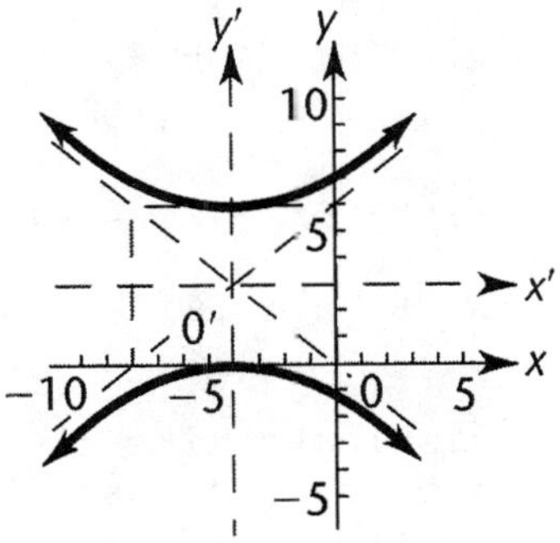

31. First find the coordinates of the foci in the translated system.

$$c'^2 = 3^2 - 2^2 = 5$$
$$c' = \sqrt{5}$$
$$-c' = -\sqrt{5}$$

Thus the coordinates in the translated system are

$F' = (-\sqrt{5}, 0)$ and $F = (\sqrt{5}, 0)$

Now use

$$x = x' + h = x' + 2$$
$$y = y' + k = y' + 2$$

to obtain

$F' = (-\sqrt{5} + 2, 2)$ and $F = (\sqrt{5} + 2, 2)$

as the coordinates of the foci in the original system.

33. First find the coordinates of the focus in the translated system. Since $a = -2$, and the parabola opens down, the coordinates are (0, -2). Now use

$x = x' + h = x' - 4 = 0 - 4$
$y = y' + k = y' + 2 = -2 + 2$

to obtain (-4, 0) as the coordinates of the focus in the original system.

35. First find the coordinates of the foci in the translated system.

$c'^2 = 3^2 + 4^2 = 25$
$c' = 5$
$-c' = -5$

Thus the coordinates in the translated system are
$F' = (0, -5)$ and $F = (0, 5)$
Now use
$x = x' + h = x' - 4$
$y = y' + k = y' + 3$
to obtain
$F' = (0 - 4, -5 + 3) = F' = (-4, -2)$ and $F = (0 - 4, 5 + 3) = F = (-4, 8)$ as the coordinates of the foci in the original system.

37.
$$x^2 - 2x + y^2 + 4y + 5 = 0$$
$$x^2 - 2x + y^2 + 4y = -5$$
$$(x^2 - 2x + ?) + (y^2 + 4y + ?) = -5 \qquad \left(\frac{-2}{1}\right)^2 = 1; \ \left(\frac{4}{2}\right)^2 = 4$$
$$(x^2 - 2x + 1) + (y^2 + 4y + 4) = -5 + 1 + 4$$
$$(x - 1)^2 + (y + 2)^2 = 0$$

This looks like the equation of a circle with center (1, -2). But the radius would be zero, so this is a degenerate circle which is really just the point (1, -2).

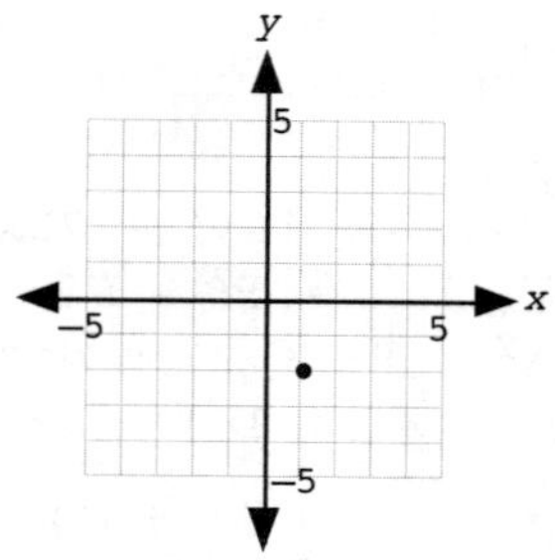

39.
$$x^2 + 8x - 4y^2 + 8y + 12 = 0$$
$$x^2 + 8x - 4y^2 + 8y = -12$$
$$x^2 + 8x - 4(y^2 - 2y) = -12$$
$$(x^2 + 8x + ?) - 4(y^2 - 2y + ?) = -12 \qquad \left(\frac{8}{2}\right)^2 = 16; \ \left(\frac{-2}{2}\right)^2 = 1$$
$$(x^2 + 8x + 16) - 4(y^2 - 2y + 1) = -12 + 16 - 4$$
$$(x + 4)^2 - 4(y - 1)^2 = 0$$

This looks like the equation of a hyperbola but it is degenerate because of the zero. To see what the graph is, we solve the equation for y:

$$4(y - 1)^2 = (x + 4)^2$$
$$2(y - 1) = \pm(x + 4)$$
$$2y - 2 = x + 4 \quad \text{or} \quad 2y - 2 = -(x + 4) = -x - 4$$
$$2y = x + 6 \qquad 2y = -x - 2$$
$$y = \frac{1}{2}x + 6 \qquad y = -\frac{1}{2}x - 1$$

The graph is 2 lines that intersect at (-4, 1) (which would have been the center of the hyperbola.)

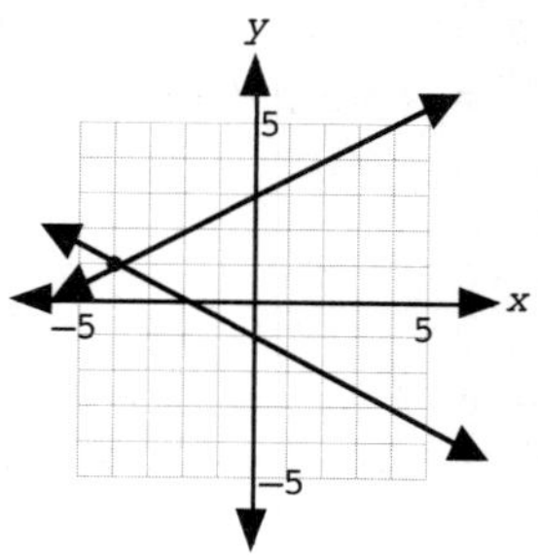

41. If $A \neq 0$, $C = 0$, and $E \neq 0$, write

$$Ax^2 + Dx + Ey + F = 0$$

Complete the square relative to x:

$$A\left(x^2 + \frac{D}{A}x + ?\right) = -Ey - F$$

$$A\left(x^2 + \frac{D}{A}x + \frac{D^2}{4A^2}\right) = -Ey - F + \frac{D^2}{4A}$$

$$\left(x + \frac{D}{2A}\right)^2 = -\frac{E}{A}y - \frac{F}{A} + \frac{D^2}{4A^2}$$

$$\left(x + \frac{D}{2A}\right)^2 = 4\cdot\left(-\frac{E}{4A}\right)\left[y + \frac{F}{E} - \frac{D^2}{4AE}\right]$$

$$\left[x - \left(-\frac{D}{2A}\right)\right]^2 = 4\left(-\frac{E}{4A}\right)\left[y - \frac{D^2 - 4AF}{4AE}\right]$$

The equations of translation are

$$x' = x - \left(-\frac{D}{2A}\right) \qquad y' = y - \frac{D^2 - 4AF}{4AE}$$

Thus, $h = -\frac{D}{2A}$, $k = \frac{D^2 - 4AF}{4AE}$

The equation would become $x'^2 = 4\left(-\frac{E}{4A}\right)y'$, that is, the equation of a parabola.

43. $x^2 + y^2 = 49$

The rotation equations are

$x = \cos 45°x' - \sin 45°y'$ $\qquad$ $y = \sin 45°x' + \cos 45°y'$

$x = \frac{\sqrt{2}}{2}x' - \frac{\sqrt{2}}{2}y'$ $\qquad$ $y = \frac{\sqrt{2}}{2}x' + \frac{\sqrt{2}}{2}y'$

Substitute into the original equation and simplify:

$$\left(\frac{\sqrt{2}}{2}x' - \frac{\sqrt{2}}{2}y'\right)^2 + \left(\frac{\sqrt{2}}{2}x' + \frac{\sqrt{2}}{2}y'\right)^2 = 49$$

$$\frac{1}{2}x'^2 - x'y' + \frac{1}{2}y'^2 + \frac{1}{2}x'^2 + x'y' + \frac{1}{2}y'^2 = 49$$

$$x'^2 + y'^2 = 49$$

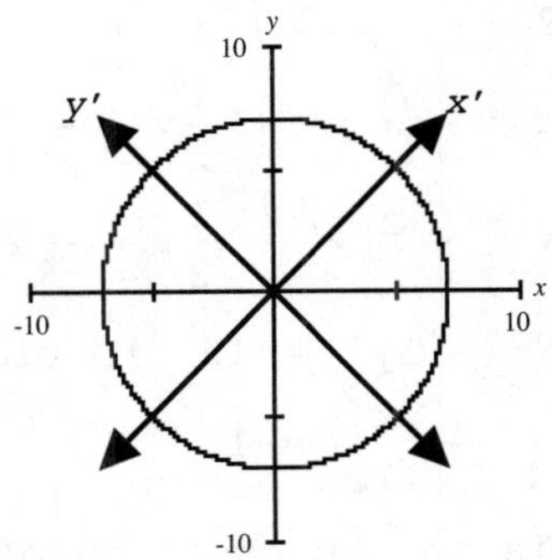

We got the original equation back! This makes sense because the graph is a circle with center (0, 0). Rotating the axes will not change the graph.

45. $2x^2 + \sqrt{3}xy + y^2 - 10 = 0$

The rotation equations are

$x = \cos 30°x' - \sin 30°y'$ $\qquad$ $y = \sin 30°x' + \cos 30°y'$

$x = \frac{\sqrt{3}}{2}x' - \frac{1}{2}y'$ $\qquad$ $y = \frac{1}{2}x' + \frac{\sqrt{3}}{2}y'$

Substitute into the original equation and simplify:

$$2\left(\frac{\sqrt{3}}{2}x' - \frac{1}{2}y'\right)^2 + \sqrt{3}\left(\frac{\sqrt{3}}{2}x' - \frac{1}{2}y'\right)\left(\frac{1}{2}x' + \frac{\sqrt{3}}{2}y'\right) + \left(\frac{1}{2}x' + \frac{\sqrt{3}}{2}y'\right)^2 - 10 = 0$$

$$2\left(\frac{3}{4}x'^2 - \frac{\sqrt{3}}{2}x'y' + \frac{1}{4}y'^2\right) + \sqrt{3}\left(\frac{\sqrt{3}}{4}x'^2 + \frac{1}{2}x'y' - \frac{\sqrt{3}}{4}y'^2\right) + \frac{1}{4}x'^2 + \frac{\sqrt{3}}{2}x'y' + \frac{3}{4}y'^2 = 10$$

$$\frac{3}{2}x'^2 - \sqrt{3}x'y' + \frac{1}{2}y'^2 + \frac{3}{4}x'^2 + \frac{\sqrt{3}}{2}x'y' - \frac{3}{4}y'^2 + \frac{1}{4}x'^2 + \frac{\sqrt{3}}{2}x'y' + \frac{3}{4}y'^2 = 10$$

$\frac{5}{2}x'^2 + \frac{1}{2}y'^2 = 10$ (Divide both sides by 10)

$$\frac{x'^2}{4} + \frac{y'^2}{20} = 1$$

This is the equation of an ellipse with intercepts (2, 0), (-2, 0), (0, $\sqrt{20}$), and (0, $-\sqrt{20}$) rotated through 30°.

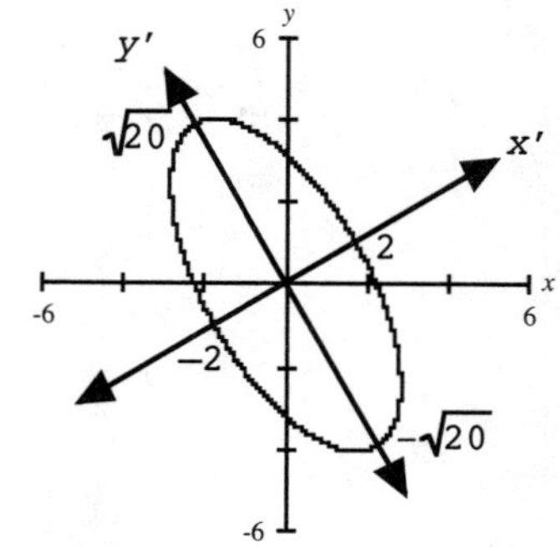

47. $x^2 - 4xy + y^2 = 12$

Use the formula $\cot 2\theta = \frac{A - C}{B}$ to find the appropriate angle.

$\cot 2\theta = \frac{1 - 1}{-4} = 0$

Since $\cot 90° = \frac{\cos 90°}{\sin 90°} = \frac{0}{1} = 0$, $2\theta = 90°$, and $\theta = 45°$.

The rotation equations are

$x = \cos 45°x' - \sin 45°y'$ $\quad$ $y = \sin 45°x' + \cos 45°y'$

$x = \frac{\sqrt{2}}{2}x' - \frac{\sqrt{2}}{2}y'$ $\quad$ $y = \frac{\sqrt{2}}{2}x' + \frac{\sqrt{2}}{2}y'$

Substitute into the original equation and simplify.

$$\left(\frac{\sqrt{2}}{2}x' - \frac{\sqrt{2}}{2}y'\right)^2 - 4\left(\frac{\sqrt{2}}{2}x' - \frac{\sqrt{2}}{2}y'\right)\left(\frac{\sqrt{2}}{2}x' - \frac{\sqrt{2}}{2}y'\right) + \left(\frac{\sqrt{2}}{2}x' + \frac{\sqrt{2}}{2}y'\right)^2 = 12$$

$$\frac{1}{2}x'^2 - x'y' + \frac{1}{2}y'^2 - 4\left(\frac{1}{2}x'^2 - \frac{1}{2}y'^2\right) + \frac{1}{2}x'^2 + x'y' + \frac{1}{2}y'^2 = 12$$

$$\frac{1}{2}x'^2 - x'y' + \frac{1}{2}y'^2 - 2x'^2 + 2y'^2 + \frac{1}{2}x'^2 + x'y' + \frac{1}{2}y'^2 = 12$$

$-x'^2 + 3y'^2 = 12$ (Divide both sides by 12)

$$\frac{y'^2}{4} - \frac{x'^2}{12} = 1$$

This is the equation of a hyperbola with transverse axis length 4 and conjugate axis length $2\sqrt{12}$ rotated through 45°.

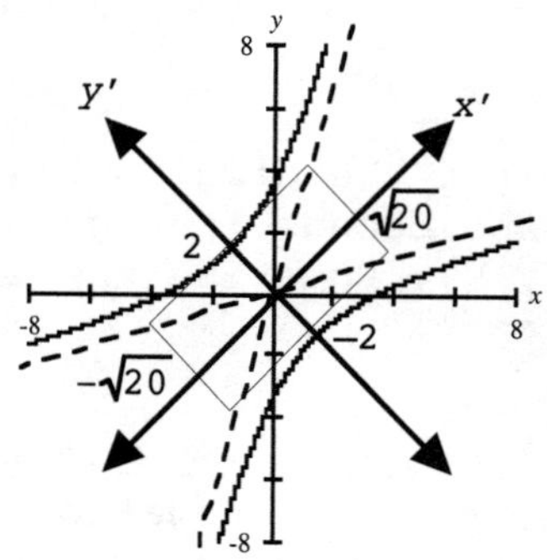

49. $8x^2 - 4xy + 5y^2 = 36$

Use the formula $\cot 2\theta = \frac{A - C}{B}$ to find the appropriate angle.

$\cot 2\theta = \frac{8 - 5}{-4} = -\frac{3}{4}$

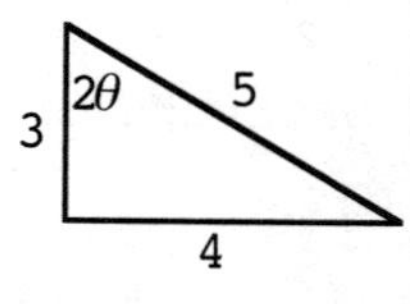

2θ is an angle in the second quadrant and, using the reference triangle at the right, we see that $\cos 2\theta = -\frac{3}{5}$.

We can find the rotation equations using the half-angle identities

$$\sin\theta = \sqrt{\frac{1-\cos 2\theta}{2}} \quad \text{and} \quad \cos\theta = \sqrt{\frac{1+\cos 2\theta}{2}}$$

$$\sin\theta = \sqrt{\frac{1-\left(-\frac{3}{5}\right)}{2}} \qquad \cos\theta = \sqrt{\frac{1+\left(-\frac{3}{5}\right)}{2}}$$

$$\sin\theta = \sqrt{\frac{8}{10}} \text{ or } \frac{2}{\sqrt{5}} \qquad \cos\theta = \sqrt{\frac{1}{5}} \text{ or } \frac{1}{\sqrt{5}}$$

The rotation equations are

$$x = \frac{1}{\sqrt{5}}x' - \frac{2}{\sqrt{5}}y' \qquad y = \frac{2}{\sqrt{5}}x' + \frac{1}{\sqrt{5}}y'$$

Substitute into the original equation and simplify:

$$8\left(\frac{1}{\sqrt{5}}x' - \frac{2}{\sqrt{5}}y'\right)^2 - 4\left(\frac{1}{\sqrt{5}}x' - \frac{2}{\sqrt{5}}y'\right)\left(\frac{2}{\sqrt{5}}x' + \frac{1}{\sqrt{5}}y'\right) + 5\left(\frac{2}{\sqrt{5}}x' + \frac{1}{\sqrt{5}}y'\right)^2 = 36$$

$$8\left(\frac{1}{5}x'^2 - \frac{4}{5}x'y' + \frac{4}{5}y'^2\right) - 4\left(\frac{2}{5}x'^2 - \frac{3}{5}x'y' - \frac{2}{5}y'^2\right) + 5\left(\frac{4}{5}x'^2 + \frac{4}{5}x'y' + \frac{1}{5}y'^2\right) = 36$$

$$\frac{8}{5}x'^2 - \frac{32}{5}x'y' + \frac{32}{5}y'^2 - \frac{8}{5}x'^2 + \frac{12}{5}x'y' + \frac{8}{5}y'^2 + 4x'^2 + 4x'y' + y'^2 = 36$$

$4x'^2 + 9y'^2 = 36$ (Divide both sides by 36)

$$\frac{x'^2}{9} + \frac{y'^2}{4} = 1$$

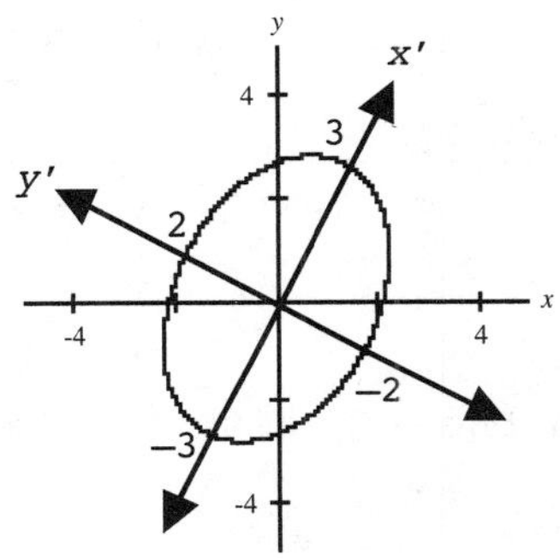

This is the equation of an ellipse with intercepts (3, 0), (-3, 0), (0, 2), and (0, -2). To find the angle of rotation, note that $\cos 2\theta = -\frac{3}{5}$, and using the inverse cosine function on a calculator, we get $2\theta = 126.87°$ so $\theta = 63.4°$.

51. $x^2 - 2\sqrt{3}xy + 3y^2 - 16\sqrt{3}x - 16y = 0$

Use the formula $\cot 2\theta = \frac{A - C}{B}$ to find the appropriate angle.

$$\cot 2\theta = \frac{1-3}{-2\sqrt{3}} = \frac{-2}{-2\sqrt{3}} = \frac{1}{\sqrt{3}}$$

Note that $\cot 60° = \frac{\cos 60°}{\sin 60°} = \frac{\frac{1}{2}}{\frac{\sqrt{3}}{2}} = \frac{1}{\sqrt{3}}$, so $2\theta = 60°$ and $\theta = 30°$.

The rotation equations are

$$x = \cos 30°x' - \sin 30°y' \qquad y = \sin 30°x' + \cos 30°y'$$

$$x = \frac{\sqrt{3}}{2}x' - \frac{1}{2}y' \qquad y = \frac{1}{2}x' + \frac{\sqrt{3}}{2}y'$$

Substitute into the original equation and simplify:

$$\left(\frac{\sqrt{3}}{2}x' - \frac{1}{2}y'\right)^2 - 2\sqrt{3}\left(\frac{\sqrt{3}}{2}x' - \frac{1}{2}y'\right)\left(\frac{1}{2}x' + \frac{\sqrt{3}}{2}y'\right) + 3\left(\frac{1}{2}x' + \frac{\sqrt{3}}{2}y'\right)^2$$
$$- 16\sqrt{3}\left(\frac{\sqrt{3}}{2}x' - \frac{1}{2}y'\right) - 16\left(\frac{1}{2}x' + \frac{\sqrt{3}}{2}y'\right) = 0$$

$$\frac{3}{4}x'^2 - \frac{\sqrt{3}}{2}x'y' + \frac{1}{4}y'^2 - 2\sqrt{3}\left(\frac{\sqrt{3}}{4}x'^2 + \frac{1}{2}x'y' - \frac{\sqrt{3}}{4}y'^2\right) + 3\left(\frac{1}{4}x'^2 + \frac{\sqrt{3}}{2}x'y' - \frac{3}{4}y'^2\right)$$
$$- 24x' + 8\sqrt{3}y' - 8x' - 8\sqrt{3}y' = 0$$

$$\frac{3}{4}x'^2 - \frac{\sqrt{3}}{2}x'y' + \frac{1}{4}y'^2 - \frac{3}{2}x'^2 - \sqrt{3}x'y' + \frac{3}{2}y'^2 + \frac{3}{4}x'^2 + \frac{3\sqrt{3}}{2}x'y' + \frac{9}{4}y'^2$$
$$- 24x' + 8\sqrt{3}y' - 8x' - 8\sqrt{3}y' = 0$$

$$4y'^2 - 32x' = 0$$
$$4y'^2 = 32x'$$
$$y'^2 = 8x'$$

This is the equation of a parabola with vertex (0, 0) rotated through 30°.

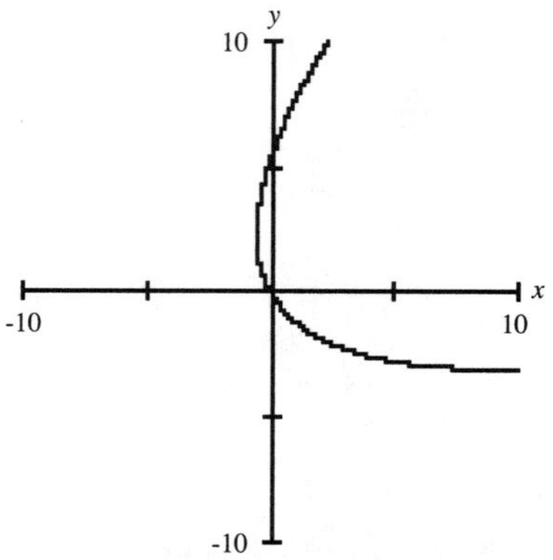

53. This is the equation of a hyperbola with center at (3, -2). The equations of translation are $x' = x - 3$, $y' = y + 2$. Making these substitutions, we obtain
$$x'^2 - y'^2 = 1$$
This equation is in standard form with $a = 1$, $b = 1$. Since the intercepts are on the x' axis, the asymptotes have equations
$$y' = \pm\frac{b}{a}x' = \pm x'$$
In the original coordinate system, these become
$$y + 2 = \pm(x - 3)$$

55. This is the equation of a hyperbola with center at (0, -1). The equations of translation are $x' = x$, $y' = y + 1$. Making these substitutions, we obtain
$$\frac{x'^2}{4} - \frac{y'^2}{25} = 1$$
This equation is in standard form with $a = 2$, $b = 5$. Since the intercepts are on the x' axis, the asymptotes have equations
$$y' = \pm\frac{b}{a}x' = \pm\frac{5}{2}x'$$
In the original coordinate system, these become
$$y + 1 = \pm\frac{5}{2}x$$

57. In standard form, this becomes
$$\frac{(y - 5)^2}{16} - \frac{(x + 2)^2}{9} = 1$$
This is the equation of a hyperbola with center at (-2, 5). The equations of translation are $x' = x + 2$, $y' = y - 5$. Making these substitutions we obtain
$$\frac{y'^2}{16} - \frac{x'^2}{9} = 1$$
This equation in standard form with $a = 4$, $b = 3$. Since the intercepts are on the y' axis, the asymptotes have equations
$$y' = \pm\frac{a}{b}x' = \pm\frac{4}{3}x'$$
In the original coordinate system, these become
$$y - 5 = \pm\frac{4}{3}(x + 2)$$

59. In standard form, this becomes
$$\frac{(y + 4)^2}{1/3} - \frac{x^2}{1} = 1$$
This is the equation of a hyperbola with center (0, -4). The equations of translation are $x' = x$, $y' = y' + 4$. Making these substitutions, we obtain
$$\frac{y'^2}{1/3} - \frac{x'^2}{1} = 1$$

This equation is in standard form with $a = \frac{1}{\sqrt{3}}$, $b = 1$. Since the intercepts are on the y' axis, the asymptotes have equations

$$y' = \pm\frac{a}{b}x' = \pm\frac{1}{\sqrt{3}}x'$$

In the original coordinate system, these become

$$y + 4 = \pm\frac{1}{\sqrt{3}}x$$

61. Applying the methods of this section (outline solution)
$A = C = 0$, $B = 1$, hence $\cot 2\theta = 0$, $\theta = 45°$.
The rotation formulas are $x = \frac{x' - y'}{\sqrt{2}}$, $y = \frac{x' + y'}{\sqrt{2}}$
In the rotated coordinate system, the equation becomes

$$\frac{x'^2 - y'^2}{2} - 9 = 0 \quad \text{or} \quad \frac{x'^2}{18} - \frac{y'^2}{18} = 1$$

This equation is in standard form and the hyperbola has asymptotes $y' = \pm x'$ in the rotated system. In the original system, applying the reverse rotation formulas,

$$x' = \frac{x + y}{\sqrt{2}} \qquad y' = -\frac{x + y}{\sqrt{2}}$$

the asymptotes become

$$-\frac{x + y}{\sqrt{2}} = \pm\frac{x + y}{\sqrt{2}}$$

which simplifies to $x = 0$, $y = 0$.
It would be much simpler, however, to note that $xy - 9 = 0$ is the equation of the rational function $y = \frac{9}{x}$, which has asymptotes $x = 0$ and $y = 0$, Section (4-4).

63. Locate the vertex and axis in the original coordinate system, then sketch the parabola and translate the origin to the vertex of the parabola. Next write the equation of the parabola in the translated system:

$$x'^2 = 4ay'$$

The origin in the translated system is at $(h, k) = (2, 5)$ and the translation formulas are

$$x' = x - h = x - 2$$
$$y' = y - k = y - 5$$

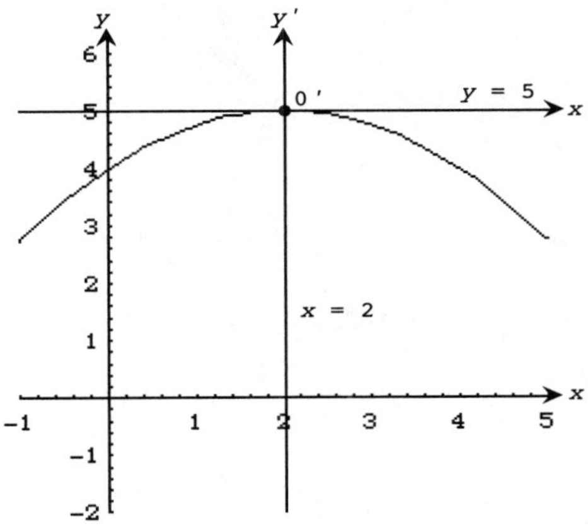

Thus, the equation of the parabola in the original system is

$$(x - 2)^2 = 4a(y - 5)$$

Since the point $(-2, 1)$ is on the parabola, its coordinates must satisfy the equation of the parabola, hence

$$[(-2) - 2]^2 = 4a(1 - 5)$$
$$16 = -16a$$
$$a = -1$$

Thus, the equation of the parabola is

$$(x - 2)^2 = -4(y - 5)$$
$$x^2 - 4x + 4 = -4y + 20$$
$$x^2 - 4x + 4y - 16 = 0$$

65. Locate the vertices in the original coordinate system, then sketch the ellipse and translate the origin to the center of the ellipse. Next write the equation of the ellipse in the translated system. Since $2a = 8$, and $2b = 4$, we know $a = 4$, $b = 2$, hence the equation is

$$\frac{x'^2}{4^2} + \frac{y'^2}{2^2} = 1$$

The origin in the translated system is at $(h, k) = (-2, -3)$ and the translation formulas are

$$x' = x - h = x - (-2) = x + 2$$
$$y' = y - k = y - (-3) = y + 3$$

Thus, the equation of the ellipse in the original system is

$$\frac{(x+2)^2}{16} + \frac{(y+3)^2}{4} = 1$$
$$(x+2)^2 + 4(y+3)^2 = 16$$
$$x^2 + 4x + 4 + 4y^2 + 24y + 36 = 16$$
$$x^2 + 4y^2 + 4x + 24y + 24 = 0$$

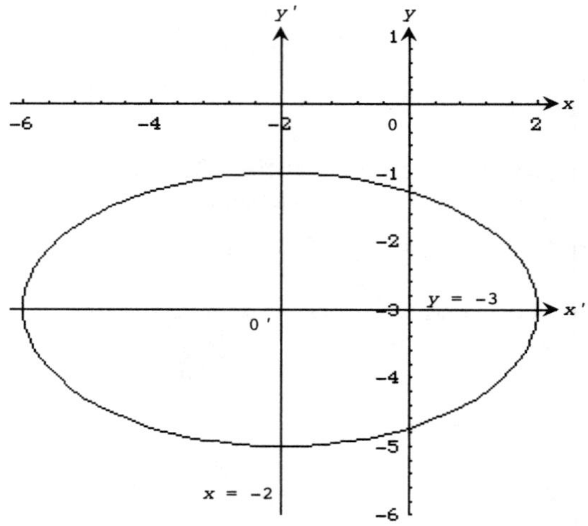

67. Locate the vertices in the original coordinate system, then sketch the ellipse and translate the origin to the center of the ellipse. Since $2a = 3 - (-7) = 10$, $a = 5$. Since $2c = 2 - (-6) = 8$, $c = 4$. Hence, $b = \sqrt{a^2 - c^2} = \sqrt{25 - 16} = 3$. The endpoints of the minor axis are symmetrically placed with respect to the center $(4, -2)$, that is, at $(1, -2)$ and $(7, -2)$.

Next write the equation of the ellipse in the translated system.

$$\frac{x'^2}{9} + \frac{y'^2}{25} = 1$$

The origin in the translated system is at $(h, k) = (4, -2)$ and the translation formulas are

$$x' = x - h = x - 4$$
$$y' = y - k = y - (-2) = y + 2$$

Thus, the equation of the ellipse in the original system is

$$\frac{(x-4)^2}{9} + \frac{(y+2)^2}{25} = 1$$
$$25(x-4)^2 + 9(y+2)^2 = 225$$
$$25(x^2 - 8x + 16) + 9(y^2 + 4y + 4) = 225$$
$$25x^2 - 200x + 400 + 9y^2 + 36y + 36 = 225$$
$$25x^2 + 9y^2 - 200x + 36y + 211 = 0$$

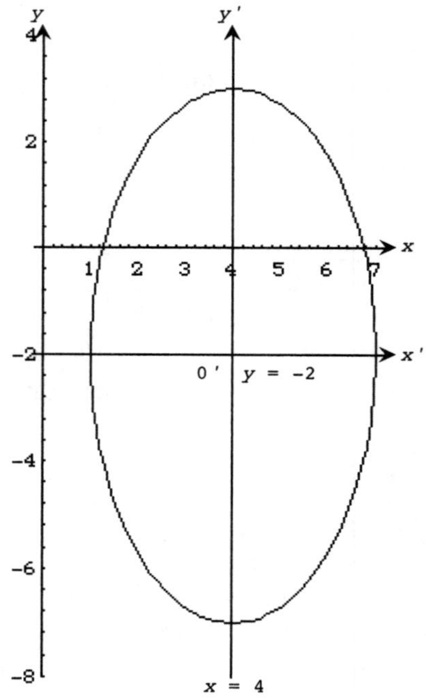

69. Locate the vertices, asymptote rectangle, and asymptotes in the original coordinate system, then sketch the hyperbola and translate the origin to the center of the hyperbola. Next write the equation of the hyperbola in the translated system.

$$\frac{y'^2}{4} - \frac{x'^2}{1} = 1$$

The origin in the translated system is at $(h, k) = (2, 3)$ and the translation formulas are

$$x' = x - h = x - 2$$
$$y' = y - k = y - 3$$

Thus, the equation of the hyperbola in the original system is

$$\frac{(y-3)^2}{4} - \frac{(x-2)^2}{1} = 1$$
$$(y-3)^2 - 4(x-2)^2 = 4$$
$$y^2 - 6y + 9 - 4(x^2 - 4x + 4) = 4$$
$$y^2 - 6y + 9 - 4x^2 + 16x - 16 = 4$$
$$-4x^2 + y^2 + 16x - 6y - 11 = 0$$
$$4x^2 - y^2 - 16x + 6y + 11 = 0$$

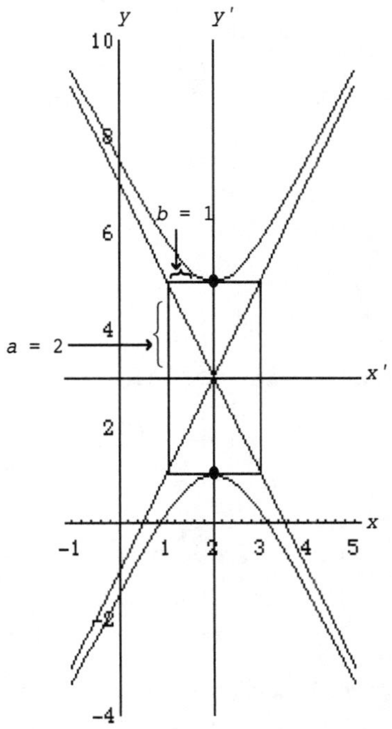

71. The center of the ellipse is at $(-2, 1) = (h, k)$. The major axis is vertical, with length $4 - (-2) = 6 = 2a$. Thus $a = 3$. The minor axis is horizontal, with length $(-1) - (-3) = 2 = 2b$. Thus $b = 1$. Hence, substituting in

$$\frac{(x-h)^2}{b^2} + \frac{(y-k)^2}{a^2} = 1$$

we obtain

$$\frac{[x-(-2)]^2}{1^2} + \frac{(y-1)^2}{3^2} = 1$$

The equation of the ellipse is thus

$$(x+2)^2 + \frac{(y-1)^2}{9} = 1$$
$$9(x+2)^2 + (y-1)^2 = 9$$
$$9(x^2 + 4x + 4) + (y^2 - 2y + 1) = 9$$
$$9x^2 + 36x + 36 + y^2 - 2y + 1 = 9$$
$$9x^2 + y^2 + 36x - 2y + 28 = 0$$

73. The center of the hyperbola is at $(1, 2) = (h, k)$. The transverse axis is horizontal, with length $2 = 2a$. Thus $a = 1$. Hence, substituting in

$$\frac{(x-h)^2}{a^2} - \frac{(y-k)^2}{b^2} = 1$$

we obtain

$$\frac{(x-1)^2}{1^2} - \frac{(y-2)^2}{b^2} = 1$$

Since the point $(4, 4)$ is on the graph, its coordinates satisfy the equation.

Substituting, we obtain

$$\frac{(4-1)^2}{1} - \frac{(4-2)^2}{b^2} = 1$$
$$9 - \frac{4}{b^2} = 1$$
$$-\frac{4}{b^2} = -8$$
$$b^2 = \frac{1}{2} \quad \text{(It is not necessary to solve for } b.)$$

The equation of the hyperbola is thus

$$\frac{(x-1)^2}{1^2} - \frac{(y-2)^2}{1/2} = 1$$
$$(x-1)^2 - 2(y-2)^2 = 1$$
$$x^2 - 2x + 1 - 2(y^2 - 4y + 4) = 1$$
$$x^2 - 2x + 1 - 2y^2 + 8y - 8 = 1$$
$$x^2 - 2y^2 - 2x + 8y - 8 = 0$$

75. $13x^2 + 10xy + 13y^2 - 72 = 0$
$B^2 - 4AC = 100 - 4(13)(13) = 100 - 676 = -576$
Since the discriminant is negative, the equation is an ellipse. To graph on a graphing calculator, we need to solve for y using the quadratic formula:

$13y^2 + 10xy + 13x^2 - 72 = 0 \quad a = 13,\ b = 10x,\ c = 13x^2 - 72$

$$y = \frac{-10x \pm \sqrt{100x^2 - 4(13)(13x^2 - 72)}}{2(13)}$$

$$y = \frac{-10x \pm \sqrt{100x^2 - 676x^2 + 3,744}}{26}$$

$$y = \frac{-10x \pm \sqrt{3,744 - 576x^2}}{26}$$

Enter $y_1 = (-10x + \sqrt{3,744 - 576x^2}) \div 26$, $y_2 = (-10x - \sqrt{3,744 - 576x^2}) \div 26$.

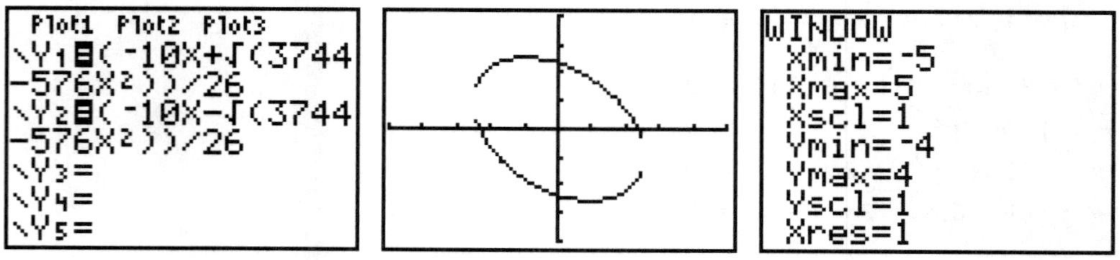

77. $x^2 - 6\sqrt{3}xy - 5y^2 - 8 = 0$
$B^2 - 4AC = (-6\sqrt{3})^2 - 4(1)(-5) = 108 + 20 = 128$

Since the discriminant is positive, the equation is a hyperbola. To graph on a graphing calculator we need to solve for y using the quadratic formula:

$-5y^2 - 6\sqrt{3}xy + x^2 - 8 = 0 \quad a = -5,\ b = -6\sqrt{3}x,\ c = x^2 - 8$

$$y = \frac{6\sqrt{3}x \pm \sqrt{\left(-6\sqrt{3}x\right)^2 - 4(-5)(x^2 - 8)}}{2(-5)}$$

$$y = \frac{6\sqrt{3}x \pm \sqrt{108x^2 + 20x^2 - 32}}{-10}$$

$$y = \frac{6\sqrt{3}x \pm \sqrt{128x^2 - 32}}{-10}$$

Enter $y_1 = (6\sqrt{3}x + \sqrt{128x^2 - 32}) \div -10$, $y_2 = (6\sqrt{3}x - \sqrt{128x^2 - 32}) \div -10$.

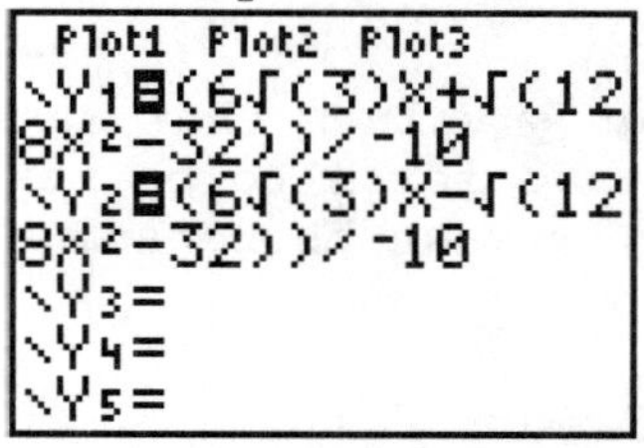

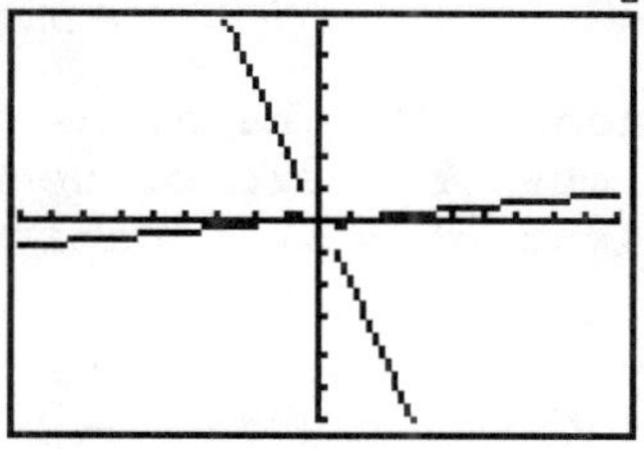
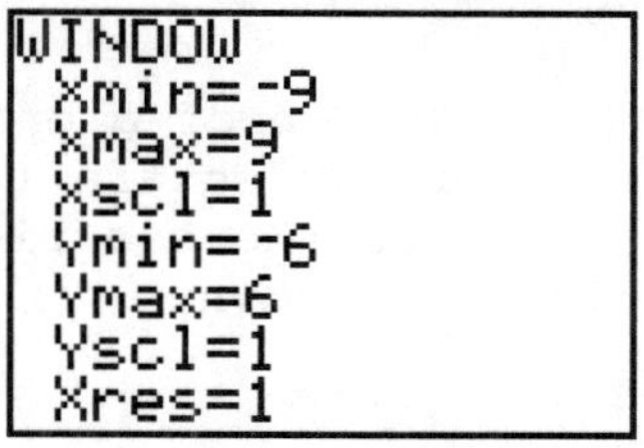

79. $16x^2 - 24xy + 9y^2 + 60x + 80y = 0$
$B^2 - 4AC = (-24)^2 - 4(16)(9) = 576 - 576 = 0$

Since the discriminant is zero, the equation is a parabola. To graph on a graphing calculator we need to solve for y using the quadratic formula:

$9y^2 + 80y - 24xy + 16x^2 + 60x = 0$

$9y^2 + (80 - 24x)y + 16x^2 + 60x = 0 \quad a = 9,\ b = 80 - 24x,\ c = 16x^2 + 60x$

$$y = \frac{-(80 - 24x) \pm \sqrt{(80 - 24x)^2 - 4(9)(16x^2 + 60x)}}{2(9)}$$

$$y = \frac{-80 + 24x \pm \sqrt{6,400 - 3,840x + 576x^2 - 576x^2 - 2,160x}}{18}$$

$$y = \frac{-80 - 24x \pm \sqrt{6,400 - 6,000x}}{18}$$

Enter $y_1 = (-80 + 24x + \sqrt{6,400 - 6,000x}) \div 18$,

$y_2 = (-80 + 24x - \sqrt{6,400 - 6,000x}) \div 18$.

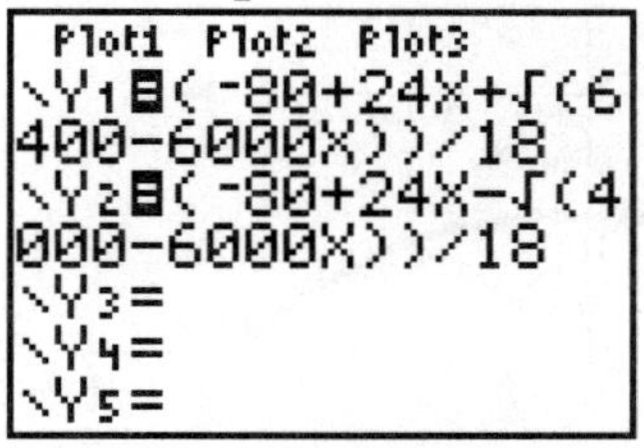

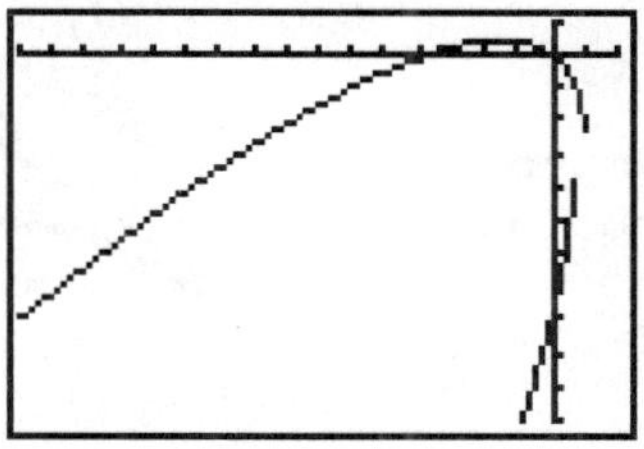
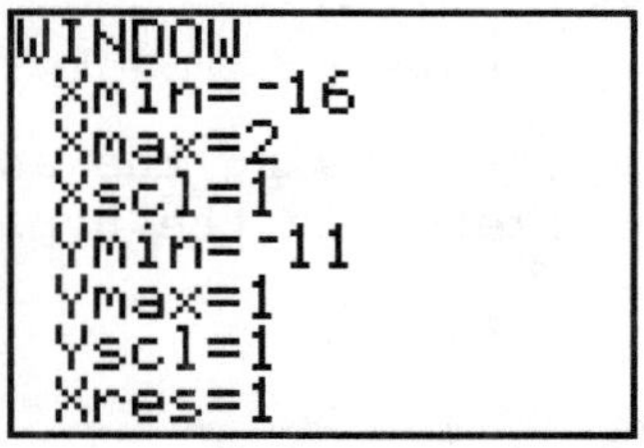

CHAPTER 9 REVIEW

1. First write the equation in standard form by dividing both sides by 225.

$9x^2 + 25y^2 = 225$

$$\frac{x^2}{25} + \frac{y^2}{9} = 1$$

In this form the equation is identifiable as that of an ellipse. Locate the intercepts.

When $y = 0$, $\frac{x^2}{25} = 1$. x intercepts: ± 5

When $x = 0$, $\frac{x^2}{9} = 1$. y intercepts: ± 3

Thus, $a = 5$, $b = 3$, and the major axis is on the x axis.

Common Error:
The relationship $c^2 = a^2 + b^2$ applies to a, b, c as defined for hyperbolas but not for ellipses.

Foci: $c^2 = a^2 - b^2$

$c^2 = 25 - 9$

$c^2 = 16$

$c = 4$

Foci: $F'(-4, 0)$, $F(4, 0)$

Major axis length = 2(5) = 10

Minor axis length = 2(3) = 6

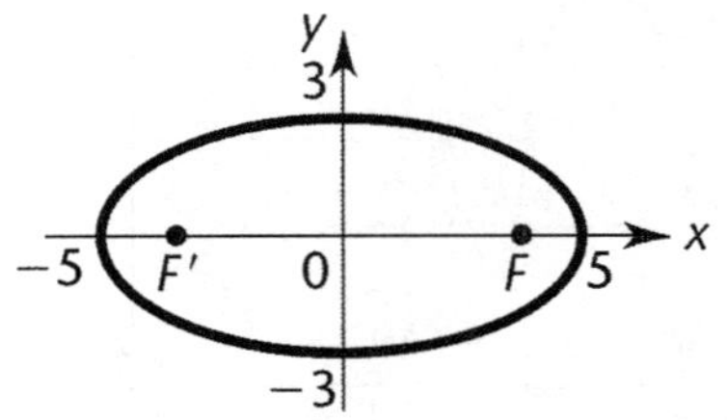

(9-2)

2. $x^2 = -12y$ is the equation of a parabola. To graph, assign y values that make the right side a perfect square (y must be zero or negative for x to be real) and solve for x. Since the coefficient of y is negative, a must be negative, and the parabola opens down.

x	0	±6	±2
y	0	−3	$-\frac{1}{3}$

To find the focus and directrix, solve

$4a = -12$

$a = -3$

Focus: (0, −3) Directrix: $y = -(-3) = 3$

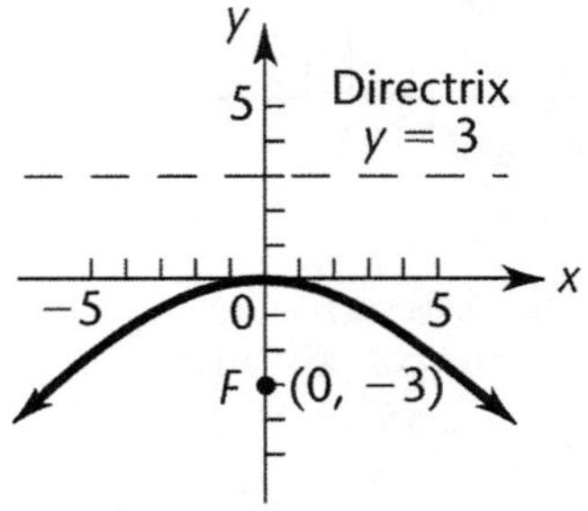

(9-1)

3. First, write the equation in standard form by dividing both sides by 225.

$25y^2 - 9x^2 = 225$

$$\frac{y^2}{9} - \frac{x^2}{25} = 1$$

In this form the equation is identifiable as that of a hyperbola.

When $y = 0$, $-\frac{x^2}{25} = 1$. There are no x intercepts, but $b = 5$.

When $x = 0$, $\frac{y^2}{9} = 1$. y intercepts: ±3

Sketch the asymptotes using the asymptote rectangle, then sketch in the hyperbola.

Foci: $c^2 = 3^2 + 5^2$

$c^2 = 34$

$c = \sqrt{34}$

Foci: $F'(0, -\sqrt{34})$, $F(0, \sqrt{34})$

Transverse axis length = 2(3) = 6

Conjugate axis length = 2(5) = 10

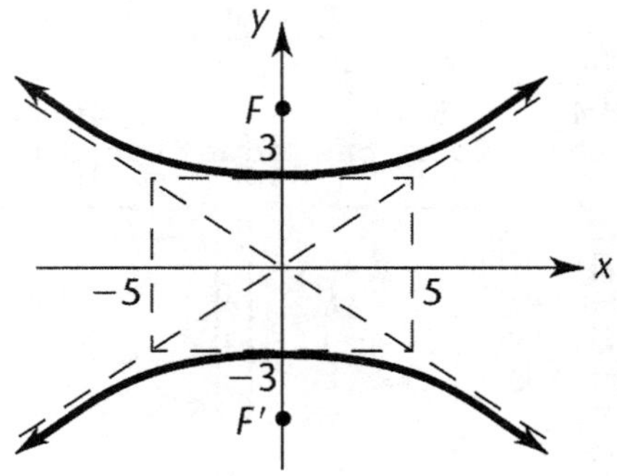

(9-3)

4. First, write the equation in standard form by dividing both sides by 16.

$x^2 - y^2 = 16$

$$\frac{x^2}{16} - \frac{y^2}{16} = 1$$

In this form the equation is identifiable as a hyperbola.

When $x = 0$, $-\frac{y^2}{16} = 1$. There are no y intercepts, but $b = 4$.

When $y = 0$, $\dfrac{x^2}{16} = 1$. x intercepts: ± 4.

Sketch the asymptotes using the asymptote rectangle, then sketch in the hyperbola.

Foci: $c^2 = 4^2 + 4^2$

$c^2 = 32$

$c = 4\sqrt{2}$

Foci: $F' = (-4\sqrt{2}, 0)$, $F = (4\sqrt{2}, 0)$

Transverse axis length = 2(4) = 8

Conjugate axis length = 2(4) = 8

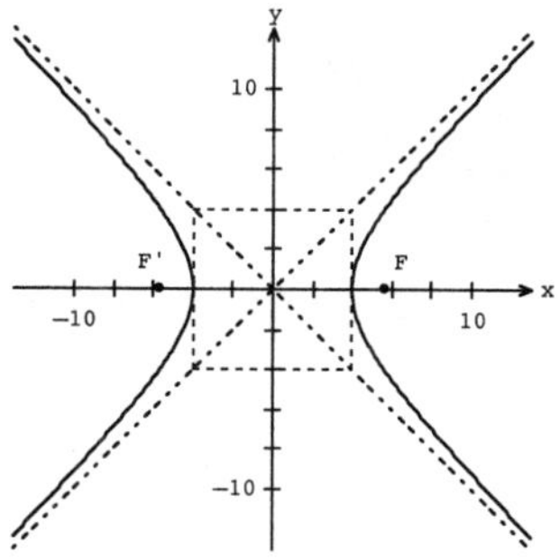

(*9-3*)

5. $y^2 = 8x$ is the equation of a parabola. To graph, assign x values that make the right side a perfect square (x must be zero or positive for y to be real) and solve for y. Since the coefficient of x is positive, a must be positive, and the parabola opens right.

x	0	2	8
y	0	±4	±8

To find the focus and directrix, solve $4a = 8$

$a = 2$.

Focus: (2, 0) Directrix: $x = -2$

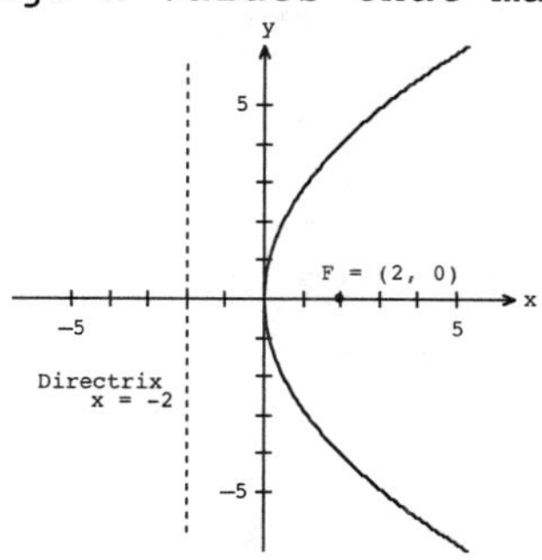

(*9-1*)

6. First, write the equation in standard form by dividing both sides by 8.

$2x^2 + y^2 = 8$

$\dfrac{x^2}{4} + \dfrac{y^2}{8} = 1$

In this form the equation is identifiable as that of an ellipse. Locate the intercepts.

When $y = 0$, $\dfrac{x^2}{4} = 1$ x intercepts: ± 2

When $x = 0$, $\dfrac{y^2}{8} = 1$ y intercepts: $\pm 2\sqrt{2}$

Thus, $a = 2\sqrt{2}$, $b = 2$, and the major axis is on the y axis.

Foci: $c^2 = a^2 - b^2$

$c^2 = 8 - 4$

$c^2 = 4$

$c = 2$

Foci: $F' = (0, -2)$, $F = (0, 2)$

Major axis length = $2(2\sqrt{2}) = 4\sqrt{2}$

Minor axis length = 2(2) = 4

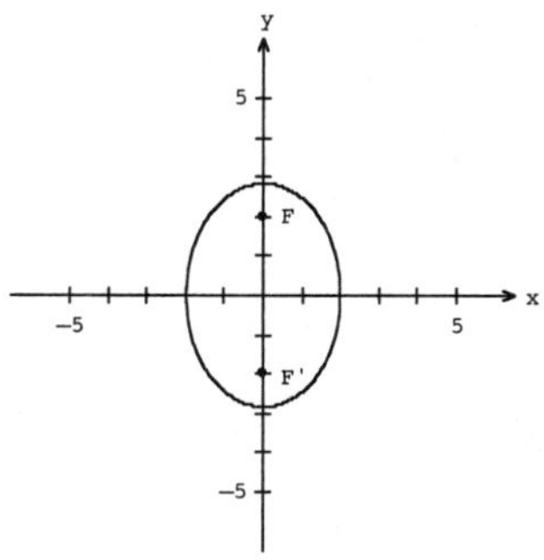

(*9-2*)

7. (A) Divide both sides by 100

$$\frac{(y + 2)^2}{25} - \frac{(x - 4)^2}{4} = 1$$

(B) This is the equation of a hyperbola. (*9-4*)

8. (A) Subtract $12(y + 4)$ from both sides.

$$(x + 5)^2 = -12(y + 4)$$

(B) This is the equation of a parabola. (*9-4*)

9. (A) Divide both sides by 144

$$\frac{(x-6)^2}{9} + \frac{(y-4)^2}{16} = 1$$

(B) This is the equation of an ellipse. (9-4)

10. Use the rotation identities $x' = x\cos\theta + y\sin\theta$, $y' = -x\sin\theta + y\cos\theta$.

(A) $x' = 3\cos 30° + 4\sin 30°$, $y' = -3\sin 30° + 4\cos 30°$

$x' = 3\cdot\frac{\sqrt{3}}{2} + 4\cdot\frac{1}{2} = \frac{3\sqrt{3}+4}{2}$ $y' = -3\cdot\frac{1}{2} + 4\cdot\frac{\sqrt{3}}{2} = \frac{-3+4\sqrt{3}}{2}$

$\left(\frac{3\sqrt{3}+4}{2}, \frac{-3+4\sqrt{3}}{2}\right)$

(B) $x' = 3\cos 45° + 4\sin 45°$, $y' = -3\sin 45° + 4\cos 45°$

$x' = 3\cdot\frac{\sqrt{2}}{2} + 4\cdot\frac{\sqrt{2}}{2} = \frac{7\sqrt{2}}{2}$ $y' = -3\cdot\frac{\sqrt{2}}{2} + 4\cdot\frac{\sqrt{2}}{2} = \frac{\sqrt{2}}{2}$

$\left(\frac{7\sqrt{2}}{2}, \frac{\sqrt{2}}{2}\right)$

(C) $x' = 3\cos 60° + 4\sin 60°$, $y' = -3\sin 60° + 4\cos 60°$

$x' = 3\cdot\frac{1}{2} + 4\cdot\frac{\sqrt{3}}{2} = \frac{3+4\sqrt{3}}{2}$ $y' = -3\cdot\frac{\sqrt{3}}{2} + 4\cdot\frac{1}{2} = \frac{-3\sqrt{3}+4}{2}$

$\left(\frac{3+4\sqrt{3}}{2}, \frac{-3\sqrt{3}+4}{2}\right)$ (9-4)

11. We need to use the rotation identities to rewrite the equations of the x axis ($y' = 0$) and the y' axis ($x' = 0$).

First, plug in $\theta = 75°$:

$x' = x\cos 75° + y\sin 75°$ $y' = -x\sin 75° + y\cos 75°$

$x' = 0.259x + 0.966y$ $y' = -0.966x + 0.259y$

The x' axis corresponds to $y' = 0$ so we set $-0.966x + 0.259y$ equal to zero and solve for y:

$$-0.966x + 0.259y = 0$$
$$0.259y = 0.966x$$
$$y = 3.73x$$

To find the equation of the y' axis set $0.259x + 0.966y$ equal to zero:

$$0.259x + 0.966y = 0$$
$$0.966y = -0.259x$$
$$y = -.268x$$

(9-4)

12. The parabola is opening either left or right and has an equation of the form $y^2 = 4ax$. Since $(-4, -2)$ is on the graph, we have:

$$(-2)^2 = 4a(-4)$$
$$4 = -16a$$
$$-\frac{1}{4} = a$$

Thus, the equation of the parabola is

$$y^2 = 4\left(-\frac{1}{4}\right)x$$
$$y^2 = -x$$

(9-1)

13. Make a rough sketch of the ellipse and compute x and y intercepts:

$$\frac{x^2}{a^2} + \frac{y^2}{b^2} = 1$$

$$a = \frac{12}{2} = 6 \qquad b = \frac{10}{2} = 5$$

$$\frac{x^2}{36} + \frac{y^2}{25} = 1$$

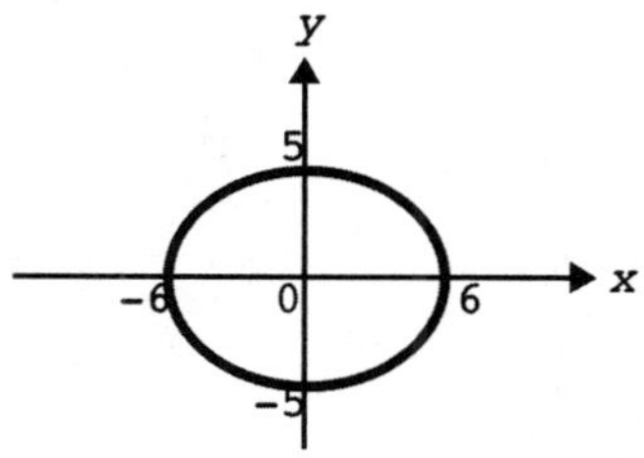

(9-2)

14. Make a rough sketch of the ellipse, locate focus and x intercepts, then determine y intercepts using the special triangle relationship.

$$\frac{x^2}{b^2} + \frac{y^2}{a^2} = 1$$

$$b = \frac{12}{2} = 6$$

$$a^2 = 6^2 + 8^2 = 100$$

$$a = 10$$

$$\frac{x^2}{36} + \frac{y^2}{100} = 1$$

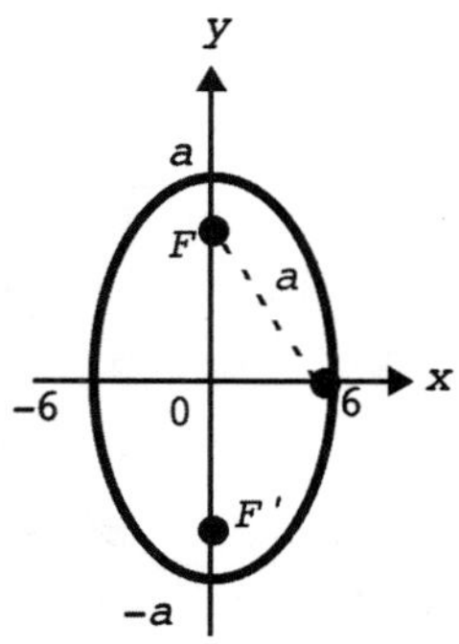

(9-2)

15. Since the transverse axis is on the y axis, start with

$$\frac{y^2}{a^2} - \frac{x^2}{b^2} = 1$$

and find a and b.

$$b = \frac{6}{2} = 3, \quad c = \frac{8}{2} = 4$$

To find a, sketch the asymptote rectangle, label known parts, and use the Pythagorean Theorem.

$$a^2 = c^2 - b^2$$

$$a^2 = 4^2 - 3^2$$

$$a^2 = 7$$

$$a = \sqrt{7}$$

Thus, the equation is

$$\frac{y^2}{7} - \frac{x^2}{9} = 1$$

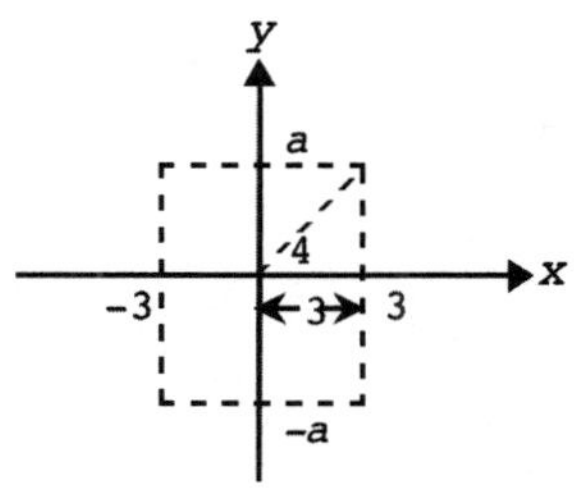

(9-3)

16. Since the transverse axis is on the x axis, start with

$$\frac{x^2}{a^2} - \frac{y^2}{b^2} = 1$$

and find a and b.

$$a = \frac{14}{2} = 7 \qquad b = \frac{16}{2} = 8$$

Thus, the equation is

$$\frac{x^2}{49} - \frac{y^2}{64} = 1$$

(9-3)

17. Comparing focus (0, -5) with the information in the chart of standard equations for a parabola, we see:
$a = -5$ Axis of symmetry: the y axis. Equation: $x^2 = 4ay$.
Thus the equation of the parabola must be $x^2 = 4(-5)y$ or $x^2 = -20y$. (9-1)

18. Since the major axis is on the x axis, the foci must be at $(\pm c, 0)$ and the equation of the ellipse is of form

$$\frac{x^2}{a^2} + \frac{y^2}{b^2} = 1$$

Since the major axis has twice the length of the minor axis, $a = 2b$. The equation then becomes

$$\frac{x^2}{4b^2} + \frac{y^2}{b^2} = 1$$

Since $(-6, 0)$ is on the graph, its coordinates satisfy the equation, hence

$$\frac{(-6)^2}{4b^2} + \frac{0^2}{b^2} = 1$$

$$\frac{36}{4b^2} = 1$$

$$b^2 = 9$$

Then $a^2 = 4b^2 = 36$ and $c^2 = a^2 - b^2 = 36 - 9 = 27$; $c = \sqrt{27}$. The foci are at $(\pm\sqrt{27}, 0)$, that is, $(-3\sqrt{3}, 0)$ and $(3\sqrt{3}, 0)$. *(9-2)*

19. Since the conjugate axis has length 4, $2b = 4$, hence $b = 2$. Since $(0, -3)$ is a focus, $c = 3$. Hence $a^2 = c^2 - b^2 = 9 - 4 = 5$ and $a = \sqrt{5}$. The y intercepts are $(0, \pm a)$, that is $(0, \sqrt{5})$ and $(0, -\sqrt{5})$. *(9-3)*

20. Since the focus is at $(-4, 0)$, the axis of symmetry is the x axis, the parabola opens left, and the directrix has equation $x = -(-4)$ or $x = 4$. *(9-1)*

21. $x^2 = 8y$
$y^2 = -x$
Solve for y in the first equation, then substitute into the second equation

$$y = \frac{x^2}{8}$$

$$\left(\frac{x^2}{8}\right)^2 = -x$$

$$\frac{x^4}{64} = -x$$

$$x^4 = -64x$$

$$x^4 + 64x = 0$$

$$x(x^3 + 64) = 0$$

$$x(x + 4)(x^2 - 4x + 16) = 0$$

$x = 0$ $\quad x + 4 = 0$ $\quad x^2 - 4x + 16 = 0$
$\quad x = -4$ $\quad$ No real solutions.

For $x = 0$ $\quad$ For $x = -4$

$y = 0$ $\quad y = \frac{(-4)^2}{8}$

$y = 2$ $\quad$ Solutions: $(0, 0)$, $(-4, 2)$ *(9-1)*

22. Since the major axis has length 14, $2a = 14$, hence $a = 7$. Since $(0, -1)$ is a focus, $c = 1$. Hence $b^2 = a^2 - c^2 = 49 - 1 = 48$; $b = \sqrt{48}$. The x intercepts are at $(\pm\sqrt{48}, 0)$, that is $(4\sqrt{3}, 0)$ and $(-4\sqrt{3}, 0)$. *(9-2)*

23. Since the transverse axis is on the y axis, the foci must be at $(0, \pm c)$ and the equation of the hyperbola is of the form

$$\frac{y^2}{a^2} - \frac{x^2}{b^2} = 1$$

Since the conjugate axis has twice the length of the transverse axis, $b = 2a$. The equation then becomes

$$\frac{y^2}{a^2} - \frac{x^2}{4a^2} = 1$$

Since (0, -4) is on the graph,, its coordinates satisfy the equation, hence

$$\frac{(-4)^2}{a^2} - \frac{0^2}{4a^2} = 1$$

$$\frac{16}{a^2} = 1$$

$$a^2 = 16$$

Then $b^2 = 4a^2 = 64$ and $c^2 = a^2 + b^2 = 16 + 64 = 80$; $c = \sqrt{80}$. The foci are at $(0, \pm\sqrt{80})$, that is $(0, 4\sqrt{5})$ and $(0, -4\sqrt{5})$. (9-3)

24.

$$\begin{aligned} 16x^2 + 4y^2 + 96x - 16y + 96 &= 0 \\ 16x^2 + 96x + 4y^2 - 16y &= -96 \\ 16(x^2 + 6x + ?) + 4(y^2 - 4y + ?) &= -96 \\ 16(x^2 + 6x + 9) + 4(y^2 - 4y + 4) &= -96 + 144 + 16 \\ 16(x + 3)^2 + 4(y - 2)^2 &= 64 \\ \frac{(x+3)^2}{4} + \frac{(y-2)^2}{16} &= 1 \end{aligned}$$

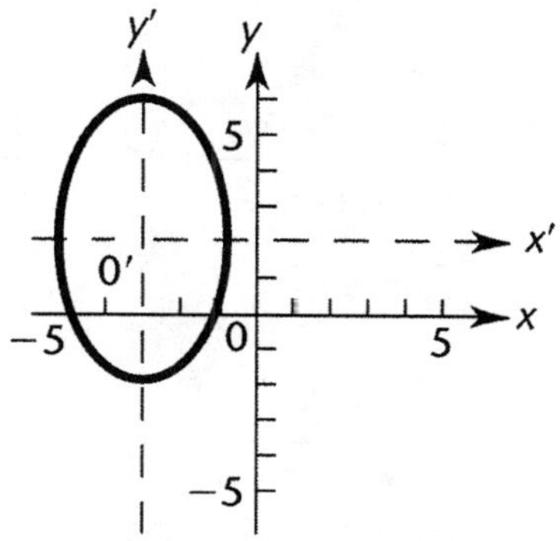

This is the equation of an ellipse with center at (-3, 2). The equations of translation are $x' = x + 3$, $y' = y - 2$. Making these substitutions, we obtain

$$\frac{x'^2}{4} + \frac{y'^2}{16} = 1$$

We graph this in the $x'y'$ system, following the process discussed in Section 9-2. (9-4)

25. $x^2 - 4x - 8y - 20 = 0$

$$\begin{aligned} x^2 - 4x &= 8y + 20 \\ x^2 - 4x + 4 &= 8y + 24 \\ (x - 2)^2 &= 4(2)(y + 3) \end{aligned}$$

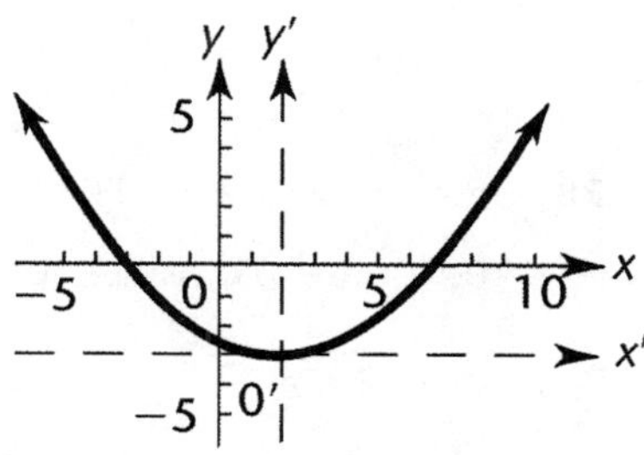

This is the equation of a parabola opening up with vertex at $(h, k) = (2, -3)$. The equations of translation are $x' = x - 2$, $y' = y + 3$. Making these substitutions, we obtain

$$x'^2 = 8y'$$

We graph this in the $x'y'$ system, following the process discussed in Section 9-1. (9-4)

26.

$$\begin{aligned} 4x^2 - 9y^2 + 24x - 36y - 36 &= 0 \\ 4x^2 + 24x - 9y^2 - 36y &= 36 \\ 4(x^2 + 6x + ?) - 9(y^2 + 4y + ?) &= 36 \\ 4(x^2 + 6x + 9) - 9(y^2 + 4y + 4) &= 36 + 36 - 36 \\ 4(x + 3)^2 - 9(y + 2)^2 &= 36 \\ \frac{(x+3)^2}{9} - \frac{(y+2)^2}{4} &= 1 \end{aligned}$$

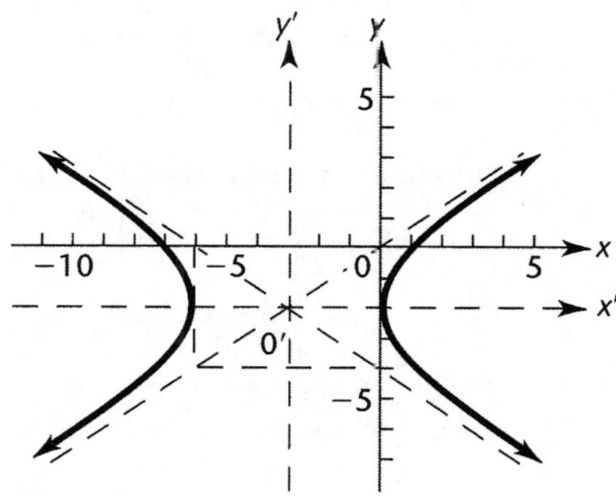

This is the equation of a hyperbola with center at (-3, -2). The equations of translation are $x' = x + 3$, $y' = y + 2$. Making these substitutions, we obtain

$$\frac{x'^2}{9} - \frac{y'^2}{4} = 1$$

We graph this in the $x'y'$ system, following the process discussed in Section 9-3. (9-4)

27. The rotation equations are

$x = \cos 30°x' - \sin 30°y'$ $\qquad$ $y = \sin 30°x' + \cos 30°y'$

$x = \frac{\sqrt{3}}{2}x' - \frac{1}{2}y'$ $\qquad$ $y = \frac{1}{2}x' + \frac{\sqrt{3}}{2}y'$

Substitute into the original equation:

$x^2 - \sqrt{3}xy + 2y^2 - 10 = 0$

$$\left(\frac{\sqrt{3}}{2}x' - \frac{1}{2}y'\right)^2 - \sqrt{3}\left(\frac{\sqrt{3}}{2}x' - \frac{1}{2}y'\right)\left(\frac{1}{2}x' + \frac{\sqrt{3}}{2}y'\right) + 2\left(\frac{1}{2}x' + \frac{\sqrt{3}}{2}y'\right)^2 - 10 = 0$$

$$\frac{3}{4}x'^2 - \frac{\sqrt{3}}{2}x'y' + \frac{1}{4}y'^2 - \sqrt{3}\left(\frac{\sqrt{3}}{4}x'^2 + \frac{3}{4}x'y' - \frac{1}{4}x'y' - \frac{\sqrt{3}}{4}y'^2\right)$$
$$+ 2\left(\frac{1}{4}x'^2 + \frac{\sqrt{3}}{2}x'y' + \frac{3}{4}y'^2\right) = 10$$

$$\frac{3}{4}x'^2 - \frac{\sqrt{3}}{2}x'y' + \frac{1}{4}y'^2 - \frac{3}{4}x'^2 - \frac{\sqrt{3}}{2}x'y' + \frac{3}{4}y'^2 + \frac{1}{2}x'^2 + \sqrt{3}x'y' + \frac{3}{2}y'^2 = 10$$

$\frac{1}{2}x'^2 + \frac{5}{2}y'^2 = 10$ (Divide both sides by 10)

$$\frac{x'^2}{20} - \frac{y'^2}{4} = 1$$

This is an ellipse with major axis length $2\sqrt{20}$ and minor axis length 4 rotated through 30°.

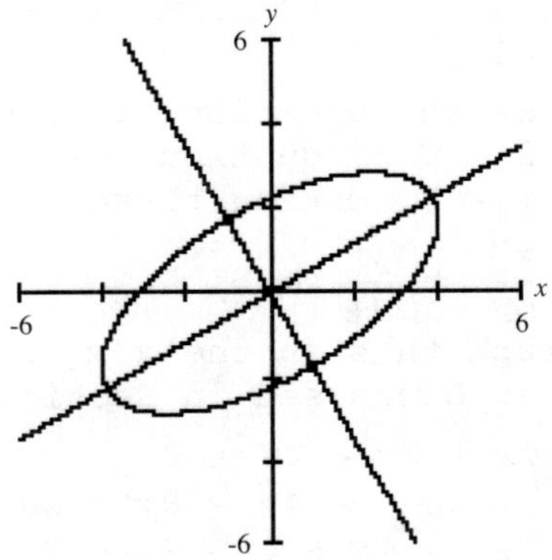

(9-4)

28. $5x^2 + 26xy + 5y^2 + 72 = 0$

Use the formula $\cot 2\theta = \frac{A - C}{B}$ to find the appropriate angle.

$\cot 2\theta = \frac{5 - 5}{26} = 0$

Since $\cot 90° = \frac{\cos 90°}{\sin 90°} = \frac{0}{1}$, $2\theta = 90°$ and $\theta = 45°$.

The rotation equations are

$x = \cos 45°x' - \sin 45°y'$ $\qquad$ $y = \sin 45°x' + \cos 45°y'$

$x = \frac{\sqrt{2}}{2}x' - \frac{\sqrt{2}}{2}y'$ $\qquad$ $y = \frac{\sqrt{2}}{2}x' + \frac{\sqrt{2}}{2}y'$

Substitute into the original equation and simplify:

$$5\left(\frac{\sqrt{2}}{2}x' - \frac{\sqrt{2}}{2}y'\right)^2 + 26\left(\frac{\sqrt{2}}{2}x' - \frac{\sqrt{2}}{2}y'\right)\left(\frac{\sqrt{2}}{2}x' + \frac{\sqrt{2}}{2}y'\right) + 5\left(\frac{\sqrt{2}}{2}x' + \frac{\sqrt{2}}{2}y'\right)^2 + 72 = 0$$

$$5\left(\frac{1}{2}x'^2 - x'y' + \frac{1}{2}y'^2\right) + 26\left(\frac{1}{2}x'^2 - \frac{1}{2}y'^2\right) + 5\left(\frac{1}{2}x'^2 + x'y' + \frac{1}{2}y'^2\right) = -72$$

$$\frac{5}{2}x'^2 - 5x'y' + \frac{5}{2}y'^2 + 13x'^2 - 13y'^2 + \frac{5}{2}x'^2 + 5x'y' + \frac{5}{2}y'^2 = -72$$

$$18 - 8y'^2 = -72 \quad \text{(Divide both sides by -72)}$$
$$\frac{y'^2}{9} - \frac{x'^2}{4} = 1$$

This is a hyperbola with intercepts (0, 3) and (0, -3) rotated through 45°.

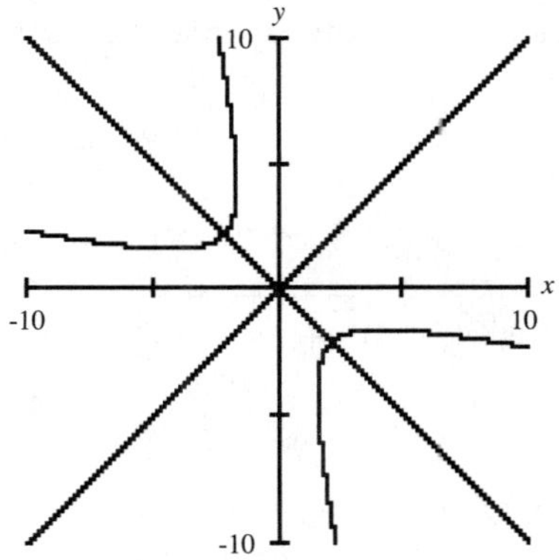

(9-4)

29. To identify the curve, we find the discriminant of the equation $3x^2 + 4xy + 2y^2 - 20 = 0$, which has $A = 3$, $B = 4$, and $C = 2$.

$$B^2 - 4AC = 16 - 4(3)(2) = -8 < 0$$

The curve is an ellipse. (9-4)

30. From the figure, we see that the point $P(x, y)$ is a point on the curve if and only if

$$\begin{aligned}
d_1 &= d_2 \\
d(F, P) &= d(M, P) \\
\sqrt{(x-2)^2 + (y-4)^2} &= \sqrt{(x-6)^2 + (y-y)^2} \\
(x-2)^2 + (y-4)^2 &= (x-6)^2 \\
x^2 - 4x + 4 + (y-4)^2 &= x^2 - 12x + 36 \\
(y-4)^2 &= -8x + 32 \\
(y-4)^2 &= -8(x-4) \text{ or} \\
y^2 - 8y + 16 &= -8x + 32 \\
y^2 - 8y + 8x - 16 &= 0
\end{aligned}$$

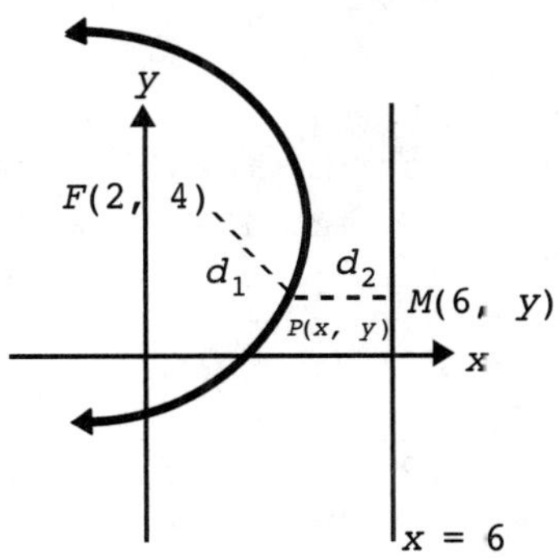

(9-1)

31. From the figure, we see that the point $P(x, y)$ is a point on the curve if and only if

$$\begin{aligned}
d_1 &= 2d_2 \\
d(F, P) &= 2d(N, P) \\
\sqrt{(x-4)^2 + (y-0)^2} &= 2\sqrt{(x-1)^2 + (y-y)^2} \\
(x-4)^2 + y^2 &= 4(x-1)^2 \\
x^2 - 8x + 16 + y^2 &= 4(x^2 - 2x + 1) \\
x^2 - 8x + 16 + y^2 &= 4x^2 - 8x + 4 \\
-3x^2 + y^2 &= -12 \\
\frac{x^2}{4} - \frac{y^2}{12} &= 1
\end{aligned}$$

This is the equation of a hyperbola.

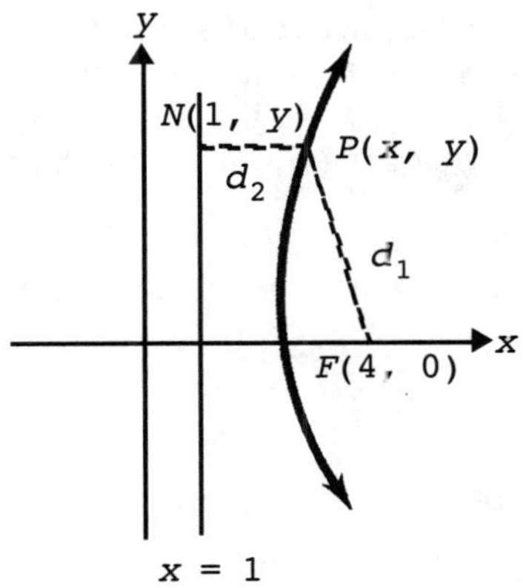

(9-3)

32. From the figure, we see that the point $P(x, y)$ is a point on the curve if and only if

$$\begin{aligned}
d_1 &= \frac{2}{3}d_2 \\
d(F, P) &= \frac{2}{3}d(M, P) \\
\sqrt{(x-4)^2 + (y-0)^2} &= \frac{2}{3}\sqrt{(x-9)^2 + (y-y)^2} \\
(x-4)^2 + y^2 &= \frac{4}{9}(x-9)^2 \\
9[(x-4)^2 + y^2] &= 4(x-9)^2 \\
9(x^2 - 8x + 16 + y^2) &= 4(x^2 - 18x + 81) \\
9x^2 - 72x + 144 + 9y^2 &= 4x^2 - 72x + 324 \\
5x^2 + 9y^2 &= 180 \\
\frac{x^2}{36} + \frac{y^2}{20} &= 1
\end{aligned}$$

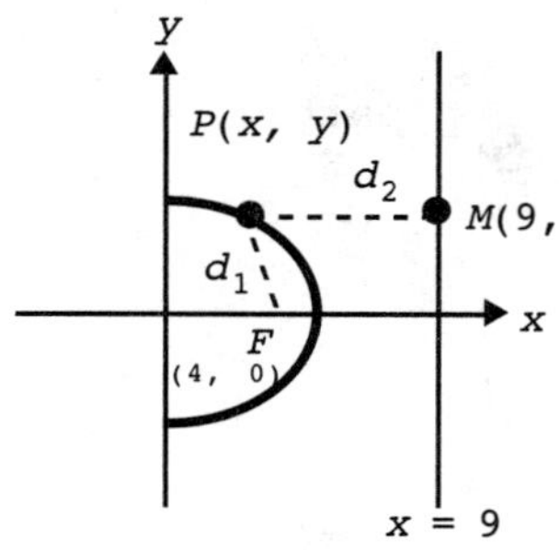

This is the equation of an ellipse. (9-2)

33. First find the coordinates of the foci in the translated system.

$c'^2 = 4^2 - 2^2 = 12$

$c' = \sqrt{12}$

$-c' = -\sqrt{12}$

Thus the coordinates in the translated system are

$F'(0, -\sqrt{12})$ and $F(0, \sqrt{12})$

Now use

$x = x' + h = x' - 3$

$y = y' + k = y' + 2$

to obtain

$F'(-3, -\sqrt{12} + 2)$ and $F(-3, \sqrt{12} + 2)$

as the coordinates of the foci in the original system. (9-4)

34. First find the coordinates of the focus in the translated system. Since $a = 2$ and the parabola opens up, they are $(0, 2)$. Now use

$x = x' + h = x' + 2 = 0 + 2 = 2$

$y = y' + k = y' - 3 = 2 - 3 = -1$

to obtain $(2, -1)$ as the coordinates of the focus in the original system. (9-4)

35. First find the coordinates of the foci in the original system.

$c'^2 = 3^2 + 2^2 = 13$

$c' = \sqrt{13}$

$-c' = -\sqrt{13}$

Thus the coordinates in the translated system are

$F'(-\sqrt{13}, 0)$ and $F(\sqrt{13}, 0)$

Now use

$x = x' + h = x' - 3$

$y = y' + k = y' - 2$

to obtain

$F'(-\sqrt{13} - 3, -2)$ and $F(\sqrt{13} - 3, -2)$ as the coordinates of the foci in the original system. (9-4)

36. The equation $\dfrac{x^2}{49} - \dfrac{y^2}{25} = 1$ is in standard form with $a^2 = 49$, $b^2 = 25$. Since the intercepts are on the x axis, the asymptotes have equations $y = \pm\dfrac{b}{a}x = \pm\dfrac{5}{7}x$. *(9-3)*

37. The equation $\dfrac{y^2}{64} - \dfrac{x^2}{4} = 1$ is in standard form with $a^2 = 64$, $b^2 = 4$. Since the intercepts are on the y axis, the asymptotes have equations $y = \pm\dfrac{a}{b}x = \pm\dfrac{8}{2}x$ or $y = \pm 4x$. *(9-3)*

38. $4x^2 - y^2 = 1$. In standard form, this becomes

$$\frac{x^2}{1/4} - \frac{y^2}{1} = 1$$

Then $a^2 = \dfrac{1}{4}$, $b^2 = 1$. Since the intercepts are on the x axis, the asymptotes have equations $y = \pm\dfrac{b}{a}x = \pm\dfrac{1}{1/2}x$ or $y = \pm 2x$. *(9-3)*

39. Applying the methods of Section 9-4:
$A = C = 0$, $B = 1$, hence $\cot 2\theta = 0$, $\theta = 45°$
The rotation formulas are $x = \dfrac{x' - y'}{\sqrt{2}}$, $y = \dfrac{x' + y'}{\sqrt{2}}$.
In the rotated coordinate system, the equation becomes

$$\frac{x'^2 - y'^2}{2} - 1 = 0 \quad \text{or} \quad \frac{x'^2}{2} - \frac{y'^2}{2} = 1$$

The equation is in standard form and the hyperbola has asymptotes $y' = \pm x'$ in the rotated system.
In the original system, applying the reverse rotation formulas.

$$x' = \frac{x + y}{\sqrt{2}} \qquad y' = \frac{-x + y}{\sqrt{2}}$$

the asymptotes become

$$-\frac{x + y}{\sqrt{2}} = \pm\frac{x + y}{\sqrt{2}}$$

which simplifies to $x = 0$, $y = 0$.
It would be much simpler, however, to note that $xy - 1 = 0$ is the equation of the rational function $y = \dfrac{1}{x}$, which has asymptotes $x = 0$ and $y = 0$ (Section 4-4). *(9-4)*

40. From the figure, we see that the parabola opens up, hence its equation must be of the form $x^2 = 4ay$. Since $(4, 1)$ is on the graph we have

$4^2 = 4a \cdot 1$
$16 = 4a$
$a = 4$
Thus a, the distance of the focus from the vertex, is 4 feet.

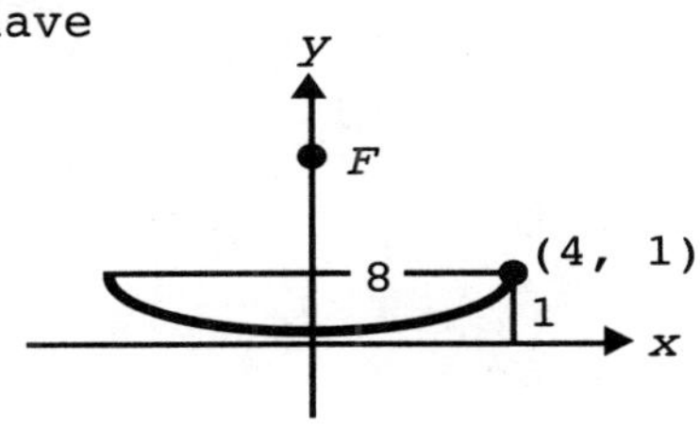

(9-1)

41. From the figure, we see that the x intercepts must be at (-5, 0) and (5, 0), the foci at (-4, 0) and (4, 0). Hence $a = 5$ and $c = 4$. We can determine the y intercepts using the special triangle relationship

$$5^2 = 4^2 + b^2$$
$$25 = 16 + b^2$$
$$b = 3$$

Hence, the equation of the ellipse is

$$\frac{x^2}{5^2} + \frac{y^2}{3^2} = 1$$

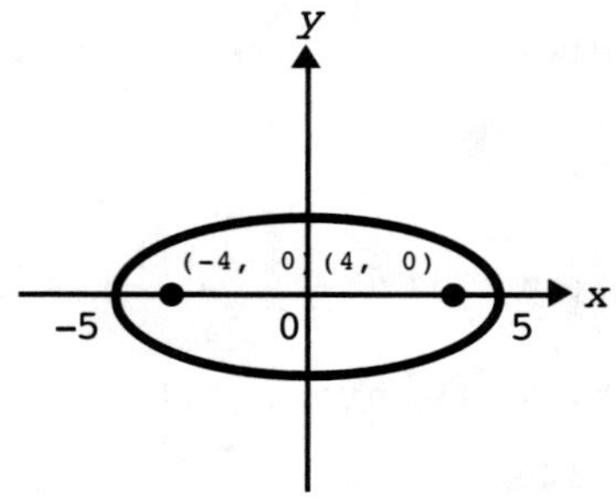

(9-2)

42. From the figure, we can see:
$d + a = y_1$ = the y coordinate of the point on the hyperbola with x coordinate 15. From the equation of the hyperbola $\frac{y^2}{40^2} - \frac{x^2}{30^2} = 1$, we have $a = 40$, hence $d + 40 = y_1$, or $d = y_1 - 40$. Since the point $(15, y_1)$ is on the hyperbola, its coordinates must satisfy the equation of the hyperbola. Thus,

$$\frac{y_1^2}{40^2} - \frac{15^2}{30^2} = 1$$
$$\frac{y_1^2}{40^2} = 1 + \frac{15^2}{30^2}$$
$$\frac{y_1^2}{40^2} = \frac{5}{4}$$
$$y_1^2 = 2000$$
$$y_1 = 44.72$$
$$\text{depth} = y_1 - 40 = 4.72 \text{ feet.}$$

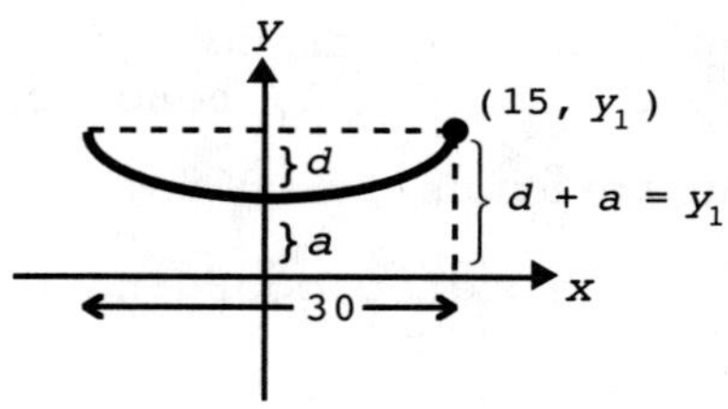

(9-3)

CHAPTER 10 Systems of Equations and Inequalities; Matrices

Section 10-1

1. Both lines in the given system are different, but they have the same slope $\left(\frac{1}{2}\right)$ and are therefore parallel. This system corresponds to (B) and has no solution.

3. In slope-intercept form, these equations are $y = 2x - 5$ and $y = -\frac{3}{2}x - \frac{3}{2}$. Thus, one has slope 2 and y intercept -5; the other has slope $-\frac{3}{2}$ and y intercept $-\frac{3}{2}$. This system corresponds to (D) and its solution can be read from the graph as (1, -3). Checking, we see that

$$2x - y = 2\cdot 1 - (-3) = 5$$
$$3x + 2y = 3\cdot 1 + 2(-3) = -3$$

5. $x + y = 7$
$x - y = 3$
If we add, we can eliminate y.
$x + y = 7$
$x - y = 3$
$2x = 10$
$x = 5$
Now substitute $x = 5$ back into the top equation and solve for y.
$5 + y = 7$
$y = 2$
(5, 2)

7. $3x - 2y = 12$
$7x + 2y = 8$
If we add, we can eliminate y.
$3x - 2y = 12$
$7x + 2y = 8$
$10x = 20$
$x = 2$
Now substitute $x = 2$ back into the bottom equation and solve for y.
$7(2) + 2y = 8$
$2y = -6$
$y = -3$
(2, -3)

9. $3u + 5v = 15$
$6u + 10v = -30$
If we multiply the top equation by -2 and add, we eliminate both u and v.
$-6u - 10v = -30$
$6u + 10v = -30$
$0 = -60$
No solution. The equations represent parallel lines.

11. $y = 2x + 3$
$y = 3x - 5$
Substitute y from the first equation into the second equation to eliminate y.
$2x + 3 = 3x - 5$
$-x + 3 = -5$
$-x = -8$
$x = 8$
Now replace x with 8 in the first equation to find y.
$y = 2 \cdot 8 + 3$
$y = 19$
Solution: $x = 8$, $y = 19$

13. $x - y = 4$
$x + 3y = 12$
Solve the first equation for x in terms of y.
$x = 4 + y$
Substitute into the second equation to eliminate x.
$(4 + y) + 3y = 12$
$4y = 8$
$y = 2$
Now replace y with 2 in the first equation to find x.
$x - 2 = 4$
$x = 6$
Solution: $x = 6$, $y = 2$

15. $3x - y = 7$
$2x + 3y = 1$
Solve the first equation for y in terms of x.
$-y = 7 - 3x$
$y = -7 + 3x$
Substitute into the second equation to eliminate y.
$2x + 3(-7 + 3x) = 1$
$2x - 21 + 9x = 1$
$11x = 22$
$x = 2$

Now replace y with 2 in the first equation to find y.

$3 \cdot 2 - y = 7$
$6 - y = 7$
$y = -1$
Solution: $x = 2$, $y = -1$

17. $4x + 3y = 26$
$3x - 11y = -7$
Solve the second equation for x in terms of y.
$3x = 11y - 7$
$x = \dfrac{11y - 7}{3}$
Substitute into the first equation to eliminate x.
$4\left(\dfrac{11y - 7}{3}\right) + 3y = 26$
$\dfrac{44y - 28}{3} + 3y = 26$
$44y - 28 + 9y = 78$
$53y = 106$
$y = 2$
$x = \dfrac{11 \cdot 2 - 7}{3}$
$x = 5$

Solution: $x = 5$, $y = 2$

19. $7m + 12n = -1$
$5m - 3n = 7$
Solve the first equation for n in terms of m.
$12n = -1 - 7m$
$n = \dfrac{-1 - 7m}{12}$
Substitute into the second equation to eliminate n.
$5m - 3\left(\dfrac{-1 - 7m}{12}\right) = 7$
$5m - \dfrac{-1 - 7m}{4} = 7$
$20m + 1 + 7m = 28$
$27m = 27$
$m = 1$
$n = \dfrac{-1 - 7(1)}{12}$
$n = -\dfrac{2}{3}$
Solution: $m = 1$, $n = -\dfrac{2}{3}$

21. $y = 0.08x$
$y = 100 + 0.04x$
Substitute y from the first equation into the second equation to eliminate y.
$0.08x = 100 + 0.04x$
$0.04x = 100$
$x = 2{,}500$
$y = 0.08(2{,}500)$
$y = 200$
Solution: $x = 2{,}500$, $y = 200$

23. $0.2u - 0.5v = 0.07$
$0.8u - 0.3v = 0.79$

For convenience, eliminate decimals by multiplying both sides of each equation by 100.

$20u - 50v = 7$
$80u - 30v = 79$

Common Error: $2u - 5v \neq 7$
$8u - 3v \neq 79$

Solve the first equation for u in terms of v and substitute into the second equation to eliminate u.

$$20u = 50v + 7$$
$$u = \frac{50v + 7}{20}$$
$$80\left(\frac{50v + 7}{20}\right) - 30v = 79$$
$$4(50v + 7) - 30v = 79$$
$$200v + 28 - 30v = 79$$
$$170v = 51$$
$$v = 0.3$$
$$u = \frac{50(0.3) + 7}{20}$$
$$u = 1.1$$

Solution: $u = 1.1$, $v = 0.3$

25. $\frac{2}{5}x + \frac{3}{2}y = 2$
$\frac{7}{3}x - \frac{5}{4}y = -5$

Eliminate fractions by multiplying both sides of the first equation by 10 and both sides of the second equation by 12.

$$10\left(\frac{2}{5}x + \frac{3}{2}y\right) = 20$$
$$4x + 15y = 20$$
$$12\left(\frac{7}{3}x - \frac{5}{4}y\right) = -60$$
$$28x - 15y = -60$$

Solve the first equation for y in terms of x and substitute into the second equation to eliminate y.

$$15y = 20 - 4x$$
$$y = \frac{20 - 4x}{15}$$
$$28x - 15\left(\frac{20 - 4x}{15}\right) = -60$$
$$28x - (20 - 4x) = -60$$
$$28x - 20 + 4x = -60$$
$$32x = -40$$
$$x = -\frac{5}{4}$$
$$y = \frac{20 - 4\left(-\frac{5}{4}\right)}{15}$$
$$y = \frac{20 + 5}{15}$$
$$y = \frac{5}{3}$$

Solution: $x = -\frac{5}{4}$, $y = \frac{5}{3}$

In Problems 27 and 29, a graphing calculator solution is presented.

27. First solve each equation for y.

$$2x - 3y = -5 \qquad 3x + 4y = 13$$
$$-3y = -2x - 5 \qquad 4y = -3x + 13$$
$$y = \frac{2}{3}x + \frac{5}{3} \qquad y = -\frac{3}{4}x + \frac{13}{4}$$

Next enter each equation, graph, and approximate the solution.

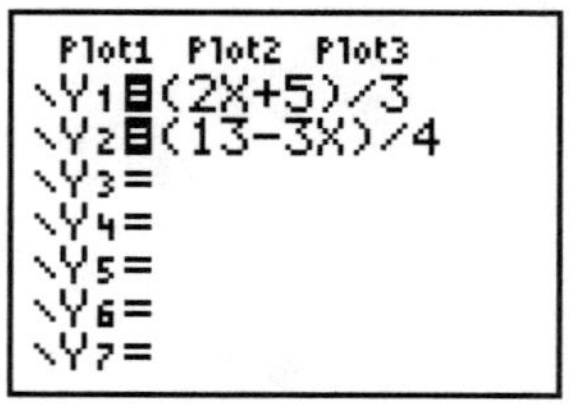

Equation definitions

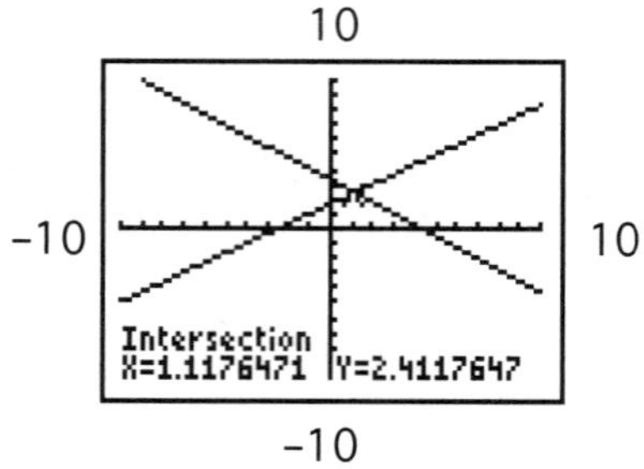

Intersection point

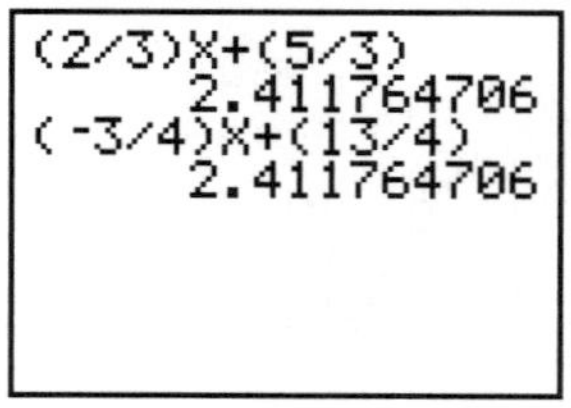

Check

To two decimal places, the solution is $x = 1.12$, $y = 2.41$ or (1.12, 2.41).

29. First solve each equation for y.

$$3.5x - 2.4y = 0.1 \qquad 2.6x - 1.7y = -0.2$$
$$-2.4y = -3.5x + 0.1 \qquad -1.7y = -2.6x - 0.2$$
$$y = \frac{3.5}{2.4}x - \frac{0.1}{2.4} \qquad y = \frac{2.6x}{1.7} + \frac{0.2}{1.7}$$

Next enter each equation, graph, and approximate the solution.

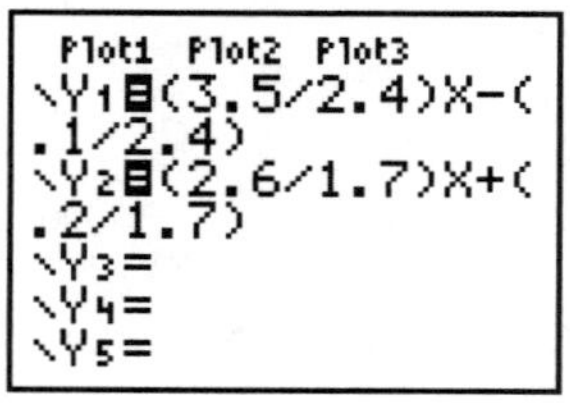

Equation definitions

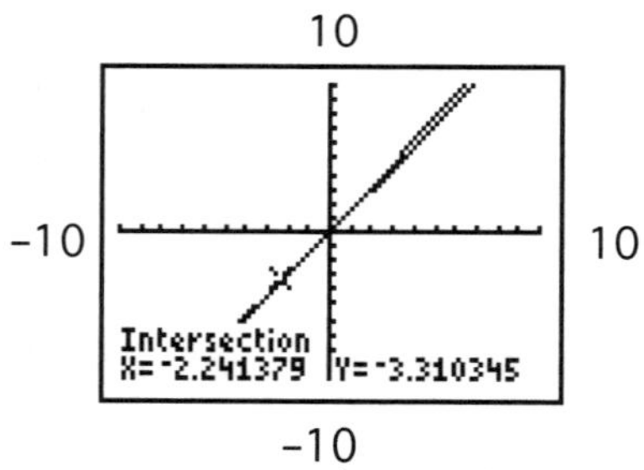

Intersection point

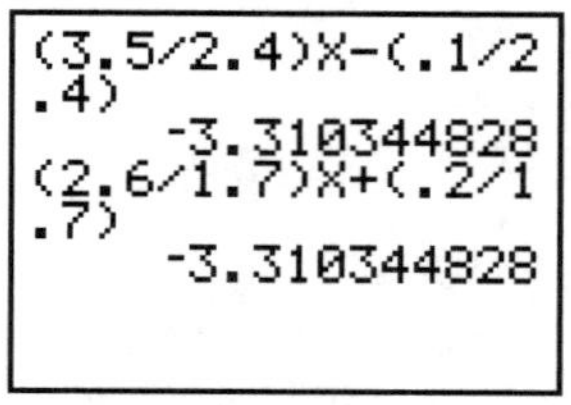

Check

To two decimal places, the solution is $x = -2.24$, $y = -3.31$ or (-2.24, -3.31).

If we add, we can eliminate y.

$$\begin{array}{r} 3x - 2y = 12 \\ 7x + 2y = 8 \\ \hline 10x \quad = 20 \\ x = 2 \end{array}$$

Now substitute $x = 2$ back into the bottom equation and solve for y.

$$7(2) + 2y = 8$$
$$2y = -6$$
$$y = -3$$

(2, -3)

31. $x = 2 + p - 2q$
$y = 3 - p + 3q$
Solve the first equation for p in terms of q, x, and y and substitute into the second equation to eliminate p, then solve for q in terms of x and y.

$$p = x - 2 + 2q$$
$$y = 3 - (x - 2 + 2q) + 3q$$
$$y = 3 - x + 2 - 2q + 3q$$
$$y = 5 - x + q$$
$$q = x + y - 5$$

Now substitute this expression for q into $p = x - 2 + 2q$ to find p in terms of x and y.

$$p = x - 2 + 2(x + y - 5)$$
$$p = x - 2 + 2x + 2y - 10$$
$$p = 3x + 2y - 12$$

Solution: $p = 3x + 2y - 12$, $q = x + y - 5$
To check this solution substitute into the original equations to see if true statements result:

$$x = 2 + p - 2q$$
$$x \stackrel{?}{=} 2 + (3x + 2y - 12) - 2(x + y - 5)$$
$$x \stackrel{?}{=} 2 + 3x + 2y - 12 - 2x - 2y + 10$$
$$x \stackrel{\checkmark}{=} x$$

$$y = 3 - p + 3q$$
$$y \stackrel{?}{=} 3 - (3x + 2y - 12) + 3(x + y - 5)$$
$$y \stackrel{?}{=} 3 - 3x - 2y + 12 + 3x + 3y - 15$$
$$y \stackrel{\checkmark}{=} y$$

33. $ax + by = h$
$cx + dy = k$
Solve the first equation for x in terms of y and the constants.

$$ax = h - by$$
$$x = \frac{h - by}{a} \quad (a \neq 0)$$

Substitute this expression into the second equation to eliminate x.

$$c\left(\frac{h - by}{a}\right) + dy = k$$
$$ac\left(\frac{h - by}{a}\right) + ady = ak$$
$$c(h - by) + ady = ak$$
$$ch - bcy + ady = ak$$
$$(ad - bc)y = ak - ch$$
$$y = \frac{ak - ch}{ad - bc} \quad ad - bc \neq 0$$

Similarly, solve the first equation for y in terms of x and the constants.

$$by = h - ax$$
$$y = \frac{h - ax}{b} \quad (b \neq 0)$$

Substitute this expression into the second equation to eliminate y.

$$cx + d\left(\frac{h - ax}{b}\right) = k$$
$$bcx + bd\left(\frac{h - ax}{b}\right) = bk$$
$$bcx + d(h - ax) = bk$$
$$bcx + dh - adx = bk$$
$$(bc - ad)x = bk - dh$$
$$x = \frac{bk - dh}{bc - ad} \quad bc - ad \neq 0$$

or, for consistency with the expression for y,

$$x = \frac{dh - bk}{ad - bc}$$

Solution: $x = \dfrac{dh - bk}{ad - bc}$, $y = \dfrac{ak - ch}{ad - bc}$ $\quad ad - bc \neq 0$

35. Let x = airspeed of the plane
y = rate at which wind is blowing
Then $x - y$ = ground speed flying from Atlanta to Los Angeles (head wind)
$x + y$ = ground speed flying from Los Angeles to Atlanta (tail wind)
Then, applying Distance = Rate ∞ Time, we have
$2{,}100 = 8.75(x - y)$
$2{,}100 = 5(x + y)$
After simplification, we have
$x - y = 240$
$x + y = 420$
Solve the first equation for x in terms of y and substitute into the second equation. $x = 240 + y$
$240 + y + y = 420$
$2y = 180$
$y = 90$ mph = wind rate
$x = 240 + y$
$x = 240 + 90$
$x = 330$ mph = airspeed

37. Let x = time rowed upstream
y = time rowed downstream
Then $x + y = \frac{1}{4}$ $\quad$ (15 min = $\frac{1}{4}$ hr.)
Since rate upstream = 20 - 2 = 18 mph and
rate downstream = 20 + 2 = 22 mph,
applying Distance = Rate × Time to the equal distances upstream and downstream, we have
$18x = 22y$
Solve the first equation for y in terms of x and substitute into the second equation.

$$y = \frac{1}{4} - x$$

$$18x = 22\left(\frac{1}{4} - x\right)$$

$18x = 5.5 - 22x$
$40x = 5.5$
$x = 0.1375$ hr.
Then the distance rowed upstream = $18x = 18(0.1375) = 2.475$ km.

39. Let x = amount of 50% solution
y = amount of 80% solution
100 milliliters are required, hence
$x + y = 100$
68% of the 100 milliliters must be acid, hence
$0.50x + 0.80y = 0.68(100)$
Solve the first equation for y in terms of x and substitute into the second equation.
$y = 100 - x$
$0.50x + 0.80(100 - x) = 0.68(100)$
$0.5x + 80 - 0.8x = 68$
$-0.3x = -12$
$x = 40$ milliliters of 50% solution
$y = 100 - x$
$y = 100 - 40 = 60$ milliliters of 80% solution

41. "Break even" means Cost = Revenue.
Let y = Cost = Revenue.
Let x = number of CDs sold

$$y = \text{Revenue} = \text{number of CDs sold} \times \text{price per CD}$$
$$y = x(8.00)$$
$$y = \text{Cost} = \text{Fixed Cost} + \text{Variable Cost}$$
$$= 17{,}680 + \text{number of CDs} \times \text{cost per CD}$$
$$y = 17{,}680 + x(4.60)$$

Substitute y from the first equation into the second equation to eliminate y.

$$8.00x = 17{,}680 + 4.60x$$
$$3.40x = 17{,}680$$
$$x = \frac{17{,}680}{3.40}$$
$$x = 5{,}200 \text{ CDs}$$

43. Let x = amount invested at 10% $0.1x$ = yield on amount invested at 10%
y = amount invested at 15% $0.15y$ = yield on amount invested at 15%
Then

$$x + y = 12{,}000 \quad \text{(amount invested)}$$
$$0.1x + 0.15y = 0.12(12{,}000) \quad \text{(total yield)}$$

Solve the first equation for y in terms of x and substitute into the second equation.

$$0.1x + 0.15(12{,}000 - x) = 0.12(12{,}000)$$
$$0.1x + 0.15(12{,}000) - 0.15x = 0.12(12{,}000)$$
$$0.1x + 1{,}800 - 0.15x = 1{,}440$$
$$-0.05x + 1{,}800 = 1{,}440$$
$$-0.05x = -360$$
$$x = \$7{,}200 \text{ invested at } 10\%$$
$$y = 12{,}000 - x = \$4{,}800 \text{ invested at } 15\%$$

45. Let x = number of hours Mexico plant is operated
y = number of hours Taiwan plant is operated
Then (Production at Mexico plant) + (Production at Taiwan plant) = (Total Production)

$$40x + 20y = 4000 \text{ (keyboards)}$$
$$32x + 32y = 4000 \text{ (screens)}$$

Solve the first equation for y in terms of x and substitute into the second equation.

$$20y = 4{,}000 - 40x$$
$$y = 200 - 2x$$
$$32x + 32(200 - 2x) = 4{,}000$$
$$32x + 6{,}400 - 64x = 4{,}000$$
$$-32x = -2{,}400$$
$$x = 75 \text{ hours Mexico plant}$$
$$y = 200 - 2x = 200 - 2(75)$$
$$= 50 \text{ hours Taiwan plant}$$

47. Let x = number of grams of Mix A
y = number of grams of Mix B
Then (Nutrition from Mix A) + (Nutrition for Mix B) = (Total Nutrition)

$$0.10x + 0.20y = 20 \text{ (Total protein)}$$
$$0.06x + 0.02y = 6 \text{ (Total fat)}$$

For convenience, eliminate decimals by multiplying both sides of the first equation by 10 and the second equation by 100.

$$x + 2y = 200$$
$$6x + 2y = 600$$

Solve the first equation for x in terms of y and substitute into the second equation.

$$\begin{aligned} x &= 200 - 2y \\ 6(200 - 2y) + 2y &= 600 \\ 1200 - 12y + 2y &= 600 \\ -10y &= -600 \\ y &= 60 \text{ grams Mix } A \\ x &= 200 - 2y = 200 - 2(60) \\ &= 80 \text{ grams Mix } B \end{aligned}$$

49. (A) If $p = 4$, then $4 = 0.007q + 3$, $q = \dfrac{1}{0.007} = 143$ T-shirts is the number that suppliers are willing to supply at this price.

$4 = -0.018q + 15$, $q = \dfrac{11}{0.018} = 611$ T-shirts is the number that consumers will purchase.

Demand exceeds supply and the price will rise.

(B) If $p = 8$, then $8 = 0.007q + 3$, $q = \dfrac{5}{0.007} = 714$ T-shirts is the number that suppliers are willing to supply.

$8 = -0.018q + 15$, $q = \dfrac{7}{0.018} = 389$ T-shirts is the number that consumers will purchase at this price.

Supply exceeds demand and the price will fall.

(C) Solve $p = 0.007q + 3$
$p = -0.018q + 315$

Substitute the expression for p in terms of q from the first equation into the second equation.

$$\begin{aligned} 0.007q + 3 &= -0.018q + 15 \\ 0.025q + 3 &= 15 \\ 0.025q &= 12 \\ q &= 480 \text{ T-shirts is the equilibrium quantity.} \\ p &= 0.007(480) + 3 = \$6.36 \text{ is the equilibrium price.} \end{aligned}$$

(D)

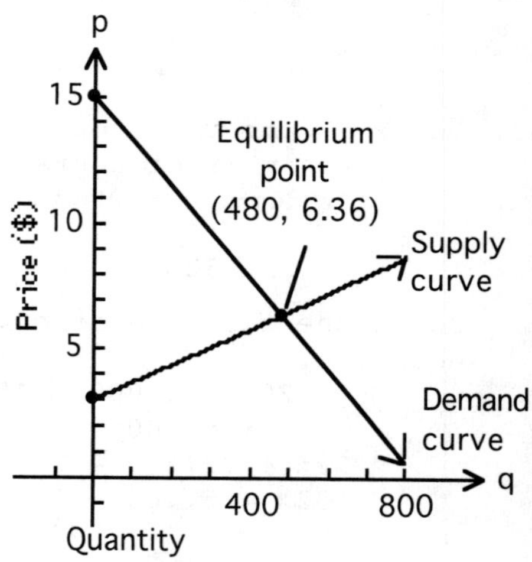

51. (A) Write $p = aq + b$.
Since $p = 0.60$ corresponds to supply $q = 450$,
$0.60 = 450a + b$
Since $p = 0.90$ corresponds to supply $q = 750$,
$0.90 = 750a + b$

Solve the first equation for b in terms of a and substitute into the second equation.

$$\begin{aligned} b &= 0.60 - 450a \\ 0.90 &= 750a + 0.60 - 450a \\ 0.30 &= 300a \\ a &= 0.001 \\ b &= 0.60 - 450a = 0.60 - 450(0.001) \\ &= 0.15 \end{aligned}$$

Thus, the supply equation is $p = 0.001q + 0.15$.

(B) Write $p = cq + d$.
Since $p = 0.60$ corresponds to demand $q = 645$,
$$0.60 = 645c + d$$
Since $p = 0.90$ corresponds to demand $q = 495$,
$$0.90 = 495c + d$$
Solve the first equation for d in terms of c and substitute into the second equation.
$$\begin{aligned} d &= 0.60 - 645c \\ 0.90 &= 495c + 0.60 - 645c \\ 0.30 &= -150c \\ c &= -0.002 \\ d &= 0.60 - 645c = 0.60 - 645(-0.002) \\ &= 1.89 \end{aligned}$$
Thus, the demand equation is $p = -0.002q + 1.89$

(C) Solve the system of equations
$$\begin{aligned} p &= 0.001q + 0.15 \\ p &= -0.002q + 1.89 \end{aligned}$$
Substitute p from the first equation into the second equation to eliminate p.
$$\begin{aligned} 0.001q + 0.15 &= -0.002q + 1.89 \\ 0.003q &= 1.74 \\ q &= 580 \text{ bushels} = \text{equilibrium quantity} \\ p &= 0.001q + 0.15 = 0.001(580) + 0.15 \\ &= \$0.73 \text{ equilibrium price} \end{aligned}$$

53. $s = a + bt^2$

(A) We are given: When $t = 1$, $s = 180$
When $t = 2$, $s = 132$

Substituting these values in the given equation, we have
$$\begin{aligned} 180 &= a + b(1)^2 \\ 132 &= a + b(2)^2 \quad \text{or} \\ 180 &= a + b \\ 132 &= a + 4b \end{aligned}$$
Solve the first equation for a in terms of b and substitute into the second equation to eliminate a.
$$\begin{aligned} a &= 180 - b \\ 132 &= 180 - b + 4b \\ 132 &= 180 + 3b \\ -48 &= 3b \\ b &= -16 \\ 180 &= a - 16 \\ a &= 196 \end{aligned}$$

(B) The height of the building is represented by s, the distance of the object above the ground, when $t = 0$. Since we now know
$$s = 196 - 16t^2$$
from part (A), when $t = 0$, $s = 196$ feet is the height of the building.

(C) The object falls until s, its distance above the ground, is zero. Since
$$s = 196 - 16t^2$$
we substitute $s = 0$ and solve for t.
$$\begin{aligned} 0 &= 196 - 16t^2 \\ 16t^2 &= 196 \\ t^2 &= \frac{196}{16} \\ t &= \frac{14}{4} \quad \text{(discarding the negative solution)} \\ t &= 3.5 \text{ seconds} \end{aligned}$$

55. Let p = time of primary wave
s = time for secondary wave
We know
$s - p = 16$ (time difference)
To find a second equation, we have to use Distance = Rate × Time
$5p$ = distance for primary wave
$3s$ = distance for secondary wave
These distances are equal, hence
$5p = 3s$

Solve the first equation for s in terms of p and substitute into the second equation to eliminate s.

$$\begin{aligned} s &= p + 16 \\ 5p &= 3(p + 16) \\ 5p &= 3p + 48 \\ 2p &= 48 \\ p &= 24 \text{ seconds} \\ s &= 24 + 16 \\ s &= 40 \text{ seconds} \end{aligned}$$

The distance traveled $= 5p = 3s = 120$ miles

Section 10-2

Note: The checking steps are not shown, but should be performed by the student.

1.
$$\begin{aligned} -2x \qquad\qquad &= 2 \quad E_1 \\ x - 3y \qquad &= 2 \quad E_2 \\ -x + 2y + 3z &= -7 \quad E_3 \end{aligned}$$

Solve E_1 for x.

$$\begin{aligned} -2x &= 2 \quad E_1 \\ x &= -1 \end{aligned}$$

Substitute $x = -1$ in E_2 and solve for y.

$$\begin{aligned} x - 3y &= 2 \quad E_2 \\ -1 - 3y &= 2 \\ y &= -1 \end{aligned}$$

Substitute $x = -1$ and $y = -1$ in E_3 and solve for z.

$$\begin{aligned} -x + 2y + 3z &= -7 \quad E_3 \\ -(-1) + 2(-1) + 3z &= -7 \\ z &= -2 \end{aligned}$$

$(-1, -1, -2)$

3.
$$\begin{aligned} 2y - z &= 2 \quad E_1 \\ -4y + 2z &= 1 \quad E_2 \\ x - 2y + 3z &= 0 \quad E_3 \end{aligned}$$

Multiply E_1 by 2 and add to E_2

$$\begin{aligned} 4y - 2z &= 4 \quad 2E_1 \\ \underline{-4y + 2z} &\underline{= 1} \quad E_2 \\ 0 &= 5 \quad E_4 \end{aligned}$$

A contradiction. No solution.

5.
$$\begin{aligned} x - 3y \qquad &= 2 \quad E_1 \\ 2y + z &= -1 \quad E_2 \\ x - y + z &= 1 \quad E_3 \end{aligned}$$

Multiply E_1 by -1 and add to E_3 to eliminate x.

$$\begin{aligned} -x + 3y \qquad &= -2 \quad (-1)E_1 \\ \underline{x - y + z} &\underline{= 1} \quad E_3 \\ 2y + z &= -1 \quad E_4 \end{aligned}$$

Equivalent system:

$$\begin{aligned} x - 3y \qquad &= 2 \quad E_1 \\ 2y + z &= -1 \quad E_2 \\ 2y + z &= -1 \quad E_4 \end{aligned}$$

If E_2 is multiplied by -1 and added to E_4, $0 = 0$ results. The system is dependent and equivalent to

$$\begin{aligned} x - 3y &= 2 \\ 2y + z &= -1 \end{aligned}$$

Let $y = s$. Then

$$2s + z = -1$$
$$z = -2s - 1$$
$$x - 3s = 2$$
$$x = 3s + 2$$

Solutions: $\{(3s + 2, s, -2s - 1) \mid s \text{ is any real number}\}$

7.
$$\begin{aligned} 4y - z &= -13 \quad E_1 \\ 3y + 2z &= 4 \quad E_2 \\ 6x - 5y - 2z &= -18 \quad E_3 \end{aligned}$$

Multiply E_1 by 2 and add to E_2 to eliminate z.

$$\begin{aligned} 8y - 2z &= -26 \quad 2E_1 \\ 3y + 2z &= 4 \quad E_2 \\ \hline 11y &= -22 \quad E_4 \\ y &= -2 \end{aligned}$$

Substitute $y = -2$ into E_2 and solve for z.

$$\begin{aligned} 3y + 2z &= 4 \quad E_2 \\ 3(-2) + 2z &= 4 \\ z &= 5 \end{aligned}$$

Substitute $y = -2$ and $z = 5$ into E_3 and solve for x.

$$\begin{aligned} 6x - 5(-2) - 2(5) &= -18 \\ x &= -3 \end{aligned}$$

$(-3, -2, 5)$

9.
$$\begin{aligned} 2x + y - z &= 5 \quad E_1 \\ x - 2y - 2z &= 4 \quad E_2 \\ 3x + 4y + 3z &= 3 \quad E_3 \end{aligned}$$

Multiply E_2 by -2 and add to E_1 to eliminate x. Also multiply E_2 by -3 and add to E_3 to eliminate x.

$$\begin{aligned} 2x + y - z &= 5 \quad E_1 \\ -2x + 4y + 4z &= -8 \quad (-2)E_2 \\ \hline 5y + 3z &= -3 \quad E_4 \end{aligned}$$

$$\begin{aligned} 3x + 4y + 3z &= 3 \quad E_3 \\ -3x + 6y + 6z &= -12 \quad (-3)E_2 \\ \hline 10y + 9z &= -9 \quad E_5 \end{aligned}$$

Equivalent system:

$$\begin{aligned} x - 2y - 2z &= 4 \quad E_2 \\ 5y + 3z &= -3 \quad E_4 \\ 10y + 9z &= -9 \quad E_5 \end{aligned}$$

Multiply E_4 by -2 and add to E_5 to eliminate y.

$$\begin{aligned} -10y - 6z &= 6 \quad (-2)E_4 \\ 10y + 9z &= -9 \quad E_5 \\ \hline 3z &= -3 \\ z &= -1 \end{aligned}$$

Substitute $z = -1$ into E_4 and solve for y.

$$\begin{aligned} 5y + 3z &= -3 \quad E_4 \\ 5y + 3(-1) &= -3 \\ y &= 0 \end{aligned}$$

Substitute $y = 0$ and $z = -1$ into E_2 and solve for x.

$$\begin{aligned} x - 2y - 2z &= 4 \quad E_2 \\ x - 2(0) - 2(-1) &= 4 \\ x &= 2 \end{aligned}$$

$(2, 0, -1)$

11.
$$\begin{aligned} x - y + z &= 1 & E_1 \\ 2x + y + z &= 6 & E_2 \\ 7x - y + 5z &= 15 & E_3 \end{aligned}$$
Multiply E_1 by -2 and add to E_2 to eliminate x. Also multiply E_1 by -7 and add to E_3 to eliminate x.
$$\begin{aligned} -2x + 2y - 2z &= -2 & (-2)E_1 \\ 2x + y + z &= 6 & E_2 \\ \hline 3y - z &= 4 & E_4 \end{aligned}$$
$$\begin{aligned} -7x + 7y - 7z &= -7 & (-7)E_1 \\ 7x - y + 5z &= 15 & E_3 \\ \hline 6y - 2z &= 8 & E_5 \end{aligned}$$
Equivalent system:
$$\begin{aligned} x - y + z &= 1 & E_1 \\ 3y - z &= 4 & E_4 \\ 6y - 2z &= 8 & E_5 \end{aligned}$$
If E_4 is multiplied by -2 and added to E_5, $0 = 0$ results. The system is dependent and equivalent to
$$\begin{aligned} x - y + z &= 1 \\ 3y - z &= 4 \end{aligned}$$
Let $y = s$. Then
$$\begin{aligned} 3s - z &= 4 \\ z &= 3s - 4 \\ x - s + (3s - 4) &= 1 \\ x &= -2s + 5 \end{aligned}$$
Solutions: $\{(-2s + 5, s, 3s - 4) \mid s \text{ is any real number}\}$

13.
$$\begin{aligned} 2a + 4b + 3c &= -6 & E_1 \\ a - 3b + 2c &= -15 & E_2 \\ a + 2b - c &= 9 & E_3 \end{aligned}$$
Add E_2 to E_3 to eliminate a. Also multiply E_2 by -2 and add to E_1 to eliminate a.
$$\begin{aligned} a - 3b + 2c &= -15 & E_2 \\ -a + 2b - c &= 9 & E_3 \\ \hline -b + c &= -6 & E_4 \end{aligned}$$
$$\begin{aligned} -2a + 6b - 4c &= 30 & (-2)E_2 \\ 2a + 4b + 3c &= -6 & E_1 \\ \hline 10b - c &= 24 & E_5 \end{aligned}$$
Equivalent system:
$$\begin{aligned} a - 3b + 2c &= -15 & E_2 \\ -b + c &= -6 & E_4 \\ 10b - c &= 24 & E_5 \end{aligned}$$
Add E_4 to E_5 to eliminate c
$$\begin{aligned} -b + c &= -6 & E_4 \\ 10b - c &= 24 & E_5 \\ \hline 9b &= 18 \\ b &= 2 \end{aligned}$$
Substitute $b = 2$ into E_4 and solve for c.
$$\begin{aligned} -b + c &= -6 & E_4 \\ -2 + c &= -6 \\ c &= -4 \end{aligned}$$
Substitute $b = 2$ and $c = -4$ into E_2 and solve for a.
$$\begin{aligned} a - 3b + 2c &= -15 & E_2 \\ a - 3(2) + 2(-4) &= -15 \\ a &= -1 \end{aligned}$$
$(-1, 2, -4)$

15. $2x - 3y + 3z = -5$ $\quad E_1$
$3x + 2y - 5z = 34$ $\quad E_2$
$5x - 4y - 2z = 23$ $\quad E_3$

Multiply E_1 by $-\frac{3}{2}$ and add to E_2 to eliminate x. Also multiply E_1 by $-\frac{5}{2}$ and add to E_3 to eliminate x.

$$\begin{array}{rcll} -3x + \frac{9}{2}y - \frac{9}{2}z &=& \frac{15}{2} & \left(-\frac{3}{2}\right)E_1 \\ 3x + 2y - 5z &=& 34 & E_2 \\ \hline \frac{13}{2}y - \frac{19}{2}z &=& \frac{83}{2} & E_4 \\ -5x + \frac{15}{2}y - \frac{15}{2}z &=& \frac{25}{2} & \left(-\frac{5}{2}\right)E_1 \\ 5x - 4y - 2z &=& 23 & E_3 \\ \hline \frac{7}{2}y - \frac{19}{2}z &=& \frac{71}{2} & E_5 \end{array}$$

Equivalent system:

$$\begin{array}{rcll} 2x - 3y + 3z &=& -5 & E_1 \\ \frac{13}{2}y - \frac{19}{2}z &=& \frac{83}{2} & E_4 \\ \frac{7}{2}y - \frac{19}{2}z &=& \frac{71}{2} & E_5 \end{array}$$

Multiply E_4 by -1 and add to E_5 to eliminate z.

$$\begin{array}{rcll} -\frac{13}{2}y + \frac{19}{2}z &=& -\frac{83}{2} & (-1)E_4 \\ \frac{7}{2}y - \frac{19}{2}z &=& \frac{71}{2} & E_5 \\ \hline -3y &=& -6 & E_6 \end{array}$$

Solve E_6 for y to obtain $y = 2$. Substitute $y = 2$ into E_4 and solve for z.

$$\begin{array}{rcll} \frac{13}{2}y - \frac{19}{2}z &=& \frac{83}{2} & E_4 \\ \frac{13}{2}(2) - \frac{19}{2}z &=& \frac{83}{2} & \\ -\frac{19}{2}z &=& \frac{57}{2} & \\ z &=& -3 & \end{array}$$

Substitute $y = 2$ and $z = -3$ into E_1 and solve for x.

$$\begin{array}{rcll} 2x - 3y + 3z &=& -5 & E_1 \\ 2x - 3(2) + 3(-3) &=& -5 & \\ x &=& 5 & \end{array}$$

(5, 2, -3)

17. $x + 2y + z = 2$ $\quad E_1$
$-2x + 3y - 2z = -3$ $\quad E_2$
$x - 5y + z = 2$ $\quad E_3$

Multiply E_1 by 2 and add to E_2 to eliminate x. Also multiply E_1 by -1 and add to E_3 to eliminate x.

$$\begin{array}{rcll} 2x + 4y + 2z &=& 4 & 2E_1 \\ -2x + 3y - 2z &=& -3 & E_2 \\ \hline 7y &=& 1 & E_4 \\ -x - 2y - z &=& -2 & (-1)E_1 \\ x - 5y + z &=& 2 & E_3 \\ \hline -7y &=& 0 & E_5 \end{array}$$

Equivalent system

$$\begin{aligned} x + 2y + z &= 2 & E_1 \\ 7y &= 1 & E_4 \\ -7y &= 0 & E_5 \end{aligned}$$

y cannot simultaneously satisfy E_4 and E_5. A contradiction. No solution.

19.

$$\begin{aligned} -x + 2y - z &= -4 & E_1 \\ 2x + 5y - 4z &= -16 & E_2 \\ x + y - z &= -4 & E_3 \end{aligned}$$

Multiply E_1 by 2 and add to E_2 to eliminate x. Also add E_1 to E_3 to eliminate x.

$$\begin{aligned} -2x + 4y - 2z &= -8 & 2E_1 \\ 2x + 5y - 4z &= -16 & E_2 \\ \hline 9y - 6z &= -24 & E_4 \\ -x + 2y - z &= -4 & E_1 \\ x + y - z &= -4 & E_3 \\ \hline 3y - 2z &= -8 & E_5 \end{aligned}$$

Equivalent system:

$$\begin{aligned} -x + 2y - z &= -4 & E_1 \\ 9y - 6z &= -24 & E_4 \\ 3y - 2z &= -8 & E_5 \end{aligned}$$

If E_5 is multiplied by -3 and added to E_4, 0 = 0 results. The system is dependent and equivalent to

$$\begin{aligned} -x + 2y - z &= -4 \\ 3y - 2z &= -8 \end{aligned}$$

Let $z = s$. Then

$$3y - 2s = -8$$

$$y = \frac{2s - 8}{3} \text{ or } \frac{2}{3}s - \frac{8}{3}$$

$$-x + 2\left(\frac{2}{3}s - \frac{8}{3}\right) - s = -4$$

$$-x = -\frac{4}{3}s + \frac{16}{3} + s - 4$$

$$x = \frac{1}{3}s - \frac{4}{3}$$

Solutions: $\left\{\left(\frac{1}{3}s - \frac{4}{3}, \frac{2}{3}s - \frac{8}{3}, s\right) \middle| s \text{ is any real number}\right\}$

21.

$$\begin{aligned} 2x + 3y + 2z &= 2 & E_1 \\ 5x + 8y + 4z &= 3 & E_2 \\ 4x + 7y + 2z &= 0 & E_3 \end{aligned}$$

Multiply E_1 by $-\frac{5}{2}$ and add to E_2 to eliminate x. Also multiply E_1 by -2 and add to E_3 to eliminate x.

$$\begin{aligned} -5x - \frac{15}{2}y - 5z &= -5 & \left(-\frac{5}{2}\right)E_1 \\ 5x + 8y + 4z &= 3 & E_2 \\ \hline \frac{1}{2}y - z &= -2 & E_4 \\ -4x - 6y - 4z &= -4 & (-2)E_1 \\ 4x + 7y + 2z &= 0 & E_3 \\ \hline y - 2z &= -4 & E_5 \end{aligned}$$

Equivalent system:

$$\begin{aligned} 2x + 3y + 2z &= 2 & E_1 \\ \frac{1}{2}y - z &= -2 & E_4 \\ y - 2z &= -4 & E_5 \end{aligned}$$

If E_4 is multiplied by -2 and added to E_5, 0 = 0 results. The system is dependent and equivalent to

$$\begin{aligned} 2x + 3y + 2z &= 2 \\ y - 2z &= -4 \end{aligned}$$

Let $z = s$. Then

$$\begin{aligned} y - 2s &= -4 \\ y &= 2s - 4 \\ 2x + 3(2s - 4) + 2s &= 2 \\ 2x &= -6s + 12 - 2s + 2 \\ x &= -4s + 7 \end{aligned}$$

Solutions: $\{(-4s + 7, 2s - 4, s) \mid s \text{ is any real number}\}$

23. If the points are on the graph, the coordinates satisfy the equation. Substituting, in turn, (-2, 9), (1, -9) and (4, 9) into $ax^2 + bx + c = y$ yields

$$\begin{aligned} 4a - 2b + c &= 9 & E_1 \\ a + b + c &= -9 & E_2 \\ 16a + 4b + c &= 9 & E_3 \end{aligned}$$

Multiply E_2 by -1 and add to E_1 to eliminate c. Also multiply E_2 by -1 and add to E_3 to eliminate c.

$$\begin{aligned} -a - b - c &= 9 & (-1)E_2 \\ \underline{4a - 2b + c} &= \underline{9} & E_1 \\ 3a - 3b &= 18 & E_4 \\ -a - b - c &= 9 & (-1)E_2 \\ \underline{16a + 4b + c} &= \underline{9} & E_3 \\ 15a + 3b &= 18 & E_5 \end{aligned}$$

Equivalent system

$$\begin{aligned} a + b + c &= -9 & E_2 \\ 3a - 3b &= 18 & E_4 \\ 15a + 3b &= 18 & E_5 \end{aligned}$$

Add E_4 to E_5 to eliminate b.

$$\begin{aligned} 3a - 3b &= 18 \\ \underline{15a + 3b} &= \underline{18} \\ 18a &= 36 \\ a &= 2 \end{aligned}$$

Substitute $a = 2$ into E_5 and solve for b.

$$\begin{aligned} 15a + 3b &= 18 & E_5 \\ 15(2) + 3b &= 18 \\ b &= -4 \end{aligned}$$

Substitute $a = 2$ and $b - 4$ into E_2 and solve for c.

$$\begin{aligned} a + b + c &= -9 & E_2 \\ 2 + (-4) + c &= -9 \\ c &= -7 \end{aligned}$$

$a = 2$, $b = -4$, $c = -7$

25. If the points are on the graph, the coordinates satisfy the equation. Substituting, in turn, (-1, 1), (5, 1), and (6, -6) into $x^2 + y^2 + ax + by + c = 0$ yields

$$\begin{aligned} 1 + 1 - a + b + c &= 0 \\ 25 + 1 + 5a + b + c &= 0 \\ 36 + 36 + 6a - 6b + c &= 0 \end{aligned}$$

Simplifying yields

$$\begin{aligned} -a + b + c &= -2 & E_1 \\ 5a + b + c &= -26 & E_2 \\ 6a - 6b + c &= -72 & E_3 \end{aligned}$$

Multiply E_1 by -1 and add to E_2 to eliminate c (and b).
Multiply E_1 by -1 and add to E_3 to eliminate c.

$$\begin{aligned} a - b - c &= 2 & (-1)E_1 \\ 5a + b + c &= -26 & E_2 \\ \hline 6a \qquad &= -24 & E_4 \end{aligned}$$

$$\begin{aligned} a - b - c &= 2 & (-1)E_1 \\ 6a - 6b + c &= -72 & E_3 \\ \hline 7a - 7b \qquad &= -70 & E_5 \end{aligned}$$

Equivalent system:

$$\begin{aligned} -a + b + c &= -2 & E_1 \\ 6a \qquad &= -24 & E_4 \\ 7a - 7b \qquad &= -70 & E_5 \end{aligned}$$

Solve E_4 for a to obtain $a = -4$. Substitute $a = -4$ into E_5 and solve for b.

$$\begin{aligned} 7a - 7b &= -70 & E_5 \\ 7(-4) - 7b &= -70 \\ b &= 6 \end{aligned}$$

Substitute $a = -4$ and $b = 6$ into E_1 and solve for c.

$$\begin{aligned} -a + b + c &= -2 & E_1 \\ -(-4) + 6 + c &= -2 \\ c &= -12 \end{aligned}$$

$a = -4$, $b = 6$, $c = -12$

27. Let

x = number of lawn mowers manufactured each week
y = number of snowblowers manufactured each week
z = number of chain saws manufactured each week

Then

E_1 $20x + 30y + 45z = 35{,}000$ Labor
E_2 $35x + 50y + 40z = 50{,}000$ Materials
E_3 $15x + 25y + 10z = 20{,}000$ Shipping

Multiply E_3 by -4.5 and add to E_1 to eliminate z. Also multiply E_3 by -4 and add to E_2 to eliminate z.

$$\begin{aligned} 20x + 30y + 45z &= 35{,}000 & E_1 \\ -67.5x - 112.5y - 45z &= -90{,}000 & (-4.5)E_3 \\ \hline -47.5x - 82.5y \qquad &= -55{,}000 & E_4 \end{aligned}$$

$$\begin{aligned} 35x + 50y + 40z &= 50{,}000 & E_2 \\ -60x - 100y - 40z &= -80{,}000 & (-4)E_3 \\ \hline -25x - 50y \qquad &= -30{,}000 & E_5 \end{aligned}$$

Equivalent system:

$$\begin{aligned} 15x + 25y + 10z &= 20{,}000 & E_3 \\ -47.5x - 82.5y \qquad &= -55{,}000 & E_4 \\ -25x - 50y \qquad &= -30{,}000 & E_5 \end{aligned}$$

Multiply E_5 by -1.9 and add to E_4 to eliminate x.

$$\begin{array}{rcrl} 47.5x + 95y & = & 57{,}000 & (-1.9)E_5 \\ 47.5x - 82.5y & = & -55{,}000 & E_4 \\ \hline 12.5y & = & 2{,}000 & \\ y & = & 160 & \end{array}$$

Substitute $y = 160$ into E_5 and solve for x.

$$\begin{array}{rcrl} -25x - 50y & = & -30{,}000 & E_5 \\ -25x - 50(160) & = & -30{,}000 & \\ x & = & 880 & \end{array}$$

Substitute $x = 880$ and $y = 160$ into E_3 and solve for z

$$\begin{array}{rcrl} 15x + 25y + 10z & = & 20{,}000 & E_3 \\ 15(880) + 25(160) + 10z & = & 20{,}000 & \\ z & = & 280 & \end{array}$$

880 lawn mowers, 160 snowblowers, 280 chain saws.

29. Let

x = number of days operating the Michigan plant
y = number of days operating the New York plant
z = number of days operating the Ohio plant

Then

$$\begin{array}{llrl} E_1 & 10x + 70y + 60z = 2{,}150 & & \text{Notebooks} \\ E_2 & 20x + 50y + 80z = 2{,}300 & & \text{Desktops} \\ E_3 & 40x + 30y + 90z = 2{,}500 & & \text{Servers} \end{array}$$

Multiply E_1 by -2 and add to E_2 to eliminate x. Also multiply E_1 by -4 and add to E_3 to eliminate x.

$$\begin{array}{rcrl} -20x - 140y - 120z & = & -4{,}300 & (-2)E_1 \\ 20x + 50y + 80z & = & 2{,}300 & E_2 \\ \hline -90y - 40z & = & -2{,}000 & E_4 \end{array}$$

$$\begin{array}{rcrl} -40x - 280y - 240z & = & -8{,}600 & (-4)E_1 \\ 40x + 30y + 90z & = & 2{,}500 & E_3 \\ \hline -250y - 150z & = & -6{,}100 & E_5 \end{array}$$

Equivalent system:

$$\begin{array}{rcrl} 10x + 70y + 60z & = & 2{,}150 & E_1 \\ -90y - 40z & = & -2{,}000 & E_4 \\ -250y - 150z & = & -6{,}100 & E_5 \end{array}$$

Multiply E_5 by -0.36 and add to E_4 to eliminate y.

$$\begin{array}{rcrl} -90y - 40z & = & -2{,}000 & E_4 \\ 90y + 54z & = & 2{,}196 & (0.36)E_5 \\ \hline 14z & = & 196 & \\ z & = & 14 & \end{array}$$

Substitute $z = 14$ into E_4 and solve for y

$$\begin{array}{rcrl} -90y - 40z & = & -2{,}000 & E_4 \\ -90y - 40(14) & = & -2{,}000 & \\ y & = & 16 & \end{array}$$

Substitute $y = 16$ and $z = 14$ into E_1 and solve for x.

$$\begin{array}{rcrl} 10x + 70y + 60z & = & 2{,}150 & E_1 \\ 10x + 70(16) + 60(14) & = & 2{,}150 & \\ x & = & 19 & \end{array}$$

19 days Michigan plant, 16 days New York plant, 14 days Ohio plant.

31. Let

x = amount invested in treasury bonds
y = amount invested in municipal bonds
z = amount invested in corporate bonds

Then

$$\begin{aligned} x + y + z &= 100{,}000 && \text{total investment} \\ .04x + .035y + .05z &= 4{,}400 && \text{total income} \\ x + y &= z && \text{risk control factor} \end{aligned}$$

Clearing of decimals and changing to standard form yields

$$\begin{aligned} x + y + z &= 100{,}000 && E_1 \\ 40x + 35y + 50z &= 4{,}400{,}000 && E_2 \\ x + y - z &= 0 && E_3 \end{aligned}$$

Multiply E_3 by -1 and add to E_1 to eliminate x and y.

$$\begin{aligned} x + y + z &= 100{,}000 && E_1 \\ \underline{-x - y + z} &= \underline{0} && (-1)E_3 \\ 2z &= 100{,}000 \\ z &= 50{,}000 \end{aligned}$$

Substituting $z = 50{,}000$ into E_1 and into E_2 yields

$$\begin{aligned} x + y + z &= 100{,}000 && E_1 \\ x + y + 50{,}000 &= 100{,}000 \\ 40x + 35y + 50z &= 4{,}400{,}000 && E_2 \\ 40x + 35y + 50(50{,}000) &= 4{,}400{,}000 \end{aligned}$$

Simplifying yields

$$\begin{aligned} x + y &= 50{,}000 && E_4 \\ 40x + 35y &= 1{,}900{,}000 && E_5 \end{aligned}$$

Multiply E_4 by -35 and add to E_5 to eliminate y.

$$\begin{aligned} -35x - 35y &= -1{,}750{,}000 && (-35)E_4 \\ \underline{40x + 35y} &= \underline{1{,}900{,}000} && E_5 \\ 5x &= 150{,}000 \\ x &= 30{,}000 \end{aligned}$$

Substituting $x = 30{,}000$ into E_4 yields $y = 20{,}000$. \$30,000 in treasury bonds, \$20,000 in municipal bonds, \$50,000 in corporate bonds.

33. Let

x = number of pounds of peanuts
y = number of pounds of cashews
z = number of pounds of walnuts

Then

$$\begin{aligned} x + y + z &= 1{,}500 && \text{total weight} \\ 1.4x + 4y + 2.9z &= 3{,}500 && \text{total value} \\ x &= y + z && \text{blend is half peanuts} \end{aligned}$$

Clearing of decimals and changing to standard form yields

$$\begin{aligned} x + y + z &= 1{,}500 && E_1 \\ 14x + 40y + 29z &= 35{,}000 && E_2 \\ x - y - z &= 0 && E_3 \end{aligned}$$

Adding equations E_1 and E_3 eliminates y and z.

$$\begin{aligned} x + y + z &= 1{,}500 && E_1 \\ \underline{x - y - z} &= \underline{0} && E_3 \\ 2x &= 1{,}500 \\ x &= 750 \end{aligned}$$

Substitute $x = 750$ into E_1 and E_2.

$$\begin{aligned} 750 + y + z &= 1{,}500 \\ 14(750) + 40y + 29z &= 35{,}000 \end{aligned}$$

Simplifying yields

$$y + z = 750 \quad E_4$$
$$40y + 29z = 24{,}500 \quad E_5$$

Multiply E_4 by -40 and add to E_5 to eliminate y

$$\begin{aligned} -40y - 40z &= -30{,}000 \quad (-40)E_4 \\ 40y + 29z &= 24{,}500 \quad E_5 \\ \hline -11z &= -5{,}500 \\ z &= 500 \end{aligned}$$

Substituting $z = 500$ into E_4 yields $y = 250$.

750 lbs. peanuts, 250 lbs. cashews, 500 lbs. walnuts.

35. Let $f(x) = ax^2 + bx + c$ be the quadratic model and evaluate f at each of the 3 data points.

$$\begin{aligned} x = 0&: & c &= 75 \quad E_1 \\ x = 50&: & 2{,}500a + 50b + c &= 150 \quad E_2 \\ x = 100&: & 10{,}000a + 100b + c &= 275 \quad E_3 \end{aligned}$$

Substitute $c = 75$ into E_2 and E_3.

$$2{,}500a + 50b + 75 = 150$$
$$10{,}000a + 100b + 75 = 275$$

Simplifying yields

$$2{,}500a + 50b = 75 \quad E_4$$
$$10{,}000a + 100b = 200 \quad E_5$$

Multiply E_4 by -2 and add to E_5 to eliminate b.

$$\begin{aligned} -5{,}000a - 100b &= -150 \quad (-2)E_4 \\ 10{,}000a + 100b &= 200 \quad E_5 \\ \hline 5{,}000a &= 50 \\ a &= 0.01 \end{aligned}$$

Substitute $a = 0.01$ into E_5 and solve for b.

$$\begin{aligned} 10{,}000a + 100b &= 200 \\ 10{,}000(0.01) + 100b &= 200 \\ b &= 1 \end{aligned}$$

The model is $f(x) = 0.01x^2 + x + 75$.

To find the predicted population in 2050 calculate $f(150)$.

$$f(150) = 0.01(150)^2 + 150 + 75 = 450 \text{ million}$$

37. Let $f(x) = ax^2 + bx + c$ be the quadratic model and evaluate f at each of the three data points.

$$\begin{aligned} x = 0&: & c &= 77.6 \quad E_1 \\ x = 5&: & 25a + 5b + c &= 78 \quad E_2 \\ x = 10&: & 100a + 10b + c &= 78.6 \quad E_3 \end{aligned}$$

Substitute $c = 77.6$ into E_2 and E_3

$$25a + 5b + 77.6 = 78$$
$$100a + 10b + 77.6 = 78.6$$

Simplifying yields

$$25a + 5b = 0.4 \quad E_4$$
$$100a + 10b = 1 \quad E_5$$

Multiply E_4 by -2 and add to E_5 to eliminate b.

$$\begin{aligned} -50a - 10b &= -0.8 \\ 100a + 10b &= 1 \\ \hline 50a &= 0.2 \\ a &= 0.004 \end{aligned}$$

Substitute $a = 0.004$ into E_5 and solve for b.

$$100a + 10b = 1 \quad E_5$$
$$100(0.004) + 10b = 1$$
$$b = 0.06$$

The model is $f(x) = 0.004x^2 + 0.06x + 77.6$.

Predicted life expectancy for females born 1995-2000:

$f(15) = 0.004(15)^2 + 0.06(15) + 77.6 = 79.4$ yrs.

Predicted life expectancy for females born 2000-2005:

$f(20) = 0.004(20)^2 + 0.06(20) + 77.6 = 80.4$ yrs.

39. The independent variable is given by the first year in the interval minus the starting year 1980, so enter 0, 5, 10, 15, 20 as L_1. The dependent variable is life expectancy, so enter the data 77.6, 78, 78.6, 79.1, 79.7 as L_2. Then use the quadratic regression routine from the STAT CALC menu and plot using the STAT PLOT screen.

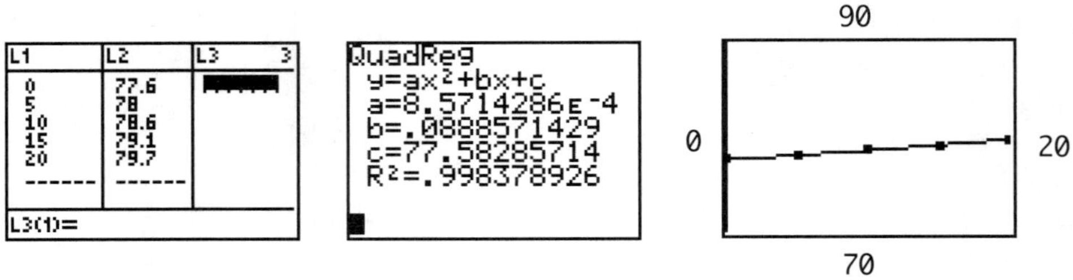

Section 10-3

1. Matrix A has 2 rows and 3 columns, so its size is 2×3.
Matrix C has 1 row and 3 columns, so its size is 1×3.

3. C is the only row matrix.

5. B is the only square matrix.

7. a_{12} in the element is row 1, column 2 of A. This is -2.
a_{23} is the element in row 2, column 3 of A. This is -6.

9. The elements on the principal diagonal are $b_{11} = -2$, $b_{22} = 6$, and $b_{33} = 0$.

11. No. Condition 2 is violated.

13. Yes

15. No. Condition 4 is violated.

17. Yes

19.
$$\begin{aligned} x_1 \qquad\qquad &= -2 \\ x_2 \qquad &= 3 \\ x_3 &= 0 \end{aligned}$$

The system is already solved.

21.
$$\begin{aligned} x_1 \qquad - 2x_3 &= 3 \\ x_2 + x_3 &= -5 \end{aligned}$$

Solution:

$x_3 = t$

$x_2 = -5 - x_3 = -5 - t$

$x_1 = 3 + 2x_3 = 3 + 2t$

Thus $x_1 = 2t + 3$, $x_2 = -t - 5$, $x_3 = t$ is the solution for t any real number.

23. $x_1 \qquad = 0$
$x_2 \quad = 0$
$0 = 1$

The system has no solution.

25. $x_1 - 2x_2 \qquad - 3x_4 = -5$
$x_3 + 3x_4 = 2$

Solution:
$x_4 = t$
$x_3 = 2 - 3x_4 = 2 - 3t$
$x_2 = s$
$x_1 = -5 + 2x_2 + 3x_4 = -5 + 2s + 3t$
Thus $x_1 = 2s + 3t - 5$, $x_2 = s$, $x_3 = -3t + 2$, $x_4 = t$ is the solution, for s and t any real numbers.

27. $R_1 \leftrightarrow R_2$ means interchange Rows 1 and 2.
$$\left[\begin{array}{rr|r} 4 & -6 & -8 \\ 1 & -3 & 2 \end{array}\right]$$

29. $-4R_1 \rightarrow R_1$ means multiply Row 1 by -4.
$$\left[\begin{array}{rr|r} -4 & 12 & -8 \\ 4 & -6 & -8 \end{array}\right]$$

31. $2R_2 \rightarrow R_2$ means multiply Row 2 by 2.
$$\left[\begin{array}{rr|r} 1 & -3 & 2 \\ 8 & -12 & -16 \end{array}\right]$$

33. $(-4)R_1 + R_2 \rightarrow R_2$ means replace Row 2 by itself plus -4 times Row 1.
$$\left[\begin{array}{rr|r} 1 & -3 & 2 \\ 4 & -6 & -8 \end{array}\right] \rightarrow \left[\begin{array}{rr|r} 1 & -3 & 2 \\ 0 & 6 & -16 \end{array}\right]$$
$\rightarrow -4 \quad 12 \quad -8$

35. $(-2)R_1 + R_2 \rightarrow R_2$ means replace Row 2 by itself plus -2 times Row 1.
$$\left[\begin{array}{rr|r} 1 & -3 & 2 \\ 4 & -6 & -8 \end{array}\right] \rightarrow \left[\begin{array}{rr|r} 1 & -3 & 2 \\ 2 & 0 & -12 \end{array}\right]$$
$\rightarrow -2 \quad 6 \quad -4$

37. $(-1)R_1 + R_2 \rightarrow R_2$ means replace Row 2 by itself plus -1 times Row 1.
$$\left[\begin{array}{rr|r} 1 & -3 & 2 \\ 4 & -6 & -8 \end{array}\right] \rightarrow \left[\begin{array}{rr|r} 1 & -3 & 2 \\ 3 & -3 & -10 \end{array}\right]$$
$\rightarrow -1 \quad 3 \quad -2$

39. $$\left[\begin{array}{rr|r} 1 & 2 & -1 \\ 0 & 1 & 3 \end{array}\right] \quad (-2)R_2 + R_1 \rightarrow R_1$$
Need a 0 here
$$\sim \left[\begin{array}{rr|r} 1 & 0 & -7 \\ 0 & 1 & 3 \end{array}\right]$$

41. $$\left[\begin{array}{rrr|r} 1 & 3 & -3 & 1 \\ 0 & 1 & 2 & 0 \\ 0 & 0 & 3 & -6 \end{array}\right] \tfrac{1}{3}R_3 \rightarrow R_3$$
↑
Need a 1 here
Need 0's here
↓
$$\sim \left[\begin{array}{rrr|r} 1 & 0 & -3 & 1 \\ 0 & 1 & 2 & 0 \\ 0 & 0 & 1 & -2 \end{array}\right] \begin{array}{l} 3R_3 + R_1 \rightarrow R_1 \\ (-2)R_3 + R_2 \rightarrow R_2 \end{array}$$
$$\sim \left[\begin{array}{rrr|r} 1 & 0 & 0 & -5 \\ 0 & 1 & 0 & 4 \\ 0 & 0 & 1 & -2 \end{array}\right]$$

43. $\left[\begin{array}{rrr|r} 1 & 2 & -2 & -1 \\ 0 & 3 & -6 & 1 \\ 0 & -1 & 2 & -\frac{1}{3} \end{array}\right] \frac{1}{3}R_2 \to R_2$

Need a 1 here

$\sim \left[\begin{array}{rrr|r} 1 & 2 & -2 & -1 \\ 0 & 1 & -2 & \frac{1}{3} \\ 0 & -1 & 2 & -\frac{1}{3} \end{array}\right] \begin{array}{l} (-2)R_2 + R_1 \to R_1 \\ R_3 + R_2 \to R_3 \end{array}$

Need 0's here

$\sim \left[\begin{array}{rrr|r} 1 & 0 & 2 & -\frac{5}{3} \\ 0 & 1 & -2 & \frac{1}{3} \\ 0 & 0 & 0 & 0 \end{array}\right]$

45. $\left[\begin{array}{rr|r} 1 & -4 & -2 \\ -2 & 1 & -3 \end{array}\right] 2R_1 + R_2 \to R_2$

Need a 0 here

2 −8 −4

$\sim \left[\begin{array}{rr|r} 1 & -4 & -2 \\ 0 & -7 & -7 \end{array}\right] -\frac{1}{7}R_2 \to R_2$

Need a 1 here

Need a 0 here

$\sim \left[\begin{array}{rr|r} 1 & -4 & -2 \\ 0 & 1 & 1 \end{array}\right] 4R_2 + R_1 \to R_1$

0 4 4

$\sim \left[\begin{array}{rr|r} 1 & 0 & 2 \\ 0 & 1 & 1 \end{array}\right]$

Therefore $x_1 = 2$ and $x_2 = 1$

47. $\left[\begin{array}{rr|r} 3 & -1 & 2 \\ 1 & 2 & 10 \end{array}\right] R_1 \leftrightarrow R_2$

Need a 1 here

$\sim \left[\begin{array}{rr|r} 1 & 2 & 10 \\ 3 & -1 & 2 \end{array}\right] (-3)R_1 + R_2 \to R_2$

Need a 0 here

−3 −6 −30

$\sim \left[\begin{array}{rr|r} 1 & 2 & 10 \\ 0 & -7 & -28 \end{array}\right] -\frac{1}{7}R_2 \to R_2$

Need a 1 here

Need a 0 here

$\sim \left[\begin{array}{rr|r} 1 & 2 & 10 \\ 0 & 1 & 4 \end{array}\right] (-2)R_2 + R_1 \to R_1$

0 −2 −8

$\sim \left[\begin{array}{rr|r} 1 & 0 & 2 \\ 0 & 1 & 4 \end{array}\right]$

Therefore $x_1 = 2$ and $x_2 = 4$.

49. $\left[\begin{array}{rr|r} 1 & 2 & 4 \\ 2 & 4 & -8 \end{array}\right] (-2)R_1 + R_2 \to R_2$

Need a 0 here

−2 −4 −8

$\sim \left[\begin{array}{rr|r} 1 & 2 & 4 \\ 0 & 0 & -16 \end{array}\right]$

This matrix corresponds to the system

$x_1 + 2x_2 = 4$

$0x_1 + 0x_2 = -16$

This system has no solution.

51. $\left[\begin{array}{rr|r} 2 & 1 & 6 \\ 1 & -1 & -3 \end{array}\right] R_1 \leftrightarrow R_2$

Need a 1 here

$\sim \left[\begin{array}{rr|r} 1 & -1 & -3 \\ 2 & 1 & 6 \end{array}\right] (-2)R_2 + R_1 \to R_1$

Need a 0 here

−2 2 6

$\sim \left[\begin{array}{rr|r} 1 & -1 & -3 \\ 0 & 3 & 12 \end{array}\right] \frac{1}{3}R_2 \to R_2$

Need a 1 here

Need a 0 here

$\sim \left[\begin{array}{rr|r} 1 & -1 & -3 \\ 0 & 1 & 4 \end{array}\right] R_2 + R_1 \to R_1$

$\sim \left[\begin{array}{rr|r} 1 & 0 & -3 \\ 0 & 1 & 4 \end{array}\right]$

Therefore $x_1 = 1$ and $x_2 = 4$.

53. $\left[\begin{array}{cc|c} 3 & -6 & -9 \\ -2 & 4 & 6 \end{array}\right]$ $\frac{1}{3}R_1 \to R_1$

Need a 1 here

$\sim \left[\begin{array}{cc|c} 1 & -2 & -3 \\ -2 & 4 & 6 \end{array}\right]$ $(-2)R_1 + R_2 \to R_2$

Need a 0 here

2   -4   -6

$\sim \left[\begin{array}{cc|c} 1 & -2 & -3 \\ 0 & 0 & 0 \end{array}\right]$

This matrix corresponds to the system

$x_1 - 2x_2 = -3$

$0x_1 + 0x_2 = 0$

Thus $x_1 = 2x_2 - 3$.

Hence there are infinitely many solutions: for any real number s, $x_2 = s$, $x_1 = 2s - 3$ is a solution.

55. $\left[\begin{array}{cc|c} 4 & -2 & 2 \\ -6 & 3 & -3 \end{array}\right]$ $\frac{1}{4}R_1 \to R_1$

Need a 1 here

$\sim \left[\begin{array}{cc|c} 1 & -\frac{1}{2} & \frac{1}{2} \\ -6 & 3 & -3 \end{array}\right]$ $6R_1 + R_2 \to R_2$

Need a 0 here

6   -3   3

$\sim \left[\begin{array}{cc|c} 1 & -\frac{1}{2} & \frac{1}{2} \\ 0 & 0 & 0 \end{array}\right]$

This matrix corresponds to the system

$x_1 - \frac{1}{2}x_2 = \frac{1}{2}$

$0x_1 + 0x_2 = 0$

Thus $x_1 = \frac{1}{2}x_2 + \frac{1}{2}$.

Hence there are infinitely many solutions: for any real number s, $x_2 = s$, $x_1 = \frac{1}{2}s + \frac{1}{2}$ is a solution.

57. $\left[\begin{array}{ccc|c} 2 & 4 & -10 & -2 \\ 3 & 9 & -21 & 0 \\ 1 & 5 & -12 & 1 \end{array}\right]$ $R_1 \leftrightarrow R_3$

Need a 1 here

$\sim \left[\begin{array}{ccc|c} 1 & 5 & -12 & 1 \\ 3 & 9 & -21 & 0 \\ -2 & 4 & -10 & -2 \end{array}\right]$ $\begin{array}{l} (-3)R_1 + R_2 \to R_2 \\ (-2)R_1 + R_3 \to R_3 \end{array}$

Need 0's here

$\sim \left[\begin{array}{ccc|c} 1 & 5 & -12 & 1 \\ 0 & -6 & 15 & -3 \\ 0 & -6 & 14 & -4 \end{array}\right]$ $-\frac{1}{6}R_2 \to R_2$

Need a 1 here

$\sim \left[\begin{array}{ccc|c} 1 & 5 & -12 & 1 \\ 0 & 1 & -\frac{5}{2} & \frac{1}{2} \\ 0 & -6 & 14 & -4 \end{array}\right]$ $\begin{array}{l} (-5)R_2 + R_1 \to R_1 \\ 6R_2 + R_3 \to R_3 \end{array}$

Need 0's here

$\sim \left[\begin{array}{ccc|c} 1 & 0 & \frac{1}{2} & -\frac{3}{2} \\ 0 & 1 & -\frac{5}{2} & \frac{1}{2} \\ 0 & 0 & -1 & -1 \end{array}\right]$ $-R_3 \to R_3$

Need a 1 here

$\sim \left[\begin{array}{ccc|c} 1 & 0 & \frac{1}{2} & -\frac{3}{2} \\ 0 & 1 & \frac{5}{2} & \frac{1}{2} \\ 0 & 0 & 1 & 1 \end{array}\right]$ $\begin{array}{l} (-\frac{1}{2})R_3 + R_1 \to R_1 \\ \frac{5}{2}R_3 + R_2 \to R_2 \end{array}$

Need 0's here

$\sim \left[\begin{array}{ccc|c} 1 & 0 & 0 & -2 \\ 0 & 1 & 0 & 3 \\ 0 & 0 & 1 & 1 \end{array}\right]$

Therefore $x_1 = -2$, $x_2 = 3$, and $x_3 = 1$.

59. $\left[\begin{array}{ccc|c} 3 & 8 & -1 & -18 \\ 2 & 1 & 5 & 8 \\ 2 & 4 & 2 & -4 \end{array}\right]$ $\begin{array}{l} \frac{1}{2}R_3 \to R_3 \\ R_3 \leftrightarrow R_1 \end{array}$

Need a 1 here

$\sim \left[\begin{array}{ccc|c} 1 & 2 & 1 & -2 \\ 2 & 1 & 5 & 8 \\ 3 & 8 & -1 & -18 \end{array}\right]$ $\begin{array}{l} (-2)R_1 + R_2 \to R_2 \\ (-3)R_1 + R_3 \to R_3 \end{array}$

Need 0's here

$\sim \left[\begin{array}{ccc|c} 1 & 2 & 1 & -2 \\ 0 & -3 & 3 & 12 \\ 0 & 2 & -4 & -12 \end{array}\right]$ $-\frac{1}{3}R_2 \to R_2$

Need a 1 here

$\sim \left[\begin{array}{ccc|c} 1 & 2 & 1 & -2 \\ 0 & 1 & -1 & -4 \\ 0 & 2 & -4 & -12 \end{array}\right]$ $\begin{array}{l} (-2)R_2 + R_1 \to R_1 \\ (-2)R_2 + R_3 \to R_3 \end{array}$

Need 0's here

$\sim \left[\begin{array}{ccc|c} 1 & 0 & 3 & 6 \\ 0 & 1 & -1 & -4 \\ 0 & 0 & -2 & -4 \end{array}\right]$ $-\frac{1}{2}R_3 \to R_3$

Need a 1 here

$\sim \left[\begin{array}{ccc|c} 1 & 0 & 3 & 6 \\ 0 & 1 & -1 & -4 \\ 0 & 0 & 1 & 2 \end{array}\right]$ $\begin{array}{l} (-3)R_3 + R_1 \to R_1 \\ R_3 + R_2 \to R_2 \end{array}$

Need 0's here

$\left[\begin{array}{ccc|c} 1 & 0 & 0 & 0 \\ 0 & 1 & 0 & -2 \\ 0 & 0 & 1 & 2 \end{array}\right]$ Therefore $x_1 = 0$, $x_2 = -2$, and $x_3 = 2$.

61. $\left[\begin{array}{rrr|r} 2 & -1 & -3 & 8 \\ 1 & -2 & 0 & 7 \end{array}\right]$ $R_1 \leftrightarrow R_2$

$\sim \left[\begin{array}{rrr|r} 1 & -2 & 0 & 7 \\ 2 & -1 & -3 & 8 \end{array}\right]$ $(-2)R_1 + R_2 \to R_2$

$\sim \left[\begin{array}{rrr|r} 1 & -2 & 0 & 7 \\ 0 & 3 & -3 & -6 \end{array}\right]$ $\frac{1}{3}R_2 \to R_2$

$\sim \left[\begin{array}{rrr|r} 1 & -2 & 0 & 7 \\ 0 & 1 & -1 & -2 \end{array}\right]$ $2R_2 + R_1 \to R_1$

$\sim \left[\begin{array}{rrr|r} 1 & 0 & -2 & 3 \\ 0 & 1 & -1 & -2 \end{array}\right]$

Let $x_3 = t$. Then

$x_2 - x_3 = -2$

$x_2 = x_3 - 2 = t - 2$

$x_1 - 2x_3 = 3$

$x_1 = 2x_3 + 3 = 2t + 3$

Solution: $x_1 = 2t + 3$, $x_2 = t - 2$, $x_3 = t$, t any real number.

63. $\left[\begin{array}{rr|r} 2 & -1 & 0 \\ 3 & 2 & 7 \\ 1 & -1 & -1 \end{array}\right]$ $R_1 \leftrightarrow R_3$

$\sim \left[\begin{array}{rr|r} 1 & -1 & -1 \\ 3 & 2 & 7 \\ 2 & -1 & 0 \end{array}\right]$ $\begin{array}{l} \\ (-3)R_1 + R_2 \to R_2 \\ (-2)R_1 + R_3 \to R_3 \end{array}$

$\sim \left[\begin{array}{rr|r} 1 & -1 & -1 \\ 0 & 5 & 10 \\ 0 & 1 & 2 \end{array}\right]$ $R_2 \leftrightarrow R_3$

$\sim \left[\begin{array}{rr|r} 1 & -1 & -1 \\ 0 & 1 & 2 \\ 0 & 5 & 10 \end{array}\right]$ $\begin{array}{l} R_2 + R_1 \to R_1 \\ \\ (-5)R_2 + R_3 \to R_3 \end{array}$

$\sim \left[\begin{array}{rr|r} 1 & 0 & 1 \\ 0 & 1 & 2 \\ 0 & 0 & 0 \end{array}\right]$

Therefore $x_1 = 1$ and $x_2 = 2$.

65. $\left[\begin{array}{rrr|r} 3 & -4 & -1 & 1 \\ 2 & -3 & 1 & 1 \\ 1 & -2 & 3 & 2 \end{array}\right]$ $R_1 \leftrightarrow R_3$

$\sim \left[\begin{array}{rrr|r} 1 & -2 & 3 & 2 \\ 2 & -3 & 1 & 1 \\ 3 & -4 & -1 & 1 \end{array}\right]$ $\begin{array}{l} \\ (-2)R_1 + R_2 \to R_2 \\ (-3)R_1 + R_3 \to R_3 \end{array}$

$\sim \left[\begin{array}{rrr|r} 1 & -2 & 3 & 2 \\ 0 & 1 & -5 & -3 \\ 0 & 2 & -10 & -5 \end{array}\right]$ $\begin{array}{l} \\ \\ (-2)R_2 + R_3 \to R_3 \end{array}$

$\sim \left[\begin{array}{rrr|r} 1 & -2 & 3 & 2 \\ 0 & 1 & -5 & -3 \\ 0 & 0 & 0 & 1 \end{array}\right]$

Since the last row corresponds to the equation $0x_1 + 0x_2 + 0x_3 = 1$, there is no solution.

67. $\left[\begin{array}{rrr|r} -2 & 1 & 3 & -7 \\ 1 & -4 & 2 & 0 \\ 1 & -3 & 1 & 1 \end{array}\right]$ $R_1 \leftrightarrow R_2$

$\sim \left[\begin{array}{rrr|r} 1 & -4 & 2 & 0 \\ -2 & 1 & 3 & -7 \\ 1 & -3 & 1 & 1 \end{array}\right]$ $\begin{array}{l} \\ 2R_1 + R_2 \to R_2 \\ (-1)R_1 + R_3 \to R_3 \end{array}$

$\sim \left[\begin{array}{rrr|r} 1 & -4 & 2 & 0 \\ 0 & -7 & 7 & -7 \\ 0 & 1 & -1 & 1 \end{array}\right]$ $R_2 \leftrightarrow R_3$

$\sim \left[\begin{array}{rrr|r} 1 & -4 & 2 & 0 \\ 0 & 1 & -1 & 1 \\ 0 & -7 & 7 & -7 \end{array}\right]$ $\begin{array}{l} 4R_2 + R_1 \to R_1 \\ \\ 7R_2 + R_3 \to R_3 \end{array}$

$\sim \left[\begin{array}{rrr|r} 1 & 0 & -2 & 4 \\ 0 & 1 & -1 & 1 \\ 0 & 0 & 0 & 0 \end{array}\right]$

Let $x_3 = t$. Then

$x_2 - x_3 = 1$

$x_2 = x_3 + 1 = t + 1$

$x_1 - 2x_3 = 4$

$x_1 = 2x_3 + 4 = 2t + 4$

Solution: $x_1 = 2t + 4$, $x_2 = t + 1$, $x_3 = t$, t any real number.

69. $\left[\begin{array}{rrr|r} 2 & -2 & -4 & -2 \\ -3 & 3 & 6 & 3 \end{array}\right]$ $\begin{array}{l} \frac{1}{2}R_1 \to R_1 \\ \frac{1}{3}R_2 \to R_2 \end{array}$

$\sim \left[\begin{array}{rrr|r} 1 & -1 & -2 & -1 \\ -1 & 1 & 2 & 1 \end{array}\right]$ $R_1 + R_2 \to R_2$

$\sim \left[\begin{array}{rrr|r} 1 & -1 & -2 & -1 \\ 0 & 0 & 0 & 0 \end{array}\right]$

Let $x_3 = t$, $x_2 = s$. Then

$x_1 - x_2 - 2x_3 = -1$

$x_1 = x_2 + 2x_3 - 1$

$= s + 2t - 1$

Solution: $x_1 = s + 2t - 1$, $x_2 = s$, $s_3 = t$, s and t any real numbers.

71. $\begin{bmatrix} 4 & -1 & 2 & | & 3 \\ -4 & 1 & -3 & | & -10 \\ 8 & -2 & 9 & | & -1 \end{bmatrix} \begin{matrix} \\ R_1 + R_2 \to R_2 \\ (-2)R_1 + R_3 \to R_3 \end{matrix}$

$\sim \begin{bmatrix} 4 & -1 & 2 & | & 3 \\ 0 & 0 & -1 & | & -7 \\ 0 & 0 & 5 & | & -7 \end{bmatrix} \begin{matrix} \\ \\ 5R_2 + R_3 \to R_3 \end{matrix}$

$\sim \begin{bmatrix} 4 & -1 & 2 & | & 3 \\ 0 & 0 & -1 & | & -7 \\ 0 & 0 & 0 & | & -42 \end{bmatrix}$

Since the last row corresponds to the equation $0x_1 + 0x_2 + 0x_3 = -42$, there is no solution.

73. $\begin{bmatrix} 2 & -5 & -3 & | & 7 \\ -4 & 10 & 2 & | & 6 \\ 6 & -15 & -1 & | & -19 \end{bmatrix} \begin{matrix} \\ 2R_1 + R_2 \to R_2 \\ (-3)R_1 + R_3 \to R_3 \end{matrix}$

$\sim \begin{bmatrix} 2 & -5 & -3 & | & 7 \\ 0 & 0 & -4 & | & 20 \\ 0 & 0 & 8 & | & -40 \end{bmatrix} \begin{matrix} \frac{1}{2} R_1 \to R_1 \\ -\frac{1}{4} R_2 \to R_2 \\ \end{matrix}$

$\sim \begin{bmatrix} 1 & -2.5 & -1.5 & | & 3.5 \\ 0 & 0 & 1 & | & -5 \\ 0 & 0 & 8 & | & -40 \end{bmatrix} \begin{matrix} 1.5R_2 + R_1 \to R_1 \\ \\ (-8)R_2 + R_3 \to R_3 \end{matrix}$

$\sim \begin{bmatrix} 1 & -2.5 & 0 & | & -4 \\ 0 & 0 & 1 & | & -5 \\ 0 & 0 & 0 & | & 0 \end{bmatrix}$

Let $x_2 = t$. Then $x_3 = -5$ and

$$x_1 - 2.5x_2 = -4$$
$$x_1 = 2.5x_2 - 4$$
$$x_1 = 2.5t - 4$$

Solution: $x_1 = 2.5t - 4$, $x_2 = t$, $x_3 = -5$, t any real number.

75. $\begin{bmatrix} 5 & -3 & 2 & | & 13 \\ 2 & -1 & -3 & | & 1 \\ 4 & -2 & 4 & | & 12 \end{bmatrix} \begin{matrix} \\ \\ \frac{1}{4} R_3 \to R_3 \end{matrix}$

$\sim \begin{bmatrix} 5 & -3 & 2 & | & 13 \\ 2 & -1 & -3 & | & 1 \\ 1 & -\frac{1}{2} & 1 & | & 3 \end{bmatrix} \; R_1 \leftrightarrow R_3$

$\sim \begin{bmatrix} 1 & -\frac{1}{2} & 1 & | & 3 \\ 2 & -1 & -3 & | & 1 \\ 5 & -3 & 2 & | & 13 \end{bmatrix} \begin{matrix} \\ (-2)R_1 + R_2 \to R_2 \\ (-5)R_1 + R_3 \to R_3 \end{matrix}$

$\sim \begin{bmatrix} 1 & -\frac{1}{2} & 1 & | & 3 \\ 0 & 0 & -5 & | & -5 \\ 0 & -\frac{1}{2} & -3 & | & -2 \end{bmatrix} \begin{matrix} (-1)R_3 + R_1 \to R_1 \\ -\frac{1}{5} R_2 \to R_2 \\ \end{matrix}$

$\sim \begin{bmatrix} 1 & 0 & 4 & | & 5 \\ 0 & 0 & 1 & | & 1 \\ 0 & -\frac{1}{2} & -3 & | & -2 \end{bmatrix} \begin{matrix} (-4)R_2 + R_1 \to R_1 \\ 3R_2 + R_3 \to R_3 \\ \end{matrix}$

$\sim \begin{bmatrix} 1 & 0 & 0 & | & 1 \\ 0 & 0 & 1 & | & 1 \\ 0 & -\frac{1}{2} & 0 & | & 1 \end{bmatrix} \begin{matrix} \\ \\ (-2)R_3 \to R_3 \end{matrix}$

$\sim \begin{bmatrix} 1 & 0 & 0 & | & 1 \\ 0 & 0 & 1 & | & 1 \\ 0 & 1 & 0 & | & -2 \end{bmatrix} \; R_2 \leftrightarrow R_3$

$\sim \begin{bmatrix} 1 & 0 & 0 & | & 1 \\ 0 & 1 & 0 & | & -2 \\ 0 & 0 & 1 & | & 1 \end{bmatrix}$

Solution: $x_1 = 1$, $x_2 = -2$, $x_3 = 1$.

77. $\left[\begin{array}{rrrr|r} 1 & 2 & -4 & -1 & 7 \\ 2 & 5 & -9 & -4 & 16 \\ 1 & 5 & -7 & -7 & 13 \end{array}\right] \begin{array}{l} \\ (-2)R_1 + R_2 \to R_2 \\ (-1)R_1 + R_3 \to R_3 \end{array}$

$\sim \left[\begin{array}{rrrr|r} 1 & 2 & -4 & -1 & 7 \\ 0 & 1 & -1 & -2 & 2 \\ 0 & 3 & -3 & -6 & 6 \end{array}\right] \begin{array}{l} (-2)R_2 + R_1 \to R_1 \\ \\ (-3)R_2 + R_3 \to R_3 \end{array}$

$\sim \left[\begin{array}{rrrr|r} 1 & 0 & -2 & 3 & 3 \\ 0 & 1 & -1 & -2 & 2 \\ 0 & 0 & 0 & 0 & 0 \end{array}\right]$

Let $x_4 = t$, $x_3 = s$. Then

$x_2 - x_3 - 2x_4 = 2$

$x_2 = s + 2t + 2 + 2$

$x_1 - 2x_3 + 3x_4 = 3$

$x_1 = 2x_3 - 3x_4 + 3$

Solution: $x_1 = 2s - 3t + 3$, $x_2 = s + 2t + 2$, $x_3 = s$, $x_4 = t$, s and t any real numbers.

79. $\left[\begin{array}{rrrr|r} 1 & -1 & 3 & -2 & 1 \\ -2 & 4 & -3 & 1 & 0.5 \\ 3 & -1 & 10 & -4 & 2.9 \\ 4 & -3 & 8 & -2 & 0.6 \end{array}\right] \begin{array}{l} \\ 2R_1 + R_2 \to R_2 \\ (-3)R_1 + R_3 \to R_3 \\ (-4)R_1 + R_4 \to R_4 \end{array}$

$\sim \left[\begin{array}{rrrr|r} 1 & -1 & 3 & -2 & 1 \\ 0 & 2 & 3 & -3 & 2.5 \\ 0 & 2 & 1 & 2 & -0.1 \\ 0 & 1 & -4 & 6 & -3.4 \end{array}\right] \begin{array}{l} \\ \\ \\ R_4 \leftrightarrow R_2 \end{array}$

$\sim \left[\begin{array}{rrrr|r} 1 & -1 & 3 & -2 & 1 \\ 0 & 1 & -4 & 6 & -3.4 \\ 0 & 2 & 1 & 2 & -0.1 \\ 0 & 2 & 3 & -3 & 2.5 \end{array}\right] \begin{array}{l} R_2 + R_1 \to R_1 \\ \\ (-2)R_2 + R_3 \to R_3 \\ (-2)R_2 + R_4 \to R_4 \end{array}$

$\sim \left[\begin{array}{rrrr|r} 1 & 0 & -1 & 4 & -2.4 \\ 0 & 1 & -4 & 6 & -3.4 \\ 0 & 0 & 9 & -10 & 6.7 \\ 0 & 0 & 11 & -15 & 9.3 \end{array}\right] \begin{array}{l} \\ \\ (-1)R_4 + R_3 \to R_3 \\ \\ \end{array}$

$\sim \left[\begin{array}{rrrr|r} 1 & 0 & -1 & 4 & -2.4 \\ 0 & 1 & -4 & 6 & -3.4 \\ 0 & 0 & -2 & 5 & -2.6 \\ 0 & 0 & 11 & -15 & 9.3 \end{array}\right] -\frac{1}{2}R_3 \leftrightarrow R_3$

$\sim \left[\begin{array}{rrrr|r} 1 & 0 & -1 & 4 & -2.4 \\ 0 & 1 & -4 & 6 & -3.4 \\ 0 & 0 & 1 & -2.5 & 1.3 \\ 0 & 0 & 11 & -15 & 9.3 \end{array}\right] \begin{array}{l} R_3 + R_1 \to R_1 \\ 4R_3 + R_2 \to R_2 \\ \\ (-11)R_3 + R_4 \to R_4 \end{array}$

$\sim \left[\begin{array}{rrrr|r} 1 & 0 & 0 & 1.5 & -1.1 \\ 0 & 1 & 0 & -4 & 1.8 \\ 0 & 0 & 1 & -2.5 & 1.3 \\ 0 & 0 & 0 & 12.5 & -5 \end{array}\right] \begin{array}{l} \\ \\ \\ \frac{1}{12.5}R_4 \to R_4 \end{array}$

$\sim \left[\begin{array}{rrrr|r} 1 & 0 & 0 & 1.5 & -1.1 \\ 0 & 1 & 0 & -4 & 1.8 \\ 0 & 0 & 1 & -2.5 & 1.3 \\ 0 & 0 & 0 & 1 & -0.4 \end{array}\right] \begin{array}{l} (-1.5)R_4 + R_1 \to R_1 \\ 4R_4 + R_2 \to R_2 \\ 2.5R_4 + R_3 \to R_3 \\ \\ \end{array}$

$\sim \left[\begin{array}{rrrr|r} 1 & 0 & 0 & 0 & -0.5 \\ 0 & 1 & 0 & 0 & 0.2 \\ 0 & 0 & 1 & 0 & 0.3 \\ 0 & 0 & 0 & 1 & -0.4 \end{array}\right]$

Solution: $x_1 = -0.5$, $x_2 = 0.2$, $x_3 = 0.3$, $x_4 = -0.4$

81. $$\left[\begin{array}{rrrrr|r} 1 & -2 & 1 & 1 & 2 & 2 \\ -2 & 4 & 2 & 2 & -2 & 0 \\ 3 & -6 & 1 & 1 & 5 & 4 \\ -1 & 2 & 3 & 1 & 1 & 3 \end{array}\right] \begin{array}{l} \\ 2R_1 + R_2 \to R_2 \\ (-3)R_1 + R_3 \to R_3 \\ R_1 + R_4 \to R_4 \end{array}$$

$$\sim \left[\begin{array}{rrrrr|r} 1 & -2 & 1 & 1 & 2 & 2 \\ -2 & 4 & 2 & 2 & -2 & 0 \\ 3 & -6 & 1 & 1 & 5 & 4 \\ -1 & 2 & 3 & 1 & 1 & 3 \end{array}\right] \quad \tfrac{1}{4}R_2 \to R_2$$

$$\sim \left[\begin{array}{rrrrr|r} 1 & -2 & 1 & 1 & 2 & 2 \\ 0 & 0 & 1 & 1 & 0.5 & 1 \\ 0 & 0 & -2 & -2 & -1 & -2 \\ 0 & 0 & 4 & 2 & 3 & 5 \end{array}\right] \begin{array}{l} (-1)R_2 + R_1 \to R_1 \\ \\ 2R_2 + R_3 \to R_3 \\ (-4)R_2 + R_4 \to R_4 \end{array}$$

$$\sim \left[\begin{array}{rrrrr|r} 1 & -2 & 0 & 0 & 1.5 & 1 \\ 0 & 0 & 1 & 1 & 0.5 & 1 \\ 0 & 0 & 0 & 0 & 0 & 0 \\ 0 & 0 & 0 & -2 & 1 & 1 \end{array}\right] \quad R_3 \leftrightarrow R_4$$

$$\sim \left[\begin{array}{rrrrr|r} 1 & -2 & 0 & 0 & 1.5 & 1 \\ 0 & 0 & 1 & 1 & 0.5 & 1 \\ 0 & 0 & 0 & -2 & 1 & 1 \\ 0 & 0 & 0 & 0 & 0 & 0 \end{array}\right] \quad \left(-\tfrac{1}{2}\right)R_3 \to R_3$$

$$\sim \left[\begin{array}{rrrrr|r} 1 & -2 & 0 & 0 & 1.5 & 1 \\ 0 & 0 & 1 & 1 & 0.5 & 1 \\ 0 & 0 & 0 & 1 & -0.5 & -0.5 \\ 0 & 0 & 0 & 0 & 0 & 0 \end{array}\right] \quad (-1)R_3 + R_2 \to R_2$$

$$\sim \left[\begin{array}{rrrrr|r} 1 & -2 & 0 & 0 & 1.5 & 1 \\ 0 & 0 & 1 & 0 & 1 & 1.5 \\ 0 & 0 & 0 & 1 & -0.5 & -0.5 \\ 0 & 0 & 0 & 0 & 0 & 0 \end{array}\right]$$

Let $x_5 = t$. Then

$$x_4 - 0.5x_5 = -0.5$$
$$x_4 = 0.5x_5 - 0.5 = 0.5t - 0.5$$
$$x_3 + x_5 = 1.5$$
$$x_3 = -x_5 + 1.5 = -t + 1.5$$

Let $x_2 = s$. Then

$$x_1 - 2x_2 + 1.5x_5 = 1$$
$$x_1 = 2x_2 - 1.5x_5 + 1 = 2s - 1.5t + 1$$

Solution: $x_1 = 2s - 1.5t + 1$, $x_2 = s$, $x_3 = -t + 1.15$, $x_4 = 0.5t - 0.5$, $x_5 = t$, s and t any real numbers.

83. (A) The reduced form matrix will have the form

$$\left[\begin{array}{rrr|r} 1 & a & b & c \\ 0 & 0 & 0 & 0 \\ 0 & 0 & 0 & 0 \end{array}\right]$$

Thus, the system has been shown equivalent to

$$x_1 + ax_2 + bx_3 = c$$
$$0 = 0$$
$$0 = 0$$

The system is dependent, and x_2 and x_3 may assume any real values. Thus, there are two parameters in the solution.

(B) The reduced form matrix will have the form

$$\left[\begin{array}{ccc|c} 1 & 0 & a & b \\ 0 & 1 & c & d \\ 0 & 0 & 0 & 0 \end{array}\right]$$

Thus, the system has been shown equivalent to

$$\begin{aligned} x_1 + ax_3 &= b \\ x_2 + cx_3 &= d \\ 0 &= 0 \end{aligned}$$

The system is dependent, with a solution for any real value of x_3. Thus, there is one parameter in the solution.

(C) The reduced form matrix will have the form

$$\left[\begin{array}{ccc|c} 1 & 0 & 0 & a \\ 0 & 1 & 0 & b \\ 0 & 0 & 1 & c \end{array}\right]$$

Thus, there is only one solution, $x_1 = a$, $x_2 = b$, $x_3 = c$, and the system is independent.

(D) This is impossible; there are only 3 equations.

85. Let x = amount of 20% solution
y = amount of 80% solution

Summarize the given information in a table:

	20% solution	80% solution	62% solution
amount of solution	x	y	100
amount of acid	$0.2x$	$0.8y$	$0.62(100) = 62$

Form equations from the information:

$$\begin{pmatrix}\text{Amount of}\\ \text{20\% solution}\end{pmatrix} + \begin{pmatrix}\text{Amount of}\\ \text{80\% solution}\end{pmatrix} = \begin{pmatrix}\text{Amount of}\\ \text{62\% solution}\end{pmatrix}$$

$$x + y = 100$$

$$\begin{pmatrix}\text{Amount of acid}\\ \text{in 20\% solution}\end{pmatrix} + \begin{pmatrix}\text{Amount of acid}\\ \text{in 80\% solution}\end{pmatrix} = \begin{pmatrix}\text{Amount of acid}\\ \text{in 62\% solution}\end{pmatrix}$$

$$0.2x + 0.8y = 62$$

Solve this system using elimination by addition.

$$\begin{aligned} -0.2x - 0.2y &= -20 \\ 0.2x + 0.8y &= 62 \\ \hline 0.6y &= 42 \\ y &= 70 \\ x + 70 &= 100 \\ x &= 30 \end{aligned}$$

30 liters of 20% solution and 70 liters of 80% solution.

87. Let x_1 = number of 500-cc containers of 10% solution
x_2 = number of 500-cc containers of 20% solution
x_3 = number of 1,000-cc containers of 50% solution

Then $500x_1 + 500x_2 + 1{,}000x_3 = 12{,}000$ (total number of cc)

$0.10(500x_1) + 0.20(500x_2) + 0.50(1{,}000x_3) = 0.30(12{,}000)$

(total amount of ingredient in solution.)

After simplification, we have

$$\begin{aligned} x_1 + x_2 + 2x_3 &= 24 \\ x_1 + 2x_2 + 10x_3 &= 72 \end{aligned}$$

We write the augmented matrix and solve by Gauss-Jordan elimination:

$$\left[\begin{array}{ccc|c} 1 & 1 & 2 & 24 \\ 1 & 2 & 10 & 72 \end{array}\right] (-1)R_1 + R_2 \rightarrow R_2$$

$$\sim \left[\begin{array}{ccc|c} 1 & 1 & 2 & 24 \\ 0 & 1 & 8 & 48 \end{array}\right] (-1)R_2 + R_1 \rightarrow R_1$$

$$\sim \left[\begin{array}{ccc|c} 1 & 0 & -6 & -24 \\ 0 & 1 & 8 & 48 \end{array}\right]$$

This augmented matrix is in reduced form. It corresponds to the system:

$x_1 - 6x_3 = -24$

$x_2 + 8x_3 = 48$

Let $x_3 = t$. Then

$x_2 = -8x_3 + 48$

$\quad = -8t + 48$

$x_1 = 6x_3 - 24$

$\quad = 6t - 24$

A solution is achieved, not for every real value of t, but for integer values of t that give rise to non-negative x_1, x_2, x_3.

$x_1 \geq 0$ means $6t - 24 \geq 0$ or $t \geq 4$

$x_2 \geq 0$ means $-8t + 48 \geq 0$ or $t \leq 6$

Thus we have the solution:

$x_1 = (6t - 24)$ 500-cc containers of 10% solution

$x_2 = (48 - 8t)$ 500-cc containers of 20% solution

$x_3 = t$ 1000-cc containers of 50% solution

where $t = 4$, 5, or 6.

89. If the curve passes through a point, the coordinates of the point satisfy the equation of the curve. Hence,

$3 = a + b(-2) + c(-2)^2$

$2 = a + b(-1) + c(-1)^2$

$6 = a + b(1) + c(1)^2$

After simplification, we have

$a - 2b + 4c = 3$

$a - b + c = 2$

$a + b + c = 6$

We write the augmented matrix and solve by Gauss-Jordan elimination.

$$\left[\begin{array}{ccc|c} 1 & -2 & 4 & 3 \\ 1 & -1 & 1 & 2 \\ 1 & 1 & 1 & 6 \end{array}\right] \begin{array}{l} \\ (-1)R_1 + R_2 \rightarrow R_2 \\ (-1)R_1 + R_3 \rightarrow R_3 \end{array}$$

$$\sim \left[\begin{array}{ccc|c} 1 & -2 & 4 & 3 \\ 0 & 1 & -3 & -1 \\ 0 & 3 & -3 & 3 \end{array}\right] \begin{array}{l} 2R_2 + R_1 \rightarrow R_1 \\ \\ (-3)R_2 + R_3 \rightarrow R_3 \end{array}$$

$$\sim \left[\begin{array}{ccc|c} 1 & 0 & -2 & 1 \\ 0 & 1 & -3 & -1 \\ 0 & 0 & 6 & 6 \end{array}\right] \begin{array}{l} \\ \\ \frac{1}{6} R_3 \rightarrow R_3 \end{array}$$

$$\sim \left[\begin{array}{ccc|c} 1 & 0 & -2 & 1 \\ 0 & 1 & -3 & -1 \\ 0 & 0 & 1 & 1 \end{array}\right] \begin{array}{l} 2R_3 + R_1 \rightarrow R_1 \\ 3R_3 + R_2 \rightarrow R_2 \\ \\ \end{array}$$

$$\sim \left[\begin{array}{ccc|c} 1 & 0 & 0 & 3 \\ 0 & 1 & 0 & 2 \\ 0 & 0 & 1 & 1 \end{array}\right] \quad \text{Thus } a = 3,\ b = 2,\ c = 1.$$

91. If the curve passes through a point, the coordinates of the point satisfy the equation of the curve. Hence,

$$6^2 + 2^2 + a(6) + b(2) + c = 0$$
$$4^2 + 6^2 + a(4) + b(6) + c = 0$$
$$(-3)^2 + (-1)^2 + a(-3) + b(-1) + c = 0$$

After simplification, we have

$$6a + 2b + c = -40$$
$$4a + 6b + c = -52$$
$$-3a - b + c = -10$$

We write the augmented matrix and solve by Gauss-Jordan elimination.

$$\left[\begin{array}{rrr|r} 6 & 2 & 1 & -40 \\ 4 & 6 & 1 & -52 \\ -3 & -1 & 1 & -10 \end{array}\right] \frac{1}{6}R_1 \to R_1$$

$$\sim \left[\begin{array}{rrr|r} 1 & \frac{1}{3} & \frac{1}{6} & -\frac{20}{3} \\ 4 & 6 & 1 & -52 \\ -3 & -1 & 1 & -10 \end{array}\right] \begin{array}{l} \\ (-4)R_1 + R_2 \to R_2 \\ 3R_1 + R_3 \to R_3 \end{array}$$

$$\sim \left[\begin{array}{rrr|r} 1 & \frac{1}{3} & \frac{1}{6} & -\frac{20}{3} \\ 0 & \frac{14}{3} & \frac{1}{3} & -\frac{76}{3} \\ -3 & 0 & \frac{3}{2} & -30 \end{array}\right] \begin{array}{l} \\ \\ \frac{2}{3}R_3 \to R_3 \end{array}$$

$$\sim \left[\begin{array}{rrr|r} 1 & \frac{1}{3} & \frac{1}{6} & -\frac{20}{3} \\ 0 & \frac{14}{3} & \frac{1}{3} & -\frac{76}{3} \\ 0 & 0 & 1 & -20 \end{array}\right] \begin{array}{l} (-\frac{1}{6})R_3 + R_1 \to R_1 \\ (-\frac{1}{3})R_3 + R_2 \to R_2 \\ \\ \end{array}$$

$$\sim \left[\begin{array}{rrr|r} 1 & \frac{1}{3} & 0 & -\frac{10}{3} \\ 0 & \frac{14}{3} & 0 & -\frac{56}{3} \\ 0 & 0 & 1 & -20 \end{array}\right] \begin{array}{l} \\ \frac{3}{14}R_2 \to R_2 \\ \\ \end{array}$$

$$\sim \left[\begin{array}{rrr|r} 1 & \frac{1}{3} & 0 & -\frac{10}{3} \\ 0 & 1 & 0 & -4 \\ 0 & 0 & 1 & -20 \end{array}\right] \begin{array}{l} (-\frac{1}{3})R_2 + R_1 \to R_1 \\ \\ \\ \end{array}$$

$$\left[\begin{array}{rrr|r} 1 & 0 & 0 & -2 \\ 0 & 1 & 0 & -4 \\ 0 & 0 & 1 & -20 \end{array}\right]$$

Thus $a = -2$, $b = -4$, and $c = -20$.

93. Let x_1 = number of one-person boats
x_2 = number of two-person boats
x_3 = number of four-person boats

We have

$0.5x_1 + 1.0x_2 + 1.5x_3 = 380$ cutting department
$0.6x_1 + 0.9x_2 + 1.2x_3 = 330$ assembly department
$0.2x_1 + 0.3x_2 + 0.5x_3 = 120$ packing department

Common Error:
The facts in this problem do not justify the equation
$0.5x_1 + 0.6x_2 + 0.2x_3 = 380$

Clearing of decimals for convenience:

$$x_1 + 2x_2 + 3x_3 = 760$$
$$6x_1 + 9x_2 + 12x_3 = 3300$$
$$2x_1 + 3x_2 + 5x_3 = 1200$$

We write the augmented matrix and solve by Gauss-Jordan elimination:

$$\left[\begin{array}{ccc|c} 1 & 2 & 3 & 760 \\ 6 & 9 & 12 & 3300 \\ 2 & 3 & 5 & 1200 \end{array}\right] \begin{array}{l} \\ (-6)R_1 + R_2 \to R_2 \\ (-2)R_1 + R_3 \to R_3 \end{array}$$

$$\sim \left[\begin{array}{ccc|c} 1 & 2 & 3 & 760 \\ 0 & -3 & -6 & -1260 \\ 0 & -1 & -1 & -320 \end{array}\right] -\tfrac{1}{3}R_2 \to R_2$$

$$\sim \left[\begin{array}{ccc|c} 1 & 2 & 3 & 760 \\ 0 & 1 & 2 & 420 \\ 0 & -1 & -1 & -320 \end{array}\right] \begin{array}{l} (-2)R_2 + R_1 \to R_1 \\ \\ R_2 + R_3 \to R_3 \end{array}$$

$$\sim \left[\begin{array}{ccc|c} 1 & 0 & -1 & -80 \\ 0 & 1 & 2 & 420 \\ 0 & 0 & 1 & 100 \end{array}\right] \begin{array}{l} R_3 + R_1 \to R_1 \\ (-2)R_3 + R_2 \to R_2 \\ \\ \end{array}$$

$$\left[\begin{array}{ccc|c} 1 & 0 & 0 & 20 \\ 0 & 1 & 0 & 220 \\ 0 & 0 & 1 & 100 \end{array}\right]$$

Therefore
x_1 = 20 one-person boats
x_2 = 220 two-person boats
x_3 = 100 four-person boats

95. This assumption discards the third equation. The system, cleared of decimals, reads

$$x_1 + 2x_2 + 3x_3 = 760$$
$$6x_1 + 9x_2 + 12x_3 = 3300$$

The augmented matrix becomes

$$\left[\begin{array}{ccc|c} 1 & 2 & 3 & 760 \\ 6 & 9 & 12 & 3300 \end{array}\right]$$

We solve by Gauss-Jordan elimination. We start by introducing a 0 into the lower left corner using $(-6)R_1 + R_2$ as in the previous problem:

$$\sim \left[\begin{array}{ccc|c} 1 & 2 & 3 & 760 \\ 0 & -3 & -6 & -1260 \end{array}\right] -\tfrac{1}{3}R_2 \to R_2$$

$$\sim \left[\begin{array}{ccc|c} 1 & 2 & 3 & 760 \\ 0 & 1 & 2 & 420 \end{array}\right] (-2)R_2 + R_1 \to R_1$$

$$\sim \left[\begin{array}{ccc|c} 1 & 0 & -1 & -80 \\ 0 & 1 & 2 & 420 \end{array}\right]$$

This augmented matrix is in reduced form. It corresponds to the system:

$$x_1 - x_3 = -80$$
$$x_2 + 2x_3 = 420$$

Let $x_3 = t$. Then

$x_2 = -2x_3 + 420$
$\quad = -2t + 420$
$x_1 = x_3 - 80$
$\quad = t - 80$

A solution is achieved, not for every real value of t, but for integer values of t that give rise to non-negative x_1, x_2, x_3.

$x_1 \ge 0$ means $t - 80 \ge 0$ or $t \ge 80$
$x_2 \ge 0$ means $-2t + 420 \ge 0$ or $210 \ge t$

Thus we have the solution
$x_1 = (t - 80)$ one-person boats
$x_2 = (-2t + 420)$ two-person boats
$x_3 = t$ four-person boats
$80 \le t \le 210$, t an integer

97. In this case we have $x_3 = 0$ from the beginning. The three equations of problem 59, cleared of decimals, read:

$$\begin{aligned} x_1 + 2x_2 &= 760 \\ 6x_1 + 9x_2 &= 3300 \\ 2x_1 + 3x_2 &= 1200 \end{aligned}$$

The augmented matrix becomes:

$$\left[\begin{array}{cc|c} 1 & 2 & 760 \\ 6 & 9 & 3300 \\ 2 & 3 & 1200 \end{array}\right]$$

Notice that the row operation

$$(-3)R_3 + R_2 \to R_2$$

transforms this into the equivalent augmented matrix:

$$\left[\begin{array}{cc|c} 1 & 2 & 760 \\ 0 & 0 & -300 \\ 2 & 3 & 1200 \end{array}\right]$$

Therefore, since the second row corresponds to the equation

$$0x_1 + 0x_2 = -300$$

there is no solution.

No production schedule will use all the work-hours in all departments.

99. Let x_1 = number of ounces of food *A*.
x_2 = number of ounces of food *B*.
x_3 = number of ounces of food *C*.

> **Common Error:**
> The facts in this problem do not justify the equation
> $30x_1 + 10x_2 + 10x_3 = 340$

Then

$$\begin{aligned} 30x_1 + 10x_2 + 20x_3 &= 340 \text{ (calcium)} \\ 10x_1 + 10x_2 + 20x_3 &= 180 \text{ (iron)} \\ 10x_1 + 30x_2 + 20x_3 &= 220 \text{ (vitamin A)} \end{aligned}$$

or

$$\begin{aligned} 3x_1 + x_2 + 2x_3 &= 34 \\ x_1 + x_2 + 2x_3 &= 18 \\ x_1 + 3x_2 + 2x_3 &= 22 \end{aligned}$$

is the system to be solved. We form the augmented matrix and solve by Gauss-Jordan elimination.

$$\left[\begin{array}{ccc|c} 3 & 1 & 2 & 34 \\ 1 & 1 & 2 & 18 \\ 1 & 3 & 2 & 22 \end{array}\right] R_1 \leftrightarrow R_2$$

$$\sim \left[\begin{array}{ccc|c} 1 & 1 & 2 & 18 \\ 3 & 1 & 2 & 34 \\ 1 & 3 & 2 & 22 \end{array}\right] \begin{array}{l} \\ (-3)R_1 + R_2 \to R_2 \\ (-1)R_1 + R_3 \to R_3 \end{array}$$

$$\sim \left[\begin{array}{ccc|c} 1 & 1 & 2 & 18 \\ 0 & -2 & -4 & -20 \\ 0 & 2 & 0 & 4 \end{array}\right] -\tfrac{1}{2}R_2 \to R_2$$

$$\sim \left[\begin{array}{ccc|c} 1 & 1 & 2 & 18 \\ 0 & 1 & 2 & 10 \\ 0 & 2 & 0 & 4 \end{array}\right] \begin{array}{l} (-1)R_2 + R_1 \to R_1 \\ \\ (-2)R_2 + R_3 \to R_3 \end{array}$$

$$\sim \left[\begin{array}{ccc|c} 1 & 0 & 0 & 8 \\ 0 & 1 & 2 & 10 \\ 0 & 0 & -4 & -16 \end{array}\right] -\tfrac{1}{4}R_3 \to R_3$$

$$\sim \left[\begin{array}{ccc|c} 1 & 0 & 0 & 8 \\ 0 & 1 & 2 & 10 \\ 0 & 0 & 1 & 4 \end{array}\right] (-2)R_3 + R_2 \rightarrow R_2$$

$$\sim \left[\begin{array}{ccc|c} 1 & 0 & 0 & 8 \\ 0 & 1 & 0 & 2 \\ 0 & 0 & 1 & 4 \end{array}\right]$$

Thus

x_1 = 8 ounces food *A*

x_2 = 2 ounces food *B*

x_3 = 4 ounces food *C*

101. In this case we have $x_3 = 0$ from the beginning. The three equations of problem 65 become

$$30x_1 + 10x_2 = 340$$
$$10x_1 + 10x_2 = 180$$
$$10x_1 + 30x_2 = 220$$

or

$$3x_1 + x_2 = 34$$
$$x_1 + x_2 = 18$$
$$x_1 + 3x_2 = 22$$

The augmented matrix becomes

$$\left[\begin{array}{cc|c} 3 & 1 & 34 \\ 1 & 1 & 18 \\ 1 & 3 & 22 \end{array}\right]$$

We solve by Gauss-Jordan elimination, starting by the row operation $R_1 \times R_2$

$$\left[\begin{array}{cc|c} 1 & 1 & 18 \\ 3 & 1 & 34 \\ 1 & 3 & 22 \end{array}\right] \begin{array}{l} \\ (-3)R_1 + R_2 \rightarrow R_2 \\ (-1)R_1 + R_3 \rightarrow R_3 \end{array}$$

$$\sim \left[\begin{array}{cc|c} 1 & 1 & 18 \\ 0 & -2 & -20 \\ 0 & 2 & 4 \end{array}\right] \begin{array}{l} \\ \\ R_2 + R_3 \rightarrow R_3 \end{array}$$

$$\sim \left[\begin{array}{cc|c} 1 & 1 & 18 \\ 0 & -2 & -20 \\ 0 & 0 & -16 \end{array}\right]$$

Since the third row corresponds to the equation

$$0x_1 + 0x_2 = -16$$

there is no solution.

103. In this case we discard the third equation. The system becomes

$$30x_1 + 10x_2 + 20x_3 = 340$$
$$10x_1 + 10x_2 + 20x_3 = 180$$

or

$$3x_1 + x_2 + 2x_3 = 34$$
$$x_1 + x_2 + 2x_3 = 18$$

The augmented matrix becomes

$$\left[\begin{array}{ccc|c} 3 & 1 & 2 & 34 \\ 1 & 1 & 2 & 18 \end{array}\right]$$

We solve by Gauss-Jordan elimination, starting by the row operation $R_1 \rightarrow R_2$.

$$\left[\begin{array}{ccc|c} 1 & 1 & 2 & 18 \\ 3 & 1 & 2 & 34 \end{array}\right] (-3)R_1 + R_2 \rightarrow R_2$$

$$\sim \left[\begin{array}{ccc|c} 1 & 1 & 2 & 18 \\ 0 & -2 & -4 & -20 \end{array}\right] -\tfrac{1}{2}R_2 \to R_2$$

$$\sim \left[\begin{array}{ccc|c} 1 & 1 & 2 & 18 \\ 0 & 1 & 2 & 10 \end{array}\right] (-1)R_2 + R_1 \to R_1$$

$$\sim \left[\begin{array}{ccc|c} 1 & 0 & 0 & 8 \\ 0 & 1 & 2 & 10 \end{array}\right]$$

This augmented matrix is in reduced form. It corresponds to the system

$x_1 = 8$

$x_2 + 2x_3 = 10$

Let $x_3 = t$

Then $x_2 = -2x_3 + 10$

$= -2t + 10$

A solution is achieved, not for every real value t, but for values of t that give rise to non-negative x_2, x_3.

$x_3 \ge 0$ means $t \ge 0$

$x_2 \ge 0$ means $-2t + 10 \ge 0$, $5 \ge t$

Thus we have the solution

$x_1 = 8$ ounces food A

$x_2 = -2t + 10$ ounces food B

$x_3 = t$ ounces food C

$0 \le t \le 5$

105. Let x_1 = number of hours company A is to be scheduled

x_2 = number of hours company B is to be scheduled

In x_1 hours, company A can handle $30x_1$ telephone and $10x_1$ house contacts.

In x_2 hours, company B can handle $20x_2$ telephone and $20x_2$ house contacts.

We therefore have:

$30x_1 + 20x_2 = 600$ telephone contacts

$10x_1 + 20x_2 = 400$ house contacts

We form the augmented matrix and solve by Gauss-Jordan elimination.

$$\left[\begin{array}{cc|c} 30 & 20 & 600 \\ 10 & 20 & 400 \end{array}\right] \begin{array}{l} \frac{1}{10}R_1 \to R_1 \\ \frac{1}{10}R_2 \to R_2 \end{array}$$

$$\sim \left[\begin{array}{cc|c} 3 & 2 & 60 \\ 1 & 2 & 40 \end{array}\right] R_1 \leftrightarrow R_2$$

$$\sim \left[\begin{array}{cc|c} 1 & 2 & 40 \\ 3 & 2 & 60 \end{array}\right] (-3)R_1 + R_2 \to R_2$$

$$\sim \left[\begin{array}{cc|c} 1 & 2 & 40 \\ 0 & -4 & -60 \end{array}\right] -\tfrac{1}{4}R_3 \to R_3$$

$$\sim \left[\begin{array}{cc|c} 1 & 2 & 40 \\ 0 & 1 & 15 \end{array}\right] (-2)R_2 + R_1 \to R_1$$

$$\sim \left[\begin{array}{cc|c} 1 & 0 & 10 \\ 0 & 1 & 15 \end{array}\right]$$

Therefore

$x_1 = 10$ hours company A

$x_2 = 15$ hours company B

107. Let x = base price
y = surcharge for each additional pound.

Since a 5-pound package costs the base price plus 4 surcharges,
$x + 4y = 27.75$

Since a 20-pound package costs the base price plus 19 surcharges,
$x + 19y = 64.50$

Solve using elimination by addition.

$$\begin{aligned} -x - 4y &= -27.75 \\ x + 19y &= 64.50 \\ \hline 15y &= 36.75 \\ y &= 2.45 \\ x + 4(2.45) &= 27.75 \\ x &= 17.95 \end{aligned}$$

The base price is \$17.95 and the surcharge per pound is \$2.45.

109. Let x = number of pounds of robust blend
y = number of pounds of mild blend

Summarize the given information in a table:

	Robust blend	Mild blend
ozs. of Columbian beans	12	6
ozs. of Brazilian beans	4	10

Form equations from the information:

$$\begin{pmatrix}\text{pounds of Columbian}\\ \text{beans needed for}\\ \text{robust blend}\end{pmatrix} + \begin{pmatrix}\text{pounds of Columbian}\\ \text{beans needed for}\\ \text{mild blend}\end{pmatrix} = \begin{pmatrix}\text{Total}\\ \text{Columbian beans}\\ \text{available}\end{pmatrix}$$

$$\frac{12}{16}x + \frac{6}{16}y = 50(132)$$

$$\begin{pmatrix}\text{pounds of Brazilian}\\ \text{beans needed for}\\ \text{robust blend}\end{pmatrix} + \begin{pmatrix}\text{pounds of Brazilian}\\ \text{beans needed for}\\ \text{mild blend}\end{pmatrix} = \begin{pmatrix}\text{Total}\\ \text{Brazilian beans}\\ \text{available}\end{pmatrix}$$

$$\frac{4}{16}x + \frac{10}{16}y = 40(132)$$

Solve using elimination by addition:

$$\begin{aligned} \frac{12}{16}x + \frac{6}{16}y &= 6{,}600 \\ -\frac{12}{16}x - \frac{30}{16}y &= -15{,}840 \\ -\frac{24}{16}y &= -9{,}240 \\ y &= 6{,}160 \\ \frac{4}{16}x + \frac{10}{16}(6{,}160) &= 40(132) \\ \frac{1}{4}x + 3{,}850 &= 5{,}280 \\ \frac{1}{4}x &= 1{,}430 \\ x &= 5{,}720 \end{aligned}$$

5,720 pounds of the robust blend and 6,160 pounds of the mild blend.

Section 10-4

1. $\begin{bmatrix} 5 & -2 \\ 3 & 0 \end{bmatrix} + \begin{bmatrix} -3 & 7 \\ 1 & -6 \end{bmatrix} = \begin{bmatrix} 5 + (-3) & (-2) + 7 \\ 3 + 1 & 0 + (-6) \end{bmatrix} = \begin{bmatrix} 2 & 5 \\ 4 & -6 \end{bmatrix}$

3. $\begin{bmatrix} 4 & 0 \\ -2 & 3 \\ 8 & 1 \end{bmatrix} + \begin{bmatrix} -1 & 2 \\ 0 & 5 \\ 4 & -6 \end{bmatrix} = \begin{bmatrix} 4 + (-1) & 0 + 2 \\ (-2) + 0 & 3 + 5 \\ 8 + 4 & 1 + (-6) \end{bmatrix} = \begin{bmatrix} 3 & 2 \\ -2 & 8 \\ 12 & -5 \end{bmatrix}$

5. These matrices have different sizes, hence the sum is not defined.

7. $\begin{bmatrix} 5 & -1 & 0 \\ 4 & 6 & 3 \end{bmatrix} - \begin{bmatrix} 2 & 4 & -6 \\ 3 & 5 & -5 \end{bmatrix} = \begin{bmatrix} 5 - 2 & (-1) - 4 & 0 - (-6) \\ 4 - 3 & 6 - 5 & 3 - (-5) \end{bmatrix} = \begin{bmatrix} 3 & -5 & 6 \\ 1 & 1 & 8 \end{bmatrix}$

9. $\begin{bmatrix} 12 & -16 & 28 \\ -8 & 36 & 20 \end{bmatrix}$

11. $[5 \quad 3]\begin{bmatrix} 4 \\ 7 \end{bmatrix} = [5 \cdot 4 + 3 \cdot 7] = [41]$

13. $\begin{bmatrix} -6 & 3 \\ 2 & -5 \end{bmatrix}\begin{bmatrix} 1 \\ 3 \end{bmatrix} = \begin{bmatrix} (-6)1 + 3 \cdot 3 \\ 2 \cdot 1 + (-5)3 \end{bmatrix} = \begin{bmatrix} 3 \\ -13 \end{bmatrix}$

15. $\begin{bmatrix} 5 & 1 \\ 4 & 6 \end{bmatrix}\begin{bmatrix} 2 & 0 \\ 3 & 8 \end{bmatrix} = \begin{bmatrix} 5 \cdot 2 + 1 \cdot 3 & 5 \cdot 0 + 1 \cdot 8 \\ 4 \cdot 2 + 6 \cdot 3 & 4 \cdot 0 + 6 \cdot 8 \end{bmatrix} = \begin{bmatrix} 13 & 8 \\ 26 & 48 \end{bmatrix}$

17. $\begin{bmatrix} 8 & -3 \\ -5 & 3 \end{bmatrix}\begin{bmatrix} 2 & 0 \\ 0 & 6 \end{bmatrix} = \begin{bmatrix} 8 \cdot 2 + (-3)0 & 8 \cdot 0 + (-3)6 \\ (-5)2 + 3 \cdot 0 & (-5)0 + 3 \cdot 6 \end{bmatrix} = \begin{bmatrix} 16 & -18 \\ -10 & 18 \end{bmatrix}$

19. $[3 \quad -6]\begin{bmatrix} -2 \\ 5 \end{bmatrix} = [3(-2) + (-6)5] = [-36]$

19. $[4 \quad -2]\begin{bmatrix} -5 \\ -3 \end{bmatrix} = [4(-5) + (-2)(-3)] = [-14]$

21. $\begin{bmatrix} -5 \\ -3 \end{bmatrix}[4 \quad -2] = \begin{bmatrix} (-5)4 & (-5)(-2) \\ (-3)4 & (-3)(-2) \end{bmatrix} = \begin{bmatrix} -20 & 10 \\ -12 & 6 \end{bmatrix}$

23. $[3 \quad -2 \quad -4]\begin{bmatrix} 1 \\ 2 \\ -3 \end{bmatrix} = [3 \cdot 1 + (-2)2 + (-4)(-3)] = [11]$

25. $\begin{bmatrix} 1 \\ 2 \\ -3 \end{bmatrix}[3 \quad -2 \quad -4] = \begin{bmatrix} 1 \cdot 3 & 1(-2) & 1(-4) \\ 2 \cdot 3 & 2(-2) & 2(-4) \\ (-3)3 & (-3)(-2) & (-3)(-4) \end{bmatrix} = \begin{bmatrix} 3 & -2 & -4 \\ 6 & -4 & -8 \\ -9 & 6 & 12 \end{bmatrix}$

27. C has 3 columns. A has 2 rows. Therefore, CA is not defined.

29. $BA = \begin{bmatrix} -3 & 1 \\ 2 & 5 \end{bmatrix}\begin{bmatrix} 2 & -1 & 3 \\ 0 & 4 & -2 \end{bmatrix} = \begin{bmatrix} (-3)2 + 1 \cdot 0 & (-3)(-1) + 1 \cdot 4 & (-3)3 + 1(-2) \\ 2 \cdot 2 + 5 \cdot 0 & 2(-1) + 5 \cdot 4 & 2 \cdot 3 + 5(-2) \end{bmatrix}$

$= \begin{bmatrix} -6 & 7 & -11 \\ 4 & 18 & -4 \end{bmatrix}$

31. $C^2 = \begin{bmatrix} -1 & 0 & 2 \\ 4 & -3 & 1 \\ -2 & 3 & 5 \end{bmatrix} \begin{bmatrix} -1 & 0 & 2 \\ 4 & -3 & 1 \\ -2 & 3 & 5 \end{bmatrix}$

$= \begin{bmatrix} (-1)(-1) + 0 \cdot 4 + 2(-2) & (-1)0 + 0(-3) + 2 \cdot 3 & (-1)2 + 0 \cdot 1 + 2 \cdot 5 \\ 4(-1) + (-3)4 + 1(-2) & 4 \cdot 0 + (-3)(-3) + 1 \cdot 3 & 4 \cdot 2 + (-3)1 + 1 \cdot 5 \\ (-2)(-1) + 3 \cdot 4 + 5(-2) & (-2)0 + 3(-3) + 5 \cdot 3 & (-2)2 + 3 \cdot 1 + 5 \cdot 5 \end{bmatrix}$

$= \begin{bmatrix} -3 & 6 & 8 \\ -18 & 12 & 10 \\ 4 & 6 & 24 \end{bmatrix}$

33. $DA = \begin{bmatrix} 3 & -2 \\ 0 & -1 \\ 1 & 2 \end{bmatrix} \begin{bmatrix} 2 & -1 & 3 \\ 0 & 4 & -2 \end{bmatrix}$

$= \begin{bmatrix} 3 \cdot 2 + (-2)0 & 3(-1) + (-2)4 & 3 \cdot 3 + (-2)(-2) \\ 0 \cdot 2 + (-1)0 & 0(-1) + (-1)4 & 0 \cdot 3 + (-1)(-2) \\ 1 \cdot 2 + 2 \cdot 0 & 1(-1) + 2 \cdot 4 & 1 \cdot 3 + 2(-2) \end{bmatrix} = \begin{bmatrix} 6 & -11 & 13 \\ 0 & -4 & 2 \\ 2 & 7 & -1 \end{bmatrix}$

$C + DA = \begin{bmatrix} -1 & 0 & 2 \\ 4 & -3 & 1 \\ -2 & 3 & 5 \end{bmatrix} + \begin{bmatrix} 6 & -11 & 13 \\ 0 & -4 & 2 \\ 2 & 7 & -1 \end{bmatrix} = \begin{bmatrix} 5 & -11 & 15 \\ 4 & -7 & 3 \\ 0 & 10 & 4 \end{bmatrix}$

35. $0.2CD = 0.2 \begin{bmatrix} -1 & 0 & 2 \\ 4 & -3 & 1 \\ -2 & 3 & 5 \end{bmatrix} \begin{bmatrix} 3 & -2 \\ 0 & -1 \\ 1 & 2 \end{bmatrix}$

$= 0.2 \begin{bmatrix} (-1)3 + 0 \cdot 0 + 2 \cdot 1 & (-1)(-2) + 0(-1) + 2 \cdot 2 \\ 4 \cdot 3 + (-3)0 + 1 \cdot 1 & 4(-2) + (-3)(-1) + 1 \cdot 2 \\ (-2)3 + 3 \cdot 0 + 5 \cdot 1 & (-2)(-2) + 3(-1) + 5 \cdot 2 \end{bmatrix}$

$= 0.2 \begin{bmatrix} -1 & 6 \\ 13 & -3 \\ -1 & 11 \end{bmatrix} = \begin{bmatrix} -0.2 & 1.2 \\ 2.6 & -0.6 \\ -0.2 & 2.2 \end{bmatrix}$

37. $DB = \begin{bmatrix} 3 & -2 \\ 0 & -1 \\ 1 & 2 \end{bmatrix} \begin{bmatrix} -3 & 1 \\ 2 & 5 \end{bmatrix} = \begin{bmatrix} 3(-3) + (-2)2 & 3 \cdot 1 + (-2)5 \\ 0(-3) + (-1)2 & 0 \cdot 1 + (-1)5 \\ 1(-3) + 2 \cdot 2 & 1 \cdot 1 + 2 \cdot 5 \end{bmatrix} = \begin{bmatrix} -13 & -7 \\ -2 & -5 \\ 1 & 11 \end{bmatrix}$

$CD = \begin{bmatrix} -1 & 6 \\ 13 & -3 \\ -1 & 11 \end{bmatrix}$ (see problem 35)

Thus, $2DB + 5CD = 2\begin{bmatrix} -13 & -7 \\ -2 & -5 \\ 1 & 11 \end{bmatrix} + 5\begin{bmatrix} -1 & 6 \\ 13 & -3 \\ -1 & 11 \end{bmatrix} = \begin{bmatrix} -26 & -14 \\ -4 & -10 \\ 2 & 22 \end{bmatrix} + \begin{bmatrix} -5 & 30 \\ 65 & -15 \\ -5 & 55 \end{bmatrix}$

$= \begin{bmatrix} -31 & 16 \\ 61 & -25 \\ -3 & 77 \end{bmatrix}$

39. $(-1)AC$ is a matrix of size 2×3. $3DB$ is a matrix of size 3×2. Hence, $(-1)AC + 3DB$ is not defined.

41. $CD = \begin{bmatrix} -1 & 6 \\ 13 & -3 \\ -1 & 11 \end{bmatrix}$ (see problem 35)

Hence $CDA = \begin{bmatrix} -1 & 6 \\ 13 & -3 \\ -1 & 11 \end{bmatrix} \begin{bmatrix} 2 & -1 & 3 \\ 0 & 4 & -2 \end{bmatrix}$

$= \begin{bmatrix} (-1)2 + 6 \cdot 0 & (-1)(-1) + 6 \cdot 4 & (-1)3 + 6(-2) \\ 13 \cdot 2 + (-3)0 & 13(-1) + (-3)4 & 13 \cdot 3 + (-3)(-2) \\ (-1)2 + 11 \cdot 0 & (-1)(-1) + 11 \cdot 4 & (-1)3 + 11(-2) \end{bmatrix} = \begin{bmatrix} -2 & 25 & -15 \\ 26 & -25 & 45 \\ -2 & 45 & -25 \end{bmatrix}$

43. $DB = \begin{bmatrix} -13 & -7 \\ -2 & -5 \\ 1 & 11 \end{bmatrix}$ (see problem 37)

Hence

$$DBA = \begin{bmatrix} -13 & -7 \\ -2 & -5 \\ 1 & 11 \end{bmatrix} \begin{bmatrix} 2 & -1 & 3 \\ 0 & 4 & -2 \end{bmatrix}$$

$$= \begin{bmatrix} (-13)2 + (-7)0 & (-13)(-1) + (-7)4 & (-13)3 + (-7)(-2) \\ (-2)2 + (-5)0 & (-2)(-1) + (-5)4 & (-2)3 + (-5)(-2) \\ 1 \cdot 2 + 11 \cdot 0 & 1(-1) + 11 \cdot 4 & 1 \cdot 3 + 11(-2) \end{bmatrix} = \begin{bmatrix} -26 & -15 & -25 \\ -4 & -18 & 4 \\ 2 & 43 & -19 \end{bmatrix}$$

45. Entering matrix B in a graphing calculator, we obtain the results

```
[B]²
    [[.28 .72]
     [.24 .76]]
[B]³
  [[.256 .744]
   [.248 .752]]
```

```
[B]^4
 [[.2512 .7488]
  [.2496 .7504]]
[B]^5
[[.25024 .74976…
 [.24992 .75008…
```

```
[B]^6
[[.250048 .7499…
 [.249984 .7500…
[B]^7
[[.2500096 .749…
 [.2499968 .750…
```

It appears that $B^n \to \begin{bmatrix} 0.25 & 0.75 \\ 0.25 & 0.75 \end{bmatrix}$

We calculate AB, AB^2, AB^3, … and obtain the results

```
[A]*[B]
     [[.26 .74]]
[A]*[B]²
   [[.252 .748]]
[A]*[B]³
 [[.2504 .7496]]
```

```
[A]*[B]^4
[[.25008 .74992…
[A]*[B]^5
[[.250016 .7499…
[A]*[B]^6
[[.2500032 .749…
```

It appears that $AB^n \to [0.25 \quad 0.75]$.

47. $\begin{bmatrix} a & b \\ c & d \end{bmatrix} + \begin{bmatrix} 2 & -3 \\ 0 & 1 \end{bmatrix} = \begin{bmatrix} a+2 & b-3 \\ c & d+1 \end{bmatrix} = \begin{bmatrix} 1 & -2 \\ 3 & -4 \end{bmatrix}$

if and only if corresponding elements are equal.

$a + 2 = 1 \qquad b - 3 = -2 \qquad c = 3 \qquad d + 1 = -4$

$a = -1 \qquad b = 1 \qquad c = 3 \qquad d = -5$

49. $\begin{bmatrix} 3x & 5 \\ -1 & 4x \end{bmatrix} + \begin{bmatrix} 2y & -3 \\ -6 & -y \end{bmatrix} = \begin{bmatrix} 3x + 2y & 2 \\ -7 & 4x - y \end{bmatrix} = \begin{bmatrix} 7 & 2 \\ -7 & 2 \end{bmatrix}$

if and only if corresponding elements are equal.

$3x + 2y = 7 \qquad 2 \overset{\checkmark}{=} 2$

$-7 \overset{\checkmark}{=} -7 \qquad 4x - y = 2$

Two conditions are already met.

To find x and y, we solve the system:

$3x + 2y = 7$

$4x - y = 2$ to obtain $x = 1$, $y = 2$. (Solution left to the student.)

51. Compute the square matrix A:

$$A^2 = \begin{bmatrix} a & b \\ c & -a \end{bmatrix} \begin{bmatrix} a & b \\ c & -a \end{bmatrix} = \begin{bmatrix} a^2 + bc & ab + (-ab) \\ ac + (-ac) & cb + a^2 \end{bmatrix} = \begin{bmatrix} a^2 + bc & 0 \\ 0 & a^2 + bc \end{bmatrix}$$

Two of the entries are already zero and the other two are both $a^2 + bc$. So if $a^2 + bc = 0$, then $A^2 = 0$.

If $a = 1$, $b = 1$, $c = -1$, then $a^2 + bc = 0$, so the matrix $A = \begin{bmatrix} 1 & 1 \\ -1 & -1 \end{bmatrix}$ will have $A^2 = 0$. If $a = 2$, $b = 4$, $c = -1$, then $a^2 + bc = 0$, so the matrix $A = \begin{bmatrix} 2 & 4 \\ -1 & -2 \end{bmatrix}$ will have $A^2 = 0$.

53. Compute the product AB:

$$AB = \begin{bmatrix} a & b \\ c & d \end{bmatrix}\begin{bmatrix} 1 & 1 \\ 1 & -1 \end{bmatrix} = \begin{bmatrix} a + b & a + b \\ c + d & c + d \end{bmatrix}$$

Two of the entries are $a + b$ and the other two are $c + d$, so if $a = -b$ and $c = -d$, then $AB = 0$.
The following are a couple of examples of matrices A that will satisfy $AB = 0$:

$$\begin{bmatrix} 2 & -2 \\ 4 & -4 \end{bmatrix} \begin{bmatrix} -5 & 5 \\ 1 & -1 \end{bmatrix}$$

55. $$\begin{bmatrix} 1 & 3 \\ -2 & -2 \end{bmatrix}\begin{bmatrix} x & 1 \\ 3 & 2 \end{bmatrix} = \begin{bmatrix} x + 9 & 7 \\ -2x - 6 & -6 \end{bmatrix} = \begin{bmatrix} y & 7 \\ y & -6 \end{bmatrix}$$

if and only if corresponding elements are equal.

$$\begin{matrix} x + 9 = y & 7 \overset{\checkmark}{=} 7 \\ -2x - 6 = y & -6 \overset{\checkmark}{=} -6 \end{matrix} \Bigg\} \text{ Two conditions are already met.}$$

To find x and y, we solve the system:

$x + 9 = y$

$-2x - 6 = y$ to obtain $x = -5$, $y = 4$. (Solution left to the student.)

57. $$\begin{bmatrix} 1 & 3 \\ 1 & 4 \end{bmatrix}\begin{bmatrix} a & b \\ c & d \end{bmatrix} = \begin{bmatrix} a + 3c & b + 3d \\ a + 4c & b + 4d \end{bmatrix} = \begin{bmatrix} 6 & -5 \\ 7 & -7 \end{bmatrix}$$

if and only if corresponding elements are equal.

$a + 3c = 6 \quad b + 3d = -5$

$a + 4c = 7 \quad b + 4d = -7$

Solving these systems we obtain $a = 3$, $b = 1$, $c = 1$, $d = -2$.
(Solution left to the student.)

59. (A) Since $\begin{bmatrix} a_1 & 0 \\ 0 & d_1 \end{bmatrix} + \begin{bmatrix} a_2 & 0 \\ 0 & d_2 \end{bmatrix} = \begin{bmatrix} a_1 + a_2 & 0 \\ 0 & d_1 + d_2 \end{bmatrix}$, the statement is true.

(B) $A + B = B + A$ is true for any matrices for which $A + B$ is defined, as it is in this case.

(C) Since $\begin{bmatrix} a_1 & 0 \\ 0 & d_1 \end{bmatrix}\begin{bmatrix} a_2 & 0 \\ 0 & d_2 \end{bmatrix} = \begin{bmatrix} a_1a_2 & 0 \\ 0 & d_1d_2 \end{bmatrix}$, the statement is true.

(D) Since $\begin{bmatrix} a_1 & 0 \\ 0 & d_1 \end{bmatrix}\begin{bmatrix} a_2 & 0 \\ 0 & d_2 \end{bmatrix} = \begin{bmatrix} a_1a_2 & 0 \\ 0 & d_1d_2 \end{bmatrix} = \begin{bmatrix} a_2a_1 & 0 \\ 0 & d_2d_1 \end{bmatrix} = \begin{bmatrix} a_2 & 0 \\ 0 & d_2 \end{bmatrix}\begin{bmatrix} a_1 & 0 \\ 0 & d_2 \end{bmatrix}$,

the statement is true.

61. $$\frac{1}{2}(A + B) = \frac{1}{2}\left(\begin{bmatrix} 30 & 25 \\ 60 & 80 \end{bmatrix} + \begin{bmatrix} 36 & 27 \\ 54 & 74 \end{bmatrix}\right) = \frac{1}{2}\begin{bmatrix} 66 & 52 \\ 114 & 154 \end{bmatrix} = \begin{array}{c} \text{Guitar} \quad \text{Banjo} \\ \begin{bmatrix} 33 & 26 \\ 57 & 77 \end{bmatrix} \end{array} \begin{array}{l} \text{Materials} \\ \text{Labor} \end{array}$$

63. If a quantity is increased b 15%, the result is a multiplication by 1.15.
If a quantity is increased by 10%, the result is a multiplication by 1.1.
Thus we must calculate $1.1N - 1.15M$. The mark-up matrix is:

$$1.1N - 1.15M = 1.1\begin{bmatrix} 13,900 & 783 & 263 & 215 \\ 15,000 & 838 & 395 & 236 \\ 18,300 & 967 & 573 & 248 \end{bmatrix} - 1.15\begin{bmatrix} 10,400 & 682 & 215 & 182 \\ 12,500 & 721 & 295 & 182 \\ 16,400 & 827 & 443 & 192 \end{bmatrix}$$

$$= \begin{bmatrix} 15,290 & 861.3 & 289.3 & 236.5 \\ 16,500 & 921.8 & 434.5 & 259.6 \\ 20,130 & 1,063.7 & 630.3 & 272.8 \end{bmatrix} - \begin{bmatrix} 11,960 & 784.3 & 247.25 & 209.3 \\ 14,375 & 829.15 & 339.25 & 209.3 \\ 18,860 & 951.05 & 509.45 & 220.8 \end{bmatrix}$$

=

	Basic Car	Air	AM/FM Radio	Cruise Control
Model A	\$3,330	\$77	\$42	\$27
Model B	\$2,125	\$93	\$95	\$50
Model C	\$1,270	\$113	\$121	\$52

= Mark up

65. (A) $[0.6 \quad 0.6 \quad 0.2]\begin{bmatrix}8\\10\\5\end{bmatrix} = (0.6)8 + (0.6)10 + (0.2)5 = 11.80$ dollars per boat

(B) $[1.5 \quad 1.2 \quad 0.4]\begin{bmatrix}9\\12\\6\end{bmatrix} = (1.5)9 + (1.2)12 + (0.4)6 = 30.30$ dollars per boat

(C) The matrix *NM* has no obvious meaning, but the matrix *MN* gives the labor costs per boat at each plant.

(D) $$MN = \begin{bmatrix}0.6 & 0.6 & 0.2\\1.0 & 0.9 & 0.3\\1.5 & 1.2 & 0.4\end{bmatrix}\begin{bmatrix}8 & 9\\10 & 12\\5 & 6\end{bmatrix}$$

$$= \begin{bmatrix}(0.6)8 + (0.6)10 + (0.2)5 & (0.6)9 + (0.6)12 + (0.2)6\\(1.0)8 + (0.9)10 + (0.3)5 & (1.0)9 + (0.9)12 + (0.3)6\\(1.5)8 + (1.2)10 + (0.4)5 & (1.5)9 + (1.2)12 + (0.4)6\end{bmatrix}$$

	Plant I	Plant II	
=	\$11.80	\$13.80	One-person boat
	\$18.50	\$21.60	Two-person boat
	\$26.00	\$30.30	Four-person boat

This matrix gives the labor costs for each type of boat at each plant.

67. (A) $$A^2 = AA = \begin{bmatrix}0&1&0&1&0\\0&0&1&0&0\\1&0&0&0&1\\0&0&1&0&0\\0&0&0&1&0\end{bmatrix}\begin{bmatrix}0&1&0&1&0\\0&0&1&0&0\\1&0&0&0&1\\0&0&1&0&0\\0&0&0&1&0\end{bmatrix} = \begin{bmatrix}0&0&2&0&0\\1&0&0&0&1\\0&1&0&2&0\\1&0&0&0&1\\0&0&1&0&0\end{bmatrix}$$

The 1 in row 2 and column 1 of A^2 indicates that there is one way to travel from Baltimore to Atlanta with one intermediate connection. The 2 in row 1 and column 3 indicates that there are two ways to travel from Atlanta to Chicago with one intermediate connection. In general, the elements in A^2 indicate the number of different ways to travel from the ith city to the jth city with one intermediate connection.

(B) $$A^3 = A^2A = \begin{bmatrix}0&0&2&0&0\\1&0&0&0&1\\0&1&0&2&0\\1&0&0&0&1\\0&0&1&0&0\end{bmatrix}\begin{bmatrix}0&1&0&1&0\\0&0&1&0&0\\1&0&0&0&1\\0&0&1&0&0\\0&0&0&1&0\end{bmatrix} = \begin{bmatrix}2&0&0&0&2\\0&1&0&2&0\\0&0&3&0&0\\0&1&0&2&0\\1&0&0&0&1\end{bmatrix}$$

The 1 in row 4 and column 2 of A^3 indicates that there is one way to travel from Denver to Baltimore with two intermediate connections. The 2 in row 1 and column 5 indicates that there are two ways to travel from Atlanta to El Paso with two intermediate connections. In general, the elements in A^3 indicate the number of different ways to travel from the ith city to the the jth city with two intermediate connections.

(C) A is given above.

$$A + A^2 = \begin{bmatrix}0&1&0&1&0\\0&0&1&0&0\\1&0&0&0&1\\0&0&1&0&0\\0&0&0&1&0\end{bmatrix} + \begin{bmatrix}0&0&2&0&0\\1&0&0&0&1\\0&1&0&2&0\\1&0&0&0&1\\0&0&1&0&0\end{bmatrix} = \begin{bmatrix}0&1&2&1&0\\1&0&1&0&1\\1&1&0&2&1\\1&0&1&0&1\\0&0&1&1&0\end{bmatrix}$$

$$A + A^2 + A^3 = \begin{bmatrix}0&1&2&1&0\\1&0&1&0&1\\1&1&0&2&1\\1&0&1&0&1\\0&0&1&1&0\end{bmatrix} + \begin{bmatrix}2&0&0&0&2\\0&1&0&2&0\\0&0&3&0&0\\0&1&0&2&0\\1&0&0&0&1\end{bmatrix} = \begin{bmatrix}2&1&2&1&2\\1&1&1&2&1\\1&1&3&2&1\\1&1&1&2&1\\1&0&1&1&1\end{bmatrix}$$

A zero element remains, so we must compute A^4.

$$A^4 = A^3A = \begin{bmatrix} 2 & 0 & 0 & 0 & 2 \\ 0 & 1 & 0 & 2 & 0 \\ 0 & 0 & 3 & 0 & 0 \\ 0 & 1 & 0 & 2 & 0 \\ 1 & 0 & 0 & 0 & 1 \end{bmatrix} \begin{bmatrix} 0 & 1 & 0 & 1 & 0 \\ 0 & 0 & 1 & 0 & 0 \\ 1 & 0 & 0 & 0 & 1 \\ 0 & 0 & 1 & 0 & 0 \\ 0 & 0 & 0 & 1 & 0 \end{bmatrix} = \begin{bmatrix} 0 & 2 & 0 & 4 & 0 \\ 0 & 0 & 3 & 0 & 0 \\ 3 & 0 & 0 & 0 & 3 \\ 0 & 0 & 3 & 0 & 0 \\ 0 & 1 & 0 & 2 & 0 \end{bmatrix}$$

$$\text{Then } A + A^2 + A^3 + A^4 = \begin{bmatrix} 2 & 1 & 2 & 1 & 2 \\ 1 & 1 & 1 & 2 & 1 \\ 1 & 1 & 3 & 2 & 1 \\ 1 & 1 & 1 & 2 & 1 \\ 1 & 0 & 1 & 1 & 1 \end{bmatrix} + \begin{bmatrix} 0 & 2 & 0 & 4 & 0 \\ 0 & 0 & 3 & 0 & 0 \\ 3 & 0 & 0 & 0 & 3 \\ 0 & 0 & 3 & 0 & 0 \\ 0 & 1 & 0 & 2 & 0 \end{bmatrix} = \begin{bmatrix} 2 & 3 & 2 & 5 & 2 \\ 1 & 1 & 4 & 2 & 1 \\ 4 & 1 & 3 & 2 & 4 \\ 1 & 1 & 4 & 2 & 1 \\ 1 & 1 & 1 & 3 & 1 \end{bmatrix}$$

This matrix indicates that it is possible to travel from any origin to any destination with at most 3 intermediate connections.

69. (A) $[1{,}000 \quad 500 \quad 5{,}000]\begin{bmatrix} \$0.80 \\ \$1.50 \\ \$0.40 \end{bmatrix} = 1{,}000(\$0.80) + 500(\$1.50) + 5{,}000(\$0.40) = \$3{,}550$

(B) $[2{,}000 \quad 800 \quad 8{,}000]\begin{bmatrix} \$0.80 \\ \$1.50 \\ \$0.40 \end{bmatrix} = 2{,}000(\$0.80) + 800(\$1.50) + 8{,}000(\$0.40) = \$6{,}000$

(C) The matrix MN has no obvious interpretations, but the matrix NM represents the total cost of all contacts in each town.

(D) $$NM = \begin{bmatrix} 1{,}000 & 500 & 5{,}000 \\ 2{,}000 & 800 & 8{,}000 \end{bmatrix}\begin{bmatrix} \$0.80 \\ \$1.50 \\ \$0.40 \end{bmatrix} = \begin{bmatrix} 1{,}000(0.80) + 500(1.50) + 5{,}000(0.40) \\ 2{,}000(0.80) + 800(1.50) + 8{,}000(0.40) \end{bmatrix}$$

$$= \begin{bmatrix} \$3{,}550 \\ \$6{,}000 \end{bmatrix} \begin{matrix} \text{Berkeley} \\ \text{Oakland} \end{matrix} = \text{cost of all contacts in each town.}$$

(E) The matrix $[1 \quad 1]N$ can be used to find the total number of each of the three types of contact:

$$[1 \quad 1]\begin{bmatrix} 1{,}000 & 500 & 5{,}000 \\ 2{,}000 & 800 & 8{,}000 \end{bmatrix} = [1{,}000 + 2{,}000 \quad 500 + 800 \quad 5{,}000 + 8{,}000]$$

[Telephone House Letter] = [3,000 1,300 13,000]

(F) The matrix $N\begin{bmatrix} 1 \\ 1 \\ 1 \end{bmatrix}$ can be used to find the total number of contacts in each town:

$$\begin{bmatrix} 1{,}000 & 500 & 5{,}000 \\ 2{,}000 & 800 & 8{,}000 \end{bmatrix}\begin{bmatrix} 1 \\ 1 \\ 1 \end{bmatrix}$$

$$= \begin{bmatrix} 1{,}000 + 500 + 5{,}000 \\ 2{,}000 + 800 + 8{,}000 \end{bmatrix} = \begin{bmatrix} 6{,}500 \\ 10{,}800 \end{bmatrix} = \begin{bmatrix} \text{Berkeley contacts} \\ \text{Oakland contacts} \end{bmatrix}$$

71. (A) Since player 1 did not defeat player 1, a 0 is placed in row 1, column 1.
Since player 1 did not defeat player 2, a 0 is placed in row 1, column 2.
Since player 1 defeated player 3, a 1 is placed in row 1, column 3.
Since player 1 defeated player 4, a 1 is placed in row 1, column 4.
Since player 1 defeated player 5, a 1 is placed in row 1, column 5.
Since player 1 did not defeat player 6, a 0 is placed in row 1, column 6.

Proceeding in this manner, we obtain

$$A = \begin{bmatrix} 0 & 0 & 1 & 1 & 1 & 0 \\ 1 & 0 & 0 & 1 & 1 & 0 \\ 0 & 1 & 0 & 1 & 0 & 0 \\ 0 & 0 & 0 & 0 & 0 & 1 \\ 0 & 0 & 1 & 1 & 0 & 1 \\ 1 & 1 & 1 & 0 & 0 & 0 \end{bmatrix}$$

(B) $A^2 = \begin{bmatrix} 0 & 0 & 1 & 1 & 1 & 0 \\ 1 & 0 & 0 & 1 & 1 & 0 \\ 0 & 1 & 0 & 1 & 0 & 0 \\ 0 & 0 & 0 & 0 & 0 & 1 \\ 0 & 0 & 1 & 1 & 0 & 1 \\ 1 & 1 & 1 & 0 & 0 & 0 \end{bmatrix} + \begin{bmatrix} 0 & 0 & 1 & 1 & 1 & 0 \\ 1 & 0 & 0 & 1 & 1 & 0 \\ 0 & 1 & 0 & 1 & 0 & 0 \\ 0 & 0 & 0 & 0 & 0 & 1 \\ 0 & 0 & 1 & 1 & 0 & 1 \\ 1 & 1 & 1 & 0 & 0 & 0 \end{bmatrix} = \begin{bmatrix} 0 & 1 & 1 & 2 & 0 & 2 \\ 0 & 0 & 2 & 2 & 1 & 2 \\ 1 & 0 & 0 & 1 & 1 & 1 \\ 1 & 1 & 1 & 0 & 0 & 0 \\ 1 & 2 & 1 & 1 & 0 & 1 \\ 1 & 1 & 1 & 3 & 2 & 0 \end{bmatrix}$

$$A + A^2 = \begin{bmatrix} 0 & 1 & 2 & 3 & 1 & 2 \\ 1 & 0 & 2 & 3 & 2 & 2 \\ 1 & 1 & 0 & 2 & 1 & 1 \\ 1 & 1 & 1 & 0 & 0 & 1 \\ 1 & 2 & 2 & 2 & 0 & 2 \\ 2 & 2 & 2 & 3 & 2 & 0 \end{bmatrix} = B$$

Check (using graphing calculator):

```
[A]
 [[0 0 1 1 1 0]
  [1 0 0 1 1 0]
  [0 1 0 1 0 0]
  [0 0 0 0 0 1]
  [0 0 1 1 0 1]
  [1 1 1 0 0 0]]
```

```
[A]²
 [[0 1 1 2 0 2]
  [0 0 2 2 1 2]
  [1 0 0 1 1 1]
  [1 1 1 0 0 0]
  [1 2 1 1 0 1]
  [1 1 1 3 2 0]]
```

```
[A]+[A]²→[B]
 [[0 1 2 3 1 2]
  [1 0 2 3 2 2]
  [1 1 0 2 1 1]
  [1 1 1 0 0 1]
  [1 2 2 2 0 2]
  [2 2 2 3 2 0]]
```

Finally $B^{-1}A^{-1}$ is calculated as $C = \begin{bmatrix} 1 \\ 1 \\ 1 \\ 1 \\ 1 \\ 1 \end{bmatrix}$ Then $BC = \begin{bmatrix} 0 & 1 & 2 & 3 & 1 & 2 \\ 1 & 0 & 2 & 3 & 2 & 2 \\ 1 & 1 & 0 & 2 & 1 & 1 \\ 1 & 1 & 1 & 0 & 0 & 1 \\ 1 & 2 & 2 & 2 & 0 & 2 \\ 2 & 2 & 2 & 3 & 2 & 0 \end{bmatrix}\begin{bmatrix} 1 \\ 1 \\ 1 \\ 1 \\ 1 \\ 1 \end{bmatrix} = \begin{bmatrix} 9 \\ 10 \\ 6 \\ 4 \\ 9 \\ 11 \end{bmatrix}$

(D) BC measures the relative strength of the players, with the larger numbers representing greater strength. Thus, player 6 is the strongest and player 4 the weakest; ranking:

Frank, Bart, Aaron & Elvis (tie), Charles, Dan.

Section 10-5

1. $\begin{bmatrix} 2 & -3 \\ 4 & 5 \end{bmatrix}$ **3.** $\begin{bmatrix} 2 & -3 \\ 4 & 5 \end{bmatrix}$ **5.** $\begin{bmatrix} -2 & 1 & 3 \\ 2 & 4 & -2 \\ 5 & 1 & 0 \end{bmatrix}$ **7.** $\begin{bmatrix} -2 & 1 & 3 \\ 2 & 4 & -2 \\ 5 & 1 & 0 \end{bmatrix}$

9. $\begin{bmatrix} 3 & -4 \\ -2 & 3 \end{bmatrix}\begin{bmatrix} 3 & 4 \\ 2 & 3 \end{bmatrix} = \begin{bmatrix} 3 \cdot 3 + (-4)2 & 3 \cdot 4 + (-4)3 \\ (-2)3 + 3 \cdot 2 & (-2)4 + 3 \cdot 3 \end{bmatrix} = \begin{bmatrix} 1 & 0 \\ 0 & 1 \end{bmatrix}$

Thus, these two matrices are inverses of each other.

11. $\begin{bmatrix} 2 & 2 \\ -1 & -1 \end{bmatrix}\begin{bmatrix} 1 & 1 \\ -1 & -1 \end{bmatrix} = \begin{bmatrix} 2 \cdot 1 + 2(-1) & 2 \cdot 1 + 2(-1) \\ (-1)1 + (-1)(-1) & (-1)1 + (-1)(-1) \end{bmatrix} = \begin{bmatrix} 0 & 0 \\ 0 & 0 \end{bmatrix}$

Thus, these two matrices are not inverses of each other.

13. $\begin{bmatrix} -5 & 2 \\ -8 & 3 \end{bmatrix}\begin{bmatrix} 3 & -2 \\ 8 & -5 \end{bmatrix} = \begin{bmatrix} (-5)3 + 2 \cdot 8 & (-5)(-2) + 2(-5) \\ (-8)3 + 3 \cdot 8 & (-8)(-2) + 3(-5) \end{bmatrix} = \begin{bmatrix} 1 & 0 \\ 0 & 1 \end{bmatrix}$

Thus, these two matrices are inverses of each other.

15. $\begin{bmatrix} 1 & 2 & 0 \\ 0 & 1 & 0 \\ -1 & -1 & 1 \end{bmatrix}\begin{bmatrix} 1 & -2 & 0 \\ 0 & 1 & 0 \\ 1 & -1 & 0 \end{bmatrix} =$

$$\begin{bmatrix} 1 \cdot 1 + 2 \cdot 0 + 0 \cdot 1 & 1(-2) + 2 \cdot 1 + 0(-1) & 1 \cdot 0 + 2 \cdot 0 + 0 \cdot 0 \\ 0 \cdot 1 + 1 \cdot 1 + 0(-1) & 0(-2) + 1 \cdot 1 + 0(-1) & 0 \cdot 0 + 1 \cdot 0 + 0 \cdot 0 \\ (-1)1 + (-1)1 + 1 \cdot 1 & (-1)(-2) + (-1)1 + 1(-1) & (-1)0 + (-1)0 + 0 \cdot 0 \end{bmatrix}$$

$$= \begin{bmatrix} 1 & 0 & 0 \\ 1 & 1 & 0 \\ -1 & 0 & 0 \end{bmatrix}$$

Thus, these two matrices are not inverses of each other.

17. $\begin{bmatrix} 1 & -1 & 1 \\ 0 & 2 & -1 \\ 2 & 3 & 0 \end{bmatrix}\begin{bmatrix} 3 & 3 & -1 \\ -2 & -2 & 1 \\ -4 & -5 & 2 \end{bmatrix}$

$= \begin{bmatrix} 1\cdot 3 + (-1)(-2) + 1(-4) & 1\cdot 3 + (-1)(-2) + 1(-5) & 1\cdot(-1) + (-1)\cdot 1 + 1\cdot 2 \\ 0\cdot 3 + 2(-2) + (-1)(-4) & 0\cdot 3 + 2(-2) + (-1)(-5) & 0(-1) + 2\cdot 1 + (-1)\cdot 2 \\ 2\cdot 3 + 3(-2) + 0(-4) & 2\cdot 3 + 3(-2) + 0(-5) & 2(-1) + 3\cdot 1 + 0\cdot 2 \end{bmatrix}$

$= \begin{bmatrix} 1 & 0 & 0 \\ 0 & 1 & 0 \\ 0 & 0 & 1 \end{bmatrix}$

Thus, these two matrices are inverses of each other.

19. $2x_1 - x_2 = 3$
$x_1 + 3x_2 = -2$

21. $-2x_1 + x_3 = 3$
$x_1 + 2x_2 + x_3 = -4$
$x_2 - x_3 = 2$

23. $\begin{bmatrix} 4 & -3 \\ 1 & 2 \end{bmatrix}\begin{bmatrix} x_1 \\ x_2 \end{bmatrix} = \begin{bmatrix} 2 \\ 1 \end{bmatrix}$

25. $\begin{bmatrix} 1 & -2 & 1 \\ -1 & 1 & 0 \\ 2 & 3 & 1 \end{bmatrix}\begin{bmatrix} x_1 \\ x_2 \\ x_3 \end{bmatrix} = \begin{bmatrix} -1 \\ 2 \\ -3 \end{bmatrix}$

27. Since $\begin{bmatrix} 3 & -2 \\ 1 & 4 \end{bmatrix}\begin{bmatrix} -2 \\ 1 \end{bmatrix} = \begin{bmatrix} 3(-2) + (-2)1 \\ 1(-2) + 4\cdot 1 \end{bmatrix}$

$= \begin{bmatrix} -8 \\ 2 \end{bmatrix}$, $\begin{bmatrix} x_1 \\ x_2 \end{bmatrix} = \begin{bmatrix} -8 \\ 2 \end{bmatrix}$ if and only if $x_1 = -8$ and $x_2 = 2$.

29. Since $\begin{bmatrix} -2 & 3 \\ 2 & -1 \end{bmatrix}\begin{bmatrix} 3 \\ 2 \end{bmatrix} = \begin{bmatrix} (-2)3 + 3\cdot 2 \\ 2\cdot 3 + (-1)2 \end{bmatrix} = \begin{bmatrix} 0 \\ 4 \end{bmatrix}$

$\begin{bmatrix} x_1 \\ x_2 \end{bmatrix} = \begin{bmatrix} 0 \\ 4 \end{bmatrix}$ if and only if $x_1 = 0$ and $x_2 = 4$.

31. $\underset{A}{\begin{bmatrix} 1 & -1 \\ 1 & -2 \end{bmatrix}}\underset{X}{\begin{bmatrix} x_1 \\ x_2 \end{bmatrix}} = \underset{B}{\begin{bmatrix} 5 \\ 7 \end{bmatrix}}$

$AX = B$ has solution $X = A^{-1}B$.
To find A^{-1}, we perform row operations on

$\left[\begin{array}{cc|cc} 1 & -1 & 1 & 0 \\ 1 & -2 & 0 & 1 \end{array}\right] (-1)R_1 + R_2 \to R_2$

$\sim \left[\begin{array}{cc|cc} 1 & -1 & 1 & 0 \\ 0 & -1 & -1 & 1 \end{array}\right] (-1)R_2 + R_1 \to R_1$

$\sim \left[\begin{array}{cc|cc} 1 & 0 & 2 & -1 \\ 0 & -1 & -1 & 1 \end{array}\right] (-1)R_2 \to R_2$

$\sim \left[\begin{array}{cc|cc} 1 & 0 & 2 & -1 \\ 0 & 1 & 1 & -1 \end{array}\right]$

Hence $A^{-1} = \begin{bmatrix} 2 & -1 \\ 1 & -1 \end{bmatrix}$

Check: $A^{-1}A = \begin{bmatrix} 2 & -1 \\ 1 & -1 \end{bmatrix}\begin{bmatrix} 1 & -1 \\ 1 & -2 \end{bmatrix}$

$= \begin{bmatrix} 1 & 0 \\ 0 & 1 \end{bmatrix}$

33. $\underset{A}{\begin{bmatrix} 1 & 1 \\ 2 & -3 \end{bmatrix}}\underset{X}{\begin{bmatrix} x_1 \\ x_2 \end{bmatrix}} = \underset{B}{\begin{bmatrix} 15 \\ 10 \end{bmatrix}}$

$AX = B$ has solution $X = A^{-1}B$.
To find A^{-1}, we perform row operations on

$\left[\begin{array}{cc|cc} 1 & 1 & 1 & 0 \\ 2 & -3 & 0 & 1 \end{array}\right] (-2)R_1 + R_2 \to R_2$

$\sim \left[\begin{array}{cc|cc} 1 & 1 & 1 & 0 \\ 0 & -5 & -2 & 1 \end{array}\right] \left(-\frac{1}{5}\right)R_2 \to R_2$

$\sim \left[\begin{array}{cc|cc} 1 & 1 & 1 & 0 \\ 0 & 1 & \frac{2}{5} & -\frac{1}{5} \end{array}\right] (-1)R_2 + R_1 \to R_1$

$\sim \left[\begin{array}{cc|cc} 1 & 0 & \frac{3}{5} & \frac{1}{5} \\ 0 & 1 & \frac{2}{5} & -\frac{1}{5} \end{array}\right]$

Hence $A^{-1} = \begin{bmatrix} \frac{3}{5} & \frac{1}{5} \\ \frac{2}{5} & -\frac{1}{5} \end{bmatrix}$

Check:

$A^{-1}A = \begin{bmatrix} \frac{3}{5} & \frac{1}{5} \\ \frac{2}{5} & -\frac{1}{5} \end{bmatrix}\begin{bmatrix} 1 & 1 \\ 2 & -3 \end{bmatrix} = \begin{bmatrix} 1 & 0 \\ 0 & 1 \end{bmatrix}$

Then

$$\overset{X}{\begin{bmatrix} x_1 \\ x_2 \end{bmatrix}} = \overset{A^{-1}}{\begin{bmatrix} 2 & -1 \\ 1 & -1 \end{bmatrix}} \overset{B}{\begin{bmatrix} 5 \\ 7 \end{bmatrix}} = \begin{bmatrix} 3 \\ -2 \end{bmatrix}$$

$x_1 = 3$, $x_2 = -2$.

Then

$$\overset{X}{\begin{bmatrix} x_1 \\ x_2 \end{bmatrix}} = \overset{A^{-1}}{\begin{bmatrix} \frac{3}{5} & \frac{1}{5} \\ \frac{2}{5} & -\frac{1}{5} \end{bmatrix}} \overset{B}{\begin{bmatrix} 15 \\ 10 \end{bmatrix}} = \begin{bmatrix} 11 \\ 4 \end{bmatrix}$$

$x_1 = 11$, $x_2 = 4$

35. $\left[\begin{array}{cc|cc} 1 & 9 & 1 & 0 \\ 0 & 1 & 0 & 1 \end{array}\right] \quad (-9)R_2 + R_1 \to R_1$

$\sim \left[\begin{array}{cc|cc} 1 & 0 & 1 & -9 \\ 0 & 1 & 0 & 1 \end{array}\right]$

Hence, $M^{-1} = \begin{bmatrix} 1 & -9 \\ 0 & 1 \end{bmatrix}$

Check: $M^{-1}M = \begin{bmatrix} 1 & -9 \\ 0 & 1 \end{bmatrix}\begin{bmatrix} 1 & 9 \\ 0 & 1 \end{bmatrix} = \begin{bmatrix} 1 \cdot 1 + (-9)0 & 1 \cdot 9 + (-9)1 \\ 0 \cdot 1 + 1 \cdot 0 & 0 \cdot 9 + 1 \cdot 1 \end{bmatrix} = \begin{bmatrix} 1 & 0 \\ 0 & 1 \end{bmatrix}$

37. $\left[\begin{array}{cc|cc} -1 & -2 & 1 & 0 \\ 2 & 5 & 0 & 1 \end{array}\right] 2R_1 + R_2 \to R_2$

$\sim \left[\begin{array}{cc|cc} -1 & -2 & 1 & 0 \\ 0 & 1 & 2 & 1 \end{array}\right] 2R_2 + R_1 \to R_1$

$\sim \left[\begin{array}{cc|cc} -1 & -2 & 1 & 0 \\ 0 & 1 & 2 & 1 \end{array}\right] 2R_2 + R_1 \to R_1$

$\sim \left[\begin{array}{cc|cc} -1 & 0 & 5 & 2 \\ 0 & 1 & 2 & 1 \end{array}\right] (-1)R_1 \to R_1$

$\sim \left[\begin{array}{cc|cc} 1 & 0 & -5 & -2 \\ 0 & 1 & 2 & 1 \end{array}\right]$

Hence, $M^{-1} = \begin{bmatrix} -5 & -2 \\ 2 & 1 \end{bmatrix}$

Check: $M^{-1}M = \begin{bmatrix} -5 & -2 \\ 2 & 1 \end{bmatrix}\begin{bmatrix} -1 & -2 \\ 2 & 5 \end{bmatrix} = \begin{bmatrix} (-5)(-1) + (-2)2 & (-5)(-2) + (-2)5 \\ 2(-1) + 1 \cdot 2 & 2(-2) + 1 \cdot 5 \end{bmatrix}$

$= \begin{bmatrix} 1 & 0 \\ 0 & 1 \end{bmatrix}$

39. $\left[\begin{array}{cc|cc} -5 & 7 & 1 & 0 \\ 2 & -3 & 0 & 1 \end{array}\right] 3R_2 + R_1 \to R_1$

$\sim \left[\begin{array}{cc|cc} 1 & -2 & 1 & 3 \\ 2 & -3 & 0 & 1 \end{array}\right] (-2)R_1 + R_2 \to R_2$

$\sim \left[\begin{array}{cc|cc} 1 & -2 & 1 & 3 \\ 0 & 1 & -2 & -5 \end{array}\right] 2R_2 + R_1 \to R_1$

$\sim \left[\begin{array}{cc|cc} 1 & 0 & -3 & -7 \\ 0 & 1 & -2 & -5 \end{array}\right]$

Hence, $M^{-1} = \begin{bmatrix} -3 & -7 \\ -2 & -5 \end{bmatrix}$

Check: $M^{-1}M = \begin{bmatrix} -3 & -7 \\ -2 & -5 \end{bmatrix}\begin{bmatrix} -5 & 7 \\ 2 & -3 \end{bmatrix} = \begin{bmatrix} (-3)(-5) + (-7)2 & (-3)7 + (-7)(-3) \\ (-2)(-5) + (-5)2 & (-2)7 + (-5)(-3) \end{bmatrix} = \begin{bmatrix} 1 & 0 \\ 0 & 1 \end{bmatrix}$

41. If the inverse existed we would find it by row operations on the following matrix:

$$= \left[\begin{array}{cc|cc} 3 & 9 & 1 & 0 \\ 2 & 6 & 0 & 1 \end{array}\right]$$

But consider what happens if we perform $(-\frac{2}{3})R_1 + R_2 \to R_2$

$$= \left[\begin{array}{cc|cc} 3 & 9 & 1 & 0 \\ 0 & 0 & -\frac{2}{3} & 1 \end{array}\right]$$

Since a row of zeros results to the left of the vertical line, no inverse exists.

43. $\left[\begin{array}{cc|cc} 2 & 3 & 1 & 0 \\ 3 & 5 & 0 & 1 \end{array}\right] \quad \frac{1}{2}R_1 \to R_1$

$$\sim \left[\begin{array}{cc|cc} 1 & 1.5 & 0.5 & 0 \\ 3 & 5 & 0 & 1 \end{array}\right] (-3)R_1 + R_2 \to R_2$$

$$\sim \left[\begin{array}{cc|cc} 1 & 1.5 & 0.5 & 0 \\ 0 & 0.5 & -1.5 & 1 \end{array}\right] 2R_2 \to R_2$$

$$\sim \left[\begin{array}{cc|cc} 1 & 1.5 & 0.5 & 0 \\ 0 & 1 & -3 & 2 \end{array}\right] (-1.5)R_2 + R_1 \to R_1$$

$$\sim \left[\begin{array}{cc|cc} 1 & 0 & 5 & -3 \\ 0 & 1 & -3 & 2 \end{array}\right]$$

The inverse is $\begin{bmatrix} 5 & -3 \\ -3 & 2 \end{bmatrix}$

The checking steps are omitted for lack of space in this and subsequent problems.

45. $\left[\begin{array}{ccc|ccc} 1 & -1 & 0 & 1 & 0 & 0 \\ -1 & 1 & -1 & 0 & 1 & 0 \\ 0 & -1 & 1 & 0 & 0 & 1 \end{array}\right] \quad R_1 + R_2 \to R_2$

$$\sim \left[\begin{array}{ccc|ccc} 1 & -1 & 0 & 1 & 0 & 0 \\ 0 & 0 & -1 & 1 & 1 & 0 \\ 0 & -1 & 1 & 0 & 0 & 1 \end{array}\right] \quad R_2 \leftrightarrow R_3$$

$$\sim \left[\begin{array}{ccc|ccc} 1 & -1 & 0 & 1 & 0 & 0 \\ 0 & -1 & 1 & 0 & 0 & 1 \\ 0 & 0 & -1 & 1 & 1 & 0 \end{array}\right] \quad (-1)\ R_2 + R_1 \to R_1$$

$$\sim \left[\begin{array}{ccc|ccc} 1 & 0 & -1 & 1 & 0 & -1 \\ 0 & -1 & 1 & 0 & 0 & 1 \\ 0 & 0 & -1 & 1 & 1 & 0 \end{array}\right] \begin{array}{l} (-1)R_3 + R_1 \to R_1 \\ R_3 + R_2 \to R_2 \end{array}$$

$$\sim \left[\begin{array}{ccc|ccc} 1 & 0 & 0 & 0 & -1 & -1 \\ 0 & -1 & 0 & 1 & 1 & 1 \\ 0 & 0 & -1 & 1 & 1 & 0 \end{array}\right] \begin{array}{l} (-1)R_2 \to R_2 \\ (-1)R_3 \to R_3 \end{array}$$

$$\sim \left[\begin{array}{ccc|ccc} 1 & 0 & 0 & 0 & -1 & -1 \\ 0 & 1 & 0 & -1 & -1 & -1 \\ 0 & 0 & 1 & -1 & -1 & 0 \end{array}\right]$$

Hence, $M^{-1} = \begin{bmatrix} 0 & -1 & -1 \\ -1 & -1 & -1 \\ -1 & -1 & 0 \end{bmatrix}$

$$\text{Check: } M^{-1}M = \begin{bmatrix} 0 & -1 & -1 \\ -1 & -1 & -1 \\ -1 & -1 & 0 \end{bmatrix}\begin{bmatrix} 1 & -1 & 0 \\ -1 & 1 & -1 \\ 0 & -1 & 1 \end{bmatrix}$$

$$= \begin{bmatrix} 0\cdot 1 + (-1)(-1) + (-1)0 & 0(-1) + (-1)1 + (-1)(-1) & 0\cdot 0 + (-1)(-1) + (-1)1 \\ (-1)1 + (-1)(-1) + (-1)0 & (-1)(-1) + (-1)1 + (-1)(-1) & (-1)0 + (-1)(-1) + (-1)1 \\ (-1)1 + (-1)(-1) + 0\cdot 0 & (-1)(-1) + (-1)1 + 0(-1) & (-1)0 + (-1)(-1) + 0\cdot 1 \end{bmatrix}$$

$$= \begin{bmatrix} 1 & 0 & 0 \\ 0 & 1 & 0 \\ 0 & 0 & 1 \end{bmatrix}$$

47. $$\left[\begin{array}{ccc|ccc} 1 & 2 & 5 & 1 & 0 & 0 \\ 3 & 5 & 9 & 0 & 1 & 0 \\ 1 & 1 & -2 & 0 & 0 & 1 \end{array}\right] \begin{matrix} \\ (-3)R_1 + R_2 \to R_2 \\ (-1)R_1 + R_3 \to R_3 \end{matrix}$$

$$\sim \left[\begin{array}{ccc|ccc} 1 & 2 & 5 & 1 & 0 & 0 \\ 0 & -1 & -6 & -3 & 1 & 0 \\ 0 & -1 & -7 & -1 & 0 & 1 \end{array}\right] \begin{matrix} 2R_2 + R_1 \to R_1 \\ \\ (-1)R_2 + R_3 \to R_3 \end{matrix}$$

$$\sim \left[\begin{array}{ccc|ccc} 1 & 0 & -7 & -5 & 2 & 0 \\ 0 & -1 & -6 & -3 & 1 & 0 \\ 0 & 0 & -1 & 2 & -1 & 1 \end{array}\right] \begin{matrix} (-7)R_3 + R_1 \to R_1 \\ (-6)R_3 + R_2 \to R_2 \\ \\ \end{matrix}$$

$$\sim \left[\begin{array}{ccc|ccc} 1 & 0 & 0 & -19 & 9 & -7 \\ 0 & -1 & 0 & -15 & 7 & -6 \\ 0 & 0 & -1 & 2 & -1 & 1 \end{array}\right] \begin{matrix} \\ (-1)R_2 \to R_2 \\ (-1)R_3 \to R_3 \end{matrix}$$

$$\sim \left[\begin{array}{ccc|ccc} 1 & 0 & 0 & -19 & 9 & -7 \\ 0 & 1 & 0 & 15 & -7 & 6 \\ 0 & 0 & 1 & -2 & 1 & -1 \end{array}\right]$$

$$\text{Hence, } M^{-1} = \begin{bmatrix} -19 & 9 & -7 \\ 15 & -7 & 6 \\ -2 & 1 & -1 \end{bmatrix}$$

$$\text{Check: } M^{-1}M = \begin{bmatrix} -19 & 9 & -7 \\ 15 & -7 & 6 \\ -2 & 1 & -1 \end{bmatrix}\begin{bmatrix} 1 & 2 & 5 \\ 3 & 5 & 9 \\ 1 & 1 & -2 \end{bmatrix}$$

$$= \begin{bmatrix} (-19)1 + 9\cdot 3 + (-7)1 & (-19)2 + 9\cdot 5 + (-7)1 & (-19)5 + 9\cdot 9 + (-7)(-2) \\ 15\cdot 1 + (-7)3 + 6\cdot 1 & 15\cdot 2 + (-7)5 + 6\cdot 1 & 15\cdot 5 + (-7)9 + 6(-2) \\ (-2)1 + 1\cdot 3 + (-1)1 & (-2)2 + 1\cdot 5 + (-1)1 & (-2)5 + 1\cdot 9 + (-1)(-2) \end{bmatrix}$$

$$= \begin{bmatrix} 1 & 0 & 0 \\ 0 & 1 & 0 \\ 0 & 0 & 1 \end{bmatrix}$$

49. $$\left[\begin{array}{ccc|ccc} 2 & 2 & -1 & 1 & 0 & 0 \\ 0 & 4 & -1 & 0 & 1 & 0 \\ -1 & -2 & 1 & 0 & 0 & 1 \end{array}\right] \begin{matrix} \\ R_1 \leftrightarrow R_3 \\ \\ \end{matrix}$$

$$\sim \left[\begin{array}{ccc|ccc} -1 & -2 & 1 & 0 & 0 & 1 \\ 0 & 4 & -1 & 0 & 1 & 0 \\ 2 & 2 & -1 & 1 & 0 & 0 \end{array}\right] \begin{matrix} \\ 2R_1 + R_3 \to R_3 \\ \\ \end{matrix}$$

$$\sim \left[\begin{array}{ccc|ccc} -1 & -2 & 1 & 0 & 0 & 1 \\ 0 & 4 & -1 & 0 & 1 & 0 \\ 0 & -2 & 1 & 1 & 0 & 2 \end{array}\right] \begin{matrix} (-1)R_3 + R_1 \to R_1 \\ 2R_3 + R_2 \to R_2 \\ \\ \end{matrix}$$

$$\sim \left[\begin{array}{ccc|ccc} -1 & 0 & 0 & -1 & 0 & -1 \\ 0 & 0 & 1 & 2 & 1 & 4 \\ 0 & -2 & 1 & -1 & 0 & 2 \end{array}\right] \begin{matrix} \\ (-1)R_2 + R_3 \to R_3 \\ \\ \end{matrix}$$

$$\sim \left[\begin{array}{ccc|ccc} -1 & 0 & 0 & -1 & 0 & -1 \\ 0 & 0 & 1 & 2 & 1 & 4 \\ 0 & -2 & 0 & -1 & -1 & -2 \end{array}\right] R_2 \leftrightarrow R_3$$

$$\sim \left[\begin{array}{ccc|ccc} -1 & 0 & 0 & -1 & 0 & -1 \\ 0 & -2 & 0 & -1 & -1 & -2 \\ 0 & 0 & 1 & 2 & 1 & 4 \end{array}\right] \begin{array}{l} (-1)R_1 \to R_1 \\ (-\frac{1}{2})R_2 \to R_2 \\ \end{array}$$

$$\sim \left[\begin{array}{ccc|ccc} 1 & 0 & 0 & 1 & 0 & 1 \\ 0 & 1 & 0 & \frac{1}{2} & \frac{1}{2} & 1 \\ 0 & 0 & 1 & 2 & 1 & 4 \end{array}\right]$$

The inverse is $\begin{bmatrix} 1 & 0 & 1 \\ \frac{1}{2} & \frac{1}{2} & 1 \\ 2 & 1 & 4 \end{bmatrix}$

51. If the inverse existed we would find it by row operations on the following matrix:

$$\left[\begin{array}{ccc|ccc} 2 & 1 & 1 & 1 & 0 & 0 \\ 1 & 1 & 0 & 0 & 1 & 0 \\ -1 & -1 & 0 & 0 & 0 & 1 \end{array}\right]$$

But consider what happens if we perform $R_2 + R_3 \to R_2$

$$\left[\begin{array}{ccc|ccc} 2 & 1 & 1 & 1 & 0 & 0 \\ 0 & 0 & 0 & 0 & 1 & 1 \\ -1 & -1 & 0 & 0 & 0 & 1 \end{array}\right]$$

Since a row of zeros results to the left of the vertical line, no inverse exists.

53. $\left[\begin{array}{ccc|ccc} 1 & 5 & 10 & 1 & 0 & 0 \\ 0 & 1 & 4 & 0 & 1 & 0 \\ 1 & 6 & 15 & 0 & 0 & 1 \end{array}\right] \begin{array}{l} \\ \\ (-1)R_1 + R_3 \to R_3 \end{array}$

$$\sim \left[\begin{array}{ccc|ccc} 1 & 5 & 10 & 1 & 0 & 0 \\ 0 & 1 & 4 & 0 & 1 & 0 \\ 0 & 1 & 5 & -1 & 0 & 1 \end{array}\right] \begin{array}{l} (-5)R_2 + R_1 \to R_1 \\ \\ (-1)R_2 + R_3 \to R_3 \end{array}$$

$$\sim \left[\begin{array}{ccc|ccc} 1 & 0 & -10 & 1 & -5 & 0 \\ 0 & 1 & 4 & 0 & 1 & 0 \\ 0 & 0 & 1 & -1 & -1 & 1 \end{array}\right] \begin{array}{l} 10R_3 + R_1 \to R_1 \\ (-4)R_3 + R_2 \to R_2 \\ \end{array}$$

$$\sim \left[\begin{array}{ccc|ccc} 1 & 0 & 0 & -9 & -15 & 10 \\ 0 & 1 & 0 & 4 & 5 & -4 \\ 0 & 0 & 1 & -1 & -1 & 1 \end{array}\right]$$

The inverse is $\begin{bmatrix} -9 & -15 & 10 \\ 4 & 5 & -4 \\ -1 & -1 & 1 \end{bmatrix}$

55. $\begin{bmatrix} -1 & -2 \\ 2 & 5 \end{bmatrix}\begin{bmatrix} x_1 \\ x_2 \end{bmatrix} = \begin{bmatrix} k_1 \\ k_2 \end{bmatrix}$

$AX = K$ has solution $X = A^{-1}K$.

We find $A^{-1}K$ for each given K. From problem 37, $A^{-1} = \begin{bmatrix} -5 & -2 \\ 2 & 1 \end{bmatrix}$

(A) $K = \begin{bmatrix} 2 \\ 5 \end{bmatrix}$ $A^{-1}K = \begin{bmatrix} -5 & -2 \\ 2 & 1 \end{bmatrix}\begin{bmatrix} 2 \\ 5 \end{bmatrix} = \begin{bmatrix} -20 \\ 9 \end{bmatrix}$ $x_1 = -20$, $x_2 = 9$

(B) $K = \begin{bmatrix} -4 \\ 1 \end{bmatrix}$ $A^{-1}K = \begin{bmatrix} -5 & -2 \\ 2 & 1 \end{bmatrix}\begin{bmatrix} -4 \\ 1 \end{bmatrix} = \begin{bmatrix} 18 \\ -7 \end{bmatrix}$ $x_1 = 18$, $x_2 = -7$

(C) $K = \begin{bmatrix} -3 \\ -2 \end{bmatrix}$ $A^{-1}K = \begin{bmatrix} -5 & -2 \\ 2 & 1 \end{bmatrix}\begin{bmatrix} -3 \\ -2 \end{bmatrix} = \begin{bmatrix} 19 \\ -8 \end{bmatrix}$ $x_1 = 19$, $x_2 = -8$

57. $\begin{bmatrix} -5 & 7 \\ 2 & -3 \end{bmatrix}\begin{bmatrix} x_1 \\ x_2 \end{bmatrix} = \begin{bmatrix} k_1 \\ k_2 \end{bmatrix}$

$AX = K$ has solution $X = A^{-1}K$.

We find $A^{-1}K$ for each given K. From problem 39, $A^{-1} = \begin{bmatrix} -3 & -7 \\ -2 & -5 \end{bmatrix}$

(A) $K = \begin{bmatrix} -5 \\ 1 \end{bmatrix}$ $A^{-1}K = \begin{bmatrix} -3 & -7 \\ -2 & -5 \end{bmatrix}\begin{bmatrix} -5 \\ 1 \end{bmatrix} = \begin{bmatrix} 8 \\ 5 \end{bmatrix}$ $x_1 = 8$, $x_2 = 5$

(B) $K = \begin{bmatrix} 8 \\ -4 \end{bmatrix}$ $A^{-1}K = \begin{bmatrix} -3 & -7 \\ -2 & -5 \end{bmatrix}\begin{bmatrix} 8 \\ -4 \end{bmatrix} = \begin{bmatrix} 4 \\ 4 \end{bmatrix}$ $x_1 = 4$, $x_2 = 4$

(C) $K = \begin{bmatrix} 6 \\ 0 \end{bmatrix}$ $A^{-1}K = \begin{bmatrix} -3 & -7 \\ -2 & -5 \end{bmatrix}\begin{bmatrix} 6 \\ 0 \end{bmatrix} = \begin{bmatrix} -18 \\ -12 \end{bmatrix}$ $x_1 = -18$, $x_2 = -12$

59. $\begin{bmatrix} 1 & -1 & 0 \\ -1 & 1 & -1 \\ 0 & -1 & 1 \end{bmatrix}\begin{bmatrix} x_1 \\ x_2 \\ x_3 \end{bmatrix} = \begin{bmatrix} k_1 \\ k_2 \\ k_3 \end{bmatrix}$

$AX = K$ has solution $X = A^{-1}K$.

We find $A^{-1}K$ for each given K. From problem 45, $A^{-1} = \begin{bmatrix} 0 & -1 & -1 \\ -1 & -1 & -1 \\ -1 & -1 & 0 \end{bmatrix}$

(A) $K = \begin{bmatrix} 1 \\ 1 \\ 2 \end{bmatrix}$ $A^{-1}K = \begin{bmatrix} 0 & -1 & -1 \\ -1 & -1 & -1 \\ -1 & -1 & 0 \end{bmatrix}\begin{bmatrix} 1 \\ 1 \\ 2 \end{bmatrix} = \begin{bmatrix} -3 \\ -4 \\ -2 \end{bmatrix}$ $x_1 = -3$, $x_2 = -4$, $x_3 = -2$

(B) $K = \begin{bmatrix} -1 \\ 0 \\ -4 \end{bmatrix}$ $A^{-1}K = \begin{bmatrix} 0 & -1 & -1 \\ -1 & -1 & -1 \\ -1 & -1 & 0 \end{bmatrix}\begin{bmatrix} -1 \\ 0 \\ -4 \end{bmatrix} = \begin{bmatrix} 4 \\ 5 \\ 1 \end{bmatrix}$ $x_1 = 4$, $x_2 = 5$, $x_3 = 1$

(C) $K = \begin{bmatrix} 3 \\ -2 \\ 0 \end{bmatrix}$ $A^{-1}K = \begin{bmatrix} 0 & -1 & -1 \\ -1 & -1 & -1 \\ -1 & -1 & 0 \end{bmatrix}\begin{bmatrix} 3 \\ -2 \\ 0 \end{bmatrix} = \begin{bmatrix} 2 \\ -1 \\ -1 \end{bmatrix}$ $x_1 = 2$, $x_2 = -1$, $x_3 = -1$

61. $\begin{bmatrix} 1 & 2 & 5 \\ 3 & 5 & 9 \\ 1 & 1 & -2 \end{bmatrix}\begin{bmatrix} x_1 \\ x_2 \\ x_3 \end{bmatrix} = \begin{bmatrix} k_1 \\ k_2 \\ k_3 \end{bmatrix}$

$AX = K$ has solution $X = A^{-1}K$.

We find $A^{-1}K$ for each given K. From problem 47, $A^{-1} = \begin{bmatrix} -19 & 9 & -7 \\ 15 & -7 & 6 \\ -2 & 1 & -1 \end{bmatrix}$

(A) $K = \begin{bmatrix} 0 \\ 1 \\ 4 \end{bmatrix}$ $A^{-1}K = \begin{bmatrix} -19 & 9 & -7 \\ 15 & -7 & 6 \\ -2 & 1 & -1 \end{bmatrix}\begin{bmatrix} 0 \\ 1 \\ 4 \end{bmatrix} = \begin{bmatrix} -19 \\ 17 \\ -3 \end{bmatrix}$ $x_1 = -19$, $x_2 = 17$, $x_3 = -3$

(B) $K = \begin{bmatrix} 5 \\ -1 \\ 0 \end{bmatrix}$ $A^{-1}K = \begin{bmatrix} -19 & 9 & -7 \\ 15 & -7 & 6 \\ -2 & 1 & -1 \end{bmatrix} \begin{bmatrix} 5 \\ -1 \\ 0 \end{bmatrix} = \begin{bmatrix} -104 \\ 82 \\ -11 \end{bmatrix}$ $x_1 = -104,\ x_2 = 82,\ x_3 = -11$

(C) $K = \begin{bmatrix} -6 \\ 0 \\ 2 \end{bmatrix}$ $A^{-1}K = \begin{bmatrix} -19 & 9 & -7 \\ 15 & -7 & 6 \\ -2 & 1 & -1 \end{bmatrix} \begin{bmatrix} -6 \\ 0 \\ 2 \end{bmatrix} = \begin{bmatrix} 100 \\ -78 \\ 10 \end{bmatrix}$ $x_1 = 100,\ x_2 = -78,\ x_3 = 10$

63. $AX = BX + C$

$AX + (-BX) = (-BX) + BX + C$ Addition property of equality

$AX - BX = 0 + C$ Additive inverse property; definition of subtraction

$AX - BX = C$

$(A - B)X = C$ Right distributive property

[To be more careful, we should write

$AX - BX = AX + -BX$ by the definition of subtraction

$= [A + (-B)]X$ by the right distributive property

$= (A - B)X$ by the definition of subtraction

but the distributive properties are generally understood as applying to subtraction also.]

$(A - B)^{-1}[(A - B)X] = (A - B)^{-1}C$ Left multiplication property of equality

$[(A - B)^{-1}(A - B)]X = (A - B)^{-1}C$ Associative property

$IX = (A - B)^{-1}C$ Multiplicative inverse property

$X = (A - B)^{-1}C$ Multiplicative identity property

65. $X = AX + C$

$X + (-AX) = (-AX) + AX + C$ Addition property of equality

$X - AX = 0 + C$ Additive inverse property; definition of subtraction

$X - AX = C$ Additive identity property

$IX - AX = C$ Multiplicative identity property

$(I - A)X = C$ Right distributive property

$(I - A)^{-1}[(I - A)X] = (I - A)^{-1} C$ Left multiplication property of equality

$[(I - A)^{-1}(I - A)]X = (I - A)^{-1} C$ Associative property

$IX = (I - A)^{-1} C$ Multiplicative inverse property

$X = (I - A)^{-1} C$ Multiplicative identity property

67. $AX + C = 3X$

$AX + (-AX) + C = 3X + (-AX)$ Addition property of equality

$0 + C = 3X - AX$ Additive inverse property; definition of subtraction

$C = 3X - AX$ Additive identity property

$C = 3IX - AX$ Multiplicative identity property

$C = (3I - A)X$ Right distributive property

$(3I - A)^{-1} C = (3I - A)^{-1}[(3I - A)X]$ Left multiplication property of equality

$(3I - A)^{-1}C = [(3I - A)^{-1}(3I - A)]X$ Associative property

$(3I - A)^{-1}C = IX$ Multiplicative inverse property

$(3I - A)^{-1}C = X$ Multiplicative inverse property

69. Try to find A^{-1}:

$$\left[\begin{array}{cc|cc} a & 0 & 1 & 0 \\ 0 & d & 0 & 1 \end{array}\right] \tfrac{1}{a}R_1 \to R_1$$

$$\sim \left[\begin{array}{cc|cc} 1 & 0 & \frac{1}{a} & 0 \\ 0 & d & 0 & 1 \end{array}\right] \tfrac{1}{d}R_2 \to R_2$$

$$\sim \left[\begin{array}{cc|cc} 1 & 0 & \frac{1}{a} & 0 \\ 0 & 1 & 0 & \frac{1}{d} \end{array}\right] \quad \text{The inverse is } A^{-1} = \begin{bmatrix} \frac{1}{a} & 0 \\ 0 & \frac{1}{d} \end{bmatrix}$$

This will exist unless either a or d is zero, which would make $\frac{1}{a}$ or $\frac{1}{d}$ undefined. So A^{-1} exists exactly when both a and d are non-zero.

71. (A) A^{-1}:

$$\left[\begin{array}{cc|cc} 3 & 2 & 1 & 0 \\ -4 & -3 & 0 & 1 \end{array}\right] R_2 + R_1 \to R_1$$

$$\sim \left[\begin{array}{cc|cc} -1 & -1 & 1 & 1 \\ -4 & -3 & 0 & 1 \end{array}\right] (-1)R_1 \to R_1$$

$$\sim \left[\begin{array}{cc|cc} 1 & 1 & -1 & -1 \\ -4 & -3 & 0 & 1 \end{array}\right] 4R_1 + R_2 \to R_2$$

$$\sim \left[\begin{array}{cc|cc} 1 & 1 & -1 & -1 \\ 0 & 1 & -4 & -3 \end{array}\right] (-1)R_2 + R_1 \to R_1$$

$$\sim \left[\begin{array}{cc|cc} 1 & 0 & 3 & 2 \\ 0 & 1 & -4 & -3 \end{array}\right] \quad A^{-1} = \begin{bmatrix} 3 & 2 \\ -4 & -3 \end{bmatrix}$$

A^2:

$$\begin{bmatrix} 3 & 2 \\ -4 & -3 \end{bmatrix}\begin{bmatrix} 3 & 2 \\ -4 & -3 \end{bmatrix}$$

$$= \begin{bmatrix} 3 \cdot 3 + 2(-4) & 3 \cdot 2 + 2(-3) \\ (-4)3 + (-3)(-4) & (-4)2 + (-3)(-3) \end{bmatrix}$$

$$= \begin{bmatrix} 1 & 0 \\ 0 & 1 \end{bmatrix}$$

(B) A^{-1}:

$$\left[\begin{array}{cc|cc} -2 & -1 & 1 & 0 \\ 3 & 2 & 0 & 1 \end{array}\right] R_2 + R_1 \to R_1$$

$$\sim \left[\begin{array}{cc|cc} 1 & 1 & 1 & 1 \\ 3 & 2 & 0 & 1 \end{array}\right] (-3)R_1 + R_2 \to R_2$$

$$\sim \left[\begin{array}{cc|cc} 1 & 1 & 1 & 1 \\ 0 & -1 & -3 & -2 \end{array}\right] R_2 + R_1 \to R_1$$

$$\sim \left[\begin{array}{cc|cc} 1 & 0 & -2 & -1 \\ 0 & -1 & -3 & -2 \end{array}\right] (-1)R_2 \to R_2$$

$$\sim \left[\begin{array}{cc|cc} 1 & 0 & -2 & -1 \\ 0 & 1 & 3 & 2 \end{array}\right] \quad A^{-1} = \begin{bmatrix} -2 & -1 \\ 3 & 2 \end{bmatrix}$$

A^2:

$$\begin{bmatrix} -2 & -1 \\ 3 & 2 \end{bmatrix}\begin{bmatrix} -2 & -1 \\ 3 & 2 \end{bmatrix}$$

$$= \begin{bmatrix} (-2)(-2) + (-1)3 & (-2)(-1) + (-1)2 \\ 3(-2) + 2 \cdot 3 & 3(-1) + 2 \cdot 2 \end{bmatrix}$$

$$= \begin{bmatrix} 1 & 0 \\ 0 & 1 \end{bmatrix}$$

Note that in both cases $A^{-1} = A$ and $A^2 = I$.

73. (A) We calculate A^{-1} by row operations on

$$\left[\begin{array}{cc|cc} 4 & 2 & 1 & 0 \\ 1 & 3 & 0 & 1 \end{array}\right] R_1 \leftrightarrow R_2$$

$$\sim \left[\begin{array}{cc|cc} 1 & 3 & 0 & 1 \\ 4 & 2 & 1 & 0 \end{array}\right] (-4)R_1 + R_2 \to R_2$$

$$\sim \left[\begin{array}{cc|cc} 1 & 3 & 0 & 1 \\ 0 & -10 & 1 & -4 \end{array}\right] \left(-\tfrac{1}{10}\right)R_2 \to R_2$$

$$\sim \left[\begin{array}{cc|cc} 1 & 3 & 0 & 1 \\ 0 & 1 & -\frac{1}{10} & \frac{2}{5} \end{array}\right] (-3)R_2 + R_1 \to R_1$$

$$\sim \left[\begin{array}{cc|cc} 1 & 0 & \frac{3}{10} & -\frac{1}{5} \\ 0 & 1 & -\frac{1}{10} & \frac{2}{5} \end{array}\right]$$

Hence $A^{-1} = \begin{bmatrix} \frac{3}{10} & -\frac{1}{5} \\ -\frac{1}{10} & \frac{2}{5} \end{bmatrix}$

We calculate $(A^{-1})^{-1}$ by row operations on

$$\left[\begin{array}{cc|cc} \frac{3}{10} & -\frac{1}{5} & 1 & 0 \\ -\frac{1}{10} & \frac{2}{5} & 0 & 1 \end{array}\right] \tfrac{10}{3} R_1 \to R_1$$

$$\sim \left[\begin{array}{cc|cc} 1 & -\frac{2}{3} & \frac{10}{3} & 0 \\ -\frac{1}{10} & \frac{2}{5} & 0 & 1 \end{array}\right] \tfrac{1}{10} R_1 + R_2 \to R_2$$

$$\sim \left[\begin{array}{cc|cc} 1 & -\frac{2}{3} & \frac{10}{3} & 0 \\ 0 & \frac{1}{3} & \frac{1}{3} & 1 \end{array}\right] 2R_2 + R_1 \to R_1$$

$$\sim \left[\begin{array}{cc|cc} 1 & 0 & 4 & 2 \\ 0 & \frac{1}{3} & \frac{1}{3} & 1 \end{array}\right] 3R_2 \to R_2$$

$$\sim \left[\begin{array}{cc|cc} 1 & 0 & 4 & 2 \\ 0 & 1 & 1 & 3 \end{array}\right]$$

Hence $(A^{-1})^{-1} = \begin{bmatrix} 4 & 2 \\ 1 & 3 \end{bmatrix}$

(B) We calculate A^{-1} by row operations on

$$\left[\begin{array}{cc|cc} 5 & 5 & 1 & 0 \\ -1 & 3 & 0 & 1 \end{array}\right] R_1 \leftrightarrow R_2$$

$$\sim \left[\begin{array}{cc|cc} -1 & 3 & 0 & 1 \\ 5 & 5 & 1 & 0 \end{array}\right] 5R_1 + R_2 \to R_2$$

$$\sim \left[\begin{array}{cc|cc} -1 & 3 & 0 & 1 \\ 0 & 20 & 1 & 5 \end{array}\right] \tfrac{1}{20} R_2 \to R_2$$

$$\sim \left[\begin{array}{cc|cc} -1 & 3 & 0 & 1 \\ 0 & 1 & \frac{1}{20} & \frac{1}{4} \end{array}\right] (-3)R_2 + R_1 \to R_1$$

$$\sim \left[\begin{array}{cc|cc} -1 & 0 & -\frac{3}{20} & \frac{1}{4} \\ 0 & 1 & \frac{1}{20} & \frac{1}{4} \end{array}\right] (-1)R_1 \to R_1$$

$$\sim \left[\begin{array}{cc|cc} 1 & 0 & \frac{3}{20} & -\frac{1}{4} \\ 0 & 1 & \frac{1}{20} & \frac{1}{4} \end{array}\right]$$

Hence $A^{-1} = \begin{bmatrix} \frac{3}{20} & -\frac{1}{4} \\ \frac{1}{20} & \frac{1}{4} \end{bmatrix}$

We calculate A^{-1} by row operations on

$$\left[\begin{array}{cc|cc} \frac{3}{20} & -\frac{1}{4} & 1 & 0 \\ \frac{1}{20} & \frac{1}{4} & 0 & 1 \end{array}\right] \tfrac{20}{3} R_1 \to R_1$$

$$\sim \left[\begin{array}{cc|cc} 1 & -\frac{5}{3} & \frac{20}{3} & 0 \\ \frac{1}{20} & \frac{1}{4} & 0 & 1 \end{array}\right] -\tfrac{1}{20} R_1 + R_2 \to R_2$$

$$\sim \left[\begin{array}{cc|cc} 1 & -\frac{5}{3} & \frac{20}{3} & 0 \\ 0 & \frac{1}{3} & -\frac{1}{3} & 1 \end{array}\right] 5R_2 + R_1 \to R_1$$

$$\sim \left[\begin{array}{cc|cc} 1 & 0 & 5 & 5 \\ 0 & \frac{1}{3} & -\frac{1}{3} & 1 \end{array}\right] 3R_2 \to R_2$$

$$\sim \left[\begin{array}{cc|cc} 1 & 0 & 5 & 5 \\ 0 & 1 & -1 & 3 \end{array}\right]$$

Hence $(A^{-1})^{-1} = \begin{bmatrix} 5 & 5 \\ -1 & 3 \end{bmatrix}$

Note that in both cases the inverse of the inverse works out to be the original matrix.

75. (A) Using a graphing calculator, we enter A and B and calculate A^{-1} and B^{-1}

```
[A]
          [[3 4]
           [2 3]]
[A]-1
         [[3  -4]
          [-2 3 ]]
```

```
[B]
          [[3 7]
           [2 5]]
[B]-1
         [[5  -7]
          [-2 3 ]]
```

Then $(AB)^{-1}$ and $A^{-1}B^{-1}$ are calculated as

```
([A]*[B])-1
       [[29   -41]
        [-12  17 ]]
[A]-1*[B]-1
       [[23   -33]
        [-16  23 ]]
```

Finally $B^{-1}A^{-1}$ is calculated as

```
[B]-1*[A]-1
      [[29   -41]
       [-12 17  ]]
```

(B) Using a graphing calculator, we enter A and B and calculate A^{-1} and B^{-1}

```
[A]
        [[1 -1]
         [2 3 ]]
[A]-1
      [[.6   .2]
       [-.4 .2]]
```

```
[B]
        [[6 2]
         [2 1]]
[B]-1
      [[.5 -1]
       [-1 3 ]]
```

Then $(AB)^{-1}$ and $A^{-1}B^{-1}$ are calculated as

```
([A]*[B])-1
     [[.7     -.1]
      [-1.8 .4  ]]
[A]-1*[B]-1
        [[.1   0]
         [-.4 1]]
```

Finally $B^{-1}A^{-1}$ is calculated as

```
[B]-1*[A]-1
     [[.7     -.1]
      [-1.8 .4  ]]
```

Notice that in each case $(AB)^{-1} = B^{-1}A^{-1}$ but $(AB)^{-1} \neq A^{-1}B^{-1}$.

77. Using the assignation numbers 1 to 27 with the letters of the alphabet and a blank as in the text, write

C	A	T		I	N		T	H	E		H	A	T
3	1	20	27	9	14	27	20	8	5	27	8	1	20

and calculate

$$\begin{bmatrix} 3 & 5 \\ 1 & 2 \end{bmatrix} \begin{bmatrix} 3 & 20 & 9 & 27 & 8 & 27 & 1 \\ 1 & 27 & 14 & 20 & 5 & 8 & 20 \end{bmatrix} = \begin{bmatrix} 14 & 195 & 97 & 181 & 49 & 121 & 103 \\ 5 & 74 & 37 & 67 & 18 & 43 & 41 \end{bmatrix}$$

The encoded message is thus

14 5 195 74 97 37 181 67 49 18 121 43 103 41

79. The inverse of matrix A is easily calculated to be

$$A^{-1} = \begin{bmatrix} 2 & -5 \\ -1 & 3 \end{bmatrix}$$

Putting the coded message into matrix form and multiplying by A^{-1} yields:

$$\begin{bmatrix} 2 & -5 \\ -1 & 3 \end{bmatrix} \begin{bmatrix} 111 & 40 & 177 & 50 & 116 & 86 & 62 & 121 & 68 \\ 43 & 15 & 68 & 19 & 45 & 29 & 22 & 43 & 27 \end{bmatrix}$$

$$= \begin{bmatrix} 7 & 5 & 14 & 5 & 7 & 27 & 14 & 27 & 1 \\ 18 & 5 & 27 & 7 & 19 & 1 & 4 & 8 & 13 \end{bmatrix}$$

This decodes to

7	18	5	5	14	27	5	7	7	19	27	1	14	4	27	8	1	13
G	R	E	E	N		E	G	G	S		A	N	D		H	A	M

81. Using the assignation of numbers 1 to 27 with the letters of the alphabet and a blank as in the text, write

D	W	I	G	H	T		D	A	V	I	D		E	I	S	E	N	H	O	W	E	R		
4	23	9	7	8	20	27	4	1	22	9	4	27	5	9	19	5	14	8	15	23	5	18	27	27

and calculate

$$\begin{bmatrix} 1 & 0 & 1 & 0 & 1 \\ 0 & 1 & 1 & 0 & 3 \\ 2 & 1 & 1 & 1 & 1 \\ 0 & 0 & 1 & 0 & 2 \\ 1 & 1 & 1 & 2 & 1 \end{bmatrix} \begin{bmatrix} 4 & 20 & 9 & 19 & 23 \\ 23 & 27 & 4 & 5 & 5 \\ 9 & 4 & 27 & 14 & 18 \\ 7 & 1 & 5 & 8 & 27 \\ 8 & 22 & 9 & 15 & 27 \end{bmatrix} = \begin{bmatrix} 21 & 46 & 45 & 48 & 68 \\ 56 & 97 & 58 & 64 & 104 \\ 55 & 94 & 63 & 80 & 123 \\ 25 & 48 & 45 & 44 & 72 \\ 58 & 75 & 59 & 69 & 127 \end{bmatrix}$$

The encoded message is thus

21 56 55 25 58 46 97 94 48 75 45 58 63 45 59 48 64
80 44 69 68 104 123 72 127

83. The inverse of B is calculated to be

$$B^{-1} = \begin{bmatrix} -2 & -1 & 2 & 2 & -1 \\ 3 & 2 & -2 & -4 & 1 \\ 6 & 2 & -4 & -5 & 2 \\ -2 & -1 & 1 & 2 & 0 \\ -3 & -1 & 2 & 3 & -1 \end{bmatrix}$$

Putting the coded message into matrix form and multiplying by B^{-1} yields

$$\begin{bmatrix} -2 & -1 & 2 & 2 & -1 \\ 3 & 2 & -2 & -4 & 1 \\ 6 & 2 & -4 & -5 & 2 \\ -2 & -1 & 1 & 2 & 0 \\ -3 & -1 & 2 & 3 & -1 \end{bmatrix} \begin{bmatrix} 41 & 25 & 43 & 44 & 68 \\ 84 & 56 & 54 & 67 & 135 \\ 82 & 67 & 89 & 86 & 136 \\ 44 & 20 & 39 & 44 & 81 \\ 74 & 54 & 102 & 90 & 149 \end{bmatrix} = \begin{bmatrix} 12 & 14 & 14 & 15 & 14 \\ 25 & 27 & 5 & 8 & 27 \\ 14 & 2 & 19 & 14 & 27 \\ 4 & 1 & 27 & 19 & 27 \\ 15 & 9 & 10 & 15 & 27 \end{bmatrix}$$

This decodes to

12	25	14	4	15	14	27	2	1	9	14	5	19	27	10	15	8	14	19	15	14	27	27	27
L	Y	N	D	O	N		B	A	I	N	E	S		J	O	H	N	S	O	N			

85. The system to be solved, for an arbitrary return, is derived as follows:
Let x_1 = number of \$4 tickets sold
x_2 = number of \$8 tickets sold
Then $x_1 + x_2 = 10{,}000$ number of seats
$4x_1 + 8x_2 = k_2$ return required
We solve the system by writing it as a matrix equation.

$$\overset{A}{\begin{bmatrix} 1 & 1 \\ 4 & 8 \end{bmatrix}} \overset{X}{\begin{bmatrix} x_1 \\ x_2 \end{bmatrix}} = \overset{B}{\begin{bmatrix} 10{,}000 \\ k_2 \end{bmatrix}}$$

If A^{-1} exists, then $X = A^{-1}B$. To find A^{-1}, we perform row operations on

$$\left[\begin{array}{cc|cc} 1 & 1 & 1 & 0 \\ 4 & 8 & 0 & 1 \end{array}\right] (-4)R_1 + R_2 \to R_2$$

$$\sim \left[\begin{array}{cc|cc} 1 & 1 & 1 & 0 \\ 0 & 4 & -4 & 1 \end{array}\right] 0.25R_2 \to R_2$$

$$\sim \left[\begin{array}{cc|cc} 1 & 1 & 1 & 0 \\ 0 & 1 & -1 & 0.25 \end{array}\right] (-1)R_2 + R_1 \to R_1$$

$$\sim \left[\begin{array}{cc|cc} 1 & 0 & 2 & -0.25 \\ 0 & 1 & -1 & 0.25 \end{array}\right]$$

Hence $A^{-1} = \begin{bmatrix} 2 & -0.25 \\ -1 & 0.25 \end{bmatrix}$

Check: $A^{-1}A = \begin{bmatrix} 2 & -0.25 \\ -1 & 0.25 \end{bmatrix} \begin{bmatrix} 1 & 4 \\ 1 & 8 \end{bmatrix} = \begin{bmatrix} 1 & 0 \\ 0 & 1 \end{bmatrix}$

We can now solve the system as

$$\overset{X}{\begin{bmatrix} x_1 \\ x_2 \end{bmatrix}} = \overset{A^{-1}}{\begin{bmatrix} 2 & -0.25 \\ -1 & 0.25 \end{bmatrix}} \overset{B}{\begin{bmatrix} 10{,}000 \\ k_2 \end{bmatrix}}$$

If $k_2 = 56{,}000$ (Concert 1),

$$\begin{bmatrix} x_1 \\ x_2 \end{bmatrix} = \begin{bmatrix} 2 & -0.25 \\ -1 & 0.25 \end{bmatrix} \begin{bmatrix} 10{,}000 \\ 56{,}000 \end{bmatrix} = \begin{bmatrix} 6{,}000 \\ 4{,}000 \end{bmatrix}$$

Concert 1: 6,000 \$4 tickets and 4,000 \$8 tickets

If k_2 = 60,000 (Concert 2),

$$\begin{bmatrix} x_1 \\ x_2 \end{bmatrix} = \begin{bmatrix} 2 & -0.25 \\ -1 & 0.25 \end{bmatrix} \begin{bmatrix} 10,000 \\ 60,000 \end{bmatrix} = \begin{bmatrix} 5,000 \\ 5,000 \end{bmatrix}$$

Concert 2: 5,000 \$4 tickets and 5,000 \$8 tickets

If k_2 = 68,000 (Concert 3),

$$\begin{bmatrix} x_1 \\ x_2 \end{bmatrix} = \begin{bmatrix} 2 & -0.25 \\ -1 & 0.25 \end{bmatrix} \begin{bmatrix} 10,000 \\ 68,000 \end{bmatrix} = \begin{bmatrix} 3,000 \\ 7,000 \end{bmatrix}$$

Concert 3: 3,000 \$4 tickets and 7,000 \$8 tickets

87. We solve the system, for arbitrary V_1 and V_2, by writing it as a matrix equation.

$$\overset{A}{\begin{bmatrix} 1 & -1 & 1 \\ 1 & 1 & 0 \\ 0 & 1 & 2 \end{bmatrix}} \overset{J}{\begin{bmatrix} I_1 \\ I_2 \\ I_3 \end{bmatrix}} = \overset{B}{\begin{bmatrix} 0 \\ V_1 \\ V_2 \end{bmatrix}}$$

If A^{-1} exists, then $J = A^{-1}B$. To find A^{-1}, we perform row operations on

$$\left[\begin{array}{ccc|ccc} 1 & -1 & 1 & 1 & 0 & 0 \\ 1 & 1 & 0 & 0 & 1 & 0 \\ 0 & 1 & 2 & 0 & 0 & 1 \end{array}\right] (-1)R_1 + R_2 \rightarrow R_2$$

$$\sim \left[\begin{array}{ccc|ccc} 1 & -1 & 1 & 1 & 0 & 0 \\ 0 & 2 & -1 & -1 & 1 & 0 \\ 0 & 1 & 2 & 0 & 0 & 1 \end{array}\right] R_2 \leftrightarrow R_3$$

$$\sim \left[\begin{array}{ccc|ccc} 1 & -1 & 1 & 1 & 0 & 0 \\ 0 & 1 & 2 & 0 & 0 & 1 \\ 0 & 2 & -1 & -1 & 1 & 0 \end{array}\right] \begin{matrix} R_2 + R_1 \rightarrow R_1 \\ \\ (-2)R_2 + R_3 \rightarrow R_3 \end{matrix}$$

$$\sim \left[\begin{array}{ccc|ccc} 1 & 0 & 3 & 1 & 0 & 1 \\ 0 & 1 & 2 & 0 & 0 & 1 \\ 0 & 0 & -5 & -1 & 1 & -2 \end{array}\right] -\tfrac{1}{5} R_3 \rightarrow R_3$$

$$\sim \left[\begin{array}{ccc|ccc} 1 & 0 & 3 & 1 & 0 & 1 \\ 0 & 1 & 2 & 0 & 0 & 1 \\ 0 & 0 & 1 & \frac{1}{5} & -\frac{1}{5} & \frac{2}{5} \end{array}\right] \begin{matrix} (-3)R_3 + R_1 \rightarrow R_1 \\ (-2)R_3 + R_2 \rightarrow R_2 \\ \\ \end{matrix}$$

$$\sim \left[\begin{array}{ccc|ccc} 1 & 0 & 0 & \frac{2}{5} & \frac{3}{5} & -\frac{1}{5} \\ 0 & 1 & 0 & -\frac{2}{5} & \frac{2}{5} & \frac{1}{5} \\ 0 & 0 & 1 & \frac{1}{5} & -\frac{1}{5} & \frac{2}{5} \end{array}\right]$$

Hence $A^{-1} = \frac{1}{5}\begin{bmatrix} 2 & 3 & -1 \\ -2 & 2 & 1 \\ 1 & -1 & 2 \end{bmatrix}$

Check: $A^{-1}A = \frac{1}{5}\begin{bmatrix} 2 & 3 & -1 \\ -2 & 2 & 1 \\ 1 & -1 & 2 \end{bmatrix}\begin{bmatrix} 1 & -1 & 1 \\ 1 & 1 & 0 \\ 0 & 1 & 2 \end{bmatrix} = \begin{bmatrix} 1 & 0 & 0 \\ 0 & 1 & 0 \\ 0 & 0 & 1 \end{bmatrix}$

We can now solve the system as

$$\overset{J}{\begin{bmatrix} I_1 \\ I_2 \\ I_3 \end{bmatrix}} = \frac{1}{5}\overset{A^{-1}}{\begin{bmatrix} 2 & 3 & -1 \\ -2 & 2 & 1 \\ 1 & -1 & 2 \end{bmatrix}} \overset{B}{\begin{bmatrix} 0 \\ V_1 \\ V_2 \end{bmatrix}}$$

(A) $V_1 = 10 \quad V_2 = 10$

$$\begin{bmatrix} I_1 \\ I_2 \\ I_3 \end{bmatrix} = \frac{1}{5}\begin{bmatrix} 2 & 3 & -1 \\ -2 & 2 & 1 \\ 1 & -1 & 2 \end{bmatrix}\begin{bmatrix} 0 \\ 10 \\ 10 \end{bmatrix} = \begin{bmatrix} 4 \\ 6 \\ 2 \end{bmatrix} \qquad I_1 = 4,\ I_2 = 6,\ I_3 = 2 \text{ (amperes)}$$

(B) $V_1 = 10 \quad V_2 = 15$

$$\begin{bmatrix} I_1 \\ I_2 \\ I_3 \end{bmatrix} = \frac{1}{5}\begin{bmatrix} 2 & 3 & -1 \\ -2 & 2 & 1 \\ 1 & -1 & 2 \end{bmatrix}\begin{bmatrix} 0 \\ 10 \\ 15 \end{bmatrix} = \begin{bmatrix} 3 \\ 7 \\ 4 \end{bmatrix} \qquad I_1 = 3,\ I_2 = 7,\ I_3 = 4 \text{ (amperes)}$$

(C) $V_1 = 15 \quad V_2 = 10$

$$\begin{bmatrix} I_1 \\ I_2 \\ I_3 \end{bmatrix} = \frac{1}{5}\begin{bmatrix} 2 & 3 & -1 \\ -2 & 2 & 1 \\ 1 & -1 & 2 \end{bmatrix}\begin{bmatrix} 0 \\ 15 \\ 10 \end{bmatrix} = \begin{bmatrix} 7 \\ 8 \\ 1 \end{bmatrix} \qquad I_1 = 7,\ I_2 = 8,\ I_3 = 1 \text{ (amperes)}$$

89. If the graph of $f(x) = ax^2 + bx + c$ passes through a point, the coordinates of the point must satisfy the equation of the graph. Hence

$k_1 = a(1)^2 + b(1) + c$

$k_2 = a(2)^2 + b(2) + c$

$k_3 = a(3)^2 + b(3) + c$

After simplification, we obtain:

$a + b + c = k_1$

$4a + 2b + c = k_2$

$9a + 3b + c = k_3$

We solve this system, for arbitrary k_1, k_2, k_3, by writing it as a matrix equation.

$$\overset{A}{\begin{bmatrix} 1 & 1 & 1 \\ 4 & 2 & 1 \\ 9 & 3 & 1 \end{bmatrix}}\overset{X}{\begin{bmatrix} a \\ b \\ c \end{bmatrix}} = \overset{B}{\begin{bmatrix} k_1 \\ k_2 \\ k_3 \end{bmatrix}}$$

If A^{-1} exists, then $X = A^{-1}B$. To find A^{-1} we perform row operations on

$$\left[\begin{array}{ccc|ccc} 1 & 1 & 1 & 1 & 0 & 0 \\ 4 & 2 & 1 & 0 & 1 & 0 \\ 9 & 3 & 1 & 0 & 0 & 1 \end{array}\right] \begin{array}{l} \\ (-4)R_1 + R_2 \to R_2 \\ (-9)R_1 + R_3 \to R_3 \end{array}$$

$$\sim \left[\begin{array}{ccc|ccc} 1 & 1 & 1 & 1 & 0 & 0 \\ 0 & -2 & -3 & -4 & 1 & 0 \\ 0 & -6 & -8 & -9 & 0 & 1 \end{array}\right] -\frac{1}{2}R_2 \to R_2$$

$$\sim \left[\begin{array}{ccc|ccc} 1 & 1 & 1 & 1 & 0 & 0 \\ 0 & 1 & \frac{3}{2} & 2 & -\frac{1}{2} & 0 \\ 0 & -6 & -8 & -9 & 0 & 1 \end{array}\right] \begin{array}{l} (-1)R_2 + R_1 \to R_1 \\ \\ 6R_2 + R_3 \to R_3 \end{array}$$

$$\sim \left[\begin{array}{ccc|ccc} 1 & 0 & -\frac{1}{2} & -1 & \frac{1}{2} & 0 \\ 0 & 1 & \frac{3}{2} & 2 & -\frac{1}{2} & 0 \\ 0 & 0 & 1 & 3 & -3 & 1 \end{array}\right] \begin{array}{l} \frac{1}{2}R_3 + R_1 \to R_1 \\ (-\frac{3}{2})R_3 + R_2 \to R_2 \\ \\ \end{array}$$

$$\sim \left[\begin{array}{ccc|ccc} 1 & 0 & 0 & \frac{1}{2} & -1 & \frac{1}{2} \\ 0 & 1 & 0 & -\frac{5}{2} & 4 & -\frac{3}{2} \\ 0 & 0 & 1 & 3 & -3 & 1 \end{array}\right]$$

Hence $A^{-1} = \frac{1}{2}\begin{bmatrix} 1 & -2 & 1 \\ -5 & 8 & -3 \\ 6 & -6 & 2 \end{bmatrix}$

Check: $A^{-1}A = \frac{1}{2}\begin{bmatrix} 1 & -2 & 1 \\ -5 & 8 & -3 \\ 6 & -6 & 2 \end{bmatrix}\begin{bmatrix} 1 & 1 & 1 \\ 4 & 2 & 1 \\ 9 & 3 & 1 \end{bmatrix} = \begin{bmatrix} 1 & 0 & 0 \\ 0 & 1 & 0 \\ 0 & 0 & 1 \end{bmatrix}$

We can now solve the system as

$$\underset{X}{\begin{bmatrix} a \\ b \\ c \end{bmatrix}} = \underset{A^{-1}}{\frac{1}{2}\begin{bmatrix} 1 & -2 & 1 \\ -5 & 8 & -3 \\ 6 & -6 & 2 \end{bmatrix}}\underset{B}{\begin{bmatrix} k_1 \\ k_2 \\ k_3 \end{bmatrix}}$$

(A) $\begin{bmatrix} a \\ b \\ c \end{bmatrix} = \frac{1}{2}\begin{bmatrix} 1 & -2 & 1 \\ -5 & 8 & -3 \\ 6 & -6 & 2 \end{bmatrix}\begin{bmatrix} -2 \\ 1 \\ 6 \end{bmatrix} = \begin{bmatrix} 1 \\ 0 \\ -3 \end{bmatrix}$ $a = 1$, $b = 0$, $c = -3$

(B) $\begin{bmatrix} a \\ b \\ c \end{bmatrix} = \frac{1}{2}\begin{bmatrix} 1 & -2 & 1 \\ -5 & 8 & -3 \\ 6 & -6 & 2 \end{bmatrix}\begin{bmatrix} 4 \\ 3 \\ -2 \end{bmatrix} = \begin{bmatrix} -2 \\ 5 \\ 1 \end{bmatrix}$ $a = -2$, $b = 5$, $c = 1$

(C) $\begin{bmatrix} a \\ b \\ c \end{bmatrix} = \frac{1}{2}\begin{bmatrix} 1 & -2 & 1 \\ -5 & 8 & -3 \\ 6 & -6 & 2 \end{bmatrix}\begin{bmatrix} 8 \\ -5 \\ 4 \end{bmatrix} = \begin{bmatrix} 11 \\ -46 \\ 43 \end{bmatrix}$ $a = 11$, $b = -46$, $c = 43$

91. The system to be solved, for an arbitrary diet, is derived as follows:
Let x_1 = amount of mix A
x_2 = amount of mix B
Then $0.20x_1 + 0.10x_2 = k_1$ (k_1 = amount of protein)
$0.02x_1 + 0.06x_2 = k_2$ (k_2 = amount of fat)
We solve the system by writing it as a matrix equation.

$$\underset{A}{\begin{bmatrix} 0.20 & 0.10 \\ 0.02 & 0.06 \end{bmatrix}}\underset{X}{\begin{bmatrix} x_1 \\ x_2 \end{bmatrix}} = \underset{B}{\begin{bmatrix} k_1 \\ k_2 \end{bmatrix}}$$

If A^{-1} exists, then $X = A^{-1}B$. To find A^{-1}, we perform row operations on

$$\left[\begin{array}{cc|cc} 0.20 & 0.10 & 1 & 0 \\ 0.02 & 0.06 & 0 & 1 \end{array}\right] \begin{array}{l} 5R_1 \to R_1 \\ 50R_2 \to R_2 \end{array}$$

$$\sim \left[\begin{array}{cc|cc} 1 & 0.5 & 5 & 0 \\ 1 & 3 & 0 & 50 \end{array}\right] (-1)R_1 + R_2 \to R_2$$

$$\sim \left[\begin{array}{cc|cc} 1 & 0.5 & 5 & 0 \\ 0 & 2.5 & -5 & 50 \end{array}\right] 0.4R_2 \to R_2$$

$$\sim \left[\begin{array}{cc|cc} 1 & 0.5 & 5 & 0 \\ 0 & 1 & -2 & 20 \end{array}\right] (-0.5)R_2 + R_1 \to R_1$$

$$\sim \left[\begin{array}{cc|cc} 1 & 0 & 6 & -10 \\ 0 & 1 & -2 & 20 \end{array}\right]$$

Hence $A^{-1} = \begin{bmatrix} 6 & -10 \\ -2 & 20 \end{bmatrix}$

Check: $A^{-1}A = \begin{bmatrix} 6 & -10 \\ -2 & 20 \end{bmatrix}\begin{bmatrix} 0.20 & 0.10 \\ 0.02 & 0.06 \end{bmatrix} = \begin{bmatrix} 1 & 0 \\ 0 & 1 \end{bmatrix}$

We can now solve the system as

$$\overset{X}{\begin{bmatrix} x_1 \\ x_2 \end{bmatrix}} = \overset{A^{-1}}{\begin{bmatrix} 6 & -10 \\ -2 & 20 \end{bmatrix}} \overset{B}{\begin{bmatrix} k_1 \\ k_2 \end{bmatrix}}$$

For Diet 1, $k_1 = 20$ and $k_2 = 6$

$$\begin{bmatrix} x_1 \\ x_2 \end{bmatrix} = \begin{bmatrix} 6 & -10 \\ -2 & 20 \end{bmatrix} \begin{bmatrix} 20 \\ 6 \end{bmatrix} = \begin{bmatrix} 60 \\ 80 \end{bmatrix}$$

Diet 1: 60 ounces Mix *A* and 80 ounces Mix *B*

For Diet 2, $k_1 = 10$ and $k_2 = 4$

$$\begin{bmatrix} x_1 \\ x_2 \end{bmatrix} = \begin{bmatrix} 6 & -10 \\ -2 & 20 \end{bmatrix} \begin{bmatrix} 10 \\ 4 \end{bmatrix} = \begin{bmatrix} 20 \\ 60 \end{bmatrix}$$

Diet 2: 20 ounces Mix *A* and 60 ounces Mix *B*

For Diet 3, $k_1 = 10$ and $k_2 = 6$

$$\begin{bmatrix} x_1 \\ x_2 \end{bmatrix} = \begin{bmatrix} 6 & -10 \\ -2 & 20 \end{bmatrix} \begin{bmatrix} 10 \\ 6 \end{bmatrix} = \begin{bmatrix} 0 \\ 100 \end{bmatrix}$$

Diet 3: 0 ounces Mix *A* and 100 ounces Mix *B*

Section 10-6

1. $x^2 + y^2 = 169$
$x = -12$
$(-12)^2 + y^2 = 169$
$y^2 = 25$
$y = \pm 5$
Solution: $(-12, 5)$, $(-12, -5)$

Check: $-12 \overset{\checkmark}{=} -12$
$(-12)^2 + (\pm 5)^2 \overset{\checkmark}{=} 169$

3. $8x^2 - y^2 = 16$
$y = 2x$
Substitute y from the second equation into the first equation.
$8x^2 - (2x)^2 = 16$
$8x^2 - 4x^2 = 16$
$4x^2 = 16$
$x^2 = 4$
$x = \pm 2$
For $x = 2$: $y = 2(2)$, $y = 4$ For $x = -2$: $y = 2(-2)$, $y = -4$
Solutions: $(2, 4)$, $(-2, -4)$

Check:
For $(2, 4)$: $4 \overset{\checkmark}{=} 2 \cdot 2$, $8(2)^2 - 4^2 \overset{\checkmark}{=} 16$
For $(-2, -4)$: $-4 \overset{\checkmark}{=} 2(-2)$, $8(-2)^2 - (-4)^2 \overset{\checkmark}{=} 16$

5. $3x^2 - 2y^2 = 25$
$x + y = 0$
Solve for y in the first degree equation
$y = -x$
Substitute into the second degree equation.
$3x^2 - 2(-x)^2 = 25$
$x^2 = 25$
$x = \pm 5$
For $x = 5$: $y = -5$ For $x = -5$: $y = 5$
Solutions: $(5, -5)$, $(-5, 5)$

Check:
For $(5, -5)$: $5 + (-5) \overset{\checkmark}{=} 0$, $3(5)^2 - 2(-5)^2 \overset{\checkmark}{=} 25$
For $(-5, 5)$: $(-5) + 5 \overset{\checkmark}{=} 0$, $3(-5)^2 - 2(5)^2 \overset{\checkmark}{=} 25$

From this point on we will not show the checking steps for lack of space. The student should perform these checking steps, however.

7. $y^2 = x$
$x - 2y = 2$

Solve for x in the first degree equation.
$x = 2y + 2$
Substitute into the second degree equation.

$$y^2 = 2y + 2$$
$$y^2 - 2y - 2 = 0$$
$$y = \frac{-b \pm \sqrt{b^2 - 4ac}}{2a} \quad a = 1,\ b = -2,\ c = -2$$
$$y = \frac{-(-2) \pm \sqrt{(-2)^2 - 4(1)(-2)}}{2(1)}$$
$$y = \frac{2 \pm \sqrt{12}}{2}$$
$$y = 1 \pm \sqrt{3}$$

For $y = 1 + \sqrt{3}$
$x = 2(1 + \sqrt{3}) + 2$
$x = 4 + 2\sqrt{3}$

For $y = 1 - \sqrt{3}$
$x = 2(1 - \sqrt{3}) + 2$
$x = 4 - 2\sqrt{3}$

Solutions: $(4 + 2\sqrt{3}, 1 + \sqrt{3})$, $(4 - 2\sqrt{3}, 1 - \sqrt{3})$

9. $2x^2 + y^2 = 24$
$x^2 - y^2 = -12$

Solve using elimination by addition. Adding, we obtain:
$3x^2 = 12$
$x^2 = 4$
$x = \pm 2$

For $x = 2$
$4 - y^2 = -12$
$-y^2 = -16$
$y^2 = 16$
$y = \pm 4$

For $x = -2$
$4 - y^2 = -12$
Similarly
$y = \pm 4$

Solutions: $(2, 4)$, $(2, -4)$, $(-2, 4)$, $(-2, -4)$

11. $x^2 + y^2 = 10$
$16x^2 + y^2 = 25$

Solve using elimination by addition. Multiply the top equation by -1 and add.

$$\begin{aligned} -x^2 - y^2 &= -10 \\ 16x^2 + y^2 &= 25 \\ \hline 15x^2 &= 15 \end{aligned}$$

$x^2 = 1$
$x = \pm 1$

For $x = 1$
$1 + y^2 = 10$
$y^2 = 9$
$y = \pm 3$

For $x = -1$
$1 + y^2 = 10$
$y = \pm 3$

Solutions: $(1, 3)$, $(1, -3)$, $(-1, 3)$, $(-1, -3)$

13. $xy - 4 = 0$
$x - y = 2$

Solve for x in the first degree equation.
$x = y + 2$
Substitute into the second degree equation
$(y + 2)y - 4 = 0$
$y^2 + 2y - 4 = 0$

$$y = \frac{-b \pm \sqrt{b^2 - 4ac}}{2a} \quad a = 1, b = 2, c = -4$$
$$y = \frac{-2 \pm \sqrt{(2)^2 - 4(1)(-4)}}{2(1)}$$
$$y = \frac{-2 \pm \sqrt{20}}{2}$$
$$y = -1 \pm \sqrt{5}$$

For $y = -1 + \sqrt{5}$
$x = -1 + \sqrt{5} + 2$
$x = 1 + \sqrt{5}$

For $y = -1 - \sqrt{5}$
$x = -1 - \sqrt{5} + 2$
$x = 1 - \sqrt{5}$

Solutions: $(1 + \sqrt{5}, -1 + \sqrt{5})$, $(1 - \sqrt{5}, -1 - \sqrt{5})$

15. $x^2 + 2y^2 = 6$
$xy = 2$

Solve for y in the second equation

$y = \frac{2}{x}$

Substitute into the first equation

$$x^2 + 2\left(\frac{2}{x}\right)^2 = 6$$
$$x^2 + \frac{8}{x^2} = 6 \qquad x \neq 0$$
$$x^2 \cdot x^2 + x^2 \cdot \frac{8}{x^2} = 6x^2$$
$$x^4 + 8 = 6x^2$$
$$x^4 - 6x^2 + 8 = 0$$
$$(x^2 - 2)(x^2 - 4) = 0$$
$$(x - \sqrt{2})(x + \sqrt{2})(x - 2)(x + 2) = 0$$
$$x = \sqrt{2}, -\sqrt{2}, 2, -2$$

For $x = \sqrt{2}$: $y = \frac{2}{\sqrt{2}}$, $y = \sqrt{2}$

For $x = -\sqrt{2}$: $y = -\frac{2}{\sqrt{2}}$, $y = -\sqrt{2}$

For $x = 2$: $y = \frac{2}{2}$, $y = 1$

For $x = -2$: $y = \frac{2}{-2}$, $y = -1$

Solutions: $(\sqrt{2}, \sqrt{2})$, $(-\sqrt{2}, -\sqrt{2})$, $(2, 1)$, $(-2, -1)$

17. $2x^2 + 3y^2 = -4$
$4x^2 + 2y^2 = 8$

Solve using elimination by addition. Multiply the second equation by $-\frac{1}{2}$ and add.

$$\begin{aligned} 2x^2 + 3y^2 &= -4 \\ -2x^2 - y^2 &= -4 \\ 2y^2 &= -8 \\ \hline y^2 &= -4 \\ y &= \pm 2i \end{aligned}$$

For $y = 2i$

$2x^2 + 3(2i)^2 = -4$
$2x^2 - 12 = -4$
$2x^2 = 8$
$x^2 = 4$
$x = \pm 2$

For $y = -2i$

$2x^2 + 3(-2i)^2 = -4$
$2x^2 - 12 = -4$
Similarly
$x = \pm 2$

Solutions: $(2, 2i)$, $(-2, 2i)$, $(2, -2i)$, $(-2, -2i)$

19. $x^2 - y^2 = 2$
$y^2 = x$

Substitute y^2 from the second equation into the first equation.

$x^2 - x = 2$
$x^2 - x - 2 = 0$
$(x - 2)(x + 1) = 0$
$x = 2, -1$

For $x = 2$: $y^2 = 2$, $y = \pm\sqrt{2}$

For $x = -1$: $y^2 = -1$, $y = \pm i$

Solutions: $(2, \sqrt{2})$, $(2, -\sqrt{2})$, $(-1, i)$, $(-1, -i)$

21. $x^2 + y^2 = 9$
$x^2 = 9 - 2y$

Substitute x^2 from the second equation into the first equation.

$9 - 2y + y^2 = 9$
$y^2 - 2y = 0$
$y(y - 2) = 0$
$y = 0, 2$

For $y = 0$: $x^2 = 9 - 2(0)$, $x^2 = 9$, $x = \pm 3$

For $y = 2$: $x^2 = 9 - 2(2)$, $x^2 = 5$, $x = \pm\sqrt{5}$

Solutions: $(3, 0)$, $(-3, 0)$, $(\sqrt{5}, 2)$, $(-\sqrt{5}, 2)$

23. $x^2 - y^2 = 3$
$xy = 2$

Solve for y in the second equation.

$y = \frac{2}{x}$

Substitute into the first equation:

$$x^2 - \left(\frac{2}{x}\right)^2 = 3$$
$$x^2 - \frac{4}{x^2} = 3 \quad x \neq 0$$
$$x^4 - 4 = 3x^2$$
$$x^4 - 3x^2 - 4 = 0$$
$$(x^2 - 4)(x^2 + 1) = 0$$

$x^2 - 4 = 0$ $\quad$ $x^2 + 1 = 0$
$x^2 = 4$ $\quad$ $x^2 = -1$
$x = \pm 2$ $\quad$ $x = \pm i$

For $x = 2$: $y = \frac{2}{2}$, $y = 1$
For $x = -2$: $y = \frac{2}{-2}$, $y = -1$
For $x = i$: $y = \frac{2}{i}$, $y = -2i$
For $x = -i$: $y = \frac{2}{-i}$, $y = 2i$

Solutions: $(2, 1)$, $(-2, -1)$, $(i, -2i)$, $(-i, 2i)$

25. $y = 5 - x^2$
$y = 2 - 2x$

Substitute y from the first equation into the second equation.

$$5 - x^2 = 2 - 2x$$
$$0 = x^2 - 2x - 3$$
$$0 = (x - 3)(x + 1)$$
$$x = 3, -1$$

For $x = 3$: $y = 2 - 2(3)$, $y = -4$
For $x = -1$: $y = 2 - 2(-1)$, $y = 4$

Solutions: $(3, -4)$, $(-1, 4)$

27. $y = x^2 - x$
$y = 2x$

Substitute y from the first equation into the second equation.

$$x^2 - x = 2x$$
$$x^2 - 3x = 0$$
$$x(x - 3) = 0$$
$$x = 0, 3$$

For $x = 0$: $y = 2(0)$, $y = 0$
For $x = 3$: $y = 2(3)$, $y = 6$

Solutions: $(0, 0)$, $(3, 6)$

29. $y = x^2 - 6x + 9$
$y = 5 - x$

Substitute y from the first equation into the second equation.

$$x^2 - 6x + 9 = 5 - x$$
$$x^2 - 5x + 4 = 0$$
$$(x - 1)(x - 4) = 0$$
$$x = 1, 4$$

For $x = 1$: $y = 5 - 1$, $y = 4$
For $x = 4$: $y = 5 - 4$, $y = 1$

Solutions: $(1, 4)$, $(4, 1)$

31. $y = 8 + 4x - x^2$
$y = x^2 - 2x$

Substitute y from the first equation into the second equation.

$$8 + 4x - x^2 = x^2 - 2x$$
$$0 = 2x^2 - 6x - 8$$
$$0 = x^2 - 3x - 4$$
$$0 = (x - 4)(x + 1)$$
$$x = 4, -1$$

For $x = 4$: $y = 4^2 - 2(4)$, $y = 8$
For $x = -1$: $y = (-1)^2 - 2(-1)$, $y = 3$

Solutions: $(4, 8)$, $(-1, 3)$

33. (A) The lines are tangent to the circle.

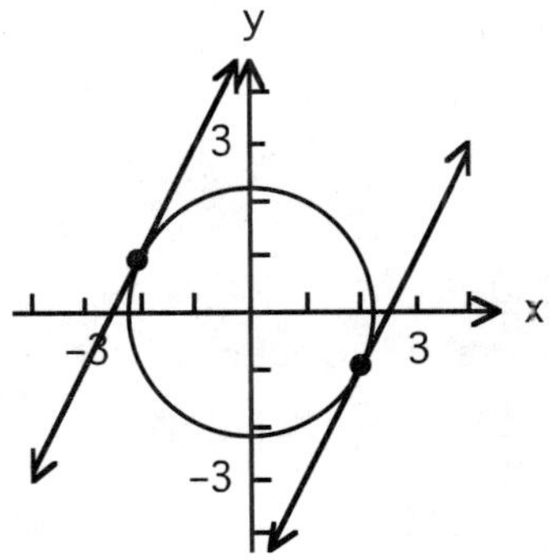

(B) To find values of b such that

$x^2 + y^2 = 5$
$2x - y = b$

has exactly one solution, we solve the system for arbitrary b. Solve for y in the second equation.

$y = 2x - b$

Substitute into the first equation:

$$x^2 + (2x - b)^2 = 5$$
$$x^2 + 4x^2 - 4bx + b^2 = 5$$
$$5x^2 - 4bx + b^2 - 5 = 0$$

This quadratic equation will have one solution if the discriminant $B^2 - 4AC = (-4b)^2 - 4(5)(b^2 - 5)$ is equal to 0.

This will occur when

$$16b^2 - 20b^2 + 100 = 0$$
$$-4b^2 + 100 = 0$$
$$b^2 = 25$$
$$b = \pm 5$$

Consider $b = 5$

Then the solution of the system

$x^2 + y^2 = 5$
$2x - y = 5$

will be given by solving

$5x^2 - 4bx + b^2 - 5 = 0$

for $b = 5$.

$$5x^2 - 4 \cdot 5x + 5^2 - 5 = 0$$
$$5x^2 - 20x + 20 = 0$$
$$5(x - 2)^2 = 0$$
$$x - 2 = 0$$
$$x = 2$$

Since $2x - y = 5$
$2 \cdot 2 - y = 5$
$y = -1$

The intersection point is $(2, -1)$ for $b = 5$.

Consider $b = -5$

Then the solution of the system

$x^2 + y^2 = 5$
$2x - y = -5$

will be given by solving

$5x^2 - 4bx + b^2 - 5 = 0$

for $b = -5$.

$$5x^2 - 4(-5)x + (-5)^2 - 5 = 0$$
$$5x^2 + 20x + 20 = 0$$
$$5(x + 2)^2 = 0$$
$$x + 2 = 0$$
$$x = -2$$

Since $2x - y = -5$
$2(-2) - y = -5$
$y = 1$

The intersection point is $(-2, 1)$ for $b = -5$.

(C) The line $x + 2y = 0$ is perpendicular to all the lines in the family and intersects the circle at the intersection points found in part B, since this line passes through the center of the circle and thus includes a diameter of the circle, which is perpendicular to the tangent line at their mutual point of intersection with the circle. Solving the system $x^2 + y^2 = 5$, $x + 2y = 0$ would determine the intersection points.

35. $2x + 5y + 7xy = 8$
$xy - 3 = 0$

Solve for y in the second equation.

$xy = 3$

$y = \frac{3}{x}$

Substitute into the first equation.

$2x + 5\left(\frac{3}{x}\right) + 7x\left(\frac{3}{x}\right) = 8$

$2x + \frac{15}{x} + 21 = 8 \quad x \neq 0$

$2x^2 + 15 + 21x = 8x$

$2x^2 + 13x + 15 = 0$

$(2x + 3)(x + 5) = 0$

$x = -\frac{3}{2}, -5$

For $x = -\frac{3}{2}$: $y = 3 \div \left(-\frac{3}{2}\right)$, $y = -2$

For $x = -5$: $y = \frac{3}{-5}$, $y = -\frac{3}{5}$

Solutions: $\left(-\frac{3}{2}, -2\right)$, $\left(-5, -\frac{3}{5}\right)$

37. $x^2 - 2xy + y^2 = 1$
$x - 2y = 2$

Solve for x in terms of y in the first-degree equation.

$x = 2y + 2$

Substitute into the second-degree equation.

$(2y + 2)^2 - 2(2y + 2)y + y^2 = 1$

$4y^2 + 8y + 4 - 4y^2 - 4y + y^2 = 1$

$y^2 + 4y + 3 = 0$

$(y + 1)(y + 3) = 0$

$y = -1, -3$

For $y = -1$: $x = 2(-1) + 2 = 0$

For $y = -3$: $x = 2(-3) + 2 = -4$

Solutions: $(0, -1)$, $(-4, -3)$

39. $2x^2 - xy + y^2 = 8$
$x^2 - y^2 = 0$

Factor the left side of the equation that has a zero constant term.

$(x - y)(x + y) = 0$

$x = y$ or $x = -y$

> **Common Error:**
> It is incorrect to replace $x^2 - y^2 = 0$ or $x^2 = y^2$ by $x = y$. This neglects the possibility $x = -y$.

Thus, the original system is equivalent to the two systems

$2x^2 - xy + y^2 = 8$, $x = y$ $\qquad$ $2x^2 - xy + y^2 = 8$, $x = -y$

These systems are solved by substitution.

First system:

$2x^2 - xy + y^2 = 8$

$x = y$

$2y^2 - yy + y^2 = 8$

$2y^2 = 8$

$y^2 = 4$

$y = \pm 2$

Second system:

$2x^2 - xy + y^2 = 8$

$x = -y$

$2(-y)^2 - (-y)y + y^2 = 8$

$2y^2 + y^2 + y^2 = 8$

$4y^2 = 8$

$y^2 = 2$

$y = \pm\sqrt{2}$

For $y = 2$: $x = 2$ $\qquad$ For $y = -2$: $x = -2$ $\qquad$ For $y = \sqrt{2}$: $x = -\sqrt{2}$ $\qquad$ For $y = -\sqrt{2}$: $x = \sqrt{2}$

Solutions: $(2, 2)$, $(-2, -2)$, $(-\sqrt{2}, \sqrt{2})$, $(\sqrt{2}, -\sqrt{2})$

41. $x^2 + xy - 3y^2 = 3$
$x^2 + 4xy + 3y^2 = 0$

Factor the left side of the equation that has a zero constant term.

$(x + y)(x + 3y) = 0$

$x = -y$ or $x = -3y$

Thus the original system is equivalent to the two systems

$x^2 + xy - 3y^2 = 3$, $x = -y$ $\qquad$ $x^2 + xy - 3y^2 = 3$, $x = -3y$

These systems are solved by substitution.

First system:

$$\begin{aligned} x^2 + xy - 3y^2 &= 3 \\ x &= -y \\ (-y)^2 + (-y)y - 3y^2 &= 3 \\ y^2 - y^2 - 3y^2 &= 3 \\ -3y^2 &= 3 \\ y^2 &= -1 \\ y &= \pm i \end{aligned}$$

Second system:

$$\begin{aligned} x^2 + xy - 3y^2 &= 3 \\ x &= -3y \\ (-3y)^2 + (-3y)y - 3y^2 &= 3 \\ 9y^2 - 3y^2 - 3y^2 &= 3 \\ 3y^2 &= 3 \\ y^2 &= 1 \\ y &= \pm 1 \end{aligned}$$

For $y = i$, $x = -i$ $\qquad$ For $y = -i$, $x = i$ $\qquad$ For $y = 1$, $x = -3$ $\qquad$ For $y = -1$, $x = 3$

Solutions: $(-i, i)$, $(i, -i)$, $(-3, 1)$, $(3, -1)$

43. Before we can enter these equations in our graphing calculator, we must solve for y:

$$\begin{aligned} -x^2 + 2xy + y^2 &= 1 \\ y^2 + 2xy - 1 - x^2 &= 0 \end{aligned} \qquad \begin{aligned} 3x^2 - 4xy + y^2 &= 2 \\ y^2 - 4xy + 3x^2 - 2 &= 0 \end{aligned}$$

Applying the quadratic formula to each equation, we have

$$y = \frac{-2x \pm \sqrt{4x^2 - 4(-1 - x^2)}}{2} \qquad y = \frac{4x \pm \sqrt{16x^2 - 4(3x^2 - 2)}}{2}$$

$$y = \frac{-2x \pm \sqrt{8x^2 + 4}}{2} \qquad y = \frac{4x \pm \sqrt{4x^2 + 8}}{2}$$

$$y = -x \pm \sqrt{2x^2 + 1} \qquad y = 2x \pm \sqrt{x^2 + 2}$$

Entering each of these four equations into a graphing calculator produces the graph shown at the right.

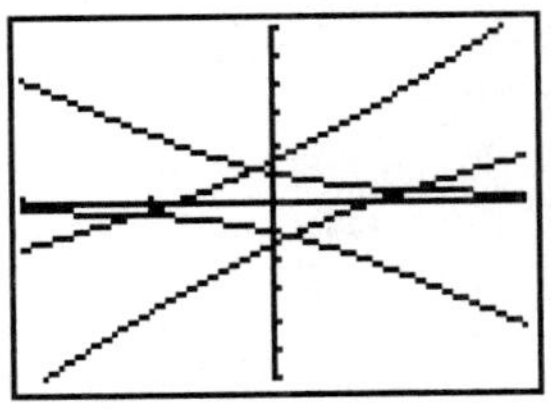

Zooming in on the four intersection points, or using a built-in intersection routine (details omitted), yields (-1.41, -0.82), (-0.13, 1.15), (0.13, -1.15), and (1.41, 0.82) to two decimal places.

45. Before we can enter these equations in our graphing calculator, we must solve for y:

$$\begin{aligned} 3x^2 - 4xy - y^2 &= 2 \\ y^2 + 4xy + 2 - 3x^2 &= 0 \end{aligned} \qquad \begin{aligned} 2x^2 + 2xy + y^2 &= 9 \\ y^2 + 2xy + 2x^2 - 9 &= 0 \end{aligned}$$

Applying the quadratic formula to each equation, we have

$$y = \frac{-4x \pm \sqrt{16x^2 - 4(2 - 3x^2)}}{2} \qquad y = \frac{-2x \pm \sqrt{4x^2 - 4(2x^2 - 9)}}{2}$$

$$y = \frac{-4x \pm \sqrt{28x^2 - 8}}{2} \qquad y = \frac{-2x \pm \sqrt{36 - 4x^2}}{2}$$

$$y = -2x \pm \sqrt{7x^2 - 2} \qquad y = -x \pm \sqrt{9 - x^2}$$

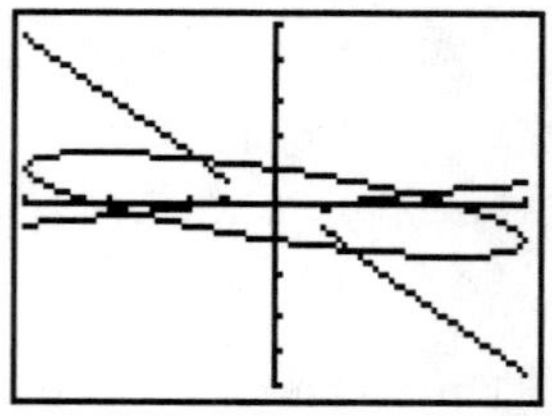

Entering each of these four equations into a graphing calculator produces the graph shown at the right.

Zooming in on the four intersection points, or using a built-in intersection routine (details omitted), yields (-1.66, -0.84), (-0.91, 3.77), (0.91, -3.77), and (1.66, 0.84) to two decimal places.

47. Before we can enter these equations in our graphing calculator, we must solve for y:

$$2x^2 - 2xy + y^2 = 9 \qquad\qquad 4x^2 - 4xy + y^2 + x = 3$$
$$y^2 - 2xy + 2x^2 - 9 = 0 \qquad\qquad y^2 - 4xy + 4x^2 + x - 3 = 0$$

Applying the quadratic formula to each equation, we have

$$y = \frac{2x \pm \sqrt{4x^2 - 4(2x^2 - 9)}}{2} \qquad\qquad y = \frac{4x \pm \sqrt{16x^2 - 4(4x^2 + x - 3)}}{2}$$
$$y = \frac{2x \pm \sqrt{36 - 4x^2}}{2} \qquad\qquad y = \frac{4x \pm \sqrt{12 - 4x}}{2}$$
$$y = x \pm \sqrt{9 - x^2} \qquad\qquad y = 2x \pm \sqrt{3 - x}$$

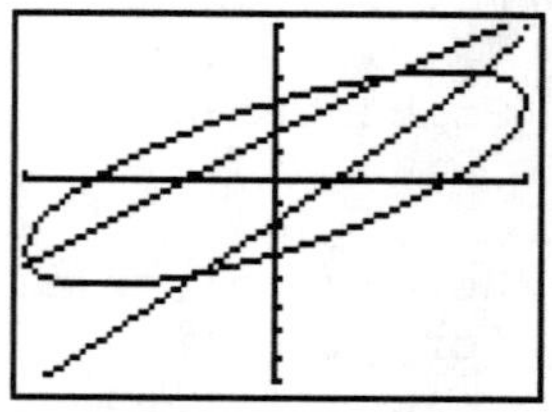

Entering each of these four equations into a graphing calculator produces the graph shown at the right.

Zooming in on the four intersection points, or using a built-in intersection routine (details omitted), yields (-2.96, -3.47), (-0.89, -3.76), (1.39, 4.05), and (2.46, 4.18) to two decimal places.

49. Let x and y equal the two numbers. We have the system

$$x + y = 3$$
$$xy = 1$$

Solve the first equation for y in terms of x, then substitute into the second degree equation.

$$y = 3 - x$$
$$x(3 - x) = 1$$
$$3x - x^2 = 1$$
$$-x^2 + 3x - 1 = 0$$
$$x^2 - 3x + 1 = 0$$

$$x = \frac{-b \pm \sqrt{b^2 - 4ac}}{2a} \qquad a = 1,\ b = -3,\ c = 1$$
$$x = \frac{-(-3) \pm \sqrt{(-3)^2 - 4(1)(1)}}{2(1)}$$
$$x = \frac{3 \pm \sqrt{5}}{2}$$

For $x = \frac{3 + \sqrt{5}}{2}$

$$y = 3 - x = 3 - \frac{3 + \sqrt{5}}{2} = \frac{6 - 3 - \sqrt{5}}{2} = \frac{3 - \sqrt{5}}{2}$$

For $x = \frac{3 - \sqrt{5}}{2}$

$$y = 3 - x = 3 - \frac{3 - \sqrt{5}}{2} = \frac{6 - 3 + \sqrt{5}}{2} = \frac{3 + \sqrt{5}}{2}$$

Thus the two numbers are $\frac{1}{2}(3 - \sqrt{5})$ and $\frac{1}{2}(3 + \sqrt{5})$.

51. Sketch a figure.

Let x and y represent the lengths of the two legs.

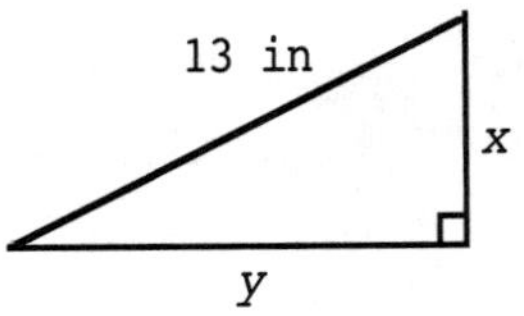

From the Pythagorean Theorem we have

$x^2 + y^2 = 13^2$

From the formula for the area of a triangle we have

$\frac{1}{2}xy = 30$

Thus the system of equations is

$$x^2 + y^2 = 169$$
$$\frac{1}{2}xy = 30$$

Solve the second equation for y in terms of x, then substitute into the first equation.

$$xy = 60$$
$$y = \frac{60}{x}$$
$$x^2 + \left(\frac{60}{x}\right)^2 = 169$$
$$x^2 + \frac{3600}{x^2} = 169 \qquad x \neq 0$$
$$x^4 + 3600 = 169x^2$$
$$x^4 - 169x^2 + 3600 = 0$$
$$(x^2 - 144)(x^2 - 25) = 0$$
$$(x - 12)(x + 12)(x - 5)(x + 5) = 0$$
$$x = \pm 12, \pm 5$$

Discarding the negative solutions, we have

$x = 12$ or $x = 5$

For $x = 12$

$$y = \frac{60}{x}$$
$$y = 5$$

For $x = 5$

$$y = \frac{60}{x}$$
$$y = 12$$

The lengths of the legs are 5 inches and 12 inches.

53. Let x = width of screen.

y = height of screen.

From the Pythagorean Theorem, we have

$x^2 + y^2 = (7.5)^2$

From the formula for the area of a rectangle we have

$xy = 27$

Thus the system of equations is:

$$x^2 + y^2 = 56.25$$
$$xy = 27$$

Solve the second equation for y in terms of x, then substitute into the first equation.

$$y = \frac{27}{x}$$
$$x^2 + \left(\frac{27}{x}\right)^2 = 56.25 \qquad x \neq 0$$
$$x^2 + \frac{729}{x^2} = 56.25 \qquad x \neq 0$$
$$x^4 + 729 = 56.25x^2$$
$$x^4 - 56.25x^2 + 729 = 0 \quad \text{quadratic in } x^2$$

$$x^2 = \frac{-b \pm \sqrt{b^2 - 4ac}}{2a} \qquad a = 1,\ b = -56.25,\ c = 729$$

$$x^2 = \frac{-(-56.25) \pm \sqrt{(-56.25)^2 - 4(1)(729)}}{2(1)}$$

$$x^2 = \frac{56.25 \pm 15.75}{2}$$

$$x^2 = 36,\ 20.25$$

$$x = 6,\ 4.5 \text{ (discarding the negative solutions)}$$

For $x = 6$

$$y = \frac{27}{6} = 4.5$$

For $x = 4.5$

$$y = \frac{27}{4.5} = 6$$

The dimensions of the screen must be 6 inches by 4.5 inches.

55.

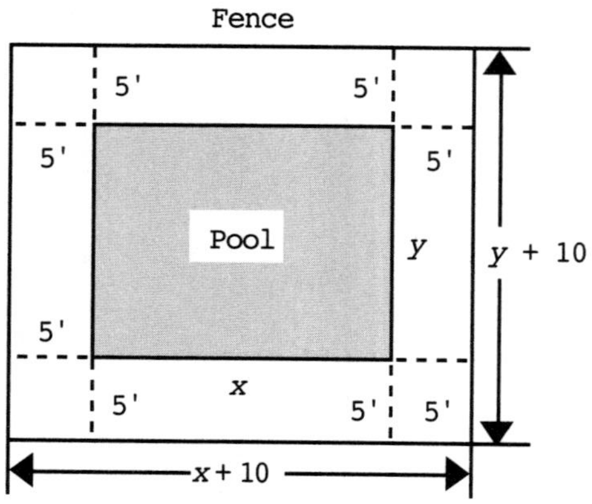

Redrawing and labeling the figure as shown, we have

Area of pool = 572

$$xy = 572$$

Area enclosed by fence = 1,152

$$(x + 10)(y + 10) = 1{,}152$$

We solve this system by solving for y in terms of x in the first equation, then substituting into the second equation.

$$y = \frac{572}{x}$$

$$(x + 10)\left(\frac{572}{x} + 10\right) = 1{,}152$$

$$572 + 10x + \frac{5{,}720}{x} + 100 = 1{,}152$$

$$10x + \frac{5{,}720}{x} - 480 = 0 \qquad x \neq 0$$

$$10x^2 + 5{,}720 - 480x = 0$$

$$x^2 - 48x + 572 = 0$$

$$(x - 26)(x - 22) = 0$$

$$x = 26,\ 22$$

For $x = 26$

$$y = \frac{572}{26}$$

$$y = 22$$

For $x = 22$

$$y = \frac{572}{22}$$

$$y = 26$$

The dimensions of the pool are 22 feet by 26 feet.

57. Let x = average speed of Boat B
Then $x + 5$ = average speed of Boat A

Let y = time of Boat B, then $y - \frac{1}{2}$ = time of Boat A

Using Distance = rate × time, we have

$75 = xy$

$75 = (x + 5)\left(y - \frac{1}{2}\right)$

Note: The *faster* boat, A, has the *shorter* time. It is a common error to confuse the signs here. Another common error: if rates are expressed in miles per hour, then $y - 30$ is not the correct time for boat A. Times must be expressed in hours.

Solve the first equation for y in terms of x, then substitute into the second equation.

$$y = \frac{75}{x}$$
$$75 = (x + 5)\left(\frac{75}{x} - \frac{1}{2}\right)$$
$$75 = 75 - \frac{1}{2}x + \frac{375}{x} - \frac{5}{2}$$
$$0 = -\frac{1}{2}x + \frac{375}{x} - \frac{5}{2} \qquad x \neq 0$$
$$2x(0) = 2x\left(-\frac{1}{2}x\right) + 2x\left(\frac{375}{x}\right) - 2x\left(\frac{5}{2}\right)$$
$$0 = -x^2 + 750 - 5x$$
$$x^2 + 5x - 750 = 0$$
$$(x - 25)(x + 30) = 0$$
$$x = 25, -30$$

Discarding the negative solution, we have

$x = 25$ mph = average speed of Boat B

$x + 5 = 30$ mph = average speed of Boat A

Section 10-7

1. Graph $2x - 3y = 6$ as a dashed line, since equality is not included in the original statement. The origin is a suitable test point.
$2x - 3y < 6$
$2(0) - 3(0) = 0 < 6$
Hence (0, 0) is in the solution set. The graph is the half-plane containing (0, 0).

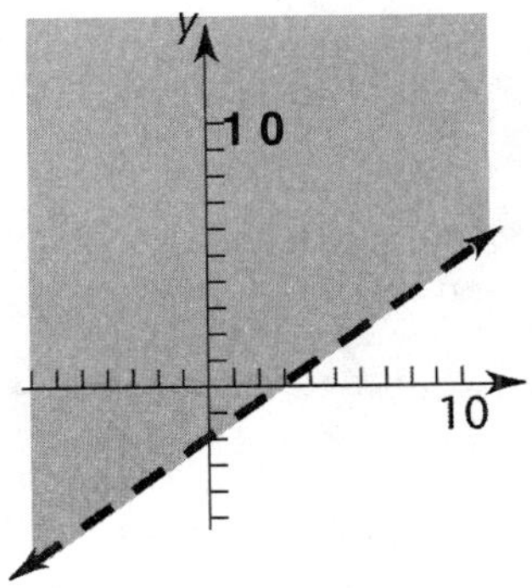

3. Graph $3x + 2y = 18$ as a solid line, since equality is included in the original statement. The origin is a suitable test point.
$3(0) + 2(0) = 0 \not\geq 18$
Hence $(0, 0)$ is not in the solution set. The graph is the line $3x + 2y = 18$ and the half-plane not containing the origin.

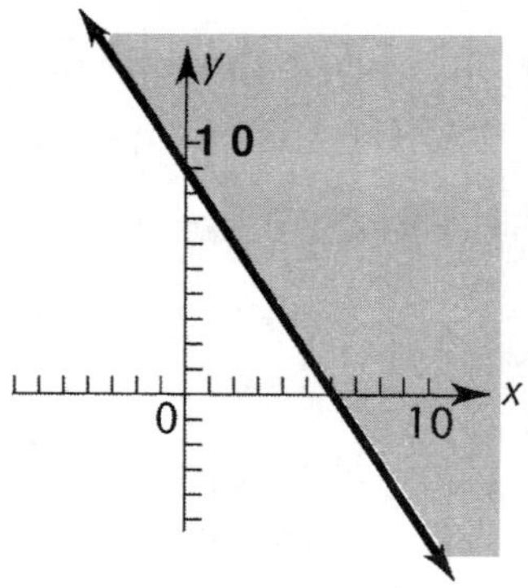

5. Graph $y = \frac{2}{3}x + 5$ as a solid line, since equality is included in the original statement. The origin is a suitable test point.
$0 \overset{?}{\leq} (0) + 5$
$0 \leq 5$
Hence $(0, 0)$ is in the solution set. The graph is the line $y = \frac{2}{3}x + 5$ and the half-plane containing the origin.

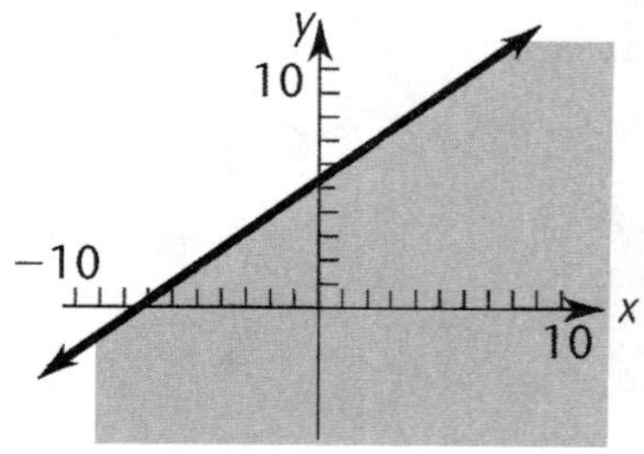

7. Graph $y = 8$ as a dashed line, since equality is not included in the original statement. Clearly the graph consists of all points whose y-coordinates are less than 8, that is, the lower half-plane.

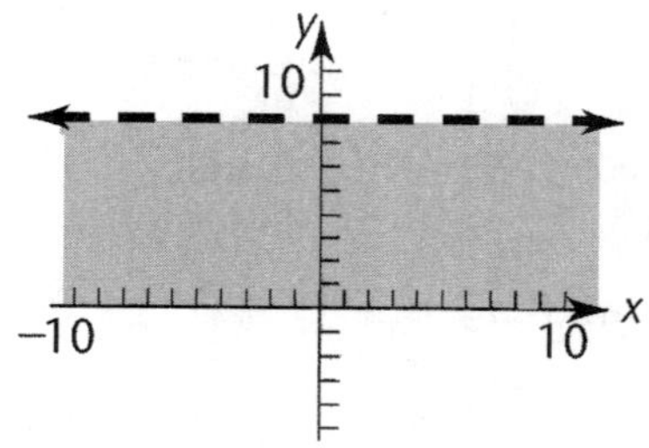

9. This system is equivalent to the system
$y \geq -3$
$y < 2$
and its graph is the intersection of the graphs of these inequalities.

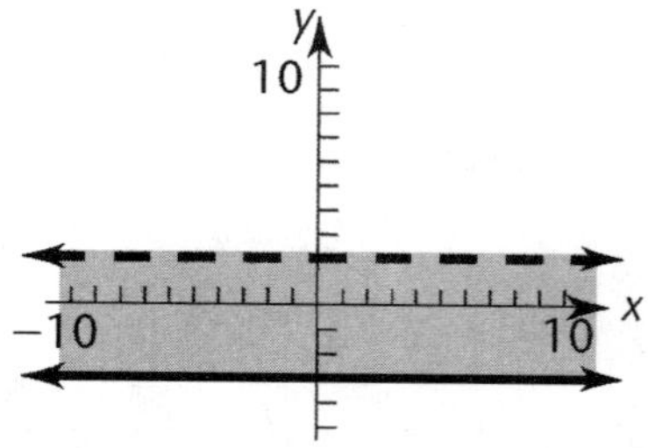

11. $x + 2y \leq 8$
$3x - 2y \geq 0$
Choose a suitable test point that lies on neither line, for example, $(2, 0)$.
$2 + 2(0) = 2 \leq 8$ Hence, the solution region is *below* the graph of $x + 2y = 8$.
$3(2) - 2(0) = 6 \geq 0$ Hence, the solution region is *below* the graph of $3x - 2y = 0$.
Thus the solution region is region IV in the diagram.

13. $x + 2y \geq 8$
$3x - 2y \geq 0$
Choose a suitable test point that lies on neither line, for example, (2, 0).
$2 + 2(0) = 2 \not\geq 8$ Hence, the solution region is *above* the graph of $x + 2y = 8$.
$3(2) - 2(0) = 6 \geq 0$ Hence, the solution region is *below* the graph of $3x - 2y = 0$.
Thus the solution region is region I in the diagram.

15.

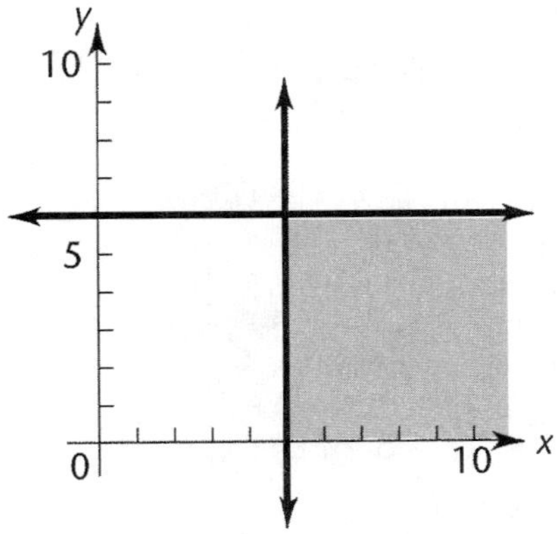

17.

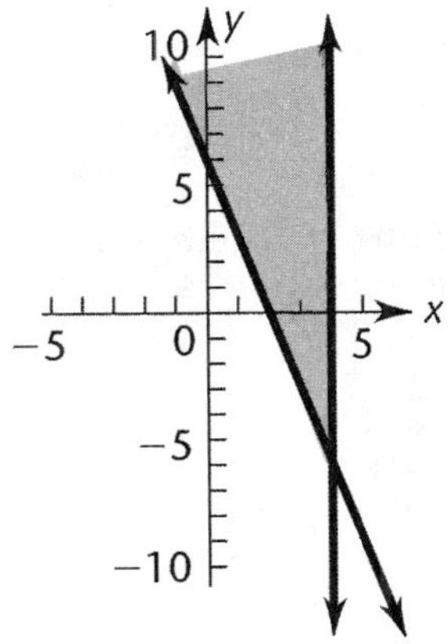

19.

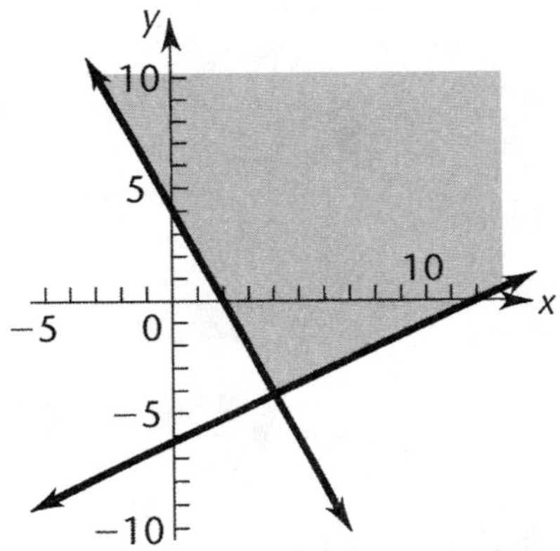

21. Choose a suitable test point that lies on none of the lines, say (5, 1).
$5 + 3(1) = 8 \leq 18$
Hence, the solution region is *below* the graph of $x + 3y = 18$.
$2(5) + 1 = 11 \not\geq 16$
Hence, the solution region is *above* the graph of $2x + y = 16$.
$5 \geq 0$
$1 \geq 0$
Thus the solution region is region IV in the diagram. The corner points are the labeled points (6, 4), (8, 0), and (18, 0).

23. Choose a suitable test point that lies on none of the lines, say (5, 1).
$5 + 3(1) = 8 \not\geq 18$ Hence, the solution region is *above* the graph of $x + 3y = 18$.
$2(5) + 1 = 11 \not\geq 16$ Hence, the solution region is *above* the graph of $2x + y = 16$.
$5 \geq 0$
$1 \geq 0$
Thus the solution region is region I in the diagram. The corner points are the labeled points (0, 16), (6, 4), and (18, 0).

25. The solution region is bounded (contained in, for example, the circle $x^2 + y^2 = 16$). The corner points are obvious from the graph: (0, 0), (0, 2), (3, 0).

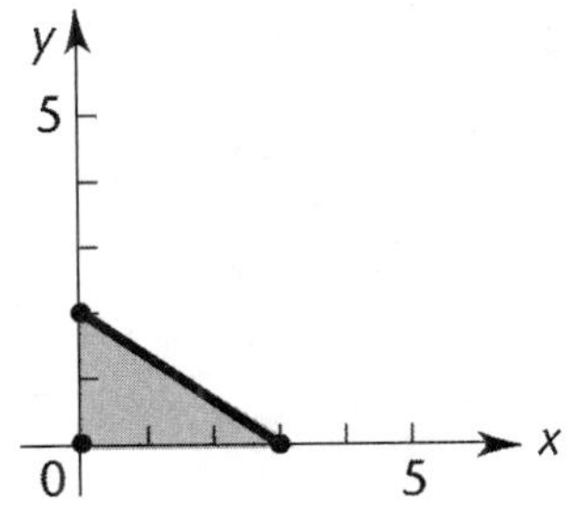

27. The solution region is unbounded. The corner points are obvious from the graph: (0, 4) and (5, 0).

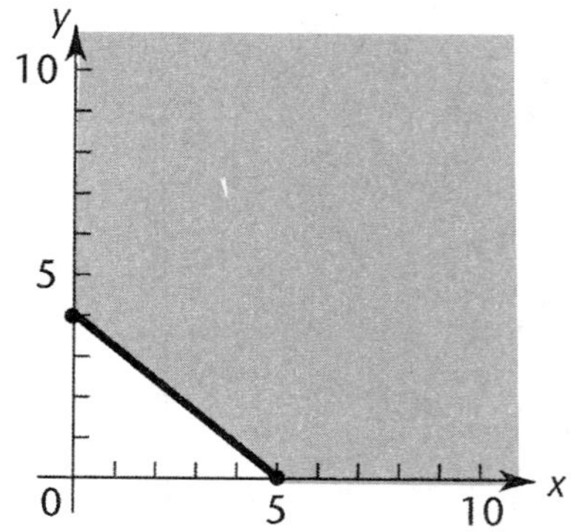

29. The solution region is bounded. Three corner points are obvious from the graph: (0, 4), (0, 0), (4, 0). The fourth corner point is obtained by solving the system

$x + 3y = 12$

$2x + y = 8$ to obtain $\left(\frac{12}{5}, \frac{16}{5}\right)$.

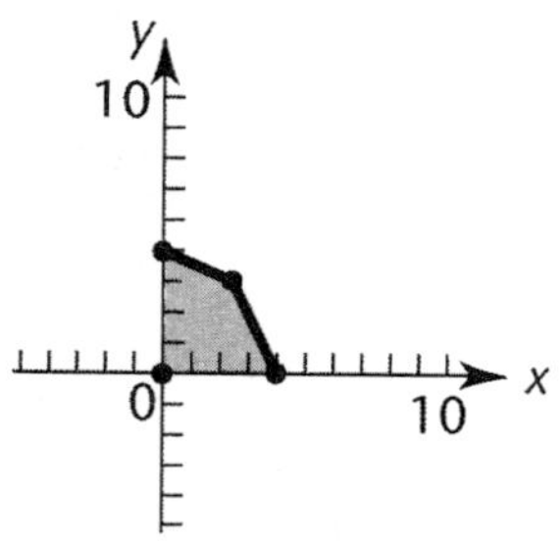

31. The solution region is unbounded. The corner points are obvious from the graph: (9, 0) and (0, 8). The third corner point is obtained by solving the system

$4x + 3y = 24$

$2x + 3y = 18$ to obtain (3, 4).

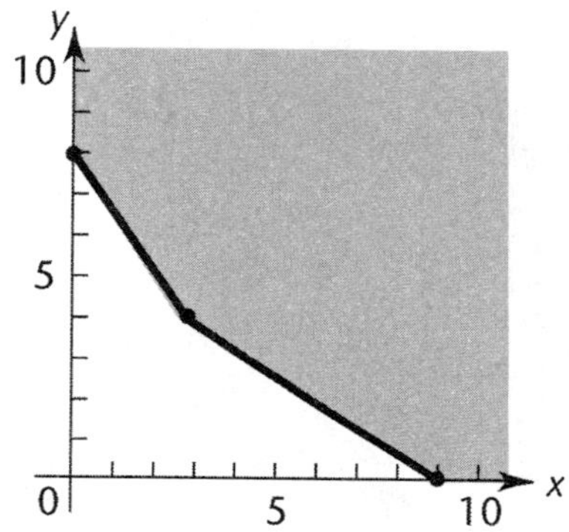

33. The solution region is bounded. Three corner points are obvious from the graph: (6, 0), (0, 0), and (0, 5). The other corner points are obtained by solving:

$2x + y = 12$ and $x + y = 7$
$x + y = 7$ to obtain (5, 2)

$x + y = 7$
$x + 2y = 10$ to obtain (4, 3)

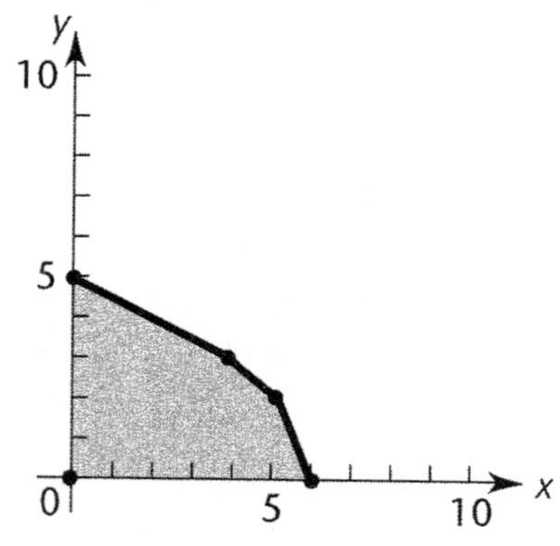

35. The solution region is unbounded. Two of the corner points are obvious from the graph: (16, 0) and (0, 14). The other corner points are obtained by solving:

$x + 2y = 16$ and $x + y = 12$ to obtain (8, 4)

$x + y = 12$
$2x + y = 14$ to obtain (2, 10)

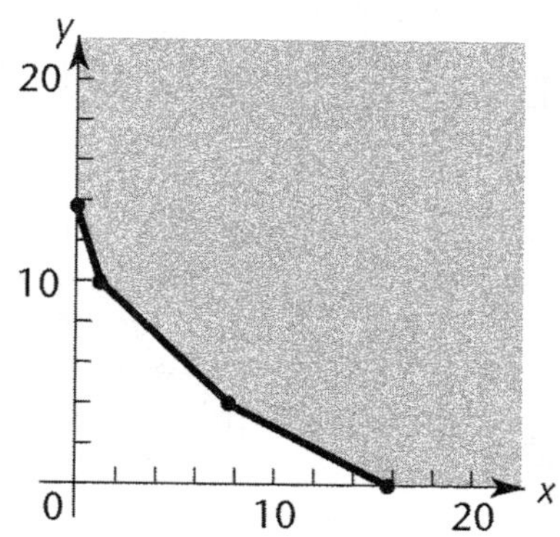

37. The solution region is bounded. The corner points are obtained by solving:

$\begin{cases} x + y = 11 \\ 5x + y = 15 \end{cases}$ to obtain (1, 10)

$\begin{cases} 5x + y = 15 \\ x + 2y = 12 \end{cases}$ to obtain (2, 5), and

$\begin{cases} x + y = 11 \\ x + 2y = 12 \end{cases}$ to obtain (10, 1)

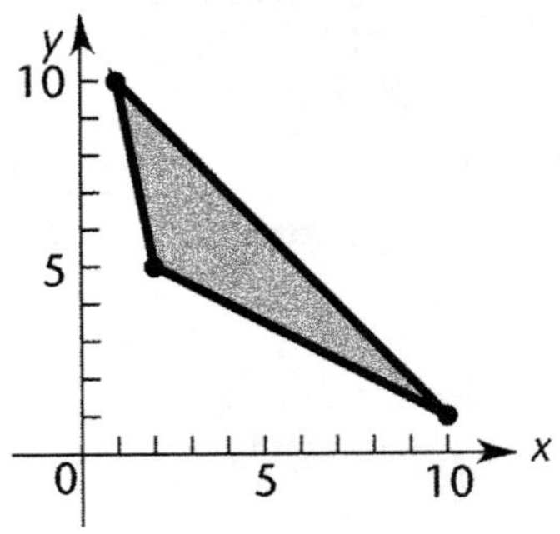

39. From the graph it should be clear that there is no point with x coordinate greater than 4 which satisfies both $3x + 2y \le 24$ (arrows pointing, roughly, northeast) and $3x + y \le 15$ (arrows pointing, roughly, southwest). The feasible region is empty.

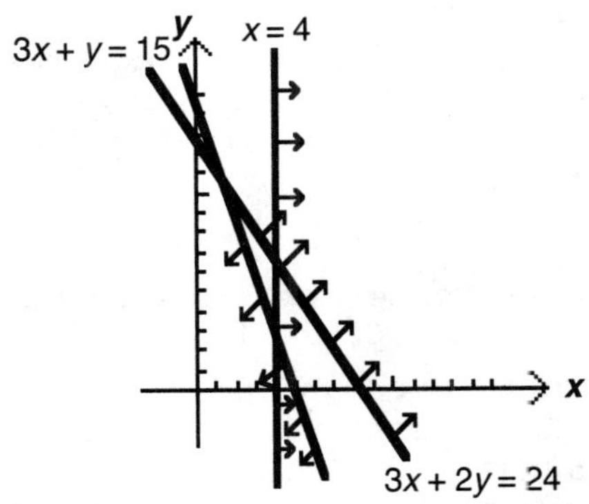

41. The feasible region is bounded. The corner points are obtained by solving:

$\begin{cases} x + y = 10 \\ 3x - 2y = 15 \end{cases}$ to obtain (7, 3)

$\begin{cases} 3x - 2y = 15 \\ 3x + 5y = 15 \end{cases}$ to obtain (5, 0),

$\begin{cases} 3x + 5y = 15 \\ -5x + 2y = 6 \end{cases}$ to obtain (0, 3), and

$\begin{cases} -5x + 2y = 6 \\ x + y = 10 \end{cases}$ to obtain (2, 8)

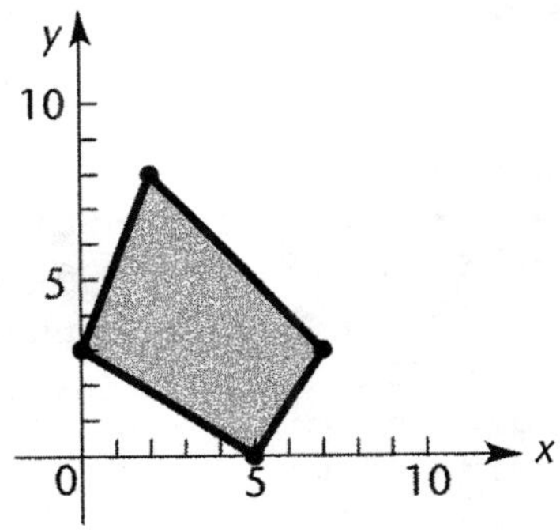

43. The feasible region is bounded. The corner points are obtained by solving:

$\begin{cases} 16x + 13y = 119 \\ 12x + 16y = 101 \end{cases}$ to obtain (5.91, 1.88)

$\begin{cases} 16x + 13y = 119 \\ -4x + 3y = 11 \end{cases}$ to obtain (2.14, 6.53)

and

$\begin{cases} 12x + 16y = 101 \\ -4x + 3y = 11 \end{cases}$ to obtain (1.27, 5.36)

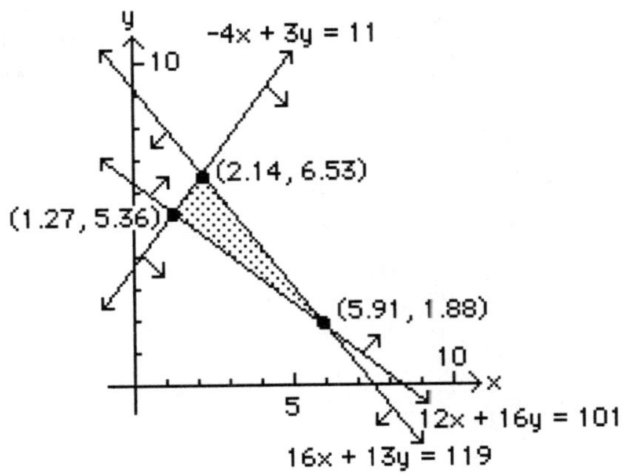

45. Let x = number of trick skis produced per day.
y = number of slalom skis produced per day

Clearly x and y must be non-negative.
Hence $x \ge 0$ (1)
$y \ge 0$ (2)

To fabricate x trick skis requires $6x$ hours. To fabricate y slalom skis requires $4y$ hours.
108 hours are available for fabricating; hence
$6x + 4y \le 108$ (3)
To finish x trick skis requires $1x$ hours.
To finish y slalom skis requires $1y$ hours.
24 hours are available for finishing, hence
$x + y \le 24$ (4)

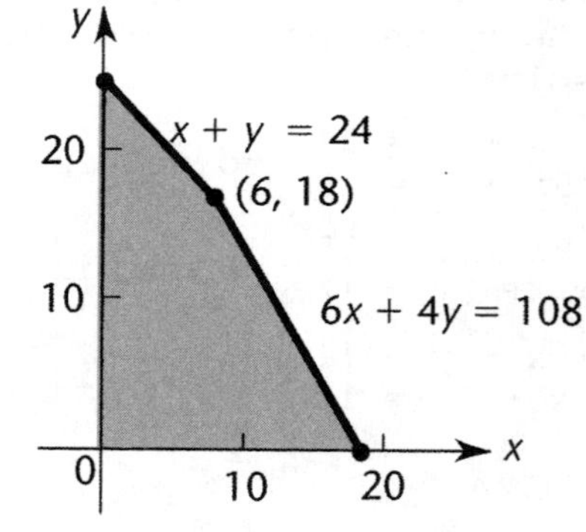

Graphing the inequality system (1), (2), (3), (4), we have the diagram.

47. (A) All production schedules in the feasible region that are on the graph of $50x + 60y = 1{,}100$ will result in a profit of \$1,100.

(B) There are many possible choices. For example, producing 5 trick and 15 slalom skis will produce a profit of \$1,150. The graph of the line $50x + 60y = 1{,}150$ includes all the production schedules in the feasible region that result in a profit of \$1,150.

(C) A graphical approach would involve drawing other lines of the type $50x + 60y = A$. The graphs of these lines include all production schedules that will result in a profit of A. Increase A until the line either intersects the feasible region only in 1 corner point or contains an edge of the feasible region. This value of A will be the maximum profit possible.

49. Clearly x and y must be non-negative.
Hence $x \ge 0$ (1)
$y \ge 0$ (2)
x cubic yards of mix A contains $20x$ pounds of phosphoric acid.
y cubic yards of mix B contains $10y$ pounds of phosphoric acid.

At least 460 pounds of phosphoric acid are required, hence
$20x + 10y \ge 460$ (3)
x cubic yards of mix A contains $30x$ pounds of nitrogen.
y cubic yards of mix B contains $30y$ pounds of nitrogen.

At least 960 pounds of nitrogen are required, hence
$30x + 30y \ge 960$ (4)
x cubic yards of mix A contains $5x$ pounds of potash.
y cubic yards of mix B contains $10y$ pounds of potash.

At least 220 pounds of potash are required, hence
$5x + 10y \ge 220$ (5)

Graphing the inequality system (1), (2), (3), (4), (5), we have the diagram:

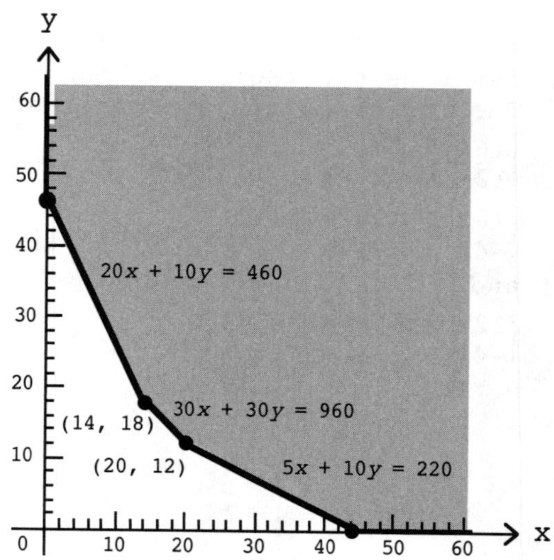

51. Clearly x and y must be non-negative.
Hence $x \geq 0$ (1)
$y \geq 0$ (2)
Each sociologist will spend 10 hours collecting data: $10x$ hours.
Each research assistant will spend 30 hours collecting data: $30y$ hours.
At least 280 hours must be spent collecting data; hence
$10x + 30y \geq 280$ (3)
Each sociologist will spend 30 hours analyzing data: $30x$ hours.
Each research assistant will spend 10 hours analyzing data: $10y$ hours.
At least 360 hours must be spent analyzing data; hence
$30x + 10y \geq 360$ (4)
Graphing the inequality system (1), (2), (3), (4), we have the diagram.

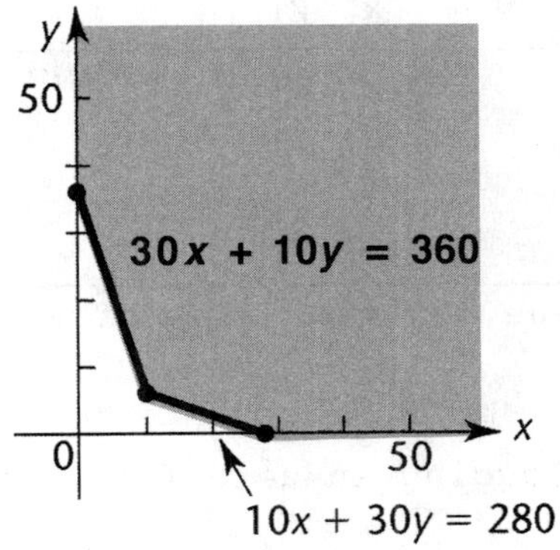

Section 10-8

1.

Corner Point (x, y)	Objective Function $z = x + y$	
(0, 12)	12	
(7, 9)	16	Maximum value
(10, 0)	10	
(0, 0)	0	

The maximum value of z on S is 16 at (7, 9).

3.

Corner Point (x, y)	Objective Function $z = 3x + 7y$		
(0, 12)	84	Minimum value	Multiple optimal solutions
(7, 9)	84	Minimum value	
(10, 0)	30		
(0, 0)	0		

The maximum value of z on S is 84 at both (0, 12) and (7, 9).

5. Plugging in zero for x, we get $z = 2y$ or $y = \frac{z}{2}$.
Plugging in zero for y, we get $z = x$. So the intercepts of the line $z = x + 2y$ are $\left(0, \frac{z}{2}\right)$ and $(z, 0)$. The feasible region is shown at the right with several constant z lines drawn in.

Sliding a straight edge parallel to these constant z lines in the direction of increasing z, we can see that the point in the feasible region that will intersect the constant z line for largest possible z is (7, 9). When $x = 7$ and $y = 9$, $z = 25$.

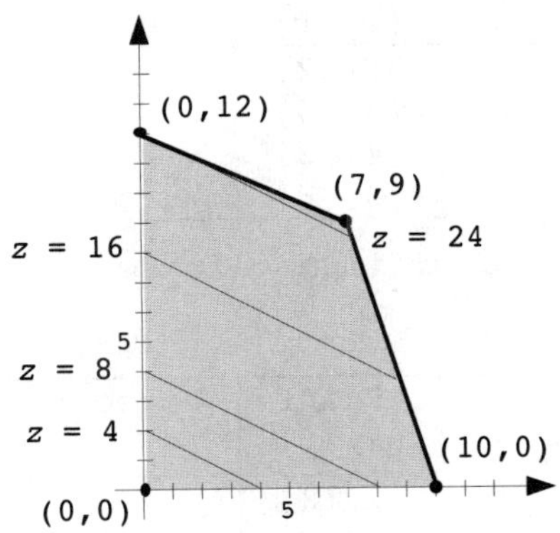

Check:

Corner Point (x, y)	Objective Function $z = x + 2y$	
(0, 12)	24	
(7, 9)	25	Maximum Value
(10, 0)	10	
(0, 0)	0	

The maximum value of z on S is 25 at (7, 9).

7. Plugging in zero for x, we get $z = 2y$ or $y = \frac{z}{2}$.

Plugging in zero for y, we get $z = 7x$ or $x = \frac{z}{7}$.

So the intercepts of the line $z = 7x + 2y$ are $\left(0, \frac{z}{2}\right)$ and $\left(\frac{z}{7}, 0\right)$. The feasible region is shown at the right with several constant z lines drawn in.

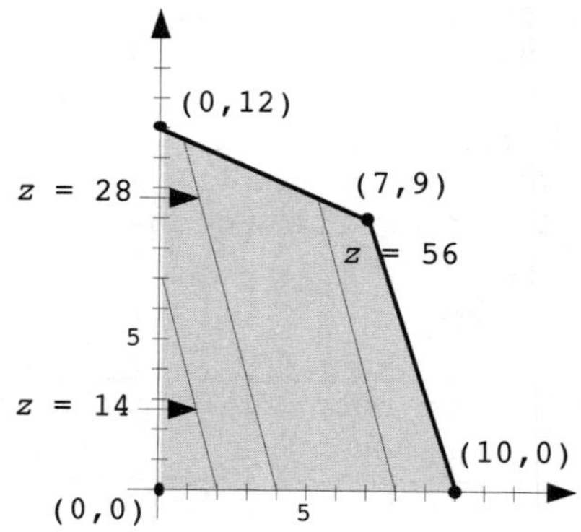

If we slide a straight edge parallel to the constant z lines as z increases, the last point in S that will intersect our lines is (10, 0). When $x = 10$ and $y = 0$, $z = 70$.

Check:

Corner Point (x, y)	Objective Function $z = 7x + 2y$	
(0, 12)	24	
(7, 9)	67	
(10, 0)	70	Maximum Value
(0, 0)	0	

The maximum value of z on S is 70 at (10, 0).

9.

Corner Point (x, y)	Objective Function $z = 7x + 4y$	
(0, 12)	48	
(12, 0)	84	
(4, 3)	40	
(0, 8)	32	Minimum value

The minimum value of z on S is 32 at (0, 8).

11.

Corner Point (x, y)	Objective Function $3x + 8y$		
(0, 12)	96		
(12, 0)	36	Minimum value	Multiple optimal solutions
(4, 3)	36	Minimum value	
(0, 8)	64		

The minimum value of z on S is 36 at both (12, 0) and (4, 3).

13. If $x = 5$ and $y = 5$, $z = 5 + 2(5) = 15$, so the constant-value line we need is $x + 2y = 15$. The feasible region is shown at the right with the constant-value line.

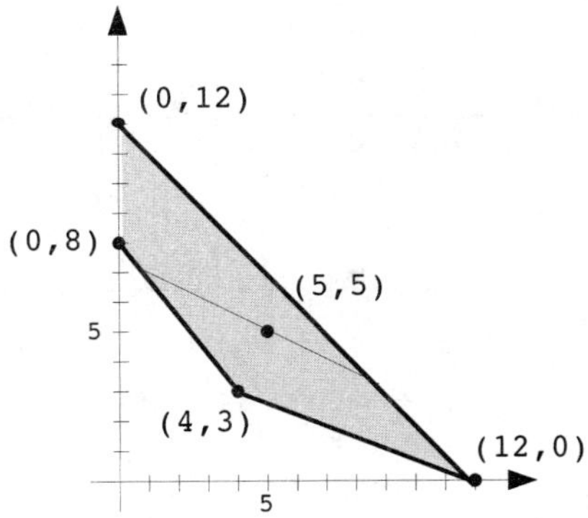

If we slide a straightedge parallel to this constant-value line in the direction of decreasing z (downward), the last point in T that will intersect our line is (4, 3). When $x = 4$ and $y = 3$, $z = 10$.

Check:

Corner Point (x, y)	Objective Function $z = x + 2y$	
(0, 12)	24	
(12, 0)	12	
(0, 8)	16	
(4, 3)	10	Minimum Value

The minimum value of z on T is 10 at (4, 3).

15. If $x = 5$ and $y = 5$, $z = 5(5) + 4(5) = 45$, so the constant-value line we need is $5x + 4y = 45$. The feasible region is shown at the right with the constant-value line.

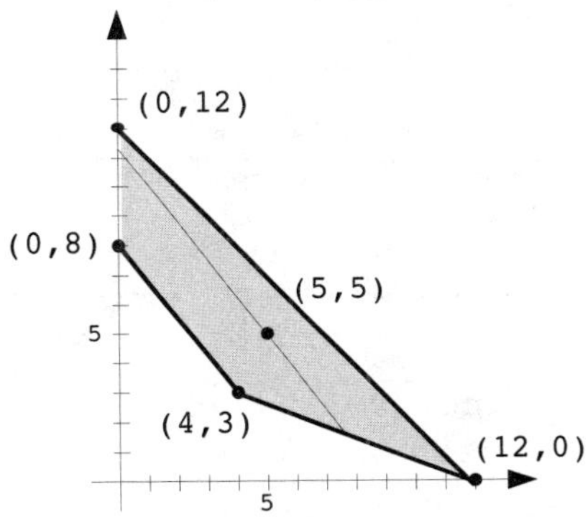

The constant-value line appears to be parallel to the edge connecting (0, 8) and (4, 3). Thus the minimum could occur at either (0, 8) or (4, 3) when $x = 0$ and $y = 8$, $z = 32$. When $x = 4$ and $y = 3$, $z = 32$ also.

Check:

Corner Point (x, y)	Objective Function $z = 5x + 4y$	
(0, 12)	48	
(12, 0)	60	
(0, 8)	32	Minimum Value
(4, 3)	32	Minimum Value

The minimum value of 32 occurs at both (0, 8) and (4, 3).

17. The feasible region is graphed as follows:
The corner points (0, 5), (5, 0) and (0, 0) are obvious from the graph. The corner point (4, 3) is obtained by solving the system
$x + 2y = 10$
$3x + y = 15$
We now evaluate the objective function at each corner point.

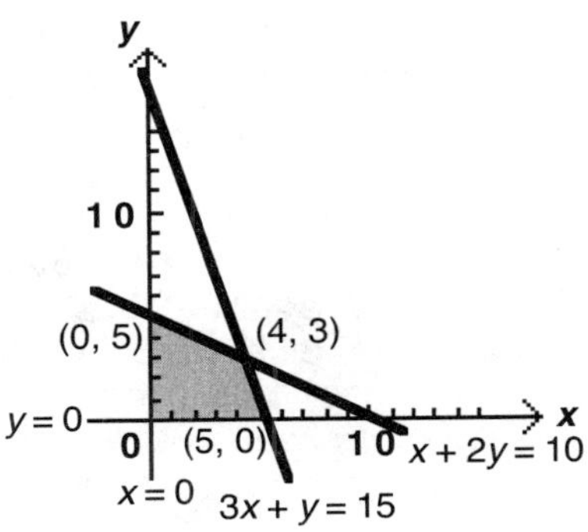

Corner Point (x, y)	Objective Function $z = 3x + 2y$	
(0, 5)	10	
(0, 0)	0	
(5, 0)	15	
(4, 3)	18	Maximum value

The maximum value of z on S is 18 at (4, 3).

19. The feasible region is graphed as follows:
The corner points (4, 0) and (10, 0) are obvious from the graph. The corner point (2, 4) is obtained by solving the system
$x + 2y = 10$
$2x + y = 8$
We now evaluate the objective function at each corner point.

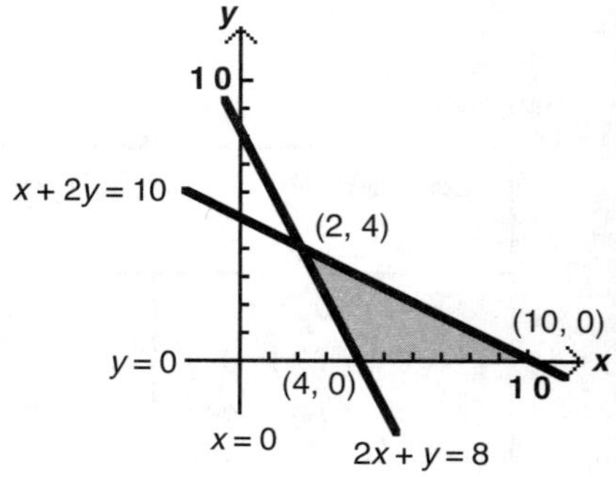

Corner Point (x, y)	Objective Function $z = 3x + 4y$	
(4, 0)	12	Minimum value
(10, 0)	30	
(2, 4)	22	

The minimum value of z on S is 12 at (4, 0).

21. The feasible region is graphed below. The corner points (0, 12), (0, 0), and (12, 0) are obvious from the graph. The other corner points are obtained by solving:
$x + 2y = 24$ and $x + y = 14$
$x + y = 14$ to obtain (4, 10) $2x + y = 24$ to obtain (10, 4)
We now evaluate the objective function at each corner point.

Corner Point (x, y)	Objective Function $z = 3x + 4y$	
(0, 12)	48	
(0, 0)	0	
(12, 0)	36	
(10, 4)	46	
(4, 10)	52	Maximum value

The maximum value of z on S is 52 at (4, 10).

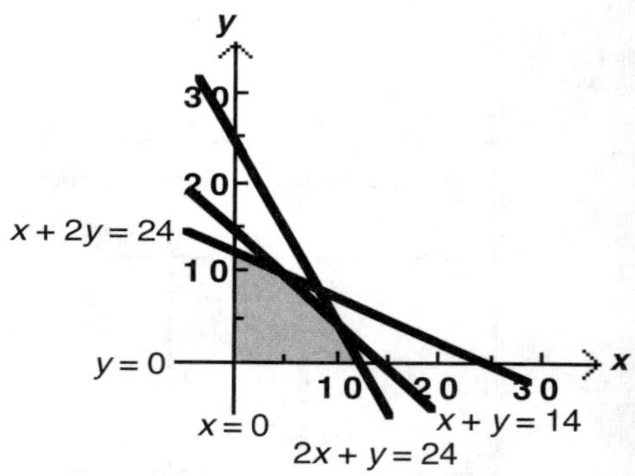

23. The feasible region is graphed as follows:
The corner points (0, 20) and (20, 0) are obvious from the graph. The third corner point is obtained by solving:
$x + 4y = 20$
$4x + y = 20$ to obtain (4, 4)
We now evaluate the objective function at each corner point.

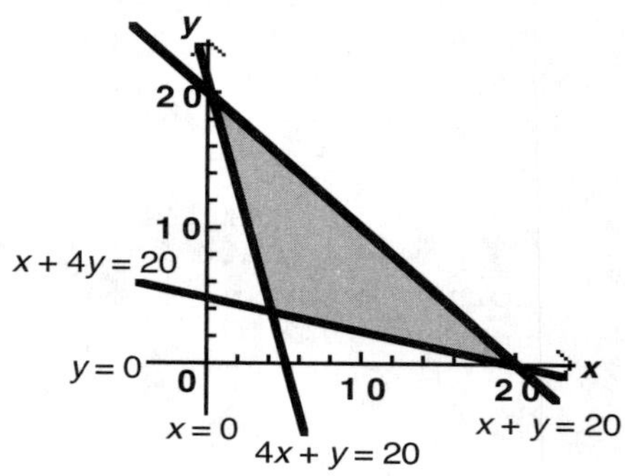

Corner Point (x, y)	Objective Function $z = 5x + 6y$	
(0, 20)	120	
(20, 0)	100	
(4, 4)	44	Minimum value

The minimum value of z on S is 44 at (4, 4).

25. The feasible region is graphed as follows: The corner points (60, 0) and (120, 0) are obvious from the graph. The other corner points are obtained by solving:

$x + y = 60$ and $x + 2y = 120$
$x - 2y = 0$ $\quad$ $x - 2y = 0$
to obtain (40, 20) $\quad$ to obtain (60, 30)

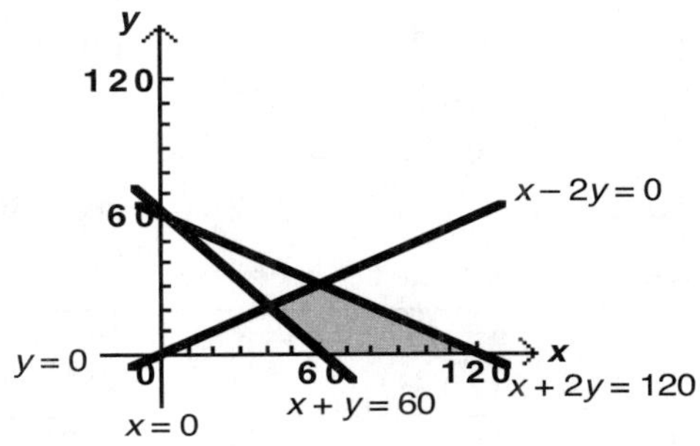

We now evaluate the objective function at each corner point.

Corner Point (x, y)	Objective Function $25x + 50y$	
(60, 0)	1,500	Minimum value
(40, 20)	2,000	
(60, 30)	3,000	Maximum value } Multiple optimal solutions
(120, 0)	3,000	Maximum value }

The minimum value of z on S is 1,500 at (60, 0). The maximum value of z on S is 3,000 at (60, 30) and (120, 0).

27. The feasible region is graphed as follows:
The corner points (0, 45), (0, 20), (25, 0), and (60, 0) are obvious from the graph shown below. The other corner points are obtained by solving:

$3x + 4y = 240$ and $3x + 4y = 240$
$y = 45$ to obtain (20, 45) $\quad$ $x = 60$ to obtain (60, 15)

We now evaluate the objective function at each corner point.

Corner Point (x, y)	Objective Function $25x + 15y$	
(0, 45)	675	
(0, 20)	300	Minimum value
(25, 0)	625	
(60, 0)	1,500	
(60, 15)	1,725	Maximum value
(20, 45)	1,175	

The minimum value of z on S is 300 at (0, 20). The maximum value of z on S is 1,725 at (60, 15).

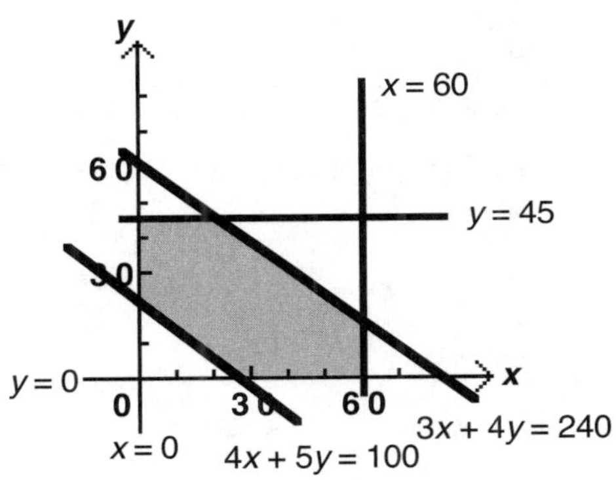

29. The feasible region is graphed as shown at the right.

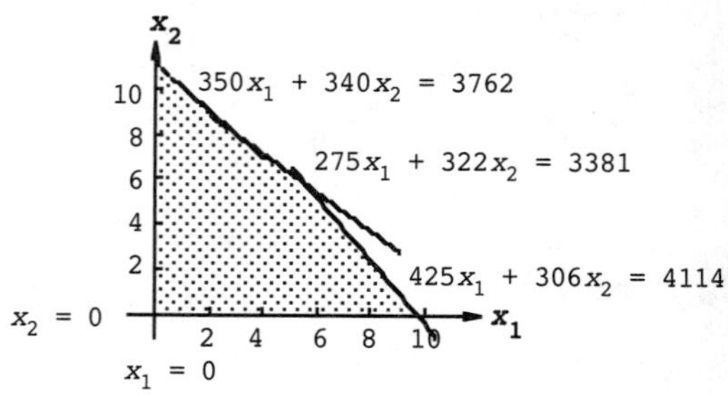

The corner point (0, 0) is obvious from the graph. The other corner points are obtained by solving:

$x_1 = 0$	$350x_1 + 340x_2 = 3762$
$275x_1 + 322x_2 = 3381$	$275x_1 + 322x_2 = 3381$
to obtain (0, 10.5)	to obtain (3.22, 7.75)
$350x_1 + 340x_2 = 3762$	$425x_1 + 306x_2 = 4114$
$425x_1 + 306x_2 = 4114$	$x_2 = 0$
to obtain (6.62, 4.25)	to obtain (9.68, 0)

We now evaluate the objective function at each corner point.

Corner Point (x_1, x_2)	Objective Function $525x_1 + 478x_2$	
(0, 0)	0	Minimum value
(0, 10.5)	5019	
(3.22, 7.75)	5395	
(6.62, 4.25)	5507	Maximum value
(9.68, 0)	5082	

The maximum value of P is 5507 at the corner point (6.62, 4.25).

31. The feasible region is graphed as follows: (heavily outlined for clarity)

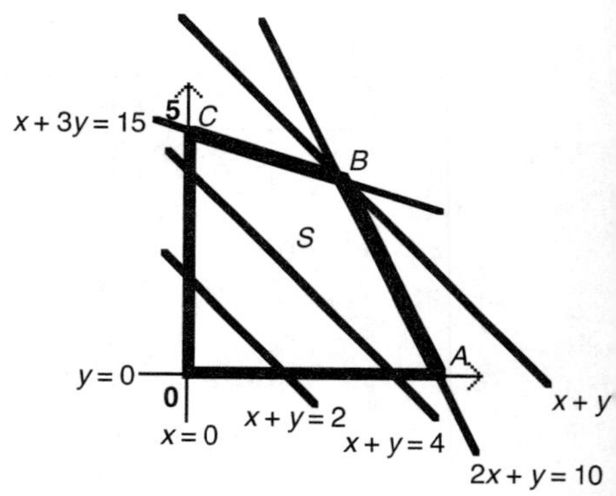

Consider the objective function $x + y$. It should be clear that it takes on the value 2 along $x + y = 2$, the value 4 along $x + y = 4$, the value 7 along $x + y = 7$, and so on. The maximum value of the objective function, then, on S, is 7, which occurs at B. Graphically this occurs when the line $x + y = c$ coincides with the boundary of S. Thus to answer questions (A)-(E) we must determine values of a and b such that the appropriate line $ax + by = c$ coincides with the boundary of S only at the specified points.

(A) The line $ax + by = c$ must have slope negative, but greater in absolute value that that of line segment AB, $2x + y = 10$. Therefore $a > 2b$.

(B) The line $ax + by = c$ must have slope negative but between that of $x + 3y = 15$ and $2x + y = 10$. Therefore $\frac{1}{3}b < a < 2b$.

(C) The line $ax + by = c$ must have slope greater than that of line segment BC, $x + 3y = 15$. Therefore $a < \frac{1}{3}b$ or $b > 3a$

(D) The line $ax + by = c$ must be parallel to line segment AB, therefore $a = 2b$.

(E) The line $ax + by = c$ must be parallel to line segment BC, therefore $b = 3a$.

33. We let x = the number of trick skis
y = the number of slalom skis
The problem constraints were
$6x + 4y \le 108$
$x + y \le 24$
The non-negative constraints were
$x \ge 0$
$y \ge 0$
The feasible region was graphed there.
(A) We note now: the linear objective function
$P = 40x + 30y$ represents the profit.

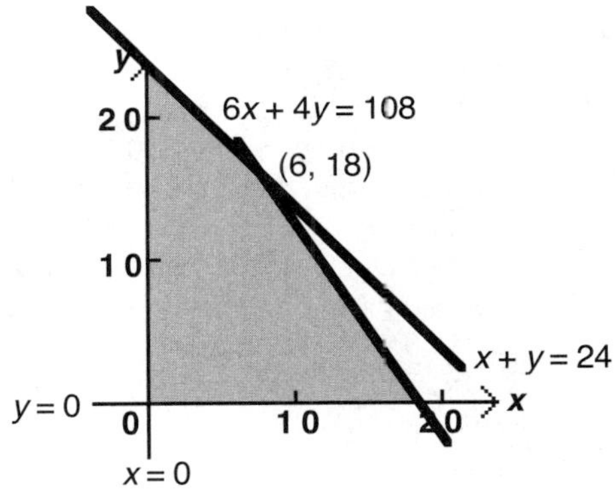

Three of the corner points are obvious from the graph: (0, 24), (0, 0), and (18, 0). The fourth corner point is obtained by solving:

$6x + 4y = 108$
$x + y = 24$ to obtain (6, 18).

Summarizing: the mathematical model for this problem is:

Maximize $P = 40x + 30y$
subject to: $6x + 4y \le 108$
$x + y \le 24$
$x, y \ge 0$

We now evaluate the objective function $40x + 30y$ at each corner point.

Corner Point (x, y)	Objective Function $40x + 30y$	
(0, 0)	0	
(18, 0)	720	
(6, 18)	780	Maximum value
(0, 24)	720	

The optimal value is 780 at the corner point (6, 18). Thus, 6 trick skis and 18 slalom skis should be manufactured to obtain the maximum profit of $780.

(B) The objective function now becomes $40x + 25y$. We evaluate this at each corner point.

Corner Point (x, y)	Objective Function $40x + 25y$	
(0, 0)	0	
(18, 0)	720	Maximum value
(6, 18)	690	
(0, 24)	600	

The optimal value is now 720 at the corner point (18, 0). Thus, 18 trick skis and no slalom skis should be produced to obtain a maximum profit of $720.

(C) The objective function now becomes $40x + 45y$. We evaluate this at each corner point.

Corner Point (x, y)	Objective Function $40x + 45y$	
(0, 0)	0	
(18, 0)	720	
(6, 18)	1,050	
(0, 24)	1,080	Maximum value

The optimal value is now 1,080 at the corner point (0, 24). Thus, no trick skis and 24 slalom skis should be produced to obtain a maximum profit of $1,080.

35. Let x = number of model A trucks
y = number of model B trucks
We form the linear objective function
$C = 15{,}000x + 24{,}000y$
We wish to minimize C, the cost of buying x trucks @ \$15,000 and y trucks @ \$24,000, subject to the constraints.
$x + y \le 15$ maximum number of trucks constraint
$2x + 3y \ge 36$ capacity constraint
$x, y \ge 0$ non-negative constraints.
Solving the system of constraint inequalities graphically, we obtain the feasible region S shown in the diagram.

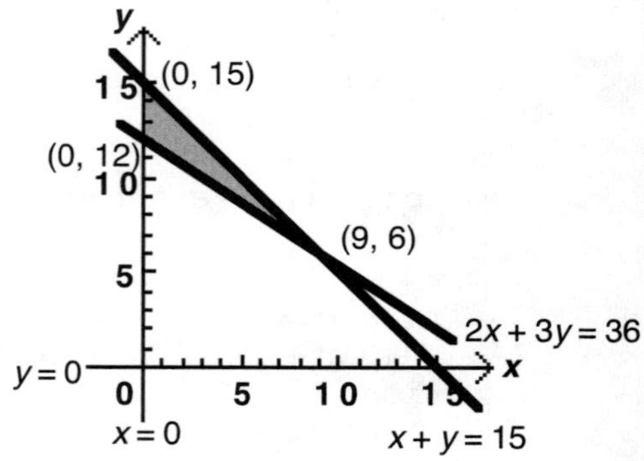

Next we evaluate the objective function at each corner point.

Corner Point (x, y)	Objective Function $C = 15{,}000x + 24{,}000y$	
(0, 12)	288,000	
(0, 15)	360,000	
(9, 6)	279,000	Minimum value

The optimal value is \$279,000 at the corner point (9, 6). Thus, the company should purchase 9 model A trucks and 6 model B trucks to realize the minimum cost of \$279,000.

37. (A) Let x = number of tables
y = number of chairs
We form the linear objective function
$P = 90x + 25y$
We wish to maximize P, the profit from x tables @ \$90 and y chairs @ \$25, subject to the constraints
$8x + 2y \le 400$ assembly department constraint
$2x + y \le 120$ finishing department constraint
$x, y \ge 0$ non-negative constraints
Solving the system of constraint inequalities graphically, we obtain the feasible region S shown in the diagram.

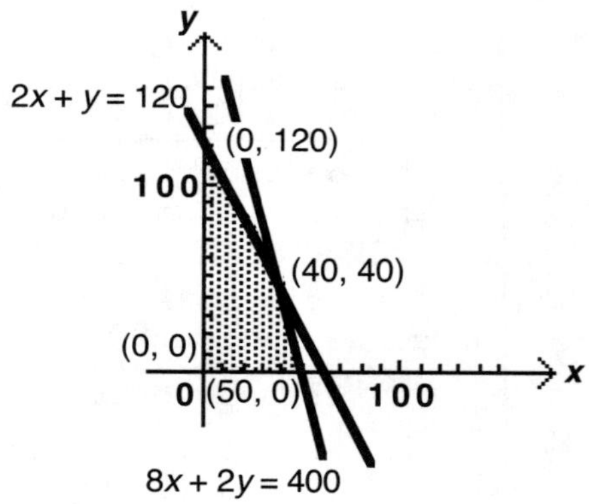

Next we evaluate the objective function at each corner point.

Corner Point (x, y)	Objective Function $P = 90x + 25y$	
(0, 0)	0	
(50, 0)	4,500	
(40, 40)	4,600	Maximum value
(0, 120)	3,000	

The optimal value is 4,600 at the corner point (40, 40). Thus, the company should manufacture 40 tables and 40 chairs for a maximum profit of \$4,600.

(B) We are faced with the further condition that $y \geq 4x$. We wish, then, to maximize $P = 90x + 25y$ under the constraints

$8x + 2y \leq 400$
$2x + y \leq 120$
$y \geq 4x$
$x, y \geq 0$

The feasible region is now S' as graphed.

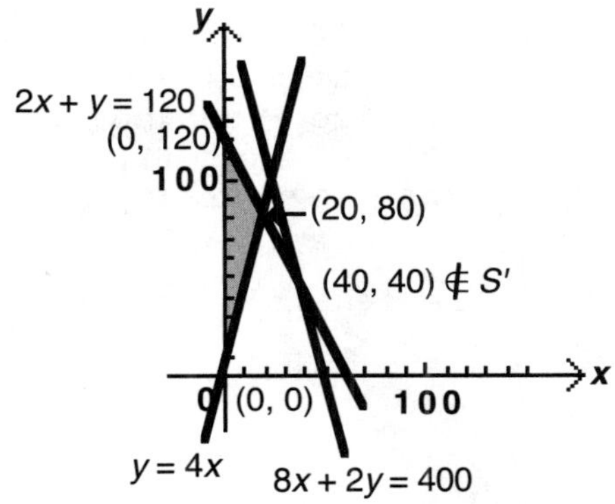

Note that the new condition has the effect of excluding (40, 40) from the feasible region.

We now evaluate the objective function at the new corner points.

Corner Point (x, y)	Objective Function $P = 90x + 25y$	
(0, 120)	3,000	
(0, 0)	0	
(20, 80)	3,800	Maximum value

The optimal value is now 3,800 at the corner point (20, 80). Thus the company should manufacture 20 tables and 80 chairs for a maximum profit of $3,800.

39. Let x = number of gallons produced using the old process
y = number of gallons produced using the new process

We form the linear objective function
$P = 0.6x + 0.2y$

(A) We wish to maximize P, the profit from x gallons using the old process and y gallons using the new process, subject to the constraints

$20x + 5y \leq 16,000$ sulfur dioxide constraint
$40x + 20y \leq 30,000$ particulate matter constraint
$x, y \geq 0$ non-negative constraints

Solving the system of constraint inequalities graphically, we obtain the feasible region S shown in the diagram. Note that no corner points are determined by this (very weak) sulfur dioxide constraint.

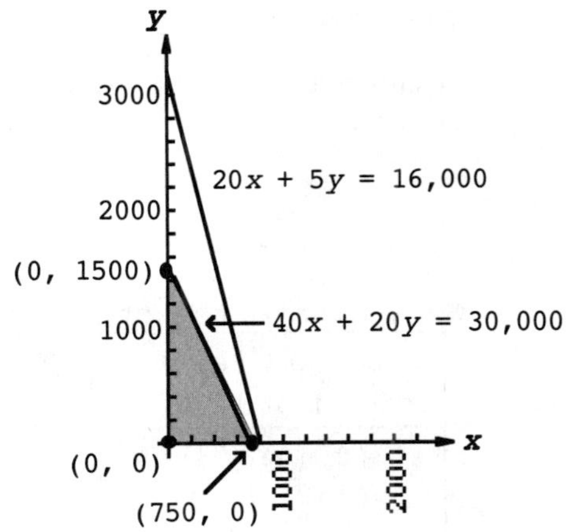

We evaluate the objective function at each corner point.

Corner Point (x, y)	Objective Function $P = 0.6x + 0.2y$	
(0, 0)	0	
(0, 1500)	300	
(750, 0)	450	Maximum value

The optimal value is 450 at the corner point (750, 0). Thus, the company should manufacture 750 gallons by the old process exclusively, for a profit of $450.

(B) The sulfur dioxide constraint is now $20x + 5y \le 11{,}500$. We now wish to maximize P subject to the constraints

$$\begin{aligned} 20x + 5y &\le 11{,}500 \\ 40x + 20y &\le 30{,}000 \\ x, y &\ge 0 \end{aligned}$$

The feasible region is now S_1 as shown.

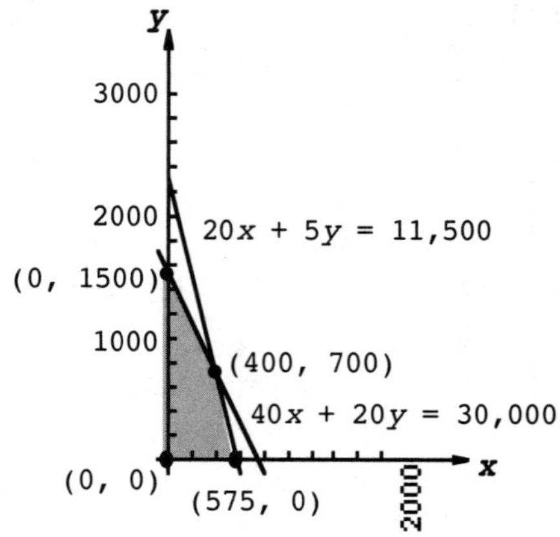

We evaluate the objective function at the new corner points.

Corner Point (x, y)	Objective Function $P = 0.6x + 0.2y$	
(0, 0)	0	
(575, 0)	345	
(400, 700)	380	Maximum value
(0, 1500)	300	

The optimal value is now 380 at the corner point (400, 700). Thus, the company should manufacture 400 gallons by the old process and 700 gallons by the new process, for a profit of $380.

(C) The sulfur dioxide constraint is now $20x + 5y \le 7{,}200$. We now wish to maximize P subject to the constraints

$$\begin{aligned} 20x + 5y &\le 7{,}200 \\ 40x + 20y &\le 30{,}000 \\ x, y &\ge 0 \end{aligned}$$

The feasible region is now S_2 as shown.

Note that now no corner points are determined by the particulate matter constraint.

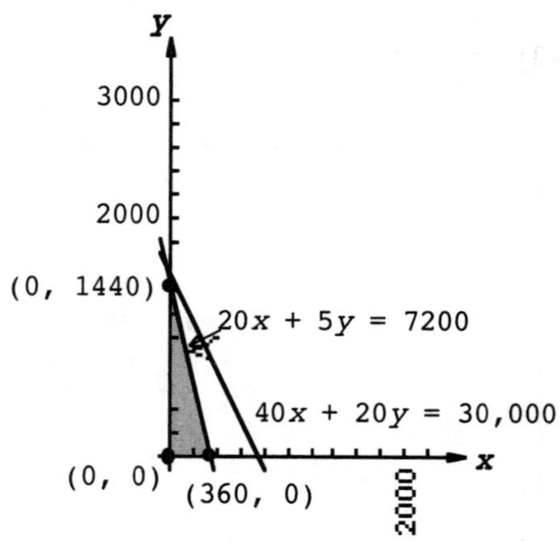

We evaluate the objective function at the new corner points.

Corner Point (x, y)	Objective Function $P = 0.6x + 0.2y$	
(0, 0)	0	
(360, 0)	216	
(0, 1440)	288	Maximum value

The optimal value is now 288 at the corner point (0, 1440). Thus, the company should manufacture 1,440 gallons by the new process exclusively, for a profit of $288.

CHAPTER 10 REVIEW

1. $2x + y = 7$
$3x - 2y = 0$
We multiply the top equation by 2 and add.

$$\begin{array}{rcl} 4x + 2y &=& 14 \\ 3x - 2y &=& 0 \\ \hline 7x &=& 14 \\ x &=& 2 \end{array}$$

Substituting $x = 2$ in the top equation, we have
$2(2) + y = 7$
$y = 3$
Solution: (2, 3) *(10-1)*

2. $3x - 6y = 5$
$-2x + 4y = 1$
We multiply the top equation by 2, the bottom by 3, and add.

$$\begin{array}{rcl} 6x - 12y &=& 10 \\ -6x + 12y &=& 3 \\ \hline 0 &=& 13 \end{array}$$

No solution *(10-1)*

3. $4x - 3y = -8$
$-2x + \frac{3}{2}y = 4$
We multiply the bottom equation by 2 and add.

$$\begin{array}{rcl} 4x - 3y &=& -8 \\ -4x + 3y &=& 8 \\ \hline 0 &=& 0 \end{array}$$

There are infinitely many solutions. For any real number t, $4t - 3y = -8$, hence,
$-3y = -4t - 8$
$y = \frac{4t + 8}{3}$
Thus, $\left(t, \frac{4t + 8}{3}\right)$ is a solution for any real number t. *(10-1)*

4.
$$\begin{array}{rcll} x - 3y + z &=& 4 & E_1 \\ -x + 4y - 4z &=& 1 & E_2 \\ 2x - y + 5z &=& -3 & E_3 \end{array}$$

Add E_1 to E_2 to eliminate x.
Also multiply E_1 by -2 and add to E_3 to eliminate x.

$$\begin{array}{rcll} x - 3y + z &=& 4 & E_1 \\ -x + 4y - 4z &=& 1 & E_2 \\ \hline y - 3z &=& 5 & E_4 \end{array}$$

$$\begin{array}{rcll} -2x + 6y - 2z &=& -8 & (-2)E_1 \\ 2x - y + 5z &=& -3 & E_3 \\ \hline 5y + 3z &=& -11 & E_5 \end{array}$$

Equivalent system:

$$\begin{array}{rcll} x - 3y + z &=& 4 & E_1 \\ y - 3z &=& 5 & E_4 \\ 5y + 3z &=& -11 & E_5 \end{array}$$

Add E_4 to E_5 to eliminate z

$$\begin{array}{rcll} y - 3z &=& 5 & E_4 \\ 5y + 3z &=& -11 & E_5 \\ \hline 6y &=& -6 & \\ y &=& -1 & \end{array}$$

Substitute $y = -1$ into E_5 and solve for z.

$$\begin{aligned} 5y + 3z &= -11 \quad E_5 \\ 5(-1) + 3z &= -11 \\ z &= -2 \end{aligned}$$

Substitute $y = -1$ and $z = -2$ into E_1 and solve for x.

$$\begin{aligned} x - 3y + z &= 4 \quad E_1 \\ x - 3(-1) + (-2) &= 4 \\ x &= 3 \end{aligned}$$

$(3, -1, -2)$ (10-2)

5. $x^2 + y^2 = 2$
$2x - y = 3$
We solve the first-degree equation for y in terms of x, then substitute into the second-degree equation.

$$\begin{aligned} 2x - y &= 3 \\ -y &= 3 - 2x \\ y &= 2x - 3 \\ x^2 + (2x - 3)^2 &= 2 \\ x^2 + 4x^2 - 12x + 9 &= 2 \\ 5x^2 - 12x + 7 &= 0 \\ (5x - 7)(x - 1) &= 0 \\ x &= \frac{7}{5}, 1 \end{aligned}$$

For $x = \frac{7}{5}$

$$y = 2\left(\frac{7}{5}\right) - 3 = -\frac{1}{5}$$

For $x = 1$

$$y = 2(1) - 3 = -1$$

Solutions: $(1, -1)$, $\left(\frac{7}{5}, -\frac{1}{5}\right)$

The checking steps are omitted for lack of space. (10-6)

6. $2x + y - z = 5 \quad E_1$
$x - 2y - 2z = 4 \quad E_2$
$3x + 4y + 3z = 3 \quad E_3$

Multiply E_2 by -2 and add to E_1 to eliminate x. Also multiply E_2 by -3 and add to E_3 to eliminate x.

$$\begin{aligned} 2x + y - z &= 5 \quad E_1 \\ -2x + 4y + 4z &= -8 \quad (-2)E_2 \\ 5y + 3z &= -3 \quad E_4 \\ 3x + 4y + 3z &= 3 \quad E_3 \\ -3x + 6y + 6z &= -12 \quad (-3)E_2 \\ 10y + 9z &= -9 \quad E_5 \end{aligned}$$

Equivalent system:

$$\begin{aligned} x - 2y - 2z &= 4 \quad E_2 \\ 5y + 3z &= -3 \quad E_4 \\ 10y + 9z &= -9 \quad E_5 \end{aligned}$$

Multiply E_4 by -2 and add to E_5 to eliminate y.

$$\begin{aligned} -10y - 6z &= 6 \quad (-2)E_4 \\ 10y + 9z &= -9 \quad E_5 \\ 3z &= -3 \\ z &= -1 \end{aligned}$$

Substitute $z = -1$ into E_4 and solve for y.

$$\begin{aligned} 5y + 3z &= -3 \quad E_4 \\ 5y + 3(-1) &= -3 \\ y &= 0 \end{aligned}$$

Substitute $y = 0$ for $z = -1$ into E_2 and solve for x.

$$\begin{aligned} x - 2y - 2z &= 4 \quad E_2 \\ x - 2(0) - 2(-1) &= 4 \\ x &= 2 \end{aligned}$$

$(2, 0, -1)$ (10-2)

7.

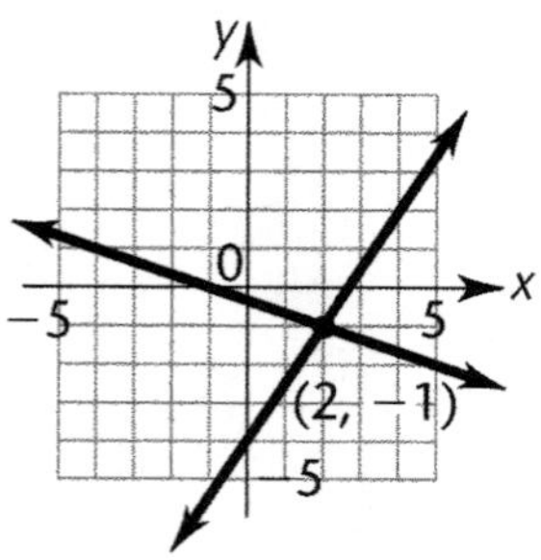

(10-1)

8. Graph $3x - 4y = 24$ as a solid line since equality is included in the original statement. (0, 0) is a suitable test point.
$3(0) - 4(0) = 0 \not\geq 24$
Therefore (0, 0) is not in the solution set. The line $3x - 4y = 24$ and the half-plane not containing (0, 0) form the graph.

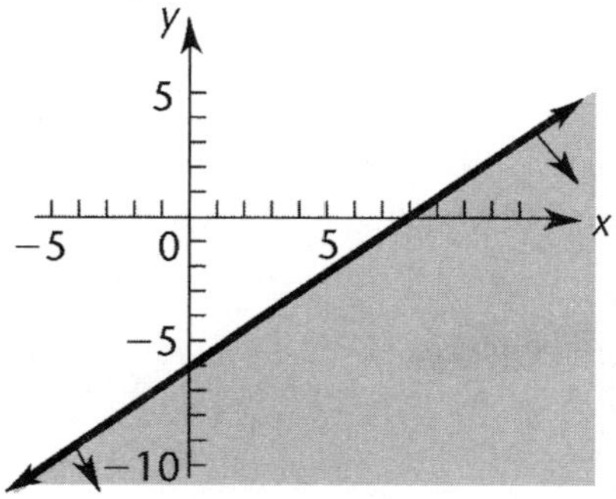

(10-7)

9.

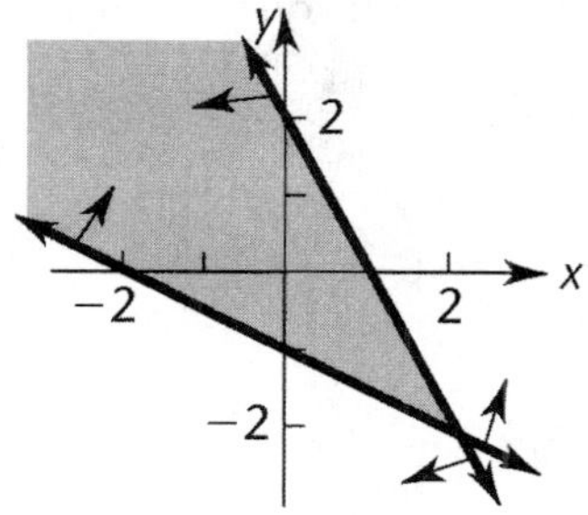

(10-7)

10. $R_1 \leftrightarrow R_2$ means interchange Rows 1 and 2.
$$\left[\begin{array}{cc|c} 3 & -6 & 12 \\ 1 & -4 & 5 \end{array}\right]$$ (10-3)

11. $\frac{1}{3}R_2 \to R_2$ means multiply Row 2 by $\frac{1}{3}$.
$$\left[\begin{array}{cc|c} 1 & -4 & 5 \\ 1 & -2 & 4 \end{array}\right]$$ (10-3)

12. $(-3)R_1 + R_2 \to R_2$ means replace Row 2 by itself plus -3 times Row 1.
$$\left[\begin{array}{cc|c} 1 & -4 & 5 \\ 0 & 6 & -3 \end{array}\right]$$ (10-3)

13. $x_1 = 4$
$x_2 = -7$
The solution is (4, -7) (10-3)

14. $x_1 - x_2 = 4$
$0 = 1$
No solution (10-3)

15. $x_1 - x_2 = 4$
$0 = 0$
Solution:
$x_2 = t$
$x_1 = x_2 + 4 = t + 4$
Thus $x_1 = t + 4$, $x_2 = t$ is the solution, for t any real number.
(10-3)

16. $$AB = \begin{bmatrix} 4 & -2 \\ 0 & 3 \end{bmatrix}\begin{bmatrix} -1 & 5 \\ -4 & 6 \end{bmatrix} = \begin{bmatrix} 4(-1) + (-2)(-4) & 4 \cdot 5 + (-2)6 \\ 0(-1) + 3(-4) & 0 \cdot 5 + 3 \cdot 6 \end{bmatrix} = \begin{bmatrix} 4 & 8 \\ -12 & 18 \end{bmatrix}$$ (10-4)

17. $$CD = [-1 \quad 4]\begin{bmatrix} 3 \\ -2 \end{bmatrix} = [(-1)3 + 4(-2)] = [-11]$$ (10-4)

18. $$CB = [-1 \quad 4]\begin{bmatrix} -1 & 5 \\ -4 & 6 \end{bmatrix} = [(-1)(-1) + 4(-4) \quad (-1)5 + 4 \cdot 6] = [-15 \quad 19]$$ (10-4)

19. $$AD = \begin{bmatrix} 4 & -2 \\ 0 & 3 \end{bmatrix}\begin{bmatrix} 3 \\ -2 \end{bmatrix} = \begin{bmatrix} 4 \cdot 3 + (-2)(-2) \\ 0 \cdot 3 + 3(-2) \end{bmatrix} = \begin{bmatrix} 16 \\ -6 \end{bmatrix}$$ (10-4)

20. $$A + B = \begin{bmatrix} 4 & -2 \\ 0 & 3 \end{bmatrix} + \begin{bmatrix} -1 & 5 \\ -4 & 6 \end{bmatrix} = \begin{bmatrix} 4 + (-1) & (-2) + 5 \\ 0 + (-4) & 3 + 6 \end{bmatrix} = \begin{bmatrix} 3 & 3 \\ -4 & 9 \end{bmatrix}$$ (10-4)

21. $C + D$ is not defined (*10-4*) **22.** $A + C$ is not defined (*10-4*)

23. $2A - 5B = 2\begin{bmatrix} 4 & -2 \\ 0 & 3 \end{bmatrix} - 5\begin{bmatrix} -1 & 5 \\ -4 & 6 \end{bmatrix} = \begin{bmatrix} 8 & -4 \\ 0 & 6 \end{bmatrix} - \begin{bmatrix} -5 & 25 \\ -20 & 30 \end{bmatrix} = \begin{bmatrix} 13 & -29 \\ 20 & -24 \end{bmatrix}$ (*10-4*)

24. $CA + C = [-1 \quad 4]\begin{bmatrix} 4 & -2 \\ 0 & 3 \end{bmatrix} + [-1 \quad 4] = [(-1)4 + 4 \cdot 0 \quad (-1)(-2) + 4 \cdot 3] + [-1 \quad 4]$

$= [-4 \quad 14] + [-1 \quad 4] = [-5 \quad 18]$ (*10-4*)

25. $\left[\begin{array}{rr|rr} 4 & 7 & 1 & 0 \\ -1 & -2 & 0 & 1 \end{array}\right] R_1 \leftrightarrow R_2$

$\sim \left[\begin{array}{rr|rr} -1 & -2 & 0 & 1 \\ 4 & 7 & 1 & 0 \end{array}\right] 4R_1 + R_2 \to R_2$

$\sim \left[\begin{array}{rr|rr} -1 & -2 & 0 & 1 \\ 0 & -1 & 1 & 4 \end{array}\right] (-2)R_2 + R_1 \to R_1$

$\sim \left[\begin{array}{rr|rr} -1 & 0 & -2 & -7 \\ 0 & -1 & 1 & 4 \end{array}\right] \begin{array}{l} (-1)R_1 \to R_1 \\ (-1)R_2 \to R_2 \end{array}$

$\sim \left[\begin{array}{rr|rr} 1 & 0 & 2 & 7 \\ 0 & 1 & -1 & -4 \end{array}\right]$

Hence, $A^{-1} = \begin{bmatrix} 2 & 7 \\ -1 & -4 \end{bmatrix}$

$A^{-1}A = \begin{bmatrix} 2 & 7 \\ -1 & -4 \end{bmatrix}\begin{bmatrix} 4 & 7 \\ -1 & -2 \end{bmatrix} = \begin{bmatrix} 2 \cdot 4 + 7(-1) & 2 \cdot 7 + 7(-2) \\ (-1)4 + (-4)(-1) & (-1)7 + (-4)(-2) \end{bmatrix} = \begin{bmatrix} 1 & 0 \\ 0 & 1 \end{bmatrix} = I$

(*10-5*)

26. As a matrix equation the system becomes

$$\underset{A}{\begin{bmatrix} 3 & 2 \\ 4 & 3 \end{bmatrix}}\underset{X}{\begin{bmatrix} x_1 \\ x_2 \end{bmatrix}} = \underset{B}{\begin{bmatrix} k_1 \\ k_2 \end{bmatrix}}$$

The solution of $AX = B$ is $X = A^{-1}B$.

Applying matrix methods, we obtain $A^{-1} = \begin{bmatrix} 3 & -2 \\ -4 & 3 \end{bmatrix}$ (details omitted). Applying this inverse, we have

$X = \begin{bmatrix} 3 & -2 \\ -4 & 3 \end{bmatrix}\begin{bmatrix} k_1 \\ k_2 \end{bmatrix}$

(A) $\begin{bmatrix} x_1 \\ x_2 \end{bmatrix} = \begin{bmatrix} 3 & -2 \\ -4 & 3 \end{bmatrix}\begin{bmatrix} 3 \\ 5 \end{bmatrix} = \begin{bmatrix} 3 \cdot 3 + (-2)5 \\ (-4)3 + 3 \cdot 5 \end{bmatrix} = \begin{bmatrix} -1 \\ 3 \end{bmatrix}$ $x_1 = -1$, $x_2 = 3$

(B) $\begin{bmatrix} x_1 \\ x_2 \end{bmatrix} = \begin{bmatrix} 3 & -2 \\ -4 & 3 \end{bmatrix}\begin{bmatrix} 7 \\ 10 \end{bmatrix} = \begin{bmatrix} 3 \cdot 7 + (-2)10 \\ (-4)7 + 3 \cdot 10 \end{bmatrix} = \begin{bmatrix} 1 \\ 2 \end{bmatrix}$ $x_1 = 1$, $x_2 = 2$

(C) $\begin{bmatrix} x_1 \\ x_2 \end{bmatrix} = \begin{bmatrix} 3 & -2 \\ -4 & 3 \end{bmatrix}\begin{bmatrix} 4 \\ 2 \end{bmatrix} = \begin{bmatrix} 3 \cdot 4 + (-2)2 \\ (-4)4 + 3 \cdot 2 \end{bmatrix} = \begin{bmatrix} 8 \\ -10 \end{bmatrix}$ $x_1 = 8$, $x_2 = -10$ (*10-5*)

27.

Corner Point (x, y)	Objective Function $z = 5x + 3y$	
(0, 10)	30	
(0, 6)	18	Minimum value
(4, 2)	26	
(6, 4)	42	Maximum value

The maximum value of z on S is 42 at (6, 4). The minimum value of z on S is 18 at (0, 6). (*10-8*)

28. We write the augmented matrix:

$$\left[\begin{array}{cc|c} 1 & -1 & 4 \\ 2 & 1 & 2 \end{array}\right] \quad \underbrace{(-2)R_1} + R_2 \to R_2$$

$-2 \quad 2 \quad -8$

Need a 0 here

$$\sim \left[\begin{array}{cc|c} 1 & -1 & 4 \\ 0 & 3 & -6 \end{array}\right] \quad \tfrac{1}{3}R_2 \to R_2 \quad \text{corresponds to the linear system} \quad \begin{aligned} x_1 - x_2 &= 4 \\ 3x_2 &= -6 \end{aligned}$$

Need a 1 here

Need a 0 here

$$\sim \left[\begin{array}{cc|c} 1 & -1 & 4 \\ 0 & 1 & -2 \end{array}\right] \quad R_2 + R_1 \to R_1 \quad \text{corresponds to the linear system} \quad \begin{aligned} x_1 - x_2 &= 4 \\ x_2 &= -2 \end{aligned}$$

$$\sim \left[\begin{array}{cc|c} 1 & 0 & 2 \\ 0 & 1 & -2 \end{array}\right] \quad \text{corresponds to the linear system} \quad \begin{aligned} x_1 &= 2 \\ x_2 &= -2 \end{aligned}$$

The solution is $x_1 = 2$, $x_2 = -2$. Each pair of lines graphed below has the same intersection point, (2, -2).

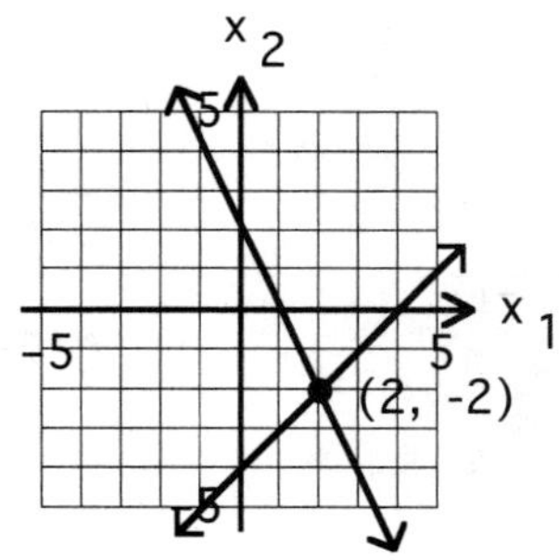

$x_1 - x_2 = 4$
$2x_1 + x_2 = 2$

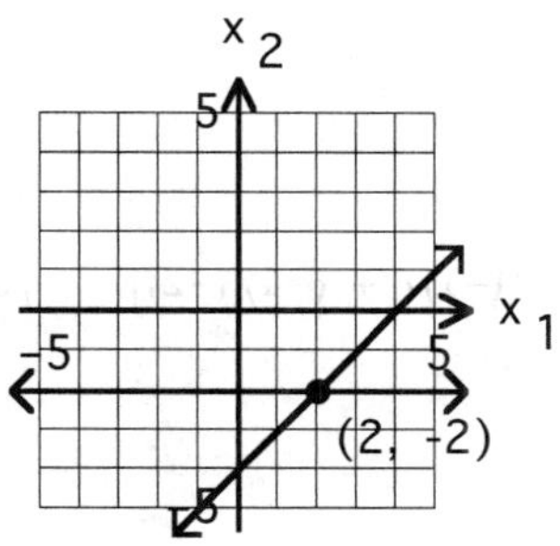

$x_1 - x_2 = 4$
$3x_2 = -6$

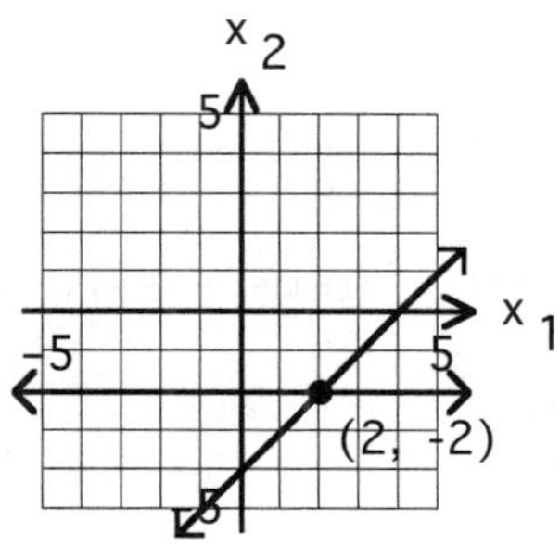

$x_1 - x_2 = 4$
$x_2 = -2$

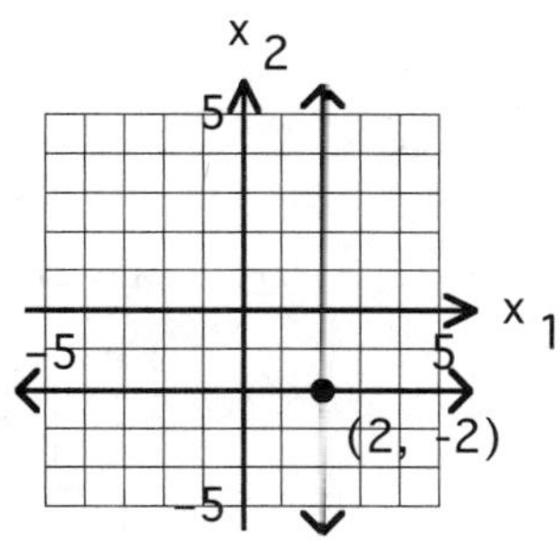

$x_1 = 2$
$x_2 = -2$

(*10-1, 10-3*)

29. Here is a computer-generated graph of the system, entered as

$$y = \frac{9 - x}{3}$$
$$y = \frac{10 + 2x}{7}$$

After zooming in (graph not shown) the intersection point is located at (2.54, 2.15) to two decimal places.

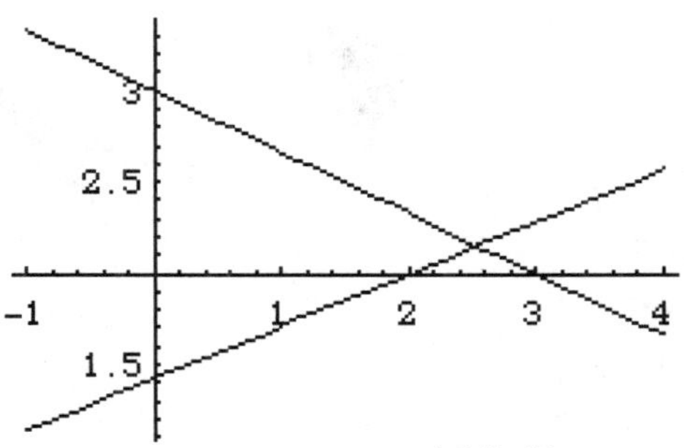

(*10-1*)

30. $$\left[\begin{array}{cc|c} 3 & 2 & 3 \\ 1 & 3 & 8 \end{array}\right] \quad R_1 \leftrightarrow R_2$$

$$\sim \left[\begin{array}{cc|c} 1 & 3 & 8 \\ 3 & 2 & 3 \end{array}\right] \quad \underbrace{(-3)R_1} + R_2 \to R_2$$

$-3 \quad -9 \quad -24$

$$\sim \left[\begin{array}{cc|c} 1 & 3 & 8 \\ 0 & -7 & -21 \end{array}\right] \quad -\tfrac{1}{7}R_2 \to R_2$$

$$\sim \left[\begin{array}{cc|c} 1 & 3 & 8 \\ 0 & 1 & 3 \end{array}\right] \quad \underbrace{(-3)R_2} + R_1 \to R_1$$

$0 \quad -3 \quad -9$

$$\sim \left[\begin{array}{cc|c} 1 & 0 & -1 \\ 0 & 1 & 3 \end{array}\right]$$

Solution: $x_1 = -1$, $x_2 = 3$ (*10-3*)

31. $\left[\begin{array}{ccc|c} 1 & 1 & 0 & 1 \\ 1 & 0 & -1 & -2 \\ 0 & 1 & 2 & 4 \end{array}\right]$ $(-1)R_1 + R_2 \to R_2$

$\sim \left[\begin{array}{ccc|c} 1 & 1 & 0 & 1 \\ 0 & -1 & -1 & -3 \\ 0 & 1 & 2 & 4 \end{array}\right]$ $(-1)\ R_2 \to R_2$

$\sim \left[\begin{array}{ccc|c} 1 & 1 & 0 & 1 \\ 0 & 1 & 1 & 3 \\ 0 & 1 & 2 & 4 \end{array}\right]$ $\begin{array}{l} (-1)R_2 + R_1 \to R_1 \\ (-1)R_2 + R_3 \to R_3 \end{array}$

$\sim \left[\begin{array}{ccc|c} 1 & 0 & -1 & -2 \\ 0 & 1 & 1 & 3 \\ 0 & 0 & 1 & 1 \end{array}\right]$ $\begin{array}{l} R_3 + R_1 \to R_1 \\ (-1)R_3 + R_2 \to R_2 \end{array}$

$\sim \left[\begin{array}{ccc|c} 1 & 0 & 0 & -1 \\ 0 & 1 & 0 & 2 \\ 0 & 0 & 1 & 1 \end{array}\right]$

Solution: $x_1 = -1$, $x_2 = 2$, $x_3 = 1$ *(10-3)*

32. $\left[\begin{array}{ccc|c} 1 & 2 & 3 & 1 \\ 2 & 3 & 4 & 3 \\ 1 & 2 & 1 & 3 \end{array}\right]$ $\begin{array}{l} (-2)R_1 + R_2 \to R_2 \\ (-1)R_1 + R_3 \to R_3 \end{array}$

$\sim \left[\begin{array}{ccc|c} 1 & 2 & 3 & 1 \\ 0 & -1 & -2 & 1 \\ 0 & 0 & -2 & 2 \end{array}\right]$ $\begin{array}{l} (-1)R_2 \to R_2 \\ -\frac{1}{2}R_3 \to R_3 \end{array}$

$\sim \left[\begin{array}{ccc|c} 1 & 2 & 3 & 1 \\ 0 & 1 & 2 & -1 \\ 0 & 0 & 1 & -1 \end{array}\right]$ $(-2)R_2 + R_1 \to R_1$

$\sim \left[\begin{array}{ccc|c} 1 & 0 & -1 & 3 \\ 0 & 1 & 2 & -1 \\ 0 & 0 & 1 & -1 \end{array}\right]$ $\begin{array}{l} R_3 + R_1 \to R_1 \\ (-2)R_3 + R_2 \to R_2 \end{array}$

$\sim \left[\begin{array}{ccc|c} 1 & 0 & 0 & 2 \\ 0 & 1 & 0 & 1 \\ 0 & 0 & 1 & -1 \end{array}\right]$

Solution: $x_1 = 2$, $x_2 = 1$, $x_3 = -1$ *(10-3)*

33. $\left[\begin{array}{ccc|c} 1 & 2 & -1 & 2 \\ 2 & 3 & 1 & -3 \\ 3 & 5 & 0 & -1 \end{array}\right]$ $\begin{array}{l} (-2)R_1 + R_2 \to R_2 \\ (-3)R_1 + R_3 \to R_3 \end{array}$

$\sim \left[\begin{array}{ccc|c} 1 & 2 & -1 & 2 \\ 0 & -1 & 3 & -7 \\ 0 & -1 & 3 & -7 \end{array}\right]$ $(-1)R_2 + R_3 \to R_3$

$\sim \left[\begin{array}{ccc|c} 1 & 2 & -1 & 2 \\ 0 & -1 & 3 & -7 \\ 0 & 0 & 0 & 0 \end{array}\right]$ $2R_2 + R_1 \to R_1$

$\sim \left[\begin{array}{ccc|c} 1 & 0 & 5 & -12 \\ 0 & -1 & 3 & -7 \\ 0 & 0 & 0 & 0 \end{array}\right]$ $(-1)R_2 \to R_2$

$\sim \left[\begin{array}{ccc|c} 1 & 0 & 5 & -12 \\ 0 & 1 & -3 & 7 \\ 0 & 0 & 0 & 0 \end{array}\right]$

This corresponds to the system

$$\begin{aligned} x_1 + \quad 5x_3 &= -12 \\ x_2 - 3x_3 &= 7 \end{aligned}$$

Let $x_3 = t$

Then $x_2 = 3x_3 + 7$

$= 3t + 7$

$x_1 = -5x_3 - 12$

$= -5t - 12$

Hence $x_1 = -5t - 12$, $x_2 = 3t + 7$, $x_3 = t$ is a solution for every real number t. There are infinitely many solutions. *(10-3)*

34. $\left[\begin{array}{cc|c} 1 & -2 & 1 \\ 2 & -1 & 0 \\ 1 & -3 & -2 \end{array}\right]$ $\begin{array}{l} (-2)R_1 + R_2 \to R_2 \\ (-1)R_1 + R_3 \to R_3 \end{array}$

$\sim \left[\begin{array}{cc|c} 1 & -2 & 1 \\ 0 & 3 & -2 \\ 0 & -1 & -3 \end{array}\right]$ $3R_3 + R_2 \to R_2$

$\sim \left[\begin{array}{cc|c} 1 & -2 & 1 \\ 0 & 0 & -11 \\ 0 & -1 & -3 \end{array}\right]$

The second row corresponds to the equation $0x_1 + 0x_2 = -11$, hence there is no solution. *(10-3)*

35. $\left[\begin{array}{ccc|c} 1 & 2 & -1 & 2 \\ 3 & -1 & 2 & -3 \end{array}\right]$ $(-3)R_1 + R_2 \to R_2$

$\sim \left[\begin{array}{ccc|c} 1 & 2 & -1 & 2 \\ 0 & -7 & 5 & -9 \end{array}\right]$ $-\frac{1}{7}R_2 \to R_2$

$\sim \left[\begin{array}{ccc|c} 1 & 2 & -1 & 2 \\ 0 & 1 & -\frac{5}{7} & \frac{9}{7} \end{array}\right]$ $(-2)R_2 + R_1 \to R_1$

$\sim \left[\begin{array}{ccc|c} 1 & 0 & \frac{3}{7} & -\frac{4}{7} \\ 0 & 1 & -\frac{5}{7} & \frac{9}{7} \end{array}\right]$

This corresponds to the system

$x_1 + \frac{3}{7}x_3 = -\frac{4}{7}$

$x_2 - \frac{5}{7}x_3 = \frac{9}{7}$

Let $x_3 = t$

Then $x_2 = \frac{5}{7}x_3 + \frac{9}{7}$

$= \frac{5}{7}t + \frac{9}{7}$

$x_1 = -\frac{3}{7}x_3 - \frac{4}{7}$

$= -\frac{3}{7}t - \frac{4}{7}$

Hence $x_1 = -\frac{3}{7}t - \frac{4}{7}$, $x_2 = \frac{5}{7}t + \frac{9}{7}$, $x_3 = t$ is a solution for every real number t. There are infinitely many solutions. (*10-3*)

36. $AD = \begin{bmatrix} 1 & 2 \\ 4 & 5 \\ -3 & -1 \end{bmatrix}\begin{bmatrix} 7 & 0 & -5 \\ 0 & 8 & -2 \end{bmatrix} = \begin{bmatrix} 1\cdot 7 + 2\cdot 0 & 1\cdot 0 + 2\cdot 8 & 1(-5) + 2(-2) \\ 4\cdot 7 + 5\cdot 0 & 4\cdot 0 + 5\cdot 8 & 4(-5) + 5(-2) \\ (-3)7 + (-1)0 & (-3)0 + (-1)8 & (-3)(-5) + (-1)(-2) \end{bmatrix}$

$= \begin{bmatrix} 7 & 16 & -9 \\ 28 & 40 & -30 \\ -21 & -8 & 17 \end{bmatrix}$ (*10-4*)

37. $DA = \begin{bmatrix} 7 & 0 & -5 \\ 0 & 8 & -2 \end{bmatrix}\begin{bmatrix} 1 & 2 \\ 4 & 5 \\ -3 & -1 \end{bmatrix} = \begin{bmatrix} 7\cdot 1 + 0\cdot 4 + (-5)(-3) & 7\cdot 2 + 0\cdot 5 + (-5)(-1) \\ 0\cdot 1 + 8\cdot 4 + (-2)(-3) & 0\cdot 2 + 8\cdot 5 + (-2)(-1) \end{bmatrix}$

$= \begin{bmatrix} 22 & 19 \\ 38 & 42 \end{bmatrix}$ (*10-4*)

38. $BC = \begin{bmatrix} 6 \\ 0 \\ -4 \end{bmatrix}[2 \quad 4 \quad -1] = \begin{bmatrix} 6\cdot 2 & 6\cdot 4 & 6(-1) \\ 0\cdot 2 & 0\cdot 4 & 0(-1) \\ (-4)2 & (-4)4 & (-4)(-1) \end{bmatrix} = \begin{bmatrix} 12 & 24 & -6 \\ 0 & 0 & 0 \\ -8 & -16 & 4 \end{bmatrix}$ (*10-4*)

39. $CB = [2 \quad 4 \quad -1]\begin{bmatrix} 6 \\ 0 \\ -4 \end{bmatrix} = [2\cdot 6 + 4\cdot 0 + (-1)(-4)] = [16]$ (*10-4*)

40. Since D has 3 columns and E has 2 rows, DE is not defined. (*10-4*)

41. $ED = \begin{bmatrix} 9 & -3 \\ -6 & 2 \end{bmatrix}\begin{bmatrix} 7 & 0 & -5 \\ 0 & 8 & -2 \end{bmatrix}$

$= \begin{bmatrix} 9\cdot 7 + (-3)0 & 9\cdot 0 + (-3)8 & 9(-5) + (-3)(-2) \\ (-6)7 + 2\cdot 0 & (-6)0 + 2\cdot 8 & (-6)(-5) + 2(-2) \end{bmatrix} = \begin{bmatrix} 63 & -24 & -39 \\ -42 & 16 & 26 \end{bmatrix}$ (*10-4*)

42. $\left[\begin{array}{ccc|ccc} 1 & 0 & 4 & 1 & 0 & 0 \\ -2 & 1 & 0 & 0 & 1 & 0 \\ 4 & -1 & 4 & 0 & 0 & 1 \end{array}\right]$ $2R_1 + R_1 \to R_2$, $(-4)R_1 + R_3 \to R_3$

$\sim \left[\begin{array}{ccc|ccc} 1 & 0 & 4 & 1 & 0 & 0 \\ 0 & 1 & 8 & 2 & 1 & 0 \\ 0 & -1 & -12 & -4 & 0 & 1 \end{array}\right]$ $R_2 + R_3 \to R_3$

$\sim \left[\begin{array}{ccc|ccc} 1 & 0 & 4 & 1 & 0 & 0 \\ 0 & 1 & 8 & 2 & 1 & 0 \\ 0 & 0 & -4 & -2 & 1 & 1 \end{array}\right]$ $R_3 + R_1 \to R_1$, $2R_3 + R_2 \to R_2$

$\sim \left[\begin{array}{ccc|ccc} 1 & 0 & 0 & -1 & 1 & 1 \\ 0 & 1 & 0 & -2 & 3 & 2 \\ 0 & 0 & -4 & -2 & 1 & 1 \end{array}\right]$ $(-\frac{1}{4})R_3 \to R_3$

$$\sim \left[\begin{array}{ccc|ccc} 1 & 0 & 0 & -1 & 1 & 1 \\ 0 & 1 & 0 & -2 & 3 & 2 \\ 0 & 0 & 1 & \frac{1}{2} & -\frac{1}{4} & -\frac{1}{4} \end{array}\right]$$

Hence

$$A^{-1} = \begin{bmatrix} -1 & 1 & 1 \\ -2 & 3 & 2 \\ \frac{1}{2} & -\frac{1}{4} & -\frac{1}{4} \end{bmatrix}$$

$$A^{-1}A = \begin{bmatrix} -1 & 1 & 1 \\ -2 & 3 & 2 \\ \frac{1}{2} & -\frac{1}{4} & -\frac{1}{4} \end{bmatrix}\begin{bmatrix} 1 & 0 & 4 \\ -2 & 1 & 0 \\ 4 & -1 & 4 \end{bmatrix}$$

$$= \begin{bmatrix} (-1)1 + 1(-2) + 1 \cdot 4 & (-1)0 + 1 \cdot 1 + 1(-1) & (-1)4 + 1 \cdot 0 + 1 \cdot 4 \\ (-2)1 + 3(-2) + 2 \cdot 4 & (-2)0 + 3 \cdot 1 + 2(-1) & (-2)4 + 3 \cdot 0 + 2 \cdot 4 \\ (\frac{1}{2})1 + (-\frac{1}{4})(-2) + (-\frac{1}{4})4 & (\frac{1}{2})0 + (-\frac{1}{4})1 + (-\frac{1}{4})(-1) & (\frac{1}{2})4 + (-\frac{1}{4})0 + (-\frac{1}{4})4 \end{bmatrix}$$

$$= \begin{bmatrix} 1 & 0 & 0 \\ 0 & 1 & 0 \\ 0 & 0 & 1 \end{bmatrix}$$ (10-5)

43. $$\underset{A}{\begin{bmatrix} 1 & 2 & 3 \\ 2 & 3 & 4 \\ 1 & 2 & 1 \end{bmatrix}} \underset{X}{\begin{bmatrix} x_1 \\ x_2 \\ x_3 \end{bmatrix}} = \underset{B}{\begin{bmatrix} k_1 \\ k_2 \\ k_3 \end{bmatrix}}$$

The solution to $AX = B$ is $X = A^{-1}B$.

Applying matrix methods, we obtain $A^{-1} = \begin{bmatrix} -\frac{5}{2} & 2 & -\frac{1}{2} \\ 1 & -1 & 1 \\ \frac{1}{2} & 0 & -\frac{1}{2} \end{bmatrix}$ (details omitted).
Applying the inverse, we have

(A) $B = \begin{bmatrix} 1 \\ 3 \\ 3 \end{bmatrix}$ $X = \begin{bmatrix} x_1 \\ x_2 \\ x_3 \end{bmatrix} = \begin{bmatrix} -\frac{5}{2} & 2 & -\frac{1}{2} \\ 1 & -1 & 1 \\ \frac{1}{2} & 0 & -\frac{1}{2} \end{bmatrix}\begin{bmatrix} 1 \\ 3 \\ 3 \end{bmatrix} = \begin{bmatrix} 2 \\ 1 \\ -1 \end{bmatrix}$ $x_1 = 2,\ x_2 = 1,\ x_3 = -1$

(B) $B = \begin{bmatrix} 0 \\ 0 \\ -2 \end{bmatrix}$ $X = \begin{bmatrix} x_1 \\ x_2 \\ x_3 \end{bmatrix} = \begin{bmatrix} -\frac{5}{2} & 2 & -\frac{1}{2} \\ 1 & -1 & 1 \\ \frac{1}{2} & 0 & -\frac{1}{2} \end{bmatrix}\begin{bmatrix} 0 \\ 0 \\ -2 \end{bmatrix} = \begin{bmatrix} 1 \\ -2 \\ 1 \end{bmatrix}$ $x_1 = 1,\ x_2 = -2,\ x_3 = 1$

(C) $B = \begin{bmatrix} -3 \\ -4 \\ 1 \end{bmatrix}$ $X = \begin{bmatrix} x_1 \\ x_2 \\ x_3 \end{bmatrix} = \begin{bmatrix} -\frac{5}{2} & 2 & -\frac{1}{2} \\ 1 & -1 & 1 \\ \frac{1}{2} & 0 & -\frac{1}{2} \end{bmatrix}\begin{bmatrix} -3 \\ -4 \\ 1 \end{bmatrix} = \begin{bmatrix} -1 \\ 2 \\ -2 \end{bmatrix}$ $x_1 = -1,\ x_2 = 2,\ x_3 = -2$ (10-5)

The checking steps are omitted in problems 44–46 for lack of space.

44. $x^2 - y^2 = 2$
$y^2 = x$

Substitute y^2 in the second equation into the first equation.

$$x^2 - x = 2$$
$$x^2 - x - 2 = 0$$
$$(x - 2)(x + 1) = 0$$
$$x = 2, -1$$

For $x = 2$: $y^2 = 2$, $y = \pm\sqrt{2}$

For $x = -1$: $y^2 = -1$, $y = \pm i$

Solutions: $(2, \sqrt{2})$, $(2, -\sqrt{2})$, $(-1, i)$, $(-1, -i)$ (10-6)

45. $x^2 + 2xy + y^2 = 1$
$xy = -2$

Solve for y in the second equation.

$y = \frac{-2}{x}$

Substitute into the first equation

$$x^2 + 2x\left(-\frac{2}{x}\right) + \left(-\frac{2}{x}\right)^2 = 1$$

$$x^2 - 4 + \frac{4}{x^2} = 1 \quad x \neq 0$$

$$x^4 - 4x^2 + 4 = x^2$$

$$x^4 - 5x^2 + 4 = 0$$

$$(x^2 - 4)(x^2 - 1) = 0$$

$$(x - 2)(x + 2)(x - 1)(x + 1) = 0$$

$x = 2, -2, 1, -1$

For $x = 2$: $y = \frac{-2}{2}$, $y = -1$

For $x = -2$: $y = \frac{-2}{-2}$, $y = 1$

For $x = 1$: $y = \frac{-2}{1}$, $y = -2$

For $x = -1$: $y = \frac{-2}{-1}$, $y = 2$

Solutions: $(2, -1)$, $(-2, 1)$, $(1, -2)$, $(-1, 2)$ *(10-6)*

46. $2x^2 + xy + y^2 = 8$
$x^2 - y^2 = 0$

We factor the left side of the equation that has a zero constant term.

$(x - y)(x + y) = 0$

$x = y$ or $x = -y$

Thus, the original system is equivalent to the two systems

$2x^2 + xy + y^2 = 8$, $x = y$ $2x^2 + xy + y^2 = 8$, $x = -y$

These systems are solved by substitution.

First System:

$$2x^2 + xy + y^2 = 8$$
$$x = y$$
$$2y^2 + yy + y^2 = 8$$
$$4y^2 = 8$$
$$y^2 = 2$$
$$y = \pm\sqrt{2}$$

Second System:

$$2x^2 + xy + y^2 = 8$$
$$x = -y$$
$$2(-y)^2 + (-y)y + y^2 = 8$$
$$2y^2 = 8$$
$$y^2 = 4$$
$$y = \pm 2$$

For $y = \sqrt{2}$: $x = \sqrt{2}$

For $y = -\sqrt{2}$: $x = -\sqrt{2}$

For $y = 2$: $x = -2$

For $y = -2$: $x = 2$

Solutions: $(\sqrt{2}, \sqrt{2})$, $(-\sqrt{2}, -\sqrt{2})$, $(2, -2)$, $(-2, 2)$ *(10-6)*

47. The solution region is bounded (contained in, for example, the circle $x^2 + y^2 = 36$). Three corner points are obvious from the graph: $(0, 4)$, $(0, 0)$, and $(4, 0)$. The fourth corner point is obtained by solving the system

$2x + y = 8$
$2x + 3y = 12$ to obtain $(3, 2)$.

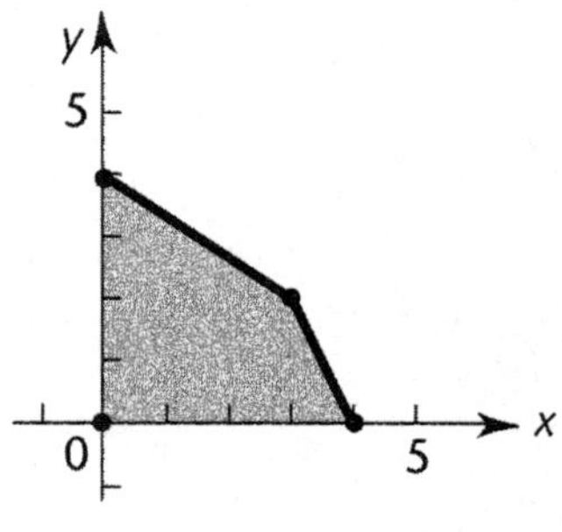

(10-7)

48. The solution region is unbounded. Two corner points are obvious from the graph: (0, 8) and (12, 0). The third corner point is obtained by solving the system

$2x + y = 8$
$x + 3y = 12$

to obtain $\left(\frac{12}{5}, \frac{16}{5}\right)$.

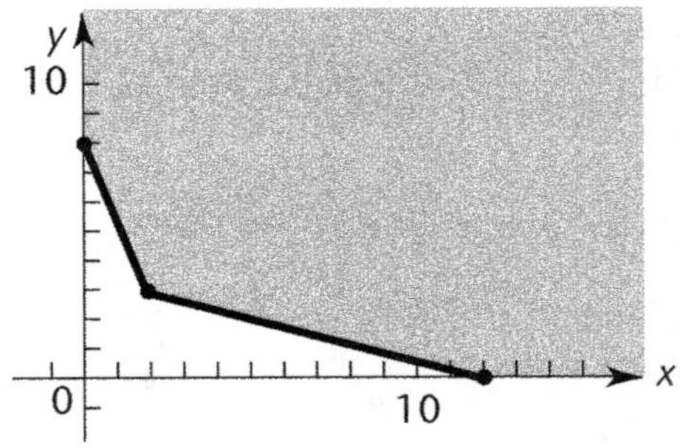

(10-7)

49. The solution region is bounded. The corner point (20, 0) is obvious from the graph. The other corner points are found by solving the system

$\begin{cases} x + y = 20 \\ x - y = 0 \end{cases}$ to obtain (10, 10)

and

$\begin{cases} x - y = 0 \\ x + 4y = 20 \end{cases}$ to obtain (4, 4)

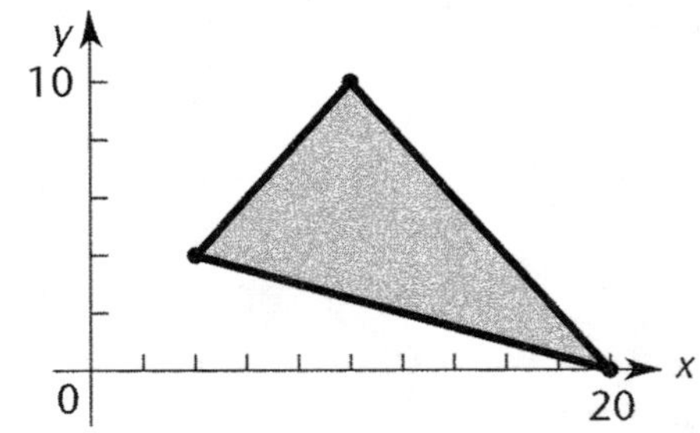

(10-7)

50. The feasible region is graphed as follows:
The corner points (0, 4), (0, 0) and (5, 0) are obvious from the graph. The corner point (4, 2) is obtained by solving the system

$x + 2y = 8$
$2x + y = 10$

We now evaluate the objective function at each corner point.

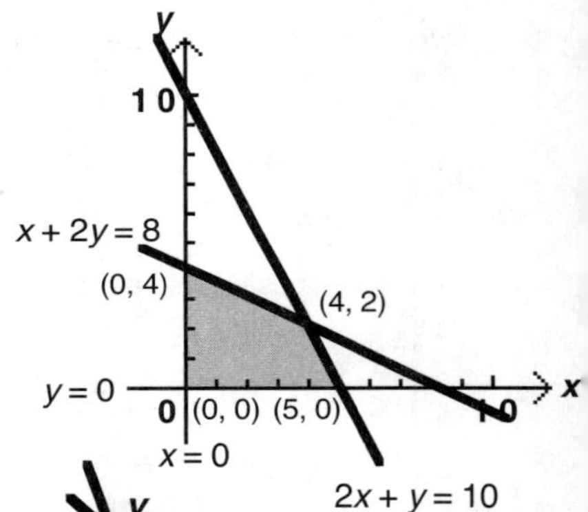

Corner Point (x, y)	Objective Function $z = 7x + 9y$	
(0, 4)	36	
(0, 0)	0	
(5, 0)	35	
(4, 2)	46	Maximum value

The maximum value of z on S is 46 at (4, 2).

(10-8)

51. The feasible region is graphed as follows:
The corner points (0, 20), (0, 15), (15, 0) and (20, 0) are obvious from the graph. The corner point (3, 6) is obtained by solving the system

$3x + y = 15$
$x + 2y = 15$

We now evaluate the objective function at each corner point.

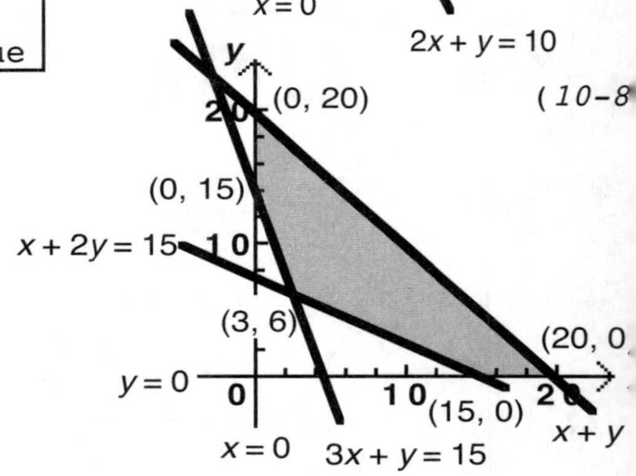

Corner Point (x, y)	Objective Function $z = 5x + 10y$	
(0, 20)	200	
(0, 15)	150	
(3, 6)	75	Minimum value } Multiple optimal solutions
(15, 0)	75	Minimum value } Multiple optimal solutions
(20, 0)	100	

The minimum value of z on S is 75 at (3, 6) and (15, 0). (10-8)

52. The feasible region is graphed below. The corner points (0, 10) and (0, 7) are obvious from the graph. The other corner points are obtained by solving the systems:

$x + 2y = 20$ and $3x + y = 15$
$3x + y = 15$ to obtain (2, 9) $x + y = 7$ to obtain (4, 3)

We now evaluate the objective function at each corner point.

Corner Point (x, y)	Objective Function $z = 5x + 8y$	
(0, 10)	80	
(0, 7)	56	
(4, 3)	44	Minimum value
(2, 9)	82	Maximum value

The minimum value of z on S is 44 at (4, 3).
The maximum value of z on S is 82 at (2, 9).

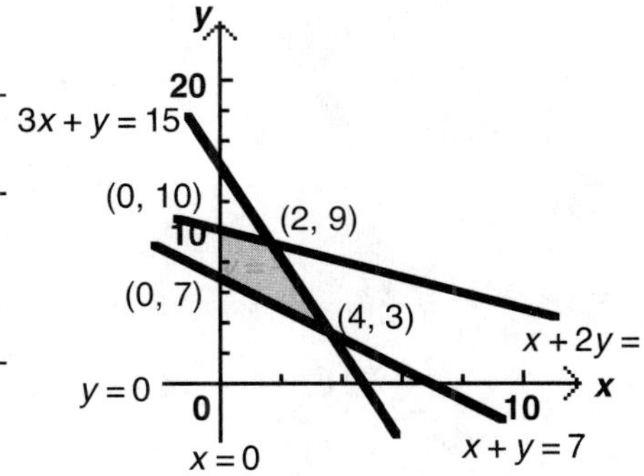

(*10-8*)

53. $$\left[\begin{array}{ccc|c} 1 & 1 & 1 & 7000 \\ 0.04 & 0.05 & 0.06 & 360 \\ 0.04 & 0.05 & -0.06 & 120 \end{array}\right] \begin{array}{l} \\ (-0.04)R_1 + R_2 \to R_2 \\ (-0.04)R_1 + R_3 \to R_3 \end{array}$$

$$\sim \left[\begin{array}{ccc|c} 1 & 1 & 1 & 7000 \\ 0 & 0.01 & 0.02 & 80 \\ 0 & 0.01 & -0.1 & -160 \end{array}\right] \quad 100R_2 \to R_2$$

$$\sim \left[\begin{array}{ccc|c} 1 & 1 & 1 & 7000 \\ 0 & 1 & 2 & 8000 \\ 0 & 0.01 & -0.1 & -160 \end{array}\right] \begin{array}{l} (-1)R_2 + R_1 \to R_1 \\ \\ (-0.01)R_2 + R_3 \to R_3 \end{array}$$

$$\sim \left[\begin{array}{ccc|c} 1 & 0 & -1 & -1000 \\ 0 & 1 & 2 & 8000 \\ 0 & 0 & -0.12 & -240 \end{array}\right] \; -\frac{25}{3} R_3 \to R_3$$

$$\sim \left[\begin{array}{ccc|c} 1 & 0 & -1 & -1000 \\ 0 & 1 & 2 & 8000 \\ 0 & 0 & 1 & 2000 \end{array}\right] \begin{array}{l} R_3 + R_1 \to R_1 \\ (-2)R_3 + R_2 \to R_2 \\ \\ \end{array}$$

$$\sim \left[\begin{array}{ccc|c} 1 & 0 & 0 & 1000 \\ 0 & 1 & 0 & 4000 \\ 0 & 0 & 1 & 2000 \end{array}\right]$$

Solution: $x_1 = 1000$, $x_2 = 4000$, $x_3 = 2000$ (*10-3*)

54. $x^2 - xy + y^2 = 4$
$x^2 + xy - 2y^2 = 0$
Factor the left side of the equation that has a zero constant term.
$(x + 2y)(x - y) = 0$
$x = -2y$ or $x = y$
Thus, the original system is equivalent to the two systems
$x^2 - xy + y^2 = 4$ $\quad$ $x^2 - xy + y^2 = 4$
$x = -2y$ $\quad$ $x = y$
These systems are solved by substitution

First System:

$$x^2 - xy + y^2 = 4$$
$$x = -2y$$
$$(-2y)^2 - (-2y)y + y^2 = 4$$
$$4y^2 + 2y^2 + y^2 = 4$$
$$7y^2 = 4$$
$$y = \pm\sqrt{\frac{4}{7}}$$
$$y = \pm\frac{2\sqrt{7}}{7}$$

Second System:

$$x^2 - xy + y^2 = 4$$
$$x = y$$
$$y^2 - yy + y^2 = 4$$
$$y^2 = 4$$
$$y = \pm 2$$

For $y = \frac{2\sqrt{7}}{7}$

$$x = -2\left(\frac{2\sqrt{7}}{7}\right) = -\frac{4\sqrt{7}}{7}$$

For $y = -\frac{2\sqrt{7}}{7}$

$$x = -2\left(-\frac{2\sqrt{7}}{7}\right) = \frac{4\sqrt{7}}{7}$$

For $y = -2$

$$x = -2$$

For $y = 2$

$$x = 2$$

Solutions: $\left(\frac{4\sqrt{7}}{7}, -\frac{2\sqrt{7}}{7}\right)$, $\left(-\frac{4\sqrt{7}}{7}, \frac{2\sqrt{7}}{7}\right)$, (2, 2), (-2, -2). The checking steps are omitted for lack of space. *(10-6)*

55. For convenience we rewrite the constraint conditions:

$1.2x + 0.6y \le 960$ becomes $2x + y \le 1{,}600$
$0.04x + 0.03y \le 36$ becomes $4x + 3y \le 3{,}600$
$0.2x + 0.3y \le 270$ becomes $2x + 3y \le 2{,}700$
$x, y \ge 0$ are unaltered.

The feasible region is graphed in the diagram. The corner points (0, 900), (0, 0), and (800, 0) are obvious from the graph. The other corner points are obtained by solving the system:

$2x + y = 1{,}600$
$4x + 3y = 3{,}600$ to obtain (600, 400)

and $4x + 3y = 3{,}600$
$2x + 3y = 2{,}700$ to obtain (450, 600)

We now evaluate the objective function at each corner point.

Corner Point (x, y)	Objective Function $z = 30x + 20y$	
(0, 900)	18,000	
(0, 0)	0	
(800, 0)	24,000	
(600, 400)	26,000	Maximum value
(450, 600)	25,500	

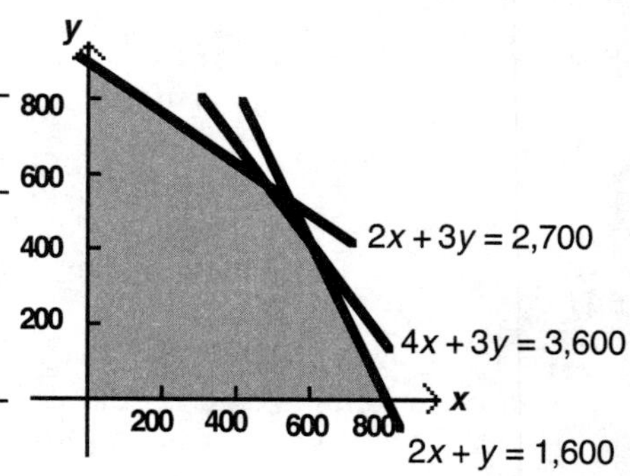

The maximum value of z on S is 26,000 at (600, 400).

(10-8)

56. Before we can enter these equations in our graphing calculator, we must solve for y:

$$x^2 + 4xy + y^2 = 8 \qquad\qquad 5x^2 + 2xy + y^2 = 25$$
$$y^2 + 4xy + x^2 - 8 = 0 \qquad\qquad y^2 + 2xy + 5x^2 - 25 = 0$$

Applying the quadratic formula to each equation, we have

$$y = \frac{-4x \pm \sqrt{16x^2 - 4(x^2 - 8)}}{2} \qquad\qquad y = \frac{-2x \pm \sqrt{4x^2 - 4(5x^2 - 25)}}{2}$$
$$y = \frac{-4x \pm \sqrt{12x^2 + 32}}{2} \qquad\qquad y = \frac{-2x \pm \sqrt{100 - 16x^2}}{2}$$
$$y = -2x \pm \sqrt{3x^2 + 8} \qquad\qquad y = -x \pm \sqrt{25 - 4x^2}$$

Entering each of these four equations into a graphing utility produces the graph shown at the right.

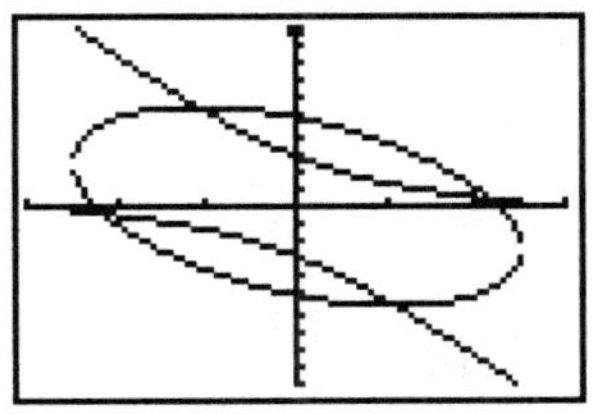

Zooming in on the four intersection points, or using a built-in intersection routine (details omitted), yields (-2.16, -0.37), (-1.09, 5.59), (1.09, -5.59), and (2.16, 0.37) to two decimal places. *(10-6)*

57. (A) The system is independent. There is one solution.

(B) The matrix is in fact

$$\left[\begin{array}{ccc|c} 1 & 0 & -3 & 4 \\ 0 & 1 & 2 & 5 \\ 0 & 0 & 0 & n \end{array}\right]$$

The third row corresponds to the equation $0x_1 + 0x_2 + 0x_3 = n$. This is impossible. The system has no solution.

(C) The matrix is in fact

$$\left[\begin{array}{ccc|c} 1 & 0 & -3 & 4 \\ 0 & 1 & 2 & 5 \\ 0 & 0 & 0 & 0 \end{array}\right]$$

Thus, there are an infinite number of solutions ($x_3 = t$, $x_2 = 5 - 2t$, $x_1 = 4 + 3t$, for t any real number.) *(10-3)*

58. (A) If the coefficient matrix has an inverse, then the system can be written as $AX = B$ and its solution can be written $X = A^{-1}B$. Thus the system has one solution.

(B) If the coefficient matrix does not have an inverse, then the system can be solved by Gauss-Jordan elimination, but it will not have exactly one solution. The other possibilities are that the system has no solution or an infinite number of solutions, and either possibility may occur. *(10-5)*

59. If we assume that A is a non-zero matrix with an inverse A^{-1}, then if $A^2 = 0$ we can write $A^{-1}A^2 = A^{-1}0$ or $A^{-1}AA = 0$ or $IA = 0$ or $A = 0$. But A was assumed non-zero, so there is a contradiction. Hence A^{-1} cannot exist for such a matrix. *(10-5)*

60.

$AX - B = CX$		
$AX - B + B - CX = CX - CX + B$	Addition property
$AX + 0 - CX = 0 + B$	$M + (-M) = 0$
$AX - CX = B$	$M + 0 = M$
$(A - C)X = B$	Right distributive property
$(A - C)^{-1}[(A - C)X] = (A - C)^{-1}B$	Left multiplication property
$[(A - C)^{-1}(A - C)]X = (A - C)^{-1}B$	Associative property
$IX = (A - C)^{-1}B$	$A^{-1}A = I$
$X = (A - C)^{-1}B$	$IX = X$

Common Errors: $(A - C)^{-1}B \neq B(A - C)^{-1}$
$(A - C)^{-1} \neq A^{-1} - C^{-1}$

(10-5)

61. $\left[\begin{array}{ccc|ccc} 4 & 5 & 6 & 1 & 0 & 0 \\ 4 & 5 & -6 & 0 & 1 & 0 \\ 1 & 1 & 1 & 0 & 0 & 1 \end{array}\right]$ $(-1)R_1 + R_2 \rightarrow R_2$

$\sim \left[\begin{array}{ccc|ccc} 4 & 5 & 6 & 1 & 0 & 0 \\ 0 & 0 & -12 & -1 & 1 & 0 \\ 1 & 1 & 1 & 0 & 0 & 1 \end{array}\right]$ $R_1 \leftrightarrow R_3$

$\sim \left[\begin{array}{ccc|ccc} 1 & 1 & 1 & 0 & 0 & 1 \\ 0 & 0 & -12 & -1 & 1 & 0 \\ 4 & 5 & 6 & 1 & 0 & 0 \end{array}\right]$ $(-4)R_1 + R_3 \rightarrow R_3$

$\sim \left[\begin{array}{ccc|ccc} 1 & 1 & 1 & 0 & 0 & 1 \\ 0 & 0 & -12 & -1 & 1 & 0 \\ 0 & 1 & 2 & 1 & 0 & -4 \end{array}\right]$ $R_2 \leftrightarrow R_3$

$\sim \left[\begin{array}{ccc|ccc} 1 & 1 & 1 & 0 & 0 & 1 \\ 0 & 1 & 2 & 1 & 0 & -4 \\ 0 & 0 & -12 & -1 & 1 & 0 \end{array}\right]$ $(-1)R_2 + R_1 \rightarrow R_1$

$\sim \left[\begin{array}{ccc|ccc} 1 & 0 & -1 & -1 & 0 & 5 \\ 0 & 1 & 2 & 1 & 0 & -4 \\ 0 & 0 & -12 & -1 & 1 & 0 \end{array}\right]$ $-\frac{1}{12}R_3 \rightarrow R_3$

$\sim \left[\begin{array}{ccc|ccc} 1 & 0 & -1 & -1 & 0 & 5 \\ 0 & 1 & 2 & 1 & 0 & -4 \\ 0 & 0 & 1 & \frac{1}{12} & -\frac{1}{12} & 0 \end{array}\right]$ $R_3 + R_1 \rightarrow R_1$, $(-2)R_3 + R_2 \rightarrow R_2$

$\sim \left[\begin{array}{ccc|ccc} 1 & 0 & 0 & -\frac{11}{12} & -\frac{1}{12} & 5 \\ 0 & 1 & 0 & \frac{10}{12} & \frac{2}{12} & -4 \\ 0 & 0 & 1 & \frac{1}{12} & -\frac{1}{12} & 0 \end{array}\right]$

Hence

$$A^{-1} = \begin{bmatrix} -\frac{11}{12} & -\frac{1}{12} & 5 \\ \frac{10}{12} & \frac{2}{12} & -4 \\ \frac{1}{12} & -\frac{1}{12} & 0 \end{bmatrix} \text{ or } \frac{1}{12}\begin{bmatrix} -11 & -1 & 60 \\ 10 & 2 & -48 \\ 1 & -1 & 0 \end{bmatrix}$$

$$A^{-1}A = \frac{1}{12}\begin{bmatrix} -11 & -1 & 60 \\ 10 & 2 & -48 \\ 1 & -1 & 0 \end{bmatrix}\begin{bmatrix} 4 & 5 & 6 \\ 4 & 5 & -6 \\ 1 & 1 & 1 \end{bmatrix}$$

$$= \frac{1}{12}\begin{bmatrix} (-11)4 + (-1)4 + 60 \cdot 1 & (-11)5 + (-1)5 + 60 \cdot 1 & (-11)6 + (-1)(-6) + 60 \cdot 1 \\ 10 \cdot 4 + 2 \cdot 4 + (-48)1 & 10 \cdot 5 + 2 \cdot 5 + (-48)1 & 10 \cdot 6 + 2(-6) + (-48)1 \\ 1 \cdot 4 + (-1)4 + 0 \cdot 1 & 1 \cdot 5 + (-1)5 + 0 \cdot 1 & 1 \cdot 6 + (-1)(-6) + 0 \cdot 1 \end{bmatrix}$$

$$= \frac{1}{12}\begin{bmatrix} 12 & 0 & 0 \\ 0 & 12 & 0 \\ 0 & 0 & 12 \end{bmatrix} = \begin{bmatrix} 1 & 0 & 0 \\ 0 & 1 & 0 \\ 0 & 0 & 1 \end{bmatrix} = I$$

(10-5)

62. Multiplying the first two equations by 100, the system becomes

$$4x_1 + 5x_2 + 6x_3 = 36{,}000$$
$$4x_1 + 5x_2 - 6x_3 = 12{,}000$$
$$x_1 + x_2 + x_3 = 7{,}000$$

As a matrix equation, we have

$$\overset{A}{\begin{bmatrix} 4 & 5 & 6 \\ 4 & 5 & -6 \\ 1 & 1 & 1 \end{bmatrix}} \overset{X}{\begin{bmatrix} x_1 \\ x_2 \\ x_3 \end{bmatrix}} = \overset{B}{\begin{bmatrix} 36{,}000 \\ 12{,}000 \\ 7{,}000 \end{bmatrix}}$$

The solution to $AX = B$ is $X = A^{-1}B$. Using A^{-1} from problem 61, we have

$$X = \begin{bmatrix} x_1 \\ x_2 \\ x_3 \end{bmatrix} = \frac{1}{12}\begin{bmatrix} -11 & -1 & 60 \\ 10 & 2 & -48 \\ 1 & -1 & 0 \end{bmatrix}\begin{bmatrix} 36{,}000 \\ 12{,}000 \\ 7{,}000 \end{bmatrix}$$

$$= \frac{1}{12}\begin{bmatrix} (-11)(36{,}000) + (-1)(12{,}000) + (60)(7{,}000) \\ (10)(36{,}000) + (2)(12{,}000) + (-48)(7{,}000) \\ 1(36{,}000) + (-1)(12{,}000) + (0)(7{,}000) \end{bmatrix} = \frac{1}{12}\begin{bmatrix} 12{,}000 \\ 48{,}000 \\ 24{,}000 \end{bmatrix} = \begin{bmatrix} 1{,}000 \\ 4{,}000 \\ 2{,}000 \end{bmatrix}$$

Hence, $x_1 = 1{,}000$, $x_2 = 4{,}000$, $x_3 = 2{,}000$ *(10-5)*

63. Let x = number of $\frac{1}{2}$-pound packages

y = number of $\frac{1}{3}$-pound packages

There are 120 packages. Hence

$x + y = 120$ (1)

Since x $\frac{1}{2}$-pound packages weigh $\frac{1}{2}x$ pounds and y $\frac{1}{3}$-pound packages weigh $\frac{1}{3}y$ pounds, we have

$\frac{1}{2}x + \frac{1}{3}y = 48$ (2)

We solve the system (1), (2) using elimination by addition. We multiply the second equation by -3 and add.

$$x + y = 120$$
$$-\frac{3}{2}x - y = -144$$
$$-\frac{1}{2}x = -24$$
$$x = 48$$

Substituting into equation (1), we have

$$48 + y = 120$$
$$y = 72$$

48 $\frac{1}{2}$-pound packages and 72 $\frac{1}{3}$-pound packages. *(10-1, 10-3)*

64. Using the formulas for perimeter and area of a rectangle, we have

$2a + 2b = 28$

$ab = 48$

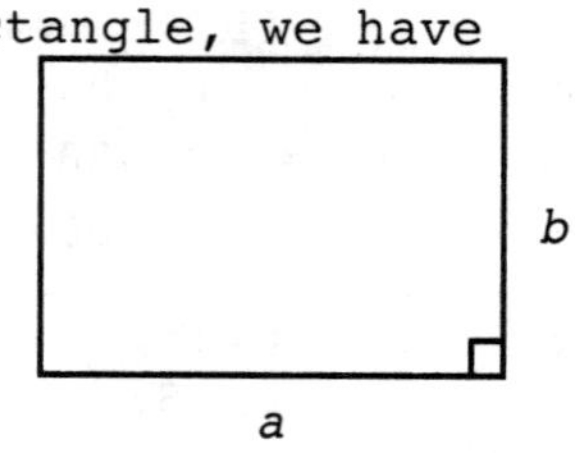

Solving the first-degree equation for b and substituting into the second-degree equation, we have

$$\begin{aligned} 2b &= 28 - 2a \\ b &= 14 - a \\ a(14 - a) &= 48 \\ -a^2 + 14a &= 48 \\ a^2 - 14a + 48 &= 0 \\ (a - 6)(a - 8) &= 0 \\ a &= 6,\ 8 \end{aligned}$$

For $a = 6$
$$b = 14 - 6 = 8$$

For $a = 8$
$$b = 14 - 8 = 6$$

Dimensions: 6 meters by 8 meters. *(10-6)*

65. Let x_1 = number of grams of mix A

x_2 = number of grams of mix B

x_3 = number of grams of mix C

We have

$0.30x_1 + 0.20x_2 + 0.10x_3 = 27$ (protein)

$0.03x_1 + 0.05x_2 + 0.04x_3 = 5.4$ (fat)

$0.10x_1 + 0.20x_2 + 0.10x_3 = 19$ (moisture)

Clearing of decimals for convenience, we have

$3x_1 + 2x_2 + x_3 = 270$

$3x_1 + 5x_2 + 4x_3 = 540$

$x_1 + 2x_2 + x_3 = 190$

Form the augmented matrix and solve by Gauss-Jordan elimination.

$$\left[\begin{array}{ccc|c} 3 & 2 & 1 & 270 \\ 3 & 5 & 4 & 540 \\ 1 & 2 & 1 & 190 \end{array}\right] R_3 \leftrightarrow R_1$$

$$\sim \left[\begin{array}{ccc|c} 1 & 2 & 1 & 190 \\ 3 & 5 & 4 & 540 \\ 3 & 2 & 1 & 270 \end{array}\right] \begin{array}{l} (-3)R_1 + R_2 \to R_2 \\ (-3)R_1 + R_3 \to R_3 \end{array}$$

$$\sim \left[\begin{array}{ccc|c} 1 & 2 & 1 & 190 \\ 0 & -1 & 1 & -30 \\ 0 & -4 & -2 & -300 \end{array}\right] (-1)R_2 \to R_2$$

$$\sim \left[\begin{array}{ccc|c} 1 & 2 & 1 & 190 \\ 0 & 1 & -1 & 30 \\ 0 & -4 & -2 & -300 \end{array}\right] \begin{array}{l} (-2)R_2 + R_1 \to R_1 \\ 4R_2 + R_3 \to R_3 \end{array}$$

$$\sim \left[\begin{array}{ccc|c} 1 & 0 & 3 & 130 \\ 0 & 1 & -1 & 30 \\ 0 & 0 & -6 & -180 \end{array}\right] -\tfrac{1}{6}R_3 \to R_3$$

$$\sim \left[\begin{array}{ccc|c} 1 & 0 & 3 & 130 \\ 0 & 1 & -1 & 30 \\ 0 & 0 & 1 & 30 \end{array}\right] \begin{array}{l} (-3)R_3 + R_1 \to R_1 \\ R_3 + R_2 \to R_2 \end{array}$$

$$\sim \left[\begin{array}{ccc|c} 1 & 0 & 0 & 40 \\ 0 & 1 & 0 & 60 \\ 0 & 0 & 1 & 30 \end{array}\right]$$

Therefore

$x_1 = 40$ grams Mix A

$x_2 = 60$ grams Mix B

$x_3 = 30$ grams Mix C *(10-5)*

66. Let x_1 = number of tons at Big Bend
x_2 = number of tons at Saw Pit

Then
$0.05x_1 + 0.03x_2$ = number of tons of nickel at both mines = k_1
$0.07x_1 + 0.04x_2$ = number of tons of copper at both mines = k_2

We solve
$0.05x_1 + 0.03x_2 = k_1$
$0.07x_1 + 0.04x_4 = k_2,$
For arbitrary k_1 and k_2, by writing the system as a matrix equation.

$$\overset{A}{\begin{bmatrix} 0.05 & 0.03 \\ 0.07 & 0.04 \end{bmatrix}} \overset{X}{\begin{bmatrix} x_1 \\ x_2 \end{bmatrix}} = \overset{B}{\begin{bmatrix} k_1 \\ k_2 \end{bmatrix}}$$

If A^{-1} exists, then $X = A^{-1}B$. To find A^{-1}, we perform row operations on

$$\left[\begin{array}{cc|cc} 0.05 & 0.03 & 1 & 0 \\ 0.07 & 0.04 & 0 & 1 \end{array}\right] 20R_1 \to R_1$$

$$\sim \left[\begin{array}{cc|cc} 1 & 0.6 & 20 & 0 \\ 0.07 & 0.04 & 0 & 1 \end{array}\right] (-0.07)R_1 + R_2 \to R_2$$

$$\sim \left[\begin{array}{cc|cc} 1 & 0.6 & 20 & 0 \\ 0 & -0.002 & -1.4 & 1 \end{array}\right] -500R_2 \to R_2$$

$$\sim \left[\begin{array}{cc|cc} 1 & 0.6 & 20 & 0 \\ 0 & 1 & 700 & -500 \end{array}\right] (-0.6)R_2 + R_1 \to R_1$$

$$\sim \left[\begin{array}{cc|cc} 1 & 0 & -400 & 300 \\ 0 & 1 & 700 & -500 \end{array}\right]$$

Hence $A^{-1} = \begin{bmatrix} -400 & 300 \\ 700 & -500 \end{bmatrix}$

Check: $A^{-1}A = \begin{bmatrix} -400 & 300 \\ 700 & -500 \end{bmatrix} \begin{bmatrix} 0.05 & 0.03 \\ 0.07 & 0.04 \end{bmatrix} = \begin{bmatrix} 1 & 0 \\ 0 & 1 \end{bmatrix}$

We can now solve the system as:

$$\overset{X}{\begin{bmatrix} x_1 \\ x_2 \end{bmatrix}} = \overset{A^{-1}}{\begin{bmatrix} -400 & 300 \\ 700 & -500 \end{bmatrix}} \overset{B}{\begin{bmatrix} k_1 \\ k_2 \end{bmatrix}}$$

(A) If $k_1 = 3.6$, $k_2 = 5$,

$$\begin{bmatrix} x_1 \\ x_2 \end{bmatrix} = \begin{bmatrix} -400 & 300 \\ 700 & -500 \end{bmatrix} \begin{bmatrix} 3.6 \\ 5 \end{bmatrix} = \begin{bmatrix} 60 \\ 20 \end{bmatrix}$$

60 tons of ore must be produced at Big Bend, 20 tons of ore at Saw Pit.

(B) If $k_1 = 3$, $k_2 = 4.1$,

$$\begin{bmatrix} x_1 \\ x_2 \end{bmatrix} = \begin{bmatrix} -400 & 300 \\ 700 & -500 \end{bmatrix} \begin{bmatrix} 3 \\ 4.1 \end{bmatrix} = \begin{bmatrix} 30 \\ 50 \end{bmatrix}$$

30 tons of ore must be produced at Big Bend, 50 tons of ore at Saw Pit.

(C) If $k_1 = 3.2$, $k_2 = 4.4$,

$$\begin{bmatrix} x_1 \\ x_2 \end{bmatrix} = \begin{bmatrix} -400 & 300 \\ 700 & -500 \end{bmatrix} \begin{bmatrix} 3.2 \\ 4.4 \end{bmatrix} = \begin{bmatrix} 40 \\ 40 \end{bmatrix}$$

40 tons of ore must be produced at Big Bend, 40 tons of ore at Saw Pit. *(10-5)*

67. (A) The labor cost of producing one printer stand at the South Carolina plant is the product of the stand row of L with South Carolina column of H.

$$[0.9 \quad 1.8 \quad 0.6]\begin{bmatrix} 10.00 \\ 8.50 \\ 4.50 \end{bmatrix} = 27 \text{ dollars}$$

(B) The matrix HL has no obvious meaning, but the matrix LH represents the total labor costs for each item at each plant.

(C) $$LH = \begin{bmatrix} 1.7 & 2.4 & 0.8 \\ 0.9 & 1.8 & 0.6 \end{bmatrix}\begin{bmatrix} 11.50 & 10.00 \\ 9.50 & 8.50 \\ 5.00 & 4.50 \end{bmatrix}$$

$$= \begin{bmatrix} (1.7)(11.50) + (2.4)(9.50) + (0.8)(5.00) & (1.7)(10.00) + (2.4)(8.50) + (0.8)(4.50) \\ (0.9)(11.50) + (1.8)(9.50) + (0.6)(5.00) & (0.9)(10.00) + (1.8)(8.50) + (0.6)(4.50) \end{bmatrix}$$

$$\begin{array}{c} \text{N.C.} \quad\quad \text{S.C.} \\ \begin{bmatrix} \$46.35 & \$41.00 \\ \$30.45 & \$27.00 \end{bmatrix} \begin{array}{l} \text{Desk} \\ \text{Stands} \end{array} \end{array}$$

(*10-4*)

68. (A) The average monthly production for the months of January and February is represented by the matrix $\frac{1}{2}(J + F)$

$$\frac{1}{2}(J + F) = \frac{1}{2}\left(\begin{bmatrix} 1,500 & 1,650 \\ 850 & 700 \end{bmatrix} + \begin{bmatrix} 1,700 & 1,810 \\ 850 & 740 \end{bmatrix}\right) = \frac{1}{2}\begin{bmatrix} 3,200 & 3,460 \\ 1,780 & 1,440 \end{bmatrix}$$

$$\begin{array}{c} \text{N.C.} \quad \text{S.C.} \\ = \begin{bmatrix} 1,600 & 1,730 \\ 890 & 720 \end{bmatrix} \begin{array}{l} \text{Desks} \\ \text{Stands} \end{array} \end{array}$$

(B) The increase in production from January to February is represented by the matrix $F - J$.

$$F - J = \begin{bmatrix} 1,700 & 1,810 \\ 930 & 740 \end{bmatrix} - \begin{bmatrix} 1,500 & 1,650 \\ 850 & 700 \end{bmatrix}$$

$$\begin{array}{c} \text{N.C.} \quad \text{S.C.} \\ = \begin{bmatrix} 200 & 160 \\ 80 & 40 \end{bmatrix} \begin{array}{l} \text{Desks} \\ \text{Stands} \end{array} \end{array}$$

(C) $$J\begin{bmatrix} 1 \\ 1 \end{bmatrix} = \begin{bmatrix} 1,500 & 1,650 \\ 850 & 700 \end{bmatrix}\begin{bmatrix} 1 \\ 1 \end{bmatrix} = \begin{bmatrix} 3,150 \\ 1,550 \end{bmatrix} \begin{array}{l} \text{Desks} \\ \text{Stands} \end{array}$$

This matrix represents the total production of each item in January.

(*10-4*)

69. The inverse of matrix B is calculated to be

$$B^{-1} = \begin{bmatrix} 1 & 1 & -1 \\ 0 & -1 & 1 \\ -1 & 0 & 1 \end{bmatrix}$$

Putting the coded message into matrix form and multiplying by B^{-1} yields

$$\begin{bmatrix} 1 & 1 & -1 \\ 0 & -1 & 1 \\ -1 & 0 & 1 \end{bmatrix}\begin{bmatrix} 25 & 24 & 21 & 41 & 21 & 52 \\ 8 & 25 & 41 & 30 & 32 & 52 \\ 26 & 33 & 48 & 50 & 41 & 79 \end{bmatrix} = \begin{bmatrix} 7 & 16 & 14 & 21 & 12 & 25 \\ 18 & 8 & 7 & 20 & 9 & 27 \\ 1 & 9 & 27 & 9 & 20 & 27 \end{bmatrix}$$

This decodes to

7 18 1 16 8 9 14 7 27 21 20 9 12 9 20 25 27 27
G R A P H I N G U T I L I T Y

(*10-5*)

70. (A) Let x_1 = number of nickels

x_2 = number of dimes

Then $x_1 + x_2 = 30$ (total number of coins)

$5x_1 + 10x_2 = 190$ (total value of coins)

We form the augmented matrix and solve by Gauss-Jordan elimination.

$$\left[\begin{array}{cc|c} 1 & 1 & 30 \\ 5 & 10 & 190 \end{array}\right] (-5)R_1 + R_2 \to R_2$$

$$\sim \left[\begin{array}{cc|c} 1 & 1 & 30 \\ 0 & 5 & 40 \end{array}\right] \tfrac{1}{5}R_2 \to R_2$$

$$\sim \left[\begin{array}{cc|c} 1 & 1 & 30 \\ 0 & 1 & 8 \end{array}\right] (-1)R_2 + R_1 \to R_1$$

$$\sim \left[\begin{array}{cc|c} 1 & 0 & 22 \\ 0 & 1 & 8 \end{array}\right]$$

The augmented matrix is in reduced form. It corresponds to the system

x_1 = 22 nickels

x_2 = 8 dimes

(B) Let x_1 = number of nickels

x_2 = number of dimes

x_3 = number of quarters

Then $x_1 + x_2 + x_3 = 30$ (total number of coins)

$5x_1 + 10x_2 + 25x_3 = 190$ (total value of coins)

We form the augmented matrix and solve by Gauss-Jordan elimination

$$\left[\begin{array}{ccc|c} 1 & 1 & 1 & 30 \\ 5 & 10 & 25 & 190 \end{array}\right] (-5)R_1 + R_2 \to R_2$$

$$\sim \left[\begin{array}{ccc|c} 1 & 1 & 1 & 30 \\ 0 & 5 & 20 & 40 \end{array}\right] \tfrac{1}{5}R_2 \to R_2$$

$$\sim \left[\begin{array}{ccc|c} 1 & 1 & 1 & 30 \\ 0 & 1 & 4 & 8 \end{array}\right] (-1)R_2 + R_1 \to R_1$$

$$\sim \left[\begin{array}{ccc|c} 1 & 0 & -3 & 22 \\ 0 & 1 & 4 & 8 \end{array}\right]$$

The augmented matrix is in reduced form. It corresponds to the system:

$$x_1 \quad - 3x_3 = 22$$
$$x_2 + 4x_3 = 8$$

Let $x_3 = t$. Then $x_2 = -4x_3 + 8$

$= -4t + 8$

$x_1 = 3x_3 + 22$

$= 3t + 22$

A solution is achieved, not for every real value of t, but for integer values of t that give rise to non-negative x_1, x_2, x_3.

$x_1 \geq 0$ means $3t + 22 \geq 0$ or $t \geq -7\frac{1}{3}$

$x_2 \geq 0$ means $-4t + 8 \geq 0$ or $t \leq 2$

$x_3 \geq 0$ means $t \geq 0$

The only integer values of t that satisfy these conditions are 0, 1, 2. Thus we have the solutions

$x_1 = 3t + 22$ nickels

$x_2 = 8 - 4t$ dimes

$x_3 = t$ quarters

where t = 0, 1, or 2 *(10-1)*

71. Let x = number of regular sails
y = number of competition sails

(A) We form the linear objective function
$P = 60x + 100y$
We wish to maximize P, the profit from x regular sails @ \$60 and y competition sails @ \$100, subject to the constraints

$x + 2y \le 140$ cutting department constraint
$3x + 4y \le 360$ sewing department constraint
$x, y \ge 0$ non-negative constraints

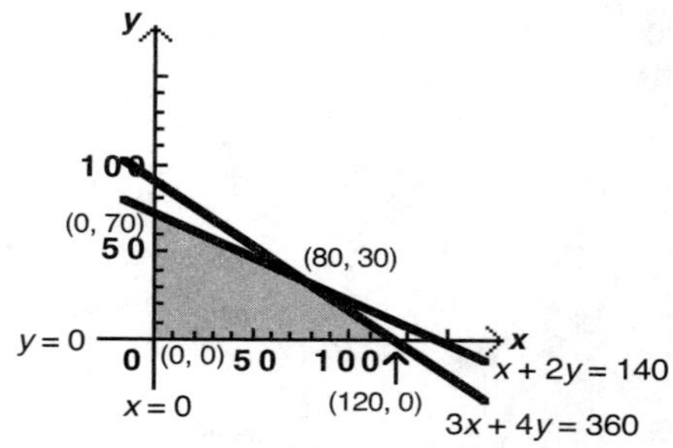

Solving the system of constraint inequalities graphically, we obtain the feasible region S shown.

Next we evaluate the objective function at each corner point.

Corner Point (x, y)	Objective Function $P = 60x + 100y$	
(0, 0)	0	
(0, 70)	7,000	
(80, 30)	7,800	Maximum value
(120, 0)	7,200	

The optimal value is 7,800 at the corner point (80, 30). Thus, the company should manufacture 80 regular sails and 30 competition sails for a maximum profit of \$7,800.

(B) The objective function now becomes $60x + 125y$. We evaluate this at each corner point.

Corner Point (x, y)	Objective Function $P = 60x + 125y$	
(0, 0)	0	
(0, 70)	8,750	Maximum value
(80, 30)	8,550	
(120, 0)	7,200	

The optimal value is now 8,750 at the corner point (0, 70). Thus, the company should manufacture 70 competition sails and no regular sails for a maximum profit of \$8,750.

(C) The objective function now becomes $60x + 75y$. We evaluate this at each corner point.

Corner Point (x, y)	Objective Function $P = 60x + 75y$	
(0, 0)	0	
(0, 70)	5,250	
(80, 30)	7,050	
(120, 0)	7,200	Maximum value

The optimal value is now 7,200 at the corner point (120, 0). Thus, the company should manufacture 120 regular sails and no competition sails for a maximum profit of \$7,200. *(10-8)*

72. Let x = number of grams of mix A
y = number of grams of mix B

(A) We form the objective function
$C = 0.07x + 0.04y$

C represents the cost of x grams at \$0.07 per gram and y grams at \$0.04 per gram. We wish to minimize C subject to

$5x + 2y \geq 800$	vitamin constraint
$2x + 4y \geq 800$	mineral constraint
$4x + 4y \leq 1300$	calorie constraint
$x, y \geq 0$	non-negative constraints

Solving the system of constraint inequalities graphically, we obtain the feasible region S shown in the diagram.

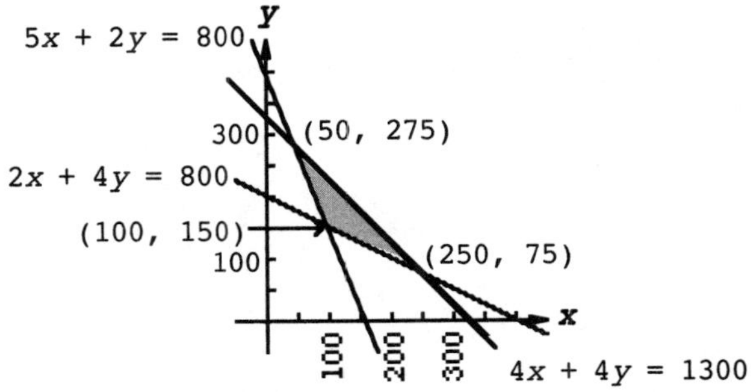

Next we evaluate the objective function at each corner point.

Corner Point (x, y)	Objective Function $C = 0.07x + 0.04y$	
(100, 150)	13	Minimum value
(250, 75)	20.5	
(50, 275)	14.5	

The optimal value is 13 at the corner point (100, 150). Thus, 100 grams of mix A and 150 grams of mix B should be used for a cost of \$13.

(B) The objective function now becomes $0.07x + 0.02y$. We evaluate this at each corner point.

Corner Point (x, y)	Objective Function $C = 0.07x + 0.02y$	
(100, 150)	10	
(250, 75)	19	
(50, 275)	9	Minimum value

The optimal value is now 9 at the corner point (50, 275). Thus, 50 grams of mix A and 275 grams of mix B should be used for a cost of \$9.

(C) The objective function now becomes $0.07x + 0.15y$. We evaluate this at each corner point.

Corner Point (x, y)	Objective Function $C = 0.07x + 0.15y$	
(100, 150)	29.5	
(250, 75)	28.75	Minimum value
(50, 275)	44.75	

The optimal value is now 28.75 at the corner point (250, 75). Thus, 250 grams of mix A and 75 grams of mix B should be used for a cost of \$28.75.

(10-8)

CHAPTER 11 Sequences, Induction, and Probability

Section 11-1

1. $a_1 = 1 - 2 = -1 \quad a_2 = 2 - 2 = 0 \quad a_3 = 3 - 2 = 1 \quad a_4 = 4 - 2 = 2$

3. $a_1 = \frac{1 - 1}{1 + 1} = 0 \quad a_2 = \frac{2 - 1}{2 + 1} = \frac{1}{3} \quad a_3 = \frac{3 - 1}{3 + 1} = \frac{2}{4} = \frac{1}{2} \quad a_4 = \frac{4 - 1}{4 + 1} = \frac{3}{5}$

5. $a_1 = (-2)^{1+1} = (-2)^2 = 4 \quad a_2 = (-2)^{2+1} = (-2)^3 = -8$
$a_3 = (-2)^{3+1} = (-2)^4 = 16 \quad a_4 = (-2)^{4+1} = (-2)^5 = -32$

7. $a_8 = 8 - 2 = 6$

9. $a_{100} = \frac{100 - 1}{100 + 1} = \frac{99}{101}$

11. $S_5 = 1 + 2 + 3 + 4 + 5$

13. $S_3 = \frac{1}{10^1} + \frac{1}{10^2} + \frac{1}{10^3} = \frac{1}{10} + \frac{1}{100} + \frac{1}{1000}$

15. $S_4 = (-1)^1 + (-1)^2 + (-1)^3 + (-1)^4 = (-1) + 1 + (-1) + 1 = -1 + 1 - 1 + 1$

17. $a_1 = (-1)^{1+1}\, 1^2 = 1 \quad a_2 = (-1)^{2+1}\, 2^2 = -4 \quad a_3 = (-1)^{3+1}\, 3^2 = 9$
$a_4 = (-1)^{4+1}\, 4^2 = -16 \quad a_5 = (-1)^{5+1}\, 5^2 = 25$

19.
$$a_1 = \frac{1}{3}\left[1 - \frac{1}{10^1}\right] = \frac{1}{3} \cdot \frac{9}{10} = \frac{3}{10} = 0.3$$
$$a_2 = \frac{1}{3}\left[1 - \frac{1}{10^2}\right] = \frac{1}{3} \cdot \frac{99}{100} = \frac{33}{100} = 0.33$$
$$a_3 = \frac{1}{3}\left[1 - \frac{1}{10^3}\right] = \frac{1}{3} \cdot \frac{999}{1,000} = \frac{333}{1,000} = 0.333$$
$$a_4 = \frac{1}{3}\left[1 - \frac{1}{10^4}\right] = \frac{1}{3} \cdot \frac{9,999}{10,000} = \frac{3,333}{10,000} = 0.3333$$
$$a_5 = \frac{1}{3}\left[1 - \frac{1}{10^5}\right] = \frac{1}{3} \cdot \frac{99,999}{100,000} = \frac{33,333}{100,000} = 0.33333$$

21.
$$a_1 = \left(-\frac{1}{2}\right)^{1-1} = \left(-\frac{1}{2}\right)^0 = 1$$
$$a_2 = \left(-\frac{1}{2}\right)^{2-1} = -\frac{1}{2}$$
$$a_3 = \left(-\frac{1}{2}\right)^{3-1} = \frac{1}{4}$$
$$a_4 = \left(-\frac{1}{2}\right)^{4-1} = -\frac{1}{8}$$
$$a_5 = \left(-\frac{1}{2}\right)^{5-1} = \frac{1}{16}$$

23. $a_1 = 7$
$a_2 = a_{2-1} - 4 = a_1 - 4 = 7 - 4 = 3$
$a_3 = a_{3-1} - 4 = a_2 - 4 = 3 - 4 = -1$
$a_4 = a_{4-1} - 4 = a_3 - 4 = -1 - 4 = -5$
$a_5 = a_{5-1} - 4 = a_4 - 4 = -5 - 4 = -9$

> **Common Error:**
> $a_2 \neq a_2 - 1 - 4$. The 1 is in the subscript:
> a_{2-1}; that is, a_1

25. $a_1 = 4$

$a_2 = \frac{1}{4}a_{2-1} = \frac{1}{4}a_1 = \frac{1}{4} \cdot 4 = 1$

$a_3 = \frac{1}{4}a_{3-1} = \frac{1}{4}a_2 = \frac{1}{4} \cdot 1 = \frac{1}{4}$

$a_4 = \frac{1}{4}a_{4-1} = \frac{1}{4}a_3 = \frac{1}{4} \cdot \frac{1}{4} = \frac{1}{16}$

$a_5 = \frac{1}{4}a_{5-1} = \frac{1}{4}a_4 = \frac{1}{4} \cdot \frac{1}{16} = \frac{1}{64}$

27. $a_1 = 1$

$a_2 = 2$

$a_3 = a_1 + 2a_2 = 1 + 2(2) = 5$

$a_4 = a_2 + 2a_3 = 2 + 2(5) = 12$

$a_5 = a_3 + 2a_4 = 5 + 2(12) = 29$

$a_6 = a_4 + 2a_5 = 12 + 2(29) = 70$

$a_7 = a_5 + 2a_6 = 29 + 2(70) = 169$

The graphic demonstrates use of the RECUR program on a TI-83.

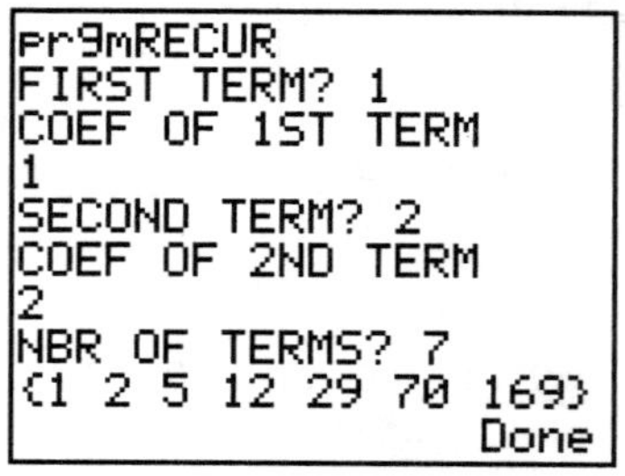

29. $a_1 = -1$

$a_2 = 2$

$a_3 = 2a_1 + a_2 = 2(-1) + 2 = 0$

$a_4 = 2a_2 + a_3 = 2(2) + 0 = 4$

$a_5 = 2a_3 + a_4 = 2(0) + 4 = 4$

$a_6 = 2a_4 + a_5 = 2(4) + 4 = 12$

$a_7 = 2a_5 + a_6 = 2(4) + 12 = 20$

The graphic demonstrates use of the RECUR program on a TI-83.

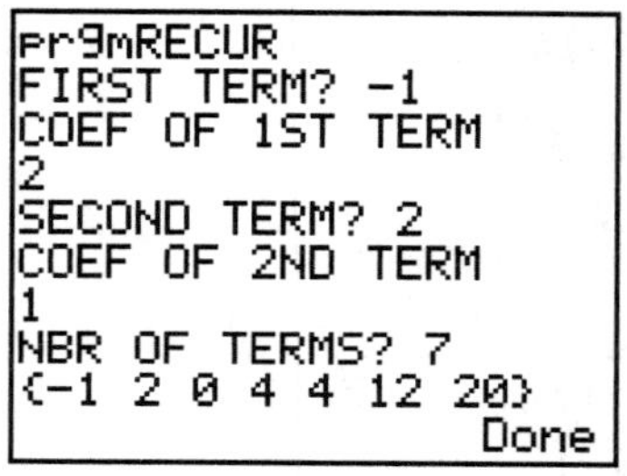

31. a_n: -2, -1, 0, 1, …
n: 1, 2, 3, 4, …
Comparing a_n with n, we see that $a_n = n - 3$.

33. a_n: 5, 7, 9, 11, …
$a_n - 1$: 4, 6, 8, 10, …
$a_n - 3$: 2, 4, 6, 8, …
n: 1, 2, 3, 4, …
Comparing $a_n - 3$ with n, we see that $a_n - 3 = 2n$, hence $a_n = 2n + 3$.

35. a_n: -1, 1, -1, 1, …
n: 1, 2, 3, 4, …
Comparing a_n with n, we see that $a_n = (-1)^n$.

37. a_n: 2, $\frac{3}{2}$, $\frac{4}{3}$, $\frac{5}{4}$, …
n: 1, 2, 3, 4, …
Comparing a_n with n, we see that $a_n = \frac{n+1}{n}$.

39. a_n: -3, 9, -27, 81, …
n: 1, 2, 3, 4, …
Comparing a_n with n, we see that a_n consists of successive nth powers of -3. $a_n = (-3)^n$

41. a_n: x, $\frac{x^2}{2}$, $\frac{x^3}{3}$, $\frac{x^4}{4}$, …
n: 1, 2, 3, 4, …
Comparing a_n with n, we see that $a_n = \frac{x^n}{n}$.

43. (A) $a_1 = 1^2 - 1 + 2 = 2$; $a_2 = 2^2 - 2 + 2 = 4$; $a_3 = 3^2 - 3 + 2 = 8$;
$a_4 = 4^2 - 4 + 2 = 14$

(B) A (simpler) sequence with the first three terms 2, 4, 8 would be $b_n = 2^n$.

45. (A) $a_1 = 6 \cdot 1^2 - 11 \cdot 1 + 6 = 1$; $a_2 = 6 \cdot 2^2 - 11 \cdot 2 + 6 = 8$;
$a_3 = 6 \cdot 3^2 - 11 \cdot 3 + 6 = 27$; $a_4 = 6 \cdot 4^2 - 11 \cdot 4 + 6 = 58$

(B) A (simpler) sequence with the first three terms 1, 8, 27 would be $b_n = n^3$.

47. (A) $a_1 = 2 \cdot 1^2 - 8 \cdot 1 + 7 = 1$; $a_2 = 2 \cdot 2^2 - 8 \cdot 2 + 7 = -1$;
$a_3 = 2 \cdot 3^2 - 8 \cdot 3 + 7 = 1$; $a_4 = 2 \cdot 4^2 - 8 \cdot 4 + 7 = 7$

(B) A (simpler) sequence with the first three terms 1, -1, 1 would be $b_n = (-1)^{n+1}$.

49.

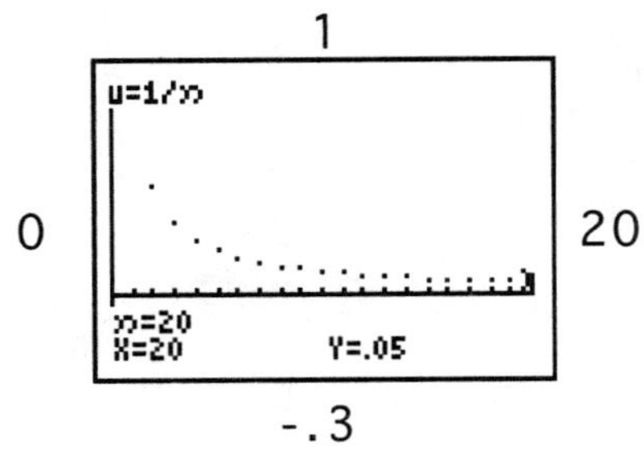

51.

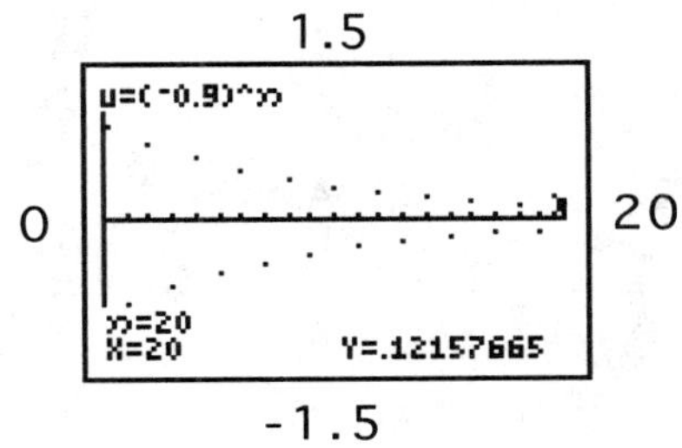

53. $S_4 = \dfrac{(-2)^{1+1}}{1} + \dfrac{(-2)^{2+1}}{2} + \dfrac{(-2)^{3+1}}{3} + \dfrac{(-2)^{4+1}}{4}$
$= \dfrac{4}{1} - \dfrac{8}{2} + \dfrac{16}{3} - \dfrac{32}{4}$

55. $S_3 = \dfrac{1}{1}x^{1+1} + \dfrac{1}{2}x^{2+1} + \dfrac{1}{3}x^{3+1}$
$= x^2 + \dfrac{x^3}{2} + \dfrac{x^4}{3}$

57. $S_5 = \dfrac{(-1)^{1+1}}{1}x^1 + \dfrac{(-1)^{2+1}}{2}x^2 + \dfrac{(-1)^{3+1}}{3}x^3 + \dfrac{(-1)^{4+1}}{4}x^4 + \dfrac{(-1)^{5+1}}{5}x^5$
$= x - \dfrac{x^2}{2} + \dfrac{x^3}{3} - \dfrac{x^4}{4} + \dfrac{x^5}{5}$

59. $S_4 = 1^2 + 2^2 + 3^2 + 4^2$
$k = 1, 2, 3, 4, \cdots$
Clearly, $a_k = k^2$, $k = 1, 2, 3, 4$
$S_4 = \sum_{k=1}^{4} k^2$

61. $S_5 = \dfrac{1}{2^1} + \dfrac{1}{2^2} + \dfrac{1}{2^3} + \dfrac{1}{2^4} + \dfrac{1}{2^5}$
$k = 1, 2, 3, 4, \cdots$
Clearly, $a_k = \dfrac{1}{2^k}$, $k = 1, 2, 3, 4, 5$
$S_5 = \sum_{k=1}^{5} \dfrac{1}{2^k}$

63. $S_n = 1 + \dfrac{1}{2^2} + \dfrac{1}{3^2} + \cdots + \dfrac{1}{n^2}$
$k = 1, 2, 3, \cdots, n$
Clearly, $a_k = \dfrac{1}{k^2}$, $k = 1, 2, 3, \cdots, n$
$S_n = \sum_{k=1}^{n} \dfrac{1}{k^2}$

65. $S_n = 1 - 4 + 9 + \cdots + (-1)^{n+1}n^2$
$k = 1, 2, 3, \cdots, n$
Clearly, $a_k = (-1)^{k+1}k^2$, $k = 1, 2, 3, \cdots, n$
$S_n = \sum_{k=1}^{n} (-1)^{k+1}k^2$

67. (A) $a_1 = 3$

$$a_2 = \frac{a_{2-1}^2 + 2}{2a_{2-1}} = \frac{a_1^2 + 2}{2a_1} = \frac{3^2 + 2}{2 \cdot 3} \approx 1.83$$

$$a_3 = \frac{a_{3-1}^2 + 2}{2a_{3-1}} = \frac{a_2^2 + 2}{2a_2} = \frac{(1.83)^2 + 2}{2(1.83)} \approx 1.46$$

$$a_4 = \frac{a_{4-1}^2 + 2}{2a_{4-1}} = \frac{a_3^2 + 2}{2a_3} = \frac{(1.46)^2 + 2}{2(1.46)} \approx 1.415$$

(B) Calculator $\sqrt{2} = 1.4142135...$

(C) $a_1 = 1$

$$a_2 = \frac{a_1^2 + 2}{2a_1} = \frac{1^2 + 2}{2 \cdot 1} = 1.5$$

$$a_3 = \frac{a_2^2 + 2}{2a_2} = \frac{(1.5)^2 + 2}{2(1.5)} \approx 1.417$$

$$a_4 = \frac{a_3^2 + 2}{2a_3} = \frac{(1.417)^2 + 2}{2(1.417)} \approx 1.414$$

69. The first ten terms of the Fibonacci sequence a_n are
1, 1, 2, 3, 5, 8, 13, 21, 34, 55
The first ten terms of b_n are
1, 3, 4, 7, 11, 18, 29, 47, 76, 123
The first ten terms of $c_n = \frac{b_n}{a_n}$ are

$$\frac{1}{1}, \frac{3}{1}, \frac{4}{2}, \frac{7}{3}, \frac{11}{5}, \frac{18}{8}, \frac{29}{13}, \frac{47}{21}, \frac{76}{34}, \frac{123}{55}$$

In decimal notation this becomes
1, 3, 2, 2.33..., 2.2, 2.25, 2.23..., 2.238..., 2.235..., 2.236...

71. $e^{0.2} = 1 + \frac{0.2}{1!} + \frac{(0.2)^2}{2!} + \frac{(0.2)^3}{3!} + \frac{(0.2)^4}{4!}$
$= 1.2214000$

$e^{0.2} = 1.2214028$ (calculator—direct evaluation)

73. $\sum_{k=1}^{n} ca_k = ca_1 + ca_2 + ca_3 + \cdots + ca_n = c(a_1 + a_2 + a_3 + \cdots a_n) = c\sum_{k=1}^{n} a_k$

Section 11-2

1. $(3 + 5)^1 = 3^1 + 5^1$ True
$(3 + 5)^2 = 3^2 + 5^2$ False
Fails at $n = 2$

3. $1^2 = 3 \cdot 1 - 2$ True
$2^2 = 3 \cdot 2 - 2$ True
$3^2 = 3 \cdot 3 - 2$ False
Fails at $n = 3$

5. P_1: $2 = 2 \cdot 1^2$ $\quad 2 = 2$
P_2: $2 + 6 = 2 \cdot 2^2$ $\quad 8 = 8$
P_3: $2 + 6 + 10 = 2 \cdot 3^2$ $\quad 18 = 18$

7. P_1: $a^5a^1 = a^{5+}$ $\quad a^6 = a^6$

P_2: $a^5a^2 = a^{5+2}$ $\quad a^5a^2 = a^5(a^1a) = (a^5a)a = a^6a = a^7 = a^{5+2}$

P_3: $a^5a^3 = a^{5+3}$ $\quad a^5a^3 = a^5(a^2a) = a^5(a^1a)a = [(a^5a)a]a = a^8 = a^{5+3}$

9. P_1: $9^1 - 1 = 8$ is divisible by 4

P_2: $9^2 - 1 = 80$ is divisible by 4

P_3: $9^3 - 1 = 728$ is divisible by 4

11. P_k: $2 + 6 + 10 + \cdots + (4k - 2) = 2k^2$

P_{k+1}: $2 + 6 + 10 + \cdots + (4k - 2) + [4(k + 1) - 2] = 2(k + 1)^2$

or $2 + 6 + 10 + \cdots + (4k - 2) + (4k + 2) = 2(k + 1)^2$

13. P_k: $a^5a^k = a^{5+k}$

P_{k+1}: $a^5a^{k+1} = a^{5+k+1}$

15. P_k: $9^k - 1 = 4r$ for some integer r

P_{k+1}: $9^{k+1} - 1 = 4s$ for some integer s

17. Prove: $2 + 6 + 10 + \cdots + (4n - 2) = 2n^2$, $n \in N$

Write: P_n: $2 + 6 + 10 + \cdots + (4n - 2) = 2n^2$

Part 1: Show that P_1 is true:

P_1: $2 = 2 \cdot 1^2$

$= 2$ $\quad P_1$ is true

Part 2: Show that if P_k is true, then P_{k+1} is true:

Write out P_k and P_{k+1}.

P_k: $2 + 6 + 10 + \cdots + (4k - 2) = 2k^2$

P_{k+1}: $2 + 6 + 10 + \cdots + (4k - 2) + (4k + 2) = 2(k + 1)^2$

We start with P_k:

$2 + 6 + 10 + \cdots + (4k - 2) = 2k^2$

Adding $(4k + 2)$ to both sides, we get

$$\begin{aligned} 2 + 6 + 10 + \cdots + (4k - 2) + (4k + 2) &= 2k^2 + 4k + 2 \\ &= 2(k^2 + 2k + 1) \\ &= 2(k + 1)^2 \end{aligned}$$

We have shown that if P_k is true, then P_{k+1} is true.

Conclusion: P_n is true for all positive integers n.

> **Common Errors:**
> P_1 is not the nonsensical statement
> $2 + 6 + 10 + \cdots + (4 \cdot 1 - 2) = 2 \cdot 1^2$
> achieved by "substitution" of 1 for n. The left side of P_n has n terms; the left side of P_1 must have 1 term. Also, $P_{k+1} \neq P_k + 1$
> Nor is P_{k+1} simply its $k + 1$st term. $4k + 2 \neq 2(k + 1)^2$

19. Prove: $a^5a^n = a^{5+n}$, $n \in N$
Write: P_n: $a^5a^n = a^{5+n}$

Part 1: Show that P_1 is true:
P_1: $a^5a^1 = a^{5+1}$
$a^5a = a^{5+1}$. True by the recursive definition of a^n.

Part 2: Show that if P_k is true, then P_{k+1} is true:
Write out P_k and P_{k+1}.
P_k: $a^5a^k = a^{5+k}$
P_{k+1}: $a^5a^{k+1} = a^{5+k+1}$

We start with P_k:
$a^5a^k = a^{5+k}$

Multiply both sides by a:
$a^5a^ka = a^{5+k}a$
$a^5a^{k+1} = a^{5+k+1}$ by the recursive definition of a^n.

We have shown that if P_k is true, then P_{k+1} is true.
Conclusion: P_n is true for all positive integers n.

21. Prove: $9^n - 1$ is divisible by 4.
Write: P_n: $9^n - 1$ is divisible by 4

Part 1: Show that P_1 is true:
P_1: $9^1 - 1 = 8 = 4 \cdot 2$ is divisible by 4. Clearly true.

Part 2: Show that if P_k is true, then P_{k+1} is true:
Write out P_k and P_{k+1}.
P_k: $9^k - 1 = 4r$ for some integer r
P_{k+1}: $9^{k+1} - 1 = 4s$ for some integer s

We start with P_k:
$9^k - 1 = 4r$ for some integer r

Now,
$$\begin{aligned} 9^{k+1} - 1 &= 9^{k+1} - 9^k + 9^k - 1 \\ &= 9^k(9 - 1) + 9^k - 1 \\ &= 9^k \cdot 8 + 9^k - 1 \\ &= 4 \cdot 9^k \cdot 2 + 4r \\ &= 4(9^k \cdot 2 + r) \end{aligned}$$
Therefore,

$9^{k+1} - 1 = 4s$ for some integer s $(= 2 \cdot 9^k + r)$
$9^{k+1} - 1$ is divisible by 4.

We have shown that if P_k is true, then P_{k+1} is true.
Conclusion: P_n is true for all positive integers n.

23. The polynomial $x^4 + 1$ is a counterexample, since it has degree 4, and no real zeros.

25. 23 is a counterexample, since there are no prime numbers p such that $23 < p < 29$.

27. P_n: $2 + 2^2 + 2^3 + \cdots + 2^n = 2^{n+1} - 2$

Part 1: Show that P_1 is true:

$$\begin{aligned} P_1: 2 &= 2^{1+1} - 2 \\ &= 2^2 - 2 \\ &= 4 - 2 \\ &= 2 \quad P_1 \text{ is true} \end{aligned}$$

Part 2: Show that if P_k is true, then P_{k+1} is true:
Write out P_k and P_{k+1}.

P_k: $2 + 2^2 + 2^3 + \cdots + 2^k = 2^{k+1} - 2$

P_{k+1}: $2 + 2^2 + 2^3 + \cdots + 2^k + 2^{k+1} = 2^{k+2} - 2$

We start with P_k:

$$2 + 2^2 + 2^3 + \cdots + 2^k = 2^{k+1} - 2$$

Adding 2^{k+1} to both sides,

$$\begin{aligned} 2 + 2^2 + 2^3 + \cdots + 2^k + 2^{k+1} &= 2^{k+1} - 2 + 2^{k+1} \\ &= 2^{k+1} + 2^{k+1} - 2 \\ &= 2 \cdot 2^{k+1} - 2 \\ &= 2^1 \cdot 2^{k+1} - 2 \\ &= 2^{k+2} - 2 \end{aligned}$$

We have shown that if P_k is true, then P_{k+1} is true.

Conclusion: P_n is true for all positive integers n.

29. P_n: $1^2 + 3^2 + 5^2 + \cdots + (2n - 1)^2 = \frac{1}{3}(4n^3 - n)$

Part 1: Show that P_1 is true:

$$\begin{aligned} P_1: 1^2 &= \frac{1}{3}(4 \cdot 1^3 - 1) \\ &= \frac{1}{3}(4 - 1) \\ &= 1 \quad \text{True.} \end{aligned}$$

Part 2: Show that if P_k is true, then P_{k+1} is true:
Write out P_k and P_{k+1}.

P_k: $1^2 + 3^2 + 5^2 + \cdots + (2k - 1)^2 = \frac{1}{3}(4k^3 - k)$

P_{k+1}: $1^2 + 3^2 + 5^2 + \cdots + (2k - 1)^2 + (2k + 1)^2 = \frac{1}{3}[4(k + 1)^3 - (k + 1)]$

We start with P_k:

$$1^2 + 3^2 + 5^2 + \cdots + (2k - 1)^2 = \frac{1}{3}(4k^3 - k)$$

Adding $(2k + 1)^2$ to both sides,

$$\begin{aligned} 1^2 + 3^2 + 5^2 + \cdots + (2k - 1)^2 + (2k + 1)^2 &= \frac{1}{3}(4k^3 - k) + (2k + 1)^2 \\ &= \frac{1}{3}[4k^3 - k + 3(2k + 1)^2] \\ &= \frac{1}{3}[4k^3 - k + 3(4k^2 + 4k + 1)] \\ &= \frac{1}{3}[4k^3 - k + 12k^2 + 12k + 3] \\ &= \frac{1}{3}[4k^3 + 12k^2 + 12k - k + 3] \end{aligned}$$

$$= \frac{1}{3}[4k^3 + 12k^2 + 12k + 4 - k - 1]$$
$$= \frac{1}{3}[4(k^3 + 3k^2 + 3k + 1) - (k + 1)]$$
$$= \frac{1}{3}[4(k + 1)^3 - (k + 1)]$$

We have shown that if P_k is true, then P_{k+1} is true.

Conclusion: P_n is true for all positive integers n.

31. P_n: $1^2 + 2^2 + 3^2 + \cdots + n^2 = \frac{n(n + 1)(2n + 1)}{6}$

Part 1: Show that P_1 is true:

$$P_1\text{: } 1^2 = \frac{1(1 + 1)(2 \cdot 1 + 1)}{6} = \frac{1 \cdot 2 \cdot 3}{6} = 1 \qquad P_1 \text{ is true.}$$

Part 2: Show that if P_k is true, then P_{k+1} is true:

Write out P_k and P_{k+1}.

$$P_k\text{: } 1^2 + 2^2 + 3^2 + \cdots + k^2 = \frac{k(k + 1)(2k + 1)}{6}$$
$$P_{k+1}\text{: } 1^2 + 2^2 + 3^2 + \cdots + k^2 + (k + 1)^2 = \frac{(k + 1)(k + 2)(2k + 3)}{6}$$

We start with P_k:

$$1^2 + 2^2 + 3^2 + \cdots + k^2 = \frac{k(k + 1)(2k + 1)}{6}$$

Adding $(k + 1)^2$ to both sides,

$$1^2 + 2^2 + 3^2 + \cdots + k^2 + (k + 1)^2 = \frac{k(k + 1)(2k + 1)}{6} + (k + 1)^2$$
$$= \frac{k(k + 1)(2k + 1)}{6} + \frac{6(k + 1)^2}{6}$$
$$= \frac{k(k + 1)(2k + 1) + 6(k + 1)^2}{6}$$
$$= \frac{(k + 1)[k(2k + 1) + 6(k + 1)]}{6}$$
$$= \frac{(k + 1)[2k^2 + 7k + 6]}{6}$$
$$= \frac{(k + 1)(k + 2)(2k + 3)}{6}$$

We have shown that if P_k is true, then P_{k+1} is true.

Conclusion: P_n is true for all positive integers n.

33. P_n: $\dfrac{a^n}{a^3} = a^{n-3} \quad n > 3$

Part 1: Show that P_4 is true:

P_4: $\dfrac{a^4}{a^3} = a^{4-3}$

But $\dfrac{a^4}{a^3} = \dfrac{a^3a}{a^3} = a = a^1 = a^{4-3}$ So, P_4 is true.

Part 2: Show that if P_k is true, then P_{k+1} is true:

Write out P_k and P_{k+1}.

P_k: $\dfrac{a^k}{a^3} = a^{k-3}$

P_{k+1}: $\dfrac{a^{k+1}}{a^3} = a^{k-2}$

We start with P_k:

$$\frac{a^k}{a^3} = a^{k-3}$$

Multiply both sides by a:

$$\frac{a^k}{a^3}a = a^{k-3}a$$

$$\frac{a^{k+1}}{a^3} = a^{k-3+1} \text{ by the recursive definition of } a^n$$

$$\frac{a^{k+1}}{a^3} = a^{k-2}$$

We have shown that if P_k is true, then P_{k+1} is true.

Conclusion: P_n is true for all integers $n > 3$.

35. Write: P_n: $a^ma^n = a^{m+n}$ m an arbitrary element of N.

Part 1: Show that P_1 is true:

P_1: $a^ma^1 = a^{m+1}$ This is true by the recursive definition of a^n.

Part 2: Show that if P_k is true, then P_{k+1} is true:

Write out P_k and P_{k+1}.

P_k: $a^ma^k = a^{m+k}$

P_{k+1}: $a^ma^{k+1} = a^{m+k+1}$

We start with P_k:

$a^ma^k = a^{m+k}$

Multiplying both sides by a:

$a^ma^ka = a^{m+k}a$

$a^ma^{k+1} = a^{m+k+1}$ by the recursive definition of a^n.

We have shown that if P_k is true, then P_{k+1} is true.

Conclusion: $a^ma^n = a^{m+n}$ is true for arbitrary m, n positive integers.

37. Write: P_n: $x^n - 1 = (x - 1)Q_n(x)$ for some polynomial $Q_n(x)$.

Part 1: Show that P_1 is true:

P_1: $x^1 - 1 = (x - 1)Q_1(x)$ for some $Q_1(x)$. This is true since we can choose $Q_1(x) = 1$.

Part 2: Show that if P_k is true, then P_{k+1} is true:
Write out P_k and P_{k+1}.

P_k: $x^k - 1 = (x - 1)Q_k(x)$ for some $Q_k(x)$
P_{k+1}: $x^{k+1} - 1 = (x - 1)Q_{k+1}(x)$ for some $Q_{k+1}(x)$

We start with P_k:

$x^k - 1 = (x - 1)Q_k(x)$ for some $Q_k(x)$

Now, $x^{k+1} - 1 = x^{k+1} - x^k + x^k - 1$

$= x^k(x - 1) + x^k - 1$

$= x^k(x - 1) + (x - 1)Q_k(x)$

$= (x - 1)(x^k + Q_k(x))$

$= (x - 1)Q_{k+1}(x)$, where $Q_{k+1}(x) = x^k + Q_k(x)$

Conclusion: $x^n - 1 = (x - 1)Q_n(x)$ for all positive integers n. Thus $x^n - 1$ is divisible by $x - 1$ for all positive integers n.

39. Write: P_n: $x^{2n} - 1 = (x - 1)Q_n(x)$ for some polynomial $Q_n(x)$.

Part 1: Show that P_1 is true:

P_1: $x^2 - 1 = (x - 1)Q_1(x)$ for some polynomial $Q_1(x)$. This is true since we can choose $Q_1(x) = x + 1$.

Part 2: Show that if P_k is true, then P_{k+1} is true:
Write out P_k and P_{k+1}.

P_k: $x^{2k} - 1 = (x - 1)Q_k(x)$ for some $Q_k(x)$
P_{k+1}: $x^{2k+2} - 1 = (x - 1)Q_{k+1}(x)$ for some $Q_{k+1}(x)$

Now, $x^{2k+2} - 1 = x^{2k+2} - x^{2k} + x^{2k} - 1$

$= x^{2k}(x^2 - 1) + x^{2k} - 1$

$= x^{2k}(x + 1)(x - 1) + (x - 1)Q_k(x)$

$= (x - 1)[x^{2k}(x + 1) + Q_k(x)]$

$= (x - 1)Q_{k+1}(x)$, where $Q_{k+1}(x) = x^{2k}(x + 1) + Q_k(x)$

Conclusion: $x^{2n} - 1 = (x - 1)Q_n(x)$ for all positive integers n. Thus $x^{2n} - 1$ is divisible by $x - 1$ for all positive integers n.

41. P_n: $1^3 + 2^3 + 3^3 + \cdots + n^3 = (1 + 2 + 3 + \cdots + n)^2$

Part 1: Show that P_1 is true:

P_1: $1^3 = 1^2$ True

Part 2: Show that if P_k is true, then P_{k+1} is true:

Write out P_k and P_{k+1}.

P_k: $1^3 + 2^3 + 3^3 + \cdots + k^3 = (1 + 2 + 3 + \cdots + k)^2$

P_{k+1}: $1^3 + 2^3 + 3^3 + \cdots + k^3 + (k + 1)^3 = (1 + 2 + 3 + \cdots + k + k + 1)^2$

We take it as proved in text Matched Problem 1 that

$$1 + 2 + 3 + \cdots + n = \frac{n(n + 1)}{2} \quad \text{for} \quad n \in N$$

Thus, $1 + 2 + 3 + \cdots + k = \dfrac{k(k + 1)}{2}$

$$1 + 2 + 3 + \cdots + k + (k + 1) = \frac{(k + 1)(k + 2)}{2} \text{ are known}$$

We start with P_k:

$$1^3 + 2^3 + 3^3 + \cdots + k^3 = (1 + 2 + 3 + \cdots + k)^2$$

Adding $(k + 1)^3$ to both sides:

$$\begin{aligned} 1^3 + 2^3 + 3^3 + \cdots + k^3 + (k + 1)^3 &= (1 + 2 + 3 + \cdots + k)^2 + (k + 1)^3 \\ &= \left[\frac{k(k + 1)}{2}\right]^2 + (k + 1)^3 \\ &= (k + 1)^2\left[\left(\frac{k}{2}\right)^2 + (k + 1)\right] \\ &= (k + 1)^2\left[\frac{k^2}{4} + \frac{4(k + 1)}{4}\right] \\ &= (k + 1)^2\left[\frac{k^2 + 4k + 4}{4}\right] \\ &= \frac{(k + 1)^2(k + 2)^2}{4} \\ &= \left[\frac{(k + 1)(k + 2)}{2}\right]^2 \\ &= [1 + 2 + 3 + \cdots + k + (k + 1)]^2 \end{aligned}$$

Conclusion: P_n is true for all positive integers n.

43. We note:

$$\begin{aligned} 2 &= 2 = 1\cdot 2 & n = 1 \\ 2 + 4 &= 6 = 2\cdot 3 & n = 2 \\ 2 + 4 + 6 &= 12 = 3\cdot 4 & n = 3 \\ 2 + 4 + 6 + 8 &= 20 = 4\cdot 5 & n = 4 \end{aligned}$$

Hypothesis: P_n: $2 + 4 + 6 + \cdots + 2n = n(n + 1)$

Proof: Part 1: Show P_1 is true

P_1: $2 = 1\cdot 2$ True

Part 2: Show that if P_k is true, then P_{k+1} is true:

Write out P_k and P_{k+1}.

P_k: $2 + 4 + 6 + \cdots + 2k = k(k + 1)$

P_{k+1}: $2 + 4 + 6 + \cdots + 2k + (2k + 2) = (k + 1)(k + 2)$

We start with P_k:

$$2 + 4 + 6 + \cdots + 2k = k(k + 1)$$

Adding $2k + 2$ to both sides:

$$\begin{aligned} 2 + 4 + 6 + \cdots + 2k + (2k + 2) &= k(k + 1) + 2k + 2 \\ &= (k + 1)k + (k + 1)2 \\ &= (k + 1)(k + 2) \end{aligned}$$

Conclusion: The hypothesis P_n is true for all positive integers n.

45. $n = 1$: no line is determined.
$n = 2$: one line is determined.

$n = 3$: three lines are determined.

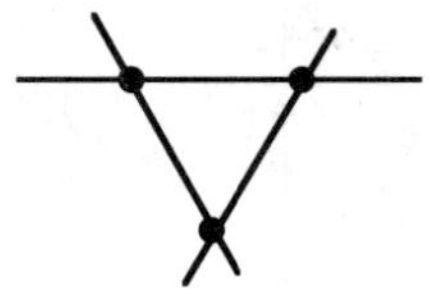

$n = 4$: six lines are determined.

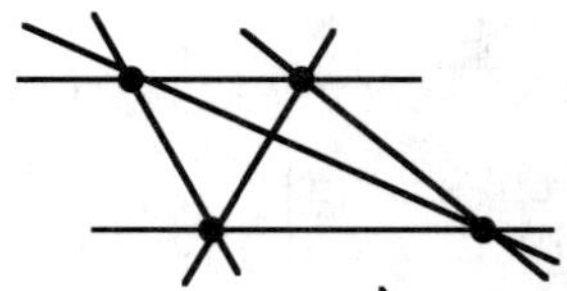

$n = 5$: ten lines are determined.

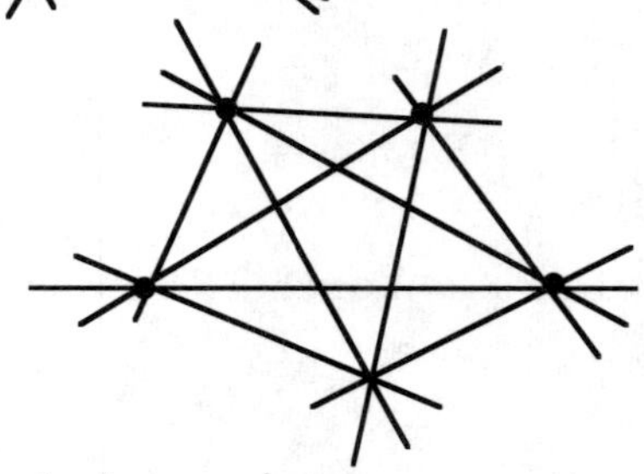

Hypothesis: P_n: n points (no three collinear) determine

$1 + 2 + 3 + \cdots + (n - 1) = \dfrac{n(n-1)}{2}$ lines, $n \ge 2$.

Proof: Part 1: Show that P_2 is true.

P_2: 2 points determine one line:

$$\begin{aligned} 1 &= \frac{2(2-1)}{2} \\ &= \frac{2 \cdot 1}{2} \\ &= 1 \text{ is true} \end{aligned}$$

Part 2: Show that if P_k is true, then P_{k+1} is true:
Write out P_k and P_{k+1}.

P_k: k points determine $1 + 2 + 3 + \cdots + (k - 1) = \dfrac{k(k-1)}{2}$ lines

P_{k+1}: $k + 1$ points determine $1 + 2 + 3 + \cdots + (k - 1) + k = \dfrac{(k+1)k}{2}$ lines

We start with P_k:

k points determine $1 + 2 + 3 + \cdots + (k - 1) = \dfrac{k(k-1)}{2}$ lines

Now, the k + 1st point will determine a total of k new lines, one with each of the previously existing k points. These k new lines will be added to the previously existing lines. Hence, $k + 1$ points determine

$$1 + 2 + 3 + \cdots + (k - 1) + k = \frac{k(k-1)}{2} + k \text{ lines}$$
$$= k\left[\frac{k-1}{2} + 1\right] \text{ lines}$$
$$= k\left[\frac{k-1+2}{2}\right] \text{ lines}$$
$$= \frac{k(k+1)}{2} \text{ lines}$$

Conclusion: The hypothesis P_n is true for $n \geq 2$.

47. P_n: $a > 1 \Rightarrow a^n > 1$, $n \in N$.

Part 1: Show that P_1 is true:

P_1: $a > 1 \Rightarrow a^1 > 1$. This is automatically true.

Part 2: Show that if P_k is true, then P_{k+1} is true:

Write out P_k and P_{k+1}.

P_k: $a > 1 \Rightarrow a^k > 1$

P_{k+1}: $a > 1 \Rightarrow a^{k+1} > 1$

We start with P_k. Further, assume $a > 1$ and try to derive $a^{k+1} > 1$. If this succeeds, we have proved P_{k+1}.

Assume $a > 1$.

From P_k we know that $a^k > 1$, also $1 > 0$, hence $a > 0$. We may therefore multiply both sides of the inequality $a^k > 1$ by a without changing the sense of the inequality. Hence,

$a^k a > 1a$

$a^{k+1} > a$

But, $a > 1$ by assumption. Hence, $a^{k+1} > 1$. We have derived this from P_k and $a > 1$. Thus, if P_k is true, then P_{k+1} is true.

Conclusion: P_n is true for all $n \in N$.

49. P_n: $n^2 > 2n$, $n \geq 3$.

Part 1: Show that P_3 is true:

P_3: $3^2 > 2 \cdot 3$

$9 > 6$ True

Part 2: Show that if P_k is true, then P_{k+1} is true:

Write out P_k and P_{k+1}.

P_k: $k^2 > 2k$

P_{k+1}: $(k + 1)^2 > 2(k + 1)$

We start with P_k:

$k^2 > 2k$

Adding $2k + 1$ to both sides: $k^2 + 2k + 1 > 2k + 2k + 1$

$(k + 1)^2 > 2k + 2 + 2k - 1$

Now, $2k - 1 > 0$, $k \in N$, hence $2k + 2 + 2k - 1 > 2k + 2$.

Therefore, $(k + 1)^2 > 2k + 2$

$(k + 1)^2 > 2(k + 1)$

Thus, if P_k is true, then P_{k+1} is true.

Conclusion: P_n is true for all integers $n \geq 3$.

51. The statement is true. Proof by mathematical induction:

P_n: $1 - 2 + 3 - \dots (2n - 1) = n$

Show P_1 is true

$1 = 1$

Show if P_k is true, P_{k+1} is true.

Write out P_k and P_{k+1}.

P_k: $1 - 2 + 3 - \dots + (2k - 1) = k$

P_{k+1}: $1 - 2 + 3 - \dots + (2k + 1) = k + 1$

We start with P_k:

$1 - 2 + 3 - \dots + (2k - 1) = k$

Now add $-2k + (2k + 1)$ to both sides:

$$\begin{aligned} 1 - 2 + 3 \dots + (2k - 1) - 2k + (2k + 1) &= k - 2k + (2k + 1) \\ &= k - 2k + 2k + 1 \\ &= k + 1 \end{aligned}$$

We have shown that if P_k is true, P_{k+1} is true.

Conclusion: $1 - 2 + 3 - \dots + (2n - 1) = n$ for all alternating sums of an odd number of positive integers.

53. The statement is false. For example, if $n = 3$

$3^4 + 4^4 + 5^4 + 6^4 = 2258$ while $7^4 = 2401$.

55. To prove $a_n = b_n$, $n \in N$, write:

P_n: $a_n = b_n$.

Proof: Part 1: Show P_1 is true.

P_1: $a_1 = b_1$ $a_1 = 1$ $b_1 = 2 \cdot 1 - 1 = 1$

Thus, $a_1 = b_1$

Part 2: Show that if P_k is true, then P_{k+1} is true:

Write out P_k and P_{k+1}.

P_k: $a_k = b_k$

P_{k+1}: $a_{k+1} = b_{k+1}$

We start with P_k:

$a_k = b_k$

$$\begin{aligned} \text{Now, } a_{k+1} = a_{k+1-1} + 2 = a_k + 2 = b_k + 2 &= 2k - 1 + 2 = 2k + 1 \\ &= 2(k + 1) - 1 = b_{k+1} \end{aligned}$$

Therefore, $a_{k+1} = b_{k+1}$

Thus, if P_k is true, then P_{k+1} is true.

Conclusion: P_n is true for all $n \in N$. Hence, $\{a_n\} = \{b_n\}$.

57. To prove: $a_n = b_n$, $n \in N$, write:
P_n: $a_n = b_n$.

Proof: Part 1: Show P_1 is true.

P_1: $a_1 = b_1 \quad a_1 = 2 \quad b_1 = 2^{2\cdot 1-1} = 2^1 = 2$

Thus, $a_1 = b_1$

Part 2: Show that if P_k is true, then P_{k+1} is true:

Write out P_k and P_{k+1}.

P_k: $a_k = b_k$

P_{k+1}: $a_{k+1} = b_{k+1}$

We start with P_k:

$a_k = b_k$

Now, $a_{k+1} = 2^2 a_{k+1-1} = 2^2 a_k = 2^2 b_k = 2^2 2^{2k-1} = 2^{2+2k-1} = 2^{2(k+1)-1} = b_{k+1}$

Therefore, $a_{k+1} = b_{k+1}$

Thus, if P_k is true, then P_{k+1} is true.

Conclusion: P_n is true for all $n \in N$. Hence, $\{a_n\} = \{b_n\}$.

Section 11-3

1. (A) Since (-16) - (-11) = -5 and (-21) - (-16) = -5, the given terms can start an arithmetic sequence with $d = -5$. The next terms are then -21 + (-5) = -26, and (-26) + (-5) = -31.

(B) Since (-4) - 2 = -6 and 8 - (-4) = 12, there is no common difference. Since (-4) ÷ 2 = -2 and 8 ÷ (-4) = -2, the given terms can start a geometric sequence with $r = -2$. The next terms are then $8\cdot(-2) = -16$, and $(-16)\cdot(-2) = 32$.

(C) Since $4 - 1 \neq 9 - 4$, there is no common difference, so the sequence is not an arithmetic sequence. Since $4 \div 1 \neq 9 \div 4$, there is no common ratio, so the sequence is not geometric either.

(D) Since $\frac{1}{6} - \frac{1}{2} = -\frac{1}{3}$ and $\frac{1}{18} - \frac{1}{6} = -\frac{1}{9}$, there is no common difference. Since $\frac{1}{6} \div \frac{1}{2} = \frac{1}{3}$ and $\frac{1}{18} \div \frac{1}{6} = \frac{1}{3}$, the given terms can start a geometric sequence with $r = \frac{1}{3}$. The next terms are then $\frac{1}{18} \cdot \frac{1}{3} = \frac{1}{54}$ and $\frac{1}{54} \cdot \frac{1}{3} = \frac{1}{162}$.

3. $a_2 = a_1 + d = -5 + 4 = -1$

$a_3 = a_2 + d = -1 + 4 = 3$

$a_4 = a_3 + d = 3 + 4 = 7$

$a_n = a_1 + (n - 1)d$

5. $a_{15} = a_1 + 14d = -3 + 14\cdot 5 = 67$

$$S_n = \frac{n}{2}[2a_1 + (n - 1)d]$$

$$S_{11} = \frac{11}{2}[2(-3) + (11 - 1)5] = \frac{11}{2}(44) = 242$$

7. $a_2 - a_1 = 5 - 1 = d = 4$

$$S_n = \frac{n}{2}[2a_1 + (n - 1)d]$$

$$S_{21} = \frac{21}{2}[2\cdot 1 + (21 - 1)4] = \frac{21}{2}(82) = 861$$

9. $a_2 - a_1 = 5 - 7 = -2 = d$

$a_n = a_1 + (n - 1)d$

$a_{15} = 7 + (15 - 1)(-2) = -21$

11. $a_2 = a_1 r = (-6)\left(-\frac{1}{2}\right) = 3$

$a_3 = a_2 r = 3\left(-\frac{1}{2}\right) = -\frac{3}{2}$

$a_4 = a_3 r = \left(-\frac{3}{2}\right)\left(-\frac{1}{2}\right) = \frac{3}{4}$

13. $a_n = a_1 r^{n-1}$

$a_{10} = 81\left(\frac{1}{3}\right)^{10-1}$

$= \frac{1}{243}$

15. $S_n = \frac{a_1 - ra_n}{1 - r}$

$S_7 = \frac{3 - 3(2,187)}{1 - 3}$

$= 3,279$

17. $a_n = a_1 + (n - 1)d$

$a_{20} = a_1 + 19d$

$117 = 3 + 19d$

$114 = 19d$

So, $d = 6$. Therefore

$a_{101} = a_1 + (100)d$

$= 3 + 100(6) = 603$

19. $S_n = \frac{n}{2}(a_1 + a_n)$

$S_{40} = \frac{40}{2}(-12 + 22)$

$= 200$

21. $a_2 - a_1 = d$

$\frac{1}{2} - \frac{1}{3} = \frac{1}{6} = d$

$a_n = a_1 + (n - 1)d$

$a_{11} = \frac{1}{3} + (11 - 1)\frac{1}{6}$

$= 2$

$S_n = \frac{n}{2}(a_1 + a_n)$

$S_{11} = \frac{11}{2}\left(\frac{1}{3} + 2\right) = \frac{77}{6}$

23. $a_n = a_1 + (n - 1)d$

$a_{10} = a_1 + 9d$

$a_3 = a_1 + 2d$

Eliminating d between these two statements by addition, we have

$2a_{10} = 2a_1 + 18d$

$-9a_3 = -9a_1 - 18d$

$2a_{10} - 9a_3 = -7a_1$

$a_1 = \frac{2a_{10} - 9a_3}{-7} = \frac{2(55) - 9(13)}{-7} = 1$

25. $a_1 = 8,\ a_2 = 2$

$a_2 = a_1 r$

$2 = 8r$

$r = \frac{1}{4}$

27. $a_1 = 120 \quad a_4 = -15$

$a_n = a_1 r^{n-1}$

$a_4 = a_1 r^3$

$-15 = 120r^3$

$r^3 = -\frac{1}{8}$

$r = -\frac{1}{2}$

29. $S_n = \frac{a_1 - a_1 r^n}{1 - r}$

$S_{10} = \frac{9 - 9(\frac{2}{3})^{10}}{1 - \frac{2}{3}}$

$= \frac{\frac{58,025}{6,561}}{\frac{1}{3}}$

$= \frac{58,025}{2,187}$

31. First find r

$a_n = a_1 r^{n-1}$

$a_8 = a_1 r^7$

$2,187 = 1 \cdot r^7$

$r = 3$

Therefore

$S_8 = \frac{a_1 - a_1 r^8}{1 - r}$

$= \frac{1 - 1 \cdot 3^8}{1 - 3}$

$= \frac{-6,560}{-2}$

$= 3,280$

33. $a_3 = a_1 r^2,\ a_6 = a_1 r^5$

$72 = a_1 r^2,\ -243 = a_1 r^5$

Dividing, we obtain

$\frac{a_1 r^5}{a_1 r^2} = r^3 = -\frac{243}{72}$

$r^3 = -\frac{27}{8}$

$r = -\frac{3}{2}$

Therefore

$72 = a_1\left(-\frac{3}{2}\right)^2$

$a_1 = 72\left(-\frac{2}{3}\right)^2$

$a_1 = 32$

35. $a_4 = a_1 r^3$

$-1 = 1 \cdot r^3$

$r^3 = -1$

$r = -1$

$a_{100} = a_1 r^{99}$

$= 1 \cdot (-1)^{99}$

$= -1$

37.
$$\begin{aligned} a_n &= a_1 + (n-1)d \\ d &= a_2 - a_1 \\ &= (3\cdot 2 + 3) - (3\cdot 1 + 3) = 3 \\ a_{51} &= a_1 + (51-1)d \\ &= (3\cdot 1 + 3) + 50\cdot 3 \\ &= 156 \\ S_n &= \frac{n}{2}(a_1 + a_n) \\ S_{51} &= \frac{51}{2}(g(1) + g(51)) \\ &= \frac{51}{2}[4 + (-46)] \\ &= -1{,}071 \end{aligned}$$

39. $S_n = \dfrac{a_1 - a_1 r^n}{1 - r}$

First, note $a_1 = (-3)^{1-1} = 1$

$$\begin{aligned} r &= \frac{a_2}{a_1} = \frac{(-3)^{2-1}}{(-3)^{1-1}} = -3 \\ S_7 &= \frac{1 - 1(-3)^7}{1 - (-3)} \\ &= \frac{2{,}188}{4} \\ &= 547 \end{aligned}$$

41.
$$\begin{aligned} g(t) &= 5 - t \\ g(1) &= 5 - 1 = 4 \\ g(51) &= 5 - 51 = -46 \end{aligned}$$

$g(1) + g(2) + g(3) + \cdots + g(51) = S_{51}$

$$\begin{aligned} S_n &= \frac{n}{2}(a_1 + a_n) \\ S_{51} &= \frac{51}{2}(g(1) + g(51)) \\ &= \frac{51}{2}[4 + (-46)] \\ &= -1{,}071 \end{aligned}$$

43. $g(1) + g(2) + \cdots + g(10)$ is a geometric series, with

$$\begin{aligned} g(1) &= a_1 = \left(\frac{1}{2}\right)^1 = \frac{1}{2} \\ r &= \frac{g(2)}{g(1)} = \frac{\left(\frac{1}{2}\right)^2}{\left(\frac{1}{2}\right)^1} = \frac{1}{2} \\ S_n &= \frac{a_1 - a_1 r^n}{1 - r} \\ S_{10} &= \frac{\frac{1}{2} - \frac{1}{2}\left(\frac{1}{2}\right)^{10}}{1 - \frac{1}{2}} \\ &= \frac{1 - \left(\frac{1}{2}\right)^{10}}{2 - 1} \\ &= \frac{1{,}023}{1{,}024} \end{aligned}$$

45. First, find n:

$$\begin{aligned} a_n &= a_1 + (n-1)d \\ 134 &= 22 + (n-1)2 \\ n &= 57 \end{aligned}$$

Now, find S_{57}

$$\begin{aligned} S_n &= \frac{n}{2}(a_1 + a_n) \\ S_{57} &= \frac{57}{2}(22 + 134) \\ &= 4{,}446 \end{aligned}$$

47. To prove:

$$1 + 3 + 5 + \cdots + (2n - 1) = n^2$$

The sequence $1, 3, 5, \cdots$ is an arithmetic sequence, with $d = 2$. We are to find S_n.

But, $S_n = \dfrac{n}{2}(a_1 + a_n)$

$$\begin{aligned} &= \frac{n}{2}[1 + (2n - 1)] \\ &= \frac{n}{2}\cdot 2n \end{aligned}$$

So, $S_n = n^2$

49. $\dfrac{a_2}{a_1} = \dfrac{a_3}{a_2} = r$ for $a_1 + a_2 + a_3$ to be a geometric series. Hence,

$$\begin{aligned} \frac{x}{-2} &= \frac{-6}{x} \\ x^2 &= 12 \\ x &= 2\sqrt{3}, \end{aligned}$$

since x is specified positive.

51. Note that $a_n - a_{n-1} = 3$. Hence, this is an arithmetic sequence, with $d = 3$.
We are to find a_n.
But, $a_n = a_1 + (n - 1)d$
So, $a_n = -3 + (n - 1)3$ or $3n - 6$

53.

Enter the sequences in the sequence editor.

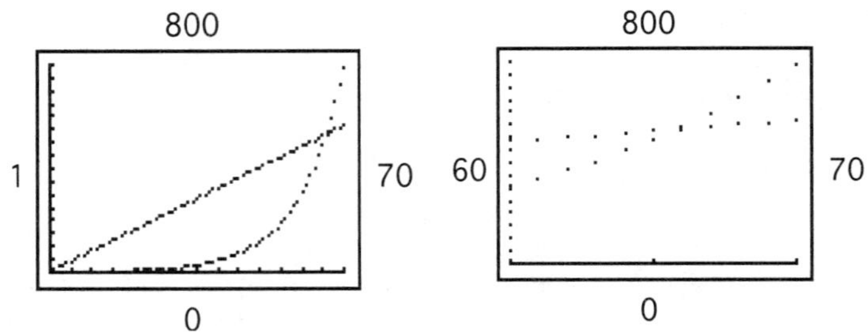

Graph the sequences.

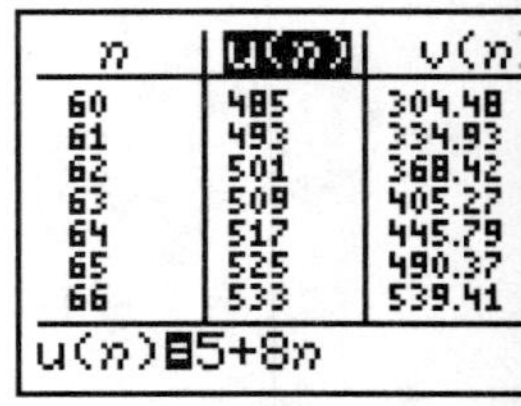

n	u(n)	v(n
60	485	304.48
61	493	334.93
62	501	368.42
63	509	405.27
64	517	445.79
65	525	490.37
66	533	539.41

u(n)=5+8n

Display the points (in the appropriat interval) in a tab

From the second graph, 66 is the least positive integer n such that $a_n < b_n$.
The table confirms this.

55.

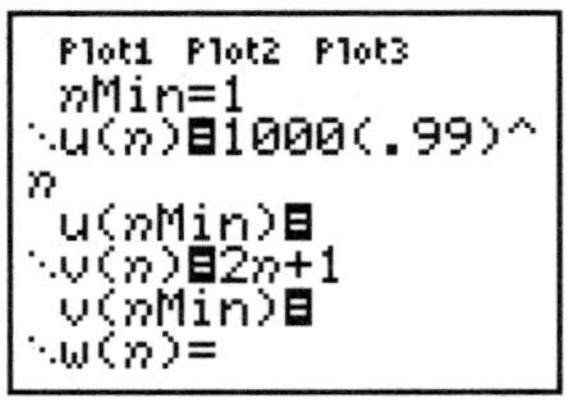

Enter the sequences in the sequence editor.

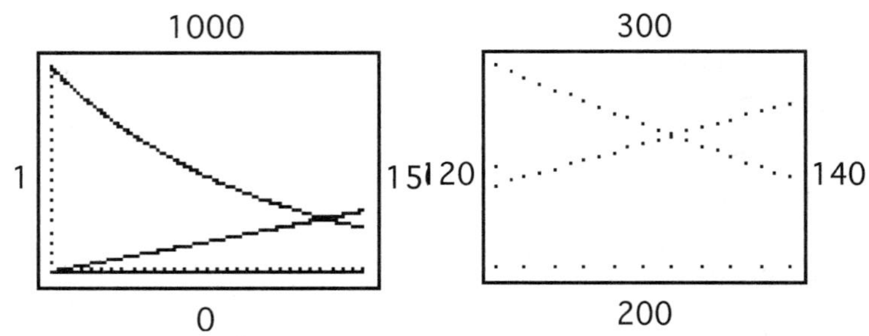

Graph the sequences.

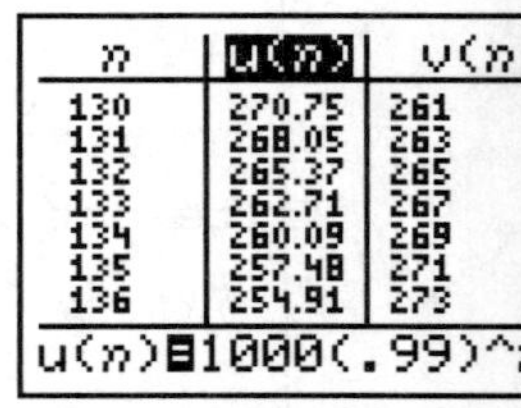

n	u(n)	v(n
130	270.75	261
131	268.05	263
132	265.37	265
133	262.71	267
134	260.09	269
135	257.48	271
136	254.91	273

u(n)=1000(.99)^

Display the points (in the appropriat interval) in a tab

From the second graph, 133 is the least positive integer n such that $a_n < b_n$.
The table confirms this.

57. $r = \dfrac{a_2}{a_1} = \dfrac{1}{2} \div 2 = \dfrac{1}{4}$

Since $|r| = \dfrac{1}{4} < 1$, the series has a sum.

$S_\infty = \dfrac{a_1}{1 - r} = \dfrac{2}{1 - \frac{1}{4}} = \dfrac{8}{3}$

59. $r = \dfrac{a_2}{a_1} = \dfrac{-1}{3}$

Since $|r| = \left|\dfrac{-1}{3}\right| = \dfrac{1}{3} < 1$, the series has a sum.

$S_\infty = \dfrac{a_1}{1 - r} = \dfrac{3}{1 - \frac{-1}{3}} = \dfrac{9}{4}$

61. $r = \dfrac{a_2}{a_1} = \dfrac{0.1}{1} = 0.1$

Since $|r| = 0.1 < 1$, the series has a sum.

$S_\infty = \dfrac{a_1}{1 - r} = \dfrac{1}{1 - 0.1} = \dfrac{10}{9}$

63. $r = \dfrac{a_2}{a_1} = \dfrac{1}{2} \div (-1) = -\dfrac{1}{2}$

Since $|r| = \left|-\dfrac{1}{2}\right| = \dfrac{1}{2} < 1$, the series has a sum.

$S_\infty = \dfrac{a_1}{1 - r} = \dfrac{-1}{1 - \left(-\frac{1}{2}\right)} = -\dfrac{2}{3}$

65. $r = \frac{a_2}{a_1} = \frac{1}{-1} = -1$

Since $|r| = |-1| = 1 \geq 1$, the series has no sum.

67. $0.\bar{7} = 0.777\ldots = 0.7 + 0.07 - 0.007 + 0.0007 + \ldots$

This is an infinite geometric series with

$a_1 = 0.7$ and $r = 0.1$.

Thus,

$$S_\infty = \frac{a_1}{1 - r} = \frac{0.7}{1 - 0.1} = \frac{0.7}{0.9} = \frac{7}{9}$$

69. $0.\overline{54}\ldots = 0.54 + 0.0054 + 0.000054 + \ldots$

This is an infinite geometric series with $a_1 = 0.54$ and $r = 0.01$. Thus,

$$S_\infty = \frac{a_1}{1 - r} = \frac{0.54}{1 - 0.01} = \frac{0.54}{0.99} = \frac{6}{11}$$

71. $3.\overline{216} = 3.216216216\ldots = 3 + 0.216 + 0.000216 + 0.000000216 + \ldots$

Therefore, we note: $0.216 + 0.000216 + 0.000000216 + \ldots$ is an infinite geometric series with $a_1 = 0.216$ and $r = 0.001$. Thus,

$$3.\overline{216} = 3 + S_\infty = 3 + \frac{a_1}{1 - r} = 3 + \frac{0.216}{1 - 0.001} = 3 + \frac{0.216}{0.999} = 3\tfrac{8}{37} \text{ or } \tfrac{119}{37}$$

73. Write: P_n: $a_n = a_1 + (n - 1)d$

Proof: Part 1: Show that P_1 is true:

P_1: $a_1 = a_1 + (1 - 1)d = a_1$ P_1 is true.

Part 2: Show that if P_k is true, then P_{k+1} is true:

Write out P_k and P_{k+1}.

P_k: $a_k = a_1 + (k - 1)d$

P_{k+1}: $a_{k+1} = a_1 + kd$

We start with P_k:

$a_k = a_1 + (k - 1)d$

Adding d to both sides:

$$\begin{aligned} a_k + d &= a_1 + (k - 1)d + d \\ a_{k+1} &= a_1 + d[(k - 1) + 1] \\ a_{k+1} &= a_1 + kd \end{aligned}$$

Thus, if P_k is true, then P_{k+1} is true.

Conclusion: P_n is true for all positive integers n.

75. This is a geometric sequence with ratio $\frac{a_n}{a_{n-1}} = -3$. Hence,

$$\begin{aligned} a_n &= a_1 r^{n-1} \\ &= (-2)(-3)^{n-1} \end{aligned}$$

77. Assume x, y, z are consecutive terms of an arithmetic progression.

Then, $y - x = d$, $z - y = d$

To show $a = x^2 + xy + y^2$, $b = z^2 + xz + x^2$, $c = y^2 + yz + z^2$, are consecutive terms of an arithmetic progression, we need only show that

$b - a = c - b$

$$\begin{aligned} b - a &= z^2 + xz + x^2 - (x^2 + xy + y^2) \\ &= z^2 + xz - xy - y^2 \\ &= z^2 + x(z - y) - y^2 \\ &= z^2 - y^2 + x(z - y) \\ &= (z - y)(z + y + x) \\ &= d(x + y + z) \end{aligned}$$

$$\begin{aligned} c - b &= y^2 + yz + z^2 - (z^2 + xz + x^2) \\ &= y^2 + yz - xz - x^2 \\ &= y^2 - x^2 + yz - xz \\ &= y^2 - x^2 + z(y - x) \\ &= (y - x)[y + x + z] \\ &= d(x + y + z) \end{aligned}$$

Hence, a, b, c, that is, $x^2 + xy + y^2$, $z^2 + xz + x^2$, $y^2 + yz + z^2$ are consecutive terms of an arithmetic progression.

79. Write: P_n: $a_n = a_1r^{n-1}$, $n \in N$

Proof: Part 1: Show that P_1 is true:

$$\begin{aligned} P_1\text{: } a_1 &= a_1r^{1-1} \\ &= a_1r^0 \\ &= a_1 \quad \text{True.} \end{aligned}$$

Part 2: Show that if P_k is true, then P_{k+1} is true:

Write out P_k and P_{k+1}.

P_k: $a_k = a_1r^{k-1}$

P_{k+1}: $a_{k+1} = a_1r^k$

We start with P_k:

$a_k = a_1r^{k-1}$

Multiply both sides by r:

$a_kr = a_1r^{k-1}r$

$a_kr = a_1r^k$

But, we are given that $a_{k+1} = ra_k = a_kr$.

Hence, $a_{k+1} = a_1r^k$

Thus, if P_k is true, then P_{k+1} is true.

Conclusion: P_n is true for all $n \in N$.

81. Given a, b, c, d, e, f is an arithmetic progression, let D be the common difference. Then, assuming $D \neq 0$:

$b = a + D$, $c = a + 2D$, $d = a + 3D$, $e = a + 4D$, $f = a + 5D$

From the solution to Exercise 7-1, problem 33, we know that there will be a solution of

$$\begin{aligned} ax + by &= c \\ dx + ey &= f \end{aligned}$$

if $ae - bd \neq 0$

In this case,
$$\begin{aligned} ae - bd &= a(a + 4D) - (a + 3D)(a + D) \\ &= a^2 + 4aD - (a^2 + 4aD + 3D^2) \\ &= -3D^2 \neq 0 \end{aligned}$$

Then, from the same problem, we have

$$\begin{aligned} x &= \frac{ce - bf}{ae - bd} = \frac{(a + 2D)(a + 4D) - (a + D)(a + 5D)}{-3D^2} \\ &= \frac{a^2 + 6aD + 8D^2 - (a^2 + 6aD + 5D^2)}{-3D^2} \\ &= \frac{3D^2}{-3D^2} \\ &= -1 \end{aligned}$$

$$y = \frac{af - dc}{ae - bd} = \frac{a(a + 5D) - (a + 3D)(a + 2D)}{-3D^2}$$
$$= \frac{a^2 + 5aD - (a^2 + 5aD + 6D^2)}{-3D^2}$$
$$= \frac{-6D^2}{-3D^2}$$
$$= 2$$

83. If $\frac{a_n}{a_{n-1}} = r$ and $a_n = a_{n-1} + d$, then

$$a_n = ra_{n-1} = a_{n-1} + d$$
$$ra_{n-1} - a_{n-1} = d$$
$$a_{n-1}(r - 1) = d$$

Thus, a_{n-1} is a constant for all terms of the sequence. Moreover, if all terms are the same constant, $d = 0$ and $r = 1$.

85. With each firm, the salaries form an arithmetic sequence. We are asked for the sum of fifteen terms, or S_{15}, given a_1 and d.

$$S_n = \frac{n}{2}[2a_1 + (n - 1)d\ ]$$
$$S_{15} = \frac{15}{2}[2a_1 + 14d\ ] = 15(a_1 + 7d)$$

Firm A: $a_1 = 25{,}000$ $d = 1{,}200$ $S_{15} = 15(25{,}000 + 7 \cdot 1{,}200) = \$501{,}000$

Firm B: $a_1 = 28{,}000$ $d = 800$ $S_{15} = 15(28{,}000 + 7 \cdot 800) = \$504{,}000$

87. We are asked for the sum of an infinite geometric series.

$a_1 = \$800{,}000$

$r = 0.8 \quad |r| \le 1,$

so the series has a sum,

$$S_\infty = \frac{a_1}{1 - r}$$
$$= \frac{\$800{,}000}{1 - 0.8}$$
$$= \frac{\$800{,}000}{0.2}$$
$$= \$4{,}000{,}000$$

89. After one year, $P(1 + r)$ is present. Hence, the geometric sequence has $a_1 = P(1 + r)$. The ratio is given as $(1 + r)$, hence

$$a_n = a_1(1 + r)^{n-1}$$
$$= P(1 + r)(1 + r)^{n-1}$$
$$= P(1 + r)^n$$ is the amount present after n years.

The time taken for P to double is represented by n years. We set $A = 2P$, $r = 0.06$, then solve

$$2P = P(1 + 0.06)^n$$
$$2 = (1.06)^n$$
$$\log 2 = n \log 1.06$$
$$n = \frac{\log 2}{\log 1.06}$$
$$\approx 12 \text{ years}$$

91. This involves an arithmetic sequence. Let d = increase in earnings each year.

$$a_1 = 7{,}000$$
$$a_{11} = 14{,}000$$
$$a_n = a_1 + (n - 1)d$$
$$14{,}000 = 7{,}000 + (11 - 1)d$$
$$d = \$700$$

The amount of money received over the 11 years is

$$S_{11} = a_1 + a_2 + \cdots + a_{11}$$
$$S_n = \frac{n}{2}(a_1 + a_n)$$
$$S_{11} = \frac{11}{2}(7{,}000 + 14{,}000) = \$115{,}500$$

93. We are asked for the sum of an infinite geometric series.

a_1 = number of revolutions in the first minute = 300

$r = \frac{2}{3}$ $\quad |r| < 1$, so the series has a sum,

$$S_\infty = \frac{a_1}{1 - r} = \frac{300}{1 - \frac{2}{3}} = 900 \text{ revolutions.}$$

95. This involves a geometric sequence. Let a_n = 2,000 calories. There are five stages: $n = 5$. We require a_1 on the assumption that $r = 20\% = \frac{1}{5}$.

$$a_n = a_1 r^{n-1}$$
$$2000 = a_1\left(\frac{1}{5}\right)^{5-1}$$
$$a_1 = 2{,}000 \cdot 5^4$$
$$= 1{,}250{,}000 \text{ calories}$$

97. We have an arithmetic sequence, 16, 48, 80, ⋯, with $a_1 = 16$, $d = 32$.

(A) This requires a_{11}: $a_n = a_1 + (n - 1)d$

$$a_{11} = 16 + (11 - 1)32$$
$$= 336 \text{ feet}$$

(B) This requires s_{11}: $s_n = \frac{n}{2}(a_1 + a_n)$

$$s_{11} = \frac{11}{2}(16 + 336)$$
$$= 1{,}936 \text{ feet}$$

(C) This requires s_t: $s_n = \frac{n}{2}[2a_1 + (n - 1)d]$

$$s_t = \frac{t}{2}[2a_1 + (t - 1)32]$$
$$= \frac{t}{2}[32 + (t - 1)32]$$
$$= \frac{t}{2} \cdot 32t$$
$$= 16t^2 \text{ feet}$$

99. This involves a geometric sequence. Let $a_1 = 2A_0$ = number present after 1 half-hour period. In t hours, $2t$ half-hours will have elapsed, hence, $n = 2t$; $r = 2$, since the number of bacteria doubles in each period.

$$a_n = a_1 r^{n-1}$$
$$a_{2t} = 2A_0 2^{2t-1}$$
$$= A_0 2^1 2^{2t-1}$$
$$= A_0 2^{2t}$$

101. If b_n is the brightness of an nth-magnitude star, we find r for the geometric progression $b_1, b_2, b_3, \ldots,$ given $b_1 = 100b_6$.

$$b_n = b_1 r^{n-1}$$
$$b_6 = b_1 r^{6-1}$$
$$b_6 = b_1 r^5$$
$$b_6 = 100 b_6 r^5$$
$$\frac{1}{100} = r^5$$
$$r = \sqrt[5]{0.01} = 10^{-0.4} = 0.398$$

103. This involves a geometric sequence. a_1 = amount of money on first square; $a_1 = \$0.01$, $r = 2$; a_n = amount of money on nth square.

$$a_n = a_1 r^{n-1}$$
$$a_{64} = 0.01 \cdot 2^{64-1}$$
$$= 9.22 \times 10^{16} \text{ dollars}$$

The amount of money on the whole board $= a_1 + a_2 + a_3 + \cdots + a_{64} = S_{64}$.

$$S_n = \frac{a_1 - ra_n}{1 - r}$$
$$S_{64} = \frac{0.01 - 2(9.223 \times 10^{16})}{1 - 2}$$
$$= 1.845 \times 10^{17} \text{ dollars}$$

105. This involves a geometric sequence. Let $a_1 = 15$ = pressure at sea level, $r = \frac{1}{10}$ = factor of decrease. We require a_5 = pressure after four 10-mile increases in altitude.

$$a_5 = 15\left(\frac{1}{10}\right)^{5-1}$$
$$= 0.0015 \text{ pounds per square inch.}$$

107. This involves an infinite geometric series. a_1 = perimeter of first triangle = 1, $r = \frac{1}{2}$, $|r| < 1$, so the series has a sum.

$$S_\infty = \frac{a_1}{1 - r} = \frac{1}{1 - \frac{1}{2}} = 2$$

109.

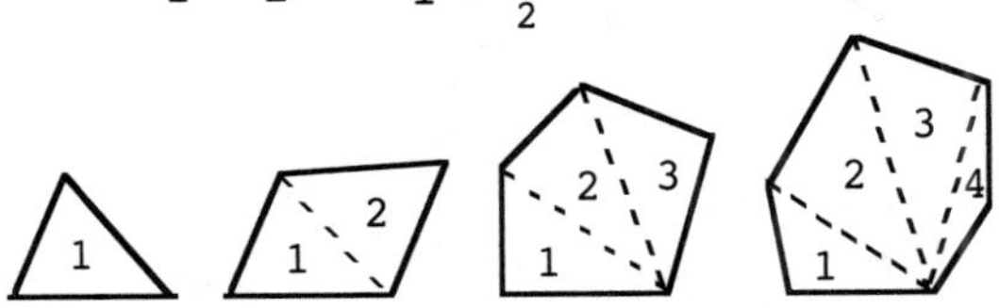

From the figure, and ordinary induction, it should be clear that the interior angles of an $n + 2$-sided polygon, $n = 1, 2, 3, \cdots$, are those of n triangles, that is, $180°n$. Thus, for the sequence of interior angles $\{a_n\}$,

$$a_n - a_{n-1} = 180°n - 180°(n - 1) = 180° = d.$$

A detailed proof by mathematical induction is omitted.

For a 21-sided polygon, we use $a_n = a_1 + (n - 1)d$ with $a_1 = 180°$, $d = 180°$, and $n + 2 = 21$, hence, $n = 19$.

$$a_{19} = 180° + (19 - 1)180° = 3{,}420°$$

Section 11-4

1. $9! = 362{,}880$

3. $11! = 39{,}916{,}800$

5. $\dfrac{11!}{8!} = \dfrac{11 \cdot 10 \cdot 9 \cdot 8!}{8!} = 990$

7. $\dfrac{5!}{2!\,3!} = \dfrac{5 \cdot 4 \cdot 3!}{2 \cdot 1 \cdot 3!} = 10$

9. $\dfrac{7!}{4!\,(7 - 4)\,!} = \dfrac{7!}{4!\,3!} = \dfrac{7 \cdot 6 \cdot 5 \cdot 4!}{4!\,3 \cdot 2 \cdot 1} = 35$

11. $\dfrac{7!}{7!\,(7 - 7)\,!} = \dfrac{7!}{7!\,0!} = \dfrac{7!}{7!\,(1)} = 1$

13. The alleged solution for number 5 is incorrect. The function "11 n Pr8" represents the permutations of 11 objects taken 8 at a time, which is calculated by

$$\frac{11!}{(11 - 8)\,!} = \frac{11!}{3!}$$

This is different from problem 5, which is 11!/8!.
The alleged solution for number 7 is incorrect also. The calculator interprets the input 5!/2!3! as $\dfrac{5!}{2!} \cdot 3!$. The correct calculation is 5!/(2!3!).
The solution for number 9 is correct.

15. $P_{8,2} = \dfrac{8!}{(8 - 2)\,!} = \dfrac{8!}{6!} = \dfrac{8 \cdot 7 \cdot 6!}{6!} = 56$

17. $C_{8,2} = \dfrac{8!}{2!\,(8 - 2)\,!} = \dfrac{8!}{2!\,6!} = \dfrac{8 \cdot 7 \cdot 6!}{2 \cdot 1 \cdot 6!} = \dfrac{56}{2} = 28$

19. $P_{8,6} = \dfrac{8!}{(8 - 6)\,!} = \dfrac{8!}{2!} = \dfrac{40{,}320}{2} = 20{,}160$

21. $C_{8,6} = \dfrac{8!}{6!\,(8 - 6)\,!} = \dfrac{8!}{6!\,2!} = \dfrac{8 \cdot 7 \cdot 6!}{6! \cdot 2 \cdot 1} = \dfrac{8 \cdot 7}{2 \cdot 1} = \dfrac{56}{2} = 28$

23. $P_{30,5} = \dfrac{30!}{(30 - 5)\,!} = \dfrac{30!}{25!} = \dfrac{30 \cdot 29 \cdot 28 \cdot 27 \cdot 26 \cdot 25!}{25!} = 30 \cdot 29 \cdot 28 \cdot 27 \cdot 26$
$= 17{,}100{,}720$

25. $C_{6,5} = \dfrac{6!}{5!\,(6 - 5)\,!} = \dfrac{6!}{5!\,1!} = \dfrac{6 \cdot 5!}{5! \cdot 1} = 6$

27. (A) Order is important. The selection is a permutation.
(B) Order is not important. The selection is a combination.

29. O_1: Selecting the color — N_1: 5 ways
O_2: Selecting the transmission — N_2: 3 ways
O_3: Selecting the interior — N_3: 4 ways
O_4: Selecting the engine — N_4: 2 ways
Applying the multiplication principle, there are $5 \cdot 3 \cdot 4 \cdot 2 = 120$ variations of the car.

31. Order is important here. We use permutations, selecting, in order, three horses out of ten:
$P_{10,3} = 10 \cdot 9 \cdot 8 = 720$ different finishes

33. For the subcommittee, order is not important. We use combinations, selecting three persons out of seven:
$C_{7,3} = \frac{7!}{3!\,(7-3)!} = \frac{7!}{3!\,4!} = \frac{7 \cdot 6 \cdot 5 \cdot 4!}{3 \cdot 2 \cdot 1 \cdot 4!} = 35$ subcommittees
In choosing a president, a vice-president, and a secretary, we can use permutations, or apply the multiplication principle.
O_1: Selecting the president N_1: 7 ways
O_2: Selecting the vice-president N_2: 6 ways (the president is not considered)
O_3: Selecting the secretary N_3: 5 ways (the president and vice-president are not considered)
Thus, there are $N_1 \cdot N_2 \cdot N_3 = 7 \cdot 6 \cdot 5$ $(= P_{7,3}) = 210$ ways.

35. For each game, we are selecting two teams out of ten to be opponents. Since the order of the opponents does not matter (this has nothing to do with the order in which the games might be played, which is not under discussion here), we use combinations.
$C_{10,2} = \frac{10!}{2!\,(10-2)!} = \frac{10!}{2!\,8!} = \frac{10 \cdot 9 \cdot 8!}{2 \cdot 1 \cdot 8!} = 45$ games

37.

	No letter can be repeated	Allowing letters to repeat
O_1: Selecting first letter	N_1: 6 ways	N_1: 6 ways
O_2: Selecting second letter	N_2: 5 ways	N_2: 6 ways
O_3: Selecting third letter	N_3: 4 ways	N_3: 6 ways
O_4: Selecting fourth letter	N_4: 3 ways	N_4: 6 ways
	$P_{6,4} = 6 \cdot 5 \cdot 4 \cdot 3 = 360$ possible code words	$6 \cdot 6 \cdot 6 \cdot 6 = 1{,}296$ possible code words

39.

	No digit can be repeated	Allowing digits to repeat
O_1: Selecting first digit	N_1: 10 ways	N_1: 10 ways
O_2: Selecting second digit	N_2: 9 ways	N_2: 10 ways
⋮	⋮	⋮
O_5: Selecting fifth digit	N_5: 6 ways	N_5: 10 ways
	$P_{10,5} = 10 \cdot 9 \cdot 8 \cdot 7 \cdot 6 =$ 30,240 lock combinations	$10 \cdot 10 \cdot 10 \cdot 10 \cdot 10 = 100{,}000$ lock combinations

41. We are selecting five cards out of the 13 hearts in the deck. The order is not important, so we use combinations.

$C_{13,5} = \frac{13!}{5!\,(13-5)!} = \frac{13!}{5!\,8!} = \frac{13 \cdot 12 \cdot 11 \cdot 10 \cdot 9 \cdot 8!}{5 \cdot 4 \cdot 3 \cdot 2 \cdot 1 \cdot 8!} = 1{,}287$

43.

	Repeats allowed	No repeats allowed
O_1: Selecting first letter	N_1: 26 ways	N_1: 26 ways
O_2: Selecting second letter	N_2: 26 ways	N_2: 25 ways
O_3: Selecting third letter	N_3: 26 ways	N_3: 24 ways
O_4: Selecting first digit	N_4: 10 ways	N_4: 10 ways
O_5: Selecting second digit	N_5: 10 ways	N_5: 9 ways
O_6: Selecting third digit	N_6: 10 ways	N_6: 8 ways
	$26\cdot 26\cdot 26\cdot 10\cdot 10\cdot 10 =$ 17,576,000 license plates	$26\cdot 25\cdot 24\cdot 10\cdot 9\cdot 8 =$ 11,232,000 license plates

45. O_1: Choosing 5 spades out of 13 possible (order is not important)
N_1: $C_{13,5}$
O_2: Choosing 2 hearts out of 13 possible (order is not important)
N_2: $C_{13,2}$
Using the multiplication principle, we have:

$$\text{Number of hands} = C_{13,5}\cdot C_{13,2} = \frac{13!}{5!(13-5)!}\cdot\frac{13!}{2!(13-2)!} = 1,287\cdot 78 = 100,386$$

47. O_1: Choosing 3 appetizers out of 8 possible (order is not important)
N_1: $C_{8,3}$
O_2: Choosing 4 main courses out of 10 possible (order is not important)
N_2: $C_{10,4}$
O_3: Choosing 2 desserts out of 7 possible (order is not important)
N_3: $C_{7,2}$
Using the multiplication principle, we have:

$$\text{Number of banquets} = C_{8,3}\cdot C_{10,4}\cdot C_{7,2} = \frac{8!}{3!(8-3)!}\cdot\frac{10!}{4!(10-4)!}\cdot\frac{7!}{2!(7-2)!} = 56\cdot 210\cdot 21 = 246,960$$

49. (A)

$P_{10,0} = 1$	$0! = 1$
$P_{10,1} = 10$	$1! = 1$
$P_{10,2} = 90$	$2! = 2$
$P_{10,3} = 720$	$3! = 6$
$P_{10,4} = 5,040$	$4! = 24$
$P_{10,5} = 30,240$	$5! = 120$
$P_{10,6} = 151,200$	$6! = 720$
$P_{10,7} = 604,800$	$7! = 5,040$
$P_{10,8} = 1,814,400$	$8! = 40,320$
$P_{10,9} = 3,628,800$	$9! = 362,880$
$P_{10,10} = 3,628,800$	$10! = 3,628,800$

Thus, $P_{10,r} \ge r!$ for $r = 0, 1, \ldots, 10$

(B) If $r = 0$, $P_{10,0} = 0! = 1$. If $r = 10$, $P_{10,10} = 10! = 3,628,800$

(C) $P_{n,r}$ and $r!$ are each the product of r consecutive integers, the largest of which is n for $P_{n,r}$ and r for $r!$. Thus if $r \le n$, $P_{n,r} \ge r!$

51. O_1: Choosing a left glove out of 12 possible
N_1: 12
O_2: Choosing a right glove out of all the right gloves that do *not* match the left glove already chosen
N_2: 11
Using the multiplication principle, we have:
Number of ways to mismatch gloves = $12 \cdot 11 = 132$.

53. (A) We are choosing 2 points out of the 8 to join by a chord. Order is not important.

$$C_{8,2} = \frac{8!}{2!(8-2)!} = 28 \text{ chords}$$

(B) No three of the points can be collinear, since no line intersects a circle at more than two points. Thus, we can select any three of the 8 to use as vertices of the triangle. Order is not important.

$$C_{8,3} = \frac{8!}{3!(8-3)!} = 56 \text{ triangles}$$

(C) We can select any four of the eight points to use as vertices of a quadrilateral. Order is not important.

$$C_{8,4} = \frac{8!}{4!(8-4)!} = 70 \text{ quadrilaterals}$$

55. To seat two people, we can seat one person, then the second person.

O_1: Seat the first person in any chair

N_1: 5 ways

O_2: Seat the second person in any remaining chair

N_2: 4 ways

Thus, applying the multiplication principle, we can seat two persons in $N_1 \cdot N_2 = 5 \cdot 4 = 20$ ways.

We can continue this reasoning for a third person.

O_3: Seat the third person in any of the three remaining chairs

N_3: 3 ways

Thus we can seat 3 persons in $5 \cdot 4 \cdot 3 = 60$ ways.

For a fourth person:

O_4: Seat the fourth person in any of the two remaining chairs

N_4: 2 ways

Thus we can seat 4 persons in $5 \cdot 4 \cdot 3 \cdot 2 = 120$ ways.

For a fifth person:

O_5: Seat the fifth person. There will be only one chair remaining.

N_5: 1 way

Thus we can seat 5 persons in $5 \cdot 4 \cdot 3 \cdot 2 \cdot 1 = 120$ ways

57. (A) Order is important, so we use permutations, selecting, in order, 5 persons out of 8:

$$P_{8,5} = 8\cdot7\cdot6\cdot5\cdot4 = 6,720 \text{ teams.}$$

(B) Order is not important, so we use combinations, selecting 5 persons out of 8

$$C_{8,5} = \frac{8!}{5!\,(8-5)!} = \frac{8!}{5!\,3!} = \frac{8\cdot7\cdot6\cdot5!}{5!\cdot3\cdot2\cdot1} = 56 \text{ teams}$$

(C) O_1: Selecting either Mike or Ken out of {Mike, Ken}

N_1: $C_{2,1}$

O_2: Selecting the 4 remaining players out of the 6 possibilities that do not include either Mike or Ken.

N_2: $C_{6,4}$

Using the multiplication principle, we have

$$\begin{aligned} N_1\cdot N_2 = C_{2,1}\cdot C_{6,4} &= \frac{2!}{1!\,(2-1)!}\cdot\frac{6!}{4!\,(6-4)!} \\ &= \frac{2!}{1!\,1!}\cdot\frac{6!}{4!\,2!} \\ &= 2\cdot15 \\ &= 30 \text{ teams} \end{aligned}$$

59. The number of ways to deal a hand containing exactly 1 king is computed as follows:

O_1: Choosing 1 king out of 4 possible (order is not important)

N_1: $C_{4,1}$

O_2: Choosing 4 cards out of 48 possible (order is not important)

N_2: $C_{48,4}$

Using the multiplication principle, we have

$$\begin{aligned} N_1\cdot N_2 = C_{4,1}\cdot C_{48,4} &= \frac{4!}{1!\,(4-1)!}\cdot\frac{48!}{4!\,(48-4)!} \\ &= \frac{4!}{1!\,3!}\cdot\frac{48!}{4!\,44!} \\ &= 4\cdot194,580 \\ &= 778,320 \text{ ways} \end{aligned}$$

The number of ways to deal a hand containing no hearts is

$$C_{39,5} = \frac{39!}{5!\,(39-5)!} = \frac{39!}{5!\,34!} = 575,757 \text{ ways}$$

Thus the hand that contains exactly one king is more likely.

Section 11-5

1. A suitable sample space is $S = \{1, 2, 3, 4, 5, 6\}$. Assigning the probability of $\frac{1}{6}$ to each simple event in S, we find that $E = \{1, 6\}$ has probability $\frac{1}{6} + \frac{1}{6} = \frac{1}{3}$.

3. Since there are 52 equally likely cards and 26 of them are red, the event of drawing a red card has probability $\frac{26}{52} = \frac{1}{2}$.

5. A suitable sample space is S = {HH, HT, TH, TT}. Assigning the probability of $\frac{1}{4}$ to each simple event in S, we find that E = {HH} has probability $\frac{1}{4}$.

7. A suitable sample space has the 36 elements shown in Example 2 of the text. Assigning the probability of $\frac{1}{36}$ to each simple event in this space, we find that the probability of $E = \{(1,1), (2,2), (3,3), (4,4), (5,5), (6,6)\}$ is $\frac{6}{36} = \frac{1}{6}$.

9. Since there are 52 equally likely cards and 40 of them are kings and 4 of them are queens, the event of drawing a king or queen has probability $\frac{4}{52} + \frac{4}{52} = \frac{2}{13}$.

11. A suitable sample space is S = {HHH, HHT, HTH, HTT, THH, THT, TTH, TTT}. Assigning the probability of $\frac{1}{8}$ to each simple event in S, we find that the probability of E = {HTT, THT, TTT} is $\frac{3}{8}$.

13. The occurrence of E is certain.

15. (A) Must be rejected. It violates condition 1; no probability can be negative.
(B) Must be rejected. It violates condition 2, since $P(R) + P(G) + P(Y) + P(B) \neq 1$.
(C) Is an acceptable probability assignment.

17. This is a compound event made up of the simple events R and Y.
$P(E) = P(R) + P(Y) = .26 + .30 = .56$

19. $P(E) \approx \frac{f(E)}{n} = \frac{25}{250} = .1$

21. $P(E) \approx \frac{f(E)}{n} = \frac{189}{420} = .45$

23. The sample space S is the set of all possible sequences of three digits with no repeats. $n(S) = P_{10,3} = 10\cdot 9\cdot 8 = 720$. The event is the simple event of one particular sequence. $n(E) = 1$. Then $P(E) = \frac{n(E)}{n(S)} = \frac{1}{720} \approx .0014$.

25. The sample space S is the set of all possible 5-card hands, chosen from a deck of 52 cards, $n(S) = C_{52,5}$. The event E is the set of all possible 5-card hands that are all black, thus, are chosen from the 26 black cards. $n(E) = C_{26,5}$.

$$P(E) = \frac{n(E)}{n(S)} = \frac{C_{26,5}}{C_{52,5}} = \frac{26!}{5!\,(26-5)!} \div \frac{52!}{5!\,(52-5)!} \approx .025$$

27. The sample space S is the set of all possible 5-card hands, chosen from a deck of 52 cards. $n(S) = C_{52,5}$. The event E is the set of all possible 5-card hands that are all face cards, thus, are chosen from the 4 face cards in each of 4 suits, or 16 face cards. $n(E) = C_{16,5}$.

$$P(E) = \frac{n(E)}{n(S)} = \frac{C_{16,5}}{C_{52,5}} = \frac{16!}{5!\,(16-5)!} \div \frac{52!}{5!\,(52-5)!} \approx .0017$$

29. Yes. There are four aces in a deck so a 5-card hand could have anywhere from zero to four aces. But it is not an equally likely sample space. Drawing four aces in a 5-card hand is far less likely than drawing none.

31. The sample space S is the set of all possible four digit numbers less than 5,000 formed from the digits 1, 3, 5, 7, 9.

$n(S) = N_1 \cdot N_2 \cdot N_3 \cdot N_4$

where

$N_1 = 2$ (digits 1 or 3 only for the first digit since the number is less than 5,000)

$N_2 = N_3 = N_4 = 5$ (digits 1, 3, 5, 7, 9)

$n(S) = 2 \cdot 5 \cdot 5 \cdot 5 = 250$

The event E is the set of those numbers in S which are divisible by 5. These must end in 5.

$n(E) = N_1 \cdot N_2 \cdot N_3 \cdot N_4'$

where

$N_4' = 1$ (digit 5 only)

$n(E) = 2 \cdot 5 \cdot 5 \cdot 1 = 50$

$$P(E) = \frac{n(E)}{n(S)} = \frac{50}{250} = .2$$

33. The sample space S is the set of all arrangements of the five notes in the five envelopes. $n(S) = P_{5,5} = 5! = 120$. The event E is the one correct arrangement.

$n(E) = 1$. $\quad P(E) = \dfrac{n(E)}{n(S)} = \dfrac{1}{120} \approx .008$

In problems 35—49, using Figure 1 of the text, we have that S = the set of all possible ordered pairs of dots showing, $n(S) = 36$.

35. The sum of the dots can be 2 in exactly one way. $E = \{(1, 1)\}$

$n(E) = 1$

$$P(E) = \frac{n(E)}{n(S)} = \frac{1}{36}$$

37. The sum of the dots can be 6 in 5 ways:

$E = \{(5, 1), (4, 2), (3, 3), (2, 4), (1, 5)\}$

$n(E) = 5$

$$P(E) = \frac{n(E)}{n(S)} = \frac{5}{36}$$

39. The sum of the dots can be less than 5 in 6 ways. (upper left-hand corner of figure) $n(E) = 6$.

$$P(E) = \frac{n(E)}{n(S)} = \frac{6}{36} = \frac{1}{6}$$

41. The sum of the dots can be 7 in 6 ways and 11 in 2 ways. Thus

$n(E) = 36 - (6 + 2) = 28$

$$P(E) = \frac{n(E)}{n(S)} = \frac{28}{36} = \frac{7}{9}$$

43. The sum of the dots cannot be 1. Thus $P(E) = 0$.

45. The sum of the dots is divisible by 3 if the sum is 3, 6, 9, or 12. 3 can occur in 2 ways, 6 in 5 ways, 9 in 4 ways, and 12 in 1 way. Thus

$n(E) = 2 + 5 + 4 + 1 = 12$.

$$P(E) = \frac{n(E)}{n(S)} = \frac{12}{36} = \frac{1}{3}$$

47. The sum of the dots can be 7 in 6 ways and 11 in 2 ways. Thus $n(E) = 6 + 2 = 8$.

$$P(E) = \frac{n(E)}{n(S)} = \frac{8}{36} = \frac{2}{9}$$

49. The sum of the dots will *not* be divisible by 2 or 3 if the sum is 5, 7, or 11. 5 can occur in 4 ways, 7 in 6 ways, and 11 in 2 ways. Otherwise the sum will be divisible by 2 or 3. Thus $n(E) = 36 - (4 + 6 + 2) = 36 - 12 = 24$.

$$P(E) = \frac{n(E)}{n(S)} = \frac{24}{36} = \frac{2}{3}$$

51. The sample space is the set of events consisting of choosing a person with birthday on day 1 through day 365 of the year. (February 29 is neglected.) $S = \{1, 2, 3, \dots, 365\}$

Since each of these days is essentially equally likely, $P(e_i) = \frac{1}{365}$.

53. (A) Use $P(E) \approx \frac{f(E)}{n}$ with $n = 500$.

$P(2) = \frac{11}{500} = .022$ $\quad P(3) = \frac{35}{500} = .07$ $\quad P(4) = \frac{44}{500} = .088$

$P(5) = \frac{50}{500} = .1$ $\quad P(6) = \frac{71}{500} = .142$ $\quad P(7) = \frac{89}{500} = .178$

$P(8) = \frac{72}{500} = .144$ $\quad P(9) = \frac{52}{500} = .104$ $\quad P(10) = \frac{36}{500} = .072$

$P(11) = \frac{26}{500} = .052$ $\quad P(12) = \frac{14}{500} = .028$

(B) Use Figure 1 of the text, with S = the set of all possible ordered pairs of dots showing, $n(S) = 36$.

$P(2) = \frac{1}{36}$ $\quad P(3) = \frac{2}{36}$ $\quad P(4) = \frac{3}{36}$ $\quad P(5) = \frac{4}{36}$

$P(6) = \frac{5}{36}$ $\quad P(7) = \frac{6}{36}$ $\quad P(8) = \frac{5}{36}$ $\quad P(9) = \frac{4}{36}$

$P(10) = \frac{3}{36}$ $\quad P(11) = \frac{2}{36}$ $\quad P(12) = \frac{1}{36}$

(C) The expected frequency of the occurrence of E is $n \cdot P(E)$. Thus, the expected frequency of the occurrence of sum 2 is $500 \cdot P(2) = 500 \cdot \frac{1}{36}$ or 13.9.

The expected frequency of the occurrence of sum 3 is $500 \cdot P(3) = 500 \cdot \frac{2}{36} = 27.8$.

The expected frequency of the occurrence of sum 4 is $500 \cdot P(4) = 500 \cdot \frac{3}{36} = 41.7$.

The expected frequency of the occurrence of sum 5 is $500 \cdot P(5) = 500 \cdot \frac{4}{36} = 55.6$.

The expected frequency of the occurrence of sum 6 is $500 \cdot P(6) = 500 \cdot \frac{5}{36} = 69.4$.

The expected frequency of the occurrence of sum 7 is $500 \cdot P(7) = 500 \cdot \frac{6}{36} = 83.3$.

The expected frequency of the occurrence of sum 8 is $500 \cdot P(8) = 500 \cdot \frac{5}{36} = 69.4$.

The expected frequency of the occurrence of sum 9 is $500 \cdot P(9) = 500 \cdot \frac{4}{36} = 55.6$.

The expected frequency of the occurrence of sum 10 is $500 \cdot P(10) = 500 \cdot \frac{3}{36} = 41.7$.

The expected frequency of the occurrence of sum 11 is $500 \cdot P(11) = 500 \cdot \frac{2}{36} = 27.8$.

The expected frequency of the occurrence of sum 12 is $500 \cdot P(12) = 500 \cdot \frac{1}{36} = 13.9$.

(D) Use the command *r* and Int (*i*, *k*, *n*). Let 1 to 6 represent the possible outcomes in tossing one die; then rand Int (1, 6, 500) will give a list of 500 possible outcomes in tossing one die. rand Int (1, 6, 500) + rand Int (1, 6, 500) will give a list of 500 possible outcomes in tossing 2 dice. A possible result is outlined here, but the student's result will be different (with a high probability!)

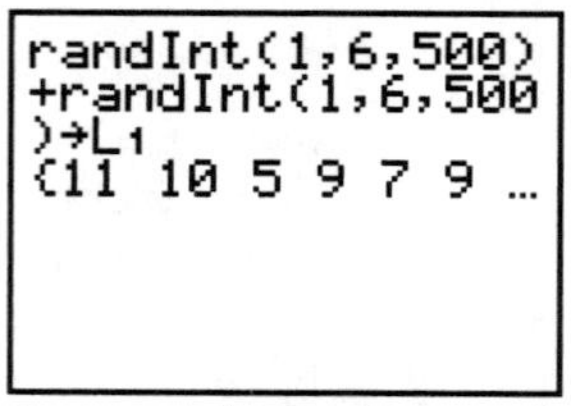

Generating the random numbers.

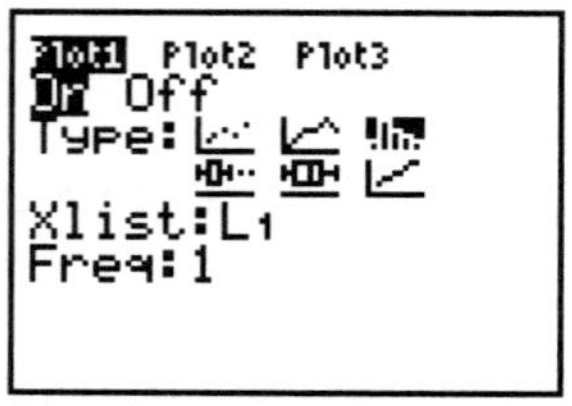

Setting up the histogram.

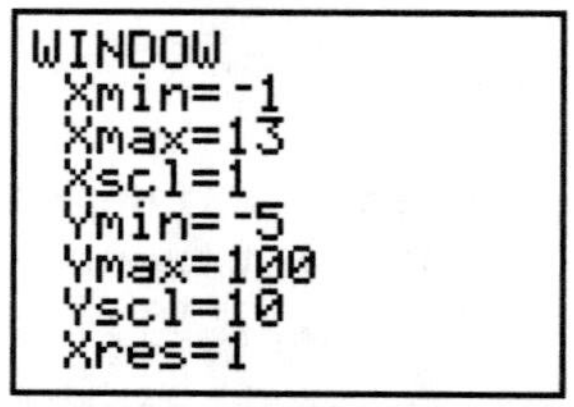

Selecting the window variables.

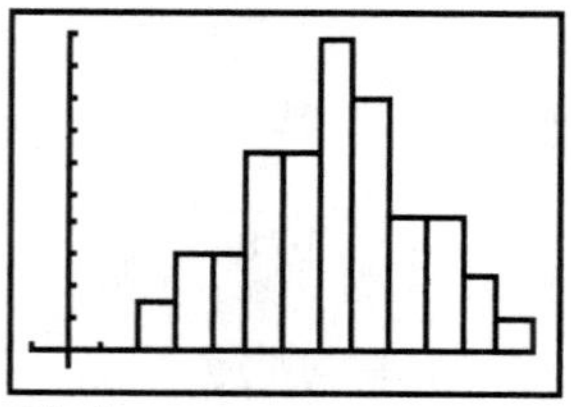
Histogram

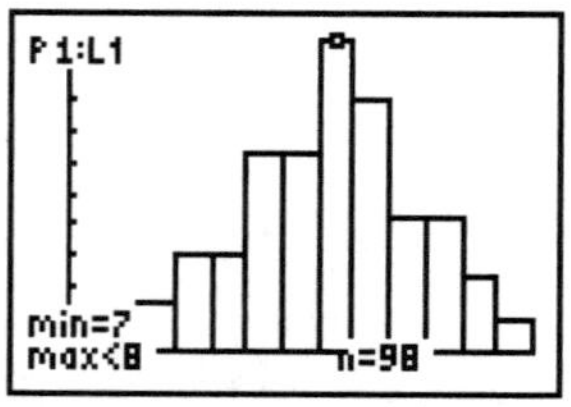

Use trace to display any result.

55. (A) It is possible, but very unlikely, to get 29 heads in 30 flips of a coin. If one thinks of the 30 flips as a sequence of words with 30 letters chosen from $\{H, T\}$, then there are 2 ways to choose each of the letters and therefore 2^{30} such words. Only 30 of these words have exactly 29 *H*'s because only 30 have exactly one *T*. Thus the probability of getting 29 heads in 30 flips of a fair coin is

$$P(E) = \frac{n(E)}{n(S)} = \frac{30}{2^{30}} \approx 2.8 \times 10^{-8}$$

(B) There is a strong suspicion that the coin is unfair, because this event is unlikely, although not as unlikely as in part A. If the coin is unfair, the empirical probabilities are

$$P(H) \approx \frac{f(H)}{n} = \frac{42}{50} = .84 \qquad P(T) \approx \frac{f(T)}{n} = \frac{8}{50} = .16$$

57. The sample space for this experiment = $\{HTH, HHH, THH, TTH\}$ assuming that the third coin is the one with a head on both sides. Thus $n(S) = 4$. E is the event $\{TTH\}$. Thus $n(E) = 1$.

$$P(E) = \frac{n(E)}{n(S)} = \frac{1}{4}$$

59. See problem 57. $n(S) = 4$. E is the event $\{HHH\}$. Thus $n(E) = 1$

$$P(E) = \frac{n(E)}{n(S)} = \frac{1}{4}$$

61. See problem 57. $n(S) = 4$. E is the event $\{HTH, HHH, THH\}$. Thus $n(E) = 3$

$$P(E) = \frac{n(E)}{n(S)} = \frac{3}{4}$$

63. Since it is equally likely that each die shows a 1, a 2 or a 3, we can draw the following figure to represent S, the set of all possibilities:

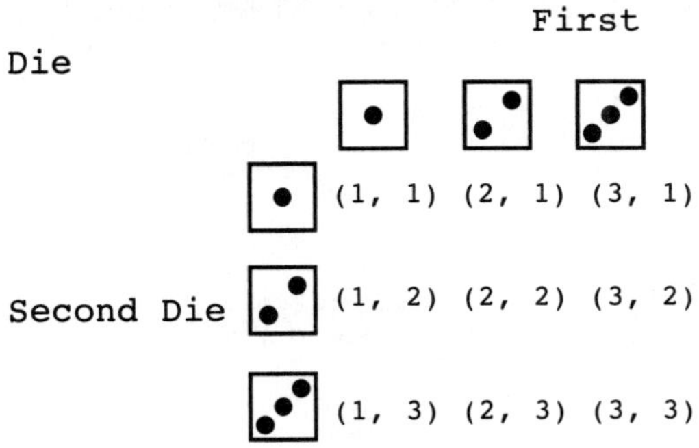

$n(S) = 9$
The sum of the dots can be 2 in exactly one way. $n(E) = 1$

$P(E) = \frac{n(E)}{n(S)} = \frac{1}{9}$

65. See figure, problem 63, $n(S) = 9$. The sum of the dots can be 4 in 3 ways:
$E = \{(1, 3), (2, 2), (3, 1)\}$
$n(E) = 3$

$P(E) = \frac{n(E)}{n(S)} = \frac{3}{9} = \frac{1}{3}$

67. See figure, problem 63. $n(S) = 9$. The sum of the dots can be 6 in exactly one way: $E = \{(3, 3)\}$. $n(E) = 1$

$P(E) = \frac{n(E)}{n(S)} = \frac{1}{9}$

69. See figure, problem 63. $n(S) = 9$. The sum of the dots will be odd if the sum is 3 or 5.
$E = \{(1, 2), (2, 1)(2, 3), (3, 2)\}$ $n(E) = 4$

$P(E) = \frac{n(E)}{n(S)} = \frac{4}{9}$

71. The sample space S is the set of all possible 5-card hands, chosen from a deck of 52 cards. $n(S) = C_{52,5} = 2{,}598{,}960$. The event E is the set of all possible 5-card hands that are all jacks through aces, thus, are chosen from the 16 cards (4 jacks, 4 queens, 4 kings, 4 aces) of this type.
$n(E) = C_{16,5} = 4{,}368$.

$P(E) = \frac{n(E)}{n(S)} = \frac{C_{16,5}}{C_{52,5}} = \frac{4,368}{2,598,960} \approx .00168$

73. The sample space S is the set of all possible 5-card hands chosen from a deck of 52 cards. $n(S) = C_{52,5}$. The event E is the set of all possible 5-card hands of the type $AAAAx$ where x is any of the $52 - 4 = 48$ cards that are not aces.
$n(E) = 48$

$P(E) = \frac{n(E)}{n(S)} = \frac{48}{C_{52,5}} = \frac{48}{2,598,960} \approx .000\,0185$

75. The sample space S is the set of all possible 5-card hands, chosen from a deck of 52 cards. $n(S) = C_{52,5}$. The event E is the set consisting of the cards ace, king, queen, jack, ten, in each of the 4 suits, abbreviated:
{club AKQJ10, diamond AKQJ10, heart AKQJ10, spade AKQJ10}. Thus $n(E) = 4$.

$P(E) = \frac{n(E)}{n(S)} = \frac{4}{C_{52,5}} = \frac{4}{2,598,960} \approx .000\,0015$

77. The sample space S is the set of all possible 5-card hands, chosen from a deck of 52 cards. $n(S) = C_{52,5}$. The event E is the set of all possible 5-card hands chosen as follows:

O_1: Choose 2 aces out of the set of 4 aces
N_1: $C_{4,2}$
O_2: Choose 3 queens out of the set of 4 queens
N_2: $C_{4,3}$
Thus, applying the multiplication principle, $n(E) = N_1N_2 = C_{4,2}C_{4,3}$

$$P(E) = \frac{n(E)}{n(S)} = \frac{C_{4,2} \cdot C_{4,3}}{C_{52,5}} = \frac{6 \cdot 4}{2,598,960} = .000\,009$$

79. (A) $P \text{ (this event)} = \dfrac{15}{1,000} = .015$

(B) $P \text{ (this event)} = \dfrac{130 + 80 + 12}{1,000} = .222$

(C) $P \text{ (this event)} = \dfrac{30 + 32 + +28 + 25 + 20 + 21 + 12 + 1}{1,000} = .169$

(D) $P \text{ (this event)} = 1 - P \text{ (owning zero television sets)}$

$$= 1 - \frac{2 + 10 + 30}{1,000} = .958$$

Section 11-6

For Problems 1—10, the first nine lines of Pascal's triangle are as follows:

```
0                 1
1               1   1
2             1   2   1
3           1   3   3   1
4         1   4   6   4   1
5       1   5  10  10   5   1
6     1   6  15  20  15   6   1
7   1   7  21  35  35  21   7   1
8 1   8  28  56  70  56  28   8   1
```

1. $\binom{6}{2}$ is the third entry in line 6 (start counting from 0) of the triangle, 15.

3. $\binom{7}{4}$ is the fifth entry in line 7 (start counting from 0) of the triangle, 35.

5. $\binom{8}{5}$ is the sixth entry in line 8 (start counting from 0) of the triangle, 56.

7. $C_{7,1}$ is the second entry in line 7 (start counting from 0) of the triangle, 7.

9. $C_{5,4}$ is the fifth entry in line 5 (start counting from 0) of the triangle, 5.

11. $\binom{26}{2} = \frac{26!}{2!(26-2)!} = \frac{26!}{2!\,24!} = \frac{26 \cdot 25 \cdot 24!}{2 \cdot 1 \cdot 24!} = \frac{26 \cdot 25}{2 \cdot 1} = 325$

13. $\binom{10}{6} = \frac{10!}{6!(10-6)!} = \frac{10!}{6!\,4!} = \frac{3,628,800}{720 \cdot 24} = 210$

15. $\binom{10}{5} = \frac{10!}{5!(10-5)!} = \frac{10!}{5!\,5!} = \frac{3,628,800}{120 \cdot 120} = 252$

17. $C_{52,5} = \frac{52!}{5!(52-5)!} = \frac{52!}{5!\,47!} = \frac{52 \cdot 51 \cdot 50 \cdot 49 \cdot 48 \cdot 47!}{5 \cdot 4 \cdot 3 \cdot 2 \cdot 1 \cdot 47!} = \frac{52 \cdot 51 \cdot 50 \cdot 49 \cdot 48}{5 \cdot 4 \cdot 3 \cdot 2 \cdot 1}$
$= 2,598,960$

19. $C_{13,9} = \frac{13!}{9!(13-9)!} = \frac{13!}{9!\,4!} = \frac{13 \cdot 12 \cdot 11 \cdot 10 \cdot 9!}{9!\,4 \cdot 3 \cdot 2 \cdot 1} = \frac{13 \cdot 12 \cdot 11 \cdot 10}{4 \cdot 3 \cdot 2 \cdot 1} = 715$

21. $(m+n)^3 = \sum_{k=0}^{3} \binom{3}{k} m^{3-k} n^k$
$= \binom{3}{0} m^3 + \binom{3}{1} m^2 n + \binom{3}{2} mn^2 + \binom{3}{3} n^3$
$= m^3 + 3m^2 n + 3mn^2 + n^3$

23. $(2x - 3y)^3 = [2x + (-3y)]^3$
$= \sum_{k=0}^{3} \binom{3}{k} (2x)^{3-k} (-3y)^k$
$= \binom{3}{0}(2x)^3 + \binom{3}{1}(2x)^2(-3y)^1 + \binom{3}{2}(2x)(-3y)^2 + \binom{3}{3}(-3y)^3$
$= 8x^3 + 3(4x^2)(-3y) + 3(2x)(9y^2) + (-27y^3)$
$= 8x^3 - 36x^2 y + 54xy^2 - 27y^3$

25. $(x-2)^4 = [x + (-2)]^4 = \sum_{k=0}^{4} \binom{4}{k} x^{4-k} (-2)^k$
$= \binom{4}{0} x^4 + \binom{4}{1} x^3 (-2)^1 + \binom{4}{2} x^2 (-2)^2 + \binom{4}{3} x(-2)^3 + \binom{4}{4} (-2)^4$
$= x^4 - 8x^3 + 24x^2 - 32x + 16$

27. $(m + 3n)^4 = \sum_{k=0}^{4} \binom{4}{k} m^{4-k} (3n)^k$
$= \binom{4}{0} m^4 + \binom{4}{1} m^3 (3n)^1 + \binom{4}{2} m^2 (3n)^2 + \binom{4}{3} m(3n)^3 + \binom{4}{4} (3n)^4$
$= m^4 + 12m^3 n + 54m^2 n^2 + 108mn^3 + 81n^4$

29. $(2x - y)^5 = [2x + (-y)]^5 = \sum_{k=0}^{5} \binom{5}{k} (2x)^{5-k} (-y)^k$
$= \binom{5}{0}(2x)^5 + \binom{5}{1}(2x)^4(-y)^1 + \binom{5}{2}(2x)^3(-y)^2 + \binom{5}{3}(2x)^2(-y)^3$
$+ \binom{5}{4}(2x)(-y)^4 + \binom{5}{5}(-y)^5$
$= 32x^5 - 80x^4 y + 80x^3 y^2 - 40x^2 y^3 + 10xy^4 - y^5$

31. $(m + 2n)^6 = \sum_{k=0}^{6} \binom{6}{k} m^{6-k}(2n)^k$

$$= \binom{6}{0}m^6 + \binom{6}{1}m^5(2n)^1 + \binom{6}{2}m^4(2n)^2 + \binom{6}{3}m^3(2n)^3 + \binom{6}{4}m^2(2n)^4 + \binom{6}{5}m(2n)^5 + \binom{6}{6}(2n)^6$$

$$= m^6 + 12m^5n + 60m^4n^2 + 160m^3n^3 + 240m^2n^4 + 192mn^5 + 64n^6$$

33. In the expansion

$$(x + 1)^7 = \sum_{k=0}^{7} \binom{7}{k} x^{7-k}1^k$$

the exponent of x is 4 when $k = 3$. Thus the term containing x^4 is

$$\binom{7}{3}x^4 1^3 = 35x^4$$

35. In the expansion

$$(2x - 1)^{11} = [2x + (-1)]^{11} = \sum_{k=0}^{11} \binom{11}{k}(2x)^{11-k}(-1)^k$$

the exponent of x is 6 when $k = 5$. Thus the term containing x^6 is

$$\binom{11}{5}(2x)^6(-1)^5 = (462)(64x^6)(-1) = -29{,}568x^6$$

37. In the expansion

$$(2x + 3)^{18} = \sum_{k=0}^{18} \binom{18}{k}(2x)^{18-k}(3)^k$$

the exponent of x is 14 when $k = 4$. Thus the term containing x^{14} is

$$\binom{18}{4}(2x)^{14}(3)^4 = 3{,}060(2^{14}x^{14})(3^4) = 4{,}060{,}938{,}240x^{14}$$

39. In the expansion

$$(x^2 - 1)^6 = \sum_{k=0}^{6} \binom{6}{k}(x^2)^{6-k}(-1)^k = \sum_{k=0}^{6} \binom{6}{k}x^{12-2k}(-1)^k$$

the exponent of x, $12 - 2k$, is 8, when $12 - 2k = 8$ and $2k = 4$, $k = 2$. Thus the term containing x^8 is

$$\binom{6}{2}x^8(-1)^2 = 15x^8$$

41. In the expansion

$$(x^2 + 1)^9 = \sum_{k=0}^{9} \binom{9}{k}(x^2)^{9-k}(1)^k = \sum_{k=0}^{9} \binom{9}{k}x^{18-2k}(1)^k$$

the exponent of x, $18 - 2k$, is even and can never equal 11.

43. In the expansion of $(a + b)^n$, the exponent of b in the rth term is $r - 1$ and the exponent of a is $n - (r - 1)$. Here, $r = 7$, $n = 15$.

$$\text{Seventh term} = \binom{15}{6}u^9v^6$$
$$= \frac{15!}{9!\,6!}u^9v^6$$
$$= \frac{15 \cdot 14 \cdot 13 \cdot 12 \cdot 11 \cdot 10}{6 \cdot 5 \cdot 4 \cdot 3 \cdot 2 \cdot 1}u^9v^6$$
$$= 5{,}005u^9v^6$$

45. In the expansion of $(a + b)^n$, the exponent of b in the rth term is $r - 1$ and the exponent of a is $n - (r - 1)$. Here, $r = 11$, $n = 12$.

$$\text{Eleventh term} = \binom{12}{10}(2m)^2n^{10}$$
$$= \frac{12!}{10!\,2!}4m^2n^{10}$$
$$= \frac{12 \cdot 11}{2 \cdot 1}4m^2n^{10}$$
$$= 264m^2n^{10}$$

47. In the expansion of $(a + b)^n$, the exponent of b in the rth term is $r - 1$ and the exponent of a is $n - (r - 1)$. Here, $r = 7$, $n = 12$.

$$\text{Seventh term} = \binom{12}{6}\left(\frac{w}{2}\right)^6(-2)^6$$
$$= \frac{12!}{6!\,6!}\,\frac{w^6}{2^6}\,2^6$$
$$= \frac{12 \cdot 11 \cdot 10 \cdot 9 \cdot 8 \cdot 7 \cdot 6!}{6 \cdot 5 \cdot 4 \cdot 3 \cdot 2 \cdot 1 \cdot 6!}w^6$$
$$= 924w^6$$

49. In the expansion of $(a + b)^n$, the exponent of b in the rth term is $r - 1$ and the exponent of a is $n - (r - 1)$. Here, $r = 6$, $n = 8$.

$$\text{Sixth term} = \binom{8}{5}(3x)^3(-2y)^5$$
$$= \frac{8!}{5!\,3!}(3x)^3(-2y)^5$$
$$= \frac{8 \cdot 7 \cdot 6 \cdot 5!}{5! \cdot 3 \cdot 2 \cdot 1}(27x^3)(-32y^5)$$
$$= -(56)(27)(32)x^3y^5$$
$$= -48{,}384x^3y^5$$

51. $$\frac{f(x+h) - f(x)}{h} = \frac{(x+h)^3 - x^3}{h} = \frac{x^3 + 3x^2h + 3xh^2 + h^3 - x^3}{h}$$
$$= \frac{3x^2h + 3xh^2 + h^3}{h} = \frac{h(3x^2 + 3xh + h^2)}{h} = 3x^2 + 3xh + h^2$$

As h approaches 0, this quantity approaches $3x^2$.

53. $\dfrac{f(x + h) - f(x)}{h} = \dfrac{(x + h)^5 - x^5}{h} = \dfrac{x^5 + 5x^4h + 10x^3h^2 + 10x^2h^3 + 5xh^4 + h^5 - x^5}{h}$

$$= \frac{5x^4h + 10x^3h^2 + 10x^2h^3 + 5xh^4 + h^5}{h}$$

$$= \frac{h(5x^4 + 10x^3h + 10x^2h^2 + 5xh^3 + h^4)}{h}$$

$$= 5x^4 + 10x^3h + 10x^2h^2 + 5xh^3 + h^4$$

As h approaches 0, this approaches $5x^4$.

55. Graphing the sequence, we obtain

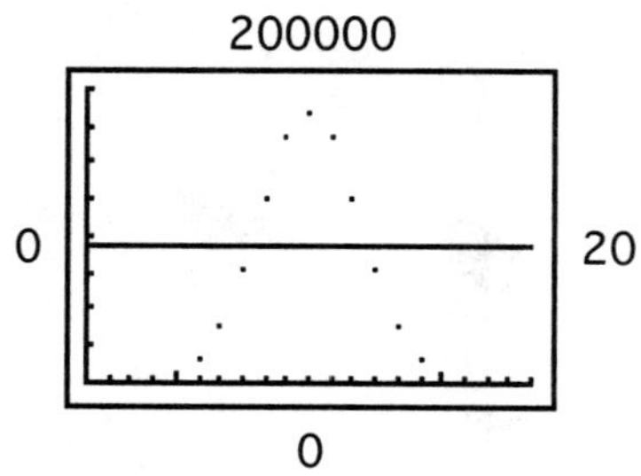

The solid line indicates $\frac{1}{2}(184{,}756) = 92{,}378$, that is, half the largest term. The graph shows five terms greater than half the largest term. The table display confirms this:

n	$u(n)$
7	77520
8	125970
9	167960
10	184756
11	167960
12	125970
13	77520

$u(n)$=20 nCr n

$\binom{20}{8} = 125{,}970$ $\binom{20}{9} = 167{,}960$ $\binom{20}{10} = 184{,}756$ $\binom{20}{11} = 167{,}960$ $\binom{20}{12} = 125{,}970$

are all greater than 92,378, but the other terms are less than 92,378.

57. (A) Graphing the sequence, we obtain the graph at the right.

From the table display we find the largest term to be $\binom{10}{4}(.6)^6(.4)^4 =$ 0.251 as displayed in the graph.

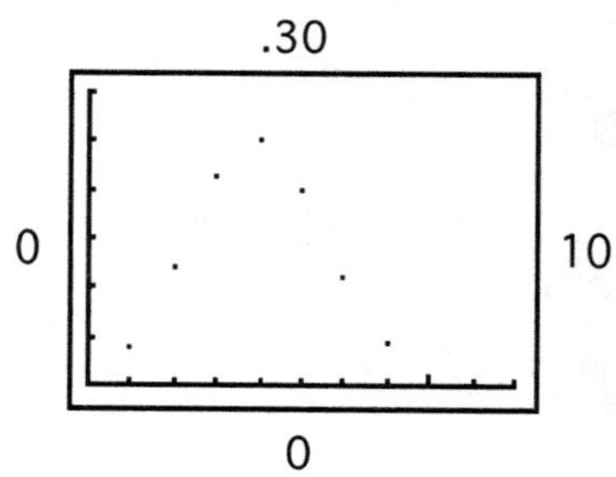

n	$u(n)$
0	.00605
1	.04031
2	.12093
3	.21499
4	.25082
5	.20066
6	.11148

n=4

(B) According to the binomial formula,

$$\sum_{k=0}^{n} \binom{n}{k} a^{n-k} b^k = (a + b)^n$$

Thus, $a_0 + a_1 + a_2 + \cdots + a_{10} = \sum_{k=0}^{10} \binom{n}{k} (0.6)^{10-k}(0.4)^k = (0.6 + 0.4)^{10} = 1^{10} = 1$

59. $(1.01)^{10} = (1 + 0.01)^{10} = \sum_{k=0}^{10} \binom{10}{k} 1^{10-k}(0.01)^k,\ 1^{10-k} = 1$

$$= \sum_{k=0}^{10} \binom{10}{k} (0.01)^k$$

$$= \binom{10}{0} + \binom{10}{1}(0.01)^1 + \binom{10}{2}(0.01)^2 + \binom{10}{3}(0.01)^3 + \binom{10}{4}(0.01)^4 + \binom{10}{5}(0.01)^5 + \binom{10}{6}(0.01)^6 + \binom{10}{7}(0.01)^7 + \binom{10}{8}(0.01)^8 + \binom{10}{9}(0.01)^9 + \binom{10}{10}(0.01)^{10}$$

$= 1 + 0.1 + 0.0045 + 0.00012 + 0.0000021 +$ (other terms with no effect in fourth decimal place)

$= 1.1046$

61. $\binom{n}{r} = \dfrac{n!}{r!(n - r)!} = \dfrac{n!}{(n - r)!r!} = \dfrac{n!}{(n - r)![n - (n - r)]!} = \binom{n}{n - r}$

63. $\binom{k}{r - 1} + \binom{k}{r} = \dfrac{k!}{(r - 1)!(k - r + 1)!} + \dfrac{k!}{r!(k - r)!}$

$$= \frac{rk! + (k - r + 1)k!}{r!(k - r + 1)!} = \frac{(r + k - r + 1)k!}{r!(k - r + 1)!}$$

$$= \frac{(k + 1)k!}{r!(k - r + 1)!} = \frac{(k + 1)!}{r!(k - r + 1)!}$$

$$= \binom{k + 1}{r}$$

65. $\binom{k}{k} = \dfrac{k!}{k!(k - k)!} = 1 = \dfrac{(k + 1)!}{(k + 1)![(k + 1) - (k + 1)]!} = \binom{k + 1}{k + 1}$

67. $2^n = (1 + 1)^n = \sum_{k=0}^{n} \binom{n}{k} 1^{n-k}1^k,\ 1^{n-k} = 1^k = 1$

$$= \sum_{k=0}^{n} \binom{n}{k}$$

$$= \binom{n}{0} + \binom{n}{1} + \binom{n}{2} + \cdots + \binom{n}{n}$$

CHAPTER 11 REVIEW

1. (A) Since $\dfrac{-8}{16} = \dfrac{4}{-8} = -\dfrac{1}{2}$, this could start a geometric sequence.

(B) Since $7 - 5 = 9 - 7 = 2$, this could start an arithmetic sequence.

(C) Since $-5 - (-8) = -2 - (-5) = 3$, this could start an arithmetic sequence.

(D) Since $\dfrac{3}{2} \neq \dfrac{5}{3}$ and $3 - 2 \neq 5 - 3$, this could start neither an arithmetic nor a geometric sequence.

(E) Since $\dfrac{2}{-1} = \dfrac{-4}{2} = -2$, this could start a geometric sequence. *(11-1, 11-3)*

2. $a_n = 2n + 3$

(A) $a_1 = 2 \cdot 1 + 3 = 5$
$a_2 = 2 \cdot 2 + 3 = 7$
$a_3 = 2 \cdot 3 + 3 = 9$
$a_4 = 2 \cdot 4 + 3 = 11$

(B) This is an arithmetic sequence with $d = 2$. Hence

$$a_n = a_1 + (n - 1)d$$
$$a_{10} = 5 + (10 - 1)2 = 23$$

(C) $$S_n = \frac{n}{2}(a_1 + a_n)$$
$$S_{10} = \frac{10}{2}(5 + 23) = 140$$

(11-1, 11-3)

3. $a_n = 32\left(\frac{1}{2}\right)^n$

(A) $a_1 = 32\left(\frac{1}{2}\right)^1 = 16$
$a_2 = 32\left(\frac{1}{2}\right)^2 = 8$
$a_3 = 32\left(\frac{1}{2}\right)^3 = 4$
$a_4 = 32\left(\frac{1}{2}\right)^4 = 2$

(B) This is a geometric sequence with $r = \frac{1}{2}$. Hence

$$a_n = a_1 r^{n-1}$$
$$a_{10} = 16\left(\frac{1}{2}\right)^{10-1} = \frac{1}{32}$$

(C) $$S_n = \frac{a_1 - ra_n}{1 - r}$$
$$S_{10} = \frac{16 - \frac{1}{2}\left(\frac{1}{32}\right)}{\frac{1}{2}} = \frac{16 - \frac{1}{64}}{\frac{1}{2}} = 31\tfrac{31}{32}$$

(11-1, 11-3)

4. $a_1 = -8$, $a_n = a_{n-1} + 3$, $n \geq 2$

(A) $a_1 = -8$
$a_2 = a_1 + 3 = -8 + 3 = -5$
$a_3 = a_2 + 3 = -5 + 3 = -2$
$a_4 = a_3 + 3 = -2 + 3 = 1$

(B) This is an arithmetic sequence with $d = 3$. Hence

$$a_n = a_1 + (n - 1)d$$
$$a_{10} = -8 + (10 - 1)3 = 19$$

(C) $$S_n = \frac{n}{2}(a_1 + a_n)$$
$$S_{10} = \frac{10}{2}(-8 + 19) = 55$$

(11-1, 11-3)

5. $a_1 = -1$, $a_n = (-2)a_{n-1}$, $n \geq 2$

(A) $a_1 = -1$
$a_2 = (-2)a_1 = (-2)(-1) = 2$
$a_3 = (-2)a_2 = (-2)2 = -4$
$a_4 = (-2)a_3 = (-2)(-4) = 8$

(B) This is a geometric sequence with $r = -2$. Hence

$$a_n = a_1 r^{n-1}$$
$$a_{10} = (-1)(-2)^{10-1} = 512$$

(C) $S_n = \dfrac{a_1 - ra_n}{1 - r}$

$S_{10} = \dfrac{-1 - (-2)(512)}{1 - (-2)}$

$= 341$ (*11-1, 11-3*)

6. This is a geometric sequence with $a_1 = 16$ and $r = \frac{1}{2} < 1$, so the sum exists:

$$S_\infty = \frac{a_1}{1 - r} = \frac{16}{1 - \frac{1}{2}} = 32$$

(*11-3*)

7. $6! = 6 \cdot 5 \cdot 4 \cdot 3 \cdot 2 \cdot 1 = 720$ (*11-4*)

8. $\dfrac{22!}{19!} = \dfrac{22 \cdot 21 \cdot 20 \cdot 19!}{19!} = 9{,}240$ (*11-4*)

9. $\dfrac{7!}{2!(7-2)!} = \dfrac{7!}{2!\,5!} = \dfrac{7 \cdot 6 \cdot 5!}{2 \cdot 1 \cdot 5!} = 21$ (*11-4*)

10. $C_{6,2} = \dfrac{6!}{2!(6-2)!} = \dfrac{6!}{2!\,4!} = \dfrac{6 \cdot 5 \cdot 4!}{2 \cdot 1 \cdot 4!} = 15 \quad P_{6,2} = 6 \cdot 5 = 30$ (*11-5*)

11. (A) The outcomes can be displayed in a tree diagram as follows:

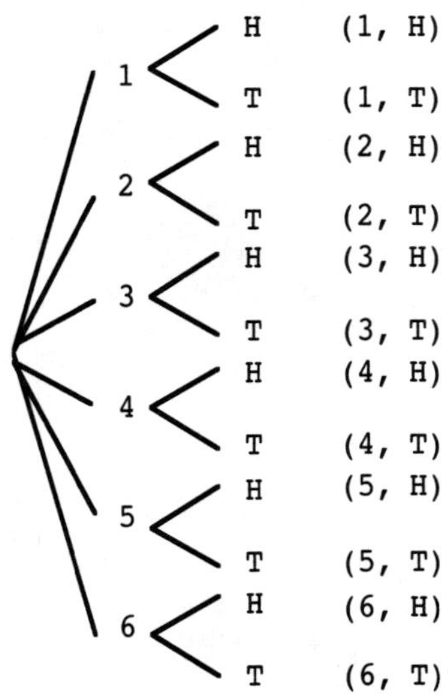

(B) O_1: Rolling the die
N_1: 6 outcomes
O_2: Flipping the coin
N_2: 2 outcomes

Applying the multiplication principle, there are $6 \cdot 2 = 12$ combined outcomes. (*11-5*)

12. O_1: Seating the first person N_1: 6 ways
O_2: Seating the second person N_2: 5 ways
O_3: Seating the third person N_3: 4 ways
O_4: Seating the fourth person N_4: 3 ways
O_5: Seating the fifth person N_5: 2 ways
O_6: Seating the sixth person N_6: 1 way

Applying the multiplication principle, there are $6 \cdot 5 \cdot 4 \cdot 3 \cdot 2 \cdot 1 = 720$ arrangements. (*11-5*)

13. Order is important here. We use permutations to determine the number of arrangements of 6 objects. $P_{6,6} = 6! = 720$ (*11-5*)

14. The sample space S is the set of all possible 5-card hands, chosen from a deck of 52 cards. $n(S) = C_{52,5}$

The event E is the set of all possible 5-card hands that are all clubs, thus, are chosen from the 13 clubs. $n(E) = C_{13,5}$

$$P(E) = \frac{n(E)}{n(S)} = \frac{C_{13,5}}{C_{52,5}} = \frac{13!}{5!(13-5)!} \div \frac{52!}{5!(52-5)!} \approx .0005$$ (*11-5*)

15. The sample space S is the set of all possible arrangements of two persons, chosen from a set of 15 persons. Order is important here, so we use permutations. $n(S) = P_{15,2}$. The event E is one of these arrangements. $n(E) = 1$.

$$P(E) = \frac{n(E)}{n(S)} = \frac{1}{P_{15,2}} = \frac{1}{15 \cdot 14} \approx .005$$ (*11-5*)

16. $P(E) \approx \dfrac{f(E)}{n} = \dfrac{50}{1,000} = .05$ (*11-5*)

17. P_1: $5 = 1^2 + 4 \cdot 1 = 5$

P_2: $5 + 7 = 2^2 + 4 \cdot 2$

$12 = 12$

P_3: $5 + 7 + 9 = 3^2 + 4 \cdot 3$

$21 = 21$ (*11-2*)

18. P_1: $2 = 2^{1+1} - 2 = 4 - 2 = 2$

P_2: $2 + 4 = 2^{2+1} - 2$

$6 = 6$

P_3: $2 + 4 + 8 = 2^{3+1} - 2$

$14 = 14$ (*11-2*)

19. P_1: $49^1 - 1$ is divisible by 6

$48 = 6 \cdot 8$ true

P_2: $49^2 - 1$ is divisible by 6

$2,400 = 6 \cdot 400$ true

P_3: $49^3 - 1$ is divisible by 6

$117,648 = 19,608 \cdot 6$ true (*11-2*)

20. P_k: $5 + 7 + 9 + \cdots + (2k + 3) = k^2 + 4k$

P_{k+1}: $5 + 7 + 9 + \cdots + (2k + 3) + (2k + 5) = (k + 1)^2 + 4(k + 1)$ (*11-2*)

21. P_k: $2 + 4 + 8 + \cdots + 2^k = 2^{k+1} - 2$

P_{k+1}: $2 + 4 + 8 + \cdots + 2^k + 2^{k+1} = 2^{k+2} - 2$ (*11-2*)

22. P_k: $49^k - 1 = 6r$ for some integer r

P_{k+1}: $49^{k+1} - 1 = 6s$ for some integer s (*11-2*)

23. Although 1 is less than 4, $1 + \frac{1}{2}$ is less than 4, $1 + \frac{1}{2} + \frac{1}{3}$ is less than 4, and so on, the statement is false. In fact,

$$1 + \frac{1}{2} + \frac{1}{3} + \ldots + \frac{1}{31} \approx 4.027245$$

hence $n = 31$ is a counterexample. (*11-2*)

24. $S_{10} = (2 \cdot 1 - 8) + (2 \cdot 2 - 8) + (2 \cdot 3 - 8) + (2 \cdot 4 - 8) + (2 \cdot 5 - 8) + (2 \cdot 6 - 8) + (2 \cdot 7 - 8) + (2 \cdot 8 - 8) + (2 \cdot 9 - 8) + (2 \cdot 10 - 8)$

$= (-6) + (-4) + (-2) + 0 + 2 + 4 + 6 + 8 + 10 + 12$

$= 30$ (*11-3*)

25. $S_7 = \frac{16}{2^1} + \frac{16}{2^2} + \frac{16}{2^3} + \frac{16}{2^4} + \frac{16}{2^5} + \frac{16}{2^6} + \frac{16}{2^7}$

$= 8 + 4 + 2 + 1 + \frac{1}{2} + \frac{1}{4} + \frac{1}{8}$

$= 15\frac{7}{8}$ (*11-3*)

26. This is an infinite geometric sequence with $a_1 = 27$.

$r = \frac{-18}{27} = -\frac{2}{3}$ $\left|-\frac{2}{3}\right| < 1$, hence the sum exists

$S_\infty = \frac{a_1}{1 - r}$

$= \frac{27}{1 - \left(-\frac{2}{3}\right)}$

$= \frac{81}{5}$ (*11-3*)

27. $S_n = \sum_{k=1}^{n} \frac{(-1)^{k+1}}{3^k}$

This geometric sequence has $a_1 = \frac{1}{3}$.

$r = \left(-\frac{1}{9}\right) \div \frac{1}{3} = -\frac{1}{3}$ $\left|-\frac{1}{3}\right| < 1$, hence the sum exists.

$S_\infty = \frac{a_1}{1 - r}$

$= \frac{\frac{1}{3}}{1 - \left(-\frac{1}{3}\right)}$

$= \frac{1}{4}$ (*11-3*)

28. The probability of an event cannot be negative, but $P(e_2)$ is given as negative. The sum of the probabilities of the simple events must be 1, but it is given as 2.5. The probability of an event cannot be greater than 1, but $P(e_4)$ is given as 2. (*11-5*)

29. We can select any three of the six points to use as vertices of the triangle. Order is not important. $C_{6,3} = \frac{6!}{3!(6-3)!} = 20$ triangles (*11-5*)

30. First, find d:

$a_n = a_1 + (n - 1)d$
$31 = 13 + (7 - 1)d$
$31 = 13 + 6d$
$d = 3$

Hence, $a_5 = 13 + (5 - 1)3$
$= 25$ (*11-3*)

31.

	Case 1	Case 2	Case 3
O_1: select the first letter	N_1: 8 ways	8 ways	8 ways
O_2: select the second letter	N_2: 7 ways	8 ways	7 ways
O_3: select the third letter	N_3: 6 ways	8 ways	7 ways
	(exclude first and second letter)		(exclude second letter.)
	$8 \cdot 7 \cdot 6 = 336$ words	$8 \cdot 8 \cdot 8 = 512$ words	$8 \cdot 7 \cdot 7 = 392$ words

(*11-4*)

32. (A) $P(\text{2 heads}) \approx \dfrac{210}{1,000} = .21$

$P(\text{1 head}) \approx \dfrac{480}{1,000} = .48$

$P(\text{0 heads}) \approx \dfrac{310}{1,000} = .31$

(B) A sample space of equally likely events is:

$S = \{HH, HT, TH, TT\}$

Let E_1 = 2 heads = $\{HH\}$

E_2 = 1 head = $\{HT, TH\}$

E_3 = 0 heads = $\{TT\}$

then $P(E_1) = \dfrac{n(E_1)}{n(S)} = \dfrac{1}{4} = .25$

$P(E_2) = \dfrac{n(E_2)}{n(S)} = \dfrac{2}{4} = .5$

$P(E_3) = \dfrac{n(E_3)}{n(S)} = \dfrac{1}{4} = .25$

(C) Expected frequency = $P(E)\cdot$number of trials

Expected frequency of 2 heads = $P(E_1)\cdot 1,000 = .25(1,000) = 250$

Expected frequency of 1 head = $P(E_2)\cdot 1,000 = .5(1,000) = 500$

Expected frequency of 0 heads = $P(E_3)\cdot 1,000 = .25(1,000) = 250$ (11-5)

33. The sample space S is the set of all possible 5-card hands chosen from a deck of 52 cards. $n(S) = C_{52,5}$.

(A) The event E is the set of all possible 4-card hands that are all diamonds, thus, are chosen from the 13 diamonds. $n(E) = C_{13,5}$

$$P(E) = \frac{n(E)}{n(S)} = \frac{C_{13,5}}{C_{52,5}}$$

(B) The event E is the set of all possible 5-card hands chosen as follows:

O_1: Choose 3 diamonds out of the 13 in the deck. N_1: $C_{13,3}$

O_2: Choose 2 spades out of the 13 in the deck. N_2: $C_{13,2}$

Applying the multiplication principle, $n(E) = N_1\cdot N_2 = C_{13,3}\cdot C_{13,2}$

$$P(E) = \frac{n(E)}{n(S)} = \frac{C_{13,3}\cdot C_{13,2}}{C_{52,5}}$$ (11-5)

34. The sample space S is the set of all possible 4-person choices out of the 10 people. $n(S) = C_{10,4}$

The event E is the set of all of those choices that include 2 particular people, thus

O_1: Choose the 2 married people N_1: 1

O_2: Choose 2 more people out of the 8 remaining N_2: $C_{8,2}$

Applying the multiplication principle, $n(E) = N_1\cdot N_2 = 1\cdot C_{8,2} = C_{8,2}$

$$P(E) = \frac{n(E)}{n(S)} = \frac{C_{8,2}}{C_{10,4}} = \frac{8!}{2!\,(8-2)!} \div \frac{10!}{4!\,(10-4)!} = 28 \div 210 = \frac{2}{15}$$ (11-5)

35. The sample space S is the set of all possible results of spinning the device twice:

$\{(1, 1), (1, 2), (1, 3), (2, 1), (2, 2), (2, 3), (3, 1), (3, 2), (3, 3)\}$.

$n(S) = 9$.

(A) The event E is the set of all possible results in which both spins are the same.

$E = \{(1, 1), (2, 2), (3, 3)\}$. $n(E) = 3$.

$$P(E) = \frac{n(E)}{n(S)} = \frac{3}{9} = \frac{1}{3}$$

(B) The event E is the set of those spin sequences that add up to 5:
$E = \{(2, 3), (3, 2)\}$. $n(E) = 2$.

$$P(E) = \frac{n(E)}{n(S)} = \frac{2}{9}$$ (11-5)

36. $0.\overline{72} = 0.72 + 0.0072 + 0.000072 + \cdots$
This is an infinite geometric sequence with $a_1 = 0.72$ and $r = 0.01$.

$$0.\overline{72} = S_\infty = \frac{a_1}{1 - r} = \frac{0.72}{1 - 0.01} = \frac{0.72}{0.99} = \frac{8}{11}$$ (11-3)

37. (A) Order is important here. We are selecting 3 digits out of 6 possible.
$P_{6,3} = 6 \cdot 5 \cdot 4 = 120$ lock combinations

(B) Order is not important here. We are selecting 2 players out of 5.
$C_{5,2} = \frac{5!}{2!(5 - 2)!} = 10$ games (11-4)

38. $$\frac{20!}{18!(20 - 18)!} = \frac{20!}{18!\,2!} = \frac{20 \cdot 19 \cdot 18!}{18!\,2 \cdot 1} = 190$$ (11-6)

39. $$\binom{16}{12} = \frac{16!}{12!(16 - 12)!} = \frac{16!}{12!\,4!} = \frac{16 \cdot 15 \cdot 14 \cdot 13 \cdot 12!}{12! \cdot 4 \cdot 3 \cdot 2 \cdot 1} = 1{,}820$$ (11-6)

40. $$\binom{11}{11} = \frac{11!}{11!(11 - 11)!} = \frac{11!}{11!\,0!} = 1$$ (11-6)

41. $$(x - y)^5 = [x + (-y)]^5 = \sum_{k=0}^{5} \binom{5}{k} (x)^{5-k}(-y)^k$$
$$= \binom{5}{0}x^5 + \binom{5}{1}x^4(-y)^1 + \binom{5}{2}x^3(-y)^2 + \binom{5}{3}x^2(-y)^3 + \binom{5}{4}x(-y)^4 + \binom{5}{5}(-y)^5$$
$$= x^5 - 5x^4y + 10x^3y^2 - 10x^2y^3 + 5xy^4 - y^5$$ (11-6)

42. In the expansion
$$(x + 2)^9 = \sum_{k=0}^{9} \binom{9}{k} x^{9-k}\, 2^k$$
the exponent of x is 6 when $k = 3$. Thus the term containing x^6 is
$$\binom{9}{3}x^6 2^3 = 84x^6 \cdot 8 = 672x^6$$ (11-6)

43. In the expansion of $(a + b)^n$, the exponent of b in the rth term is $r - 1$ and the exponent of a is $n - (r - 1)$. Here, $r = 10$, $n = 12$.

$$\begin{aligned}\text{Tenth term} &= \binom{12}{9}(2x)^3(-y)^9\\ &= \frac{12!}{9!\,3!}(8x^3)(-y^9)\\ &= \frac{12\cdot 11\cdot 10\cdot 9!}{9!\cdot 3\cdot 2\cdot 1}(-8x^3y^9)\\ &= -1760x^3y^9\end{aligned}$$

(11-6)

44. Write: P_n: $5 + 7 + 9 + \cdots + (2n + 3) = n^2 + 4n$

Proof: Part 1: Show that P_1 is true:

P_1: $5 = 1^2 + 4\cdot 1$

$= 1 + 4$ Clearly true.

Part 2: Show that if P_k is true, then P_{k+1} is true:

Write out P_k and P_{k+1}.

P_k: $5 + 7 + 9 + \cdots + (2k + 3) = k^2 + 4k$

P_{k+1}: $5 + 7 + 9 + \cdots + (2k + 3) + (2k + 5) = (k + 1)^2 + 4(k + 1)$

We start with P_k:

$5 + 7 + 9 + \cdots + (2k + 3) = k^2 + 4k$

Adding $2k + 5$ to both sides:

$$\begin{aligned}5 + 7 + 9 + \cdots + (2k + 3) + (2k + 5) &= k^2 + 4k + 2k + 5\\ &= k^2 + 6k + 5\\ &= k^2 + 2k + 1 + 4k + 4\\ &= (k + 1)^2 + 4(k + 1)\end{aligned}$$

We have shown that if P_k is true, then P_{k+1} is true.

Conclusion: P_n is true for all positive integers n. (11-2)

45. Write: P_n: $2 + 4 + 8 + \cdots + 2^n = 2^{n+1} - 2$

Proof: Part 1: Show that P_1 is true:

P_1: $2 = 2^{1+1} - 2$

$= 4 - 2$ Clearly true.

Part 2: Show that if P_k is true, then P_{k+1} is true:

Write out P_k and P_{k+1}.

P_k: $2 + 4 + 8 + \cdots + 2^k = 2^{k+1} - 2$

P_{k+1}: $2 + 4 + 8 + \cdots + 2^k + 2^{k+1} = 2^{k+2} - 2$

We start with P_k:

$2 + 4 + 8 + \cdots + 2^k = 2^{k+1} - 2$

Adding 2^{k+1} to both sides:

$$\begin{aligned}2 + 4 + 8 + \cdots + 2^k + 2^{k+1} &= 2^{k+1} - 2 + 2^{k+1}\\ &= 2^{k+1} + 2^{k+1} - 2\\ &= 2\cdot 2^{k+1} - 2\\ &= 2^{k+2} - 2\end{aligned}$$

We have shown that if P_k is true, then P_{k+1} is true.

Conclusion: P_n is true for all positive integers n. (11-2)

46. Write: P_n: $49^n - 1 = 6r$ for some r in N.

Part 1: Show that P_1 is true:

P_1: $49^1 - 1 = 6r$ is true ($r = 8$).

Part 2: Show that if P_k is true, then P_{k+1} is true:

Write out P_k and P_{k+1}.

P_k: $49^k - 1 = 6r$, for some integer r

P_{k+1}: $49^{k+1} - 1 = 6s$ for some integer s

We start with P_k:

$49^k - 1 = 6r$ for some integer r

$$\begin{aligned} \text{Now, } 49^{k+1} - 1 &= 49^{k+1} - 49^k + 49^k - 1 \\ &= 49^k(49 - 1) + 49^k - 1 \\ &= 49^k \cdot 8 \cdot 6 + 6r \\ &= 6(49^k \cdot 8 + r) \\ &= 6s \text{ with } s = 49^k \cdot 8 + r \end{aligned}$$

We have shown that if P_k is true, then P_{k+1} is true.

Conclusion: P_n is true for all positive integers n. *(11-2)*

47.

Enter the sequences in the sequence editor.

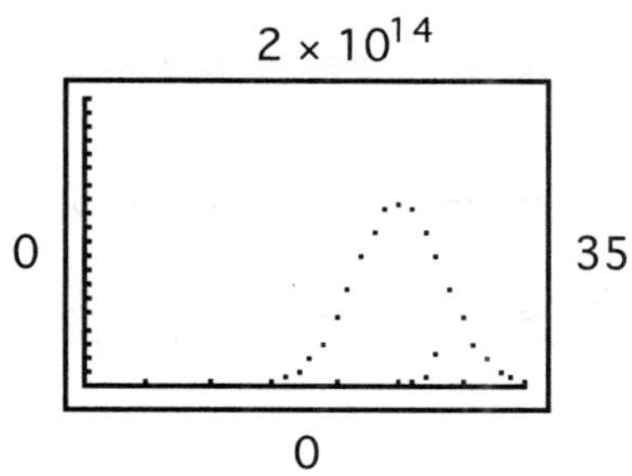

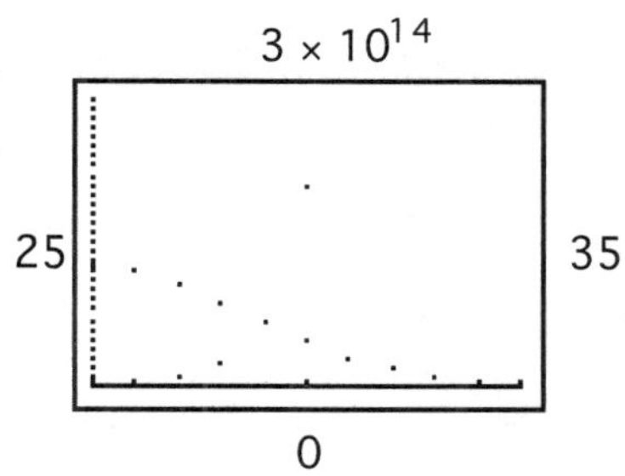

Graph the sequences.

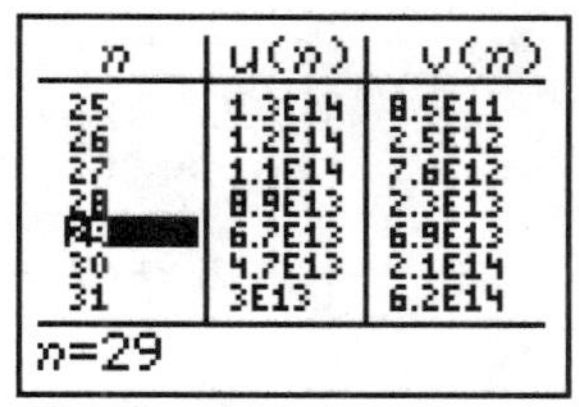

Display the points (in the appropriate interval) in a table.

From the second graph, 29 is the least positive integer n such that $a_n < b_n$. The table confirms this. *(11-6)*

48.

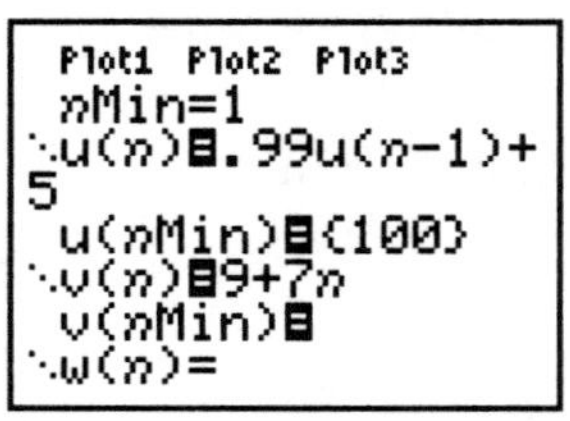

Enter the sequences in the sequence editor.

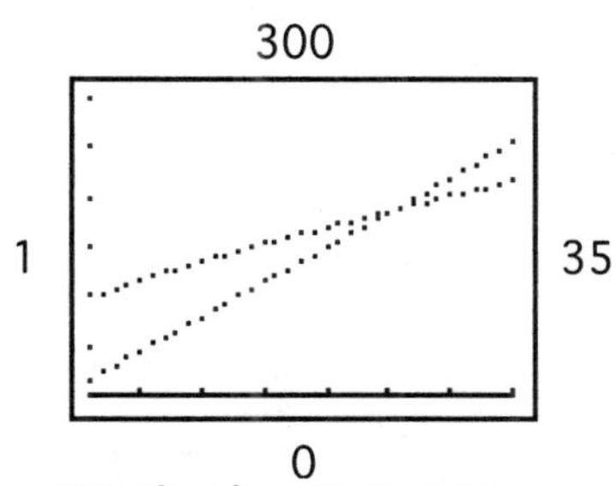

Graph the sequences.

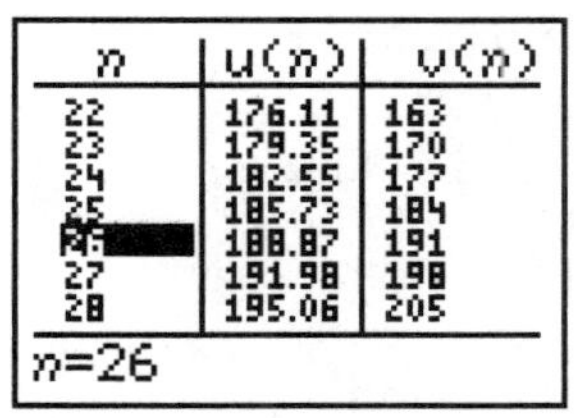

Display the points in a table.

From the graph, the least positive integer n such that $a_n < b_n$ is between 25 and 30. The table confirms that $n = 26$. *(11-1)*

49. In the first case, the order matters. We have five successive events, each of which can happen two ways (girl or boy). Applying the multiplication principle, there are $2 \cdot 2 \cdot 2 \cdot 2 \cdot 2 = 32$ possible families. In the second case, the possible families can be listed as {0 girls, 1 girl, 2 girls, 3 girls, 4 girls, 5 girls}, thus there are 6 possibilities. *(11-4)*

50. An arithmetic sequence is involved, with $a_1 = \frac{g}{2}$, $d = \frac{3g}{2} - \frac{g}{2} = g$.
Distance fallen during the twenty-fifth second $= a_{25}$.

$$a_n = a_1 + (n - 1)d$$
$$a_{25} = \frac{g}{2} + (25 - 1)g$$
$$= \frac{49g}{2} \text{ feet}$$

Total distance fallen after twenty-five seconds $= a_1 + a_2 + a_3 + \cdots + a_{25} = S_{25}$

$$S_n = \frac{n}{2}(a_1 + a_n)$$
$$S_{25} = \frac{25}{2}\left(\frac{g}{2} + \frac{49g}{2}\right)$$
$$= \frac{625g}{2} \text{ feet}$$

(*11-3*)

51. To seat two people, we can seat one person, then the second person.
O_1: Seat the first person in any chair.
N_1: 4 ways
O_2: Seat the second person in any remaining chair.
N_2: 3 ways
Thus, applying the multiplication principle, there are $4 \cdot 3 = 12$ ways to seat two persons. (*11-4*)

52.
$$(x + i)^6 = \sum_{k=0}^{6} \binom{6}{k} x^{6-k} i^k$$
$$= \binom{6}{0}x^6 + \binom{6}{1}x^5 i^1 + \binom{6}{2}x^4 i^2 + \binom{6}{3}x^3 i^3 + \binom{6}{4}x^2 i^4 + \binom{6}{5}x i^5 + \binom{6}{6}i^6$$
$$= x^6 + 6ix^5 - 15x^4 - 20ix^3 + 15x^2 + 6ix - 1$$

(*11-6*)

53. Since there are several ways in which at least one woman can be selected, it is simplest to note: probability of selecting at least one woman =
1 - probability of selecting 0 women. To calculate the probability of selecting 0 women, that is, three men, we note:
The sample space S is the set of all 3-person subsets, chosen from the ten people. $n(S) = C_{10,3}$
The event E is the set of all 3-man subsets, chosen from the 7 men.
$n(E) = C_{7,3}$

Then $P(E) = \frac{C_{7,3}}{C_{10,3}}$

Hence $P(\text{at least one woman}) = 1 - P(0 \text{ women}) = 1 - P(E)$

$$= 1 - \frac{C_{7,3}}{C_{10,3}} = 1 - \frac{7}{24} = \frac{17}{24}$$

(*11-5*)

54. (A) $P(E) \approx \frac{f(E)}{n} = \frac{350}{1,000} = .350$

(B) A sample space of equally likely events is:
$S = \{HHH, HHT, HTH, HTT, THH, THT, TTH, TTT\}$
Let E = 2 heads $= \{HHT, HTH, THH\}$
then $P(E) = \frac{n(E)}{n(S)} = \frac{3}{8} = .375$

(C) Expected frequency $= P(E) \cdot$ number of trials $= .375(1,000) = 375$ (*11-5*)

55. Write: P_n: $\sum_{k=1}^{n} k^3 = \left(\sum_{k=1}^{n} k\right)^2$

Proof: Part 1: Show that P_1 is true.

$$P_1: \sum_{k=1}^{1} k^3 = 1^3 = 1^2 = \left(\sum_{k=1}^{1} k\right)^2$$

Part 2: Show that if P_j is true, then P_{j+1} is true.
Write out P_j and P_{j+1}.

$$P_j: \sum_{k=1}^{j} k^3 = \left(\sum_{k=1}^{j} k\right)^2$$

$$P_{j+1}: \sum_{k=1}^{j+1} k^3 = \left(\sum_{k=1}^{j+1} k\right)^2$$

We start with P_j:

$$\sum_{k=1}^{j} k^3 = \left(\sum_{k=1}^{j} k\right)^2$$

Adding $(j + 1)^3$ to both sides:

$$\sum_{k=1}^{j} k^3 + (j + 1)^3 = \left(\sum_{k=1}^{j} k\right)^2 + (j + 1)^3$$

$$\sum_{k=1}^{j+1} k^3 = (1 + 2 + 3 + \dots + j)^2 + (j + 1)^3$$

$$= \left[\frac{j(j + 1)}{2}\right]^2 + (j + 1)^3 \text{ using Matched Problem 1, Section 11-2}$$

$$= (j + 1)^2 \frac{j^2}{4} + (j + 1)^2(j + 1)$$

$$= (j + 1)^2\left[\frac{j^2}{4} + j + 1\right]$$

$$= (j + 1)^2\left[\frac{j^2 + 4j + 4}{4}\right]$$

$$= (j + 1)^2 \frac{(j + 2)^2}{2^2}$$

$$= \left[\frac{(j + 1)(j + 2)}{2}\right]^2$$

$$= [1 + 2 + 3 + \dots + (j + 1)]^2 \text{ using Matched Problem 1, Section 11-2}$$

$$= \left(\sum_{k=1}^{j+1} k\right)^2$$

We have shown that if P_j is true, then P_{j+1} is true.

Conclusion: P_n is true for all positive integers n. *(11-2)*

56. Write: P_n: $x^{2n} - y^{2n} = (x - y)Q_n(x, y)$, where $Q_n(x, y)$ denotes some polynomial in x and y.

Proof: Part 1: Show that P_1 is true.

P_1: $x^{2\cdot 1} - y^{2\cdot 1} = (x - y)(x + y) = (x - y)Q_1(x, y)$ $\quad P_1$ is true.

Part 2: Show that if P_k is true, then P_{k+1} is true.

Write out P_k and P_{k+1}.

P_k: $x^{2k} - y^{2k} = (x - y)Q_k(x, y)$

P_{k+1}: $x^{2k+2} - y^{2k+2} = (x - y)Q_{k+1}(x, y)$

We start with P_k:

$x^{2k} - y^{2k} = (x - y)Q_k(x, y)$

Now,
$$\begin{aligned} x^{2k+2} - y^{2k+2} &= x^{2k+2} - x^{2k}y^2 + x^{2k}y^2 - y^{2k+2} \\ &= x^{2k}(x^2 - y^2) + y^2(x^{2k} - y^{2k}) \\ &= (x - y)x^{2k}(x + y) + y^2(x - y)Q_k(x, y) \quad \text{by } P_k \\ &= (x - y)[x^{2k}(x + y) + y^2Q_k(x, y)] \\ &= (x - y)Q_{k+1}(x, y) \end{aligned}$$

We have shown that if P_k is true, then P_{k+1} is true.

Conclusion: P_n is true for all positive integers n. (*11-2*)

57. Write: P_n: $\dfrac{a^n}{a^m} = a^{n-m}$, m an arbitrary positive integer, $n > m$.

Proof: Part 1: Show P_{m+1} is true.

$$P_{m+1}: \quad \frac{a^{m+1}}{a^m} = a^{m+1-m}$$

$$\frac{a^m a}{a^m} = a^1 \quad \text{by the recursive definition of } a^n$$

$$a = a \quad P_{m+1} \text{ is true.}$$

Part 2: Show that if P_k is true, then P_{k+1} is true.

Write out P_k and P_{k+1}.

$$P_k: \quad \frac{a^k}{a^m} = a^{k-m}$$

$$P_{k+1}: \quad \frac{a^{k+1}}{a^m} = a^{k-m+1}$$

We start with P_k:

$$\frac{a^k}{a^m} = a^{k-m}$$

Multiplying both sides by a:

$$\frac{a^k}{a^m}a = a^{k-m}a$$

$$\frac{a^k a}{a^m} = a^{k-m+1}$$

$$\frac{a^{k+1}}{a^m} = a^{k-m+1}$$

We have show that if P_k is true, then P_{k+1} is true, for m an arbitrary positive integer.

Conclusion: P_n is true for all positive integers m, n. (*11-2*)

58. To prove $a_n = b_n$, n a positive integer, write:

P_n: $a_n = b_n$

Proof: Part 1: Show P_1 is true.

$a_1 = -3 \qquad b_1 = -5 + 2 \cdot 1 = -3.$

Thus, $a_1 = b_1$

Part 2: Show that if P_k is true, then P_{k+1} is true.

Write out P_k and P_{k+1}.

P_k: $a_k = b_k$

P_{k+1}: $a_{k+1} = b_{k+1}$

We start with P_k:

$a_k = b_k$

$$\begin{aligned} \text{Now, } a_{k+1} = a_k + 2 &= b_k + 2 \\ &= -5 + 2k + 2 \\ &= -5 + 2(k + 1) \\ &= b_{k+1} \end{aligned}$$

Therefore, $a_{k+1} = b_{k+1}$.

Thus, if P_k is true, then P_{k+1} is true.

Conclusion: P_n is true for all $n \in N$. Hence, $\{a_n\} = \{b_n\}$ (*11-2*)

59. Write: P_n: $(1!)1 + (2!)2 + (3!)3 + \cdots + (n!)n = (n + 1)! - 1$.

Proof: Part 1: Show that P_1 is true.

P_1: $(1!)1 = 1 = 2 - 1 = 2! - 1$ is true.

Part 2: Show that if P_k is true, then P_{k+1} is true.

Write out P_k and P_{k+1}.

P_k: $(1!)1 + (2!)2 + (3!)3 + \cdots + (k!)k = (k + 1)! - 1$

P_{k+1}: $(1!)1 + (2!)2 + (3!)3 + \cdots + (k!)k + (k + 1)!(k + 1) = (k + 2)! - 1$

We start with P_k:

$(1!)1 + (2!)2 + (3!)3 + \cdots + (k!)k = (k + 1)! - 1$

Adding $(k + 1)!(k + 1)$ to both sides:

$$\begin{aligned} &(1!)1 + (2!)2 + (3!)3 + \cdots + (k!)k + (k + 1)!(k + 1) \\ &\qquad = (k + 1)! - 1 + (k + 1)!(k + 1) \\ &\qquad = (k + 1)!(1 + k + 1) - 1 \\ &\qquad = (k + 2)(k + 1)! - 1 \\ &\qquad = (k + 2)! - 1 \end{aligned}$$

Thus, if P_k is true, then P_{k+1} is true.

Conclusion: P_n is true for all positive integers n. (*11-2*)

60. The unpaid balance starts at \$7,200, and decreases each month by \$300. Thus the interest starts at 7,200(0.01) and decreases each month by 300(0.01). This is an arithmetic sequence with

$a_1 = 7{,}200(0.01)$, $a_{24} = 300(0.01)$, $n = 24$.

Then the total interest paid is

$$S_n = \frac{n}{2}(a_1 + a_n)$$

$$\begin{aligned} S_{24} &= \frac{24}{2}[7{,}200(0.01) + 300(0.01)] \\ &= \$900 \end{aligned}$$

(*11-3*)

61. This involves an infinite geometric series. $a_1 = (0.75)2,400$. $r = 0.75$.

$|r| < 1$, so the series has a sum.

$$S_\infty = \frac{a_1}{1 - r} = \frac{(0.75)2400}{1 - 0.75} = \$7,200$$ (11-3)

62. Since $A(n)$ is a geometric sequence with common ratio 1.06, we can write

$A(n) = a_n = (1.06)^n\,500$

If $n = 10$, $a_{10} = (1.06)^{10}\,500 = \$895.42$

If $n = 20$, $a_{10} = (1.06)^{20}\,500 = \$1,603.57$ (11-3)

63. A route plan can be regarded as a series of choices of stores, thus an arrangement of the 5 stores. Since the order matters, we use permutations: $P_{5,5} = 5! = 120$ route plans. (11-4)

64. (A) $P(\text{this event}) = \dfrac{40}{1,000} = .04$

(B) $P(\text{this event}) = \dfrac{100 + 60}{1,000} = .16$

(C) $P(\text{this event}) = 1 - P$ (not 12 - 18 and buys 0 or 1 cassette annually)

$$= 1 - \frac{60 + 70 + 70 + 110 + 100 + 50}{1,000} = 1 - \frac{460}{1,000} = .54$$ (11-5)

65. P (shipment returned) = 1 - P (no substandard part found).
The sample space S is the set of all possible 4-part subsets of the 12 part set.
$n(S) = C_{12,4}$
The event, no substandard part found, is the set of all possible 4-part subsets of the 10 acceptable parts. $n(E) = C_{10,4}$.

Then $P(\text{shipment returned}) = 1 - \dfrac{n(E)}{n(S)} = 1 - \dfrac{C_{10,4}}{C_{12,4}} = 1 - \dfrac{14}{33} \approx .576.$ (11-5)

Chapters 9, 10, & 11 Cumulative Review

1. We choose elimination by addition. We multiply the top equation by 3, the bottom by 5, and add.

$$\begin{aligned} 9x - 15y &= 33 \\ 10x + 15y &= 5 \\ \hline 19x \quad &= 38 \\ x &= 2 \end{aligned}$$

Substituting $x = 2$ in the bottom equation, we have

$$\begin{aligned} 2(2) + 3y &= 1 \\ 4 + 3y &= 1 \\ 3y &= -3 \\ y &= -1 \end{aligned}$$

Solution: (2, -1) *(10-1)*

2.

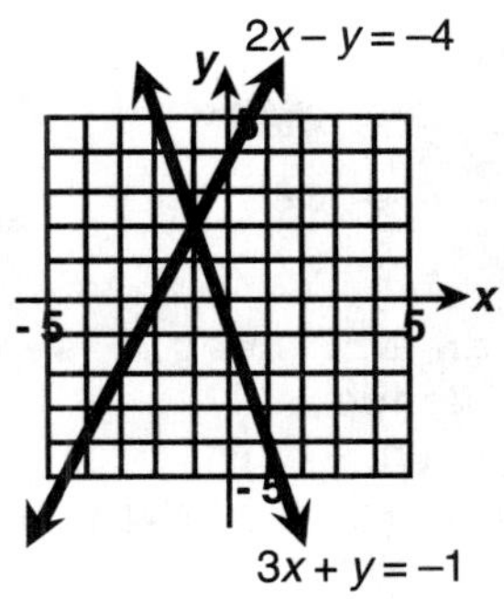

(-1, 2) *(10-1)*

3.

$$\begin{aligned} -6x + 3y &= 2 \\ 2x - y &= 1 \end{aligned}$$

Solve the second equation for y and substitute into the first.

$$\begin{aligned} 2x - y &= 1 \\ 2x - 1 &= y \\ -6x + 3(2x - 1) &= 2 \\ -6x + 6x - 3 &= 2 \\ -3 &= 2 \end{aligned}$$

This is a contradiction so the system has no solution. *(10-1)*

4.

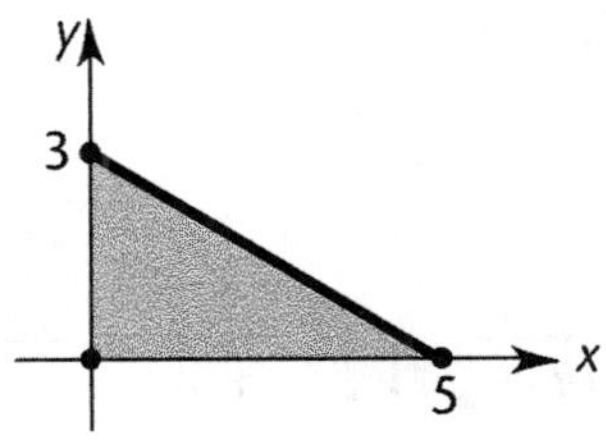

(10-7)

5. (A) Since $15 - 20 = 10 - 15 = -5$, this could start an arithmetic sequence.

(B) Since $\frac{25}{5} = \frac{125}{25} = 5$, this could start a geometric sequence.

(C) Since $\frac{25}{5} \neq \frac{50}{25}$ and $25 - 5 \neq 50 - 25$, this could start neither an arithmetic nor a geometric sequence.

(D) Since $\frac{-9}{27} = \frac{3}{-9} = -\frac{1}{3}$, this could start a geometric sequence.

(E) Since $(-6) - (-9) = (-3) - (-6) = 3$, this could start an arithmetic sequence. *(11-3)*

6. $a_n = 2 \cdot 5^n$

(A)
$$\begin{aligned} a_1 &= 2 \cdot 5^1 = 10 \\ a_2 &= 2 \cdot 5^2 = 50 \\ a_3 &= 2 \cdot 5^3 = 250 \\ a_4 &= 2 \cdot 5^4 = 1{,}250 \end{aligned}$$

(B) This is a geometric sequence with $r = 5$. Hence,

$$\begin{aligned} a_n &= a_1 r^{n-1} \\ a_8 &= 10(5)^{8-1} \\ &= 781{,}250 \end{aligned}$$

(C)
$$\begin{aligned} S_n &= \frac{a_1 - ra_n}{1 - r} \\ S_8 &= \frac{10 - 5(781{,}250)}{1 - 5} \\ &= 976{,}560 \end{aligned}$$

(11-3)

7. $a_n = 3n - 1$

(A)
$$\begin{aligned} a_1 &= 3 \cdot 1 - 1 = 2 \\ a_2 &= 3 \cdot 2 - 1 = 5 \\ a_3 &= 3 \cdot 3 - 1 = 8 \\ a_4 &= 3 \cdot 4 - 1 = 11 \end{aligned}$$

(B) This is an arithmetic sequence with $d = 3$. Hence,

$$\begin{aligned} a_n &= a_1 + (n - 1)d \\ a_8 &= 2 + (8 - 1)3 \\ &= 23 \end{aligned}$$

(C)
$$\begin{aligned} S_n &= \frac{n}{2}(a_1 + a_n) \\ S_8 &= \frac{8}{2}(2 + 23) \\ &= 100 \end{aligned}$$

(11-3)

8. $a_1 = 100 \quad a_n = a_{n-1} - 6 \quad n \geq 2$

(A) $a_1 = 100$

$a_2 = a_1 - 6 = 94$

$a_3 = a_2 - 6 = 88$

$a_4 = a_3 - 6 = 82$

(B) This is an arithmetic sequence with $d = -6$. Hence,

$a_n = a_1 + (n - 1)d$

$a_8 = 100 + (8 - 1)(-6)$

$= 58$

(C) $S_n = \frac{n}{2}(a_1 + a_n)$

$S_8 = \frac{8}{2}(100 + 58)$

$= 632$ (*11-3*)

9. (A) $8! = 8 \cdot 7 \cdot 6 \cdot 5 \cdot 4 \cdot 3 \cdot 2 \cdot 1 = 40{,}320.$

(B) $\frac{32!}{30!} = \frac{32 \cdot 31 \cdot 30!}{30!} = 992$

(C) $\frac{9!}{3!(9-3)!} = \frac{9!}{3!\,6!}$

$= \frac{9 \cdot 8 \cdot 7 \cdot 6!}{3 \cdot 2 \cdot 1 \cdot 6!} = 84$ (*11-4*)

10. (A) $\binom{7}{2} = \frac{7!}{2!(7-2)!} = \frac{7!}{2!\,5!} = \frac{7 \cdot 6 \cdot 5!}{2 \cdot 1 \cdot 5!} = 21$ (B) $C_{7,2} = \frac{7!}{2!(7-2)!} = 21$

(C) $P_{7,2} = \frac{7!}{(7-2)!} = \frac{7!}{5!} = \frac{7 \cdot 6 \cdot 5!}{5!} = 42$ (*11-4, 11-5*)

11. First, write the equation in standard form by dividing both sides by 900.

$25x^2 - 36y^2 = 900$

$\frac{x^2}{36} - \frac{y^2}{25} = 1$

In this form the equation is identifiable as that of a hyperbola.

When $x = 0$, $-\frac{y^2}{25} = 1$. There are no y intercepts, but $b = 5$.

When $y = 0$, $\frac{x^2}{36} = 1$. x intercepts: ± 6.

Sketch the asymptotes using the asymptote rectangle, then sketch in the hyperbola.

Foci: $c^2 = 5^2 + 6^2$

$c^2 = 61$

$c = \sqrt{61}$

Foci: $F' = (-\sqrt{61}, 0)$, $F = (\sqrt{61}, 0)$

Transverse axis length = 2(6) = 12

Conjugate axis length = 2(5) = 10

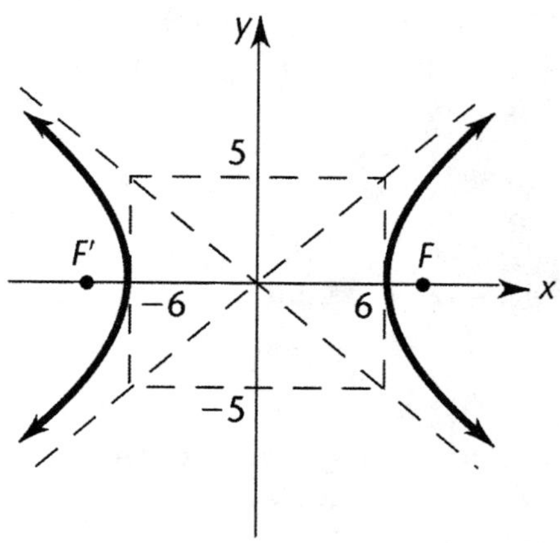

(*9-3*)

12. First, write the equation in standard form by dividing both sides by 900.

$$25x^2 + 36y^2 = 900$$

$$\frac{x^2}{36} + \frac{y^2}{25} = 1$$

In this form the equation is identifiable as that of an ellipse. Locate the intercepts:

When $y = 0$, $\frac{x^2}{36} = 1$. x intercepts: ± 6.

When $x = 0$, $\frac{y^2}{25} = 1$. y intercepts: ± 5

Thus, $a = 6$, $b = 5$, and the major axis is on the x axis.

Foci: $c^2 = a^2 - b^2$

$c^2 = 36 - 25$

$c^2 = 11$

$c = \sqrt{11}$

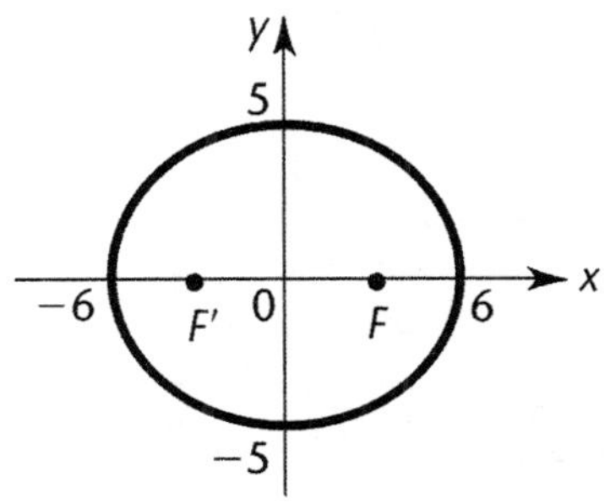

Foci: $F' = (-\sqrt{11}, 0)$, $F = (\sqrt{11}, 0)$

Major axis length = 2(6) = 12; Minor axis length = 2(5) = 10 (9-2)

13. $25x^2 - 36y = 0$ is the equation of a parabola. For convenience, we rewrite this as $25x^2 = 36y$. To graph, assign y values that make $25x^2$ a perfect square (y must be positive or zero for x to be real) and solve for x. Since the coefficient of x is positive, a must be positive, and the parabola opens up.

x	0	$\pm\frac{6}{5}$	$\pm\frac{12}{5}$
y	0	1	4

To find the focus and directrix, solve

$$4a = \frac{36}{25}$$

$$a = \frac{9}{25}$$

Focus: $\left(0, \frac{9}{25}\right)$. Directrix: $y = -\frac{9}{25}$

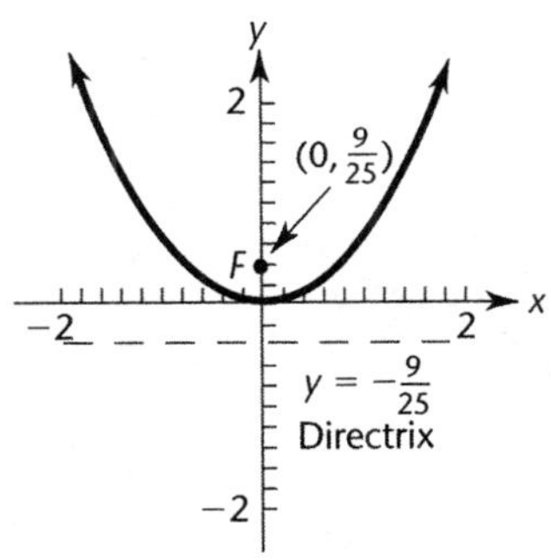

(9-1)

14. Find the discriminant with $A = 3$, $B = -4$, $C = 2$:

$$B^2 - 4AC = 16 - 4(3)(2) = -8$$

Since the discriminant is negative, this is an ellipse. (9-4)

15. $x^2 + y^2 = 2$

$2x - y = 1$

We choose substitution, solving the first-degree equation for y in terms of x, then substituting into the second-degree equation.

$$2x - y = 1$$

$$-y = 1 - 2x$$

$$y = 2x - 1$$

$$x^2 + (2x - 1)^2 = 2$$

$$x^2 + 4x^2 - 4x + 1 = 2$$

$$5x^2 - 4x - 1 = 0$$

$$(5x + 1)(x - 1) = 0$$

$$x = -\tfrac{1}{5}, 1$$

For $x = -\frac{1}{5}$

$$y = 2\left(-\tfrac{1}{5}\right) - 1 = -\tfrac{7}{5}$$

For $x = 1$

$$y = 2(1) - 1 = 1$$

Solutions: $\left(-\frac{1}{5}, -\frac{7}{5}\right)$, $(1, 1)$ The checking steps are omitted for lack of space. (10-6)

16.

Corner Point (x, y)	Objective Function $z = 2x + 3y$	
(0, 4)	12	
(5, 0)	10	Minimum value
(6, 7)	33	Maximum value
(0, 10)	30	

The minimum value of z on S is 10 at (5, 0).
The maximum value of z on S is 33 at (6, 7). (*10-8*)

17. (A) $M - 2N = \begin{bmatrix} 2 & 1 \\ 1 & -3 \end{bmatrix} - 2\begin{bmatrix} 1 & 2 \\ -1 & 3 \end{bmatrix} = \begin{bmatrix} 2 & 1 \\ 1 & -3 \end{bmatrix} - \begin{bmatrix} 2 & 4 \\ -2 & 6 \end{bmatrix} = \begin{bmatrix} 0 & -3 \\ 3 & -9 \end{bmatrix}$

(B) $P + Q$ is not defined

(C) $PQ = [1 \quad 2]\begin{bmatrix} -1 \\ 2 \end{bmatrix} = [1(-1) + 2 \cdot 2] = [3]$

(D) $MN = \begin{bmatrix} 2 & 1 \\ 1 & -3 \end{bmatrix}\begin{bmatrix} 1 & 2 \\ -1 & 3 \end{bmatrix} = \begin{bmatrix} 2 \cdot 1 + 1(-1) & 2 \cdot 2 + 1 \cdot 3 \\ 1 \cdot 1 + (-3)(-1) & 1 \cdot 2 + (-3)3 \end{bmatrix} = \begin{bmatrix} 1 & 7 \\ 4 & -7 \end{bmatrix}$

(E) $PN = [1 \quad 2]\begin{bmatrix} 1 & 2 \\ -1 & 3 \end{bmatrix} = [1 \cdot 1 + 2(-1) \quad 1 \cdot 2 + 2 \cdot 3] = [-1 \quad 8]$

(F) QM is not defined (*10-4*)

18. (A) The outcomes can be displayed in a tree diagram as follows:

HHH
HHT
HTH
HTT
THH
THT
TTH
TTT

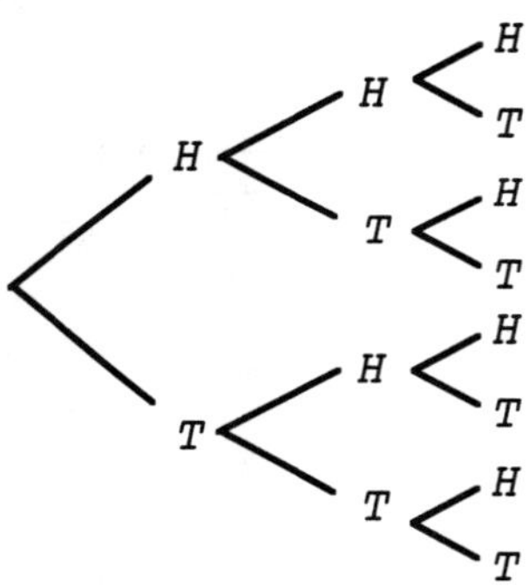

(B) O_1: First flip of the coin
N_1: 2 outcomes
O_2: Second flip of the coin
N_2: 2 outcomes
O_3: Third flip of the coin
N_3: 2 outcomes

Applying the multiplication principle, there are $2 \cdot 2 \cdot 2 = 8$ possible outcomes. (*11-5*)

19. (A) O_1: Place first book N_1: 4 ways
O_2: Place second book N_2: 3 ways
O_3: Place third book N_3: 2 ways
O_4: Place fourth book N_4: 1 way

Applying the multiplication principle, there are $4 \cdot 3 \cdot 2 \cdot 1$, or 24, arrangements.

(B) Order is important here. We use permutations to determine the number of arrangements of 4 objects. $P_{4,4} = 4! = 24$. (*11-5*)

20. The sample space S is the set of all possible 3-card hands, chosen from a deck of 52 cards. $n(S) = C_{52,3}$.
The event E is the set of all possible 3-card hands that are all diamonds, thus, are chosen from the 13 diamonds. $n(E) = C_{13,3}$

$$P(E) = \frac{n(E)}{n(S)} = \frac{C_{13,3}}{C_{52,3}} = \frac{13!}{3!(13-3)!} \div \frac{52!}{3!(52-3)!} \approx .0129$$ (*11-5*)

21. In the first case, order is important. The sample space is the set of all possible arrangements of four objects, drawn from a set of 10 objects. $n(S) = P_{10,4}$.
The event E is one of these arrangements. $n(E) = 1$.

$$P(E) = \frac{n(E)}{n(S)} = \frac{1}{P_{10,4}} = \frac{1}{10 \cdot 9 \cdot 8 \cdot 7} \approx .0002$$

In the second case, order is not important. The sample space is the set of all possible ways of drawing four objects from a set of 10 objects $n(S) = C_{10,4}$.
The event E is one of these ways. $n(E) = 1$.

$$P(E) = \frac{n(E)}{n(S)} = \frac{1}{C_{10,4}} = 1 \div \frac{10!}{4!(10-4)!} \approx .0048$$ *(11-5)*

22. $P(E) \approx \dfrac{f(E)}{n} = \dfrac{100 - 38}{100} = .62$ *(11-5)*

23. (A) $x_1 = 3$
$x_2 = -4$

(B) $x_1 - 2x_2 = 3$
Let $x_2 = t$
Then $x_1 = 2x_2 + 3$
$= 2t + 3$
Hence, $x_1 = 2t + 3$, $x_2 = t$ is a solution for every real number t.

(C) $x_1 - 2x_2 = 3$
$0x_1 + 0x_2 = 1$
No solution. *(10-3)*

24. (A) $\left[\begin{array}{rr|r} 1 & 1 & 3 \\ -1 & 1 & 5 \end{array}\right]$

(B) $\left[\begin{array}{rr|r} 1 & 1 & 3 \\ -1 & 1 & 5 \end{array}\right] R_1 + R_2 \to R_2$

$\sim \left[\begin{array}{rr|r} 1 & 1 & 3 \\ 0 & 2 & 8 \end{array}\right] \frac{1}{2}R_2 \to R_2$

$\sim \left[\begin{array}{rr|r} 1 & 1 & 3 \\ 0 & 1 & 4 \end{array}\right] (-1)R_2 + R_1 \to R_1$

$\sim \left[\begin{array}{rr|r} 1 & 0 & -1 \\ 0 & 1 & 4 \end{array}\right]$

(C) Solution: $x_1 = -1$, $x_2 = 4$ *(10-3)*

25. (A) As a matrix equation the system becomes

$$\underset{A}{\begin{bmatrix} 1 & -3 \\ 2 & -5 \end{bmatrix}} \underset{X}{\begin{bmatrix} x_1 \\ x_2 \end{bmatrix}} = \underset{B}{\begin{bmatrix} k_1 \\ k_2 \end{bmatrix}}$$

(B) To find A^{-1}, we perform row operations on

$\left[\begin{array}{rr|rr} 1 & -3 & 1 & 0 \\ 2 & -5 & 0 & 1 \end{array}\right] (-2)R_1 + R_2 \to R_2$

$\sim \left[\begin{array}{rr|rr} 1 & -3 & 1 & 0 \\ 0 & 1 & -2 & 1 \end{array}\right] 3R_2 + R_1 \to R_1$

$\sim \left[\begin{array}{rr|rr} 1 & 0 & -5 & 3 \\ 0 & 1 & -2 & 1 \end{array}\right]$

Hence $A^{-1} = \begin{bmatrix} -5 & 3 \\ -2 & 1 \end{bmatrix}$

Check: $A^{-1}A = \begin{bmatrix} -5 & 3 \\ -2 & 1 \end{bmatrix}\begin{bmatrix} 1 & -3 \\ 2 & -5 \end{bmatrix} = \begin{bmatrix} (-5)1 + 3 \cdot 2 & (-5)(-3) + 3(-5) \\ (-2)1 + 1 \cdot 2 & (-2)(-3) + 1(-5) \end{bmatrix} = \begin{bmatrix} 1 & 0 \\ 0 & 1 \end{bmatrix}$

(C) The solution of $AX = B$ is $X = A^{-1}B$. Using the result of (B), we have

$$X = \begin{bmatrix} -5 & 3 \\ -2 & 1 \end{bmatrix}\begin{bmatrix} k_1 \\ k_2 \end{bmatrix}$$

If $k_1 = -2$, $k_2 = 1$, we have

$$\begin{bmatrix} x_1 \\ x_2 \end{bmatrix} = \begin{bmatrix} -5 & 3 \\ -2 & 1 \end{bmatrix}\begin{bmatrix} -2 \\ 1 \end{bmatrix} = \begin{bmatrix} (-5)(-2) + 3 \cdot 1 \\ (-2)(-2) + 1 \cdot 1 \end{bmatrix} = \begin{bmatrix} 13 \\ 5 \end{bmatrix} \quad x_1 = 13,\ x_2 = 5$$

(D) If $k_1 = 1$, $k_2 = -2$, reasoning as in (C) we have

$$\begin{bmatrix} x_1 \\ x_2 \end{bmatrix} = \begin{bmatrix} -5 & 3 \\ -2 & 1 \end{bmatrix}\begin{bmatrix} 1 \\ -2 \end{bmatrix} = \begin{bmatrix} (-5)1 + 3(-2) \\ (-2)1 + 1(-2) \end{bmatrix} = \begin{bmatrix} -11 \\ -4 \end{bmatrix} \quad x_1 = -11,\ x_2 = -4$$

(10-5)

26. We write the augmented matrix:

$$\left[\begin{array}{cc|c} 1 & 3 & 10 \\ 2 & -1 & -1 \end{array}\right] \quad (-2)R_1 + R_2 \rightarrow R_2$$

$$-2 \quad -6 \quad -20$$

↑
Need a 0 here

$$\sim \left[\begin{array}{cc|c} 1 & 3 & 10 \\ 0 & -7 & -21 \end{array}\right] \quad -\tfrac{1}{7}R_2 \rightarrow R_2 \quad \text{corresponds to the linear system} \quad \begin{aligned} x_1 + 3x_2 &= 10 \\ -7x_2 &= -21 \end{aligned}$$

↑
Need a 1 here

Need a 0 here
↓

$$\left[\begin{array}{cc|c} 1 & 3 & 10 \\ 0 & 1 & 3 \end{array}\right] \quad (-3)R_2 + R_1 \rightarrow R_1 \quad \text{corresponds to the linear system} \quad \begin{aligned} x_1 + 3x_2 &= 10 \\ 7x_2 &= 3 \end{aligned}$$

$$0 \quad -3 \quad -9$$

$$\sim \left[\begin{array}{cc|c} 1 & 0 & 1 \\ 0 & 1 & 3 \end{array}\right] \quad \text{corresponds to the linear system} \quad \begin{aligned} x_1 &= 1 \\ x_2 &= 3 \end{aligned}$$

The solution is $x_1 = 1$, $x_2 = 3$. Each pair of lines graphed below has the same intersection point, (1, 3).

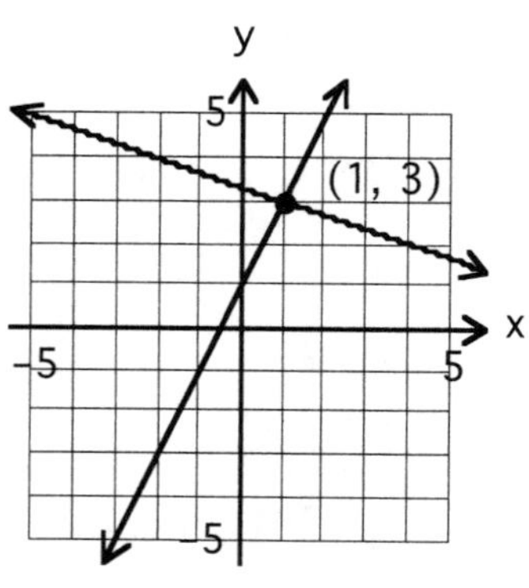

$x_1 + 3x_2 = 10$
$2x_1 - x_2 = -1$

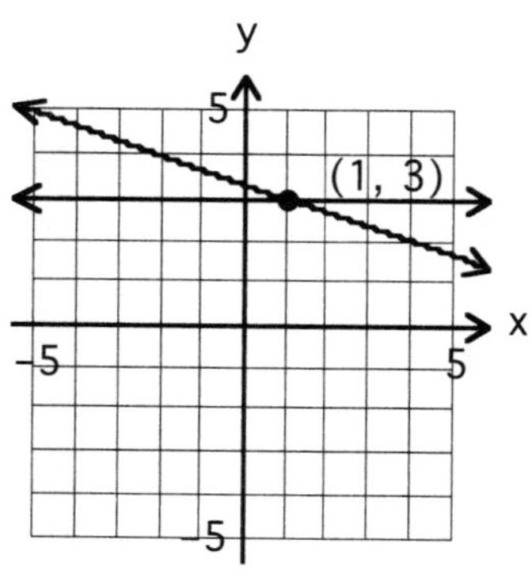

$x_1 + 3x_2 = 10$
$-7x_2 = -21$

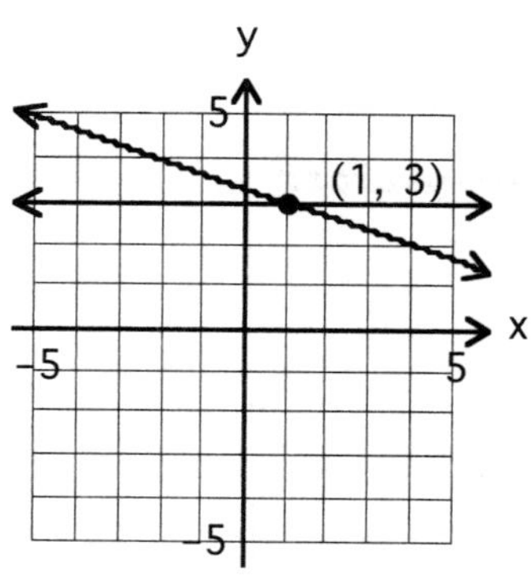

$x_1 + 3x_2 = 10$
$x_2 = 3$

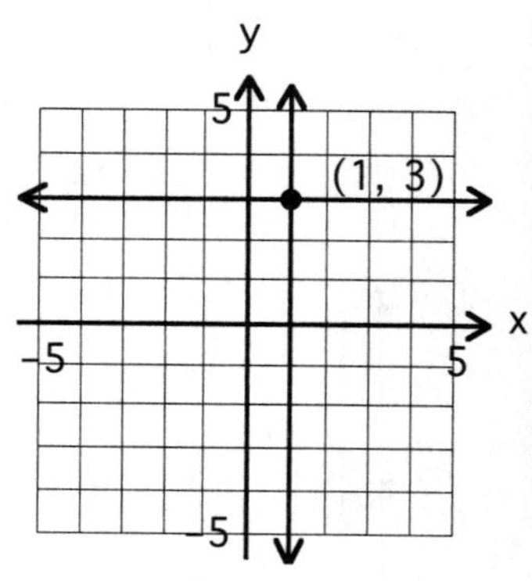

$x_1 = 1$
$x_2 = 3$

(10-1)

27. Here is a computer-generated graph of the system, entered as

$$y = \frac{7 + 2x}{3}$$
$$y = \frac{18 - 3x}{4}$$

After zooming in (graph not shown) the intersection point is located at (1.53, 3.35) to two decimal places.

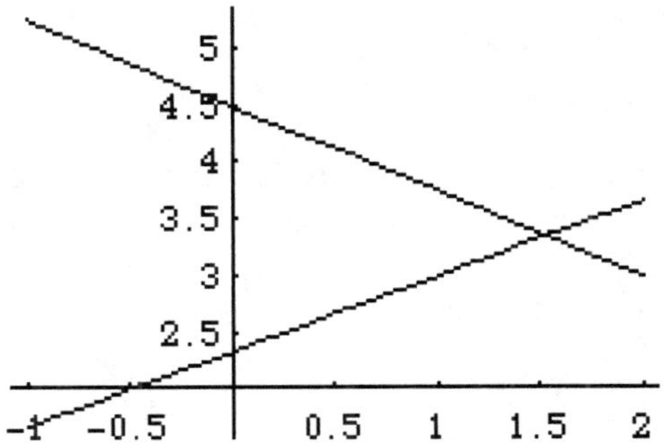

(10-1)

28. P_1: $1 = 1(2\cdot1 - 1) = 1\cdot1 = 1$
P_2: $1 + 5 = 2(2\cdot2 - 1)$
$6 = 6$
P_3: $1 + 5 + 9 = 3(2\cdot3 - 1)$
$15 = 15$ (11-2)

29. P_1: $1^2 + 1 + 2$ is divisible by 2
$4 = 2\cdot2$ true
P_2: $2^2 + 2 + 2$ is divisible by 2
$8 = 2\cdot4$ true
P_3: $3^2 + 3 + 2$ is divisible by 2
$14 = 2\cdot7$ true (11-2)

30. P_k: $1 + 5 + 9 + \cdots + (4k - 3) = k(2k - 1)$
P_{k+1}: $1 + 5 + 9 + \cdots + (4k - 3) + (4k + 1) = (k + 1)(2k + 1)$ (11-2)

31. P_k: $k^2 + k + 2 = 2r$ for some integer r
P_{k+1}: $(k + 1)^2 + (k + 1) + 2 = 2s$ for some integer s (11-2)

32. The parabola is opening either up or down and has an equation of the form $x^2 = 4ay$. Since (2, -8) is on the graph, we have:

$$2^2 = 4a(-8)$$
$$4 = -32a$$
$$a = -\frac{1}{8}$$

Thus, the equation of the parabola is

$$x^2 = 4\left(-\frac{1}{8}\right)y$$
$$x^2 = -\frac{1}{2}y$$
$$y = -2x^2$$

(9-1)

33. Make a rough sketch of the ellipse, locate focus and x intercepts, then determine y intercepts using the special triangle relationship.

$$\frac{x^2}{a^2} + \frac{y^2}{b^2} = 1$$
$$a = \frac{10}{2} = 5$$
$$b^2 = a^2 - c^2 = 5^2 - 3^2 = 25 - 9 = 16$$
$$b = 4$$
$$\frac{x^2}{25} + \frac{y^2}{16} = 1$$

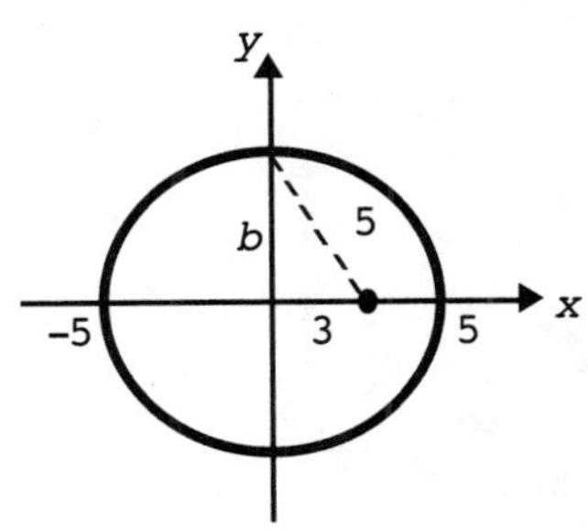

(9-2)

34. Start with $\frac{x^2}{a^2} - \frac{y^2}{b^2} = 1$ and find a and b.

$a = \frac{16}{2} = 8$

To find b, sketch the asymptote rectangle, label known parts, and use the Pythagorean Theorem.

$$b^2 = (\sqrt{89})^2 - 8^2$$
$$= 89 - 64$$
$$b^2 = 25$$
$$b = 5$$

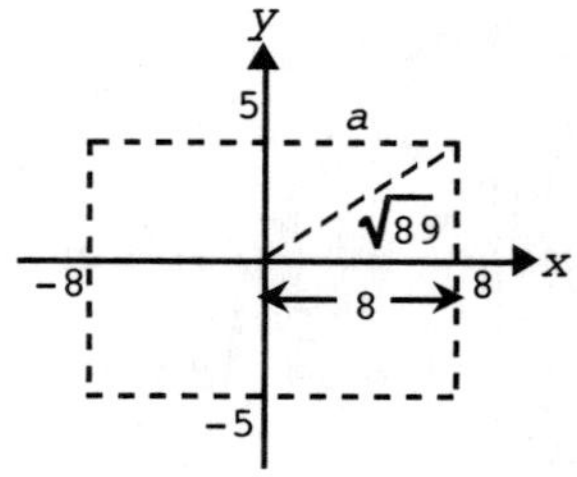

Thus, the equation is $\frac{x^2}{64} - \frac{y^2}{25} = 1$

(9-3)

35. $\left[\begin{array}{rrr|r} 1 & 2 & -1 & 3 \\ 0 & 1 & 1 & -2 \\ 2 & 3 & 1 & 0 \end{array}\right] (-2)R_1 + R_3 \rightarrow R_3$

$\sim \left[\begin{array}{rrr|r} 1 & 2 & -1 & 3 \\ 0 & 1 & 1 & -2 \\ 0 & -1 & 3 & -6 \end{array}\right] \begin{array}{l} (-2)R_2 + R_1 \rightarrow R_1 \\ \\ R_2 + R_3 \rightarrow R_3 \end{array}$

$\sim \left[\begin{array}{rrr|r} 1 & 0 & -3 & 7 \\ 0 & 1 & 1 & -2 \\ 0 & 0 & 4 & -8 \end{array}\right] \frac{1}{4} R_3 \rightarrow R_3$

$\sim \left[\begin{array}{rrr|r} 1 & 0 & -3 & 7 \\ 0 & 1 & 1 & -2 \\ 0 & 0 & 1 & -2 \end{array}\right] \begin{array}{l} 3R_3 + R_1 \rightarrow R_1 \\ (-1)R_3 + R_2 \rightarrow R_2 \\ \end{array}$

$\sim \left[\begin{array}{rrr|r} 1 & 0 & 0 & 1 \\ 0 & 1 & 0 & 0 \\ 0 & 0 & 1 & -2 \end{array}\right]$

Solution: $x_1 = 1$, $x_2 = 0$, $x_3 = -2$

(10-3)

36. $\left[\begin{array}{rrr|r} 1 & 1 & -2 & 2 \\ 0 & 4 & 6 & -1 \\ 0 & 6 & 9 & 0 \end{array}\right] (-\tfrac{2}{3})R_2 + R_3 \rightarrow R_3$

$\sim \left[\begin{array}{rrr|r} 1 & 1 & -2 & 2 \\ 0 & 4 & 6 & -1 \\ 0 & 0 & 0 & \frac{2}{3} \end{array}\right]$

The last row corresponds to the equation

$$0x_1 + 0x_2 + 0x_3 = \frac{2}{3},$$

hence there is no solution. (10-3)

37. $\left[\begin{array}{rrr|r} 1 & -2 & 1 & 1 \\ 3 & -2 & -1 & -5 \end{array}\right] (-3)R_1 + R_2 \rightarrow R_2$

$\sim \left[\begin{array}{rrr|r} 1 & -2 & 1 & 1 \\ 0 & 4 & -4 & -8 \end{array}\right] \frac{1}{4} R_2 \rightarrow R_2$

$\sim \left[\begin{array}{rrr|r} 1 & -2 & 1 & 1 \\ 0 & 1 & -1 & -2 \end{array}\right] 2R_2 + R_1 \rightarrow R_1$

$\sim \left[\begin{array}{rrr|r} 1 & 0 & -1 & -3 \\ 0 & 1 & -1 & -2 \end{array}\right]$

This corresponds to the system

$$x_1 - x_3 = -3$$
$$x_2 - x_3 = -2$$

Let $x_3 = t$

Then $x_2 = x_3 - 2$
$= t - 2$

$x_1 = x_3 - 3$
$= t - 3$

Hence $x_1 = t - 3$, $x_2 = t - 2$, $x_3 = t$ is a solution for every real number t.

(10-3)

38. (A) $MN = [1 \quad 2 \quad -1]\begin{bmatrix}1\\-1\\2\end{bmatrix} = [1 \cdot 1 + 2(-1) + (-1)2] = [-3]$

(B) $NM = \begin{bmatrix}1\\-1\\2\end{bmatrix}[1 \quad 2 \quad -1] = \begin{bmatrix}1 \cdot 1 & 1 \cdot 2 & 1(-1)\\(-1)1 & (-1)2 & (-1)(-1)\\2 \cdot 1 & 2 \cdot 2 & 2(-1)\end{bmatrix} = \begin{bmatrix}1 & 2 & -1\\-1 & -2 & 1\\2 & 4 & -2\end{bmatrix}$ *(10-4)*

39. (A) $LM = \begin{bmatrix}2 & -1 & 0\\1 & 2 & 1\end{bmatrix}\begin{bmatrix}1 & 2\\-1 & 0\\1 & 1\end{bmatrix} = \begin{bmatrix}2 \cdot 1 + (-1)(-1) + 0 \cdot 1 & 2 \cdot 2 + (-1)0 + 0 \cdot 1\\1 \cdot 1 + 2(-1) + 1 \cdot 1 & 1 \cdot 2 + 2 \cdot 0 + 1 \cdot 1\end{bmatrix}$

$LM - 2N = \begin{bmatrix}3 & 4\\0 & 3\end{bmatrix} - 2\begin{bmatrix}2 & 1\\-1 & 0\end{bmatrix} = \begin{bmatrix}3 & 4\\0 & 3\end{bmatrix} - \begin{bmatrix}4 & 2\\-2 & 0\end{bmatrix} = \begin{bmatrix}-1 & 2\\2 & 3\end{bmatrix}$

(B) Since ML is a 3×3 matrix and N is a 2×2 matrix, $ML + N$ is not defined. *(10-4)*

40. $x^2 - 3xy + 3y^2 = 1$
$xy = 1$

Solve for y in the second equation.

$$y = \frac{1}{x}$$

$$x^2 - 3x\left(\frac{1}{x}\right) + 3\left(\frac{1}{x}\right)^2 = 1$$

$$x^2 - 3 + \frac{3}{x^2} = 1 \quad x \neq 0$$

$$x^4 - 3x^2 + 3 = x^2$$

$$x^4 - 4x^2 + 3 = 0$$

$$(x^2 - 1)(x^2 - 3) = 0$$

$x^2 - 1 = 0$ $\qquad$ $x^2 - 3 = 0$
$x^2 = 1$ $\qquad$ $x^2 = 3$
$x = \pm 1$ $\qquad$ $x = \pm\sqrt{3}$

For $x = 1$: $y = \frac{1}{1}$, $y = 1$

For $x = -1$: $y = \frac{1}{-1}$, $y = -1$

For $x = \sqrt{3}$: $y = \frac{1}{\sqrt{3}}$, $y = \frac{\sqrt{3}}{3}$

For $x = -\sqrt{3}$: $y = \frac{1}{-\sqrt{3}}$, $y = -\frac{\sqrt{3}}{3}$

Solutions: $(1, 1)$, $(-1, -1)$, $\left(\sqrt{3}, \frac{\sqrt{3}}{3}\right)$, $\left(-\sqrt{3}, -\frac{\sqrt{3}}{3}\right)$ *(10-6)*

41. $x^2 - 3xy + y^2 = -1$
$x^2 - xy = 0$

Factor the left side of the equation that has a zero constant term.

$x(x - y) = 0$
$x = 0$ or $x = y$

Thus the original system is equivalent to the two systems

$x^2 - 3xy + y^2 = -1$ $\qquad$ $x^2 - 3xy + y^2 = -1$
$x = 0$ $\qquad$ $x = y$

These systems are solved by substitution.

First system:

$$\begin{aligned} x^2 - 3xy + y^2 &= -1 \\ x &= 0 \\ 0^2 - 3(0)y + y^2 &= -1 \\ y^2 &= -1 \\ y &= \pm i \\ x &= 0 \end{aligned}$$

Second system:

$$\begin{aligned} x^2 - 3xy + y^2 &= -1 \\ x &= y \\ y^2 - 3yy + y^2 &= -1 \\ -y^2 &= -1 \\ y^2 &= 1 \\ y &= \pm 1 \end{aligned}$$

For $y = 1$, $x = 1$ For $y = -1$, $x = -1$

Solutions: $(1, 1)$, $(-1, -1)$, $(0, i)$, $(0, -i)$ *(10-6)*

42. $2\sqrt{3}xy + 2y^2 + 3 = 0$

Use the formula $\cot 2\theta = \dfrac{A - C}{B}$ to find the appropriate angle.

$$\cot 2\theta = \frac{0 - 2}{2\sqrt{3}} = \frac{-2}{2\sqrt{3}} = \frac{-1}{\sqrt{3}}$$

Note that $\cot 120° = \dfrac{\cos(120°)}{\sin(120°)} = \dfrac{-\frac{1}{2}}{\frac{\sqrt{3}}{2}} = -\dfrac{1}{\sqrt{3}}$, so $2\theta = 120°$, and $\theta = 60°$.

The rotation formulas are:

$$x = \cos 60°x' - \sin 60°y' \qquad y = \sin 60°x' + \cos 60°y'$$

$$x = \frac{1}{2}x' - \frac{\sqrt{3}}{2}y' \qquad y = \frac{\sqrt{3}}{2}x' + \frac{1}{2}y'$$

Substituting these into the original equation, we get:

$$2\sqrt{3}\left(\frac{1}{2}x' - \frac{\sqrt{3}}{2}y'\right)\left(\frac{\sqrt{3}}{2}x' + \frac{1}{2}y'\right) + 2\left(\frac{\sqrt{3}}{2}x' + \frac{1}{2}y'\right)^2 + 3 = 0$$

$$2\sqrt{3}\left(\frac{\sqrt{3}}{4}x'^2 + \frac{1}{4}x'y' - \frac{3}{4}x'y' - \frac{\sqrt{3}}{4}y'^2\right) + 2\left(\frac{3}{4}x'^2 + \frac{\sqrt{3}}{4}x'y' + \frac{\sqrt{3}}{4}x'y' + \frac{1}{4}y'^2\right) + 3 = 0$$

$$2\sqrt{3}\left(\frac{3}{4}x'^2 - \frac{1}{2}x'y' - \frac{\sqrt{3}}{4}y'^2\right) + 2\left(\frac{3}{4}x'^2 + \frac{\sqrt{3}}{2}x'y' + \frac{1}{4}y'^2\right) + 3 = 0$$

$$\frac{3}{2}x'^2 - \sqrt{3}x'y' - \frac{3}{2}y'^2 + \frac{3}{2}x'^2 + \sqrt{3}x'y' + \frac{1}{2}y'^2 + 3 = 0$$

$$3x'^2 - y'^2 + 3 = 0$$

This can be rewritten as:

$$y'^2 - 3x'^2 = 3 \quad \text{or}$$

$$\frac{y'^2}{3} - x'^2 = 1$$

which is the equation of a hyperbola.

To find the graph on a graphing utility, we solve the original equation for y using the quadratic formula:

$2y^2 + 2\sqrt{3}xy + 3 = 0$ $a = 2$, $b = 2\sqrt{3}x$, $c = 3$

$$y = \frac{-2\sqrt{3}x \pm \sqrt{12x^2 - 4(2)(3)}}{2(2)} = \frac{-2\sqrt{3}x \pm \sqrt{12x^2 - 24}}{4}$$

Enter $y_1 = (-2\sqrt{3}x + \sqrt{12x^2 - 24}) \div 4$ and $y_2 = (-2\sqrt{3}x - \sqrt{12x^2 - 24}) \div 4$.

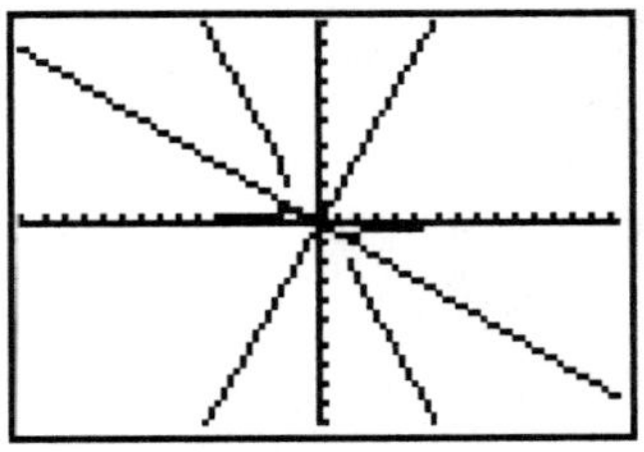

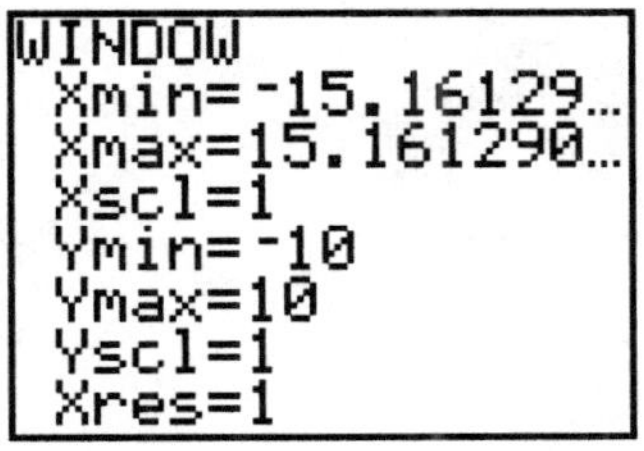

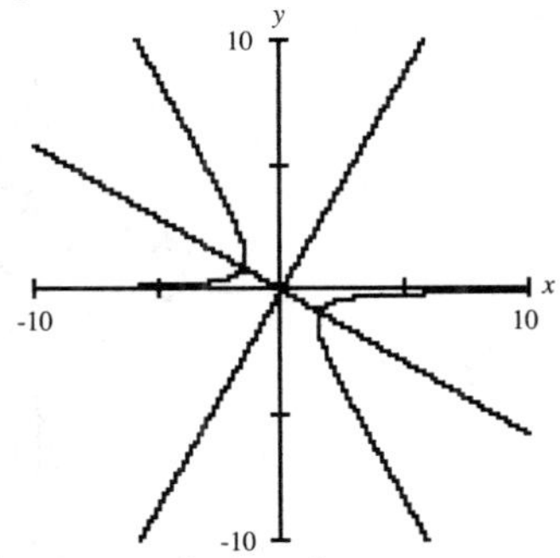

The graph is shown in a square viewing window with the rotated $x' + y'$ axes drawn in as well. (9-4)

43. $x^2 + 2xy + y^2 + 4\sqrt{2}x - 4\sqrt{2}y = 0$

Use the formula $\cot 2\theta = \dfrac{A - C}{B}$ to find the appropriate angle.

$$\cot 2\theta = \frac{1 - 1}{2} = 0$$

Since $\cot 90° = \dfrac{\cos 90°}{\sin 90°} = \dfrac{0}{1} = 0$, $2\theta = 90°$, and $\theta = 45°$.

The rotation formulas are:

$$x = \cos 45°x' - \sin 45°y' \qquad y = \sin 45°x' + \cos 45°y'$$

$$x = \frac{\sqrt{2}}{2}x' - \frac{\sqrt{2}}{2}y' \qquad y = \frac{\sqrt{2}}{2}x' + \frac{\sqrt{2}}{2}y'$$

Substituting these into the original equation, we get:

$$\left(\frac{\sqrt{2}}{2}x' - \frac{\sqrt{2}}{2}y'\right)^2 + 2\left(\frac{\sqrt{2}}{2}x' + \frac{\sqrt{2}}{2}y'\right) + \left(\frac{\sqrt{2}}{2}x' + \frac{\sqrt{2}}{2}y'\right)^2 + 4\sqrt{2}\left(\frac{\sqrt{2}}{2}x' - \frac{\sqrt{2}}{2}y'\right) - 4\sqrt{2}\left(\frac{\sqrt{2}}{2}x' + \frac{\sqrt{2}}{2}y'\right) = 0$$

$$\frac{1}{2}x'^2 - \frac{1}{2}x'y' - \frac{1}{2}x'y' + \frac{1}{2}y'^2 + 2\left(\frac{1}{2}x'^2 - \frac{1}{2}y'^2\right) + \frac{1}{2}x'^2 + \frac{1}{2}x'y' + \frac{1}{2}x'y' + \frac{1}{2}y'^2 + 4x' - 4y' - 4x' - 4y' = 0$$

$$\frac{1}{2}x'^2 - x'y' + \frac{1}{2}y'^2 + x'^2 - y'^2 + \frac{1}{2}x'^2 + x'y' + \frac{1}{2}y'^2 + 4x' - 4y' - 4x' - 4y' = 0$$

$$2x'^2 - 8y' = 0$$
$$8y' = 2x'^2$$
$$y = \frac{1}{4}x'^2$$

This is the equation of a parabola with vertex at the origin rotated through 45°. To graph, find the intercepts:

x intercept: $x^2 + 4\sqrt{2}x = 0$
$x(x + 4\sqrt{2}) = 0$
$x = 0, -4\sqrt{2}$

y intercept: $y^2 - 4\sqrt{2}y = 0$
$y(y - 4\sqrt{2}) = 0$
$y = 0, 4\sqrt{2}$

The graph is shown below with the rotated axes.

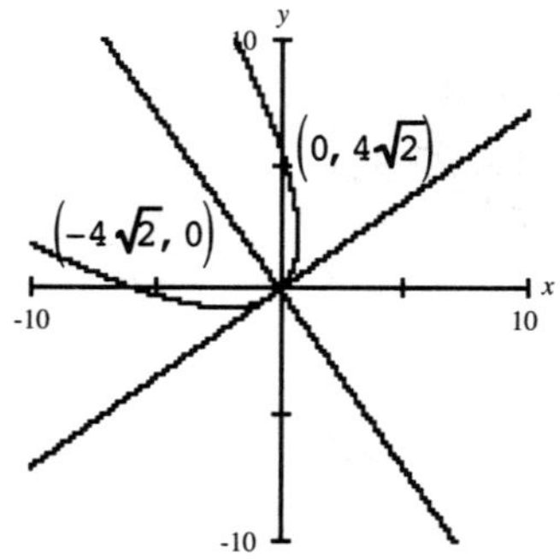

(9-4)

44. Before we can enter these equations in our graphing utility, we must solve for y:

$$x^2 + 2xy - y^2 = 1 \qquad\qquad 9x^2 + 4xy + y^2 = 15$$
$$y^2 - 2xy - x^2 + 1 = 0 \qquad\qquad y^2 + 4xy + 9x^2 - 15 = 0$$

Applying the quadratic formula to each equation, we have

$$y = \frac{2x \pm \sqrt{4x^2 - 4(-x^2 + 1)}}{2} \qquad\qquad y = \frac{-4x \pm \sqrt{16x^2 - 4(9x^2 - 15)}}{2}$$
$$y = \frac{2x \pm \sqrt{8x^2 - 4}}{2} \qquad\qquad y = \frac{-4x \pm \sqrt{60 - 20x^2}}{2}$$
$$y = x \pm \sqrt{2x^2 - 1} \qquad\qquad y = -2x \pm \sqrt{15 - 5x^2}$$

Entering each of these four equations into a graphing utility produces the graph shown at the right.

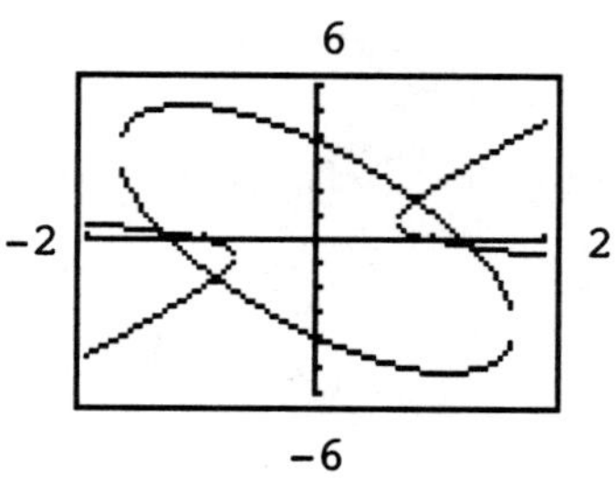

Zooming in on the four intersection points, or using a built-in intersection routine (details omitted), yields (-1.35, 0.28), (-0.87, -1.60), (0.87, 1.60), and (1.35, -0.28) to two decimal places.

(10-6)

45. $\sum_{k=1}^{5} k^k = 1^1 + 2^2 + 3^3 + 4^4 + 5^5 = 1 + 4 + 27 + 256 + 3{,}125 = 3{,}413$ (11-1)

46. $S_6 = \frac{2}{2!} - \frac{2^2}{3!} + \frac{2^3}{4!} - \frac{2^4}{5!} + \frac{2^5}{6!} - \frac{2^6}{7!}$

$k = 1, 2, 3, 4, 5, 6$

Noting that the terms alternate in sign, we can rewrite as follows:

$$S_6 = (-1)^2\frac{2}{2!} + (-1)^3\frac{2^2}{3!} + (-1)^4\frac{2^3}{4!} + (-1)^5\frac{2^4}{5!} + (-1)^6\frac{2^5}{6!} + (-1)^7\frac{2^6}{7!}$$

Clearly, $a_k = (-1)^{k+1}\frac{2^k}{(k+1)!}$

$$S_6 = \sum_{k=1}^{6} (-1)^{k+1}\frac{2^k}{(k+1)!}$$ (11-1)

47. $\frac{a_2}{a_1} = \frac{-36}{108} = -\frac{1}{3} = r.\ |r| < 1.$

Therefore, this infinite geometric series has a sum.

$$S_\infty = \frac{a_1}{1 - r} = \frac{108}{1 - \left(-\frac{1}{3}\right)} = 81$$ (11-3)

48. The solution region is unbounded. Two corner points are obvious from the graph: (0, 6) and (8, 0). The third corner point is obtained by solving the system
$3x + 2y = 12$
$x + 2y = 8$ to obtain (2, 3).

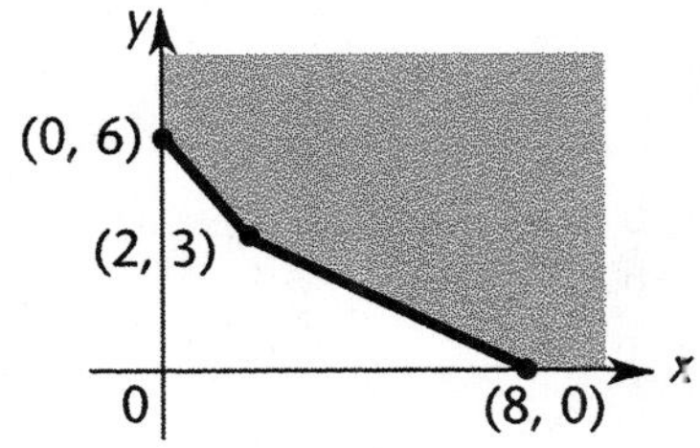

(10-7)

49. The feasible region is graphed as follows: The corner points (0, 7), (0, 0) and (8, 0) are obvious from the graph. The corner point (6, 4) is obtained by solving the system
$x + 2y = 14$
$2x + y = 16$
We now evaluate the objective function at each corner point.

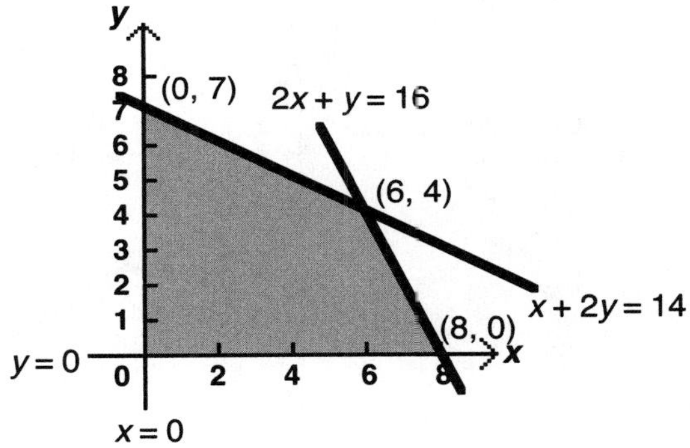

Corner Point (x, y)	Objective Function $z = 4x + 9y$	
(0, 7)	63	Maximum value
(0, 0)	0	
(8, 0)	32	
(6, 4)	60	

(10-8)

50. (A) As a matrix equation the system becomes

$$\underset{A}{\begin{bmatrix} 1 & 4 & 2 \\ 2 & 6 & 3 \\ 2 & 5 & 2 \end{bmatrix}} \underset{X}{\begin{bmatrix} x_1 \\ x_2 \\ x_3 \end{bmatrix}} = \underset{B}{\begin{bmatrix} k_1 \\ k_2 \\ k_3 \end{bmatrix}}$$

(B) To find A^{-1}, we perform row operations on

$$\left[\begin{array}{ccc|ccc} 1 & 4 & 2 & 1 & 0 & 0 \\ 2 & 6 & 3 & 0 & 1 & 0 \\ 2 & 5 & 2 & 0 & 0 & 1 \end{array}\right] \begin{matrix} \\ (-2)R_1 + R_2 \rightarrow R_2 \\ (-2)R_1 + R_3 \rightarrow R_3 \end{matrix}$$

$$\sim \left[\begin{array}{ccc|ccc} 1 & 4 & 2 & 1 & 0 & 0 \\ 0 & -2 & -1 & -2 & 1 & 0 \\ 0 & -3 & -2 & -2 & 0 & 1 \end{array}\right] \begin{matrix} \\ -\frac{1}{2}R_2 \rightarrow R_2 \\ \\ \end{matrix}$$

$$\sim \left[\begin{array}{ccc|ccc} 1 & 4 & 2 & 1 & 0 & 0 \\ 0 & 1 & \frac{1}{2} & 1 & -\frac{1}{2} & 0 \\ 0 & -3 & -\frac{1}{2} & -2 & 0 & 1 \end{array}\right] \begin{matrix} (-4)R_2 + R_1 \rightarrow R_1 \\ \\ 3R_2 + R_3 \rightarrow R_3 \end{matrix}$$

$$\sim \left[\begin{array}{ccc|ccc} 1 & 0 & 0 & -3 & 2 & 0 \\ 0 & 1 & \frac{1}{2} & 1 & -\frac{1}{2} & 0 \\ 0 & 0 & -\frac{1}{2} & 1 & -\frac{3}{2} & 1 \end{array}\right] \begin{matrix} \\ R_3 + R_2 \rightarrow R_2 \\ \\ \end{matrix}$$

$$\sim \left[\begin{array}{ccc|ccc} 1 & 0 & 0 & -3 & 2 & 0 \\ 0 & 1 & 0 & 2 & -2 & 1 \\ 0 & 0 & -\frac{1}{2} & 1 & -\frac{3}{2} & 1 \end{array}\right] -2R_3 \to R_3$$

$$\sim \left[\begin{array}{ccc|ccc} 1 & 0 & 0 & -3 & 2 & 0 \\ 0 & 1 & 0 & 2 & -2 & 1 \\ 0 & 0 & 1 & -2 & 3 & -2 \end{array}\right]$$

Hence $A^{-1} = \begin{bmatrix} -3 & 2 & 0 \\ 2 & -2 & 1 \\ -2 & 3 & -2 \end{bmatrix}$

Check: $A^{-1}A = \begin{bmatrix} -3 & 2 & 0 \\ 2 & -2 & 1 \\ -2 & 3 & -2 \end{bmatrix} \begin{bmatrix} 1 & 4 & 2 \\ 2 & 6 & 3 \\ 2 & 5 & 2 \end{bmatrix}$

$$= \begin{bmatrix} (-3)1 + 2 \cdot 2 + 0 \cdot 2 & (-3)4 + 2 \cdot 6 + 0 \cdot 5 & -3 \cdot 2 + 2 \cdot 3 + 0 \cdot 2 \\ 2 \cdot 1 + (-2)2 + 1 \cdot 2 & 2 \cdot 4 + (-2)6 + 1 \cdot 5 & 2 \cdot 2 + (-2)3 + 1 \cdot 2 \\ (-2)1 + 3 \cdot 2 + (-2)2 & (-2)4 + 3 \cdot 6 + (-2)5 & (-2)2 + 3 \cdot 3 + (-2)2 \end{bmatrix}$$

$$= \begin{bmatrix} 1 & 0 & 0 \\ 0 & 1 & 0 \\ 0 & 0 & 1 \end{bmatrix}$$

(C) The solution of $AX = B$ is $X = A^{-1}B$. Using the result of (B), we have

$$X = \begin{bmatrix} -3 & 2 & 0 \\ 2 & -2 & 1 \\ -2 & 3 & -2 \end{bmatrix} \begin{bmatrix} k_1 \\ k_2 \\ k_3 \end{bmatrix}$$

If $k_1 = -1$, $k_2 = 2$, $k_3 = 1$, we have

$$\begin{bmatrix} x_1 \\ x_2 \\ x_3 \end{bmatrix} = \begin{bmatrix} -3 & 2 & 0 \\ 2 & -2 & 1 \\ -2 & 3 & -2 \end{bmatrix} \begin{bmatrix} -1 \\ 2 \\ 1 \end{bmatrix} = \begin{bmatrix} 7 \\ -5 \\ 6 \end{bmatrix} \quad x_1 = 7,\ x_2 = -5,\ x_3 = 6$$

(D) If $k_1 = 2$, $k_2 = 0$, $k_3 = -1$, reasoning as in (C) we have

$$\begin{bmatrix} x_1 \\ x_2 \\ x_3 \end{bmatrix} = \begin{bmatrix} -3 & 2 & 0 \\ 2 & -2 & 1 \\ -2 & 3 & -2 \end{bmatrix} \begin{bmatrix} 2 \\ 0 \\ -1 \end{bmatrix} = \begin{bmatrix} -6 \\ 3 \\ -2 \end{bmatrix} \quad x_1 = -6,\ x_2 = 3,\ x_3 = -2$$

(10-5)

51.

	Case 1	Case 2	Case 3
O_1: Select the first letter	N_1: 6 ways	6 ways	6 ways
O_2: Select the second letter	N_2: 5 ways	6 ways	5 ways (exclude first letter)
O_3: Select the third letter	N_3: 4 ways	6 ways	5 ways (exclude second letter)
O_4: Select the fourth letter	N_4: 3 ways	6 ways	5 ways (exclude third letter)
	6·5·4·3 = 360 words	6·6·6·6 = 1,296 words	6·5·5·5 = 750 words

(11-4)

52. The sample space is the set of all possible ways to choose 5 players out of 12.
$n(S) = C_{12,5}$

The event E is the set of all those choices that include 2 particular people, thus

O_1: Choose the two centers — N_1: 1

O_2: Choose 3 more people out of the 10 remaining — N_2: $C_{10,3}$

Applying the multiplication principle, $n(E) = N_1 \cdot N_2 = 1 \cdot C_{10,3} = C_{10,3}$

$$P(E) = \frac{n(E)}{n(S)} = \frac{C_{10,3}}{C_{12,5}} = \frac{10!}{3!(10-3)!} \div \frac{12!}{5!(12-5)!} = 120 \div 792 = \frac{5}{33} = .\overline{15}$$

(11-5)

53. (A) P (this event) $= \dfrac{195 + 170}{1000} = .365$

(B) A sample space of equally likely events is:

$S = \{1, 2, 3, 4, 5, 6\} \quad n(S) = 6$

The event E is the set of those outcomes that are divisible by 3.

$E = \{3, 6\}.\ n(E) = 2$

Then $P(E) = \dfrac{n(E)}{n(S)} = \dfrac{2}{6} = \dfrac{1}{3}$ (11-5)

54.

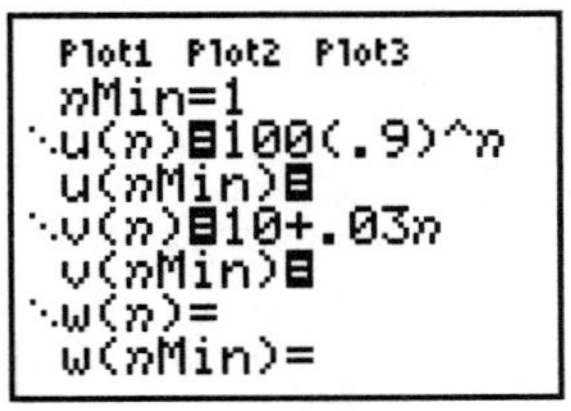

Enter the sequences in the sequence editor.

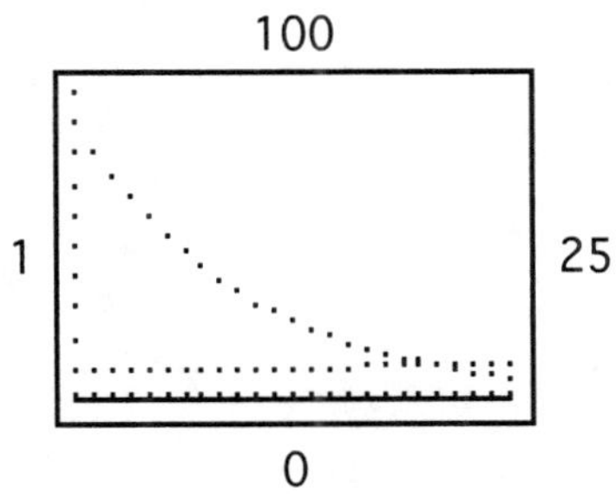

Graph the sequences.

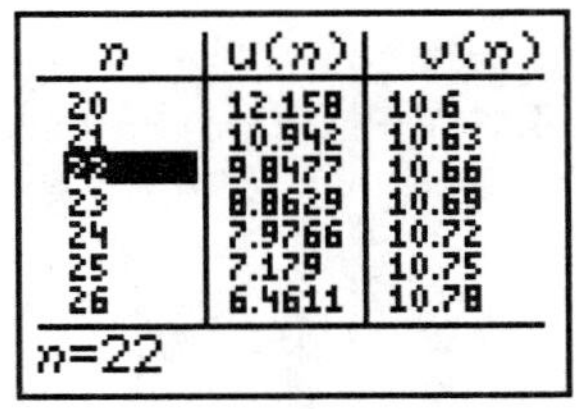

Display the points in a table.

From the graph, 22 is the least positive integer n such that $a_n < b_n$. The table confirms this. (11-3)

55. (A) $P_{25,5} = 25 \cdot 24 \cdot 23 \cdot 22 \cdot 21 = 6,375,600$

(B) $C_{25,5} = \dfrac{25!}{5!(25-5)!} = \dfrac{25!}{5!\,20!} = \dfrac{25 \cdot 24 \cdot 23 \cdot 22 \cdot 21 \cdot 20!}{5 \cdot 4 \cdot 3 \cdot 2 \cdot 1 \cdot 20!} = 53,130$

(C) $\dbinom{25}{5} = \dfrac{25!}{5!(25-5)!} = 53,130$ (11-4, 11-6)

56.

$$\left(a + \frac{1}{2}b\right)^6 = \sum_{k=0}^{6} \binom{6}{k} a^{6-k}\left(\frac{1}{2}b\right)^k$$

$$= \binom{6}{0}a^6 + \binom{6}{1}a^5\left(\frac{1}{2}b\right)^1 + \binom{6}{2}a^4\left(\frac{1}{2}b\right)^2 + \binom{6}{3}a^3\left(\frac{1}{2}b\right)^3 + \binom{6}{4}a^2\left(\frac{1}{2}b\right)^4 + \binom{6}{5}a\left(\frac{1}{2}b\right)^5 + \binom{6}{6}\left(\frac{1}{2}b\right)^6$$

$$= a^6 + 3a^5b + \frac{15}{4}a^4b^2 + \frac{5}{2}a^3b^3 + \frac{15}{16}a^2b^4 + \frac{3}{16}ab^5 + \frac{1}{64}b^6$$ (11-6)

57. In the expansion of $(a + b)^n$, the exponent of b in the rth term is $r - 1$ and the exponent of a is $n - (r - 1)$. Here, in the first case, $r = 5$, $n = 10$.

Fifth term $= \dbinom{10}{4}(3x)^6(-y)^4$

$= 153,090x^6y^4$

In the second case, $r = 8$, n is still 10.

Eighth term $= \dbinom{10}{7}(3x)^3(-y)^7$

$= -3,240x^3y^7$ (11-6)

58. P_n: $1 + 5 + 9 + \cdots + (4n - 3) = n(2n - 1)$

Part 1: Show that P_1 is true.

P_1: $1 = 1(2\cdot 1 - 1) = 1$ True

Part 2: Show that if P_k is true, then P_{k+1} is true.

Write out P_k and P_{k+1}.

P_k: $1 + 5 + 9 + \cdots + (4k - 3) = k(2k - 1)$

P_{k+1}: $1 + 5 + 9 + \cdots + (4k - 3) + (4k + 1) = (k + 1)(2k + 1)$

We start with P_k:

$1 + 5 + 9 + \cdots + (4k - 3) = k(2k - 1)$

Adding $4k + 1$ to both sides:

$$\begin{aligned} 1 + 5 + 9 + \cdots + (4k - 3) + (4k + 1) &= k(2k - 1) + 4k + 1 \\ &= 2k^2 - k + 4k + 1 \\ &= 2k^2 + 3k + 1 \\ &= (k + 1)(2k + 1) \end{aligned}$$

We have shown that if P_k is true, then P_{k+1} is true.

Conclusion: P_n is true for all positive integers n. *(10-2)*

59. P_n: $n^2 + n + 2 = 2p$ for some integer p.

Part 1: Show that P_1 is true.

P_1: $1^2 + 1 + 2 = 4 = 2\cdot 2$ is true.

Part 2: Show that if P_k is true, then P_{k+1} is true.

Write out P_k and P_{k+1}.

P_k: $k^2 + k + 2 = 2r$ for some integer r

P_{k+1}: $(k + 1)^2 + (k + 1) + 2 = 2s$ for some integer s

We start with P_k:

$k^2 + k + 2 = 2r$ for some integer r

$$\begin{aligned} \text{Now, } (k + 1)^2 + (k + 1) + 2 &= k^2 + 2k + 1 + k + 1 + 2 \\ &= k^2 + k + 2 + 2k + 2 \\ &= 2r + 2k + 2 \\ &= 2(r + k + 2) \end{aligned}$$

Therefore,

$(k + 1)^2 + (k + 1) + 2 = 2s$ for some integer s $(= r + k + 2)$

$(k + 1)^2 + (k + 1) + 2$ is divisible by 2.

We have shown that if P_k is true, then P_{k+1} is true.

Conclusion: P_n is true for all positive integers n. *(10-2)*

60. We are to find $51 + 53 + \cdots + 499$. This is the sum of an arithmetic sequence, S_n, with $d = 2$.

First, find n:

$$\begin{aligned} a_n &= a_1 + (n - 1)d \\ 499 &= 51 + (n - 1)2 \\ n &= 225 \end{aligned}$$

Now, find S_{225}

$$\begin{aligned} S_n &= \frac{n}{2}(a_1 + a_n) \\ S_{225} &= \frac{225}{2}(51 + 499) \\ &= 61{,}875 \end{aligned}$$

(11-3)

61. $2.\overline{45} = 2.454545\ldots$
$= 2 + 0.454545\ldots$

$0.454545\ldots = 0.45 + 0.0045 + 0.000045 + \cdots$

This is an infinite geometric series with $a_1 = 0.45$ and $r = 0.01$. Thus

$$S_\infty = \frac{a_1}{1 - r} = \frac{0.45}{1 - 0.01} = \frac{0.45}{0.99} = \frac{5}{11}$$

Hence, $2.\overline{45} = 2 + \frac{5}{11} = \frac{27}{11}$ (11-3)

62. Graphing the sequence, we obtain the graph at the right. The solid line indicates 0.01. The largest term of the sequence is a_{27}. From the table display,

$$a_{27} = \binom{30}{27}(0.1)^{30-27}(0.9)^{27} = 0.236088$$

There are 8 terms larger than 0.01, as can be seen from the graph or the table display (details omitted).

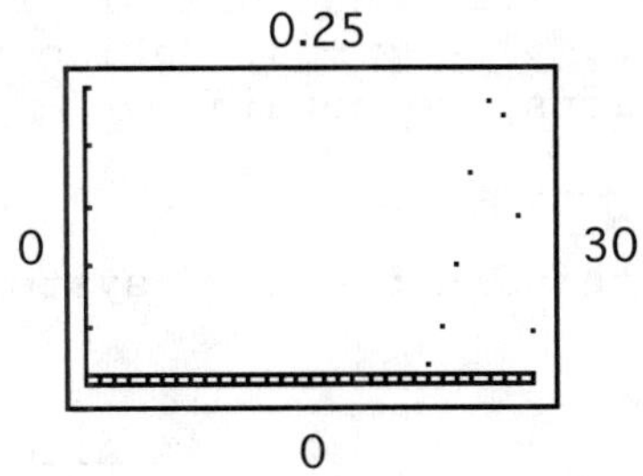

(11-6)

63. $4x + 4y - y^2 + 8 = 0$

$$4x + 8 = y^2 - 4y$$
$$4x + 8 + 4 = y^2 - 4y + 4$$
$$4x + 12 = (y - 2)^2$$
$$4(x + 3) = (y - 2)^2$$

This is the equation of a parabola opening right with vertex at $(h, k) = (-3, 2)$.
The equations of translation are $x' = x + 3$, $y' = y - 2$. Making these substitutions, we obtain

$$4x' = y'^2$$

We graph this in the $x'y'$ system.

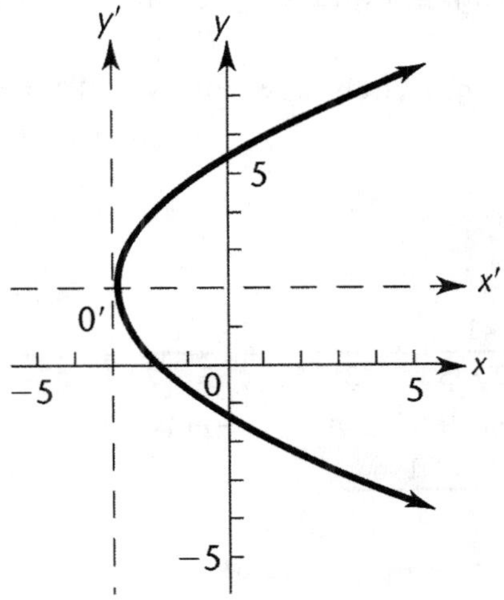

(9-1, 9-4)

64.

$$x^2 + 2x - 4y^2 - 16y + 1 = 0$$
$$x^2 + 2x - 4y^2 - 16y = -1$$
$$x^2 + 2x + ? - 4(y^2 + 4y + ?) = -1$$
$$x^2 + 2x + 1 - 4(y^2 + 4y + 4) = -1 + 1 - 16$$
$$(x + 1)^2 - 4(y + 2)^2 = -16$$
$$\frac{(y + 2)^2}{4} - \frac{(x + 1)^2}{16} = 1$$

This is the equation of a hyperbola with center at $(-1, -2)$. The equations of translation are $x' = x + 1$, $y' = y + 2$. Making these substitutions, we obtain

$$\frac{y'^2}{4} - \frac{x'^2}{16} = 1$$

We graph this in the $x'y'$ system.

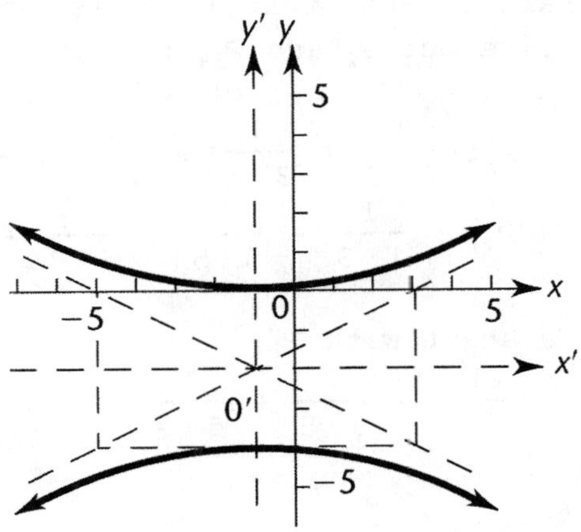

(9-3, 9-4)

65.

$$4x^2 - 16x + 9y^2 + 54y + 61 = 0$$
$$4x^2 - 16x + 9y^2 + 54y = -61$$
$$4(x^2 - 4x + ?) + 9(y^2 + 6y + ?) = -61$$
$$4(x^2 - 4x + 4) + 9(y^2 + 6y + 9) = -61 + 16 + 81$$
$$4(x - 2)^2 + 9(y + 3)^2 = 36$$
$$\frac{(x - 2)^2}{9} + \frac{(y + 3)^2}{4} = 1$$

This is the equation of an ellipse with center at (2, -3). The equations of translation are $x' = x - 2$, $y' = y + 3$. Making these substitutions, we obtain

$$\frac{x'^2}{9} + \frac{y'^2}{4} = 1$$

We graph this in the $x'y'$ system.

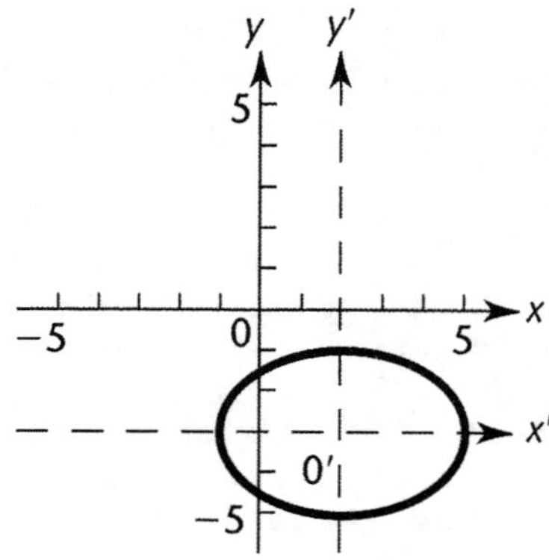

(9-2, 9-4)

66.

	Allowing digits to repeat	No digit can be repeated
O_1: Selecting first digit	N_1: 10 ways	N_1: 10 ways
O_2: Selecting second digit	N_2: 10 ways	N_2: 9 ways
⋮	⋮	⋮
O_9: Selecting ninth digit	N_9: 10 ways	N_9: 2 ways
	$10\cdot10\cdot10\cdot10\cdot10\cdot10\cdot10\cdot10\cdot10$	$P_{10,9} = 10\cdot9\cdot8\cdot7\cdot6\cdot5\cdot4\cdot3\cdot2$
	$= 10^9$ zip codes	$= 3{,}628{,}800$ zip codes

(11-1)

67. Prove P_n: $\frac{1}{1 \cdot 3} + \frac{1}{3 \cdot 5} + \frac{1}{5 \cdot 7} + \cdots + \frac{1}{(2n - 1)(2n + 1)} = \frac{n}{2n + 1}$

Part 1: Show that P_1 is true.

$$P_1: \frac{1}{1 \cdot 3} = \frac{1}{2 \cdot 1 + 1}$$
$$\frac{1}{3} = \frac{1}{3} \quad P_1 \text{ is true}$$

Part 2: Show that if P_k is true, then P_{k+1} is true.
Write out P_k and P_{k+1}:

$$P_k: \frac{1}{1 \cdot 3} + \frac{1}{3 \cdot 5} + \frac{1}{5 \cdot 7} + \cdots + \frac{1}{(2k - 1)(2k + 1)} = \frac{k}{2k + 1}$$
$$P_{k+1}: \frac{1}{1 \cdot 3} + \frac{1}{3 \cdot 5} + \frac{1}{5 \cdot 7} + \cdots + \frac{1}{(2k - 1)(2k + 1)} + \frac{1}{(2k + 1)(2k + 3)} = \frac{k + 1}{2k + 3}$$

We start with P_k:

$$\frac{1}{1 \cdot 3} + \frac{1}{3 \cdot 5} + \frac{1}{5 \cdot 7} + \cdots + \frac{1}{(2k - 1)(2k + 1)} = \frac{k}{2k + 1}$$

Adding $\dfrac{1}{(2k + 1)(2k + 3)}$ to both sides, we get

$$\frac{1}{1 \cdot 3} + \frac{1}{3 \cdot 5} + \frac{1}{5 \cdot 7} + \cdots + \frac{1}{(2k - 1)(2k + 1)} + \frac{1}{(2k + 1)(2k + 3)}$$
$$= \frac{k}{2k + 1} + \frac{1}{(2k + 1)(2k + 3)}$$
$$= \frac{k(2k + 3)}{(2k + 1)(2k + 3)} + \frac{1}{(2k + 1)(2k + 3)}$$
$$= \frac{k(2k + 3) + 1}{(2k + 1)(2k + 3)}$$
$$= \frac{2k^2 + 3k + 1}{(2k + 1)(2k + 3)}$$
$$= \frac{(2k + 1)(k + 1)}{(2k + 1)(2k + 3)}$$
$$= \frac{k + 1}{2k + 3}$$

We have shown that if P_k is true, then P_{k+1} is true.

Conclusion: P_n is true for positive integers n. *(10-2)*

68. Case 1: No repetition.
The sample space S is the set of all possible sequences of three digits, chosen from {1, 2, 3, 4, 5} with no repeats.
$n(S) = P_{5,3} = 5 \cdot 4 \cdot 3 = 60$
The event E is the set of those sequences that end in 2 or 4.
$n(E) = N_1 \cdot N_2 \cdot N_3$
where
$N_1 = 2$ (digits 2 or 4 only for the last digit)
$N_2 = 4$ (digits remaining for middle digit after last digit is chosen)
$N_3 = 3$ (digits remaining for first digit after others are chosen)
$n(E) = 2 \cdot 4 \cdot 3 = 24$
$P(E) = \dfrac{n(E)}{n(S)} = \dfrac{24}{60} = \dfrac{2}{5}$

Case 2: Allowing repetition
The sample space S is the set of all possible three digit numbers chosen from {1, 2, 3, 4, 5}.
$n(S) = 5 \cdot 5 \cdot 5 = 125$
The event E is the set of those numbers that end in 2 or 4.
$n(E) = 5 \cdot 5 \cdot 2 = 50$
$P(E) = \dfrac{n(E)}{n(S)} = \dfrac{50}{125} = \dfrac{2}{5}$ *(11-5)*

69. (A) The matrix is in fact

$$\left[\begin{array}{ccc|c} 1 & 0 & 5 & 2 \\ 0 & 1 & 3 & 6 \\ 0 & 0 & 0 & 0 \end{array}\right]$$

Thus there are an infinite number of solutions ($x_3 = t$, $x_2 = 6 - 3t$, $x_1 = 2 + 5t$, for t any real number.)

(B) The matrix is in fact

$$\left[\begin{array}{ccc|c} 1 & 0 & -5 & 2 \\ 0 & 1 & 3 & 6 \\ 0 & 0 & 0 & n \end{array}\right]$$

The third row corresponds to the equation $0x_1 + 0x_2 + 0x_3 = n$. This is impossible. The system has no solution.

(C) The system is independent. There is one solution. (10-3)

70. If $A^2 = A$ and A^{-1} exists, then $A^{-1}A^2 = A^{-1}A$ or $A^{-1}AA = I$ or $IA = I$. Thus, $A = I$, the $n \times n$ identity matrix. (10-5)

71. L, M, and P are in reduced form.
N is not in reduced form; it violates condition 1 of the definition of reduced matrix, Section 10-3: there is a row of 0's above rows having non-zero elements.
(10-3)

72. True. For example,

$$\begin{bmatrix} a_{11} & a_{12} & a_{13} \\ 0 & a_{22} & a_{23} \\ 0 & 0 & a_{33} \end{bmatrix} + \begin{bmatrix} b_{11} & b_{12} & b_{13} \\ 0 & b_{22} & b_{23} \\ 0 & 0 & b_{33} \end{bmatrix} = \begin{bmatrix} a_{11} + b_{11} & a_{12} + b_{12} & a_{13} + b_{13} \\ 0 & a_{22} + b_{22} & a_{23} + b_{23} \\ 0 & 0 & a_{33} + b_{33} \end{bmatrix}$$

proves the statement for 3 ∞ 3 matrices. A similar proof can be given for any order matrix. (10-4)

73. True. For example, $\begin{bmatrix} a_{11} & 0 \\ a_{21} & a_{22} \end{bmatrix}\begin{bmatrix} b_{11} & 0 \\ b_{21} & b_{22} \end{bmatrix} = \begin{bmatrix} a_{11}b_{11} & 0 \\ a_{21}b_{11} + a_{22}b_{21} & a_{22}b_{22} \end{bmatrix}$ proves the statement for 2 × 2 matrices. A similar proof can be given for any order matrix. (10-4)

74. False. For example, $\begin{bmatrix} 1 & 1 \\ 0 & 1 \end{bmatrix} + \begin{bmatrix} 1 & 0 \\ 1 & 1 \end{bmatrix} = \begin{bmatrix} 2 & 1 \\ 1 & 2 \end{bmatrix}$ which is not a diagonal matrix.
(10-4)

75. False. For example, $\begin{bmatrix} 1 & 1 \\ 0 & 1 \end{bmatrix}\begin{bmatrix} 1 & 0 \\ 1 & 1 \end{bmatrix} = \begin{bmatrix} 2 & 1 \\ 1 & 1 \end{bmatrix}$ which is not a diagonal matrix.
(10-4)

76. True. If all elements above and all elements below the principal diagonal are zero, then all elements not on the principal diagonal are zero. (10-5)

77. True. For example, the inverse of $\begin{bmatrix} a_{11} & 0 & 0 \\ 0 & a_{22} & 0 \\ 0 & 0 & a_{33} \end{bmatrix}$ is $\begin{bmatrix} \frac{1}{a_{11}} & 0 & 0 \\ 0 & \frac{1}{a_{22}} & 0 \\ 0 & 0 & \frac{1}{a_{33}} \end{bmatrix}$.

This is defined if none of elements on the principal diagonal are zero. A similar situation holds for any order matrix. (10-5)

78.
$$\begin{aligned}(x - 2i)^6 = [x + (-2i)]^6 &= \sum_{k=0}^{6} \binom{6}{k} x^{6-k}(-2i)^k \\ &= \binom{6}{0}x^6 + \binom{6}{1}x^5(-2i)^1 + \binom{6}{2}x^4(-2i)^2 + \binom{6}{3}x^3(-2i)^3 \\ &\quad + \binom{6}{4}x^2(-2i)^4 + \binom{6}{5}x(-2i)^5 + \binom{6}{6}(-2i)^6 \\ &= x^6 - 12ix^5 - 60x^4 + 160ix^3 + 240x^2 - 192ix - 64\end{aligned}$$
(11-6)

79. Let $P(x, y)$ be a point on the parabola.

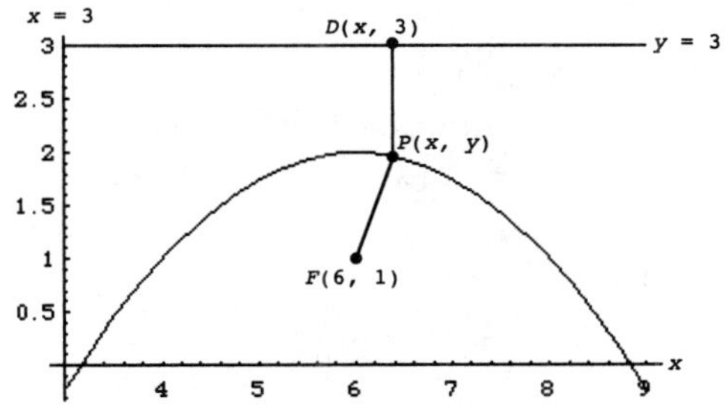

Then by the definition of the parabola, the distance from $P(x, y)$ to the focus $F(6, 1)$ must equal the perpendicular distance from P to the directrix at $D(x, 3)$. Applying the distance formula, we have

$$\begin{aligned} d(P, F) &= d(P, D) \\ \sqrt{(x-6)^2 + (y-1)^2} &= \sqrt{(x-x)^2 + (y-3)^2} \\ (x-6)^2 + (y-1)^2 &= (x-x)^2 + (y-3)^2 \\ x^2 - 12x + 36 + y^2 - 2y + 1 &= 0 + y^2 - 6y + 9 \\ x^2 - 12x + 4y + 28 &= 0 \end{aligned}$$

(9-1)

80. Make a rough sketch of the ellipse, locate focus and x intercepts, then determine y intercepts using the special triangle relationship.

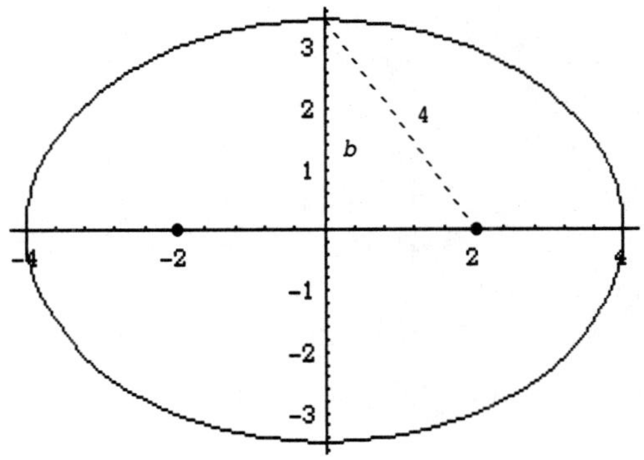

$$\begin{aligned} a &= 4 \\ b^2 &= a^2 - c^2 \\ &= 4^2 - 2^2 \\ &= 12 \\ b &= \sqrt{12} = 2\sqrt{3} \end{aligned}$$

y intercepts: $\pm 2\sqrt{3}$

(9-2)

81. Sketch the asymptote rectangle, label known parts, and use the Pythagorean Theorem.

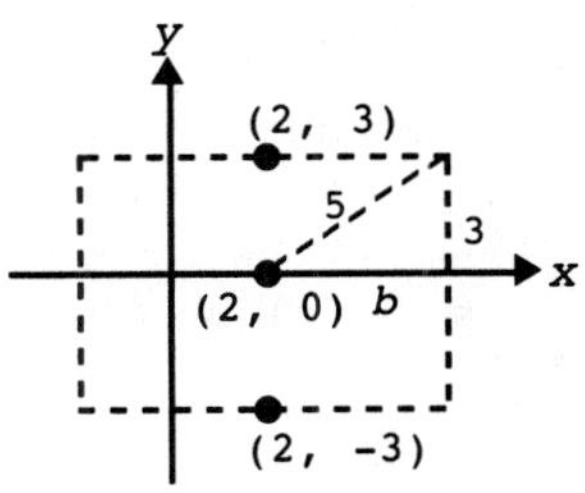

$a = 3 \quad c = 5$

$$\begin{aligned} b^2 &= c^2 - a^2 \\ &= 5^2 - 3^2 \\ &= 16 \\ b &= 4 \end{aligned}$$

Thus, the length of the conjugate axis is $2b = 8$.

(9-3)

82. We can select any three of the seven points to use as vertices of the triangle. Order is not important.

$$C_{7,3} = \frac{7!}{3!(7-3)!} = 35 \text{ triangles}$$

(11-4)

83. P_n: $2^n < n!$ $\quad n \geq 4$ (for n an integer, this is identical with the condition $n > 3$)

Part 1: Show that P_4 is true

P_4: $2^4 < 4!$

$16 < 24$ True.

Part 2: Show that if P_k is true, then P_{k+1} is true.

Write out P_k and P_{k+1}:

P_k: $2^k < k!$

P_{k+1}: $2^{k+1} < (k+1)!$

We start with P_k:

$2^k < k!$

Multiplying both sides by 2: $2 \cdot 2^k < 2 \cdot k!$

Now, $k > 3$, thus $1 < k$, $2 < k + 1$, hence $2 \cdot k! < (k + 1)k!$

Therefore, $2^{k+1} = 2 \cdot 2^k < 2 \cdot k! < (k + 1)k! = (k + 1)!$

$2^{k+1} < (k + 1)!$

Thus, if P_k is true, then P_{k+1} is true.

Conclusion: P_n is true for all $n \geq 4$. *(10-2)*

84. To prove $a_n = b_n$ for all positive integers n, write:

P_n: $a_n = b_n$

Proof: Part 1: Show P_1 is true.

P_1: $a_1 = b_1$. $a_1 = 3$. $b_1 = 2^1 + 1 = 3$.

Thus, $a_1 = b_1$

Part 2: Show that if P_k is true, then P_{k+1} is true.

Write out P_k and P_{k+1}.

P_k: $a_k = b_k$

P_{k+1}: $a_{k+1} = b_{k+1}$

We start with P_k:

$a_k = b_k$

Now, $a_{k+1} = 2a_{k+1-1} - 1 = 2a_k - 1 = 2b_k - 1 = 2(2^k + 1) - 1$

$= 2 \cdot 2^k + 2 - 1 = 2^1 \cdot 2^k + 1 = 2^{k+1} + 1 = b_{k+1}$

Therefore, $a_{k+1} = b_{k+1}$

Thus, if P_k is true, then P_{k+1} is true.

Conclusion: P_n is true for all positive integers n. Hence, $\{a_n\} = \{b_n\}$. *(10-2)*

85. Make a rough sketch of the situation.

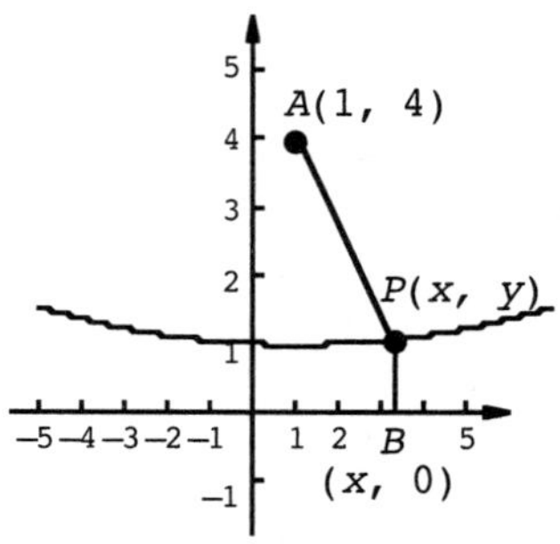

We are given $d(P, A) = 3d(P, B)$. Applying the distance formula, we have

$$\sqrt{(x - 1)^2 + (y - 4)^2} = 3\sqrt{(x - x)^2 + (y - 0)^2}$$
$$(x - 1)^2 + (y - 4)^2 = 9[(x - x)^2 + (y - 0)^2]$$
$$x^2 - 2x + 1 + y^2 - 8y + 16 = 9y^2$$
$$x^2 - 2x - 8y^2 - 8y + 17 = 0$$

is the equation of the curve. Completing the square relative to x and y, we have

$$x^2 - 2x + 1 - 8\left(y^2 + y + \frac{1}{4}\right) - 1 + 2 + 17 = 0$$

$$(x - 1)^2 - 8\left(y + \frac{1}{2}\right)^2 = -18$$

$$\frac{\left(y+\frac{1}{2}\right)^2}{\frac{18}{8}} - \frac{(x - 1)^2}{18} = 1$$

This is the equation of a hyperbola. *(9-3)*

86. The sample space S is the set of all possible ways of selecting 3 bulbs out of 12.
$n(S) = C_{12,3}$
P(at least one defective bulb) = 1 - P(no defective bulbs chosen)
The event, no defective bulbs chosen, is the set of all possible 3-bulb subsets of the 8 acceptable bulbs. $n(E) = C_{8,3}$

$$\text{Then } P(\text{at least one defective bulb}) = 1 - \frac{n(E)}{n(S)} = 1 - \frac{C_{8,3}}{C_{12,3}}$$
$$= 1 - \frac{56}{220} = \frac{41}{55} = .7\overline{45} \qquad (11\text{-}5)$$

87. We are asked for the sum of an infinite geometric series.
$a = \$2{,}000{,}000(0.75)$
$r = 0.75 \quad |r| \le 1$, so the series has a sum,

$$S_\infty = \frac{a_1}{1 - r}$$
$$= \frac{\$2{,}000{,}000(0.75)}{1 - 0.75}$$
$$= \frac{\$2{,}000{,}000(0.75)}{0.25}$$
$$= \$6{,}000{,}000 \qquad (11\text{-}3)$$

88. Let x = the length and y = the width.
The perimeter is $2x + 2y$ so $2x + 2y = 24$.
The area is xy so $xy = 32$.

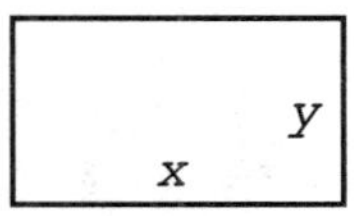

We solve the second equation for y and substitute into the first:

$$xy = 32 \qquad\qquad 2x + 2\left(\frac{32}{x}\right) = 24$$
$$y = \frac{32}{x} \qquad\qquad 2x + \frac{64}{x} = 24 \quad \text{(Multiply both sides by } x.)$$
$$2x^2 + 64 = 24x$$
$$2x^2 - 24x + 64 = 0$$
$$2(x^2 - 12x + 32) = 0$$
$$2(x - 8)(x - 4) = 0$$
$$x = 8 \quad \text{or} \quad x = 4$$

If $x = 8$, $y = \frac{32}{8} = 4$. If $x = 4$, $y = \frac{32}{4} = 8$. The dimensions are 4m × 8m. $(10\text{-}6)$

89.

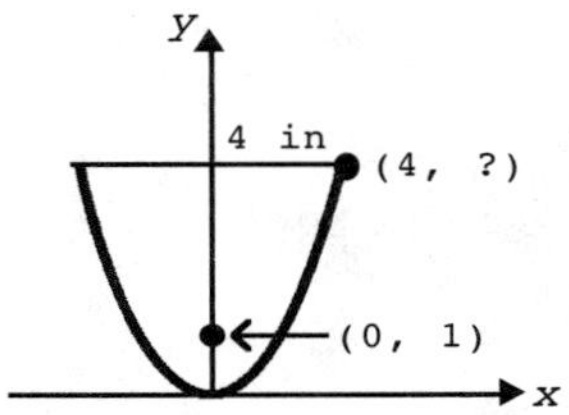

From the figure, we see that the parabola can be positioned so that it is opening up with axis the y axis, hence it has an equation of the form $x^2 = 4ay$. Since the focus is at $(0, a) = (0, 1)$, $a = 1$ and the equation of the parabola is $x^2 = 4y$.

Since the depth represents the y coordinate y_1 of a point on the parabola with $x = 4$, we have
$4^2 = 4y_1$
y_1 = depth = 4 in. $(9\text{-}1)$

90.

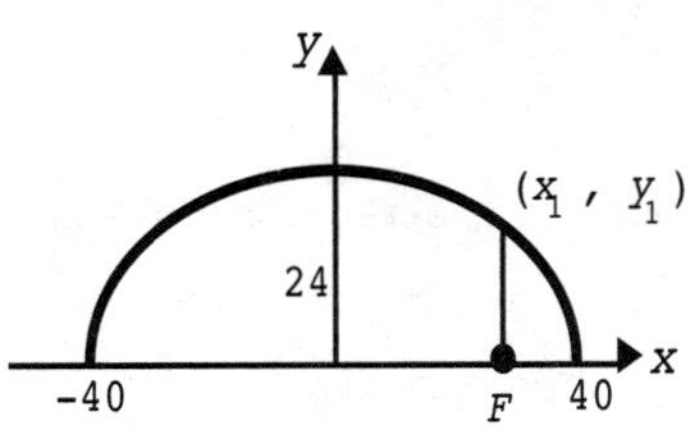

From the figure, we see that the x and y intercepts of the ellipse must be 40 and 24 respectively. Hence the equation of the ellipse must be

$$\frac{x^2}{(40)^2} + \frac{y^2}{(24)^2} = 1$$
$$\text{or} \quad \frac{x^2}{1{,}600} + \frac{y^2}{576} = 1$$

To find the distance c of each focus from the center of the arch, we use the special triangle relationship. $a = 40$, $b = 24$

$$\begin{aligned} c^2 &= a^2 - b^2 \\ &= 40^2 - 24^2 \\ c^2 &= 1024 \\ c &= 32 \text{ ft.} \end{aligned}$$

To find the height of the arch above each focus, we need the y coordinate y_1 of the point P whose x coordinate is 32. Since P is on the ellipse, we have

$$\frac{32^2}{1{,}600} + \frac{y_1^2}{576} = 1$$

$$\frac{1{,}024}{1{,}600} + \frac{y_1^2}{576} = 1$$

$$0.64 + \frac{y_1^2}{576} = 1$$

$$\frac{y_1^2}{576} = 0.36$$

$$y_1^2 = 576(0.36)$$

$$y_1^2 = 207.36$$

$$y_1 = 14.4 \text{ ft.}$$

(9-2)

91. Let x = amount invested at 8%
y = amount invested at 14%
Then $x + y = 12{,}000$ (total amount invested) (1)
$0.08x + 0.14y = 0.10(12{,}000)$ (total yield on investment) (2)

We solve the system of equations (1), (2) using elimination by addition.

$$\begin{aligned} -0.08x - 0.08y &= -0.08(12{,}000) \quad && -0.08[\text{equation (1)}] \\ 0.08x + 0.14y &= 0.10(12{,}000) && \text{equation (2)} \\ \hline 0.06y &= -0.08(12{,}000) + 0.10(12{,}000) \\ 0.06y &= 240 \\ y &= 4{,}000 \\ x + y &= 12{,}000 \\ x &= 8{,}000 \end{aligned}$$

\$8,000 at 8% and \$4,000 at 14%. *(10-1)*

92. Let x_1 = number of ounces of mix A.
x_2 = number of ounces of mix B.
x_3 = number of ounces of mix C.

Then

$$\begin{aligned} 0.2x_1 + 0.1x_2 + 0.15x_3 &= 23 \quad \text{(protein)} \\ 0.02x_1 + 0.06x_2 + 0.05x_3 &= 6.2 \quad \text{(fat)} \\ 0.15x_1 + 0.1x_2 + 0.05x_3 &= 16 \quad \text{(moisture)} \end{aligned}$$

or

$$\begin{aligned} 4x_1 + 2x_2 + 3x_3 &= 460 \\ 2x_1 + 6x_2 + 5x_3 &= 620 \\ 3x_1 + 2x_2 + x_3 &= 320 \end{aligned}$$

is the system to be solved. We form the augmented matrix and solve by Gauss-Jordan elimination.

$$\left[\begin{array}{ccc|c} 4 & 2 & 3 & 460 \\ 2 & 6 & 5 & 620 \\ 3 & 2 & 1 & 320 \end{array}\right] \begin{array}{l} \frac{1}{4}R_1 \to R_1 \\ \\ \\ \end{array}$$

$$\sim \left[\begin{array}{ccc|c} 1 & \frac{1}{2} & \frac{3}{4} & 115 \\ 2 & 6 & 5 & 620 \\ 3 & 2 & 1 & 320 \end{array}\right] \begin{array}{l} \\ (-2)R_1 + R_2 \to R_2 \\ (-3)R_1 + R_3 \to R_3 \end{array}$$

$$\sim \left[\begin{array}{ccc|c} 1 & \frac{1}{2} & \frac{3}{4} & 115 \\ 0 & 5 & \frac{7}{2} & 390 \\ 0 & \frac{1}{2} & -\frac{5}{4} & -25 \end{array}\right] R_2 \leftrightarrow R_3$$

$$\sim \left[\begin{array}{ccc|c} 1 & \frac{1}{2} & \frac{3}{4} & 115 \\ 2 & \frac{1}{2} & -\frac{5}{4} & -25 \\ 3 & 5 & \frac{7}{2} & 390 \end{array}\right] \begin{array}{l} (-1)R_2 + R_1 \to R_1 \\ \\ (-10)R_2 + R_3 \to R_3 \end{array}$$

$$\sim \left[\begin{array}{ccc|c} 1 & 0 & 2 & 140 \\ 0 & \frac{1}{2} & -\frac{5}{4} & -25 \\ 0 & 0 & 16 & 640 \end{array}\right] \begin{array}{l} \\ 2R_2 \to R_2 \\ \frac{1}{16}R_3 \to R_3 \end{array}$$

$$\sim \left[\begin{array}{ccc|c} 1 & 0 & 2 & 140 \\ 0 & 1 & -\frac{5}{2} & -50 \\ 0 & 0 & 1 & 40 \end{array}\right] \begin{array}{l} (-2)R_3 + R_1 \to R_1 \\ \left(-\frac{5}{2}\right)R_3 + R_2 \to R_2 \\ \\ \end{array}$$

$$\sim \left[\begin{array}{ccc|c} 1 & 0 & 0 & 60 \\ 0 & 1 & 0 & 50 \\ 0 & 0 & 1 & 40 \end{array}\right]$$

Thus, x_1 = 60g Mix *A*, x_2 = 50g Mix *B*, x_3 = 40g Mix *C* (*10-2, 10-3*)

93. Let x_1 = number of model *A* trucks
x_2 = number of model *B* trucks
x_3 = number of model *C* trucks

Then $x_1 + x_2 + x_3 = 12$ (total number of trucks)
$18{,}000x_1 + 22{,}000x_2 + 30{,}000x_3 = 300{,}000$ (total funds needed)

We form the augmented matrix and solve by Gauss-Jordan elimination.

$$\left[\begin{array}{ccc|c} 1 & 1 & 1 & 12 \\ 18{,}000 & 22{,}000 & 30{,}000 & 300{,}000 \end{array}\right] \begin{array}{l} \\ \frac{1}{2{,}000}R_2 \to R_2 \end{array}$$

$$\sim \left[\begin{array}{ccc|c} 1 & 1 & 1 & 12 \\ 9 & 11 & 15 & 150 \end{array}\right] \quad (-9)R_1 + R_2 \to R_2$$

$$\sim \left[\begin{array}{ccc|c} 1 & 1 & 1 & 12 \\ 0 & 2 & 6 & 42 \end{array}\right] \quad -\tfrac{1}{2}R_2 + R_1 \to R_1$$

$$\sim \left[\begin{array}{ccc|c} 1 & 0 & -2 & -9 \\ 0 & 2 & 6 & 42 \end{array}\right] \quad \tfrac{1}{2}R_2 \to R_2$$

$$\sim \left[\begin{array}{ccc|c} 1 & 0 & -2 & -9 \\ 0 & 1 & 3 & 21 \end{array}\right]$$

Let $x_3 = t$. Then

$$\begin{aligned} x_2 &= -3x_3 + 21 \\ &= -3t + 21 \\ x_1 &= 2x_3 - 9 \\ &= 2t - 9 \end{aligned}$$

A solution is achieved, not for every value of t, but for integer values of t that give rise to non-negative x_1, x_2, x_3.

$x_1 \ge 0$ means $2t - 9 \ge 0$ or $t \ge 4\frac{1}{2}$
$x_2 \ge 0$ means $-3t + 21 \ge 0$ or $t \le 7$

The only integer values of t that satisfy these conditions are 5, 6, 7. Thus we have the solutions

$x_1 = 2t - 9$ model A trucks

$x_2 = -3t + 21$ model B trucks

$x_3 = t$ model C trucks

where $t = 5, 6,$ or 7. Thus the distributor can purchase

1 model A truck, 6 model B trucks, and 5 model C trucks, or

3 model A trucks, 3 model B trucks, and 6 model C trucks, or

5 model A trucks and 7 model C trucks. (10-3)

94. Let x = number of standard day packs
y = number of deluxe day packs

(A) We form the linear objective function

$P = 8x + 12y$

We wish to maximize P, the profit from x standard packs @ \$8 and y deluxe packs @ \$12, subject to the constraints

$0.5x + 0.5y \le 300$ fabricating constraint

$0.3x + 0.6y \le 240$ sewing constraint

$x, y \ge 0$ non-negative constraints

Solving the system of constraint inequalities graphically, we obtain the feasible region S shown in the diagram.

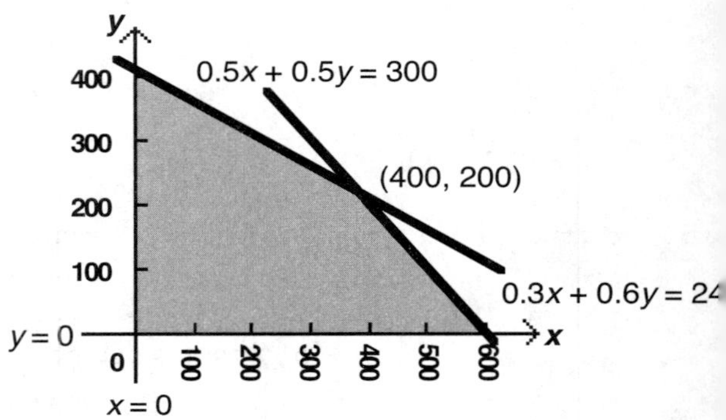

The three corner points (0, 400), (0, 0), and (600, 0) are obvious from the diagram. The corner point (400, 200) is found by solving the system

$0.5x + 0.5y = 300$

$0.3x + 0.6y = 240$

Next we evaluate the objective function at each corner point.

Corner Point (x, y)	Objective Function $P = 5x + 15y$	
(0, 0)	0	
(600, 0)	4,800	
(400, 200)	5,600	Maximum value
(0, 400)	4,800	

The optimal value is 5,600 at the corner point (400, 200). Thus, the company should manufacture 400 standard packs and 200 deluxe packs for a maximum profit of \$5,600.

(B) The objective function now becomes $5x + 15y$. We evaluate this at each corner point.

Corner Point (x, y)	Objective Function $P = 5x + 15y$	
(0, 0)	0	
(600, 0)	3,000	
(400, 200)	5,000	
(0, 400)	6,000	Maximum value

The optimal value is now 6,000 at the corner point (0, 400). Thus, the company should manufacture no standard packs and 400 deluxe packs for a maximum profit of \$6,000.

(C) The objective function now becomes $11x + 9y$. We evaluate this at each corner point.

Corner Point (x, y)	Objective Function $P = 5x + 15y$	
(0, 0)	0	
(600, 0)	6,600	Maximum value
(400, 200)	6,200	
(0, 400)	3,600	

The optimal value is now 6,600 at the corner point (600, 0). Thus, the company should manufacture 600 standard packs and no deluxe packs for a maximum profit of $6,600. *(10-8)*

95. (A) $M\begin{bmatrix}0.25\\0.25\\0.25\\0.25\end{bmatrix} = \begin{bmatrix}78 & 84 & 81 & 86\\91 & 65 & 84 & 92\\95 & 90 & 92 & 91\\75 & 82 & 87 & 91\\83 & 88 & 81 & 76\end{bmatrix}\begin{bmatrix}0.25\\0.25\\0.25\\0.25\end{bmatrix}$

$$= \begin{bmatrix}0.25(78 + 84 + 81 + 86)\\0.25(91 + 65 + 84 + 92)\\0.25(95 + 90 + 92 + 91)\\0.25(75 + 82 + 87 + 91)\\0.25(83 + 88 + 81 + 76)\end{bmatrix} = \begin{bmatrix}82.25\\83\\92\\83.75\\82\end{bmatrix}\begin{matrix}\text{Ann}\\\text{Bob}\\\text{Carol}\\\text{Dan}\\\text{Eric}\end{matrix}$$

(B) $M\begin{bmatrix}0.2\\0.2\\0.2\\0.4\end{bmatrix} = \begin{bmatrix}78 & 84 & 81 & 86\\91 & 65 & 84 & 92\\95 & 90 & 92 & 91\\75 & 82 & 87 & 91\\83 & 88 & 81 & 76\end{bmatrix}\begin{bmatrix}0.2\\0.2\\0.2\\0.4\end{bmatrix}$

$$= \begin{bmatrix}0.2(78 + 84 + 81) + 0.4 \cdot 86\\0.2(91 + 65 + 84) + 0.4 \cdot 92\\0.2(95 + 90 + 92) + 0.4 \cdot 91\\0.2(75 + 82 + 87) + 0.4 \cdot 91\\0.2(83 + 88 + 81) + 0.4 \cdot 76\end{bmatrix}\begin{bmatrix}83\\84.8\\91.8\\85.2\\80.8\end{bmatrix}\begin{matrix}\text{Ann}\\\text{Bob}\\\text{Carol}\\\text{Dan}\\\text{Eric}\end{matrix}$$

(C) $[0.2\ \ 0.2\ \ 0.2\ \ 0.2\ \ 0.2]M = [0.2\ \ 0.2\ \ 0.2\ \ 0.2\ \ 0.2]\begin{bmatrix}78 & 84 & 81 & 86\\91 & 65 & 84 & 92\\95 & 90 & 92 & 91\\75 & 82 & 87 & 91\\83 & 88 & 81 & 76\end{bmatrix}$

$= [0.2(78 + 91 + 95 + 75 + 83)\ \ 0.2(84 + 65 + 90 + 82 + 88)\ \ 0.2(81 + 84 + 92 + 87 + 81)\ \ 0.2(86 + 92 + 91 + 91 + 76)]$

$\begin{matrix} & \text{Test 1} & \text{Test 2} & \text{Test 3} & \text{Test 4}\\ = [& 84.4 & 81.8 & 85 & 87.2] \end{matrix}$ *(10-4)*

96. (A) $P\text{ (this event)} = \dfrac{130}{1,000} = .13$

(B) $P\text{ (this event)} = \dfrac{80 + 90}{1,000} = .17$

(C) $P\text{ (this event)} = P\text{ (being an independent)} + P\text{ (being over 59)} - P\text{ (being both)}$

$= \dfrac{120}{1,000} + \dfrac{230}{1,000} - \dfrac{30}{1,000} = .32$ *(11-5)*

CHAPTER 12

Section 12-1

1. $\lim_{x \to 5}(3x - 8)$

x	4	4.1	4.9	4.99	5.01	5.1	5.5	6
$3x - 8$	4	4.3	6.7	6.97	7.03	7.3	8.5	10

From the table, we conclude that the limit is 7.

3. $\lim_{x \to -2} \frac{x^2 - 4}{x + 2}$

x	-3	-2.5	-2.1	-2.01	-2	-1.99	-1.9	-1.5	-1.1
$\frac{x^2 - 4}{x + 2}$	-5	-4.5	-4.1	-4.01	Undefined	-3.99	-3.9	-3.5	-3.1

From the table, we conclude that the limit is -4.

5. $\lim_{x \to -3} \frac{4}{(x + 3)^2}$

x	-4	-3.5	-3.1	-3.01	-3	-2.99	-2.9	-2.5	-2
$\frac{4}{(x + 3)^2}$	4	16	400	40,000	Undefined	40,000	400	16	4

From the table, we conclude that the limit is ∞.

7. $\lim_{x \to 12} \frac{x}{12 - x}$

x	11	11.5	11.9	11.99	12	12.01	12.1	12.5	13
$\frac{x}{12 - x}$	11	23	119	1199	Undefined	-1201	-121	-25	-13

From the table, we conclude that the limit does not exist.

9. $\lim_{x \to -2} (4 - 7x)$

x	-3	-2.5	-2.1	-2.01	-1.99	-1.9	-1.5	-1
$4 - 7x$	25	21.5	18.7	18.07	17.93	17.3	14.5	11

From the table, we conclude that the limit is 18.

11. $\lim_{x \to 0} (2x^2 - 3x + 1)$

x	-1	-0.5	-0.1	-0.01	0.01	0.1	0.5	1
$2x^2 - 3x + 1$	6	3	1.32	1.0302	0.9702	0.72	0	0

From the table, we conclude that the limit is 1.

13. $\lim_{x \to 4} \frac{\sqrt{x} - 2}{x - 4}$

x	3	3.5	3.9	3.99	4	4.01	4.1	4.5	5
$\frac{\sqrt{x} - 2}{x - 4}$	0.268	0.258	0.252	0.250	Undefined	0.250	0.248	0.243	0.236

From the table, we conclude that the limit is 0.25 or $\frac{1}{4}$.

15. $\lim_{x \to 0} \frac{(3 + x)^2 - 9}{x}$

x	-1	-0.5	-0.1	-0.01	0	0.01	0.1	0.5	1
$\frac{(3 + x)^2 - 9}{x}$	5	5.5	5.9	5.99	Undefined	6.01	6.1	6.5	7

From the table, we conclude that the limit is 6.

17. $\lim_{x \to 0} \frac{2e^x - 2}{x}$

x	-1	-0.5	-0.1	-0.01	0	0.01	0.1	0.5	1
$\frac{2e^x - 2}{x}$	1.26	1.57	1.90	1.99	Undefined	2.01	2.10	2.59	3.44

From the table, we conclude that the limit is 2.

19. $\lim_{x \to -2} \frac{-4x}{(x + 2)^2}$

x	-3	-2.5	-2.1	-2.01	-2	-1.99	-1.9	-1.5	-1
$\frac{-4x}{(x + 2)^2}$	12	40	840	80,400	Undefined	79,600	760	24	4

From the table, we conclude that the limit is ∞.

21. $\lim_{x \to -10} \frac{x + 4}{x + 10}$

x	-11	-10.5	-10.1	-10.01	-10	-9.99	-9.9	-9.5	-9
$\frac{x + 4}{x + 10}$	7	13	61	601	Undefined	-599	-59	-11	-5

From the table, we conclude that the limit does not exist.

23. $\lim_{x \to 5} f(x)$, where $f(x) = \begin{cases} 2x + 7 & x < 5 \\ x^2 - 8 & x > 5 \end{cases}$

x	4	4.5	4.9	4.99	5	5.01	5.1	5.5	6
$f(x)$	15	16	16.8	16.98	Undefined	17.1	18.01	22.25	28

From the table, we conclude that the limit is 17.

25. 2 **27.** ∞ **29.** −∞ **31.** Does not exist

33.

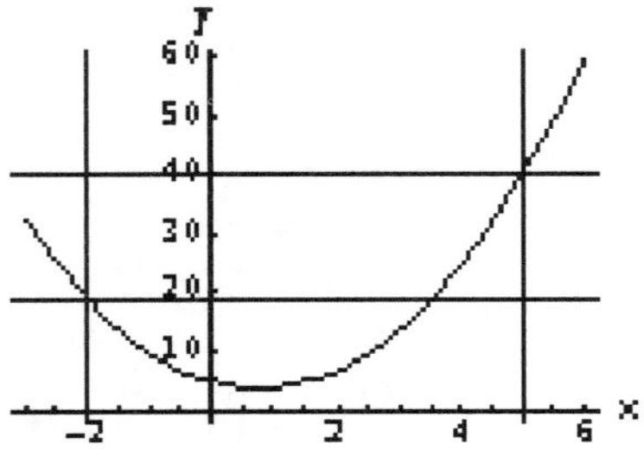

From the graph, we conclude that $\lim_{x \to -2} f(x) = 19$, $\lim_{x \to 5} f(x) = 40$.

35.

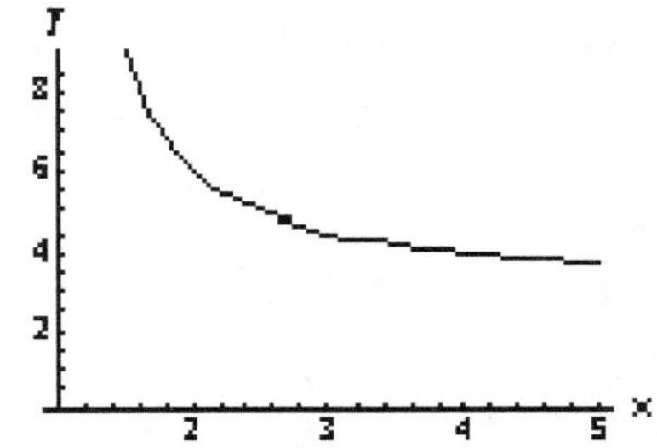

From the graph, we conclude that $\lim_{x \to 2} f(x) = 6$, $\lim_{x \to 4} f(x) = 4$.

37.

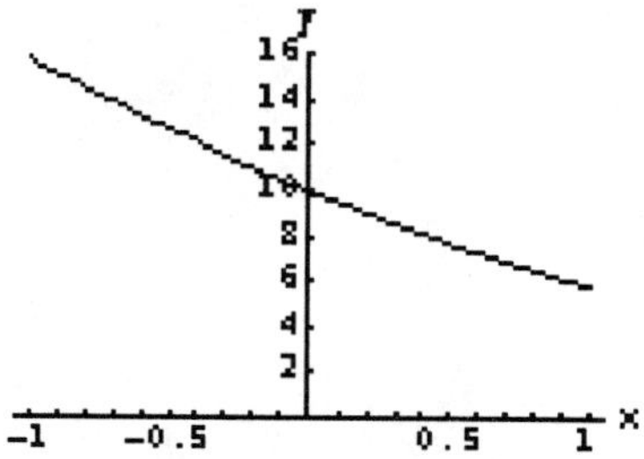

From the graph, we conclude that $\lim_{x \to 0} f(x) = 10$.

39.

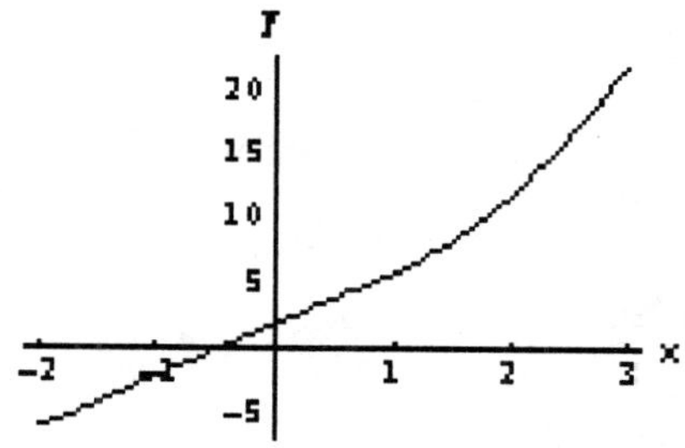

From the graph, we conclude that $\lim_{x \to 1} f(x) = 6$.

41.

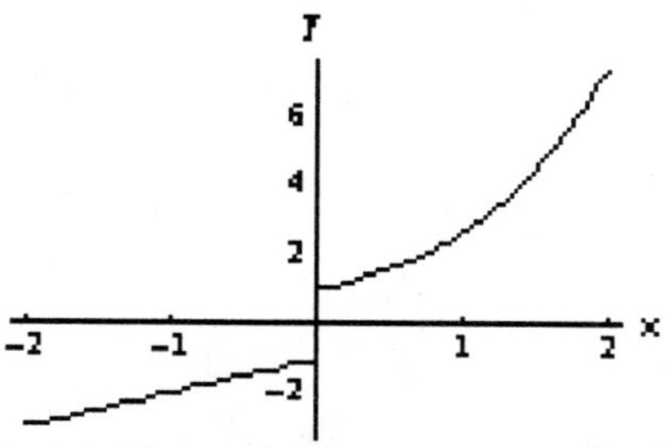

From the graph, we conclude that $\lim_{x \to 0} f(x)$ does not exist.

43. $\lim_{x \to 4} \frac{2}{x - 5}$

x	3	3.5	3.9	3.99	4.01	4.1	4.5	5
$\frac{2}{x-5}$	-1	-1.333	-1.818	-1.98	-2.02	-2.222	-4	Undefined

From the table, we conclude that the limit is -2.

45. $\lim_{x \to 2} e^{x-1}$

x	1	1.5	1.9	1.99	2.01	2.1	2.5	3
e^{x-1}	1	1.649	2.460	2.691	2.746	3.004	4.482	7.389

From the table, we conclude that the limit is approximately 2.7. (The exact limit is e.)

47. $\lim_{x \to 3^-} (2x - 19)$. We substitute in values less than 3.

x	2	2.5	2.9	2.99
$2x - 19$	-15	-14	-13.2	-13.02

From the table, we conclude that the limit is -13.

49. $\lim_{x \to 0^+} \frac{x^2 + 4}{x}$. We substitute in values greater than 0.

x	1	0.5	0.1	0.01
$\frac{x^2 + 4}{x}$	5	8.5	40.1	400.01

From the table, we conclude that the limit is ∞.

51. To find $\lim_{x \to -1^-} f(x)$, we substitute in only x values less than -1, so only the formula $\frac{1}{2}x + \frac{5}{2}$ is relevant.

x	-2	-1.5	-1.1	-1.01
$\frac{1}{2}x + \frac{5}{2}$	1.5	1.75	1.95	1.995

We conclude that $\lim_{x \to -1^-} f(x) = 2$.

To find $\lim_{x \to -1^+} f(x)$, we substitute in only x values greater than -1, so only the formula $\frac{5x}{x + 2}$ is relevant.

x	0	-0.5	-0.9	-0.99
$\frac{5x}{x + 2}$	0	-1.667	-4.091	-4.901

We conclude that $\lim_{x \to -1^+} f(x) = -5$.

53. -2 **55.** Does not exist **57.** $-\infty$ **59.** 0

61. 0 **63.** $-\infty$

65. $\lim_{t \to 220} (22{,}000 - 100t)$

t	200	210	219	219.9	220.1	221	230	240
$22{,}000 - 100t$	2000	1000	100	10	-10	-100	-1000	-2000

We conclude that $\lim_{t \to 220} V(t) = 0$. As the time since pumping began approaches 220 minutes, the basement empties completely.

67. To find $\lim_{x \to 10{,}000^-} T(x)$, we substitute in only x values less than 10,000, so only the formula $0.03x$ is relevant.

x	9,000	9,900	9,990	9,999
$0.03x$	270	297	299.7	299.97

We conclude that $\lim_{x \to 10{,}000^-} T(x) = 300$.

To find $\lim_{x \to 10{,}000^+} T(x)$, we substitute in only x values greater than 10,000, so only the formula $0.05x - 200$ is relevant.

x	11,000	10,100	10,010	10,001
$0.05x - 200$	350	305	300.5	300.05

We conclude that $\lim_{x \to 10{,}000^+} T(x) = 300$. There is no jump as you pass $10,000 in income.

Section 12-2

1. Since $\lim_{x \to a} c = c$, $\lim_{x \to 5} 12 = 12$

3. Since $\lim_{x \to a} c = c$, $\lim_{t \to \pi} -14 = -14$

5. Since $\lim_{x \to a} x = a$, $\lim_{x \to -4} x = -4$

7. Since $\lim_{x \to a} x = a$, $\lim_{z \to 111} z = 111$

9.
$$\begin{aligned}\lim_{x \to 4}(7 - 5x) &= \lim_{x \to 4} 7 - \lim_{x \to 4} 5x \\ &= \lim_{x \to 4} 7 - 5\lim_{x \to 4} x \\ &= 7 - 5(4) = -13\end{aligned}$$

11.
$$\begin{aligned}\lim_{y \to 10} (5y + 2) &= \lim_{y \to 10} 5y + \lim_{y \to 10} 2 \\ &= 5 \lim_{y \to 10} y + \lim_{y \to 10} 2 \\ &= 5(10) + 2 = 52\end{aligned}$$

13. $\lim_{t\to-3}(t^2+2t) = \lim_{t\to-3} t^2 + \lim_{t\to-3} 2t$
$= (\lim_{t\to-3} t)^2 + 2\lim_{t\to-3} t$
$= (-3)^2 + 2(-3) = 3$

15. $\lim_{x\to-1}(2x^2+3x-4) = \lim_{x\to-1} 2x^2 + \lim_{x\to-1} 3x - \lim_{x\to-1} 4$
$= 2(\lim_{x\to-1} x)^2 + 3\lim_{x\to-1} x - \lim_{x\to-1} 4$
$= 2(-1)^2 + 3(-1) - 4 = -5$

17. $\lim_{z\to-2}(2z^3+3z^2+5) = \lim_{z\to-2} 2z^3 + \lim_{z\to-2} 3z^2 + \lim_{z\to-2} 5$
$= 2(\lim_{z\to-2} z)^3 + 3(\lim_{z\to-2} z)^2 + \lim_{z\to-2} 5$
$= 2(-2)^3 + 3(-2)^2 + 5 = 1$

19. $\lim_{x\to-1}\dfrac{x^2-5}{2x+3} = \dfrac{\lim_{x\to-1}(x^2-5)}{\lim_{x\to-1}(2x+3)}$
$= \dfrac{(\lim_{x\to-1} x)^2 - \lim_{x\to-1} 5}{2\lim_{x\to-1} x + \lim_{x\to-1} 3}$
$= \dfrac{(-1)^2-5}{2(-1)+3} = -4$

21. $\lim_{y\to5}\dfrac{6y-9}{y+2} = \dfrac{\lim_{y\to5}(6y-9)}{\lim_{y\to5}(y+2)}$
$= \dfrac{6\lim_{y\to5} y - \lim_{y\to5} 9}{\lim_{y\to5} y + \lim_{y\to5} 2}$
$= \dfrac{6(5)-9}{5+2} = 3$

23. $\lim_{t\to\frac{1}{2}}\dfrac{4t-6}{8t^2+4} = \dfrac{\lim_{t\to\frac{1}{2}}(4t-6)}{\lim_{t\to\frac{1}{2}}(8t^2+4)}$
$= \dfrac{4\lim_{t\to\frac{1}{2}} t - \lim_{t\to\frac{1}{2}} 6}{8\left(\lim_{t\to\frac{1}{2}} t\right)^2 + \lim_{t\to\frac{1}{2}} 4}$
$= \dfrac{4\left(\frac{1}{2}\right)-6}{8\left(\frac{1}{2}\right)^2+4} = -\dfrac{2}{3}$

25. $\lim_{x\to5}\sqrt{2x+6} = \sqrt{\lim_{x\to5}(2x+6)}$
$= \sqrt{2\lim_{x\to5} x + \lim_{x\to5} 6}$
$= \sqrt{2(5)+6} = 4$

27. $\lim_{y\to-1}\sqrt{y^2+6} = \sqrt{\lim_{y\to-1}(y^2+6)}$
$= \sqrt{(\lim_{y\to-1} y)^2 + \lim_{y\to-1} 6}$
$= \sqrt{(-1)^2+6} = \sqrt{7}$

29. $\lim_{x\to-2}(x+3)\sqrt{x+6} = \lim_{x\to-2}(x+3)\ \lim_{x\to-2}\sqrt{x+6}$
$= \lim_{x\to-2}(x+3)\sqrt{\lim_{x\to-2}(x+6)}$
$= (\lim_{x\to-2} x + \lim_{x\to-2} 3)\sqrt{\lim_{x\to-2} x + \lim_{x\to-2} 6}$
$= (-2+3)\sqrt{(-2)+6} = 2$

31. Since $3x^2+5x$ is a polynomial, $\lim_{x\to2}(3x^2+5x) = 3(2)^2 + 5(2) = 22.$

33. Since $5+2y^2-y^3$ is a polynomial,
$\lim_{y\to-4}(5+2y^2-y^3) = 5 + 2(-4)^2 - (-4)^3 = 101.$

35. Since $\dfrac{2t + 16}{t^2 - 8}$ is a rational function, $\lim_{t \to 4} \dfrac{2t + 16}{t^2 - 8} = \dfrac{2(4) + 16}{4^2 - 8} = 3$.

37. Since $\dfrac{18 - x^2}{x^2 + x}$ is a rational function, $\lim_{x \to -6} \dfrac{18 - x^2}{x^2 + x} = \dfrac{18 - (-6)^2}{(-6)^2 + 6} = -\dfrac{3}{5}$.

39. Since $4 - y^2$ is negative near $y = 3$, the limit does not exist.

41. Since $\dfrac{t^2 - 2}{12 - t^2}$ is negative near $t = 4$, the limit does not exist.

43.
$$\lim_{y \to 4} \frac{y - 4}{y^2 - 16} = \lim_{y \to 4} \frac{y - 4}{(y - 4)(y + 4)} = \lim_{y \to 4} \frac{1}{y + 4} = \frac{1}{4 + 4} = \frac{1}{8}$$

45.
$$\lim_{x \to -3} \frac{x^2 + x - 6}{x + 3} = \lim_{x \to -3} \frac{(x + 3)(x - 2)}{(x + 3)} = \lim_{x \to -3} (x - 2) = (-3) - 2 = -5$$

47.
$$\lim_{x \to 25} \frac{\sqrt{x} - 5}{x - 25} = \lim_{x \to 25} \frac{\sqrt{x} - 5}{x - 25} \frac{\sqrt{x} + 5}{\sqrt{x} + 5} = \lim_{x \to 25} \frac{x - 25}{(x - 25)(\sqrt{x} + 5)} = \lim_{x \to 25} \frac{1}{\sqrt{x} + 5} = \frac{1}{\sqrt{25} + 5} = \frac{1}{10}$$

49.
$$\lim_{x \to 0} \frac{(x + 3)^2 - 9}{x} = \lim_{x \to 0} \frac{x^2 + 6x + 9 - 9}{x} = \lim_{x \to 0} \frac{x^2 + 6x}{x} = \lim_{x \to 0} \frac{x(x + 6)}{x} = \lim_{x \to 0} (x + 6) = 0 + 6 = 6$$

51.
$$\lim_{h \to 0} \frac{(1 + h)^3 - 1}{h} = \lim_{h \to 0} \frac{1 + 3h + 3h^2 + h^3 - 1}{h} = \lim_{h \to 0} \frac{3h + 3h^2 + h^3}{h} = \lim_{h \to 0} \frac{h(3 + 3h + h^2)}{h} = \lim_{h \to 0} (3 + 3h + h^2) = 3 + 3(0) + 0^2 = 3$$

53. Since $\lim_{x \to 5^-} \dfrac{7}{x - 5} = -\infty$ and $\lim_{x \to 5^+} \dfrac{7}{x - 5} = \infty$, the limit does not exist.

55. Since $\lim_{y \to 11^-} \dfrac{y - 10}{y - 11} = -\infty$ and $\lim_{y \to 11^+} \dfrac{y - 10}{y + 10} = \infty$, the limit does not exist.

57. Since $\lim_{k \to -5^-} \dfrac{2k}{(k + 5)^2} = \lim_{k \to -5^+} \dfrac{2k}{(k + 5)^2} = -\infty$, $\lim_{k \to -5} \dfrac{2k}{(k + 5)^2} = -\infty$.

59. Since $\lim_{x \to 6^-} \dfrac{x + 10}{(x - 6)^4} = \lim_{x \to 6^+} \dfrac{x + 10}{(x - 6)^4} = \infty$, $\lim_{x \to 6} \dfrac{x + 10}{(x - 6)^4} = \infty$.

61. 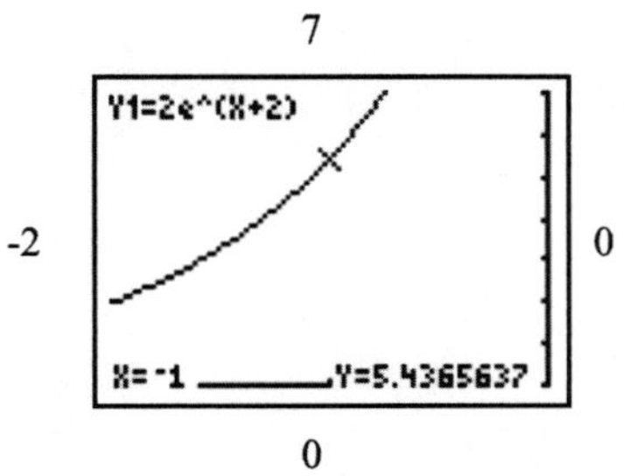

From the graph, we conclude that the limit is approximately 5.4.
(The exact limit is $2e \approx 5.4365637$.)

63. 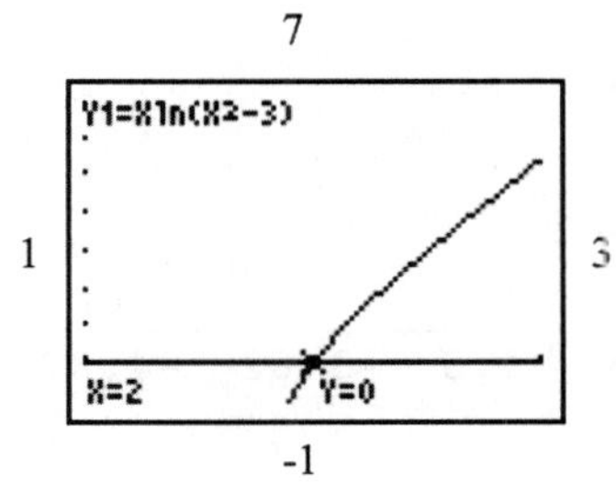

From the graph, we conclude that the limit is 0.

65. 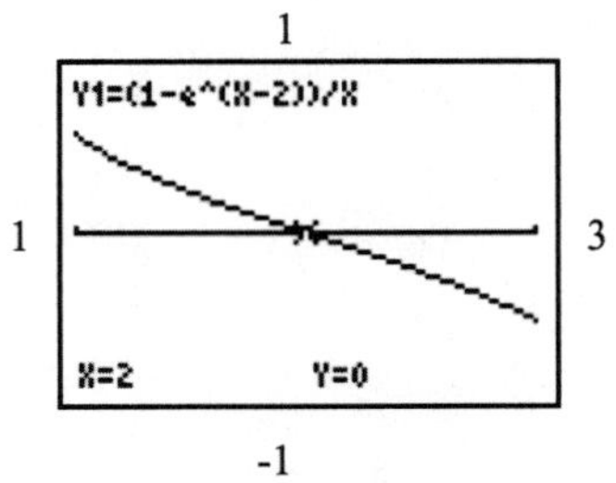

From the graph, we conclude that the limit is 0.

67. (A) 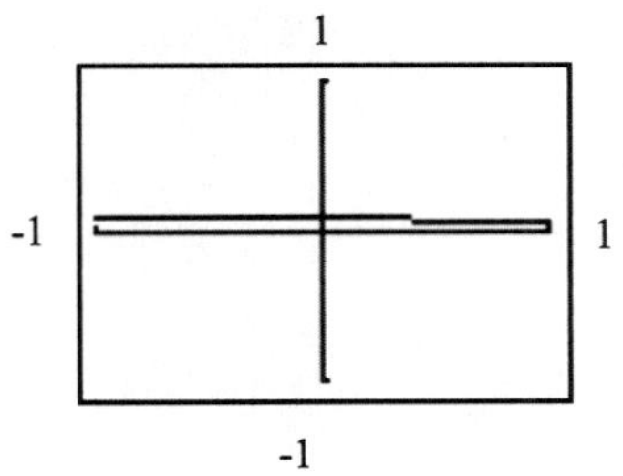

The limit appears to be very close to 0.

(B)

x	11	11.5	11.9	11.99	12	12.01	12.1	12.5	13
$\frac{x-12}{x^2-144}$	0.0435	0.0426	0.0418	0.0417	Undefined	0.0416	0.0415	0.0408	0.04

The limit appears to be about 0.04.

(C) $$\lim_{x\to 12}\frac{x-12}{x^2-144} = \lim_{x\to 12}\frac{x-12}{(x-12)(x+12)}$$
$$= \lim_{x\to 12}\frac{1}{x+12}$$
$$= \frac{1}{12+12} = \frac{1}{24} = 0.41\overline{66}$$

69. The limit of a difference is the difference of the limits.

71. The limit of a quotient is the quotient of the limits, provided the limit of the denominator is not zero.

73. The limit of a positive integral root is the root of the limit of the radicand, provided, for even integers, that the limit of the radicand is positive.

Section 12-3

1. $\lim_{x \to \infty} \frac{8 - 5x}{x + 1}$

x	1	10	100	1,000	10,000
$\frac{8 - 5x}{x + 1}$	1.5	-3.82	-4.87	-4.987	-4.999

From the table, we conclude that the limit is -5.

3. $\lim_{x \to -\infty} \frac{e^x + 10}{x}$

x	-1	-10	-100	-1,000	-10,000
$\frac{e^x + 10}{x}$	-10.37	-1	-0.1	-0.01	-0.001

From the table, we conclude that the limit is 0.

5. $\lim_{x \to \infty} \frac{3x^2}{x^2 + x}$

x	1	10	100	1,000	10,000
$\frac{3x^2}{x^2 + x}$	1.5	2.7273	2.9703	2.997	2.9997

From the table, we conclude that the limit is 3.

7. $\lim_{x \to \infty} \left(\frac{1}{2} x^2 + 3x\right)$

x	1	10	100	1,000	10,000
$\frac{1}{2}x^2 + 3x$	3.5	80	5,300	503,000	5×10^7

From the table, we conclude that the limit is ∞.

9. $\lim_{x \to -\infty} (2x^3 - x^2)$

x	-1	-10	-100	-1,000	-10,000
$2x^3 - x^2$	-3	-2,100	-2×10^6	-2×10^9	-2×10^{12}

From the table, we conclude that the limit is -∞.

11. $\lim_{x \to \infty} f(x) = 4, \quad \lim_{x \to -\infty} f(x) = 4$

13. $\lim_{x \to \infty} f(x) = \infty, \quad \lim_{x \to -\infty} f(x) = -\infty$

15.

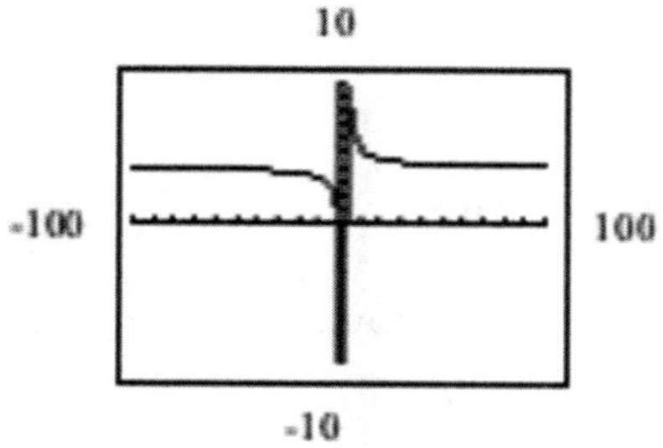

From the graph, we conclude that $\lim_{x \to \infty} f(x) = \lim_{x \to -\infty} f(x) = 4$.

17.

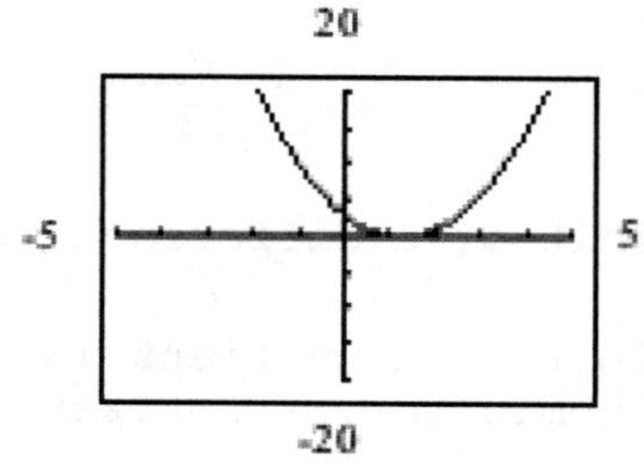

From the graph, we conclude that $\lim_{x \to \infty} f(x) = \lim_{x \to -\infty} f(x) = \infty$.

19. 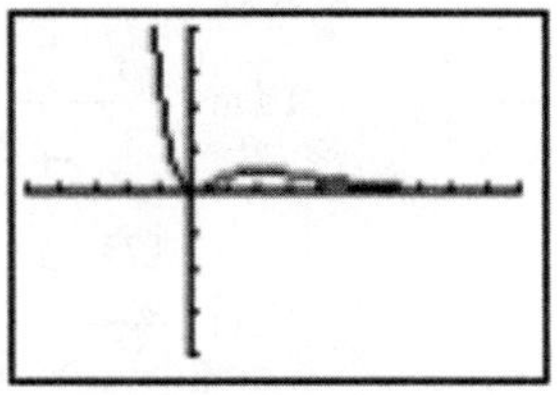

From the graph, we conclude that $\lim_{x\to\infty} f(x) = 0$ and $\lim_{x\to-\infty} f(x) = \infty$.

21. ∞

23. ∞

25. ∞

27. 0

29. 0

31. 0

33. $\lim_{x\to\infty}(x^2 - 3x + 5) = \lim_{x\to\infty} x^2 = \infty$

35. $\lim_{x\to-\infty}(2x^3 + x^2) = \lim_{x\to-\infty} 2x^3 = -\infty$

37. $\lim_{x\to\infty}(-4x + 7x^2 - 2x^4) = \lim_{x\to\infty}(-2x^4) = -\infty$

39. $\lim_{x\to-\infty}(7x - 7x^2 - 7x^3) = \lim_{x\to-\infty}(-7x^3) = \infty$

41. $$\lim_{x\to\infty}\frac{4x}{x^2+1} = \lim_{x\to\infty}\frac{\frac{4x}{x^2}}{\frac{x^2}{x^2}+\frac{1}{x^2}} = \lim_{x\to\infty}\frac{\frac{4}{x}}{1+\frac{1}{x^2}} = \frac{\lim_{x\to\infty}\frac{4}{x}}{\lim_{x\to\infty}1 + \lim_{x\to\infty}\frac{1}{x^2}} = \frac{0}{1+0} = 0$$

43. $$\lim_{x\to-\infty}\frac{2x^2}{x-x^2} = \lim_{x\to-\infty}\frac{\frac{2x^2}{x^2}}{\frac{x}{x^2}+\frac{x^2}{x^2}} = \lim_{x\to-\infty}\frac{2}{\frac{1}{x}-1} = \frac{\lim_{x\to\infty}2}{\lim_{x\to-\infty}\frac{1}{x} - \lim_{x\to-\infty}1} = \frac{2}{0-1} = -2$$

45. $$\lim_{x\to\infty}\frac{x^3-x+3}{2x^2+1} = \lim_{x\to\infty}\frac{\frac{x^3}{x^3}-\frac{x}{x^3}+\frac{3}{x^3}}{\frac{2x^2}{x^3}+\frac{1}{x^3}} = \lim_{x\to\infty}\frac{1-\frac{1}{x^2}+\frac{3}{x^3}}{\frac{2}{x}+\frac{1}{x^3}}$$

$= \infty$, since the limit of the numerator is positive and the limit of the denominator is 0, approached through positive values.

47. $$\lim_{x\to\infty}\frac{2x^3-7x^2+x}{4+3x-7x^3}=\lim_{x\to\infty}\frac{\frac{2x^3}{x^3}-\frac{7x^2}{x^3}+\frac{x}{x^3}}{\frac{4}{x^3}+\frac{3x}{x^3}-\frac{7x^3}{x^3}}=\lim_{x\to\infty}\frac{2-\frac{7}{x}+\frac{1}{x^2}}{\frac{4}{x^3}+\frac{3}{x^2}-7}=\frac{\lim\limits_{x\to\infty}2-\lim\limits_{x\to\infty}\frac{7}{x}+\lim\limits_{x\to\infty}\frac{1}{x^2}}{\lim\limits_{x\to\infty}\frac{4}{x^3}+\lim\limits_{x\to\infty}\frac{3}{x^2}-\lim\limits_{x\to\infty}7}=\frac{2-0+0}{0+0-7}=-\frac{2}{7}$$

49. $$\lim_{x\to-\infty}\frac{171+x}{3-x^2}=\lim_{x\to-\infty}\frac{\frac{171}{x^2}+\frac{x}{x^2}}{\frac{3}{x^2}-\frac{x^2}{x^2}}=\lim_{x\to-\infty}\frac{\frac{171}{x^2}+\frac{1}{x}}{\frac{3}{x^2}-1}=\frac{\lim\limits_{x\to-\infty}\frac{171}{x^2}+\lim\limits_{x\to-\infty}\frac{1}{x}}{\lim\limits_{x\to-\infty}\frac{3}{x^2}-\lim\limits_{x\to-\infty}1}=\frac{0+0}{0-1}=0$$

51. $$\lim_{x\to-\infty}\frac{2x^2+3x-1}{x+10}=\lim_{x\to-\infty}\frac{\frac{2x^2}{x^2}+\frac{3x}{x^2}-\frac{1}{x^2}}{\frac{x}{x^2}+\frac{10}{x^2}}=\lim_{x\to-\infty}\frac{2+\frac{3}{x}-\frac{1}{x^2}}{\frac{1}{x}+\frac{10}{x^2}}$$

$= -\infty$, since the limit of the numerator is positive and the limit of the denominator is 0, approached through negative values.

53. $$\lim_{x\to\infty}\sqrt{\frac{4x^2+3x-5}{2+2x^2}}=\sqrt{\lim_{x\to\infty}\frac{4x^2+3x-5}{2+2x^2}}=\sqrt{\lim_{x\to\infty}\frac{\frac{4x^2}{x^2}+\frac{3x}{x^2}-\frac{5}{x^2}}{\frac{2}{x^2}+\frac{2x^2}{x^2}}}=\sqrt{\lim_{x\to\infty}\frac{4+\frac{3}{x}-\frac{5}{x^2}}{\frac{2}{x^2}+2}}=\sqrt{\frac{\lim\limits_{x\to\infty}4+\lim\limits_{x\to\infty}\frac{3}{x}-\lim\limits_{x\to\infty}\frac{5}{x^2}}{\lim\limits_{x\to\infty}\frac{2}{x^2}+\lim\limits_{x\to\infty}2}}=\sqrt{\frac{4+0-0}{0+2}}=\sqrt{2}$$

55. Since $2x$ is negative when $x < 0$, $\lim\limits_{x\to-\infty}\sqrt{2x}$ does not exist.

57. $\lim\limits_{t\to\infty}(25 - 25e^{-0.21t})$

t	1	10	100	1,000	10,000
$25 - 25e^{-0.21t}$	4.7	21.9	25	25	25

From the table, we conclude that the limit is 25. In the long term, experienced workers will be able to produce 25 game consoles per day.

59. $$\lim_{t\to\infty} p(t) = \lim_{t\to\infty} \frac{8t^3 - t + 22}{t^3 + 8}$$
$$= \lim_{t\to\infty} \frac{\frac{8t^3}{t^3} - \frac{t}{t^3} + \frac{22}{t^3}}{\frac{t^3}{t^3} + \frac{8}{t^3}}$$
$$= \lim_{t\to\infty} \frac{8 - \frac{1}{t^2} + \frac{22}{t^3}}{1 + \frac{8}{t^3}}$$
$$= \frac{\lim_{t\to\infty} 8 - \lim_{t\to\infty} \frac{1}{t^2} + \lim_{t\to\infty} \frac{22}{t^3}}{\lim_{t\to\infty} 1 + \lim_{t\to\infty} \frac{8}{t^3}}$$
$$= \frac{8 - 0 + 0}{1 + 0} = 8$$
The long range population of the town is therefore 8,000.

61. $$\lim_{x\to\infty} V(x) = \lim_{x\to\infty} \left(\frac{42,300}{0.6x + 2} + 300\right)$$
$$= \lim_{x\to\infty} \left(\frac{\frac{42,300}{x}}{\frac{0.6x}{x} + \frac{2}{x}} + 300\right)$$
$$= \lim_{x\to\infty} \left(\frac{\frac{42,300}{x}}{0.6 + \frac{2}{x}} + 300\right)$$
$$= \frac{\lim_{x\to\infty} \frac{42,300}{x}}{\lim_{x\to\infty} 0.6 + \lim_{x\to\infty} \frac{2}{x}} + \lim_{x\to\infty} 300$$
$$= \frac{0}{0.6 + 0} + 300$$
$$= \$300$$

Section 12-4

1. $$\bar{v} = \frac{s(10) - s(5)}{10 - 5} = \frac{[8(10) - 3] - [8(5) - 3]}{10 - 5} = \frac{77 - 37}{5} = 8\,\frac{\text{ft}}{\text{sec}}$$

3. $$\bar{v} = \frac{s(2) - s(1)}{2 - 1} = \frac{[16(2)^2 + 100] - [16(1)^2 + 100]}{2 - 1} = \frac{164 - 116}{1} = 48\,\frac{\text{ft}}{\text{sec}}$$

5. On [1, 2], $\bar{v} = \frac{s(2) - s(1)}{2 - 1} = \frac{[2^2 + 12(2)] - [1^2 + 12(1)]}{2 - 1} = \frac{28 - 13}{1} = 15$ mi/hr.

On [1, 1.5], $\bar{v} = \frac{s(1.5) - s(1)}{1.5 - 1} = \frac{20.25 - 13}{0.5} = 14.5$ mi/hr.

On [1, 1.1], $\bar{v} = \frac{s(1.1) - s(1)}{1.1 - 1} = \frac{14.41 - 13}{0.1} = 14.1$ mi/hr.

On [1, 1.01], $\bar{v} = \frac{s(1.01) - s(1)}{1.01 - 1} = \frac{13.1401 - 13}{0.01} = 14.01$ mi/hr.

We can estimate the instantaneous velocity as 14 mi/hr.

7. On $[3, 4]$, $\bar{v} = \dfrac{s(4) - s(3)}{4 - 3} = \dfrac{[2(4)^3 - 8] - [2(3)^3 - 8]}{4 - 3} = \dfrac{120 - 46}{1} = 74$ mi/hr.

On $[3, 3.5]$, $\bar{v} = \dfrac{s(3.5) - s(3)}{3.5 - 3} = \dfrac{77.75 - 46}{0.5} = 63.5$ mi/hr.

On $[3, 3.1]$, $\bar{v} = \dfrac{s(3.1) - s(3)}{3.1 - 3} = \dfrac{51.582 - 46}{0.1} = 55.82$ mi/hr.

On $[3, 3.01]$, $\bar{v} = \dfrac{s(3.01) - s(3)}{3.01 - 3} = \dfrac{46.541802 - 46}{0.01} = 54.1802$ mi/hr.

We can estimate the instantaneous velocity as 54 mi/hr.

9. On $[1, 2]$, $\bar{v} = \dfrac{s(2) - s(1)}{2 - 1} = \dfrac{\frac{18(2)}{2+2} - \frac{18(1)}{1+2}}{2 - 1} = \dfrac{9 - 6}{1} = 3$ m/sec.

On $[1, 1.5]$, $\bar{v} = \dfrac{s(1.5) - s(1)}{1.5 - 1} = \dfrac{\frac{54}{7} - 6}{\frac{1}{2}} = 3\frac{3}{7}$ m/sec.

On $[1, 1.1]$, $\bar{v} = \dfrac{s(1.1) - s(1)}{1.1 - 1} = \dfrac{6.387 - 6}{0.1} = 3.87$ m/sec.

On $[1, 1.01]$, $\bar{v} = \dfrac{s(1.01) - s(1)}{1.01 - 1} = \dfrac{6.039867 - 6}{0.01} = 3.9867$ m/sec.

We can estimate the instantaneous velocity as 4 m/sec.

11. On $[2, 3]$, $\bar{v} = \dfrac{s(3) - s(2)}{3 - 2} = \dfrac{5e^{3-2} - 5e^{2-2}}{3 - 2} = 8.59$ m/sec.

On $[2, 2.5]$, $\bar{v} = \dfrac{s(2.5) - s(2)}{2.5 - 2} = 6.487$ m/sec.

On $[2, 2.1]$, $\bar{v} = \dfrac{s(2.1) - s(2)}{2.1 - 2} = 5.2585$ m/sec.

On $[2, 2.01]$, $\bar{v} = \dfrac{s(2.01) - s(2)}{2.01 - 2} = 5.025$ m/sec.

We can estimate the instantaneous velocity as 5 m/sec.

13. On $[10, 10 + h]$,
$$\bar{v} = \frac{s(10 + h) - s(10)}{10 + h - 10} = \frac{[3(10 + h) - 5] - [3(10) - 5]}{h} = \frac{30 + 3h - 5 - 25}{h} = \frac{3h}{h} = 3$$

$\lim\limits_{h \to 0} 3 = 3$ m/min = the exact instantaneous velocity.

15. On $[2, 2 + h]$,
$$\bar{v} = \frac{s(2 + h) - s(2)}{2 + h - 2} = \frac{s(2 + h)^2 - s(2)^2}{h} = \frac{8 + 8h + 2h^2 - 8}{h} = \frac{h(8 + 2h)}{h} = 8 + 2h$$

$\lim\limits_{h \to 0}(8 + 2h) = 8$ m/min = the exact instantaneous velocity.

17. On $[4, 4 + h]$,
$$\begin{aligned}\bar{v} &= \frac{s(4 + h) - s(4)}{4 + h - 4} \\ &= \frac{[(4 + h)^2 - 3(4 + h) + 10] - [4^2 - 3(4) + 10]}{h} \\ &= \frac{16 + 8h + h^2 - 12 - 3h + 10 - 14}{h} \\ &= \frac{h(5 + h)}{h} \\ &= 5 + h\end{aligned}$$

$\lim_{h\to 0}(5 + h) = 5$ m/min = the exact instantaneous velocity.

19.
$$\begin{aligned}m &= \lim_{h\to 0} \frac{f(10 + h) - f(10)}{h} \\ &= \lim_{h\to 0} \frac{[8(10 + h) - 4] - [8(10) - 4]}{h} \\ &= \lim_{h\to 0} \frac{80 + 8h - 4 - 76}{h} \\ &= \lim_{h\to 0} \frac{8h}{h} \\ &= \lim_{h\to 0} 8 = 8\end{aligned}$$

21.
$$\begin{aligned}m &= \lim_{h\to 0} \frac{g(-1 + h) - g(-1)}{h} \\ &= \lim_{h\to 0} \frac{[2(-1 + h)^2 + 5] - [2(-1)^2 + 5]}{h} \\ &= \lim_{h\to 0} \frac{2 - 4h + 2h^2 + 5 - 7}{h} \\ &= \lim_{h\to 0} \frac{h(-4 + 2h)}{h} \\ &= \lim_{h\to 0}(-4 + 2h) = -4\end{aligned}$$

23.
$$\begin{aligned}m &= \lim_{h\to 0} \frac{k(-2 + h) - k(-2)}{h} \\ &= \lim_{h\to 0} \frac{[(-2 + h)^3 - 2] - [(-2)^3 - 2]}{h} \\ &= \lim_{h\to 0} \frac{-8 + 12h - 6h^2 + h^3 - 2 + 10}{h} \\ &= \lim_{h\to 0} \frac{h(12 - 6h + h^2)}{h} \\ &= \lim_{h\to 0}(12 - 6h + h^2) = 12\end{aligned}$$

25. First find the slope of the tangent line.
$$\begin{aligned}m &= \lim_{h\to 0} \frac{q(2 + h) - q(2)}{h} \\ &= \lim_{h\to 0} \frac{[6 - (2 + h)^2] - (6 - 2^2)}{h} \\ &= \lim_{h\to 0} \frac{6 - 4 - 4h - h^2 - 2}{h} \\ &= \lim_{h\to 0} \frac{h(-4 - h)}{h} \\ &= \lim_{h\to 0}(-4 - h) = -4\end{aligned}$$
Now find the equation of the line through $(2, q(2)) = (2, 2)$ with slope -4, using the point-slope formula.
$$\begin{aligned}y - 2 &= -4(x - 2) \\ y - 2 &= -4x + 8 \\ y &= -4x + 10\end{aligned}$$

27. First find the slope of the tangent line.
$$\begin{aligned}m &= \lim_{h\to 0} \frac{p(-1 + h) - p(-1)}{h} \\ &= \lim_{h\to 0} \frac{[2(-1 + h)^2 - 4(-1 + h)] - [2(-1)^2 - 4(-1)]}{h} \\ &= \lim_{h\to 0} \frac{2 - 4h + 2h^2 + 4 - 4h - 6}{h} \\ &= \lim_{h\to 0} \frac{h(-8 + 2h)}{h} \\ &= \lim_{h\to 0}(-8 + 2h) = -8\end{aligned}$$
Now find the equation of the line through $(-1, p(-1)) = (-1, 6)$ with slope -8, using the point-slope formula.
$$\begin{aligned}y - 6 &= -8[x - (-1)] \\ y - 6 &= -8x - 8 \\ y &= -8x - 2\end{aligned}$$

29. $$f'(x) = \lim_{h\to 0}\frac{f(x+h)-f(x)}{h} = \lim_{h\to 0}\frac{[8-3(x+h)]-(8-3x)}{h} = \lim_{h\to 0}\frac{8-3x-3h-8+3x}{h} = \lim_{h\to 0}\frac{-3h}{h} = \lim_{h\to 0}(-3) = -3$$

31. $$g'(x) = \lim_{h\to 0}\frac{g(x+h)-g(x)}{h} = \lim_{h\to 0}\frac{[(x+h)^2-7]-(x^2-7)}{h} = \lim_{h\to 0}\frac{x^2+2xh+h^2-7-x^2+7}{h} = \lim_{h\to 0}\frac{h(2x+h)}{h} = \lim_{h\to 0}(2x+h) = 2x$$

33. $$g'(x) = \lim_{h\to 0}\frac{g(x+h)-g(x)}{h} = \lim_{h\to 0}\frac{[2(x+h)^2-3(x+h)]-(2x^2-3x)}{h} = \lim_{h\to 0}\frac{2x^2+4xh+2h^2-3x-3h-2x^2+3x}{h} = \lim_{h\to 0}\frac{h(4x+2h-3)}{h} = \lim_{h\to 0}(4x+2h-3) = 4x-3$$

35. $$f'(x) = \lim_{h\to 0}\frac{f(x+h)-f(x)}{h} = \lim_{h\to 0}\frac{[(x+h)^3+(x+h)]-(x^3+x)}{h} = \lim_{h\to 0}\frac{x^3+3x^2h+3xh^2+h^3+x+h-x^3-x}{h} = \lim_{h\to 0}\frac{h(3x^2+3xh+h^2+1)}{h} = \lim_{h\to 0}(3x^2+3xh+h^2+1) = 3x^2+1$$

37. $$r'(x) = \lim_{h\to 0}\frac{r(x+h)-r(x)}{h} = \lim_{h\to 0}\frac{\sqrt{x+h-5}-\sqrt{x-5}}{h} = \lim_{h\to 0}\frac{\sqrt{x+h-5}-\sqrt{x-5}}{h}\cdot\frac{\sqrt{x+h-5}+\sqrt{x-5}}{\sqrt{x+h-5}+\sqrt{x-5}} = \lim_{h\to 0}\frac{(x+h-5)-(x-5)}{h(\sqrt{x+h-5}+\sqrt{x-5})} = \lim_{h\to 0}\frac{h}{h(\sqrt{x+h-5}+\sqrt{x-5})} = \lim_{h\to 0}\frac{1}{\sqrt{x+h-5}+\sqrt{x-5}} = \frac{1}{2\sqrt{x-5}}$$

39. $$f'(x) = \lim_{h\to 0} \frac{f(x+h) - f(x)}{h}$$

$$= \lim_{h\to 0} \frac{3\sqrt{2(x+h)+1} - 3\sqrt{2x+1}}{h}$$

$$= \lim_{h\to 0} \frac{3\sqrt{2(x+h)+1} - 3\sqrt{2x+1}}{h} \cdot \frac{3\sqrt{2(x+h)+1} + 3\sqrt{2x+1}}{3\sqrt{2(x+h)+1} + 3\sqrt{2x+1}}$$

$$= \lim_{h\to 0} \frac{9[2(x+h)+1] - 9(2x+1)}{h(3\sqrt{2(x+h)+1} + 3\sqrt{2x+1})}$$

$$= \lim_{h\to 0} \frac{18x + 18h + 9 - 18x - 9}{h(3\sqrt{2(x+h)+1} + 3\sqrt{2x+1})}$$

$$= \lim_{h\to 0} \frac{18h}{3h(\sqrt{2(x+h)+1} + \sqrt{2x+1})}$$

$$= \lim_{h\to 0} \frac{6}{\sqrt{2(x+h)+1} + \sqrt{2x+1}} = \frac{3}{\sqrt{2x+1}}$$

41. $$y'(x) = \lim_{h\to 0} \frac{y(x+h) - y(x)}{h}$$

$$= \lim_{h\to 0} \frac{\frac{3}{x+h} - \frac{3}{x}}{h}$$

$$= \lim_{h\to 0} \frac{\frac{3x}{x(x+h)} - \frac{3(x+h)}{x(x+h)}}{h}$$

$$= \lim_{h\to 0} \frac{\frac{3x-3(x+h)}{x(x+h)}}{h}$$

$$= \lim_{h\to 0} \frac{\frac{-3h}{x(x+h)}}{h}$$

$$= \lim_{h\to 0} \frac{-3}{x(x+h)} = -\frac{3}{x^2}$$

43. $$y'(x) = \lim_{h\to 0} \frac{y(x+h) - y(x)}{h}$$

$$= \lim_{h\to 0} \frac{\frac{x+h}{x+h+1} - \frac{x}{x+1}}{h}$$

$$= \lim_{h\to 0} \frac{\frac{(x+h)(x+1)}{(x+h+1)(x+1)} - \frac{x(x+h+1)}{(x+h+1)(x+1)}}{h}$$

$$= \lim_{h\to 0} \frac{\frac{x^2+x+hx+h-x^2-hx-x}{(x+h+1)(x+1)}}{h}$$

$$= \lim_{h\to 0} \frac{\frac{h}{(x+h+1)(x+1)}}{h}$$

$$= \lim_{h\to 0} \frac{1}{(x+h+1)(x+1)} = \frac{1}{(x+1)^2}$$

45. $y'(x) = \lim\limits_{h \to 0} \dfrac{y(x+h) - y(x)}{h}$

$$= \lim_{h \to 0} \frac{\frac{1}{2\sqrt{x+h}} - \frac{1}{2\sqrt{x}}}{h}$$

$$= \lim_{h \to 0} \frac{\frac{\sqrt{x}}{2\sqrt{x}\sqrt{x+h}} - \frac{\sqrt{x+h}}{2\sqrt{x}\sqrt{x+h}}}{h}$$

$$= \lim_{h \to 0} \frac{\frac{\sqrt{x} - \sqrt{x+h}}{2\sqrt{x}\sqrt{x+h}}}{h}$$

$$= \lim_{h \to 0} \frac{\frac{\sqrt{x} - \sqrt{x+h}}{2\sqrt{x}\sqrt{x+h}} \cdot \frac{\sqrt{x} + \sqrt{x+h}}{\sqrt{x} + \sqrt{x+h}}}{h}$$

$$= \lim_{h \to 0} \frac{\frac{x-(x+h)}{2\sqrt{x}\sqrt{x+h}(\sqrt{x} + \sqrt{x+h})}}{h}$$

$$= \lim_{h \to 0} \frac{\frac{-h}{2\sqrt{x}\sqrt{x+h}(\sqrt{x} + \sqrt{x+h})}}{h}$$

$$= \lim_{h \to 0} \frac{-1}{2\sqrt{x}\sqrt{x+h}(\sqrt{x} + \sqrt{x+h})}$$

$$= \frac{-1}{2\sqrt{x}\sqrt{x}(\sqrt{x} + \sqrt{x})} = \frac{-1}{2x(2\sqrt{x})} = \frac{-1}{4\sqrt{x^3}}$$

47. The derivative of $g(x)$ is $g'(x) = 2x$. At $x = -2$, $g'(-2) = 2(-2) = -4$. At $x = -2$, $g(x) = g(-2) = -3$. We use the point-slope formula to find the equation of the line through $(-2, g(-2)) = (-2, -3)$ with slope $g'(-2) = -4$.

$$y - (-3) = -4[x - (-2)]$$
$$y + 3 = -4x - 8$$
$$y = -4x - 11$$

49. The derivative of $r(x)$ is $r'(x) = \dfrac{1}{2\sqrt{x-5}}$. At $x = 9$, $r'(9) = \dfrac{1}{2\sqrt{9-5}} = \dfrac{1}{4}$.
At $x = 9$, $r(x) = r(9) = \sqrt{9-5} = 2$. We use the point-slope formula to find the equation of the line through $(9, r(9)) = (9, 2)$ with slope $r'(9) = \dfrac{1}{4}$.

$$y - 2 = \frac{1}{4}(x - 9)$$
$$y - 2 = \frac{1}{4}x - \frac{9}{4}$$
$$y = \frac{1}{4}x - \frac{1}{4}$$

51. (A) First find the average velocity on the interval $[t, t + h]$.

$$\bar{v} = \frac{s(t+h) - s(t)}{t + h - t}$$
$$= \frac{[-16(t+h)^2 + 70(t+h)] - (-16t^2 + 70t)}{4}$$
$$= \frac{-16t^2 - 32ht - 16h^2 + 70t + 70h + 16t^2 - 70t}{h}$$
$$= \frac{h(-32t - 16h + 70)}{h}$$
$$= -32t - 16h + 70$$

Then the instantaneous velocity of time t is

$$v(t) = \lim_{h \to 0}(-32t - 16h + 70) = -32t + 70$$

At $t = 2$, $v(2) = -32(2) + 70 = 6$ ft/s.
At $t = 4$, $v(4) = -32(4) + 70 = -58$ ft/s.

(B) The positive velocity tells us that the ball is moving upward after 2 seconds, while the negative velocity tells us it is moving downward after 4 seconds. The ball must have reached maximum height and changed direction at some time between 2 and 4 seconds.

53. (A) $v(a) = s'(a) = \lim_{h \to 0} \frac{s(a + h) - s(a)}{h}$

$$= \lim_{h \to 0} \frac{[-6.08(a + h)^2 + 70(a + h)] - (-6.08a^2 + 70a)}{h}$$
$$= \lim_{h \to 0} \frac{-6.08a^2 - 12.16ah - 6.08h^2 + 70a + 70h + 6.08a^2 - 70a}{h}$$
$$= \lim_{h \to 0} \frac{h(-12.16a - 6.08h + 70)}{h}$$
$$= \lim_{h \to 0} (-12.16a - 6.08h + 70) = -12.16a + 70$$

(B) $v(3) = -12.16(3) + 70 = 33.52$ ft/s
$v(7) = -12.16(7) + 70 = -15.12$ ft/s

(C) The object reaches the ground when $s(t) = 0$.
Solve $-6.08t^2 + 70t = 0$
$t(-6.08t + 70) = 0$
$t = 0$ (discarded) or $-6.08t + 70 = 0$
$t = \frac{70}{6.08} \approx 11.5$ sec.

(D) Substitute the value of t obtained in part (C) to obtain
$v(t) = v(11.5) = -12.16(11.5) + 70 \approx -70$ ft/s.

55.

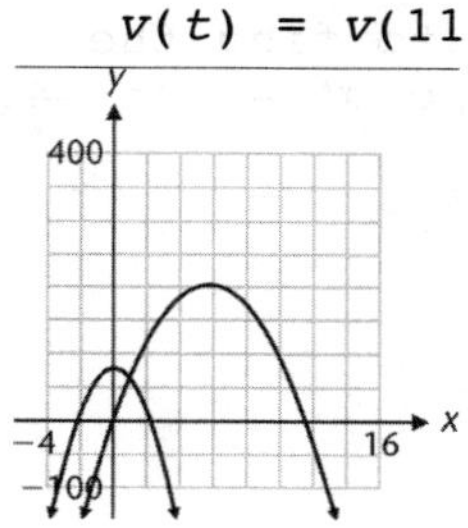

Applying the maximum routine to $-6.08t^2 + 70t$ yields 201.48026 feet on Mars. Applying the maximum routine to $-16t^2 + 70t$ yields 76.5625 feet on Earth. The difference is approximately 124.9 feet higher on Mars.

57. $v'(5) = \lim_{h \to 0} \frac{v(5 + h) - v(5)}{h}$

$$= \lim_{h \to 0} \frac{[-41(5 + h)^2 + 1,680(5 + h) + 7,400] - [-41(5)^2 + 1,680(5) + 7,400]}{h}$$
$$= \lim_{h \to 0} \frac{-1,025 - 410h - 41h^2 + 8,400 + 1,680h + 7,400 - 14,775}{h}$$
$$= \lim_{h \to 0} \frac{h(-410 - 41h + 1,680)}{h}$$
$$= \lim_{h \to 0} (1,270 - 41h) = \$1,270 \text{ dollars/yr.}$$

Five years after the artist dies, the painting is gaining value at $1,270 per year.

59. (A) $p'(3) = -0.000348(3)^2 + 0.111(3) - 1.02 \approx -0.69$ dollars/month
$p'(24) = -0.000348(24)^2 + 0.111(24) = 1.02 \approx 1.44$ dollars/month
In September 2002, the value of the stock was decreasing at the rate of 69 cents per month. In June 2004, it was increasing at $1.44 per month.

(B)

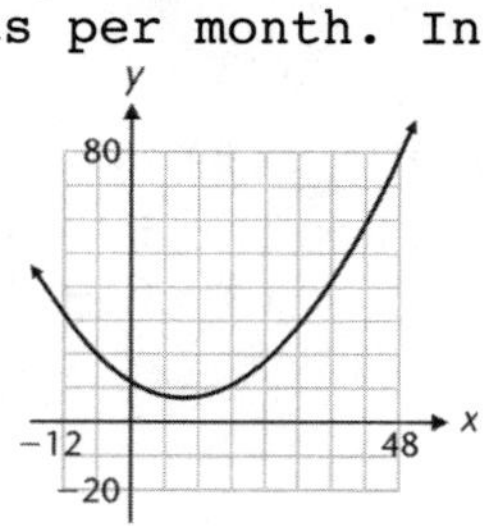

At $x = 3$, the height of the graph is decreasing, so a negative change in value is appropriate. At $x = 24$, the height is increasing fairly rapidly, so a larger positive change in value makes sense.

Section 12-5

1. The region is a rectangle with width 4 units and height 5 units. The area is width × height, so $A = 4 \cdot 5 = 20$ square units.

3. The region is a rectangle with width $2 - (-2) = 4$ units and height π units. The area is width × height, so $A = 4\pi$ square units.

5. Rectangle 1: $A = \text{width} \cdot \text{height} = 2\left(\frac{1}{2} \cdot 2 + 10\right) = 22$

   Rectangle 2: $A = \text{width} \cdot \text{height} = 2\left(\frac{1}{2} \cdot 4 + 10\right) = 24$

   Rectangle 3: $A = \text{width} \cdot \text{height} = 2\left(\frac{1}{2} \cdot 6 + 10\right) = 26$

   Rectangle 4: $A = \text{width} \cdot \text{height} = 2\left(\frac{1}{2} \cdot 8 + 10\right) = 28$

   $A \approx 22 + 24 + 26 + 28 = 100$

7. Rectangle 1: $A = \text{width} \cdot \text{height} = \frac{1}{2}\left[-\left(\frac{3}{2}\right)^2 + 4\left(\frac{3}{2}\right) + 1\right] = \frac{19}{8}$

   Rectangle 2: $A = \text{width} \cdot \text{height} = \frac{1}{2}[(-2)^2 + 4(2) + 1] = \frac{5}{2}$

   Rectangle 3: $A = \text{width} \cdot \text{height} = \frac{1}{2}\left[-\left(\frac{5}{2}\right)^2 + 4\left(\frac{5}{2}\right) + 1\right] = \frac{19}{8}$

   Rectangle 4: $A = \text{width} \cdot \text{height} = \frac{1}{2}[(-3)^2 + 4(3) + 1] = 2$

   $A \approx \frac{19}{8} + \frac{5}{2} + \frac{19}{8} + 2 = \frac{37}{4}$

9. 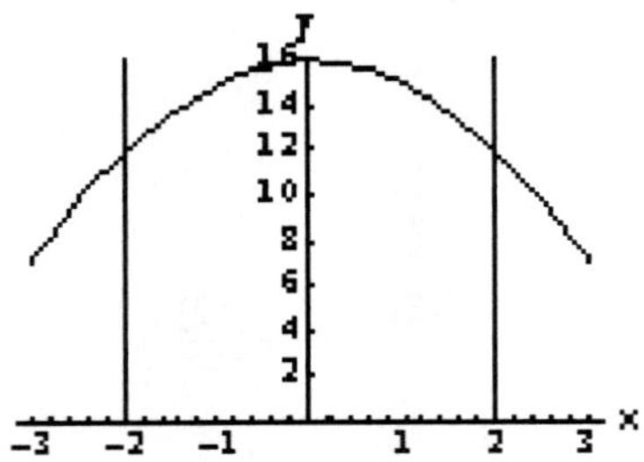

   $\Delta x = \frac{2 - (-2)}{4} = 1$

   $S_4 = f(-1)\Delta x + f(0)\Delta x + f(1)\Delta x + f(2)\Delta x$
   $= [16 - (-1)^2]1 + [16 - 0^2]1 + [16 - 1^2]1 + [16 - 2^2]1$
   $= 15 + 16 + 15 + 12 = 58$

11. 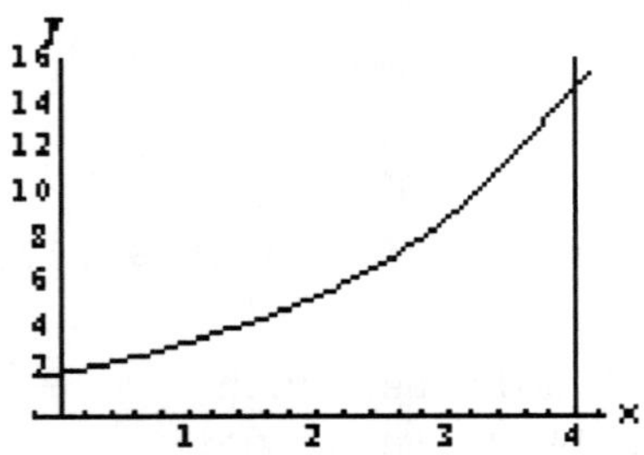

   $\Delta x = \frac{4 - 0}{4} = 1$

   $S_4 = f(1)\Delta x + f(2)\Delta x + f(3)\Delta x + f(4)\Delta x$
   $= 2e^{1/2} \cdot 1 + 2e^{2/2} \cdot 1 + 2e^{3/2} \cdot 1 + 2e^{4/2} \cdot 1$
   $= 32.48$

13. 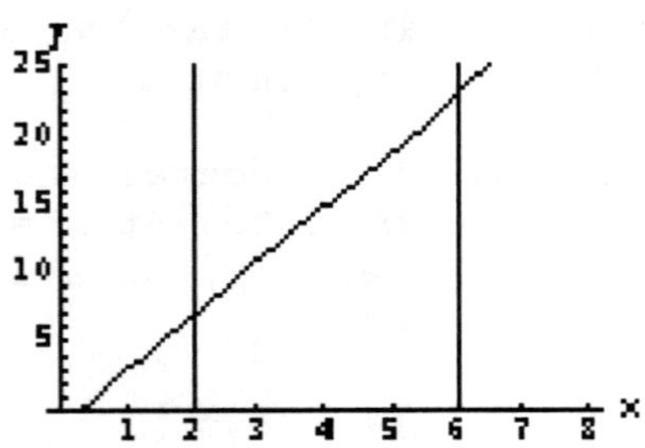

   For 4 rectangles, $\Delta x = \frac{6 - 2}{4} = 1.$

   $S_4 = f(3)\Delta x + f(4)\Delta x + f(5)\Delta x + f(6)\Delta x$
   $= (4 \cdot 3 - 1)1 + (4 \cdot 4 - 1)1 + (4 \cdot 5 - 1)1 + (4 \cdot 6 - 1)1$
   $= 11 + 15 + 19 + 23 = 68,$

For 8 rectangles, $\Delta x = \dfrac{6 - 2}{8} = 0.5$.

$$S_8 = f(2.5)\Delta x + f(3)\Delta x + f(3.5)\Delta x + f(4)\Delta x + f(4.5)\Delta x + f(5)\Delta x + f(5.5)\Delta x + f(6)\Delta x$$

$$= [4(2.5) - 1](0.5) + [4 \cdot 3 - 1](0.5) + [4(3.5) - 1](0.5) + [4 \cdot 4 - 1](0.5) + [4(4.5) - 1]0.5 + [4 \cdot 5 - 1](0.5) + [4(5.5) - 1](0.5) + [4 \cdot 6 - 1](0.5)$$

$$= 64$$

15. 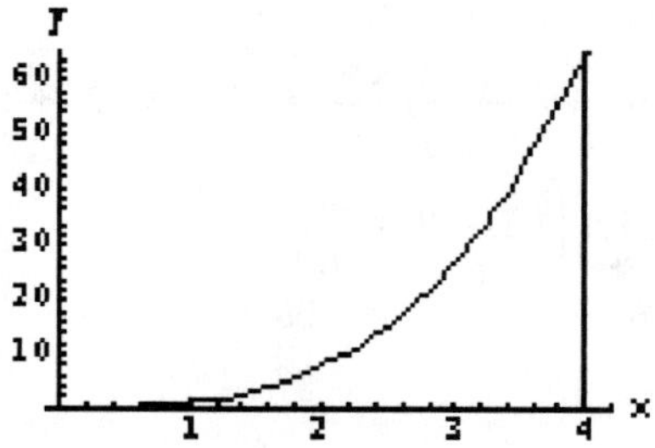

For 4 rectangles, $\Delta x = \dfrac{4 - 0}{4} = 1$.

$$S_4 = f(1)\Delta x + f(2)\Delta x + f(3)\Delta x + f(4)\Delta x$$
$$= 1^3 \cdot 1 + 2^3 \cdot 1 + 3^3 \cdot 1 + 4^3 \cdot 1$$
$$= 1 + 8 + 27 + 64 = 100,$$

For 8 rectangles, $\Delta x = \dfrac{4 - 0}{8} = 0.5$.

$$S_8 = f(0.5)\Delta x + f(1)\Delta x + f(1.5)\Delta x + f(2)\Delta x + f(2.5)\Delta x + f(3)\Delta x + f(3.5)\Delta x + f(4)\Delta x$$

$$= (0.5)^3(0.5) + 1^3(0.5) + (1.5)^3(0.5) + 2^3(0.5) + (2.5)^3(0.5) + 3^3(0.5) + (3.5)^3(0.5) + 4^3(0.5)$$

$$= 81$$

17. $\Delta x = \dfrac{8 - 2}{n} = \dfrac{6}{n}$, $x_k = 2 + k\Delta x = 2 + k \cdot \dfrac{6}{n}$

19. $\Delta x = \dfrac{4 - 0}{4} = \dfrac{4}{n}$, $x_k = 0 + k\Delta x = k \cdot \dfrac{4}{n}$

21. $\Delta x = \dfrac{4 - 2}{n} = \dfrac{2}{n}$, $x_k = 2 + k\Delta x = 2 + k \cdot \dfrac{2}{n}$

$$S_n = \sum_{k=1}^{n} f(x_k)\Delta x$$

$$= \sum_{k=1}^{n} f\left(2 + k \cdot \frac{2}{n}\right)\left(\frac{2}{n}\right)$$

$$= \sum_{k=1}^{n} \left(2 + k \cdot \frac{2}{n} + 6\right)\left(\frac{2}{n}\right)$$

$$= \sum_{k=1}^{n} \left(8 + k \cdot \frac{2}{n}\right)\left(\frac{2}{n}\right)$$

$$= \sum_{k=1}^{n} \left(\frac{16}{n} + k \cdot \frac{4}{n^2}\right)$$

23. $\Delta x = \dfrac{2-0}{n} = \dfrac{2}{n}$,

$$x_k = 0 + k\Delta x = k \cdot \frac{2}{n}$$

$$\begin{aligned} S_n &= \sum_{k=1}^{n} g(x_k)\Delta x \\ &= \sum_{k=1}^{n} g\left(k \cdot \frac{2}{n}\right)\left(\frac{2}{n}\right) \\ &= \sum_{k=1}^{n} \left[2\left(k \cdot \frac{2}{n}\right)^2 + 1\right]\left(\frac{2}{n}\right) \\ &= \sum_{k=1}^{n} \left(k^2 \cdot \frac{8}{n^2} + 1\right)\left(\frac{2}{n}\right) \\ &= \sum_{k=1}^{n} \left(k^2 \cdot \frac{16}{n^3} + \frac{2}{n}\right) \end{aligned}$$

25. $$\begin{aligned} S_n &= \sum_{k=1}^{n} \left(\frac{16}{n} + k \cdot \frac{4}{n^2}\right) \\ &= \sum_{k=1}^{n} \frac{16}{n} + \sum_{k=1}^{n} k \cdot \frac{4}{n^2} \\ &= \sum_{k=1}^{n} \frac{16}{n} + \frac{4}{n^2}\sum_{k=1}^{n} k \\ &= \frac{16}{n} \cdot n + \frac{4}{n^2}\,\frac{n(n+1)}{2} \\ &= 16 + \frac{4n^2 + 4n}{2n^2} = 16 + 2 + \frac{2}{n} \\ A &= \lim_{n\to\infty}\left(16 + 2 + \frac{2}{n}\right) \\ &= 16 + 2 + 0 = 18 \end{aligned}$$

27. $$\begin{aligned} S_n &= \sum_{k=1}^{n} \left(k^2 \cdot \frac{16}{n^3} + \frac{2}{n}\right) \\ &= \sum_{k=1}^{n} k^2 \cdot \frac{16}{n^3} + \sum_{k=1}^{n} \frac{2}{n} \\ &= \frac{16}{n^3}\sum_{k=1}^{n} k^2 + \sum_{k=1}^{n} \frac{2}{n} \\ &= \frac{16}{n^3}\,\frac{n(n+1)(2n+1)}{6} + n \cdot \frac{2}{n} \\ &= \frac{16(2n^3 + 3n^2 + n)}{6n^3} + 2 \\ &= \frac{32}{6} + \frac{48}{6n} + \frac{16}{6n^2} + 2 \\ A &= \lim_{n\to\infty}\left(\frac{32}{6} + \frac{48}{6n} + \frac{16}{6n^2} + 2\right) \\ &= \frac{16}{3} + 0 + 0 + 2 = \frac{22}{3} \end{aligned}$$

29. $\Delta x = \dfrac{0-(-2)}{n} = \dfrac{2}{n}$,

$$x_k = -2 + k\Delta x = -2 + k \cdot \frac{2}{n}$$

$$\begin{aligned} S_n &= \sum_{k=1}^{n} f(x_k)\Delta x \\ &= \sum_{k=1}^{n} 3\left(-2 + k \cdot \frac{2}{n}\right)^2\left(\frac{2}{n}\right) \\ &= \sum_{k=1}^{n} \left(4 - k \cdot \frac{8}{n} + k^2 \cdot \frac{4}{n^2}\right)\left(\frac{6}{n}\right) \\ &= \sum_{k=1}^{n} \left(\frac{24}{n} - k \cdot \frac{48}{n^2} + k^2 \cdot \frac{24}{n^3}\right) \\ &= \sum_{k=1}^{n} \frac{24}{n} - \frac{48}{n^2}\sum_{k=1}^{n} k + \frac{24}{n^3}\sum_{k=1}^{n} k^2 \\ &= \frac{24}{n} \cdot n - \frac{48}{n^2}\,\frac{n(n+1)}{2} \\ &\quad + \frac{24}{n^3}\,\frac{n(n+1)(2n+1)}{6} \\ &= 24 - \frac{48n^2 + 48n}{2n^2} \\ &\quad + \frac{24(2n^3 + 3n^2 + n)}{6n^3} \\ &= 24 - 24 - \frac{24}{n} + 8 + \frac{12}{n} + \frac{4}{n^2} \\ A &= \lim_{n\to\infty} S_n = 24 - 24 + 8 = 8 \end{aligned}$$

31. $\Delta x = \dfrac{4-0}{n} = \dfrac{4}{n}$, $x_k = 0 + k\Delta x = k \cdot \dfrac{4}{n}$

$$S_n = \sum_{k=1}^{n} g(x_k)\Delta x$$
$$= \sum_{k=1}^{n} \left[\left(k \cdot \frac{4}{n}\right)^2 + 2\left(k \cdot \frac{4}{n}\right)\right]\left(\frac{4}{n}\right)$$
$$= \sum_{k=1}^{n} \left(k^2 \cdot \frac{16}{n^2} + k \cdot \frac{8}{n}\right)\left(\frac{4}{n}\right)$$
$$= \sum_{k=1}^{n} \left(k^2 \cdot \frac{64}{n^3} + k \cdot \frac{32}{n^2}\right)$$
$$= \frac{64}{n^3}\sum_{k=1}^{n} k^2 + \frac{32}{n^2}\sum_{k=1}^{n} k$$
$$= \frac{64}{n^3}\,\frac{n(n+1)(2n+1)}{6} + \frac{32}{n^2}\,\frac{n(n+1)}{2}$$
$$= \frac{64(2n^3+3n^2+n)}{6n^3} + \frac{32n^2+32n}{2n^2}$$
$$= \frac{64}{3} + \frac{192}{6n} + \frac{64}{6n^2} + 16 + \frac{16}{n}$$
$$A = \lim_{n\to\infty} S_n = \frac{64}{3} + 16 = \frac{112}{3}$$

33. Using the hint, we calculate the area first.

$A \approx 14 \cdot 4 + 16 \cdot 4 + 16 \cdot 4 + 14 \cdot 4 + 12 \cdot 4 + 14 \cdot 4 + 10 \cdot 4$

$= 384$ square feet

Then volume = Area × average depth

$= 384 \cdot 5 = 1{,}920$ cubic feet.

CHAPTER 12 REVIEW

1. $\lim_{x\to 3}(4 - 5x + x^2)$

x	2	2.5	2.9	2.99	3.01	3.1	3.5	4
$4 - 5x + x^2$	-2	-2.25	-2.09	-2.01	-1.99	-1.89	-1.25	0

From the table, we conclude that the limit is -2. *(12-1)*

2. $\lim_{x\to 10}\dfrac{x^2-100}{x-10}$

x	9	9.5	9.9	9.99	10	10.01	10.1	10.5	11
$\dfrac{x^2-100}{x-10}$	19	19.5	19.9	19.99	↑ Undefined	20.01	20.1	20.5	21

From the table, we conclude that the limit is 20. *(12-1)*

3. $\lim_{x\to -1}\dfrac{8}{(x+1)^2}$

x	-2	-1.5	-1.1	-1.01	-1	-0.99	-0.9	-0.5	0
$\dfrac{8}{(x+1)^2}$	8	32	800	80,000	↑ Undefined	80,000	800	32	8

From the table, we conclude that the limit is ∞. *(12-1)*

4. $\lim\limits_{x \to 3} \dfrac{x}{x-3}$

x	2	2.5	2.9	2.99	3	3.01	3.1	3.5	4
$\dfrac{x}{x-3}$	-2	-5	-29	-299	↑ Undefined	301	31	7	4

From the table, we conclude that the limit does not exist. (*12-1*)

5. (A) $-\infty$ (B) $-\infty$ (C) Does not exist (D) 0 (E) -3 (*12-1*)

6. 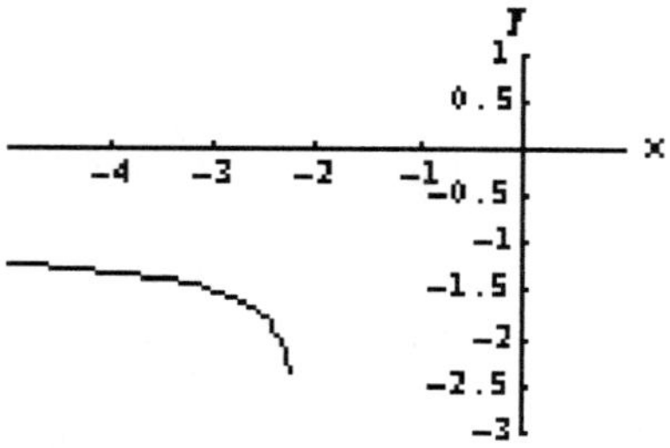

From the graph, we conclude that $\lim\limits_{x \to -3} f(x) = -1.5$. (*12-1*)

7. 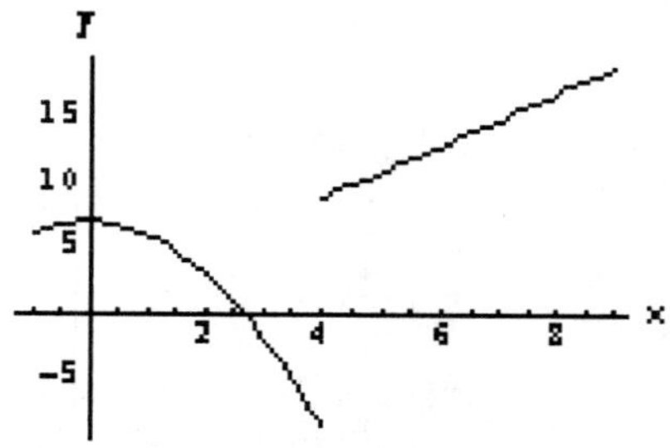

From the graph, we conclude that $\lim\limits_{x \to 4} g(x)$ does not exist. (*12-1*)

8. (A) $\lim\limits_{x \to 1^+} (3 - 12x)$. We substitute in values greater than 1.

x	2	1.5	1.1	1.01	1.001
$3 - 12x$	-21	-15	-10.2	-9.12	-9.012

From the table, we conclude that the limit is -9.

(B) To find $\lim\limits_{x \to 2^+} f(x)$, we substitute in only x values greater than 2, so only the formula $x^2 - 3$ is relevant.

x	3	2.5	2.1	2.01
$x^2 - 3$	6	3.25	1.41	1.0401

We conclude that $\lim\limits_{x \to 2^+} f(x) = 1$.

To find $\lim\limits_{x \to 2^-} f(x)$, we substitute in only x values less than 2, so only the formula $\ln(x - 1)$ is relevant.

x	1.5	1.9	1.99	1.999
$\ln(x-1)$	-0.6931	-0.1054	-0.0101	-0.001

We conclude that $\lim\limits_{x \to 2^-} f(x) = 0$. (*12-1*)

9. (A) ∞ (B) $-\infty$ (C) -10 (D) 10 (*12-1*)

10. If you substitute x values that get arbitrarily close to 5 into $f(x)$, the corresponding outputs will get arbitrarily close to 8. (*12-1*)

11.
$$\begin{aligned}\lim_{x \to 5}(-3x + 4) &= \lim_{x \to 5}(-3x) + \lim_{x \to 5} 4 \\ &= -3\lim_{x \to 5} x + \lim_{x \to 5} 4 \\ &= -3(5) + 4 = -11\end{aligned}$$
(*12-2*)

12.
$$\begin{aligned}\lim_{x \to -2}(3x^2 + 7x - 5) &= \lim_{x \to -2}(3x^2) + \lim_{x \to -2}(7x) - \lim_{x \to -2} 5 \\ &= 3(\lim_{x \to -2} x)^2 + 7\lim_{x \to -2} x - \lim_{x \to -2} 5 \\ &= 3(-2)^2 + 7(-2) - 5 = -7\end{aligned}$$
(*12-2*)

13. $\lim_{x \to -1} \frac{x + x^3}{10x + 2} = \frac{\lim_{x \to -1}(x + x^3)}{\lim_{x \to -1}(10x + 2)}$

$= \frac{\lim_{x \to -1} x + (\lim_{x \to -1} x)^3}{10 \lim_{x \to -1} x + \lim_{x \to -1} 2}$

$= \frac{-1 + (-1)^3}{10(-1) + 2} = \frac{-2}{-8} = \frac{1}{4}$ (12-2)

14. $\lim_{x \to 2}(x - 1)\sqrt{x + 7} = \lim_{x \to 2}(x - 1) \lim_{x \to 2}\sqrt{x + 7}$

$= \lim_{x \to 2}(x - 1) \sqrt{\lim_{x \to 2}(x + 7)}$

$= (\lim_{x \to 2} x - \lim_{x \to 2} 1)\sqrt{\lim_{x \to 2} x + \lim_{x \to 2} 7}$

$= (2 - 1)\sqrt{2 + 7} = \sqrt{9} = 3$ (12-2)

15. Since $x^4 - 7$ is a polynomial, $\lim_{x \to -2}(x^4 - 7) = (-2)^4 - 7 = 9$. (12-2)

16. Since $\frac{x + 1}{x^2 - 2}$ is a rational function, $\lim_{x \to 1} \frac{x + 1}{x^2 - 2} = \frac{1 + 1}{1^2 - 2} = -2$. (12-2)

17. Since $3x^2 - 4$ is negative near $x = 0$, the limit does not exist. (12-2)

18. $\lim_{x \to 2} \frac{x - 2}{x^2 + 3x - 10} = \lim_{x \to 2} \frac{x - 2}{(x - 2)(x + 5)}$

$= \lim_{x \to 2} \frac{1}{x + 5}$

$= \frac{1}{2 + 5} = \frac{1}{7}$ (12-2)

19. $\lim_{x \to 0} \frac{\sqrt{9 + x} - 3}{x} = \lim_{x \to 0} \frac{\sqrt{9 + x} - 3}{x} \cdot \frac{\sqrt{9 + x} + 3}{\sqrt{9 + x} + 3}$

$= \lim_{x \to 0} \frac{9 + x - 9}{x(\sqrt{9 + x} + 3)}$

$= \lim_{x \to 0} \frac{x}{x(\sqrt{9 + x} + 3)}$

$= \lim_{x \to 0} \frac{1}{\sqrt{9 + x} + 3}$

$= \frac{1}{\sqrt{9 + 0} + 3} = \frac{1}{6}$ (12-2)

20. $\lim_{x \to 0} \frac{(-4 - x)^2 - 16}{x} = \lim_{x \to 0} \frac{16 + 8x + x^2 - 16}{x}$

$= \lim_{x \to 0} \frac{8x + x^2}{x}$

$= \lim_{x \to 0} \frac{x(8 + x)}{x}$

$= \lim_{x \to 0}(8 + x)$

$= 8 + 0 = 8$ (12-2)

21. Since $\lim_{x \to -4^-} \frac{3}{x + 4} = -\infty$ and $\lim_{x \to -4^+} \frac{3}{x + 4} = \infty$, the limit does not exist. (12-2)

22. Since $\lim_{x\to 1^-} \frac{5x}{(x-1)^4} = \lim_{x\to 1^+} \frac{5x}{(x-1)^4} = \infty$, $\lim_{x\to 1} \frac{5x}{(x-1)^4} = \infty$. (12-2)

23. $\lim_{x\to\infty} \frac{7x^2+2}{2x^2}$

x	1	10	100	1,000	10,000
$\frac{7x^2+2}{2x^2}$	4.5	3.51	3.5001	3.5	3.5

From the table, we conclude that the limit is 3.5 $\left(= \frac{7}{2}\right)$. (12-3)

24. $\lim_{x\to\infty} (3 - x + x^2)$

x	1	10	100	1,000	10,000
$3 - x + x^2$	3	93	9,903	999,003	10^8

From the table, we conclude that the limit is ∞. (12-3)

25. $\lim_{x\to -\infty} (2x^2 - 6x)$

x	-1	-10	-100	-1,000	-10,000
$2x^2 - 6x$	8	260	20,600	2.01×10^6	2×10^8

From the table, we conclude that the limit is ∞. (12-3)

26. $\lim_{x\to\infty} f(x) = -\infty$, $\lim_{x\to -\infty} f(x) = 2$ (12-3)

27. (A)

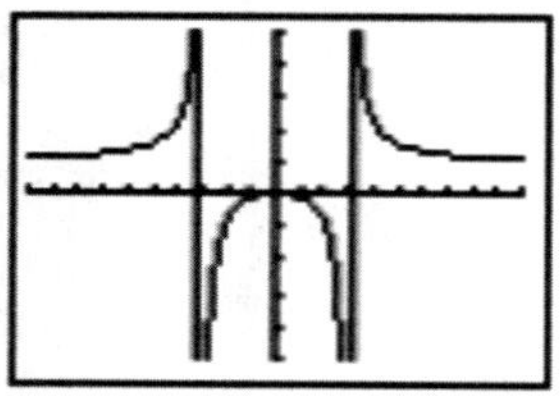

From the graph, we conclude that

$$\lim_{x\to\infty} \frac{x^2}{x^2-9} = \lim_{x\to -\infty} \frac{x^2}{x^2-9} = 1$$

(B)

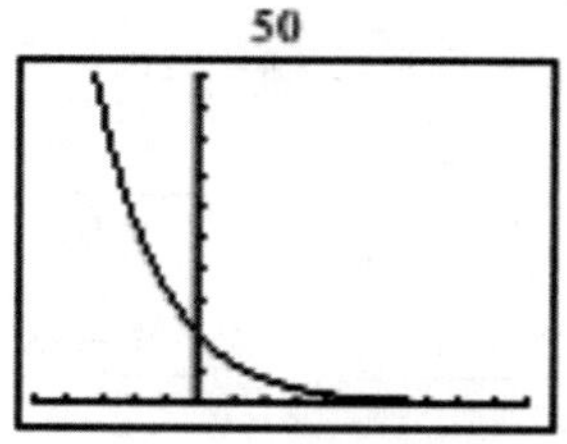

From the graph, we conclude that $\lim_{x\to\infty} \frac{10}{e^{0.5x}} = 0$ and $\lim_{x\to -\infty} \frac{10}{e^{0.5x}} = \infty$. (12-3)

28. ∞ (12-3)

29. 0 (12-3)

30. $\lim_{x\to\infty} (-3x^4 + 10x^2 - 11x) = \lim_{x\to\infty} (-3x^4) = -\infty$ (12-3)

31.
$$\lim_{x\to\infty} \frac{x - x^2}{2x^3+7} = \lim_{x\to\infty} \frac{\frac{x}{x^3} - \frac{x^2}{x^3}}{\frac{2x^3}{x^3} + \frac{7}{x^3}}$$
$$= \lim_{x\to\infty} \frac{\frac{1}{x^2} - \frac{1}{x}}{2 + \frac{7}{x^3}}$$
$$= \frac{\lim_{x\to\infty} \frac{1}{x^2} - \lim_{x\to\infty} \frac{1}{x}}{\lim_{x\to\infty} 2 + \lim_{x\to\infty} \frac{7}{x^3}}$$
$$= \frac{0 - 0}{2 + 0} = 0$$
(12-3)

32. $$\lim_{x\to-\infty}\frac{5x^3+2x}{1+x+2x^3} = \lim_{x\to-\infty}\frac{\frac{5x^3}{x^3}+\frac{2x}{x^3}}{\frac{1}{x^3}+\frac{x}{x^3}+\frac{2x^3}{x^3}}$$
$$= \lim_{x\to-\infty}\frac{5+\frac{2}{x^2}}{\frac{1}{x^3}+\frac{1}{x^2}+2}$$
$$= \frac{\lim_{x\to-\infty}5+\lim_{x\to-\infty}\frac{2}{x^2}}{\lim_{x\to-\infty}\frac{1}{x^3}+\lim_{x\to-\infty}\frac{1}{x^2}+\lim_{x\to-\infty}2}$$
$$= \frac{5+0}{0+0+2} = \frac{5}{2}$$ (*12-3*)

33. $$\lim_{x\to\infty}\frac{x^4-3x^3+1}{10x^2+20x} = \lim_{x\to\infty}\frac{\frac{x^4}{x^4}-\frac{3x^3}{x^4}+\frac{1}{x^4}}{\frac{10x^2}{x^4}+\frac{20x}{x^4}}$$
$$= \lim_{x\to\infty}\frac{1-\frac{3}{x}+\frac{1}{x^4}}{\frac{10}{x^2}+\frac{20}{x^3}}$$
$= \infty$ since the limit of the numerator is positive and the limit of the denominator is 0, approaching through positive values. (*12-3*)

34. On [3, 4], $\bar{v} = \frac{s(4)-s(3)}{4-3} = \frac{[-4(4)^2+60\cdot4]-[-4(3)^2+60\cdot3]}{4-3} = \frac{176-144}{1} = 32$ ft/s

On [3, 3.5], $\bar{v} = \frac{s(3.5)-s(3)}{3.5-3} = \frac{161-144}{0.5} = 34$ ft/s

On [3, 3.1], $\bar{v} = \frac{s(3.1)-s(3)}{3.1-3} = \frac{147.56-144}{0.1} = 35.6$ ft/s

On [3, 3.01], $\bar{v} = \frac{s(3.01)-s(3)}{3.01-3} = \frac{144.3596-144}{0.01} = 35.96$ ft/s

We can estimate the instantaneous velocity as 36 ft/s. (*12-4*)

35. On [3, 4], $\bar{v} = \frac{s(4)-s(3)}{4-3} = \frac{\frac{4\cdot4}{4+1}-\frac{4\cdot3}{3+1}}{4-3} = \frac{\frac{16}{5}-3}{1} = \frac{1}{5}$ yd/min

On [3, 3.5], $\bar{v} = \frac{s(3.5)-s(3)}{3.5-3} = \frac{\frac{28}{9}-3}{\frac{1}{2}} = \frac{2}{9}$ yd/min

On [3, 3.1], $\bar{v} = \frac{s(3.1)-s(3)}{3.1-3} = \frac{\frac{124}{41}-3}{\frac{1}{10}} = \frac{10}{41}$ yd/min

On [3, 3.01], $\bar{v} = \frac{s(3.01)-s(3)}{3.01-3} = \frac{\frac{1204}{401}-3}{\frac{1}{100}} = \frac{100}{401}$ yd/min

We can estimate the instantaneous velocity as $\frac{1}{4}$ yd/min. (*12-4*)

36. (A) On $[5, 5 + h]$, $\bar{v} = \frac{s(5 + h) - s(5)}{5 + h - 5}$

$$= \frac{[8(5 + h) + 12] - [8(5) + 12]}{h}$$
$$= \frac{40 + 8h + 12 - 52}{h}$$
$$= \frac{8h}{h}$$
$$= 8$$

$\lim_{h\to 0} 8 = 8$ ft/s = the exact instantaneous velocity

(B) On $[2, 2 + h]$, $\bar{v} = \frac{s(2 + h) - s(2)}{2 + h - 2}$

$$= \frac{[2(2 + h)^2 + (2 + h)] - [2(2)^2 + 2]}{h}$$
$$= \frac{8 + 8h + 2h^2 + 2 + h - 10}{h}$$
$$= \frac{9h + 2h^2}{h}$$
$$= \frac{h(9 + 2h)}{h}$$
$$= 9 + 2h$$

$\lim_{h\to 0} (9 + 2h) = 9$ ft/s = the exact instantaneous velocity (12-4)

37. $m = \lim_{h\to 0} \frac{f(2 + h) - f(2)}{h}$

$$= \lim_{h\to 0} \frac{[10 - 6(2 + h)] - [10 - 6(2)]}{h}$$
$$= \lim_{h\to 0} \frac{10 - 12 - 6h + 2}{h}$$
$$= \lim_{h\to 0} \frac{-6h}{h}$$
$$= \lim_{h\to 0} (-6) = -6$$

(12-4)

38. $m = \lim_{h\to 0} \frac{g(-1 + h) - g(-1)}{h}$

$$= \lim_{h\to 0} \frac{[(-1 + h)^2 + 9(-1 + h)] - [(-1)^2 + 9(-1)]}{h}$$
$$= \lim_{h\to 0} \frac{1 - 2h + h^2 - 9 + 9h + 8}{h}$$
$$= \lim_{h\to 0} \frac{h(7 + h)}{h}$$
$$= \lim_{h\to 0} (7 + h) = 7$$

(12-4)

39. First find the slope of the tangent line.

$$m = \lim_{h\to 0} \frac{y(3 + h) - y(3)}{h}$$
$$= \lim_{h\to 0} \frac{[2(3 + h)^2 + 4] - [2(3)^2 + 4]}{h}$$
$$= \lim_{h\to 0} \frac{18 + 12h + 2h^2 + 4 - 22}{h}$$
$$= \lim_{h\to 0} \frac{h(12 + 2h)}{h}$$
$$= \lim_{h\to 0} (12 + 2h) = 12$$

Now find the equation of the line through $(3, y(3)) = (3, 22)$ with slope 12, using the point-slope formula.

$$y - 22 = 12(x - 3)$$
$$y - 22 = 12x - 36$$
$$y = 12x - 14$$

(12-4)

40. $f'(x) = \lim_{h\to 0} \frac{f(x+h) - f(x)}{h}$

$= \lim_{h\to 0} \frac{[9 - \frac{1}{2}(x+h)] - [9 - \frac{1}{2}x]}{h}$

$= \lim_{h\to 0} \frac{9 - \frac{1}{2}x - \frac{1}{2}h - 9 + \frac{1}{2}x}{h}$

$= \lim_{h\to 0} \frac{-\frac{1}{2}h}{h}$

$= \lim_{h\to 0}\left(-\frac{1}{2}\right) = -\frac{1}{2}$ (12-4)

41. $g'(x) = \lim_{h\to 0} \frac{g(x+h) - g(x)}{h}$

$= \lim_{h\to 0} \frac{[8 - 5(x+h)^2] - (8 - 5x^2)}{h}$

$= \lim_{h\to 0} \frac{8 - 5x^2 - 10hx - 5h^2 - 8 + 5x^2}{h}$

$= \lim_{h\to 0} \frac{h(-10x - 5h)}{h}$

$= \lim_{h\to 0}(-10x - 5h) = -10x$ (12-4)

42. $k'(x) = \lim_{h\to 0} \frac{k(x+h) - k(x)}{h}$

$= \lim_{h\to 0} \frac{\sqrt{x+h+11} - \sqrt{x+11}}{h}$

$= \lim_{h\to 0} \frac{\sqrt{x+h+11} - \sqrt{x+11}}{h} \cdot \frac{\sqrt{x+h+11} + \sqrt{x+11}}{\sqrt{x+h+11} + \sqrt{x+11}}$

$= \lim_{h\to 0} \frac{(x+h+11) - (x+11)}{h(\sqrt{x+h+11} + \sqrt{x+11})}$

$= \lim_{h\to 0} \frac{h}{h(\sqrt{x+h+11} + \sqrt{x+11})}$

$= \lim_{h\to 0} \frac{1}{\sqrt{x+h+11} + \sqrt{x+11}} = \frac{1}{2\sqrt{x+11}}$ (12-4)

43. $r'(x) = \lim_{h\to 0} \frac{r(x+h) - r(x)}{h}$

$= \lim_{h\to 0} \frac{\frac{7}{x+h+1} - \frac{7}{x+1}}{h}$

$= \lim_{h\to 0} \frac{\frac{7(x+1)}{(x+1)(x+h+1)} - \frac{7(x+h+1)}{(x+1)(x+h+1)}}{h}$

$= \lim_{h\to 0} \frac{\frac{7(x+1) - 7(x+h+1)}{(x+1)(x+h+1)}}{h}$

$= \lim_{h\to 0} \frac{\frac{-7h}{(x+1)(x+h+1)}}{h}$

$= \lim_{h\to 0} \frac{-7}{(x+1)(x+h+1)} = \frac{-7}{(x+1)^2}$ (12-4)

44. The derivative of $g(x)$ is $g'(x) = -10x$. At $x = 0$, $g'(0) = -10(0) = 0$. At $x = 0$, $g(x) = g(0) = 8$. The equation of the line with slope 0 through (0, 8) is $y = 8$. (12-4)

45. The region is a rectangle with width $3 - (-3) = 6$ units and height 25 units. The area is width × height, so $A = 6 \cdot 25 = 150$ square units. (12-5)

46.

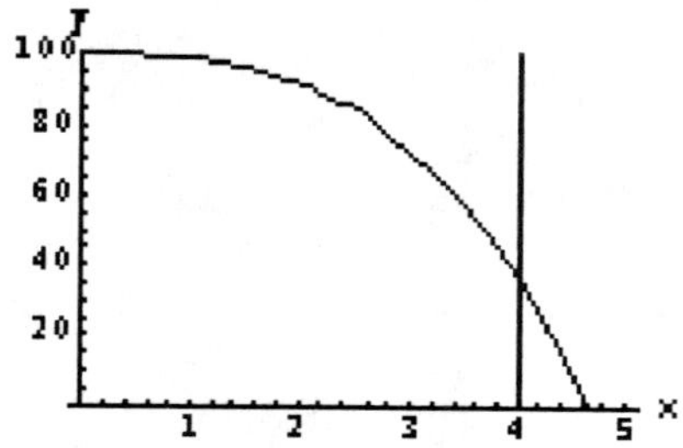

$$\Delta x = \frac{4 - 0}{4} = 1$$

$$\begin{aligned} S_4 &= f(1)\Delta x + f(2)\Delta x + f(3)\Delta x + f(4)\Delta x \\ &= (100 - 1^3) + (100 - 2^3)1 + (100 - 3^3)1 \\ &\quad + (100 - 4^3)1 \\ &= 99 + 92 + 73 + 36 = 300 \end{aligned}$$

(12-5)

47.

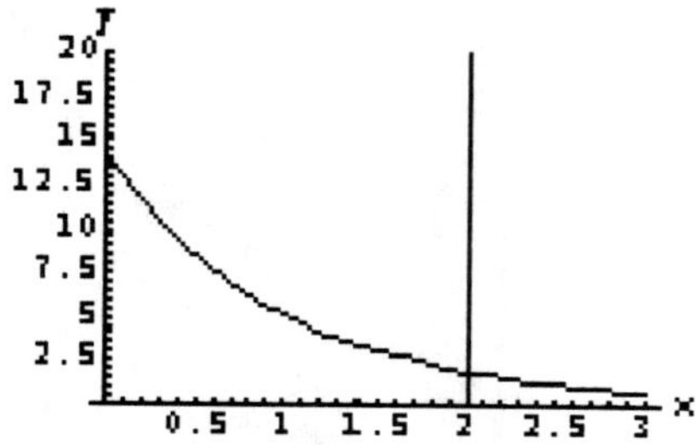

$$\Delta x = \frac{2 - 0}{4} = 0.5$$

$$\begin{aligned} S_4 &= f(0.5)\Delta x + f(1)\Delta x + f(1.5)\Delta x + f(2)\Delta x \\ &= 14e^{-0.5}(0.5) + 14e^{-1}(0.5) + 14e^{-1.5}(0.5) \\ &\quad + 14e^{-2}(0.5) \\ &= 9.33 \end{aligned}$$

(12-5)

48. $\Delta x = \dfrac{10 - 5}{n} = \dfrac{5}{n}$, $x_k = 5 + k\Delta x = 5 + k \cdot \dfrac{5}{n}$ (12-5)

49. (A) $\Delta x = \dfrac{5 - 1}{n} = \dfrac{4}{n}$, $x_k = 1 + k\Delta x = 1 + k \cdot \dfrac{4}{n}$

$$\begin{aligned} S_n &= \sum_{k=1}^{n} f(x_k)\Delta x \\ &= \sum_{k=1}^{n} f\left(1 + k \cdot \frac{4}{n}\right)\left(\frac{4}{n}\right) \\ &= \sum_{k=1}^{n} \left[6\left(1 + k \cdot \frac{4}{n}\right) - 3\right]\left(\frac{4}{n}\right) \\ &= \sum_{k=1}^{n} \left(6 + k \cdot \frac{24}{n} - 3\right)\left(\frac{4}{n}\right) \\ &= \sum_{k=1}^{n} \left(\frac{12}{n} + k \cdot \frac{96}{n^2}\right) \end{aligned}$$

(B) $\Delta x = \dfrac{2 - 0}{n} = \dfrac{2}{n}$, $x_k = 0 + k\Delta x = k \cdot \dfrac{2}{n}$

$$\begin{aligned} S_n &= \sum_{k=1}^{n} g(x_k)\Delta x \\ &= \sum_{k=1}^{n} \left[\left(k \cdot \frac{2}{n}\right)^2 + 3\right]\left(\frac{2}{n}\right) \\ &= \sum_{k=1}^{n} \left(k^2 \cdot \frac{4}{n^2} + 3\right)\left(\frac{2}{n}\right) \\ &= \sum_{k=1}^{n} \left(k^2 \cdot \frac{8}{n^3} + \frac{6}{n}\right) \end{aligned}$$

(12-5)

50. For 49(A)

$$S_n = \sum_{k=1}^{n}\left(\frac{12}{n} + k \cdot \frac{96}{n^2}\right)$$

$$= \sum_{k=1}^{n}\frac{12}{n} + \sum_{k=1}^{n} k \cdot \frac{96}{n^2}$$

$$= \sum_{k=1}^{n}\frac{12}{n} + \frac{96}{n^2}\sum_{k=1}^{n} k$$

$$= \frac{12}{n} \cdot n + \frac{96}{n^2}\,\frac{n(n+1)}{2}$$

$$= 12 + \frac{96n^2 + 96n}{2n^2}$$

$$= 12 + 48 + \frac{48}{n}$$

$$A = \lim_{n \to \infty}\left(12 + 48 + \frac{48}{n}\right) = 12 + 48 + 0 = 60$$

For 49(B)

$$S_n = \sum_{k=1}^{n}\left(k^2 \cdot \frac{8}{n^3} + \frac{6}{n}\right)$$

$$= \sum_{k=1}^{n} k^2 \cdot \frac{8}{n^3} + \sum_{k=1}^{n}\frac{6}{n}$$

$$= \frac{8}{n^3}\sum_{k=1}^{n} k^2 + \sum_{k=1}^{n}\frac{6}{n}$$

$$= \frac{8}{n^3}\,\frac{n(n+1)(2n+1)}{6} + \frac{6}{n} \cdot n$$

$$= \frac{8(2n^3 + 3n^2 + n)}{6n^3} + 6$$

$$= \frac{16}{6} + \frac{24}{6n} + \frac{8}{6n^2} + 6$$

$$A = \lim_{n \to \infty}\left(\frac{16}{6} + \frac{24}{6n} + \frac{8}{6n^2} + 6\right) = \frac{16}{6} + 0 + 0 + 6 = \frac{26}{3}$$ (12-5)

51. $\lim_{x \to 1500} (1{,}500{,}000 - 1{,}000x)$

x	1490	1499	1499.9	1499.99	1500.01	1500.1	1501	1510
$1{,}500{,}000 - 100x$	10,000	1,000	100	10	-10	-100	-1,000	-10,000

We conclude that $\lim_{x \to 1500} R(x) = 0$. As the time approaches 1,500 minutes after filling began, the tank will fill completely. (12-1)

52. To find $\lim_{t \to 30^-} C(t)$, we substitute in only t values less than 30, so only the formula $3 + 0.3t$ is relevant.

t	20	25	28	29
$3 + 0.3t$	9	10.5	11.4	11.7

We conclude that $\lim_{t \to 30^-} C(t) = 12$.

To find $\lim_{t \to 30^+} C(t)$, we substitute in only t values greater than 30, so only the formula $7.5 + 0.15t$ is relevant.

t	40	35	32	31
$7.5 + 0.15t$	13.5	12.75	12.3	12.15

We conclude that $\lim_{t \to 30^+} C(t) = 12$. There is no jump as you pass 30 hours in usage.

(12-1)

53. $\lim_{t \to \infty} (80 - 55e^{-0.16t})$

t	1	10	100	1,000	10,000
$80 - 55e^{-0.16t}$	33.1	68.9	80	80	80

From the table, we conclude that the limit is 80. As sorters gain experience, the number of packages they can process per hour will approach 80. (12-3)

54. $$\lim_{t \to \infty} \frac{3.2t^2 - 0.9t + 7.1}{t^2 + 1.4} = \lim_{t \to \infty} \frac{\frac{3.2t^2}{t^2} - \frac{0.9t}{t^2} + \frac{7.1}{t^2}}{\frac{t^2}{t^2} + \frac{1.4}{t^2}}$$

$$= \lim_{t \to \infty} \frac{3.2 - \frac{0.9t}{t} + \frac{7.1}{t^2}}{1 + \frac{1.4}{t^2}}$$

$$= \frac{\lim_{t \to \infty} 3.2 - \lim_{t \to \infty} \frac{0.9t}{t} + \lim_{t \to \infty} \frac{7.1}{t^2}}{\lim_{t \to \infty} 1 + \lim_{t \to \infty} \frac{1.4}{t^2}}$$

$$= \frac{3.2 - 0 + 0}{1 + 0} = 3.2 \text{ (thousand)}$$

The long-range population of voles is therefore 3,200. (12-3)

55. (A) The velocity is given by the derivative.

$$v(t) = s'(t) = \lim_{h \to 0} \frac{s(t + h) - s(t)}{h}$$

$$= \lim_{h \to 0} \frac{[-9.8(t + h)^2 + 240] - (-9.8t^2 + 240)}{h}$$

$$= \lim_{h \to 0} \frac{-9.8t^2 - 19.6ht - 9.8h^2 + 240 + 9.8t^2 - 240}{h}$$

$$= \lim_{h \to 0} \frac{h(-19.6t - 9.8h)}{h}$$

$$= \lim_{h \to 0} (-19.6t - 9.8h)$$

$$= -19.6t$$

After 1 second, $s'(1) = -19.6(1) = -19.6$ m/sec.
After 3 seconds, $s'(3) = -19.6(3) = -58.8$ m/sec.

(B) He reaches the river when $s(t) = 0$.

Solve $-9.8t^2 + 240 = 0$

$$-9.8t^2 = -240$$

$$t^2 = \frac{240}{9.8}$$

$$t = \sqrt{\frac{240}{9.8}} \approx 4.95 \text{ seconds}$$

(C) Substitute the value of t obtained in part (B) to obtain
$v(t) = v(4.95) = -19.6(4.95) = -97.02$ m/sec. (12-4)

56. $V'(x) = \lim_{h\to 0} \frac{V(x + h) - V(x)}{h}$

$= \lim_{h\to 0} \frac{[-0.15(x + h)^2 + 7(x + h) + 160] - (-0.15x^2 + 7x + 160)}{h}$

$= \lim_{h\to 0} \frac{-0.15x - 0.3xh - 0.15h^2 + 7x + 7h + 160 + 0.15x^2 + 7x - 160}{h}$

$= \lim_{h\to 0} \frac{h(-0.3x - 0.15h + 7)}{h}$

$= \lim_{h\to 0}(-0.3x - 0.15h + 7)$

$= -0.3x + 7$

$V'(5) = -0.3(5) + 7 = 5.5$ (thousand)
$V'(30) = -0.3(30 + 7 = -2$ (thousand)

5 years after purchase, the property is gaining value at the rate cf \$5,500 per year. 30 years after purchase, it is losing value at \$2,000 per year. *(12-4)*

57. Estimate the area:

$A \approx 34 \cdot 6 + 30 \cdot 6 + 28 \cdot 6 + 30 \cdot 6 + 31 \cdot 6 + 33 \cdot 6$

= 1,116 square feet.

Cost = Area × \$2.20 per square foot

= (1,116)(2.20)

= \$2,455.20 *(12-5)*

APPENDIX A Special Topics

Section A-1

1. 123,005 **3.** 20,040 **5.** 6.0 **7.** 80.000 **9.** 0.012 **11.** 0.000 960

13. 3.08 **15.** 924,000 **17.** 23.7 **19.** 2.82×10^3 **21.** 6.78×10^{-4}

23.

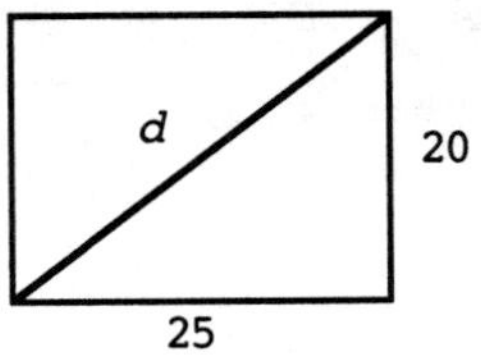

Using the Pythagorean Theorem,

$$25^2 + 20^2 = d^2$$
$$625 + 400 = d^2$$
$$1{,}025 = d^2$$
$$32.0156 = d$$

Rounded to 1 significant digit, which is the smallest number of significant digits in the original measurements, the diagonal is 30 ft.

Section A-2

1. $$\frac{7x - 14}{(x - 4)(x + 3)} = \frac{A}{x - 4} + \frac{B}{x + 3} = \frac{A(x + 3) + B(x - 4)}{(x - 4)(x + 3)}$$

Thus, for all x

$7x - 14 = A(x + 3) + B(x - 4)$

Graphical Solution:

$7x - 14 = A(x + 3) + B(x - 4) \Rightarrow 7x - 14 = (A + B)x + 3A - 4B$

$$\begin{aligned} A + B &= 7 \\ 3A - 4B &= -14 \end{aligned} \quad \Rightarrow \quad \begin{aligned} B &= 7 - A \\ B &= \frac{14 + 3A}{4} \end{aligned}$$

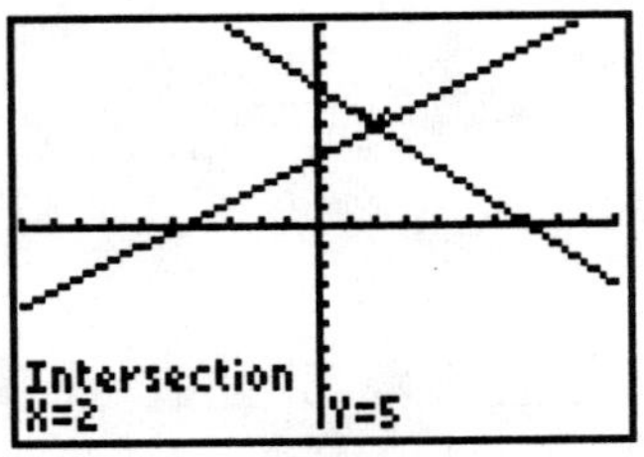

$A = 2,\ B = 5$

Algebraic Solution:

If $x = -3$, then

$-35 = -7B$

$B = 5$

If $x = 4$, then

$14 = 7A$

$A = 2$

$A = 2,\ B = 5$

3. $\dfrac{17x - 1}{(2x - 3)(3x - 1)} = \dfrac{A}{2x - 3} + \dfrac{B}{3x - 1} = \dfrac{A(3x - 1) + B(2x - 3)}{(2x - 3)(3x - 1)}$

Thus, for all x

$17x - 1 = A(3x - 1) + B(2x - 3)$

Graphical Solution:

$17x - 1 = A(3x - 1) + B(2x - 3) \Rightarrow 17x - 1 = (3A + 2B)x - A - 3B$

$$\begin{aligned} 3A + 2B &= 17 \\ -A - 3B &= -1 \end{aligned} \quad \Rightarrow \quad \begin{aligned} B &= \frac{17 - 3A}{2} \\ B &= \frac{1 - A}{3} \end{aligned}$$

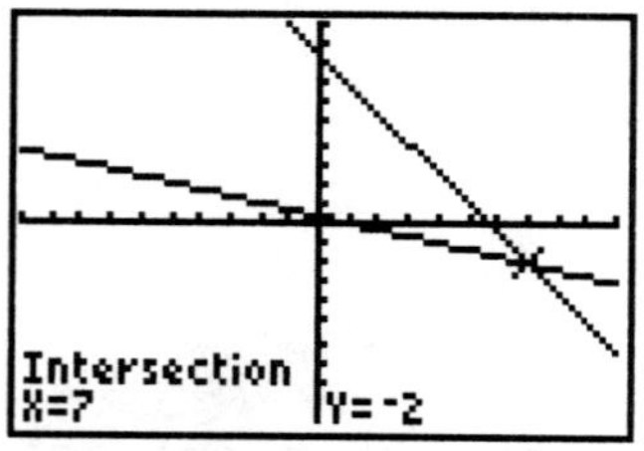

$A = 7,\ B = -2$

Algebraic Solution: If $x = \frac{1}{3}$, then

$$\begin{aligned} 17\left(\frac{1}{3}\right) - 1 &= B\left[2\left(\frac{1}{3}\right) - 3\right] \\ \frac{14}{3} &= -\frac{7}{3}B \\ B &= -2 \end{aligned}$$

If $x = \frac{3}{2}$, then

$$\begin{aligned} 17\left(\frac{3}{2}\right) - 1 &= A\left[3\left(\frac{3}{2}\right) - 1\right] \\ \frac{49}{2} &= \frac{7}{2}A \\ A &= 7 \end{aligned}$$

$A = 7,\ B = -2$

5. $\dfrac{3x^2 + 7x + 1}{x(x + 1)^2} = \dfrac{A}{x} + \dfrac{B}{x + 1} + \dfrac{C}{(x + 1)^2} = \dfrac{A(x + 1)^2 + Bx(x + 1) + Cx}{x(x + 1)^2}$

$3x^2 + 7x + 1 = A(x + 1)^2 + Bx(x + 1) + Cx$

$3x^2 + 7x + 1 = A(x^2 + 2x + 1) + Bx^2 + Bx + Cx$

$3x^2 + 7x + 1 = Ax^2 + 2AX + A + Bx^2 + Bx + Cx$

$3x^2 + 7x + 1 = (A + B)x^2 + (2A + B + C)x + A$

Equating coefficients, we know $A = 1$. Also, $A + B = 3$, so $1 + B = 3$ and $B = 2$. Finally, $2A + B + C = 7$, so $2 + 2 + C = 7$ and $C = 3$.

Check by addition:

$$\begin{aligned} \frac{1}{x} + \frac{2}{x + 1} + \frac{3}{(x + 1)^2} &= \frac{(x + 1)^2}{x(x + 1)^2} + \frac{2x(x + 1)}{x(x + 1)^2} + \frac{3x}{x(x + 1)^2} \\ &= \frac{x^2 + x + 1 + 2x^2 + 2x + 3x}{x(x + 1)^2} \\ &= \frac{3x^2 + 7x + 1}{x(x + 1)^2} \end{aligned}$$

Check with graphing calculator:

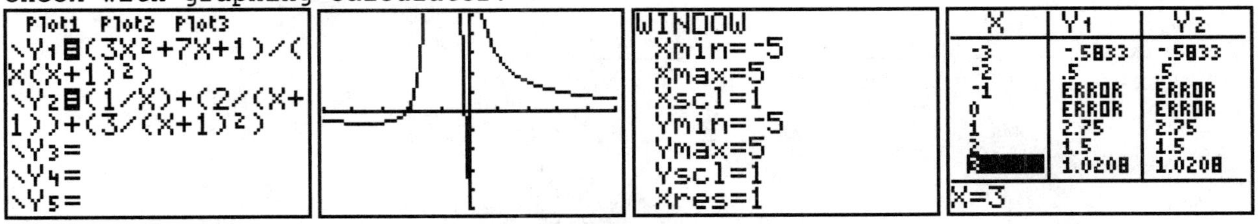

The graphs appear to be identical and the tables are the same.

7. $$\frac{3x^2 + x}{(x - 2)(x^2 + 3)} = \frac{A}{x - 2} + \frac{Bx + C}{x^2 + 3} = \frac{A(x^2 + 3) + (Bx + C)(x - 2)}{(x - 2)(x^2 + 3)}$$

$3x^2 + x = A(x^2 + 3) + (Bx + C)(x - 2)$

First, plug in $x = 2$:

$$3(2)^2 + 2 = A(4 + 3) + 0$$
$$14 = 7A$$
$$A = 2$$
$$3x^2 + x = 2(x^2 + 3) + (Bx + C)(x - 2)$$
$$3x^2 + x = 2x^2 + 6 + Bx^2 - 2Bx + Cx - 2C$$
$$3x^2 + x = (2 + B)x^2 + (C - 2B)x + 6 - 2C$$

Equating coefficients,

$2 + B = 3$, so $B = 1$

$6 - 2C = 0$, so $6 = 2C$ and $C = 3$.

Check by addition:

$$\frac{x}{x - 2} + \frac{x + 3}{x^2 + 3} = \frac{2(x^2 + 3) + (x + 3)(x - 2)}{(x - 2)(x^2 + 3)}$$
$$= \frac{2x^2 + 6 + x^2 + x - 6}{(x - 2)(x^2 + 3)}$$
$$= \frac{3x^2 + x}{(x - 2)(x^2 + 3)}$$

Check with graphing calculator:

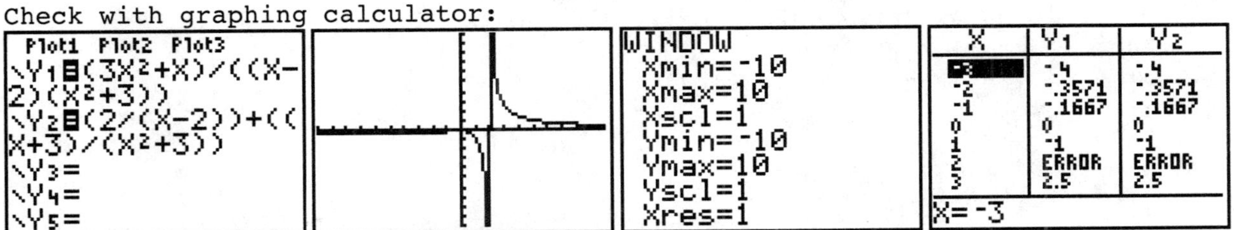

The graphs appear to be identical and the tables are the same.

9. $$\frac{2x^2 + 4x - 1}{(x^2 + x + 1)^2} = \frac{Ax + B}{x^2 + x + 1} + \frac{Cx + D}{(x^2 + x + 1)^2} = \frac{(Ax + B)(x^2 + x + 1) + (Cx + D)}{(x^2 + x + 1)^2}$$

$2x^2 + 4x - 1 = (Ax + B)(x^2 + x + 1) + Cx + D$

$2x^2 + 4x - 1 = Ax^3 + Ax^2 + Ax + Bx^2 + Bx + B + Cx + D$

$2x^2 + 4x - 1 = Ax^3 + (A + B)x^2 + (A + B + C)x + B + D$

Equating coefficients,

$A = 0$

$A + B = 2$, so $B = 2$

$A + B + C = 4$, so $0 + 2 + C = 4$, and $C = 2$

$B + D = -1$, so $2 + D = -1$, and $D = -3$

Check by addition:

$$\frac{2}{x^2 + x + 1} + \frac{2x - 3}{(x^2 + x + 1)^2} = \frac{2(x^2 + x + 1)}{(x^2 + x + 1)^2} + \frac{2x - 3}{(x^2 + x + 1)^2}$$

$$= \frac{2x^2 + 2x + 2 + 2x - 3}{(x^2 + x + 1)^2}$$

$$= \frac{2x^2 + 4x - 1}{(x^2 + x + 1)^2}$$

Check with graphing calculator:

```
Plot1 Plot2 Plot3
\Y1■(2X²+4X-1)/(X²+X+1)²
\Y2■(2/(X²+X+1))+((2X-3)/(X²+X+1)²)
\Y3=
\Y4=
```

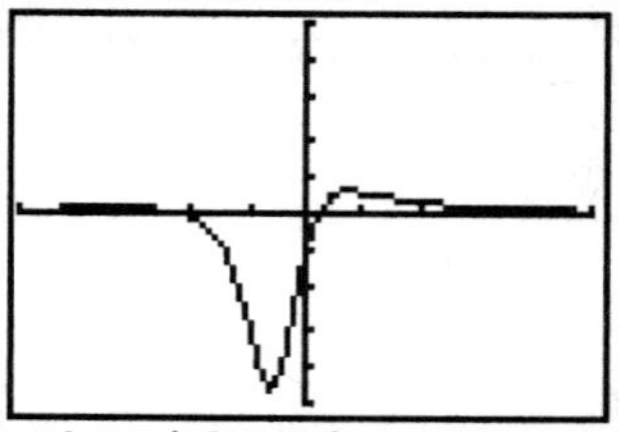

```
WINDOW
 Xmin=-5
 Xmax=5
 Xscl=1
 Ymin=-5
 Ymax=5
 Yscl=1
 Xres=1
```

X	Y1	Y2
-3	.10204	.10204
-2	-.1111	-.1111
-1	-3	-3
0	-1	-1
1	.55556	.55556
2	.30612	.30612
3	.1716	.1716

X=-3

The graphs appear to be identical and the tables are the same.

11. Since $x^2 - 2x - 8 = (x + 2)(x - 4)$, we write

$$\frac{-x + 22}{x^2 - 2x - 8} = \frac{A}{x + 2} + \frac{B}{x - 4} = \frac{A(x - 4) + B(x + 2)}{(x + 2)(x - 4)}$$

Thus, for all x

$-x + 22 = A(x - 4) + B(x + 2)$

If $x = 4$

$18 = 6B$

$B = 3$

If $x = -2$

$24 = -6A$

$A = -4$

So $\dfrac{-x + 22}{x^2 - 2x - 8} = \dfrac{-4}{x + 2} + \dfrac{3}{x - 4}$.

13. Since $6x^2 - x - 12 = (3x + 4)(2x - 3)$, we write

$$\frac{3x - 13}{6x^2 - x - 12} = \frac{A}{3x + 4} + \frac{B}{2x - 3} = \frac{A(2x - 3) + B(3x + 4)}{(3x + 4)(2x - 3)}$$

Thus, for all x

$3x - 13 = A(2x - 3) + B(3x + 4)$

If $x = \frac{3}{2}$

$$3\left(\frac{3}{2}\right) - 13 = B\left[3\left(\frac{3}{2}\right) + 4\right]$$

$$-\frac{17}{2} = \frac{17}{2}B$$

$$B = -1$$

If $x = -\frac{4}{3}$

$$3\left(-\frac{4}{3}\right) - 13 = A\left[2\left(-\frac{4}{3}\right) - 3\right]$$

$$-17 = -\frac{17}{3}A$$

$$A = 3$$

So $\dfrac{3x - 13}{6x^2 - x - 12} = \dfrac{3}{3x + 4} - \dfrac{1}{2x - 3}$

15. Since $x^3 - 6x^2 + 9x = x(x^2 - 6x + 9) = x(x - 3)^2$, we write

$$\frac{x^2 - 12x + 18}{x^3 - 6x^2 + 9x} = \frac{A}{x} + \frac{B}{x - 3} + \frac{C}{(x - 3)^2} = \frac{A(x - 3)^2 + Bx(x - 3) + Cx}{x(x - 3)^2}$$

Thus, for all x

$x^2 - 12x + 18 = A(x - 3)^2 + Bx(x - 3) + Cx$

If $x = 3$
$-9 = 3C$
$C = -3$

If $x = 0$
$18 = 9A$
$A = 2$

If $x = 2$
$-2 = A - 2B + 2C$
$-2 = 2 - 2B - 6$ using $A = 2$ and $C = -3$
$2 = -2B$
$B = -1$

So $\dfrac{x^2 - 12x + 18}{x^3 - 6x^2 + 9x} = \dfrac{2}{x} - \dfrac{1}{x - 3} - \dfrac{3}{(x - 3)^2}$

17. Since $x^3 + 2x^2 + 3x = x(x^2 + 2x + 3)$, we write

$$\frac{5x^2 + 3x + 6}{x^3 + 2x^2 + 3x} = \frac{A}{x} + \frac{Bx + C}{x^2 + 2x + 3} = \frac{A(x^2 + 2x + 3) + (Bx + C)x}{x(x^2 + 2x + 3)}$$

> **Common Error:**
> Writing $\frac{B}{x^2+2x+3}$. Since $x^2 + 2x + 3$ is quadratic, the numerator must be first degree.

Thus, for all x

$5x^2 + 3x + 6 = A(x^2 + 2x + 3) + (Bx + C)x = (A + B)x^2 + (2A + C)x + 3A$

Equating coefficients of like terms, we have
$5 = A + B$
$3 = 2A + C$
$3A = 6$

Hence $A = 2$, $C = -1$, $B = 3$

So $\dfrac{5x^2 + 3x + 6}{x^3 + 2x^2 + 3x} = \dfrac{2}{x} + \dfrac{3x - 1}{x^2 + 2x + 3}$

19. Since $x^4 + 4x^2 + 4 = (x^2 + 2)^2$, we write

$$\frac{2x^3 + 7x + 5}{x^4 + 4x^2 + 4} = \frac{Ax + B}{x^2 + 2} + \frac{Cx + D}{(x^2 + 2)^2} = \frac{(Ax + B)(x^2 + 2) + Cx + D}{(x^2 + 2)^2}$$

Thus, for all x

$2x^3 + 7x + 5 = (Ax + B)(x^2 + 2) + Cx + D = Ax^3 + Bx^2 + (2A + C)x + 2B + D$

Equating coefficients of like terms, we have
$2 = A$
$0 = B$
$7 = 2A + C$
$5 = 2B + D$

Hence $A = 2$, $B = 0$, $C = 3$, $D = 5$

$$\frac{2x^3 + 7x + 5}{x^4 + 4x^2 + 4} = \frac{2x}{x^2 + 2} + \frac{3x + 5}{(x^2 + 2)^2}$$

21. First we divide to obtain a polynomial plus a proper fraction.

$$\begin{array}{r} x - 2 \\ x^2 - 5x + 6 \overline{\smash{)}\, x^3 - 7x^2 + 17x - 17} \\ \underline{x^3 - 5x^2 + 6x} \\ -2x^2 + 11x - 17 \\ \underline{-2x^2 + 10x - 12} \\ x - 5 \end{array}$$

So, $\dfrac{x^3 - 7x^2 + 17x - 17}{x^2 - 5x + 6} = x - 2 + \dfrac{x - 5}{x^2 - 5x + 6}$

To decompose the proper fraction, we note $x^2 - 5x + 6 = (x - 2)(x - 3)$ and we write:

$$\frac{x - 5}{x^2 - 5x + 6} = \frac{A}{x - 2} + \frac{B}{x - 3} = \frac{A(x - 3) + B(x - 2)}{(x - 2)(x - 3)}$$

Thus for all x

$x - 5 = A(x - 3) + B(x - 2)$

If $x = 3$

$-2 = B$

If $x = 2$

$-3 = -A$

$A = 3$

So $\dfrac{x^3 - 7x^2 + 17x - 17}{x^2 - 5x + 6} = x - 2 + \dfrac{3}{x - 2} - \dfrac{2}{x - 3}$

23. First, we must factor $x^3 - 6x - 9$. Possible rational zeros of this polynomial are ±1, ±3, ±9. Forming a synthetic division table, we see

	1	0	-6	-9
1	1	1	-5	-14
3	1	3	3	0

Thus $x^3 - 6x - 9 = (x - 3)(x^2 + 3x + 3)$

$x^2 + 3x + 3$ cannot be factored further in the real numbers.

So $\dfrac{4x^2 + 5x - 9}{x^3 - 6x - 9} = \dfrac{A}{x - 3} + \dfrac{Bx + C}{x^2 + 3x + 3} = \dfrac{A(x^2 + 3x + 3) + (Bx + C)(x - 3)}{(x - 3)(x^2 + 3x + 3)}$

Thus, for all x

$$\begin{aligned} 4x^2 + 5x - 9 &= A(x^2 + 3x + 3) + (Bx + C)(x - 3) \\ &= Ax^2 + 3Ax + 3A + Bx^2 - 3Bx + Cx - 3C \\ &= (A + B)x^2 + (3A - 3B + C)x + 3A - 3C \end{aligned}$$

Before equating coefficients of like terms, we note that if $x = 3$

$$\begin{aligned} 4(3)^2 + 5(3) - 9 &= A(3^2 + 3 \cdot 3 + 3) \\ \text{So } 42 &= 21A \\ A &= 2 \\ \text{Since } 4 &= A + B,\ B = 2 \\ 5 &= 3A - 3B + C = 6 - 6 + C, \text{ so } C = 5 \end{aligned}$$

We have $A = 2$, $B = 2$, $C = 5$

So $\dfrac{4x^2 + 5x - 9}{x^3 - 6x - 9} = \dfrac{2}{x - 3} + \dfrac{2x + 5}{x^2 + 3x + 3}$

25. First, we must factor $x^3 + 2x^2 - 15x - 36$. The possible rational zeros are ±1, ±2, ±3, ±4, ±6, ±9, ±12, ±18, ±36. Forming a synthetic division table, we see

	1	2	-15	-36
1	1	3	-12	-48
2	1	4	-7	-50
3	1	5	0	-36
4	1	6	9	0

Hence $x^3 + 2x^2 - 15x - 36 = (x - 4)(x^2 + 6x + 9) = (x - 4)(x + 3)^2$.

$$\text{So } \frac{x^2 + 16x + 18}{x^3 + 2x^2 - 15x - 36} = \frac{A}{x - 4} + \frac{B}{x + 3} + \frac{C}{(x + 3)^2}$$
$$= \frac{A(x + 3)^2 + B(x - 4)(x + 3) + C(x - 4)}{(x - 4)(x + 3)^2}$$

Thus, for all x

$x^2 + 16x + 18 = A(x + 3)^2 + B(x - 4)(x + 3) + C(x - 4)$

If $x = -3$

$-21 = -7C$

$C = 3$

If $x = 4$

$98 = 49A$

$A = 2$

If $x = 5$

$123 = 64A + 8B + C$

$= 128 + 8B + 3$ using $A = 2$ and $C = 3$

$-8 = 8B$

$B = -1$

$$\text{So } \frac{x^2 + 16x + 18}{x^3 + 2x^2 - 15x - 36} = \frac{2}{x - 4} - \frac{1}{x + 3} + \frac{3}{(x + 3)^2}$$

27. First, we must factor $x^4 - 5x^3 + 9x^2 - 8x + 4$. The possible rational zeros are ±1, ±2, ±4. We form a synthetic division table:

	1	-5	9	-8	4
1	1	-4	5	-3	1
2	1	-3	3	-2	0

We examine $x^3 - 3x^2 + 3x - 2$ to see if 2 is a double zero.

	1	-3	3	-2
2	1	-1	1	0

Thus $x^4 - 5x^3 + 9x^2 - 8x + 4 = (x - 2)^2(x^2 - x + 1)$. $x^2 - x + 1$ cannot be factored further in the real numbers, so

$$\frac{-x^2 + x - 7}{x^4 - 5x^3 + 9x^2 - 8x + 4} = \frac{A}{x - 2} + \frac{B}{(x - 2)^2} + \frac{Cx + D}{x^2 - x + 1}$$
$$= \frac{A(x - 2)(x^2 - x + 1) + B(x^2 - x + 1) + (Cx + D)(x - 2)^2}{(x - 2)^2(x^2 - x + 1)}$$

Thus, for all x

$-x^2 + x - 7 = A(x - 2)(x^2 - x + 1) + B(x^2 - x + 1) + (Cx + D)(x - 2)^2$

If $x = 2$

$-9 = 3B$

$B = -3$

$$-x^2 + x - 7 = A(x^3 - 3x^2 + 3x - 2) + B(x^2 - x + 1) + (Cx + D)(x^2 - 4x + 4)$$
$$= (A + C)x^3 + (-3A + B - 4C + D)x^2 + (3A - B + 4C - 4D)x - 2A + B + 4D$$

We have already $B = -3$, so equating coefficients of like terms,

$0 = A + C$
$-1 = -3A - 3 - 4C + D$
$1 = 3A + 3 + 4C - 4D$
$-7 = -2A - 3 + 4D$

Since $C = -A$, we can write

$-1 = -3A - 3 + 4A + D$
$1 = 3A + 3 - 4A - 4D$
$-7 = -2A - 3 + 4D$

So $D = 0$ (adding the first equations), $A = 2$, $C = -2$.

Hence $\dfrac{-x^2 + x - 7}{x^4 - 5x^3 + 9x^2 - 8x + 4} = \dfrac{2}{x - 2} - \dfrac{3}{(x - 2)^2} - \dfrac{2x}{x^2 - x + 1}$

29. First we divide to obtain a polynomial plus a proper fraction.

$$\begin{array}{r|l} & \quad x + 2 \\ 4x^4 + 4x^3 - 5x^2 + 5x - 2 & 4x^5 + 12x^4 - x^3 + 7x^2 - 4x + 2 \\ & \underline{4x^5 + 4x^4 - 5x^3 + 5x^2 - 2x} \\ & \qquad 8x^4 + 4x^3 + 2x^2 - 2x + 2 \\ & \qquad \underline{8x^4 + 8x^3 - 10x^2 + 10x - 4} \\ & \qquad\quad - 4x^3 + 12x^2 - 12x + 6 \end{array}$$

Now we must decompose $\dfrac{-4x^3 + 12x^2 - 12x + 6}{4x^4 + 4x^3 - 5x^2 + 5x - 2}$, starting by factoring $4x^4 + 4x^3 - 5x^2 + 5x - 2$.

Possible rational zeros are ± 1, ± 2, $\pm\frac{1}{2}$, $\pm\frac{1}{4}$. We form a synthetic division table:

	4	4	-5	5	-2	
1	4	8	3	8	6	1 is an upper bound, eliminating 2
-1	4	0	-5	10	-12	
-2	4	-4	3	-1	0	-2 is a zero

We investigate $4x^3 - 4x^2 + 3x - 1$; the only remaining rational zeros are $\pm\frac{1}{2}$, $\pm\frac{1}{4}$.

We form a synthetic division table:

	4	-4	3	-1
$\frac{1}{2}$	4	-2	2	0

So $4x^4 + 4x^3 - 5x^2 + 5x - 2 = (x + 2)\left(x - \frac{1}{2}\right)(4x^2 - 2x + 2)$

$= (x + 2)\left(x - \frac{1}{2}\right)2(2x^2 - x + 1)$

$= (x + 2)(2x - 1)(2x^2 - x + 1)$

$2x^2 - x + 1$ cannot be factored further in the real numbers, so we write:

$$\frac{-4x^3 + 12x^2 - 12x + 6}{4x^4 + 4x^3 - 5x^2 + 5x - 2}$$
$$= \frac{A}{x + 2} + \frac{B}{2x - 1} + \frac{Cx + D}{2x^2 - x + 1}$$

$$= \frac{A(2x-1)(2x^2-x+1) + B(x+2)(2x^2-x+1) + (Cx+D)(x+2)(2x-1)}{(x+2)(2x-1)(2x^2-x+1)}$$

Thus, for all x

$-4x^3 + 12x^2 - 12x + 6$

$= A(2x - 1)(2x^2 - x + 1) + B(x + 2)(2x^2 - x + 1) + (Cx + D)(x + 2)(2x - 1)$

If $x = -2$

$$-4(-2)^3 + 12(-2)^2 - 12(-2) + 6 = A(-5)[2(-2)^2 - (-2) + 1]$$
$$32 + 48 + 24 + 6 = -55A$$
$$110 = -55A$$
$$A = -2$$

If $x = \frac{1}{2}$

$$-4\left(\frac{1}{2}\right)^3 + 12\left(\frac{1}{2}\right)^2 - 12\left(\frac{1}{2}\right) + 6 = B\left(\frac{5}{2}\right)\left[2\left(\frac{1}{2}\right)^2 - \frac{1}{2} + 1\right]$$
$$-\frac{1}{2} + 3 - 6 + 6 = B\left(\frac{5}{2}\right)(1)$$
$$B = 1$$

If $x = 0$

$6 = A(-1)(1) + B(2)(1) + D(2)(-1)$

$6 = -A + 2B - 2D$

$6 = 2 + 2 - 2D$

$D = -1$

If $x = 1$

$$-4 + 12 - 12 + 6 = A(1)(2) + B(3)(2) + (C + D)(3)(1)$$
$$2 = 2A + 6B + 3C + 3D$$
$$= -4 + 6 + 3C - 3$$
$$C = 1$$

So $\dfrac{4x^5 + 12x^4 - x^3 + 7x^2 - 4x + 2}{4x^4 + 4x^3 - 5x^2 + 5x - 2} = x + 2 - \dfrac{2}{x+2} + \dfrac{1}{2x-1} + \dfrac{x-1}{2x^2 - x + 1}$.

Section A-3

1. Note that if you eliminate the parameter from the parametric equations by substituting x for t^2 in $y = t^2 - 2$, the result is the equation $y = x - 2$. Since $x = t^2$, x must be non-negative. Thus, the graph of the parametric equations will consist only of a ray, the portion of the line $y = x - 2$ for $x \geq 0$.

3. Construct a table and graph.

t	0	1	2	3	−1	−2	−3
x	0	−1	−2	−3	1	2	3
y	−2	0	2	4	−4	−6	−8

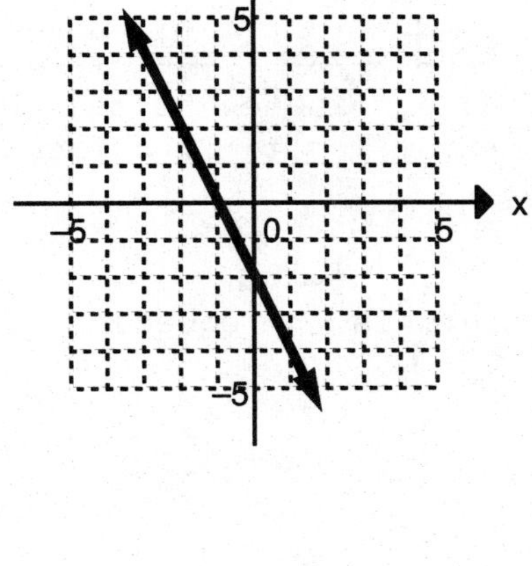

To eliminate the parameter t we solve $x = -t$ for t to obtain

$t = -x$

Then we substitute the expression for t into $y = 2t - 2$ to obtain

$y = 2(-x) - 2$

$y = -2x - 2$

This is the equation of a straight line.

5. Construct a table and graph.

t	0	1	2	3	−1	−2	−3	
x	0	−1	−4	−9	−1	−4	−9	Note: $x \le 0$
y	−2	0	6	16	0	6	16	

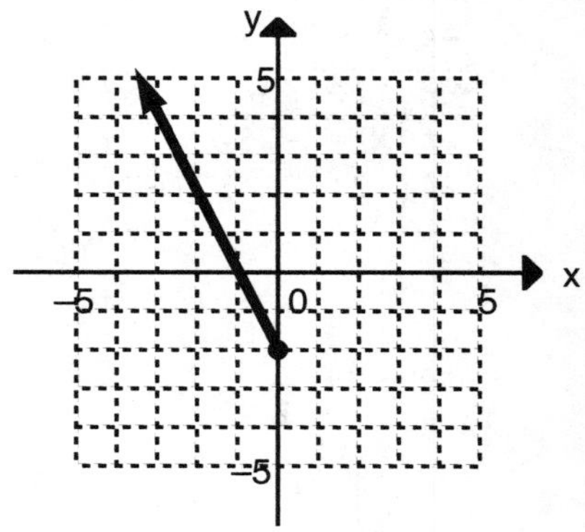

To eliminate the parameter t we solve $x = -t^2$ for t^2 (we do not need t) to obtain

$$t^2 = -x$$

Then we substitute the expression for t^2 into $y = 2t^2 - 2$ to obtain

$$y = 2(-x) - 2$$
$$y = -2x - 2$$

This is the equation of a straight line. However, since $x = -t^2$, x is restricted so that $x \le 0$. Therefore the equation of the curve is actually

$$y = -2x - 2, \quad x \le 0$$

This is the equation of a ray (part of a straight line).

7. Construct a table and graph.

t	0	1	2	3	−1	−2	−3
x	0	3	6	9	−3	−6	−9
y	0	−2	−4	−6	2	4	6

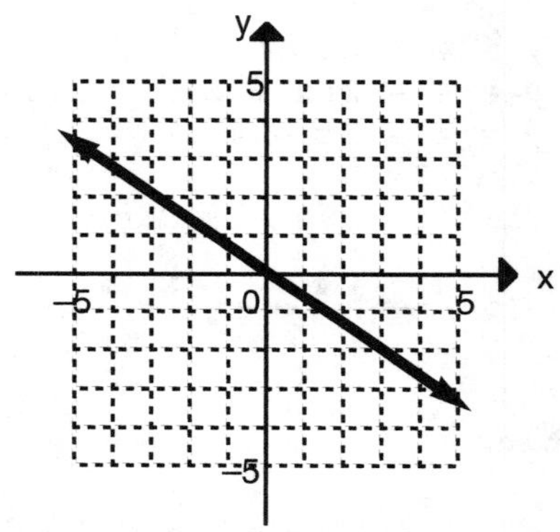

To eliminate the parameter t we solve $x = 3t$ for t to obtain

$$t = \frac{x}{3}$$

Then we substitute the expression for t into $y = -2t$ to obtain

$$y = -2\left(\frac{x}{3}\right)$$
$$y = -\frac{2}{3}x$$

This is the equation of a straight line.

9. Construct a table of values and graph.

t	0	1	2	3	−1	−2	−3
x	0	$\frac{1}{4}$	1	$\frac{9}{4}$	$\frac{1}{4}$	1	$\frac{9}{4}$
y	0	1	2	3	−1	−2	−3

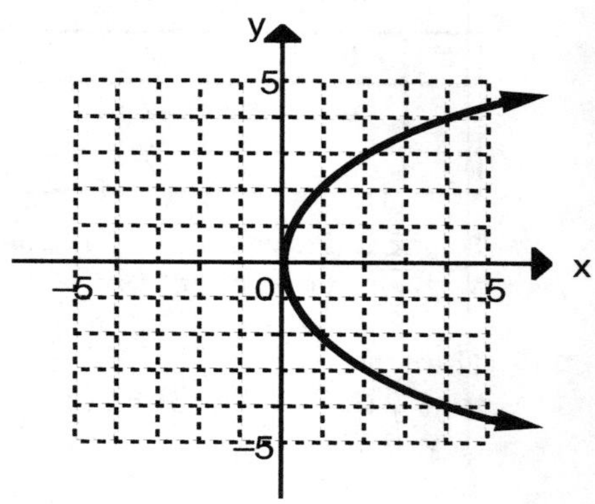

To eliminate the parameter t we substitute $t = y$ (from $y = t$) into the expression for x to obtain

$$x = \frac{1}{4}y^2$$
$$y^2 = 4x$$

This is the equation of a parabola.

11. Construct a table of values and graph.

t	0	1	2	3	−1	−2	−3
x	0	$\frac{1}{4}$	4	$\frac{81}{4}$	$\frac{1}{4}$	4	$\frac{81}{4}$
y	0	1	4	9	1	4	9

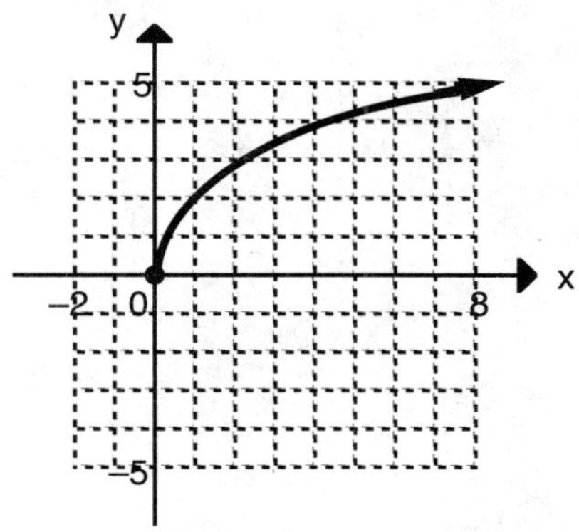

To eliminate the parameter we substitute $t^2 = y$ (from $y = t^2$) into the expression for x to obtain

$$x = \frac{1}{4}t^4$$
$$x = \frac{1}{4}(t^2)^2$$
$$x = \frac{1}{4}y^2$$
$$y^2 = 4x$$

This is the equation of a parabola. However, since $y = t^2$, y is restricted to nonnegative values only. Hence the equation of the curve is actually

$y^2 = 4x,\ y \geq 0$

This is the equation of the upper half of a parabola.

13. We eliminate the parameter t as follows:

$$x = t - 2$$
$$x + 2 = t$$

Substituting this expression for t into the expression for y, we obtain

$$y = 4 - 2t$$
$$y = 4 - 2(x + 2)$$
$$y = 4 - 2x - 4$$
$$y = -2x$$

This is the equation of a line.

15. We eliminate the parameter t as follows:
Solve $x = t - 1$ for t to obtain $t = x + 1$. Substitute the expression for t into $y = \sqrt{t}$ to obtain

$y = \sqrt{x + 1}$ or

$y^2 = x + 1,\ y \geq 0.$

Since t is restricted so that $t \geq 0$, we also have $x + 1 \geq 0$ or $x \geq -1$. This graph is a parabola with vertex at $(-1, 0)$, opening right. Since $y \geq 0$ only the upper half of the parabola is traced out.

17. We eliminate the parameter t as follows:
Solve $x = \sqrt{t}$ for t to obtain $t = x^2$.
Substitute the expression for t into $y = 2\sqrt{16 - t}$ to obtain

$$y = 2\sqrt{16 - x^2} \text{ or}$$
$$y^2 = 4(16 - x^2)$$
$$y^2 = 64 - 4x^2$$
$$4x^2 + y^2 = 64$$
$$\frac{x^2}{16} + \frac{y^2}{64} = 1 \quad \text{(ellipse in standard form)}$$

Since $0 \leq t \leq 16$ and $x = \sqrt{t}$, it follows that $0 \leq x \leq 4$.
Since $0 \leq t \leq 16$ and $t = x^2$, it follows that $0 \leq x^2 \leq 16$, $0 \leq 16 - x^2 \leq 16$, $0 \leq \sqrt{16 - x^2} \leq 4$, and $0 \leq 2\sqrt{16 - x^2} \leq 8$. Thus $0 \leq y \leq 8$.

The graph of $\frac{x^2}{16} + \frac{y^2}{64} = 1$, $0 \leq x \leq 4$ and $0 \leq y \leq 8$, is the portion of an ellipse lying in the first quadrant.

19. We eliminate the parameter t as follows:
Solve $x = -\sqrt{t+1}$ for t to obtain

$$x^2 = t + 1$$
$$t = x^2 - 1$$

Substitute the expression for t into $y = -\sqrt{t-1}$ to obtain

$$y = -\sqrt{x^2 - 1 - 1}$$
$$y = -\sqrt{x^2 - 2}$$
$$y^2 = x^2 - 2$$
$$x^2 - y^2 = 2 \quad \text{(hyperbola)}$$

Since $t \geq 1$ it follows that $t + 1 \geq 2$, $\sqrt{t+1} \geq \sqrt{2}$ and $x = -\sqrt{t+1} \leq -\sqrt{2}$.
Also it follows that $t - 1 \geq 0$, $\sqrt{t-1} \geq 0$, and $y = -\sqrt{t-1} \leq 0$.
The graph of $x^2 - y^2 = 2$, $x \leq -\sqrt{2}$ and $y \leq 0$, is the portion of a hyperbola lying in the third quadrant.

21. We eliminate the parameter θ as follows:

$$\frac{x}{3} = \sin\theta \qquad \frac{y}{4} = \cos\theta$$
$$\left(\frac{x}{3}\right)^2 + \left(\frac{y}{4}\right)^2 = \sin^2\theta + \cos^2\theta$$
$$\frac{x^2}{9} + \frac{y^2}{16} = 1$$

This is the equation of an ellipse in standard form.

23. We eliminate the parameter θ as follows:

$$x - 2 = 2\sin\theta \qquad y - 3 = 2\cos\theta$$
$$\frac{x-2}{2} = \sin\theta \qquad \frac{y-3}{2} = \cos\theta$$
$$\left(\frac{x-2}{2}\right)^2 + \left(\frac{y-3}{2}\right)^2 = \sin^2\theta + \cos^2\theta$$
$$\frac{(x-2)^2}{4} + \frac{(y-3)^2}{4} = 1$$
$$(x-2)^2 + (y-3)^2 = 4$$

This is the equation of a circle with center (2, 3) and radius 2.

25. If $A \neq 0$, $C = 0$, and $E \neq 0$, write

$$Ax^2 + Dx + Ey + F = 0$$

There are many possible parametric equations for this curve. A simple approach is to set $x = t$ and solve for y in terms of t.

$$At^2 + Dt + Ey + F = 0$$
$$At^2 + Dt + F = -Ey$$
$$\frac{At^2 + Dt + F}{-E} = y$$

Then $x = t$, $y = \dfrac{At^2 + Dt + F}{-E}$, $-\infty < t < \infty$, are parametric equations for this curve. The curve is a parabola.

27. We eliminate the parameter t as follows:

$$x - 3 = 6\cos t \qquad y - 2 = 4\sin t$$
$$\frac{x-3}{6} = \cos t \qquad \frac{y-2}{4} = \sin t$$
$$\left(\frac{x-3}{6}\right)^2 + \left(\frac{y-2}{4}\right)^2 = \cos^2 t + \sin^2 t$$
$$\frac{(x-3)^2}{36} + \frac{(y-2)^2}{16} = 1$$

The curve is an ellipse with center (3, 2).

29. We eliminate the parameter t as follows:

$$x + 3 = 2\tan t \qquad y + 1 = 5\sec t$$
$$\frac{x+3}{2} = \tan t \qquad \frac{y+1}{5} = \sec t$$
$$\left(\frac{y+1}{5}\right)^2 - \left(\frac{x+3}{2}\right)^2 = \sec^2 t - \tan^2 t$$
$$\frac{(y+1)^2}{25} - \frac{(x+3)^2}{4} = 1$$

The curve is a hyperbola with center (-3, -1).

31. We eliminate the parameter t as follows:
Square the expressions for x and y.

$$x^2 = t^2 + 1 \qquad y^2 = t^2 + 9$$

Solve the first equation for t^2 and substitute the result into the second equation.

$$t^2 = x^2 - 1$$
$$y^2 = x^2 - 1 + 9$$
$$y^2 = x^2 + 8$$
$$y^2 - x^2 = 8$$

This is the equation of a hyperbola.

Since $t^2 \ge 0$, $t^2 + 1 \ge 1$, $\sqrt{t^2+1} \ge 1$. It follows that $x \ge 1$.
Since $t^2 \ge 0$, $t^2 + 9 \ge 9$, $\sqrt{t^2+9} \ge 3$. It follows that $y \ge 3$.
The curve is the part of the hyperbola $y^2 - x^2 = 8$ for $x \ge 1$ and $y \ge 3$.

33. We eliminate the parameter t as follows:
Square the expressions for x and y.

$$x^2 = \frac{4}{t^2+1} \qquad y^2 = \frac{4}{t^2+1}$$
$$x^2 + y^2 = \frac{4}{t^2+1} + \frac{4}{t^2+1}$$
$$x^2 + y^2 = \frac{4+4t^2}{t^2+1}$$
$$x^2 + y^2 = \frac{4(1+t^2)}{t^2+1}$$
$$x^2 + y^2 = 4$$

This is the equation of a circle.
Since $t^2 \ge 0$, $t^2 + 1 \ge 1$, $\sqrt{t^2+1} \ge 1$, $0 < \frac{1}{\sqrt{t^2+1}} \le 1$, $0 < \frac{2}{\sqrt{t^2+1}} \le 2$.
It follows that $0 < x \le 2$. Therefore $0 < x^2 \le 4$, $0 > -x^2 \ge -4$, $4 > 4 - x^2 \ge 0$, $2 > \sqrt{4-x^2} \ge 0$, $-2 < -\sqrt{4-x^2} \le 0$. Since $y = \pm\sqrt{4-x^2}$ on the circle $x^2 + y^2 = 4$, it follows that $-2 < y < 2$ (since $x \ne 0$). Thus the curve is the portion of the circle $x^2 + y^2 = 4$ for which $0 < x \le 2$, $-2 < y < 2$. This is the right-hand semicircle, excluding the points $(0, \pm 2)$.

35. We eliminate the parameter t as follows: First note that since $x = \frac{8}{t^2 + 4}$, $(t^2 + 4)x = 8$ or $t^2 + 4 = \frac{8}{x}$. Therefore x cannot be 0. Now square the expressions for x and y.

$$x^2 = \frac{64}{(t^2 + 4)^2}$$

$$y^2 = \frac{16t^2}{(t^2 + 4)^2}$$

$$x^2 + y^2 = \frac{64}{(t^2 + 4)^2} + \frac{16t^2}{(t^2 + 4)^2}$$

$$x^2 + y^2 = \frac{64 + 16t^2}{(t^2 + 4)^2}$$

$$x^2 + y^2 = \frac{16(4 + t^2)}{(t^2 + 4)^2}$$

$$x^2 + y^2 = \frac{16}{t^2 + 4}$$

Since $x = \frac{8}{t^2 + 4}$, $2x = \frac{16}{t^2 + 4}$, so

$$x^2 + y^2 = 2x \quad x \neq 0$$

This is the equation of a circle. Completing the square we see

$$x^2 - 2x + y^2 = 0$$
$$x^2 - 2x + 1 + y^2 = 1$$
$$(x - 1)^2 + y^2 = 1, \ x \neq 0$$

Hence the circle has center (1, 0) and radius 1. The origin is not part of the curve since x cannot equal zero, hence there is a hole in the graph there.

37. $25x^2 - 200x - 9y^2 - 18y + 616 = 0$

Complete the square relative to both x and y.

$$25x^2 - 200x \quad - 9y^2 - 18y \quad = -616$$
$$25(x^2 - 8x \quad) - 9(y^2 + 2y \quad) = -616$$
$$25(x^2 - 8x + 16) - 9(y^2 + 2y + 1) = -616 + 400 - 9$$
$$25(x - 4)^2 - 9(y + 1)^2 = -225$$
$$\frac{(y + 1)^2}{25} - \frac{(x - 4)^2}{9} = 1$$

The curve is a hyperbola with center (4, -1).

We introduce parameter t in analogy with problem 35:

$$\frac{y + 1}{5} = \sec t \qquad \frac{x - 4}{3} = \tan t$$
$$y + 1 = 5 \sec t \qquad x - 4 = 3 \tan t$$
$$x = 4 + 3 \tan t \qquad y = -1 + 5 \sec t, \ 0 \le t \le 2\pi,$$
$$t \neq \frac{\pi}{2}, \frac{3\pi}{2}$$

39. $4x^2 - 24x + 49y^2 + 392y + 624 = 0$

Complete the square relative to both x and y.

$$4x^2 - 24x \quad + 49y^2 + 392y \quad = -624$$
$$4(x^2 - 6x \quad) + 49(y^2 + 8y \quad) = -624$$
$$4(x^2 - 6x + 9) + 49(y^2 + 8y + 16) = -624 + 36 + 784$$
$$4(x - 3)^2 + 49(y + 4)^2 = 196$$
$$\frac{(x-3)^2}{49} + \frac{(y+4)^2}{4} = 1$$

The curve is an ellipse with center (3, -4).

We introduce parameter t in analogy with problem 33:

$$\frac{x-3}{7} = \cos t \qquad \frac{y+4}{2} = \sin t$$
$$x - 3 = 7\cos t \qquad y + 4 = 2\sin t$$
$$x = 3 + 7\cos t \qquad y = -4 + 2\sin t,\ 0 \le t \le 2\pi$$

41. Here are the graphs in a squared viewing window (shown together with the line $y = x$).

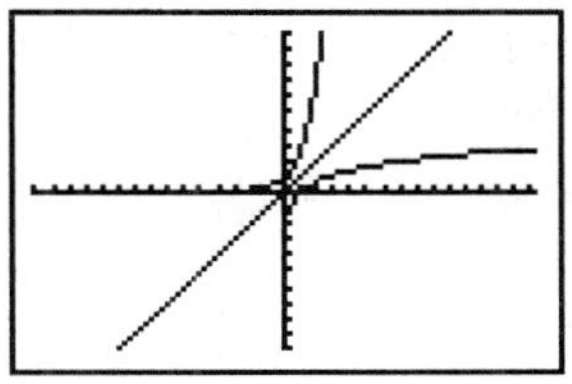

$x_1 = t,\ y_1 = e^t$
$x_2 = e^t,\ y_2 = t$

(A) The graphs are reflections of each other in the line $y = x$.

(B) Since $t = x_1$, $y_1 = e^{x_1}$. Since $t = y_2$, $x_2 = e^{y_2}$, $y_2 = \ln x_2$. The functions are inverse functions of each other.

43. At the instant the supplies are dropped, the velocity is 125 meters per second, the angle with the horizontal is 0, and the altitude is 1,000 meters. Substituting these values in

$$x = (v_0 \cos \alpha)t$$
$$y = a_0 + (v_0 \sin \alpha)t - 4.9t^2$$

the parametric equations for the path of the supplies are

$$x = 125t$$
$$y = 1000 - 4.9t^2$$

The supplies strike the ground when $y = 0$. Using the parametric equations for y, we obtain

$$y = 1{,}000 - 4.9t^2 = 0$$
$$-4.9t^2 = -1{,}000$$
$$t = \sqrt{\frac{-1{,}000}{-4.9}} \approx 14.2857 \text{ sec.}$$

The distance inland required is the value of x for this value of t. Substituting $t = 14.2857$ in $x = 125t$, the distance is
$x = 125(14.2857) \approx 1{,}786$ meters

45. Using equations (6) in the text, we have $v_0 = 300$, $\alpha = 45°$.

$x = (300 \cos 45°)t$
$y = (300 \sin 45°)t - 4.9t^2$

(A) The time of impact is the positive value of t for which $y = 0$.
Solving $0 = (300 \sin 45°)t - 4.9t^2$
we have $t = 0$ (which we discard) or

$$0 = 300 \sin 45° - 4.9t$$

$$t = \frac{300 \sin 45°}{4.9} = 43.292 \text{ seconds}$$

(B) The range is the value of x which corresponds to this value of t. Using the exact value for t we obtain $x = 300 \cos 45°\left(\frac{300 \sin 45°}{4.9}\right) = 9{,}183.674$ meters or, using the rounded value for t, we have $x = 300 \cos 45°(43.292) = 9{,}183.620$ meters. To convert meters to kilometers we divide either by 1,000 to obtain 9.184 kilometers.

(C) To find the maximum height attained by the projectile, we compute the height of the vertex of the parabola above the ground.
Since $y = (v_0 \sin \alpha)t - 4.9t^2$ or $-4.9t^2 + v_0 \sin \alpha t$ is the equation of the parabola, the height of the vertex of the parabola is the value of y found by completing the square relative to t.

$$y = -4.9\left(t^2 - \frac{v_0 \sin \alpha}{4.9}t\right)$$

$$= -4.9\left[t^2 - \frac{v_0 \sin \alpha}{4.9}t + \left(\frac{v_0 \sin \alpha}{9.8}\right)\right] + 4.9\left(\frac{v_0 \sin \alpha}{9.8}\right)^2$$

$$= -4.9\left(t - \frac{v_0 \sin \alpha}{9.8}\right)^2 + 4.9\left(\frac{v_0 \sin \alpha}{9.8}\right)^2$$

Thus the maximum height of the projectile occurs when $t = \frac{v_0 \sin \alpha}{9.8}$ and is given by

$$y = 4.9\left(\frac{v_0 \sin \alpha}{9.8}\right)^2$$

$$= 4.9\left(\frac{300 \sin 45°}{9.8}\right)^2$$

$$= 2{,}295.918 \text{ meters}$$

Notes

Notes

Notes

Notes